T0075511

Studies in Computational Intelligence

Volume 809

The series "Studies in Computational Intelligence" (SCI) publishes new developments and advances in the various areas of computational intelligence—quickly and with a high quality. The intent is to cover the theory, applications, and design methods of computational intelligence, as embedded in the fields of engineering, computer science, physics and life sciences, as well as the methodologies behind them. The series contains monographs, lecture notes and edited volumes in computational intelligence spanning the areas of neural networks, connectionist systems, genetic algorithms, evolutionary computation, artificial intelligence, cellular automata, self-organizing systems, soft computing, fuzzy systems, and hybrid intelligent systems. Of particular value to both the contributors and the readership are the short publication timeframe and the world-wide distribution, which enable both wide and rapid dissemination of research output.

More information about this series at http://www.springer.com/series/7092

Vladik Kreinovich · Nguyen Ngoc Thach
Nguyen Duc Trung · Dang Van Thanh
Editors

Beyond Traditional Probabilistic Methods in Economics

Editors
Vladik Kreinovich
Department of Computer Science
University of Texas at El Paso
El Paso, TX, USA

Nguyen Ngoc Thach
Banking University HCMC
Ho Chi Minh City, Vietnam

Nguyen Duc Trung
Banking University HCMC
Ho Chi Minh City, Vietnam

Dang Van Thanh
TTC Group
Ho Chi Minh City, Vietnam

ISSN 1860-949X ISSN 1860-9503 (electronic)
Studies in Computational Intelligence
ISBN 978-3-030-04199-1 ISBN 978-3-030-04200-4 (eBook)
https://doi.org/10.1007/978-3-030-04200-4

Library of Congress Control Number: 2018960912

This Springer imprint is published by the registered company Springer Nature Switzerland AG
The registered company address is: Gewerbestrasse 11, 6330 Cham, Switzerland

Preface

Economics is a very important and, at the same, a very difficult discipline. It is very difficult to predict how an economy will evolve, and it is very difficult to find out which measures we should undertake to make economy prosper. One of the main reasons for this difficulty is that in economics, there is a lot of uncertainty: Different difficult-to-predict events can influence the future economic behavior. To make good predictions, to make reasonable recommendations, we need to take this uncertainty into account.

In the past, most related research results were based on using traditional techniques from probability and statistics, such as p-value-based hypothesis testing and the use of normal distributions. These techniques led to many successful applications, but in the last decades, many examples emerged showing the limitations of these traditional techniques: Often, these techniques lead to non-reproducible results and to unreliable and inaccurate predictions. It is therefore necessary to come up with new techniques for processing the corresponding uncertainty, techniques that go beyond the traditional probabilistic techniques.

Such techniques and their economic applications are the main focus of this book. This book contains both related theoretical developments and practical applications to various economic problems. The corresponding techniques range from more traditional methods—such as methods based on Bayesian approach—to innovative methods utilizing ideas and techniques from quantum physics. A special section is devoted to fixed point techniques—mathematical techniques corresponding to the important economic notions of stability and equilibrium. And, of course, there are still many remaining challenges and many open problems.

We hope that this volume will help practitioners to learn how to apply various uncertainty techniques to economic problems, and help researchers to further improve the existing techniques and to come up with new techniques for dealing with uncertainty in economics.

We want to thank all the authors for their contributions and all anonymous referees for their thorough analysis and helpful comments.

The publication of this volume is partly supported by the Banking University of Ho Chi Minh City, Vietnam. Our thanks to the leadership and staff of the Banking University, for providing crucial support. Our special thanks to Prof. Hung T. Nguyen for his valuable advice and constant support.

We would also like to thank Prof. Janusz Kacprzyk (Series Editor) and Dr. Thomas Ditzinger (Senior Editor, Engineering/Applied Sciences) for their support and cooperation in this publication.

January 2019

Vladik Kreinovich
Nguyen Duc Trung
Nguyen Ngoc Thach
Dang Van Thanh

Contents

General Theory

Beyond Traditional Probabilistic Methods in Econometrics

Hung T. Nguyen[1,2](✉), Nguyen Duc Trung[3], and Nguyen Ngoc Thach[3]

[1] Department of Mathematical Sciences, New Mexico State University, Las Cruces, NM 88003, USA
hunguyen@nmsu.edu

[2] Faculty of Economics, Chiang Mai University, Chiang Mai 50200, Thailand

[3] Banking University of Ho-Chi-Minh City, 36 Ton That Dam Street, District 1, Ho-Chi-Minh City, Vietnam
{trungnd,thachnn}@buh.edu.vn

Abstract. We elaborate on various uncertainty calculi in current research efforts to improve empirical econometrics. These consist essentially of considering appropriate non additive (and non commutative) probabilities, as well as taking into account economic data which involved economic agents' behavior. After presenting a panorama of well-known non traditional probabilistic methods, we focus on the emerging effort of taking the analogy of financial econometrics with quantum mechanics to exhibit the promising use of quantum probability for modeling human behavior, and of Bohmian mechanics for modeling economic data.

Keywords: Fuzzy sets · Kolmogorov probability
Machine learning · Neural networks · Non-additive probabilities
Possibility theory · Quantum probability

1 Introduction

The purpose of this paper is to give a survey of research methodologies extending traditional probabilistic methods in economics. For a general survey on "new directions in economics", we refer the reader to [25].

In economics (e.g., consumers' choices) and econometrics (e.g., modeling of economic dynamics), it is all about uncertainty. Specifically, it is all about foundational questions such as what are possible sources (types) of uncertainty?, how to quantify a given type of uncertainty?. This is so since, depending upon which uncertainty we face, and how we quantify it, that we proceed to conduct our economic research.

The so-called traditional probabilistic methodology refers to the "standard" one based upon the thesis that uncertainty is taken as "chance/randomness", and we quantify it by additive set functions (subjectively/Bayes or objectively/Kolmogorov). This is exemplified by von Neumann's expected utility theory and stochastic models (resulting in using statistical methods for "inference"/predictions).

V. Kreinovich et al. (Eds.): ECONVN 2019, SCI 809, pp. 3–21, 2019.
https://doi.org/10.1007/978-3-030-04200-4_1

Thus, first, by non-traditional (probabilistic) methods, we mean those which are based upon uncertainty measures that are not "conventional", i.e., not "additive". Secondly, not using methods based on Kolmogorov probability can be completely different than just replacing an uncertainty quantification by another one. Thus, non probabilistic methods in machine learning, such as neural networks, are also considered as non traditional probabilistic methods.

In summary, we will discuss non traditional methods such as non-additive probabilities, possibility theory based on fuzzy sets, quantum probability, and then machine learning methods such as neural networks. Intensive references given at the end of the paper should provide a comprehensive picture of all probabilistic methods in economics so far.

2 Machine Learning

Let's start out by looking at traditional (or standard) methods (model-based) in economics in general, and econometrics in particular, to contrast with what can be called "model-free approaches" in machine learning.

Recall that uncertainty enters economic analysis at two main places: consumers' choice and economic equilibrium in micro economics [22,23,35,54], and stochastic modells in econometrics. At both places, even observed data are in general affected by economic agents (such as in finance), their dynamics (fluctuations over time), which are model-based, are modeled as stochastics processes in the standard theory of (Kolmogorov) probability theory (using also Ito stochastic calculus). And this is based on the "assumption" that the observed data can be viewed as a realization of a stochastic process, such as a random walk, or more generally a martingale. At the "regression" level, stochastic relations between economic variables are suggested by models, taking into account economic knowledge. Roughly speaking, we learn, teach and do research as follows. Having a problem of interest, e.g., predicting future economic states, we collect relevant (observed) data, pick a "suitable" model from our toolkit, such as a GARCH model, then use statistical methods to "identify" that model from data (e.g., estimating model parameters), then arguing that the chosen model is "good" (i.e., representing faithfully the data/data fitting, so that people can trust our derived conclusions). The last step can be done by "statistical tests" or by model selection procedures. The whole "program" is model-based [12,24]. The data is used after a model has been chosen! That is why econometrics is not quite an empirical science [25].

Remark. It has been brought to our attention in the research literature that, in fact, to achieve the main goal of econometrics, namely making forecasts, we do not need "significant tests". And this is consistent with the successful practice in physics, namely forecasting methods should be judged by their predictive ability. This will avoid the actual "crisis of p-value in science"! [7,13,26,27,43,55].

At the turn of the century, Breiman [6] called our attention to two cultures in statistical modeling (in the context of regression). In fact, a statistical model-based culture of 98% of statisticians, and a model-free (or really data-driven modeling) culture of 2% of the rest, while the main common goal is prediction. Note that, as explained in [51], we should distinguish clearly between statistical modeling towards "explaining" and/or "prediction".

After pointing out limitations of the statistical modeling culture, Breiman called our attention to the "algorithmic modeling" culture, from computer science, where the methodology is direct and data-driven: by passing the explanation step, and getting directly to prediction, using algorithms tuning for predictive ability. Perhaps, the most familiar algorithmic modeling to us is neural networks (one tool in machine learning among other such as decision trees, support vector machines, and recently, deep learning, data mining, big data and data science).

Before saying few words about the rationale of these non probabilistic methods, it is "interesting" to note that Breiman [6] classified"prediction in financial markets" in the category of "complex prediction problems where it was obvious that data model (i.e., statistical model) were not applicable" (p. 205). See also [9].

The learning capability of neural networks (see e.g., [42]), via backpropagation algorithms, is theoretically justified by the so- called "universal approximation property" which is formulated as a problem of approximating for functions (algorithms connecting inputs to outputs). As such, it is simply the well-known Stone-Weierstrass theorem, namely

Stone-Weierstrass Theorem. Let (X, d) be a compact metric space, and $C(X)$ be the space of continuous real-valued functions on X. If $H \subseteq C(X)$ such that

(i) H is a subalgebra of $C(X)$,
(ii) H vanishes at no point of X,
(iii) H separates points of X, then H is dense in $C(X)$.

Note that in practice we also need to know how much training data is needed to obtain a good approximation. This clearly depends on the complexity on the neural network considered. It turns out that, just like for support vector machines (in supervised machine learning), a measure of the complexity of neural networks is given as the Vapnik-Chervonenkis dimension (of the class of functions computable by neural networks).

3 Non Additive Probabilities

Roughly speaking, in view of Ellsberg "paradox" [19] (also [1]) in von Neumann's expected utility [54], the problem of quantifying uncertainty became central in social sciences, especially in economics. While standard probability calculus (Kolmogorov) is natural for roulette wheels, see [17] for a recent account, its basic additivity axiom seems not natural for the kind of uncertainty faced by humans in making decisions. In fact, it is precisely the additivity axiom (of probability measures) which is responsible to Ellsberg's paradox. This phenomenon triggered immediately the search for non-additive set functions to replace Kolmogorov probability in economics.

Before embarking on a brief review of efforts in the literature concerning non additive probabilities, it seems useful, at least to avoid of possible confusions among empirical econometricians, to say few words about the Bayesian approach to risk and uncertainty. In the Bayesian approach to uncertainty (which is also applied to economic analysis), there is no distinction between risk (uncertainty with known objective probabilities, e.g., in games of chance) and Knight's uncertainty (uncertainty with unknown probabilities, e.g., epistemic uncertainty, or caused by nature): When you face Knight's uncertainty, just use your own subjective probabilities to proceed, and treat your problems in the same framework as standard probability, i.e., using the additivity axiom to arrive as things such as the "law of total probability", the"Bayes updating rule" (leading to "conditional models" in econometrics). Without asking how reliable a subjective probability could be, let's ask "Can all types of uncertainty be quantified as additive probabilities, subjective or objective?". Philosophical debate aside (nobody can win!), let's look at real situations, e.g., experiments performed by psychologists to see whether, even if it is possible, additive probabilities are "appropriate" for quantitatively modeling human uncertainty. Bayesians like A. Gelman, M. Betancourt [28] recognized that "Does quantum uncertainty have a place in everyday applied statistics?" (noting that, see later, quantum uncertainty is quantified as a non additive probability). In fact, as we will see, as a Bayesian, A. Dempster [14] pioneered in modeling subjective probabilities (beliefs) by non additive set functions, which means simply that not all types on uncertainties can be modeled as additive probabilities.

Is there really a probability "measure" which is non additive? Well, it does! That was exactly what Richard Feynman told us in 1951 [21]: although the concept of chance is the same, the context of quantum mechanics (the way particles behave) only allows physicists to compute it in another way so that the additive axiom is violated. Thus, we do have a concrete calculus which does not follow standard Kolmogorov probability calculus, and yet it leads to successful physical results as we all knew. This illustrates an extremely important thing to focus on, and that is, whenever we face an uncertainty (for making decisions or predictions), we cannot force a calculus on it, but instead, we need to find out not only how to quantify it, but also how the context dictates its quantitative modeling. We will elaborate on this when we come to human decision-making under risk.

Inspired by Dempster's work [14], Shafer [50] proposed a non additive measure of uncertainty (called a "belief function") to model "generalized prior/subjective probability" (called "evidence"). In his formulation on a finite set U, a belief function is a set function $F : 2^U \to [0,1]$ satisfying a weaken form of Poincare's equality (making it non additive): $F(\varnothing) = 0, F(\Omega) = 1$, and, for any $k \geq 2$, and $A_1, A_2, ..., A_k$, subsets of U (denoting $|I|$ the cardinality of the set I):

$$F(\cup_{j=1}^{k} A_j) \geq \sum_{\varnothing \neq I \subseteq \{1,2,...,k\}} (-1)^{|I|+1} F(\cap_{i \in I} A_i)$$

But it was quickly pointed out [39] that such a set function is precisely the "probability distribution function" of a *random set* (see [41]), i.e., $F(A) = P(\omega : S(\omega) \subseteq A)$, where $S : \Omega \to 2^U$ is a random set (a random element) defined on a standard probability space $(\Omega, \mathscr{A}, P)$ and taking subsets of U as values. It is so since

$$f : 2^U \to [0,1], \qquad f(A) = \sum_{B \subseteq A} (-1)^{|A \setminus B|} F(B)$$

is a bona fide probability density function of 2^U, and $F(A) = \sum_{B \subseteq A} f(B)$.

As such, as a set function, it is non additive, but it does not really model another kind of uncertainty calculus. It just raises the uncertainty to a higher level, say, for coarse data. See also [20].

Other non additive probabilities arises in, say, robust Bayesian statistics, as "imprecise probabilities" [56], or in economics as "ambiguity" [29,30,37,47], or in general mathematics [15]. A general and natural way at arrive at non additive uncertainty measures is to consider Choquet capacity in Potential Theory, such as for statistics [33], for financial risk analysis [53]. For a favor of using non additive uncertainty measures in decision-making, see, e.g., [40]. For a behavioral approach to economics, see e.g., [34].

Remark on Choquet Capacities. Capacities are non additive set functions in potential theory, investigated by Gustave Choquet. They happened to generalize (additive) probability measures, and hence are imported into the area of uncertainty analysis with applications in social sciences, including economics. What is "interesting" for econometricians to learn from Choquet's work on the theory of capacities is not this mathematical theory itself, but from "how he achieved it?". He revealed it in the following paper "The birth of the theory of capacity: Reflexion on a personal experience" in *La vie des Sciences, Comptes Rendus 3(4), 385–397 (1986):* He solved a problem considered as difficult by specialists because he is not a specialist! A fresh look at a problem (such as "how to provide a model for a set of observed economic data?") without being an econometrician, and hence without constraints by previous knowledge of model-based approaches, may lead to a better model (i.e., closer to reality). Here is what Gustave Choquet wrote:

"Voila le probleme que Marcel Brelot et Henri Cartan signalaient vers 1950 comme un probleme difficile (et important) et pour lequel je finis par me passinonner en me persuadant que sa reponse devrait etre positive (pourquoi cette passion? C'est la le mistere des atomes crochus). Or je ne connaissais alors pratiquement rien de la theorie du potentiel. A la reflexion, je pense maintenant que ce fut cette raison qui me parmit de resoudre un probleme qui arretait les specialists. C'est la un point interessant pour les philosophes; aussi vais - je y insister un peu. Mon ignorance m'evitait en effet des prejuges: elle m'ecartait d'outils potentialistes trop sophistiques".

4 Possibility and Fuzziness

We illustrate now the question "Are there different kinds of uncertainty than randomness?". In economics, ambiguity is a kind of uncertainty. Another popular type of uncertainty is fuzziness [44,57]. Mathematically, fuzzy sets were considered to enlarge ordinary events (represented as sets) to events with no sharply defined boundaries. Originally, they are used in various situations in engineering and artificial intelligence, such as for representing imprecise information, coarsening information, building rule-based systems (e.g., in fuzzy neural control [42]). There is a large research community using fuzzy sets and logics in economics. What we are talking about here is a type of uncertainty which is built from the concept of fuzziness, called possibility theory [57]. It is a non additive uncertainty measure, and is also called an *idempotent probability* [46]. Mathematically, possibility measures arise as limits in the study of large deviations in Kolmogorov probability theory. Its definition is this. For any set Ω, a possibility measure is a set function $\mu(.) : 2^{\Omega} \to [0,1]$ such that $\mu(\varnothing) = 0$, $\mu(\Omega) = 1$, and for any family of subsets of Ω, $A_i, i \in I$, we have $\mu(\cup_{i \in I} A_i) = \sup\{\mu(A_i) : i \in I\}$.

Like all other non additive probabilities, possibility measures remain commutative and monotone increasing. As such, they might be useful for situations where events, information are consistent with their calculi, e.g., for economic data having no "thinking participants" involved. See [52] for a discussion about economic data in which a distinction between "natural economic data" (e.g., data fluctuating because of, say, weather; or data from industrial quality control of machines), and "data arising from free will of economic agents" is made. This distinction seems important for modeling of their dynamics, not only because these are different sources of dynamics (factors which create data fluctuations), but also the different types of uncertainty associated with them.

5 Quantum Probability and Mechanics

We have just seen a panorama of non traditional probabilistic tools which are developed either to improve conventional studies in economics (e.g., von Neumann's expected utility in social choice and economic equilibria) or to handle more complex situations (e.g., imprecise information). They are all centered around modeling (quantifying) various types of uncertainty, i.e., developing uncertainty calculi. Two things need to be noted. First, even with the specific goal of modeling how humans (economic agents) behave, say, under uncertainty (in making decisions), these non additive probabilities only capture one aspect of human behavior, namely non additivity! Secondly, although some analyses based on these non additive measures (i.e., associated integral calculi) were developed [15,47,48,53], namely Choquet integral, non additive integrals (which are useful for investigating financial risk measures), they are not appropriate to model economic data, i.e., not for proposing better models in econometrics. For example, Ito stochastic calculus is still used in financial econometrics. This is due to the fact that a connection between cognitive decision-making and economic data involving "thinking participants" was not yet discovered. This is, in fact, a delicate (and very important) issue, as stated earlier.

The latest research effort that we discuss now is precisely about these two things: improving cognitive decision modeling and economic data modeling. Essentially, we will elaborate on rationale and techniques to arrive at uncertainty measures capturing, not only non additivity of human behavior, but also other aspects such as non-monotonicity and non- commutativity, which were missing from previous studies. Note that these "aspects" in cognition were discovered by psychologists, see e.g. [8,31,34]. But the most important, and novel thing in economic research is the recognition that, even when using a model-based approach ("traditional"), the "nature" of data should be examined more "carefully" than just postulate that they are realizations of a (traditional) stochastic process! from which "better" models (which could be a "law", i.e., an useful model in the sense of Box [4,5]).

The above "program" was revealed partly in [52], and thanks to Hawking [32] for calling our attention to the analogy with mechanics. Of course, we have followed and borrowed concepts and techniques from natural sciences (e.g., physics, mechanics), such as "entropy", to conduct research in social sciences, especially in economics, but not "all the way"!, i.e., stopping at Newtonian mechanics (not go all the way to quantum mechanics).

First, what is "quantum probability?". The easy answer is "It is a calculus, i.e., a way to measure chance, in the subatomic world" which is used in quantum mechanics (motion of particles). Note that, at this junction, econometricians do not really need to "know" quantum mechanics (or, as a matter of fact, physics in general!). We will come to the "not-easy answer" shortly, but before that, it is important to "see" the following. As excellently emphasizing in the recent book [17], while the concept of "chance" is somewhat understood for everybody, but only qualitatively, it is useful in science only if we understand its "quantitative" face. While this book addressed only the notion of chance as uncertainty, and not other types of uncertainty such as fuzziness ("ambiguity" is included in the context of quantum mechanics as any path is a plausible path taken by a moving particle), it digged deeply into how uncertainty is quantified from various points of view. And this is important in science (natural or social) because, for example, decision-making under uncertainty is based on how we get its measure. When we put down a (mathematical) definition of an uncertainty measure (for chance), we actually put down "axioms", i.e., basic properties of such a measure (in other words, a specific calculus). The fundamental "axiom" of standard probability calculus (for both frequentist and Bayesian) is additivity because of the way we think we can "measure" chances of events, say by ratios of favorable cases over possible cases. When it was discovered that quantum mechanics is intrinsically unpredictable, the only way to observe nature at the subatomic world is computing probabilities of quantum events. Can we use standard probability theory for this purpose? Well, we can, but we will get the wrong probabilities we seek! The simple and well-known two-slit experiment says it all [21]. It all depends on how we can "measure" chance in a specific situation, here, motion of particles.

And this should be refered back to experiments performed by psychologists, not only violating standard probability calculus used in von Neumann's expected utility, leading to the considerations of non additive probabilities [19,20,34], but also bringing out the fact that it is the quantitative aspect of uncertainty which is important in science. As for quantum probability, i.e., how physicists measure probabilities of quantum events, the evidence in the two-slit experiment is this. The state of a particle in quantum mechanics is determined by its wave function $\psi(x,t)$, solution of the Schrodinger's equation (counterpart of Newton's second law of motion):

$$ih\frac{\partial\psi(x,t)}{\partial t} = -\frac{h^2}{2m}\Delta_x\psi(x,t) + V(x)\psi(x,t)$$

where Δ_x is the Laplacian, i complex unit, and h is the Planck's constant, with the meaning that the wave function $\psi(x,t)$ is the "probability amplitude" of position x at time t, i.e., $x \to |\psi(x,t)|^2$ is the probability density function for the particle position at time t, so that the probability of finding the particle, at time t, in a region $A \subseteq \mathbb{R}^2$ is $\int_A |\psi(x,t)|^2 dx$. That is how physicists predict quantum events. Thus, in the experiment where particles travel through two slits A, B, we have $|\psi_{A \cup B}|^2 = |\psi_A + \psi_B|^2 \neq |\psi_A|^2 + |\psi_B|^2$ implying that "quantum probability" is not additive. It turns out that other experiments reveal that $QP(A \text{ and } B) \neq QP(B \text{ and } A)$, i.e., quantum probabilities are not commutative (of course the connective "and" here should be specified mathematically). It is a "nice" coincidence that the same phenomena appeared in cognition, see e.g., [31]. Whether there is some "similarity" between particles and economic agents with free will is a matter of debate. What econometricians should be aware to take advantage of is there is a mathematical language (called functional analysis) available to construct a non commutative probability, see e.g., [38,45].

Let's turn now to the second important point for econometricians, namely how to incorporate economic agents' free will (affecting economic dynamics) into the "art" of economic model building? remembering that, traditionally, our model-based approach to econometrics does not take this fundamental and obvious information into account.

It is about a careful data analysis towards the most important step in modeling dynamics of economic data for prediction, remembering that, as an effective theory, econometrics at present is only "moderately successful", as opposed to "totally successful of quantum mechanics" [32]. Moreover, at clearly stated in [25], present econometrics is not quite an empirical science. Is it because of the fact that we did not examine carefully the data we see? Are there other sources causing the fluctuations of our data that we missed (to incorporate into our modeling process)?. Should we use the "bootstrap spirit": Get more out of the data?

One direction of research using quantum mechanic formalism to finance, e.g., [2], is to replace Kolmogorov probability calculus by quantum stochastic calculus, as well as using Feynman's path integral. Basically, this seems because of assertions such as "A natural explanation of extreme irregularities in the evolution of prices in financial markets is provided by quantum effects", [49]. See also [11,16].

Remark on Path Integral. For those who wish to have a quick look at what is path integral. Here it is.

How to obtain probabilities for "quantum events"? This question was answered by the main approach to quantum mechanics, namely, by the famous Schrodinger's equation (playing the role of "law of quantum mechanics", counterpart of Newton's second law in classical mechanics). The solution $\psi(x,t)$ to the Schrodinger's equation is a probability amplitude for (x,t), i.e., $|\psi(x,t)|^2$ is the probability you seek. Beautiful!

But why it is so? Lots of physical justifications are needed to arrive at the above conclusion, but they are nothing to do with classical mechanics, just like there is no connections between the two kinds of mechanics. However, see later for Bohmian mechanics.

It was right here that Richard Feynman came in. Can we find the above quantum probability amplitude without solving the (PDE) Schrodinger's equation, and yet connecting quantum mechanics with classical mechanics? If the answer is yes, then, at least, from a technical viewpoint, we have a new technique to solve difficult PDE, at least for PDE related to physics!

Technically speaking, the above question is somewhat similar to what giant mathematicians like Lagrange, Euler and Hamilton have asked within the context of classical mechanics. And that is "can we study mechanics by another, but equivalent, way than solving Newton's differential equation?". The answer is Lagrangian mechanics. Rather than solving Newton's differential equation (his second law), we optimize a functional (on paths) called "action" which is an integral of the Lagrangian of the dynamical system: $S(x) = \int L(x,x')dt$. Note that Newton's law is expressed in term of force. Now motion is also caused by energy. The Lagrangian is the difference between kinetic energy and potential energy (which is not conserved, as opposed to the Hamiltonian of the system, which is the sum of these energies). It turns out that the extremum of the action provides solution to the Newton's equation, the so-called the Least Action Principle (LAP) in classical mechanics (but you need "calculus of variations" to solve this *functional optimization*!).

With LAP in mind, Feynman proceeded as follows. From an initial condition $(x(0) = a)$ of an emitting particle, we know that, for it to be at $(T, x(T) = b)$, it must take a path (a continuous function) joining point a to point b. There are lots of such paths, denoted as $\mathscr{P}([a,b])$. Unlike Newtonian mechanics where the object (here a particle) can take one path which is determined either by solving Newton's equation, or by LAP, a particle can take any path $x(t), t \in [0,T]$, each with some probability. Thus, a "natural" question is "how much each possible path contributes to the global probability amplitude of being at $(T, x(T) = b)$?

If p_x is a probability amplitude, contributed by the path $x(.) \in \mathscr{P}([a,b])$, then their *sum* over all paths, informally $\sum_{x\in\mathscr{P}([a,b])} p_x$, could be the probability amplitude we seek (this is what Feynman called "sum over histories"). But how to "sum" $\sum_{x\in\mathscr{P}([a,b])} p_x$ when the set of summation indices $\mathscr{P}([a,b])$ is uncountable? Well, that is so familiar in mathematics, and we know how to handle it: Use integral! But what kinds of integral? None of the integrals you knew so far (Stieltjes, Lebesgue integrals) "fits" our need here, since the integration domain $\mathscr{P}([a,b])$ is a function space, i.e., an uncountable, infinitely dimensional set (similar to the concept of "derivative with respect to a function", i.e., functional derivatives, leading to the development of the Calculs of Variations). We are facing the problem of *functional integration*. What do we mean by an expression like $\int_{\mathscr{P}([a,b])} \Psi(x)Dx$, where the integration variable x is a function? Well, we might proceed as follows. Except Riemann integral, all other integrals arrive after we have a measure on the integration domain (measure theory is in fact an integration theory: measures are used to construct associated integrals). Note that, historically, Lebesgue developed his integral (later extended to an abstract setting) in this spirit. A quick search on literature reveals that N. Wiener (The average value of a functional, *Proc. London Math. Soc.* (22), 454–467, 1924) has defined a measure on the space of continuous functions (paths of Brownian motion) and from it constructed a functional integral. Unfortunately, we cannot use his functional integral (based on his measure) to interprete $\int_{\mathscr{P}([a,b])} \Psi(x)Dx$ here, since, as far as quantum mechanics is concerned, the integrand $\Psi(x) = \exp\{\frac{i}{\hbar}S(x)\}$, where i is the imaginary unit, so that, in order to use Wiener measure, we need to replace it by a complex measure involving a Gaussian distribution with a complex variance (!), and no such ($\sigma-$) additive measure exists, as shown by R. H. Cameron ("A family of integrals serving to connect the Wiener and Feynman integrals", *J. Math. and Phys* (39), 126–140, 1960). To date, there is no possible measure-theoretic definition of Feynman's path integral.

So how Feynman managed to define his "path integral" to represent $\int_{\mathscr{P}([a,b])} \exp\{\frac{i}{\hbar}S(x)\}Dx$?

Clearly, without the existence of a complex measure on $\mathscr{P}([a,b])$, we have to construct integral without it! The only way to do that is to follow Riemann!!!!

Thus, Feynman's path integral is a Riemann-based approach, as I will elaborate now.

Once the integral $\int_{\mathscr{P}([a,b])} \exp\{\frac{i}{\hbar}S(x)\}Dx$ is defined, we still need to show that it does provide the correct probability amplitude. How? Well, just verify that it is precisely the solution for the initial value problem of the PDE Schrodinger's equation! In fact, more can be proved: the Schrodinger's equation came from the path integral formalism, i.e., Feynman's approach to quantum mechanics, via his path integral concept, is equivalent to Schrodinger's formalism (which is in fact, equivalent to Heinsenberg's matrix formalism, via representation theory in mathematics), constituting a third equivalent formalism for quantum mechanics.

The Principle of Least Action

How to study (classical) mechanics? Well, easy, just use and solve Newton's equation (Newton's Second law)! 150 years after Newton, giant mathematicians like Lagrange, Euler and Hamilton reformulated it for good reasons:

(i) More elegant!
(ii) More powerful: providing new methods to solve hard problems in a straightforward way,
(iii) Universal, and providing a framework that can be extended to other laws of physics, and revealing a relationship with quantum mechanics (that we will explore in this Lecture).

Solving Newton's equation, we should get the trajectory of the moving object under study. Is there another way for obtaining the same result? Yes, the following one will also lead to the equations of motion of that object.

Let the moving object have (total) mass m, subject to a force F, then according to Newton, the trajectory of it $x(t) \in \mathbb{R}$ (for simplicity) is solution of $F = m\frac{dx(t)^2}{dt^2} = mx''(t)$. Here, we need to solve a second order differential equation (with initial condition: $x(t_o), x'(t_o)$). Note that trajectories are differentiable functions (paths).

Now, instead of force, let's use energy of the system. There are two kinds of energy. The Kinetic energy K (inherent in motion, e.g., energy emitted by light photon), which is a function of the object's velocity $K(x')$ (e.g., $K(x') = \frac{1}{2}m(x')^2$), and potential energy $V(x)$, function of position x, which depends on the configuration of the system (e.g., force: $F = -\nabla V(x)$). The sum $H = K + V$ is called the Hamiltonian of the system, whereas the difference $L(x, x') = K(x') - V(x)$ is called the Lagrangian, which is a function of x and x'. The Lagrangian L summarizes the dynamics of the system. In this setting, instead of specifying the initial condition as $x(t_o), x'(t_o)$, we specify initial and final positions, say, $x(t_1), x(t_2)$, and ask "how the object moves from $x(t_1)$ to $x(t_2)$?". More specifically, among all possible paths connecting $x(t_1)$ to $x(t_2)$, what path does the object actually take?

For each such (differentiable) path, assign a number, which we call an "action"

$$S(x) = \int_{t_1}^{t_2} L(x(t), x'(t))dt$$

The map $S(.)$ is a functional on differentiable paths.

Theorem. The path taken by the moving object is an extremum of the action S.

This theorem is referred to as "The Principle of Least Action" in Lagrangian Mechanics.

The optimization is over all paths $x(.)$ joining $x(t_1)$ to $x(t_2)$. The action $S(.)$ is a functional. To show that such an extremum is indeed the trajectory of the moving object, it suffices to show that it satisfies Newton's equation! For example, with $L = \frac{1}{2}m(x')^2 - V(x)$, then $\delta S = 0$ when $m(x'')^2 = -\nabla V$ which is precisely the Newton's equation.

As we will see shortly, physics will also lead us to an integral (i.e., a way to express summation in continuous context) unfamiliar to standard mathematics: *a functional integral,* i.e., an integral over an infinitely dimensional domain (function spaces). It is a perfect example of "where fancy mathematics came from?"!

In studying Brownian motion of a particle (caused by chocs of surrounding particles, as explained by Einstein in 1905) modeled according to Kolmogorov probability theory (note that Einstein contributed to quantum physics/structures of matter/particles, but not really to quantum mechanics), N. Wiener, in 1922, introduced a measure on the space of continuous functions (paths of Brownian motion) from which he considered a functional integral with respect to that measure. As we will see, for the need of quantum mechanics, Feynman was led to consider also a functional integral, but in a quantum world. Feynman's path integral is different than Wiener's integral and was constructed without first constructing a measure, using the old Riemann's method of constructing integral without the need of a measure.

Recall also the basic problem in quantum mechanics: From a starting known position x_o, how the particle will travel? In view of the random nature of its travels, the realistic question to ask is "what is the chance it will pass through a point $x \in \mathbb{R}$ (in one dimension for simplicity/possibly extended to $\mathbb{R}^d$) at a later time t?". In the Schrodinger's formalism, the answer to this question is $|\psi(x,t)|^2$, where the wave function satisfies the Schrodinger' s equation (noting that, the wave function, as solution of Schrodinger's equation, "describes" the particle motion in the sense that it provides a probability amplitude). As you can realize, this formalism came from examining the nature of particles, and not from any attempt to "extending" classical mechanics to the quantum context (from macroobjects to microobjects). Of course, any such attempts cannot be based upon "extending" Newton's laws of motion to quantum laws. But for the fundamental question above, namely "what is the probability for a particle to be in some given position?", an "extension" is possible, although not "directly".

As we have seen above, Newton's laws are "equivalent" to the Least Action Principle. The question is "Can we use the Least Action Principle to find quantum probabilities?", i.e., solving Schrodinger's equation without actually "solving" it! i.e., just get its solution from some place else!

Having the two-slit experiment in the back of our mind, consider the situation where a particle is starting its voyage from a point (emission source) $(t = 0, x(0) = a)$ to a point $(t = T, x(T) = b)$. To star from a and arrive at b, clearly the particle must take some "path" (a *continuous* function $t \in [0,T] \to x(t)$, such that $x(0) = a, x(T) = b$) joining a and b. But unlike Newtonian mechanics (where the moving object will certainty take only one path, among all such paths, which is determined by the Least Action Principle/LAP), in the quantum world, the particle can take any paths (sometimes it takes this path, sometimes it takes another path), each one with some probability. In view of this, it seems natural to think that the "overall" probability amplitude should be the sum of all "local" probability amplitude, i.e., contributed by each path. The crucial question is "what is the probability amplitude contributed by a given path?". The great idea of Richard Feynman, inspired from LAP in classical mechanics, via Paul Dirac's remark "the transition amplitude is governed by the value of the classical action", is to take (of course, from physical considerations) the local contribution (called the "propagator") to be $\exp\{\frac{i}{h}S(x)\}$, where

$S(x)$ is the action on the path $x(.)$, namely, $S(x) = \int_0^T L(x, x')dt$, where L is the Lagrangian of the system (Recall that, in Schrodinger's formalism, it was the Hamiltonian which was used). Each path contributes a transition amplitude, a (complex) number, proportional to $e^{\frac{i}{\hbar}S(x)}$, to the total probability amplitude of getting from a to b.

Feynman claimed that the "sum over histories", an informal expression (a "functional" integral form) of the form $\int_{\text{all paths}} e^{\frac{i}{\hbar}S(x)}Dx$, could be the total probability amplitude that the particle, staring at a, will be at b. Specifically, the probability that the particle will go from a to b is

$$|\int_{\text{all paths}} e^{\frac{i}{\hbar}S(x)}Dx|^2$$

Note that here, {all paths} means paths joining a to b. and Dx denotes "informally" the "measure" on the space of paths $x(.)$.

It should be noted that, while the probability amplitude in Shrodinger's formalism is associated with the position of the particle, at a given time t, namely $\psi(x, t)$, Feynman's probability amplitude is associated with an entire motion of the particle as a function of time (paths).

Moreover, just like the LAP is equivalent to Newton's law, this path integral formalism to quantum mechanics is equivalent to Schrodinger's formalism, in the sense that the path integral can be used to represent the solution of initial value problem for the Schrodinger equation.

Thus, first, we need to define rigorously the "path integral" $\int_{\{path\ x\}} f(x)Dx$, of a functional $f : \{pathx\} \to \mathbb{C}$, over the integration domain $\{pathx\}$, a functional space.

Note that the space of paths from a to b, denoted as $\mathscr{P}([a, b])$, is the set of all continuous functions. Technically speaking, the Lagrangian $L(.,.)$ operates only on differentiable paths, so that the integrand $e^{\frac{i}{\hbar}S(x)}$ is defined also only for differentiable paths. We will need to extend the action $S(x) = \int_{t_a}^{t_b} L(x, x')dt$ to continuous paths. The path integral of interest in quantum mechanics is $\int_{\mathscr{P}([a,b])} e^{\frac{i}{\hbar}S(x)}Dx$, where Dx stands for "summation symbol" of path integral.

In general, a path integral is of the form $\int_{\mathscr{C}} \Psi(x)Dx$, where $\mathscr{C}$ is a set of continuous functions, and $\Psi : \mathscr{C} \to \mathbb{C}$ a functional. The construction (definition) of such an integral starts with replacing $\Psi(x)$ by an approximating Riemann sum, then using a limiting procedure for a multiple ordinary integrals. Let's illustrate it with the specific $\int_{\mathscr{P}([a,b])} e^{\frac{i}{\hbar}S(x)}Dx$.

We have, noting that $L(x, x') = \frac{(mv)^2}{2m} - V(x) = \frac{m}{2}(\frac{dx}{dt})^2 - V(x)$, so that

$$S(x) = \int_0^T L(x, x')dt = \int_0^T [\frac{m}{2}(\frac{dx}{dt})^2 - V(x)]dt$$

For $x(t)$ continuous, we represent $\frac{dx(t)}{dt}$ by a difference quotient, and represent the integral by an approximate sum.

For that purpose, dividing the time interval $[0, T]$ into n equal subintervals, each of length $\Delta t = \frac{T}{n}$, and let $t_j = j\Delta t$, $j = 0, 1, 2, ..., n$ and $x_j = x(t_j)$

Now, for each fixed t_j, we vary the paths $x(.)$, so that at t_j, we have the set of values $\{x(t_j) = x_j : x(.) \in \mathscr{P}([a,b])\}$, so dx_j denotes the integration over all $\{x_j : x(.) \in \mathscr{P}([a,b])\}$. Put it differently, $x_j(.) : \mathscr{P}([a,b]) \to \mathbb{R}$: $x_j(x) = x(t_j)$.

Then, approximate $S(x)$ by

$$\sum_{j=1}^{n}[\frac{m}{2}(\frac{x_{j+1}-x_j}{\Delta t})^2 - V(x_{j+1})]\Delta t = \sum_{j=1}^{n}[\frac{m(x_{j+1}-x_j)^2}{2\Delta t} - V(x_{j+1})\Delta t]$$

Integrating with respect to $x_1, x_2, ..., x_{n-1}$,

$$\int_{-\infty}^{\infty}...\int_{-\infty}^{\infty}\exp\{\frac{i}{h}[\sum_{j=1}^{n}[\frac{m(x_{j+1}-x_j)^2}{2\Delta t} - V(x_{j+1})\Delta t]dx_1...dx_{n-1}$$

By physical considerations, the normalizing factor $(\frac{mn}{2\pi ihT})^{\frac{n}{2}}$ is used before taking the limit. Thus, the path integral $\int_{\mathscr{P}([a,b])} e^{\frac{i}{h}S(x)}Dx$ is defined as

$$\int_{\mathscr{P}([a,b])} e^{\frac{i}{h}S(x)}Dx$$

$$= \lim_{n\to\infty}(\frac{mn}{2\pi ihT})^{\frac{n}{2}}\int_{-\infty}^{\infty}...\int_{-\infty}^{\infty}\exp\{\frac{i}{h}[\sum_{j=1}^{n}[\frac{m(x_{j+1}-x_j)^2}{2\Delta t} - V(x_{j+1})\Delta t]dx_1...dx_{n-1}$$

Remark. Similarly to the normalizing factor $\Delta t = \frac{T}{n}$ in the Riemann integral

$$S(x) = \int_0^T[\frac{m}{2}(\frac{dx}{dt})^2 - V(x)]dt = \lim_{n\to\infty}(\Delta t)\sum_{j=1}^{n}[\frac{m}{2}(\frac{x_{j+1}-x_j}{\Delta t})^2 - V(x_{j+1})]$$

a suitable normalizing factor $A(n)$ is needed in path integral to ensure that the limit exists:

$$\int_{\mathscr{C}}\Psi(x)Dx = \lim_{n\to\infty}\frac{1}{A}\int_{\mathbb{R}^{n-1}}\Psi(x)\frac{dx_1}{A}...\frac{dx_{n-1}}{A}$$

The factor $A(n)$ is calculated on a case by case basis. For example, for $\int_{\mathscr{P}([a,b])} e^{\frac{i}{h}S(x)}Dx$, the normalizing factor is found to be

$$A(n) = (\frac{2\pi ih\Delta t}{m})^{\frac{1}{2}} = (\frac{2\pi ihT}{mn})^{\frac{1}{2}}$$

Finally, let $T = t$, and $b = x$ (a position), then $\psi(x,t) = \int_{\mathscr{P}([a,x])} e^{\frac{i}{h}S(z)}Dz$, defined as above, can be shown to be the solution of the initial value Schrodinger's equation

$$ih\frac{\partial\psi}{\partial t} = -\frac{h^2}{2m}\frac{\partial^2\psi}{\partial x^2} + V(x)\psi(x,t)$$

Moreover, it can be shown that Schrodinger 's equation follows from Feynman's path integral formalism. Thus, Feynman's path integral is an equivalent formalism for quantum mechanics.

Some Final Notes

(i) The connection between classical and quantum mechanics is provided by the concept of "action" from classical mechanics. Specifically, in classical mechanics, the trajectory of a moving object is the path making its action $S(x)$ stationary. In quantum mechanics, the probability amplitude is a path integral of the integrand $\exp\{\frac{i}{h}S(x)\}$. Both procedures are based upon the notion of "action" in classical mechanics (in Lagrange's formulation).

(ii) Once $\psi(b,T) = \int_{\mathscr{P}([a,b])} e^{\frac{i}{h}S(x)} Dx$ is defined (known theoretically, for each (b,T)), all the rest of quantum analysis can be carried out, from the quantum probability density for the particle position, at each time, $b \to |\int_{\mathscr{P}([a,b])} e^{\frac{i}{h}S(x)} Dx|^2$. Thus, for applications, computational algorithms for path integrals are needed.

But as mentioned in [10], even path integral in quantum mechanics is equivalent to the formalism of stochastic (Ito) calculus [2], a model for stock market of the form

$$dS_t = \mu S_t dt + \sigma S_t dW_t$$

does not contain terms describing the *behavior of agents* of the market. Thus, recognizing that any financial data is a result of natural randomness ("hard" effect) and of decisions of investors ("soft" effect), we have to consider these two sources of uncertainties causing its dynamics. And this is for "explaining" the data, recalling that "explaining" modeling is different than "predictive" modeling [51]. Since, obviously, we are interested in prediction, the predictive modeling, based on the available data, should be proceeded in the same spirit. Specifically, we need to "identify" or formulate the "soft effect" which is related to things such as expectations (of investors) and the market psychology, as well as a stochastic process representing the "hard effect". Again, as pointed out in [10], an additional stochastic process, to the above Ito stochastic equation, to represent behavior of investors, is not appropriate since it cannot describe the "mental state of the market" which is of infinite complexity, requiring an infinitely dimensional representation, not suitable in classical probability theory.

The crucial problem becomes: How to formulate and put these two "effects" into our modeling process leading to a more faithfull representation of the data, for purpose of prediction? We think this is a challenge for econometricians in this century. At present, here is the state-of-the-art of the research efforts in the literature.

Since we are talking about modeling of dynamics of financial data, we should think about mechanics! Dynamics is caused by forces, and forces are derived from energies or potentials. Since we have in mind two types of "potentials" soft and hard which could correspond to two types of energies in classical mechanics, namely potential energy (dues to position) and kinetic energy (due to motion), we could think about Hamiltonian formalism of classical mechanics. On the other hand, not only human decision-making seems to carry out in the context of non commutative probability (which has a formalism in quantum mechanics), but also, as stated above, the stochastic part should be infinitely dimensional, again a known situation in quantum mechanics! As such, the analogies with quantum mechanics seems obvious. However, in the standard formalism of quantum mechanics (the so-called Copenhagen interpretation), the state of a particle is "described" by Schrodinger's wave function (with a probabilist interpretation, leading, in fact, to successful predictions, as we all know), and as such (in view of Heisenberg's uncertainty principle) there is no trajectories of dynamics. So how can we use (an analogy with) quantum mechanics to portray economic dynamics? Well, while standard formalism is popular among physicists, there is another interpretation of quantum mechanics which relates quantum mechanics with classical mechanics, called Bohmian mechanics, see e.g. [31], in which we can talk about the classical concept of trajectories of particles, although their randomness (caused by subjective probability/imperfect knowledge of initial conditions) is due to initial conditions.

Remark on Bohmian Mechanics
The choice of Bohmian interpretation of quantum mechanics [3] for econometrics is dictated by econometric needs, and not by Ockham's razor (a heuristic concept to decide between several feasible interpretations or physical theories). Since Bohmian interpretation is currently proposed to construct financial models from data which exhibit both natural randomness and investors' behavior, let's elaborate a bit on it.

Recall that the "standard" (Copenhaven) interpretation of quantum mechanics is this [18]. Roughly speaking the "state" of a quantum system (say, of a particle with mass m, in $\mathbb{R}^3$) is "described" by its wave function $\psi(x,t)$, solution of the Schrodinger's equation, in the sense that $x \to |\psi(x,t)|^2$ is the probability density function of the position x at time t. This randomness (about particle's positions) is intrinsic, i.e., due to nature itself, in other words, quantum mechanic is a (objective) probability theory, so that the notion of trajectory (of a particle) is not defined, as opposed to classical mechanics. Essentially, the wave function is a tool for prediction purposes. The main point of this interpretation is the objectivity of the probabilities (of quantum events) based soly on the wave function.

Another "empirically equivalent" interpretation of quantum mechanics is Bohmian interpretation which indicates that classical mechanics is a limiting case of quantum mechanics (when the Planck constant $h \to 0$). Although the interpretation leads to the consideration of classical notion of trajectories (which is good for economics when we will take, say, stock prices as analogues of particles!), these trajectories remain random (by our lack of knowledge about initial conditions/by our ignorance), characterized by wave functions, but "subjectively" instead (i.e., epistemic). Specifically, the Bohmian interpretation considers two ingredients: the wave function, and the particles. Its connection with classical mechanics manifests in its Hamiltonian formalism of classical mechanics, derived from Schrodinger's equation, which makes the applications to economic modeling plausible, especially, as potential induces force (source of dynamics), one can "store" (or extract) mental energy in potential energy expression, for explaining (or for prediction) purposes. Roughly speaking, with the Bohmian formalism of quantum mechanics, econometricians should be in position to carry out a new approach to economic modeling, in which the human factor is taken into account. A final note is this. We are mentioning the classical context of quantum mechanics, and not just classical mechanics because classical mechanics is deterministic, whereas quantum mechanics, even in Bohmian formalism, is stochastic with a probability calculus (quantum probability) exhibiting the uncertainty calculus in cognition, as spelled out in the first point (quantum probability for human decision-making).

References

1. Allais, M.: Le comportement de l'homme rationnel devant le risque: Critique des postulats et axiomes de l'ecole americaine. Econometrica **21**(4), 503–546 (1953)
2. Baaquie, B.E.: Quantum Finance: Path Integrals and Hamiltonians for Options and Interest Rates. Cambridge University Press, Cambridge (2007)
3. Bohm, D.: Quantum Theory. Prentice Hall, Englewood Cliffs (1951)
4. Box, G.E.P.: Science and statistics. J. Am. Stat. Assoc. **71**(356), 791–799 (1976)
5. Box, G.E.P.: Robustness in the strategy of scientific model building. In: Launer, R.L., Wilkinson, G.N. (eds.) Robustness in Statistics, pp. 201–236. Academic Press, New York (1979)
6. Breiman, L.: Statistical modeling: the two cultures. Stat. Sci. **16**(3), 199–215 (2001)
7. Briggs, W.: Uncertainty: The Soul of Modeling, Probability and Statistics. Springer, New York (2016)
8. Busemeyer, J.R., Bruza, P.D.: Quantum Models of Cognitive and Decision. Cambridge University Press, Cambridge (2012)
9. Campbell, J.Y., Lo, A.W., Mackinlay, A.C.: The Econometrics of Financial Markets. Princeton University Press, Princeton (1997)
10. Choustova, O.: Quantum Bohmian model for financial markets. Phys. A **347**, 304–314 (2006)
11. Darbyshire, P.: Quantum physics meets classical finance. Phys. World **18**(5), 25–29 (2005)
12. Dejong, D.N., Dave, C.: Structural Macroeconometrics. Princeton University Press, Princeton (2007)
13. De Saint Exupery, A.: The Little Prince. Penguin Books (1995)
14. Dempster, A.: Upper and lower probabilities induced by a multivalued mapping. Ann. Math. Stat. **38**, 325–339 (1967)
15. Denneberg, D.: Non-additive Measure and Integral. Kluwer Academic Press, Dordrecht (1994)
16. Derman, D.: My life as a Quant: Reflections on Physics and Finance. Wiley, Hoboken (2004)
17. Diaconis, P., Skyrms, B.: Ten Great Ideas About Chance. Princeton University Press, Princeton and Oxford (2018)
18. Dirac, D.: The Principles of Quantum Mechanics. Clarendon Press, Oxford (1947)
19. Ellsberg, D.: Risk, ambiguity, and the savage axioms. Q. J. Econ. **75**(4), 643–669 (1961)
20. Fegin, R., Halpern, J.Y.: Uncertainty, belief and probability. Comput. Intell. **7**, 160–173 (1991)
21. Feynman, R.: The concept of probability in quantum mechanics. In: Berkeley Symposium on Mathematical Statistics and Probability, pp. 533–541 (1951)

22. Fishburn, P.C.: Non Linear Preference and Utility Theory. Wheatsheaf Books, Sussex (1988)
23. Fishburn, P.C.: Utility Theory for Decision Making. Wiley, New York (1970)
24. Florens, J.P., Marimoutou, V., Peguin-Feissolle, A.: Econometric Modeling and Inference. Cambridge University Press, Cambridge (2007)
25. Focardi, S.M.: Is economics an empirical science? If not, can it become one? Front. Appl. Math. Stat. **1**, 7 (2015)
26. Freedman, D., Pisani, R., Purves, R.: Statistics, 4th edn. W.W. Norton, New York (2007)
27. Gale, R.P., Hochhaus, A., Zhang, M.J.: What is the (p-) value of the p-value? Leukemia **30**, 1965–1967 (2016)
28. Gelman, A., Betancourt, M.: Does quantum uncertainty have a place in everyday applied statistics? Behav. Brain Sci. **36**(3), 285 (2013)
29. Gilboa, I., Marinacci, M.: Ambiguity and the Bayesian paradigm. In: Acemoglu, D. (ed.) Advances in Economics and Econometrics, pp. 179–242. Cambridge University Press, Cambridge (2013)
30. Gilboa, I., Postlewaite, A.W., Schmeidler, D.: Probability and uncertainty in economic modeling. J. Econ. Perspect. **22**(3), 173–188 (2008)
31. Haven, E., Khrennikov, A.: Quantum Social Science. Cambridge University Press, Cambridge (2013)
32. Hawking, S., Mlodinow, L.: The Grand Design. Bantam Books, London (2010)
33. Huber, P.J.: The use of Choquet capacities in statistics. Bull. Inst. Intern. Stat. **4**, 181–188 (1973)
34. Kahneman, D., Tversky, A.: Prospect theory: an analysis of decision under risk. Econometrica **47**, 263–292 (1979)
35. Kreps, D.M.: Notes on the Theory of Choice. Westview Press, Boulder (1988)
36. Lambertini, L.: John von Neumann between physics and economics: a methodological note. Rev. Econ. Anal. **5**, 177–189 (2013)
37. Marinacci, M., Montrucchio, L.: Introduction to the mathematics of ambiguity. In: Gilboa, I. (ed.) Uncertainty in Economic Theory, pp. 46–107. Routledge, New York (2004)
38. Meyer, P.A.: Quantum Probability for Probabilists. Lecture Notes in Mathematics. Springer, Heidelberg (1995)
39. Nguyen, H.T.: On random sets and belief functions. J. Math. Anal. Appl. **65**(3), 531–542 (1978)
40. Nguyen, H.T., Walker, A.E.: On decision making using belief functions. In: Yager, R., Kacprzyk, J., Pedrizzi, M. (eds.) Advances the Dempster-Shafer Theory of Evidence, pp. 311–330. Wiley, New York (1994)
41. Nguyen, H.T.: An Introduction to Random Sets. Chapman and Hall/CRC Press, Boca Raton (2006)
42. Nguyen, H.T., Prasad, N.R., Walker, C.L., Walker, E.A.: A first Course in Fuzzy and Neural Control. Chapman and Hall/CRC Press, Boca Raton (2003)
43. Nguyen, H.T.: On evidence measures of support for reasoning with integrated uncertainty: a lesson from the ban of p-values in statistical inference. In: Huynh, V.N., et al. (eds.) Integrated Uncertainty in Knowledge Modeling and Decision Making. Lecture Notes in Artificial Intelligence, vol. 9978, pp. 3–15. Springer, Cham (2016)
44. Nguyen, H.T., Walker, E.A.: A First Course in Fuzzy Logic, 3rd edn. Chapman and Hall/CRC Press, Boca Raton (2006)
45. Parthasarathy, K.R.: An Introduction to Quantum Stochastic Calculus. Springer, Basel (1992)

46. Puhalskii, A.: Large Deviations and Idempotent Probability. Chapman and Hall/CRC Press, Boca Raton (2001)
47. Schmeidler, D.: Integral representation without additivity. Proc. Am. Math. Soc. **97**, 255–261 (1986)
48. Schmeidler, D.: Subjective probability and expected utility without additivity. Econometrica **57**(3), 571–587 (1989)
49. Segal, W., Segal, I.E.: The Black-Scholes pricing formula in the quantum context. Proc. Natl. Acad. Sci. **95**, 4072–4075 (1998)
50. Shafer, G.: A Mathematical Theory of Evidence. Princeton University Press, Princeton (1976)
51. Shmueli, G.: To explain or TP predict. Stat. Sci. **25**(3), 289–310 (2010)
52. Soros, J.: The Alchemy of Finance: Reading of Mind of the Market. Wiley, New York (1987)
53. Sriboonchitta, S., Wong, W.K., Dhompongsa, S., Nguyen, H.T.: Stochastic Dominance and Applications to Finance, Risk and Economics. Chapman and Hall/CRC Press, Boca Raton (2010)
54. Von Neumann, J., Morgenstern, O.: The Theory of Games and Economic Behavior. Princeton University Press, Princeton (1944)
55. Wasserstein, R.L., Lazar, N.A.: The ASA's statement on p-values: context, process and purpose. Am. Stat. **70**, 129–133 (2016)
56. Walley, P.: Statistical Reasoning with Imprecise Probabilities. Chapman and Hall, London (1991)
57. Zadeh, L.A.: Fuzzy sets as a basis for a theory of possibility. J. Fuzzy Sets Syst. **1**, 3–28 (1978)

Everything Wrong with P-Values Under One Roof

William M. Briggs(✉)

340 E. 64th Apt 9A, New York, USA
matt@wmbriggs.com

Abstract. P-values should not be used. They have no justification under frequentist theory; they are pure acts of will. Arguments justifying p-values are fallacious. P-values are not used to make all decisions about a model, where in some cases judgment overrules p-values. There is no justification for this in frequentist theory. Hypothesis testing cannot identify cause. Models based on p-values are almost never verified against reality. P-values are never unique. They cause models to appear more real than reality. They lead to magical or ritualized thinking. They do not allow the proper use of decision making. And when p-values seem to work, they do so because they serve a loose proxies for predictive probabilities, which are proposed as the replacement for p-values.

Keywords: Causation · P-values · Hypothesis testing
Model selection · Model validation · Predictive probability

1 The Beginning of the End

It is past time for p-values to be retired. They do not do what is claimed, there are better alternatives, and their use has led to a pandemic of over-certainty. All these claims will be proved here.

Criticisms of p-values are as old as the measures themselves. None was better than Jerzy Neyman's original, however, who called decisions made conditional on p-values "acts of will"; see [1,2]. This criticism is fundamental: once the force of it is understood, as I hope readers agree, it is seen there is no justification for p-values.

Many are calling for an end to p-value-drive hypothesis testing. An important recent paper is [3] which concludes that given the many flaws with p-values "it is sensible to dispense with significance testing altogether." The book *The Cult of Statistical Significance* [4] has had some influence. The shift away from formal testing, and parameter-based inference, is also called for in [5].

There are scores of critical articles. Here is an incomplete, small, but representative list: [6–18]. The mood that was once uncritical is changing, best demonstrated by the critique by [19], which leads with the modified harsh words of Sir Thomas Beecham, "One should try everything in life except incest, folk

V. Kreinovich et al. (Eds.): ECONVN 2019, SCI 809, pp. 22–44, 2019.
https://doi.org/10.1007/978-3-030-04200-4_2

dancing and calculating a P-value." A particularly good resource of p-value criticisms is the web page "A Litany of Problems With p-values" compiled and routinely updated by Harrell [20].

Replacements, tweaks, manipulations have all been proposed to save p-values, such as lowering the magic number. Prominent among these is Benjamin et al. [21], who would divide the magic number by 10. There are many others suggestions which seek to put p-values in their "proper" but still respected place. Yet none of the proposed fixes solve the underlying problems with p-values, which I hope to demonstrate below.

Why are p-values used? To say something about a theory's or hypothesis's truth or goodness. But the relationship between a theory's truth and p-values is non-existent by design. Frequentist theory forbids speaking of the probability of a theory's truth. The connection between a theory's truth and Bayes factors is more natural, e.g. [22], but because Bayes factors focus on unobservable parameters, and rely just as often on "point nulls" as do p-values, they too exaggerate evidence for or against a theory. It is also unclear in both frequentist and Bayesian theory what *precisely* a hypothesis or theory is. The definition is usually taken to mean non-zero value of a parameter, but that parameter, attached to a certain measurable in a model (the "X"), does not say how the observable (the "Y") itself changes in any causal sense. It only says how our *uncertainty* in the observable changes. Probability theories and hypotheses, then, are epistemic and not ontic statements; i.e., they speak of our knowledge of the observable, given certain conditions, and not on what causes the observable.

This means probability models are only needed when causes are unknown (at least in some degree; there are rare exceptions). Though there is some disagreement on the topic, e.g. [23–25], there is no ability for a wholly statistical model to identify cause. Everybody agrees models can, and do, find correlations. And because correlations are not causes, hypothesis testing cannot find causes, nor does it claim to in theory. At best, hypothesis testing highlights possibly interesting relationships. So that finding a correlation is all a p-values or Bayes factor, of indeed any measure, can do. But correlations exist whether or not they are identified as "significant" by these measures. And that identification, as I show below, is rife with contradictions and fallacies. Accepting that, it appears the only solution is to move from purely a hypothesis testing (frequentist or Bayes) scheme to a predictive one in which the model claimed to be good or true or useful can be verified and tested against reality. See the latter chapters of [26] for a complete discussion of this.

Now every statistician knows about at least these limitations of p-values (and Bayes factors), and all agree with them to varying extent (most disputes are about the nature of cause, e.g. contrast [25,26]). But the "civilians" who use our tools do not share our caution. P-values, as we all know, work like magic for most civilians. This explains the overarching desire for p-value hacking and the like. The result is massive over-certainty and a much-lamented reproducibility crisis; e.g. see among many others [27,28]; see too [13].

The majority—which includes all users of statistical models, not just careful academics—treat p-values like ritual, e.g. [8]. If the p-value is less than the magic number, a theory has been proved, or taken to be proved, or almost proved. It does not matter that frequentist statistical theory insists that this is not so. It is what everybody believes. And the belief is impossible to eradicate. For that reason alone, it's time to retire p-values.

Some definitions are in order. I take probability to be everywhere conditional, and nowhere causal, in the same manner as [26,29–31]. Accepting this is not strictly necessary for understanding the predictive position, which is compared with hypothesis testing below, but understanding the conditional nature of all probability required is for a complete philosophical explanation. Predictive philosophy's emphasis on observables and measurable values which only inform uncertainty in observables is the biggest point of departure between hypothesis testing, which assumes probability is real and, at times, even causal.

Predictive probabilities make an apt, easy, and verifiable replacement for p-values; see [26,32] for fuller explanations. Predictive probability is demonstrated in the schematic equation:

$$\Pr(\mathrm{Y}|\text{new X}, \mathrm{DMA}), \tag{1}$$

where Y is the proposition of interest. For example, Y = "$y > 0$", Y = "yellow", Y = "$y < -1$ or $y > 1$ but not $y = 0$ if x_3 = 'Detroit'"; basically, Y is any proposition that can be asked (and answered!). D is the old data, i.e. prior measures X and the observable Y (where the dimension of all is clear from the context), both of which may have been measured or merely assumed. The model characterizing uncertainty in Y is M, usually parameterized, and A is a list of assumptions probative to M and Y. Everything thought about Y goes into A, even if it is not quantifiable. For instance, in A is information on the priors of the parameters, or *whatever* other information that is relevant to Y. The new X are those values of the measures that must be assumed or measured each time the probability of Y is computed. They are necessary because they are in D, and modeled in M.

A book could be written summarizing all of the literature for and against p-values. Here I tackle only the major arguments against p-values. The first arguments are those showing they have no or sketchy justification, that their use reflects, as Neyman originally said, acts of will; that their use is even fallacious. These will be less familiar to most readers. The second set of arguments assume the use of p-values, but show the severe limitations arising from that use. These are more common. Why p-values seem to work is also addressed. When they do seem to work it is because they are related to or proxies for the more natural predictive probabilities.

The emphasis in this paper is philosophical not mathematical. Technical mathematical arguments and formula, though valid and of interest, must always assume, tacitly or explicitly, a philosophy. If the philosophy on which a mathematical argument is based is shown to be in error, the "downstream" mathematical arguments supposing this philosophy are thus not independent evidence for or against p-values, and, whatever mathematical interest they may have, become irrelevant.

2 Arguments Against P-Values

2.1 Fisher's Argument

A version of an argument given first by Fisher appears in every introductory statistics book. The original argument is this, [33]:

> Belief in a null hypothesis as an accurate representation of the population sampled is confronted by a logical disjunction: Either the null hypothesis is false, or the p-value has attained by chance an exceptionally low value.

A logical disjunction would be a proposition of the type "Either it is raining or it is not raining." Both parts of the proposition relate to the state of rain. The proposition "Either it is raining or the soup is cold" is a disjunction, but not a logical one because the first part relates to rain and the second to soup. Fisher's "logical disjunction" is evidently not a logical disjunction because the first part relates to the state of the null hypothesis and the second to the p-value.

Fisher's argument can be made into a logical disjunction, however, by a simple fix. Restated: Either the null hypothesis is false and we see a small p-value, or the null hypothesis is true and we see a small p-value. Stated another way, "Either the null hypothesis is true or it is false, and we see a small p-value." The first clause of this proposition, "Either the null hypothesis is true or it is false", is a tautology, a necessary truth, which transforms the proposition to (loosely) "TRUE and we see a small p-value." Adding a logical tautology to a proposition does not change its truth value; it is like multiplying a simple algebraic equation by 1. So, in the end, Fisher's dictum boils down to: "We see a small p-value."

In other words, in Fisher's argument a small p-value has no bearing on any hypothesis (any hypothesis unrelated to the p-value itself, of course). Making a decision about a parameter or data because the p-value takes any particular value is thus always fallacious: it is not justified by Fisher's argument, which is a non sequitur. The decision made using p-values may be serendipitously correct, of course, as indeed any decision based on any criterion might be. Decisions made by researchers are often likely correct because experimenters are good at controlling their experiments, and because (as we will see) the p-value is a proxy for the predictive probability, but if the final decision is dependent on a p-value it is reached by a fallacy. It becomes a pure act of will.

2.2 All P-Values Support the Null?

Frequentist theory claims that, assuming the truth of the null, we can equally likely see any p-value whatsoever, i.e. the p-value under the null is uniformly distributed. That is, assuming the truth of the null, we deduce we can see any p-value between 0 and 1. It is thus asserted the following proposition is true:

$$\text{If the null is true, then } p \in (0, 1). \tag{2}$$

where the bounds may or may not be not sharp, depending on one's definition of probability.

We always do see any value between 0 and 1, and so it might seem that *any* p-value confirms the null. But it is not a formal argument to then say that the null is true, which would be the fallacy of affirming the consequent.

Assume the bounds on the p-value's possibilities are sharp, i.e. $p \in [0, 1]$. Now it is not possible to observe a p-value *except* in the interval $[0, 1]$. So that if the null hypothesis is judged true a fallacy of affirming the consequent is committed, and if the null is rejected, i.e. judged false, a non sequitur fallacy is committed. It does not follow from the premise (2) that any particular p-value confirms the falsity (or unlikelihood) of the null.

If the bounds were not sharp, and a p-value *not* in $(0, 1)$ was observed, then it would logically follow that the null would be false, from the classic modus tollens argument. That is, if either $p = 0$ or $p = 1$, which can occur in practice (given obvious trivial data sets), then it is not true that the null is true, which is to say, the null would be false. But that means an observed $p = 1$ would declare the null false! The only way to validly declare the null false, to repeat, would be if $p = 0$ or $p = 1$, but as mentioned, this doesn't happen except in trivial cases. Using any other value to reject the null does not follow, and thus any decision is again fallacious.

Other than those two extreme cases, then, any observed $p \in (0, 1)$ says nothing logically about the null hypothesis. At no point in frequentist theory is it proved that

$$\text{If the null is false, then } p \text{ is wee.} \tag{3}$$

Indeed, as just mentioned, all frequentist theory states is (2). Yet practice, and not theory, insists small p-value are evidence the null is false. Yet not quite "not false", but "not true". It is said the null "has not been falsified." This is because of Fisher's reliance on the then popular theory of Karl Popper that propositions could never be affirmed but only falsified; see [34] for a discussion of Popper's philosophy, which is now largely discredited among philosophers of science, e.g. [35].

2.3 Probability Goes Missing

Holmes [36] wrote "Data currently generated in the fields of ecology, medicine, climatology, and neuroscience often contain tens of thousands of measured variables. If special care is not taken, the complexity associated with statistical analysis of such data can lead to publication of results that prove to be irreproducible." These words every statistician will recognize as true. They are true because of the use of p-values and hypothesis testing.

Holmes defines the use of p-values in the following very useful and illuminating way:

> Statisticians are willing to pay "some chance of error to extract knowledge" (J.W. Tukey) using induction as follows.
> "If, given A $\implies$ B, then the existence of a small ϵ such that $P(B) < \epsilon$ tells us that A is probably not true."
> This translates into an inference which suggests that if we observe data X, which is very unlikely if A is true (written $P(\mathrm{X}|\mathrm{A}) < \epsilon$), then A is not plausible.

The last sentence had the following footnote: "We do not say here that the probability of A is low; as we will see in a standard frequentist setting, either A is true or not and fixed events do not have probabilities. In the Bayesian setting we would be able to state a probability for A."

We have just seen in (2) (A $\implies$ B in Holmes's notation) that because the probability of B (conditional on what?) is low, it most certainly does not tell us A is probably not true. Nevertheless, let us continue with this example.

In my notation, Holmes's statement translates to this:

$$\Pr\left(\mathrm{A}|\mathrm{X}\,\&\,\Pr(\mathrm{X}|\mathrm{A}) = \text{small}\right) = \text{small}. \tag{4}$$

This equation is equally fallacious. First, under the theory of frequentism the statement "fixed events do not have probabilities" is true. Under objective Bayes and logical probability anything can have a probability: under these systems, the probability of *any* proposition is always conditional on assumed premises. Yet every frequentist acts as if fixed events *do* have probabilities when they say things like "A is not plausible." *Not plausible* is a synonym for *not likely*, which is a synonym for *of low probability*. In other words, every time a frequentist uses a p-value, he makes a probability judgment, which is forbidden by the theory he claims to hold. In frequentist theory A has to believed or rejected with certainty. Any uncertainty in A, quantified or not, is, as Holmes said, forbidden.

Frequentists may believe, if they like, that singular events like A cannot have probabilities, but then they cannot, via a back door trick using imprecise language, give A a (non-quantified) probability after all. This is an inconsistency.

Let that pass and consider more closely (4). It helps to have an example. Let A be the theory "There is a six-sided object that when activated must show one of the six sides, just one of which is labeled 6." And, for fun, let X = "6 6s in a row." We are all tired of dice examples, but there is still some use in them (and here we do not have to envisage a real die, merely a device which takes one of six states). Given these facts, Pr(X|A) = small, where the value of "small" is much weer than the magic number (it's about 2×10^{-5}). We want

$$\Pr\left(\mathrm{A}|6\ 6\text{s on six-sided device}\,\&\,\Pr(6\ 6\text{s}|\mathrm{A}) = 2\times 10^{-5}\right) = ? \tag{5}$$

It should be obvious there is no (direct) answer to (5). That is, unless we magnify some implicit premise, or add new ones entirely.

The right-hand-side (the givens) tell us that if we accept A as true, then 6 6s are a possibility; and so when we see 6 6s, if anything, it is evidence in favor of A's truth. After all, something that A said could happen did happen. An implicit premise might be that in noticing we just rolled 6 6s in a row, there were other possibilities beside A we should consider. Another implicitly premise is that we notice we can't identify the precise *causes* of the 6s showing (this is just some mysterious device), but we understand the causes must be there and are, say, related to standard physics. These implicit premises can be used to infer A. But they cannot reject it.

We now come to the classic objection, which is that no alternative to A is given. A is the only thing going. Unless we add new implicit premises to (5) that give us a hint about something beside A. Whatever this premise is, it cannot be "Either A is true or something else is", because that is a tautology, and in logic adding a tautology to the premises changes nothing about the truth status of the conclusion.

Now if you told a frequentist that you were rejecting A because you just saw 6 6s in the row, because "another number is due", he'd probably (rightly) accuse you of falling prey to the gambler's fallacy. The gambler's fallacy can only be judged were we to add more information to the right hand side of (5). This is the key. *Everything* we are using as evidence for or against A goes on the right hand side of (5). Even if it is not written, it is there. This is often forgotten in the rush to make everything mathematical and quantitative.

In our case, to have any evidence of the gambler's fallacy would entail adding evidence to the RHS of (5) that is similar to "We're in a casino, where I'm sure they're careful about the dice, replacing worn and even 'lucky' ones; plus, the way they make you throw the dice make it next to impossible to physically control the outcome." That, of course, is only a small summary of a large thought. All evidence that points to A or away from it that we consider is there on the right hand side, even if it is, I stress again, not formalized.

For instance, suppose we're on 34th street in New York City at the famous Tannen's Magic Store and we've just seen the 6 6s, or even 20 6s, or however many you like, by some dice labeled "magic". What of the probability then? The RHS of (5) in *that* situation changes dramatically, adding possibilities other than A, by implicit premise.

In short, it is not the observations alone in (5) that get you anywhere. It is the extra information we add that does the trick, as it were. Most important of all—and this cannot be overstated—*whatever* is added to (5), then (5) *is no longer* (5), *but something else*! That is because (5) specifies all the information it needs. If we add to the right hand side, we change (5) into a new equation.

Once again it is shown there is no justification for p-values, except the appeal to authority which states wee p-values cause rejection.

2.4 An Infinity of Null Hypotheses

An ordinary regression model is written $\mu = \beta_0 x_1 + \cdots + \beta_0 x_p$, where μ is the central parameter of the normal distribution used to quantify uncertainty in the observable. Hypothesis tests help hone the eventual list of measures appearing on the right hand side. The point here is not about regression *per se*, but about all probability models; regression is a convenient, common, and easy example.

For every measure included in a model, an infinity of measures have been tacitly excluded, exclusions made without benefit of hypothesis tests. Suppose in a regression the observable is patient weight loss, and the measures the usual list of medical and demographic states. One potential measure is the preferred sock color of the third nearest neighbor from the patient's main residence. It is a silly measure because, we judge using outside common-sense knowledge, that this neighbor's sock color cannot have any causal bearing on our patient's weight loss. The point is not that nobody would add such a measure—nobody would—but that it could have been but was excluded without the use of hypothesis testing.

Sock color *could* have been measured and incorporated into the model. That it wasn't proves two things: (1) that inclusion and exclusion of measures in models can and are made without guidance of p-values and hypothesis tests, and (2) since there are an infinity of possible measures for every model, we always must make many judgments without p-values. There is no guidance in frequentist (or Bayesian) theory that says use p-values here, but use your judgment there. One man will insist on p-values for a certain X, and another will use judgment. Who is right? Why not use p-values everywhere? Or judgment everywhere? (The predictive method uses judgment aided by probability and decision.)

The only measures put into models are those which are at least suspected to be in the "causal path" of the observable. Measures which may, in part, be directly involved with the efficient and material cause of the observable are obvious, such as adding sex to medical observable models, because it is known differences in biological sex cause different things to happen to many observables. But those measures which might cause a change in the direct partial cause, or a change in the change and so on, like income in the weight loss model, also naturally find homes (income does not directly cause weight loss, but might cause changes which in turn cause others etc. which cause weight loss). Sock color belongs to this chain only if we can tell ourselves a just-so story of how this sock color can cause changes in other causes etc. of eventual causes of the observable. This can *always* be done: it only takes imagination.

The (initial) knowledge or surmise of material or efficient causes comes from *outside* the model, or the evidence of the model. Models begin with the assumption of measures included in the causal chain. A wee p-value does not, however, confirm a cause (or cause of a cause etc.) because non-causal correlations happen. Think of seeing a rabbit in a cloud. P-values, at best (see the Sect. 3 below) highlight large correlations.

It is also common that measures with small correlations, i.e. with large p-values, where there are known, or highly suspected, causal chains between the X and Y are not expunged from models; i.e. they are kept regardless what they p-value said. These are yet more cases where p-values are ignored.

The predictive approach is agnostic about cause: it accepts conditional hypotheses and surmises and outside knowledge of cause. The predictive approach simply says the best model is that which makes the best verified predictions.

2.5 Non-unique Adjustments

This criticism is similar to the infinity of hypotheses. P-values are often adjusted for multiple tests using methods like Bonferroni corrections. There are no corrections for those hypotheses rejected out of hand without the benefit of hypothesis tests.

Corrections are not used consistently. For instance, in model selection and in interim analyses, which is often informal. How many working statisticians have heard the request, "How much more data do I need to get significance?" It is, of course, except under the most controlled situations, impossible to police abuse. This is contrasted with the predictive method, which reports the model in a form which can be verified by (theoretically) anybody. So that even if abuse, such as confirmation bias, was used in building the model, it can still be checked. Confirmation bias using p-values is easier to hide. The predictive method does not assume a true model in the frequentist senses: instead, all models are conditional on the premises, evidence, and data assumed.

Harrell [20] says, "There remains controversy over the choice of 1-tailed vs. 2-tailed tests. The 2-tailed test can be thought of as a multiplicity penalty for being potentially excited about either a positive effect or a negative effect of a treatment. But few researchers want to bring evidence that a treatment harms patients... So when one computes the probability of obtaining an effect larger than that observed if there is no true effect, why do we too often ignore the sign of the effect and compute the (2-tailed) p-value?"

The answer is habit married to the fecundity of two-tailed tests at producing wee p-values.

2.6 P-Values Cannot Identify Cause

Often when a wee p-value is seen in accord with some hypothesis, it will be taken as implying that the cause, or one of the causes, of the observable has been verified. But p-values cannot identify cause; see [37] for a full discussion. This is because parameters inside probability models are not (or almost never) representations of cause, thus any decision based upon parameters cannot confirm nor deny any cause. Regression model parameters in particular are not representations of cause.

It helps to have a semi-fictional example. Third-hand smoking, which is not fictional [38], is when items touched by second-hand smokers, who have touched things by first-hand smokers, are in turn touched by others, who become "third-hand smokers". There is no reason this chain cannot be continued indefinitely. One gathers data from x-hand smokers (which are down the touched-smoke chain somewhere) and non-x-hand smokers and the presence or absence of a list of maladies. If in some parameterized model relating these a wee p-value is found for one of the maladies, x-hand smoking will be said to have been "linked to" the malady. This "linked to" only means a "statistically significant result" was found, which in turn only means wee p-value was seen.

Those keen on promoting x-hand smoking as causing the disease will take the "linked to" as statistical validation of cause. Careful statisticians won't, but stopping the causal interpretation from being used is by now an impossible task. This is especially so when even statisticians use "linked to" without carefully defining it.

Now if x-hand smoking caused the particular disease, then it would always do so, and statistical testing would scarcely be needed to ascertain this because each individual exposed to the cause would be always contract the disease—unless the cause were blocked. What blocks this cause could be various, such as a person's particular genetic makeup, or state of hand calluses (to block absorption of x-hand smoke), or whether a certain vegetable was eaten (that somehow cancels out the effect of x-hand smoke), and so on. If these blocking causes were known (the blocks are also causes), again statistical models would not be needed, because all we would need know is whether any x-hand-smoke-exposed individual had the relevant blocking mechanism. Each individual would get the disease for certain unless he had (for certain) a block.

Notice that (and also see below the criticism that p-values are not always believed) models are only tested when the causes or blocks are not known. If causes were known, then models would not be needed. In many physical cases, cause or block can be demonstrated by "bench" science, and then the cause or block becomes known with certainty. It may not be known how this cause or block interacts or behaves in the face of multiple other potential causes or blocks, of course. Statistical models can be used to help quantify this kind of uncertainty, given appropriate experiments. But then this cause or block would not be added or expunged from a model regardless of the size of its p-value.

It can be claimed hypothesis tests are only used where causes or blocks are unknown, but testing cannot confirm unknown causes or blocks.

2.7 P-Values Aren't Verified

One reason for the reproducibility crisis is the presumed finality of p-values. Once a "link" has been "validated" with a wee p-value, it is taken by most to mean the "link" definitely exists. This thinking is enforced since frequentist theory forbids assigning a probability measure to any "link's" veracity. The wee-p-confirmed "link" enters the vocabulary of the field. This thinking is especially rife in purely statistically driven fields, like sociology, education, and so forth, where direct experimentation to identify cause is difficult or impossible.

Given the ease of finding wee p-values, it is no surprise that popular theories are not re-validated when in rare instances they are attempted to be replicated. And then not every finding can be replicated at least because of the immense cost and time involved. So, many spurious "links" are taken as true or causal.

Using Bayes factors, or adjusting the magic number lower, would not solve the inherent problem. Only verifying models can, i.e. testing them against reality. When a civil engineer proposes a new theory for bridge construction, testing via simulation and incorporating outside causal knowledge provides guidance whether the new bridge built using the theory will stand or fall. But even given a positive judgment from this process does not mean the new bridge will stand. The only way to know with any certainty is to build the bridge and see. And, as readers will know, not every new bridge does stand. Even the best considered models fail.

What is true for bridges is true for probability models. P-value-based models are never verified against reality using new, never before seen or used in any way data. The predictive approach makes predictions that can, and must, be verified. Whatever measures are assumed results in probabilistic predictions about the observable. These predictions can be checked in theory by anybody, even without having the data which built the model, in the same way even a novice driver can understand whether the bridge under him is collapsing or not. How verification is done is explained elsewhere. e.g. [26,32,39–41].

A change in practice is needed. Models should only be taken as preliminary and unproved until they can be verified using outside, never-before-seen or used data. Every paper which uses statistical results should announce "This model has not yet been verified using outside data and is therefore unproven." The practice of printing wee p-values, announcing "links", and then moving on to the next model must end. This would move statistics into the realm of the harder sciences, like physics and chemistry, which take pains to verify all proposed models.

2.8 P-Values Are Not Unique

We now begin the more familiar arguments against p-values, with some added insight. As all know, the p-value is never unique, and is dependent on *ad hoc* statistics. Statistics themselves are not unique. The models on which the statistics are computed are, with very rare exceptions in practice, also *ad hoc*; thus, they are not unique. The rare exceptions are when the model is deduced from first principles, and are therefore parameter-free, obviating the need for hypothesis testing. The simplest examples of fully deduced models are found in introductory probability books. Think of dice or urn examples. But then nobody suggests using p-values on these models.

If in any parameterized model the resulting p-value is not wee, or otherwise has not met the criteria for publishing, then different statistics can be sought to remedy the "problem." An amusing case found its way into the *Wall Street Journal*, [42].

The paper reported that Boston Scientific (BS) introduced a new stent called the Taxus Liberte. The company did the proper experiments and analyzed their data using a Wald test. This give them a p-value that was just under the magic number, a result which is looked upon with favor by the Food and Drug Administration. But a competitor charged that the Wald statistic is not one they would have used. So they hired their own statistician to reevaluate their rival's data. This statisticians computed p-values for several other statistics and discovered each of these were a fraction larger than the magic number. This is when the lawyers entered the story, and where we exit it.

Now the critique that the model and statistic is not unique must be qualified. Under frequentism, probability is said to exist unconditionally; which is to say, the moment a parameterized model is written—somehow, somewhere—at "the limit" the "actual" or "true" probability is created. This theory is believed even though alternate parameterized models for the same observable may be created, which in turn create their own "true" values of parameters. All rival models and parameters are thus "true" (at the limit), which is a contradiction. This is further confused if probability is believed to be ontic, i.e. actually existing as apples or pencils exist. It would seem that rival models battle over probability somehow, picking one which is the truly true or really true model (at the limit).

Contrast this with the predictive approach, which accepts all probability is conditional. Probability at the limit may never need be referenced. All is allowed to remain finite (asymptotics can of course be used as convenient approximations). Changing any assumptions changes the model by definition, and all probability is epistemic. Different people using different models, or even using the same models, would come to different conclusions quite naturally.

2.9 The Deadly Sin of Reification

If in some collection of data a difference in means between two groups is seen, this difference is certain (assuming no calculation mistakes). We do not need to do any tests to verify whether the difference is real. It was seen: it is real. Indeed, any question that can be asked of the observed data can be answered with a simple yes or no. Probability models are not needed.

Hypothesis testing acknowledges the observed difference, but then asks whether this difference is "really real". If the p-value is wee, it is; if not, the observed real difference is declared not really real. It will even be announced (by most) "No difference was found", a very odd thing to say. If it does not sound odd to your ears, it shows how successful frequentist theory is. The attitude that actual difference is not really real comes from assuming probability is ontic, that we have only sampled from an infinite reality where the model itself is larger and realer than the observed data. The model is said to have "generated" the value in some vague way, where the notion of the causal means by which the model does this forever recedes into the distance the more it is pursued. The model is reified. It becomes better than reality.

The predictive method is, as said, agnostic about cause. It takes the observed difference as real and given and then calculates the chance that such differences will be seen in *new* observations. Predictive models can certainly err and can be fooled by spurious correlations just as frequentist ones can (though far less frequently). But the predictive model asks to be verified: if it says differences will persist, this can be checked. Hypothesis tests *declare* they will be seen (or not), end of story.

If the difference is observed but the p-value not wee, it is declared that chance or randomness *caused* the observed difference; other verbiage is to say the observed difference is "due to" chance, etc. This is causal language, but it is false. Chance and randomness do not exist. They are purely epistemic. They therefore cannot cause anything. Some thing or things caused the observed difference. But it cannot have been chance. The reification of chance comes, I believe, from the reluctance of researchers to say, "I have no idea what happened."

If *all*—and I mean this word in its strictest sense—we allow is X as the potential cause (or in the causal path) of an observed difference, then we must accept that X is the cause regardless of what a p-value says to do with X (usually, of course, the parameter associated with X). We can say "Either X is the cause or something else is", but this will always be true, even in the face of knowledge X is not a cause. This argument is only to reinforce the idea that knowledge of cause must come from outside the probability model. Also that chance is never a cause. And that any probability model that gives non-extreme predictive probabilities is always an admission that we do not know all the causes of the observable. This is true (and for chance and randomness, too) even for quantum mechanical observations, the discussion of which would take us too far afield here. But see [26], Chap. 5 for a discussion.

2.10 P-Values Are Magic

Every working statistician will have a client who has been reduced to grief after receiving the awful news that the p-value for their hypothesis was larger than the magic number, and therefore unpublishable. "What can we do to make it smaller?" ask many clients (I have had this happen many times). All statisticians know the tricks to oblige this request. Some do oblige.

Gigerenzer [8] calls p-value hunting a ritualized approach to doing science. As long as the proper (dare we say magic) formulas are used and the p-values are wee, science is said to have been done. Yet is there any practical, scientific difference between a p-value of 0.49 and 0.051? Are the resulting post-model decisions made always so finely tuned and hair-breadth crucial that the tiny step between 0.49 and 0.51 throws everything off balance? Most scientists, and all statisticians, will say no. But most will act as if the answer is yes. A wee p-value is mesmerizing.

The counter-argument to abandoning p-values in the fact of this criticism is better education. But that education would have to overcome decades of beliefs and actions that the magic number is in fact magic. The word preferred is not *magic*, of course, but *significant*. Anyway, this educational initiative would have to cleanse all books and material that bolsters this belief, which is not possible.

2.11 P-Values Are Not Believed When Convenient

In any given set of data, with some parameterized model, its p-value are assumed true, and thus the decisions based upon them sound. Theory insists on this. The decisions "work", whether the p-value is wee or not wee.

Suppose a wee p-value. The null is rejected, and the "link" between the measure and the observable is taken as proved, or supported, or believable, or whatever it is "significance" means. We are then directed to act as if the hypothesis is true. Thus if it is shown that per capita cheese consumption and the number of people who died tangled in their bed sheets are "linked" via a wee p, we are to believe this. And we are to believe all of the links found at the humorous web site Spurious Correlations, [43].

I should note that we can either accept that grief of loved ones strangulated in their beds drives increased cheese eating, *or* that cheese eating causes sheet strangulation. This is joke, but also a valid criticism. The direction of causal link is not mandated by the p-value, which is odd. That means the direction comes from outside the hypothesis test itself. Direction is thus (always) a form of prior information. But prior information like this is forbidden in frequentist theory. Everybody dismisses, as they should, these spurious correlations, but they do so using prior information. They are thus violating frequentist theory.

Suppose next a non-wee p-value. The null has been "accepted" in any practical sense. There is the idea, started by Fisher, that if the p-value was not wee that one should collect more data, and that the null is not accepted but that we have failed to reject it. Collecting more data will lead to a wee p-value eventually, even when the correlations are spurious (this is a formal criticism, given below). Fisher did not have in mind spurious correlations, but genuine effects, where he took it the parameter represented something real in the causal chain of the observable. But this is a form of prior information, which is forbidden because it is independent (I use this word in its philosophical not mathematical sense) of the p-value. The p-value then becomes a self-fulfilling prophecy. It must be, because we started by declaring the effect was real. This practice does not make any finding false, as Cohen pointed out [9]. But if we knew the effect was real before the p-value was calculated, we know it even after. And we reject the p-values that do not conform to our prior knowledge. This, again, goes against frequentist theory.

2.12 P-Values Base Decisions on What Did Not Occur

P-values calculate the probability of what did not happen on the assumption that what did not happen should be rare. As Jefferys [44] famously said: "What the use of P[-value] implies, therefore, is that a hypothesis that may be true may be rejected because it has not predicted observable results that have not occurred."

Decisions should instead be conditioned of what did happen and on uncertainty in the observable itself, and not on parameters (or functions of them) inside models.

2.13 P-Values Are Not Decisions

If the p-value is wee, a decision is made to reject the null hypothesis, and vice versa (ignoring the verbiage "fail to reject"). Yet the consequences of this decision are not quantified using the p-value. The decision to reject is just the same, and therefore just as consequential, for a p-value of 0.05 as one of 0.0005. Some have the habit of calling especially wee p-values as "highly significant", and so forth, but this does not accord with frequentist theory, and is in fact forbidden by that theory because it seeks a way around the proscription of applying probability to hypotheses. The p-value, as frequentist theory admits, is not related in any way to the probability the null is true or false. Therefore the size of the p-value does not matter. Any level chosen as "significant" is, as proved above, an act of will.

A consequence of the frequentist idea that probability is ontic and that true models exist (at the limit) is the idea that the decision to reject or accept some hypothesis should be the same for all. Steve Goodman calls this idea "naive inductivism", which is "a belief that all scientists seeing the same data should come to the same conclusions," [45]. That this is false should be obvious enough. Two men do not always make the same bets even when the probabilities are deduced from first principles, and are therefore true. We should not expect all to come to agreement on believing a hypothesis based on tests concocted from *ad hoc* models. This is true, and even stronger, in a predictive sense, where conditionality is insisted upon.

Two (or more) people can come to completely different predictions, and therefore difference decisions, even when using the same data. Incorporating decision in the face of uncertainty implied by models is only partly understood. New efforts along these lines using quantum probability calculus, especially in economic decisions, are bound to pay off, see e.g. [46].

A striking and in-depth example of how using the same model and same data can lead people to *opposite* beliefs and decisions is given by Jaynes in his chapter "Queer uses for probability theory", [30].

2.14 No One Remembers the Definition of P-Values

The p-value is (usually) the conditional probability an *ad hoc* test statistic being larger (in absolute value) than the observed statistic, assuming the null hypothesis is true, given the values of the observed data, and assuming the truth of the model. The probability of exceeding the test statistic assuming the alternate hypothesis is true, or given the null hypothesis is false, given the other conditions, is not known. Nor is the second-most important probability known: whether or not the null hypothesis is true.

It is the second-most important probability because most null hypotheses are "point nulls", because continuous parameters take fixed single values, which because parameters live on the continuum, "points" have a probability of 0. The most important probability, or rather probabilities, is that of Y given X, and Y given X's absence, where it is assumed (as with p-values) X is part of the model. This is a direct measure of relevance of X. If the conditional probability of Y given X (in the model) is a, and the probability of Y given X's absence is also a, then X is irrelevant, conditional on the model and other information listed in (1). If X is relevant, the difference in probabilities because a matter of individual decision, not a mandated universal judgment, as with p-values.

Now frequentists do not accept the criticism of the point null having zero probability, because according to frequentist theory parameters (the uncertainty in them) do not have probabilities. Again, once any model is written, parameters come into existence (somehow) as some sort of Platonic form at the limit. They take "true" values there; it is inappropriate in the theory to use probability to express uncertainty in their unknown values. Why? It is not, after all, thought wrong to express uncertainty in unknown observables using probability. The restriction to probability only on observables has no satisfactory explanation: the difference just exists by declaration. See [47–49] for these and other unanswerable criticisms of frequentist theories (including those in the following paragraphs) well known to philosophers, but somehow more-or-less unknown to statisticians.

Rival models, i.e. those with different parameterizations (Normal versus Weibull model, say) somehow create parameters, too, which are also "true". Which set of parameters are the truest? Are all equally true? Or are all models merely crude approximations to *the* true model which nobody knows or can know? Frequentists might point to central limit theorems to answer these questions, but it is not the case all rival models converge to the same limit, so the problem is not solved.

Here is one of a myriad of examples showing failing memories, from a paper whose intent is to teach proper p-value use: [50] says, "The p value is the probability to obtain an effect equal to or more extreme than the one observed presuming the null hypothesis of no effect is true; it gives researchers a measure of the strength of evidence against the null hypothesis."

The p-value is mute on the size of an effect (and also on what an effect is; see above). And though it is widely believed, this conclusion is false, accepting the frequentist theory in which p-values are embedded. "Strength" is not a measure of probability, so just what is it? It is never defined formally inside frequentist theory. The discussion below on why p-values sometimes seem to work is relevant here.

2.15 Increasing the Sample Size Lowers P-Values

Large and increasing sample sizes show low and lowering p-values. Even small differences become "significant" eventually. This is so well known there are routine discussions warning people to, for instance, not conflate clinical versus statistical "significance", e.g. [51]. What is statistical significance? A wee p-value. And what is a wee p-value? Statistical significance.

Suppose the uncertainty in some observable y_0 in a group 0 is characterized by a normal distribution with parameters $\theta_0 = a$ and with a σ also known; and suppose the same for the observable y_1 in a group 1, but with $\theta_1 = a + 0.00001$. The groups represent, say, the systolic blood pressure measures of people who live on the same block but with even (group 0) and odd (group 1) street addresses. We are in this case *certain* of the values of the parameters. Obviously, $\theta_1 - \theta_0 = 0.00001$ with certainty. P-values are only calculated with observed measures, and here there are none, but since there is a certain difference, we would expect the "theoretical" p-value to be precisely 0. As it would be for *any* sized difference in the θs.

This by itself is not especially interesting, except that it confirms low p-values can be found for small differences, which here flows from the knowledge of the true difference in the parameters. The p-value would (or should) in these cases always be "significant".

Now a tradition has developed to call the difference in parameters the "effect size", borrowing language used by physicists. In physics (and similar fields) parameters are often written as direct or proxy causes and can then be taken as effects. This isn't the case for the vast, vast majority of statistical models. Parameters are not ontic or causal effects. They represent only changes in our epistemic knowledge.

This is a small critique, but the use of p-values, since they are parameter-centric, encourages this false view of effect. Parameter-focused analyses of any kind always exaggerates the certainty we have in any measure and its epistemic influence on the observable. We can have absolute certainty of parameter values, as in the example just given, but that does not translate into large differences in the probability of new differences in the observable. If that example, $\Pr(\theta_1 > \theta_0|\mathrm{DMA}) = 1$, but for most scenarios $\Pr(\mathrm{Y}_1 > \mathrm{Y}_0|\mathrm{DMA}) \approx 0.5$. That means frequentist point estimates bolstered by wee p-values, or Bayesians parameter posteriors, all exaggerate evidence. Given that nearly all analyses are parameter-centric, we do not only have a reproducibility crisis, we have an over-certainty crisis.

2.16 It Ain't Easy

Tests for complicated decisions do not always exist; the further we venture from simple models and hypotheses, the more this is true. For instance, how to test whether groups 3 or 4 exceed some values but not group 1 when there is indifference about group 2, and where the values depend in some way on the state of other measures (say, these other measures being in some range)?

This is no problem at all for predictive statistics. Any question that can be conceived, and can theoretically be measured, can be formulated in probability in a predictive model.

P-values also make life too easy for modelers. Data is "submitted" to software (a not uncommon phrase), and if wee p-values are found, after suitable tweaking, everybody believes their job is done. I don't mean that researchers don't call for "future work", which they will always do, but the belief that the model has been sufficiently proved. That the model just proposed for, say, this small set of people existing in one location for a small time out of history, and having certain attributes, somehow then applies to all people everywhere. This is not *per se* a p-value criticism, but p-values do make this kind of thinking easy.

2.17 The P-Value for What?

Neyman fixed "test level", which is practically identical with p-values fixed at the magic number, are for tests on the whole, and not for the test at hand, which is itself in no way guaranteed to have a Type I or even Type II error level. These numbers (whatever they might mean) apply to infinite sets of tests. And we haven't got there yet.

2.18 Frequentists Become Secret Bayesians

That is because people argue: For most small p-values I have seen in the past, I believe the null has been false (and vice versa); I now see a new small p-value, therefore the null hypothesis in this new problem is likely false. That argument works, but it has no place in frequentist theory (which anyway has innumerable other difficulties). It is the Bayesian-like interpretation. Newman's method is to accept with finality the decisions of the tests as certainty. But people, even ardent frequentists, cannot help but put probability, even if unquantified, on the truth value of hypotheses. They may believe that by omitting the quantification and only speaking of the truth of the hypothesis as "likely", "probable" or other like words, that they have not violated frequentist theory. If you don't write it down as math, it doesn't count! This is, of course, false.

3 If P-Values Are Bad, Why Do They Sometimes Work?

3.1 P-Values Can Be Approximations to Predictive Probability

Perhaps the most-used statistic is the t (and I make this statement without benefit of a formal hypothesis test, you notice, and you understood it without one, too), which is in its numerator the mean of one measure minus the mean of a second. The more the means of measures under different groups differ, the smaller the p-value will in general be, with the caveats about standard deviations and sample sizes understood.

Now consider the objective Bayesian or logical probability interpretation of the same observations, taken in a predictive sense. The probability the measure with the larger observed mean exhibits in new data larger values than the measure with the smaller mean increases the larger t is (with similar caveats). That is, loosely,

$$\text{As } t \to \infty, \quad \Pr(Y_2 > Y_1 | \text{DMA}, t) \to 1, \tag{6}$$

where D is the old data, M is a parameterized model with its host of assumptions (such as about the priors) A, and t the t-statistic for the two groups Y_2 and Y_1, assuming the group 2 has the larger observed mean. As t increases, so does in general the probability Y_2 will be larger than Y_1, again with the caveats understood (most models will converge not to 1, but to some number larger than 0.5 less than 1). Since this is a predictive interpretation, the parameters have been "integrated out." (In the observed data, it will be *certain* if the mean of one group was larger than the other.) This is an abuse of notation, since t is derived from D. It is also a cartoon equation meant only to convey a general idea; it is, as is obvious enough, true in the normal case (assuming finite variance and conjugate or flat priors).

What (6) says is that the p-value in this sense is a proxy for the predictive probability. And it's the predictive probability all want, since again there is no uncertainty in the past data. When p-values work, they do so because they are representing reasonable predictions about future values of the observables.

This is only rough because those caveats become important. Small p-values, as mentioned above, are had just by increasing sample size. With a fixed standard deviation, and miniscule difference between observed means, a small p-value can be got by increasing the sample size, but the probability the observables differ won't budge much beyond 0.5.

Taking these caveats into consideration, why not use p-values, since they, at least in the case of t- and other similar statistics, can do a reasonable job approximating the magnitude of the predictive probability? The answer is obvious: since it's easy to get, and it is what is desired, calculate the predictive probability instead of the p-value. Even better, with predictive probabilities none of the caveats must be worried about: they take care of themselves in the modeling. There will be no need of any discussions about clinical versus statistical significance. Wee p-values can lead to small or large predictive probability differences. And all we need are the predictive probability differences.

The interpretation of predictive probabilities is also natural and easy to grasp, a condition which is certainly false with p-values. If you tell a civilian, "Given the experiment, the probability your blood pressure will be lower if you take this new drug rather than the old is 70%", he'll understand you. But if you tell him that if the experiment were repeated an infinite number of times, and if we assume the new drug is no different than the old, then a certain test statistic in each of these infinite experiments will be larger than the one observed in the experiment 5% of the time, he won't understand you.

Decisions are easier and more natural—and verifiable—using predictive probability.

3.2 Natural Appeal of Some P-Values

There is a natural and understandable appeal to some p-values. An example is in tests of psychic abilities, [52]. An experiment will be designed, say guessing numbers from 1 to 100. On the hypothesis that no psychic ability is present, and the only information the would-be psychic has is that the numbers will be in a certain set, and where knowledge of successive numbers is irrelevant (each time it's 1–100, and it's not numbered balls in urns), then the probability of guessing correctly can be deduced as 0.01. The would-be psychic will be asked to guess more than once, and his total correct out of n is his score.

Suppose conditional on this information the probability of the would-be psychic's score assuming he is only guessing is some small number, say, much lower than the magic number. The lower this probability is, the more likely, it is thought, of the fellow having genuine psychic powers. Interestingly, a probability at or near the magic number in psychic would be taken by no one as conclusive evidence. The reason is that cheating and sloppy and misleading experiments are far from unknown. But those suspicions, while true, do not accord with p-value theory, which has no way to incorporate anything but quantifiable hypotheses (see the discussion above about incorporating prior information).

But never mind that. Let's assume no cheating. This probability of the score assuming guessing, or the probability of scores at least as large as the one observed, functions as a p-value. Wee ones are taken as indicating psychic ability, or at least as indicating psychic ability is likely. Saying ability is "likely" is forbidden under frequentist theory, as discussed above, so when people do this they are acting as predictivists. Nor can we say the small p-value confirms psychic powers are the cause of the results. Nor chance.

So what do the scores mean? Same thing batting averages do in baseball. Nobody bats a thousand, nor do we expect psychics to guess correctly 100% of the time. Abilities differ. Now a high batting average, say from Spring Training, is taken as a predictive of a high batting average in the regular season. This often does not happen—the prediction does not verify—and when it doesn't Spring Training is taken as a fluke. The excellent performance during Spring Training will be put down to a variety of causes. One of these won't be good hitting ability.

A would-be psychic's high score is the same thing. Looks good. Something caused the hits. What? Could have been genuine ability. Let's get to the big leagues and really put him to the test. Let magicians watch him. If the would-be psychic doesn't make it there, and so far none have, then the prior performance just like in baseball will be ascribed to any number of causes, one of which may be cheating.

In other words, even when a p-value seems natural, it is again a proxy for a predictive probability or an estimate of ability assuming cause (but not proving it).

4 What Are the Odds of That?

As should be clear, many of the arguments used against p-values could for the most part also be used against Bayes factors. This is especially so if probability is taken as subjective (where a bad burrito can shift probabilities in any direction), where the notion of cause becomes murky. Many of the arguments against p-values can also be marshaled against using point (parameter) estimation. As said, parameter-based analyses exaggerates evidence, often to extent that is surprising, especially if one is unfamiliar with predictive output. Parameters are too often reified as "the" effects, when all they are, in nearly all probability models, are expressions of uncertainty in how the measure X affects the uncertainty in the observable Y. Why not then speak directly of the how changes in X, and not in some *ad hoc* uninteresting parameter, relate to changes in the uncertainty of Y? About the mechanics of how to decide which X are relevant and important in a model, I leave to other sources, as mentioned above.

People often quip, when seeing something curious, "What are the odds of that?" The probability of any observed thing is 1, conditional on its occurence. It happened. There is therefore no need to discuss its probability—*unless* one wanted to make predictions of future possibilities. Then the conditions on which the curious thing are stated dictate the probability. Different people can come to different conditions, and therefore come to different probabilities. As often happens. This isn't so with frequentist theory, which must embed every event in some unique not-debatable infinite sequence in which, at the limit, probability becomes real and unchangeable. But nothing is actually infinite, only potentially infinite. It is these fundamental differences in philosophy that drive many of the criticisms of p-values, and therefore of frequentism itself. Most statisticians will not have read these arguments, given by authors like Hájek [47,49], Franklin [29,53], and Stove [54] (the second half of this reference). They are therefore urged to review them. The reader does not now have to believe frequentism is false, as these authors argue, to grasp the arguments against p-values above. But if frequentism is false, then p-values are ruled moot *tout court.*

A common refrain in the face of criticisms like these is to urge caution. "Use p-values wisely," it will be said, or use them "in the proper way." But there is no wise or proper use of p-values. They are not justified in any instance.

Some think p-values are justified by simulations which purport to show p-values behave as expected when probabilities are known. But those who make those arguments forget that there is nothing in a simulation that was not first put there. All simulations are self-fulfilling. The simulation said, in some lengthy path, that the p-value should look like this, and, lo, it did. There is also, in most cases, reification of probability in these simulations. Probability is taken as real, ontic. When all simulations do is manipulate known formulas given known and fully expected input. That it, simulations begin by stating that given an input u produce via this long path p. Except that semi-blind eyes are turned to u, which makes it "random", and therefore makes p ontic. This is magical thinking. I do not expect readers to be convinced by this telegraphic and wholly unfamiliar argument, given how common simulations are, so see Chap. 5 in [26] for a full explication. This argument will seem more shocking the more one is convinced probability is real.

Predictive probability takes the model not as true or real as in hypothesis testing, but as the best summary of knowledge available to the modeler (some models can be deduced from first principles, and thus have no parameters, and are thus true). Statements made about the model are therefore more naturally cautious. Predictive probability is no panacea. People can cheat and fool themselves just as easily as before, but the exposure of the model in a form that can be checked by anybody will propel and enhance caution. P-value-based models say 'Here is the result, which you must accept.' Rather, that is what theory directs. Actual interpretation often departs from theory dogma, which is yet another reason to abandon p-values.

Future work is not needed. The totality of all arguments insists that p-values should be retired immediately.

References

1. Neyman, J.: Philos. Trans. R. Soc. Lond. A **236**, 333 (1937)
2. Lehman, E.: Jerzy Neyman, 1894–1981. Technical report, Department of Statistics, Berkeley (1988)

3. Trafimow, D., Amrhein, V., Areshenkoff, C.N., Barrera-Causil, C.J., Beh, E.J., Bilgiç, Y.K., Bono, R., Bradley, M.T., Briggs, W.M., Cepeda-Freyre, H.A., Chaigneau, S.E., Ciocca, D.R., Correa, J.C., Cousineau, D., de Boer, M.R., Dhar, S.S., Dolgov, I., Gómez-Benito, J., Grendar, M., Grice, J.W., Guerrero-Gimenez, M.E., Gutiérrez, A., Huedo-Medina, T.B., Jaffe, K., Janyan, A., Karimnezhad, A., Korner-Nievergelt, F., Kosugi, K., Lachmair, M., Ledesma, R.D., Limongi, R., Liuzza, M.T., Lombardo, R., Marks, M.J., Meinlschmidt, G., Nalborczyk, L., Nguyen, H.T., Ospina, R., Perezgonzalez, J.D., Pfister, R., Rahona, J.J., Rodríguez-Medina, D.A., Romão, X., Ruiz-Fernández, S., Suarez, I., Tegethoff, M., Tejo, M., van de Schoot, R., Vankov, I.I., Velasco-Forero, S., Wang, T., Yamada, Y., Zoppino, F.C.M., Marmolejo-Ramos, F.: Front. Psychol. **9**, 699 (2018). https://doi.org/10.3389/fpsyg.2018.00699
4. Ziliak, S.T., McCloskey, D.N.: The Cult of Statistical Significance. University of Michigan Press, Ann Arbor (2008)
5. Greenland, S.: Am. J. Epidemiol. **186**, 639 (2017)
6. McShane, B.B., Gal, D., Gelman, A., Robert, C., Tackett, J.L.: The American Statistician (2018, forthcoming)
7. Berger, J.O., Selke, T.: JASA **33**, 112 (1987)
8. Gigerenzer, G.: J. Socio-Econ. **33**, 587 (2004)
9. Cohen, J.: Am. Psychol. **49**, 997 (1994)
10. Trafimow, D.: Philos. Psychol. **30**(4), 411 (2017)
11. Nguyen, H.T.: Integrated Uncertainty in Knowledge Modelling and Decision Making, pp. 3–15. Springer (2016)
12. Trafimow, D., Marks, M.: Basic Appl. Soc. Psychol. **37**(1), 1 (2015)
13. Nosek, B.A., Alter, G., Banks, G.C., et al.: Science **349**, 1422 (2015)
14. Ioannidis, J.P.: PLoS Med. **2**(8), e124 (2005)
15. Nuzzo, R.: Nature **526**, 182 (2015)
16. Colquhoun, D.: R. Soc. Open Sci. **1**, 1 (2014)
17. Greenland, S., Senn, S.J., Rothman, K.J., Carlin, J.B., Poole, C., Goodman, S.N., Altman, D.G.: Eur. J. Epidemiol. **31**(4), 337 (2016). https://doi.org/10.1007/s10654-016-0149-3
18. Greenwald, A.G.: Psychol. Bull. **82**(1), 1 (1975)
19. Hochhaus, R.G.A., Zhang, M.: Leukemia **30**, 1965 (2016)
20. Harrell, F.: A litany of problems with p-values (2018). http://www.fharrell.com/post/pval-litany/
21. Benjamin, D., Berger, J., Johannesson, M., Nosek, B., Wagenmakers, E., Berk, R., et al.: Nat. Hum. Behav. **2**, 6 (2018)
22. Mulder, J., Wagenmakers, E.J.: J. Math. Psychol. **72**, 1 (2016)
23. Hitchcock, C.: The Stanford Encyclopedia of Philosophy (Winter 2016 Edition) (2016). https://plato.stanford.edu/archives/win2016/entries/causation--probabilistic
24. Breiman, L.: Stat. Sci. **16**(3), 199 (2001)
25. Pearl, J.: Causality: Models, Reasoning, and Inference. Cambridge University Press, Cambridge (2000)
26. Briggs, W.M.: Uncertainty: The Soul of Probability, Modeling & Statistics. Springer, New York (2016)
27. Nuzzo, R.: Nature **506**, 50 (2014)
28. Begley, C.G., Ioannidis, J.P.: Circ. Res. **116**, 116 (2015)
29. Franklin, J.: Erkenntnis **55**, 277 (2001)
30. Jaynes, E.T.: Probability Theory: The Logic of Science. Cambridge University Press, Cambridge (2003)

31. Keynes, J.M.: A Treatise on Probability. Dover Phoenix Editions, Mineola (2004)
32. Briggs, W.M., Nguyen, H.T., Trafimow, D.: Structural Changes and Their Econometric Modeling. Springer (2019, forthcoming)
33. Fisher, R.: Statistical Methods for Research Workers, 14th edn. Oliver and Boyd, Edinburgh (1970)
34. Briggs, W.M.: arxiv.org/pdf/math.GM/0610859 (2006)
35. Stove, D.: Popper and After: Four Modern Irrationalists. Pergamon Press, Oxford (1982)
36. Holmes, S.: Bull. Am. Math. Soc. **55**, 31 (2018)
37. Briggs, W.M.: arxiv.org/abs/1507.07244 (2015)
38. Protano, C., Vitali, M.: Environ. Health Perspect. **119**, a422 (2011)
39. Briggs, W.M.: JASA **112**, 897 (2017)
40. Gneiting, T., Raftery, A.E., Balabdaoui, F.: J. R. Stat. Soc. Ser. B Stat. Methodol. **69**, 243 (2007)
41. Gneiting, T., Raftery, A.E.: JASA **102**, 359 (2007)
42. Winstein, K.J.: Wall Str. J. (2008). https://www.wsj.com/articles/SB121867148093738861
43. Vigen, T.: Spurious correlations (2018). http://www.tylervigen.com/spurious-correlations
44. Jeffreys, H.: Theory of Probability. Oxford University Press, Oxford (1998)
45. Goodman, S.N.: Epidemiology **12**, 295 (2001)
46. Nguyen, H.T., Sriboonchitta, S., Thac, N.N.: Structural Changes and Their Econometric Modeling. Springer (2019, forthcoming)
47. Hájek, A.: Erkenntnis **45**, 209 (1997)
48. Hájek, A.: Uncertainty: Multi-disciplinary Perspectives on Risk. Earthscan (2007)
49. Hájek, A.: Erkenntnis **70**, 211 (2009)
50. Biau, D.J., Jolles, B.M., Porcher, R.: Clin. Orthop. Relat. Res. **468**(3), 885 (2010)
51. Sainani, K.L.: Phys. Med. Rehabil. **4**, 442 (2012)
52. Briggs, W.M.: So, You Think You're Psychic? Lulu, New York (2006)
53. Campbell, S., Franklin, J.: Synthese **138**, 79 (2004)
54. Stove, D.: The Rationality of Induction. Clarendon, Oxford (1986)

Mean-Field-Type Games for Blockchain-Based Distributed Power Networks

Boualem Djehiche[1](✉), Julian Barreiro-Gomez[2], and Hamidou Tembine[2]

[1] Department of Mathematics, KTH Royal Institute of Technology, Stockholm, Sweden
boualem@math.kth.se

[2] Learning and Game Theory Laboratory, New York University in Abu Dhabi, Abu Dhabi, UAE
{jbarreiro,tembine}@nyu.edu

Abstract. In this paper we examine mean-field-type games in blockchain-based distributed power networks with several different entities: investors, consumers, prosumers, producers and miners. Under a simple model of jump-diffusion and regime switching processes, we identify risk-aware mean-field-type optimal strategies for the decision-makers.

Keywords: Blockchain · Bond · Cryptocurrency · Mean-field game
Oligopoly · Power network · Stock

1 Introduction

This paper introduces mean-field-type games for blockchain-based smart energy systems. The cryptocurrency system consists in a peer to peer electronic payment platform in which the transactions are made without the need of a centralized entity in charge of authorizing them. Therefore, the aforementioned transactions are validated/verified by means of a coded scheme called *blockchain* [1]. In addition, the blockchain is maintained by its participants, which are called *miners*.

Blockchain or distributed ledger technology is an emerging technology for peer-to-peer transaction platforms that uses decentralized storage to record all transaction data [2]. One of the first blockchain applications was developed in the e-commerce sector to serve as the basis for the cryptocurrency "Bitcoin" [3]. Since then, several other altcoins and cryptocurrencies including Ethereum, Litecoin, Dash, Ripple, Solarcoin, Bitshare etc have been widely adopted and are all based on blockchain. More and more new applications have recently been emerging that add to the technology's core functionality - decentralized storage of transaction data - by integrating mechanisms that allow for the actual transactions to be implemented on a decentralized basis. The lack of a centralized entity, that could have control over the security of transactions, requires

V. Kreinovich et al. (Eds.): ECONVN 2019, SCI 809, pp. 45–64, 2019.
https://doi.org/10.1007/978-3-030-04200-4_3

the development of a sophisticated verification procedure to validate transactions. Such task is known as *Proof-of-Work*, which brings new technological and algorithmic challenges as presented in [4]. For instance, [5] discusses the sustainability of bitcoin and blockchain in terms of the needed energy in order to perform the verification procedure. In [6], algorithms to validate transactions are studied by considering propagation delays. On the other hand, alternative directions are explored in order to enhance the blockchain, e.g., [7] discusses how the blockchain-based identity and access management systems can be improved by using an *Internet of Things* security approach.

In this paper the possibility of implementing distributed power networks on the blockchain and its pros and contras are presented. The core model (Fig. 1) uses a Bayesian mean-field-type game theory on the blockchain. The base interaction model considers producers, consumers and a new important element of distributed power networks called prosumers. A prosumer (producer-consumer) is a user that not only consumes electricity, but can also produce and store electricity [8,9]. We identify and formulate the key interactions between consumers, prosumers and producers on the blockchain. Based on forecasted demand generated from the blockchain, each producer determines its production quantity, its mismatch cost, and engages an auction mechanism to the prosumer market on the blockchain. The resulting supply is completed by the prosumers auction market. This determines a market price, and the consumers react to the offers and the price and generate a certain demand. The consistency relationship between demand and supply provides a fixed-point system, whose solution is a mean-field-type equilibrium [10].

The rest of paper is organized as follows. The next subsection presents the emergence of decentralized platform. Section 3 focuses on the game model. Section 4 presents risk-awareness and price stability analysis. Section 5 focuses on consumption-insurance and investment tradeoffs.

2 Towards a Decentralized Platform

The distributed ledger technology is a peer-to-peer transaction platform that integrates mechanisms that allow decentralized transactions or decentralized and distributed exchange system. These mechanisms, called "smart contracts", operate on the basis of individually defined rules (e.g. specifications as to quantity, quality, price, location) that enable an autonomous matching of distributed producers and their prospective customers. Recently the energy sector is also moving towards a semi-decentralized platform with the integration of prosumers' market and aggregators to the power grid. Distributed power is a power generated at or near the point of use. This includes technologies that supply both electric power and mechanical power. In electrical applications, distributed power systems stand in contrast to central power stations that supply electricity from a centralized location, often far from users. The rise of distributed power is being driven by broader decentralization movement of smarter cities. With blockchain transaction, every participant in a network can transact directly with every other network participant without involving a third-party intermediary (aggregator, operator). In other words, aggregators and the third parties are replaced by the blockchain. All transaction data is stored on a distributed blockchain, with all relevant information being stored identically on the computers of all participants, all transactions are made on the basis of smart contracts, i.e., based on predefined individual rules concerning quality, price, quantity, location, feasibility etc.

2.1 A Blockchain for Underserved Areas

One of the first questions that rises in blockchain is the service to Society. An authentication service offering to make environment-friendly (solar/wind/hydro) energy certificates available via a blockchain. The new service works by connecting solar panels and wind farms to an Internet of Things (IoT)-enabled device that measures the quality (of the infrastructure), quantity and the location of the power produced and fed into the grid. Certificates supporting PV growth and wind power can be bought and sold anonymously via a blockchain platform. Then, solar and wind energy produced by prosumers in undeserved areas can be transmitted to end-users. SolarCoin [11] was developed following that idea, with blockchain technology to generate an additional reward for solar electricity producers. Solar installation owners registering to the SolarCoin network receive one SolarCoin for each MWh of solar electricity that they produce. This digital asset will allow solar electricity producers to receive an additional reward for their contribution to the energy transition, which will develop itself through network effect. SolarCoin is freely distributed to any owner of a solar installation owner. Participating in the SolarCoin program can be done online, directly on the SolarCoin website. As of October 2017, more than 2,134,893 MWh of solar energy have been incentivized through SolarCoin across 44 countries. The ElectriCChain aims to provide the bulk of Blockchain recording for the solar installation owners in order to micro-finance the solar installation, incentivize it (through the SolarCoin tool), and monitor the install production. The idea of Wattcoin is to build this scheme for other renewable energies such as wind, thermo, hydro power plants to incentivize global electricity generation from several renewable energy sources. The incentive scheme influences the prosumers decision because they will be rewarded in WattCoins as an additional incentive to initiate the energy transition and possibly to compensate a fraction of the peak-hours energy demand.

2.2 Security, Energy Theft and Regulation Issues

If fully adopted, blockchain-based distributed power networks (b-DIPONET) is not without challenge. One of the challenges is security. This includes not only network security but also robustness, double spending and false/fake accounts. Stokens are regulated securities tokens built on the blockchain using smart contracts. They provide a way for accredited investors to interact with regulated companies through a digital ecosystem. Currently, the cryptocurrency industry has enormous potential - but it needs to be accompanied properly.

The blockchain technology can be used to reduce energy theft and unpaid bills by means of the automation of the prosumers who are connected to the power grid and their produced energy data is monitored in the network.

3 Mean-Field-Type Game Analysis

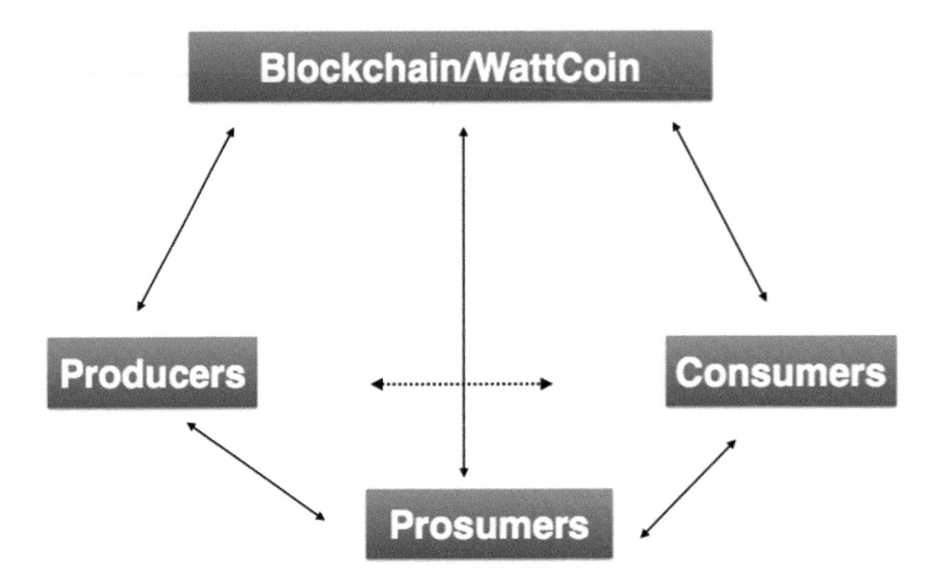

Fig. 1. Interaction blocks for blockchain-based distributed power networks.

This section presents the base mean-field-type game model. We identify and formulate the key interactions between consumers, prosumers and producers (see Fig. 1). Based on the forecasted demand from the blockchain-based history matching, each prosumer determines its production quantity, its mismatch cost, and use the blockchain to respond directly to consumers. All the energy producers together are engaged in a competitive energy market share. The resulting supply is completed by the prosumers energy market. This determines a market price, and the consumers react to the price and generate a demand. The consistency relationship between demand and supply of the three components provides a fixed-point system, whose solution is a mean-field equilibrium.

3.1 The Game Setup

Consumer i can decide to install a solar panel on her roof or a wind power station. Depending on sunlight or wind speed consumer i may produce surplus energy. She is no longer just an energy consumer but a prosumer. *A prosumer can decide to participate or not to the blockchain.* If the prosumer decides to participate to the blockchain to sell her surplus energy, the energy produced by this prosumer is measured by a dedicated meter which is connected and linked to the blockchain. The measurement and the validation is done ex-post from the quality-of-experience of the consumers of prosumer i. The characteristics and the bidding price of the energy produced by the prosumer are registered in the blockchain. This allows to give a certain score or Wattcoin to that prosumer for incentivization and participation level. This data is public if in the public blockchain's distributed register. All the transactions are verified and validated by the users of the blockchain ex-post. If the energy transaction does not happen in the blockchain platform, the proof-of-validation is simply an ex-post quality-experience measurement and therefore it does not need to use the heavy proof-of-work used by some crypto-currencies. The adoption of energy transactions to be blockchain requires a significantly reduction of the energy consumption of the proof-of-work itself. If the proof-of-work is energy consuming (and costly) then the energy transactions is kept to the traditional channel and only proof-of-validation is used as a recommendation system to monitor and to incentivize the prosumers. The blockchain technology makes it public and more transparent. If j and k are neighbors of the location of where i produced the energy, j and k can buy electricity off him and the consumption needs recorded in the blockchain ex-post. The transactions need to be technically secure and automated. Once prosumer i reaches a total of 1 MWh of energy sold to its neighbors, consumer i gets an equivalent of a certain unit of blockchain cryptocurrency such as Wattcoin, WindCoin, Solarcoin etc. It is an extra reward to the revenue of the prosumer. This scheme incentivizes prosumers to participate and promotes environment-friendly energy. Instead of a digitally mined product (transaction), the WattCoin proof-of-validity happens in the physical world, and those who have wind/thermo/photovoltaic arrays can earn Wattcoin just for generating electricity and serving it successfully. It is essentially a global rewarding/loyalty program, and is designed to help incentivize more renewable electricity production, while also serving as a lower-carbon cryptocurrency than Bitcoin and similar alternative currencies.

Each entity can

- Purchase and supply energy and have automated and verifiable proof of the amounts of green energy purchased/supplied via the information stored on the blockchain.
- Ensure that local generation (and feasibility) is supported, as it becomes possible to track the exact geographical origin of each energy MWh produced. For example, it becomes possible to pay additional premiums for green energy if it is generated locally, to promote further local energy generation capacity. Since the incentive reward is received only ex-post by the prosumer after checking the quality-of-experience, the proof-of-validity will improve the feasibility status of the energy supply and demand.

- Spatial energy price (price field) is publicly available to the consumers and prosumers who would like to purchase. This includes production cost and migration/distribution fee for moving energy from its point of production to its point of use.
- Each producer can supply energy on the platform and make smart contract for the delivery.
- Miners can decide to mine environment-friendly energy blocks. Honest miners are entities or people who validate the proof-of-work or proof-of-stakes (or other scheme). This can be individual, a pool or a coalition. There should be an incentive for them to mine. Selfish miners are those who may aim to pool their effort to maximize their own-interest. This can be individual, a pool or a coalition. Deviators or Malicious miners are entities or people who buy tokens for market and vote to impose their version of blockchain (different assigns at different block).

The game is described by the following four key elements:

- Platform: A Blockchain
- Players: Investors, consumers, prosumers, producers, miners.
- Decisions: Each player can decide and act via the blockchain.
- Outcomes: The outcome is given by gain minus loss for each participant.

Note that in this model, there is no energy trading option on the blockchain. However, the model can be modified to include trading at some part of the private blockchain. The electricity price dynamics regulation and stability will be discussed below.

3.2 Analysis

How can blockchain improve the penetration rate of renewable energy?
Thanks to the blockchain-based incentive, a non-negligible portion of prosumers will participate to the program. This will increase the produced renewable energy volumes. A basic rewarding scheme is that simple and easy to implement is a Tullock-like scheme, where probabilities to win a winner-take-all contest are considered, defining some *constest success functions* [12–14]. It consists of taking a spatial rewarding scheme to be added to the prosumers if a certain number of criteria are satisfied. In terms of incentives, a prosumer producing energy from location x will be rewarded ex-post $R(x)$ with probability $\frac{h_j(x,a_j)}{\sum_{i=1}^n h_i(x,a_i)}$ if $\sum_{i=1}^n h_i(x,a_i) > R(x) > 0$, where h_i is non-decreasing in its second component. Clearly, with this incentive scheme, a non-negligible portion of producers can reinvest more funds in the renewable energy production.

Implementation Cost
We identify basic costs for the blockchain-based energy system need to be implemented properly with largest coverage. As the next generation wireless communication and internet-of-everything is moving toward advanced devices with

high-speed, well-connected and more security and reliability than the previous version, blockchain technology should take advantage of it to decentralized operation. The wireless communication devices can be used as hotspots to connect to the blockchain as mobile calls are using wireless access points and hotspots as relays. Thus, a large coverage of the technology as related to the wireless coverage and connectivity of the location. Thus, the cost is reflected to the consumers and to the producers from their internet subscription fees. In addition to that cost, miners operations consume energy and powers. Supercomputers (CPUs, GPUs) and operating machines cost should be added to.

Demand-Supply Mismatch Cost

Let $\mathscr{T} := [t_0, t_1]$ be the time horizon with $t_0 < t_1$. In presence of blockchain, prosumers aim to anticipate their production strategies by solving the following problem:

$$\left\{\begin{array}{c} \inf_s \mathbb{E}L(s, e, \mathscr{T}) \\ L(s, e, \mathscr{T}) = l_{t_1}(e(t_1)) + \int_{t_0}^{t_1} l(t, D(t) - S(t))\, dt \\ \frac{d}{dt}e_{jk}(t) = x_{jk}(t)\mathbb{1}_{\{k \in A_j(t)\}} - s_{jk}(t), \\ n \geq 1, \\ j \in \{1, \ldots, n\}, \\ k \in \{1, \ldots, K_j\}, \\ K_j \geq 1, \\ x_{jk}(t) \geq 0, \\ s_{jk}(t) \in [0, \bar{s}_{jk}],\ \forall j, k, t \\ \bar{s}_{jk} \geq 0, \\ e_{jk}(t_0) \text{ given}, \end{array}\right. \tag{1}$$

where

- the instant loss is $l(t, D(t) - S(t))$, l_{t_1} is the terminal loss function.
- the energy supply at time t is

$$S(t) = \sum_{j=1}^{n} \sum_{k=1}^{K_j} s_{jk}(t),$$

  $s_{jk}(t)$ is the production rate of power plant/generator k of prosumer j at time t, $\bar{s}_{jk}$ is an upper bound for s_{jk} which will be used as a control action.
- The stock of energy $e_{jk}(t)$ of prosumer j at power plant k at time t is given by the following classical motion dynamics:

$$\frac{d}{dt}e_{jk}(t) = \text{incoming flow}_{jk}(t) - \text{outgoing flow}_{jk}(t), \tag{2}$$

  The incoming flow happens only when the power station is active. In that case, the arrival rate is $x_{jk}(t)\mathbb{1}_{\{k \in A_j(t)\}}$ where $x_{jk}(t) \geq 0$, and the set of active power plant of j is defined by $A_j(t)$, the set of all active power plants is $A(t) = \cup_j A_j(t)$. $D(t)$ is the demand on the blockchain at time t.

In general, the demand needs to be anticipated/estimated/predicted so that the produced quantity is enough to serve the consumers. If the supply S is less than

D some of the consumers will not be served, hence it is costly for the operator. If the supply S is greater that D then the operator needs to store the exceed amount of energy. It will be lost if the storage is enough. Thus, it is costly in both cases, and the cost is represented by $l(\cdot, D - S)$. The demand-supply mismatch cost is determined by solving (1).

3.3 Oligopoly with Incomplete Information

There are $n \geq 2$ potential interacting energy producers over the horizon $\mathcal{T}$. At time $t \in \mathcal{T}$, producer i's output is $u_i(t) \geq 0$. The dynamics of the log-price, $p(t) :=$ logarithm of the price of energy at time t, is given by $p(t_0) = p_0$ and

$$dp(t) = \eta[a - D(t) - p(t)]dt + \left(\sigma dB(t) + \int_{\theta\in\Theta} \mu(\theta)\tilde{N}(dt, d\theta)\right) + \sigma_o dB_o(t), \quad (3)$$

where

$$D(t) := \sum_{i=1}^{n} u_i(t),$$

is the supply at time $t \in \mathcal{T}$, and B_o is standard Brownian motion representing a global uncertainty observed by all participant to the market. The processes B and N describe local uncertainties or noises. B is a standard Brownian motion, N is a jump process with Lévy measure $\nu(\mathrm{d}\theta)$ defined over Θ. It is assumed that ν is a Radon measure over Θ (the jump space) which is subset of $\mathbb{R}^m$. The process

$$\tilde{N}(\mathrm{dt}, \mathrm{d}\theta) = N(\mathrm{dt}, \mathrm{d}\theta) - \nu(\mathrm{d}\theta)\mathrm{dt}$$

is the compensated martingale. We assume that all these processes are mutually independent. Denote by $\mathcal{F}_t^{B,N,B_o}$ the natural filtration generated by the union of events $\{B, N, B_o\}$ up to time t, and by $(\mathcal{F}_t^{B_o}, t \in \mathcal{T})$ the natural filtration generated by the observed common noise, where $\mathcal{F}_t^{B_o} = \sigma(B_0(s), s \leq t)$ is the smallest σ-field generated by the process B_0 up to time t (see e.g. [15]).

The number η is positive. For larger values of the real number η the market price adjusts quicker along the inverse demand, all in the logarithmic scale. The terms a, σ, σ_o are fixed constant parameters. The jump rate size $\mu(\cdot)$ is in $L^2_\nu(\Theta, \mathbb{R})$ i.e.

$$\int_\Theta \mu^2(\theta)\nu(\mathrm{d}\theta) < +\infty.$$

The initial distribution of $p(0)$ is square integrable: $\mathbb{E}[p_0^2] < \infty$.

Producers know only their own types $(c_i, r_i, \bar{r}_i)$ but not the types of the others $(c_j, r_j, \bar{r}_j)_{j\neq i}$. We define a game with incomplete information denoted by G_ξ. The game G_ξ has n producers. A strategy for producer j is a map $\tilde{u}_j : I_j \to U_j$ prescribing an action for each possible type of producer j. We denote the set of actions of producer j by $\tilde{\mathcal{U}}_j$. Let ξ_j denote the distribution on the type vector $(c_j, r_j, \bar{r}_j)$ from the perspective of the jth producer. Given ξ_j, producer j can compute the conditional distribution $\xi_{-j}(c_{-j}, r_{-j}, \bar{r}_{-j}|c_j, r_j, \bar{r}_j)$, where

$$c_{-j} = (c_1, \ldots, c_{j-1}, c_{j+1}, \ldots, c_n) \in \mathbb{R}^{n-1}.$$

Producer j can then evaluate her expected payoff based on the expected types of other producers. We call a Nash equilibrium of G_ξ Bayesian equilibrium as.

At time $t \in \mathcal{T}$, producer i receives $\hat{p}(t)u_i - C_i(u_i)$ where $C_i : \mathbb{R} \to \mathbb{R}$, given by

$$C_i(u_i) = c_i u_i + \frac{1}{2} r_i u_i^2 + \frac{1}{2} \bar{r}_i \hat{u}_i^2,$$

is the instant cost function of i. The term $\hat{u}_i = \mathbb{E}[u_i \mid \mathcal{F}_t^{B_o}]$ is the conditional expectation of producer i's output given the global uncertainty B_o observed in the market. The last term $\frac{1}{2}\bar{r}_i \hat{u}_i^2$, in the expression of the instant cost C_i, aims to capture the risk-sensitivity of producer i. The conditional expectation of the price given the global uncertainty B_o up to time t is $\hat{p}(t) = \mathbb{E}[p(t) \mid \mathcal{F}_t^{B_o}]$. At the terminal time t_1 the revenue is $-\frac{q}{2}e^{-\lambda_i t_1}\left(p(t_1) - \hat{p}(t_1)\right)^2$. The long-term revenue of producer i is

$$R_{i,\mathcal{T}}(p_0, u) = -\frac{q}{2}e^{-\lambda_i t_1}\left(p(t_1) - \hat{p}(t_1)\right)^2 + \int_{t_0}^{t_1} e^{-\lambda_i t}\left[\hat{p}u_i - C_i(u_i)\right] dt,$$

where λ_i is a discount factor of producer i. Finally, each producer optimizes her long-term expected revenue. The case of deterministic complete information was investigated in [16,17]. Extension of the complete information to the stochastic case with mean-field term was done recently in [18]. Below, we investigate the equilibrium solution under incomplete information.

3.3.1 Bayesian Mean-Field-Type Equilibria

A Bayesian-Nash Mean-Field-Type Equilibrium is defined as a strategy profile and beliefs specified for each producer about the types of the other producers that minimizes the expected performance functional for each producer given their beliefs about the other producers' types and given the strategies played by the other producers. We compute the generic expression of the Bayesian mean-field-type equilibria.

Any strategy $u_i^* \in \tilde{\mathcal{U}_i}$ satisfying the maximum in

$$\begin{cases} \max_{u_i \in \tilde{\mathcal{U}_i}} \mathbb{E}\left[R_{i,\mathcal{T}}(p_0, u) \mid c_i, r_i, \bar{r}_i, \xi\right], \\ dp(t) = \eta\left[a - D(t) - p(t)\right] dt + \left(\sigma dB(t) + \int_\Theta \mu(\theta)\tilde{N}(dt, d\theta)\right) \\ \qquad + \sigma_o dB_o(t), \\ p(t_0) = p_0, \end{cases} \tag{4}$$

is called a Bayesian best-response strategy of producer i to the other producers strategy $u_{-i} \in \prod_{j \neq i} \tilde{\mathcal{U}_j}$.

Generically, Problem (4) has the following interior solution:

The Bayesian equilibrium strategy in state-and-conditional mean-field feedback form and is given by

$$\tilde{u}_i^*(t) = -\frac{\eta\hat{\alpha}_i(t)}{r_i}(p(t) - \hat{p}(t)) + \frac{\hat{p}(t)(1 - \eta\hat{\beta}_i(t)) - (c_i + \eta\hat{\gamma}_i)(t)}{r_i + \bar{r}_i},$$

where the conditional equilibrium price $\hat{p}$ is

$$\begin{cases} d\hat{p}(t) = \eta \Big\{ a + \frac{c_i+\eta\hat{\gamma}_i(t)}{r_i+\bar{r}_i} + \int \sum_{j\neq i} \frac{c_j+\eta\hat{\gamma}_j(t)}{r_j+\bar{r}_j} d\xi_{-i}(.|c_i,r_i,\bar{r}_i) \\ -\hat{p}(t)\left(1+\frac{1-\eta\hat{\beta}_i(t)}{r_i+\bar{r}_i} + \int \sum_{j\neq i} \frac{1-\eta\hat{\beta}_j(t)}{r_j+\bar{r}_j} d\xi_{-i}(.|c_i,r_i,\bar{r}_i)\right)\Big\} dt + \sigma_o \mathrm{d}B_o(t), \\ \hat{p}(t_0) = \hat{p}_0, \end{cases}$$

and the random parameters $\hat{\alpha}, \hat{\beta}, \hat{\gamma}, \hat{\delta}$ solve the stochastic Bayesian Riccati system:

$$\begin{cases} d\hat{\alpha}_i(t) = \Big\{(\lambda_i+2\eta)\hat{\alpha}_i(t) - \frac{\eta^2}{r_i}\hat{\alpha}_i^2(t) - 2\eta^2\hat{\alpha}_i(t)\int\sum_{j\neq i}\frac{\hat{\alpha}_j(t)}{r_j}d\xi_{-i}(.|c_i,r_i,\bar{r}_i)\Big\}dt \\ \quad + \hat{\alpha}_{i,o}(t)dB_o(t), \\ \hat{\alpha}_i(t_1) = -q, \\ d\hat{\beta}_i(t) = \Big\{(\lambda_i+2\eta)\hat{\beta}_i(t) - \frac{(1-\eta\hat{\beta}_i(t))^2}{r_i+\bar{r}_i} + 2\eta\hat{\beta}_i(t)\int\sum_{j\neq i}\frac{1-\eta\hat{\beta}_j(t)}{r_j+\bar{r}_j}d\xi_{-i}(.|c_i,r_i,\bar{r}_i)\Big\}dt \\ \quad + \hat{\beta}_{i,o}(t)dB_o(t), \\ \hat{\beta}_i(t_1) = 0, \\ d\hat{\gamma}_i(t) = \Big\{(\lambda_i+\eta)\hat{\gamma}_i(t) - \eta a\hat{\beta}_i(t) - \hat{\beta}_{i,o}(t)\sigma_o + \frac{(1-\eta\hat{\beta}_i(t))(c_i+s\hat{\gamma}_i(t))}{r_i+\bar{r}_i} \\ + \eta\hat{\gamma}_i(t)\int\sum_{j\neq i}\frac{1-\eta\hat{\beta}_j(t)}{r_j+\bar{r}_j}d\xi_{-i}(.|c_i,r_i,\bar{r}_i) - \eta\hat{\beta}_i(t)\int\sum_{j\neq i}\frac{c_j+\eta\hat{\gamma}_j(t)}{r_j+\bar{r}_j}d\xi_{-i}(.|c_i,r_i,\bar{r}_i)\Big\}dt \\ \quad - \hat{\beta}_i(t)\sigma_o dB_o(t), \\ \hat{\gamma}_i(0) = 0, \\ d\hat{\delta}_i(t) = -\Big\{-\lambda_i\hat{\delta}_i(t) + \frac{1}{2}\sigma_o^2\hat{\beta}_i(t) + \frac{1}{2}\hat{\alpha}_i(t)\left(\sigma^2 + \int_\Theta \mu^2(\theta)\nu(d\theta)\right) + \eta a\hat{\gamma}_i(t) \\ \quad + \hat{\gamma}_{i,o}(t)\sigma_o + \frac{1}{2}\frac{(c_i+\eta\hat{\gamma}_i(t))^2}{r_i+\bar{r}_i} + \eta\hat{\gamma}_i(t)\int\sum_{j\neq i}\frac{c_j+s\hat{\gamma}_j(t)}{r_j+\bar{r}_j}d\xi_{-i}(.|c_i,r_i,\bar{r}_i)\Big\}dt \\ \quad -\sigma_o\hat{\gamma}_i(t)\mathrm{d}B_o(t), \\ \hat{\delta}_i(t_1) = 0, \end{cases}$$

and the equilibrium revenue of producer i is

$$\mathbb{E}\left[\frac{1}{2}\hat{\alpha}_i(t_0)(p(t_0)-\hat{p}_0)^2 + \frac{1}{2}\hat{\beta}_i(t_0)\hat{p}_0^2 + \hat{\gamma}_i(t_0)\hat{p}_0 + \hat{\delta}_i(t_0)\right].$$

The proof of the Bayesian Riccati system follows from a Direct Method by conditioning on the type $(c_i, r_i, \bar{r}_i, \xi)$.

Noting that the Riccati system of the Bayesian mean-field-type game is different from the Riccati system of mean-field-type game, it follows that the Bayesian equilibrium costs are different. They become equal when $\xi_{-j} = \delta_{(c_{-j}, r_{-j}, \bar{r}_{-j})}$. This also shows that there is a value of information in this game. Note that the equilibrium supply is

$$\sum_i \tilde{u}_i^*(t) = -\eta(p(t)-\hat{p}(t))\sum_i \frac{\hat{\alpha}_i(t)}{r_i} + \sum_i \frac{\hat{p}(t)(1-\eta\hat{\beta}_i(t)) - (c_i+s\hat{\gamma}_i(t))}{r_i+\bar{r}_i}.$$

3.3.2 Ex-Post Resilience

Definition 1. We define a strategy profile $\tilde{u}$ as ex-post resilient if for every type profile $(c_j, r_j, \bar{r}_j)_j$, and for each producer i,

$$\begin{aligned}\text{argmax}_{\tilde{u}_i \in \tilde{\mathscr{U}}_i} \mathbb{E} \int R_{i,\mathscr{T}}(p_0, c_i, , r_i, \bar{r}_i, \tilde{u}_i, \tilde{u}_{-i}) \xi_{-i}(dc_{-i} dr_{-i} d\bar{r}_{-i} \mid c_i, r_i, \bar{r}_i) \\ = \text{argmax}_{\tilde{u}_i \in \tilde{\mathscr{U}}_i} \mathbb{E} R_{i,\mathscr{T}}(p_0, \tilde{u}_i, \tilde{u}_{-i}).\end{aligned}$$

We show that generically the Bayesian equilibrium is not ex-post resilient. An $n-$tuple of strategies is said to be ex-post resilient if each producer's strategy is a best response to the other producers' strategies, under all possible realizations of the others' types. An ex-post resilient strategy must be an equilibrium of every game with the realized type profile $(c, r, \bar{r})$. Thus, any ex-post resilient strategy is a robust strategy of the game in which all the parameters $(c, r, \bar{r})$ are taken. Here, each producer makes her ex-ante decision based on ex-ante information, that is, distribution and expectation, which is not necessarily identical to her ex-post information, that is, the realized actions and types of other producers. Thus, ex-post, or after the producer observes the actually produced quantities of energy of all the other producers, she may prefer to alter her ex-ante optimal production decision.

4 Price Stability and Risk-Awareness

This section examines the price stability of a stylized blockchain-based market under regulation designs. As a first step we design a target price dynamics that allows a high volume of transactions while fulfilling the regulation requirement. However, the target price is not the market price. In a second step, we propose and examine a simple price market dynamics under jump-diffusion process. The market price model builds on the market demand, supply and token quantity. We use three different token supply strategies to evaluate the proposed market price motion. The first strategy designs a supply of tokens to the market more frequently balancing the mismatch between market supply and market demand. The second strategy is a mean-field control strategy. The third strategy is a mean-field-type control strategy that incorporates the risk of deviating from the regulation bounds.

4.1 Unstable and High Variance Market

As an illustration of high variance price, we take the fluctuations of bitcoin price between December 2017 and February 2018. The data is from coindesk (https://www.coindesk.com/price/). The price went from 10 K USD to 20 K USD and back to 7 K USD within 3 months. The variance was extremely high within that period, which implied very high risks in the market (Fig. 1). This extremely high variance and unstable market is far beyond the risk-sensitivity index distributions of users and investors. Therefore the market needs to be re-designed to fit investors and users risk-sensitivity distributions.

Fig. 2. Coindesk database: the price of bitcoin went from 10K USD to 20 K USD and back to below 7 K USD within 2–3 months in 2017–2018.

4.2 Fully Stable and Zero Variance

We have seen that the above example is too risky and is beyond the risk-sensitivity index of the many users. Thus, it is important to have a more stable market price in the blockchain.

A fully stable situation is the case of constant price. For that case the variance is zero and there is no risk on that market. However, this case may not be interesting for producers, and investors: if they know that the price will not vary they will not buy. Thus, the volume of transactions will be significantly reduced which is not convenient for the blockchain technology which aims to be a place of innovations and investments.

Electricity market price cannot be constant because demand is variable on a daily basis or from one season to another within the same year. Peak hours price may be different from off-peak hours price as it is already the case in most countries.

Below we propose a price dynamics that is somehow in between the two scenarios: it is of relatively low variance and it allows several transaction opportunities.

4.3 What Is a More Stable Price Dynamics?

An example of a more stable cryptocurrency within similar time frame as the bitcoin is the tether USD (USDT) which oscillates between 0.99 and 1.01 but with an important volume of transactions (see Fig. 2). The maximum magnitude variation of the price remains very small while the number oscillations in between is large, allowing several investment, buying/selling opportunities (Fig. 3).

Is token supply possible in the blockchain?
Tokens in blockchain-based cryptocurrencies are generated by blockchain algorithms. Token supply is a decision process that can be incorporated in the algorithm. Thus, token supply can be used to influence the market price. In our model below we will use it as a control action variable.

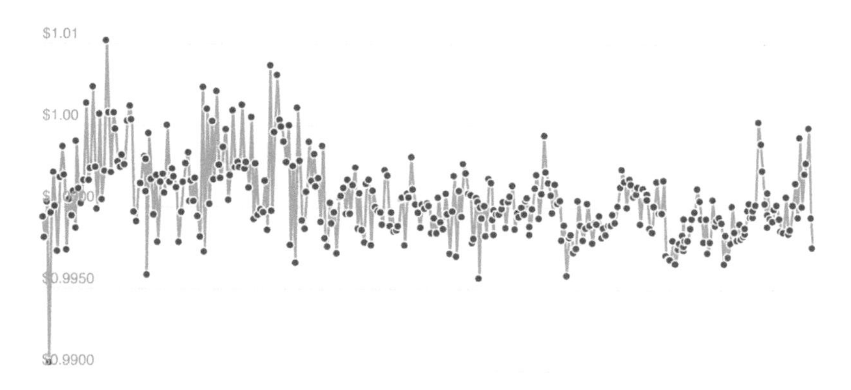

Fig. 3. Coindesk database: the price of tether USD went from 0.99 USD to 1.01 USD

4.4 A More Stable and Regulated Market Price

Let $\mathscr{T} := [t_0, t_1]$ be the time horizon with $t_0 < t_1$. There are n potential interacting regulated blockchain-based technologies over the horizon $\mathscr{T}$. The regulation authority of each blockchain-based technology has to choose the regulation bounds: the price of cryptocurrency i should be between $[\underline{p}_i, \bar{p}_i], \quad \underline{p}_i < \bar{p}_i$. We construct a target price $p_{tp,i}$ from an historical data-driven price dynamics of i. The target price should stay within the interval $[\underline{p}_i, \bar{p}_i]$ target range. The market price $p_{mp,i}$ depends on the quantity of token supplied, demanded and is given by a simple price adjustment dynamics obtained from Roos 1925 (see [16,17]). The idea of the Roos's model is very simple: Suppose that the cryptocurrency authority supplies a very small number of token in total, it will result in high prices and if the authorities expect these high price conditions not to continue in the following period, they will raise the number of tokens and, as a result, the market price will decrease a bit. If low prices are expected to continue, the authorities will decrease the number of token, resulting again in higher prices. Thus, oscillating between periods of low number of tokens with high prices and high number of tokens with low prices, the set price-quantity traces out an oscillatory phenomenon (which will allow large volume of transactions).

4.4.1 Designing a Regulated Price Dynamics

For any given $\underline{p}_i < \bar{p}_i$ one can choose the coefficients $c, \hat{c}$ such that the target price $p_{tp,i}(t) \in [\underline{p}_i, \bar{p}_i]$ for all time t. An example of such an oscillatory function is as follows:

$$p_{tp,i}(t) = c_{i0} + \sum_{k=1}^{2} c_{ik} \cos(2\pi kt) + \hat{c}_{ik} \sin(2\pi kt),$$

with $c_{ik}, \hat{c}_{ik}$ to be designed to fulfill the regulation requirement. Let $c_{i0} := \frac{\underline{p}_i + \bar{p}_i}{2}$, $c_1 := \frac{\bar{p}_i - \underline{p}_i}{100}$, $\hat{c}_{i1} := \frac{\bar{p}_i - \underline{p}_i}{150}$, $c_{12} := \frac{\bar{p}_i - \underline{p}_i}{200}$, $\hat{c}_{12} := \frac{\bar{p}_i - \underline{p}_i}{250}$. We want the target function to stay between 0.98 USD and 1.02 USD we set $\underline{p}_i = 0.98, \bar{p}_i = 1.02$. Figure 4 plots such a target function.

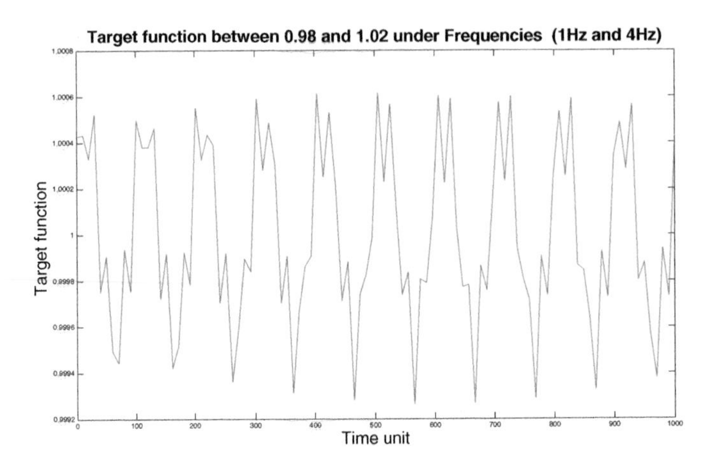

Fig. 4. Target price function $p_{tp,i}(t)$ between 0.98 and 1.02 under Frequencies (1 Hz and 4 Hz)

Note that this target price is not the market price. In order to incorporate a more realistic market behavior we introduce a dependence on demand and supply of tokens.

4.4.2 Proposed Price Model for Regulated Monopoly

We propose a market price dynamics that takes into consideration the market demand and the market supply. The blockchain-based market log-price (i.e. the logarithm of the price) dynamics is given by $p_i(t_0) = p_0$ and

$$\begin{aligned} dp_i(t) &= \eta_i[D_i(t) - p_i(t) - (S_i(t) + u_i(t))]dt \\ &\quad + \left(\sigma_i dB_i(t) + \int_{\theta\in\Theta} \mu_i(\theta)\tilde{N}_i(dt, d\theta)\right) + \sigma_o dB_o(t), \end{aligned} \tag{5}$$

where $u_i(t)$ is the total token injected to the market at time t, B_o is standard Brownian motion representing a global uncertainty observed by all participant to the market. As above, the processes B and N are local uncertainty or noise. B is a standard Brownian motion, N is a jump process with Lévy measure $\nu(d\theta)$ defined over Θ. It is assumed that ν is a Radon measure over Θ (the jump space). The process

$$\tilde{N}(dt, d\theta) = N(dt, d\theta) - \nu(d\theta)dt,$$

is the compensated martingale. We assume that all these processes are mutually independent. Denote by $(\mathscr{F}_t^{B_o}, t \in \mathscr{T})$ the filtration generated by the observed common noise B_0 (see Sect. 3.3). The number η_i is positive. For larger values of

η_i the market price adjusts quicker along the inverse demand. a, σ, σ_o are fixed constant parameters. The jump rate size $\mu(.)$ is in $L^2_\nu(\Theta, \mathbb{R})$ i.e.

$$\int_\Theta \mu^2(\theta)\nu(\mathrm{d}\theta) < +\infty.$$

The initial distribution p_0 is square integrable: $\mathbb{E}[p_0^2] < \infty$.

4.4.3 A Control Design that Tracks the Past Price

We formulate a basic control design that tracks the past price and the trend. A typical example is to choose the control action $u_{ol,i}(t) = -p_{tp,i}(t) + D_i(t) - S_i(t)$. This is an open-loop control strategy if D_i and S_i are explicit functions of time. Then the price dynamics becomes

$$\begin{aligned} dp_i(t) &= \eta_i[p_{tp,i}(t) - p_i(t)]dt \\ &\quad + \sigma_i dB_i(t) + \int_{\theta\in\Theta} \mu_i(\theta)\tilde{N}_i(dt, d\theta) + \sigma_o dB_o(t). \end{aligned} \tag{6}$$

Figure 5 illustrates an example of real price evolution from prosumer electricity markets in which we have incorporated a simulation of a regulated price dynamics as a continuation of real market. We observe that the open-loop control action $u_{ol,i}(t)$ decreases the magnitude of the fluctuations under similar circumstances.

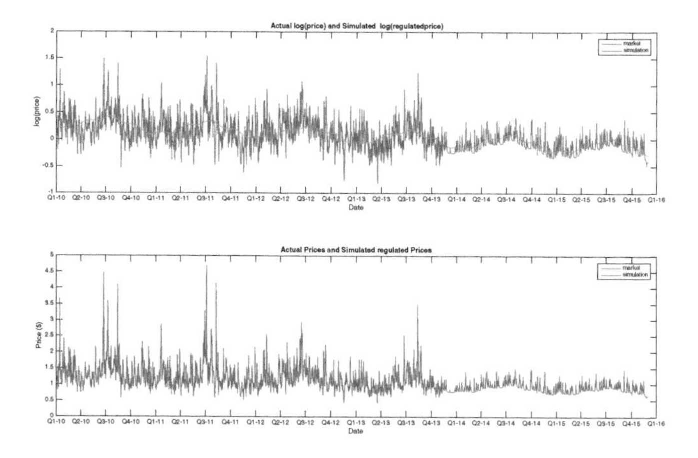

Fig. 5. Real market price and simulation of the regulated price dynamics as a continuation price under open-loop strategy.

4.4.4 An LQR Control Design

We formulate a basic LQR problem to a control strategy. Choose the control action that minimize $\mathbb{E}\{(p_i(t_1) - p_{tp,i}(t_1))^2 + \int_{t_0}^{t_1} (p_i(t) - p_{tp,i}(t))^2 dt\}$. Then the price dynamics becomes

$$\begin{aligned} dp_i(t) &= \eta_i[D_i(t) - p_i(t) - (S_i(t) + u_i(t))]dt \\ &\quad + \sigma_i dB_i(t) + \int_{\theta\in\Theta} \mu_i(\theta)\tilde{N}_i(dt, d\theta) + \sigma_o dB_o(t). \end{aligned} \tag{7}$$

4.4.5 A Mean-Field Game Strategy

The mean-field game strategy is obtained by freezing the mean-field term $\mathbb{E}p_i(t) := m(t)$ resulting from other cryptocurrencies and choosing the control action that minimizes

$$\begin{aligned}&\mathbb{E}q(t_1)(p_i(t_1) - f(t_1))^2 + \bar{q}(t_1)[m(t_1) - f(t_1)]^2 \\ &\qquad + \mathbb{E}\int_{t_0}^{t_1} q(t)(p_i(t) - f(t))^2 + \bar{q}(t)[m(t) - f(t)]^2 dt.\end{aligned} \tag{8}$$

The mean-field term $\mathbb{E}p_i(t) := m(t)$ is a frozen quantity and does not depend on the individual control action $u_{mfg,i}$. Then, the price dynamics becomes

$$\begin{aligned}dp_i(t) &= \eta[D_i(t) - p_i(t) - (S_i(t) + u_{mfg,i}(t))]dt \\ &\qquad + \sigma_i dB_i(t) + \int_{\theta\in\Theta} \mu_i(\theta)\tilde{N}_i(dt, d\theta) + \sigma_o dB_o(t).\end{aligned} \tag{9}$$

4.4.6 A Mean-Field-Type Game Strategy

A mean-field-type game strategy consists of a choice of a control action $u_{mftg,i}$ that minimizes

$$\begin{aligned}L_{mftg} &= \mathbb{E}q_i(t_1)(p_i(t_1) - p_{tp,i}(t_1))^2 + \bar{q}_i(t_1)[\mathbb{E}(p_i(t_1) - p_{tp,i}(t_1))]^2 \\ &\quad + \mathbb{E}\int_{t_0}^{t_1} q_i(t)(p_i(t) - p_{tp,i}(t))^2 + \bar{q}_i(t)[\mathbb{E}p_i(t) - p_{tp,i}(t)]^2 \, dt.\end{aligned} \tag{10}$$

Note that here the mean-field-type term $\mathbb{E}p_i(t)$ is not a frozen quantity. It depends significantly on the control action $u_{mftg,i}$. The performance index can be rewritten in terms of variance as

$$\begin{aligned}L_{mftg} &= \mathbb{E}q_i(t_1)var(p_i(t_1) - p_{tp,i}(t_1)) + [q_i(t_1) + \bar{q}_i(t_1)][\mathbb{E}p_i(t_1) - p_{tp,i}(t_1)]^2 \\ &\quad + \int_{t_0}^{t_1} q_i(t)\mathrm{Var}(p_i(t) - p_{tp,i}(t))dt \\ &\quad + \mathbb{E}\int_{t_0}^{t_1} [q_i(t) + \bar{q}_i(t)][\mathbb{E}p_i(t) - p_{tp,i}(t)]^2 \, dt.\end{aligned} \tag{11}$$

Then the price dynamics becomes

$$\begin{aligned}dp_i(t) &= \eta_i[D_i(t) - p_i(t) - (S_i(t) + u_{mftg,i})(t)]dt \\ &\qquad + \left(\sigma_i dB_i(t) + \int_{\theta\in\Theta} \mu_i(\theta)\tilde{N}_i(dt, d\theta)\right) + \sigma_o dB_o(t),\end{aligned} \tag{12}$$

The cost to be paid to the regulation authority if the price does not stay within $[\underline{p}_i, \bar{p}_i]$ is $\bar{c}_i(1 - \mathbb{1}_{[\underline{p}_i, \bar{p}_i]}(p_i(t)))$, $\bar{c}_i > 0$. Since the market price is stochastic due to demand, exchange and random events, there is still a probability to be out of the regulation range $[\underline{p}_i, \bar{p}_i]$. The outage probabilities under the three strategies $u_{ol,i}, u_{mfg,i}, u_{mftg,i}$ can be computed and used as a decision-support with respect to the regulation bounds. However, these continuous time strategies may not be convenient.

Very often, the supply of tokens decision is made in fixed times τ_i and not continuously. We look for a simpler strategy that is piecewise constant and takes a finite number of values within the horizon $\mathcal{T}$. Since the price may fluctuates very quickly due the jump terms, we propose an adjustment based on the recent moving average called the trend: $y(t) = \int_{t-\tau_i}^{t} x(t')\phi(t, t')\lambda(dt')$, implemented at different discrete time block units.

Different regulated blockchain technologies may choose different ranges $[\underline{p}_i, \bar{p}_i]$, so that investors and users can diversify their portfolios depending on their risk-sensitivity index distribution across the assets. This means that there will be an interaction between the cryptocurrencies and the altcoins. For example, the demand $D = \sum_{i=1}^{n} D_i$ will be shared between them. Users may exchange between coins and switch into another altcoins. The payoff of the blockchain-based technology i is $R_i = \hat{p}_i D_i - \bar{c}_i(1 - 1\!\!1_{[\underline{p}_i, \bar{p}_i]}(\hat{p}_i(t)))$, where $\hat{p}_i(t) = \mathbb{E}[p_i(t) \mid \mathscr{F}_t^{B_o}]$ is the conditional expectation of the market price with respect to $\mathscr{F}_t^{B_o}$.

4.5 Handling Positive Constraints

The price of the energy asset under cryptocurrency k is $x_k = e^{p_k} \geq 0$. The wealth of decision-maker i is $x = \sum_{k=0}^{d} \kappa_k x_k$.

Set $u_k^I = \kappa_k x_k$ to get the state dynamics. The sum of all the u_k is x. The variation is

$$
\begin{aligned}
dx = & [\kappa_0(r_0 + \hat{\mu}_0)x + \textstyle\sum_{k=1}^{d}[\hat{\mu}_k - (r_0 + \hat{\mu}_0)\kappa_0]u_k^I]dt \\
& + \textstyle\sum_{k=1}^{d} u_k^I\{Drift_k + Diffusion_k + Jump_k\},
\end{aligned}
\tag{13}
$$

where

$$
\begin{aligned}
& Drift_k = \eta_k[D_k - p_k - (S_k + u_{mftg,k})]dt + \tfrac{1}{2}(\sigma_i^2 + \sigma_o^2)dt \\
& \qquad + \textstyle\int_{\Theta}[e^{\gamma_k} - 1 - \gamma_k]\nu(d\theta)dt, \\
& Diffusion_k = (\sigma_k dB_k + \sigma_o dB_o), \\
& Jump_k = \textstyle\int_{\Theta}[e^{\gamma_k} - 1]\tilde{N}_k(dt, d\theta).
\end{aligned}
\tag{14}
$$

5 Consumption-Investment-Insurance

A generic agent wants to decide between consumption-Investment-Insurance [19–21] when the blockchain market is constituted of a bond with price p_0 and several stocks with prices $p_k, k > 0$ and is under different switching regime defined over a complete probability space $(\Omega, \mathscr{F}, \mathbb{P})$ in which a standard Brownian motion B, a jump process N, an observable Brownian motion B_o and an observable continuous-time finite-state Markov chain $\tilde{s}(t)$ representing a regime switching, with $\tilde{S}$ being the set of regimes, and $\tilde{q}_{\tilde{s}\tilde{s}'}$ a generator (intensity matrix) of $\tilde{s}(t)$. The log-price processes are the ones given above. The total wealth of the generic agent follows the dynamics

$$
\begin{aligned}
dx = & \kappa_0(r_0(\tilde{s}) + \hat{\mu}_0(\tilde{s}))x dt \\
& + \textstyle\sum_{k=1}^{d}[\hat{\mu}_k - (r_0(\tilde{s}) + \hat{\mu}_0(\tilde{s}))\kappa_0 + \mathrm{Drift}_k(\tilde{s})]u_k^I dt - u^c dt \\
& - \bar{\lambda}(\tilde{s})(1 + \bar{\theta}(\tilde{s}))E[u^{ins}]dt + \textstyle\sum_{k=1}^{d} u_k^I \mathrm{Diffusion}_k(\tilde{s}) \\
& + \textstyle\sum_{k=1}^{d} u_k^I \mathrm{Jump}_k(\tilde{s}) - (L - u^{ins})dN,
\end{aligned}
\tag{15}
$$

where $L = l(\tilde{s})x$.

In the dynamics (15) we have considered per-claim insurance of u^{ins}. That is, if the agent suffers a loss L at time t, the indemnity pays $u^{ins}(L)$. Such indemnity arrangements are common in private insurance at the individual level, among others. Motivated by new blockchain-based insurance products, we allow not only the cryptocurrency market but also the insurable loss to depend on the regime of the cryptocurrency economy and mean-field terms.

The payoff functional of the generic agent is

$$R = -qe^{-\lambda t_1}\{\hat{x}(t_1) - [x(t_1) - \hat{x}(t_1)]^2\} + \int_{t_0}^{t_1} e^{-\lambda t} \log u^c(t)\, dt,$$

where the process $\hat{x}$ denotes $\hat{x}(t) = \mathbb{E}[x(t) \mid \mathscr{F}_t^{\tilde{s}_0, B_o}]$. The generic agent seeks for a strategy $u = (u^c, u^I, u^{ins})$ that optimizes the expected value of R given $x(t_0), \tilde{s}(t_0)$ and the filtration generated by the common noise B_o.

For $q = 0$ an explicit solution can be found. To prove it, we choose a guess functional of the form

$$f = \alpha_1(t, \tilde{s}(t)) \log x(t) + \alpha_2(t, \tilde{s}(t)).$$

Applying Itô's formula for jump-diffusion-regime switching yields

$$\begin{array}{l} f(t,x,\tilde{s}) = f(t_0,x_0,\tilde{s}_0) + \int_{t_0}^t \{\dot{\alpha}_1 \log x + \dot{\alpha}_2 + \frac{\alpha_1}{x}\kappa_0(r_0(\tilde{s}) + \hat{\mu}_0(\tilde{s}))x \\ +\frac{\alpha_1}{x}\sum_{k=1}^d[\hat{\mu}_k - (r_0(\tilde{s}) + \hat{\mu}_0(\tilde{s}))\kappa_0 + \text{Drift}_k(\tilde{s})]u_k^I \\ -\frac{\alpha_1}{x}u^c - \frac{\alpha_1}{x}\bar{\lambda}(\tilde{s})(1+\bar{\theta}(\tilde{s}))E[u^{ins}] - \frac{\alpha_1}{x^2}\frac{1}{2}\sum_{k=1}^d\{(u_k^I\sigma_k)^2 + (u_k^I\sigma_o)^2\} \\ +\sum_{k=1}^d \int_\Theta \alpha_1 \log\{x + u_k^I(e^{\gamma_k}-1)\} - \alpha_1 \log x - \frac{\alpha_1}{x}u_k^I(e^{\gamma_k}-1)\nu(d\theta) \\ +\bar{\lambda}[\alpha_1 \log(x - (L - u^{ins})) - \alpha_1 \log x + \frac{\alpha_1}{x}(L - u^{ins})] \\ \sum_{\tilde{s}'}[\alpha_1(t,\tilde{s}') - \alpha_1(t,\tilde{s})]\log x + \sum_{\tilde{s}'}\alpha_2(t,\tilde{s}') - \alpha_2(t,\tilde{s}) \;\}\, dt + \int_{t_0}^t d\tilde{\varepsilon}, \end{array} \tag{16}$$

where $\tilde{\varepsilon}$ is a martingale. The term $\bar{\theta}(\tilde{s})$ represents $\frac{\theta(\tilde{s})}{1+\bar{m}(t)}$ where $\bar{m}(t)$ the average amount invested by other agents for insurance.

$$\begin{array}{l} R - f(t_0,x_0,\tilde{s}_0) = -f(t_1,x(t_1),\tilde{s}(t_1)) - qe^{-\lambda t_1}[x(t_1) - \hat{x}(t_1)]^2 \\ +\int_{t_0}^{t_1}\{\dot{\alpha}_1 \log x + \dot{\alpha}_2 + \frac{\alpha_1}{x}\kappa_0(r_0(\tilde{s}) + \hat{\mu}_0(\tilde{s}))x + e^{-\lambda t}\log u^c - \frac{\alpha_1}{x}u^c \\ +\frac{\alpha_1}{x}\sum_{k=1}^d[\hat{\mu}_k - (r_0(\tilde{s}) + \hat{\mu}_0(\tilde{s}))\kappa_0 + \text{Drift}_k(\tilde{s})]u_k^I \\ -\frac{\alpha_1}{x^2}\frac{1}{2}\sum_{k=1}^d\{(u_k^I\sigma_k)^2 + (u_k^I\sigma_o)^2\} \\ +\sum_{k=1}^d \int_\Theta \alpha_1 \log\{x + u_k^I(e^{\gamma_k}-1)\} - \alpha_1 \log x - \frac{\alpha_1}{x}u_k^I(e^{\gamma_k}-1)\nu(d\theta) \\ -\frac{\alpha_1}{x}\bar{\lambda}(\tilde{s})(1+\bar{\theta}(\tilde{s}))E[u^{ins}] \\ +\bar{\lambda}[\alpha_1 \log(x - (L - u^{ins})) - \alpha_1 \log x + \frac{\alpha_1}{x}(L - u^{ins})] \\ +\sum_{\tilde{s}'}[\alpha_1(t,\tilde{s}') - \alpha_1(t,\tilde{s})]\log x + \sum_{\tilde{s}'}\alpha_2(t,\tilde{s}') - \alpha_2(t,\tilde{s}) \;\}\, dt + \int_{t_0}^{t_1} d\tilde{\varepsilon}. \end{array} \tag{17}$$

The optimal u^c is obtained by direct optimization of $e^{-\lambda t}\log u^c - \frac{\alpha_1}{x}u^c$. This is a strictly concave function and its maximum is achieved at $u^c = \frac{e^{-\lambda t}}{\alpha_1}x$, provided that $\alpha_1(t,\cdot) > 0$ and $x(\cdot) > 0$.

This latter result can be interpreted as follows. The optimal consumption strategy process is proportional to the wealth process, i.e., the ratio $\frac{u^c(t)}{x^*(t)} > 0$.

This means that the blockchain-based cryptocurrency investors will consume proportionally more when they become wealthier in the market. Similarly, the insurance strategy u^{ins} can be obtained by optimizing

$$-\frac{1}{x}(1+\bar{\theta}(\tilde{s}))\mathbb{E}[u^{ins}(\tilde{s})] + \log(x - (L(\tilde{s}) - u^{ins}(\tilde{s}))) + \frac{1}{x}(L(\tilde{s}) - u^{ins}(\tilde{s}))],$$

which yields that

$$\frac{1}{x - L + u^{ins}} = \frac{1}{x}(2+\bar{\theta}).$$

Thus, noting that we have set $L(\tilde{s}) = l(\tilde{s})x$, we obtain

$$u^{ins}(\tilde{s}) = \left[l(\tilde{s}) - \frac{1+\bar{\theta}(\tilde{s})}{2+\bar{\theta}(\tilde{s})}\right]^{+} x = \max\left\{0, l(\tilde{s}) - \frac{1+\bar{\theta}(\tilde{s})}{2+\bar{\theta}(\tilde{s})}\right\} x.$$

We observe that, for each fixed regime $\tilde{s}$, the optimal insurance is proportional to the blockchain investor's wealth x. We note that it is optimal to buy insurance only if $l(\tilde{s}) > \frac{1+\bar{\theta}(\tilde{s})}{2+\bar{\theta}(\tilde{s})}$. When this condition is satisfied, the insurance strategy is $u^{ins}(\tilde{s}) := \left[l(\tilde{s}) - \frac{1+\bar{\theta}(\tilde{s})}{2+\bar{\theta}(\tilde{s})}\right] x$ which is a decreasing and convex function of $\bar{\theta}$. This monotonicity property means that, as the premium loading $\bar{\theta}$ increases, it is optimal to reduce the purchase of insurance.

The optimal investment strategy u_k^I can be found explicitly by mean-field-type optimization. Incorporating all together, a system of backward ordinary differential equations can be found for the coefficient functions $\{\alpha(t,\tilde{s})\}_{\tilde{s}\in\tilde{S}}$. Lastly, a fixed-point problem is solved by computing the total wealth invested in insurance to match with $\bar{m}$.

6 Concluding Remarks

In this paper we have examined mean-field-type games in blockchain-based distributed power networks with several different entities: investors, consumers, prosumers, producers and miners. We have identified a simple class of mean-field-type strategies under a rather simple model of jump-diffusion and regime switching processes. In our future work, we plan to extend these works to higher moments and predictive strategies.

References

1. Di Pierro, M.: What is the blockchain? Comput. Sci. Eng. **19**(5), 92–95 (2017)
2. Mansfield-Devine, S.: Beyond bitcoin: using blockchain technology to provide assurance in the commercial world. Comput. Fraud. Secur. **2017**(5), 14–18 (2017)
3. Nakamoto, S.: Bitcoin: A peer-topeer electronic cash system (2008)
4. Henry, R., Herzberg, A., Kate, A.: Blockchain access privacy: challenges and directions. IEEE Secur. Privacy **16**(4), 38–45 (2018)
5. Vranken, H.: Sustainability of bitcoin and blockchains. Curr. Opin. Environ. Sustain. **28**, 1–9 (2017)

6. Göbel, J., Keeler, H.P., Krzesinki, A.E., Taylor, P.G.: Bitcoin blockchain dynamics: the selfish-mine strategy in the presence of propagation delay. Perform. Eval. **104**, 23–41 (2016)
7. Kshetri, N.: Can blockchain strengthen the internet of things? IT Prof. **19**(4), 68–72 (2017)
8. Zafar, R., Mahmood, A., Razzaq, S., Ali, W., Naeem, U., Shehzad, K.: Prosumer based energy management and sharing in smart grid. Renew. Sustain. Energy Rev. **82**(2018), 1675–1684 (2018)
9. Dekka, A., Ghaffari, R., Venkatesh, B., Wu, B.: A survey on energy storage technologies in power systems. In: IEEE Electrical Power and Energy Conference (EPEC), pp. 105–111, Canada (2015)
10. Djehiche, B., Tcheukam, A., Tembine, H.: Mean-field-type games in engineering. AIMS Electron. Electr. Eng. **1**(2017), 18–73 (2017)
11. SolarCoin at https://solarcoin.org/en
12. Tullock, G.: Efficient rent seeking. Texas University Press, College Station, TX, USA pp. 97–112 (1980)
13. Kafoglis, M.Z., Cebula, R.J.: The buchanan-tullock model: some extensions. Public Choice **36**(1), 179–186 (1981)
14. Chowdhury, S.M., Sheremeta, R.M.: A generalized tullock contest. Public Choice **147**(3), 413–420 (2011)
15. Karatzas, I., Shreve, S.E.: Brownian Motion and Stochastic Calculus, 2nd edn. Springer, New York (1991)
16. Roos, C.F.: A mathematical theory of competition. Am. J. Math. **47**, 163–175 (1925)
17. Roos, C.F.: A dynamic theory of economics. J. Polit. Econ. **35**, 632–656 (1927)
18. Djehiche, B., Barreiro-Gomez, J., Tembine, H.: Electricity price dynamics in the smart grid: a mean-field-type game perspective. In: 23rd International Symposium on Mathematical Theory of Networks and Systems (MTNS), pp. 631–636, Hong Kong (2018)
19. Mossin, J.: Aspects of rational insurance purchasing. J. Polit. Econ. **79**, 553–568 (1968)
20. Van Heerwaarden, A.: Ordering of risks. Thesis, Tinbergen Institute, Amsterdam (1991)
21. Moore, K.S., Young, V.R.: Optimal insurance in a continuous-time model. Insur. Math. Econ. **39**, 47–68 (2006)

Finance and the Quantum Mechanical Formalism

Emmanuel Haven[1,2](✉)

[1] Memorial University, St. John's, Canada
ehaven@mun.ca
[2] IQSCS, Leicester, UK

Abstract. This contribution tries to sketch how we may want to embed formalisms from the exact sciences (more precisely physics) into social science. We begin to answer why such an endeavour may be necessary. We then consider more specifically how some formalisms of quantum mechanics can aid in possibly extending some finance formalisms.

1 Introduction

It is very enticing to think that a new avenue of research should almost instantaneously command respect, just by the mere fact that it is 'new'. We often hear, what I would call 'feeling' statements such as "since we have never walked the new path, there must be promise". The popular media does aid in furthering such a feeling. New flagship titles do not help much in dispelling such sort of myth that 'new', by definition must be good.

The title of this contribution attempts to introduce how some elements of the formalism of quantum mechanics may aid in extending our knowledge in finance. This is a very difficult objective to realize within the constraint of a few pages. In what follows, we will try to sketch some of the contributions, first starting from classical (statistical) mechanics for then to move towards showing how some of the quantum formalism may be contributing to a better understanding of some finance theories.

2 New Movements...

It is probably not incorrect to state that about 15 years ago, work was started in the area of using quantum mechanics in macroscopic environments. This is important to stress. Quantum mechanics, is formally residing at inquiries which take place on incredibly small scales. Maybe some of you have heard about the Planck constant and the atomic scale. Quantum mechanics works on those scales and a very quick question may arise in your minds: why would one want to be interested in analyzing the macroscopic world with such a formalism? Why? The answer is resolutely NOT because we believe that the macroscopic world would exhibit traces of quantum mechanics. Very few researchers will claim this.

V. Kreinovich et al. (Eds.): ECONVN 2019, SCI 809, pp. 65–75, 2019.
https://doi.org/10.1007/978-3-030-04200-4_4

Before we discuss how we can rationalize the quantum mechanical formalism in macroscopic applications, I would like to first, very briefly, sketch, with the aid of some historical notes, what we need to be careful of when we think of 'new' movements of research.

The academic world is sometimes very conservative. There is a very good reason for this. One must carefully investigate new avenues. Hence, progress is piece-wise and very often subject to many types and levels of critique. When a new avenue of research is being opened like, what we henceforth will call, quantum social science (QSS), one of the almost immediate 'tasks' (so to speak) is to test how the proposed new theories shall be embedded in the various existing social science theories. One way to test progress on this goal is to check how output can be successfully published in the host discipline. This embedding is progressive albeit moving sometimes at a very slow pace. Quantum social science (QSS) initially published much work in the physics area. Thereafter, work began to be published in psychology. Much more recently, research output started penetrating into mainstream journals in economics and finance. This is to show that the QSS movement is still extremely new. There is a lot which still needs doing.

For those who are very critical about anything 'new' in the world of knowledge, it is true that the wider academy is replete with examples of new movements. However, being 'new' does not need to presage anything negative. Fuzzy set theory, the theory which applies multivalued logic to a set of engineering problems (and other problems), came onto the world scene in a highly publicized way in the 1990's and although it is less noticeable nowadays, this theory has still a lot of relevance. But we need to realize that with whatever is 'new', whether it is a new product or a new idea, there are 'cycles' which trace out time dependent evolutions of levels of exposure. Within our very setting of economics and finance, fuzzy set theory actually contributed to augmenting models in finance and economics. Key work on fuzzy set theory is by Nguyen and Walker [1], Nguyen et al. [2] and also Billot [3].

A contender, from the physics world, which also applies ideas from physics to social science, especially economics and finance, is the so called 'econophysics' movement. Econophysics is mostly interested in applying formalisms from statistical mechanics to social science. From the outset, we can not pretend there are no connections between classical mechanics and quantum mechanics. For those of you who know a little more about physics, there are beautiful connections. I hint for instance at how a Poisson bracket has a raison d'être in both classical and quantum mechanics.

Quantum mechanics in macroscopic environments is probably still too new to write its history....I think this is true. The gist of this section of the paper is to keep in mind that knowledge expands and contracts according to cycles, and quantum social science will not be an exception to this observation.

3 And 'Quantum-Like' Is What Precisely?

Our talk at the ECONVN2019 conference in Vietnam will center around how quantum mechanics is paving new avenues of research in economics and finance. After this first section of the paper, which I hope, guards you against too much exuberance, it is maybe time to whet the appetite a little. We used, very loosely, the terminology 'quantum social science (QSS)' to mean that we apply elements of the quantum mechanical formalism to social science. We could equally have called it 'quantum-like research' for instance. Again, we repeat: we never mean that by using the toolkit from quantum mechanics to a world where '1 m' makes more sense to a human than 10^{-10} m (the atomic scale), we therefore have proven that the '1 m' world is quantum mechanical. To convince yourself, a very good starting point is the work by Khrennikov [4]. This paper sets the tone of what is to come (back in 1999). I recommend this paper to any novice in the field. I also recommend the short course by Nguyen [5] which also gives an excellent overview.

If you want to start reading papers, without further reading this paper, I recommend some other work, albeit it is much more technical than what will appear in this conference paper. Here are some key references if you really want to whet your appetite. I have made it somewhat symmetrical. The middle paper in the list below, is very short, and should be the first paper to read. Then, if your appetite is really of a technical flavour, go on to read either Baaquie or Segal. Here they are: Baaquie [6]; Shubik (a very short paper) [7] and Segal and Segal [8].

To conclude this brief section, please keep one premise in mind if you decide to continue reading the sequel of this paper. 'Quantum-like' when we pose it as a paradigm, shall mean first and foremost that the concept of 'information' is the key driver. I hope that you have some idea what we mean with 'information'. You may recall that information can be measured: Shannon entropy and Fisher information are examples of such measurement formalisms. Quantum-like then essentially means this: information is an integral part of any system[1] and information can be measured. If we accept that the wave function (in quantum mechanics) is purely informational in nature then we claim that we can use (elements) of the formalism of quantum mechanics to formalize the processing of information, *and* we claim we can use this formalism outside of its natural remit (i.e. outside of the scale of objects where quantum mechanical processes happen, such as the 10^{-10} m scale). One immediate critique to our approach is this: but why a quantum mechanical wave function? Engineers know all to well that one can work with wave functions which have no connection at all with quantum mechanics. Let us clarify a little more. At least two consequences follow from our paradigm. One consequence is more or less expected, and the other one is quite more subtle. Consequence one is as follows: we do *not*, by any means, claim that the macroscopic world is quantum mechanical. We already hinted to this

[1] A society is an example of a system; cell re-generation is another example of a system etc.

in the beginning of this paper. Consequence 2, is more subtle: the wave function of quantum mechanics is chosen for a very precise reason! In the applications of the quantum mechanical formalism in decision making one will see this consequence pops up all the time. Why? Because the wave function in quantum mechanics is in effect a probability amplitude. This amplitude is a key component in the formation of the so called probability interference rule. There are currently important debates forming on whether this type of probability forms part of classical probability; *or* whether it provides for a departure of the so called law of total probability (which is classical probability). For those who are interested in the interpretations of probability, please do have a look at Andrei Khrennikov's [9] work. We give a more precise definition of what we mean with quantum-like in our Handbook (see Haven and Khrennikov [10], p. v).

At this point in your reading, I would dare to believe that some of you will say very quietly: 'but why this connection between physics and social science. Why?' It is an excellent question and a difficult one to answer. First, it is surely not unreasonable to propose that the physics formalism, whatever guise it takes (classical; quantum; statistical), was developed to theorize about physical processes *not* societal processes. Nobody can make an argument against such point of view. Second, even if there is reason to believe that societal processes could be formalized with physics models, there are difficult hurdles to jump. I list five difficult hurdles (and I explain each one of them below). The list is non-exhaustive, unfortunately.

1. Equivalent data needs
2. The notion of time
3. Conservation principle
4. Social science works with other tools
5. Integration issues within social science
   - Hurdle 1, equivalent data needs, sounds haughty but it is a very good point. In physics, we have devices which can measure events which contain an enormous amount of information. If we import the physics formalism in social science, do we have tools at our disposal to amalgamate the same sort of massive information into one measurement? As an example: a gravitational wave is the outcome of a huge amount of data points which lead to the detection of such wave. What we mean with equivalent data needs is this. A physics formalism would require, in many instances, samples of a size which in social science are unheard of. So, naively, we may say: if you import the edifice of physics in social science can you comply, in social science, with the same data needs that physics uses? The answer is 'no'. Is this an issue? The answer is again 'no'. Why should we think that the whole edifice of physics is to be imported in social science. We use 'bits and pieces' of physics to advance knowledge in social science. Can we do this without consequence? Where is the limit? Those two questions need to be considered very carefully.

- Hurdle 2, the notion of time in physics may not at all be the same as the notion of time used in decision making or finance for instance. As an example, if we were to think of 'trading time' as the minimum time needed to make a new trade. Then in the beginning of the twentieth century that minimum time would several times be a multiple of the minimum trading time needed to make a trade nowadays. There is a subjective value to the notion of time in social science. Surely, we can consider a time series on prices of a stock. But time in a time series, in terms of the time reference used, is different. A time series from stocks traded in the 1960's has a different time reference than a time series from stocks traded in the 1990's (trading times were different for starters). This is quite different from physics: in the 1960's the time used for a ball of lead to fall from a skyscraper will be the same - exactly the same - as the time used for the ball of lead to fall from that same skyscraper in the 1990's. We may argue that time has an objective value in physics, whilst this may not be the case in social science. There is also the added issue of time reversibility in classical mechanics which we need to consider.
- Hurdle 3, there are many processes in social science which are not conserved. Conservation is a key concept in physics. Energy conservation for instance is intimately connected to Newton's second law (we come back to this law below). Gallegati et al. [11] remarked that "....income is not, like energy in physics, conserved by economic processes."
- Hurdle 4, comes, of course, as no surprise. The formalism used in social science, surely is very different from physics. As an example, there is very little use of differential equations in economics (although in finance, the Black-Scholes theory [12] has a partial differential equation which has very clear links with physics). Another example: the formalism underpinning mathematical economics is measure-theoretic for a large part. This is very different from physics.
- Hurdle 5, mentions integration issues within social science. This can pose additional resistance to having physics being used in social science. As an example, in Black-Scholes option pricing theory (a finance theory), one does not need any 'preference modelling'. The physics formalism which is maybe allied best with finance, therefore integrates badly with economics.

A question now becomes: how much of the physics edifice needs going into social science? There are no definite answers at all (as would be expected). In fact, I strongly believe that the (humble) stance one wants to take is this: 'why just not borrow tool X or Y from physics and see if it furthers knowledge in social science?' But are there pitfalls? As an example: when one uses probability interference from quantum mechanics (in social science) should we assume that orthogonal states need to remain orthogonal throughout time (as quantum physics requires it)? The answer should be no: i.e. not when we consider social science applications. Hence, taking the different view, i.e. that the social world is physics based, is I think, wrong. That one can uncover power laws in financial data does not mean that finance is physics based. That one emulates time dependent (and random) stock price behavior with Brownian motion does not mean that stocks are basic building blocks from physics. In summary, I do believe that there are insurmountable barriers to import the full physics edifice in social science. It is futile, I think, to argue to the contrary. There is a lot of work written on this. If you are interested check out Georgescu-Roegen [13] for instance.

4 Being 'Formal' About 'Quantum-Like'

An essential idea we need to take into account when introducing the quantum-like approach is that, besides[2] the paradigm (i.e. that the wave function is information and that we capture probability amplitude), there is a clear distinction in quantum mechanics between a state and a measurement. It is this distance between state and measurement which leaves room to interpret decision making as the result of what we could call 'contextual interaction'. I notice that I use terms which have a very precise meaning in quantum mechanics. 'Context' is such an example. In your future (or past) readings you will (you may have) come across other terms such as 'non-locality' or also 'entanglement' and 'no-signalling'. Those terms have very precise definitions in quantum mechanics and we must really thread very carefully when using them in a macroscopic environment.

In this paper we are interested in finance and the quantum mechanical formalism. From the outset it is essential to note that classical quantum mechanics does not allow for paths in its formalism. The typical finance formalism will have paths (such as stock price paths). What we have endeavoured to do with our quantum-like approach, within finance per sé, is to consider:

- (i) quantum mechanics via the quantum-like paradigm (thus centering our efforts on the concept of information) and;
- (ii) try to use a path approach within this quantum mechanical setting

In Baaquie [6] (p. 99) we can read this important statement: "The random evolution of the stock price $S(t)$ implies that if one knows the value of the stock price, then one has no information regarding its velocity..." This statement encapsulates the idea of the uncertainty principle from quantum mechanics. The above two points (i) and (ii), are important to bear in mind as in fact, if one uses (ii), one connects quite explicitly with (i). Let me explain. The path approach, if one can use this terminology, does not mean that quantum mechanics can be formulated with the notion of path in mind. However, it gets close: there are multiplicity of paths under a non-zero Planck constant and when one wants to approach the classical world, the multiplicity of paths reduces to one path. For those of you who are really interested in knowing what this is all about, it is important to properly set the contributions of this type of approach towards quantum mechanics in its context. In the 1950's David Bohm did come up with,

[2] It is not totally 'besides' though...

what one could call, a semi-classical approach to quantum mechanics. The key readings are Bohm [14], [15] and Bohm and Hiley [16].

The essential contribution which we think is characterizing Bohmian mechanics to an area like finance (for which it was certainly not developed), is that it provides for a re-interpretation of the second law of Newton (now embedded within a finance context) *and* it gives an information approach to finance which is squarely embedded within the argument that point (ii) is explicitly connected to point (i) above.

Let us explain this a little more formally. We follow Choustova [17] (see also Haven and Khrennikov [18] (p. 102–) and Haven et al. [19] (p. 143)).

The first thing to consider is the so called polar form of the wave function: $\psi(q,t) = R(q,t)e^{i\frac{S(q,t)}{h}}$; where $R(q,t)$ is the amplitude and $S(q,t)$ is the phase. Note that h is the Planck constant [3] and i is a complex number, q is position and t is time. Now plug $\psi(q,t)$ into the Schrödinger equation. Hold on though! How can we begin to intuitively grasp this equation? There is a lot of background to be given to the Schrödinger equation and there are various ways to approach this equation. In a nutshell, two basic building blocks are needed[4]: (i) a Hamiltonian[5] and (ii) an operator on that Hamiltonian. The Hamiltonian can be thought of as the sum of potential[6] and kinetic energy. When an operator is applied on that Hamiltonian, one essentially uses the momentum operator on the kinetic part of the Hamiltonian. The Schrödinger equation is a partial differential equation[7] which, in the time dependent format, shows us the evolution of the wave function - when not disturbed. The issue of disturbance and non-disturbance has much to do with the issue of collapse of the wave function. We do not discuss it here. If you want an analogy with classical mechanics, you can think of the equation which portrays the time dependent evolution of a probability density function over a particle. This equation is known as the Fokker-Planck equation. Note that the wave function here, is a probability amplitude and NOT a probability. The transition towards probability occurs via so called complex conjugation of the amplitude function.

This is now the Schrödinger equation: $ih\frac{\partial\psi}{\partial t} = -\frac{h^2}{2m}\frac{\partial^2\psi}{\partial q^2} + V(q,t)\psi(q,t)$; where V denotes the real potential and m denotes mass. You can see that the operator on momentum is contained in the $\frac{\partial^2}{\partial q^2}$ term. When $\psi(q,t) = R(q,t)e^{i\frac{S(q,t)}{h}}$ is plugged into that equation, one can separate out the real and imaginary part (recall we have a complex number here) and one of the equations which are generated is: $\frac{\partial S}{\partial t} + \frac{1}{2m}\left(\frac{\partial S}{\partial q}\right)^2 + \left(V - \frac{h^2}{2mR}\frac{\partial^2 R}{\partial q^2}\right) = 0$. Note that if $\frac{h^2}{2m} \ll 1$ then

[3] Note that in the sequel h will be set to one. In physics this constant is essential to have the left and right hand sides of the Schrödinger partial differential equation to have units which agree.

[4] This is one way to look at this equation. There are other ways.

[5] Not to be confused with the so called Lagrangian!.

[6] Contrary to the idea of energy conservation we mentioned above, potential energy need not be conserved.

[7] Yes: physics is replete with differential equations (see our discussion above).

the term $\frac{h^2}{2mR}\frac{\partial^2 R}{\partial q^2}$ becomes negligible. Now assume, we set $\frac{h^2}{2m} = 1$, i.e. we are beginning preparatory work to use the formalism in a macroscopic setting.

The term, $Q(q,t) = -\frac{h^2}{2mR}\frac{\partial^2 R}{\partial q^2}$ with its Planck constant is called the 'quantum potential'. This is a subtle concept and I would recommend to go back to the work of Bohm and Hiley [16] for a proper interpretation. A typical question which arises is this one: how does this quantum potential compare to the real potential? This is not an easy question. From this approach, one can write a revised second law of Newton, as follows: $m\frac{d^2q(t)}{dt^2} = -\frac{\partial V(q,t)}{\partial q} - \frac{\partial Q(q,t)}{\partial q}$ with initial conditions. We note that $Q(q,t)$ depends on the wave function which itself follows the Schr ödinger equation. Paths can be traced out of this differential equation. We mentioned above, that the Bohmian mechanics approach gives an information approach to finance where the paths are connected to information. So where does this notion of information come from? It can be shown that the quantum potential is related to a measure of information known as 'Fisher information'. See Reginatto [21]. Finally, we would also want to note that Edward Nelson obtains a quantum potential, but via a different route. See Nelson [22].

As we remarked in Haven, Khrennikov and Robinson [19], the issue with the Bohmian trajectories is that they do not reflect the idea (well founded in finance) of so called non-zero quadratic variation. One can remedy this problem to some extent with constraining conditions on the mass parameter. See Choustova [20] and Khrennikov [9].

5 What Now...?

Now that we have been attempting to begin to be a little formal about 'quantum-like', the next, and very logical, question is: 'what can we now really do with all this?' I do want to refer the interested reader to some more references if they want to get much more of a background. Besides Khrennikov [9] and Haven and Khrennikov [18] we need to cite the work of Busemeyer and Bruza [23], which focusses heavily on successful applications in psychology.

With regard to the applications of the quantum potential in finance, we want to make some mention of how this new tool can be estimated from financial data and what the results are, if we compare both potentials with each other. As we mentioned above, it is a subtle debate, in which we will not enter in this paper, on how both potentials can be compared, from a purely physics based point of view. But we have attempted to compare them in applied work. More on this now.

It may come as a surprise that the energy concepts from physics do have social science traction. This is quite a recent phenomenon. We mentioned at the beginning of this paper that one hurdle (amongst the many hurdles one needs jumping when physics formalisms are to be applied to social science) says that social science uses different tools altogether. A successful example of work which has overcome that hurdle is the work by Baaquie [24]. This is work which firmly plants a classical physics formalism, where the Hamiltonian (i.e. the sum of potential and kinetic energy) plays a central role, into one of the most basic frameworks of economic theory, i.e. the framework from which equilibrium prices are found. In his paper potential energy is defined for the very first time as being the sum of the demand and supply of a good. From the minimization of that potential one can find the equilibrium prices (which coincide with the equilibrium price one would have found by finding the intersection of supply and demand functions). This work shows how the Hamiltonian can give an enriched view of a very basic economics based framework. Not only does the minimization of the real potential allow to trace out more information around the minimum of that potential, it also allows to bring in dynamics via the kinetic energy term.

To come back now to furthering the argument that energy concepts from physics have traction in social science, we can mention that in a recent paper by Shen and Haven [25] some estimates were provided on the quantum potential from financial data. This paper follows in line of another paper by Tahmasebi et al. [26].

Essentially, for the estimation of the quantum potential, one sources R from the probability density function on daily returns on a set of commodities. In the paper, returns on the prices of several commodities are sourced from Bloomberg. The real potential V was sourced from: $f(q) = N \exp(-\frac{2V(q)}{Q})$, Q is a diffusion coefficient and N a constant. An interesting result is that the real potential exhibits an equilibrium value (reflective of the mean return of the prices (depending on the time frame they have been sampled on). The quantum potential, however does not have such an equilibrium. Both potentials clearly show that if returns try to jump out of range, a strong negative reaction force will pull those returns back and such forces may well be reflective of some sort of sort of efficiency mechanism. We also report in the Shen and Haven paper that when forces are considered (i.e. the negative gradient of the potentials), the gradient of the force associated with the real potential is higher than the gradient of the force associated with the quantum potential. This may indicate that the potentials may well pick up different types of information. More work is warranted in this area. But the argument was made before, that the quantum and real potential, when connected to financial data may pick up soft (psychologically based) information and hard (finance based only) information. This was already laid out in Khrennikov [9].

6 Conclusion

If you have read until this section then you may wonder what the next steps are. The quantum formalism in the finance area is currently growing out of three different research veins. The Bohmian mechanics approach we alluded to in this paper is one of them. The path integration approach is another one and mainly steered by Baaquie. A third vein, which we have not discussed in this paper consists of applications of quantum field theory to finance. Quantum field theory regards the wave function now as a field and fields are operators. This allows for the creation and destruction of different energy levels (via so called eigenvectors). Again, the idea of energy can be noticed. The first part of the book by Haven, Khrennikov and Robinson [19] goes into much depth on the field theory approach. A purely finance application which uses quantum field theory principles is by Bagarello and Haven [27]. More to come!!

References

1. Nguyen, H.T., Walker, E.A.: A First Course in Fuzzy Logic, 3rd edn. Chapman and Hall/CRC Press, Boca Raton (2006)
2. Nguyen, H.T., Prasad, N.R., Walker, C.L., Walker, E.A.: A First Course in Fuzzy and Neural Control. Chapman and Hall/CRC Press, Boca Raton (2003)
3. Billot, A.: Economic Theory of Fuzzy Equilibria: An Axiomatic Analysis. Springer, Heidelberg (1995)
4. Khrennikov, A.Y.: Classical and quantum mechanics on information spaces with applications to cognitive, psychological, social and anomalous phenomena. Found. Phys. **29**, 1065–1098 (1999)
5. Nguyen, H.T.: Quantum Probability for Behavioral Economics. Short Course at BUH. New Mexico State University (2018)
6. Baaquie, B.: Quantum Finance. Cambridge University Press, Cambridge (2004)
7. Shubik, M.: Quantum economics, uncertainty and the optimal grid size. Econ. Lett. **64**(3), 277–278 (1999)
8. Segal, W., Segal, I.E.: The Black-Scholes pricing formula in the quantum context. Proc. Natl. Acad. Sci. USA **95**, 4072–4075 (1998)
9. Khrennikov, A.: Ubiquitous Quantum Structure: From Psychology to Finance. Springer, Heidelberg (2010)
10. Haven, E., Khrennikov, A.Y.: The Palgrave Handbook of Quantum Models in Social Science, p. v. Springer - Palgrave MacMillan, Heidelberg (2017)
11. Gallegati, M., Keen, S., Lux, T., Ormerod, P.: Worrying trends in econophysics. Physica A **370**, 1–6 (2006). page 5
12. Black, F., Scholes, M.: The pricing of options and corporate liabilities. J. Polit. Econ. **81**, 637–659 (1973)
13. Georgescu-Roegen, N.: The Entropy Law and the Economic Process. Harvard University Press (2014, Reprint)
14. Bohm, D.: A suggested interpretation of the quantum theory in terms of hidden variables. Phys. Rev. **85**, 166–179 (1952a)
15. Bohm, D.: A suggested interpretation of the quantum theory in terms of hidden variables. Phys. Rev. **85**, 180–193 (1952b)
16. Bohm, D., Hiley, B.: The Undivided Universe: An Ontological Interpretation of Quantum Mechanics. Routledge and Kegan Paul, London (1993)
17. Choustova, O.: Quantum Bohmian model for financial market. Department of Mathematics and System Engineering. International Center for Mathematical Modelling. Växjö University (Sweden) (2007)
18. Haven, E., Khrennikov, A.: Quantum Social Science. Cambridge University Press (2013)
19. Haven, E., Khrennikov, A., Robinson, T.: Quantum Methods in Social Science: A First Course. World Scientific, Singapore (2017)
20. Choustova, O.: Quantum model for the price dynamics: the problem of smoothness of trajectories. J. Math. Anal. Appl. **346**, 296–304 (2008)
21. Reginatto, M.: Derivation of the equations of nonrelativistic quantum mechanics using the principle of minimum fisher information. Phys. Rev. A **58**(3), 1775–1778 (1998)

22. Nelson, E.: Stochastic mechanics of particles and fields. In: Atmanspacher, H., Haven, E., Kitto, K., Raine, D. (eds.) Quantum Interaction: 7th International Conference, QI 2013. Lecture Notes in Computer Science, vol. 8369, pp. 1–5 (2013)
23. Busemeyer, J.R., Bruza, P.: Quantum Models of Cognition and Decision. Cambridge University Press, Cambridge (2012)
24. Baaquie, B.: Statistical microeconomics. Physica A **392**(19), 4400–4416 (2013)
25. Shen, C., Haven, E.: Using empirical data to estimate potential functions in commodity markets: some initial results. Int. J. Theor. Phys. **56**(12), 4092–4104 (2017)
26. Tahmasebi, F., Meskinimood, S., Namaki, A., Farahani, S.V., Jalalzadeh, S., Jafari, G.R.: Financial market images: a practical approach owing to the secret quantum potential. Eur. Lett. **109**(3), 30001 (2015)
27. Bagarello, F., Haven, E.: Toward a formalization of a two traders market with information exchange. Phys. Scr. **90**(1), 015203 (2015)

Quantum-Like Model of Subjective Expected Utility: A Survey of Applications to Finance

Polina Khrennikova(✉)

School of Business, University of Leicester, Leicester LE1 7RH, UK
pk228@le.ac.uk

Abstract. In this survey paper we review the potential financial applications of quantum probability (QP) framework of subjective expected utility formalized in [2]. The model serves as a generalization to the classical probability (CP) scheme and relaxes the core axioms of commutativity and distributivity of events. The agents form subjective beliefs via the rules of projective probability calculus and make decisions between prospects or lotteries by employing utility functions and some additional parameters given by a so called 'comparison operator'. Agents' comparison between lotteries involves interference effects that denote their risk perceptions from the ambiguity about prospect realisation when making a lottery selection. The above framework that builds upon the assumption of *non-commuting* lottery observables can have a wide class of applications to finance and asset pricing. We review here a case of an investment in two complementary risky assets about which the agent possesses non-commuting price expectations that give raise to a state dependence in her trading preferences. We summarise by discussing some other behavioural finance applications of the QP based selection behaviour framework.

Keywords: Subjective expected utility · Quantum probability
Belief state · Decision operator · Interference effects
Complementary of observables · Behavioural finance

1 Introduction

Starting with the seminal paradoxes revealed in thought experiments by [1,10] the classical neo-economic theory was preoccupied with modelling of the impact of ambiguity and risk upon agent's probabilistic belief formation and preference formation. In classical decision theories due to [43,54] there are two core components of a decision making process: (i) probabilistic processing of information via Bayesian scheme, and formation of subjective beliefs; (ii) preference formation that is based on an attachment of utility to each (monetary) outcome. The domain of behavioural economics and finance, starting among others with the early works by [22–26,35,45,46] as well as works based on aggregate

V. Kreinovich et al. (Eds.): ECONVN 2019, SCI 809, pp. 76–89, 2019.
https://doi.org/10.1007/978-3-030-04200-4_5

finance data, [47,49,50] laid the foundation to a further exploration and modeling of human belief and preference evolution under ambiguity and risk. The revealed deviations from rational reasoning (with some far reaching implications for the domains of asset pricing, corporate finance, agents' reaction to important economic news etc.) suggested that human mental capabilities, as well as environmental conditions, can shape belief and preference formation in an context specific mode. The interplay between human mental variables and the surrounding decision-making environment is often alluded to in the above literature as mental biases or 'noise' that are perceived as a manifestation of a deviation from the normative rules of probabilistic information processing and preference formation, [9,22,25].[1] More specifically, these biases create fallacious probabilistic judgments and 'colour' information update in a non-classical mode, where a context of ambiguity or a experienced decision state (e.g. a previous gain and loss, framing, order of decision making task) can affect: (a) beliefs about the probabilities, (b) tolerance to risk and ambiguity and hence, the perceived value of the prospects. The prominent Prospect Theory by [23,53], approaches these effects via functionals that have an 'inflection point' corresponding to an agent's 'status quo' state. In different decision making situations a switch in beliefs or risk attitudes is captured via the different probability weighting functionals or value function. The models by [32,37] tackle preference reversals under ambiguity through a different perspective by assuming a different utility between risky and ambiguous prospects to incorporate agents' ambiguity premiums. Other works also tackle the non-linearity of human probability judgements that are identified in the literature as causes of preference reversals over lotteries and ambiguous prospects, [13,14,35,45]. Agents can also update the probabilities in a non-Bayesian mode under ambiguity and risk, see experimental findings in [46,53] and recently [19,51].

Ambiguity impact on the formation of subjective beliefs and preferences, as well as uncertain information processing, has been also successfully formalized through the notion of quantum probability (QP) wave interference, starting with early works by [27,28]. In the recent applications of QP in economics and decision theory contributions by [7,8,17,18,30,38,56] tackle the emergence beliefs and preferences under non-classical ambiguity that describe well the violation of classical Bayesian updating scheme in 'Savage Sure Thing principle' problems and the 'agree to disagree' paradox. The authors in [19] non-consequential preferences in risky investment choices are modelled in via generalized operator projectors. A QP model for order effects that accounts for specific QP regularity in preference frequency from non-commutativity is devised [55] and further explored in [29]. Ellsberg and Machina paradox-type behaviour from context

[1] A deviation from classical information processing and other instances of 'non-optimization' in a vNM sense are not universally considered as an exhibition of 'low intelligence', but as a mode of a faster and more efficient decision making process that is built upon using mental shortcuts and heuristics, in a given decision making situation, also known through Herbert Simon's notion of 'bounded rationality' that is reinforced in the work by [12].

dependence and ambiguous beliefs is explained in [18] through positive and negative interference effects. A special ambiguity sensitive probability weighting function is derived with an special parameter from the interference term λ in [2]. The existence of the 'zero prior paradox' that challenges the Bayesian updating from uninformative priors is solved in [5] with the aid of quantum transition probabilities that follow the Born rule of state transition and probability computation. The recent work by [6] serves as an endeavour to generalise the process of lottery ranking, based on their utility and risk combined with other internal decision making processes and agent's preference 'fluctuations'.

The remainder of this survey is organized as follows: in the next Sect. 2 we present a non-technical introduction to the neo-classical utility theories under uncertainty and risk. In Sect. 3 we discuss the main causes of non-rational behaviour in finance, pertaining among other to inflationary and deflationary asset prices that deviate from a fundamental valuation of assets. In Sect. 4 we summarize assumptions of the proposed QP based model of subjective expected utility and define the core mathematical rules pertaining to lottery selection from an agent's (indefinite) comparison state. In Sect. 5, we outline a simple QP rule of belief formation, when evaluating the price dynamics of two complimentary risky assets. Finally, in Sect. 6 we conclude and consider some possible future venues of research in the domain of QP based preference formation in asset trading.

2 VNM Framework of Preferences over Risky Lotteries

The most well-known and debated theory of choice in modern economics, the expected utility theory for preferences under risk, (henceforth vNM utility theory) was derived by von Neumann and Morgenstern, [54]. Similar axiomatics for subjective probability judgements over uncertain states of the world and expected utility preferences over outcomes was conceived by Savage in 1954 [43], and is mostly familiar to the public through the key axiom of rational behaviour, the "Sure Thing Principle". These theories served as a benchmark in social science (primarily in modern economics and finance) in respect to how an individual, confronted with different choice alternatives in situations involving risk and uncertainty should act, as to maximise her perceived benefits. Due to their prescriptive appeal and reliance on employment of the canons of formal logic, the above theories were coined as normative decision theories.[2] The notion of maximization of personal utility that quantifies the moral expectations associated with a decision outcome together with the possibility of quantifying risk and uncertainty through objective and subjective probabilities, allowed to

[2] Johnson-Laird and Shafir, [20], separate choice theories into three categories: normative, descriptive and prescriptive. The descriptive accounts have as their goal to capture the real process of decision formation, see e.g. Prospect Theory and its advances. Prescriptive theories are not easy to fit into either category (normative, or descriptive). In a sense, prescriptive theories would provide a prognosis on how a decision maker ought to reason in different contexts.

establish a simple optimization technique that each decision maker ought to follow by computing the expectation values of lotteries or state outcomes in terms of the level of utility, to always choose a lottery with highest expected utility.

According to Karni [21], the main premises of vNM utility theory that relate to risk attitude are based on: (i) separability in evaluation of mutually exclusive outcomes; (b) the evaluations of outcomes may be quantified by the cardinal utility U; (c) utilities may be obtained by firstly computing the expectations of each outcome with respect to the risk encoded in the objective probabilities; and finally d) the utilities of the considered outcomes are aggregated. These assumptions imply that utilities of outcomes are context independent and the agents can form joint probabilistic picture of the consequences of all considered lotteries.[3] We stress that agents ought to evaluate the objective probabilities associated with the prospects following the rules of classical probability theory and employ a Bayesian updating scheme to obtain posterior probabilities, following [34].

3 Anomalies in Preference Formation and Some Financial Market Implications

The deviations from classical probability based information processing hinged by the state dependence of economic agents' valuation of payoffs has far reaching implications for their trading on the finance market, fuelling disequilibrium prices of the traded risky assets. In this section we provide a compressed review of the mispricing of financial assets combined with the failure of classical models, such as Capital Asset Pricing Model to incorporate agents' risk evaluation of the traded assets. The mispricing of assets from agents' trading behaviour can be attributed to their non-classical beliefs, characterised by optimism in some trading periods that gives raise to instances of overpricing that surface in financial bubbles, see foundational works by [16,44]. Such disequilibrium market prices can also be observed for specific classes of assets, as well as exhibit intertemporal patterns, cf. the seminal works by [3,4]. The former work attributes mispricing of some classes of assets to informational incompleteness of markets (put differently, the findings show a non-reflection of all information in the asset prices of classes of assets with a high P/E ratio that is not in accord with the semi-strong form of efficiency), while the latter work explores under-pricing of small companies' shares, and stipulates that agents demand a higher risk premium for these types of assets. Banz [3] brings forwards an important argument about the mispricing causes, by attributing the under-valuation of small companies' assets to the possible ambiguous information content about the fundamentals.[4] The notion of

[3] This assumption is also central for a satisfaction of the independence axiom and the reduction axiom of compound lotteries, in addition to other axioms establishing the preference rule, such as completeness and transitivity.

[4] A theoretical analysis in [36] in a similar vein shows an existence of a negative welfare effect from agents' ambiguity averse beliefs about the idiosyncratic risk component of some asset classes that also yields under-pricing of these assets and a reduced diversification with these assets.

informational ambiguity and its impact upon agents' trading decisions attracted a large wave of attention in finance literature, with theoretical contributions, as well as experimental studies, looking into possible deviations from the rational expectations equilibrium and the corresponding welfare implications. We can mention among others the stream of 'ambiguity aversion' centered frameworks by Epstein and his colleagues, [11], as well as model [36] on specific type for ambiguity in respect to asset specific risks and related experimental findings by [42,51]. Investors can have a heterogeneous attitude towards ambiguity, and also, exhibit state dependent shifts in their attitude towards some kinds of uncertainties. For instance, 'ambiguity seeking' expectations, manifest in an overweighting of uncertain probabilities can also take place under specific agent states, [41], and references herein. The notion of *state dependence* that we attached a more outspread meaning in the above discussion is formalized more precisely via an inflection of the functionals related to preferences and expectations: (i) the value function that captures an attitude towards the risk has a dual shape around this point; ii) probability weighting function that depicts individual beliefs about the risky and ambiguous probabilities of prospects in the Prospect Theory formalisation by [23,53].[5] The notion of loss aversion and its impact on asset trading is also widely explored in the literature. Agents can similarly exhibit a discrepancy in their valuation of the already owned assets and the ones they did not yet invest in, known as a manifestation of endowment effect introduces in [24]. The work by [?] shows the reference point dependence of investors' perception of the positive and negative return, supported by related experimental findings with other types of payoffs by [19,46,48] in investment setting. Loss aversion gives raise to investors' unwillingness to sell an asset, if they treat the purchase price as a reference point, and a negative return as a sure loss. The agents exhibit a high level of disutility from losing this change in the price, which feeds into a sticky asset holding behaviour on their side, in a hope to break even in respect to the reference point. This trading behaviour clearly shows that trading behaviour and previous gains and losses can affect the subsequent investment behaviour of the agents, even in the absence of important news. The proposed QP based subjective expected utility theory has the potential to describe some of the above reviewed investment 'anomalies' from the viewpoint of rational decision making. We provide a short summary of the model in the next Sect. 4.

[5] We note that 'state dependence' that we can also allude to as 'context dependence', as coined in [26], indicates that agents can be affected by other factors besides, e.g., previous losses or levels of risk in the process of their preference and belief formation. As we indicated earlier, agents beliefs and value perception can be interconnected in their mind, whereby shifts in their welfare level can also transform their beliefs. This more broad based type of impact of the current decision making state of the agent upon her beliefs and risk preferences is well addressed by the 'mental state' wave function in QP models see, e.g., detailed illustration in [8,17,39].

4 QP Lottery Selection from an Ambiguous State

The QP lottery selection theory can be considered a generalization of Prospect theory that captures a state dependence in lottery evaluation, where utilities and beliefs about lottery realizations are dependent on the riskiness of the set of lotteries that are considered.

The lottery evaluation and comparison process devised in [2] and generalized to a multiple lottery comparison in [6] is in nutshell based on the following premises:

- The choice lotteries L_A and L_B are treated by the decision maker as complimentary, and she does not perform a joint probability evaluation of the outcomes of these lotteries. The initial comparison state, ψ, is an undetermined preference state, for which interference effects are present that encode agent's attitude to the risk of each lottery separately. This attitude is quantified by the *degree of evaluation of risk* (DER). The attitude to risk is different from the classical risk attitude measure (based on the shape of the utility function), and is related to the fear of the agent of getting an undesirable lottery outcome. The interference parameter, λ, serves as an input in the probability weighting function (i.e. the interference of probability amplitudes corresponds well to the probability weights in the Prospect Theory value function, [53]. Another source of indeterminacy are preference reflections between the desirability of the two lotteries that are given by *non-commuting* lottery operators.
- The utilities that are attached to each lottery's eigenvalue correspond to the individual benefit from some monetary outcome (e.g. \$100 or \$−50) and are given by classical vNM utility functions that are computed via mappings from each observed lottery eigenstate to a real number associated with a specific utility value. We should note that the utilities $u(x_i)$ are attached to the outcome of a specific lottery. With other words the utilities are 'lottery dependent' and can change, when the lottery setting (lottery observable) changes. If the lotteries to be compared are sharing the same basis then their corresponding observables are said to be compatible and the same amounts of each lottery payoffs would correspond the equivalent utilities as in the classical vNM formalization, e.g., $u(L_A; 100) = u(L_B; 100)$.
- The comparisons of utilities between the lottery outcomes are driven by a special *comparison operator D*, coined in the earlier work by [2]. This operator induces sequential comparison between the utilities obtained from lottery outcomes, such as L_1^A and L_2^B. Mathematically this operator consists of two 'sub-operators' that induce comparisons of the relative utility from switching the preferences between the two lotteries. State transition driven by $D_{B \to A}$ component generates the positive utility from selection of the L_A and negative utility from foregoing the L_B. The component $D_{A \to B}$ triggers a reverse state dynamics of the agents' comparison state. Hence, the composite comparison operator D allows to compute the difference in relative utility from the above comparisons, mathematically given as $D = D_{B \to A} - D_{A \to B}$. If the value is positive, then a preference rule for L_A is established.

- The indeterminacy in respect to the lottery realization is given by interference term associated with the beliefs about the outcomes of each lottery. More precisely the beliefs of the representative agents about the lottery realizations are affected by the interference of the complex probability amplitudes and therefore, can deviate from the objectively given lottery probability distributions. The QP based subjective probabilities are closely reproducing specific type of probability weighting function that captures ambiguity attraction to low probabilities and ambiguity aversion to high (>> 1) probabilities, cf. concrete probability weighting functionals estimated in [15,40,53].[6]
This function is of the form:

$$w_{\lambda,\delta}(x) = \frac{\delta x^{\lambda}}{\delta x^{\lambda} + (1-x)^{\lambda}}, \tag{1}$$

The parameters λ and δ control the curvature and elevation of the function 1, see for instance [15]. The smaller the value of the above concavity/convexity parameter the more 'curved' is the probability weighting function. The derivation of such a curvature of the probability weighting function from the QP amplitudes corresponds to one specific type of parameter function with $\lambda = 1/2$.

4.1 A Basic Outline of the QP Selection Model

In classical vNM mode we assume that an agent evaluates some ordinary risky lotteries L_A and L_B. Every lot contains $n =$ outcomes, with $i = 1, 2, 3..n$ each of them given with an objective probability p. Probabilities across lots sum up to one, and all outcomes are different, whereby no lottery stochastically dominates the other. We denote the lots by their outcomes and probabilities, $L_A = (x_i; p_i)$, $L_B = (y_i; q_i)$, where x_i, y_i are some random outcomes and p_i, q_i are the corresponding probabilities. The outcomes of both lots can be associated with a specific utility, e.g. assume that $x_1 = 100$ we can get $u(x_1) = u(100)$.[7] The comparison state is given in a simplest mode as a superposition state ψ in respect to the orthonormal bases associated with each lottery. In a two lot example, they are given by Hermitian operators that do not commute. Mathematically they posses different basis vectors. We denote these lots as L_A and L_B, each of them consisting of n eigenvectors, $|i_a\rangle$, respective $|i_b\rangle$ that form two orthonormal bases in the complex Hilbert space H. Each eigenvector $|i_a\rangle$ corresponds to a realization of a lottery specific monetary consequence given by the same eigenvalue. The agent forms her preferences by mapping from eigenvalues (x_i or y_i) to some numerical utilities, $|i_a\rangle \rightarrow u(x_i)$, $|j_b\rangle \rightarrow u(y_j)$. The utility values can be context specific in respect to: (a) L_A and L_B outcomes and their probabilistic composition; (b) correlation between the set of lotteries to be selected. The difference in

[6] Some psychological factors that can contribute to the particular parameter values are further explored in [57].

[7] We stress one important distinction of the utility computation in the QP framework, where utility value is depending on the particular lottery observable, and not only to the monetary outcome.

coordinates that determine the corresponding bases gives rise to a variance in the mapping from the eigenvalues to utilities.

The comparison state ψ can be represented with respect to the basis of the lottery operators, denoted as A or B, $\psi = \sum_i c_i|i_a\rangle$, where c_i are complex coordinates satisfying the normalization condition via: $\sum_i |c_i|^2 = 1$. This is a linear superposition representation of an agent's comparison state, when an evaluation of the consequences of L_A given by corresponding operator takes place. The comparison state can be fixed in a similar mode with respect to the basis of the operator L_B. The squared absolute values of the complex coefficients, c_i, provide a classical probability measure for obtaining the outcome i, $p_i = |c_i|^2$, given by the Born Rule.

An important feature of complex probability amplitude calculus that each c_i is associated with a phase that is due to oscillations of these probability amplitudes. For detailed representation consult an earlier work by [6] and monographs by [8,17]. Without going into mathematical details in this survey, we emphasise the importance of the phases between the basis vectors that quantify the interference effects of the probability amplitudes that correspond to underweighting (destructive interference), respective overweighting (constructive interference) of subjective probabilities. The non-classical effects cause deviations of agents' probabilistic beliefs from the objectively given odds as derived in Eq. (1).

The selection process of an agent is complicated by the need to carry out comparisons between several lots (limit the discussion to two lots L_A and L_B without the loss of generalisability). These comparisons are sequential since the agent cannot measure two of the corresponding observables jointly. The composite comparison operator D that serves to generate preference fluctuations of the agent between the lotteries is given by two comparison operators $D_{B \to A}$ and $D_{A \to B}$ that describe the relative utility of transiting from a preference for one lottery to the other.[8] The sub-operator, $D_{B \to A}$, represents the utility of a selection of the lottery A relative to the utility of the lottery B. This is the net utility the agent gets, after accounting in utility gain from L_A and utility loss by abandoning L_B. Formally this difference can be represented as: $u_{ij} = u(x_i) - u(y_j)$, where $u(x_n)$ is utility of the potential outcome x_i of L_A and $u(y_j)$ is the utility of a potential outcome y_j part of L_B.

In the same way the transition operator $D_{A \to B}$ provides a relative utility of the selection of the lottery L_B relatively to the utility of a selection of the lottery L_A. The comparison state of the agent fluctuates between preferring the outcomes of the A-lottery to outcomes of the B-lottery (formally represented by the operator $D_{B \to A}$) and inverse preference (formally represented by the operator component $D_{A \to B}$). Finally, an agent is computing the average utility from preferring L_A to L_B in comparison with choosing L_B over L_A that is given by a difference in the net utilities in the above described preference transition scheme. A comparison operator based judgment of the agent is in essence a comparison of

[8] The splitting of the composite comparison operator into two sub-operators that generate the reflection dynamics of the agents' indeterminate preference state is a mathematical construct that aims to illustrate the *process* behind lottery evaluation.

two relative utilities represented by the sub-operators $D_{B \to A}$ and $D_{A \to B}$ establishing a preference rule that gives $L_A \geq L_B$ *iff* the average utility computed by the composite comparison operator D is positive, i.e. the average of the comparison operator is higher than zero. Finally, on the composite state space level of lottery selection, the interference effects between the probability amplitudes, denoted by λ occur depending on the lottery payoff composition. The parameter gives a measure of an agent's DER (degree of evaluation of risk), associated with a preference for a particular lottery that is psychologically associated with a fear to obtain an 'undesirable' outcome, such as a loss.

5 Selection of Complimentary Financial Assets

On the level of the composite finance market agents are often influenced by order effects when forming the beliefs about the traded risky assets' price realizations. These effects are often coined 'overreaction' in behavioural finance literature [47,49], and can be considered as a manifestation of state dependence in agents' belief formation that affect their selling and buying preferences. We also refer to some experimental studies on the effect of previous gains and losses upon agents' investment behaviour, see for instance, [19,33,49].

Based on the assumptions made in [31], about the non-classical correlations that assets' returns can exhibit, we present here a simple QP model of an agent's asset evaluation process with an example of two risky assets, k and n as she observes the price dynamics. The agent is uncertain about the price dynamics of these assets and does not possess a joint probability evaluation of their price outcomes. Hence, interference effects exist in respect to the price realizations beliefs of these assets. In other words, asset observable are complimentary, and order effects in respect to the final evaluation of the price dynamics of these assets emerge. The asset price variables are depicted through non-commuting operators following the QP models of order effects, [52,55].

By making a decision $\alpha = \pm 1$ or the asset k, an agent's state ψ is projected onto the eigenvector $|\alpha_i\rangle$ that corresponds to an eigenstate for a particular price realization for that asset.[9] After the next trading period price realization belief about the asset k, the agent proceeds by forming a belief about the possible price behaviour of the asset n and she performs a measurement of the corresponding expectation observable, but for the updated belief-state $|+_i\rangle$ and she obtains the eigenvalues of the price behaviour observable of asset n with $\beta = \pm 1$ given by the transition probabilities:

$$p_{k \to n}(\alpha \to \beta) = |\langle \alpha_k | \beta_n \rangle|^2. \tag{2}$$

[9] In the simple setup with two types of discrete price movements, we fix only two eigenvectors $|\alpha_+\rangle$ and $|\alpha_-\rangle$, corresponding to eigenvalues $a = \pm 1$.

The eigenvalues correspond to the possible price realizations of the respective assets.[10]

The above exposition of state transition allows to obtain the quantum transition probabilities that denote agents beliefs about the asset n prices when she has observed the asset k price realization. The transition probabilities have also an objective interpretation. Consider an ensemble of agents in the same state ψ, who made a decision α, with respect to the price behavior of the kth asset. As a next step, the agents form preferences about the nth asset and we choose only those, whose firm decision is β. In this way it is possible to find the frequency-probability $p_{k\to n}(\alpha \to \beta)$. Following the classical tradition, we can consider these quantum probabilities as analogues of the conditional probabilities, $p_{k\to n}(\alpha \to \beta) \equiv p_{n|k}(\beta|\alpha)$. We remark that the belief formation about asset prices in this setup takes place under informational ambiguity. Hence, in each of the subsequent belief states about the price behaviour the agent is in a superposition in respect price behaviour of the complementary asset, and interference effects exist for each agent's pure belief state (that can be approximated by a notion of a representative agent).

Given the probabilities, in (2) we can define a quantum joint probability distribution for forming beliefs about both of the two assets k and n.

$$p_{kn}(\alpha, \beta) = p_k(\alpha)p_{n|k}(\beta|\alpha). \tag{3}$$

This joint probability respects the order structure, as such:

$$p_{kn}(\alpha, \beta) \neq p_{nk}(\beta, \alpha), \tag{4}$$

This is a manifestation of order effects, or state dependence in belief formation that is not in accord with the classical Bayesian probability update, see e.g., analysis in [39,51,55]. Order effect imply a non-satisfaction of the joint probability distribution and bring a violation of the commutativity principle, as pointed out earlier.[11]

The obtained results with the QP formula can be also interpreted as subjective probabilities or an agent' degree of belief about the distribution of asset prices. As an example, the agent in the belief-state ψ considers two possibilities for the dynamics of the kth price. She speculates: suppose that kth asset would

[10] The model can be generalized to include the actual trading behaviour, i.e., where the agent does not only observe the price dynamics of the assets between the trading periods that feeds back into her beliefs about the complimentary assets' future price realizations, but also actually trades the assets, based on the perceived utility of each portfolio holding. In this setting the agent's mental state in relation to the future price expectations is also affected by the realized losses and gains.

[11] Order effects can exist for: (i) information processing related to the order effect for the observation of some sequences of signals; (ii) preference formation related to the sequence of asset evaluation or actual asset trading that we described now. Non-commuting observables allow to depict agents' state dependence in preference formation. As noted, when state dependence is absent, the observable operators are commuting.

demonstrate the $\alpha(=\pm1)$ behavior. Under this assumption (which is a type of 'counter-factual' update of her state ψ), she forms her beliefs about a possible outcome for the nth asset price. Starting with the counterfactually updated state $|\alpha_k\rangle$, she generates subjective probabilities for the price outcomes of both of these assets. These probabilities give the conditional expectations of the asset n price value $\beta = \pm$, after observing price behaviour of asset k, with a price value $\alpha = \pm1$.

We remark that following the QP setup the operators for the asset k and n price behaviour do not commute, i.e., $[\pi_k, \pi_n] \neq 0$. This means that these price observables are complementary in the same mode, as the lotteries that we considered in the Sect. 4. As a consequence, it is impossible to define a family of random variables $\xi_i : \Omega \rightarrow \{\pm1\}$ on the same classical probability space, $(\Omega, \mathcal{F}; P)$, which would reproduce the quantum probabilities $p_i(\pm1) = |\langle\pm_i|\psi\rangle|^2$ as $P(\xi_i = \pm)$ and quantum transition probabilities $p_{k\rightarrow n}(\alpha \rightarrow \beta) = |\langle\alpha_k|\beta_n\rangle|^2$, $\alpha, \beta = \pm$, as classical conditional probabilities $P(\xi_n = \beta|\xi_k = \alpha)$. If it were possible, then in the process of asset trading the agent's decision making state would be able to define sectors $\Omega(\alpha_1, ..., \alpha_N) = \{\omega \in \Omega : \xi_1(\omega) = \alpha_1, ..., \xi_N(\omega) = \alpha_N\}, \alpha_j = \pm$ and form firm probabilistic measures associated with the realization of the price of each asset, part of the N financial assets.

QP frameworks aids to depict agents' non-definite opinions about the prices behavior for traded 'complementary assets' and their ambiguity in respect to the vague probabilistic composition of the price state realizations of such set of assets. In the case of such assets, an agent forms her beliefs *sequentially*, and not jointly as is the case in the standard finance portfolio theory. She firstly resolves her uncertainty about the asset k, and only with this knowledge can she resolve the uncertainty about other assets (in our simple example the asset n.) The quantum probability belief formation scheme based on non-commuting asset price-observables can be applied to describe subjective belief formation of a representative agent by exploring the 'bets' or price observations of an ensemble of agents and approximate the frequencies by probabilities, see also an analysis in other information processing settings, [8,17,19,38].

6 Concluding Remarks

We presented a short summary of the advances of QP based decision theory with an example of lottery selection under risk, based on classical vNM expect utility function, [54]. The core premise of the presented framework is that *non-commutativity of lottery observables* can give raise to agents' belief ambiguity in respect to the subjective probability evaluation, in a similar mode, as captured by the probability weighing function presented in [2] based on the original weighting function from Prospect Theory in [53], followed by advances in [15,40]. In particular, the interference effects that are present in an agent's ambiguous comparison state, translate into over-, or underweighting of objective probabilities associated with the riskiness of the lots. The interference term and its size allows to quantify an agent's fear to obtain an undesirable outcome that is a part of her ambiguous comparison state. The agent compares the relative utilities of the lottery outcomes that are given by the eigenstates associated with the lottery specific orthonormal bases in the complex Hilbert space. This setup creates a lottery dependence of an agent's utility, where the lottery payoffs and probability composition play a role in her preference formation.

We also aimed to set the ground for broader application of QP based utility theory in financial applications, given the wide range of revealed behavioural anomalies that are often associated with non-classical information processing by investors and a state dependence in their trading preferences. The main motivation for the application of QP mathematical framework as a mechanism of probability calculus under non-neutral ambiguity attitudes among agents coupled with a state dependence of their utility perception derived from its ability to generalise the rules of classical probability theory, and capture the indeterminacy state before a preference is formed through the notion a *superposition*, as elaborated in a thorough synthesis provided in reviews by [18,39], and monographs by [8,17].

References

1. Allais, M.: Le comportement de l'homme rationnel devant le risque: critique des postulats et axiomes de l'Ecole americaine. Econometrica **21**, 503–536 (1953)
2. Asano, M., Basieva, I., Khrennikov, A., Ohya, M., Tanaka, Y.: A quantum-like model of selection behavior. J. Math. Psych. **78**, 2–12 (2017)
3. Banz, R.W.: The relationship between return and market value of common stocks. J. Fin. Econ. **9**(1), 3–18 (1981)
4. Basu, S.: Investment performance of common stocks in relation to their price-earning ratios: a test of the Efficient Market Hypothesis. J. Financ. **32**(3), 663–682 (1977)
5. Basieva, I., Pothos, E., Trueblood, J., Khrennikov, A., Busemeyer, J.: Quantum probability updating from zero prior (by-passing Cromwell's rule). J. Math. Psych. **77**, 58–69 (2017)
6. Basieva, I., Khrennikova, P., Pothos, E., Asano, M., Khrennikov, A.: Quantum-like model of subjective expected utility. J. Math. Econ. (2018). https://doi.org/10.1016/j.jmateco.2018.02.001
7. Busemeyer, J.R., Wang, Z., Townsend, J.T.: Quantum dynamics of human decision making. J. Math. Psych. **50**, 220–241 (2006)
8. Busemeyer, J., Bruza, P.: Quantum models of Cognition and Decision. Cambridge University Press (2012)
9. Costello, F., Watts, P.: Surprisingly rational: probability theory plus noise explains biases in judgment. Psych. Rev. **121**(3), 463–480 (2014)
10. Ellsberg, D.: Risk, ambiguity and the Savage axioms. Q. J. Econ. **75**, 643–669 (1961)
11. Epstein, L.G., Schneider, M.: Ambiguity, information quality and asset pricing. J. Finance **LXII**(1), 197–228 (2008)
12. Gigerenzer, G., Selten, R.: Bounded Rationality: The Adaptive Toolbox. MIT Press (2002)
13. Gilboa, I., Schmeidler, D.: Maxmin expected utility with non-unique prior. J. Math. Econ. **18**, 141–153 (1989)

14. Gilboa, I.: Theory of decision under uncertainty. Econometric Society Monographs (2009)
15. Gonzales, R., Wu, G.: On the shape of the probability weighting function. Cogn. Psych. **38**, 129–166 (1999)
16. Harrison, M., Kreps, D.: Speculative investor behaviour in a stock market with heterogeneous expectations. Q. J. Econ. **89**, 323–336 (1978)
17. Haven, E., Khrennikov, A.: Quantum Social Science. Cambridge University Press, Cambridge (2013)
18. Haven, E., Sozzo, S.: A generalized probability framework to model economic agents' decisions under uncertainty. Int. Rev. Financ. Anal. **47**, 297–303 (2016)
19. Haven, E., Khrennikova, P.: A quantum probabilistic paradigm: non-consequential reasoning and state dependence in investment choice. J. Math. Econ. (2018). https://doi.org/10.1016/j.jmateco.2018.04.003
20. Johnson-Laird, P.M., Shafir, E.: The interaction between reasoning and decision making: an introduction. In: Johnson-Laird, P.M., Shafir, E.: Reasoning and Decision Making. Blackwell Publishers, Cambridge (1994)
21. Karni, E.: Axiomatic foundations of expected utility and subjective probability. In: Machina, M.J., Kip Viscusi, W. (eds.) Handbook of Economics of Risk and Uncertainty, pp. 1–39. Oxford, North Holland (2014)
22. Kahneman, D., Tversky, A.: Subjective probability: a judgement of representativeness. Cogn. Psych. **3**(3), 430–454 (1972)
23. Kahneman, D., Tversky, A.: Prospect theory: an analysis of decision under risk. Econometrica **47**, 263–291 (1979)
24. Kahneman, D., Knetch, J.L., Thaler, R.H.: Experimental tests of the endowment effect and the coarse theorem. J. Polit. Econ. **98**(6), 1325–1348 (1990)
25. Kahneman, D.: Maps of bounded rationality: psychology for behavioral economics. Am. Econ. Rev. **93**(5), 1449–1475 (2003)
26. Kahneman, D., Thaler., R.: Utility maximization and experienced utility. J. Econ. Persp. **20**, 221–234 (2006)
27. Khrennikov, A.: Classical and quantum mechanics on information spaces with applications to cognitive, psychological, social and anomalous phenomena. Found. Phys. **29**, 1065–1098 (1999)
28. Khrennikov, A.: Quantum-like formalism for cognitive measurements. Biosystems **70**, 211–233 (2003)
29. Khrennikov, A., Basieva, I., Dzhafarov, E.N., Busemeyer, J.R.: Quantum models for psychological measurements : An unsolved problem. PLoS ONE **9** (2014). Article ID: e110909
30. Khrennikov, A.: Quantum version of Aumann's approach to common knowledge: sufficient conditions of impossibility to agree on disagree. J. Math. Econ. **60**, 89–104 (2015)
31. Khrennikova, P.: Application of quantum master equation for long-term prognosis of asset-prices. Physica A **450**, 253–263 (2016)
32. Klibanoff, P., Marinacci, M., Mukerji, S.: A smooth model of decision making under ambiguity. Econometrica **73**, 1849–1892 (2005)
33. Knutson, B., Samanez-Larkin, G.R., Kuhnen, C.M.: Gain and loss learning differentially contribute to life financial outcomes. PLoS ONE **6**(9), e24390 (2011)
34. Kolmogorov, A.N.: Grundbegriffe der Warscheinlichkeitsrechnung, Springer, Berlin (1933). English translation: Foundations of the Probability Theory. Chelsea Publishing Company, New York (1956)
35. Machina, M.J.: Choice under uncertainty: problems solved and unsolved. J. Econ. Perspect. **1**(1), 121–154 (1987)

36. Mukerji, S., Tallan, J.M.: Ambiguity aversion and incompleteness of financial markets. Rev. Econ. Stud. **68**, 883–904 (2001)
37. Nau, R.F.: Uncertainty aversion with second-order utilities and probabilities. Manag. Sci. **52**, 136–145 (2006)
38. Pothos, M.E., Busemeyer, J.R.: A quantum probability explanation for violations of rational decision theory. Proc. Roy. Soc. B **276**(1665), 2171–2178 (2009)
39. Pothos, E.M., Busemeyer, J.R.: Can quantum probability provide a new direction for cognitive modeling? Behav. Brain Sc. **36**(3), 255–274 (2013)
40. Prelec, D.: The probability weighting function. Econometrica **60**, 497–528 (1998)
41. Roca, M., Hogarth, R.M., Maule, A.J.: Ambiguity seeking as a result of the status quo bias. J. Risk and Uncertainty **32**, 175–194 (2006)
42. Sarin, R.K., Weber, M.: Effects of ambiguity in market experiments. Manag. Sci. **39**, 602–615 (1993)
43. Savage, L.J.: The Foundations of Statistics. Wiley, US (1954)
44. Scheinkman, J., Xiong, W.: Overconfidence and speculative bubbles. J. Polit. Econ. **111**, 1183–1219 (2003)
45. Schemeidler, D.: Subjective probability and expected utility without additivity. Econometrica **57**(3), 571–587 (1989)
46. Shafir, E.: Uncertainty and the difficulty of thinking through disjunctions. Cognition **49**, 11–36 (1994)
47. Shiller, R.: Speculative asset prices. Amer. Econ. Rev. **104**(6), 1486–1517 (2014)
48. Thaler, R.H., Johnson, E.J.: Gambling with the house money and trying to break even: the effects of prior outcomes on risky choice. Manag. Sci. **36**(6), 643–660 (1990)
49. Thaler, R.: Misbehaving. W.W. Norton & Company (2015)
50. Thaler, R.: Quasi-Rational Economics. Russel Sage Foundations (1994)
51. Trautman, S.T.: Shunning uncertainty: the neglect of learning opportunities. Games Econ. Behav. **79**, 44–55 (2013)
52. Trueblood, J.S., Busemeyer, J.R.: A quantum probability account of order effects in inference. Cogn. Sci. **35**, 1518–1552 (2011)
53. Tversky, D., Kahneman, D.: Advances in prospect theory: cumulative representation of uncertainty. J. Risk Uncertainty **5**, 297–323 (1992)
54. von Neumann, J., Morgenstern, O.: Theory of Games and Economic Behaviour. Princeton University Press, Princeton (1944)
55. Wang, Z., Busemeyer, J.R.: A quantum question order model supported by empirical tests of an a priori and precise prediction. Topics in Cogn. Sci. **5**, 689–710 (2013)
56. Yukalov, V.I., Sornette, D.: Decision Theory with prospect inference and entanglement. Theory Dec. **70**, 283–328 (2011)
57. Wu, G., Gonzales, R.: Curvature of the probability weighting function. Manag. Sci. **42**(12), 1676–1690 (1996)

Agent-Based Artificial Financial Market

Akira Namatame(✉)

Department of Computer Science, National Defense Academy, Yokosuka, Japan
akiranamatame@gmail.com

Abstract. In this paper, we study the agent modelling in an artificial stock market. In an artificial stock market, we consider two broad types of agents, "rational traders" and "imitators". Rational traders trade to optimize their short-term profit and imitators invest based on the trend follow strategy. We examine how the coexistence of rational and irrational traders affect stock prices and their long run performance. We show the performances of these traders depend on their ratio in the market. In the region where rational traders are in the minority, they can come to win the market, in that they eventually have a high share of wealth. On the other hand, in the region where rational traders are in the majority, imitators can come to win the market. We conclude that the survival in a finance market is a kind of the minority game, and mimic traders (noise traders) might survive and come to win.

1 Introduction

Economists have long asked whether traders who misperceive the future price can survive in a competitive market such as a stock or a currency market. The classic answer, given by Friedman (1953), is that they cannot. Friedman argued that mistaken investors buy high and sell low, as a result lose money to rational trader, and eventually lose all their wealth. Therefore, in the long run irrational investors cannot survive as they tend to lose wealth and disappear from the market. Offering an operational definition of rational investors, however, presents conceptual difficulties as all investors are boundedly rational. No agent can realistically claim to have the kind of supernatural knowledge needed to formulate rational expectations. The fact that different populations of agents with different strategies prone to forecast errors can coexist in the long run is a fact that still requires an explanation.

De Long et al. (1991) questioned the presumption that traders who misperceive returns do not survive. Since noise traders who are on average bullish bear more risk than do rational investors holding rational expectations, as long as the market rewards risk-taking such noise traders can earn a higher expected return even though they buy high and sell low on average. Because Friedmanś argument does not take account of the possibility that some patterns of noise traders' misperceptions might lead them to take on more risk, it cannot be correct as stated. But this objection to Friedman does not settle the matter, for

V. Kreinovich et al. (Eds.): ECONVN 2019, SCI 809, pp. 90–99, 2019.
https://doi.org/10.1007/978-3-030-04200-4_6

expected returns are not an appropriate measure of long run survival. To adequately analyze whether irrational (noise) traders are likely to persist in an asset market, one must describe the long-run distribution of their wealth, not just the level of expected returns.

In recent economic and finance research, there is a growing interest in marrying the two viewpoints, that is, in incorporating ideas from social sciences to account for the facts that markets reflect the thoughts, emotions, and actions of real people as opposed to the idealized economic investors who under lies the efficient markets and random walk hypotheses (Le Baron 2000). A real investors may intend to be rational and may try to optimize his or her actions, but that rationality tends to be hampered by cognitive biases, emotional quirks, and social influences. The behaviours of financial markets is thought to result from varying attitudes towards risk, the heterogeneity in the framing of information, cognitive errors, self-control and lack thereof, regret in financial decision making, and the influence of mass psychology. There is also growing empirical evidence of the existence of herd or crowd behaviour in markets. Herd behaviour is often said to occur when many traders take the same action, because they mimic the actions of others.

The question whether or not there are winning and losing market strategies, and what determines their characteristics have been discussed from the practical point of view (Cinocotti 2003). If a consistently winning market strategy exists, the losing trading strategies will disappear with the force of natural selection in the long run. Understanding if there are winning and losing market strategies and determine their characteristics is an important question. On one side, it seems obvious that different investors exhibit different investing behaviour which is, at least partially, responsible for the time evolution of market prices. On the other side, it is difficult to reconcile the regular functioning of financial markets with the coexistence of different populations of investors. If there is a consistently winning market strategy than it is reasonable to assume that the losing populations disappear in the long run.

In the past, several researchers tried to explain the stylized facts as the macroscopic outcome of an assemble of heterogeneous interacting agents (Cont 2000, Le Baron 2001). According this view, the market is populated by agents with different characteristics such as differences in access to and interpretation of available information, different expectations, or different trading strategies. The agents interact by changing information or they trade imitating the behaviour of other traders. Then, the market possesses an endogenous dynamics, and the universality of the statistical regularities is seen as an emergent property of this endogenous dynamics which is governed by the interactions of agents.

Boswijk et al. estimated the model to annual US stock price data from 1871 to 2003 (Boswijk 2007). The estimation results support the existence of two expectation regimes. One regime can be characterized as a fundamentalist regime, where agents believe in mean reversion of stock prices toward the benchmark fundamental value. The second regime can be characterized as a chartist, trend following regime where agents expect the deviations from the fundamental to trend. The fraction of agents using the fundamentalists and trend following forecasting rules show substantial time variation and switching between two regimes. It is suggested that behavioural heterogeneity is significant and that there are two different regimes: A mean reversion regime and a trend following regime. To each regime, there are corresponds a different investor type: fundamentalists and trend followers. These two investors types coexist and their fraction show considerable fluctuation over time. The mean-reversion regime corresponds to the situation when the market is dominated by the fundamentalists who recognize the asset and expect the stock price to move back towards its fundamental value. The other trend following regime represents a situation when the market is dominated by trend followers, expecting continuation of good news in the near future and expect positive stock returns. They also allow the coexistence of different types of investors with heterogeneous expectations about future pay-offs.

2 Efficient Market Hypothesis vs Interacting Agent Hypothesis

Rationality is one of the major assumptions behind many economic theories. Here we shall examine the efficient market hypothesis (EMH), which is behind most economic analysis of financial markets. In conventional economics, markets are assumed efficient if all available information is fully reflected in current market prices. Depending on the information set available, there are different forms of the EMH. It suggests that the information set includes only the history of prices or returns themselves. If the weak form of EMH holds in a market, abnormal profits cannot be acquired from analysis of historical stock prices or volume. In other words, analysing charts of past price movements, is a waste of time. The weak form of EMH is associated with the term random walk hypothesis. Random walk hypothesis suggests that investment returns are serially independent. That means the next period's return is not a function of previous returns. Prices only changes as a result of new information, such as the company has new, significant personnel changes, being made available.

A large number of empirical tests have been conducted to test the weak form of EMH. Recent work illustrated many anomalies, which are events or patterns that may offer investors opportunities to earn abnormal return. Those anomalies could not be explained by the form of EMH. To explain the empirical anomalies, many believe that new theories for explaining market efficiency remain to be discovered. Alfarano et al. (2005) estimated an EMH with fundamentalists and chartists to exchange rates and found considerable fluctuations of the market impact of fundamentalists. Their research suggests that behavioural heterogeneity is significant and that there are two different regimes: "A mean reversion regime" and "a trend following regime". To each regime, there corresponds a different investor type: fundamentalists and followers. These two investor types co-exist and their fractions show considerable fluctuations over time. The mean-version-reversion regime corresponds to the situation when the market is dominated by fundamentalists who recognize over or under pricing of the asset and expect the stock price to move back towards its fundamental value. The other trend following regime represents a situation when the market is dominated by trend followers, expecting continuation of good news in the near future and positive stock returns.

We may distinguish two competing hypotheses: One derive from the traditional Efficient Market Hypothesis (EMH) and a recent alternative which we might call Interacting Agent Hypothesis (IAH) (Tesfatsion 2002). The EMH states that the price fully and instantaneously reflects any new information: Therefore, the market is efficient in aggregating available information with its invisible hand. The traders (agents) are assumed to be rational and homogeneous with respect to the access and their assessment of information, and as a consequence, interactions among them can be neglected. Advances in computing give rise to a whole new area of research in the study of economics and social sciences. From an academic point of view, advances in computing give many challenges in economics. Some researchers attempt to gain better insight into the behaviour of markets. Agent-based research plays an important role in understanding the market behaviour. The design of the behaviour of the agents that participate in an agent-based model is very important. The type of agents can vary from very simple agents to very sophisticated ones. The mechanisms by which the agents learn can be based on many techniques like genetic algorithms, learning classifier systems, genetic programming, etc. Agent-based methods have been applied in many different economic environments. For instance, a price increase may induce agents to buy more or less depending on whether they believe there is new information carried in this change.

3 Agent-Based Modelling of an Artificial Market

One way to study properties of a market is to build artificial markets, whose dynamics are solely determined by agents that model various behaviours of humans. Some of these programs may attempt to model naive behaviour, others may attempt to exhibit intelligence. Since the behaviour of agents is completely under the designers' control, the experimenters have means to control various experimental factors and relate market behaviour to observed phenomena. The enormous degrees of freedom that one faces when one designs an agent-based market make the process very complex. The work by Arthur opened a new way of thinking about the use of artificial agents that behave like humans in financial markets simulations (Tesfasion 2002).

One of the most important part of agent based markets is the actual mechanism that governs the trading of assets. In most agent based markets they assume a simple price response to excess demand. Most markets of this type poll traders for their current demands, sum the market demands, and if there is an excess demand, increase the price. If there is an excess supply they decrease the price. Simple form of this rule would be where D(t) and S(t) are the demand and supply at time t respectively. The agent is maintaining the stock and the capital in the artificial market model in this research. The agent loses the capital by obtaining the stock and gets it by selling off the stock.

The basic model is to assume that the stock price reflect the excess demand, which is governed as

$$P(t) = P(t-1) + k[N_1(t) - N_2(t)] \tag{1}$$

where $P(t)$ is stock prices at time t, $N_1(t)$ is a number of agents to buy and $N_2(t)$ is a number of agents to sell respectively at time t, k is a constant. This expression implies that the stock price is a function of the excess demand, and the price rises when there are more agents to buy, and it descend when more agents to sell it. The price volatility as

$$v(t) = (P(t) - P(t-1))/P(t-1) \tag{2}$$

The stock one agent can buy and sell in one trading is one unit. We introduce a notional wealth $W_i(t)$ of agent i as:

$$W_i(t) = P(t)\Phi_i(t) + C_i(t) \tag{3}$$

where Φi is the number of assets held and C_i is the amount of cash held by agent i.

It is clear from equation that an exchange of cash for assets at any price does not in any way affect the agent's notional wealth. However, the point is in the terminology: the wealth $W_i(t)$ is only notional and not real in any sense. The only real measure of wealth $C_i(t)$, the amount of capital the agent has available to spend. Thus, it is evident that an agent has to do a round trip: buy (sell) an asset then sell (buy) it back to discover whether a real profit is made.

The profit rate of agent i at time t is given as

$$\gamma = W_i(t)/W_i(0) \tag{4}$$

4 Formulation of Trading Rules

In this paper, traders are segmented into two types depending on their trading behaviours: rational traders (chartist) and imitators. We address the important issue of the existence both types of traders.

(1) Rational traders (Chartists)

For modelling purposes, we have rational traders who make rational decision in the following stylized behaviour: If they expect the price goes up, then they will buy, and if they expect the stock price goes down then they will sell right now. Rational traders observe the trend of the market and trade so that their short-term pay-off will be improved. Therefore if the trend of the markets is "buy", then this agent's attitude is "sell". On the other hand, if the trend of the markets is "sell", then this agent's attitude is "buy". As can be seen, trading with the minority decision creates wealth for the agent on performing the necessary trip, whereas trading with majority decision loses wealth. However, if the agent had held the asset for a length of time between buying it and selling it back, his/her wealth would also depend on the rise and fall of the stock price over the holding period. However, the property that the purchaser (or seller) can be put in a single deal and bought (clearance) is one unit, so the agent who cannot buy and sell it when the number of the buyer and seller is different.

(i) When buyers are minority The agent cannot sell it even if it is selected to sell it exists. Because the price falls in the buyer's market still, it is an agent that sells who is maintaining a lot of properties. The agent who is maintaining the property more is enabled the clearance it.
(ii) When buyers are majority The agent cannot buy it even if it is selected to buy it exists. Because the price rises, being able to buy is still an agent who is maintaining a lot of capitals. The agent who is maintaining the more capital is able to purchase it.

We use the following terminology:

- N: Number of agent who participate in markets.
- $N_1(t)$: Number of agent who buy at time t.
- $R(t)$: The rate of buying agents at time t

$$R(t) = N_1(t)/N \tag{5}$$

We also denote $R_F(t)$ as the estimated value of R(t) by the rational trader i, which is defined as

$$R_F(t) = R(t-1) + \varepsilon_i \tag{6}$$

where $\varepsilon_i(-0.5 < \varepsilon_i < 0.5)$ is the rate of bullishness and timidity of agent i. If ε_i is large, this agent has tendency to "buy", and it is small, the tendency to "sell" is high. In a population of rational traders, ε is normally distributed.

<A trading rule of a rational trader>

$$\begin{aligned} &if\ R_F(t)\ < 0.5,\ then\ sell \\ &if\ R_F(t)\ > 0.5,\ then\ buy \end{aligned} \tag{7}$$

(2) Imitators

Imitators observe the behaviours of rational traders. If the majority of rational traders "buy", then imitators also "buy", on the other hand, if the majority of rational traders "sell" then they also "sell". We can formulate the imitator's behaviour as follows. $R_F(t)$: The ratio of rational traders to buy at time t $R_I(t)$: The estimated value of $R_F(t)$ by imitator j

$$R_I(t) = R_F(t-1) + \varepsilon_j \tag{8}$$

where $\varepsilon_j(-0.5 < \varepsilon_j < 0.5)$ is the rate of bullishness and timidity of imitator j which differs depending by each imitator. In a population of imitators ε is also normally distributed.

<A trading rule of imitator>

$$\begin{aligned} &if\ P_I(t)\ > 0.5,\ then\ buy \\ &if\ P_I(t)\ < 0.5,\ then\ sell \end{aligned} \tag{9}$$

5 Simulation Results

We consider a artificial stock market consists of 2,500 traders and simulate markets behaviour by varying the ratio of rational traders. We also obtain the long-run accumulation of wealth of each type of traders.

(Case 1) The ratio of rational traders: 20%

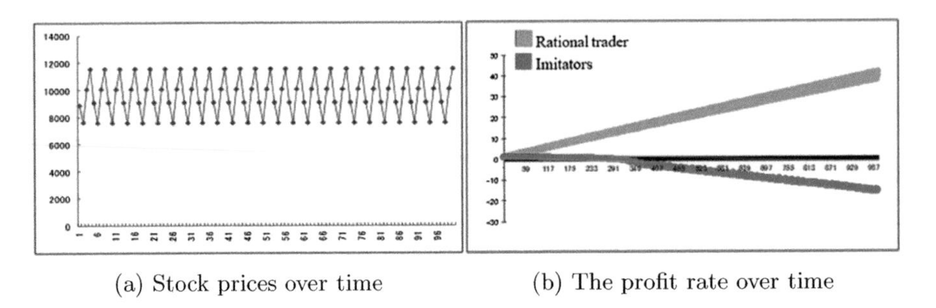

(a) Stock prices over time (b) The profit rate over time

Fig. 1. The stock price changes (a), and the profit rates of rational traders and imitators (b). The ratio of rational traders is 20%, and the ratio of imitators is 80%.

In Fig. 1(a) we show transition of the price when the ratio of the rational traders is 20%. Figure 1(b) show the transition of the average profit rate of the rational traders and imitators over time. In this case where the rational traders are in the minority, the average wealth of the rational traders is increasing over time and that of the imitator decreasing. When a majority of the traders are imitators, the stock price changes drastically. When stock prices goes up, a large number of traders buy then the stock price goes down next time period. Imitators mimic the movement of the small number of rational traders. If rational traders start to raise the stock price, imitators also move towards raising the stock price. If rational traders start to lower stock price, imitators also lower the stock price further. Therefore the movement of a large number of imitators amplifies the

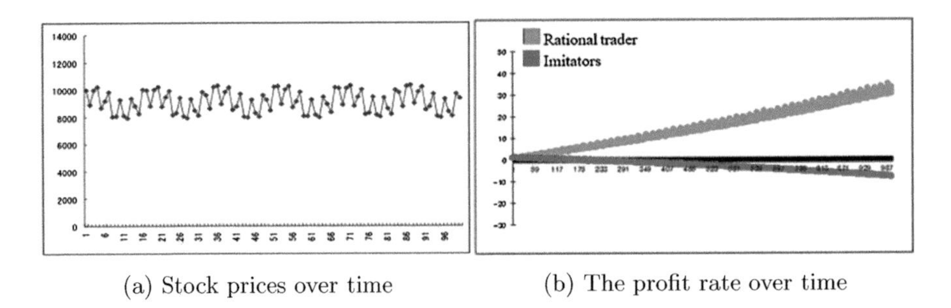

(a) Stock prices over time (b) The profit rate over time

Fig. 2. The stock price changes (a), and the profit rates of rational traders and imitators (b). The ratio of rational traders and imitators are the same: 50%.

movement of price caused by the rational traders causing a big fluctuation in stock prices. The profit rate of imitators is declining and that of the rational trader keeps to rise (Fig. 2).

(Case 2) The ratio of rational traders: 50%

In Case 2, the fluctuation of stock price is small compared with Case 1. The co-existence of the rational traders and imitators who mimic the behaviour of rational traders offset the fluctuation. The increase of the ratio of the rational traders stabilizes the market. About the rate of profit, rational trader is raising their profit but it is smaller compared with Case 1 (Fig. 3).

(Case 3) The ratio of rational traders: 80%

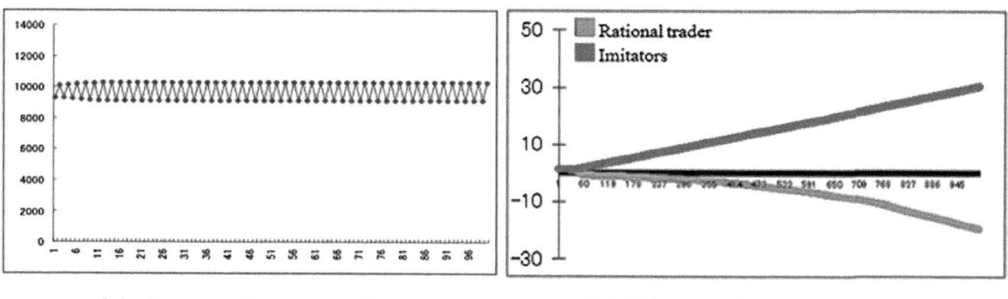

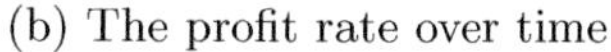

(a) Stock prices over time

(b) The profit rate over time

Fig. 3. The stock price changes (a), and the profit rates of rational traders and imitators (b). The ratio of rational traders is 80%, and that of imitators is 20%.

In Case 3, the fluctuation of stock prices becomes much smaller. Because there are a lot of rational traders, the market becomes efficient, the price change becomes to be small. In such an efficient market, case rational traders cannot raise the profit but imitators can raise their profit. In the region where the

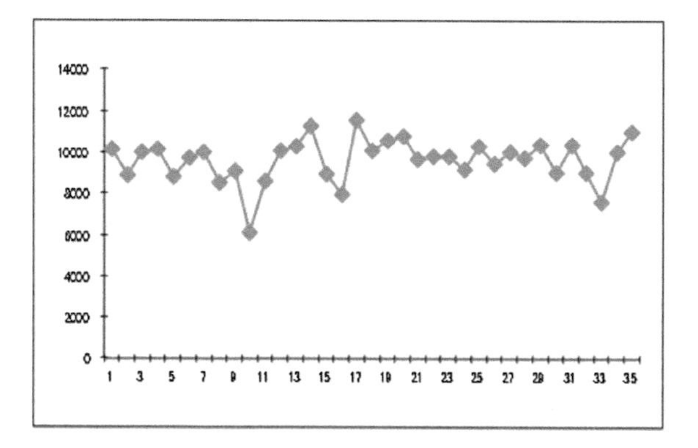

Fig. 4. The stock price changes when the ratio of rational traders is chosen randomly between 20% and 80%

rational traders are in the majority, and the imitators are in the minority, the average wealth of the imitator is increasing over time and that of the rational traders is decreasing. Therefore, in the region where imitators are in the minority, they are better off and their success in accumulating the wealth is due to the loss of the rational traders.

(Case 4) The ratio of rational traders: random between 20% and 80%

In Fig. 4, we show the change of the stock price when ratio of rational traders is changed randomly between 20%–80%. Because trader's ratio changes every five times, price fluctuations become random.

6 Summary

The computational experiments performed using the agent-based modelling show a number of important results. First, they demonstrate that the average price level and the trends are set by the amount of cash present and eventually injected in the market. In a market with a fixed amount of stocks, a cash injection creates an inflation pressure on prices. The other important finding of this work is that different populations of traders characterized by simple but fixed trading strategies cannot coexist in the long run. One population prevails and the other progressively lose weight and disappear. Which population will prevail and which will lose cannot be decided on the basis of the strategies alone. Trading strategies yield different results in different market conditions. In real life, different populations of traders with different trading strategies do coexist. These strategies are boundedly rational and thus one cannot really invoke rational expectations in any operational sense. Though market price processes in the absence of arbitrage can always be described as the rational activity of utility maximizing agents, the behaviour of these agents cannot be operationally defined.

This work shows that the coexistence of different trading strategies is not a trivial fact but requires explanation. One could randomize strategies imposing that traders statistically shift from one strategy to another. It is however difficult to explain why a trader embracing a winning strategy should switch to a losing strategy. Perhaps market change continuously and make trading strategies randomly more or less successful. More experimental work is necessary to gain an understanding of the conditions that allow the coexistence of different trading populations.

References

Alfarano, S., Lux, T.: A noise trader model as a generator of apparent financial power laws and long memory, Economics working paper, University of Kiel (2005)

Boswijk, H, Hommes, C.H., and Manzan, S.: Behavioral heterogeneity in Stock price. J. Econ. Dyn. Control **31**(6), 1938–1970 (2007)

Cincotti, S., Focardi, S., Marchesi, M., Raberto, M.: Who wins? Study of long-run trader survival in an artificial stock market. Physica A **324**, 227–233 (2003)

Cont, R., Bouchaud, J.P.: Herd behavior and aggregate fluctuations in financial markets. Macroeconomic Dyn. **4**(2), 170–196 (2000)

De Long, J.B., Shleifer, A., Summers, A., Waldmann, R.J.: The survival of noise traders in financial markets. J. Bus. **64**(1), 1–19 (1991)
Friedman, M.: Essays in Positive Economics. University of Chicago Press (1953)
LeBaron, B.: Agent based computational finance: suggested readings and early research. J. Econ. Dyn. Control **24**, 679–702 (2000)
LeBaron, B.: A builder's guide to agent-based financial markets. Quant. Finance **1**(2), 254–261 (2001)
Levy, H., Levy, M., Solomon, L.: Microscopic Simulation of Financial Markets. From Investor Behaviour to Market Phenomena. Academic Press, San Diego (2000)
Lux, T., Marchesi, L.: Scaling and criticality in a stochastic multi-agent model of a financial market. Nature **397**, 498–500 (2000)
Raberto, M., Cincotti, S., Focardi, S.M., Marchesi, M.: Agent-based simulation of a financial market. Physica A **299**(1-2), 320–328 (2001)
Sornette, D.: Why Stock Markets Crash. Princeton University Press (2003)
Tesfatsion, L.: Agent-based computational economics: growing economies from the bottom up. Artif. Life **8**, 55–82 (2002)
Palmer, R.G., Arthur, W.B., Holland, J., LeBaron, P.T.: Artificial economic life: a simple model of a stock market. Physica D **75**(1–3), 264–274 (1994)

A Closer Look at the Modeling of Economics Data

Hung T. Nguyen[1,2](✉) and Nguyen Ngoc Thach[3](✉)

[1] Department of Mathematical Sciences, New Mexico State University, Las Cruces, NM 88003, USA
hunguyen@nmsu.edu
[2] Faculty of Economics, Chiang Mai University, Chiang Mai 50200, Thailand
[3] Banking University of Ho-Chi-Minh City, 36 Ton That Dam Street, District 1, Ho-Chi-Minh City, Vietnam
thachnn@buh.edu.vn

Abstract. By taking a closer look at the traditional way we used to proceed to conduct empirical research in economics, especially in using "traditional" proposed models for economical dynamics, we elaborate on current efforts to improve its research methodology. This consists essentially of focusing on the possible use of quantum mechanics formalism to derive dynamical models for economic variables, as well as the use of quantum probability as an appropriate uncertainty calculus in human decision process (under risk). This approach is not only in line with the recent emerging approach of behavioral economics, but also should provide an improvement upon it. For practical purposes, we will elaborate a bit on the concrete road map for applying this "quantum-like" approach to financial data.

Keywords: Behavioral econometrics · Bohmian mechanics
Financial models · Quantum mechanics · Quantum probability

1 Introduction

A typical text book in economics, such as [9], is about using a proposed class of models, namely "dynamic stochastic general equilibrium" (DSGE), to conduct macroeconomic empirical research, before seeing the data! Moreover, as in almost all other texts, there is no distinction (with respect to the sources of fluctuation/dynamics) between data arising from "physical" sources and data "created" by economic agents (humans), e.g., data from industrial quality control area or stock prices, as far as (stochastic) modeling of dynamics is concerned.

When we view econometrics as a combination of economic theories, statistics and mathematics, we proceed as follows. There is a number of issues in economics to be investigated, such as prediction of asset prices. For such an issue, economic considerations (theories?), such as the well-known Efficient Market Hypothesis (EMH), dictates the model (e.g., martingales) for data to be seen! Of course,

V. Kreinovich et al. (Eds.): ECONVN 2019, SCI 809, pp. 100–112, 2019.
https://doi.org/10.1007/978-3-030-04200-4_7

given a time series, what we need to start (solidly) the analysis is a model of its dynamics. The economic theory gives us a model, in fact, many possible models (but we just pick one and rarely comparing it with another one!). From a given model, we need, among other things, to specifying it, e.g., estimating its parameters. It is only here that the data is used with statistical methods. The model "exists" before we see the data. Is this an empirical approach? See [13] for a clear explanation: Economics is not an empirical science if we proceed this way, since the data does not really suggest the model (to capture its dynamics). Perhaps the practice is based upon the argument that "it is the nature of the economic issue which already reveals a reasonable model for it (i.e., using economic theory)". But even so, what we mean by an empirical science is some procedure to arrive at a model "using" the data. We all known that for observational data, like time series, it is not easy to "figure out" its dynamics (true model), that is why proposed models are not only necessary but famous! As we will see, the point of insisting on "data-driven modeling" is more important than just for terminology!

In awarding the Prize in Economic Sciences in Memory of Alfred Nobel 2017 to Richard H. Thaler for his foundational works on behavioral economics (integrating economics with psychology), the Nobel Committee stated "Economists aim to develop models of human behavior and interactions in markets and other economic settings. But we humans behave in *complex ways*". As clearly explained in [13], economies are "complex systems" made up of human agents, and as such their behavior (in making decisions affecting economic data that we see and use to model its dynamics/model) must be taken into account. But a complex system is somewhat "similar" to a "quantum system", at least at a formalism level (of course, humans with their free will in making choices are not quite like particles!). According to [18], behavior of traders at financial markets, due to their free will, produces an additional "stochasticity" (to the "non mental", classical random fluctuations) and could not be reduced to it.

On the other hand, as Stephen Hawking reminded us [16], psychology was created precisely to study human's free will. Recent advances in psychological studies seem to indicate that quantum probability is appropriate to describe cognitive decision-making. Thus, in both aspects (for economics) of a theory of (consumer) choice and economic modeling of dynamics, quantum mechanic formalism is present. This paper will offer precisely an elaboration on the need of quantum mechanics in psychology, economics and finance. The point is this. Empirically, a new look at data is necessary to come up with better economic models.

The paper is organized as follows. In Sect. 2, we briefly recall how we get economic models so far, to emphasize the fact that we did not take into account the "human factor" in the data we observed. In Sect. 3, we talk about behavioral economics to emphasize the psychological integration into economics where cognitive decision-making could be improved with quantum probability calculus. In Sect. 4, we focus on our main objective, namely, why and how quantum mechanics formalism could help improving economic modeling. Finally, Sect. 5 presents a road map for applications.

2 How Models in Economics Were Obtained?

As clearly explained in the Preface of [6], financial economics (a subfield of econometrics), while highly empirical, is traditionally studied using a "model-based" approach. Specifically, [12], economic theories (i.e., knowledge from economic subject, they are "models" that link observations/ to be observed, without any pretense of being descriptive) bring out models, for possible relations between economic variables, or of their dynamics, such as regression models and stochastic dynamics models (e.g., common time series models, GARCH models, structural models). Given that it is a model-based approach (i.e., when facing a "real" economic problem, we just look at our toolkit to pick out a model to use), we need to identify a chosen model (in fact, we should "justify" why this model and not another). And then we use the observed data for that purpose (e.g., estimating model parameters) after "viewing" that our observed data is a realization of a stochastic process (where the probability theory in the "background" is the standard one, i.e., Kolmogorov), allowing us to use statistical theory to accept or reject the model.

Of course, new models could be suggested to, say, improve old ones. For example, in finance, volatility might not be constant over time, but it is a hidden variable (unobservable). The ARCH/GARCH models were proposed to improve models for stock prices. Note that GARCH models are used to "measure" volatility, once a concept of volatility is specified. At present, GARCH models are Kolmogorov stochastic models, i.e., based on standard probability theory. We say this because, GARCH models are models for stochastic dynamics of volatility (models for a non-observable "object") which is treated as a random variable. But what is the "source" of its "random variations"? The volatility (of a stock price) is high or low is clearly due to investors' behavior!. Should economic agents' behavior (in making decisions) be taken into account in the process to build a more coherent dynamic model for volatility? Perhaps, it is easy said than done! But here is the light: If volatility varies "randomly" (like in a game of chance) then Kolmogorov probability is appropriate for modeling it, but if volatility is due to "free will" of traders, then it is another matter: as we will see, the quantitative modeling of this type of uncertainty could be quantum probability instead.

Remark on "closer looks". We need closer looks at lots of things in sciences! A typical case is "A closer look at tests of significance" which is the whole last chapter of [17] with the final conclusion:

"Nowadays, tests of significant are extremely popular. One reason is that the tests are part on an impressive and well-developed mathematical theory. Another reason is that many investigators just cannot be bothered to set up chance models. The language of testing makes it easy to bypass the model, and talk about "statistically significant" results. This sounds so impressive, and there is so much

mathematical machinery clanking around in the background, that tests seem truly scientific - even when they are complete nonsense, St Exupery understood this kind of problem very well: when a mystery is too overwhelming, you do not dare to question it ([10]*, page 8).*

3 Behavioral Economic Approach

Standard economic practices are exposed in texts such as [6], [12]. Important aspects (for modeling) such as "individual behavior", "nature of economic data", were spelled out, but only on the surface, rather than taking a "closer look" at them! A closer look at them is what behavioral economics is all about.

Roughly speaking, the distinction between "economics" and "behavioral economics" (say, in microeconomics or financial econometrics) is the addition of human factors into the way we model stochastic models of observed economic data. More specifically, "fluctuations" of economic phenomena are explained by "free will" of economic agents (using psychology) and incorporating it into the search for better representation of dynamic models of economic data.

At present, by behavioral economics, we refer it to methodology pursued by economists like Richard Thaler (considered as the founder of behavioral finance). Specially, the focus is on investigating how human behavior affecting prices in financial markets. It all boils down to how to quantitatively model the uncertainty "considered" by economic agents when they make decisions. Psychological experiments have revealed that von Neumann 's expected utility and Bayes' updating procedure are both violated. As such, non additive uncertainty measures, as well as psychological-oriented theories (such as prospect theory) should be used instead. This seems to be in the right direction to improve standard practices in econometrics, in general. However, the Nobel Committee, while recognizing that "humans behave in complex ways", did not go all the way to elaborate on "what is a complex system?". This issue is clearly explained in [13]. The point is this. It is true that economic agents, with their free will (in choosing economic strategies) behave and interact in a complex fashion, but the complexity is not yet fully analyzed. Thus, a closer look at behavioral economics is desirable.

4 Quantum Probability and Mechanics

When taking into account "human factors" (in the data) to arrive at "better" dynamical models, we see that quantum mechanics exhibits two main "things" which seem to be useful:

(i) At the "micro" level, it "explains" how human factors affect the dynamics of observed data (by quantum probability calculus),

(ii) At the "macro" level, it provides a dynamical "law" (from Schrodinger's wave equation), i.e., a unique model for the fluctuations in the data.

So let's us elaborate a bit on these two things.

4.1 Quantum Probability

At the cognitive decision-making level, recall what we used to do. There are different types of uncertainty involved in social sciences, exemplified by the distinction by Frank Knight (1921): "risk" as a situation in which (standard/ additive) probabilities are known or knowable, i.e., they can be estimated from past data and calculated from the usual axioms of Kolmogorov probability theory; "uncertainty" as a situation in which "probabilities" are neither known, nor can they be calculated in an objective way. The Bayesian approach ignores this distinction by saying this: when you face Knight uncertainty, just model it by your own "subjective" probability (beliefs)! How you get your own subjective beliefs and how reliable they are another matter, what to be emphasized is that the subjective probability in the Bayesian approach is an additive set function (besides how you get it, its calculus is the same as objective probability measures), from it the law of total probability follows (as well as the so-called Bayesian updating rule). As another note, rather than ask whether any kind of uncertainty can be probabilistically quantified, it seems more useful to look at actually how humans make decisions under uncertainty.

In psychological experiments, see e.g., [5, 15], the intuitive notion of "likelihood" used by humans exhibits non-additivity, non monotone increasing and non-commutativity (so that non-additivity alone of an uncertainty measure is not enough to capture the source of uncertainty in cognitive decision-making). We are thus looking for an uncertainty measure having all these properties, to be used in behavioral economics.

It turns out that we already have precisely such an uncertainty measure used in quantum physics! It is simply a generalization of Kolmogorov probability measures, from a commutative one to a noncommutative one. The following is a tutorial on how to extend a commutative theory to a noncommutative one.

The cornerstone of Kolmogorov's theory is a probability space $(\Omega, \mathscr{A}, P)$ describing the source of uncertainty for derived variables. For example, if X is a real-valued random variable, then "under P", it has a probability law given by $P_X = PX^{-1}$ on $(\mathbb{R}, \mathscr{B}(\mathbb{R}))$. Random variables can be observed (or measured) directly.

Let's generalize the triple $(\Omega, \mathscr{A}, P)$! Ω is just a set, for example $\mathbb{R}^d$, a separable, finitely dimensional Hilbert space, which plays precisely the role of a "sampling space" (the space where we collect data). While the counterpart of a sampling space in classical mechanics is the "phase space" $\mathbb{R}^6$, the space of "states" in quantum mechanics is a complex, separable, infinitely dimensional Hilbert space H. So let's extend $\mathbb{R}^d$ to H (or take Ω to be H). Next, the Boolean ring $\mathscr{B}(\mathbb{R})$ (or $\mathscr{A}$) is replaced by a more general structure, namely by the bounded (non-distributive) lattice $\mathscr{P}(H)$ of projectors on H (we consider this since "quantum events" are represented by projectors). The "measurable" space $(\mathbb{R}, \mathscr{B}(\mathbb{R}))$ is thus replaced by the "observable" space $(H, \mathscr{P}(H))$. Kolmogorov probability measure $P(.)$ is defined on the boolean ring $\mathscr{A}$ with properties: $P(\Omega) = 1$, and $\sigma-$ additive. It is replaced by a map $Q : \mathscr{P}(H) \to [0, 1]$, with similar properties, in the language of operators: $Q(I) = 1$, $\sigma-$additive for mutually orthogonal projectors. All such maps arise from positive operators ρ on H (hence self adjoint) with unit trace. Specifically, P is replaced by $Q_\rho(.) : \mathscr{P}(H) \to [0,1]$, $Q_\rho(A) = tr(\rho A)$. Note that ρ plays the role of a probability density function.

In summary, a *quantum probability space* is a triple $(H, \mathscr{P}(H), Q_\rho)$, or simple $(H, \mathscr{P}(H), \rho)$, where H is a complex, separable, infinitely dimensional Hilbert space; $\mathscr{P}(H)$ is the set of all (orthogonal) projections on H; and ρ is a positive operator on H with unit trace (called a *density operator* , or matrix). For more details on quantum stochastic calculus, see Parthasarathy [17].

The quantum probability space describes the source of quantum uncertainty on the dynamics of particles, since, as we will see, the density matrix ρ arises from the fundamental law of quantum mechanics, the Schrodinger's equation (counterpart of Newton's law in classical mechanics), in view of the intrinsic randomness of particles motion, together with the so-called wave/particle duality.

Random variables in quantum mechanics are physical quantities associated with particles' motion, such as position, momentum, energy.

What is a "quantum random variable?" It is called an "observable". An observable is a (bounded) self adjoint operator on H with the following interpretation: A self adjoint operator A_Q "represents" a physical quantity Q in the sense that the range of Q (i.e., the set of its possible values) is the spectrum $\sigma(A_Q)$ of A_Q (i.e., the set of $\lambda \in \mathbb{C}$ such that $A_Q - \lambda I$ is not a $1-1$ map from H to H). Note that physical quantities are real-valued, and self adjoint A_Q has $\sigma(A_Q) \subseteq \mathbb{R}$.

Projections (i.e., self adjoint operators p such that $p = p^2$) represent special Q-random variables which take only two values 0, and 1 (just like indicator functions of Boolean events). Moreover, projections are in bijective correspondence with closed subspaces of H. Thus, events in classical setting can be identified with the closed subspaces of H. Boolean operations are: intersection of subspaces corresponds to event intersection; closed subspace generated by union of subspaces corresponds to event union; and orthogonal subspace corresponds to set complement. Note however, the *non-commutativity* of operators!

The probability measure of Q, on $(\mathbb{R}, \mathscr{B}(\mathbb{R}))$ is given by $P(Q \in B) = tr(\rho \zeta_{A_Q}(B))$, where $\zeta_{A_Q}(.)$ is the spectral measure of A_Q (a $\mathscr{P}(H)$ -valued measure).

In view of its intrinsic randomness, we can no longer talk about trajectories of moving objects (like in Newtonian mechanics), i.e., about "phase spaces", but instead, we should consider probability distributions of quantum states (i.e., positions of the moving particle, at each given time). In other words, quantum states are probabilistic. How to describe probabilistic behavior of quantum states, i.e., discover "quantum law of motion" (counterpart of Newton's laws)? Well, just like Newton where his laws were not "proved" but just "good guesses", i.e., confirmed by experiments (making good predictions, i.e. it "works"!), Schrodinger in 1927 got it. The random law governing the particle dynamics (with mass m, in a potential $V(x)$) is a wave-like function $\psi(x,t)$, solution of the complex PDE, known as the Schrodinger's equation

$$ih\frac{\partial\psi(x,t)}{\partial t} = -\frac{h^2}{2m}\Delta_x\psi(x,t) + V(x)\psi(x,t)$$

where Δ_x is the Laplacian, i complex unit, and h is the Planck's constant, with the meaning that the wave function $\psi(x,t)$ is the "probability amplitude" of position x at time t, i.e., $x \to |\psi(x,t)|^2$ is the probability density function for the particle position at time t.

Now, having the Schrodinger's equation as the quantum law, we obtain "quantum state" $\psi(x,t)$ at each time t, i.e., for given t, we have the probability density for the position $x \in \mathbb{R}^3$ which allows us to compute, for example, the probability that the particle will land in a neighborhood of a given position x.

Let us now specify the setting of quantum probability space $(H, \mathscr{P}(H), \rho)$. First, it can be shown that the complex functions $\psi(x,t)$ live on the complex, separable, infinitely dimensional Hilbert space $H = L^2(\mathbb{R}^3, \mathscr{B}(\mathbb{R}^3), d\mu)$. Without going into details, we write $\psi(x,t) = \varphi(x)\eta(t)$ (separation of variables), with $\eta(t) = e^{-iEt/h}$, and using Fourier transform, we can choose $\varphi \in H$ with $||\varphi|| = 1$. Let φ_n be a (countable) orthonormal basis of H, we have $\varphi = \sum_n < \varphi_n, \varphi > \varphi_n = \sum_n c_n\varphi_n$ with $\sum_n |c_n|^2 = 1$.

Then

$$\rho = \sum_n c_n|\varphi_n >< \varphi_n|$$

is a positive operator on H with

$$tr(\rho) = \sum_n < \varphi_n|\rho|\varphi_n > = \sum_n \int \varphi_n^*\rho\varphi_n = 1$$

Remark. In Diract's notation, Dirac [11], for $\tau, \alpha, \beta \in H$, $|\alpha >< \beta|$ is the operator sending τ to $< \beta, \tau > \alpha = (\int \beta^*\tau dx)\alpha$. If A is a self adjoint operator on H, then

$$tr(\rho A) = < \varphi|A|\varphi > = \sum_n c_n < \varphi_n|A|\varphi_n >$$

Thus, the "state" $\varphi \in H$ determines the density matrix ρ in $(H, \mathscr{P}(H), \rho)$. In other words, ρ is the density operator of the state ψ.

4.2 Quantum Mechanics

Let's be clear on "how to use quantum probability outside of quantum mechanics?" before entering application domains.

First of all, quantum systems are random systems with "known" probability distributions, just like "games of chance", with the exception that their probability distributions "behave" differently, such as the additivity property is violated (entailing everything which follow from it, such as the commonly use of "the law of total probability", so that Bayesian conditioning cannot be used).

Having a known probability distribution avoids the problem of "choosing models".

When we postulate that general random phenomena are like games of chance except that their probability distributions are unknown, we need to propose models as their possible candidates. Carrying out this process, we need to remember what G. Box has said "All models are wrong, but some are useful". Several questions arise immediately, such as "what is a useful model?", "how to get such a model?".

Box [3,4] already had this vision:

"Since all models are wrong, the scientist cannot obtain a "correct" one by excessive elaboration. On the contrary, following William of Occam, he should seek an economical description of natural phenomenon. Just as the ability to devise simple but evocative models is the signature of the great scientist so over elaboration and over parametrization is often the mark of mediocrity".

"Now it would be very remarkable if any system existing in the real world could be exactly represented by any simple model. However, cunningly chosen parsimonious models often do provide remarkably useful approximations. For example, the law $PV=RT$ relating pressure P, volume V and temperature T of an "ideal" gas via a constant R is not exactly true for any real gas, but it frequently provides a useful approximation and furthermore its structure is informative since it springs from a physical view of the behavior of gas molecules".

"For such models, there is no need to ask the question "Is the model true?". If "truth" is to be the "whole truth", the answer is "no". The only question of interest is "Is the model illuminating and useful?"

Usually, we rely on past data to suggest "good models". Once a suggested model is established, how do we "validate" it so that we can have enough "confidence" to "pretend" that it is our best guess of the true (but unknown) probability law generating the observed data, and then use it to predict the future. How did we validate our chosen model?

Recall that, in a quantum system, the probability law is completely determined: we know the game of nature. We can't tell where the electron will be, but we know its probability, exactly like when rolling a die, we cannot predict which number it will show, but we know the probability distribution of its states. We discover the law of "nature". The way to this information is systematic, so that "quantum machanics is an information theory": it gives us the information needed to predict future.

Imagine if we can discover the "theory" (something like Box's useful model) of the fluctuations of stock returns? where "useful" means "capable of making good predictions".

You can see that, if a random phenomenon can be modeled as a quantum system, then we can get a useful model (which we should call it, a theory, and not a model)! Moreover, in such a modeling, we may explain, or discover patterns that are hidden in traditional statistics, such as interference as opposed to correlation of variables.

Are there any things wrong with traditional statistical methodology? Well, as pointed out in Haven and Khrennikov [15].

"Consider the recent financial crisis. Are we comfortable to propose that physics should now lend a helping hand to the social sciences?"

Quantum mechanics is a science of prediction, and is one of the most successful theories humans ever devised. No existing theory in economics can come close to the predictive power of quantum physics. Note that there is no "testing" in physics! Physicists got their theories by *confirmation by experiments*, not by statistical testing.

As such, there is no doubt that when a random system can be modeled as a quantum system (by analogy), we do not need "models" anymore, we have a theory (i.e., a "useful" model). An example in finance is this. The position of a moving "object" is a price vector $x(t) \in \mathbb{R}^n$ where component $x_j(t)$ is the price of the share of the j corporation. The dynamics of the prices is the "velocity" $v(t)$, the change of prices. The analogy with quantum mechanics: mass as number of shares of stock j (m_j); kinetic energy as $\frac{1}{2}\sum_{j=1}^{n} m_j v_j^2$; potential energy as $V(x(t))$, describing interactions between traders and other macroeconomic factors.

For more concrete applications to finance with emphasis on the use of path integral, see Baaquie [1] A short summary of actual developments of *quantum pricing of options* is in Darbyshire [8] in which the rationale was spelled out clearly, since, e.g., "*The value of a financial derivative depends on the path followed by the underlying asset*". In any case, while keeping in mind the successful predictive power of quantum mechanics, the research efforts towards applying it to social sciences should be welcome.

5 How to Apply Quantum Mechanics to Building Financial Models?

When citing economics as an effective theory, Hawking [16] gave an example similar to quantum mechanics in view of the free will of humans, as a counterpart of the intrinsic randomness of particles. Now, as we have seen, the "official" view of quantum mechanics is that dynamics of particles is provided by a "quantum law" (via the Schrodinger's wave equation), thus it is expected that some "counterpart" of the quantum law (of motion) could be found to describe economic dynamics, based upon the fact that under the same type of uncertainty (quantified by noncommutative probability) the behavior of subatomic particles is similar to that of firms and consumers.

With all "clues" above, it is time to get to work! As suggested by current research, e.g. [7,15], we are going to talk about a (non conventional) version of quantum theory which seems suitable for modeling of economic dynamics, namely *Bohmian mechanics*, [2,15]. Pedagogically, every time we face a new thing, we investigate it in this logical order: What? Why? and then How?

But upfront, what we have in mind is this. Taking finance as the setting, we seek to model the dynamics of prices in a more comprehensive way than traditionally done. Specifically, as explained above, besides "classical" fluctuations, the price dynamics is also "caused" by mental factors of economic agents in the market (by their free will which can be described as "quantum stochastic"). As such, we seek a dynamical model having these both uncertainty components. It will be about the dynamics of prices, so that we are going to "view" a price as a "particle", so that price dynamics will be studied as quantum mechanics (the price at a time is its position, and the change in price is its speed).

So let's see what quantum mechanics can offer? Without going into to details of quantum mechanics, it suffices to note the following. In the "conventional" view, unlike macro objects (in Newtonian mechanics), particles in motion do not have trajectories (in their phase space), or put it more specifically, their motion cannot be described (mathematically) by trajectories (because of the Heisenberg's uncertainty principle). The dynamics of a particle with mass m is "described" by a wave function $\psi(x,t)$, where $x \in \mathbb{R}^3$ is the particle position at time t, which is the solution of the Schrodinger's equation (counterpart of Newton's law of motion of macro objects):

$$ih\frac{\partial\psi(x,t)}{\partial t} = -\frac{h^2}{2m}\Delta_x\psi(x,t) + V(x)\psi(x,t)$$

and where $f_t(x) = |\psi(x,t)|^2$ is the probability density function of the particle position X at time t, i.e., $P_t(X \in A) = \int_A |\psi(x,t)|^2 dx$.

But, our price variable does have trajectories! Its is "interesting" to note that, we used to display financial prices fluctuations (data) which look like paths of a (geometric) Brownian motion. But Brownian motions, while having continuous paths, are nowhere differentiable, and as such, there are no derivatives to represent velocities (the second component of a "state" in the phase space)!

Well, we are lucky since there exists a non-conventional formulation of quantum mechanics, called Bohmian mechanics [2] (see also [7]) in which it is possible to consider trajectories for particles! The following is sufficient for our discussions here.

Remark. Before deriving Bohmian mechanics and using it for financial applications, the following should be kept in mind. For physicists, Schrodinger's equation is everything: the state of a particle is "described" by the wave function $\psi(x,t)$ in the sense that the probability to find it in a region A, at time t, is given by $\int_A |\psi(x,t)|^2 dx$. As we will see, Bohmian mechanics is related to Schrodinger's equation, but presents a completely different interpretation of the quantum world, namely, it is possible to consider trajectories of particles, just like in classical, deterministic mechanics. This quantum formalism is not shared by the majority of physicists. Thus, using Bohmian mechanics in statistics should not mean that statisticians "endorse" Bohmian mechanics as the appropriate formulation of quantum mechanics! We use it since, by analogy, we can formulate (and derive) dynamics (trajectories) of economic variables.

The following leads to a new interpretation of Schrodinger's equation.

The wave function $\psi(x,t)$ is complex-valued, so that, in polar form, $\psi(x,t) = R(x,t)\exp\{\frac{i}{h}S(x,t)\}$, with $R(x,t)$, $S(x,t)$ being real-valued. The above Schrodinger's equation becomes

$$ih\frac{\partial}{\partial t}[R(x,t)\exp\{\frac{i}{h}S(x,t)\}]$$

$$= -\frac{h^2}{2m}\Delta_x[R(x,t)\exp\{\frac{i}{h}S(x,t)\}] + V(x)[R(x,t)\exp\{\frac{i}{h}S(x,t)\}]$$

from it partial derivatives (with respect to time t) of $R(x,t)$, $S(x,t)$ can be derived. Not only that x will play the role of our price, but for simplicity, we take x as one dimentional variable, i.e., $x \in \mathbb{R}$ (so that the Laplacian Δ_x is simply $\frac{\partial^2}{\partial x^2}$) in the derivation below.

Differentiating

$$ih\frac{\partial}{\partial t}[R(x,t)\exp\{\frac{i}{h}S(x,t)\}]$$

$$= -\frac{h^2}{2m}\frac{\partial^2}{\partial x^2}[R(x,t)\exp\{\frac{i}{h}S(x,t)\}] + V(x)[R(x,t)\exp\{\frac{i}{h}S(x,t)\}]$$

and identifying real and imaginary parts of both sides, we get, respectively

$$\frac{\partial S(x,t)}{\partial t} = -\frac{1}{2m}(\frac{\partial S(x,t)}{\partial x})^2 + V(x) - \frac{h^2}{2mR(x,t)}\frac{\partial^2 R(x,t)}{\partial x^2}$$

$$\frac{\partial R(x,t)}{\partial t} = -\frac{1}{2m}[R(x,t)\frac{\partial^2 S(x,t)}{\partial x^2} + 2\frac{\partial R(x,t)}{\partial x}\frac{\partial S(x,t)}{\partial x}]$$

The equation for $\frac{\partial R(x,t)}{\partial t}$ gives rise to the dynamical equation for the probability density function $f_t(x) = |\psi(x,t)|^2 = R^2(x,t)$. Indeed,

$$\frac{\partial R^2(x,t)}{\partial t} = 2R(x,t)\frac{\partial R(x,t)}{\partial t}$$

$$= 2R(x,t)\{-\frac{1}{2m}[R(x,t)\frac{\partial^2 S(x,t)}{\partial x^2} + 2\frac{\partial R(x,t)}{\partial x}\frac{\partial S(x,t)}{\partial x}]\}$$

$$= -\frac{1}{m}[R^2(x,t)\frac{\partial^2 S(x,t)}{\partial x^2} + 2R(x,t)\frac{\partial R(x,t)}{\partial x}\frac{\partial S(x,t)}{\partial x}]$$

$$= -\frac{1}{m}\frac{\partial}{\partial x}[R^2(x,t)\frac{\partial S(x,t)}{\partial x}]$$

If we stare at the equation for $\frac{\partial S(x,t)}{\partial t}$ (corresponding to the real part of the wave function in Schrodinger's equation), then we see some analogy with classical mechanics in Hamiltonian formalism.

Recall that in Newtonian mechanics, the state of a moving object of mass m , at time t, is described as $(x, m\dot{x})$ (position $x(t)$, and momentum $p(t) = mv(t)$, with velocity $v(t) = \frac{dx}{dt} = \dot{x}(t)$). The Hamiltonian of the system is the sum of the kinetic energy and potential energy $V(x)$, namely $H(x,p) = \frac{1}{2m}v^2 + V(x) = \frac{mp^2}{2} + V(x)$. From it, $\frac{\partial H(x,p)}{\partial p} = mp$, or $\dot{x}(t) = \frac{1}{m}\frac{\partial H(x,p)}{\partial p}$. Thus, if we look at

$$\frac{\partial S(x,t)}{\partial t} = -\frac{1}{2m}(\frac{\partial S(x,t)}{\partial x})^2 + V(x) - \frac{h^2}{2mR(x,t)}\frac{\partial^2 R(x,t)}{\partial x^2}$$

ignoring the term $\frac{h^2}{2mR(x,t)}\frac{\partial^2 R(x,t)}{\partial x^2}$ for the moment, i.e., the Hamiltonian $\frac{1}{2m}(\frac{\partial S(x,t)}{\partial x})^2 - V(x)$, then the velocity of this system is $v(t) = \frac{dx}{dt} = \frac{1}{m}\frac{\partial S(x,t)}{\partial x}$.

Now the full equation has the term $Q(x,t) = \frac{h^2}{2mR(x,t)}\frac{\partial^2 R(x,t)}{\partial x^2}$, coming from Schrodinger's equation, and which we call it a "quantum potential", we follow Bohm to interprete it similarly., leading to the Bohm-Newton equation

$$m\frac{dv(t)}{dt} = m\frac{d^2x(t)}{dt^2} = -(\frac{\partial V(x,t)}{\partial x} - \frac{\partial Q(x,t)}{\partial x})$$

giving rise to the concept of "trajectory" for the "particle".

Remark. As you can guess, Bohmian mechanics (also called "pilot wave theory") is "appropriate" for modeling financial dynamics. Roughly speaking, Bohmian mechanics is this. While fundamental to all is the wave function coming out from Schrodinger's equation, the wave function itself provides only a partial description of the dynamics. This description is completed by the specification of the actual positions of the particle, which evolve according to $v(t) = \frac{dx}{dt} = \frac{1}{m}\frac{\partial S(x,t)}{\partial x}$, called the "guiding equation" (expressing the velocities of the particle in terms of the wave function). In other words, the state is specified as (ψ, x). Regardless of the debate in physics about this formalism of quantum mechanics, Bohmian mechanics is useful for economics! Note right away that the quantum potential (field) $Q(x,t)$, giving rise to the "quantum force" $-\frac{\partial Q(x,t)}{\partial x}$, disturbing the "classical" dynamics, will play the role of "mental factor" (of economic agents) when we apply Bohmian formalism to economics.

With the fundamentals of Bohmian mechanics in place, you are surely interested in a road map to economic applications! Perhaps, [7] provided the best road map.

The "Bohmian program" for applications is this. With all economic quantities analogous to those in quantum mechanics, we seek to solve the Schrodinger' s equation to obtain the (pilot) wave function $\psi(x,t)$ (representing expectation of traders in the market), where $x(t)$ is, say, the stock price at time t; from which we obtain the mental (quantum) potential $Q(x,t) = \frac{h^2}{2mR(x,t)}\frac{\partial^2 R(x,t)}{\partial x^2}$ producing the associated mental force $-\frac{\partial Q(x,t)}{\partial x}$; solve the Bohm-Newton's equation to obtain the "trajectory" for $x(t)$. Note that, the quantum randomness is encoded in the wave function via the way quantum probability is calculated, namely, $P(X(t) \in A) = \int_A |\psi(x,t)|^2 dx$. Of course, economic counterparts of quantities such as m (mass), h (the Planck constant) should be spelled out (e.g., number of shares, price scaling parameter, i.e., the unit in which we measure price change). The potential energy describes the interactions among traders (e.g., competition) together with external conditions (e.g., price of oil, weather, etc....) whereas the kinetic energy represents the efforts of economic agents to change prices. Finally, note that the amplitude $R(x,t)$ of the wave function $\psi(x,t)$ is the square root of the probability density function $x \to |\psi(x,t)|^2$, and satisfies the "continuity equation"

$$\frac{\partial R^2(x,t)}{\partial t} = -\frac{1}{m}\frac{\partial}{\partial x}[R^2(x,t)\frac{\partial S(x,t)}{\partial x}].$$

References

1. Baaquie, B.E.: Quantum Finance: Path Integrals and Hamiltonians for Options and Interest Rates. Cambridge University Press, Cambridge (2007)
2. Bohm, D.: Quantum Theory. Prentice Hall, Englewood Cliffs (1951)
3. Box, G.E.P.: Science and statistics. J. Am. Stat. Assoc. **71**(356), 791–799 (1976)
4. Box, G.E.P.: Robustness in the strategy of scientific model building. In: Launer, R.L., Wilkinson, G.N. (eds.) Robustness in Statistics, pp. 201–236. Academic Press, New York (1979)
5. Busemeyer, J.R., Bruza, P.D.: Quantum Models of Cognitive and Decision. Cambridge University Press, Cambridge (2012)
6. Campbell, J.Y., Lo, A.W., Mackinlay, A.C.: The Econometrics of Financial Markets. Princeton University Press, Princeton (1997)
7. Choustova, O.: Quantum Bohmian model for financial markets. Phys. A **347**, 304–314 (2006)
8. Darbyshire, P.: Quantum physics meets classical finance. Phys. World, 25–29 (2005)
9. Dejong, D.N., Dave, C.: Structural Macroeconometrics. Princeton University Press, Princeton (2007)
10. De Saint Exupery, A.: The Little Prince. Penguin Books, London (1995)
11. Dirac, D.: The Principles of Quantum Mechanics. Clarendon Press, Oxford (1947)
12. Florens, J.P., Marimoutou, V., Peguin-Feissolle, A.: Econometric Modeling and Inference. Cambridge University Press, Cambridge (2007)
13. Focardi, S.M.: Is economics an empirical science? If not, can it become one?. Front. Appl. Math. Stat. 1(7) (2015)
14. Freedman, D., Pisani, R., Purves, R.: Statistics, 4th edn. W.W. Norton, New York (2007)
15. Haven, E., Khrennikov, A.: Quantum Social Science. Cambridge University Press, Cambridge (2013)
16. Hawking, S., Mlodinow, L.: The Grand Design. Bantam Books, London (2011)
17. Parthasarathy, K.R.: An Introduction to Quantum Stochastix Calculus. Springer, Basel (1992)
18. Soros, J.: The Alchemy of Finance: Reading of Mind of the Market. Wiley, New York (1987)

What to Do Instead of Null Hypothesis Significance Testing or Confidence Intervals

David Trafimow(✉)

Department of Psychology, New Mexico State University, MSC 3452, P. O. Box 30001, 88003-8001 Las Cruces, NM, USA
dtrafimo@nmsu.edu

Abstract. Based on the banning of null hypothesis significance testing and confidence intervals in *Basic and Applied Psychology* (2015), this presentation focusses on alternative ways for researchers to think about inference. One section reviews literature on the a priori procedure. The basic idea, here, is that researchers can perform much inferential work before the experiment. Furthermore, this possibility changes the scientific philosophy in important ways. A second section moves to what researchers should do after they have collected their data, with an accent on obtaining a better understanding of the obtained variance. Researchers should try out a variety of summary statistics, instead of just one type (such as means), because seemingly conceptually similar summary statistics nevertheless can imply very different qualitative stories. Also, rather than engage in the typical bipartite distinction between variance due to the independent variable and variance not due to the independent variable; a tripartite distinction is possible that divides variance not due to the independent variable into variance due to systematic or random factors, with important positive consequences for researchers. Finally, the third major section focusses on how researchers should or should not draw causal conclusions from their data. This section features a discussion of within-participants causation versus between-participants causation, with an accent on whether the type of causation specified in the theory is matched or mismatched by the type of causation tested in the experiment. There also is a discussion of causal modeling approaches, with criticisms. The upshot is that researchers could do much more a priori work, and much more a posteriori work too, to maximize the scientific gains they obtain from their empirical research.

1 What to Do Instead of Null Hypothesis Significance Testing or Confidence Intervals

In a companion piece to the present one (Trafimow (2018) at TES2019), I argued against null hypothesis significance testing and confidence intervals (also see Trafimow 2014; Trafimow and Earp 2017; Trafimow and Marks 2015; 2016; Trafimow et al. 2018a).[1] In contrast to the TES2019 piece, the present work is designed to answer the question, "What should we do instead?" There are many alternatives, such as not performing inferential statistics and focusing on descriptive statistics (e.g., Trafimow

[1] Nguyen (2016) provided an informative theoretical perspective on the ban.

V. Kreinovich et al. (Eds.): ECONVN 2019, SCI 809, pp. 113–128, 2019.
https://doi.org/10.1007/978-3-030-04200-4_8

2019), including visual displays for better understanding the data (Valentine et al. 2015); Bayesian procedures (Gillies 2000 reviewed and criticized different Bayesian methods); quantum probability (Trueblood and Busemeyer 2011; 2012); and others. Rather than comparing or contrasting different alternatives, my goal is to provide alternatives that I personally like, admitting beforehand that my liking may be due to my history of personal involvement.

Many scientists fail to do sufficient thinking prior to data collection. A longer document than I can provide here is needed to describe all the types of a priori thinking researchers should do, and my present focus is limited to a priori inferential work. In addition, it is practically a truism among statisticians that many science researchers fail to look at their data with sufficient care, and so there is much a posteriori work to be performed too. Thus, the two subsequent sections concern a priori inferential work and a posteriori data analyses, respectively. Finally, as most researchers wish to draw causal conclusions from their data, the final section includes some thoughts on causation, including distinguishing within-participants and between-participants causation, and the (de)merits of causal modeling.

2 The a Priori Procedure

Let us commence by considering why researchers often collect as much data as they can afford to collect, rather than collecting only a single participant. Most statisticians would claim that under the usual assumption that participants are randomly selected from a population, the larger the sample size, the more the sample resembles the population. Thus, for example, if the researcher obtains a sample mean to estimate the population mean, the larger the sample, the more *confident* the researcher can be that the sample mean will be *close* to the population mean. I have pointed out that this statement raises two questions (Trafimow 2017a).

- How close is *close*?
- How confident is *confident*?

It is possible to write an equation that gives the necessary sample size to reach a priori specifications for confidence and closeness. This will be discussed in more detail later, but right now it is more important to explain the philosophical changes implied by this thinking.

First, the foregoing thinking assumes that the researcher wishes to use sample statistics to estimate population parameters. In fact, practically any statistical procedure that uses the concept of a population assumes—at least tacitly—that the researcher cares about the population. Whether the researcher really does care about the population may depend on the type of research being conducted. It is not mandatory that the researcher care about the population from which the sample is taken, but that will be the guiding premise, for now.

A second point to consider is that the goal of using sample statistics to estimate population parameters is very different from the goal implied by the null hypothesis significance testing procedure, which is to test (null) hypotheses. At this point, it is worth pausing to consider the potential argument that the goal of testing hypotheses is a better goal than that of estimating population parameters.[2] Thus, the reader already has a reason to ignore the present section of this document. But appearances can be deceiving.

To see the main issues quickly, imagine that you have access to Laplace's Demon who knows everything and always speaks truthfully. The Demon informs you that sample statistics have absolutely nothing to do with population parameters. With this extremely inconvenient pronouncement in mind, suppose a researcher randomly assigns participants to experimental and control conditions to test a hypothesis about whether a drug lowers blood pressure. Here is the question: no matter how the data come out, does it matter given the Demon's pronouncement? Even supposing the means in the two conditions differ in accordance with the researcher's hypothesis, this is irrelevant if the researcher has no reason to believe that the sample means are relevant to the larger potential populations of people who could have been assigned to the two conditions. The point of the example, and of invoking the Demon, is to illustrate that the ability to estimate population parameters from sample statistics is a prerequisite for hypothesis testing. Put another way, hypothesis testing means nothing if the researcher has no reason whatsoever to believe that similar results likely would happen again if the experiment were replicated or if the researcher has no reason to believe the sample data pertain to the relevant population or populations. And furthermore, much research is not about hypothesis testing, but rather about establishing empirical facts about relevant populations, establishing a proper foundation for subsequent theorizing, exploration, application, and so on.

Now that we see that the parameters really do matter, and matter extremely, let us continue to consider the philosophical implications of asking the bullet-listed questions. Researchers in different scientific areas may have different theories, goals, applications, and many other differences. A consequence of these many differences is that there can be different answers to the bullet-listed questions. For example, one researcher might be satisfied to be confident that the sample statistics are within four-tenths of a standard deviation of the corresponding population parameters whereas another researcher might insist on being confident that the sample statistics are within one-tenth of a standard deviation of the corresponding population parameters. Obviously, the latter researcher will need to collect a larger sample size than the former one, all else being equal.

Now suppose that, whatever the researcher's specifications for the degree of closeness and the degree of confidence, she collects a sufficiently large sample size to meet them. After computing the sample statistics of interest, what should she then do? Although recommendations will be forthcoming in the subsequent section, for right now, it is reasonable to argue that the researcher can simply stop, satisfied in the knowledge that the sample statistics are good estimates of their corresponding population parameters. How does the researcher know that this is so? The answer is that the researcher has performed the requisite a priori inferential work. Let us consider a specific example.

---

[2] Of course, the null hypothesis significance testing procedure does not test the hypothesis of interest but rather the null hypothesis that is not of interest, which is one of the many criticisms to which the procedure has been subjected. But as the present focus is on what to do instead, I will not focus on these criticisms. The interested reader can consult Trafimow and Earp (2017).

Suppose that a researcher wishes to be 95% confident that the sample mean to be obtained from a one-group experiment is within four-tenths of a standard deviation of the population mean. Equation 1 shows how to obtain the necessary sample size n to meet specifications where Z_C is the z-score that corresponds to the desired confidence level and f is the desired closeness, in standard deviation units:

$$n = \left(\frac{Z_C}{f}\right)^2. \tag{1}$$

As 1.96 is the z-score that corresponds to 95% confidence, instantiating this value for Z_C, as well as .4 for f, results in the following: $n = \left(\frac{Z_C}{f}\right)^2 = 24.01$. Rounding up to the nearest whole number, then, implies that the researcher needs to obtain 25 participants to meet specifications for closeness and confidence. Based on the many admonitions for researchers to collect increased samples sizes, 25 may seem a low number. But remember that 25 is the result from a very liberal assumption that it only is necessary for the sample mean to be within four-tenths of a standard deviation of the population mean; had we specified something more stringent, such as one-tenth, the result would have been much more extreme: $n = \left(\frac{Z_C}{f}\right)^2 = \left(\frac{1.96}{.1}\right)^2 = 384.16$.

Equation 1 is limited in a variety of ways. One limitation is that it only works for a single mean. To overcome this limitation, Trafimow and MacDonald (2017) derived more general equations that work for any number of means. Another limitation is that the Equations in Trafimow (2017a) and Trafimow and MacDonald (2017) assume random selection from normally distributed populations. However, most distributions are not normal but rather are skewed (Blanca et al. 2013; Cain et al. 2017; Ho and Yu 2015; Micceri 1989). Trafimow et al. (in press) showed how to expand the a priori procedure for the family of skew-normal distributions. Skew-normal distributions are interesting for many reasons, one of which is that they are defined by three parameters rather than two of them. Instead of the mean μ and standard deviation σ parameters, skew-normal distributions are defined by the location ξ, scale ω, and shape λ parameters. When using the Trafimow et al. skew-normal equations, it is ξ rather than μ which is of interest, and the researcher learns the sample size needed to be confident that the sample location statistic is close to the population location parameter.[3] Contrary to many people's intuition, as distributions become increasingly skewed, it takes fewer, rather than more, participants to meet specifications. For example, to be 95% confident that the sample location is within .1 of a scale unit of the population location, we saw earlier that it takes 385 participants when the distribution is normal, and the mean and location are the same ($\mu = \xi$). In contrast, when the shape parameter is mildly different from 0, such as .5, the number of participants necessary to meet specifications drops dramatically to 158. Thus, at least from a precision standpoint,

[3] In addition, ω is of more interest than σ though this is not of great importance yet.

skewness is an advantage and researchers who perform data transformations to reduce skewness are making a mistake.[4]

To expand the a priori procedure further, my colleagues and I also have papers "submitted" concerning differences in locations for skewed distributions across matched samples or independent samples (Wang 2018a; 2018b). Finally, we expect also to have equations concerning proportions, correlations, and standard deviations in the future.

To summarize, when using the a priori procedure, the researcher commits, before collecting data, to specifications for closeness and confidence. The researcher then uses appropriate a priori equations to find the necessary sample size. Once the required sample size is collected, the researcher can compute the sample statistics of interest and trust that these are good estimates of their corresponding population parameters, with "good" having been defined by the a priori specifications. There is thus no need to go on to perform significance tests, compute confidence intervals, or any of the usual sorts of inferential statistics that researchers routinely perform on already collected data. As a bonus, instead of skewness being a problem, as it is for traditional significance tests that assume normality or at least that the data are symmetric; skewness is an advantage, and a large one, from the point of view of a priori equations.

Before moving on, however, there are two issues that are worth mentioning. The first issue is that the a priori procedure may seem, at first glance, as merely another way to perform power analysis. But this is not so and two points should make this clear. First, power analysis depends on one's threshold for statistical significance. The more stringent the threshold, the greater the necessary sample size. In contrast, there is no statistical significance threshold for the a priori procedure, and so a priori calculations are not influenced by significance thresholds. Second, a priori calculations are strongly influenced by the desired closeness of sample statistics to corresponding population parameters, whereas power calculations are not. For both reasons, a priori calculations and power calculations render different values.

A second issue pertains to the replication crisis. The Open Science Collaboration (2015) showed that well over 60% of published findings in top journals failed to replicate, and matters may well be worse in other sciences, such as in medicine. The a priori procedure suggests an interesting way to address the replication crisis Trafimow (2018). Consider that a priori equations can be algebraically rearranged to yield probabilities under specifications for f and n. Well, then, imagine the ideal case where an experiment really is performed the same way twice, with the only difference between the original and replication experiments being randomness. Of course, in real research, this is impossible, as there will be systematic differences with respect to dates, times, locations, experimenters, background conditions, and so on. Thus, the probability of replicating in real research conditions is less than the probability of replicating under ideal conditions. But by merely expanding a priori equations to account for two

[4] The reader may wonder why skewness increases precision. For a quantitative answer, see Trafimow et al. (in press). For a qualitative answer, simply look up pictures of skew-normal distributions (contained in Trafimow et al., among other places). Observe that as the absolute magnitude of skewness increases, the bulk of the distributions become taller and narrower. Hence, sampling precision increases.

experiments, as opposed to only one experiment, it is possible to calculate the probability of replication under ideal conditions, and before collecting any data under whatever sample sizes the researcher contemplates collecting. In turn, this calculation can serve as an upper bound for the probability of replication under real conditions. Consequently, if the a priori calculations for replicating under ideal conditions are unfavorable, and I showed that this is so under typical sample sizes Trafimow (2018), they are even more unfavorable under real conditions. Therefore, we have an explanation of the replication crisis, as well as a procedure to calculate, a priori, the minimal conditions necessary to give the researcher a reasonable chance at conducting a replicable experiment. This solution to the replication crisis was an unexpected benefit of a priori thinking.

3 After Data Collection

Once data have been collected, researchers typically compute the sample statistics of interest (means, correlations, and so on) and perform null hypothesis significance tests or compute confidence intervals. But there is much more that researchers can do to understand their data as completely as possible. For example, Valentine et al. (2015) showed how a variety of visual displays can be useful for helping researchers gain a more complete understanding of their data. And there is more.

3.1 Consider Different Summary Statistics

Researchers who perform experiments typically use means and standard deviations. If the distribution is normal, this makes sense, but few distributions are normal (Blanca et al. 2013; Cain et al. 2017; Ho and Yu 2015; Micceri 1989). In fact, there are other summary statistics researchers could use such as medians, percentile cutoffs, and many more. A particularly interesting alternative, given the foregoing focus on skew-normal distributions, is to use the location. To reiterate, for normal distributions the mean and location are the same, but for skew-normal distributions they are different. But why should you care?

To use one of my own examples (Trafimow et al. 2018), imagine a researcher performs an experiment to test whether a new blood pressure medicine really does reduce blood pressure. In addition, suppose that the means in the two conditions differ in the hypothesized direction. According to appearances, the data support that the blood pressure medicine "works." But consider the possibility that the blood pressure medicine merely changed the shape of the distribution, say by introducing negative skewness. In that case, even if the location of the two distributions is the same, the means would necessarily differ, and in the hypothesized direction too. If the locations are the same, though the means are different, it would be difficult to argue that the medicine works, though in the absence of a location computation, this would be the seemingly obvious conclusion.

Alternatively, it is possible for an impressive difference in locations to be masked by a lack of difference in means. In this case, based on the difference in locations, the experiment worked but based on the lack of differences in means, it did not. Yet more dramatically, it is possible for there to be a difference in means and a difference in locations, but in opposite directions. Returning to the example of blood pressure medicine, it could easily happen that the difference in means indicates that the medicine reduces blood pressure whereas the difference in locations indicates that the blood pressure medicine increases blood pressure. More generally, Trafimow et al. 2018 showed that mean effects and location effects can (a) be in the same direction, (b) be in opposite directions, (c) be impressive for means but not for locations, or (d) be impressive for locations but not for means.

Lest the reader believe the foregoing is too dramatic and that skewness is not really that big an issue, it is worth pointing out that impressive differences can occur even at low skews, such as .5, which is well under criteria of .8 or 1.0 that authorities have set as thresholds for deciding whether a distribution should be considered normal or skewed. We saw earlier, during the discussion of the a priori procedure with normal or skew-normal distributions, that a skew of only .5 is sufficient to reduce the number of participants needed for the same sampling precision of .1 from 385 to only 158. Dramatic effects also can occur with effect sizes. One demonstration from Trafimow et al. (2018) shows that even when the effect size is zero using locations, a difference in skew of only .5 between the two conditions leads to $d = .37$ using means, which would be considered reasonably successful by most researchers.

To drive these points home consider Figs. 1 and 2. To understand Fig. 1, imagine an experiment where the control group population is normal, $\mu = \xi = 0$ and $\sigma = \omega = 1$; and there is an experimental group population with a skew-normal distribution with the same values for location and scale ($\xi = 0$ and $\omega = 1$). Clearly, the experiment does not support that the manipulation influences the location. And yet, we can imagine that the experimental manipulation does influence the shape of the distribution, and Fig. 1 allows the shape parameter of the experimental condition to vary between 0 and 1 along the horizontal axis, with the resultant effect size along the vertical axis. The three curves in Fig. 1 illustrate three ways to calculate the effect size. Because skewness decreases the standard deviation, relative to the scale, it follows that if the standard deviation of the experimental group is used in the effect size calculation, the standard deviation used is at its lowest, and so the effect size is at its largest magnitude, though in the negative direction, consistent with the blood pressure example. Alternatively, a pooled standard deviation can be used, as is typical in calculations of Cohen's *D*. And yet another alternative is to use the standard deviation of the control condition, as is typical in calculations of Glass's Δ. No matter how the effect size is calculated, though, Fig. 1 shows that seemingly impressive effect sizes can be generated by changing the shape of the distribution, even when the locations and scales are unchanged. Figure 1 illustrates the importance of not depending just on means and standard deviations, but of performing location, scale, and shape computations too (see Trafimow et al. 2018; in press; for relevant equations).

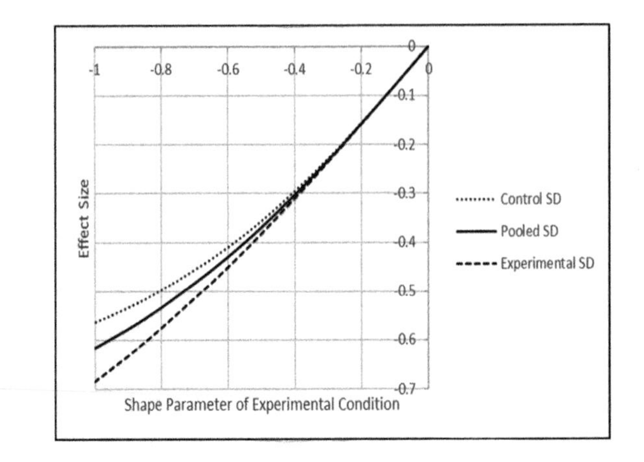

Fig. 1. The effect size is represented along the vertical axis as a function of the shape parameter along the horizontal axis, with effect size calculations based on the control group, pooled, or experimental group standard deviations.

Figure 2 might be considered even more dramatic than Fig. 1 for driving home the importance of location, scale, and shape; in addition to mean and standard deviation. In Fig. 2, the control group again is normal, with $\mu = \xi = 0$ and $\sigma = \omega = 1$. In contrast, the experimental group location is $\xi = -1$. Thus, based on a difference in locations, it should be clear that the manipulation decreased scores on the dependent variable. But will comparing means render a qualitatively similar or different story than comparing locations? Interestingly, the answer depends both on the shape and scale of the experimental condition. In Fig. 2, the shape parameter of the experimental condition varied along the horizontal axis, from -2 to 2. In addition, the scale value was set at 1, 2, 3, or 4. In the scenario modeled by Fig. 2, the difference in means is always negative, regardless of the shape, when the scale is set at 1. Thus, in this case, although the quantitative implications of comparing means versus comparing locations differ, the qualitative implications are similar. In contrast, as the scale increases to 2, 3, or 4, the difference in means can be positive, depending on the shape parameter. And in fact, especially when the scale value is 4, a substantial proportion of the curve is in positive territory. Thus, Fig. 2 dramatizes the disturbing possibility that location differences and mean differences can go in opposite directions. There is no way for researchers who neglect to calculate location, scale, and shape statistics to be aware of the possibility that a comparison of locations might suggest implications opposite to those suggested by the typical comparison of means. Thus, I cannot stress too strongly the importance of researchers not settling just for means and standard deviations; but rather that they should calculate location, scale, and shape statistics too.

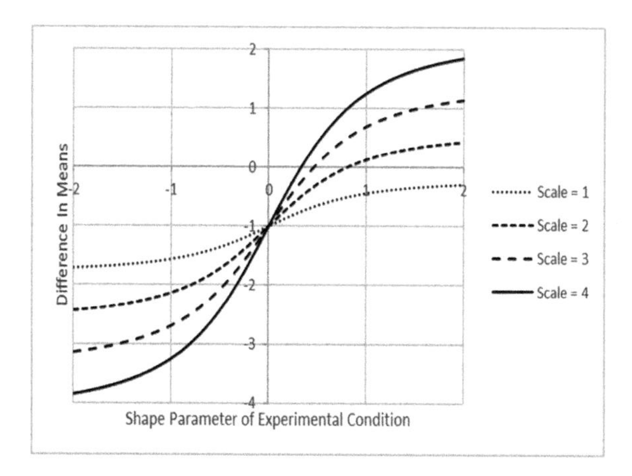

Fig. 2. The difference in means is represented along the vertical axis as a function of the shape parameter of the experimental condition, with curves representing four experimental condition scale levels.

3.2 Consider a Tripartite Division of Variance

Whatever the direction of differences in means, locations, and so on; or whatever the size of obtained correlations or statistics based on correlations; there is the issue of variance to consider.[5] Typically, researchers mainly care about variance in the context of inferential statistics. That is, researchers are used to parsing variance into "good" variance due to the independent variable of interest and "bad" variance due to everything else. The more the good variance, and the less the bad variance, the lower the *p*-value. And lower *p*-values are generally favored, especially if they pass the $p < .05$ bar needed for declarations of "statistical significance." But I have shown recently that it is possible to parse variance into three components rather than the usual two (Trafimow 2018). Provided that the researcher has measured the reliability of the dependent variable, it is possible to parse variance into that which is due to the independent variable, that which is random, and that which is systematic but due to variables unknown to the researcher; that is, a tripartite parsing. In Eq. 2, σ_{IV}^2 is the variance due to the independent variable, σ_X^2 is the total variance, and T is the population level *t*-score:

$$\sigma_{IV}^2 = \frac{T^2}{T^2 + df}\sigma_X^2. \tag{2}$$

[5] For skew-normal distributions it makes more sense to consider the square of the scale than to consider the square of the standard deviation, known as the variance. But researchers are used to variance and variance is sufficient to make the necessary points in this section.

Alternatively, in a correlational study, σ_{IV}^2 can be calculated more straightforwardly using the square of the correlation coefficient ρ_{YX}^2, as Eq. 3 shows:

$$\sigma_{IV}^2 = \rho_{YX}^2 \sigma_X^2. \tag{3}$$

Equation 4 provides the amount of random variance σ_R^2, where $\rho_{XX'}$ is the reliability of the depending variable:

$$\sigma_R^2 = \sigma_X^2 - \rho_{XX'}\sigma_X^2 = (1 - \rho_{XX'})\sigma_X^2 \tag{4}$$

Finally, because of the tripartite split of total variance into three variance components, Eq. 5 gives the systematic variance not due to the independent variable; that is, the variance due to "other" systematic factors σ_O^2.

$$\sigma_O^2 = \sigma_X^2 - \sigma_R^2 - \sigma_{IV}^2 \tag{5}$$

The equations for performing the sample-level versions of Eqs. 2–5 are presented in Trafimow (2018) and need not be repeated here. The important point for now is that it is possible, and not particularly difficult, to estimate the three types of variance. But what is the gain in doing so?

To see the gain, consider a reasonably typical case where a researcher collects data on a set of variables and finds that she can account for 10% of the variance in the variable of interest with the other variables that were included in the study. An important question, then, is whether the researcher should search for additional variables to improve on the original 10% figure. Based on the usual partition of variance into good versus bad variance, there is no straightforward way to address this important question. In contrast, by using tripartite variance parsing, the researcher can garner important clues. Suppose that the researcher finds that much of the 90% of the variance that is unaccounted for is due to systematic factors. In this case, the search for additional variables makes a lot of sense because those variables are out there to be discovered. In contrast, suppose that the variance that is unaccounted for is mostly due to random measurement error. In this case, the search for more variables makes very little sense; it would make much more sense to devote research efforts towards improving the measurement device to decrease measurement error.

Or to use an experiment as the example, suppose the researcher had obtained an effect of an experimental manipulation on the dependent variable, with the independent variable accounting for 10% of the variance in the dependent variable. Clearly, 90% of the variance in the dependent variable is due to other stuff, but to what extent is that other stuff systematic or random? If it is mostly systematic, it makes sense to search for the relevant variables and attempt to manipulate them. But if it is mostly random, the researcher cannot expect such a search likely to be worth the investment; as in the correlational example, it would be better to invest in obtaining a dependent variable less subject to random measurement error.

4 Causation

In this section, I consider two important causation issues. First, there is the issue of whether the theory pertains to within-participants or between-participants causation and whether the experimental design pertains to within-participants or between-participants causation. If there is a mismatch, empirical findings hardly can be said to provide strong evidence with respect to the theory. Second, there are causal modeling approaches, that are very popular, but nevertheless problematic. The following subsections discuss each, respectively.

4.1 Within-Participants and Between-Participants Causation

It is a truism that researchers wish to draw causal conclusions from their data. In this connection, most methodology textbooks tout the excellence of true experimental designs, with random assignment of participants to conditions. Nor do I disagree but with a discrepancy. Specifically, what most methodology textbooks do not say is that there is a difference between within-person and between-person causation. Consider the textbook case where participants are randomly assigned to experimental and control conditions, there is a difference between the means in the two conditions, and the researcher concludes that the manipulation caused the difference. Even pretending the ideal experiment, where there are zero differences between conditions other than the manipulation, and even imagining the ideal case where both distributions are normal, there nevertheless remains an issue. To see the issue, let us include some theoretical material. Let us imagine that the researcher performed an attitude manipulation to test the effect on intentions to wear seat belts. Theoretically, then, the causation is from attitudes to intentions and here is the rub. At the level of attitude theories in social psychology (see Fishbein and Ajzen 2010 for a review), each person's attitude allegedly causes his or her intention to wear or not wear a seat belt; that is, at the theoretical level the causation is within-participants. But empirically, the researcher uses a between-participants design, so all that is known is that the mean is different in the two conditions. Thus, although the researcher is safe (in our idealized setting) in concluding that the manipulation caused seat belt intentions, the empirical causation is between-participants. There is no way to know the extent to which, or whether at all, attitudes cause intentions at the theorized within-participants level.

What can be done about it? The most obvious solution is to use within-participants designs. Suppose, for example, that participants' attitudes and intentions are measured prior to a manipulation designed to influence attitudes in either the positive or negative direction; but subsequently too. In that case, according to attitude theories, participants whose attitude changes in the positive direction after the manipulation also should have corresponding intention change in the positive direction. Participants whose attitude changes in the negative direction also should have corresponding intention change in the negative direction. Those participants with matching attitude and intention changes support the theory whereas those participants with mismatching attitude and intention changes (e.g., attitude becomes more positive but intentions do not) disconfirm the theory. One option for the researcher, though far from the only option, is to simply count the number of participants who support or disconfirm the theory to gain an idea of the proportion of participants for whom the theorized within-participants causation manifests. Alternatively, if the frequency of participants with attitude changes or intention changes differs substantially from 50% in the positive or negative direction, the researcher can supplement the frequency count by computing the adjusted success rate, which takes chance matching into account and has nicer properties than alternatives, such as the phi coefficient, the odds ratio, and the difference between conditional proportions (Trafimow 2017b).[6]

4.2 Causal Modeling

It often happens that researchers wish to draw causal conclusions from correlational data via mediation, moderation, or some other kind of causal analysis. I am very skeptical of these sorts of analyses. The main reason is what Spirtes et al. (2000) termed the statistical indistinguishability problem. When a statistical analysis cannot distinguish between alternative causal pathways, which is generally the case with correlational research, then there is no way to strongly support one hypothesized causal pathway over another. A recent special issue of *Basic and Applied Social Psychology* (2015) contains articles that discuss this and related problems (Grice et al. 2015; Kline 2015; Tate 2015; Thoemmes 2015; Trafimow 2015).

But there is an additional way to criticize causal analysis as applied to correlational data that does not depend on an understanding of the philosophical issues that pertain to causation, but rather on simple arithmetic (Trafimow 2017c). Consider the case where there are only two variables and a single correlation coefficient is computed. One could create a causal model but as only two variables are considered, the causal model would be very simple as it depends on only a single underlying correlation coefficient. In contrast, suppose there are three variables, and the researcher wishes to support that A causes C, mediated by B. In that case, there are three relevant correlations: r_{AB}, r_{AC}, and r_{BC}. Note that in the case of only two variables, only a single correlation must be for the "right" reason for the model to be true. In contrast, when there are three variables, there are three correlations, and all of them must be for the right reason for the model to be true. In the case where there are four variables, there are six underlying correlations: $r_{AB}, r_{AC}, r_{AD}, r_{BC}, r_{BD}$, and r_{CD}. When there are 5 variables, there are ten underlying correlations, and matters continue to worsen as the causal model becomes increasingly complex.

Well, then, suppose that we generously assume that the probability that a correlation is for the right reason (caused by what it is supposed to be caused by and not caused by what it is not supposed to be caused by) is .7. In that case, when there are only two variables, the probability of the causal model being true is .7. But when there are three variables and three underlying correlation coefficients, the probability of the causal model being true is $.7^3 = .343$—well under a coin toss. And matters continue to worsen as more variables are included in the model. Under less optimistic scenarios, where the probability that a correlation is for the right reason is less than .7, and where

[6] I provide all the equations necessary to calculate the adjusted success rate in Trafimow (2017b).

more variables are included in the model, Table 1 shows how low model probabilities can go. And it is worth stressing that all of this is under the generous assumption that all obtained correlations are consistent with the researcher's model.

Table 1. Model probabilities when the probability for each correlation being for the right reason is .4, .5, .6, or .7; and when there are 1, 2, 3, 4, 5, 6, or 7 variables in the causal model.

# Variables	Number of correlations	Correlation Probability			
		.4	.5	.6	.7
2	1	.4	.5	.6	.7
3	3	.064	.125	.216	.343
4	6	.004	.016	.047	.118
5	10	1.04E-4	9.77E-4	6.05E-3	.028
6	15	1.07E-6	3.05E-5	4.70E-4	4.75E-3
7	21	4.40E-9	4.77E-7	2.19E-5	5.59E-4

Yet another problem with causal analysis is reminiscent of what already has been covered; the level of analysis of causal modeling articles is between-participants whereas most theories specify within-participants causation. To see this, consider another attitude instance. According a portion of the theory of reasoned action (see Fishbein and Ajzen 2010 for a review), attitudes cause intentions which, in turn, cause behaviors. The theory is clearly a within-participants theory; that is, the causal chain is supposed to happen for everyone. Although there have been countless causal modeling articles, these have been at the between-participants level and consequently fail to adequately test the theory. This is not to say that the theory is wrong; in fact, when within-participants analyses have been used they have tended to support the theory (e.g., Trafimow and Finlay 1996; Trafimow et al. 2010). Rather, the point is that thousands of empirical articles pertaining to the theory failed to adequately test it because of, among other issues, a failure to understand the difference between causation that is within versus between-participants. It is worth stressing that between-participants and within-participants analyses can suggest very different, and even contradictory, causal conclusions (Trafimow et al. 2004). Thus, there is no way to know whether this is so with respect to the study under consideration except to perform both types of analyses.

In summary, those researchers who are interested in finding causal relations between variables should ask at least two kinds of questions. First, what kind of causation—within-participants or between participants? Once this question is answered it is then possible to design an experiment more suited to the type of causation of interest. If the type of causation, at the level of the theory, really is between-participants, there is no problem with researchers using between-participants designs and comparing summary statistics across between-participant conditions. However, it is rare that theorized causation is between-participants; it is usually within-participants. In that case, although between-participants designs accompanied by a comparison of summary statistics across between-participants conditions can still yield some useful information; much more useful information is yielded by within-participants designs that allow the researcher to keep track of whether each participant's responses support or disconfirm the theorized causation. Even if the responses on one or more variables is highly imbalanced, thereby rendering chance matching of variables problematic, the problem can be handled well by using the adjusted success rate. Keeping track of participants who support or disconfirm the theorized causation, accompanied by an adjusted success rate computation, constitutes a combination that facilitates the ability of researchers to draw stronger within-participants causal conclusions than they otherwise would be able to draw.

The second causation question is specific to researchers who use causal modeling: that is, how many variables are included in the causal model and how many underlying correlations does this number imply? Aside from the statistical indistinguishability problem that plagues researchers who wish to infer causation from a set of correlations, simple arithmetic also is problematic. Table 1 shows that as the number of variables increases, the number of underlying correlations increases even more, and the probability that the model is correct decreases accordingly. The values in Table 1 show that researchers are on thin ice when they use causal modeling to support causal models based on correlational evidence. (And I urge causal modelers also not to forget to consider the issue of within-participants causation at the level of theory not matched by between-participants causation at the level of the correlations that underlie the causal analysis.) If researchers continue to use causal modeling, at least they should take the trouble to count the number of variables and underlying correlations, to arrive at probabilities such as those presented in Table 1. To my knowledge, no causal modelers do this, but they clearly should to appropriately qualify the strength of their support for proposed models.

5 Conclusion

All three sections, on a priori procedures, a posteriori analyses, and causation, imply that researchers could, and should, do much more before and after collecting their data. By using a priori procedures, researchers can assure themselves of collecting sufficient data to meet a priori specifications for closeness and confidence. They also can meet a priori specifications for replicability for ideal experiments, remembering that if the sample size is too low for good ideal replicability, it certainly is too low for good replicability in the real scientific universe. Concerning a posteriori analyses, researchers can try out different summary statistics, such as means and locations, to see if they imply similar, different, or even opposing qualitative stories (see Figs. 1 and 2). Researchers also can engage in the tripartite parsing of variance, as opposed to the currently typical bipartite parsing, to gain a much better understanding of their data and the direction future research efforts should follow.

The comments pertaining to causation do not fall neatly into the category of a priori procedures or a posteriori analyses. This is because these comments imply the necessity for careful thinking before and after obtaining data. Before conducting the research, it is useful to consider whether the type of causation tested in the research matches or mismatches the type of causation specified by the theory under investigation. And after the data have been collected, there are analyses that can be done in addition to merely comparing means (or locations) to test between-participants causation. Provided a within-participants design has been used, or at least that there is a within-participants component of the research paradigm, it is possible to investigate frequencies of participants that support or disconfirm the hypothesized within-participants causation. It is even possible to use the adjusted success rate to obtain a formal evaluation of the causal mechanism under investigation. Finally, with respect to causal modeling, the researcher can do much a priori thinking by using Table 1 and counting the number of variables to be included in the final causal model. If the count indicates a sufficiently low probability of the model, even under the very favorable assumption that all correlations work out as the researcher desires, the researcher should consider not performing that research. And if the researcher does so anyway, the findings should be interpreted with the caution that Table 1 implies is appropriate.

Compared to what researchers could be doing, what they currently are doing is blatantly underwhelming. My hope and expectation is that this paper, as well as TES2019 and ECONVN2-019 more generally, persuade researchers to dramatically increase the quality of their research with respect to a priori procedures and a posteriori analyses. As explained here, much improvement is possible. It only remains to be seen whether researchers will do it.

References

Blanca, M.J., Arnau, J., López-Montiel, D., Bono, R., Bendayan, R.: Skewness and kurtosis in real data samples. Methodol. Eur. J. Res. Methods Behav. Soc. Sci. **9**(2), 78–84 (2013)

Cain, M.K., Zhang, Z., Yuan, K.H.: Behav. Res. Methods **49**(5), 1716–1735 (2017)

Earp, B.D., Trafimow, D.: Replication, falsification, and the crisis of confidence in social psychology. Front. Psychol. **6**(621), 1–11 (2015)

Fishbein, M., Ajzen, I.: Predicting and changing behavior: The Reasoned Action Approach. Psychology Press (Taylor & Francis), New York (2010)

Gillies, D.: Philosophical theories of probability. Routledge, London (2000)

Grice, J.W., Cohn, A., Ramsey, R.R., Chaney, J.M.: On muddled reasoning and mediation modeling. Basic Appl. Soc. Psychol. **37**(4), 214–225 (2015)

Gulliksen, H.: Theory of Mental Tests. Lawrence Erlbaum Associates Publishers, Hillsdale (1987)

Ho, A.D., Yu, C.C.: Descriptive statistics for modern test score distributions: Skewness, kurtosis, discreteness, and ceiling effects. Educ. Psychol. Measur. **75**(3), 365–388 (2015)

Kline, R.B.: The mediation myth. Basic Appl. Soc. Psychol. **37**(4), 202–213 (2015)

Lord, F.M., Novick, M.R.: Statistical theories of mental test scores. Addison-Wesley, Reading (1968)

Micceri, T.: The unicorn, the normal curve, and other improbable creatures. Psychol. Bull. **105** (1), 156–166 (1989)

Nguyen, H.T.: On evidential measures of support for reasoning with integrated uncertainty: a lesson from the ban of P-values in statistical inference. In: Huynh, V.N. et al. (Eds.) Integrated Uncertainty in Knowledge Modeling and Decision Making, Lecture notes in Artificial Intelligence, vol, 9978, pp. 3–15. Springer, Cham (2016)

Spirtes, P., Glymour, C., Scheines, R.: Causation, Prediction, and Search. The MIT Press, Cambridge (2000)

Tate, C.U.: On the overuse and misuse of mediation analysis: it may be a matter of timing. Basic Appl. Soc. Psychol. **37**(4), 235–246 (2015)

Thoemmes, F.: Reversing arrows in mediation models does not distinguish plausible models. Basic Appl. Soc. Psychol. **37**(4), 226–234 (2015)

Trafimow, D.: Editorial. Basic Appl. Soc. Psychol. **36**(1), 1–2 (2014)

Trafimow, D.: Introduction to special issue: what if planetary scientists used mediation analysis to infer causation? Basic Appl. Soc. Psychol. **37**(4), 197–201 (2015)

Trafimow, D.: Using the coefficient of confidence to make the philosophical switch from a posteriori to a priori inferential statistics. Educ. Psychol. Measur. **77**(5), 831–854 (2017a)

Trafimow, D.: Comparing the descriptive characteristics of the adjusted success rate to the phi coefficient, the odds ratio, and the difference between conditional proportions. Int. J. Stat. Adv. Theory Appl. **1**(1), 1–19 (2017b)

Trafimow, D.: The probability of simple versus complex causal models in causal analyses. Behav. Res. Methods **49**(2), 739–746 (2017c)

Trafimow, D.: Some implications of distinguishing between unexplained variance that is systematic or random. Educ. Psychol. Measur. **78**(3), 482–503 (2018)

Trafimow, D.: My ban on null hypothesis significance testing and confidence intervals. Studies in Computational Intelligence (in press a)

Trafimow, D.: An *a priori* solution to the replication crisis. Philos. Psychol. **31**(8), 1188–1214 (2018)

Trafimow, D., Amrhein, V., Areshenkoff, C.N., Barrera-Causil, C.J., Beh, E.J., Bilgiç, Y.K., Bono, R., Bradley, M.T., Briggs, W.M., Cepeda-Freyre, H.A., Chaigneau, S.E., Ciocca, D.R., Correa, J.C., Cousineau, D., de Boer, M.R., Dhar, S.S., Dolgov, I., Gómez-Benito, J., Grendar, M., Grice, J.W., Guerrero-Gimenez, M.E., Gutiérrez, A., Huedo-Medina, T.B., Jaffe, K., Janyan, A., Karimnezhad, A., Korner-Nievergelt, F., Kosugi, K., Lachmair, M., Ledesma, R.D., Limongi, R., Liuzza, M.T., Lombardo, R., Marks, M.J., Meinlschmidt, G., Nalborczyk, L., Nguyen, H.T., Ospina, R., Perezgonzalez, J.D., Pfister, R., Rahona, J.J., Rodríguez-Medina, D.A., Romão, X., Ruiz-Fernández, S., Suarez, I., Tegethoff, M., Tejo, M., van de Schoot, R., Vankov, I.I., Velasco-Forero, S., Wang, T., Yamada, Y., Zoppino, F.C.M., Marmolejo-Ramos, F.: Manipulating the alpha level cannot cure significance testing. Front. Psychol. **9**, 699 (2018a)

Trafimow, D., Clayton, K.D., Sheeran, P., Darwish, A.-F.E., Brown, J.: How do people form behavioral intentions when others have the power to determine social consequences? J. Gen. Psychol. **137**, 287–309 (2010)

Trafimow, D., Kiekel, P.A., Clason, D.: The simultaneous consideration of between-participants and within-participants analyses in research on predictors of behaviors: the issue of dependence. Eur. J. Soc. Psychol. **34**, 703–711 (2004)

Trafimow, D., MacDonald, J.A.: Performing inferential statistics prior to data collection. Educ. Psychol. Measur. **77**(2), 204–219 (2017)

Trafimow, D., Marks, M.: Editorial. Basic Appl. Soc. Psychol. **37**(1), 1–2 (2015)

Trafimow, D., Marks, M.: Editorial. Basic Appl. Soc. Psychol. **38**(1), 1–2 (2016)

Trafimow, D., Wang, T., Wang, C.: Means and standard deviations, or locations and scales? That is the question! New Ideas Psychol. **50**, 34–37 (2018b)

Trafimow, D., Wang, T., Wang, C.: From a sampling precision perspective, skewness is a friend and not an enemy! Educ. Psychol. Meas. (in press)

Trueblood, J.S., Busemeyer, J.R.: A quantum probability account of order effects in inference. Cogn. Sci. **35**, 1518–1552 (2011)

Trueblood, J.S., Busemeyer, J.R.: A quantum probability model of causal reasoning. Front. Psychol. **3**, 138 (2012)

Valentine, J.C., Aloe, A.M., Lau, T.S.: Life after NHST: How to describe your data without "p-ing" everywhere. Basic Appl. Soc. Psychol. **37**(5), 260–273 (2015)

Why Hammerstein-Type Block Models Are so Efficient: Case Study of Financial Econometrics

Thongchai Dumrongpokaphan[1], Afshin Gholamy[2], Vladik Kreinovich[2(✉)], and Hoang Phuong Nguyen[3]

[1] Department of Mathematics, Faculty of Science, Chiang Mai University, Chiang Mai, Thailand
tcd43@hotmail.com

[2] University of Texas at El Paso, El Paso, TX 79968, USA
afshingholamy@gmail.com, vladik@utep.edu

[3] Division Informatics, Math-Informatics Faculty, Thang Long University, Nghiem Xuan Yem Road, Hoang Mai District, Hanoi, Vietnam
nhphuong2008@gmail.com

Abstract. In the first approximation, many economic phenomena can be described by linear systems. However, many economic processes are non-linear. So, to get a more accurate description of economic phenomena, it is necessary to take this non-linearity into account. In many economic problems, among many different ways to describe non-linear dynamics, the most efficient turned out to be Hammerstein-type block models, in which the transition from one moment of time to the next consists of several consequent blocks: linear dynamic blocks and blocks describing static non-linear transformations. In this paper, we explain why such models are so efficient in econometrics.

1 Formulation of the Problem

Linear models and need to go beyond them. In the first approximation, the dynamics of an economic system can be often well described by a linear model, in which the values $y_1(t), \ldots, y_n(t)$ of the desired quantities at the current moment of time linearly depend:

- on the values of these quantities at the previous moments of time, and
- on the values of related quantities $x_1(t), \ldots, x_m(t)$ at the current and previous moments of time:

$$y_i(t) = \sum_{j=1}^{n} \sum_{s=1}^{S} C_{ijs} \cdot y_j(t-s) + \sum_{p=1}^{m} \sum_{s=0}^{S} D_{ips} \cdot x_p(t-s) + y_{i0}. \quad (1)$$

In practice, however, many real-life processes are non-linear. To get a more accurate description of real-life economic processes, it is therefore desirable to take this non-linearity into account.

V. Kreinovich et al. (Eds.): ECONVN 2019, SCI 809, pp. 129–136, 2019.
https://doi.org/10.1007/978-3-030-04200-4_9

Hammerstein-type block models for nonlinear dynamics are very efficient in econometrics. There are many different ways to describe non-linearity. In many econometric applications, the most accurate and the most efficient models turned out to be models which in control theory are known as *Hammerstein-type block models*, i.e., models that combine linear dynamic equations like (1) with non-linear static transformations; see, e.g., [5,9,10].

To be more precise, in such models, the transition from the state at one moment of time to the state at the next moment of time consists of several sequential transformations:

- some of which are linear dynamical transformations of the type (1), and
- some correspond to static non-linear transformations, i.e., nonlinear transformations that take into account only the current values of the corresponding quantities.

A toy example of a block model. To illustrate the idea of a Hammerstein-type block model, let us consider the simplest case, when:

- the state of the system is described by a single quantity y_1,
- the state $y_1(t)$ at the current moment of time is uniquely determined only by its previous state $y_1(t-1)$ (so there is no need to take into account earlier values like $y_1(t-2)$), and
- no other quantities affect the dynamics.

In the linear approximation, the dynamics of such a system is described by a linear dynamic equation

$$y_1(t) = C_{111} \cdot y_1(t-1) + y_{10}.$$

The simplest possible non-linearity here will be an additional term which is quadratic in $y_1(t)$:

$$y_1(t) = C_{111} \cdot y_1(t-1) + c \cdot (y_1(t-1))^2 + y_{10}.$$

The resulting non-linear system can be naturally reformulated in Hammerstein-type block terms if we introduce an auxiliary variable $s(t) \stackrel{\text{def}}{=} (y_1(t))^2$. In terms of this auxiliary variable, the above system can be described in terms of two blocks:

- a linear dynamical block described by a linear dynamic equation

$$y_1(t) = C_{111} \cdot y_1(t-1) + c \cdot s(t-1) + y_{10}, \text{ and}$$

- a nonlinear block described by the following non-linear static transformation

$$s(t) = (y(t))^2.$$

Comment. In this simple case, we use a quadratic non-linear transformation. In econometrics, other non-linear transformations are often used: e.g., logarithms and exponential functions that transform a multiplicative relation $z = x \cdot y$ between quantities into a linear relation between their logarithms: $\ln(z) = \ln(x) + \ln(y)$.

Formulation of the problem. The above example shows that in many cases, a non-linear dynamical system can indeed be represented in the Hammerstein-type block form, but the question remains why necessarily such models often work the best in econometrics – while there are many other techniques for describing non-linear dynamical systems (see, e.g., [1,7]), such as:

- Wiener models, in which the values $y_i(t)$ are described as Taylor series in terms of $y_j(t-s)$ and $x_p(t-s)$,
- models that describe the dynamics of wavelet coefficients,
- models that formulate the non-linear dynamics in terms of fuzzy rules, etc.

What we do in this paper. In this paper, we provide an explanation of why such block models are indeed empirically efficient in econometrics, especially in financial econometrics.

2 Analysis of the Problem and the Resulting Explanation

Specifics of computations related to econometrics, especially to financial econometrics. In many economics-related problems, it is important not only to predict future values of the corresponding quantities, but also to predict them as fast as possible. This need for speed is easy to explain.

For example, an investor who is the first to finish computation of the future stock price will have an advantage of knowing in what direction this price will go. If his or her computations show that the price will go up, the investor will buy the stock at the current price, before everyone else realizes that this price will go up – and thus gain a lot. Similarly, if the investor's computations show that the price will go down, the investor will sell his/her stock at the current price and thus avoid losing money.

Similarly, an investor who is the first to predict the change in the ratio of two currencies will gain a lot. In all these cases, fast computations are extremely important. Thus, the nonlinear models that we use in these predictions must be appropriate for the fastest possible computations.

How can we speed up computations: need for parallel computations. If a task takes a lot of time for a single person, a natural way to speed it up is to have someone else help, so that several people can perform this task in parallel. Similarly, if a task takes too much time on a single computer processor, a natural way to speed it up is to have several processors work in parallel on different parts of this general task.

Need to consider the simplest possible computational tasks for each processor. For a massively parallel computation, the overall computation time is determined by the time during which each processor finishes its task. Thus, to make the overall computations as fast as possible, it is necessary to make the elementary tasks assigned to each processor as fast – and thus, as simple – as possible.

Each computational task involves processing numbers. Since we are talking about the transition from linear to nonlinear models, it makes sense to consider linear versus nonlinear transformations. Clearly, linear transformations are much faster than nonlinear ones.

However, if we only use linear transformations, then we only get linear models. To take nonlinearity into account, we need to have some nonlinear transformations as well. A nonlinear transformation can mean:

- having one single input number and transforming it into another,
- it can mean having two input numbers and applying a nonlinear transformation to these two numbers,
- it can mean having three input numbers, etc.

Clearly, in general, the fewer numbers we process, the faster the data processing. Thus, to make computations as fast as possible, it is desirable to restrict ourselves to the fastest possible nonlinear transformations: namely, the transformations of one number into one number.

Thus, to make computations as fast as possible, it is desirable to make sure that on each computation stage, each processor performs one of the fastest possible transformations:

- either a linear transformation
- or the simplest possible nonlinear transformation $y = f(x)$.

Need to minimize the number of computational stages. Now that we agreed how to minimize the computation time needed to perform each computation stage, the overall computation time is determined by the number of computational stages. To minimize the overall computation time, we thus need to minimize the overall number of such computational stages.

In principle, we can have all kinds of nonlinearities in economic systems. Thus, we need to select the smallest number of computational stages that would still allow us to consider all possible nonlinearities. How many stages do we need?

One stage is not sufficient. One stage is clearly not enough. Indeed, during one single stage, we can compute:

- either a linear function $Y = c_0 + \sum\limits_{i=1}^{N} c_i \cdot X_i$ of the inputs $X_1, \ldots, X_N$,
- or a nonlinear function of one of these inputs $Y = f(X_i)$,
- but not, e.g., a simple nonlinear function of two inputs, such as $Y = X_1 \cdot X_2$.

What about two stages? Can we use two stages?

- If both stages are linear, all we get is a composition of two linear functions which is also linear.
- Similarly, if both stages are nonlinear, all we get is compositions of functions of one variable – which is also a function of one variable.

Thus, we need to consider two different stages.

If:

- on the first stage we use nonlinear transformations $Y_i = f_i(X_i)$, and
- on the second stage, we use a linear transformation $Y = \sum_{i=1}^{N} c_i \cdot Y_i + c_0$,

we get the expression

$$Y = \sum_{i=1}^{N} c_i \cdot f_i(X_i) + c_0.$$

For this expression, the partial derivative

$$\frac{\partial Y}{\partial X_1} = c_1 \cdot f_1'(X_1)$$

does not depend on X_2 and thus,

$$\frac{\partial^2 Y}{\partial X_1 \partial X_2} = 0,$$

which means that we cannot use such a scheme to describe the product $Y = X_1 \cdot X_2$ for which

$$\frac{\partial^2 Y}{\partial X_1 \partial X_2} = 1.$$

But what if:

- we use linear transformation on the first stage, getting

$$Z = \sum_{i=1}^{N} c_i \cdot X_i + c_0,$$

  and then
- we apply a nonlinear transformation $Y = f(Z)$.

This would result in

$$Y(X_1, X_2, \ldots) = f\left(\sum_{i=1}^{N} c_i \cdot X_i + c_0\right).$$

In this case, the level set $\{(X_1, X_2, \ldots) : Y(X_1, X_2, \ldots) = \text{const}\}$ of thus computed function is described by the equation

$$\sum_{i=1}^{N} c_i \cdot X_i = \text{const},$$

and is, thus, a plane. In particular, in the 2-D case when $N = 2$, this level set is a straight line. Thus, a 2-stage function cannot describe or approximate multiplication $Y = X_1 \cdot X_2$, because for multiplication, the level sets are hyperbolas $X_1 \cdot X_2 = \text{const}$ – and not straight lines.

So, two computational stages are not sufficient, we need at least three.

Are three computational stages sufficient? The positive answer to this equation comes from the fact that an arbitrary function can be represented as a Fourier transform and thus, can be approximated, with any given accuracy, as a linear combination of trigonometric functions:

$$Y(X_1, \ldots, X_N) \approx \sum_{k} c_k \cdot \sin\left(\omega_{k1} \cdot X_1 + \ldots + \omega_{kN} \cdot X_N + \omega_{k0}\right).$$

The right-hand side expression can be easily computed in three simple computational stages of one of the above types:

- first, we have a linear stage where we compute the linear combinations
$$Z_k = \omega_{k1} \cdot X_1 + \ldots + \omega_{kN} \cdot X_N + \omega_{k0},$$
- then, we have a nonlinear stage at which we compute the values $Y_k = \sin(Z_k)$, and
- finally, we have another linear stage at which we combine the values Y_k into a single value $Y = \sum\limits_{k} c_k \cdot Y_k$.

Thus, three stages are indeed sufficient – and so, in our computations, we should use three stages, e.g., linear-nonlinear-linear as above.

Relation to traditional 3-layer neural networks. The same three computational stages form the basis of the traditional 3-layer neural networks (see, e.g., [2,4,6,8]):

- on the first stage, we compute a linear combination of the inputs
$$Z_k = \sum_{i=1}^{N} w_{ki} \cdot X_i - w_{k0};$$
- then, we apply a nonlinear transformation $Y_k = s_0(Z_k)$; the corresponding *activation function* $s_0(z)$ usually has either the form $s_0(z) = \dfrac{1}{1 + \exp(-z)}$ or the rectified linear form $s_0(z) = \max(z, 0)$ [3,6];
- finally, a linear combination of the values Y_k is computed: $Y = \sum\limits_{k=1}^{K} W_k \cdot Y_k - W_0$.

Comments

- It should be mentioned that in neural networks, the first two stages are usually merged into a single stage in which we compute the values

$$Y_k = s_0\left(\sum_{i=1}^{N} w_{ki} \cdot X_i - w_{k0}\right).$$

The reason for this merger is that in the biological neural networks, these two stages are performed within the same neuron:
  - first, the signals X_i from different neurons come together, forming a linear combination $Z_k = \sum_{i=1}^{N} w_{ki} \cdot X_i - w_{k0}$, and
  - then, within the same neuron, the nonlinear transformation $Y_k = s_0(Z_k)$ is applied.
- Instead of using the same activation function $s_0(z)$ for all the neurons, it is sometimes beneficial to use different functions in different situations, i.e., take $Y_k = s_k(Z_k)$ for several different functions $s_k(z)$; see, e.g., [6] and references therein.

How all this applies to non-linear dynamics. In non-linear dynamics, as we have mentioned earlier, to predict each of the desired quantities $y_i(t)$, we need to take into account the previous values $y_j(t-s)$ of the quantities $y_1, \ldots, y_n$, and the current and previous values $x_p(t-s)$ of the related quantities $x_1, \ldots, x_m$. In line with the above-described 3-stage computation scheme, the corresponding prediction of each value $y_i(t)$ consists of the following three stages:

- first, there is a linear stage, at which we form appropriate linear combinations of all the inputs; we will denote the values of these linear combinations by $\ell_{ik}(t)$:

$$\ell_{ik}(t) = \sum_{j=1}^{n}\sum_{s=1}^{S} w_{ikjs} \cdot y_j(t-s) + \sum_{p=1}^{m}\sum_{s=0}^{S} v_{ikps} \cdot x_p(t-s) - w_{ik0}; \quad (2)$$

- then, there is a non-linear stage when we apply the appropriate nonlinear functions $s_{ik}(z)$ to the values ℓ_{ik}; the results of this application will be denoted by $a_{ik}(t)$:

$$a_{ik}(t) = s_{ik}(\ell_{ik}(t)); \quad (3)$$

- finally, we again apply a linear stage, at which we estimate $y_i(t)$ as a linear combination of the values $a_{ik}(t)$ computed on the second stage:

$$y_i(t) = \sum_{k=1}^{K} W_{ik} \cdot a_{ik}(t) - W_{i0}. \quad (4)$$

We thus have the desired Hammerstein-type block structure:

- a linear dynamical part (2) is combined with
- static transformations (3) and (4), in which we only process values corresponding to the same moment of time t.

Thus, *the desire to perform computations as fast as possible indeed leads to the Hammerstein-type block models.* We have therefore explained the efficiency of such models in econometrics.

Comment. Since, as we have mentioned, 3-layer models of the above type are universal approximators, we can conclude that:

- not only Hammesterin-type models compute as fast as possible,
- these models also allow us to approximate any possible nonlinear dynamics with as much accuracy as we want.

Acknowledgments. This work was supported by Chiang Mai University. It was also partially supported by the US National Science Foundation via grant HRD-1242122 (Cyber-ShARE Center of Excellence).

The authors are greatly thankful to Hung T. Nguyen for valuable discussions.

References

1. Billings, S.A.: Nonlinear System Identification: NARMAX Methods in the Time, Frequency, and Spatio-Temporal Domains. Wiley, Chichester (2013)
2. Bishop, C.M.: Pattern Recognition and Machine Learning. Springer, New York (2006)
3. Fuentes, O., Parra, J., Anthony, E., Kreinovich, V.: Why rectified linear neurons are efficient: a possible theoretical explanations. In: Kosheleva, O., Shary, S., Xiang, G., Zapatrin, R. (eds.) Beyond Traditional Probabilistic Data Processing Techniques: Interval, Fuzzy, etc. Methods and Their Applications. Springer, Cham (to appear)
4. Gholamy, A., Parra, J., Kreinovich, V., Fuentes, O., Anthony, E.: How to best apply deep neural networks in geosciences: towards optimal 'Averaging' in dropout training. In: Watada, J., Tan, S.C., Vasant, P., Padmanabhan, E., Jain, L.C. (eds.) Smart Unconventional Modelling, Simulation and Optimization for Geosciences and Petroleum Engineering. Springer (to appear)
5. Giri, F., Bai, E.-W. (eds.): Block-oriented Nonlinear System Identification. Lecture Notes in Control and Information Sciences, vol. 404. Springer, Berlin (2010)
6. Goodfellow, I., Bengio, Y., Courville, A.: Deep Learning. MIT Press, Cambridge (2016)
7. Nelles, O.: Nonlinear System Identification: From Classical Approaches to Neural Networks and Fuzzy Models. Springer, Berlin (2010)
8. Nguyen, H.T., Kreinovich, V.: Applications of Continuous Mathematics to Computer Science. Kluwer, Dordrecht (1997)
9. Strmcnik, S., Juricic, D. (eds.): Case Studies in Control: Putting Theory to Work. Springer, London (2013)
10. van Drongelen, W.: Signal Processing for Neuroscientists. London, UK (2018)

Why Threshold Models: A Theoretical Explanation

Thongchai Dumrongpokaphan[1], Vladik Kreinovich[2(✉)], and Songsak Sriboonchitta[3]

[1] Department of Mathematics, Faculty of Science, Chiang Mai University, Chiang Mai, Thailand
tcd43@hotmail.com

[2] University of Texas at El Paso, El Paso, TX 79968, USA
vladik@utep.edu

[3] Faculty of Economics, Chiang Mai University, Chiang Mai, Thailand
songsakecon@gmail.com

Abstract. Many economic phenomena are well described by linear models. In such models, the predicted value of the desired quantity – e.g., the future value of an economic characteristic – linearly depends on the current values of this and related economic characteristic and on the numerical values of external effects. Linear models have a clear economic interpretation: they correspond to situations when the overall effect does not depend, e.g., on whether we consider a loose federation as a single country or as several countries. While linear models are often reasonably accurate, to get more accurate predictions, we need to take into account that real-life processes are nonlinear. To take this nonlinearity into account, economists use piece-wise linear (*threshold*) models, in which we have several different linear dependencies in different domains. Surprisingly, such piece-wise linear models often work better than more traditional models of non-linearity – e.g., models that take quadratic terms into account. In this paper, we provide a theoretical explanation for this empirical success.

1 Formulation of the Problem

Linear models are often successful in econometrics. In econometrics, often, linear models are efficient, when the values $q_{1,t}, \ldots, q_{k,t}$ of quantities of interest $q_1, \ldots, q_k$ at time t can be predicted as linear functions of the values of these quantities at previous moments of time $t-1$, $t-2$, ..., and of the current (and past) values $e_{m,t}, e_{m,t-1}, \ldots$ of the external quantities $e_1, \ldots, e_n$ that can influence the values of the desired characteristics:

$$q_{i,t} = a_i + \sum_{j=1}^{k} \sum_{\ell=1}^{\ell_0} a_{i,j,\ell} \cdot q_{j,t-\ell} + \sum_{m=1}^{n} \sum_{\ell=0}^{\ell_0} b_{i,m,\ell} \cdot e_{m,t-\ell}; \tag{1}$$

see, e.g., [3,4,7] and references therein.

V. Kreinovich et al. (Eds.): ECONVN 2019, SCI 809, pp. 137–145, 2019.
https://doi.org/10.1007/978-3-030-04200-4_10

At first glance, this ubiquity of linear models is in line with general ubiquity of linear models in science and engineering. At first glance, the ubiquity of linear models in econometrics is not surprising, since linear models are ubiquitous in science and engineering in general; see, e.g., [5].

Indeed, we can start with a general dependence

$$q_{i,t} = f_i\left(q_{1,t}, q_{1,t-1}, \ldots, q_{k,t-\ell_0}, e_{1,t}, e_{1,t-1}, \ldots, e_{n,t-\ell_0}\right). \tag{2}$$

In science and engineering, the dependencies are usually smooth [5]. Thus, we can expand the dependence in Taylor series and keep the first few terms in this expansion. In particular, in the first approximation, when we only keep linear terms, we get a linear model.

Linear models in econometrics are applicable way beyond the Taylor series explanation. In science and engineering, linear models are effective in a small vicinity of each state, when the deviations from a given state are small and we can therefore safely ignore terms which are quadratic (or of higher order) in terms of these deviations.

However, in econometrics, linear models are effectively even when deviations are large and quadratic terms cannot be easily ignored; see, e.g., [3,4,7]. How can we explain this unexpected efficiency?

Why linear models are ubiquitous in econometrics. A possible explanation for the ubiquity of linear models in econometrics was proposed in [7]. Let us illustrate this explanation on the example of formulas for predicting how the country's Gross Domestic Product (GDP) $q_{1,t}$ changes with time t. To estimate the current year's GDP, it is reasonable to use:

- GDP values in the past years, and
- different characteristics that affect the GDP, such as the population size, the amount of trade, the amount of minerals extracted in a given year, etc.

In many cases, the corresponding description is un-ambiguous. However, in many other cases, there is an ambiguity in what to consider a country. Indeed, in many cases, countries form a loose federation: European Union is a good example. Most of European countries have the same currency, there are no barriers for trade and for movement of people between different countries, so, from the economic viewpoint, it make sense to treat the European Union as a single country. On the other hand, there are still differences between individual members of the European Union, so it is also beneficial to view each country from the European Union on its own.

Thus, we have two possible approaches to predicting the European Union's GDP:

- we can treat the whole European Union as a single country, and apply the formula (2) to make the desired prediction;
- alternatively, we can apply the general formula (2) to each country $c = 1, \ldots, C$ independently

$$q_{i,t}^{(c)} = f_i\left(q_{1,t}^{(c)}, q_{1,t-1}^{(c)}, \ldots, q_{k,t-\ell_0}^{(c)}, e_{1,t}^{(c)}, e_{1,t-1}^{(c)}, \ldots, e_{n,t-\ell_0}^{(c)}\right). \quad (3)$$

and then add up the resulting predictions.

The overall GDP $q_{1,t}$ is the sum of GDPs of all the countries:

$$q_{1,t} = q_{1,t}^{(1)} + \ldots + q_{1,t}^{(C)}.$$

Similarly, the overall population, the overall trade, etc., can be computed as the sum of the values corresponding to individual countries:

$$e_{m,t} = e_{m,t}^{(1)} + \ldots + e_{m,t}^{(C)}.$$

Thus, the prediction of $q_{1,t}$ based on applying the formula (2) to the whole European Union takes the form

$$f_i\left(q_{1,t}^{(1)} + \ldots + q_{1,t}^{(C)}, \ldots, e_{n,t-\ell_0}^{(1)} + \ldots + e_{n,t-\ell_0}^{(C)}\right),$$

while the sum of individual predictions takes the form

$$f_i\left(q_{1,t}^{(1)}, \ldots, e_{n,t-\ell_0}^{(1)}\right) + \ldots + f_i\left(q_{1,t}^{(C)}, \ldots, e_{n,t-\ell_0}^{(C)}\right).$$

Thus, the requirement that these two predictions return the same result means that

$$f_i\left(q_{1,t}^{(1)} + \ldots + q_{1,t}^{(C)}, \ldots, e_{n,t-\ell_0}^{(1)} + \ldots + e_{n,t-\ell_0}^{(C)}\right)$$
$$= f_i\left(q_{1,t}^{(1)}, \ldots, e_{n,t-\ell_0}^{(1)}\right) + \ldots + f_i\left(q_{1,t}^{(C)}, \ldots, e_{n,t-\ell_0}^{(C)}\right).$$

In mathematical terms, this means that the function f_i should be *additive*.

It also makes sense to require that very small changes in q_i and e_m lead to small changes in the predictions, i.e., that the function f_i be continuous. It is known that every continuous additive function is linear (see, e.g., [1]) – thus the above requirement explains the ubiquity of linear econometric models.

Need to go beyond linear models. While linear models are reasonably accurate, the actual econometric processes are often non-linear. Thus, to get more accurate predictions, we need to go beyond linear models.

A seemingly natural idea: take quadratic terms into account. As we have mentioned earlier, linear models correspond to the case when we expand the original dependence in Taylor series and keep only linear terms in this expansion. From this viewpoint, if we want to get a more accurate model, a natural idea is to take into account next order terms in the Taylor expansion – i.e., quadratic terms.

The above seemingly natural idea works well in science and engineering, but in econometrics, threshold models are often better. Quadratic models are indeed very helpful in science and engineering [5]. However, surprisingly, in econometrics, different types of models turn out to be more empirically successful: namely, so-called *threshold models* in which the expression f_i in the formula (2) is piece-wise linear; see, e.g., [2,6,8–10].

Terminological comment. Piece-wise linear models are called *threshold models* since in the simplest case of a dependence on a single variable $q_{1,t} = f_1(q_{1,t-1})$, such models can be described by listing:

- thresholds $T_0 = 0, T_1, \ldots, T_S, T_{S+1} = \infty$ separating different linear expressions, and
- linear expressions corresponding to each of the intervals $[0, T_1]$, $[T_1, T_2]$, ..., $[T_{S-1}, T_S]$, $[T_S, \infty)$:

$$q_{1,t} = a^{(s)} + a_1^{(s)} \cdot q_{1,t-1} \text{ when } T_s \le q_{1,t-1} \le T_{s+1}.$$

Problem and what we do in this paper. The challenge is how to explain the surprising efficiency of partial-linear models in econometrics.

In this paper, we provide such an explanation.

2 Our Explanation

Main assumption behind linear models: reminder. As we have mentioned in the previous section, the ubiquity of linear models can be explained if we assume that for loose federations, we get the same results whether we consider the whole federation as a single country or whether we view it as several separate countries.

A similar assumption can be made if we have a company consisting of several reasonable independent parts, etc.

This assumption needs to be made more realistic. If we always require the above assumption, then we get exactly linear models. The fact that in practice, we encounter some non-linearities means that the above assumption is not always satisfied.

Thus, to take into account non-linearities, we need to replace the above too-strong assumption with a more realistic one.

How can we make the above assumption more realistic: analysis of the problem. It should not matter that much if inside a loose federation, we move an area from one country to another – so that one becomes slightly bigger and another slightly smaller – as long as the overall economy remains the same.

However, from the economic sense, it makes sense to expect somewhat different results from a "solid" country – in which the economics is tightly connected – and a loose federation of sub-countries, in which there is a clear separation between different regions. Thus:

- instead of requiring that the results of applying (2) to the whole country lead to the same prediction as results of applying (2) to sub-countries,

- we make a weaker requirement: that the sum of the result of applying (2) to sub-countries should not change if we slightly change the values within each sub-country – as long as the sum remains the same.

The crucial word here is "slightly". There is a difference between a loose federation of several economies of about the same size – as in the European Union – and an economic union of, say, France and Monaco, in which Monaco's economy is orders of magnitude smaller.

To take this difference into account, it makes sense to divide the countries into finitely many groups by size, so that the above the-same-prediction requirement be applicable only when by changing the values, we keep each country within the same group.

These groups should be reasonable from the topological viewpoint – e.g., we should require that each of the corresponding domains D of possible values is contained in a closure of its interior: $D \subseteq \overline{\text{Int}\,(D)}$, i.e., that each point on its boundary is a limit of some interior points.

Each domain should be strongly connected – in the sense that each two points in each interior should be connected by a curve which lies fully inside this interior.

Let us describe the resulting modified assumption in precise terms.

A precise description of the modified assumption. We assume that the set of all possible values of the input

$$v = (q_{1,t}, \ldots, e_{n,t-\ell_0})$$

to the function f_i is divided into a finite number of non-empty non-intersecting strongly connected domains $D^{(1)}, \ldots, D^{(S)}$. We require that each of these domains is contained in a closure of its interior $D^{(s)} \subseteq \overline{\text{Int}\left(D^{(s)}\right)}$. We then require that if the following conditions are satisfied for the fours inputs $v^{(1)}$, $v^{(2)}$, $u^{(1)}$, and $u^{(2)}$:

- the inputs $v^{(1)}$ and $u^{(1)}$ belong to the same domain,
- the inputs $v^{(2)}$ and $u^{(2)}$ also belong to the same domain (which may be different from the domain containing $v^{(1)}$ and $u^{(1)}$), and
- we have $v^{(1)} + v^{(2)} = u^{(1)} + u^{(2)}$,

then we should have

$$f_i\left(v^{(1)}\right) + f_i\left(v^{(2)}\right) = f_i\left(u^{(1)}\right) + f_i\left(u^{(1)}\right).$$

Our main result. Our main result – proven in the next section – is that under the above assumption, the function $f_i\,(v)$ is piece-wise linear.

Discussion. This result explains why piece-wise linear models are indeed ubiquitous in econometrics.

Comment. Since the functions f_i are continuous, on the border between two zones with different linear expressions E and E', these two linear expressions should attain the same value. Thus, the border between two zones can be described by the equation $E = E'$, i.e., equivalently, $E - E' = 0$. Since both expressions are linear, the equation $E - E' = 0$ is also linear, and thus, describes a (hyper-)plane in the space of all possible inputs.

So, the zones are separated by hyper-planes.

3 Proof of the Main Result

1°. We want to prove that the function f_i is linear on each domain $D^{(s)}$. To prove this, let us first prove that this function is linear in the vicinity of each point $v^{(0)}$ from the interior of the domain $D^{(s)}$.

1.1°. Indeed, by definition of the interior, it means that there exists a neighborhood of the point $v^{(0)}$ that fully belongs to the domain $D^{(s)}$. To be more precise, there exists an $\varepsilon > 0$ such that if $|d_q| \le \varepsilon$ for all components d_q of the vector d, then the vector $v^{(0)} + d$ also belongs to the domain $D^{(s)}$.

Thus, because of our assumption, if for two vectors d and d', we have

$$|d_q| \le \varepsilon, \quad |d'_q| \le \Delta, \text{ and } |d_q + d'_q| \le \varepsilon \text{ for all } q, \tag{4}$$

then we have

$$f_i\left(v^{(0)} + d\right) + f_i\left(v^{(0)} + d'\right) = f_i\left(v^{(0)}\right) + f\left(v^{(0)} + d + d'\right). \tag{5}$$

Subtracting $2f_i\left(v^{(0)}\right)$ from both sides of the equality (5), we conclude that for the auxiliary function

$$F(v) \stackrel{\text{def}}{=} f_i\left(v^{(0)} + v\right) - f_i\left(v^{(0)}\right), \tag{6}$$

we have

$$F(d + d') = F(d) + F(d'), \tag{7}$$

as long as the inequalities (4) are satisfied.

1.2°. Each vector $d = (d_1, d_2, \ldots)$ can be represented as

$$d = (d_1, 0, \ldots) + (0, d_2, 0, \ldots) + \ldots \tag{8}$$

If $|d_q| \le \varepsilon$ for all q, then the same inequalities are satisfied for all the terms in the right-hand side of the formula (8). Thus, due to the property (6), we have

$$F(d) = F_1(d_1) + F_2(d_2) + \ldots, \tag{9}$$

where we denoted

$$F_1(d_1) \stackrel{\text{def}}{=} F(d_1, 0, \ldots), \quad F_2(d_2) \stackrel{\text{def}}{=} F(0, d_2, 0, \ldots), \ldots \tag{10}$$

1.3°. For each of the functions $F_q(d_q)$, the formula (6) implies that

$$F_q\left(d_q + d'_q\right) = F_q(d_q) + F_q\left(d'_q\right). \tag{11}$$

In particular, when $d_q = d'_q = 0$, we conclude that $F_q(0) = 2F_q(0)$, hence that

$$F_q(0) = 0.$$

Now, for $d'_q = -d_q$, formula (11) implies that

$$F_q(-d_q) = -F_q(d_q). \tag{12}$$

So, to find the values of $F_q(d_q)$ for all d_q for which $|d_q| \le \varepsilon$, it is sufficient to consider the positive values d_q.

1.4°. For every natural number N, formula (11) implies that

$$F_q\left(\frac{1}{N}\cdot\varepsilon\right) + \ldots + F_q\left(\frac{1}{N}\cdot\varepsilon\right)(N \text{ times}) = F_q(\varepsilon), \tag{13}$$

thus

$$F_q\left(\frac{1}{N}\cdot\varepsilon\right) = \frac{1}{N}\cdot F_q(\varepsilon). \tag{14}$$

Similarly, for every natural number M, we have

$$F_q\left(\frac{M}{N}\cdot\varepsilon\right) = F_q\left(\frac{1}{N}\cdot\varepsilon\right) + \ldots + F_q\left(\frac{1}{N}\cdot\varepsilon\right)(M \text{ times}),$$

thus

$$F_q\left(\frac{M}{N}\cdot\varepsilon\right) = M\cdot F_q\left(\frac{1}{N}\cdot\varepsilon\right) = M\cdot\frac{1}{N}\cdot F_q(\varepsilon) = \frac{M}{N}\cdot F_q(\varepsilon).$$

So, for every rational number $r = \dfrac{M}{N} \le 1$, we have

$$F_q(r\cdot\varepsilon) = r\cdot F_q(\varepsilon). \tag{15}$$

Since the function f_i is continuous, the functions F and F_q are continuous too. Thus, we can conclude that the equality (15) holds for all real values $r \le 1$.

By using formula (12), we can conclude that the same formula holds for all real values r for which $|r| \le 1$.

Now, each d_q for which $|d_q| \le \varepsilon$ can be represented as $d_q = r\cdot\varepsilon$, where $r \stackrel{\text{def}}{=} \dfrac{d_q}{\varepsilon}$. Thus, formula (15) takes the form

$$F_q(d_q) = \frac{d_q}{\varepsilon}\cdot F_q(\varepsilon),$$

i.e., the form

$$F_q(d_q) = a_q\cdot d_q, \tag{16}$$

where we denoted $a_q \stackrel{\text{def}}{=} \dfrac{F_q(\varepsilon)}{\varepsilon}$. Formula (9) now implies that

$$F(d) = a_1\cdot d_1 + a_2\cdot d_2 + \ldots \tag{17}$$

By definition (6) of the auxiliary function $F(v)$, we have

$$f_i\left(v^{(0)}+d\right)=f_i\left(v^{(0)}\right)+F(d),$$

so for any v, if we take $d \stackrel{\text{def}}{=} v-v^{(0)}$, we would get

$$f_i(v)=f_i\left(v^{(0)}\right)+F\left(v-v^{(0)}\right). \tag{18}$$

The first term is a constant, the second term, due to (17), is a linear function of v, so indeed the function $f_i(v)$ is linear in the ε-vicinity of the given point $v^{(0)}$.

2°. To complete the proof, we need to prove that the function $f_i(v)$ is linear on the whole domain. Indeed, since the domain $D^{(s)}$ is strongly connected, any two points are connected by a finite chain of intersecting open neighborhood.

In each neighborhood, the function $f_i(v)$ is linear, and when two linear function coincide in the whole open region, their coefficients are the same. Thus, by following the chain, we can conclude that the coefficients that describe $f_i(v)$ as a locally linear function are the same for all points in the interior of the domain.

Our result is thus proven.

Acknowledgments. This work was supported by Chiang Mai University, Thailand. We also acknowledge the partial support of the Center of Excellence in Econometrics, Faculty of Economics, Chiang Mai University, Thailand, and of the US National Science Foundation via grant HRD-1242122 (Cyber-ShARE Center of Excellence).

The authors are greatly thankful to Professor Hung T. Nguyen for his help and encouragement.

References

1. Aczél, J., Dhombres, J.: Functional Equations in Several Variables. Cambridge University Press, Cambridge (2008)
2. Bollerslev, T., Chou, R.Y., Kroner, K.F.: ARCH modeling in finance: a review of the theory and empirical evidence. J. Econ. **52**, 5–59 (1992)
3. Brockwell, P.J., Davis, R.A.: Time Series: Theories and Methods. Springer, New York (2009)
4. Enders, W.: Applied Econometric Time Series. Wiley, New York (2014)
5. Feynman, R., Leighton, R., Sands, M.: The Feynman Lectures on Physics. Addison Wesley, Boston (2005)
6. Glosten, L.R., Jagannathan, R., Runkle, D.E.: On the relation between the expected value and the volatility of the nominal excess return on stocks. J. Financ. **48**, 1779–1801 (1993)
7. Nguyen, H.T., Kreinovich, V., Kosheleva, O., Sriboonchitta, S.: Why ARMAX-GARCH linear models successfully describe complex nonlinear phenomena: a possible explanation. In: Huynh, V.-N., Inuiguchi, M., Denoeux, T. (eds.) Integrated Uncertainty in Knowledge Modeling and Decision Making, Proceedings of The Fourth International Symposium on Integrated Uncertainty in Knowledge Modelling and Decision Making IUKM 2015. Lecture Notes in Artificial Intelligence, Nha Trang, Vietnam, 15–17 October 2015, vol. 9376, pp. 138–150. Springer (2015)

8. Tsay, R.S.: Analysis of Financial Time Series. Wiley, New York (2010)
9. Zakoian, J.M.: Threshold heteroskedastic models. Technical report, Institut National de la Statistique et des Études Économiques (INSEE) (1991)
10. Zakoian, J.M.: Threshold heteroskedastic functions. J. Econ. Dyn. Control **18**, 931–955 (1994)

The Inference on the Location Parameters Under Multivariate Skew Normal Settings

Ziwei Ma[1], Ying-Ju Chen[2], Tonghui Wang[1(✉)], and Wuzhen Peng[3]

[1] Department of Mathematical Sciences, New Mexico State University, Las Cruces, USA
{ziweima,twang}@nmsu.edu

[2] Department of Mathematics, University of Dayton, Dayton, USA
ychen4@udayton.edu

[3] Dongfang College Zhejiang Unversity of Finance and Economics, Hangzhou, China
pengwuzhen@163.com

Abstract. In this paper, the sampling distributions of multivariate skew normal distribution are studied. Confidence regions of the location parameter, $\boldsymbol{\mu}$, with known scale parameter and shape parameter are obtained by the pivotal method, Inferential Models (IMs), and robust method, respectively. The hypothesis test is proceeded based on the pivotal method and the power of the test is studied using non-central skew Chi-square distribution. For illustration of these results, the graphs of confidence regions and the power of the test are presented for combinations of various values of parameters. A group of Monte Carlo simulation studies is proceeded to verify the performance of the coverage probabilities at last.

Keywords: Multivariate skew-normal distributions
Confidence regions · Inferential Models
Non-central skew chi-square distribution · Power of the test

1 Introduction

The skew normal (SN) distribution was proposed by Azzalini [5,8] to cope with departures from normality. Later on, the studies on multivariate skew normal distribution are considered in Azzalini and Arellano-Valle [7], Azzalini and Capitanio [6], Branco and Dey [11], Sahu et al. [22], Arellano-Valle et al. [1], Wang et al. [25] and references therein. A k-dimensional random vector $\mathbf{Y}$ follows a skew normal distribution with location vector $\boldsymbol{\mu} \in \mathbb{R}^k$, dispersion matrix Σ (a $k \times k$ positive definite matrix), and skewness vector $\boldsymbol{\lambda} \in \mathbb{R}^k$, if its pdf is given by

$$f_{\boldsymbol{Y}}(\boldsymbol{y}) = 2\phi_k(\boldsymbol{y}; \boldsymbol{\mu}, \Sigma)\, \Phi\left(\boldsymbol{\lambda}' \Sigma^{-1/2}(\boldsymbol{y} - \boldsymbol{\mu})\right), \qquad \boldsymbol{y} \in \mathbb{R}^k, \tag{1}$$

which is denoted by $\boldsymbol{Y} \sim SN_k(\boldsymbol{\mu}, \Sigma, \boldsymbol{\lambda})$, where $\phi_k(\boldsymbol{y}; \boldsymbol{\mu}, \Sigma)$ is the k dimensional multivariate normal density (pdf) with mean $\boldsymbol{\mu}$ and covariance matrix Σ, and

V. Kreinovich et al. (Eds.): ECONVN 2019, SCI 809, pp. 146–162, 2019.
https://doi.org/10.1007/978-3-030-04200-4_11

$\Phi(u)$ is the cumulative distribution function (cdf) of the standard normal distribution. Note that $\boldsymbol{Y} \sim SN_k(\boldsymbol{\lambda})$ if $\boldsymbol{\mu} = 0$ and $\Sigma = I_k$, the k-dimensional identity matrix.

In many practical cases, a skew normal model is suitable for the analysis of data which is unimodal empirical distributed but with some skewness, see Arnold et al. [3] and Hill and Dixon [14]. For more details on the family of skew normal distributions, readers are referred to the monographs such as Genton [13] and Azzalini [9].

Making statistical inference about the parameters of a skew normal distribution is challenging. Some issues raise when using maximum likelihood (ML) based approach, such as the ML estimator for the skewness parameter could be infinite with a positive probability, and the Fisher information matrix is singular when $\lambda = 0$, even there may exist local maximum. Lots of scholars have been working on solving this issue, readers are referred to Azzalini [5,6], Pewsey [21], Liseo and Loperfido [15], Sartori [23], Bayes and Branco [10], Dey [12], Mameli et al. [18] and Zhu et al. [28] and references therein for further details. In this paper, several methods are used to construct the confidence regions for location parameter under multivariate skew normal setting and the hypothesis testing on location parameter is established as well.

The remainder of this paper is organized as follows. In Sect. 2, we discuss some properties of multivariate and matrix variate skew normal distributions, and corresponding statistical inference. In Sect. 3, confidence regions and hypothesis tests for location parameter are developed. Section 4 presents simulation studies for illustrations of our main results.

2 Preliminaries

We first introduce the basic notations and terminology which will be used throughout this article. Let $M_{n\times k}$ be the set of all $n \times k$ matrices over the real field $\mathbb{R}$ and $\mathbb{R}^n = M_{n\times 1}$. For any $B \in M_{n\times k}$, use B' to denote the transpose of B. Specifically, let I_n be the $n \times n$ identity matrix, $\mathbf{1}_n = (1, \ldots, 1)' \in \mathbb{R}^n$ and $\overline{J}_n = \frac{1}{n}\mathbf{1}_n'\mathbf{1}_n$. For $B = (\boldsymbol{b}_1, \boldsymbol{b}_2, \ldots, \boldsymbol{b}_n)'$ with $\boldsymbol{b}_i \in \mathbb{R}^k$, let $P_B = B\,(B'B)^{-}\,B'$ and $\mathrm{Vec}\,(B) = (\boldsymbol{b}_1', \boldsymbol{b}_2', \ldots, \boldsymbol{b}_n')'$. For any non negatively definite matrix $T \in M_{n\times n}$ and $m > 0$, use $\mathrm{tr}(T)$, $\mathrm{etr}(T)$ to denote the trace, exponential trace of T, respectively, and use $T^{1/2}$ and $T^{-1/2}$ to denote the square root of T and T^{-1}, respectively. For $B \in M_{m\times n}$, $C \in M_{n\times p}$ and $D \in M_{p\times q}$, use $B \otimes C$ to denote the Kronecker product of B and C, $\mathrm{Vec}\,(BCD) = (B \otimes D')\,\mathrm{Vec}\,(C)$. In addition to the notations introduced above, we use $N(0,1)$, $U(0,1)$ and χ^2_k to represent the standard normal distribution, standard uniform distribution and Chi-square distribution with degrees of freedom k, respectively. Also, bold phase letters are used to represent vectors.

2.1 Some Useful Properties of Multivariate and Matrix Variate Skew Normal Distributions

In this subsection, we introduce some fundamental properties of skew normal distributions for both multivariate and matrix variate cases, which will be used in developing the main results.

Suppose a k-dimensional random vector $\mathbf{Z} \sim SN_k(\boldsymbol{\lambda})$, i.e. its pdf is given by (1). Here, we list some useful properties of multivariate skew normal distributions that will be needed for the proof of the main results.

Lemma 1 (Arellano-Valle et al. [1]). *Let* $\mathbf{Y} = \boldsymbol{\mu} + \Sigma^{1/2}\mathbf{Z}$ *where* $\mathbf{Z} \sim SN_k(0, I_k, \boldsymbol{\lambda})$. *Then* $\mathbf{Y} \sim SN_k(\boldsymbol{\mu}, \Sigma, \boldsymbol{\lambda})$.

Lemma 2 (Wang et al. [25]). *Let* $\boldsymbol{Y} \sim SN_k(\boldsymbol{\mu}, I_k, \boldsymbol{\lambda})$. *Then* $\boldsymbol{Y}$ *has the following properties.*

(a) The moment generating function (mgf) of $\boldsymbol{Y}$ *is given by*

$$M_{\boldsymbol{Y}}(\boldsymbol{t}) = 2\exp\left(\boldsymbol{t}'\boldsymbol{\mu} + \frac{\boldsymbol{t}'\boldsymbol{t}}{2}\right)\Phi\left\{\frac{\boldsymbol{\lambda}'\boldsymbol{t}}{(1+\boldsymbol{\lambda}'\boldsymbol{\lambda})^{1/2}}\right\}, \qquad \text{for } \boldsymbol{t} \in \mathbb{R}^k, \tag{2}$$

and

(b) Two linear functions of $\boldsymbol{Y}$, $A'\boldsymbol{Y}$ *and* $B'\boldsymbol{Y}$ *are independent if and only if (i)* $A'B = 0$ *and (ii)* $A'\boldsymbol{\lambda} = 0$ *or* $B'\boldsymbol{\lambda} = 0$.

Lemma 3 (Wang et al. [25]). *Let* $Y \sim SN_k(\boldsymbol{\nu}, I_k, \boldsymbol{\lambda}_0)$, *and let* A *be a* $k \times p$ *matrix with full column rank, then the linear function of* $\boldsymbol{Y}$, $A'Y \sim SN_p(\boldsymbol{\mu}, \Sigma, \boldsymbol{\lambda})$, *where*

$$\boldsymbol{\mu} = A'\boldsymbol{\nu}, \qquad \Sigma = A'A, \qquad \text{and} \quad \boldsymbol{\lambda} = \frac{(A'A)^{-1/2}A'\boldsymbol{\lambda}_0}{\sqrt{1+\boldsymbol{\lambda}_0'\left(I_k - A(A'A)^{-1}A'\right)\boldsymbol{\lambda}_0}}. \tag{3}$$

To proceed statistical inference on multivariate skew normal population based on observed sample vectors, we need to consider the random matrix obtained from a sample of random vectors. The definition and features of matrix variate skew normal distributions are presented in the following part.

Definition 1. The $n \times p$ random matrix Y is said to have a skew-normal matrix variate distribution with location matrix μ, scale matrix $V \otimes \Sigma$, with known V and skewness parameter matrix $\boldsymbol{\gamma} \otimes \boldsymbol{\lambda}'$, denoted by $Y \sim SN_{n\times p}(\mu, V \otimes \Sigma, \boldsymbol{\gamma} \otimes \boldsymbol{\lambda}')$, if $\boldsymbol{y} \equiv \mathrm{Vec}(Y) \sim SN_{np}(\boldsymbol{\mu}, V \otimes \Sigma, \boldsymbol{\gamma} \otimes \boldsymbol{\lambda})$, where $\mu \in M_{n\times p}$, $V \in M_{n\times n}$, $\Sigma \in M_{p\times p}$, $\boldsymbol{\mu} = \mathrm{Vec}(\mu)$, $\boldsymbol{\gamma} \in \mathbb{R}^n$, and $\boldsymbol{\lambda} \in \mathbb{R}^p$.

Lemma 4 (Ye et al. [27]). *Let* $Z = (\boldsymbol{Z}_1, \ldots, \boldsymbol{Z}_k)' \sim SN_{k\times p}(0, I_{kp}, \mathbf{1}_k \otimes \boldsymbol{\lambda}')$ *with* $\mathbf{1}_k = (1, \ldots, 1)' \in \mathbb{R}^k$ *where* $\boldsymbol{Z}_i \in \mathbb{R}^p$ *for* $i = 1, \ldots, k$. *Then*

(i) The pdf of Z is

$$f(Z) = 2\phi_{k\times p}(Z)\,\Phi(\mathbf{1}_k' Z\boldsymbol{\lambda}), \qquad Z \in M_{k\times p}, \tag{4}$$

where $\phi_{k\times p}(Z) = (2\pi)^{-kp/2}\, etr(-Z'Z/2)$ and $\Phi(\cdot)$ is the standard normal distribution function.

(ii) The mgf of Z is

$$M_Z(T) = 2etr(T'T/2)\,\Phi\left\{\frac{\mathbf{1}_k' T\boldsymbol{\lambda}}{(1+k\boldsymbol{\lambda}'\boldsymbol{\lambda})^{1/2}}\right\}, \qquad T \in M_{k\times p}. \tag{5}$$

(iii) The marginals of Z, $\boldsymbol{Z}_i$ is distributed as

$$\boldsymbol{Z}_i \sim SN_p(0, I_p, \boldsymbol{\lambda}^*) \qquad \text{for} \quad i = 1, \ldots, k \tag{6}$$

with $\boldsymbol{\lambda}^* = \frac{\boldsymbol{\lambda}}{\sqrt{1+(k-1)\boldsymbol{\lambda}'\boldsymbol{\lambda}}}$.

(iv) For $i = 1, 2$, let $Y_i = \mu_i + A_i' Z \Sigma_i^{1/2}$ with $\mu_i, A_i \in M_{k\times n_i}$ and $\Sigma_i \in M_{p\times p}$, then Y_1 and Y_2 are independent if and only if

(a) $A_1' A_2 = 0$, and

(b) either $(A_1'\mathbf{1}_k) \otimes \boldsymbol{\lambda} = 0$ or $(A_2'\mathbf{1}_k) \otimes \boldsymbol{\lambda} = 0$.

2.2 Non-central Skew Chi-Square Distribution

We will make use of other related distributions to make inference on parameters for multivariate skew normal distribution, which, specifically refers to non-central skew chi-square distribution in this study.

Definition 2. Let $\boldsymbol{Y} \sim SN_m(\boldsymbol{\nu}, I_m, \boldsymbol{\lambda})$. The distribution of $\boldsymbol{Y}'\boldsymbol{Y}$ is defined as the ***noncentral skew chi-square distribution*** with degrees of freedom m, the noncentrality parameter $\xi = \boldsymbol{\nu}'\boldsymbol{\nu}$, and the skewness parameters $\delta_1 = \boldsymbol{\lambda}'\boldsymbol{\nu}$ and $\delta_2 = \boldsymbol{\lambda}'\boldsymbol{\lambda}$, denoted by $\boldsymbol{Y}'\boldsymbol{Y} \sim S\chi_m^2(\xi, \delta_1, \delta_2)$.

Lemma 5 (Ye et al. [26]). *Let $\boldsymbol{Z}_0 \sim SN_k(0, I_k, \boldsymbol{\lambda})$, $\boldsymbol{Y}_0 = \boldsymbol{\mu} + B'\boldsymbol{Z}_0$, $Q_0 = \boldsymbol{Y}_0' A \boldsymbol{Y}_0$, where $\boldsymbol{\mu} \in \mathbb{R}^n$, $B \in M_{k\times n}$ with full column rank, and A is nonnegative definite in $M_{n\times n}$ with rank m. Then the necessary and sufficient conditions under which $Q_0 \sim S\chi_m^2(\xi, \delta_1, \delta_2)$, for some $\delta_1 \in \mathbb{R}$ including $\delta_1 = 0$, are:*

(a) BAB' is idempotent of rank m,

(b) $\xi = \boldsymbol{\mu}' A \boldsymbol{\mu} = \boldsymbol{\mu}' AB'BA\boldsymbol{\mu}$,

(c) $\delta_1 = \boldsymbol{\lambda}' BA\boldsymbol{\mu}/d$,

(d) $\delta_2 = \boldsymbol{\lambda}' P_1 P_1' \boldsymbol{\lambda}/d^2$, where $d = (1 + \boldsymbol{\lambda}' P_2 P_2' \boldsymbol{\lambda})^{1/2}$, and $P = (P_1, P_2)$ is an orthogonal matrix in $M_{n\times n}$ such that

$$BAB' = P \begin{pmatrix} I_m & 0 \\ 0 & 0 \end{pmatrix} P' = P_1 P_1'.$$

Lemma 6 (Ye et al. [27]). *Let* $Z \sim SN_{k\times p}(0, I_{kp}, \mathbf{1}_k \otimes \boldsymbol{\lambda}')$, $Y = \mu + A'Z\Sigma^{1/2}$, *and* $Q = Y'WY$ *with nonnegative definite* $W \in M_{n\times n}$. *Then the necessary and sufficient conditions under which* $Q \sim SW_p(m, \Sigma, \xi, \delta_1, \delta_2)$ *for some* $\delta_1 \in M_{p\times p}$ *including* $\delta_1 = 0$, *are:*

(a) AWA' *is idempotent of rank* m,
(b) $\xi = \mu'W\mu = \mu'WVW\mu = \mu'WVWVW\mu$,
(c) $\delta_1 = \boldsymbol{\lambda}\mathbf{1}_k'AW\mu/d$, *and*
(d) $\delta_2 = \mathbf{1}_k'P_1P_1'\mathbf{1}_k\boldsymbol{\lambda}\boldsymbol{\lambda}'/d^2$, *where* $V = A'A$, $d = \sqrt{1 + \mathbf{1}_k'P_2P_2'\mathbf{1}_k\boldsymbol{\lambda}'\boldsymbol{\lambda}}$ *and* $P = (P_1, P_2)$ *is an orthogonal matrix in* $M_{k\times k}$ *such that*

$$AWA' = P\begin{pmatrix} I_m & 0 \\ 0 & 0 \end{pmatrix}P' = P_1P_1'.$$

3 Inference on Location Parameters of Multivariate Skew Normal Population

Let $Y = (\mathbf{Y}_1, \ldots, \mathbf{Y}_n)'$ be a sample of p-dimension skew normal population with sample size n such that

$$Y \sim SN_{n\times p}(\mathbf{1}_n \otimes \boldsymbol{\mu}', I_n \otimes \Sigma, \mathbf{1}_n \otimes \boldsymbol{\lambda}'), \tag{7}$$

where $\boldsymbol{\mu}, \boldsymbol{\lambda} \in \mathbb{R}^p$ and $\Sigma \in M_{p\times p}$ is positive definite. In this study, We focus on the case when the scale matrix Σ and shape parameter $\boldsymbol{\lambda}$ are known.

Based on the joint distribution of the observed sample defined by (7), we study the sampling distributions of sample mean, $\overline{\mathbf{Y}}$, and sample covariance matrix, S, respectively.

Let

$$\overline{\mathbf{Y}} = \left(\frac{1}{n}\mathbf{1}_n'Y\right)' \tag{8}$$

and

$$S = \frac{1}{n-1}\sum_{i=1}^{n}\left(\mathbf{Y}_i - \overline{\mathbf{Y}}\right)\left(\mathbf{Y}_i - \overline{\mathbf{Y}}\right)'. \tag{9}$$

The matrix form for S is

$$S = \frac{1}{n-1}Y'\left(I_n - \bar{J}_n\right)Y.$$

Theorem 1. *Let the sample matrix* $Y \sim SN_{n\times p}(\mathbf{1}_n \otimes \boldsymbol{\mu}', I_n \otimes \Sigma, \mathbf{1}_n \otimes \boldsymbol{\lambda}')$, *and* $\overline{\mathbf{Y}}$ *and* S *be defined by (8) and (9), respectively. Then*

$$\overline{\mathbf{Y}} \sim SN_p\left(\boldsymbol{\mu}, \frac{\Sigma}{n}, \sqrt{n}\boldsymbol{\lambda}\right) \tag{10}$$

and

$$(n-1)S \sim W_p(n-1, \Sigma) \tag{11}$$

are independently distributed where $W_p(n-1, \Sigma)$ *represents the p-dimensional Wishart distribution with degrees of freedom* $n-1$ *and scale matrix* Σ.

Proof. To derive the distribution of $\overline{\boldsymbol{Y}}$, consider the mgf of $\overline{\boldsymbol{Y}}$

$$M_{\overline{\boldsymbol{Y}}}(\boldsymbol{t}) = E\left(\exp\left\{\overline{\boldsymbol{Y}}'\boldsymbol{t}\right\}\right) = E\left(\operatorname{etr}\left(\overline{\boldsymbol{Y}}'\boldsymbol{t}\right)\right) = E\left(\operatorname{etr}\left(\frac{1}{n}\mathbf{1}_n'Y\boldsymbol{t}\right)\right)$$
$$= 2\operatorname{etr}\left(\boldsymbol{t}'\boldsymbol{\mu} + \frac{\frac{1}{n}\boldsymbol{t}\Sigma\boldsymbol{t}'}{2}\right)\Phi\left(\frac{\boldsymbol{t}'\Sigma^{1/2}\boldsymbol{\lambda}'}{(1+n\boldsymbol{\lambda}'\boldsymbol{\lambda})^{1/2}}\right).$$

Then the desired result follows by combining Lemmas 1 and 2.

To obtain the distribution of S, let $Q = (n-1)S = Y'\left(I_n - \overline{J}_n\right)Y$. We apply Lemma 6 to Q with $W = I_n - \overline{J}_n$, $A = I_n$ and $V = I_n$, and check conditions (a)–(d) as follows. For (a), $AWA' = I_nWI_n' = W = I_n - \overline{J}_n$ which is idempotent of rank $n-1$. For (b), from the facts $\mathbf{1}_n' \otimes \boldsymbol{\mu} = \boldsymbol{\mu}\mathbf{1}_n'$ and $\mathbf{1}_n'\left(I_n - \overline{J}_n\right) = 0$, we obtain

$$\mu' W \mu = (\mathbf{1}_n \otimes \boldsymbol{\mu}')'\left(I_n - \overline{J}_n\right)(\mathbf{1}_n \otimes \boldsymbol{\mu}') = (\mathbf{1}_n' \otimes \boldsymbol{\mu})\left(I_n - \overline{J}_n\right)(\mathbf{1}_n \otimes \boldsymbol{\mu}')$$
$$= \boldsymbol{\mu}\mathbf{1}_n'\left(I_n - \overline{J}_n\right)(\mathbf{1}_n \otimes \boldsymbol{\mu}') = 0$$

Therefore, $\xi = \mu'W\mu = \mu'WVW\mu = \mu'WVWVW\mu = 0$. For (c) and (d), we compute

$$\delta_1 = \lambda\mathbf{1}_n'\left(I_n - \overline{J}_n\right)\mu/d = 0 \qquad \text{and} \qquad \delta_2 = \mathbf{1}_n'AWA'\mathbf{1}_n\boldsymbol{\lambda}\boldsymbol{\lambda}'/d = 0$$

where $d = \sqrt{1 + n\boldsymbol{\lambda}'\boldsymbol{\lambda}}$. Therefore, we obtain that

$$Q = (n-1)\,S \sim SW_p\,(n-1, \Sigma, 0, 0, 0) = W_p\,(n-1, \Sigma)\,.$$

Now, we show that $\overline{\boldsymbol{Y}}$ and S are independent, we apply Lemma 4 part (iv) with $A_1 = \frac{1}{n}\mathbf{1}_n$ and $A_2 = I_n - \overline{J}_n$, then check the conditions (a) and (b) in Lemma 4 part (iv). For condition (a), we have

$$A_1'A_2 = \frac{1}{n}\mathbf{1}_n'(I_n - \overline{J}_n) = \mathbf{0}'.$$

For condition (b), we have

$$(A_2'\mathbf{1}_n) = (I_n - \overline{J}_n)'\mathbf{1}_n = \mathbf{0}.$$

Thus condition (b) follows automatically. Therefore the desired result follows immediately. □

3.1 Inference on Location Parameter μ When Σ and λ Are Known

After studying the sampling distributions of sample mean and covariance matrix, the inference on location parameters for a multivariate skew normal random variable defined in (7) will be performed.

3.1.1 Confidence Regions for μ

Method 1: Pivotal Method. Pivotal method is a basic method to construct confidence intervals when a pivotal quantity for the parameter of interest is available. We consider the pivotal quantity

$$P = n\left(\overline{\boldsymbol{Y}} - \boldsymbol{\mu}\right)' \Sigma^{-1} \left(\overline{\boldsymbol{Y}} - \boldsymbol{\mu}\right). \tag{12}$$

From Eq. (10) in Theorem 1 and Lemma 5, we obtain the distribution of the pivotal quantity P as follow

$$P = n\left(\overline{\boldsymbol{Y}} - \boldsymbol{\mu}\right)' \Sigma^{-1} \left(\overline{\boldsymbol{Y}} - \boldsymbol{\mu}\right) \sim \chi^2_p. \tag{13}$$

Thus we obtain the first confidence regions for the location parameter $\boldsymbol{\mu}$.

Theorem 2. *Suppose that a sample matrix Y follows the distribution (7) and Σ and $\boldsymbol{\lambda}$ are known. The confidence regions for $\boldsymbol{\mu}$ is given by*

$$C^P_{\mu}(\alpha) = \left\{\boldsymbol{\mu} : n\left(\overline{\boldsymbol{Y}} - \boldsymbol{\mu}\right)' \Sigma^{-1} \left(\overline{\boldsymbol{Y}} - \boldsymbol{\mu}\right) < \chi^2_p(1-\alpha)\right\}, \tag{14}$$

where $\chi^2_p(1-\alpha)$ represents the $1-\alpha$ quantile of χ^2_p distribution.

Remark 1. The confidence regions, given by Theorem 2, is independent with the skewness parameter, because the distribution of pivotal quantity P is free of skewness parameter $\boldsymbol{\lambda}$.

Method 2: Inferential Models (IMs). Inferential Model is a novel method proposed by Martin and Liu [19,20] recently. And Zhu et al. [28] and Ma et al. [16] applied IMs to univariate Skew normal distribution successfully. Here, we extend some of their results to multivariate skew normal distribution case. The detail derivation for creating confidence regions of the location $\boldsymbol{\mu}$ using MIs is reported in Appendix. Here, we just present the resulted theorem.

Theorem 3. *Suppose that a sample matrix Y follows the distribution (7) and Σ and $\boldsymbol{\lambda}$ are known, for the singleton assertion $B = \{\boldsymbol{\mu}\}$ at plausibility level $1-\alpha$, the plausibility region (the counter part of confidence region) for $\boldsymbol{\mu}$ is given by*

$$\Pi_{\mu}(\alpha) = \{\boldsymbol{\mu} : \mathsf{pl}(\boldsymbol{\mu}; \mathscr{S}) > \alpha\}, \tag{15}$$

where $\mathsf{pl}(\boldsymbol{\mu}; \mathscr{S}) = 1 - \left(\max\left|2G\left(A'\Sigma^{-1/2}\left(\overline{\boldsymbol{y}} - \boldsymbol{\mu}\right)\right) - 1\right|\right)^p$ is the plausibility function for the singleton assertion $B = \{\boldsymbol{\mu}\}$. The details of notations and derivation are presented in Appendix.

Method 3: Robust Method. By Theorem 1 Eq. (10), the distribution of sample mean

$$f_{\overline{\boldsymbol{Y}}}(\boldsymbol{y}) = 2\phi_p(\boldsymbol{y}; \boldsymbol{\mu}, \frac{\Sigma}{n})\Phi(n\boldsymbol{\lambda}\Sigma^{-1/2}(\boldsymbol{y} - \boldsymbol{\mu})) \qquad \text{for} \qquad \boldsymbol{y} \in \mathbb{R}^p.$$

For a given sample, we can treat above function as a confidence distribution function [24] on parameter space Θ, i.e.

$$f\left(\boldsymbol{\mu}|\overline{\boldsymbol{Y}}=\overline{\boldsymbol{y}}\right)=2\phi_p\left(\boldsymbol{\mu};\overline{\boldsymbol{y}},\frac{\Sigma}{n}\right)\Phi\left(n\boldsymbol{\lambda}\Sigma^{-1/2}\left(\overline{\boldsymbol{y}}-\boldsymbol{\mu}\right)\right)\qquad\text{for}\qquad\boldsymbol{\mu}\in\Theta\subset\mathbb{R}^p.$$

Thus, we can construct the confidence regions for $\boldsymbol{\mu}$ based on above confidence distribution of $\boldsymbol{\mu}$. Particularly, We can obtain the robust confidence regions following the talk given by Ayivor et al. [4] as follows (see details in Appendix)

$$C_{\boldsymbol{\mu}}^{R}\left(\alpha\right)=\left\{\boldsymbol{y}:\int_{\mathscr{S}}f_{\overline{\boldsymbol{Y}}}\left(\boldsymbol{y}|\boldsymbol{\mu}=\overline{\boldsymbol{y}}\right)\mathrm{d}\boldsymbol{y}=1-\alpha\right\},\tag{16}$$

where for $\boldsymbol{y}\in\partial\mathscr{S}$, $f_{\overline{\boldsymbol{Y}}}\left(\boldsymbol{y}|\boldsymbol{\mu}=\overline{\boldsymbol{y}}\right)\equiv c_0$, here $c_0>0$ is a constant value associated with the confidence distribution satisfying the condition in Eq. (16).

For comparison of these three confidence regions graphically, we draw the confidence regions $C_{\boldsymbol{\mu}}^{P}$, $\Pi_{\boldsymbol{\mu}}(\alpha)$ and $C_{\boldsymbol{\mu}}^{R}$ when $p=2$, sample size $n=5,10,30$ and $\Sigma=\begin{pmatrix}1&\rho\\\rho&1\end{pmatrix}$ where $\rho=0.1$ and 0.5.

From Figs. 1, 2 and 3, it is clear to see all these three methods can capture the location information properly. The values of ρ determine the directions of the confidence regions. The larger a sample size is, the more accurate estimation on the location could be archived.

3.1.2 Hypothesis Test on $\boldsymbol{\mu}$

In this subsection, we consider the problem of determining whether a given p-dimension vector $\boldsymbol{\mu}_0\in\mathbb{R}^p$ is a plausibility vector for the location parameter $\boldsymbol{\mu}$

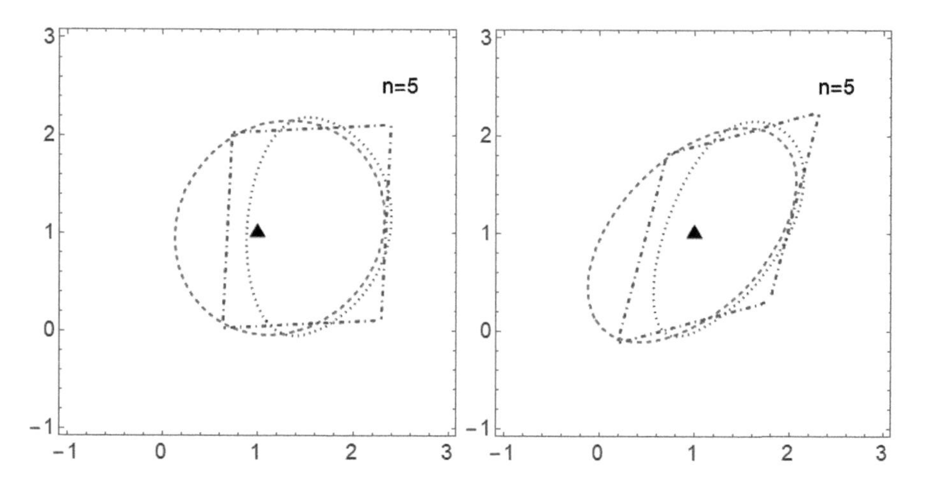

Fig. 1. Confidence regions of $\boldsymbol{\mu}$ when $\boldsymbol{\mu}=(1,1)'$, $\rho=0.1,0.5$ (left, right) and $\boldsymbol{\lambda}=(1,0)$ for sample size $n=5$. The red dashed, blue dashdotted and black dotted curves enclosed the confidence regions for $\boldsymbol{\mu}$ based on pivotal, IMs and robust methods, respectively.

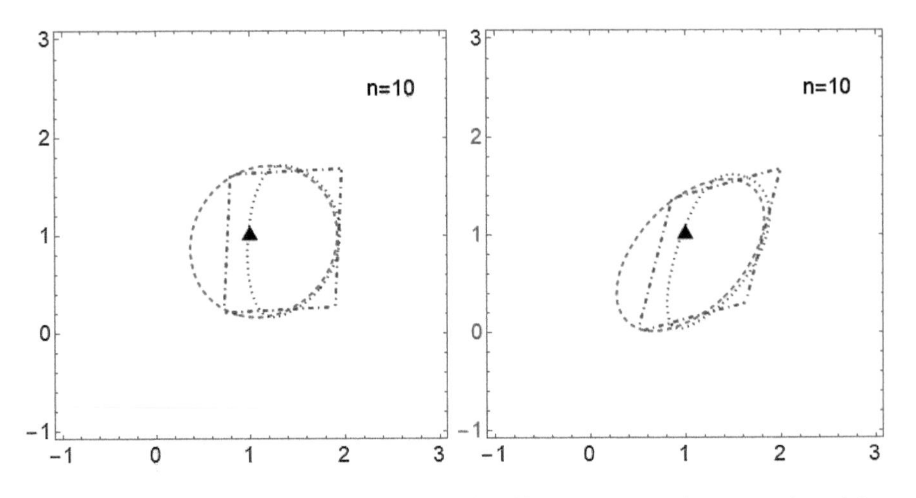

Fig. 2. Confidence regions of $\boldsymbol{\mu}$ when $\boldsymbol{\mu} = (1,1)'$, $\rho = 0.1, 0.5$ (left, right) and $\boldsymbol{\lambda} = (1,0)$ for sample size $n = 10$. The red dashed, blue dashdotted and black dotted curves enclosed the confidence regions for $\boldsymbol{\mu}$ based on pivotal, IMs and robust methods, respectively.

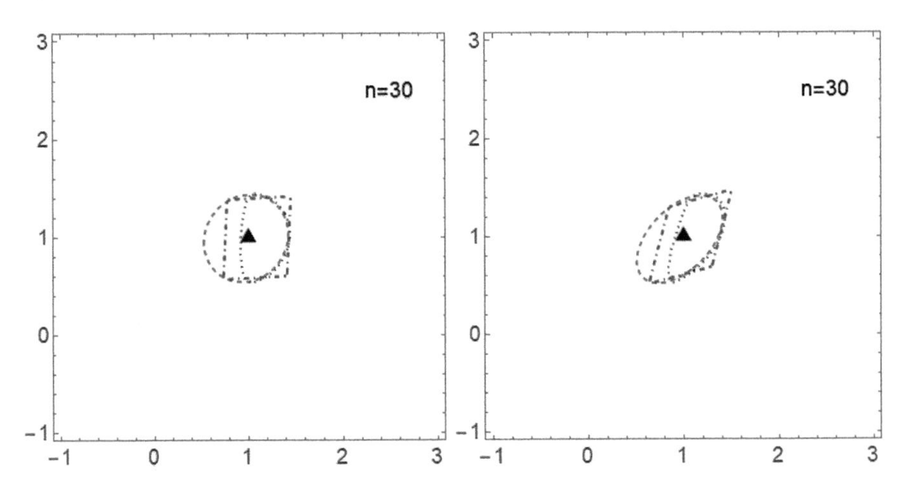

Fig. 3. Confidence regions of $\boldsymbol{\mu}$ when $\boldsymbol{\mu} = (1,1)'$, $\rho = 0.1, 0.5$ (left, right) and $\boldsymbol{\lambda} = (1,0)$ for sample size $n = 30$. The red dashed, blue dashdotted and black dotted curves enclosed the confidence regions for $\boldsymbol{\mu}$ based on pivotal, IMs and robust methods, respectively.

of a multivariate skew normal distribution. We have the hypotheses

$$\mathbf{H_0} : \boldsymbol{\mu} = \boldsymbol{\mu_0} \qquad \text{v.s.} \qquad \mathbf{H_A} : \boldsymbol{\mu} \neq \boldsymbol{\mu_0}.$$

For the case when Σ is known, we use the test statistics

$$q = n\left(\overline{\mathbf{Y}} - \boldsymbol{\mu}_0\right)' \Sigma^{-1} \left(\overline{\mathbf{Y}} - \boldsymbol{\mu}_0\right). \tag{17}$$

For the distribution of test statistic q, under the null hypothesis, i.e. $\boldsymbol{\mu} = \boldsymbol{\mu}_0$, we have

$$q = n\left(\overline{\mathbf{Y}} - \boldsymbol{\mu}_0\right)' \Sigma^{-1} \left(\overline{\mathbf{Y}} - \boldsymbol{\mu}_0\right) \sim \chi_p^2.$$

Thus, at significance level α, we reject $\mathbf{H_0}$ if $q > \chi_p^2(1-\alpha)$. To obtain the power of this test, we need to derive the distribution of q under alternative hypothesis. By the Definition 2, we obtain

$$q = n\left(\overline{\mathbf{Y}} - \boldsymbol{\mu}_0\right)' \Sigma^{-1} \left(\overline{\mathbf{Y}} - \boldsymbol{\mu}_0\right) \sim S\chi_p^2\left(\xi, \delta_1, \delta_2\right) \tag{18}$$

with $\boldsymbol{\mu}_* = \sqrt{n}\Sigma^{-1/2}(\boldsymbol{\mu} - \boldsymbol{\mu}_0)$, $\xi = \boldsymbol{\mu}_*'\boldsymbol{\mu}_*$, $\delta_1 = \boldsymbol{\mu}_*'\boldsymbol{\lambda}$ and $\delta_2 = \boldsymbol{\lambda}'\boldsymbol{\lambda}$. Therefore, we obtain the power of this test

$$\text{Power} = 1 - F(\chi_p^2(1-\alpha)), \tag{19}$$

where $F(\cdot)$ represents the cdf of $S\chi_p^2\left(\xi, \delta_1, \delta_2\right)$.

To illustrate the performance of the above hypothesis test, we calculate the power values of above test for different combinations of ξ, δ_1, δ_2 and degrees of freedom df. The results are presented in Tables 1, 2 and 3.

Table 1. Power values for hypothesis testing when Σ and $\boldsymbol{\lambda}$ are known with $\mu \in \mathbb{R}^p, p = 5$, and $\xi = n(\boldsymbol{\mu} - \boldsymbol{\mu}_0)'\Sigma^{-1}(\boldsymbol{\mu} - \boldsymbol{\mu}_0)$.

Nominal level		$1-\alpha=0.9$				$1-\alpha=0.95$			
ξ		3	5	10	20	3	5	10	20
$\delta_2=0$	$\delta_1=0$	0.33	0.49	0.78	0.98	0.22	0.36	0.68	0.95
$\delta_2=5$	$\delta_1=-\sqrt{\xi\delta_2}$	0.17	0.21	0.58	0.95	0.09	0.11	0.41	0.90
	$\delta_1=\sqrt{\xi\delta_2}$	0.50	0.77	0.98	1.00	0.35	0.62	0.95	1.00
$\delta_2=10$	$\delta_1=-\sqrt{\xi\delta_2}$	0.13	0.19	0.57	0.95	0.06	0.10	0.39	0.90
	$\delta_1=\sqrt{\xi\delta_2}$	0.54	0.79	0.99	1.00	0.38	0.63	0.97	1.00
$\delta_2=20$	$\delta_1=-\sqrt{\xi\delta_2}$	0.12	0.18	0.57	0.95	0.06	0.09	0.38	0.90
	$\delta_1=\sqrt{\xi\delta_2}$	0.54	0.80	1.00	1.00	0.38	0.64	0.97	1.00

Table 2. Power values for hypothesis testing when Σ and $\boldsymbol{\lambda}$ are known with $\mu \in \mathbb{R}^p, p = 10$, and $\xi = n(\boldsymbol{\mu} - \boldsymbol{\mu}_0)'\Sigma^{-1}(\boldsymbol{\mu} - \boldsymbol{\mu}_0)$.

Nominal level		$1-\alpha=0.9$				$1-\alpha=0.95$			
ξ		3	5	10	20	3	5	10	20
$\delta_2=0$	$\delta_1=0$	0.26	0.39	0.67	0.94	0.17	0.27	0.54	0.89
$\delta_2=5$	$\delta_1=-\sqrt{\xi\delta_2}$	0.15	0.17	0.42	0.88	0.08	0.09	0.27	0.78
	$\delta_1=\sqrt{\xi\delta_2}$	0.38	0.60	0.91	1.00	0.25	0.45	0.81	1.00
$\delta_2=10$	$\delta_1=-\sqrt{\xi\delta_2}$	0.12	0.16	0.40	0.88	0.06	0.08	0.25	0.78
	$\delta_1=\sqrt{\xi\delta_2}$	0.41	0.61	0.93	1.00	0.27	0.45	0.83	1.00
$\delta_2=20$	$\delta_1=-\sqrt{\xi\delta_2}$	0.12	0.16	0.40	0.88	0.06	0.08	0.24	0.78
	$\delta_1=\sqrt{\xi\delta_2}$	0.41	0.62	0.94	1.00	0.27	0.46	0.84	1.00

Table 3. Power values for hypothesis testing when Σ and $\boldsymbol{\lambda}$ are known with $\boldsymbol{\mu} \in \mathbb{R}^p, p = 20$, and $\xi = n(\boldsymbol{\mu} - \boldsymbol{\mu}_0)'\Sigma^{-1}(\boldsymbol{\mu} - \boldsymbol{\mu}_0)$.

Nominal level		$1-\alpha=0.9$				$1-\alpha=0.95$			
ξ		3	5	10	20	3	5	10	20
$\delta_2 = 0$	$\delta_1 = 0$	0.21	0.30	0.53	0.86	0.13	0.19	0.40	0.78
$\delta_2 = 5$	$\delta_1 = -\sqrt{\xi\delta_2}$	0.13	0.15	0.31	0.73	0.07	0.08	0.19	0.59
	$\delta_1 = \sqrt{\xi\delta_2}$	0.29	0.45	0.76	0.99	0.18	0.31	0.62	0.96
$\delta_2 = 10$	$\delta_1 = -\sqrt{\xi\delta_2}$	0.11	0.14	0.29	0.73	0.06	0.08	0.17	0.58
	$\delta_1 = \sqrt{\xi\delta_2}$	0.31	0.46	0.77	0.99	0.19	0.31	0.63	0.97
$\delta_2 = 20$	$\delta_1 = -\sqrt{\xi\delta_2}$	0.11	0.14	0.29	0.72	0.06	0.07	0.17	0.57
	$\delta_1 = \sqrt{\xi\delta_2}$	0.31	0.46	0.78	1.00	0.19	0.31	0.63	0.98

Since there are three parameters regulate the distribution of the test statistic shown in Eq. (18) and the relations among those parameters is complicated, we need to address how to properly interpret the values in Tables 1, 2 and 3. Among three parameters, ξ, δ_1 and δ_2, the values of ξ and δ_1 are related to the location parameter $\boldsymbol{\mu}$.

For ξ, it is the square of (a kind of) "Mahalanobis distance" between $\boldsymbol{\mu}$ and $\boldsymbol{\mu}_0$, so the power of the test is a strictly increasing function of ξ when other parameters are fixed. Furthermore, the power of the test approaches 1 in most cases when $\xi = 20$ which indicates the test based on the test statistic (17) is consistent.

We note that δ_1 is essentially the inner product of $\boldsymbol{\mu} - \boldsymbol{\mu}_0$ and $(\Sigma/n)^{-1/2}\boldsymbol{\lambda}$. When $\delta_1 = 0$, the distribution of the test statistic is free of the shape parameter $\boldsymbol{\lambda}$, and it follows the non-central chi-square distribution with non-centrality ξ under the alternative hypothesis which means the test is based on the normality assumption. For the case $\delta_1 \neq 0$, we only list the power of the test for $\delta_1 = \pm\sqrt{\xi\delta_2}$ because the tail of distribution of the test statistic is monotonically increasing with the increasing value of δ_1 for $\delta_1^2 \leq \xi\delta_2$ [17,26]. So it is clear to see the power of the test is highly influenced by δ_1. For example, for $p = 5, \xi = 3, \delta_2 = 5$, the power varies from 0.17 to 0.50 when δ_1 changes from $-\sqrt{15}$ to $\sqrt{15}$. But when ξ is large, the power of the test does not change too much. For example, when $p = 5, \xi = 20$, the power values of the test are between 0.95 and 1 at significance level $\alpha = 0.1$ for $\delta_2 = 0, 5, 10, 20$ and $\delta_1^2 \leq \sqrt{\xi\delta_2}$.

For δ_2, it is also easy to see the power values of the test have larger variation when δ_2 increases and p, ξ are fixed. For example, when $p = 5, \xi = 3$ the power values of the test are varied from 0.17 to 0.50 for $\delta_2 = 5$, but the range of the power of the test is from 0.13 to 0.54 for $\delta_2 = 10$. It makes sense since δ_2 is the measure of the skewness [2], say the larger δ_2 indicates the distribution is far away from the normal distribution. This also serves an evidence to support our study on skew normal distribution. The flexibility of the skew normal model may provide more accurate information or further understanding of the statistical inference result.

4 Simulations

In this section, a Monte Carlo simulation study is provided to study the performance of coverage rates for location parameter $\boldsymbol{\mu}$ when Σ and $\boldsymbol{\lambda}$ take different values for $p = 2$.

Set $\boldsymbol{\mu} = (1,1)'$, $\Sigma = \begin{pmatrix} 1 & \rho \\ \rho & 1 \end{pmatrix}$ with $\rho = \pm 0.1, \pm 0.5, \pm 0.8$, $\boldsymbol{\lambda} = (1,0)', (1,-1)'$ and$(3,5)'$, we simulated 10,000 runs for sample size $n = 5, 10, 30$. The coverage probabilities of all combinations of ρ, $\boldsymbol{\lambda}$ and sample size n are given in Tables 4, 5 and 6.

From the simulation results shown in Tables 4, 5 and 6, all these three methods can capture the correct location information with the coverage probabilities around the nominal confidence level. But comparing with IMs and robust method, the pivotal method gives less accurate inference in the sense of the area of confidence region. The reason is the pivotal quantity we employed is free of shape parameter which means it does not fully use the information. But the advantage of pivotal method is it is easy to proceed and just based on the

Table 4. Simulation results of coverage probabilities of the 95% coverage regions for $\boldsymbol{\mu}$ when $\boldsymbol{\lambda} = (1,0)'$ using pivotal method, IMs method and robust method.

		n=5			n=10			n=30	
	Pivotal	IM	Robust	Pivotal	IM	Robust	Pivotal	IM	Robust
$\rho = 0.1$	0.9547	0.9628	0.9542	0.9466	0.9595	0.9519	0.9487	0.9613	0.9499
$\rho = 0.5$	0.9533	0.9636	0.9524	0.9447	0.9566	0.9443	0.9508	0.9608	0.9510
$\rho = 0.8$	0.9500	0.9607	0.9493	0.9501	0.9621	0.9490	0.9493	0.9545	0.9496
$\rho = -0.1$	0.9473	0.9528	0.9496	0.9490	0.9590	0.9481	0.9528	0.9651	0.9501
$\rho = -0.5$	0.9495	0.9615	0.9466	0.9495	0.9603	0.9492	0.9521	0.9567	0.9516
$\rho = -0.8$	0.9541	0.9586	0.9580	0.9552	0.9599	0.9506	0.9563	0.9533	0.9522

Table 5. Simulation results of coverage probabilities of the 95% coverage regions for $\boldsymbol{\mu}$ when $\boldsymbol{\lambda} = (1,-1)'$ using pivotal method, IMs method and robust method.

		n=5			n=10			n=30	
	Pivotal	IM	Robust	Pivotal	IM	Robust	Pivotal	IM	Robust
$\rho = 0.1$	0.9501	0.9644	0.9558	0.9505	0.9587	0.9537	0.9500	0.9611	0.9491
$\rho = 0.5$	0.9529	0.9640	0.9565	0.9464	0.9622	0.9552	0.9515	0.9635	0.9537
$\rho = 0.8$	0.9471	0.9592	0.9538	0.9512	0.9623	0.9479	0.9494	0.9614	0.9556
$\rho = -0.1$	0.9511	0.9617	0.9530	0.9511	0.9462	0.9597	0.9480	0.9623	0.9532
$\rho = -0.5$	0.9517	0.9544	0.9469	0.9517	0.9643	0.9526	0.9496	0.9537	0.9510
$\rho = -0.8$	0.9526	0.9521	0.9464	0.9511	0.9576	0.9575	0.9564	0.9610	0.9532

Table 6. Simulation results of coverage probabilities of the 95% coverage regions for μ when $\lambda = (3, 5)'$ using pivotal method, IMs method and robust method.

		n=5			n=10			n=30	
	Pivotal	IM	Robust	Pivotal	IM	Robust	Pivotal	IM	Robust
$\rho = 0.1$	0.9497	0.9647	0.9558	0.9511	0.9636	0.9462	0.9457	0.9598	0.9495
$\rho = 0.5$	0.9533	0.9644	0.9455	0.9475	0.9597	0.9527	0.9521	0.9648	0.9535
$\rho = 0.8$	0.9500	0.9626	0.9516	0.9496	0.9653	0.9534	0.9569	0.9625	0.9506
$\rho = -0.1$	0.9525	0.9533	0.9434	0.9518	0.9573	0.9488	0.9500	0.9651	0.9502
$\rho = -0.5$	0.9508	0.9553	0.9556	0.9491	0.9548	0.9475	0.9514	0.9614	0.9518
$\rho = -0.8$	0.9489	0.9626	0.9514	0.9520	0.9613	0.9531	0.9533	0.9502	0.9492

chi-square distribution. The simulation results from IMs and robust method are similar but robust method is more straightforward than IMs since there is no extra concepts or algorithm introduced. But to determine the level set, i.e. the value of c_0, is computational inefficient and time consuming.

5 Discussion

In this study, the confidence regions of location parameters are constructed based on three different methods, pivotal method, IMs and robust method. All of these methods are verified by the simulation studies of coverage probabilities for the combination of various values of parameters and sample sizes.

From the confidence regions constructed by those methods shown in Figs. 1, 2, and 3, the pivot used in pivotal method is independent of the shape parameter so that the confidence regions constructed by pivotal method can not effectively use the information of the known shape parameter. On the contrary, both IMs and robust method give more accurate confidence regions for location parameter than pivotal method. Further more, the power values of the test presented in Tables 1, 2 and 3 show clearly how the shape parameters impact on the power of the test. It provides not only a strong motivation for practitioners to apply skewed distributions to model their data when the empirical distribution is away from normal, like skew normal distribution, but also clarifies and deepens the understanding of how the skewed distributions affect the statistical inference for statisticians, specifically how the shape parameters involved into the power of the test on location parameters.

The value of the shape information is shown in Tables 1, 2 and 3, which clearly suggests that the skewness influences the power of the test on the location parameter based on the pivotal method.

Appendix

Inferential Models (IMs) for Location Parameter $\boldsymbol{\mu}$ When Σ Is Known

In general, IMs consist three steps, association step, predict step and combination step. We will follow this three steps to set up an IM for the location parameter $\boldsymbol{\mu}$.

Association Step. Based on the sample matrix Y which follows the distribution (7), we use the sample mean $\overline{Y}$ defined by (8) following the distribution (10). Thus we obtain the potential association

$$\overline{\mathbf{Y}} = a(\boldsymbol{\mu}, \mathbf{W}) = \boldsymbol{\mu} + \mathbf{W},$$

where the auxiliary random vector $\mathbf{W} \sim SN_p(0, \Sigma/n, \sqrt{n}\boldsymbol{\lambda})$ but the components of $\boldsymbol{W}$ are not independent. So we use transformed IMs as follow, (see Martin and Liu [20] Sect. 4.4 for more detail on validity of transformed IMs). By Lemmas 1 and 3, we use linear transformations

$$\boldsymbol{V} = A'\Sigma^{-1/2}\boldsymbol{W}$$

where A is an orthogonal matrix with the first column is $\boldsymbol{\lambda}/||\boldsymbol{\lambda}||$, then

$$\boldsymbol{V} \sim SN_p(0, I_p, \boldsymbol{\lambda}^*)$$

where $\boldsymbol{\lambda}^* = (\lambda^*, 0, \ldots, 0)'$ with $\lambda^* = ||\boldsymbol{\lambda}||$. Thus each component of V are independent. To be concrete, let $\boldsymbol{V} = (V_1, \ldots, V_p)'$, $V_1 \sim SN(0, 1, \lambda^*)$ and $V_i \sim N(0, 1)$ for $i = 2, \ldots, p$. Therefore, we obtain a new association

$$\begin{aligned} A'\Sigma^{-1/2}\overline{\mathbf{Y}} &= A'\Sigma^{-1/2}\boldsymbol{\mu} + \boldsymbol{V} \\ &= A'\Sigma^{-1/2}\boldsymbol{\mu} + G^{-1}(\boldsymbol{U}) \end{aligned}$$

where $\boldsymbol{U} = (U_1, U_2, \ldots, U_p)'$, $G^{-1}(\boldsymbol{U}) = \left(G_1^{-1}(U_1), G_2^{-1}(U_2), \ldots, G_p^{-1}(U_p)\right)'$ with $G_1(\cdot)$ is the cdf of $SN(0, 1, \lambda^*)$, $G_i(\cdot)$ is the cdf of $N(0,1)$ for $i = 2, \ldots, p$, and U_i's follow $U(0,1)$ independently for $i = 1, \ldots, p$.

To make the association to be clearly presented, we write down the component wise associations as follows

$$\begin{aligned} \left(A'\Sigma^{-1/2}\overline{\mathbf{Y}}\right)_1 &= \left(A'\Sigma^{-1/2}\boldsymbol{\mu}\right)_1 + G_1^{-1}(U_1) \\ \left(A'\Sigma^{-1/2}\overline{\mathbf{Y}}\right)_2 &= \left(A'\Sigma^{-1/2}\boldsymbol{\mu}\right)_2 + G_2^{-1}(U_2) \\ &\vdots \\ \left(A'\Sigma^{-1/2}\overline{\mathbf{Y}}\right)_p &= \left(A'\Sigma^{-1/2}\boldsymbol{\mu}\right)_p + G_p^{-1}(U_p) \end{aligned}$$

where $\left(A'\Sigma^{-1/2}\overline{\mathbf{Y}}\right)_i$ and $\left(A'\Sigma^{-1/2}\boldsymbol{\mu}\right)_i$ represents the ith component of $A'\Sigma^{-1/2}\overline{\mathbf{Y}}$ and $A'\Sigma^{-1/2}\boldsymbol{\mu}$, respectively. $G_1(\cdot)$ represents the cdf of $SN(0, 1, \lambda^*)$

and $G_i(\cdot)$ represents the cdf of $N(0,1)$ for $i = 2, \ldots, p$, and $U_i \sim U(0,1)$ are independently distributed for $i = 1, \ldots, p$.

Thus for any observation $\overline{\boldsymbol{y}}$, and $u_i \in (0,1)$ for $i = 1, \ldots, p$, we have the solution set

$$\begin{aligned}\Theta_{\overline{y}}(\boldsymbol{\mu}) &= \left\{\boldsymbol{\mu} : A'\Sigma^{-1/2}\overline{\boldsymbol{y}} = A'\Sigma^{-1/2}\boldsymbol{\mu} + G^{-1}(\boldsymbol{U})\right\}\\ &= \left\{\boldsymbol{\mu} : G\left(A'\Sigma^{-1/2}(\overline{\boldsymbol{y}} - \boldsymbol{\mu})\right) = \boldsymbol{U}\right\}\end{aligned}$$

Predict Step. To predict the auxiliary vector $\boldsymbol{U}$, we use the default predictive random set for each components

$$S(U_1, \ldots, U_p) = \left\{(u_1, \ldots, u_p) : \max_{i=1,..,p}\{|u_i - 0.5|\} \le \max_{i=1,..,p}\{|U_i - 0.5|\}\right\}.$$

Combine Step. By the above two steps, we have the combined set

$$\Theta_{\overline{\mathbf{Y}}}(S) = \left\{\boldsymbol{\mu} : \max\left\{|G\left(A'\Sigma^{-1/2}(\overline{\boldsymbol{y}} - \boldsymbol{\mu})\right) - 0.5|\right\} \le \max\{|\mathbf{U} - 0.5|\}\right\}.$$

where

$$\max\left\{\left|G\left(A'\Sigma^{-1/2}(\overline{\boldsymbol{y}} - \boldsymbol{\mu})\right) - 0.5\right|\right\} = \max_{i=1,\ldots,p}\left\{\left|\left(G\left(A'\Sigma^{-1/2}(\overline{\boldsymbol{y}} - \boldsymbol{\mu})\right)\right)_i - 0.5\right|\right\}$$

and

$$max\{|\boldsymbol{U} - 0.5|\} = \max_{i=1,\ldots,p}\{|U_i - 0.5|\}.$$

Thus, apply above IM, for any singleton assertion $A = \{\boldsymbol{\mu}\}$, by definition of believe function and plausibility function, we obtain

$$\mathsf{bel}_{\overline{\mathbf{Y}}}(A; \mathscr{S}) = P\left(\Theta_{\overline{\mathbf{Y}}}(\mathscr{S}) \subseteq A\right) = 0$$

since $\left\{\Theta_{\overline{\mathbf{Y}}}(\mathscr{S}) \subseteq A\right\} = \emptyset$, and

$$\begin{aligned}\mathsf{pl}_{\overline{\mathbf{Y}}}(A; \mathscr{S}) &= 1 - \mathsf{bel}_{\overline{\mathbf{Y}}}\left(A^C; \mathscr{S}\right) = 1 - P_{\mathscr{S}}\left(\Theta_{\overline{\mathbf{Y}}}(\mathscr{S}) \subseteq A^C\right)\\ &= 1 - \left(\max\left\{|2G\left(A'\Sigma^{-1/2}(\overline{\boldsymbol{y}} - \boldsymbol{\mu})\right) - 1|\right\}\right)^p.\end{aligned}$$

Then the Theorem 3 follows by above computations.

Robust Method for Location Parameter $\boldsymbol{\mu}$ When Σ and $\boldsymbol{\lambda}$ Are Known

Based on the distribution of $\overline{\boldsymbol{Y}} \sim SN_p(\boldsymbol{\mu}, \frac{\Sigma}{n}, \sqrt{n}\boldsymbol{\lambda})$, we obtain the confidence distribution of $\boldsymbol{\mu}$ given $\overline{\boldsymbol{y}}$ has pdf

$$f(\boldsymbol{\mu}|\overline{\boldsymbol{Y}} = \overline{\boldsymbol{y}}) = 2\phi(\boldsymbol{\mu}; \overline{\boldsymbol{y}}, \frac{\Sigma}{n})\Phi(n\boldsymbol{\lambda}\Sigma^{-1/2}(\overline{\boldsymbol{y}} - \boldsymbol{\mu})).$$

At confidence level $1-\alpha$, it is natural to construct the confidence set $\mathscr{S}$, i.e. a set $\mathscr{S}$ such that

$$P(\boldsymbol{\mu} \in \mathscr{S}) = 1-\alpha. \tag{20}$$

To choose one set out of infinity many possible sets satisfying condition (20), we follow the idea of the most robust confidence set discussed by Kreinovich [4], for any connected set $\mathscr{S}$, defines the measure of robustness of the set $\mathscr{S}$

$$r(\mathscr{S}) \equiv \max_{\boldsymbol{y} \in \partial \mathscr{S}} f_{\overline{\boldsymbol{Y}}}(\boldsymbol{y}).$$

Then at confidence level $1-\alpha$, we obtain the most robust confidence set

$$\mathscr{S} = \{\boldsymbol{y} : f_{\overline{\boldsymbol{Y}}}(\boldsymbol{y}) \geq c_0\},$$

where c_0 is uniquely determined by the conditions $f_{\overline{\boldsymbol{Y}}}(\boldsymbol{y}) \equiv c_0$ and $\int_{\mathscr{S}} f_{\overline{\boldsymbol{Y}}}(\boldsymbol{y})\,\mathrm{d}\boldsymbol{y} = 1-\alpha$.

Remark 2. As mentioned by Kreinovich in [4], for Gaussian distribution, such an ellipsoid is indeed selected as a confidence set.

References

1. Arellano-Valle, R.B., Bolfarine, H., Lachos, V.H.: Skew-normal linear mixed models. J. Data Sci. **3**(4), 415–438 (2005)
2. Arevalillo, J.M., Navarro, H.: A stochastic ordering based on the canonical transformation of skew-normal vectors. TEST, 1–24 (2018)
3. Arnold, B.C., Beaver, R.J., Groeneveld, R.A., Meeker, W.Q.: The nontruncated marginal of a truncated bivariate normal distribution. Psychometrika **58**(3), 471–488 (1993)
4. Ayivor, F., Govinda, K.C., Kreinovich, V.: Which confidence set is the most robust? In: 21st Joint UTEP/NMSU Workshop on Mathematics, Computer Science, and Computational Sciences (2017)
5. Azzalini, A.: A class of distributions which includes the normal ones. Scand. J. Stat. **12**(2), 171–178 (1985)
6. Azzalini, A., Capitanio, A.: Statistical applications of the multivariate skew normal distribution. J. R. Stat. Soc. Ser. B (Stat. Methodol.) **61**(3), 579–602 (1999)
7. Azzalini, A., Dalla Valle, A.: The multivariate skew-normal distribution. Biometrika **83**(4), 715–726 (1996)
8. Azzalini, A.: Further results on a class of distributions which includes the normal ones. Statistica **46**(2), 199–208 (1986)
9. Azzalini, A.: The Skew-Normal and Related Families, vol. 3. Cambridge University Press, Cambridge (2013)
10. Bayes, C.L., Branco, M.D.: Bayesian inference for the skewness parameter of the scalar skew-normal distribution. Braz. J. Probab. Stat. **21**(2), 141–163 (2007)
11. Branco, M.D., Dey, D.K.: A general class of multivariate skew-elliptical distributions. J. Multivar. Anal. **79**(1), 99–113 (2001)
12. Dey, D.: Estimation of the parameters of skew normal distribution by approximating the ratio of the normal density and distribution functions. University of California, Riverside (2010)

13. Genton, M.G.: Skew-Elliptical Distributions and Their Applications: A Journey Beyond Normality. CRC Press, London (2004)
14. Hill, M.A., Dixon, W.J.: Robustness in real life: a study of clinical laboratory data. Biometrics **38**(2), 377–396 (1982)
15. Liseo, B., Loperfido, N.: A note on reference priors for the scalar skew-normal distribution. J. Stat. Plan. Inference **136**(2), 373–389 (2006)
16. Ma, Z., Zhu, X., Wang, T., Autchariyapanitkul, K.: Joint plausibility regions for parameters of skew normal family. In: International Conference of the Thailand Econometrics Society, pp. 233–245. Springer, Cham (2018)
17. Ma, Z., Tian, W., Li, B., Wang, T.: The decomposition of quadratic forms under skew normal settings. In: International Conference of the Thailand Econometrics Society, pp. 222–232. Springer, Cham (2018)
18. Mameli, V., Musio, M., Sauleau, E., Biggeri, A.: Large sample confidence intervals for the skewness parameter of the skew-normal distribution based on fisher's transformation. J. Appl. Stat. **39**(8), 1693–1702 (2012)
19. Martin, R., Liu, C.: Inferential models: a framework for prior-free posterior probabilistic inference. J. Am. Stat. Assoc. **108**(501), 301–313 (2013)
20. Martin, R., Liu, C.: Inferential Models: Reasoning with Uncertainty, vol. 145. CRC Press, New York (2015)
21. Pewsey, A.: Problems of inference for Azzalini's skewnormal distribution. J. Appl. Stat. **27**(7), 859–870 (2000)
22. Sahu, S.K., Dey, D.K., Branco, M.D.: A new class of multivariate skew distributions with applications to Bayesian regression models. Can. J. Stat. **31**(2), 129–150 (2003)
23. Sartori, N.: Bias prevention of maximum likelihood estimates for scalar skew normal and skew t distributions. J. Stat. Plan. Inference **136**(12), 4259–4275 (2006)
24. Schweder, T., Hjort, N.L.: Confidence and likelihood. Scand. J. Stat. **29**(2), 309–332 (2002)
25. Wang, T., Li, B., Gupta, A.K.: Distribution of quadratic forms under skew normal settings. J. Multivar. Anal. **100**(3), 533–545 (2009)
26. Ye, R.D., Wang, T.H.: Inferences in linear mixed models with skew-normal random effects. Acta Math. Sin. Engl. Ser. **31**(4), 576–594 (2015)
27. Ye, R., Wang, T., Gupta, A.K.: Distribution of matrix quadratic forms under skew-normal settings. J. Multivar. Anal. **131**, 229–239 (2014)
28. Zhu, X., Ma, Z., Wang, T., Teetranont, T.: Plausibility regions on the skewness parameter of skew normal distributions based on inferential models. In: Kreinovich, V., Sriboonchitta, S., Huynh, V.N. (eds.) Robustness in Econometrics, pp. 267–286. Springer, Cham (2017)

Blockchains Beyond Bitcoin: Towards Optimal Level of Decentralization in Storing Financial Data

Thach Ngoc Nguyen[1], Olga Kosheleva[2], Vladik Kreinovich[2](✉), and Hoang Phuong Nguyen[3]

[1] Banking University of Ho Chi Minh City, 56 Hoang Dieu 2, Quan Thu Duc, Thu Duc, Ho Chi Minh City, Vietnam
Thachnn@buh.edu.vn

[2] University of Texas at El Paso, 500 W. University, El Paso, TX 79968, USA
{olgak,vladik}@utep.edu

[3] Division Informatics, Math-Informatics Faculty, Thang Long University, Nghiem Xuan Yem Road, Hoang Mai District, Hanoi, Vietnam
nhphuong2008@gmail.com

Abstract. In most current financial transactions, the record of each transaction is stored in three places: with the seller, with the buyer, and with the bank. This currently used scheme is not always reliable. It is therefore desirable to introduce duplication to increase the reliability of financial records. A known absolutely reliable scheme is blockchain – originally invented to deal with bitcoin transactions – in which the record of each financial transaction is stored at every single node of the network. The problem with this scheme is that, due to the enormous duplication level, if we extend this scheme to all financial transactions, it would require too much computation time. So, instead of sticking to the current scheme or switching to the blockchain-based full duplication, it is desirable to come up with the optimal duplication scheme. Such a scheme is provided in this paper.

1 Formulation of the Problem

How Financial Information is Currently Stored. At present, usually, the information about each financial transaction is stored in three places:

- with the buyer,
- with the seller, and
- with the bank.

This Arrangement is not Always Reliable. In many real-life financial transactions, a problem later appears, so it becomes necessary to recover the information about the sale. From this viewpoint, the current system of storing information is not fully reliable: if a buyer has a problem, and his/her computer crashes

V. Kreinovich et al. (Eds.): ECONVN 2019, SCI 809, pp. 163–167, 2019.
https://doi.org/10.1007/978-3-030-04200-4_12

and deletes the original record, the only neutral source of information is then the bank – but the bank may have gone bankrupt since then.

It is therefore desirable to incorporate more duplication, so as to increase the reliability of storing financial records.

Blockchain as an Absolutely Reliable – But Somewhat Wasteful – Scheme for Storing Financial Data. The known reliable alternative to the usual scheme of storing financial data is the *blockchain* scheme, originally designed to keep track of bitcoin transactions; see, e.g., [1–12].

In this scheme, the record of each transaction is stored at every single node, i.e., at the location of every single participant. This extreme duplication makes blockchains a very reliable way of storing financial data. On the other hand, in this scheme, every time anyone performs a financial transaction, this information needs to be transmitted to all the nodes. This takes a lot of computation time, so, from this viewpoint, this scheme – while absolutely reliable – is very wasteful.

Formulation of the Problem. What scheme should we select to store the financial data?

It would be nice to have our data stored in an absolutely reliable way. Thus, it may seem reasonable to use blockchain for all financial transactions, not just for ones involving bitcoins. The problem is that:

- Already for bitcoins – which at present participate in a very small percentage of financial transactions – the world-wide update corresponding to each transaction takes about 10 seconds.
- If we apply the same technique to all financial transactions, this delay would increase drastically – and the resulting hours of delay will make the system completely impractical.

So, instead of using no duplication at all (as in the traditional scheme) or using absolute duplication (as in bitcoin), it is desirable to find the *optimal* level of duplication for each financial transaction.

This level may be different for different transactions:

- When a customer buys a relatively cheap product, too much duplication probably does not make sense, since the risk is small but the need for additional storage would increase the cost.
- On the other hand, for an expensive purchase, we may want to spend a little more to decrease the risk – just like we buy insurance when we buy a house or a car.

Good news is that the blockchain scheme itself – with its encryptions etc. – does not depend on whether we store each transaction at every single node or only in some selected nodes. In this sense, the technology is there, no matter what level of duplication we choose. The only problem is to find the optimal duplication level.

What We Do in This Paper. In this paper, we show how to find the optimal level of duplication for each type of financial transaction.

2 What Is the Optimal Level of Decentralization in Financial Transactions: Towards Solving the Problem

Notations. Let us start with some notations.

- Let d denote the level of duplication of a given transaction, i.e., the number of copies of the original transaction record that will be independently stored.
- Let p be the probability that each copy can be lost.

  This probability can be estimated based on experience.

- Let c denote the total cost of storing one copy of the transaction record.
- Finally, let L be the expected financial loss that will happen if a problem emerges related to the original sale, and all the copies of the corresponding record have disappeared.

  This expected financial loss L can estimated by multiplying the cost of the transaction by the probability that the bought item will turn out to be faulty.

Comments.

- The cost c of storing a copy is about the same for all the transactions, whether they are small or large.
- On the other hand, the potential loss L depends on the size of the transaction – and on the corresponding risk.

Analysis of the Problem. Since the cost of storing one copy of the financial transaction is c, the cost of storing d copies is equal to $d \cdot c$.

To this cost, we need to add the expected loss in the situation in which all copies of the transaction are accidentally deleted. For each copy, the probability that it will be accidentally deleted is p. The copies are assumed to be independent. Since we have d copies, the probability that all d of them will be accidentally deleted is therefore equal to the product of the d probabilities p corresponding to each copy, i.e., is equal to p^d.

So, we have the loss L with probability p^d – and, correspondingly, zero loss with the remaining probability. Thus, the expected loss from losing all the copies of the record is equal to the product $p^d \cdot L$.

Hence, once we have selected the number d of copies, the overall expected loss E is equal to the sum of the above two values, i.e., to

$$E = d \cdot c + p^d \cdot L. \tag{1}$$

We need to find the value d for which this overall loss is the smallest possible.

Let us Find the Optimal Level of Duplication, i.e., the Optimal d. To find the optimal value d, we can differentiate the expression (1) with respect to d and equate the derivative to 0. As a result, we get the following equation:

$$\frac{dE}{dd} = c + \ln(p) \cdot p^d \cdot L = 0, \tag{2}$$

hence

$$p^d = \frac{c}{L \cdot |\ln(p)|}.$$

By taking logarithms of both sides of this formula, we get

$$d \cdot \ln(p) = \ln\left(\frac{c}{L \cdot |\ln(p)|}\right).$$

Since $p < 1$, the logarithm $\ln(p)$ is negative, so it is convenient to change the sign of both sides of this formula. By taking into account that for all possible a and b, we have $-\ln\left(\frac{a}{b}\right) = \ln\left(\frac{b}{a}\right)$, we conclude that

$$d \cdot |\ln(p)| = \ln\left(\frac{L \cdot |\ln(p)|}{c}\right),$$

thus

$$d = \frac{\ln\left(\frac{L \cdot |\ln(p)|}{c}\right)}{|\ln(p)|}. \quad (3)$$

When p and c are fixed, then we transform this expression into an equivalent form in which we explicitly describe the dependence of the optimal duplication level on the expected loss L:

$$d = \frac{1}{|\ln(p)|} \cdot \ln(L) + \frac{\ln|\ln(p)| - \ln(c)}{|\ln(p)|}. \quad (4)$$

Comments.

- As one can easily see, the larger the expected loss L, the more duplications we need. In general, as we see from the formula (4), the number of duplications is proportional to the logarithm of the expected loss.
- The value d computed by using the formulas (3) and (4) may be not an integer. However, as we can see from the formula (2), the derivative of the overall loss E is first decreasing then increasing. Thus, to find the optimal integer value d, it is sufficient to consider and compare two integers which are on the two sides of the value (3)–(4): namely,
  - its floor $\lfloor d \rfloor$ and
  - its ceiling $\lceil d \rceil$.

  Out of these two values, we need to find the one for which the overall loss E attains the smallest possible value.

Acknowledgments. This work was supported in part by the US National Science Foundation via grant HRD-1242122 (Cyber-ShARE Center of Excellence).

The authors are thankful to Professor Hung T. Nguyen for valuable discussions.

References

1. Antonopoulos, A.M.: Mastering Bitcoin: Programming the Open Blockchain. O'Reilly, Sebastopol (2017)
2. Bambara, J.J., Allen, P.R., Iyer, K., Lederer, S., Madsen, R., Wuehler, M.: Blockchain: A Practical Guide to Developing Business, Law, and Technology Solutions. McGraw Hill Education, New York (2018)
3. Bashir, I.: Mastering Blockchain. Packt Publishing, Birmingham (2017)
4. Connor, M., Collins, M.: Blockchain: Ultimate Beginner's Guide to Blockchain Technology - Cryptocurrency, Smart Contracts, Distributed Ledger, Fintech and Decentralized Applications. CreateSpace Independent Publishing Platform, Scotts Valley (2018)
5. Drescher, D.: Blockchain Basics: A Non-Technical Introduction in 25 Steps. Apress, New York (2017)
6. Gates, M.: Blockchain: Ultimate Guide to Understanding Blockchain, Bitcoin, Cryptocurrencies, Smart Contracts and the Future of Money. CreateSpace Independent Publishing Platform, Scotts Valley (2017)
7. Laurence, T.: Blockchain For Dummies. John Wiley, Hoboken (2017)
8. Norman, A.T.: Blockchain Technology Explained: The Ultimate Beginner's Guide About Blockchain Wallet, Mining, Bitcoin, Ethereum, Litecoin, Zcash, Monero, Ripple, Dash, IOTA And Smart Contracts. CreateSpace Independent Publishing Platform, Scotts Valley (2017)
9. Swan, M.: Blockchain: Blueprint for a New Economy. O'Reilly, Sebastopol (2015)
10. Tapscott, D., Tapscott, A.: Blockchain Revolution: How the Technology Behind Bitcoin is Changing Money, Business, and the World Hardcover. Penguin Random House, New York (2016)
11. Vigna, P., Casey, M.J.: The Truth Machine: The Blockchain and the Future of Everything. St. Martin's Press, New York (2018)
12. White, A.K.: Blockchain: Discover the Technology behind Smart Contracts, Wallets, Mining and Cryptocurrency (Including Bitcoin, Ethereum, Ripple, Digibyte and Others). CreateSpace Independent Publishing Platform, Scotts Valley (2018)

Why Quantum (Wave Probability) Models Are a Good Description of Many Non-quantum Complex Systems, and How to Go Beyond Quantum Models

Miroslav Svítek[1], Olga Kosheleva[2], Vladik Kreinovich[2](✉), and Thach Ngoc Nguyen[3]

[1] Faculty of Transportation Sciences, Czech Technical University in Prague, Konviktska 20, 110 00 Prague 1, Czech Republic
svitek@fd.cvut.cz

[2] University of Texas at El Paso, 500 W. University, El Paso, TX 79968, USA
{olgak,vladik}@utep.edu

[3] Banking University of Ho Chi Minh City, 56 Hoang Dieu 2, Quan Thu Duc, Thu Duc, Ho Chi Minh City, Vietnam
Thachnn@buh.edu.vn

Abstract. In many practical situations, it turns out to be beneficial to use techniques from quantum physics in describing non-quantum complex systems. For example, quantum techniques have been very successful in econometrics and, more generally, in describing phenomena related to human decision making. In this paper, we provide a possible explanation for this empirical success. We also show how to modify quantum formulas to come up with an even more accurate descriptions of the corresponding phenomena.

1 Formulation of the Problem

Quantum Models are Often a Good Description of Non-quantum Systems: A Surprising Phenomenon. Quantum physics has been designed to describe quantum objects, i.e., objects – mostly microscopic but sometimes macroscopic as well – that exhibit quantum behavior. Somewhat surprisingly, however, it turns out that quantum-type techniques – techniques which are called *wave probability* techniques in [16,17] – can also be useful in describing non-quantum complex systems, in particular, economic systems and other systems involving human behavior, etc.; see, e.g., [1,5,9,16,17] and references therein.

Why quantum techniques can help in non-quantum situations is largely a mystery.

Natural Questions. The first natural question is why? Why quantum models are often a good description of non-quantum systems.

V. Kreinovich et al. (Eds.): ECONVN 2019, SCI 809, pp. 168–175, 2019.
https://doi.org/10.1007/978-3-030-04200-4_13

The next natural question is related to the fact that while quantum models provide a good description of non-quantum systems, this description is not perfect. So, a natural question: how to get a better approximation?

What We Do in This Paper. In this paper, we provide answers to the above two questions.

2 Towards an Explanation

Ubiquity of multi-D Normal Distributions. To describe the state of a complex system, we need to describe the values of the quantities $x_1, \ldots, x_n$ that form this state.

In many cases, the system consists of a large number of reasonably independent parts. In this case, each of the quantities x_i describing the system is approximately equal to the sum of the values of the corresponding quantity that describes these parts. For example:

- The overall trade volume of a country can be described as the sum of the trades performed by all its companies and all its municipal units.
- Similarly, the overall number of unemployed people in a country is equal to the sum of numbers of unemployed folks in different regions, etc.

It is known that the distribution of the sum of a large number of independent random variables is – under certain reasonable conditions – close to Gaussian (normal); this result is known as the *Central Limit Theorem*; see, e.g., [15].

Thus, with reasonable accuracy, we can assume that the vectors $x = (x_1, \ldots, x_n)$ formed by all the quantities that characterize the system as a whole are normally distributed.

Let us Simplify the Description of the multi-D Normal Distribution. A multi-D normal distribution is uniquely characterized by its means $\mu = (\mu_1, \ldots, \mu_n)$, where $\mu_i \stackrel{\text{def}}{=} E[x_i]$, and by its covariance matrix $\sigma_{ij} \stackrel{\text{def}}{=} E[(x_i - \mu_i) \cdot (x_j - \mu_j)]$.

By observing the values of the characteristics x_i corresponding to different systems, we can estimate the mean values μ_i and thus, instead of the original values x_i, consider deviations $\delta_i \stackrel{\text{def}}{=} x_i - \mu_i$ from these values.

For these deviations, the description is simpler. Indeed, their means are 0s, so to fully describe the distribution of the corresponding vector $\delta = (\delta_1, \ldots, \delta_n)$, it is sufficient to know the covariance matrix σ_{ij}.

An additional simplification is that since the means are all 0s, the formula for the covariance matrix has a simplified form $\sigma_{ij} = E[\delta_i \cdot \delta_j]$.

For Complex Systems, With a Large Number of Parameters, a Further Simplification is Needed. After the above simplification, to fully describe the corresponding distribution, we need to describe all the values of the $n \times n$ covariance matrix σ_{ij}. In general, an $n \times n$ matrix contains n^2 elements, but since the covariance matrix is symmetric, we only need to describe

$$\frac{n \cdot (n+1)}{2} = \frac{n^2}{2} + \frac{n}{2}$$

parameters – slightly more than half as many.

The big question is: can we determine all these parameters from the observations?

In general in statistics, if we want to find a reasonable estimate for a parameter, we need to have a certain number of observations. Based on N observations, we can find the value of each quantity with accuracy $\approx \frac{1}{\sqrt{N}}$; see, e.g., [15]. Thus, to be able to determine a parameter with a reasonable accuracy of 20%, we need to select N for which $\frac{1}{\sqrt{N}} \approx 20\% = 0.2$, i.e., $N = 25$. So, to find the value of one parameter, we need approximately 25 observations. By the same logic, for any integer k, to find the values of k parameters, we need to have $25k$ observations. In particular, to determine $\frac{n \cdot (n+1)}{2} \approx \frac{n^2}{2}$ parameters, we need to have $25 \cdot \frac{n^2}{2}$ observations.

Each fully detailed observation of a system leads to n numbers $x_1, \ldots, x_n$ and thus, to n numbers $\delta_1, \ldots, \delta_n$. So, to estimate $25 \cdot \frac{n^2}{2} = 12.5 \cdot n^2$ parameters, we need to have $12.5 \cdot n$ different systems. And we often do not have that many system to observe.

For example, to have a detailed analysis of a country's economics, we need to have at least several dozen parameters, at least $n \cdot 30$. By the above logic, to fully describe the joint distribution of all these parameters, we will need at least

$$12.5 \cdot 30 \approx 375$$

countries – and on the Earth, we do not have that many of them.

This problem occurs not only in econometrics, it is even more serious, e.g., in medical applications of bioinformatics: there are thousands of genes, and not enough data to be able to determine all the correlations between them.

Since we cannot determine the covariance matrix σ_{ij} exactly, we therefore need to come up with an approximate description, a description that would require fewer parameters.

Need for a Geometric Description. What does it means to have a good approximation? Intuitively, approximations means having a model which is, in some reasonable sense, close to the original one – i.e., is at a small distance from the original model. Thus, to come up with an understanding of what is a good approximation, it is desirable to have a geometric representation of the corresponding problem, a representation in which different objects would be represented by points in a certain space – so that we could easily understand what is the distance between different objects.

From this viewpoint, to see how we can reasonably approximate multi-D normal distributions, it is desirable to use an appropriate geometric representation of such distributions. Good news is that such a representation is well known. Let us recall this representation.

Geometric Description of multi-D Normal Distribution: Reminder. It is well known that a 1D normally distributed random variable x with 0 mean and standard deviation σ can be presented as $\sigma \cdot X$, where X is "standard" normal distribution, with 0 mean and standard deviation 1.

Similarly, it is known that any normally distributed n-dimensional random vector $\delta = (\delta_1, \ldots, \delta_n)$ can be represented as linear combinations $\delta_i = \sum\limits_{j=1}^{n} a_{ij} \cdot X_j$ of n independent standard random variables $X_1, \ldots, X_n$. These variables can be found, e.g., as eigenvectors of the covariance matrix divided by the corresponding eigenvalues.

This way, each of the original quantities δ_i is represented by the n-dimensional vector $a_i = (a_{i1}, \ldots, a_{in})$. The known geometric feature of this representation is that for every two linear combinations $\delta' = \sum\limits_{i=1}^{n} c'_i \cdot \delta_i$ and $\delta'' = \sum\limits_{i=1}^{n} c''_i \cdot \delta_i$ of the quantities δ_i:

- the standard deviation $\sigma[\delta' - \delta'']$ of the difference between these linear combinations is equal to
- the (Euclidean) distance $d(a', a'')$ between the corresponding n-dimensional vectors $a' = \sum\limits_{i=1} c'_i \cdot a_i$ and $a'' = \sum\limits_{i=1} c''_i \cdot a_i$, with components $a'_j = \sum\limits_{i=1}^{n} c'_i \cdot a_{ij}$ and $a''_j = \sum\limits_{i=1}^{n} c''_i \cdot a_{ij}$:

$$\sigma[\delta' - \delta''] = d(a', a'').$$

Indeed, since $\delta_i = \sum\limits_{j=1}^{n} a_{ij} \cdot X_j$, we conclude that

$$\delta' = \sum_{i=1}^{n} c'_i \cdot \delta_i = \sum_{i=1}^{n} c'_i \cdot \sum_{j=1}^{n} a_{ij} \cdot X_j.$$

By combining together all the coefficients at X_j, we conclude that

$$\delta' = \sum_{j=1}^{n} \left(\sum_{i=1}^{n} c'_i \cdot a_{ij} \right) \cdot X_j,$$

i.e., by using the formula for a'_j, that

$$\delta' = \sum_{j=1}^{n} a'_j \cdot X_j.$$

Similarly, we can conclude that

$$\delta'' = \sum_{j=1}^{n} a''_j \cdot X_j,$$

thus

$$\delta' - \delta'' = \sum_{j=1}^{n} (a'_j - a''_j) \cdot X_j.$$

Since the mean of the difference $\delta' - \delta''$ is thus equal to 0, the square of its standard deviation is simply equal to $\sigma^2[\delta' - \delta''] = E\left[(\delta' - \delta'')^2\right]$. In our case,

$$(\delta' - \delta'')^2 = \sum_{i=1}^{n} (a'_j - a''_j)^2 \cdot X_j^2 + \sum_{i \neq j} (a'_i - a''_i) \cdot (a'_j - a''_j) \cdot X_i \cdot X_j.$$

Thus,

$$\sigma^2[\delta' - \delta''] = E[(\delta' - \delta'')^2]$$

$$= \sum_{i=1}^{n} (a'_j - a''_j)^2 \cdot E[X_j^2] + \sum_{i \neq j} (a'_i - a''_i) \cdot (a'_j - a''_j) \cdot E[X_i \cdot X_j].$$

The variables X_j are independent and have 0 mean, so for $i \neq j$, we have $E[X_i \cdot X_j] = E[X_i] \cdot E[X_j] = 0$. For each i, since X_i are standard normal distributions, we have $E[X_j^2] = 1$. Thus, we conclude that

$$\sigma^2[\delta' - \delta''] = \sum_{i=1}^{n} (a'_j - a''_j)^2,$$

i.e., indeed, $\sigma^2[\delta' - \delta''] = d^2(a', a'')$ and thus, $\sigma[\delta' - \delta''] = d(\delta', \delta'')$.

How Can We Use This Geometric Description to Find a Fewer-Parameters ($k \ll n$) Approximation to the Corresponding Situation. We have n quantities $x_1, \ldots, x_n$ that describe the complex system. By subtracting the mean values μ_i from each of the quantities, we get shifted values $\delta_1, \ldots, \delta_n$. To absolutely accurately describe the joint distribution of these n quantities, we need to describe n n-dimensional vectors $a_1, \ldots, a_n$ corresponding to each of these quantities.

In our approximate description, we still want to keep all n quantities, but we cannot keep them as n-dimensional vectors – this would require too many parameters to determine, and, as we have mentioned earlier, we do not have that many observations to be able to experimentally determine all these parameters. Thus, the natural thing to do is to decrease their dimension.

In other words:

- instead of representing each quantity δ_i as an n-dimensional vector $a_i = (a_{i1}, \ldots, a_{in})$ corresponding to $\delta_i = \sum\limits_{j=1}^{n} a_{ij} \cdot X_j$,
- we select some value $k \ll n$ and represent each quantity δ_i as a k-dimensional vector $a_i = (a_{i1}, \ldots, a_{ik})$ corresponding to $\delta_i = \sum\limits_{j=1}^{k} a_{ij} \cdot X_j$.

For $k = 2$, the Above Approximation Idea Leads to a Quantum-Type Description. In one of the simplest cases $k = 2$, each quantity δ_i is represented by a 2-D vector $a_i = (a_{i1}, a_{i2})$. Similarly to the above full-dimensional case, for every two linear combinations $\delta' = \sum_{i=1}^{n} c'_i \cdot \delta_i$ and $\delta'' = \sum_{i=1}^{n} c''_i \cdot \delta_i$ of the quantities δ_i,

- the standard deviation $\sigma[\delta' - \delta'']$ of the difference between these linear combinations is equal to
- the (Euclidean) distance $d(a', a'')$ between the corresponding 2-dimensional vectors $a' = \sum_{i=1}^{n} c'_i \cdot a_i$ and $a'' = \sum_{i=1}^{n} c''_i \cdot a_i$, with components $a'_j = \sum_{i=1}^{n} c'_i \cdot a_{ij}$ and $a''_j = \sum_{i=1}^{n} c''_i \cdot a_{ij}$:

$$\sigma[\delta' - \delta''] = d(a', a'') = \sqrt{(a'_1 - a''_1)^2 + (a'_2 - a''_2)^2}.$$

However, in the 2-D case, we can alternatively represent each 2-D vector $a_i = (a_{i1}, a_{i2})$ as a complex number

$$a_i = a_{i1} + \mathrm{i} \cdot a_{i2},$$

where, as usual, $\mathrm{i} \stackrel{\text{def}}{=} \sqrt{-1}$. In this representation, the modulus (absolute value)

$$|a' - a''|$$

of the difference

$$a' - a'' = (a'_1 - a''_1) + \mathrm{i} \cdot (a'_2 - a''_2)$$

is equal to $\sqrt{(a'_1 - a''_1)^2 + (a'_2 - a''_2)^2}$, i.e., exactly the distance between the original points.

Thus, in this approximation:

- each quantity is represented by a complex number, and
- the standard deviation of the difference between different quantities is equal to the modulus of the difference between the corresponding complex numbers – and thus, the variance is equal to the square of this modulus,
- in particular, the standard deviation of each linear combination is equal to the modulus of the corresponding complex number – and thus, the variance is equal to the square of this modulus.

This is exactly what happens when we use quantum-type formulas. *Thus, we have indeed explained the empirical success of quantum-type formulas as a reasonable approximation to the description of complex systems.*

Comment. Similar argument explain why, in fuzzy logic (see, e.g., [2,6,10,12,13, 18]) complex-valued quantum-type techniques have also been successfully used – see, e.g., [4,7,8,11,14].

What Can We Do to Get a More Accurate Description of Complex Systems? As we have mentioned earlier, while quantum-type descriptions are often reasonably accurate, quantum formulas often do not provide the exact description of the corresponding complex systems. So, how can we extend and/or modify these formulas to get a more accurate description?

Based on the above arguments, a natural way to do is to switch from complex-valued 2-dimensional ($k = 2$) approximate descriptions to higher-dimensional ($k = 3$, $k = 4$, etc.) descriptions, where:

- each quantity is represented by a k-dimensional vector, and
- the standard deviation of each linear combination is equal to the length of the corresponding linear combination of vectors.

In particular:

- for $k = 4$, we can geometrically describe this representation in terms of *quaternions* [3] $a + b \cdot \mathrm{i} + c \cdot \mathrm{j} + d \cdot \mathrm{k}$, where:

$$\mathrm{i}^2 = \mathrm{j}^2 = \mathrm{k}^2 = -1, \quad \mathrm{i} \cdot \mathrm{j} = \mathrm{k}, \quad \mathrm{j} \cdot \mathrm{k} = \mathrm{i}, \quad \mathrm{k} \cdot \mathrm{i} = \mathrm{j},$$

$$\mathrm{j} \cdot \mathrm{i} = -\mathrm{k}, \quad \mathrm{k} \cdot \mathrm{j} = -\mathrm{i}, \quad \mathrm{i} \cdot \mathrm{k} = -\mathrm{j};$$

- for $k = 8$, we can represent it in terms of *octonions* [3], etc.

Similar representations are possible for multi-D generalizations of complex-valued fuzzy logic.

Acknowledgments. This work was supported by the Project AI & Reasoning CZ.02.1.01/0.0/0.0/15003/0000466 and the European Regional Development Fund. It was also supported in part by the US National Science Foundation grant HRD-1242122 (Cyber-ShARE Center). This work was performed when M. Svítek was a Visiting Professor at the University of Texas at El Paso.

The authors are thankful to Vladimir Marik and Hung T. Nguyen for their support and valuable discussions.

References

1. Baaquie, B.E.: Quantum Finance: Path Integrals and Hamiltonians for Options and Interest Rates. Camridge University Press, New York (2004)
2. Belohlavek, R., Dauben, J.W., Klir, G.J.: Fuzzy Logic and Mathematics: A Historical Perspective. Oxford University Press, New York (2017)
3. Conway, J.H., Smith, D.A.: On Quaternions and Octonions: Their Geometry, Arithmetic, and Symmetry. A. K. Peters, Natick (2003)
4. Dick, S.: Towards complex fuzzy logic. IEEE Trans. Fuzzy Syst. **13**(3), 405–414 (2005)
5. Haven, E., Khrennikov, A.: Quantum Social Science. Cambridge University Press, Cambridge (2013)
6. Klir, G., Yuan, B.: Fuzzy Sets and Fuzzy Logic. Prentice Hall, Upper Saddle River (1995)

7. Kosheleva, O., Kreinovich, V.: Approximate nature of traditional fuzzy methodology naturally leads to complex-valued fuzzy degrees. In: Proceedings of the IEEE World Congress on Computational Intelligence WCCI 2014, Beijing, China, 6–11 July 2014
8. Kosheleva, O., Kreinovich, V., Ngamsantivong, T.: Why complex-valued fuzzy? Why complex values in general? A computational explanation. In: Proceedings of the Joint World Congress of the International Fuzzy Systems Association and Annual Conference of the North American Fuzzy Information Processing Society IFSA/NAFIPS 2013, Edmonton, Canada, pp. 1233–1236, 24–28 June 2013
9. Kreinovich, V., Nguyen, H.T., Sriboonchitta, S.: Quantum ideas in economics beyond quantum econometrics. In: Anh, L.Y., Dong, L.S., Kreinovich, V., Thach, N.N. (eds.) Econometrics for Financial Applications, pp. 146–151. Springer, Cham (2018)
10. Mendel, J.M.: Uncertain Rule-Based Fuzzy Systems: Introduction and New Directions. Springer, Cham (2017)
11. Nguyen, H.T., Kreinovich, V., Shekhter, V.: On the possibility of using complex values in fuzzy logic for representing inconsistencies. Int. J. Intell. Syst. **13**(8), 683–714 (1998)
12. Nguyen, H.T., Walker, E.A.: A First Course in Fuzzy Logic. Chapman and Hall/CRC, Boca Raton (2006)
13. Novák, V., Perfilieva, I., Močkoř, J.: Mathematical Principles of Fuzzy Logic. Kluwer, Boston, Dordrecht (1999)
14. Servin, C., Kreinovich, V., Kosheleva, O.: From 1-D to 2-D fuzzy: a proof that interval-valued and complex-valued are the only distributive options. In: Proceedings of the Annual Conference of the North American Fuzzy Information Processing Society NAFIPS'2015 and 5th World Conference on Soft Computing, Redmond, Washington, 17–19 August 2015
15. Sheskin, D.J.: Handbook of Parametric and Nonparametric Statistical Procedures. Chapman and Hall/CRC, Boca Raton (2011)
16. Svítek, M.: Quantum System Theory: Principles and Applications. VDM Verlag, Saarbrucken (2010)
17. Svítek, M.: Towards complex system theory. Neural Netw. World **15**(1), 5–33 (2015)
18. Zadeh, L.A.: Fuzzy sets. Inf. Control **8**, 338–353 (1965)

Decision Making Under Interval Uncertainty: Beyond Hurwicz Pessimism-Optimism Criterion

Tran Anh Tuan[1], Vladik Kreinovich[2(✉)], and Thach Ngoc Nguyen[3]

[1] Ho Chi Minh City Institute of Development Studies, 28, Le Quy Don Street, District 3, Ho Chi Minh City, Vietnam
at7tran@gmail.com

[2] Department of Computer Science, University of Texas at El Paso, El Paso, TX 79968, USA
vladik@utep.edu

[3] Banking University of Ho Chi Minh City, 56 Hoang Dieu 2, Quan Thu Duc, Thu Duc, Ho Chi Minh City, Vietnam
Thachnn@buh.edu.vn

Abstract. In many practical situations, we do not know the exact value of the quantities characterizing the consequences of different possible actions. Instead, we often only known lower and upper bounds on these values, i.e., we only know intervals containing these values. To make decisions under such interval uncertainty, the Nobelist Leo Hurwicz proposed his optimism-pessimism criterion. It is known, however, that this criterion is not perfect: there are examples of actions which this criterion considers to be equivalent but which for which common sense indicates that one of them is preferable. These examples mean that Hurwicz criterion must be extended, to enable us to select between alternatives that this criterion classifies as equivalent. In this paper, we provide a full description of all such extensions.

1 Formulation of the Problem

Decision Making in Economics: Ideal Case. In the ideal case, when we know the exact consequence of each action, a natural idea is to select an action that will lead to the largest profit.

Need for Decision Making Under Interval Uncertainty. In real life, we rarely know the exact consequence of each action. In many cases, all we know are the lower and upper bound on the quantities describing such consequences, i.e., all we know is an interval $[\underline{a}, \overline{a}]$ that contains the actual (unknown) value a.

How can make a decision under such interval uncertainty? If we have several alternatives a for each of which we only have an interval estimate $[\underline{u}(a), \overline{u}(a)]$, which alternative should we select?

Hurwicz Optimism-Pessimism Criterion. The problem of decision making under interval uncertainty was first handled by a Nobelist Leo Hurwicz; see, e.g., [2,4,5].

V. Kreinovich et al. (Eds.): ECONVN 2019, SCI 809, pp. 176–184, 2019.
https://doi.org/10.1007/978-3-030-04200-4_14

Hurwicz's main idea was as follows. We know how to make decisions when for each alternative, we know the exact value of the resulting profit. So, to help decision makers make decisions under interval uncertainty, Hurwicz proposed to assign, to each interval $\mathbf{a} = [\underline{a}, \overline{a}]$, an equivalent value $u_H(\mathbf{a})$, and then select an alternative with the largest equivalent value.

Of course, for the case when we know the exact consequence a, i.e., when the interval is degenerate $[a, a]$, the equivalent value should be just a: $u_H([a, a]) = a$.

There are several natural requirements on the function $u_H(\mathbf{a})$. The first is that since all the values a from the interval $[\underline{a}, \overline{a}]$ are larger than (thus better than) or equal to the lower endpoint $\underline{a}$, the equivalent value must also be larger than or equal to $\underline{a}$. Similarly, since all the values a from the interval $[\underline{a}, \overline{a}]$ are smaller than (thus worse than) or equal to the upper endpoint $\overline{a}$, the equivalent value must also be smaller than or equal to $\overline{a}$:

$$\underline{a} \le u_H([\underline{a}, \overline{a}]) \le \overline{a}.$$

The second natural requirement on this function is that the equivalent value should not change if we change a monetary unit: what was better when we count in dollars should also be better when we use Vietnamese Dongs instead. A change from the original monetary unit to a new unit which is k times smaller means that all the numerical values are multiplied by k. Thus, if we have $u_H(\underline{a}, \overline{a}) = a_0$, then, for all $k > 0$, we should have

$$u_H([k \cdot \underline{a}, k \cdot \overline{a}]) = k \cdot a_0.$$

The third natural requirement is related to the fact that if have two separate independent situations with interval uncertainty, with possible profits $[\underline{a}, \overline{a}]$ and $[\underline{b}, \overline{b}]$, then we can do two different things:

- first, we can take into account that the overall profit of these two situations can take any value from $\underline{a} + \underline{b}$ to $\overline{a} + \overline{b}$, and compute the equivalent value of the corresponding interval

$$\mathbf{a} + \mathbf{b} \stackrel{\text{def}}{=} [\underline{a} + \underline{b}, \overline{a} + \overline{b}],$$

- second, we can first find equivalent values of each of the intervals and then add them up.

It is reasonable to require that the resulting value should be the same in both cases, i.e., that we should have

$$u_H([\underline{a} + \underline{b}, \overline{a} + \overline{b}]) = u_H([\underline{a}, \overline{a}]) + h_H([\underline{b}, \overline{b}]).$$

This property is known as *additivity*.

These three requirements allow us to find an explicit formula for the equivalent value $h_H(\mathbf{a})$. Namely, let us denote $\alpha_H \stackrel{\text{def}}{=} u_H([0, 1])$. Due to the first natural requirement, the value α_H is itself between 0 and 1: $0 \le \alpha_H \le 1$. Now, due to scale-invariance, for every value $a > 0$, we have $u_H([0, a]) = \alpha_H \cdot a$. For $a = 0$, this is also true, since in this case, we have $u_H([0,0]) = 0$. In particular, for every two values $\underline{a} \le \overline{a}$, we have $u_H([0, \overline{a} - \underline{a}]) = \alpha_H \cdot (\overline{a} - \underline{a})$.

Now, we also have $u_H([\underline{a}, \underline{a}]) = \underline{a}$. Thus, by additivity, we get

$$u_H([\underline{a}, \overline{a}]) = (\overline{a} - \underline{a}) \cdot \alpha_H + \underline{a},$$

i.e., equivalently, that

$$u_H([\underline{a}, \overline{a}]) = \alpha_H \cdot \overline{a} + (1 - \alpha_H) \cdot \underline{a}.$$

This is the formula for which Leo Hurwicz got his Nobel prize. The meaning of this formula is straightforward:

- When $\alpha_H = 1$, this means that the equivalent value is equal to the largest possible value $\overline{a}$. So, when making a decision, the person only takes into account the best possible scenario and ignores all other possibilities. In real life, such a person is known as an *optimist.*
- When $\alpha_H = 0$, this means that the equivalent value is equal to the smallest possible value $\underline{a}$. So, when making a decision, the person only takes into account the worst possible scenario and ignores all other possibilities. In real life, such a person is known as an *pessimist.*
- When $0 < \alpha_H < 1$, this means that a person takes into account both good and bad possibilities.

Because of this interpretation, the coefficient α_H is called *optimism-pessimism coefficient*, and the whole procedure is known as *optimism-pessimism criterion.*

Need to go Beyond Hurwicz Criterion. While Hurwicz criterion is reasonable, it leaves several options equivalent which should not be equivalent. For example, if $\alpha_H = 0.5$, then, according to Hurwicz criterion, the interval $[-1, 1]$ should be equivalent to 0. However, in reality:

- A risk-averse decision maker will definitely prefer status quo (0) to a situation $[-1, 1]$ in which he/she can lose.
- Similarly, a risk-prone decision maker would probably prefer an exciting gambling-type option $[-1, 1]$ in which he/she can gain.

To take this into account, we need to go beyond assigning a numerical value to each interval. We need, instead, to describe possible orders on the class of all intervals.

This is what we do in this paper.

2 Analysis of the Problem, Definitions, and the Main Result

For every two alternatives a and b, we want to provide the decision maker with one of the following three recommendations:

- select the first alternative; we will denote this recommendation by $b < a$;
- select the second alternative; we will denote this recommendation by $a < b$; or
- treat these two alternatives as equivalent ones; we will denote this recommendation by $a \sim b$.

Our recommendations should be consistent: e.g.,

- if we recommend that b is preferable to a and that c is preferable to b,
- then we should also recommend that c is preferable to a.

Such consistency can be described by the following definition:

Definition 1. *For every set A, by a* linear pre-order, *we mean a pair of relations $(<, \sim)$ for which the following properties are satisfied:*

- *for every a and b, exactly one of the three possibilities must be satisfied: $a < b$, or $b < a$, or $a \sim b$;*
- *for all a, we have $a \sim a$;*
- *for all a and b, if $a \sim b$, then $b \sim a$;*
- *for all a, b, and c, if $a \sim b$ and $b \sim c$, then $a \sim c$;*
- *for all a, b, and c, if $a < b$ and $b < c$, then $a < c$;*
- *for all a, b, and c, if $a < b$ and $b \sim c$, then $a \sim c$; and*
- *for all a, b, and c, if $a \sim b$ and $b < c$, then $a < c$.*

Discussion

- To fully describe a linear pre-order, it is sufficient to describe when $a < b$: indeed, by definition, once we know the relation $<$, we can uniquely reconstruct $a \sim b$ since
$$a \sim b \Leftrightarrow (a \not< b \,\&\, b \not< a).$$
- We want to describe all possible linear pre-orders on the set of all possible intervals. Of course, when the intervals are degenerate – i.e., are, in effect, exact real numbers – this pre-order must coincide with the usual order on the set of real numbers. Also, similarly to the Hurwicz case, an interval $[\underline{a}, \overline{a}]$ cannot be worse than $\underline{a}$ and cannot be better than $\overline{a}$. Thus, we arrive at the following definition.

Definition 2. *A linear pre-order on the set of all possible intervals $\mathbf{a} = [\underline{a}, \overline{a}]$ is called* natural *if the following two properties are satisfied:*

- *for every two numbers a and b, we have*
$$[a, a] < [b, b] \Leftrightarrow a < b;$$
- *for every $\underline{a} \le \overline{a}$, we have $[\underline{a}, \overline{a}] \not< [\underline{a}, \underline{a}]$ and $[\overline{a}, \overline{a}] \not< [\underline{a}, \overline{a}]$.*

Discussion. It is reasonable to require that our linear pre-order does not change if we change a monetary unit.

Definition 3. *A linear pre-order on the set of all possible intervals is called* scale-invariant *if for every two intervals* $\mathbf{a} = [\underline{a}, \overline{a}]$ *and* $\mathbf{b} = [\underline{b}, \overline{b}]$ *and for all real numbers* $k > 0$*, the following two implications hold:*

- *if* $[\underline{a}, \overline{a}] < [\underline{b}, \overline{b}]$*, then* $[k \cdot \underline{a}, k \cdot \overline{a}] < [k \cdot \underline{b}, k \cdot \overline{b}]$*;*
- *if* $[\underline{a}, \overline{a}] \sim [\underline{b}, \overline{b}]$*, then* $[k \cdot \underline{a}, k \cdot \overline{a}] \sim [k \cdot \underline{b} + \ell, k \cdot \overline{b} + \ell]$*.*

Discussion. Our next property is additivity.

Definition 4. *A linear pre-order on the set of all possible intervals is called* additive *if for every three intervals* $\mathbf{a}$, $\mathbf{b}$, *and* $\mathbf{c}$, *the following two implications hold:*

- *if* $\mathbf{a} < \mathbf{b}$*, then* $\mathbf{a} + \mathbf{c} < \mathbf{b} + \mathbf{c}$*;*
- *if* $\mathbf{a} \sim \mathbf{b}$*, then* $\mathbf{a} + \mathbf{c} \sim \mathbf{b} + \mathbf{c}$*.*

Now, we are ready to formulate our main result.

Proposition. *For every natural scale-invariant additive linear pre-order on the set of all possible intervals, there exists a number* α_H *for which the pre-order has one of the following three forms:*

- $[\underline{a}, \overline{a}] < [\underline{b}, \overline{b}]$ *if and only if*

$$\alpha_H \cdot \overline{a} + (1 - \alpha_H) \cdot \underline{a} < \alpha_H \cdot \overline{b} + (1 - \alpha_H) \cdot \underline{b}; \tag{1}$$

- *for* $\alpha_H < 1$, $\mathbf{a} = [\underline{a}, \overline{a}] < \mathbf{b} = [\underline{b}, \overline{b}]$ *if and only if*
  - *either we have an inequality (1)*
  - *or we have an equality*

$$\alpha_H \cdot \overline{a} + (1 - \alpha_H) \cdot \underline{a} = \alpha_H \cdot \overline{b} + (1 - \alpha_H) \cdot \underline{b}, \tag{2}$$

  and $\mathbf{a}$ *is wider than* $\mathbf{b}$*, i.e.,* $\overline{a} - \underline{a} > \overline{b} - \underline{b}$*;*
- *for* $\alpha_H > 0$, $\mathbf{a} = [\underline{a}, \overline{a}] < \mathbf{b} = [\underline{b}, \overline{b}]$ *if and only if:*
  - *either we have the inequality (1)*
  - *or we have the equality (2) and* $\mathbf{a}$ *is narrower than* $\mathbf{b}$*, i.e.,* $\overline{a} - \underline{a} < \overline{b} - \underline{b}$*.*

Vice versa, for each $\alpha_H \in [0, 1]$*, all three relations are natural scale-invariant consistent pre-orders on the set of all possible intervals.*

Discussion

- The first relation describes a risk-neutral decision maker, for whom all intervals with the same Hurwicz equivalent value are indeed equivalent.
- The second relation describes a risk-averse decision maker, who from all the intervals with the same Hurwicz equivalent value selects the one which is the narrowest, i.e., for which the risk is the smallest.
- Finally, the third relation describes a risk-prone decision maker, who from all the intervals with the same Hurwicz equivalent value selects the one which is the widest, i.e., for which the risk is the largest.

Interesting Fact. All three cases can be naturally described in yet another way: in terms of the so-called *non-standard analysis* (see, e.g., [1,3,6,7]), where, in addition to usual ("standard") real numbers, we have *infinitesimal* real numbers, i.e., e.g., objects ε which are positive but which are smaller than all positive standard real numbers.

We can perform usual arithmetic operations on all the numbers, standard and others ("non-standard"). In particular, for every real number x, we can consider non-standard numbers $x+\varepsilon$ and $x-\varepsilon$, where $\varepsilon > 0$ is a positive infinitesimal number – and, vice versa, every non-standard real number which is bounded from below and from above by some standard real numbers can be represented in one of these two forms.

From the above definition, we can conclude how to compare two non-standard numbers obtained by using the same infinitesimal $\varepsilon > 0$, i.e., to be precise, how to compare the numbers $x+k\cdot\varepsilon$ and $x'+k'\cdot\varepsilon$, where x, k, x', and k' are standard real numbers. Indeed, the inequality

$$x + k \cdot \varepsilon < x' + k' \cdot \varepsilon \tag{3}$$

is equivalent to

$$(k - k') \cdot \varepsilon < (x' - x).$$

- If $x' > x$, then this inequality is true since any infinitesimal number (including the number $(k-k')\cdot\varepsilon$) is smaller than any standard positive number – in particular, smaller than the standard real number $x' - x$.
- If $x' < x$, then this inequality is not true, because we will then similarly have $(k'-k)\cdot\varepsilon < (x-x')$, and thus, $(k-k')\cdot\varepsilon > (x'-x)$.
- Finally, if $x = x'$, then, since $\varepsilon > 0$, the above inequality is equivalent to $k < k'$.

Thus, the inequality (3) holds if and only if:

- either $x < x'$,
- or $x = x'$ and $k < k'$.

If we use non-standard numbers, then all three forms listed in the Proposition can be described in purely Hurwicz terms:

$$(\mathbf{a} = [\underline{a}, \overline{a}] < \mathbf{b} = [\underline{b}, \overline{b}]) \Leftrightarrow (\alpha_{NS}\cdot\overline{a} + (1-\alpha_{NS})\cdot\underline{a} < \alpha_{NS}\cdot\overline{b} + (1-\alpha_{NS})\cdot\underline{b}), \tag{4}$$

for some $\alpha_{NS} \in [0,1]$; the only difference from the traditional Hurwicz approach is that now the value α_{NS} can be non-standard. Indeed:

- If α_{NS} is a standard real number, then we get the usual Hurwicz ordering – which is the first form from the Proposition.
- If α_{NS} has the form $\alpha_{NS} = \alpha_H - \varepsilon$ for some standard real number α_H, then the inequality (4) takes the form

$$(\alpha_H - \varepsilon)\cdot\overline{a} + (1 - (\alpha_H - \varepsilon))\cdot\underline{a} < (\alpha_H - \varepsilon)\cdot\overline{b} + (1 - (\alpha_H - \varepsilon))\cdot\underline{b},$$

i.e., separating the standard and infinitesimal parts, the form

$$(\alpha_H \cdot \overline{a} + (1 - \alpha_H) \cdot \underline{a}) - (\overline{a} - \underline{a}) \cdot \varepsilon < (\alpha_H \cdot \overline{b} + (1 - \alpha_H) \cdot \underline{b}) - (\overline{b} - \underline{b}) \cdot \varepsilon.$$

Thus, according to the above description of how to compare non-standard numbers, we conclude that for $\alpha_{NS} = \alpha_H - \varepsilon$, we have $\mathbf{a} < \mathbf{b}$ if and only if:
- either we have the inequality (1)
- or we have the equality (2) and **a** is wider than **b**, i.e., $\overline{a} - \underline{a} > \overline{b} - \underline{b}$.

This is exactly the second form from our Proposition.

- Finally, if α_{NS} has the form $\alpha_{NS} = \alpha_H + \varepsilon$ for some standard real number α_H, then the inequality (4) takes the form

$$(\alpha_H + \varepsilon) \cdot \overline{a} + (1 - (\alpha_H + \varepsilon)) \cdot \underline{a} < (\alpha_H + \varepsilon) \cdot \overline{b} + (1 - (\alpha_H + \varepsilon)) \cdot \underline{b},$$

i.e., separating the standard and infinitesimal parts, the form

$$(\alpha_H \cdot \overline{a} + (1 - \alpha_H) \cdot \underline{a}) + (\overline{a} - \underline{a}) \cdot \varepsilon < (\alpha_H \cdot \overline{b} + (1 - \alpha_H) \cdot \underline{b}) + (\overline{b} - \underline{b}) \cdot \varepsilon.$$

Thus, according to the above description of how to compare non-standard numbers, we conclude that for $\alpha_{NS} = \alpha_H + \varepsilon$, we have $\mathbf{a} < \mathbf{b}$ if and only if:
- either we have the inequality (1)
- or we have the equality (2) and **a** is narrower than **b**, i.e., $\overline{a} - \underline{a} < \overline{b} - \underline{b}$.

This is exactly the third form from our Proposition.

3 Proof

1°. Let us start with the same interval $[0, 1]$ as in the above derivation of the Hurwicz criterion.

1.1°. If the interval $[0, 1]$ is equivalent to some real number α_H – i.e., strictly speaking, to the corresponding degenerate interval $[0, 1] \sim [\alpha_H, \alpha_H]$, then, similarly to that derivation, we can conclude that every interval $[\underline{a}, \overline{a}]$ is equivalent to its Hurwicz equivalent value $\alpha_H \cdot \overline{a} + (1 - \alpha_H) \cdot \underline{a}$. Here, because of naturalness, we have $\alpha_H \in [0, 1]$.

This is the first option from the formulation of our Proposition.

1.2°. To complete the proof, it is thus sufficient to consider the case when the interval $[0, 1]$ is *not* equivalent to any real number. Since we consider a linear pre-order, this means that for every real number r, the interval $[0, 1]$ is either smaller or larger.

- If for some real number a, we have $a < [0, 1]$, then, due to transitivity and naturalness, we have $a' < [0, 1]$ for all $a' < a$.
- Similarly, if for some real number b, we have $[0, 1] < b$, then we have $[0, 1] < b'$ for all $b' > b$.

Thus, there is a threshold value

$$\alpha_H = \sup\{a : a < [0, 1]\} = \inf\{b : [0, 1] < b\}$$

such that:

- for $a < \alpha_H$, we have $a < [0,1]$, and
- for $a > \alpha_H$, we have $[0,1] < a$.

Because of naturalness, we have $\alpha_H \in [0,1]$.

Since we consider the case when the interval $[0,1]$ is not equivalent to any real number, we this have either $[0,1] < \alpha_H$ or $\alpha_H < [0,1]$.

Let us first consider the first option.

2°. In the first option, due to scale-invariance and additivity with $\mathbf{c} = [\underline{a}, \underline{a}]$, similarly to the above derivation of the Hurwicz criterion, for every interval $[\underline{a}, \overline{a}]$, we have:

- when $a < \alpha_H \cdot \overline{a} + (1 - \alpha_H) \cdot \underline{a}$, then $a < [\underline{a}, \overline{a}]$; and
- when $a \geq \alpha_H \cdot \overline{a} + (1 - \alpha_H) \cdot \underline{a}$, then $[\underline{a}, \overline{a}] \leq a$.

Thus, if the Hurwicz equivalent value $u_H(\mathbf{a})$ of a non-degenerate interval $\mathbf{a}$ is smaller than the Hurwicz equivalent value $u_H(\mathbf{a})$ of a non-degenerate interval $\mathbf{b}$, we can conclude that

$$\mathbf{a} < \frac{u_H(\mathbf{a}) + u_H(\mathbf{b})}{2} < \mathbf{b}$$

and hence, that $\mathbf{a} < \mathbf{b}$. So, to complete the description of the desired linear pre-order, it is sufficient to be able to compare the intervals with the same Hurwicz equivalent value.

3°. One can easily check that for every $k > 0$, the Hurwicz equivalent value of the interval $[-k \cdot \alpha_H, k \cdot (1 - \alpha_H)]$ is 0.

Thus, in the first option, we have $[-k \cdot \alpha_H, k \cdot (1 - \alpha_H)] < 0$. So, for every $k' > 0$, by using additivity with $\mathbf{c} = [-k' \cdot \alpha_H, k' \cdot (1 - \alpha_H)]$, we conclude that

$$[-(k + k') \cdot \alpha_H, (k + k') \cdot (1 - \alpha_H)] < [-k \cdot \alpha_H, k \cdot (1 - \alpha_H)].$$

Hence, for two intervals with the same Hurwicz equivalent value 0, the narrower one is better.

By applying additivity with $\mathbf{c}$ equal to Hurwicz value, we conclude that the same is true for all possible Hurwicz equivalent values.

This is the second case in the formulation of our proposition.

4°. Similarly to Part 2 of this proof, in the second option, when $\alpha_H < [0,1]$, we can also conclude that if the Hurwicz equivalent value $u_H(\mathbf{a})$ of a non-degenerate interval $\mathbf{a}$ is smaller than the Hurwicz equivalent value $u_H(\mathbf{a})$ of a non-degenerate interval $\mathbf{b}$, then $\mathbf{a} < \mathbf{b}$.

Then, similarly to Part 3 of this proof, we can prove that for two intervals with the same Hurwicz equivalent value, the wider one is better.

This is the third option as described in the Proposition.

The Proposition is thus proven.

Acknowledgments. This work was supported by Chiang Mai University. It was also partially supported by the US National Science Foundation via grant HRD-1242122 (Cyber-ShARE Center of Excellence).

The authors are greatly thankful to Hung T. Nguyen for valuable discussions.

References

1. Gordon, E.I., Kutateladze, S.S., Kusraev, A.G.: Infinitesimal Analysis. Kluwer Academic Publishers, Dordrecht (2002)
2. Hurwicz, L.: Optimality Criteria for Decision Making Under Ignorance, Cowles Commission Discussion Paper, Statistics, No. 370 (1951)
3. Keisler, H.J.: Elementary Calculus: An Infinitesimal Approach. Dover, New York (2012)
4. Kreinovich, V.: Decision making under interval uncertainty (and beyond). In: Guo, P., Pedrycz, W. (eds.) Human-Centric Decision-Making Models for Social Sciences, pp. 163–193. Springer (2014)
5. Luce, R.D., Raiffa, R.: Games and Decisions: Introduction and Critical Survey. Dover, New York (1989)
6. Robinson, A.: Non-Standard Analysis. Princeton University Press, Princeton (1974)
7. Robinson, A.: Non-Standard Analysis. Princeton University Press, Princeton (1996). Revised edition

Comparisons on Measures of Asymmetric Associations

Xiaonan Zhu[1], Tonghui Wang[1](✉), Xiaoting Zhang[2], and Liang Wang[3]

[1] Department of Mathematical Sciences, New Mexico State University, Las Cruces, USA
{xzhu,twang}@nmsu.edu

[2] Department of Information System, College of Information Engineering, Northwest A & F University, Yangling, China
zxt@nwsuaf.edu.cn

[3] School of Mathematics and Statistics, Xidian University, Xian, China
liang610112@163.com

Abstract. In this paper, we review some recent contributions to multivariate measures of asymmetric associations, i.e., associations in an n-dimension random vector, where $n > 1$. Specially, we pay more attention on measures of complete dependence (or functional dependence). Nonparametric estimators of several measures are provided and comparisons among several measures are given.

Keywords: Asymmetric association · Mutually complete dependence
Functional dependence · Association measures · Copula

1 Introduction

Complete dependence (or functional dependence) is an important concept in many aspects of our life, such as econometrics, insurance, finance, etc. Recently, measures of (mutually) complete dependence have been defined and studied by many authors, e.g. [2,6,7,9–11,13–15], etc. In this paper, measures defined in above works are reviewed. Comparisons among measures are obtained.

Also nonparametric estimators of several measures are provided.

This paper is organized as follows. Some necessary concepts and definitions are reviewed briefly in Sect. 2. Measures of (mutually) complete dependence are summarized in Sect. 3. Estimators and comparisons of measures are provided in Sects. 4 and 5.

2 Preliminaries

Let $(\Omega, \mathscr{A}, P)$ be a probability space, where Ω is a sample space, $\mathscr{A}$ is a σ-algebra of Ω and P is a probability measure on $\mathscr{A}$. A *random variable* is a measurable function from Ω to the real line $\mathbb{R}$, and for any integer $n \geq 2$, an *n-dimensional*

V. Kreinovich et al. (Eds.): ECONVN 2019, SCI 809, pp. 185–197, 2019.
https://doi.org/10.1007/978-3-030-04200-4_15

random vector is a measurable function from Ω to $\mathbb{R}^n$. For any $a = (a_1, \cdots, a_n)$ and $b = (b_1, \cdots, b_n) \in \mathbb{R}^n$, we say $a \leq b$ if and only if $a_i \leq b_i$ for all $i = 1, \cdots, n$. Let X and Y be random vectors defined on the same probability space. X and Y are said to be *independent* if and only if $P(X \leq x, Y \leq y) = P(X \leq x)P(Y \leq y)$ for all x and y. Y is *completely dependent* (CD) on X if Y is a measurable function of X almost surely, i.e., there is a measurable function ϕ such that $P(Y = \phi(X)) = 1$. X and Y are said to be *mutually completely dependent* (MCD) if X and Y are completely dependent on each other.

Let $E_1, \cdots, E_n$ be nonempty subsets of $\mathbb{R}$ and Q a real-valued function with the domain $Dom(Q) = E_1 \times \cdots \times E_n$. Let $[a, b] = [a_1, b_1] \times \cdots \times [a_n, b_n]$ such that all vertices of $[a, b]$ belong to $Dom(Q)$. The *Q-volume of* $[a, b]$ is defined by

$$\mathscr{V}_Q([a, b]) = \sum sgn(c)Q(c),$$

where the sum is taken over all vertices $c = (c_1, \cdots, c_n)$ of $[a, b]$ and

$$sgn(c) = \begin{cases} 1, & \text{if } c_i = a_i \text{ for an even number of } i's, \\ -1, & \text{if } c_i = a_i \text{ for an odd number of } i's. \end{cases}$$

An *n-dimensional subcopula* (or *n-subcopula* for short) is a function C with the following properties [5].

(i) The domain of C is $Dom(C) = D_1 \times \cdots \times D_n$, where $D_1, \cdots, D_n$ are nonempty subsets of the unit interval $I = [0, 1]$ containing 0 and 1;
(ii) C is *grounded*, i.e., for any $u = (u_1, \cdots, u_n) \in Dom(C)$, $C(u) = 0$ if at least one $u_i = 0$;
(iii) For any $u_i \in D_i$, $C(1, \cdots, 1, u_i, 1, \cdots, 1) = u_i$, $i = 1, \cdots, n$;
(iv) C is *n-increasing*, i.e., for any $u, v \in Dom(C)$ such that $u \leq v$, $\mathscr{V}_C([u, v]) \geq 0$.
For any n random variables $X_1, \cdots, X_n$, by Sklar's Theorem [8], there is a unique n-subcopula such that

$$H(x_1, \cdots, x_n) = C(F_1(x_1), \cdots, F_n(x_n)), \quad \text{for all } (x_1, \cdots, x_n) \in \bar{\mathbb{R}}^n,$$

where $\bar{\mathbb{R}} = \mathbb{R} \cup \{-\infty, \infty\}$, H is the joint cumulative distribution function (c.d.f.) of $X_1, \cdots, X_n$, and F_i is the marginal c.d.f. of X_i, $i = 1, \cdots, n$. In addition, if $X_1, \cdots, X_n$ are continuous, then $Dom(C) = I^n$ and the unique C is called the *n-copula* (or *copula*) of $X_1, \cdots, X_n$. For more details about the copula theory, see [5] and [3].

3 Measures of Mutual Complete Dependence

3.1 Measures for Continuous Cases

In 2010, Siburg and Stoimenov [7] defined an MCD measure for continuous random variables as

$$\omega(X, Y) = \left(3\|C\|^2 - 2\right)^{\frac{1}{2}}, \tag{1}$$

where X and Y are continuous random variables with the copula C and $\|\cdot\|$ is the Sobolev norm of bivariate copulas given by

$$\|C\| = \left(\int\int |\nabla C(u,v)|^2 \, dudv\right)^{\frac{1}{2}},$$

where $\nabla C(u,v)$ is the gradient of $C(u,v)$.

Theorem 1. *[7] Let X and Y be random variables with continuous distribution functions and copula C. Then $\omega(X,Y)$ has the following properties:*

(i) $\omega(X,Y) = \omega(Y,X)$.
(ii) $0 \leq \omega(X,Y) \leq 1$.
(iii) $\omega(X,Y) = 0$ *if and only if X and Y are independent.*
(iv) $\omega(X,Y) = 1$ *if and only if X and Y are MCD.*
(v) $\omega(X,Y) \in (\sqrt{2}/2, 1]$ *if Y is completely dependent on X (or vice versa).*
(vi) If $f, g : R \to R$ are strictly monotone functions, then $\omega(f(X), g(Y)) = \omega(X,Y)$.
(vii) If $(X_n, Y_n)_{n\in N}$ is a sequence of pairs of random variables with continuous marginal distribution functions and copulas $(C_n)_{n\in N}$ and if $\lim_{n\to\infty} \|C_n - C\| = 0$, *then* $\lim_{n\to\infty} \omega(X_n, Y_n) = \omega(X,Y)$.

In 2013, Tasena and Dhompongsa [9] generalized Siburg and Stoimenov's measure to multivariate cases as follows. Let $X_1, \cdots, X_n$ be continuous variables with the n-copula C. Define

$$\delta_i(X_1, \cdots, X_n) = \delta_i(C) = \frac{\int \cdots \int [\partial_i C(u_1, \cdots, u_n) - \pi_i C(u_1, \cdots, u_n)]^2 \, du_1 \cdots du_n}{\int \cdots \int \pi_i C(u_1, \cdots, u_n)(1 - \pi_i C(u_1, \cdots, u_n)) du_1 \cdots du_n},$$

where $\partial_i C$ is the partial derivative on the ith coordinate of C and $\pi_i C : I^{n-1} \to I$ is defined by $\pi_i C(u_1, \cdots, u_{n-1}) = C(u_1, \cdots, u_{i-1}, 1, u_i, \cdots, u_{n-1})$, $i = 1, 2, \cdots, n$. Let

$$\delta(X_1, \cdots, X_n) = \delta(C) = \frac{1}{n} \sum_{i=1}^{n} \delta_i(C). \tag{2}$$

Then δ is an MCD measure of $X_1, \cdots, X_n$.

The measure δ has the following properties.

Theorem 2. *[9] For any random variables $X_1, \cdots, X_n$,*

(i) $0 \leq \delta(X_1, \cdots, X_n) \leq 1$.
(ii) $\delta(X_1, \cdots, X_n) = 0$ *if and only if all $X_i, i = 1, \cdots, n$, are independent.*
(iii) $\delta(X_1, \cdots, X_n) = 1$ *if and only if $X_1, \cdots, X_n$ are mutually completely dependent.*
(iv) $\delta(X_1, \cdots, X_n) = \delta(X_{\sigma(1)}, \cdots, X_{\sigma(n)})$ *for any permutation σ.*
(v) $\lim_{k\to\infty} \delta(X_1^k, \cdots, X_n^k) = \delta(X_1, \cdots, X_n)$ *whenever the copulas associated to $(X_1^k, \cdots, X_n^k)$ converge to the copula associated to $(X_1, \cdots, X_n)$ under the modified Sobolev norm defined by* $\|C\| = \sum_i \int |\partial_i C|^2$.

(vi) *If* X_{n+1} *and* $(X_1, \cdots, X_n)$ *are independent, then* $\delta(X_1, \cdots, X_{n+1}) < \frac{2}{3}\delta(X_1, \cdots, X_n)$.
(vii) *If* $\delta(X_1, \cdots, X_n) \geq \frac{2n-2}{3n}$, *then none of* X_i *is independent from the rest.*
(viii) $\delta^{(n)}$ *is not a function of* $\delta^{(2)}$ *for any* $n > 2$.

In 2016 Tasena and Dhompongsa [10] defined a measure of CD for random vectors. Let X and Y be two random vectors. Define

$$\omega_k(Y|X) = \left[\int\int \left|F_{Y|X}(y|x) - \frac{1}{2}\right|^k dF_X(x)dF_Y(y)\right]^{\frac{1}{k}},$$

where $k \geq 1$. The measure of Y CD on X is given by

$$\bar{\omega}_k(Y|X) = \left[\frac{\omega_k^k(Y|X) - \omega_k^k(Y'|X')}{\omega_k^k(Y|Y) - \omega_k^k(Y'|X')}\right]^{\frac{1}{k}}, \tag{3}$$

where X' and Y' are independent random vectors with the same distributions as X and Y, respectively.

Theorem 3. *[10]* ω_k *and* $\bar{\omega}_k$ *have following properties:*

(i) $\omega_k(Y|X) \geq \omega_k(Y|f(X))$ *for all measurable function* f *and all random vectors* X *and* Y.
(ii) $\omega_k(Y'|X') \leq \omega_k(Y|X) \leq \omega_k(Y|Y)$ *where* (Y', X') *have the same marginals as* (Y, X) *but* X' *and* Y' *are independent.*
(iii) $\omega_k(Y'|X') = \omega_k(Y|X)$ *if and only if* X *and* Y *are independent.*
(iv) $\omega_k(Y|X) = \omega_k(Y|Y)$ *if and only if* Y *is a function of* X.
(v) $\omega_k(Y, Y, Z|X) = \omega_k(Y, Z|X)$ *for all random vectors* X, Y, *and* Z.
(vi) $\bar{\omega}_2(Y, Z|X) \leq \bar{\omega}_2(Y|X)$ *for any random vectors* X, Y, *and* Z *in which* Z *is independent of* X *and* Y.

In the same period, Boonmee and Tasena [2] defined a measure of CD for continuous random vectors by using *linkages* which were introduced by Li et al. [4]. Let X and Y be two continuous random vectors with the linkage C. The measure of Y being completely dependent on X is defined by

$$\zeta_p(Y|X) = \left[\int\int \left|\frac{\partial}{\partial u}C(u, v) - \Pi(v)\right|^p dudv\right]^{\frac{1}{p}}, \tag{4}$$

where $\Pi(v) = \prod_{i=1}^{n} v_i$ for all $v = (v_1, \cdots, v_n) \in I^n$.

Theorem 4. *[2] The measure* ζ_p *has the following properties:*

(i) *For any random vectors* X *and* Y *and any measurable function* f *in which* $f(X)$ *has absolutely continuous distribution function,* $\zeta_p(Y|f(X)) \leq \zeta_p(Y|X)$.
(ii) *For any random vectors* X *and* Y, $\zeta_p(Y|X) = 0$ *if and only if* X *and* Y *are independent.*

(iii) *For any random vectors X and Y, $0 \leq \zeta_p(Y|X) \leq \zeta_p(Y|Y)$.*

(iv) *For any random vectors X and Y, the three following properties are equivalent.*

(a) *Y is a measurable function of X,*

(b) *$\Psi_{F_Y}(Y)$ is a measurable function of $\Psi_{F_X}(X)$, where*

$$\Psi_{F_X}(x_1, \cdots, x_n) \\ = \left(F_{X_1}(x_1), F_{X_2|X_1}(x_2|x_1), \cdots, F_{X_n|(X_1,\cdots,X_{n-1})}(x_n|(x_1, \cdots, x_{n-1}))\right).$$

(c) *$\zeta_p(Y|X) = \zeta_p(Y|Y)$.*

(v) *For any random vectors X, Y, and Z in which Z has dimension k and (X, Y) and Z are independent, $\zeta_p(Y, Z|X) = \left(\frac{1}{p+1}\right)^{\frac{k}{p}} \zeta_p(Y|X)$. In particular $\zeta_p(Y, Z|X) < \zeta_p(Y|X)$.*

(vi) *For any $\varepsilon > 0$, there are random vectors X and Y of arbitrary marginals but with the same dimension such that Y is completely dependent on X but $\zeta_p(X|Y) \leq \varepsilon$.*

3.2 Measures for Discrete Cases

In 2015, Shan et al. [6] considered discrete random variables. Let X and Y be two discrete random variables with the subcopula C. Measures $\mu_t(Y|X)$ and $\mu_t(X|Y)$ for Y completely depends on X and X completely depends on Y, respectively, are defined by

$$\mu_t(Y|X) = \left(\frac{\sum_i \sum_j C_{\Delta i,j} \Delta u_i \Delta v_j - L_t^{(2)}}{U_t^{(2)} - L_t^{(2)}}\right)^{\frac{1}{2}} \tag{5}$$

and

$$\mu_t(X|Y) = \left(\frac{\sum_i \sum_j C_{i,\Delta j} \Delta u_i \Delta v_j - L_t^{(1)}}{U_t^{(1)} - L_t^{(1)}}\right)^{\frac{1}{2}}. \tag{6}$$

An MCD measure of X and Y is given by

$$\mu_t(X, Y) = \left(\frac{\|C\|_t^2 - L_t}{U_t - L_t}\right)^{\frac{1}{2}}, \tag{7}$$

where $t \in [0, 1]$ and $\|C\|_t^2$ is the discrete norm of C defined by

$$\|C\|_t^2 = \sum_i \sum_j \left\{ \left(tC_{\Delta i,j}^2 + (1-t)C_{\Delta i,j+1}^2\right) \frac{\Delta v_j}{\Delta u_i} + \left(tC_{i,\Delta j}^2 + (1-t)C_{i+1,\Delta j}^2\right) \frac{\Delta u_i}{\Delta v_j} \right\},$$

$$C_{\Delta i,j} = C(u_{i+1}, v_j) - C(u_i, v_j), \quad C_{i,\Delta j} = C(u_i, v_{j+1}) - C(u_i, v_j), \\ \Delta u_i = u_{i+1} - u_i, \quad \Delta v_j = v_{j+1} - v_j,$$

$$L_t = L_t^{(1)} + L_t^{(2)} = \sum_i (tu_i^2 + (1-t)u_{i+1}^2)\Delta u_i + \sum_j (tv_j^2 + (1-t)v_{j+1}^2)\Delta v_j,$$

and

$$U_t = U_t^{(1)} + U_t^{(2)} = \sum_i (tu_i + (1-t)u_{i+1})\Delta u_i + \sum_j (tv_j + (1-t)v_{j+1})\Delta v_j.$$

Theorem 5. *[6] For any discrete random variables X and Y, measures $\mu_t(Y|X)$, $\mu_t(X|Y)$ and $\mu_t(X,Y)$ have the following properties:*

(i) $0 \le \mu_t(Y|X), \mu_t(X|Y), \mu_t(X,Y) \le 1$.
(ii) $\mu_t(X,Y) = \mu_t(Y,X)$.
(iii) $\mu_t(Y|X) = \mu_t(X|Y) = \mu_t(X,Y) = 0$ *if and only if X and Y are independent.*
(iv) $\mu_t(X,Y) = 1$ *if and only if X and Y are MCD.*
(v) $\mu_t(Y|X) = 1$ *if and only if Y is complete dependent on X.*
(vi) $\mu_t(X|Y) = 1$ *if and only if X is complete dependent on Y.*

In 2017, Wei and Kim [11] defined a *measure of subcopula-based asymmetric association* of discrete random variables. Let X and Y be two discrete random variables with I and J categories having the supports S_0 and S_1, where $S_0 = \{x_1, x_2, \cdots, x_I\}$, and $S_1 = \{y_1, y_2, \cdots, y_J\}$, respectively. Denote the marginal distributions of X and Y be $F(x), G(y)$, and the joint distribution of (X,Y) be $H(x,y)$, respectively. Let $U = F(X)$ and $V = G(Y)$. The supports of U and V are $D_0 = F(S_0) = \{u_1, u_2, \cdots, u_I\}$ and $D_1 = G(S_1) = \{v_1, v_2, \cdots, v_J\}$, respectively. Let $P = \{p_{ij}\}$ be the matrix of the joint cell proportions in the $I \times J$ contingency table of X and Y, where $i = 1, \cdots, I$ and $j = 1, \cdots, J$, i.e., $u_i = \sum_{s=1}^{i} p_{s\cdot}$ and $v_j = \sum_{t=1}^{j} p_{\cdot t}$. A measure of subcopula-based asymmetric association of Y on X is defined by

$$\rho^2_{X\to Y} = \frac{\sum_{i=1}^{I}\left(\sum_{j=1}^{J} v_j p_{j|i} - \sum_{j=1}^{J} v_j p_{\cdot j}\right)^2 p_{i\cdot}}{\sum_{j=1}^{J}\left(v_j - \sum_{j=1}^{J} v_j p_{\cdot j}\right)^2 p_{\cdot j}}, \tag{8}$$

where $p_{j|i} = \frac{p_{ij}}{p_{i\cdot}}$ and $p_{i|j} = \frac{p_{ij}}{p_{\cdot j}}$. A measure $\rho^2_{Y\to X}$ of asymmetric association of X on Y can be similarly defined as (8) by interchanging X and Y

The properties of $\rho^2_{X\to Y}$ is given by following theorem.

Theorem 6. *[11] Let X and Y be two variables with subcopula $C(u,v)$ in an $I \times J$ contingency table, and let $U = F(X)$ and $V = G(Y)$. Then*

(i) $0 \le \rho^2_{X\to Y} \le 1$.
(ii) If X and Y are independent, then $\rho^2_{X\to Y} = 0$; Furthermore, if $\rho^2_{X\to Y} = 0$, then the correlation of U and V is 0.

(iii) $\rho^2_{X\to Y} = 1$ *if and only if* $Y = g(X)$ *almost surely for some measurable function* g.
(iv) *If* $X_1 = g_1(X)$, *where* g_1 *is an injective function of* X, *then* $\rho^2_{X_1\to Y} = \rho^2_{X\to Y}$.
(v) *If* X *and* Y *are both dichotomous variables with only* 2 *categories, then* $\rho^2_{X\to Y} = \rho^2_{Y\to X}$.

In 2018, Zhu et al. [15] generalized Shan's measure μ_t to multivariate case. Let X and Y be two discrete random vectors with the subcopula C. Suppose that the domain of C is $Dom(C) = \mathscr{L}_1' \times \mathscr{L}_2'$, where $\mathscr{L}_1' \subseteq I^n$ and $\mathscr{L}_2' \subseteq I^m$. The *measure of* Y *being completely dependent on* X *based on* C is given by

$$\mu_C(Y|X) = \left[\frac{\omega^2(Y|X)}{\omega^2_{max}(Y|X)}\right]^{\frac{1}{2}}$$

$$= \left[\frac{\sum\limits_{v\in\mathscr{L}_2'}\sum\limits_{u\in\mathscr{L}_1'}\left[\frac{\mathscr{V}_C([(u_L,v),(u,v)])}{\mathscr{V}_C([(u_L,1_m),(u,1_m)])} - C(1_n,v)\right]^2 \mathscr{V}_C([(u_L,1_m),(u,1_m)])\mathscr{V}_C([(1_n,v_L),(1_n,v)])}{\sum\limits_{v\in\mathscr{L}_2'}\left[C(1_n,v) - (C(1_n,v))^2\right]\mathscr{V}_C([(1_n,v),(1_n,v_L])}\right]^{\frac{1}{2}}. \quad (9)$$

The *MCD measure of* X *and* Y is defined by

$$\mu_C(X,Y) = \left[\frac{\omega^2(Y|X) + \omega^2(X|Y)}{\omega^2_{max}(Y|X) + \omega^2_{max}(X|Y)}\right]^{\frac{1}{2}}, \quad (10)$$

where $\omega^2(X|Y)$ and $\omega^2_{max}(X|Y)$ are similarly defined as $\omega^2(Y|X)$ and $\omega^2_{max}(Y|X)$ by interchanging X and Y

Theorem 7. *[15] Let* X *and* Y *be two discrete random vectors with the subcopula* C. *The measures* $\mu_C(Y|X)$ *and* $\mu_C(X,Y)$ *have following properties:*

(i) $\mu_C(X,Y) = \mu_C(Y,X)$.
(ii) $0 \le \mu_C(X,Y), \mu_C(Y|X) \le 1$.
(iii) $\mu_C(X,Y) = \mu_C(Y|X) = 0$ *if and only if* X *and* Y *are independent.*
(iv) $\mu_C(Y|X) = 1$ *if and only if* Y *is a function of* X.
(v) $\mu_C(X,Y) = 1$ *if and only if* X *and* Y *are MCD.*
(vi) $\mu_C(X,Y)$ *and* $\mu_C(Y|X)$ *are invariant under strictly increasing transformations of* X *and* Y.

4 Estimators of Measures

In section, we consider estimators of measures $\mu_0(Y|X)$ and $\mu_0(X,Y)$ given by (5) and (7), $\mu(Y|X)$ and $\mu(X,Y)$ given by (9) and (10) and $\rho^2_{X\to Y}$ given by (8).

First, let $X \in \mathscr{L}_1$ and $Y \in \mathscr{L}_2$ be two discrete random vectors and $[n_{xy}]$ be their observed multi-way contingency table. Suppose that the total number of observation is n. For every $x \in \mathscr{L}_1$ and $y \in \mathscr{L}_2$, let n_{xy}, $n_{x\cdot}$ and $n_{\cdot y}$ be numbers of observations of (x,y), x and y, respectively, i.e., $n_{x\cdot} = \sum\limits_{y\in\mathscr{L}_2} n_{xy}$ and

$n_{\cdot y} = \sum_{x \in \mathscr{L}_1} n_{xy}$. If we define $\hat{p}_{xy} = n_{xy}/n$, $\hat{p}_{x\cdot} = n_{x\cdot}/n$, $\hat{p}_{\cdot y} = n_{\cdot y}/n$, $\hat{p}_{y|x} = \hat{p}_{xy}/\hat{p}_{x\cdot} = n_{xy}/n_{x\cdot}$ and $\hat{p}_{x|y} = \hat{p}_{xy}/\hat{p}_{\cdot y} = n_{xy}/n_{\cdot y}$, then estimators of measures $\mu(Y|X)$, $\mu(X|Y)$ and $\mu(X,Y)$ given by (9) and (10) can be defined as follows.

Proposition 1. *[15] Let $X \in \mathscr{L}_1$ and $Y \in \mathscr{L}_2$ be two discrete random vectors with a multi-way contingency table $[n_{xy}]$. Estimators of $\mu(Y|X)$ and $\mu(X,Y)$ are given by*

$$\hat{\mu}(Y|X)\left[\frac{\hat{\omega}^2(Y|X)}{\hat{\omega}^2_{max}(Y|X)}\right]^{\frac{1}{2}} \quad \text{and} \quad \hat{\mu}(X|Y)\left[\frac{\hat{\omega}^2(X|Y)}{\hat{\omega}^2_{max}(X|Y)}\right]^{\frac{1}{2}}, \tag{11}$$

and

$$\hat{\mu}(X,Y) = \left[\frac{\hat{\omega}^2(Y|X) + \hat{\omega}^2(X|Y)}{\hat{\omega}^2_{max}(Y|X) + \hat{\omega}^2_{max}(X|Y)}\right]^{\frac{1}{2}}, \tag{12}$$

where

$$\hat{\omega}^2(Y|X) = \sum_{y \in \mathscr{L}_2,}\sum_{x \in \mathscr{L}_1}\left[\sum_{y' \le y,}\left(\hat{p}_{y'|x} - \hat{p}_{\cdot y'}\right)\right]^2 \hat{p}_{x\cdot}\hat{p}_{\cdot y},$$

$$\hat{\omega}^2_{max}(Y|X) = \sum_{y \in \mathscr{L}_2}\left[\sum_{y' \le y,}\hat{p}_{\cdot y'} - \left(\sum_{y' \le y,}\hat{p}_{\cdot y'}\right)^2\right]\hat{p}_{\cdot y},$$

and $\hat{\omega}^2(X|Y)$ and $\hat{\omega}^2_{max}(X|Y)$ are similarly defined as $\hat{\omega}^2(Y|X)$ and $\hat{\omega}^2_{max}(Y|X)$ by interchanging X and Y.

Note that measures $\mu(Y|X)$ and $\mu(X,Y)$ given by (9) and (10) are multivariate versions of measures $\mu_0(Y|X)$ and $\mu_0(X,Y)$ given by (5) and (7). Thus, when X and Y are discrete random variables, estimators of $\mu_0(Y|X)$ and $\mu_0(X,Y)$ can be obtained similarly. By using above notations, the estimator of $\rho^2_{X\to Y}$ given by (8) is given as follows.

Proposition 2. *[11] The estimator of $\rho^2_{X\to Y}$ is given by*

$$\hat{\rho}^2_{X\to Y} = \frac{\sum_x\left(\sum_y \hat{v}_y - \sum_y \hat{v}_y\hat{p}_{\cdot y}\right)^2 \hat{p}_{i\cdot}}{\sum_y\left(\hat{v}_y - \sum_y \hat{v}_y\hat{p}_{\cdot y}\right)^2 \hat{p}_{\cdot y}} \tag{13}$$

where $\hat{v}_y = \sum_y \hat{p}_{\cdot y}$. The estimator of $\rho^2_{Y\to X}$ can be similarly obtained.

In order to make comparison of measures, we need the concept of the *functional chi-square statistic* defined by Zhang and Song [13]. Let the $r \times s$ matrix

$[n_{ij}]$ be an observed contingency table of discrete random variables X and Y. The functional chi-square statistic of X and Y is defined by

$$\chi^2(f : X \to Y) = \sum_x \sum_y \frac{(n_{xy} - n_{x\cdot}/s)^2}{n_{x\cdot}/s} - \sum_y \frac{(n_{\cdot y} - n/s)^2}{n/s} \tag{14}$$

Theorem 8. *[13] For the functional chi-square defined above, the following properties can be obtained:*

(i) If X and Y are empirically independent, then $\chi^2(f : X \to Y) = 0$.
(ii) $\chi^2(f : X \to Y) \geq 0$ for any contingency table.
(iii) The functional chi-square is asymmetric, that is, $\chi^2(f : X \to Y)$ does not necessarily equal to $\chi^2(f : Y \to X)$ for a given contingency table.
(iv) $\chi^2(f : X \to Y)$ is asymptotically chi-square distributed with $(r-1)(s-1)$ degrees of freedom under the null hypothesis that Y is uniformly distributed conditioned on X.
(v) $\chi^2(f : X \to Y)$ attains maximum if and only if the column variable Y is a function of the row variable X in the case that a contingency table is feasible. Moreover, the maximum of the functional chi-square is given by $ns\left(1 - \sum_y (n_{\cdot y}/n)^2\right)$.

Also Wongyang et al. [12] proved that the functional chi-square statistic has following additional property.

Proposition 3. *For any injective function $\phi : supp(X) \to \mathbb{R}$ and $\psi : supp(Y) \to \mathbb{R}$,*

$$\chi^2(f : \phi(X) \to Y) = \chi^2(f : X \to Y) \quad \textit{and} \quad \chi^2(f : X \to \psi(Y)) = \chi^2(f : X \to Y),$$

where $supp(\cdot)$ is the support of the random variable.

5 Comparisons of Measures

From above summaries we can see that measures given by (1), (2) and (4) are defined for continuous random variables or vectors. The measures defined by (7), (8), (9) and (10) work for discrete random variables or vectors. The measure given by (3) relies on marginal distributions of random vectors. Specifically, we have the following relations.

Proposition 4. *[6] For the measure $\mu_t(X, Y)$ given by (7), if both X and Y are continuous random variables, i.e.,* $\max\{u - u_L, v - v_L\} \to 0$*, then it can be show that*

$$\mu_t(X,Y) = \left\{3 \int\int \left[\left(\frac{\partial C}{\partial u}\right)^2 + \left(\frac{\partial C}{\partial v}\right)^2\right] dudv - 2\right\}^{\frac{1}{2}},$$

So, $\mu_t(X, Y)$ is the discrete version of the measure given by (1).

Proposition 5. *[15] For the measure $\mu_C(X,Y)$ given by (10), if both X and Y are discrete random variables with the 2-subcopula C, then we have*

$$\omega^2(Y|X) = \sum_{v\in\mathscr{L}_2'}\sum_{u\in\mathscr{L}_1'}\left[\frac{C(u,v)-C(u_L,v)^2}{u-u_L} - v\right]^2 (u-u_L)(v-v_L),$$

$$\omega^2(X|Y) = \sum_{u\in\mathscr{L}_1'}\sum_{v\in\mathscr{L}_2'}\left[\frac{C(u,v)-C(u,v_L)^2}{v-v_L} - u\right]^2 (u-u_L)(v-v_L),$$

$$\omega^2_{max}(Y|X) = \sum_{v\in\mathscr{L}_2'}(v-v^2)(v-v_L) \quad and \quad \omega^2_{max}(X|Y) = \sum_{u\in\mathscr{L}_1'}(u-u^2)(u-u_L).$$

In this case, the measure $\mu_C(X,Y) = \left[\frac{\omega^2(Y|X)+\omega^2(X|Y)}{\omega^2_{max}(Y|X)+\omega^2_{max}(X|Y)}\right]^{\frac{1}{2}}$ *is identical to the measure μ_t given by (7) with $t=0$.*

In addition, note that measures $\mu_t(Y|X)$ given by (5) and $\rho^2_{X\to Y}$ given by (8), and the functional chi-square statistic $\chi^2(f: X\to Y)$ are defined for discrete random variables. Let's compare three measures by the following examples.

Example 1. Consider the contingency table of two discrete random variables X and Y given by Table 1.

Table 1. Contingency table of X and Y.

Y	X			$n_{y\cdot}$
	1	2	3	
10	50	10	50	110
20	10	50	10	70
30	10	0	10	20
$n_{\cdot x}$	70	60	70	200

By calculation, we have

(i)

$$\hat{\omega}_0^2(Y|X) = 0.0361, \qquad \hat{\omega}^2_{0,max}(Y|X) = 0.1676,$$

and

$$\hat{\omega}_0^2(X|Y) = 0.0151, \qquad \hat{\omega}^2_{0,max}(X|Y) = 0.1479.$$

So

$$\hat{\mu}_0(Y|X) = 0.4643 \quad \text{and} \quad \hat{\mu}_0(X|Y) = 0.3198.$$

(ii)

$$\hat{\chi}^2(f : X \to Y) = 10.04, \qquad \hat{\chi}^2_{max}(f : X \to Y) = 33.9,$$

and

$$\hat{\chi}^2(f : Y \to X) = 8.38, \qquad \hat{\chi}^2_{max}(f : Y \to X) = 33.9.$$

So

$$\hat{\chi}^2_{nor}(f : X \to Y) = \frac{\hat{\chi}^2(f : X \to Y)}{\hat{\chi}^2_{max}(f : X \to Y)} = 0.2962,$$

and

$$\hat{\chi}^2_{nor}(f : Y \to X) = \frac{\hat{\chi}^2(f : Y \to X)}{\hat{\chi}^2_{max}(f : Y \to X)} = 0.2100.$$

(iii)

$$\hat{\rho}^2_{X \to Y} = 0.1884 \qquad \text{and} \qquad \hat{\rho}^2_{Y \to X} = 0.0008.$$

All measures indicate that the functional dependence of Y on X is stronger than the functional dependence of X on Y. The difference of the measure $\hat{\rho}^2$ on two directions is more significant than differences of $\hat{\mu}_0$ and $\hat{\chi}^2_{nor}$.

Example 2. Consider the contingency table of two discrete random variables X and Y given by Table 2.

Table 2. Contingency table of X and Y.

Y	X			$n_{y\cdot}$
	1	2	3	
1	10	65	5	80
2	10	5	35	50
3	50	5	15	70
$n_{\cdot x}$	70	75	55	200

By calculation, we have

(i)

$$\hat{\omega}^2_0(Y|X) = 0.0720, \qquad \hat{\omega}^2_{0,max}(Y|X) = 0.1529,$$

and

$$\hat{\omega}^2_0(X|Y) = 0.0495, \qquad \hat{\omega}^2_{0,max}(X|Y) = 0.1544.$$

So

$$\hat{\mu}_0(Y|X) = 0.6861 \qquad \text{and} \qquad \hat{\mu}_0(X|Y) = 0.5662.$$

(ii)

$$\hat{\chi}^2(f : X \to Y) = 160.17, \qquad \hat{\chi}^2_{max}(f : X \to Y) = 393,$$

and

$$\hat{\chi}^2(f : Y \to X) = 158.73, \qquad \hat{\chi}^2_{max}(f : Y \to X) = 396.75.$$

So

$$\hat{\chi}^2_{nor}(f : X \to Y) = \frac{\hat{\chi}^2(f : X \to Y)}{\hat{\chi}^2_{max}(f : X \to Y)} = 0.4075,$$

and

$$\hat{\chi}^2_{nor}(f : Y \to X) = \frac{\hat{\chi}^2(f : Y \to X)}{\hat{\chi}^2_{max}(f : Y \to X)} = 0.4001.$$

(iii)

$$\hat{\rho}^2_{X \to Y} = 0.4607 \qquad \text{and} \qquad \hat{\rho}^2_{Y \to X} = 0.2389.$$

All measures indicate that the functional dependence of Y on X is stronger than the functional dependence of X on Y.

Next, let's use one real example to illustrate the measures for discrete random vectors defined by (9) and (10).

Example 3. Table 3 is based on automobile accident records in 1988 [1], supplied by the state of Florida Department of Highway Safety and Motor Vehicles. Subjects were classified by whether they were wearing a seat belt, whether ejected, and whether killed. Denote the variables by S for wearing a seat belt, E for ejected, and K for killed. By Pearson's Chi-squared test (S, E) and K are not independent. The estimations of functional dependence between (S, E) and K are $\hat{\mu}(K|(S, E)) = 0.7081$, $\hat{\mu}((S, E)|K) = 0.2395$ and $\hat{\mu}((S, E), K) = 0.3517$.

Table 3. Automobile accident records in 1988.

Safety equipment in use	Whether ejected	Injury	
		Nonfatal	Fatal
Seat belt	Yes	1105	14
	No	411111	483
None	Yes	462	4987
	No	15734	1008

References

1. Agresti, A.: An Introduction to Categorical Data Analysis, vol. 135. Wiley, New York (1996)
2. Boonmee, T., Tasena, S.: Measure of complete dependence of random vectors. J. Math. Anal. Appl. **443**(1), 585–595 (2016)
3. Durante, F., Sempi, C.: Principles of Copula Theory. CRC Press, Boca Raton (2015)
4. Li, H., Scarsini, M., Shaked, M.: Linkages: a tool for the construction of multivariate distributions with given nonoverlapping multivariate marginals. J. Multivar. Anal. **56**(1), 20–41 (1996)
5. Nelsen, R.B.: An Introduction to Copulas. Springer, New York (2007)
6. Shan, Q., Wongyang, T., Wang, T., Tasena, S.: A measure of mutual complete dependence in discrete variables through subcopula. Int. J. Approx. Reason. **65**, 11–23 (2015)
7. Siburg, K.F., Stoimenov, P.A.: A measure of mutual complete dependence. Metrika **71**(2), 239–251 (2010)
8. Sklar, M.: Fonctions de répartition á n dimensions et leurs marges. Université Paris 8 (1959)
9. Tasena, S., Dhompongsa, S.: A measure of multivariate mutual complete dependence. Int. J. Approx. Reason. **54**(6), 748–761 (2013)
10. Tasena, S., Dhompongsa, S.: Measures of the functional dependence of random vectors. Int. J. Approx. Reason. **68**, 15–26 (2016)
11. Wei, Z., Kim, D.: Subcopula-based measure of asymmetric association for contingency tables. Stat. Med. **36**(24), 3875–3894 (2017)
12. Wongyang, T.: Copula and measures of dependence. Resarch notes, New Mexico State University (2015)
13. Zhang, Y., Song, M.: Deciphering interactions in causal networks without parametric assumptions. arXiv preprint arXiv:1311.2707 (2013)
14. Zhong, H., Song, M.: A fast exact functional test for directional association and cancer biology applications. IEEE/ACM Trans. Comput. Biol. Bioinform. (2018)
15. Zhu, X., Wang, T., Choy, S.B., Autchariyapanitkul, K.: Measures of mutually complete dependence for discrete random vectors. In: International Conference of the Thailand Econometrics Society, pp. 303–317. Springer (2018)

Fixed-Point Theory

Proximal Point Method Involving Hybrid Iteration for Solving Convex Minimization Problem and Common Fixed Point Problem in Non-positive Curvature Metric Spaces

Plern Saipara[1], Kamonrat Sombut[2(✉)], and Nuttapol Pakkaranang[3]

[1] Division of Mathematics, Department of Science, Faculty of Science and Agricultural Technology, Rajamangala University of Technology Lanna Nan, 59/13 Fai Kaeo, Phu Phiang 55000, Nan, Thailand
splernn@gmail.com

[2] Department of Mathematics and Computer Science, Faculty of Science and Technology, Rajamangala University of Technology Thanyaburi (RMUTT), 39 Rungsit-Nakorn Nayok Rd., Klong 6, Khlong Luang 12110, Thanyaburi, Pathumthani, Thailand
kamonrat_s@rmutt.ac.th

[3] Department of Mathematics, Faculty of Science, King Mongkut's University of Technology Thonburi (KMUTT), 126 Pracha-Uthit Road, Bang Mod, Thung Khru, Bangkok 10140, Thailand
nuttapol.pak@mail.kmutt.ac.th

Abstract. In this paper, we introduce a proximal point algorithm involving hybrid iteration for nonexpansive mappings in non-positive curvature metric spaces, namely CAT(0) spaces and also prove that the sequence generated by proposed algorithms converges to a minimizer of a convex function and common fixed point of such mappings.

Keywords: Proximal point algorithm · CAT(0) spaces
Convex function · Picard-S hybrid iteration

1 Introduction

Let C be a non-empty subset of a metric space (X, d). The mapping $T : C \to C$ is said to be nonexpansive if for each $x, y \in C$,

$$d(Tx, Ty) \leq d(x, y).$$

A point $x \in C$ is said to be a fixed point of T if $Tx = x$. The set of all fixed points of a mapping T will be denote by $F(T)$. There are many approximation methods for the fixed point of T, for examples, Mann iteration process, Ishikawa

V. Kreinovich et al. (Eds.): ECONVN 2019, SCI 809, pp. 201–214, 2019.
https://doi.org/10.1007/978-3-030-04200-4_16

iteration process and S-iteration process etc. More details of their iteration process can see as follows.

The Mann iteration process is defined as follows: $x_1 \in C$ and

$$x_{n+1} = (1 - \alpha_n)x_n + \alpha_n T x_n \tag{1}$$

for each $n \in \mathbb{N}$, where $\{\alpha_n\}$ is a sequence in (0,1).

The Ishikawa iteration process is defined as follows: $x_1 \in C$ and

$$\begin{cases} x_{n+1} = (1 - \alpha_n)x_n + \alpha_n T y_n, \\ y_n = (1 - \beta_n)x_n + \beta_n T x_n \end{cases} \tag{2}$$

for each $n \in \mathbb{N}$, where $\{\alpha_n\}$ and $\{\beta_n\}$ are sequences in (0,1).

Recently, the S-iteration process was introduced by Agarwal, O'Regan and Sahu [1] in a Banach space as follow:

$$\begin{cases} x_1 \in C, \\ x_{n+1} = (1 - \alpha_n)T x_n + \alpha_n T(y_n), \\ y_n = (1 - \beta_n)x_n + \beta_n T(x_n), \end{cases} \tag{3}$$

for each $n \in \mathbb{N}$, where $\{\alpha_n\}$ and $\{\beta_n\}$ are sequences in (0, 1). Pragmatically, we have to consider the rate of convergence of course, we want to fastest convergence.

The initials of CAT are in honor for three mathematicians include E. Cartan, A.D. Alexandrov and V.A. Toponogov, who have made important contributions to the understanding of curvature via inequalities for the distance function. A metric space X is a CAT(0) space if it is geodesically connected and if every geodesic triangle in X is at least as "*thin*" as its comparison triangle in the Euclidean plane. It is well known that any complete, simply connected Riemannian manifold having non-positive sectional curvature is a CAT(0) space. Kirk ([2,3]) first studied the theory of fixed point in CAT(κ) spaces. Later on, many authors generalized the notion of CAT(κ) given in [2,3], mainly focusing on CAT(0) spaces (see e.g., [4–13]). In CAT(0) spaces, they also modified the process (3) and studied strong and Δ-convergence of the S-iteration as follows: $x_1 \in C$ and

$$\begin{cases} x_{n+1} = (1 - \alpha_n)T x_n \oplus \alpha_n T y_n, \\ y_n = (1 - \beta_n)x_n \oplus \beta_n T x_n \end{cases} \tag{4}$$

for each $n \in \mathbb{N}$, where $\{\alpha_n\}$ and $\{\beta_n\}$ are sequences in (0,1).

For the case of some generalized nonexpansive mappings, Kumam, Saluja and Nashine [14] introduced modified S-iteration process and proved existence and convergence theorems in CAT(0) spaces for two mappings which is wider than that of asymptotically nonexpansive mappings as follows:

$$\begin{cases} x_1 \in K, \\ x_{n+1} = (1-\alpha_n)T^n x_n \oplus \alpha_n S^n(y_n), \\ y_n = (1-\beta_n)x_n \oplus \beta_n T^n(x_n), n \in \mathbb{N}, \end{cases} \tag{5}$$

where the sequences $\{\alpha_n\}$ and $\{\beta_n\}$ are in [0, 1], for all $n \geq 1$.

Very recently, Kumam et al. [15] introduce new type iterative scheme called a modified Picard-S hybrid iterative algorithm as follows

$$\begin{cases} x_1 \in C, \\ w_n = (1-\alpha_n)x_n \oplus \alpha_n T^n(x_n), \\ y_n = (1-\beta_n)T^n x_n \oplus \beta_n T^n(w_n), \\ x_{n+1} = T^n y_n \end{cases} \tag{6}$$

for all $n \geq 1$, where $\{\alpha_n\}$ and $\{\beta_n\}$ are real appropriate sequences in the interval $[0, 1]$. They prove Δ-convergence and strong convergence of the iterative (6) under suitable conditions for total asymptotically nonexpansive mappings in CAT(0) spaces. Various results for solving a fixed point problem of some nonlinear mappings in the CAT(0) spaces can also be found, for examples, in [16–27].

On the other hand, let (X, d) be a geodesic metric space and f be a proper and convex function from the set X to $(-\infty, \infty]$. The major problem in optimization is to find $x \in X$ such that

$$f(x) = \min_{y \in X} f(y).$$

The set of minimizers of f was denoted by $\arg\min_{y \in X} f(y)$. In 1970, Martinet [28] first introduced the effective tool for solving this problem which is the proximal point algorithm (for short term, the PPA). Later in 1976, Rockafellar [29] found that the PPA converges to the solution of the convex problem in Hilbert spaces.

Let f be a proper, convex, and lower semi-continuous function on a Hilbert space H which attains its minimum. The PPA is defined by $x_1 \in H$ and

$$x_{n+1} = \arg\min_{y \in H} \left\{ f(y) + \frac{1}{2\lambda_n} \parallel y - x_n \parallel^2 \right\}$$

for each $n \in \mathbb{N}$, where $\lambda_n > 0$ for all $n \in \mathbb{N}$. It was proved that the sequence $\{x_n\}$ converges weakly to a minimizer of f provided $\sum_{n=1}^{\infty} \lambda_n = \infty$. However, as shown by Guler [30], the PPA does not necessarily converges strongly in general. In 2000, Kamimura-Takahashi [31] combined the PPA with Halpern's algorithm [32] so that the strong convergence is guaranteed (see also [33–36]).

In 2013, Bačák [37] introduced the PPA in a CAT(0) space (X, d) as follows: $x_1 \in X$ and

$$x_{n+1} = \arg\min_{y \in X} \left\{ f(y) + \frac{1}{2\lambda_n} d^2(y, x_n) \right\}$$

for each $n \in \mathbb{N}$, where $\lambda_n > 0$ for all $n \in \mathbb{N}$. Based on the concept of the Fejér monotonicity, it was shown that, if f has a minimizer and $\Sigma_{n=1}^{\infty} \lambda_n = \infty$, then the sequence $\{x_n\}$ Δ-converges to its minimizer (see also [37]). Recently, in 2014,

Bačák [38] employed a split version of the PPA for minimizing a sum of convex functions in complete CAT(0) spaces. Other interesting results can also be found in [37,39,40].

Recently, many convergence results by the PPA for solving optimization problems have been extended from the classical linear spaces such as Euclidean spaces, Hilbert spaces and Banach spaces to the setting of manifolds [40–43]. The minimizers of the objective convex functionals in the spaces with nonlinearity play a crucial role in the branch of analysis and geometry. Numerous applications in computer vision, machine learning, electronic structure computation, system balancing and robot manipulation can be considered as solving optimization problems on manifolds (see in [44–47]).

Very recently, Cholamjiak et al. [48] introduce a new modified proximal point algorithm involving fixed point iteration of nonexpansive mappings in CAT(0) spaces as follows

$$\begin{cases} z_n = \arg\min_{y\in X}\{f(y) + \frac{1}{2\lambda_n}d^2(y, x_n)\}, \\ y_n = (1-\beta_n)x_n \oplus \beta_n T_1 z_n, \\ x_{n+1} = (1-\alpha_n)T_1 \oplus \alpha_n T_2 y_n \end{cases} \tag{7}$$

for all $n \geq 1$, where $\{\alpha_n\}$ and $\{\beta_n\}$ are real sequences in the interval $[0, 1]$.

Motivated and inspired by (6) and (7), we introduce a new type iterative scheme called modified Picard-S hybrid which is defined by the following manner:

$$\begin{cases} z_n = \arg\min_{y\in X}\{f(y) + \frac{1}{2\lambda_n}d^2(y, x_n)\}, \\ w_n = (1-a_n)x_n \oplus a_n R z_n, \\ y_n = (1-b_n)R x_n \oplus b_n S w_n, \\ x_{n+1} = S y_n \end{cases} \tag{8}$$

for all $n \geq 1$, where $\{a_n\}$ and $\{b_n\}$ are real appropriate sequences in the interval $[0, 1]$.

The propose in this paper, we introduce a proximal point algorithm involving hybrid iteration (8) for nonexpansive mappings in non-positive curvature metric spaces namely CAT(0) spaces and also prove that the sequence generated by this algorithm converges to a minimizer of a convex function and common fixed point of such mappings.

2 Preliminaries

Let (X, d) be a metric space. A geodesic path joining $x \in X$ to $y \in X$ is a mapping γ from $[0, l] \subset \mathbb{R}$ to X such that $\gamma(0) = x, \gamma(l) = y$, and $d(\gamma(t), \gamma(t')) = |t - t'|$ for all $t, t' \in [0, l]$. Especially, γ is an isometry and $d(x, y) = l$. The image $\gamma([0, l])$ of γ is called a geodesic segment joining x and y.

A geodesic triangle $\Delta(x_1, x_2, x_3)$ in a geodesic metric (X, d) consist of three points x_1, x_2, x_3 in X and a geodesic segment between each pair of vertices. A comparison triangle for the geodesic triangle $\Delta(x_1, x_2, x_3)$ in (X, d)

is a triangle $\bar{\Delta}(x_1, xx_2, x_3) := \Delta(\bar{x_1}, \bar{x_2}, \bar{x_3})$ is Euclidean space $\mathbb{R}^2$ such that $d_{\mathbb{R}^2}(\bar{x_i}, \bar{x_j}) = d(x_i, x_j)$ for each $i, j \in \{1, 2, 3\}$. A geodesic space is called a CAT(0) space if, for each geodesic triangle $\Delta(x_1, x_2, x_3)$ in X and its comparison triangle $\bar{\Delta}(x_1, x_2, x_3) := \Delta(\bar{x_1}, \bar{x_2}, \bar{x_3})$ in $\mathbb{R}^2$, the CAT(0) inequality

$$d(x, y) \leq d_{\mathbb{R}^2}(\bar{x}, \bar{y})$$

is satisfied for all $x, y \in \Delta$ and comparison points $\bar{x}, \bar{y} \in \bar{\Delta}$. A subset C of a CAT(0) space is called convex if $[x, y] \subset C$ for all $x, y \in C$. For more details, the readers may consult [49]. A geodesic space X is a CAT(0) space if and only if

$$d^2((1-\alpha))x \oplus \alpha y, z) \leq (1-\alpha)d^2(x, z) + \alpha d^2(y, z) - t(1-\alpha)d^2(x, y) \tag{9}$$

for all $x, y, z \in X$ and $\alpha \in [0, 1]$ [50]. In particular, if x, y, z are points in X and $\alpha \in [0, 1]$, then we have

$$d((1-\alpha)x \oplus \alpha y, z) \leq (1-\alpha)d(x, z) + \alpha d(y, z). \tag{10}$$

The examples of CAT(0) spaces are Euclidean spaces $\mathbb{R}^n$, Hilbert spaces, simply connected Riemannian manifolds of nonpositive sectional curvature, hyperbolic spaces and R-trees.

Let C be a nonempty closed and convex subset of a complete CAT(0) space. Then, for each point $x \in X$, there exists a unique point of C denoted by $P_c x$, such that

$$d(x, P_c x) = \inf_{y \in C} d(x, y).$$

A mapping P_c is said to be the metric projection from X onto C.

Let $\{x_n\}$ be a bounded sequence in the set C. For any $x \in X$, we set

$$r(x, \{x_n\}) = \limsup_{n \to \infty} d(x, x_n).$$

The asymptotic radius $r(\{x_n\})$ of $\{x_n\}$ is given by

$$r(\{x_n\}) = \inf\{r(x, \{x_n\}) : x \in X\}$$

and the asymptotic center $A(\{x_n\})$ of $\{x_n\}$ is the set

$$A(\{x_n\}) = \{x \in X : r(\{x_n\}) = r(x, \{x_n\})\}.$$

In CAT(0) space, $A(\{x_n\})$ consists of exactly one point (see in [51]).

Definition 1. A sequence $\{x_n\}$ in a CAT(0) space X is called Δ-convergent to a point $x \in X$ if x is the unique asymptotic center of $\{u_n\}$ for every subsequence $\{u_n\}$ of $\{x_n\}$.

We can write $\Delta - \lim_{n\to\infty} x_n = x$ and call x the Δ-limit of $\{x_n\}$. We denote $w_\Delta(x_n) := \cup\{A(\{u_n\})\}$, where the union is taken over all subsequences $\{u_n\}$ of $\{x_n\}$.

Recall that a bounded sequence $\{x_n\}$ in X is called regular if $r(\{x_n\}) = r(\{u_n\})$ for every subsequence $\{u_n\}$ of $\{x_n\}$. Every bounded sequence in X has a Δ-convergent subsequence [7].

Lemma 1. *[16] Let C be a closed and convex subset of a complete CAT(0) space X and $T : C \to C$ be a nonexpansive mapping. Let $\{x_n\}$ be a bounded sequence in C such that $\lim_{n\to\infty} d(x_n, Tx_n) = 0$ and $\Delta - \lim_{n\to\infty} x_n = x$. Then $x = Tx$.*

Lemma 2. *[16] If $\{x_n\}$ is a bounded sequence in a complete CAT(0) space with $A(\{x_n\}) = \{x\}, \{u_n\}$ is a sequence of $\{x_n\}$ with $A(\{u_n\}) = \{u\}$ and the sequence $\{d(x_n, u)\}$ converges, then $x = u$.*

Recall that a function $f : C \to (-\infty, \infty]$ define on the set C is convex if, for any geodesic $\gamma : [a, b] \to C$, the function $f \circ \gamma$ is convex. We say that a function f defined on C is lower semi-continuous at a point $x \in C$ if

$$f(x) \leq \liminf_{n\to\infty} f(x_n)$$

for each sequence $x_n \to x$. A function f is called lower semi-continuous on C if it is lower semi-continuous at any point in C.
For any $\lambda > 0$, define the Moreau-Yosida resolvent of f in CAT(0) spaces as

$$J_\lambda(x) = \arg\min_{y\in X}\{f(y) + \frac{1}{2\lambda} d^2(y, x)\} \tag{11}$$

for all $x \in X$. The mapping J_λ is well define for all $\lambda > 0$ (see in [52,53]).

Let $f : X \to (-\infty, \infty]$ be a proper convex and lower semi-continuous function. It was shown in [38] that the set $F(j_\lambda)$ of fixed points of the resolvent associated with f coincides with the set $\arg\min_{y\in X} f(y)$ of minimizers of f.

Lemma 3. *[52] Let (X, d) be a complete CAT(0) space and $f : X \to (-\infty, \infty]$ be proper convex and lower semi-continuous. For any $\lambda > 0$, the resolvent J_λ of f is nonexpansive.*

Lemma 4. *[54] Let (X, d) be a complete CAT(0) space and $f : X \to (-\infty, \infty]$ be proper convex and lower semi-continuous. Then, for all $x, y \in X$ and $\lambda > 0$, we have*

$$\frac{1}{2\lambda} d^2(J_\lambda x, y) - \frac{1}{2\lambda} d^2(x, y) + \frac{1}{2\lambda} d^2(x, J_\lambda x) + f(J_\lambda x) \leq f(y).$$

Proposition 1. *[52, 53] (The resolvent identity) Let (X, d) be a complete CAT(0) space and $f : X \to (-\infty, \infty]$ be proper convex and lower semi-continuous. Then the following identity holds:*

$$J_\lambda x = J_\mu(\frac{\lambda - \mu}{\lambda} J_\lambda x \oplus \frac{\mu}{\lambda} x)$$

for all $x \in X$ and $\lambda > \mu > 0$.

For more results in CAT(0) spaces, refer to [55].

3 The Main Results

We now establish and prove our main results.

Theorem 1. *Let (X, d) be a complete CAT(0) space and $f : X \to (-\infty, \infty]$ be a proper, convex and lower semi-continuous function. Let R, S are two nonexpansive mappings such that $\omega = F(R) \cap F(S) \cap argmin_{y\in X} f(y) \neq \emptyset$. Suppose $\{a_n\}$ and $\{b_n\}$ are sequences that $0 < a \leq a_n, b_n \leq b < 1$ for all $n \in \mathbb{N}$ and for some a, b, $\{\lambda_n\}$ be a sequence that $\lambda_n \geq \lambda > 0$ for all $n \in \mathbb{N}$ and for some λ. Let sequence $\{x_n\}$ is defined by (8) for each $n \in \mathbb{N}$. Then the sequence $\{x_n\}$ Δ-converges to common element of ω.*

Proof. Let $q^* \in \omega$. Then $Rq^* = Sq^* = Tq^* = q^*$ and $f(q^*) \leq f(y)$ for all $y \in X$. It follows that

$$f(q^*) + \frac{1}{2\lambda_n} d^2(q^*, q^*) \leq f(y) + \frac{1}{2\lambda_n} d^2(y, q^*) \ \forall y \in X$$

thus $q^* = J_{\lambda_n} q^*$ for all $n \geq 1$.

First, we will prove that $\lim_{n\to\infty} d(x_n, q^*)$ exists. Setting $z_n = J_{\lambda_n} x_n$ for all $n \geq 1$, by Lemma 2.4,

$$d(z_n, q^*) = d(J_{\lambda_n} x_n, J_{\lambda_n} q^*) \leq d(x_n, q^*). \tag{12}$$

Also,it follows form (10) and (12) we have

$$\begin{aligned} d(w_n, q^*) &= d((1 - a_n)x_n \oplus a_n R z_n, q^*) \\ &\leq (1 - a_n)d(x_n, q^*) + a_n d(Rz_n, q^*) \\ &\leq (1 - a_n)d(x_n, q^*) + a_n d(z_n, q^*) \\ &\leq d(x_n, q^*), \end{aligned} \tag{13}$$

and

$$\begin{aligned} d(y_n, q^*) &= d((1 - b_n)R_{x_n} \oplus b_n S w_n, q^*) \\ &\leq (1 - b_n)d(Rx_n, q^*) + b_n d(Sw_n, q^*) \\ &\leq (1 - b_n)d(x_n, q^*) + b_n d(w_n, q^*) \\ &\leq (1 - b_n)d(x_n, q^*) + b_n d(x_n, q^*) \\ &= d(x_n, q^*). \end{aligned} \tag{14}$$

Hence, by (13) and (14), we get

$$\begin{aligned} d(x_{n+1}, q^*) &= d(Sy_n, q^*) \\ &\leq d(y_n, q^*) \\ &\leq d(w_n, q^*) \\ &\leq d(x_n, q^*). \end{aligned} \tag{15}$$

This shows that $\lim_{n\to\infty} d(x_n, q^*)$ exists. Therefore $\lim_{n\to\infty} d(x_n, q^*) = k$ for some k.

Next, we will prove that $\lim_{n\to\infty} d(x_n, z_n) = 0$. By Lemma 2.5, we see that

$$\frac{1}{2\lambda_n} d^2(z_n, q^*) - \frac{1}{2\lambda_n} d^2(x_n, q^*) + \frac{1}{2\lambda_n} d^2(x_n, z_n) \leq f(q^*) - f(z_n).$$

Since $f(q) \leq f(z_n)$ for all $n \geq 1$, it follows that

$$d^2(x_n, z_n) \leq d^2(x_n, q^*) - d^2(z_n, q^*).$$

In order to show that $\lim_{n\to\infty} d(x_n, z_n) = 0$, it suffices to prove that

$$\lim_{n\to\infty} d(z_n, q^*) = k.$$

In fact, from (15), we have

$$\begin{aligned} d(x_{n+1}, q^*) &\leq d(y_n, q^*) \\ &\leq (1 - b_n) d(x_n, q^*) + b_n d(w_n, q^*), \end{aligned}$$

which implies that

$$\begin{aligned} d(x_n, q^*) &\leq \frac{1}{b_n}(d(x_n, q^*) - d(x_{n+1}, q^*)) + d(w_n, q^*) \\ &\leq \frac{1}{b}(d(x_n, q^*) - d(x_{n+1}, q^*)) + d(w_n, q^*), \end{aligned}$$

since $d(x_{n+1}, q^*) \leq d(x_n, q^*)$ and $b_n \geq b > 0$ for all $n \geq 1$. Thus we have

$$k = \liminf_{n\to\infty} d(x_n, q^*) \leq \liminf_{n\to\infty} d(w_n, q^*).$$

On the other hand, by (13), we observe that

$$\limsup_{n\to\infty} d(w_n, q^*) \leq \limsup_{n\to\infty} d(x_n, q^*) = k.$$

So, we get $\lim_{n\to\infty} d(w_n, q^*) = c$. Also, by (13), we have

$$\begin{aligned} d(x_n, q^*) &\leq \frac{1}{a_n}(d(x_n, q^*) - d(w_n, q^*)) + d(z_n, q^*) \\ &\leq \frac{1}{a}(d(x_n, q^*) - d(w_n, q^*)) + d(z_n, q^*), \end{aligned}$$

which yields

$$k = \liminf_{n\to\infty} d(x_n, q^*) \leq \liminf_{n\to\infty} d(z_n, q^*).$$

From (12) and (15), we obtain

$$\lim_{n\to\infty} d(z_n, q^*) = k.$$

We conclude that

$$\lim_{n\to\infty} d(x_n, z_n) = 0. \tag{16}$$

Next, we will prove that

$$\lim_{n\to\infty} d(x_n, Rx_n) = \lim_{n\to\infty} d(x_n, Sx_n) = 0.$$

We observe that

$$\begin{aligned} d^2(w_n, q^*) &= d^2((1-a_n)x_n \oplus a_n Rz_n, q^*) \\ &\leq (1-a_n)d^2(x_n, q^*) + a_n d^2(Rz_n, q^*) - a_n(1-a_n)d^2(x_n, Rz_n) \\ &\leq d^2(x_n, q^*) - a(1-b)d^2(x_n, Sz_n), \end{aligned}$$

which implies that

$$\begin{aligned} d^2(x_n, Rz_n) &\leq \frac{1}{a(1-b)}(d^2(x_n, q^*) - d^2(w_n, q^*)) \\ &\to 0 \ as \ n \to \infty. \end{aligned} \tag{17}$$

Thus,

$$\lim_{n\to\infty} d(x_n, Rz_n) = 0.$$

It follows from (16) and (17) that

$$\begin{aligned} d(x_n, Rx_n) &\leq d(x_n, Rz_n) + d(Rz_n, Rx_n) \\ &\leq d(x_n, Rz_n) + d(z_n, x_n) \\ &\to 0 \ as \ n \to \infty. \end{aligned} \tag{18}$$

In the same way, it follows from

$$\begin{aligned} d^2(y_n, q^*) &= d^2((1-b_n)Rx_n \oplus b_n Sw_n, q^*) \\ &\leq (1-b_n)d^2(Rx_n, q^*) + b_n d^2(Sw_n, q^*) - b_n(1-b_n)d^2(Rx_n, Sw_n) \\ &\leq d^2(x_n, q^*) - a(1-b)d^2(Rx_n, Sw_n) \end{aligned}$$

which implies

$$\begin{aligned} d^2(Rx_n, Sw_n) &\leq \frac{1}{a(1-b)}(d^2(x_n, q^*) - d^2(y_n, q^*)) \\ &\to 0 \ as \ n \to \infty. \end{aligned}$$

Hence

$$\lim_{n\to\infty} d(Rx_n, Sw_n) = 0. \tag{19}$$

We get

$$\begin{aligned} d(w_n, x_n) &= a_n d(Rz_n, x_n) \\ &\to 0 \ as \ n \to \infty. \end{aligned} \tag{20}$$

By (19) and (20), we obtain

$$\begin{aligned} d(x_n, Sx_n) &\leq d(x_n, Rx_n) + d(Rx_n, Sw_n) + d(Sw_n, Sx_n) \\ &\leq d(x_n, Rx_n) + d(Rx_n, Sw_n) + d(w_n, x_n) \\ &\to 0 \ as \ n \to \infty. \end{aligned}$$

Next, we will show that $\lim_{n\to\infty} d(x_n, J_{\lambda_n} x_n) = 0$.
Since $\lambda_n \geq \lambda > 0$, by (16) and Proposition 2.6,

$$\begin{aligned} d(J_\lambda x_n, J_{\lambda_n} x_n) &= d(J_\lambda x_n, J_\lambda(\frac{\lambda_n - \lambda}{\lambda_n} J_{\lambda_n} x_n \oplus \frac{\lambda}{\lambda_n} x_n)) \\ &\leq d(x_n, (1 - \frac{\lambda}{\lambda_n}) J_{\lambda_n} x_n \oplus \frac{\lambda}{\lambda_n} x_n) \\ &= (1 - \frac{\lambda}{\lambda_n}) d(x_n, z_n) \\ &\to 0 \end{aligned}$$

as $n \to \infty$.

Next, we show that $W_\Delta(x_n) \subset \omega$.

Let $u \in W_\Delta(x_n)$. Then there exists a subsequence $\{u_n\}$ of $\{x_n\}$ such that asymptotic center of $A(\{u_n\}) = \{u\}$. From Lemma 2.2, there exists a subsequence $\{v_n\}$ of $\{u_n\}$ such that $\Delta - \lim_{n\to\infty} v_n = v$ for some $v \in \omega$. So, $u = v$ by Lemma 2.3. This shows that $W_\Delta(x_n) \subset \omega$.

Finally, we will show that the sequence $\{x_n\}$ Δ-converges to a point in ω. It need to prove that $W_\Delta(x_n)$ consists of exactly one point. Let $\{u_n\}$ be a subsequence of $\{x_n\}$ with $A(\{u_n\}) = \{u\}$ and let $A(\{x_n\}) = \{x\}$. Since $u \in W_\Delta(x_n) \subset \omega$ and $\{d(x_n, u)\}$ converges, by Lemma 2.3, we have $x = u$. Hence $w_\Delta(x_n) = \{x\}$. This completes the proof.

If $R = S$ in Theorem 1 we obtain the following result.

Corollary 1. *Let (X, d) be a complete CAT(0) space and $f : X \to (-\infty, \infty]$ be a proper, convex and lower semi-continuous function. Let R be a nonexpansive mappings such that $\omega = F(R) \cap argmin_{y\in X} f(y) \neq \emptyset$. Suppose $\{a_n\}$ and $\{b_n\}$ are sequences that $0 < a \leq a_n, b_n \leq b < 1$ for all $n \in \mathbb{N}$ and for some a, b, $\{\lambda_n\}$ be a sequence that $\lambda_n \geq \lambda > 0$ for all $n \in \mathbb{N}$ and for some λ. Let sequence $\{x_n\}$ is defined by (8) for each $n \in \mathbb{N}$. Then the sequence $\{x_n\}$ Δ-converges to common element of ω.*

Since every Hilbert space is a complete CAT(0) space, we obtain following result immediately.

Corollary 2. *Let H be a Hilbert space and $f : H \to (-\infty, \infty]$ be a proper, convex and lower semi-continuous function. Let R, S are two nonexpansive mappings such that $\omega = F(R \cap S) \cap argmin_{y\in H} f(y) \neq \emptyset$. Suppose $\{a_n\}$ and $\{b_n\}$ are sequences that $0 < a \leq a_n, b_n \leq b < 1$ for all $n \in \mathbb{N}$ and for some a, b, $\{\lambda_n\}$*

be a sequence that $\lambda_n \geq \lambda > 0$ *for all* $n \in \mathbb{N}$ *and for some* λ. *Let sequence* $\{x_n\}$ *is defined by:*

$$\begin{cases} z_n = \arg\min_{y \in H}\{f(y) + \frac{1}{2\lambda_n} \parallel y - x_n \parallel^2\}, \\ w_n = (1 - a_n)x_n + a_n R z_n, \\ y_n = (1 - b_n)Rx_n + b_n S w_n, \\ x_{n+1} = S y_n \end{cases}$$

for each $n \in \mathbb{N}$. *Then the sequence* $\{x_n\}$ *weakly converges to common element of* ω.

Next, Under mild condition, we establish strong convergence theorem.

A self mapping T is said to be *semi-compact* if any sequence $\{x_n\}$ satisfying $d(x_n, Tx_n) \to 0$ has a convergent subsequence.

Theorem 2. *Let* (X, d) *be a complete CAT(0) space and* $f : X \to (-\infty, \infty]$ *be a proper, convex and lower semi-continuous function. Let* R, S *are two nonexpansive mappings such that* $\omega = F(R \cap S) \cap argmin_{y \in X} f(y) \neq \emptyset$. *Suppose* $\{a_n\}$ *and* $\{b_n\}$ *are sequences that* $0 < a \leq a_n, b_n \leq b < 1$ *for all* $n \in \mathbb{N}$ *and for some* a, b, $\{\lambda_n\}$ *be a sequence that* $\lambda_n \geq \lambda > 0$ *for all* $n \in \mathbb{N}$ *and for some* λ. *If* R *or* S, *or* J_λ *is semi-compact, then the sequence* $\{x_n\}$ *generated by* (8) *strongly converges to a common element of* ω.

Proof. Suppose that R is semi-compact. By step 3 of Theorem 1, we have

$$d(x_n, Rx_n) \to 0$$

as $n \to \infty$. Thus, there exists a subsequence $\{x_{n_k}\}$ of $\{x_n\}$ such that $x_{n_k} \to \hat{x} \in X$. Again by Theorem 1, we have $d(\hat{x}, J_\lambda \hat{x}) = 0$, and $d(\hat{x}, R\hat{x}) = d(\hat{x}, S\hat{x}) = 0$, which shows that $\hat{x} \in \omega$. For other cases, we can prove the strong convergence of $\{x_n\}$ to a common element of ω. This completes the proof.

Acknowledgements. The first author was supported by Rajamangala University of Technology Lanna (RMUTL). The second author was financial supported by RMUTT annual government statement of expenditure in 2018 and the National Research Council of Thailand (NRCT) for fiscal year of 2018 (Grant no. 2561A6502439) was gratefully acknowledged.

References

1. Agarwal, R.P., O'Regan, D., Sahu, D.R.: Iterative construction of fixed points of nearly asymptotically nonexpansive mappings. J. Nonlinear Convex. Anal. **8**(1), 61–79 (2007)
2. Kirk, W.A.: Geodesic geometry and fixed point theory In: Seminar of Mathematical Analysis (Malaga/Seville,2002/2003). Colecc. Abierta. Univ. Sevilla Secr. Publ. Seville., vol. 64, pp. 195–225 (2003)
3. Kirk, W.A.: Geodesic geometry and fixed point theory II. In: International Conference on Fixed Point Theory and Applications, pp. 113–142. Yokohama Publications, Yokohama (2004)

4. Dhompongsa, S., Kaewkhao, A., Panyanak, B.: Lim's theorems for multivalued mappings in CAT(0) spaces. J. Math. Anal. Appl. **312**, 478–487 (2005)
5. Chaoha, P., Phon-on, A.: A note on fixed point sets in CAT(0) spaces. J. Math. Anal. Appl. **320**, 983–987 (2006)
6. Leustean, L.: A quadratic rate of asymptotic regularity for CAT(0) spaces. J. Math. Anal. Appl. **325**, 386–399 (2007)
7. Kirk, W.A., Panyanak, B.: A concept of convergence in geodesic spaces. Nonlinear Anal. **68**, 3689–3696 (2008)
8. Shahzad, N., Markin, J.: Invariant approximations for commuting mappings in CAT(0) and hyperconvex spaces. J. Math. Anal. Appl. **337**, 1457–1464 (2008)
9. Saejung, S.: Halpern's iteration in CAT(0) spaces, Fixed Point Theory Appl. (2010). Article ID 471781
10. Cho, Y.J., Ciric, L., Wang, S.: Convergence theorems for nonexpansive semigroups in CAT(0) spaces. Nonlinear Anal. **74**, 6050–6059 (2011)
11. Abkar, A., Eslamian, M.: Common fixed point results in CAT(0) spaces. Nonlinear Anal. **74**, 1835–1840 (2011)
12. Shih-sen, C., Lin, W., Heung, W.J.L., Chi-kin, C.: Strong and Δ-convergence for mixed type total asymptotically nonexpansive mappings in CAT(0) spaces. Fixed Point Theory Appl. 122 (2013)
13. Jinfang, T., Shih-sen, C.: Viscosity approximation methods for two nonexpansive semigroups in CAT(0) spaces. Fixed Point Theory Appl. 122 (2013)
14. Kumam, P., Saluja, G.S., Nashine, H.K.: Convergence of modified S-iteration process for two asymptotically nonexpansive mappings in the intermediate sense in CAT(0) spaces. J. Inequalities Appl. 368 (2014)
15. Kumam, W., Pakkaranang, N., Kumam, P., Cholamjiak, P.: Convergence analysis of modified Picard-S hybrid iterative algorithms for total asymptotically nonexpansive mappings in Hadamard spaces. Int. J. Comput. Math. (2018). https://doi.org/10.1080/00207160.2018.1476685
16. Dhompongsa, S., Panyanak, B.: On Δ-convergence theorems in CAT(0) spaces. Comput. Math. Appl. **56**, 2572–2579 (2008)
17. Khan, S.H., Abbas, M.: Strong and Δ-convergence of some iterative schemes in CAT(0) spaces. Comput. Math. Appl. **61**, 109–116 (2011)
18. Chang, S.S., Wang, L., Lee, H.W.J., Chan, C.K., Yang, L.: Demiclosed principle and Δ-convergence theorems for total asymptotically nonexpansive mappings in CAT(0) spaces. Appl. Math. Comput. **219**, 2611–2617 (2012)
19. Cho, Y.J., Ćirić, L., Wang, S.: Convergence theorems for nonexpansive semigroups in CAT(0) spaces. Nonlinear Anal. **74**, 6050–6059 (2011)
20. Cuntavepanit, A., Panyanak, B.: Strong convergence of modified Halpern iterations in CAT(0) spaces. Fixed Point Theory Appl. (2011). Article ID 869458
21. Fukhar-ud-din, H.: Strong convergence of an Ishikawa-type algorithm in CAT(0) spaces. Fixed Point Theory Appl. 207 (2013)
22. Laokul, T., Panyanak, B.: Approximating fixed points of nonexpansive mappings in CAT(0) spaces. Int. J. Math. Anal. **3**, 1305–1315 (2009)
23. Laowang, W., Panyanak, B.: Strong and Δ-convergence theorems for multivalued mappings in CAT(0) spaces. J. Inequal. Appl. (2009). Article ID 730132
24. Nanjaras, B., Panyanak, B.: Demiclosed principle for asymptotically nonexpansive mappings in CAT(0) spaces. Fixed Point Theory Appl. (2010). Article ID 268780
25. Phuengrattana, W., Suantai, S.: Fixed point theorems for a semigroup of generalized asymptotically nonexpansive mappings in CAT(0) spaces. Fixed Point Theory Appl. **2012**, 230 (2012)

26. Saejung, S.: Halpern's iteration in CAT(0) spaces. Fixed Point Theory Appl. (2010). Article ID 471781
27. Shi, L.Y., Chen, R.D., Wu, Y.J.: Δ-Convergence problems for asymptotically nonexpansive mappings in CAT(0) spaces. Abstr. Appl. Anal. (2013). Article ID 251705
28. Martinet, B.: Régularisation d'inéuations variationnelles par approximations successives. Rev. Fr. Inform. Rech. Oper. **4**, 154–158 (1970)
29. Rockafellar, R.T.: Monotone operators and the proximal point algorithm. SIAM J. Control Optim. **14**, 877–898 (1976)
30. Guler, O.: On the convergence of the proximal point algorithm for convex minimization. SIAM J. Control Optim. **29**, 403–419 (1991)
31. Kamimura, S., Takahashi, W.: Approximating solutions of maximal monotone operators in Hilbert spaces. J. Approx. Theory **106**, 226–240 (2000)
32. Halpern, B.: Fixed points of nonexpanding maps. Bull. Am. Math. Soc. **73**, 957–961 (1967)
33. Boikanyo, O.A., Morosanu, G.: A proximal point algorithm converging strongly for general errors. Optim. Lett. **4**, 635–641 (2010)
34. Marino, G., Xu, H.K.: Convergence of generalized proximal point algorithm. Commun. Pure Appl. Anal. **3**, 791–808 (2004)
35. Xu, H.K.: A regularization method for the proximal point algorithm. J. Glob. Optim. **36**, 115–125 (2006)
36. Yao, Y., Noor, M.A.: On convergence criteria of generalized proximal point algorithms. J. Comput. Appl. Math. **217**, 46–55 (2008)
37. Bacak, M.: The proximal point algorithm in metric spaces. Isr. J. Math. **194**, 689–701 (2013)
38. Ariza-Ruiz, D., Leuştean, L., López, G.: Firmly nonexpansive mappings in classes of geodesic spaces. Trans. Am. Math. Soc. **366**, 4299–4322 (2014)
39. Bacak, M.: Computing medians and means in Hadamard spaces. SIAM J. Optim. **24**, 1542–1566 (2014)
40. Ferreira, O.P., Oliveira, P.R.: Proximal point algorithm on Riemannian manifolds. Optimization **51**, 257–270 (2002)
41. Li, C., López, G., Martín-Márquez, V.: Monotone vector fields and the proximal point algorithm on Hadamard manifolds. J. Lond. Math. Soc. **79**, 663–683 (2009)
42. Papa Quiroz, E.A., Oliveira, P.R.: Proximal point methods for quasiconvex and convex functions with Bregman distances on Hadamard manifolds. J. Convex Anal. **16**, 49–69 (2009)
43. Wang, J.H., L ápez, G.: Modified proximal point algorithms on Hadamard manifolds. Optimization **60**, 697–708 (2011)
44. Adler, R., Dedieu, J.P., Margulies, J.Y., Martens, M., Shub, M.: Newton's method on Riemannian manifolds and a geometric model for human spine. IMA J. Numer. Anal. **22**, 359–390 (2002)
45. Smith, S.T.: Optimization techniques on Riemannian manifolds, Hamiltonian and Gradient Flows, Algorithms and Control. Fields Inst. Commun. **3**, 113–136 (1994). Am. Math. Soc., Providence
46. Udriste, C.: Convex Functions and Optimization Methods on Riemannian Manifolds. 297. Mathematics and Its Applications. Kluwer Academic, Dordrecht (1994)
47. Wang, J.H., Li, C.: Convergence of the family of Euler-Halley type methods on Riemannian manifolds under the γ-condition. Taiwan. J. Math. **13**, 585–606 (2009)
48. Cholamjiak, P., Abdou, A., Cho, Y.J.: Proximal point algorithms involving fixed points of nonexpansive mappings in CAT(0) spaces. Fixed Point Theory Appl. 227 (2015)

49. Bridson, M.R., Haefliger, A.: Metric Spaces of Non-positive Curvature. Grundelhren der Mathematischen. Springer, Heidelberg (1999)
50. Bruhat, M., Tits, J.: Groupes réductifs sur un corps local: I. Données radicielles valuées. Publ. Math. Inst. Hautes Études Sci. **41**, 5–251 (1972)
51. Dhompongsa, S., Kirk, W.A., Sims, B.: Fixed points of uniformly Lipschitzian mappings. Nonlinear Anal. **65**, 762–772 (2006)
52. Jost, J.: Convex functionals and generalized harmonic maps into spaces of nonpositive curvature. Comment. Math. Helv. **70**, 659–673 (1995)
53. Mayer, U.F.: Gradient flows on nonpositively curved metric spaces and harmonic maps. Commun. Anal. Geom. **6**, 199–253 (1998)
54. Ambrosio, L., Gigli, N., Savare, G.: Gradient Flows in Metric Spaces and in the Space of Probability Measures. Lectures in Mathematics ETH Zurich, 2nd edn. Birkhauser, Basel (2008)
55. Bacak, M.: Convex Analysis and Optimization in Hadamard Spaces. de Gruyter, Berlin (2014)

New Ciric Type Rational Fuzzy F-Contraction for Common Fixed Points

Aqeel Shahzad[1], Abdullah Shoaib[1], Konrawut Khammahawong[2,3], and Poom Kumam[2,3](✉)

[1] Department of Mathematics and Statistics, Riphah International University, Islamabad 44000, Pakistan
aqeel4all84@gmail.com, abdullahshoaib15@yahoo.com

[2] KMUTTFixed Point Research Laboratory, Department of Mathematics, Room SCL 802 Fixed Point Laboratory, Science Laboratory Building, Faculty of Science, King Mongkut's University of Technology Thonburi (KMUTT), 126 Pracha-Uthit Road, Bang Mod, Thrung Khru, Bangkok 10140, Thailand
k.konrawut@gmail.com, poom.kum@kmutt.ac.th

[3] KMUTT-Fixed Point Theory and Applications Research Group (KMUTT-FPTA), Theoretical and Computational Science Center (TaCS), Science Laboratory Building, Faculty of Science, King Mongkut's University of Technology Thonburi (KMUTT), 126 Pracha-Uthit Road, Bang Mod, Thrung Khru, Bangkok 10140, Thailand

Abstract. In this article, common fixed point theorems for a pair of fuzzy mappings satisfying a new Ciric type rational F-contraction in complete dislocated metric spaces have been established. An example has been constructed to illustrate this result. Our results combine, extend and infer several comparable results in the existing literature.

Mathematics Subject Classification: 46S40 · 47H10 · 54H25

1 Introduction and Mathematical Preliminaries

Let $R : X \to X$ be a mapping. If $u = Ru$ then u in X is called a *fixed point* of R. In various fields of applied mathematical analysis Banach's fixed point theorem [7] plays an important role. Its importance can be seen as several authors have obtained many interesting extensions of his result in various metric spaces ([1–29]). The idea of dislocated topology has been applied in the field of logic programming semantics [11]. Dislocated metric space (metric-like space) [11] is a generalization of partial metric space [18].

A new type of contraction called F-contraction was introduced by Wardowski [29] and proved a new fixed point theorem about F-contraction. Many fixed point results were generalized in different ways. Afterwards, Secelean [22] proved fixed point theorems about of F-contractions by iterated function systems. Piri et al. [20] proved a fixed point result for F-Suzuki contractions for some weaker conditions on the self map in a complete metric spaces. Acar et al. [3] introduced the concept of generalized multivalued F-contraction mappings and extended the

V. Kreinovich et al. (Eds.): ECONVN 2019, SCI 809, pp. 215–229, 2019.
https://doi.org/10.1007/978-3-030-04200-4_17

multivalued F-contraction with δ-Distance and established fixed point results in complete metric space [2]. Sgroi et al. [23] established fixed point theorems for multivalued F-contractions and obtained the solution of certain functional and integral equations, which was a proper generalization of some multivalued fixed point theorems including Nadler's theorem [19]. Many other useful results on F-contractions can be seen in [4,5,13,17].

Zadeh was the first who presented the idea of fuzzy sets [31]. Later on Weiss [30] and Butnariu [8] gave the idea of a fuzzy mapping and obtained many fixed point results. Afterward, Heilpern [10] initiated the idea of fuzzy contraction mappings and proved a fixed point theorem for fuzzy contraction mappings which is a fuzzy analogue of Nadler's [19] fixed point theorem for multivalued mappings.

In this paper, by the concept of F-contraction we obtain some common fixed point results for fuzzy mappings satisfying a new Ciric type rational F -contraction in the context of complete dislocated metric spaces. An example is also given which supports the our proved results.

Now, we give the following definitions and results which will be needed in the sequel. In this paper, we denote $\mathbb{R}$ and $\mathbb{R}^+$ by the set of real numbers and the set of non-negative real numbers, respectively.

Definition 1. [11] Let X be a nonempty set. A mapping $d_l : X \times X \to [0, \infty)$ is called a *dislocated metric* (or simply d_l-metric) if the following conditions hold, for any $x, y, z \in X$:

(i) If $d_l(x, y) = 0$, then $x = y$;
(ii) $d_l(x, y) = d_l(y, x)$;
(iii) $d_l(x, y) \leq d_l(x, z) + d_l(z, y)$.

Then, (X, d_l) is called *dislocated metric space* or d_l metric space.

It is clear that if $d_l(x, y) = 0$, then from (i), $x = y$. But if $x = y$, $d_l(x, y)$ may not be 0.

Example 1. [11] If $X = R^+ \cup \{0\}$, then $d_l(x, y) = x + y$ defines a dislocated metric d_l on X.

Definition 2. [11] Let (X, d_l) be a dislocated metric space, then

(i) A sequence $\{x_n\}$ in (X, d_l) is called a *Cauchy sequence* if given $\varepsilon > 0$, there exists $n_0 \in N$ such that for all $n, m \geq n_0$ we have $d_l(x_m, x_n) < \varepsilon$ or $\lim_{n,m \to \infty} d_l(x_n, x_m) = 0$.

(ii) A sequence $\{x_n\}$ dislocated-converges (for short d_l-converges) to x if $\lim_{n \to \infty} d_l(x_n, x) = 0$. In this case x is called a d_l-*limit* of $\{x_n\}$.

(iii) (X, d_l) is called *complete* if every Cauchy sequence in X converges to a point $x \in X$ such that $d_l(x, x) = 0$.

Definition 3. [25] Let K be a nonempty subset of dislocated metric space X and let $x \in X$. An element $y_0 \in K$ is called a *best approximation* in K if

$$d_l(x, K) = d_l(x, y_0), \text{where } d_l(x, K) = \inf_{y \in K} d_l(x, y).$$

If each $x \in X$ has at least one best approximation in K, then K is called a *proximinal set.* We denote $P(X)$ be the set of all closed proximinal subsets of X.

Definition 4. [25] The function $H_{d_l} : P(X) \times P(X) \to R^+$, defined by

$$H_{d_l}(A, B) = \max\{\sup_{a \in A} d_l(a, B), \sup_{b \in B} d_l(A, b)\}$$

is called *dislocated Hausdorff metric* on $P(X)$.

Definition 5. [29] Let (X, d_l) be a metric space. A mapping $T : X \to X$ is said to be an F*-contraction* if there exists $\tau > 0$ such that

$$d(Tx, Ty) > 0 \Rightarrow \tau + F(d(Tx, Ty)) \leq F(d(x, y)), \text{ for all } x, y \in X, \quad (1)$$

where $F : \mathbb{R}^+ \to \mathbb{R}$ is a mapping satisfying the following conditions:

(F1) F is strictly increasing, i.e. for all $x, y \in \mathbb{R}^+$ such that $x < y$, $F(x) < F(y)$;
(F2) For each sequence $\{\alpha_n\}_{n=1}^{\infty}$ of positive numbers, $\lim_{n \to \infty} \alpha_n = 0$ if and only if $\lim_{n \to \infty} F(\alpha_n) = -\infty$;
(F3) There exists $k \in (0, 1)$ such that $\lim_{\alpha \to 0^+} \alpha^k F(\alpha) = 0$.

We denote by $\triangle_F$, the set of all functions satisfying the conditions (F1)–(F3).

Example 2. [29] The family of $\triangle_F$ is not empty.

(1) $F(x) = \ln(x)$; for $x > 0$.
(2) $F(x) = x + \ln(x)$; for $x > 0$.
(3) $F(x) = \frac{-1}{\sqrt{x}}$; for $x > 0$.

A fuzzy set in X is a function with domain X and value in $[0, 1]$, $F(X)$ is the collection of all fuzzy sets in X. If A is a fuzzy set and $x \in X$, then the function value $A(x)$ is called the *grade of membership* of x in A. The α-level set of fuzzy set A, is denoted by $[A]_\alpha$, and defined as:

$$[A]_\alpha = \{x : A(x) \geq \alpha\} \text{ where } \alpha \in (0, 1],$$
$$[A]_0 = \overline{\{x : A(x) > 0\}}.$$

Let X be any nonempty set and Y be a metric space. A mapping T is called a *fuzzy mapping*, if T is a mapping from X into $F(Y)$. A fuzzy mapping T is a fuzzy subset on $X \times Y$ with membership function $T(x)(y)$. The function $T(x)(y)$ is the grade of membership of y in $T(x)$. For convenience, we denote the α-level set of $T(x)$ by $[Tx]_\alpha$ instead of $[T(x)]_\alpha$ [28].

Definition 6. [28] A point $x \in X$ is called a *fuzzy fixed point* of a fuzzy mapping $T: X \to F(X)$ if there exists $\alpha \in (0,1]$ such that $x \in [Tx]_\alpha$.

Lemma 1. [28] *Let A and B be nonempty proximal subsets of a dislocated metric space (X, d_l). If $a \in A$, then $d_l(a, B) \leq H_{d_l}(A, B)$.*

Lemma 2. [25] *Let (X, d_l) be a dislocated metric space. Let $(P(X), H_{d_l})$ is a dislocated Hausdorff metric space on $P(X)$. If for all $A, B \in P(X)$ and for each $a \in A$ there exists $b_a \in B$ satisfies $d_l(a, B) = d_l(a, b_a)$ then $H_{d_l}(A, B) \geq d_l(a, b_a)$.*

2 Main Result

Let (X, d_l) be a dislocated metric space and $x_0 \in X$ with $A, B : X \to \hat{W}(X)$ be two fuzzy mappings on X. Let $x_1 \in [Ax_0]_{\alpha(x_0)}$ be an element such that $d_l(x_0, [Ax_0]_{\alpha(x_0)}) = d_l(x_0, x_1)$. Let $x_2 \in [Bx_1]_{\alpha(x_1)}$ be an element such that $d_l(x_1, [Bx_1]_{\alpha(x_1)}) = d_l(x_1, x_2)$. Continuing this process, we construct a sequence x_n of points in X such that $x_{2n+1} \in [Ax_{2n}]_{\alpha(x_{2n})}$ and $x_{2n+2} \in [Bx_{2n+1}]_{\alpha(x_{2n+1})}$, for $n \in \mathbb{N} \cup \{0\}$. Also $d_l(x_{2n}, [Ax_{2n}]_{\alpha(x_{2n})}) = d_l(x_{2n}, x_{2n+1})$ and $d_l(x_{2n+1}, [Bx_{2n+1}]_{\alpha(x_{2n+1})}) = d_l(x_{2n+1}, x_{2n+2})$. We denote this iterative sequence by $\{BA(x_n)\}$. We say that $\{BA(x_n)\}$ is a sequence in X generated by x_0.

Theorem 1. *Let (X, d_l) be a complete dislocated metric space and (A, B) be a pair of new Ciric type rational fuzzy F-contraction, if for all $x, y \in \{BA(x_n)\}$, we have*

$$\tau + F(H_{d_l}([Ax]_{\alpha(x)}, [By]_{\alpha(y)})) \leq F(D_l(x, y)) \tag{2}$$

where $F \in \triangle_F$, $\tau > 0$, and

$$D_l(x, y) = \max \left\{ \begin{array}{c} d_l(x, y), d_l(x, [Ax]_{\alpha(x)}), d_l(y, [By]_{\alpha(y)}), \\ \dfrac{d_l\left(x, [Ax]_{\alpha(x)}\right) . d_l\left(y, [By]_{\alpha(y)}\right)}{1 + d_l\left(x, y\right)} \end{array} \right\}. \tag{3}$$

Then, $\{BA(u_n)\} \to u \in X$. Moreover, if (2) also holds for u, then A and B have a common fixed point u in X and $d_l(u, u) = 0$.

Proof. If $D_l(x, y) = 0$, then clearly $x = y$ is a common fixed point of A and B. Then, proof is finished. Let $D_l(y, x) > 0$ for all $x, y \in \{BA(x_n)\}$ with $x \neq y$. Then, by (2), and Lemma 2 we get

$$\begin{aligned} F(d_l(x_{2i+1}, x_{2i+2})) &\leq F(H_{d_l}([Ax_{2i}]_{\alpha(x_{2i})}, [Bx_{2i+1}]_{\alpha(x_{2i+1})})) \\ &\leq F(D_l(x_{2i}, x_{2i+1})) - \tau \end{aligned}$$

for all $i \in \mathbb{N} \cup \{0\}$, where

$$\begin{aligned} D_l(x_{2i}, x_{2i+1}) &= \max \left\{ \begin{array}{c} d_l(x_{2i}, x_{2i+1}), d_l(x_{2i}, [Ax_{2i}]_{\alpha(x_{2i})}), d_l(x_{2i+1}, [Bx_{2i+1}]_{\alpha(x_{2i+1})}), \\ \dfrac{d_l\left(x_{2i}, [Ax_{2i}]_{\alpha(x_{2i})}\right) . d_l\left(x_{2i+1}, [Bx_{2i+1}]_{\alpha(x_{2i+1})}\right)}{1 + d_l\left(x_{2i}, x_{2i+1}\right)} \end{array} \right\} \\ &= \max \left\{ \begin{array}{c} d_l(x_{2i}, x_{2i+1}), d_l(x_{2i}, x_{2i+1}), d_l(x_{2i+1}, x_{2i+2}), \\ \dfrac{d_l\left(x_{2i}, x_{2i+1}\right) . d_l\left(x_{2i+1}, x_{2i+2}\right)}{1 + d_l\left(x_{2i}, x_{2i+1}\right)} \end{array} \right\} \\ &= \max\{d_l(x_{2i}, x_{2i+1}), d_l(x_{2i+1}, x_{2i+2})\}. \end{aligned}$$

If, $D_l(x_{2i}, x_{2i+1}) = d_l(x_{2i+1}, x_{2i+2})$, then

$$F(d_l(x_{2i+1}, x_{2i+2})) \leq F(d_l(x_{2i+1}, x_{2i+2})) - \tau,$$

which is a contradiction due to (F1). Therefore,

$$F(d_l(x_{2i+1}, x_{2i+2})) \leq F(d_l(x_{2i}, x_{2i+1})) - \tau, \text{ for all } i \in \mathbb{N} \cup \{0\}. \tag{4}$$

Similarly, we have

$$F(d_l(x_{2i}, x_{2i+1})) \leq F(d_l(x_{2i-1}, x_{2i})) - \tau, \text{ for all } i \in \mathbb{N}. \tag{5}$$

Using (4) in (5), we have

$$F(d_l(x_{2i+1}, x_{2i+2})) \leq F(d_l(x_{2i-1}, x_{2i})) - 2\tau.$$

Continuing the same way, we get

$$F(d_l(x_{2i+1}, x_{2i+2})) \leq F(d_l(x_0, x_1)) - (2i+1)\tau. \tag{6}$$

Similarly, we have

$$F(d_l(x_{2i}, x_{2i+1})) \leq F(d_l(x_0, x_1)) - 2i\tau, \tag{7}$$

So, by (6) and (7) we have

$$F(d_l(x_n, x_{n+1})) \leq F(d_l(x_0, x_1)) - n\tau. \tag{8}$$

On taking limit $n \to \infty$, both sides of (8), we have

$$\lim_{n\to\infty} F(d_l(x_n, x_{n+1})) = -\infty. \tag{9}$$

As, $F \in \triangle_F$, then

$$\lim_{n\to\infty} d_l(x_n, x_{n+1}) = 0. \tag{10}$$

By (8), for all $n \in \mathbb{N} \cup \{0\}$, we obtain

$$(d_l(x_n, x_{n+1}))^k (F(d_l(x_n, x_{n+1})) - F(d_l(x_0, x_1))) \leq -(d_l(x_n, x_{n+1}))^k n\tau \leq 0. \tag{11}$$

Considering (9), (10) and letting $n \to \infty$ in (11), we have

$$\lim_{n\to\infty} (n(d_l(x_n, x_{n+1}))^k) = 0. \tag{12}$$

Since (12) holds, there exists $n_1 \in \mathbb{N}$, such that $n(d_l(x_n, x_{n+1}))^k \leq 1$ for all $n \geq n_1$ or,

$$d_l(x_n, x_{n+1}) \leq \frac{1}{n^{\frac{1}{k}}} \text{ for all } n \geq n_1. \tag{13}$$

Using (13), we get form $m > n > n_1$,

$$\begin{aligned} d_l(x_n, x_m) &\leq d_l(x_n, x_{n+1}) + d_l(x_{n+1}, x_{n+2}) + \ldots + d_l(x_{m-1}, x_m) \\ &= \sum_{i=n}^{m-1} d_l(x_i, x_{i+1}) \leq \sum_{i=n}^{\infty} d_l(x_i, x_{i+1}) \leq \sum_{i=n}^{\infty} \frac{1}{i^{\frac{1}{k}}}. \end{aligned}$$

The convergence of the series $\sum_{i=n}^{\infty} \frac{1}{i^{\frac{1}{k}}}$ implies that $\lim_{n,m\to\infty} d_l(x_n, x_m) = 0$. Hence, $\{BA(x_n)\}$ is a Cauchy sequence in (X, d_l). Since (X, d_l) is a complete dislocated metric space, so there exists $u \in X$ such that $\{BA(x_n)\} \to u$ that is

$$\lim_{n\to\infty} d_l(x_n, u) = 0. \tag{14}$$

Now, by Lemma 2, we have

$$\tau + F(d_l(x_{2n+1}, [Bu]_{\alpha(u)})) \le \tau + F(H_{d_l}([Ax_{2n}]_{\alpha(x_{2n})}, [Bu]_{\alpha(u)})), \tag{15}$$

As inequality (2) also holds for u, then we have

$$\tau + F(d_l(x_{2n+1}, [Bu]_{\alpha(u)})) \le F(D_l(x_{2n}, u)), \tag{16}$$

where,

$$\begin{aligned} D_l(x_{2n}, u) &= \max \left\{ \begin{array}{c} d_l(x_{2n}, u), d_l(x_{2n}, [Ax_{2n}]_{\alpha(x_{2n})}), d_l(u, [Bu]_{\alpha(u)}), \\ \dfrac{d_l\left(x_{2n}, [Ax_{2n}]_{\alpha(x_{2n})}\right) .d_l\left(u, [Bu]_{\alpha(u)}\right)}{1 + d_l\left(x_{2n}, u\right)} \end{array} \right\} \\ &= \max \left\{ \begin{array}{c} d_l(x_{2n}, u), d_l(x_{2n}, x_{2n+1}), d_l(u, [Bu]_{\alpha(u)}), \\ \dfrac{d_l\left(x_{2n}, x_{2n+1}\right) .d_l\left(u, [Bu]_{\alpha(u)}\right)}{1 + d_l\left(x_{2n}, u\right)} \end{array} \right\}. \end{aligned}$$

Taking $\lim_{n\to\infty}$ and by using (14), we get

$$\lim_{n\to\infty} D_l(x_{2n}, u) = d_l(u, [Bu]_{\alpha(u)}). \tag{17}$$

Since F is strictly increasing, then (16) implies

$$d_l(x_{2n+1}, [Bu]_{\alpha(u)}) < D_l(x_{2n}, u).$$

By taking $\lim_{n\to\infty}$ and using (17), we get

$$d_l(u, [Bu]_{\alpha(u)}) < d_l(u, [Bu]_{\alpha(u)}).$$

Which is a contradiction. So, $d_l(u, [Bu]_{\alpha(u)}) = 0$ or $u \in [Bu]_{\alpha(u)}$. Similarly by using (14) and Lemma 2 and the inequality

$$\tau + F(d_l(x_{2n+2}, [Au]_{\alpha(u)})) \le \tau + F(H_{d_l}([Bx_{2n+1}]_{\alpha(x_{2n+1})}, [Au]_{\alpha(u)})),$$

we can show that $d_l(u, [Au]_{\alpha(u)}) = 0$ or $u \in [Au]_{\alpha(u)}$. Hence A and B have a common fixed point u in X. Now,

$$d_l(u, u) \le d_l(u, [Bu]_{\alpha(u)}) + d_l([Bu]_{\alpha(u)}, u) \le 0.$$

This implies that $d_l(u, u) = 0$.

Example 3. Let $X = [0,1]$ and $d_l(x,y) = x + y$. Then, (X, d_l) is a complete dislocated metric space. Define a pair of fuzzy mappings $A, B : X \to \hat{W}(X)$ as follows:

$$A(x)(t) = \begin{cases} \alpha & \text{if } \frac{x}{6} \leq t < \frac{x}{4} \\ \frac{\alpha}{2} & \text{if } \frac{x}{4} \leq t \leq \frac{x}{2} \\ \frac{\alpha}{4} & \text{if } \frac{x}{2} < t < x \\ 0 & \text{if } x \leq t \leq \infty \end{cases}$$

and

$$B(x)(t) = \begin{cases} \beta & \text{if } \frac{x}{8} \leq t < \frac{x}{6} \\ \frac{\beta}{4} & \text{if } \frac{x}{6} \leq t \leq \frac{x}{4} \\ \frac{\beta}{6} & \text{if } \frac{x}{4} < t < x \\ 0 & \text{if } x \leq t \leq \infty. \end{cases}$$

Define the function $F : \mathbb{R}^+ \to \mathbb{R}$ by

$$F(x) = \ln(x) \text{ for all } x \in R^+ \text{ and } F \in \triangle_F.$$

Consider,

$$[Ax]_{\alpha/2} = \left[\frac{x}{6}, \frac{x}{2}\right] \text{ and } [By]_{\beta/4} = \left[\frac{x}{8}, \frac{x}{4}\right]$$

for $x \in X$, we define the sequence $\{BA(x_n)\} = \left\{1, \frac{1}{6}, \frac{1}{48}, \cdots\right\}$ generated by $x_0 = 1$ in X. We have

$$\begin{aligned} H_{d_l}([Ax]_{\alpha/2}, [By]_{\beta/4}) &= \max\left\{\sup_{a\in Sx} d_l(a, [By]_{\beta/4}), \sup_{b\in Ty} d_l([Ax]_{\alpha/2}, b)\right\} \\ &= \max\left\{\sup_{a\in Sx} d_l\left(a, \left[\frac{y}{8}, \frac{y}{4}\right]\right), \sup_{b\in Ty} d_l\left(\left[\frac{x}{6}, \frac{x}{2}\right], b\right)\right\} \\ &= \max\left\{d_l\left(\frac{x}{6}, \frac{y}{8}\right), d_l\left(\frac{x}{6}, \frac{y}{4}\right)\right\} \\ &= \max\left\{\frac{x}{6} + \frac{y}{8}, \frac{x}{6} + \frac{y}{4}\right\} \end{aligned}$$

where

$$\begin{aligned} D_l(x,y) &= \max\left\{\begin{array}{c} d_l(x,y), \dfrac{d_l\left(x, \left[\frac{x}{6}, \frac{x}{2}\right]\right) \cdot d_l(y, \left[\frac{y}{8}, \frac{y}{4}\right])}{1 + d_l(x,y)}, d_l\left(x, \left[\frac{x}{6}, \frac{x}{2}\right]\right), \\ d_l\left(y, \left[\frac{y}{8}, \frac{y}{4}\right]\right) \end{array}\right\} \\ &= \max\left\{d_l(x,y), \frac{d_l\left(x, \frac{x}{6}\right).d_l\left(y, \frac{y}{8}\right)}{1 + d_l(x,y)}, d_l\left(x, \frac{x}{6}\right), d_l\left(y, \frac{y}{8}\right)\right\} \\ &= \max\left\{x + y, \frac{27xy}{16(1 + x + y)}, \frac{7x}{6}, \frac{9y}{8}\right\} \\ &= x + y. \end{aligned}$$

Case (i). If, $\max\left\{\frac{x}{6}+\frac{y}{8},\frac{x}{6}+\frac{y}{4}\right\}=\frac{x}{6}+\frac{y}{8}$, and $\tau=\ln(\frac{8}{3})$, then we have

$$\begin{aligned} 16x+12y &\le 36x+36y \\ \frac{8}{3}\left(\frac{x}{6}+\frac{y}{8}\right) &\le x+y \\ \ln\left(\frac{8}{3}\right)+\ln\left(\frac{x}{6}+\frac{y}{8}\right) &\le \ln(x+y). \end{aligned}$$

which implies that,

$$\tau+F(H_{d_l}([Ax]_{\alpha/2},[By]_{\beta/4})\le F(D_l(x,y)).$$

Case (ii). Similarly, if $\max\left\{\frac{x}{6}+\frac{y}{8},\frac{x}{6}+\frac{y}{4}\right\}=\frac{x}{6}+\frac{y}{4}$, and $\tau=\ln(\frac{8}{3})$, then we have

$$\begin{aligned} 16x+24y &\le 36x+36y \\ \frac{8}{3}\left(\frac{x}{6}+\frac{y}{4}\right) &\le x+y \\ \ln\left(\frac{8}{3}\right)+\ln\left(\frac{x}{6}+\frac{y}{4}\right) &\le \ln(x+y). \end{aligned}$$

Hence,

$$\tau+F(H_{d_l}([Ax]_{\alpha/2},[By]_{\beta/4})\le F(D_l(x,y)).$$

Hence all the hypothesis of Theorem 1 are satisfied. So, (A,B) have a common fixed point.

Let (X,d_l) be a dislocated metric space and $x_0\in X$ with $A:X\to\hat{W}(X)$ be a fuzzy mappings on X. Let $x_1\in[Ax0]_{\alpha(x_0)}$ be an element such that $d_l(x_0,[Ax0]_{\alpha(x_0)})=d_l(x_0,x_1)$. Let $x_2\in[Ax1]_{\alpha(x_1)}$ be an element such that $d_l(x_1,[Ax1]_{\alpha(x_1)})=d_l(x_1,x_2)$. Continuing this process, we construct a sequence x_n of points in X such that $x_{n+1}\in[Ax_n]_{\alpha(x_n)}$, for $n\in\mathbb{N}\cup\{0\}$. We denote this iterative sequence by $\{AA(x_n)\}$. We say that $\{AA(x_n)\}$ is a sequence in X generated by x_0.

Corollary 1. *Let (X,d_l) be a complete dislocated metric space and $A:X\to\hat{W}(X)$ be a fuzzy mapping such that*

$$\tau+F(H_{d_l}([Ax]_{\alpha(x)},[Ay]_{\alpha(y)}))\le F(D_l(x,y)) \tag{18}$$

for all $x,y\in\{AA(x_n)\}$, for some $F\in\Delta_F$, $\tau>0$, where

$$D_l(x,y)=\max\left\{\begin{array}{c} d_l(x,y),d_l(x,[Ax]_{\alpha(x)}),d_l(y,[Ay]_{\alpha(y)}), \\ \dfrac{d_l\left(x,[Ax]_{\alpha(x)}\right).d_l\left(y,[Ay]_{\alpha(y)}\right)}{1+d_l\left(x,y\right)} \end{array}\right\}.$$

Then, $\{AA(x_n)\}\to u\in X$. Moreover, if (18) also holds for u, then A has a fixed point u in X and $d_l(u,u)=0$.

Remark 1. By setting the following different values of $D_l(x,y)$ in (3), we can obtain different results on fuzzy $F-$contractions as corollaries of Theorem 1

(1) $D_l(x,y) = d_l(x,y)$

(2) $D_l(x,y) = \dfrac{d_l\left(x,[Ax]_{\alpha(x)}\right)\cdot d_l\left(y,[By]_{\alpha(y)}\right)}{1+d_l\left(x,y\right)}$

(3) $D_l(x,y) = \max\left\{d_l(x,y), \dfrac{d_l\left(x,[Ax]_{\alpha(x)}\right)\cdot d_l\left(y,[By]_{\alpha(y)}\right)}{1+d_l\left(x,y\right)}\right\}.$

Theorem 2. *Let (X,d_l) be a complete dislocated metric space and $A,B : X \to \hat{W}(X)$ be the two fuzzy mappings. Assume that if $F \in \triangle_F$ and $\tau \in \mathbb{R}^+$ such that*

$$\tau+F(H_{d_l}([Ax]_{\alpha(x)},[By]_{\alpha(y)})) \le F\left(\begin{array}{c} a_1 d_l(x,y)+a_2 d_l(x,[Ax]_{\alpha(x)})+a_3 d_l(y,[By]_{\alpha(y)}) \\ +a_4\dfrac{d_l^2(x,[Ax]_{\alpha(x)}).d_l(y,[By]_{\alpha(y)})}{1+d_l^2(x,y)}\end{array}\right) \tag{19}$$

for all $x,y \in \{BA(x_n)\}$, with $x \neq y$ where $a_1,a_2,a_3,a_4 > 0$, $a_1+a_2+a_3+a_4 = 1$ and $a_3+a_4 \neq 1$. Then, $\{BA(x_n)\} \to u \in X$. Moreover, if (19) also holds for u, then A and B have a common fixed point u in X and $d_l(u,u) = 0$.

Proof. As, $x_1 \in [Ax_0]_{\alpha(x_0)}$ and $x_2 \in [Bx_1]_{\alpha(x_1)}$, by using (19) and Lemma 2

$$\begin{aligned}
\tau+F(d_l(x_1,x_2)) &= \tau+F(d_l(x_1,[Bx_1]_{\alpha(x_1)})) \\
&\le \tau+F(H_{d_l}([Ax_0]_{\alpha(x_0)},[Bx_1]_{\alpha(x_1)})) \\
&\le F\left(\begin{array}{c} a_1 d_l(x_0,x_1)+a_2 d_l(x_0,[Ax_0]_{\alpha(x_0)})+a_3 d_l(x_1,[Bx_1]_{\alpha(x_1)}) \\ +\,a_4\dfrac{d_l^2(x_0,[Ax_0]_{\alpha(x_0)})\cdot d_l(x_1,[Bx_1]_{\alpha(x_1)})}{1+d_l^2(x_0,x_1)}\end{array}\right) \\
&\le F\left(\begin{array}{c} a_1 d_l(x_0,x_1)+a_2 d_l(x_0,x_1)+a_3 d_l(x_1,x_2) \\ +\,a_4 d_l(x_1,x_2)\left(\dfrac{d_l^2(x_0,x_1)}{1+d_l^2(x_0,x_1)}\right)\end{array}\right) \\
&\le F((a_1+a_2)d_l(x_0,x_1)+(a_3+a_4)d_l(x_1,x_2)).
\end{aligned}$$

Since F is strictly increasing, we have

$$\begin{aligned}
d_l(x_1,x_2) &< (a_1+a_2)d_l(x_0,x_1)+(a_3+a_4)d_l(x_1,x_2) \\
&< \left(\frac{a_1+a_2}{1-a_3-a_4}\right) d_l(x_0,x_1).
\end{aligned}$$

From $a_1+a_2+a_3+a_4 = 1$ and $a_3+a_4 \neq 1$, we deduce $1-a_3-a_4 > 0$ and so

$$d_l(x_1,x_2) < d_l(x_0,x_1).$$

Consequently

$$F(d_l(x_1,x_2)) \le F(d_l(x_0,x_1)) - \tau.$$

As we have $x_{2i+1} \in [Ax_{2i}]_{\alpha(x_{2i})}$ and $x_{2i+2} \in [Bx_{2i+1}]_{\alpha(x_{2i+1})}$ then, by (19) and Lemma 2 we get

$$\begin{aligned}
\tau + F(d_l(x_{2i+1}, x_{2i+2})) &= \tau + F(d_l(x_{2i+1}, [Bx_{2i+1}]_{\alpha(x_{2i+1})})) \\
&\leq \tau + F(H_{d_l}([Ax_{2i}]_{\alpha(x_{2i})}, [Bx_{2i+1}]_{\alpha(x_{2i+1})})) \\
&\leq F\left(\begin{array}{c} a_1 d_l(x_{2i}, x_{2i+1}) + a_2 d_l(x_{2i}, [Ax_{2i}]_{\alpha(x_{2i})}) \\ + a_3 d_l(x_{2i+1}, [Bx_{2i+1}]_{\alpha(x_{2i+1})}) \\ + a_4 \dfrac{d_l^2(x_{2i}, [Ax_{2i}]_{\alpha(x_{2i})}) \cdot d_l(x_{2i+1}, [Bx_{2i+1}]_{\alpha(x_{2i+1})})}{1 + d_l^2(x_{2i}, x_{2i+1})} \end{array} \right) \\
&\leq F(a_1 d_l(x_{2i}, x_{2i+1}) + a_2 d_l(x_{2i}, x_{2i+1}) + a_3 d_l(x_{2i+1}, x_{2i+2}) \\
&\quad + a_4 d_l(x_{2i+1}, x_{2i+2}) \frac{d_l^2(x_{2i}, x_{2i+1})}{1 + d_l^2(x_{2i}, x_{2i+1})}) \\
&\leq F(a_1 d_l(x_{2i}, x_{2i+1}) + a_2 d_l(x_{2i}, x_{2i+1}) + a_3 d_l(x_{2i+1}, x_{2i+2}) \\
&\quad + a_4 d_l(x_{2i+1}, x_{2i+2})).
\end{aligned}$$

Since F is strictly increasing, and $a_1 + a_2 + a_3 + a_4 = 1$ where $a_3 + a_4 \neq 1$, we deduce $1 - a_3 - a_4 > 0$ so we obtain

$$\begin{aligned}
d_l(x_{2i+1}, x_{2i+2}) &< a_1 d_l(x_{2i}, x_{2i+1}) + a_2 d_l(x_{2i}, x_{2i+1}) + a_3 d_l(x_{2i+1}, x_{2i+2}) \\
&\quad + a_4 d_l(x_{2i+1}, x_{2i+2})) \\
&< (a_1 + a_2) d_l(x_{2i}, x_{2i+1}) + (a_3 + a_4) d_l(x_{2i+1}, x_{2i+2}) \\
d_l(x_{2i+1}, x_{2i+2}) &< \left(\frac{a_1 + a_2}{1 - a_3 - a_4} \right) d_l(x_{2i}, x_{2i+1}) \\
&< d_l(x_{2i}, x_{2i+1}).
\end{aligned}$$

This implies that,

$$F(d_l(x_{2i+1}, x_{2i+2})) \leq F(d_l(x_{2i}, x_{2i+1})) - \tau$$

Following similar arguments as given in Theorem 1, we have $\{BA(x_n)\} \to u$ that is

$$\lim_{n \to \infty} d_l(x_n, u) = 0. \tag{20}$$

Now, by Lemma 2, we have

$$\tau + F(d_l(x_{2n+1}, [Bu]_{\alpha(u)})) \leq \tau + F(H_{d_l}([Ax_{2n}]_{\alpha(x_{2n})}, [Bu]_{\alpha(u)})),$$

By using (19), we have

$$\begin{aligned}
\tau + F(d_l(x_{2n+1}, [Bu]_{\alpha(u)})) &\leq F(a_1 d_l(x_{2n}, u) + a_2 d_l(x_{2n}, [Ax_{2n}]_{\alpha(x_{2n})}) + a_3 d_l(u, [Bu]_{\alpha(u)}) \\
&\quad + a_4 \frac{d_l^2(x_{2n}, [Ax_{2n}]_{\alpha(x_{2n})}) \cdot d_l(u, [Bu]_{\alpha(u)})}{1 + d_l^2(x_{2n}, u)}) \\
&\leq F(a_1 d_l(x_{2n}, u) + a_2 d_l(x_{2n}, x_{2n+1}) + a_3 d_l(u, [Bu]_{\alpha(u)}) \\
&\quad + a_4 \frac{d_l^2(x_{2n}, x_{2n+1}).d_l(u, [Bu]_{\alpha(u)})}{1 + d_l^2(x_{2n}, u)}).
\end{aligned}$$

Since F is strictly increasing, we have

$$d_l(x_{2n+1}, [Bu]_{\alpha(u)}) < a_1 d_l(x_{2n}, u) + a_2 d_l(x_{2n}, x_{2n+1}) + a_3 d_l(u, [Bu]_{\alpha(u)}) + a_4 \frac{d_l^2(x_{2n}, x_{2n+1}) \cdot d_l(u, [Bu]_{\alpha(u)})}{1 + d_l^2(x_{2n}, u)}.$$

Taking limit $n \to \infty$, and by using (20), we get

$$d_l(u, [Bu]_{\alpha(u)}) < a_3 d_l(u, [Bu]_{\alpha(u)}).$$

Which is a contradiction. So, $d_l(u, [Bu]_{\alpha(u)}) = 0$ or $u \in [Bu]_{\alpha(u)}$. Similarly by (19), (20), Lemma 2 and the inequality

$$\tau + F(d_l(x_{2n+2}, [Au]_{\alpha(u)})) \leq \tau + F(H_{d_l}([Bx_{2n+1}]_{\alpha(x_{2n+1})}, [Au]_{\alpha(u)}))$$

we can show that $d_l(u, [Au]_{\alpha(u)}) = 0$ or $u \in [Au]_{\alpha(u)}$. Hence the A and B have a common fixed point u in (X, d_l). Now,

$$d_l(u, u) \leq d_l(u, [Bu]_{\alpha(u)}) + d_l([Bu]_{\alpha(u)}, u) \leq 0.$$

This implies that $d_l(u, u) = 0$.

If, we take $A = B$ in Theorem 2, then we have the following result.

Corollary 2. *Let (X, d_l) be a complete dislocated metric space and $A : X \to \hat{W}(X)$ be a fuzzy mapping. Assume that $F \in \triangle_F$ and $\tau \in \mathbb{R}^+$ such that*

$$\tau + F(H_{d_l}([Ax]_{\alpha(x)}, [Ay]_{\alpha(y)})) \leq F\left(\begin{array}{c} a_1 d_l(x, y) + a_2 d_l(x, [Ax]_{\alpha(x)}) + a_3 d_l(y, [Ay]_{\alpha(y)}) \\ + a_4 \dfrac{d_l^2(x, [Ax]_{\alpha(x)}) \cdot d_l(y, [Ay]_{\alpha(y)})}{1 + d_l^2(x, y)} \end{array}\right) \tag{21}$$

for all $x, y \in \{AA(x_n)\}$, with $x \neq y$ for some $a_1, a_2, a_3, a_4 > 0$, $a_1 + a_2 + a_3 + a_4 = 1$ where $a_3 + a_4 \neq 1$. Then $\{AA(x_n)\} \to u \in X$. Moreover, if (21) also holds for u, then A has a fixed point u in X and $d_l(u, u) = 0$.

If, we take $a_2 = 0$ in Theorem 2, then we have the following result.

Corollary 3. *Let (X, d_l) be a complete dislocated metric space and $A, B : X \to \hat{W}(X)$ be the two fuzzy mappings. Assume that $F \in \triangle_F$ and $\tau \in \mathbb{R}^+$ such that*

$$\tau + F(H_{d_l}([Ax]_{\alpha(x)}, [By]_{\alpha(y)})) \leq F\left(\begin{array}{c} a_1 d_l(x, y) + a_3 d_l(y, [By]_{\alpha(y)}) + \\ a_4 \dfrac{d_l^2(x, [Ax]_{\alpha(x)}) \cdot d_l(y, [By]_{\alpha(y)})}{1 + d_l^2(x, y)} \end{array}\right) \tag{22}$$

for all $x, y \in \{BA(x_n)\}$, with $x \neq y$ where $a_1, a_3, a_4 > 0$, $a_1 + a_3 + a_4 = 1$ and $a_3 + a_4 \neq 1$. Then $\{BA(x_n)\} \to u \in X$. Moreover, if (22) also holds for u, then A and B have a common fixed point u in X and $d_l(u, u) = 0$.

If, we take $a_3 = 0$ in Theorem 2, then we have the following result.

Corollary 4. *Let (X, d_l) be a complete dislocated metric space and $A, B : X \to \hat{W}(X)$ be the two fuzzy mappings. Assume that $F \in \triangle_F$ and $\tau \in \mathbb{R}^+$ such that*

$$\tau + F(H_{d_l}([Ax]_{\alpha(x)}, [By]_{\alpha(y)})) \leq F\left(\begin{array}{c} a_1 d_l(x,y) + a_2 d_l(x, [Ax]_{\alpha(x)}) + \\ a_4 \dfrac{d_l^2(x, [Ax]_{\alpha(x)}) \cdot d_l(y, [By]_{\alpha(y)})}{1 + d_l^2(x,y)} \end{array}\right) \quad (23)$$

for all $x, y \in \{BA(x_n)\}$, with $x \neq y$ where $a_1, a_2, a_4 > 0$, $a_1 + a_2 + a_4 = 1$ and $a_4 \neq 1$. Then $\{BA(x_n)\} \to u \in X$. Moreover, if (23) also holds for u, then A and B have a common fixed point u in X and $d_l(u, u) = 0$.

If, we take $a_4 = 0$ in Theorem 2, then we have the following result.

Corollary 5. *Let (X, d_l) be a complete dislocated metric space and $A, B : X \to \hat{W}(X)$ be the two fuzzy mappings. Assume that if $F \in \triangle_F$ and $\tau \in \mathbb{R}^+$ such that*

$$\tau + F(H_{d_l}([Ax]_{\alpha(x)}, [By]_{\alpha(y)})) \leq F\left(a_1 d_l(x,y) + a_2 d_l(x, [Ax]_{\alpha(x)}) + a_3 d_l(y, [By]_{\alpha(y)})\right) \quad (24)$$

for all $x, y \in \{BA(x_n)\}$, with $x \neq y$ where $a_1, a_2, a_3 > 0$, $a_1 + a_2 + a_3 = 1$ and $a_3 \neq 1$. Then $\{BA(x_n)\} \to u \in X$. Moreover, if (24) also holds for u, then A and B have a common fixed point u in X and $d_l(u, u) = 0$.

If, we take $a_1 = a_2 = a_3 = 0$ in Theorem 2, then we have the following result.

Corollary 6. *Let (X, d_l) be a complete dislocated metric space and $A, B : X \to \hat{W}(X)$ be the two fuzzy mappings. Assume that if $F \in \triangle_F$ and $\tau \in \mathbb{R}^+$ such that*

$$\tau + F(H_{d_l}([Ax]_{\alpha(x)}, [By]_{\alpha(y)}))) \leq F\left(\frac{d_l^2(x, [Ax]_{\alpha(x)}) \cdot d_l(y, [By]_{\alpha(y)})}{1 + d_l^2(x,y)}\right) \quad (25)$$

for all $x, y \in \{BA(x_n)\}$, with $x \neq y$. Then, $\{BA(x_n)\} \to u \in X$. Moreover, if (25) also holds for u, then A and B have a common fixed point u in X and $d_l(u, u) = 0$.

3 Applications

In this section, we prove that fixed point for multivalued mappings can be derived by utilizing Theorems 1 and 2 in a dislocated metric spaces.

Theorem 3. *Let (X, d_l) be a complete dislocated metric space and (R, S) be a pair of new Ciric type rational multivalued F-contraction if for all $x, y \in \{SR(x_n)\}$, we have*

$$\tau + F(H_{d_l}(Rx, Sy)) \leq F(D_l(x,y)) \quad (26)$$

where $F \in \triangle_F$, $\tau > 0$, and

$$D_l(x,y) = \max\left\{d_l(x,y), d_l(x, Rx), d_l(y, Sy), \frac{d_l\,(x, Rx)\,.d_l\,(y, Sy)}{1 + d_l\,(x,y)}\right\}. \quad (27)$$

Then, $\{SR(x_n)\} \to x^* \in X$. Moreover, if (2) also holds for x^*, then R and S have a common fixed point x^* in X and $d_l(x^*, x^*) = 0$.

Proof. Consider an arbitrary mapping $\alpha : X \to (0, 1]$. Consider two fuzzy mappings $A, B : X \to \hat{W}(X)$ defined as

$$(Ax)(t) = \begin{cases} \alpha(x), & \text{if } t \in Rx \\ 0, & \text{if } t \notin Rx \end{cases}$$

and

$$(Bx)(t) = \begin{cases} \alpha(x), & \text{if } t \in Rx \\ 0, & \text{if } t \notin Rx \end{cases}$$

we obtain that

$$[Ax]_{\alpha(x)} = \{t : Ax(t) \geq \alpha(x)\} = Rx$$

and

$$[Bx]_{\alpha(x)} = \{t : Bx(t) \geq \alpha(x)\} = Sx.$$

Hence, the condition (26) becomes the condition (2) of Theorem 1 So, there exists $x^* \in [Ax]_{\alpha(x)} \cap [Bx]_{\alpha(x)} = Rx \cap Sx$.

Theorem 4. *Let (X, d_l) be a complete dislocated metric space and $R, S : X \to P(X)$ be the two multivalued mappings. Assume that if $F \in \triangle_F$ and $\tau \in \mathbb{R}^+$ such that*

$$\tau + F(H_{d_l}(Rx, Sy)) \leq F\left(\begin{array}{c} a_1 d_l(x, y) + a_2 d_l(x, Rx) + a_3 d_l(y, Sy) \\ + a_4 \dfrac{d_l^2(x, Rx).d_l(y, Sy)}{1 + d_l^2(x, y)} \end{array}\right) \tag{28}$$

for all $x, y \in \{SR(x_n)\}$, with $x \neq y$ where $a_1, a_2, a_3, a_4 > 0$, $a_1 + a_2 + a_3 + a_4 = 1$ and $a_3 + a_4 \neq 1$. Then, $\{SR(x_n)\} \to x^* \in X$. Moreover, if (28) also holds for x^*, then R and S have a common fixed point x^* in X and $d_l(x^*, x^*) = 0$.

Proof. Consider an arbitrary mapping $\alpha : X \to (0, 1]$. Consider two fuzzy mappings $A, B : X \to \hat{W}(X)$ defined as

$$(Ax)(t) = \begin{cases} \alpha(x), & \text{if } t \in Rx \\ 0, & \text{if } t \notin Rx \end{cases}$$

and

$$(Bx)(t) = \begin{cases} \alpha(x), & \text{if } t \in Rx \\ 0, & \text{if } t \notin Rx \end{cases}$$

we obtained that

$$[Ax]_{\alpha(x)} = \{t : Ax(t) \geq \alpha(x)\} = Rx$$

and

$$[Bx]_{\alpha(x)} = \{t : Bx(t) \geq \alpha(x)\} = Sx.$$

Hence, the condition (28) becomes the condition (18) of Theorem 2 So, there exists $x^* \in [Ax]_{\alpha(x)} \cap [Bx]_{\alpha(x)} = Rx \cap Sx$.

Acknowledgements. This project was supported by the Theoretical and Computational Science (TaCS) Center under Computational and Applied Science for Smart Innovation Cluster (CLASSIC), Faculty of Science, KMUTT. The third author would like to thank the Research Professional Development Project Under the Science Achievement Scholarship of Thailand (SAST) for financial support.

References

1. Abbas, M., Ali, B., Romaguera, S.: Fixed and periodic points of generalized contractions in metric spaces. Fixed Point Theory Appl. **243**, 11 pages (2013)
2. Acar, Ö., Altun, I.: A fixed point theorem for multivalued mappings with δ-distance. Abstr. Appl. Anal. Article ID 497092, 5 pages (2014)
3. Acar, Ö., Durmaz, G., Minak, G.: Generalized multivalued $F-$contractions on complete metric spaces. Bull. Iran. Math. Soc. **40**, 1469–1478 (2014)
4. Ahmad, J., Al-Rawashdeh, A., Azam, A.: Some new fixed point theorems for generalized contractions in complete metric spaces. Fixed Point Theory Appl. **80**, 18 pages (2015)
5. Arshad, M., Khan, S.U., Ahmad, J.: Fixed point results for F-contractions involving some new rational expressions. JP J. Fixed Point Theory Appl. **11**(1), 79–97 (2016)
6. Azam, A., Arshad, M.: Fixed points of a sequence of locally contractive multivalued maps. Comp. Math. Appl. **57**, 96–100 (2009)
7. Banach, S.: Sur les opèrations dans les ensembles abstraits et leur application aux equations itegrales. Fund. Math. **3**, 133–181 (1922)
8. Butnariu, D.: Fixed point for fuzzy mapping. Fuzzy Sets Syst. **7**, 191–207 (1982)
9. Ćirić, L.B.: A generalization of Banach's contraction principle. Proc. Am. Math. Soc. **45**, 267–273 (1974)
10. Heilpern, S.: Fuzzy mappings and fixed point theorem. J. Math. Anal. Appl. **83**(2), 566–569 (1981)
11. Hitzler, P., Seda, A.K.: Dislocated topologies. J. Electr. Eng. **51**(12/s), 3–7 (2000)
12. Hussain, N., Ahmad, J., Ciric, L., Azam, A.: Coincidence point theorems for generalized contractions with application to integral equations. Fixed Point Theory Appl. **78**, 13 pages (2015)
13. Hussain, N., Ahmad, J., Azam, A.: On Suzuki-Wardowski type fixed point theorems. J. Nonlinear Sci. Appl. **8**, 1095–1111 (2015)
14. Hussain, N., Salimi, P.: Suzuki-Wardowski type fixed point theorems for α-GF-contractions. Taiwanese J. Math. **18**(6), 1879–1895 (2014)
15. Hussain, A., Arshad, M., Khan, S.U.: $\tau-$Generalization of fixed point results for F-contraction. Bangmod Int. J. Math. Comput. Sci. **1**(1), 127–137 (2015)
16. Hussain, A., Arshad, M., Nazam, M., Khan, S.U.: New type of results involving closed ball with graphic contraction. J. Inequalities Spec. Funct. **7**(4), 36–48 (2016)
17. Khan, S.U., Arshad, M., Hussain, A., Nazam, M.: Two new types of fixed point theorems for F-contraction. J. Adv. Stud. Topology **7**(4), 251–260 (2016)
18. Matthews, S.G.: Partial metric topology. Ann. New York Acad. Sci. **728**, 183–197 (1994) In: Proceedings of 8th Summer Conference on General Topology and Applications
19. Nadler, S.: Multivalued contraction mappings. Pac. J. Math. **30**, 475–488 (1969)
20. Piri, H., Kumam, P.: Some fixed point theorems concerning F-contraction in complete metric spaces. Fixed Point Theory Appl. **210**, 11 pages (2014)
21. Rashid, M., Shahzad, A., Azam, A.: Fixed point theorems for L-fuzzy mappings in quasi-pseudo metric spaces. J. Intell. Fuzzy Syst. **32**, 499–507 (2017)
22. Secelean, N.A.: Iterated function systems consisting of F-contractions. Fixed Point Theory Appl. **277**, 13 pages (2013)
23. Sgroi, M., Vetro, C.: Multi-valued F-contractions and the solution of certain functional and integral equations. Filomat **27**(7), 1259–1268 (2013)

24. Shahzad, A., Shoaib, A., Mahmood, Q.: Fixed point theorems for fuzzy mappings in b- metric space. Ital. J. Pure Appl. Math. **38**, 419–427 (2017)
25. Shoaib, A., Hussain, A., Arshad, M., Azam, A.: Fixed point results for α_*-ψ-Ciric type multivalued mappings on an intersection of a closed ball and a sequence with graph. J. Math. Anal. **7**(3), 41–50 (2016)
26. Shoaib, A.: Fixed point results for α_*-ψ-multivalued mappings. Bull. Math. Anal. Appl. **8**(4), 43–55 (2016)
27. Shoaib, A., Ansari, A.H., Mahmood, Q., Shahzad, A.: Fixed point results for complete dislocated G_d-metric space via C-class functions. Bull. Math. Anal. Appl. **9**(4), 1–11 (2017)
28. Shoaib, A., Kumam, P., Shahzad, A., Phiangsungnoen, S., Mahmood, Q.: Fixed point results for fuzzy mappings in a b-metric space. Fixed Point Theory Appl. **2**, 12 pages (2018)
29. Wardowski, D.: Fixed point theory of a new type of contractive mappings in complete metric spaces. Fixed Point Theory Appl. **201**, 6 pages (2012). Article ID 94
30. Weiss, M.D.: Fixed points and induced fuzzy topologies for fuzzy sets. J. Math. Anal. Appl. **50**, 142–150 (1975)
31. Zadeh, L.A.: Fuzzy sets. Inf. Control **8**(3), 338–353 (1965)

Common Fixed Point Theorems for Weakly Generalized Contractions and Applications on G-metric Spaces

Pasakorn Yordsorn[1,2], Phumin Sumalai[3], Piyachat Borisut[1,2], Poom Kumam[1,2](✉), and Yeol Je Cho[4,5]

[1] KMUTTFixed Point Research Laboratory, Department of Mathematics, Room SCL 802 Fixed Point Laboratory, Science Laboratory Building, Faculty of Science, King Mongkut's University of Technology Thonburi (KMUTT), 126 Pracha-Uthit Road, Bang Mod, Thrung Khru, Bangkok 10140, Thailand
ryotarokung@gmail.com, piyachat.b@hotmail.com, poom.kum@kmutt.ac.th

[2] KMUTT-Fixed Point Theory and Applications Research Group (KMUTT-FPTA), Theoretical and Computational Science Center (TaCS), Science Laboratory Building, Faculty of Science, King Mongkut's University of Technology Thonburi (KMUTT), 126 Pracha-Uthit Road, Bang Mod, Thrung Khru, Bangkok 10140, Thailand

[3] Department of Mathematics, Faculty of Science and Technology, Muban Chombueng Rajabhat University, 46 M.3, Chombueng 70150, Ratchaburi, Thailand
phumin.su28@gmail.com

[4] Department of Mathematics Education and the RINS, Gyeongsang National University, Jinju 660-701, Korea
yjchomath@gmail.com

[5] School of Mathematical Sciences, University of Electronic Science and Technology of China, Chengdu 611731, Sichuan, People's Republic of China

Abstract. In this paper, we introduce weakly generalized contraction conditions on G-metric space and prove some common fixed point theorems for the proposed contractions. The results in this paper differ from the recent corresponding results given by some authors in literature.

Mathematics Subject Classification: 47H10 · 54H25

1 Introduction and Preliminaries

It is well known that Banach's Contraction Principle [3] has been generalized in various directions. Especially, in 1997, Alber and Guerre-Delabrere [18] introduced the concept of weak contraction in Hilbert spaces and proved the corresponding fixed point result for this contraction. In 2001, Rhoades [14] has shown that the result of Alber and Guerre-Delabrere [18] is also valid in complete metric spaces.

V. Kreinovich et al. (Eds.): ECONVN 2019, SCI 809, pp. 230–250, 2019.
https://doi.org/10.1007/978-3-030-04200-4_18

On the other hand, in 2005, Mustafa and Sims [13] introduced a new class of a generalized metric space, which is called a *G-metric space*, as a generalization of a metric space. Subsequently, Since this G-metric space, many authors have proved a lot of fixed and common fixed point results for generalized contractions in G-metric spaces (see [1,2,8,9,11,12,15–17]).

Recently, Hongqing and Gu [4,6,7] proved some common fixed point theorems for twice, third and fourth power type contractive condition in metric space. In 2017, Gu and Ye [5] proved some common fixed point theorems for three self-mappings satisfying various new contractive conditions in complete G-metric spaces.

Motivated by the recent works mentioned above, in this paper, we introduce a weakly generalized contraction condition on G-metric spaces and prove some new common fixed point theorems for our generalized contraction conditions. The results obtained in this paper differ from the recent corresponding results given by some authors in literature.

Now, we give some definitions and some propositions for our main results.

Let $a \in (0, \infty]$ and $R_a^+ = [0, a)$ and consider a function $F : R_a^+ \to \mathbb{R}$ satisfying the following conditions:

(a) $F(0) = 0$ and $f(t) > 0$ for all $t \in (0, a)$;
(b) F is nondecreasing on R_a^+;
(c) F is continuous;
(d) $F(\alpha t) = \alpha F(t)$ for all $t \in R_a^+$ and $\alpha \in [0, 1)$.

Let $\mathscr{F}[0, a)$ be the set of all the functions $F : R_a^+ \to \mathbb{R}$ satisfying the conditions (a)–(d).

Also, let $\varphi : R_a^+ \to \mathbb{R}_+$ be a function satisfying the following conditions:

(e) $\varphi(0) = 0$ and $\varphi(t) > 0$ for all $t \in (0, a)$;
(f) φ is right lower semi-continuous, i.e., for any nonnegative nonincreasing sequence $\{r_n\}$,

$$\liminf_{n\to\infty} \varphi(r_n) \geq \varphi(r)$$

provided that $\lim_{n\to\infty} r_n = r$;
(g) for any sequence $\{r_n\}$ with $\lim_{n\to\infty} r_n = 0$, there exist $b \in (0, 1)$ and $n_0 \in \mathbb{N}$ such that

$$\varphi(r_n) \geq b r_n$$

for each $n \geq n_0$;

Let $\Phi[0, a)$ be the set of all the functions $\varphi : R_a^+ \to \mathbb{R}_+$ satisfying the conditions (e)–(g).

Definition 1. [13] Let E be a metric space. Let $F \in \mathscr{F}[0, a)$, $\varphi \in \Phi[a, 0)$ and $d = \sup\{d(x, y) : x, y \in E\}$. Set $a = d$ if $d = \infty$ and $a > d$ if $d < \infty$. A multivalued mapping $G : E \to 2^E$ is called a *weakly generalized contraction* with respect to F and φ if

$$F(H_d(Gx, Gy)) \leq F(d(x, y)) - \varphi(F(d(x, y)))$$

for all $x, y \in E$ with x and y comparable.

Definition 2. [13] Let X be a nonempty set. A mapping $G : X \times X \times X \to R^+$ is called a *generalized metric* or *G-metric* if the following conditions are satisfied:

(G1) $G(x, y, z) = 0$ if $x = y = z$;
(G2) $0 < G(x, x, y)$ for all $x, y \in X$ with $x \neq y$;
(G3) $G(x, x, y) \leq G(x, y, z)$ for all $x, y, z \in X$ with $z \neq y$;
(G4) $G(x, y, z) = G(x, z, y) = G(y, z, x) = \cdots$ (symmetry in all three variables);
(G5) $G(x, y, z) \leq G(x, a, a) + G(a, y, z)$ for all $x, y, z, a \in X$ (rectangle inequality).

The pair (X, G) is called a *G-metric space.*

Every G-metric on X defines a metric d_G on X given by

$$d_G(x, y) = G(x, y, y) + G(y, x, x)$$

for all $x, y \in X$.

Recently, Kaewcharoen and Kaewkhao [10] introduced the following concepts:

Let X be a G-metric space. We denote $CB(X)$ the family of all nonempty closed bounded subsets of X. Then the Hausdorff G-distance $H(\cdot, \cdot, \cdot)$ on $CB(X)$ is defined as follows:

$$H_G(A, B, C) = \max\{\sup_{x \in A} G(x, B, C), \sup_{x \in A} G(x, C, A), \sup_{x \in A} G(x, A, B)\},$$

where

$$G(x, B, C) = d_G(x, B) + d_G(B, C) + d_G(x, C),$$
$$d_G(x, B) = \inf\{d_G(x, y) : y \in B\},$$
$$d_G(A, B) = \inf\{d_G(a, b) : a \in A, b \in B\}.$$

Recall that $G(x, y, C) = \inf\{G(x, y, z), z \in C\}$ and a point $x \in X$ is called a *fixed point* of a multi-valued mapping $T : X \to 2^X$ if $x \in Tx$.

Definition 3. [13] Let (X, G) be a G-metric space and $\{x_n\}$ be a sequence of points in X. A point $x \in X$ is called the *limit* of the sequence $\{x_n\}$ (shortly, $x_n \to x$) if

$$\lim_{m,n \to \infty} G(x, x_n, x_m) = 0,$$

which says that a sequence $\{x_n\}$ is G-convergent to a point $x \in X$.

Thus, if $x_n \to x$ in a G-metric space (X, G), then, for any $\varepsilon > 0$, there exists $n_0 \in \mathbb{N}$ such that

$$G(x, x_n, x_m) < \varepsilon$$

for all $n, m \geq n_0$.

Definition 4. [13] Let (X, G) be a G-metric space. A sequence $\{x_n\}$ is called a *G-Cauchy sequence* in X if, for any $\varepsilon > 0$, there exists $n_0 \in \mathbb{N}$ such that

$$G(x_n, x_m, x_l) < \varepsilon$$

for all $n, m, l \geq n_0$, that is,

$$G(x_n, x_m, x_l) \to 0$$

as $n, m, l \to \infty$.

Definition 5. [13] A G-metric space (X, G) is said to be *G-complete* if every G-Cauchy sequence in (X, G) is G-convergent in X.

Proposition 1. [13] *Let (X, G) be a G-metric space. Then the followings are equivalent:*

(1) $\{x_n\}$ is G-convergent to x.
(2) $G(x_n, x_n, x) \to 0$ as $n \to \infty$.
(3) $G(x_n, x, x) \to 0$ as $n \to \infty$.
(4) $G(x_n, x_m, x) \to 0$ as $n, m \to \infty$.

Proposition 2. [13] *Let (X, G) be a G-metric space. Then the following are equivalent:*

(1) The sequence $\{x_n\}$ is a G-Cauchy sequence.
(2) For any $\varepsilon > 0$, there exists $n_0 \in \mathbb{N}$ such that

$$G(x_n, x_m, x_m) < \varepsilon$$

for all $n, m \geq n_0$.

Proposition 3. [13] *Let (X, G) be a G-metric space. Then the function $G(x, y, z)$ is jointly continuous in all three of its variables.*

Definition 6. [13] Let (X, G) and (X', G') be G-metric space.

(1) A mapping $f : (X, G) \to (X', G')$ is said to be *G-continuous* at a point $a \in X$ if, for any $\varepsilon > 0$, there exists $\delta > 0$ such that

$$x, y \in X,\ G(a, x, y) < \delta \implies G'(f(a), f(x), f(y)) < \varepsilon.$$

(2) A function f is said to be *G-continuous* on X if it is G-continuous at every $a \in X$.

Proposition 4. [13] *Let (X, G) and (X', G') be G-metric space. Then a mapping $f : X \to X'$ is G-continuous at a point $x \in X$ if and only if it is G-sequentially continuous at x, that is, whenever $\{x_n\}$ is G-convergent to x, $\{f(x_n)\}$ is G-convergent to $f(x)$.*

Proposition 5. [13] *Let (X, G) be a G-metric space. Then, for any x, y, z, a in X, it follows that:*

(1) If $G(x, y, z) = 0$, then $x = y = z$.
(2) $G(x, y, z) \leq G(x, x, y) + G(x, x, z)$.
(3) $G(x, y, y) \leq 2G(y, x, x)$.
(4) $G(x, y, z) \leq G(x, a, z) + G(a, y, z)$.
(5) $G(x, y, z) \leq \frac{2}{3}(G(x, y, a) + G(x, a, z) + G(a, y, z))$.
(6) $G(x, y, z) \leq G(x, a, a) + G(y, a, a) + G(z, a, a)$.

2 Main Results

Now, we give the main results in this paper.

Theorem 1. *Let (X, G) be a complete G-metric space and G is weakly generalized contractive with respect to F and φ. Suppose the three self-mappings $f, g, h : X \to X$ satisfy the following condition:*

$$\begin{aligned} F(H_G^\theta(fx, gy, hz)) \leq F(&qH_G^\alpha(x, y, z)H_G^\beta(x, fx, fx)H_G^\gamma(y, gy, gy) \\ &H_G^\delta(z, hz, hz)) - \varphi(F(qH_G^\alpha(x, y, z)H_G^\beta(x, fx, fx) \\ &H_G^\gamma(y, gy, gy)H_G^\delta(z, hz, hz))) \end{aligned} \tag{1}$$

for all $x, y, z \in X$, where $0 \leq q < 1$, $\alpha, \beta, \gamma, \delta \in [0, +\infty)$ and $\theta = \alpha + \beta + \gamma + \delta$. Then f, g and h have a unique common fixed point (say u) and f, g, h are all G-continuous at u.

Proof. We will proceed in two steps: first we prove any fixed point of f is a fixed point of g and h. Assume that $p \in X$ is such that $fp = p$. Now, we prove that $p = gp = hp$. In fact, by using (1), we have

$$\begin{aligned} F(H_G^\theta(fp, gp, hp)) &\leq F(qH_G^\alpha(p, p, p)H_G^\beta(p, fp, fp)H_G^\gamma(p, gp, gp) \\ &\quad H_G^\delta(p, hp, hp)) - \varphi(F(qH_G^\alpha(p, p, p)H_G^\beta(p, fp, fp) \\ &\quad H_G^\gamma(p, gp, gp)H_G^\delta(p, hp, hp))) \\ &= 0. \end{aligned}$$

It follows that $F(H_G^\theta(p, gp, hp)) = 0$, hence $F(H_G^\theta(p, gp, hp) = 0$, implie $p = gp = hp$. So p is a common fixed point of f, g and h. The same conclusion holds if $p = gp$ or $p = hp$.

Now, we prove that f, g and h have a unique common fixed point. Suppose x_0 is an arbitrary point in X. Define $\{x_n\}$ by $x_{3n+1} = fx_{3n}$, $x_{3n+2} = gx_{3n+1}$, $x_{3n+3} = hx_{3n+2}$, $n = 0, 1, 2, \cdots$. If $x_n = x_{n+1}$, for some n, with $n = 3m$, then $p = x_{3m}$ is a fixed point of f, and by the first step, p is a common fixed point for f, g and h. The same holds if $n = 3m + 1$ or $n = 3m + 2$. Without loss of generality, we can assume that $x_n \neq x_{n+1}$, for all $n \in \mathbb{N}$.

Next we prove sequence $\{x_n\}$ is a G-Cauchy sequence. In fact, by (1) and $(G3)$, we have
$F(H_G^\theta(x_{3n+1}, x_{3n+2}, x_{3n+3})) = F(H_G^\theta(fx_{3n}, gx_{3n+1}, hx_{3n+2}))$

$$\begin{aligned}
&\leq F(qH_G^\alpha(x_{3n}, x_{3n+1}, x_{3n+2})H_G^\beta(x_{3n}, fx_{3n}, fx_{3n})H_G^\gamma(x_{3n+1}, gx_{3n+1}, gx_{3n+1})\\
&\quad H_G^\delta(x_{3n+2}, hx_{3n+2}, hx_{3n+2})) - \varphi(F(qH_G^\alpha(x_{3n}, x_{3n+1}, x_{3n+2})\\
&\quad H_G^\beta(x_{3n}, fx_{3n}, fx_{3n})H_G^\gamma(x_{3n+1}, gx_{3n+1}, gx_{3n+1})H_G^\delta(x_{3n+2}, hx_{3n+2}, hx_{3n+2})))\\
&= F(qH_G^\alpha(x_{3n}, x_{3n+1}, x_{3n+2})H_G^\beta(x_{3n}, x_{3n+1}, x_{3n+1})H_G^\gamma(x_{3n+1}, x_{3n+2}, x_{3n+2})\\
&\quad H_G^\delta(x_{3n+2}, x_{3n+3}, x_{3n+3})) - \varphi(F(qH_G^\alpha(x_{3n}, x_{3n+1}, x_{3n+2})H_G^\beta(x_{3n}, x_{3n+1}, x_{3n+1})\\
&\quad H_G^\gamma(x_{3n+1}, x_{3n+2}, x_{3n+2})H_G^\delta(x_{3n+2}, x_{3n+3}, x_{3n+3})))\\
&\leq F(qH_G^\alpha(x_{3n}, x_{3n+1}, x_{3n+2})H_G^\beta(x_{3n}, x_{3n+1}, x_{3n+2})H_G^\gamma(x_{3n+1}, x_{3n+2}, x_{3n+3})\\
&\quad H_G^\delta(x_{3n+2}, x_{3n+3}, x_{3n+4})) - \varphi(F(qH_G^\alpha(x_{3n}, x_{3n+1}, x_{3n+2})H_G^\beta(x_{3n}, x_{3n+1}, x_{3n+2})\\
&\quad H_G^\gamma(x_{3n+1}, x_{3n+2}, x_{3n+3})H_G^\delta(x_{3n+2}, x_{3n+3}, x_{3n+4}))).
\end{aligned}$$

Combining $\theta = \alpha + \beta + \gamma + \delta$, we have

$$\begin{aligned}
F(H_G^\theta(x_{3n+1}, x_{3n+2}, x_{3n+3})) &\leq F(qH_G^{\alpha+\beta}(x_{3n}, x_{3n+1}, x_{3n+2})H_G^{\gamma+\delta}(x_{3n+1}, x_{3n+2}, x_{3n+3}))\\
&\leq F(qH_G^{\alpha+\beta}(x_{3n}, x_{3n+1}, x_{3n+2})H_G^{\gamma+\delta}(x_{3n}, x_{3n+1}, x_{3n+2}))\\
&\leq F(qH_G^{\alpha+\beta+\gamma+\delta}(x_{3n}, x_{3n+1}, x_{3n+2}))\\
&\leq F(qH_G^\theta(x_{3n}, x_{3n+1}, x_{3n+2}))
\end{aligned}$$

which implies that

$$H_G(x_{3n+1}, x_{3n+2}, x_{3n+3}) \leq qH_G(x_{3n}, x_{3n+1}, x_{3n+2}). \tag{2}$$

On the other hand, from the condition (1) and $(G3)$ we have
$F(H_G^\theta(x_{3n+2}, x_{3n+3}, x_{3n+4})) = F(H_G^\theta(fx_{3n+1}, gx_{3n+2}, hx_{3n+3}))$

$$\begin{aligned}
&\leq F(qH_G^\alpha(x_{3n+1}, x_{3n+2}, x_{3n+3})H_G^\beta(x_{3n+1}, fx_{3n+1}, fx_{3n+1})H_G^\gamma(x_{3n+2}, gx_{3n+2}, gx_{3n+2})\\
&\quad H_G^\delta(x_{3n+3}, hx_{3n+3}, hx_{3n+3})) - \varphi(F(qH_G^\alpha(x_{3n+1}, x_{3n+2}, x_{3n+3})H_G^\beta(x_{3n+1}, fx_{3n+1}, fx_{3n+1})\\
&\quad H_G^\gamma(x_{3n+2}, gx_{3n+2}, gx_{3n+2})H_G^\delta(x_{3n+3}, hx_{3n+3}, hx_{3n+3}))\\
&= F(qH_G^\alpha(x_{3n+1}, x_{3n+2}, x_{3n+3})H_G^\beta(x_{3n+1}, x_{3n+2}, x_{3n+2})H_G^\gamma(x_{3n+2}, x_{3n+3}, x_{3n+3})\\
&\quad H_G^\delta(x_{3n+3}, x_{3n+4}, x_{3n+4})) - \varphi(F(qH_G^\alpha(x_{3n+1}, x_{3n+2}, x_{3n+3})H_G^\beta(x_{3n+1}, x_{3n+2}, x_{3n+2})\\
&\quad H_G^\gamma(x_{3n+2}, x_{3n+3}, x_{3n+3})H_G^\delta(x_{3n+3}, x_{3n+4}, x_{3n+4}))\\
&\leq F(qH_G^\alpha(x_{3n+1}, x_{3n+2}, x_{3n+3})H_G^\beta(x_{3n+1}, x_{3n+2}, x_{3n+3})H_G^\gamma(x_{3n+2}, x_{3n+3}, x_{3n+4})\\
&\quad H_G^\delta(x_{3n+2}, x_{3n+3}, x_{3n+4})) - \varphi(F(qH_G^\alpha(x_{3n+1}, x_{3n+2}, x_{3n+3})H_G^\beta(x_{3n+1}, x_{3n+2}, x_{3n+3})\\
&\quad H_G^\gamma(x_{3n+2}, x_{3n+3}, x_{3n+4})H_G^\delta(x_{3n+2}, x_{3n+3}, x_{3n+4})).
\end{aligned}$$

Combining $\theta = \alpha + \beta + \gamma + \delta$, we have

$$\begin{aligned}
F(H_G^\theta(x_{3n+2}, x_{3n+3}, x_{3n+4})) &\leq F(qH_G^{\alpha+\beta}(x_{3n+1}, x_{3n+2}, x_{3n+3})H_G^{\gamma+\delta}(x_{3n+2}, x_{3n+3}, x_{3n+4}))\\
&\leq F(qH_G^{\alpha+\beta}(x_{3n+1}, x_{3n+2}, x_{3n+3})H_G^{\gamma+\delta}(x_{3n+1}, x_{3n+2}, x_{3n+3}))\\
&\leq F(qH_G^{\alpha+\beta+\gamma+\delta}(x_{3n+1}, x_{3n+2}, x_{3n+3}))\\
&\leq F(qH_G^\theta(x_{3n+1}, x_{3n+2}, x_{3n+3}))
\end{aligned}$$

which implies that

$$H_G(x_{3n+2}, x_{3n+3}, x_{3n+4}) \leq qH_G(x_{3n+1}, x_{3n+2}, x_{3n+3}). \tag{3}$$

Again, using (1) and (G3), we can get

$$F(G^\theta(x_{3n+3}, x_{3n+4}, x_{3n+5})) = F(H_G^\theta(fx_{3n+2}, gx_{3n+3}, hx_{3n+4}))$$

$$\begin{aligned}
&\leq F(qH_G^\alpha(x_{3n+2}, x_{3n+3}, x_{3n+4})H_G^\beta(x_{3n+2}, fx_{3n+2}, fx_{3n+2})H_G^\gamma(x_{3n+3}, gx_{3n+3}, gx_{3n+3})\\
&\quad H_G^\delta(x_{3n+4}, hx_{3n+4}, hx_{3n+4})) - \varphi(F(qH_G^\alpha(x_{3n+2}, x_{3n+3}, x_{3n+4})H_G^\beta(x_{3n+2}, fx_{3n+2}, fx_{3n+2})\\
&\quad H_G^\gamma(x_{3n+3}, gx_{3n+3}, gx_{3n+3})H_G^\delta(x_{3n+4}, hx_{3n+4}, hx_{3n+4}))\\
&= F(qH_G^\alpha(x_{3n+2}, x_{3n+3}, x_{3n+4})H_G^\beta(x_{3n+2}, x_{3n+3}, x_{3n+3})H_G^\gamma(x_{3n+3}, x_{3n+4}, x_{3n+4})\\
&\quad H_G^\delta(x_{3n+4}, x_{3n+5}, x_{3n+5})) - \varphi(F(qH_G^\alpha(x_{3n+2}, x_{3n+3}, x_{3n+4})H_G^\beta(x_{3n+2}, x_{3n+3}, x_{3n+3})\\
&\quad H_G^\gamma(x_{3n+3}, x_{3n+4}, x_{3n+4})H_G^\delta(x_{3n+4}, x_{3n+5}, x_{3n+5}))\\
&\leq F(qH_G^\alpha(x_{3n+2}, x_{3n+3}, x_{3n+4})H_G^\beta(x_{3n+2}, x_{3n+3}, x_{3n+4})H_G^\gamma(x_{3n+3}, x_{3n+4}, x_{3n+5})\\
&\quad H_G^\delta(x_{3n+3}, x_{3n+4}, x_{3n+5})) - \varphi(F(qH_G^\alpha(x_{3n+2}, x_{3n+3}, x_{3n+4})H_G^\beta(x_{3n+2}, x_{3n+3}, x_{3n+4})\\
&\quad H_G^\gamma(x_{3n+3}, x_{3n+4}, x_{3n+5})H_G^\delta(x_{3n+3}, x_{3n+4}, x_{3n+5})).
\end{aligned}$$

Combining $\theta = \alpha + \beta + \gamma + \delta$, we have

$$\begin{aligned}
F(H_G^\theta(x_{3n+3}, x_{3n+4}, x_{3n+5})) &\leq F(qH_G^{\alpha+\beta}(x_{3n+2}, x_{3n+3}, x_{3n+4})H_G^{\gamma+\delta}(x_{3n+3}, x_{3n+4}, x_{3n+5}))\\
&\leq F(qH_G^{\alpha+\beta}(x_{3n+2}, x_{3n+3}, x_{3n+4})H_G^{\gamma+\delta}(x_{3n+2}, x_{3n+3}, x_{3n+4}))\\
&\leq F(qH_G^{\alpha+\beta+\gamma+\delta}(x_{3n+2}, x_{3n+3}, x_{3n+4}))\\
&\leq F(qH_G^\theta(x_{3n+2}, x_{3n+3}, x_{3n+4}))
\end{aligned}$$

which implies that

$$H_G(x_{3n+3}, x_{3n+4}, x_{3n+5}) \leq qH_G(x_{3n+2}, x_{3n+3}, x_{3n+4}). \tag{4}$$

Combining (2), (3) and (4), we have

$$H_G(x_n, x_{n+1}, x_{n+2}) \leq qH_G(x_{n-1}, x_n, x_{n+1}) \leq ... \leq q^n H_G(x_0, x_1, x_2).$$

Thus, by $(G3)$ and $(G5)$, for every $m, n \in \mathbb{N}$, $m > n$, we have

$$\begin{aligned}
H_G(x_n, x_m, x_m) &\leq H_G(x_n, x_{n+1}, x_{n+1}) + H_G(x_{n+1}, x_{n+2}, x_{n+2}) + ... + H_G(x_{m-1}, x_m, x_m)\\
&\leq H_G(x_n, x_{n+1}, x_{n+2}) + H_G(x_{n+1}, x_{n+2}, x_{n+3}) + ... + H_G(x_{m-1}, x_m, x_{m+1})\\
&\leq (q^n + q^{n+1} + ... + q^{m-1})H_G(x_0, x_1, x_2)\\
&\leq \frac{q^n}{1-q} H_G(x_0, x_1, x_2) \longrightarrow 0 (n \longrightarrow \infty)
\end{aligned}$$

which implies that $H_G(x_n, x_m, x_m) \to 0$, as $n, m \to \infty$. Thus $\{x_n\}$ is a Cauchy sequence. Due to the G-completeness of X, there exists $u \in X$, such that $\{x_n\}$ is G-convergent to u.

Now we prove u is a common fixed point of f, g and h. By using (1), we have

$$\begin{aligned}
&F(H_G^\theta(fu, x_{3n+2}, x_{3n+3})) = F(H_G^\theta(fu, gx_{3n+1}, hx_{3n+2}))\\
&\leq F(qH_G^\alpha(u, x_{3n+1}, x_{3n+2})H_G^\beta(u, fu, fu)H_G^\gamma(x_{3n+1}, gx_{3n+1}, gx_{3n+1})\\
&\quad H_G^\delta(x_{3n+2}, hx_{3n+2}, hx_{3n+2})) - \varphi(F(qH_G^\alpha(u, x_{3n+1}, x_{3n+2})H_G^\beta(u, fu, fu)\\
&\quad H_G^\gamma(x_{3n+1}, gx_{3n+1}, gx_{3n+1})H_G^\delta(x_{3n+2}, hx_{3n+2}, hx_{3n+2})).
\end{aligned}$$

Letting $n \to \infty$, and using the fact that G is continuous in its variables, we can get

$$H_G^\theta(fu, u, u) = 0.$$

Which gives that $fu = u$, hence u is a fixed point of f.
Similarly it can be shown that $gu = u$ and $hu = u$. Consequently, we have $u = fu = gu = hu$, and u is a common fixed point of f, g and h.

To prove the uniqueness, suppose that v is another common fixed point of f, g and h, then by (1), we have

$$\begin{aligned} F(H_G^\theta(u, u, v)) &= F(H_G^\theta(fu, gu, hv)) \\ &\leq F(qH_G^\alpha(u, u, v)H_G^\beta(u, fu, fu)H_G^\gamma(u, gu, gu)H_G^\delta(v, hv, hv)) \\ &\quad -\varphi(F(qH_G^\alpha(u, u, v)H_G^\beta(u, fu, fu)H_G^\gamma(u, gu, gu)H_G^\delta(v, hv, hv)) \\ &= 0. \end{aligned}$$

Then $F(H_G^\theta(u, u, v)) = 0$, implies that $(H_G^\theta(u, u, v)) = 0$. Hence $u = v$. Thus u is a unique common fixed point of f, g and h.

To show that f is G-continuous at u, let $\{y_n\}$ be any sequence in X such that $\{y_n\}$ is G-convergent to u. For $n \in \mathbb{N}$, from (1) we have

$$\begin{aligned} F(H_G^\theta(f_{y_n}, u, u)) &= F(H_G^\theta(fy_n, gu, hu)) \\ &\leq F(qH_G^\alpha(y_n, u, u)H_G^\beta(y_n, fy_n, fy_n)H_G^\gamma(u, gu, gu)H_G^\delta(u, hu, hu)) \\ &\quad -\varphi(F(qH_G^\alpha(y_n, u, u)H_G^\beta(y_n, fy_n, fy_n)H_G^\gamma(u, gu, gu)H_G^\delta(u, hu, hu)) \\ &= 0. \end{aligned}$$

Then $F(H_G^\theta(f_{y_n}, u, u)) = 0$. Therefore, we get $\lim_{n\to\infty} H_G(fy_n, u, u) = 0$, that is, $\{fy_n\}$ is G-convergent to $u = fu$, and so f is G-continuous at u. Similarly, we can also prove that g, h are G-continuous at u. This completes the proof of Theorem 1.

Corollary 1. *Let (X, G) be a complete G-metric space and G is weakly generalized contractive with respect to F and φ. Suppose the three self-mappings $f, g, h : X \to X$ satisfy the following condition:*

$$\begin{aligned} F(H_G^\theta(f^p x, g^s y, h^r z)) &\leq F(qH_G^\alpha(x, y, z)H_G^\beta(x, f^p x, f^p x)H_G^\gamma(y, g^s y, g^s y)H_G^\delta(z, h^r z, h^r z)) \\ &\quad -\varphi(F(qH_G^\alpha(x, y, z)H_G^\beta(x, f^p x, f^p x)H_G^\gamma(y, g^s y, g^s y)H_G^\delta(z, h^r z, h^r z))) \end{aligned} \tag{5}$$

for all $x, y, z \in X$, where $0 \leq q < 1$, $p, s, r \in \mathbb{N}$, $\alpha, \beta, \gamma, \delta \in [0, +\infty)$ and $\theta = \alpha + \beta + \gamma + \delta$; then f, g and h have a unique common fixed point (say u) and f^p, g^s and h^r are all G-continuous at u.

Proof. From Theorem 1 we know that f^p, g^s, h^r have a unique common fixed point (say u), that is, $f^p u = g^s u = h^r u = u$, and f^p, g^s and h^r are G-continuous at u. Since $fu = ff^p u = f^{p+1}u = f^p fu$, so fu is another fixed point of f^p,

$gu = gg^s u = g^{s+1}u = g^s gu$, so gu is another fixed point of g^s, and $hu = hh^r u = h^{r+1}u = h^r hu$, so hu is another fixed point of h^r. By the condition (5), we have

$$\begin{aligned}
&F(H_G^{\theta}(f^p fu, g^s fu, h^r fu) \\
&\leq F(qH_G^{\alpha}(fu, fu, fu)H_G^{\beta}(fu, f^p fu, f^p fu)H_G^{\gamma}(fu, g^s fu, g^s fu)H_G^{\delta}(fu, h^r fu, h^r fu)) \\
&\quad -\varphi(F(qH_G^{\alpha}(fu, fu, fu)H_G^{\beta}(fu, f^p fu, f^p fu)H_G^{\gamma}(fu, g^s fu, g^s fu)H_G^{\delta}(fu, h^r fu, h^r fu))) \\
&= 0.
\end{aligned}$$

Which implies that $H_G^{\theta}(f^p fu, g^s fu, h^r fu) = 0$, that is $fu = f^p fu = g^s fu = h^r fu$, hence fu is another common fixed point of f^p, g^s and h^r. Since the common fixed point of f^p, g^s and h^r is unique, we deduce that $u = fu$. By the same argument, we can prove $u = gu, u = fu$. Thus, we have $u = fu = gu = hu$. Suppose v is another common fixed point of f, g and h, then $v = f^p v$, and by using the condition (5) again, we have

$$\begin{aligned}
&F(H_G^{\theta}(v, u, u) = F(H_G^{\theta}(f^p v, g^s u, h^r u) \\
&\leq F(qH_G^{\alpha}(v, u, u)H_G^{\beta}(v, f^p v, f^p v)H_G^{\gamma}(u, g^s u, g^s u)H_G^{\delta}(u, h^r u, h^r u)) \\
&\quad -\varphi(F(qH_G^{\alpha}(v, u, u)H_G^{\beta}(v, f^p v, f^p v)H_G^{\gamma}(u, g^s u, g^s u)H_G^{\delta}(u, h^r u, h^r u))) \\
&= 0.
\end{aligned}$$

Which implies that $H_G^{\theta}(v, u, u) = 0$, hence $v = u$. So the common fixed point of f, g and h is unique.

Corollary 2. *Let (X, G) be a complete G-metric space and G is weakly generalized contractive with respect to F and φ. Suppose self-mapping $T : X \to X$ satisfies the condition:*

$$\begin{aligned}
F(H_G^{\theta}(Tx, Ty, Tz)) &\leq F(qH_G^{\alpha}(x, y, z)H_G^{\beta}(x, Tx, Tx)H_G^{\gamma}(y, Ty, Ty)H_G^{\delta}(z, Tz, Tz)) \\
&\quad -\varphi(F(qH_G^{\alpha}(x, y, z)H_G^{\beta}(x, Tx, Tx)H_G^{\gamma}(y, Ty, Ty)H_G^{\delta}(z, Tz, Tz)))
\end{aligned}$$

for all $x, y, z \in X$, where $0 \leq q < 1$, $\alpha, \beta, \gamma, \delta \in [0, +\infty)$ and $\theta = \alpha + \beta + \gamma + \delta$; then T has a unique fixed point (say u) and T is G-continuous at u.

Proof. Let $T = f = g = h$ in Theorem 1, we can know that the Corollary 2 holds.

Corollary 3. *Let (X, G) be a complete G-metric space and G is weakly generalized contractive with respect to F and φ. Suppose self-mapping $T : X \to X$ satisfies the condition:*

$$\begin{aligned}
F(H_G^{\theta}(T^p x, T^p y, T^p z)) &\leq F(qH_G^{\alpha}(x, y, z)H_G^{\beta}(x, T^p x, T^p x)H_G^{\gamma}(y, T^p y, T^p y)H_G^{\delta}(z, T^p z, T^p z)) \\
&\quad -\varphi(F(qH_G^{\alpha}(x, y, z)H_G^{\beta}(x, T^p x, T^p x)H_G^{\gamma}(y, T^p y, T^p y)H_G^{\delta}(z, T^p z, T^p z)))
\end{aligned}$$

for all $x, y, z \in X$, where $0 \leq q < 1$, $p \in \mathbb{N}$, $\alpha, \beta, \gamma, \delta \in [0, +\infty)$ and $\theta = \alpha + \beta + \gamma + \delta$; then T has a unique fixed point (say u) and T^p is G-continuous at u.

Proof. Let $T = f = g = h$ and $p = s = r$ in Corollary 1, we can get this condition holds.

Corollary 4. *Let (X, G) be a complete G-metric space and G is weakly generalized contractive with respect to F and φ. Suppose f, g and h are three mappings of X into itself. If one of the following conditions is satisfied*

(1) $F(H_G(fx, gy, hz)) \leq F(qH_G(x, y, z)) - \varphi(F(qH_G(x, y, z)))$;
(2) $F(H_G(fx, gy, hz)) \leq F(qH_G(x, fx, fx)) - \varphi(F(qH_G(x, fx, fx)))$;
(3) $F(H_G(fx, gy, hz)) \leq F(qH_G(y, gy, gy)) - \varphi(F(qH_G(y, gy, gy)))$;
(4) $F(H_G(fx, gy, hz)) \leq F(qH_G(z, hz, hz)) - \varphi(F(qH_G(z, hz, hz)))$
for all $x, y, z \in X$, where $0 \leq q < 1$; then f, g and h have a unique common fixed point (say u) and f, g, h are all G-continuous at u.

Proof. Taking (1) $\alpha = 1$ and $\beta = \gamma = \delta = 0$; (2) $\beta = 1$ and $\alpha = \gamma = \delta = 0$; (3) $\gamma = 1$ and $\alpha = \beta = \delta = 0$; (4) $\delta = 1$ and $\alpha = \beta = \gamma = 0$ in Theorem 1, respectively, then the conclusion of Corollary 4 can be obtained from Theorem 1 immediately.

Corollary 5. *Let (X, G) be a complete G-metric space and G is weakly generalized contractive with respect to F and φ. Suppose f, g and h are three mappings of X into itself. If one of the following conditions is satisfied*

$$(1)\quad \begin{aligned} F(H_G^2(fx, gy, hz)) &\leq F(qH_G(x, y, z)H_G(x, fx, fx)) \\ &\quad - \varphi(F(qH_G(x, y, z)H_G(x, fx, fx))); \end{aligned}$$

$$(2)\quad \begin{aligned} F(H_G^2(fx, gy, hz)) &\leq F(qH_G(x, y, z)H_G(y, gy, gy)) \\ &\quad - \varphi(F(qH_G(x, y, z)H_G(y, gy, gy))); \end{aligned}$$

$$(3)\quad \begin{aligned} F(H_G^2(fx, gy, hz)) &\leq F(qH_G(x, y, z)H_G(z, hz, hz)) \\ &\quad - \varphi(F(qH_G(x, y, z)H_G(z, hz, hz))); \end{aligned}$$

$$(4)\quad \begin{aligned} F(H_G^2(fx, gy, hz)) &\leq F(qH_G(x, fx, fx)H_G(y, gy, gy)) \\ &\quad - \varphi(F(qH_G(x, fx, fx)H_G(y, gy, gy))); \end{aligned}$$

$$(5)\quad \begin{aligned} F(H_G^2(fx, gy, hz)) &\leq F(qH_G(y, gy, gy)H_G(z, hz, hz)) \\ &\quad - \varphi(F(qH_G(y, gy, gy)H_G(z, hz, hz))); \end{aligned}$$

$$(6)\quad \begin{aligned} F(H_G^2(fx, gy, hz)) &\leq F(qH_G(z, hz, hz)H_G(x, fx, fx)) \\ &\quad - \varphi(F(qH_G(z, hz, hz)H_G(x, fx, fx))) \end{aligned}$$

for all $x, y, z \in X$, where $0 \leq q < 1$; then f, g and h have a unique common fixed point (say u) and f, g and h are all G-continuous at u.

Proof. Taking (1) $\alpha = \beta = 1$ and $\gamma = \delta = 0$; (2) $\alpha = \gamma = 1$ and $\beta = \delta = 0$; (3) $\alpha = \delta = 1$ and $\beta = \gamma = 0$; (4) $\beta = \delta = 1$ and $\alpha = \gamma = 0$; (5) $\gamma = \delta = 1$ and $\alpha = \beta = 0$; (6) $\beta = \gamma = 1$ and $\alpha = \delta = 0$ in Theorem 1, respectively, then the conclusion of Corollary 5 can be obtained from Theorem 1 immediately.

Corollary 6. *Let (X, G) be a complete G-metric space and G is weakly generalized contractive with respect to F and φ. Suppose f, g and h are three mappings of X into itself. If one of the following conditions is satisfied*

$$\text{(1)}\quad \begin{aligned} F(H_G^3(fx,gy,hz)) &\le F(qH_G(x,y,z)H_G(x,fx,fx)H_G(y,gy,gy)) \\ &\quad -\varphi(F(qH_G(x,y,z)H_G(x,fx,fx)H_G(y,gy,gy))); \end{aligned}$$

$$\text{(2)}\quad \begin{aligned} F(H_G^3(fx,gy,hz)) &\le F(qH_G(x,y,z)H_G(x,fx,fx)H_G(z,hz,hz)) \\ &\quad -\varphi(F(qH_G(x,y,z)H_G(x,fx,fx)H_G(z,hz,hz))); \end{aligned}$$

$$\text{(3)}\quad \begin{aligned} F(H_G^3(fx,gy,hz)) &\le F(qH_G(x,y,z)H_G(y,gy,gy)H_G(z,hz,hz)) \\ &\quad -\varphi(F(qH_G(x,y,z)H_G(y,gy,gy)H_G(z,hz,hz))); \end{aligned}$$

$$\text{(4)}\quad \begin{aligned} F(H_G^3(fx,gy,hz)) &\le F(qH_G(x,fx,fx)H_G(y,gy,gy)H_G(z,hz,hz)) \\ &\quad -\varphi(F(qH_G(x,fx,fx)H_G(y,gy,gy)H_G(z,hz,hz))) \end{aligned}$$

for all $x, y, z \in X$, where $0 \le q < 1$; then f, g and h have a unique common fixed point (say u) and f, g, h are all G-continuous at u.

Proof. Taking (1) $\delta = 0$ and $\alpha = \beta = \gamma = 1$; (2) $\gamma = 0$ and $\alpha = \beta = \delta = 1$; (3) $\beta = 0$ and $\alpha = \gamma = \delta = 1$; (4) $\alpha = 0$ and $\beta = \gamma = \delta = 1$ in Theorem 1, respectively, then the conclusion of Corollary 6 can be obtained from Theorem 1 immediately.

Corollary 7. *Let (X, G) be a complete G-metric space and G is weakly generalized contractive with respect to F and φ. Suppose the three self-mappings $f, g, h : X \to X$ satisfy the following condition:*

$$\begin{aligned} F(H_G^4(fx,gy,hz)) &\le F(qH_G(x,y,z)H_G(x,fx,fx)H_G(y,gy,gy)H_G(z,hz,hz)) \\ &\quad -\varphi(F(qH_G(x,y,z)H_G(x,fx,fx)H_G(y,gy,gy)H_G(z,hz,hz))) \end{aligned}$$

for all $x, y, z \in X$, where $0 \le q < 1$; then f, g and h have a unique common fixed point (say u) and f, g, h are all G-continuous at u.

Proof. Taking $\alpha = \beta = \gamma = \delta = 1$ in Theorem 1, then the conclusion of Corollary 7 can be obtained from Theorem 1 immediately.

Theorem 2. *Let (X, G) be a complete G-metric space and G is weakly generalized contractive with respect to F and φ. Suppose $f, g, h : X \to X$ be three self-mappings in X, which satisfy the following condition*

$$\begin{aligned} F(H_G^\theta(fx,gy,hz)) &\le F(qH_G^\alpha(x,y,z)H_G^\beta(x,fx,gy)H_G^\gamma(y,gy,hz)H_G^\delta(z,hz,fx)) \\ &\quad -\varphi(F(qH_G^\alpha(x,y,z)H_G^\beta(x,fx,gy)H_G^\gamma(y,gy,hz)H_G^\delta(z,hz,fx))) \end{aligned} \tag{6}$$

for all $x, y, z \in X$, where $0 \le q < 1$, $\theta = \alpha + \beta + \gamma + \delta$ and $\alpha, \beta, \gamma, \delta \in [0, +\infty)$. Then f, g and h have a unique common fixed point (say u), and f, g, h are all G-continuous at u.

Proof. We will proceed in two steps: first we prove any fixed point of f is a fixed point of g and h. Assume that $p \in X$ such that $fp = p$, by the condition (6), we have

$$\begin{aligned} F(H_G^\theta(fp,gp,hp)) &\le F(qH_G^\alpha(p,p,p)H_G^\beta(p,fp,gp)H_G^\gamma(p,gp,hp)H_G^\delta(p,hp,fp)) \\ &\quad -\varphi(F(qH_G^\alpha(p,p,p)H_G^\beta(p,fp,gp)H_G^\gamma(p,gp,hp)H_G^\delta(p,hp,fp))) \\ &= 0. \end{aligned}$$

It follows that $F(H_G^\theta(p,gp,hp))=0$, hence $H_G^\theta(p,gp,hp)=0$, implies $p=fp=gp=hp$. So p is a common fixed point of f,g and h. The same conclusion holds if $p=gp$ or $p=hp$.

Now, we prove that f, g and h have a unique common fixed point. Suppose x_0 is an arbitrary point in X. Define $\{x_n\}$ by $x_{3n+1}=fx_{3n}$, $x_{3n+2}=gx_{3n+1}$, $x_{3n+3}=hx_{3n+2}$, $n=0,1,2,\cdots$. If $x_n=x_{n+1}$, for some n, with $n=3m$, then $p=x_{3m}$ is a fixed point of f and, by the first step, p is a common fixed point for f, g and h. The same holds if $n=3m+1$ or $n=3m+2$. Without loss of generality, we can assume that $x_n\neq x_{n+1}$, for all $n\in\mathbb{N}$.

Next we prove the sequence $\{x_n\}$ is a G-Cauchy sequence. In fact, by (6) and $(G3)$, we have

$$\begin{aligned}
&F(H_G^\theta(x_{3n+1},x_{3n+2},x_{3n+3}))=F(H_G^\theta(fx_{3n},gx_{3n+1},hx_{3n+2}))\\
&\le F(qH_G^\alpha(x_{3n},x_{3n+1},x_{3n+2})H_G^\beta(x_{3n},fx_{3n},gx_{3n+1})H_G^\gamma(x_{3n+1},gx_{3n+1},hx_{3n+2})\\
&\quad H_G^\delta(x_{3n+2},hx_{3n+2},fx_{3n}))-\varphi(F(qH_G^\alpha(x_{3n},x_{3n+1},x_{3n+2})H_G^\beta(x_{3n},fx_{3n},gx_{3n+1})\\
&\quad H_G^\gamma(x_{3n+1},gx_{3n+1},hx_{3n+2})H_G^\delta(x_{3n+2},hx_{3n+2},fx_{3n})))\\
&=F(qH_G^\alpha(x_{3n},x_{3n+1},x_{3n+2})H_G^\beta(x_{3n},x_{3n+1},x_{3n+2})H_G^\gamma(x_{3n+1},x_{3n+2},x_{3n+3})\\
&\quad H_G^\delta(x_{3n+2},x_{3n+3},x_{3n+1}))-\varphi(F(qH_G^\alpha(x_{3n},x_{3n+1},x_{3n+2})H_G^\beta(x_{3n},x_{3n+1},x_{3n+2})\\
&\quad H_G^\gamma(x_{3n+1},x_{3n+2},x_{3n+3})H_G^\delta(x_{3n+2},x_{3n+3},x_{3n+1})))\\
&\le F(qH_G^\alpha(x_{3n},x_{3n+1},x_{3n+2})H_G^\beta(x_{3n},x_{3n+1},x_{3n+2})H_G^\gamma(x_{3n+1},x_{3n+2},x_{3n+3})\\
&\quad H_G^\delta(x_{3n+1},x_{3n+2},x_{3n+3}))-\varphi(F(qH_G^\alpha(x_{3n},x_{3n+1},x_{3n+2})H_G^\beta(x_{3n},x_{3n+1},x_{3n+2})\\
&\quad H_G^\gamma(x_{3n+1},x_{3n+2},x_{3n+3})H_G^\delta(x_{3n+1},x_{3n+2},x_{3n+3}))).
\end{aligned}$$

Which gives that

$$H_G(x_{3n+1},x_{3n+2},x_{3n+3})\le qH_G(x_{3n},x_{3n+1},x_{3n+2}).$$

By the same argument, we can get

$$H_G(x_{3n+2},x_{3n+3},x_{3n+4})\le qH_G(x_{3n+1},x_{3n+2},x_{3n+3}).$$

$$H_G(x_{3n+3},x_{3n+4},x_{3n+5})\le qH_G(x_{3n+2},x_{3n+3},x_{3n+4}).$$

Then for all $n\in\mathbb{N}$, we have

$$H_G(x_n,x_{n+1},x_{n+2})\le qH_G(x_{n-1},x_n,x_{n+1})\le\cdots\le q^nH_G(x_0,x_1,x_2).$$

Thus, by $(G3)$ and $(G5)$, for every $m,n\in\mathbb{N}$, $m>n$, we have

$$\begin{aligned}
H_G(x_n,x_m,x_m)&\le H_G(x_n,x_{n+1},x_{n+1})+H_G(x_{n+1},x_{n+2},x_{n+2})+\cdots+H_G(x_{m-1},x_m,x_m)\\
&\le H_G(x_n,x_{n+1},x_{n+2})+G(x_{n+1},x_{n+2},x_{n+3})+\cdots+H_G(x_{m-1},x_m,x_{m+1})\\
&\le(q^n+q^{n+1}+\cdots+q^{m-1})H_G(x_0,x_1,x_2)\\
&\le\frac{q^n}{1-q}H_G(x_0,x_1,x_2)\to 0\ (n\to\infty).
\end{aligned}$$

Which gives that $G(x_n, x_m, x_m) \to 0$, as $n, m \to \infty$. Thus $\{x_n\}$ is G-Cauchy sequence. Due to the completeness of X, there exists $u \in X$, such that $\{x_n\}$ is G-convergent to u. Next we prove u is a common fixed point of f, g and h. It follows from (6) that

$$\begin{aligned}
&F(H_G^\theta(fu, x_{3n+2}, x_{3n+3})) = F(H_G^\theta(fu, gx_{3n+1}, hx_{3n+2}))\\
&\le F(qH_G^\alpha(u, x_{3n+1}, x_{3n+2})H_G^\beta(u, fu, gx_{3n+1})H_G^\gamma(x_{3n+1}, gx_{3n+1}, hx_{3n+2})\\
&\quad H_G^\delta(x_{3n+2}, hx_{3n+2}, fu)) - \varphi(F(qH_G^\alpha(u, x_{3n+1}, x_{3n+2})H_G^\beta(u, fu, gx_{3n+1})\\
&\quad H_G^\gamma(x_{3n+1}, gx_{3n+1}, hx_{3n+2})H_G^\delta(x_{3n+2}, hx_{3n+2}, fu)))\\
&= F(qH_G^\alpha(u, x_{3n+1}, x_{3n+2})H_G^\beta(u, fu, x_{3n+2})H_G^\gamma(x_{3n+1}, x_{3n+2}, x_{3n+3})\\
&\quad H_G^\delta(x_{3n+2}, x_{3n+3}, fu)) - \varphi(F(qH_G^\alpha(u, x_{3n+1}, x_{3n+2})H_G^\beta(u, fu, x_{3n+2})\\
&\quad H_G^\gamma(x_{3n+1}, x_{3n+2}, x_{3n+3})H_G^\delta(x_{3n+2}, x_{3n+3}, fu))).
\end{aligned}$$

Letting $n \to \infty$, and using the fact that G is continuous on its variables, we get that

$$H_G^\theta(fu, u, u) = 0.$$

Similarly, we can obtain that $H_G^\theta(u, gu, u) = 0, H_G^\theta(u, u, hu) = 0$, Hence, we get $u = fu = gu = hu$, and u is a common fixed point of f, g and h. Suppose v is another common fixed point of f, g and h, then by (6) we have

$$\begin{aligned}
&F(H_G^\theta(u, u, v) = G^\theta(fu, gu, hv))\\
&\le F(qH_G^\alpha(u, u, v)H_G^\beta(u, fu, gu)H_G^\gamma(u, gu, hv)H_G^\delta(v, hv, fu))\\
&\quad -\varphi(F(qH_G^\alpha(u, u, v)H_G^\beta(u, fu, gu)H_G^\gamma(u, gu, hv)H_G^\delta(v, hv, fu)))\\
&= 0.
\end{aligned}$$

Thus, $u = v$. Then we know that the common fixed point of f, g and h is unique.

To show that f is G-continuous at u, let $\{y_n\}$ be any sequence in X such that $\{y_n\}$ is G-convergent to u. For $n \in \mathbb{N}$, from (6) we have

$$\begin{aligned}
&F(H_G^\theta(fy_n, u, u) = G^\theta(fy_n, gu, hu))\\
&\le F(qH_G^\alpha(y_n, u, u)H_G^\beta(y_n, fy_n, gu)H_G^\gamma(u, gu, hu)H_G^\delta(u, hu, fy_n))\\
&\quad -\varphi(F(qH_G^\alpha(y_n, u, u)H_G^\beta(y_n, fy_n, gu)H_G^\gamma(u, gu, hu)H_G^\delta(u, hu, fy_n)))\\
&= 0.
\end{aligned}$$

Then $F(H_G^\theta(fy_n, u, u) = 0$, which implies that $\lim_{n\to\infty} G^\theta(fy_n, u, u) = 0$. Hence $\{fy_n\}$ is G-convergent to $u = fu$. So f is G-continuous at u. Similarly, we can also prove that g, h are G-continuous at u. This completes the proof of Theorem 2.

Corollary 8. *Let (X, G) be a complete G-metric space and G is weakly generalized contractive with respect to F and φ. Suppose $f, g, h : X \to X$ be three self-mappings in X, which satisfy the following condition*

$$F(H_G^{\theta}(f^m x, g^n y, h^l z)) \leq F(qH_G^{\alpha}(x,y,z)H_G^{\beta}(x, f^m x, g^n y)H_G^{\gamma}(y, g^n y, h^l z) H_G^{\delta}(z, h^l z, f^m x)) - \varphi(F(qH_G^{\alpha}(x,y,z)H_G^{\beta}(x, f^m x, g^n y) H_G^{\gamma}(y, g^n y, h^l z)H_G^{\delta}(z, h^l z, f^m x)))$$

for all $x, y, z \in X$, where $0 \leq q < 1$, $m, n, l \in \mathbb{N}$, $\alpha, \beta, \gamma, \delta \in [0, +\infty)$ and $\theta = \alpha + \beta + \gamma + \delta$; then f, g and h have a unique common fixed point (say u), and f^m, g^n, h^l are all G-continuous at u.

Corollary 9. *Let (X, G) be a complete G-metric space and G is weakly generalized contractive with respect to F and φ. Suppose $T : X \to X$ be a self-mapping in X, which satisfies the following condition*

$$F(H_G^{\theta}(Tx, Ty, Tz)) \leq F(qH_G^{\alpha}(x,y,z)H_G^{\beta}(x, Tx, Ty)H_G^{\gamma}(y, Ty, Tz)H_G^{\delta}(z, Tz, Tx)) - \varphi(F(qH_G^{\alpha}(x,y,z)H_G^{\beta}(x, Tx, Ty)H_G^{\gamma}(y, Ty, Tz)H_G^{\delta}(z, Tz, Tx)))$$

for all $x, y, z \in X$, where $0 \leq q < 1$, $\alpha, \beta, \gamma, \delta \in [0, +\infty)$ and $\theta = \alpha + \beta + \gamma + \delta$; then T has a unique fixed point (say u), and T is G-continuous at u.

Now, we list some special cases of Theorem 2, and we get some Corollaries in the sequel.

Corollary 10. *Let (X, G) be a complete G-metric space and G is weakly generalized contractive with respect to F and φ. Suppose f, g and h are three mappings of X into itself. If one of the following conditions is satisfied*

(1) $F(H_G(fx, gy, hz)) \leq F(qH_G(x,y,z)) - \varphi(F(qH_G(x,y,z)))$;
(2) $F(H_G(fx, gy, hz)) \leq F(qH_G(x, fx, gy)) - \varphi(F(qH_G(x, fx, gy)))$;
(3) $F(H_G(fx, gy, hz) \leq F(qH_G(y, gy, hz)) - \varphi(F(qH_G(y, gy, hz)))$;
(4) $F(H_G(fx, gy, hz) \leq F(qH_G(z, hz, fx)) - \varphi(F(qH_G(z, hz, fx)))$
for all $x, y, z \in X$, where $0 \leq q < 1$; then f, g and h have a unique common fixed point (say u) and f, g, h are all G-continuous at u.

Corollary 11. *Let (X, G) be a complete G-metric space and G is weakly generalized contractive with respect to F and φ. Suppose f, g and h are three mappings of X into itself. If one of the following conditions is satisfied*

(1) $F(H_G^2(fx, gy, hz)) \leq F(qH_G(x,y,z)H_G(x, fx, gy)) - \varphi(F(qH_G(x,y,z) H_G(x, fx, gy)))$;
(2) $F(H_G^2(fx, gy, hz)) \leq F(qH_G(x,y,z)H_G(y, gy, hz)) - \varphi(F(qH_G(x,y,z) H_G(y, gy, hz)))$;
(3) $F(H_G^2(fx, gy, hz)) \leq F(qH_G(x,y,z)H_G(z, hz, fx)) - \varphi(F(qG(x,y,z) H_G(z, hz, fx)))$;
(4) $F(H_G^2(fx, gy, hz)) \leq F(qH_G(x, fx, gy)G(y, gy, hz)) - \varphi(F(qH_G(x, fx, gy) H_G(y, gy, hz)))$;

(5) $F(H_G^2(fx,gy,hz)) \leq F(qH_G(y,gy,hz)G(z,hz,fx)) - \varphi(F(qH_G(y,gy,hz) H_G(z,hz,fx)))$;

(6) $F(H_G^2(fx,gy,hz)) \leq F(qH_G(x,fx,gy)G(z,hz,fx)) - \varphi(F(qH_G(x,fx,gy) H_G(z,hz,fx)))$

for all $x,y,z \in X$, *where* $0 \leq q < 1$; *then* f,g *and* h *have a unique common fixed point (say* u*) and* f,g,h *are all* G*-continuous at* u.

Corollary 12. *Let* (X,G) *be a complete* G*-metric space and* G *is weakly generalized contractive with respect to* F *and* φ. *Suppose* f,g *and* h *are three mappings of* X *into itself. If one of the following conditions is satisfied*

(1)
$$\begin{aligned} F(H_G^3(fx,gy,hz)) &\leq F(qH_G(x,y,z)H_G(x,fx,gy)H_G(y,gy,hz)) \\ &-\varphi(F(qH_G(x,y,z)H_G(x,fx,gy)H_G(y,gy,hz))); \end{aligned}$$

(2)
$$\begin{aligned} F(H_G^3(fx,gy,hz)) &\leq F(qH_G(x,y,z)H_G(x,fx,gy)H_G(z,hz,fx)) \\ &-\varphi(F(qH_G(x,y,z)H_G(x,fx,gy)H_G(z,hz,fx))); \end{aligned}$$

(3)
$$\begin{aligned} F(H_G^3(fx,gy,hz)) &\leq F(qH_G(x,y,z)H_G(y,gy,hz)H_G(z,hz,fx)) \\ &-\varphi(F(qH_G(x,y,z)H_G(y,gy,hz)H_G(z,hz,fx))); \end{aligned}$$

(4)
$$\begin{aligned} F(H_G^3(fx,gy,hz)) &\leq F(qH_G(x,fx,gy)H_G(y,gy,hz)H_G(z,hz,fx)) \\ &-\varphi(F(qH_G(x,fx,gy)H_G(y,gy,hz)H_G(z,hz,fx))) \end{aligned}$$

for all $x,y,z \in X$, *where* $0 \leq q < 1$; *then* f,g *and* h *have a unique common fixed point (say* u*) and* f,g,h *are all* G*-continuous at* u.

Corollary 13. *Let* (X,G) *be a complete* G*-metric space and* G *is weakly generalized contractive with respect to* F *and* φ. *Suppose* f,g *and* h *are three mappings of* X *into itself. If one of the following conditions is satisfied*

$$\begin{aligned} F(H_G^4(fx,gy,hz)) &\leq F(qH_G(x,y,z)H_G(x,fx,gy)H_G(y,gy,hz)H_G(z,hz,fx)) \\ &-\varphi(F(qH_G(x,y,z)H_G(x,fx,gy)H_G(y,gy,hz)H_G(z,hz,fx))) \end{aligned}$$

for all $x,y,z \in X$, *where* $0 \leq q < 1$; *then* f,g *and* h *have a unique common fixed point (say* u*) and* f,g *and* h *are all* G*-continuous at* u.

Now, we introduce an example to support the validity of our results.

Example 1. Let $X = \{0,1,2\}$ be a set with G-metric defined by (Table 1)

Table 1. The definition of G-metric on X.

(x,y,z)	$G(x,y,z)$
$(0,0,0),(1,1,1),(2,2,2)$,	0
$(1,2,2),(2,1,2),(2,2,1)$,	1
$(0,0,1),(0,1,0),(1,0,0),(0,1,1),(1,0,1),(1,1,0)$,	2
$(0,0,2),(0,2,0),(2,0,0),(0,2,2),(2,0,2),(2,2,0)$,	3
$(1,1,2),(1,2,1),(2,1,1),(0,1,2),(0,2,1),(1,0,2),(1,2,0),(2,0,1),(2,1,0)$	4

Note that G is non-symmetric as $H_G(1,2,2) \neq H_G(1,1,2)$. Define $F(t) = I, \varphi(t) = (1-q)t$. Let $f,g,h : X \to X$ be define by (Table 2)

Table 2. The definition of maps f, g and h on X.

x	$f(x)$	$g(x)$	$h(x)$
0	2	1	2
1	2	2	2
2	2	2	2

Case 1. If $y \neq 0$, have $fx = gy = hz = 2$, then

$$F(H_G^2(fx, gy, hz)) = F(H_G^2(2,2,2)) = F(0) = 0$$
$$\leq F(\frac{1}{2}H_G(x, fx, gy)H_G(y, gy, hz))$$
$$-\varphi(F(\frac{1}{2}H_G(x, fx, gy)H_G(y, gy, hz))).$$

Case 2. If $y = 0$, then $fx = hz = 2$ and $gy = 1$, hence

$$F(H_G^2(fx, gy, hz)) = F(H_G^2(2,1,2)) = F(1) = 1.$$

We divide the study in three sub-cases:

(a) If $(x, y, z) = (0, 0, z), z \in \{0, 1, 2\}$, then we have

$$\begin{aligned} F(H_G^2(fx, gy, hz)) &= 1 \\ &\leq F(\frac{1}{2}H_G(0,2,1)H_G(0,1,2)) - \varphi(F(\frac{1}{2}H_G(0,2,1)H_G(0,1,2))) \\ &\leq F(\frac{1}{2} \cdot 4 \cdot 4) - \varphi(F(\frac{1}{2} \cdot 4 \cdot 4)) \\ &\leq F(8) - \varphi(F(8) = 8 - \varphi(8) = 8 - (1 - \frac{1}{2})8 = 4 \end{aligned}$$

(b) If $(x, y, z) = (1, 0, z), z \in \{0, 1, 2\}$, then we have

$$\begin{aligned} F(H_G^2(fx, gy, hz)) &= 1 \\ &\leq F(\frac{1}{2}H_G(1,2,1)H_G(0,1,2)) - \varphi(F(\frac{1}{2}H_G(1,2,1)H_G(0,1,2))) \\ &\leq F(\frac{1}{2} \cdot 4 \cdot 4) - \varphi(F(\frac{1}{2} \cdot 4 \cdot 4)) \\ &\leq F(8) - \varphi(F(8) = 8 - \varphi(8) = 8 - (1 - \frac{1}{2})8 = 4 \end{aligned}$$

(c) If $(x, y, z) = (2, 0, z), z \in \{0, 1, 2\}$, then we have

$$\begin{aligned} F(H_G^2(fx, gy, hz)) &= 1 \\ &\leq F(\frac{1}{2}H_G(2,2,1)H_G(0,1,2)) - \varphi(F(\frac{1}{2}H_G(2,2,1)H_G(0,1,2))) \\ &\leq F(\frac{1}{2} \cdot 1 \cdot 4) - \varphi(F(\frac{1}{2} \cdot 1 \cdot 4)) \\ &\leq F(2) - \varphi(F(2) = 2 - \varphi(2) = 2 - (1 - \frac{1}{2})2 = 1. \end{aligned}$$

In all above cases, inequality (4) of Corollary 11 is satisfied for $q = \frac{1}{2}$. Clearly, 2 is the unique common fixed point for all of the three mappings f, g and h.

3 Applications

Throughout this section, we assume that $X = C([0,T])$ be the set of all continuous functions defined on $[0,T]$. Define $G : X \times X \times X \to \mathbb{R}^+$ by

$$H_G(x,y,z) = \sup_{t\in[0,T]} |x(t) - y(t)| + \sup_{t\in[0,T]} |y(t) - z(t)| + \sup_{t\in[0,T]} |z(t) - x(t)| . \quad (7)$$

Then (X, G) is a G-complete metric spaces. And let G is weakly generalized contractive with respect to F and φ.

Consider the integral equations:

$$\begin{aligned} x(t) &= p(t) + \int_0^T K_1(t,s,x(s))ds, \ \ t \in [0,T], \\ y(t) &= p(t) + \int_0^T K_2(t,s,y(s))ds, \ \ t \in [0,T], \\ z(t) &= p(t) + \int_0^T K_3(t,s,z(s))ds, \ \ t \in [0,T], \end{aligned} \quad (8)$$

where $T > 0$, $K_1, K_2, K_3 : [0,T] \times [0,T] \times \mathbb{R} \to \mathbb{R}$.

The aim of this section is to give an existence theorem for a solution of the above integral equations by using the obtained result given by Corollary 4.

Theorem 3. *Suppose the following conditions hold:*

(i) $K_1, K_2, K_3 : [0,T] \times [0,T] \times \mathbb{R} \to \mathbb{R}$ *are all continuous,*
(ii) There exist a continuous function $H : [0,T] \times [0,T] \to \mathbb{R}^+$ *such that*

$$|K_i(t,s,u) - K_j(t,s,v)| \le H(t,s) |u - v| , \ \ i,j = 1,2,3 \quad (9)$$

for each comparable $u, v \in \mathbb{R}$ *and each* $t, s \in [0,T]$,

(iii) $\sup_{t\in[0,T]} \int_0^T H(t,s)ds \le q$ *for some* $q < 1$. *Then the integral equations (8) has a unique common solution* $u \in C([0,T])$.

Proof. Define $f, g, h : C([0,T]) \to C([0,T])$ by

$$\begin{aligned} fx(t) &= p(t) + \int_0^T K_1(t,s,x(s))ds, \ \ t \in [0,T], \\ gy(t) &= p(t) + \int_0^T K_2(t,s,y(s))ds, \ \ t \in [0,T], \\ hz(t) &= p(t) + \int_0^T K_3(t,s,z(s))ds, \ \ t \in [0,T]. \end{aligned} \quad (10)$$

For all $x, y, z \in C([0,T])$, from (7), (9), (10) and the condition (iii), we have

$$
\begin{aligned}
F(H_G(fx, gy, hz)) &= F(\sup_{t\in[0,T]} |fx(t) - gy(t)| + \sup_{t\in[0,T]} |gy(t) - hz(t)| \\
&\quad + \sup_{t\in[0,T]} |hz(t) - fx(t)|) - \varphi(F(\sup_{t\in[0,T]} |fx(t) - gy(t)| \\
&\quad + \sup_{t\in[0,T]} |gy(t) - hz(t)| + \sup_{t\in[0,T]} |hz(t) - fx(t)|)) \\
&\le F\Big(\sup_{t\in[0,T]} \Big|\int_0^T (K_1(t,s,x(s)) - K_2(t,s,y(s)))\,ds\Big| \\
&\quad + \sup_{t\in[0,T]} \Big|\int_0^T (K_2(t,s,y(s)) - K_3(t,s,z(s)))\,ds\Big| \\
&\quad + \sup_{t\in[0,T]} \Big|\int_0^T (K_3(t,s,z(s)) - K_1(t,s,x(s)))\,ds\Big|\Big) \\
&\quad -\varphi\Big(F\Big(\sup_{t\in[0,T]} \Big|\int_0^T (K_1(t,s,x(s)) - K_2(t,s,y(s)))\,ds\Big| \\
&\quad + \sup_{t\in[0,T]} \Big|\int_0^T (K_2(t,s,y(s)) - K_3(t,s,z(s)))\,ds\Big| \\
&\quad + \sup_{t\in[0,T]} \Big|\int_0^T (K_3(t,s,z(s)) - K_1(t,s,x(s)))\,ds\Big|\Big)\Big) \\
&\le F\Big(\sup_{t\in[0,T]} \int_0^T |K_1(t,s,x(s)) - K_2(t,s,y(s))|\,ds \\
&\quad + \sup_{t\in[0,T]} \int_0^T |K_2(t,s,y(s)) - K_3(t,s,z(s))|\,ds \\
&\quad + \sup_{t\in[0,T]} \int_0^T |K_3(t,s,z(s)) - K_1(t,s,x(s))|\,ds\Big) \\
&\quad -\varphi\Big(F\Big(\sup_{t\in[0,T]} \int_0^T |K_1(t,s,x(s)) - K_2(t,s,y(s))|\,ds \\
&\quad + \sup_{t\in[0,T]} \int_0^T |K_2(t,s,y(s)) - K_3(t,s,z(s))|\,ds \\
&\quad + \sup_{t\in[0,T]} \int_0^T |K_3(t,s,z(s)) - K_1(t,s,x(s))|\,ds\Big)\Big) \\
&\le F\Big(\sup_{t\in[0,T]} \int_0^T H(t,s)|x(s) - y(s)|ds + \sup_{t\in[0,T]} \int_0^T H(t,s)|y(s) - z(s)|ds \\
&\quad + \sup_{t\in[0,T]} \int_0^T H(t,s)|z(s) - x(s)|ds\Big) \\
&\quad -\varphi\Big(F\Big(\sup_{t\in[0,T]} \int_0^T H(t,s)|x(s) - y(s)|ds \\
&\quad + \sup_{t\in[0,T]} \int_0^T H(t,s)|y(s) - z(s)|ds
\end{aligned}
$$

$$
\begin{aligned}
&+\sup_{t\in[0,T]}\int_0^T H(t,s)|z(s)-x(s)|ds\Big)\Big)\\
&\leq F\Bigg(\Big(\sup_{t\in[0,T]}\int_0^T H(t,s)ds\Big)\Big(\sup_{t\in[0,T]}|x(t)-y(t)|\Big)\\
&+\Big(\sup_{t\in[0,T]}\int_0^T H(t,s)ds\Big)\Big(\sup_{t\in[0,T]}|y(t)-z(t)|\Big)\\
&+\Big(\sup_{t\in[0,T]}\int_0^T H(t,s)ds\Big)\Big(\sup_{t\in[0,T]}|z(t)-x(t)|\Big)\Bigg)\\
&-\varphi\Bigg(F\Bigg(\Big(\sup_{t\in[0,T]}\int_0^T H(t,s)ds\Big)\Big(\sup_{t\in[0,T]}|x(t)-y(t)|\Big)\\
&+\Big(\sup_{t\in[0,T]}\int_0^T H(t,s)ds\Big)\Big(\sup_{t\in[0,T]}|y(t)-z(t)|\Big)\Bigg)\Bigg)\\
&+\Big(\sup_{t\in[0,T]}\int_0^T H(t,s)ds\Big)\Big(\sup_{t\in[0,T]}|z(t)-x(t)|\Big)\Bigg)\Bigg)
\end{aligned}
$$

$$
\begin{aligned}
&\leq F\Bigg(\Big(\sup_{t\in[0,T]}\int_0^T H(t,s)ds\Big)\Big(\sup_{t\in[0,T]}|x(t)-y(t)|+\sup_{t\in[0,T]}|y(t)-z(t)|+\sup_{t\in[0,T]}|z(t)-x(t)|\Big)\Bigg)\\
&-\varphi\Bigg(F\Bigg(\Big(\sup_{t\in[0,T]}\int_0^T H(t,s)ds\Big)\Big(\sup_{t\in[0,T]}|x(t)-y(t)|+\sup_{t\in[0,T]}|y(t)-z(t)|+\sup_{t\in[0,T]}|z(t)-x(t)|\Big)\Bigg)\Bigg)\\
&\leq F(qG(x,y,z))-\varphi(F(qG(x,y,z))).
\end{aligned}
$$

This proves that the operators f, g, h satisfies the contractive condition (1) appearing in Corollary 4, and hence f, g, h have a unique common fixed point $u \in C([0,T])$, that is, u is a unique common solution to the integral equations (7).

Corollary 14. *Suppose the following hypothesis hold:*

(i) $K : [0,T] \times [0,T] \times \mathbb{R} \to \mathbb{R}$ *are all continuous,*
(ii) There exist a continuous function $H : [0,T] \times [0,T] \to \mathbb{R}^+$ *such that*

$$|K(t,s,u) - K(t,s,v)| \leq H(t,s)\,|u-v| \tag{11}$$

for each comparable $u, v \in \mathbb{R}$ *and each* $t, s \in [0,T]$,
(iii) $\sup_{t\in[0,T]} \int_0^T H(t,s)ds \leq q$ *for some* $q < 1$.
Then the integral equation

$$x(t) = p(t) + \int_0^T K(t,s,x(s))ds, \quad t \in [0,T], \tag{12}$$

has a unique common solution $u \in C([0,T])$.

Proof. Taking $K_1 = K_2 = K_3 = K$ in Theorem 3, then the conclusion of Corollary 14 can be obtained from Theorem 3 immediately.

Acknowledgements. First author would like to thank the research professional development project under scholarship of Rajabhat Rajanagarindra University (RRU) financial support. Second author was supported by Muban Chombueng Rajabhat University. Third author thank for Theoretical and Computational Science Center (TaCS), Science Laboratory Building, Faculty of Science, King Mongkut's University of Technology Thonburi (KMUTT), Bangkok, Thailand, and guidance of the fifth author, Gyeongsang National University, Jinju 660-701, Korea.

References

1. Abbas, M., Nazir, T., Radenović, S.: Some periodic point results in generalized metric spaces. Appl. Math. Comput. **217**, 4094–4099 (2010)
2. Abbas, M., Rhoades, B.E.: Common fixed point results for non-commuting mappings without continuity in generalized metric spaces. Appl. Math. Comput. **215**, 262–269 (2009)
3. Banach, S.: Sur les opérations dans les ensembles abstraits et leur application aux équations integrals. Fund. Math. **3**, 133–181 (1922)
4. Gu, F., Ye, H.: Fixed point theorems for a third power type contraction mappings in G-metric spaces. Hacettepe J. Math. Stats. **42**(5), 495–500 (2013)
5. Gu, F., Ye, H.: Common fixed point for mappings satisfying new contractive condition and applications to integral equations. J. Nonlinear Sci. Appl. **10**, 3988–3999 (2017)
6. Ye, H., Gu, F.: Common fixed point theorems for a class of twice Power type contraction maps in G-metric spaces. Abstr. Appl. Anal. Article ID 736214, 19 pages (2012)
7. Ye, H., Gu, F.: A new common fixed point theorem for a class of four power type contraction mappings. J. Hangzhou Normal Univ. (Nat. Sci. Ed.) **10**(6), 520–523 (2011)
8. Jleli, M., Samet, B.: Remarks on G-metric spaces and fixed point theorems. Fixed Point Theory Appl. **210**, 7 pages (2012)
9. Karapinar, E., Agarwal, R.: A generalization of Banach's contraction principle. Fixed Point Theory Appl. **154**, 14 pages (2013)
10. Kaewcharoen, A., Kaewkhao, A.: Common fixed points for single-valued and multi-valued mappings in G-metric spaces. Int. J. Math. Anal. **5**, 1775–1790 (2011)
11. Mustafa, Z., Aydi, H., Karapinar, E.: On common fixed points in G-metric spaces using $(E.A)$-property. Comput. Math. Appl. **64**(6), 1944–1956 (2012)
12. Mustafa, Z., Obiedat, H., Awawdeh, H.: Some fixed point theorem for mappings on complete G-metric spaces. Fixed Point Theory Appl. Article ID 189870, 12 pages (2008)
13. Mustafa, Z., Sims, B.: A new approach to generalized metric spaces. J. Nonlinear Convex Anal. **7**(2), 289–297 (2006)
14. Rhoades, B.E.: Some theorems on weakly contractive maps. Nonlinear Anal. **47**, 2683–2693 (2001)
15. Samet, B., Vetro, C., Vetro, F.: Remarks on G-metric spaces. Internat. J. Anal. Article ID 917158, 6 pages (2013)

16. Shatanawi, W.: Fixed point theory for contractive mappings satisfying Φ-maps in G-metric spaces. Fixed Point Theory Appl. Article ID 181650 (2010)
17. Tahat, N., Aydi, H., Karapinar, E., Shatanawi, W.: Common fixed points for single-valued and multi-valued maps satisfying a generalized contraction in G-metric spaces. Fixed Point Theory Appl. **48**, 9 pages (2012)
18. Alber, Y.I., Guerre-Delabriere, S.: Principle of weakly contractive maps in Hilbert spaces. New Results Oper. Theory Appl. **98**, 7–22 (1997)

A Note on Some Recent Strong Convergence Theorems of Iterative Schemes for Semigroups with Certain Conditions

Phumin Sumalai[1], Ehsan Pourhadi[2], Khanitin Muangchoo-in[3,4], and Poom Kumam[3,4](✉)

[1] Department of Mathematics, Faculty of Science and Technology, Muban Chombueng Rajabhat University, 46 M.3, Chombueng 70150, Ratchaburi, Thailand
phumin.su28@gmail.com

[2] School of Mathematics, Iran University of Science and Technology, Narmak, 16846-13114 Tehran, Iran
epourhadi@alumni.iust.ac.ir

[3] KMUTTFixed Point Research Laboratory, Department of Mathematics, Room SCL 802 Fixed Point Laboratory, Science Laboratory Building, Faculty of Science, King Mongkut's University of Technology Thonburi (KMUTT), 126 Pracha-Uthit Road, Bang Mod, Thrung Khru, Bangkok 10140, Thailand
kanitin22@gmail.com

[4] KMUTT-Fixed Point Theory and Applications Research Group (KMUTT-FPTA) Theoretical and Computational Science Center (TaCS), Science Laboratory Building, Faculty of Science, King Mongkut's University of Technology Thonburi (KMUTT), 126 Pracha-Uthit Road, Bang Mod, Thrung Khru, Bangkok 10140, Thailand
poom.kum@kmutt.ac.th

Abstract. In this note, suggesting an alternative technique we partially modify and fix the proofs of some recent results focused on the strong convergence theorems of iterative schemes for semigroups including a specific error observed frequently in several papers during the last years. Moreover, it is worth mentioning that there is no new constraint invloved in the modification process presented throughout this note.

Keywords: Nonexpansive semigroups · Strong convergence
Variational inequality · Strict pseudo-contraction
Strictly convex Banach spaces · Fixed point

1 Introduction

Throughout this note, we suppose that E is a real Banach space, E^* is the dual space of E, C is a nonempty closed convex subset of E, and $\mathbb{R}^+$ and $\mathbb{N}$ are the set

V. Kreinovich et al. (Eds.): ECONVN 2019, SCI 809, pp. 251–261, 2019.
https://doi.org/10.1007/978-3-030-04200-4_19

of nonnegative real numbers and positive integers, respectively. The normalized duality mapping $J : E \to 2^{E^*}$ is defined by

$$J(x) = \{x^* \in E^* : \langle x, x^* \rangle = ||x||^2 = ||x^*||^2\}, \ \forall x \in E$$

where $\langle \cdot, \cdot \rangle$ denotes the generalized pairing. It is well-known that if E is smooth, then J is single-valued, which is denoted by j.

Let $T : C \to C$ be a mapping. We use $F(T)$ to denote the set of fixed points of T. If $\{x_n\}$ is a sequence in E, we use $x_n \to x$ ($x_n \rightharpoonup x$) to denote strong (weak) convergence of the sequence $\{x_n\}$ to x.

Recall that a mapping $f : C \to C$ is called a contraction on C if there exists a constant $\alpha \in (0, 1)$ such that

$$||f(x) - f(y)|| \leq \alpha ||x - y||, \ \forall x, y \in C.$$

We use $\prod_C$ to denote the collection of mappings f satisfying the above inequality.

$$\prod_C = \{f : C \to C \mid f \text{ is a contraction with some constant } \alpha\}.$$

Note that each $f \in \prod_C$ has a unique fixed point in C, (see [1]). And note that if $\alpha = 1$ we call nonexpansive mapping.

Let H be a real Hilbert space, and assume that A is a strongly positive bounded linear operator (see [2]) on H, that is, there is a constant $\overline{\gamma} > 0$ with the property

$$\langle Ax, J(x) \rangle \geq \overline{\gamma} ||x||^2, \ \forall x, y \in H. \tag{1}$$

Then we can construct the following variational inequality problem with viscosity. Find $x^* \in C$ such that

$$\langle (A - \gamma f)x^*, x - x^* \rangle \geq 0, \ \forall x \in F(T), \tag{2}$$

which is the optimality condition for the minimization problem

$$\min_{x \in F(T)} \{\frac{1}{2} \langle Ax, x \rangle - h(x)\},$$

where h is a potential function for γf (i.e., $h'(x) = \gamma f(x)$ for $x \in H$), and γ is a suitable positive constant.

Recall that a mapping $T : K \to K$ is said to be a strict pseudo-contraction if there exists a constant $0 \leq k < 1$ such that

$$||Tx - Ty||^2 \leq ||x - y||^2 + k||(I - T)x - (I - T)y||^2 \tag{3}$$

for all $x, y \in K$ (if (3) holds, we also say that T is a k-strict pseudo-contraction).

The concept of strong convergence of iterative schemes for family of mapping and study on variational inequality problem have been argued extensively. Recently, some results with a special flaw in the step of proof to reach (2) have been observed which needs to be reconsidered and corrected. The existence of this error which needs a meticulous look to be seen motivates us to fix it and also warn the researchers to take another path when arriving at the mentioned step of proof.

2 Some Iterative Processes for a Finite Family of Strict Pseudo-contractions

In this section, focusing on the strong convergence theorems of iterative process for a finite family of strict pseudo-contractions, we list the main results of some recent articles which all utilized a same procedure (with a flaw) in a part of the proof. In order to amend the observed flaw we ignore some paragraphs in the corresponding proofs and fill them by the computations extracted by our simple technique.

In 2009, Qin et al. [3] presented the following nice result. They obtained a strong convergence theorem of modified Mann iterative process for strict pseudo-contractions in Hilbert space H. The sequence $\{x_n\}$ was defined by

$$\begin{cases} x_1 = x \in K, \\ y_n = P_k[\beta_n x_n + (1-\beta_n)Tx_n], \\ x_{n+1} = \alpha_n \gamma f(x_n) + (I - \alpha_n A)y_n, \ \forall n \geq 1. \end{cases} \tag{4}$$

Theorem 1 ([3]). *Let K be a closed convex subset of a Hilbert space H such that $K+K \subset K$ and $f \in \prod_K$ with the coefficient $0 < \alpha < 1$. Let A be a strongly positive linear bounded operator with the coefficient $\bar{\gamma} > 0$ such that $0 < \gamma < \frac{\bar{\gamma}}{\alpha}$ and let $T : K \to H$ be a k-strictly pseudo-contractive non-selfmapping such that $F(T) \neq \emptyset$. Given sequences $\{\alpha_n\}_{n=0}^{\infty}$ and $\{\beta_n\}_{n=0}^{\infty}$ in $[0,1]$, the following control conditions are satisfied*

(i) $\sum_{n=0}^{\infty} \alpha_n = \infty$, $\lim_{n\to\infty} \alpha_n = 0$;
(ii) $k \leq \beta_n \leq \lambda < 1$ *for all* $n \geq 1$;
(iii) $\sum_{n=1}^{\infty} |\alpha_{n+1} - \alpha_n| < \infty$ *and* $\sum_{n=1}^{\infty} |\beta_{n+1} - \beta_n| < \infty$.

Let $\{x_n\}_{n=1}^{\infty}$ be the sequence generated by the composite process (4) Then $\{x_n\}_{n=1}^{\infty}$ converges strongly to $q \in F(T)$, which also solves the following variational inequality

$$\langle \gamma f(q) - Aq, p - q \rangle \leq 0, \quad \forall p \in F(T).$$

In the proof of Theorem 1, in order to prove

$$\limsup_{t\to 0} \limsup_{n\to\infty} \langle Ax_t - \gamma f(x_t), x_t - x_n \rangle \leq 0, \quad \text{(see (2.15) in [3])}, \tag{5}$$

where x_t solves the fixed point equation $x_t = t\gamma f(x_t) + (I - tA)P_K Sx_t$, using (1) the authors obtained the following inequality

$$((\overline{\gamma}t)^2 - 2\overline{\gamma}t)\|x_t - x_n\|^2 \leq (\overline{\gamma}t^2 - 2t)\langle A(x_t - x_n), x_t - x_n \rangle$$

which is obviously impossible for $0 < t < \frac{2}{\bar{\gamma}}$. We remark that t is supposed to be vanished in the next step of proof. Here, by ignoring the computations (2.10)–(2.14) in [3] we suggest a new way to show (5) without any new condition. First let us recall the following concepts.

Definition 1. Let (X, d) be a metric space and K be a nonempty subset of X. For every $x \in K$, the distance between the point x and K is denoted by $d(x, K)$ and is defined by the following minimization problem:

$$d(x, K) := \inf d(x, y).$$

The *metric projection operator*, also said to be the nearest point mapping onto the set K is the mapping $P_K : X \to 2^K$ defined by

$$P_K(x) := \{z \in K : d(x, z) = d(x, K)\}, \quad \forall x \in X.$$

If $P_K(x)$ is singleton for every $x \in X$, then K is said to be a *Chebyshev set.*

Definition 2 ([4])**.** We say that a metric space (X, d) has property (P) if the metric projection onto any Chebyshev set is a nonexpansive mapping.

For example, any CAT(0) space has property (P). Bring in mind that Hadamard space (i.e., complete CAT(0) space) is a non-linear generalization of a Hilbert space. In the literature they are also equivalently defined as complete CAT(0) spaces.

Now, we are in a position to prove (5).

Proof. To prove inequality (5) we first find an upper bound for $\|x_t - x_n\|^2$ as follows.

$$\begin{aligned} \|x_t - x_n\|^2 &= \langle x_t - x_n, x_t - x_n\rangle \\ &= \langle t\gamma f(x_t) + (I - tA)P_K Sx_t - x_n, x_t - x_n\rangle \\ &= \langle t(\gamma f(x_t) - Ax_t) + t(Ax_t - AP_K Sx_t) \\ &\quad + (P_K Sx_t - P_K Sx_n) + (P_K Sx_n - x_n), x_t - x_n\rangle \\ &\le t\langle \gamma f(x_t) - Ax_t, x_t - x_n\rangle + t\|A\| \cdot \|x_t - P_K Sx_t\| \cdot \|x_t - x_n\| \\ &\quad + \|x_t - x_n\|^2 + \|P_K Sx_n - x_n\| \cdot \|x_t - x_n\|. \end{aligned} \tag{6}$$

We remark that following argument in the proof [3, Theorem 2.1] S is nonexpansive, on the other hand, since H has property (P) hence P_K is nonexpansive and $P_K S$ is so. Now, (6) implies that

$$\begin{aligned} \langle Ax_t - \gamma f(x_t), x_t - x_n\rangle &\le \|A\| \cdot \|x_t - P_K Sx_t\| \cdot \|x_t - x_n\| \\ &\quad + \frac{1}{t}\|P_K Sx_n - x_n\| \cdot \|x_t - x_n\| \\ &= t\|A\| \cdot \|\gamma f(x_t) - AP_K Sx_t\| \cdot \|x_t - x_n\| \\ &\quad + \frac{1}{t}\|P_K Sx_n - x_n\| \cdot \|x_t - x_n\| \\ &\le tM\|A\| \cdot \|\gamma f(x_t) - AP_K Sx_t\| + \frac{M}{t}\|P_K Sx_n - x_n\| \end{aligned} \tag{7}$$

where $M > 0$ is an appropriate constant such that $M \ge \|x_t - x_n\|$ for all $t \in (0, \|A\|^{-1})$ and $n \ge 1$ (we underline that according to [5, Proposition 3.1], the map $t \mapsto x_t$, $t \in (0, \|A\|^{-1})$ is bounded).

Therefore, firstly, utilizing (2.8) in [3], taking upper limit as $n \to \infty$, and then as $t \to 0$ in (7), we obtain that

$$\limsup_{t \to 0} \limsup_{n \to \infty} \langle Ax_t - \gamma f(x_t), x_t - x_n \rangle \leq 0. \tag{8}$$

and the claim is proved.

In what follows we concentrate on a novel result of Marino et al. [6]. They derived a strong convergence theorem of the modified Mann iterative method for strict pseudo-contractions in Hilbert space H as follows.

Theorem 2 ([6]). *Let H be a Hilbert space and let T be a k-strict pseudo-contraction on H such that $F(T) \neq \emptyset$ and f be an α-contraction. Let A be a strongly positive linear bounded self-adjoint operator with coefficient $\bar{\gamma} > 0$. Assume that $0 < \gamma < \frac{\bar{\gamma}}{\alpha}$. Given the initial guess $x_0 \in H$ chosen arbitrarily and given sequences $\{\alpha_n\}_{n=0}^{\infty}$ and $\{\beta_n\}_{n=0}^{\infty}$ in $[0,1]$, satisfying the following conditions*

(i) $\sum_{n=0}^{\infty} \alpha_n = \infty$, $\lim_{n \to \infty} \alpha_n = 0$;
(ii) $\sum_{n=1}^{\infty} |\alpha_{n+1} - \alpha_n| < \infty$ *and* $\sum_{n=1}^{\infty} |\beta_{n+1} - \beta_n| < \infty$;
(iii) $0 \leq k \leq \beta_n \leq \beta < 1$ *for all* $n \geq 1$;

let $\{x_n\}_{n=1}^{\infty}$ and $\{y_n\}_{n=0}^{\infty}$ be the sequences defined by the composite process

$$\begin{cases} y_n = \beta_n x_n + (1 - \beta_n) T x_n, \\ x_{n+1} = \alpha_n \gamma f(x_n) + (I - \alpha_n A) y_n, \ \forall n \geq 1. \end{cases}$$

Then $\{x_n\}_{n=0}^{\infty}$ and $\{y_n\}_{n=0}^{\infty}$ strongly converge to the fixed point q of T which solves the following variational inequality

$$\langle \gamma f(q) - Aq, p - q \rangle \leq 0, \quad \forall p \in F(T).$$

Similar to the arguments for Theorem 1, by ignoring the parts (2.10)–(2.14) in the proof of Theorem 2 we easily obtain the following conclusion.

Proof. Since x_t solves the fixed point equation $x_t = t\gamma f(x_t) + (I - tA)Bx_t$ we get

$$\begin{aligned} \|x_t - x_n\|^2 &= \langle x_t - x_n, x_t - x_n \rangle \\ &= \langle t\gamma f(x_t) + (I - tA)Bx_t - x_n, x_t - x_n \rangle \\ &= \langle t(\gamma f(x_t) - Ax_t) + t(Ax_t - ABx_t) \\ &\quad + (Bx_t - Bx_n) + (Bx_n - x_n), x_t - x_n \rangle \\ &\leq t\langle \gamma f(x_t) - Ax_t, x_t - x_n \rangle + t\|A\| \cdot \|x_t - Bx_t\| \cdot \|x_t - x_n\| \\ &\quad + \|x_t - x_n\|^2 + \|Bx_n - x_n\| \cdot \|x_t - x_n\| \end{aligned} \tag{9}$$

where here we used the fact that $B = kI + (1-k)T$ is a nonexpansive mapping (see [7, Theorem 2]). Now, (9) implies that

$$\begin{aligned}\langle Ax_t - \gamma f(x_t), x_t - x_n\rangle &\leq \|A\|\cdot\|x_t - Bx_t\|\cdot\|x_t - x_n\| \\ &\quad + \frac{1}{t}\|Bx_n - x_n\|\cdot\|x_t - x_n\| \\ &= t\|A\|\cdot\|\gamma f(x_t) - ABx_t\|\cdot\|x_t - x_n\| \\ &\quad + \frac{1}{t}\|Bx_n - x_n\|\cdot\|x_t - x_n\| \\ &\leq tM\|A\|\cdot\|\gamma f(x_t) - ABx_t\| + \frac{M}{t}\|Bx_n - x_n\|\end{aligned} \tag{10}$$

where $M > 0$ is an appropriate constant such that $M \geq \|x_t - x_n\|$ for all $t \in (0, \|A\|^{-1})$ and $n \geq 1$. On the other hand since $\|Bx_n - x_n\| = (1-k)\|Tx_n - x_n\|$, by using (2.8) in [6] and taking upper limit as $n \to \infty$ at first, and then as $t \to 0$ in (10), we arrive at (8) and again the claim is proved.

In 2010, Cai and Hu [8] obtained a nice strong convergence theorem of a general iterative process for a finite family of λ_i-strict pseudo-contractions in q-uniformly smooth Banach space as follows.

Theorem 3 ([8]). *Let E be a real q-uniformly smooth, strictly convex Banach space which admits a weakly sequentially continuous duality mapping J from E to E^* and C is a closed convex subset E which is also a sunny nonexpansive retraction of E such that $C + C \subset C$ with the coefficient $0 < \alpha < 1$. Let A be a strongly positive linear bounded operator with the coefficient $\bar{\gamma} > 0$ such that $0 < \gamma < \frac{\bar{\gamma}}{\alpha}$ and $T_i : C \to E$ be λ_i-strictly pseudo-contractive non-self-mapping such that $F = \cap_{i=1}^N F(T_i) \neq \emptyset$. Let $\lambda = \min\{\lambda_i : 1 \leq i \leq N\}$. Let $\{x_n\}$ be a sequence of C generated by*

$$\begin{cases} x_1 = x \in C, \\ y_n = P_C\left[\beta_n x_n + (1-\beta_n)\sum_{i=1}^{N}\eta_i^{(n)}T_i x_n\right], \\ x_{n+1} = \alpha_n\gamma f(x_n) + \gamma_n x_n + ((1-\gamma_n)I - \alpha_n A)y_n, \ \forall n \geq 1, \end{cases}$$

where f is a contraction, the sequences $\{\alpha_n\}_{n=0}^\infty$, $\{\beta_n\}_{n=0}^\infty$ and $\{\gamma_n\}_{n=0}^\infty$ are in $[0,1]$, assume for each n, $\{\eta_i^{(n)}\}_{i=1}^N$ is a finite sequence of positive numbers such that $\sum_{i=1}^N \eta_i^{(n)} = 1$ for all n and $\eta_i^{(n)} > 0$ for all $1 \leq i < N$. They satisfy the conditions (i)–(iv) of [8, Lemma 2.1] and add to the condition (v) $\gamma_n = O(\alpha_n)$. Then $\{x_n\}$ converges strongly to $z \in F$, which also solves the following variational inequality

$$\langle \gamma f(z) - Az, J(p - z)\rangle \leq 0, \quad \forall p \in F.$$

Proof. Ignoring (2.8)–(2.12) in the proof of Theorem 3 (i.e., [8, Theorem 2.2]) and using the same technique as before we see

$$\begin{aligned}\|x_t - x_n\|^2 &= \langle x_t - x_n, J(x_t - x_n)\rangle \\ &= \langle t\gamma f(x_t) + (I - tA)P_C S x_t - x_n, J(x_t - x_n)\rangle \\ &= \langle t(\gamma f(x_t) - Ax_t) + t(Ax_t - AP_C S x_t) \\ &\quad + (P_C S x_t - P_C S x_n) + (P_C S x_n - x_n), J(x_t - x_n)\rangle \\ &\le t\langle \gamma f(x_t) - Ax_t, J(x_t - x_n)\rangle + t\|A\| \cdot \|x_t - P_C S x_t\| \cdot \|x_t - x_n\| \\ &\quad + \|x_t - x_n\|^2 + \|P_C S x_n - x_n\| \cdot \|x_t - x_n\|\end{aligned} \tag{11}$$

where x_t solves the fixed point equation $x_t = t\gamma f(x_t) + (I - tA)P_C S x_t$. Again, we remark that $P_C S$ is nonexpansive and hence

$$\begin{aligned}&\langle Ax_t - \gamma f(x_t), J(x_t - x_n)\rangle \\ &\le \|A\| \cdot \|x_t - P_C S x_t\| \cdot \|x_t - x_n\| \\ &\quad + \frac{1}{t}\|P_C S x_n - x_n\| \cdot \|x_t - x_n\| \\ &= t\|A\| \cdot \|\gamma f(x_t) - AP_C S x_t\| \cdot \|x_t - x_n\| \\ &\quad + \frac{1}{t}\|P_C S x_n - x_n\| \cdot \|x_t - x_n\| \\ &\le tM\|A\| \cdot \|\gamma f(x_t) - AP_C S x_t\| + \frac{M}{t}\|P_C S x_n - x_n\|\end{aligned} \tag{12}$$

where $M > 0$ is a proper constant such that $M \ge \|x_t - x_n\|$ for $t \in (0, \|A\|^{-1})$ and $n \ge 1$. Thus, taking upper limit as $n \to \infty$ at first, and then as $t \to 0$ in (12), the following yields

$$\limsup_{t\to 0} \limsup_{n\to\infty} \langle Ax_t - \gamma f(x_t), J(x_t - x_n)\rangle \le 0. \tag{13}$$

Finally, in the last part of this section we focus on the main result of Kangtunyakarn and Suantai [9].

Theorem 4 **([9]).** *Let H be a Hilbert space, let f be an α-contraction on H and let A be a strongly positive linear bounded self-adjoint operator with coefficient $\bar{\gamma} > 0$. Assume that $0 < \gamma < \frac{\bar{\gamma}}{\alpha}$. Let $\{T_i\}_{i=1}^N$ be a finite family of κ_i-strict pseudo-contraction of H into itself for some $\kappa_i \in [0, 1)$ and $\kappa = \max\{\kappa_i : i = 1, 2, \cdots, N\}$ with $\bigcap_{i=1}^N F(T_i) \ne \emptyset$. Let S_n be the S-mappings generated by $T_1, T_2, \cdots, T_N$ and $\alpha_1^{(n)}, \alpha_2^{(n)}, \cdots, \alpha_N^{(n)}$, where $\alpha_j^{(n)} = (\alpha_1^{n,j}, \alpha_2^{n,j}, \alpha_3^{n,j}) \in I \times I \times I$, $I = [0, 1]$, $\alpha_1^{n,j} + \alpha_2^{n,j} + \alpha_3^{n,j} = 1$ and $\kappa < a \le \alpha_1^{n,j}$, $\alpha_3^{n,j} \le b < 1$ for all $j = 1, 2, \cdots, N-1$, $\kappa < c \le \alpha_1^{n,N} \le 1$, $\kappa \le \alpha_3^{n,N} \le d < 1$, $\kappa \le \alpha_2^{n,j} \le e < 1$ for all $j = 1, 2, \cdots, N$. For a point $u \in H$ and $x_1 \in H$, let $\{x_n\}$ and $\{y_n\}$ be the sequences defined iteratively by*

$$\begin{cases} y_n = \beta_n x_n + (1 - \beta_n) S_n x_n, \\ x_{n+1} = \alpha_n \gamma(a_n u + (1 - a_n) f(x_n)) + (I - \alpha_n A) y_n, \ \forall n \ge 1, \end{cases}$$

where $\{\alpha_n\}$, $\{\beta_n\}$ *and* $\{a_n\}$ *are the sequences in* $[0,1]$. *Assume that the following conditions hold:*

(i) $\sum_{n=0}^{\infty} \alpha_n = \infty$, $\lim_{n\to\infty} \alpha_n = \lim_{n\to\infty} a_n = 0$;
(ii) $\sum_{n=1}^{\infty} |\alpha_1^{n+1,j} - \alpha_1^{n,j}| < \infty$, $\sum_{n=1}^{\infty} |\alpha_3^{n+1,j} - \alpha_3^{n,j}| < \infty$ *for all* $j \in \{1,2,\cdots,N\}$, $\sum_{n=1}^{\infty} |\alpha_{n+1} - \alpha_n| < \infty$, $\sum_{n=1}^{\infty} |\beta_{n+1} - \beta_n| < \infty$ *and* $\sum_{n=1}^{\infty} |a_{n+1} - a_n| < \infty$;
(iii) $0 \le \kappa \le \beta_n < \theta < 1$ *for all* $n \ge 1$ *and some* $\theta \in (0,1)$.

Then both $\{x_n\}$ *and* $\{y_n\}$ *strongly converge to* $q \in \bigcap_{i=1}^N F(T_i)$, *which solves the following variational inequality*

$$\langle \gamma f(q) - Aq, p - q\rangle \le 0, \quad \forall p \in \bigcap_{i=1}^{N} F(T_i).$$

Proof. In the proof of Theorem 4 (i.e., [9, Theorem 3.1]), leaving the inequlities (3.9)–(3.10) behind and applying the same technique as mentioned before we derive

$$\begin{aligned}\|x_t - x_n\|^2 &= \langle x_t - x_n, x_t - x_n\rangle \\ &= \langle t\gamma f(x_t) + (I - tA)S_n x_t - x_n, x_t - x_n\rangle \\ &= \langle t(\gamma f(x_t) - Ax_t) + t(Ax_t - AS_n x_t) \\ &\quad +(S_n x_t - S_n x_n) + (S_n x_n - x_n), x_t - x_n\rangle \\ &\le t\langle \gamma f(x_t) - Ax_t, x_t - x_n\rangle + t\|A\| \cdot \|x_t - S_n x_t\| \cdot \|x_t - x_n\| \\ &\quad + \|x_t - x_n\|^2 + \|S_n x_n - x_n\| \cdot \|x_t - x_n\|\end{aligned} \tag{14}$$

where x_t solves the fixed point equation $x_t = t\gamma f(x_t) + (I - tA)S_n x_t$. Here, we notify that S_n is nonexpansive and hence

$$\begin{aligned}&\langle Ax_t - \gamma f(x_t), x_t - x_n\rangle \\ &\le \|A\| \cdot \|x_t - S_n x_t\| \cdot \|x_t - x_n\| + \frac{1}{t}\|S_n x_n - x_n\| \cdot \|x_t - x_n\| \\ &= t\|A\| \cdot \|\gamma f(x_t) - AS_n x_t\| \cdot \|x_t - x_n\| \\ &\quad + \frac{1}{t}\|S_n x_n - x_n\| \cdot \|x_t - x_n\| \\ &\le tM\|A\| \cdot \|\gamma f(x_t) - AS_n x_t\| + \frac{M}{t}\|S_n x_n - x_n\|\end{aligned} \tag{15}$$

where $M > 0$ is a proper constant such that $M \ge \|x_t - x_n\|$ for $t \in (0, \|A\|^{-1})$ and $n \ge 1$. Thus, following (3.8) in [9], taking upper limit as $n \to \infty$ at first, and then as $t \to 0$ in (15), the following yields

$$\limsup_{t\to 0} \limsup_{n\to\infty} \langle Ax_t - \gamma f(x_t), x_t - x_n\rangle \le 0$$

and the claim is proved.

3 General Iterative Scheme for Semigroups of Uniformly Asymptotically Regular Nonexpansive Mappings

Throughout this section, we focus on the main result of Yang [10] as follows. First, we recall that a continuous operator of the semigroup $\mathscr{T} = \{T(t) : 0 \leq t < \infty\}$ is said to be uniformly asymptotically regular (u.a.r.) on K if for all $h \geq 0$ and any bounded subset C of K, $\lim_{t\to\infty} \sup_{x\in C} \|T(h)T(t)x - T(t)x\| = 0$.

Theorem 5 **([10]).** *Let K be a nonempty closed convex subset of a reflexive, smooth and strictly convex Banach space E with a uniformly Gâteaux differentiable norm. Let $\mathscr{T} = \{T(t) : t \geq 0\}$ be a uniformly asymptotically regular nonexpansive semigroup on K such that $F(\mathscr{T}) \neq \emptyset$, and $f \in \Pi_K$. Let A be a strongly positive linear bounded self-adjoint operator with coefficient $\bar{\gamma} > 0$. Let $\{x_n\}$ be a sequence generated by*

$$x_{n+1} = \alpha_n \gamma f(x_n) + \delta_n x_n + ((1-\delta_n)I - \alpha_n A)T(t_n)x_n,$$

such that $0 < \gamma < \frac{\bar{\gamma}}{\alpha}$, the given sequences $\{x_n\}$ and $\{\delta_n\}$ are in $(0,1)$ satisfying the following conditions:

(i) $\sum_{n=0}^{\infty} \alpha_n = \infty$, $\lim_{n\to\infty} \alpha_n = 0$;
(ii) $0 < \liminf_{n\to\infty} \delta_n \leq \limsup_{n\to\infty} \delta_n < 1$;
(iii) $h, t_n \geq 0$ *such that* $t_{n+1} - t_n = h$ *and* $\lim_{n\to\infty} t_n = \infty$.

Then $\{x_n\}$ converges strongly to q, as $n \to \infty$, q is the element of $F(\mathscr{T})$ such that q is the unique solution in $F(\mathscr{T})$ to the variational inequality

$$\langle (A - \gamma f)q, j(q-z)\rangle \leq 0, \quad \forall z \in F(\mathscr{T}).$$

Proof. Ignoring (3.15)–(3.17) in the proof of [10, Theorem 3.5] and using the same technique as before we see that

$$\begin{aligned} \|u_m - x_n\|^2 =& \langle u_m - x_n, j(u_m - x_n)\rangle \\ =& \langle \alpha_m \gamma f(u_m) + (I - \alpha_m A)S(t_m)u_m - x_n, j(u_m - x_n)\rangle \\ =& \langle \alpha_m(\gamma f(u_m) - Au_m) + \alpha_m(Au_m - AS(t_m)u_m) \\ &+ (S(t_m)u_m - S(t_m)x_n) + (S(t_m)x_n - x_n), j(u_m - x_n)\rangle \\ \leq& \alpha_m \langle \gamma f(u_m) - Au_m, j(u_m - x_n)\rangle + \alpha_m \|A\| \\ &\cdot \|u_m - S(t_m)u_m\| \cdot \|u_m - x_n\| + \|u_m - x_n\|^2 \\ &+ \|S(t_m)x_n - x_n\| \cdot \|u_m - x_n\| \end{aligned} \tag{16}$$

where $u_m \in K$ is the unique solution of the fixed point problem $u_m = \alpha_m \gamma f(u_m) + (I - \alpha_m A)S(t_m)u_m$. It is worth mentioning that $\mathscr{S} := \{S(t) : t \geq 0\}$ is a strongly continuous semigroup of nonexpansive mapping and this helped us to find the upper bound of (16). Furthermore,

$$\begin{aligned}\langle Au_m - \gamma f(u_m), j(u_m - x_n)\rangle &\leq \|A\| \cdot \|u_m - S(t_m)u_m\| \cdot \|u_m - x_n\| \\ &\quad + \frac{1}{\alpha_m}\|S(t_m)x_n - x_n\| \cdot \|u_m - x_n\| \\ &= \alpha_m\|A\| \cdot \|\gamma f(u_m) - AS(t_m)u_m\| \cdot \|u_m - x_n\| \\ &\quad + \frac{1}{\alpha_m}\|S(t_m)x_n - x_n\| \cdot \|u_m - x_n\| \\ &\leq \alpha_m M\|A\| \cdot \|\gamma f(u_m) - AS(t_m)u_m\| \\ &\quad + \frac{M}{\alpha_m}\|S(t_m)x_n - x_n\|\end{aligned} \tag{17}$$

where $M > 0$ is a proper constant such that $M \geq \|u_m - x_n\|$ for $m, n \in \mathbb{N}$. Thus, following (i), (3.14) in [10], taking upper limit as $n \to \infty$ at first, and then as $m \to \infty$ in (17), the following yields

$$\limsup_{m\to\infty}\limsup_{n\to\infty}\langle Au_m - \gamma f(u_m), j(u_m - x_n)\rangle \leq 0 \tag{18}$$

which again proves our claim.

Remark 1. In view of the technique of the proof as above and the ones in the former section, one can easily see that we did not utilize (1) as an important property of the strongly positive bounded linear operator A. It is worth pointing out this property is crucial for the aforementioned results and we reduced the dependence of results to the property (1); we refer reader to see, for instance, (2.12) in [3], (2.10) in [8], (2.12) in [6], (3.16) in [10] and the inequalities right after (3.9) in [9].

References

1. Banach, S.: Sur les operations dans les ensembles abstraits et leur applications aux equations integrales. Fund. Math. **3**, 133–181 (1922)
2. Marino, G., Xu, H.K.: A general iterative method for nonexpansive mappings in Hilbert spaces. J. Math. Anal. Appl. **318**, 43–52 (2006)
3. Qin, X., Shang, M., Kang, S.M.: Strong convergence theorems of modified Mann iterative process for strict pseudo-contractions in Hilbert spaces. Nonlinear Anal. **70**, 1257–1264 (2009)
4. Phelps, R.R.: Convex sets and nearest points. Proc. Am. Math. Soc. **8**, 790–797 (1957)
5. Marino, G., Xu, H.K.: Weak and strong convergence theorems for strict pseudo-contractions in Hilbert spaces. J. Math. Anal. Appl. **329**, 336–346 (2007)
6. Marino, G., Colao, V., Qin, X., Kang, S.M.: Strong convergence of the modified Mann iterative method for strict pseudo-contractions. Comput. Math. Appl. **57**, 455–465 (2009)
7. Browder, F.E., Petryshyn, W.V.: Construction of fixed points of nonlinear mappings in Hilbert space. J. Math. Anal. Appl. **20**, 197–228 (1967)
8. Cai, G., Hu, C.: Strong convergence theorems of a general iterative process for a finite family of λ_i-strict pseudo-contractions in q-uniformly smooth Banach spaces. Comput. Math. Appl. **59**, 149–160 (2010)

9. Kangtunyakarn, A., Suantai, S.: Strong convergence of a new iterative scheme for a finite family of strict pseudo-contractions. Comput. Math. Appl. **60**, 680–694 (2010)
10. Yang, L.: The general iterative scheme for semigroups of nonexpansive mappings and variational inequalities with applications. Math. Comput. Model. **57**, 1289–1297 (2013)

Fixed Point Theorems of Contractive Mappings in A-cone Metric Spaces over Banach Algebras

Isa Yildirim[1], Wudthichai Onsod[2], and Poom Kumam[2,3](✉)

[1] Department of Mathematics, Faculty of Science, Ataturk University, 25240 Erzurum, Turkey
isayildirim@atauni.edu.tr

[2] KMUTT-Fixed Point Research Laboratory, Department of Mathematics, Room SCL 802 Fixed Point Laboratory, Science Laboratory Building, Faculty of Science, King Mongkut's University of Technology Thonburi (KMUTT), Bangkok, Thailand

[3] KMUTT-Fixed Point Theory and Applications Research Group (KMUTT-FPTA), Theoretical and Computational Science Center (TaCS), Science Laboratory Building, Faculty of Science, King Mongkut's University of Technology Thonburi (KMUTT), Bangkok, Thailand
wudthichai.ons@mail.kmutt.ac.th, poom.kum@kmutt.ac.th

Abstract. In this study, we prove some fixed point theorems for self-mappings satisfying certain contractive principles in **A**-cone metric spaces over Banach algebras. Our results improve and extend some main results in [8].

Keywords: **A**-cone metric space over Banach algebra · c-sequence
Generalized Lipschitz mapping

1 Introduction

Metric structure is an important tool in the study of fixed point. That is why many researchers studied to establish new classes of metric spaces, such as 2-metric space, D-metric space, D^*-metric space, G-metric space, S-metric space, partial metric space, cone metric space, etc., as a generalization of the usual metric space. In 2007, Huang and Zhang [1] introduced a new metric structure by defining the distance of two elements as a vector in an ordered Banach space and defined cone metric spaces. After that, in 2010, Du [2] showed that any cone metric space is equivalent to a usual metric space. In order to generalize and to overcome these flaws, in 2013, Liu and Xu [3] established the concept of cone

V. Kreinovich et al. (Eds.): ECONVN 2019, SCI 809, pp. 262–270, 2019.
https://doi.org/10.1007/978-3-030-04200-4_20

metric space over a Banach algebra as a proper generalization. Then, Xu and Radenovic [4] proved the results of [3] by removing the condition of normality in a solid cone. Furthermore, in 2015, **A**-metric space was introduced by Abbas et al. In the article [7], the relationship between some generalized metric spaces was given the following as:

$$G\text{-metric space} \Rightarrow D^{*}\text{-metric space} \Rightarrow S\text{-metric space} \Rightarrow \mathbf{A}\text{-metric space.}$$

Moreover, inspired by the notion of cone metric spaces over Banach algebras, Fernandez et al. [8] defined **A**-cone metric structure over Banach algebra.

2 Preliminary

A Banach algebra A is a Banach space over $\mathbf{F} = \{\mathbb{R}, \mathbb{C}\}$ which at the same time has an operation of multiplication such that it meets the following conditions:

1. $(xy)z = x(yz)$,
2. $x(y+z) = xy + xz$ and $(x+y)z = xz + yz$,
3. $\alpha(xy) = (\alpha x)y = x(\alpha y)$,
4. $||xy|| \leq ||x|| ||y||$,

for all $x, y, z \in A, \alpha \in \mathbf{F}$.

Throughout this paper, the Banach algebra has a unit element e for the multiplication that is $ex = xe = x$ for all $x \in A$. An element $x \in A$ is called invertible if there exists an element $y \in A$ such that $xy = yx = e$ and the inverse of x is denoted by x^{-1}. For more details, we refer the reader to Rudin [9]. Now let's give the concepts of cone in order to establish a semi-order on A. The cone P is a subset of A satisfied the following properties:

1. P is non-empty closed and $\{\theta, e\} \subset P$;
2. $\alpha P + \beta P \subset P$ for all non-negative real numbers α, β;
3. $P^2 = PP \subset P$;
4. $P \cap (-P) = \{\theta\}$,

where θ denotes the null of the Banach algebra A. The order relation of the elements in A is defined as $x \preceq y$ if and only if $y - x \in P$. We will indicate that

$$x \prec y \text{ iff } x \preceq y \text{ and } x \neq y,$$
$$x \ll y \text{ iff } y - x \in intP,$$

where $intP$ denotes the interior of P. A cone P is called a solid cone if $intP \neq \emptyset$, and it is called a normal cone if there is a positive real number K such that

$$\theta \preceq x \preceq y \text{ implies } ||x|| \leq K||y||$$

for all $x, y \in A$ [1].

Now, we briefly recall the spectral radius which is essential for main results. Let A be Banach algebra with a unit e and for all $x \in A$, $\lim_{n\to\infty} ||x^n||^{\frac{1}{n}}$ exists. The spectral radius of $x \in A$ satisfies

$$\rho(x) = \lim_{n\to\infty} ||x^n||^{\frac{1}{n}}.$$

If $\rho(x) < |\lambda|$, then $\lambda e - x$ is invertible and the inverse of $\lambda e - x$ is given by

$$(\lambda e - x)^{-1} = \sum_{i=0}^{\infty} \frac{x^i}{\lambda^{i+1}},$$

where λ is a complex constant [9].

From now, we always suppose that A is a real Banach algebra with unit e, P is a solid cone in A, and $\preceq$ is a semi-order with respect to P.

Lemma 1. *[4] Let u, v be vectors in A with $uv = vu$, then the following holds:*

1. $\rho(uv) \leq \rho(u)\rho(v)$,
2. $\rho(u + v) \leq \rho(u) + \rho(v)$.

Definition 1. [8] Let X be nonempty set. Suppose a mapping $d : X^t \to A$ satisfies the following conditions:

1. $\theta \preceq d(x_1, x_2, \ldots, x_{t-1}, x_t)$,
2. $d(x_1, x_2, \ldots, x_{t-1}, x_t) = \theta$ if and only if $x_1 = x_2 = \cdots = x_{t-1} = x_t$
3. $d(x_1, x_2, \ldots, x_{t-1}, x_t) \preceq d(x_1, x_1, \ldots, (x_1)_{t-1}, y) + d(x_2, x_2, \ldots, (x_2)_{t-1}, y) + \cdots + d(x_{t-1}, x_{t-1}, \ldots, (x_{t-1})_{t-1}, y) + d(x_t, x_t, \ldots, (x_t)_{t-1}, y)$

for any $x_i, y \in X$, $(i = 1, 2, \ldots, t)$. Then, (X, d) is called an **A**-cone metric space over Banach algebra.

Note that cone metric space over Banach algebra is a special case of an **A**-cone metric space over Banach algebra when $t = 2$.

Example 1. Let $X = \mathbb{R}$, $A = C[a, b]$ with the supremum norm and $P = \{x \in A | x = x(t) \geq 0 \text{ for all } t \in [a, b]\}$. Define multiplication in the usual way. Consider a mapping $d : X^3 \to A$ by

$$d(x_1, x_2, x_3)(t) = \max\{|x_1 - x_2|, |x_1 - x_3|, |x_2 - x_3|\}e^t$$

Then, (X, d) is an A-cone metric space over Banach algebra.

Lemma 2. *[8] Let (X, d) be an **A**-cone metric space over Banach algebra. Then,*

1. $d(x, x, \ldots, x, y) = d(y, y, \ldots, y, x)$,
2. $d(x, x, \ldots, x, z) \preceq (t - 1)d(x, x, \ldots, x, y) + d(y, y, \ldots, y, z)$.

Definition 2. [8] Let (X, d) be an **A**-cone metric space over Banach algebra A, $x \in X$ and let $\{x_n\}$ be sequence in X. Then:

1. $\{x_n\}$ convergence to x whenever for each $\theta \ll c$ there is a naturel number N such that for all $n \geq N$ we have $d(x_n, x_n, \dots, x_n, x) \ll c$. We denote this by $\lim_{n \to \infty} x_n = x$ or $x_n \to x, n \to \infty$.
2. $\{x_n\}$ is a Cauchy sequence whenever for each $\theta \ll c$ there is a naturel number N such that for all $n, m \geq N$ we have $d(x_n, x_n, \dots, x_n, x_m) \ll c$.
3. (X, d) said to be complete if every Cauchy sequence $\{x_n\}$ in X is convergent.

Definition 3. [4] A sequence $\{u_n\} \subset P$ is a c-sequence if for each $\theta \ll c$ there exists $n_0 \in \mathbb{N}$ such that $u_n \ll c$ for $n > n_0$.

Lemma 3. *[5] If $\rho(u) < 1$, then $\{u^n\}$ is a c-sequence.*

Lemma 4. *[4] Suppose that $\{u_n\}$ is a c-sequence in P and $k \in P$. Then, $\{ku_n\}$ is a c-sequence.*

Lemma 5. *[4] Suppose that $\{u_n\}$ and $\{v_n\}$ are c -sequences in P and $\alpha, \beta > 0$. Then, $\{\alpha u_n + \beta v_n\}$ is a c-sequence.*

Lemma 6. *[6] The following conditions are satisfied.*

1. *If $u \preceq v$ and $v \ll w$, then $u \ll w$.*
2. *If $\theta \preceq u \ll c$ for each $\theta \ll c$, then $u = \theta$.*

3 Main Results

Lemma 7. *Let (X, d) be an **A**-cone metric space over Banach algebra A and P be solid cone in A. Suppose that $\{z_n\}$ is a sequence in X satisfying the following condition:*

$$d(z_n, z_n, \dots, z_n, z_{n+1}) \preceq hd(z_{n-1}, z_{n-1}, \dots, z_{n-1}, z_n), \tag{1}$$

for all n, where for some $h \in A$ which $\rho(h) < 1$. Then, $\{z_n\}$ is a Cauchy sequence in X.

Proof. Using the inequality of (1), we have

$$\begin{aligned} d(z_n, z_n, \dots, z_n, z_{n+1}) &\preceq hd(z_{n-1}, z_{n-1}, \dots, z_{n-1}, z_n) \\ &\preceq h^2 d(z_{n-2}, z_{n-2}, \dots, z_{n-2}, z_{n-1}) \\ &\vdots \\ &\preceq h^n d(z_0, z_0, \dots, z_0, z_1). \end{aligned}$$

Since $\rho(h) < 1$, it is satisfied that $(e-h)$ is invertible and $(e-h)^{-1} = \sum_{i=0}^{\infty} h^i$. Hence, for any $m > n$, we obtain

$$\begin{aligned}
d(z_n, z_n, \ldots, z_n, z_m) &\preceq (t-1)d(z_n, z_n, \ldots, z_n, z_{n+1}) \\
&\quad + d(z_{n+1}, z_{n+1}, \ldots, z_{n+1}, z_m) \\
&\preceq (t-1)d(z_n, z_n, \ldots, z_n, z_{n+1}) \\
&\quad + (t-1)d(z_{n+1}, z_{n+1}, \ldots, z_{n+1}, z_{n+2}) \\
&\quad + \cdots + (t-1)d(z_{m-2}, z_{m-2}, \ldots, z_{m-2}, z_{m-1}) \\
&\quad + d(z_{m-1}, z_{m-1}, \ldots, z_{m-1}, z_m) \\
&\preceq (t-1)h^n d(z_0, z_0, \ldots, z_0, z_1) \\
&\quad + (t-1)h^{n+1} d(z_0, z_0, \ldots, z_0, z_1) \\
&\quad + \cdots + (t-1)h^{m-2} d(z_0, z_0, \ldots, z_0, z_1) \\
&\quad + h^{m-1} d(z_0, z_0, \ldots, z_0, z_1) \\
&\preceq (t-1)[h^n + h^{n+1} + \cdots + h^{m-1}] d(z_0, z_0, \ldots, z_0, z_1) \\
&= (t-1)h^n [e + h + \cdots + h^{m-n-1}] d(z_0, z_0, \ldots, z_0, z_1) \\
&\preceq (t-1)h^n (e-h)^{-1} d(z_0, z_0, \ldots, z_0, z_1).
\end{aligned}$$

Let $g_n = (t-1)h^n(e-h)^{-1}d(z_0, z_0, \ldots, z_0, z_1)$. By Lemmas 3 and 4, it is clear that the sequence $\{g_n\}$ is a c-sequence. Therefore, for each $\theta \ll c$, there exists $N \in \mathbb{N}$ such that $d(z_n, z_n, \ldots, z_n, z_m) \preceq g_n \ll c$ for all $n > N$. So, by using Lemma 6, $d(z_n, z_n, \ldots, z_n, z_m) \ll c$ whenever $m > n > N$. It is meaning that $\{z_n\}$ is a Cauchy sequence.

Theorem 1. *Let (X, d) be a complete* **A***-cone metric space over A and P be a solid cone in A. Let $T : X \to X$ be a map satisfying the following condition:*

$$\begin{aligned}
d(Tx, Tx, \ldots, Tx, Ty) \preceq\ & k_1 d(x, x, \ldots, x, y) + k_2 d(x, x, \ldots, x, Tx) + k_3 d(y, y, \ldots, y, Ty) \\
& + k_4 d(x, x, \ldots, x, Ty) + k_5 d(y, y, \ldots, y, Tx)
\end{aligned}$$

for all $x, y \in X$, where $k_i \in P$ $(i = 1, 2, \ldots, 5)$ are generalized Lipschitz constant vectors with $\rho(k_1) + \rho(k_2 + k_3 + k_4 + k_5) < 1$. If k_1 commutes with $k_2 + k_3 + k_4 + k_5$, then T has a unique fixed point.

Proof. Let $x_0 \in X$ be arbitrary and $\{x_n\}$ be a Picard iteration defined by $x_{n+1} = Tx_n$. Then, we get

$$\begin{aligned}
d(x_n, x_n, \ldots, x_n, x_{n+1}) &= d(Tx_{n-1}, Tx_{n-1}, \ldots, Tx_{n-1}, Tx_n) \\
&\preceq k_1 d(x_{n-1}, x_{n-1}, \ldots, x_{n-1}, x_n) + k_2 d(x_{n-1}, x_{n-1}, \ldots, x_{n-1}, x_n) \\
&\quad + k_3 d(x_n, x_n, \ldots, x_n, x_{n+1}) + k_4 d(x_{n-1}, x_{n-1}, \ldots, x_{n-1}, x_{n+1}) \\
&\quad + k_5 d(x_n, x_n, \ldots, x_n, x_n) \\
&\preceq (k_1 + k_2 + k_4) d(x_{n-1}, x_{n-1}, \ldots, x_{n-1}, x_n) \\
&\quad + (k_3 + k_4) d(x_n, x_n, \ldots, x_n, x_{n+1}),
\end{aligned}$$

which implies that

$$(e - k_3 - k_4) d(x_n, x_n, \ldots, x_n, x_{n+1}) \preceq (k_1 + k_2 + k_4) d(x_{n-1}, x_{n-1}, \ldots, x_{n-1}, x_n). \tag{2}$$

Also, we get

$$\begin{aligned}
d(x_n, x_n, \ldots, x_n, x_{n+1}) &= d(x_{n+1}, x_{n+1}, \ldots, x_{n+1}, x_n) \\
&= d(Tx_n, Tx_n, \ldots, Tx_n, Tx_{n-1}) \\
&\preceq k_1 d(x_n, x_n, \ldots, x_n, x_{n-1}) + k_2 d(x_n, x_n, \ldots, x_n, x_{n+1}) \\
&\quad + k_3 d(x_{n-1}, x_{n-1}, \ldots, x_{n-1}, x_n) + k_4 d(x_n, x_n, \ldots, x_n, x_n) \\
&\quad + k_5 d(x_{n-1}, x_{n-1}, \ldots, x_{n-1}, x_{n+1}) \\
&\preceq (k_1 + k_3 + k_5) d(x_{n-1}, x_{n-1}, \ldots, x_{n-1}, x_n) \\
&\quad + (k_2 + k_5) d(x_n, x_n, \ldots, x_n, x_{n+1}),
\end{aligned}$$

which means that

$$(e - k_2 - k_5) d(x_n, x_n, \ldots, x_n, x_{n+1}) \preceq (k_1 + k_3 + k_5) d(x_{n-1}, x_{n-1}, \ldots, x_{n-1}, x_n). \tag{3}$$

Add up (2) and (3) yields that

$$(2e - k) d(x_n, x_n, \ldots, x_n, x_{n+1}) \preceq (2k_1 + k) d(x_{n-1}, x_{n-1}, \ldots, x_{n-1}, x_n), \tag{4}$$

where $k = k_2 + k_3 + k_4 + k_5$. Since $\rho(k) \leq \rho(k_1) + \rho(k) < 1 < 2$, $(2e - k)$ is invertible and also

$$(2e - k)^{-1} = \sum_{i=0}^{\infty} \frac{k^i}{2^{i+1}}.$$

Multiplying in both sides of (4) by $(2e - k)^{-1}$, one can write

$$d(x_n, x_n, \ldots, x_n, x_{n+1}) \preceq (2e - k)^{-1}(2k_1 + k) d(x_{n-1}, x_{n-1}, \ldots, x_{n-1}, x_n). \tag{5}$$

Moreover, using that k_1 commutes with k, we can obtain that

$$\begin{aligned}
(2e - k)^{-1}(2k_1 + k) &= \left(\sum_{i=0}^{\infty} \frac{k^i}{2^{i+1}}\right)(2k_1 + k) = 2\left(\sum_{i=0}^{\infty} \frac{k^i}{2^{i+1}}\right)k_1 + \sum_{i=0}^{\infty} \frac{k^{i+1}}{2^{i+1}} \\
&= 2k_1\left(\sum_{i=0}^{\infty} \frac{k^i}{2^{i+1}}\right) + k \sum_{i=0}^{\infty} \frac{k^i}{2^{i+1}} \\
&= (2k_1 + k)\left(\sum_{i=0}^{\infty} \frac{k^i}{2^{i+1}}\right) = (2k_1 + k)(2e - k)^{-1},
\end{aligned}$$

that is, $(2e - k)^{-1}$ commutes with $(2k_1 + k)$. Let $h = (2e - k)^{-1}(2k_1 + k)$. Then, according to Lemma 1, we can conclude that

$$\begin{aligned}
\rho(h) &= \rho((2e - k)^{-1}(2k_1 + k)) \leq \rho((2e - k)^{-1}) \rho(2k_1 + k) \\
&\leq \rho\left(\sum_{i=0}^{\infty} \frac{k^i}{2^{i+1}}\right)[\rho(2k_1) + \rho(k)] \leq \left(\sum_{i=0}^{\infty} \frac{\rho(k)^i}{2^{i+1}}\right)[2\rho(k_1) + \rho(k)] \\
&= \frac{1}{2 - \rho(k)}[2\rho(k_1) + \rho(k)] < 1.
\end{aligned}$$

Considering (5) with $\rho(h) < 1$ together, we can easily say that $\{x_n\}$ is a Cauchy sequence by Lemma 7. The completeness of X indicates that there exists $x \in X$ such that $\{x_n\}$ convergence to x. Now, we will show that x is the fixed point of T. In accordance with this purpose, for one thing,

$$\begin{aligned} d(x,x,\ldots,x,Tx) &\preceq (t-1)d(x,x,\ldots,x,Tx_n) + d(Tx,Tx,\ldots,Tx,Tx_n) \\ &\preceq (t-1)d(x,x,\ldots,x,x_{n+1}) + k_1 d(x,x,\ldots,x,x_n) \\ &\quad + k_2 d(x,x,\ldots,x,Tx) + k_3 d(x_n,x_n,\ldots,x_n,x_{n+1}) \\ &\quad + k_4 d(x,x,\ldots,x,x_{n+1}) + k_5 d(x_n,x_n,\ldots,x_n,Tx) \\ &\preceq [k_1 + (t-1)(k_3+k_5)] d(x,x,\ldots,x,x_n) \\ &\quad + [(t-1)e + k_3 + k_4] d(x,x,\ldots,x,x_{n+1}) \\ &\quad + (k_2+k_5) d(x,x,\ldots,x,Tx), \end{aligned}$$

which implies that

$$\begin{aligned} (e-k_2-k_5)d(x,x,\ldots,x,Tx) &\preceq [k_1+(t-1)(k_3+k_5)]d(x,x,\ldots,x,x_n) \\ &\quad + [(t-1)e+k_3+k_4]d(x,x,\ldots,x,x_{n+1}). \end{aligned} \tag{6}$$

For another thing,

$$\begin{aligned} d(x,x,\ldots,x,Tx) &\preceq (t-1)d(x,x,\ldots,x,Tx_n) + d(Tx_n,Tx_n,\ldots,Tx_n,Tx) \\ &\preceq (t-1)d(x,x,\ldots,x,x_{n+1}) + k_1 d(x_n,x_n,\ldots,x_n,x) \\ &\quad + k_2 d(x_n,x_n,\ldots,x_n,x_{n+1}) + k_3 d(x,x,\ldots,x,Tx) \\ &\quad + k_4 d(x_n,x_n,\ldots,x_n,Tx) + k_5 d(x,x,\ldots,x,x_{n+1}) \\ &\preceq [k_1 + (t-1)(k_2+k_4)] d(x_n,x_n,\ldots,x_n,x) \\ &\quad + [(t-1)e + k_2 + k_4] d(x,x,\ldots,x,x_{n+1}) \\ &\quad + (k_3+k_4) d(x,x,\ldots,x,Tx), \end{aligned}$$

which means that

$$\begin{aligned} (e-k_3-k_4)d(x,x,\ldots,x,Tx) &\preceq [k_1+(t-1)(k_2+k_4)]d(x_n,x_n,\ldots,x_n,x) \\ &\quad + [(t-1)e+k_2+k_4]d(x,x,\ldots,x,x_{n+1}). \end{aligned} \tag{7}$$

Combining (6) and (7), we obtain

$$\begin{aligned} (2e-k)d(x,x,\ldots,x,Tx) &\preceq [2k_1+2(t-1)k]d(x,x,\ldots,x,x_n) \\ &\quad + [2(t-1)e+k]d(x,x,\ldots,x,x_{n+1}), \end{aligned} \tag{8}$$

which follows immediately from (8) that

$$\begin{aligned} d(x,x,\ldots,x,Tx) &\preceq (2e-k)^{-1}[(2k_1+2(t-1)k)d(x,x,\ldots,x,x_n) \\ &\quad + (2(t-1)e+k)d(x,x,\ldots,x,x_{n+1})]. \end{aligned}$$

Since $d(x,x,\ldots,x,x_n)$ and $d(x,x,\ldots,x,x_{n+1})$ are c-sequences, then by Lemmas 3, 4, 5 and 6, we arrive $x = Tx$. Then, x is a fixed point of T.

Finally, we prove the uniqueness of the fixed point. Suppose that y is another fixed point, then

$$d(x, x, \ldots, x, y) = d(Tx, Tx, \ldots, Tx, Ty) \preceq \alpha d(x, x, \ldots, x, y). \tag{9}$$

where $\alpha = k_1+k_2+k_3+k_4+k_5$. Note that, $\rho(\alpha) \leq \rho(k_1)+\rho(k_2+k_3+k_4+k_5) < 1$, then by Lemmas 3 and 4, $\{\alpha^n d(x, x, \ldots, x, y)\}$ is a c-sequence. The condition of (9) leads to

$$d(x, x, \ldots, x, y) \preceq \alpha^n d(x, x, \ldots, x, y).$$

Therefore, by Lemma 6, it follows that $x = y$.

Putting $k_1 = k$ and $k_2 = k_3 = k_4 = k_5 = \theta$ in Theorem 1, we can obtain the following result.

Corollary 1. *(Theorem 6.1, [8]) Let (X, d) be a complete* **A***-cone metric space over A and P be a solid cone in A. Suppose the mapping $T : X \to X$ satisfies the following condition:*

$$d(Tx, Tx, \ldots, Tx, Ty) \preceq kd(x, x, \ldots, x, y)$$

for all $x, y \in X$, where $k \in P$ with $\rho(k) < 1$. Then, T has a unique fixed point.

Choosing $k_1 = k_4 = k_5 = \theta$ and $k_2 = k_3 = k$ in Theorem 1, the following result is obvious.

Corollary 2. *(Theorem 6.3, [8]) Let (X, d) be a complete* **A** *-cone metric space over A and P be a solid cone in A. Suppose the mapping $T : X \to X$ satisfies the following condition:*

$$d(Tx, Tx, \ldots, Tx, Ty) \preceq k[d(Tx, Tx, \ldots, Tx, y) + d(Ty, Ty, \ldots, Ty, x)]$$

for all $x, y \in X$, where $k \in P$ with $\rho(k) < \frac{1}{2}$. Then, T has a unique fixed point.

Taking $k_1 = k_2 = k_3 = \theta$ and $k_4 = k_5 = k$ in Theorem 1, the following result is clear.

Corollary 3. *(Theorem 6.4, [8]) Let (X, d) be a complete* **A** *-cone metric space over A and P be a solid cone in A. Suppose the mapping $T : X \to X$ satisfies the following condition:*

$$d(Tx, Tx, \ldots, Tx, Ty) \preceq k[d(Tx, Tx, \ldots, Tx, x) + d(Ty, Ty, \ldots, Ty, y)]$$

for all $x, y \in X$, where $k \in P$ with $\rho(k) < \frac{1}{2}$. Then, T has a unique fixed point.

Remark 1. Clearly, Kannan and Chattergee type mappings in **A**-cone metric spaces over Banach algebras are not depend on t-dimension.

Remark 2. Note that Theorems 6.3 and 6.4 in [8] accept respectively the assumptions of $\rho(k) < (\frac{1}{n})^2$ and $\rho(k) < \frac{1}{n}$, which are depend on n-dimension, but Corallary 2 and 3 given above have the assumption $\rho(k) < \frac{1}{2}$. That is obviously generalize Theorems 6.3 and 6.4 in [8].

Acknowledgments. This project was supported by the Theoretical and Computational Science (TaCS) Center under Computational and Applied Science for Smart Innovation Research Cluster (CLASSIC), Faculty of Science, KMUTT.

Author contributions. All authors read and approved the final manuscript.

Competing Interests. The authors declare that they have no competing interests.

References

1. Guang, H.L., Xian, Z.: Cone metric spaces and fixed point theorems of contractive mappings. J. Math. Anal. Appl. **332**, 1468–1476 (2007)
2. Du, W.S.: A note on cone metric fixed point theory and its equivalence. Nonlinear Anal. **72**, 2259–2261 (2010)
3. Liu, H., Xu, S.: Cone metric spaces with Banach algebras and fixed point theorems of generalized Lipschitz mappings. Fixed Point Theory Appl. 320 (2013)
4. Xu, S., Radenovic, S.: Fixed point theorems of generalized Lipschitz mappings on cone metric spaces over Banach algebras without assumption of normality. Fixed Point Theory Appl. 102 (2014)
5. Huang, H., Radenovic, S.: Common fixed point theorems of generalized Lipschitz mappings in cone b-metric spaces over Banach algebras and applications. J. Non Sci. Appl. **8**, 787–799 (2015)
6. Radenovic, S., Rhoades, B.E.: Fixed point theorem for two non-self mappings in cone metric spaces. Comput. Math. Appl. **57**, 1701–1707 (2009)
7. Abbas, M., Ali, B., Suleiman, Y.I.: Generalized coupled common fixed point results in partially ordered A-metric spaces. Fixed Point Theory Appl. 64 (2015)
8. Fernandez, J., Saelee, S., Saxena, K., Malviya, N., Kumam, P.: The A-cone metric space over Banach algebra with applications. Cogent Math. 4 (2017)
9. Rudin, W.: Functional Analysis, 2nd edn. McGraw-Hill, New York (1991)

Applications

The Relationship Among Education Service Quality, University Reputation and Behavioral Intention in Vietnam

Bui Huy Khoi[1(✉)], Dang Ngoc Dai[2], Nguyen Huu Lam[2], and Nguyen Van Chuong[2]

[1] Industrial University of Ho Chi Minh City, 12 Nguyen Van Bao Street, Govap District, Ho Chi Minh City, Vietnam
buihuykhoi@iuh.edu.vn

[2] University of Economics Ho Chi Minh City, 59C Nguyen Dinh Chieu Street, District 3, Ho Chi Minh City, Vietnam

Abstract. The aim of this research was to explore the relationship among education service quality, university reputation and behavioral intention in Vietnam. Survey data was collected from 550 people graduated in HCM City. The research model was proposed from the study of education service quality, university reputation and behavioral intention of some authors in domestic and abroad. The reliability and validity of the scale were tested by Cronbach's Alpha, Average Variance Extracted (Pvc) and Composite Reliability (Pc). The analysis results of structural equation model (SEM) showed that education service quality, university reputation and behavioral intention have relationships with each other.

Keywords: Vietnam · Smartpls 3.0 · SEM · Education service quality University reputation · Behavioral intention

1 Introduction

When Vietnam entered ASEAN economic community (AEC), it gradually integrated into economies in the AEC, many foreign companies have chosen Vietnam as one of the top attractive investment location, training and applying high-quality human resources for Vietnam labor market was an urgent requirement for the period AEC integration with major economies. Many universities was established to meet the needs of integration into the AEC. Vietnam universities were facing new challenges is to improve the quality of education in order to participate international environment. With limited resources, but managers and trainers were trying to gradually improve the reputation, educational quality to gradually integration into the AEC.

In the ASEAN region, there were 11 criteria assessing the quality of education of the region (ASEAN University Network - Quality Assurance, stand for AUN-QA). The evaluation criteria of quality education stopped just above the university is considered

V. Kreinovich et al. (Eds.): ECONVN 2019, SCI 809, pp. 273–281, 2019.
https://doi.org/10.1007/978-3-030-04200-4_21

to meet targets set by the school. At the same time the purpose of the standard is a tool for the university self-assessment and to explain to the authorities about the actual quality of education, no assessment of rating agencies as a basis independently verified improved indicators of quality.

Currently, researchers and educational administrators in favor of Vietnam was the notion that education was a commodity, and students as customers. Thus, the assessment of learners on service quality of a university was increasingly managers valued education. The strong competition in the field of higher education took place between public universities, between public and private, between private and private with giving rise to the question: "Reputation and service quality of university acted as how the school intended to select students in the context of international integration?". Therefore, the article on building a service quality, reputation and behavioral intention based on standpoint of university' students to be able to contribute to the understanding of the university's service quality, reputation and behavioral intention of learners in a competitive environment and development higher education system in Vietnam gradually integration into AEC.

2 Literature Review

The quality of higher education was a multidimensional concept covering all functions and activities: teaching and training, research and academics, staff, students, housing, facilities material, equipment, community services for the and the learning environment [1]. Research by Ahmad et al. had developed four components of the quality of education services, which were seniority factor, courses factor, cultural factor and gender factor [2]. Firdaus had been shown that the measurement of the quality of higher education services with six components were: Academic Aspects, Non-Academic Aspects, Reputation, Access, Programmes issues and understanding [3]. Hence, we proposed five hypotheses:

"**Hypothesis 1 (H1).** *There was a positive impact of* Academic aspects **(ACA)** and Service quality **(SER)**"
"**Hypothesis 2 (H2).** *There was a positive impact of* Program issues **(PRO)** and Service quality **(SER)**"
"**Hypothesis 3 (H3).** *There was a positive impact of* Facilities **(FAC)** and Service quality **(SER)**"
"**Hypothesis 4 (H4).** *There was a positive impact of* Non-academic aspects **(NACA)** and Service quality **(SER)**"
"**Hypothesis 5 (H5).** *There was a positive impact of* Access **(ACC)** and Service quality **(SER)**"

Reputation was acutely aware of the individual organization. It was formed over a long period of understanding and evaluation of the success of that organization [4]. Alessandri et al. (2006) had demonstrated a relationship between the university reputation that is favored with academic performance, external performance and emotional engagement [5]. Nguyen and Leblance investigated the role of institutional image and institutional reputation in the formation of customer loyalty. The results indicated that the degree of loyalty has a tendency to be higher when perceptions of both institutional reputation and service quality are favorable [6]. Thus, we proposed five hypotheses:

"**Hypothesis 6 (H6).** *There was a positive impact of* Academic aspects (**ACA**) and Reputation (**REP**)"
"**Hypothesis 7 (H7).** *There was a positive impact of* Program issues (**PRO**) and Reputation (**REP**)"
"**Hypothesis 8 (H8).** *There was a positive impact of* Facilities (**FAC**) and Reputation (**REP**)"
"**Hypothesis 9 (H9).** *There was a positive impact of* Non-academic aspects (**NACA**) and Reputation (**REP**)"
"**Hypothesis 10 (H10).** *There was a positive impact of* Access (**ACC**) and Reputation (**REP**)"

Dehghan et al. had a significant and positive relationship between service quality and educational reputation [7]. Wang et al. found that providing high quality products and services would enhance the reputation [8]. Thus, we proposed a hypothesis:

"**Hypothesis 11 (H11).** *There was a positive impact of* Service quality (**SER**) and Reputation (**REP**)"

Walsh argued that reputation had a positive impact on customer [9]. Empirical research had shown that a company with a good reputation could reinforce customer trust in buying product and service [6]. So, we proposed a hypothesis:

"**Hypothesis 12 (H12).** *There was a positive impact of* Reputation (**REP**) and Behavior Intention (**BEIN**)"

Behaviors were actions that individuals perform to interact with service. Customer participation in the process demonstrated the best behavior in the service. Customer behavior depended heavily on their systems, service processes, and cognitive abilities. So, with a service, it could exist with different behaviors among different customers. Pratama, Sutter and Paulson gave the relationship between Service quality and Behavioral Intention [10, 11]. So we proposed a hypothesis:

"**Hypothesis 13 (H13).** *There was a positive impact of* Service quality (**SER**) and Behavioral Intention (**BEIN**)"

Finally, all hypotheses, factors and observations are modified as Fig. 1.

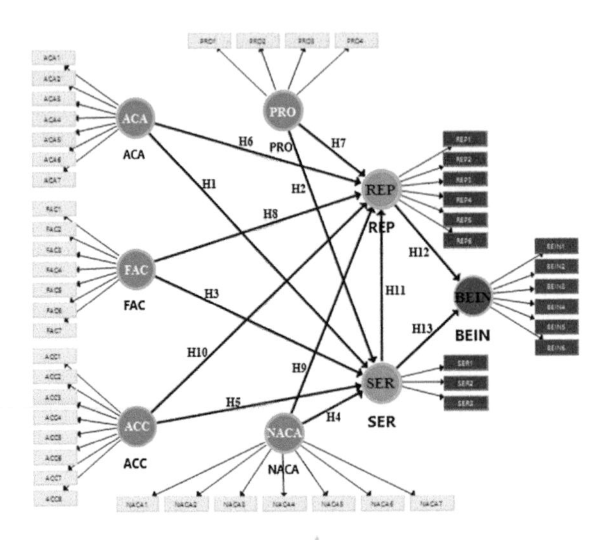

Fig. 1. Research model. **ACA:** Academic aspects, **PRO:** Program issues, **FAC:** Facilities, **NACA:** Non-academic aspects, **ACC:** Access, **REP:** Reputation, **SER:** Service quality, **BEIN:** Behavioral Intention. **Source:** *Designed by author*

3 Research Method

We followed the methods of Anh, Dong, Kreinovich, and Thach [12]. Research methodology was implemented through two steps: qualitative research and quantitative research. Qualitative research was conducted with a sample of 52 people. First period 1 was tested on a small sample to discover the flaws of the questionnaire. The questionnaire was written by Vietnamese. Second period of the official research was carried out as soon as the question was edited from the test results. Respondents were selected by convenient methods with a sample size of 550 people graduated but there were 493 people filling the correct form. There were 126 males and 367 females in this survey. Their graduated years were from 1997 to 2016. They graduated 10 universities in Vietnam as Table 1:

Table 1. Sample statistics

University graduated	Amount	Percent (%)	Year graduated	Amount	Percent (%)
AGU	16	3.2	1997	17	3.4
BDU	17	3.4	2006	17	3.4
DNTU	34	6.9	2009	51	10.3
FPTU	32	6.5	2012	51	10.3
HCMUAF	17	3.4	2013	82	16.6
IUH	279	56.6	2014	97	19.7
SGU	17	3.4	2015	82	16.6
TDTU	16	3.2	2016	96	19.5
UEH	49	9.9	**Total**	493	100.0
VNU	16	3.2			
Total	**493**	100.0			

Source: *Calculated by author*

The questionnaire answered by respondents was the main tool to collect data. The questionnaire contained questions about their graduated university and year. The survey was conducted on March 29, 2018. Data processing and statistical analysis software is used by Smartpls 3.0 developed by SmartPLS GmbH Company in Germany. The reliability and validity of the scale were tested by Cronbach's Alpha, Average Variance Extracted (Pvc) and Composite Reliability (Pc). Followed by a linear structural model SEM was used to test the research hypotheses [15].

4 Results

4.1 Consistency and Reliability

In this reflective model convergent validity was tested through composite reliability or Cronbach's alpha. Composite reliability and Average Variance Extracted were the measure of reliability since Cronbach's alpha sometimes underestimates the scale reliability [13].

Table 2 showed that composite reliability varied from 0.851 to 0.921, Cronbach's alpha from 0.835 to 0.894 and Average Variance Extracted from 0.504 to 0.795 which were above preferred value of 0.5. This proved that model was internally consistent. To check whether the indicators for variables display convergent validity, Cronbach's alpha were used. From Table 2, it can be observed that all the factors are reliable (>0.60) and Pvc > 0.5 [14].

Table 2. Cronbach's alpha, composite reliability (Pc) and AVE values (Pvc)

Factor	Cronbach's alpha	Average Variance Extracted (Pvc)	Composite Reliability (Pc)	P	Findings
ACA	0.875	0.572	0.903	0.000	Supported
ACC	0.874	0.540	0.902	0.000	Supported
BEIN	0.886	0.639	0.913	0.000	Supported
FAC	0.835	0.504	0.876	0.000	Supported
NACA	0.849	0.529	0.886	0.000	Supported
PRO	0.767	0.589	0.851	0.000	Supported
REP	0.894	0.657	0.919	0.000	Supported
SER	0.870	0.795	0.921	0.000	Supported

$$\alpha = \frac{k}{k-1}\left[1 - \frac{\sum \sigma^2(x_i)}{\sigma_x^2}\right] \quad \rho_C = \frac{\left(\sum_{i=1}^{p} \lambda_i\right)^2}{\left(\sum_{i=1}^{p} \lambda_i\right) + \sum_{i=1}^{p} (1-\lambda_i^2)} \quad \rho_{VC} = \frac{\sum_{i=1}^{p} \lambda_i^2}{\sum_{i=1}^{p} \lambda_i^2 + \sum_{i=1}^{p} (1-\lambda_i^2)}$$

k: factor, xi: observations, λ_i is a normalized weight of observation variable, σ^2: Square of Variance, i; 1- λi^2 – the variance of the observed variable i. **Source:** *Calculated by Smartpls software 3.0*

4.2 Structural Equation Modeling (SEM)

Structural Equation Modeling (SEM) was used on the theoretical framework. Partial Least Square method could handle many independent variables, even when multicollinearity exists. PLS could be implemented as a regression model, predicting one or more dependent variables from a set of one or more independent variables or it could be implemented as a path model. Partial Least Square (PLS) method could associate with the set of independent variables to multiple dependent variables [15].

SEM results in the Fig. 2 showed that the model was compatible with data research [14]. The behavioral intention was affected by quality service and reputation about 58.9%. The quality service was affected by Academic aspects, Program issues, Facilities, Non-academic aspects and Access about 54.8%. The reputation was affected by Academic aspects, Program issues, Facilities, Non-academic aspects and Access about 53.6%.

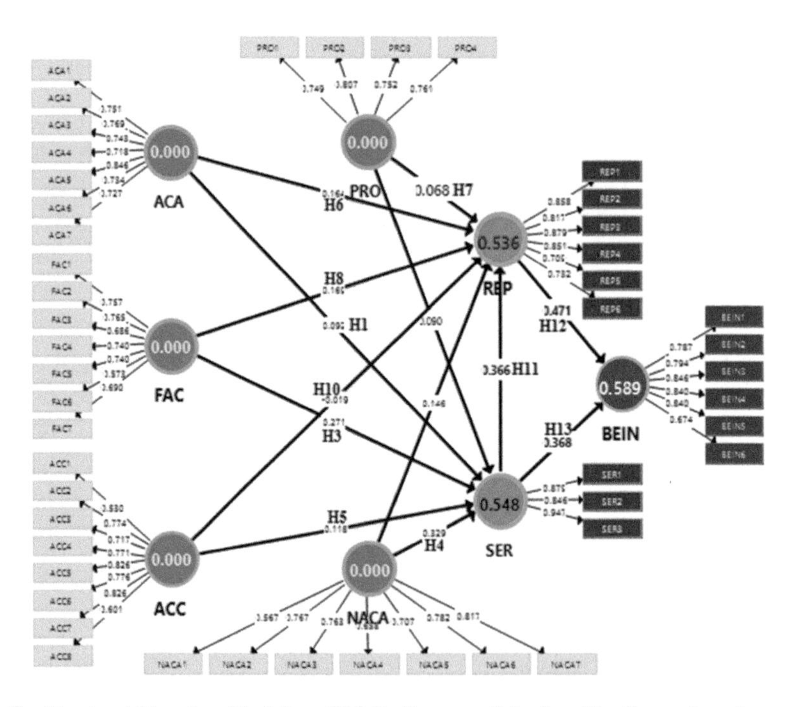

Fig. 2. Structural Equation Modeling (SEM). **Source:** *Calculated by Smartpls software 3.0*

In the SEM analysis in Table 3, the variables that associated with Behavior Intention ($p < 0.05$). The Academic aspects and Program issues were not relative with reputation as Table 3. The most important factor for service quality was Non-academic aspects with the Beta equals to 0.329. The most important factor for Reputation was Facilities with the Beta equals to 0.169. The most important factor for Behavioral Intention was Reputation with the Beta equals to 0.169.

Table 3. Structural Equation Modeling (SEM)

Relation	Beta	SE	T-value	P	Findings
ACA -> REP	0.164	0.046	3.547	0.000	Supported
ACA -> SER	0.092	0.038	2.381	0.018	Supported
ACC -> REP (H7)	−0.019	0.060	0.318	0.750	Unsupported
ACC -> SER	0.118	0.048	2.473	0.014	Supported
FAC -> REP	0.169	0.050	3.376	0.001	Supported
FAC -> SER	0.271	0.051	5.311	0.000	Supported
NACA -> REP	0.146	0.060	2.443	0.015	Supported
NACA -> SER	0.329	0.053	6.214	0.000	Supported
PRO -> REP (H10)	0.068	0.044	1.569	0.117	Unsupported
PRO -> SER	0.090	0.043	2.105	0.036	Supported
REP -> BEIN	0.471	0.040	11.918	0.000	Supported
SER -> BEIN	0.368	0.042	8.814	0.000	Supported
SER -> REP	0.366	0.055	6.706	0.000	Supported

Beta (r): SE = SQRT(1 − r2)/(n − 2); CR = (1 − r)/SE; P-value = TDIST(CR, n − 2, 2). **Source:** *Calculated by Smartpls software 3.0*

SEM results showed that the model was compatible with data research: SRMR has P-value $\leq$ 0.001 (<0.05) [13, 15] in Table 4.

Table 4. Standard of model SEM

Standard	Beta	SE	T-value	P	Findings
SRMR	0.052	0.003	15.627	0.000	Supported

Source: *Calculated by Smartpls software 3.0*

In bootstrapping, resampling methods were used to compute the significance of PLS coefficients. Output of significance levels can be retrieved from bootstrapping option. Table 3 shows the results of hypotheses testing; all the t values above 1.96 are significant at the 0.05 level. Hypotheses H1, H2, H3, H4, H5, H6, H8, H9, H11, H12 and H13 were supported. Hypotheses H7 and H10 were unsupported. The results indicated a positive (ß = 0.366, 0.368, 0.471) and significant ($p < 0.05$) association among quality service, reputation and behavioral intention.

5 Conclusion and Discussion

As the result of data analysis in results and discussion, eleven hypotheses were accepted. Hypotheses H7 and H10 were unsupported. The study results had met the research objectives and fully support the eleven hypotheses set out. Specifically, the results had confirmed the three components of the university's reputation included Academic aspects, Facilities and Non-academic aspect. There were three components contribute significantly to the university reputation in order of importance: (1) Facilities, (2) Academic aspects and (3) Non-academic aspect. All five components contributed importantly to the university service quality in order of importance: (1) Non-academic aspect, (2) Facilities, (3) Access, (4) Academic aspects and (5) Program issues. In addition, the good service quality and reputation of the university student's minds had led to their intended positive behavior in the future.

Limitations and Suggestions for Further Research

The construction of the relationship service quality, reputation and Behavioral Intention should be based on respects for Academic aspects, Program issues, Facilities, Non-academic aspects and Access. Obviously service quality and reputation were very important to the university such as competitive advantage, satisfactory and met society's expectations and only reputation was responsible for maintenance throughout the whole organization. Research results indicate that Academic aspects, Program issues, Facilities, Non-academic aspects and Access are important factors for students when determining the service and three of the five elements have a strong influence on the reputation of organizations.

Students were attracted by superficial factors and convenience. Aspect played an important role in affect the university service quality and reputation. The school should be interested in the standardization of procedures for providing services ranging from recruitment, admissions, course registration procedures, tuition, lookup points in the learning process, and so on…

Limitation of research has two hypotheses in model to be canceled so we would try to complete next researches.

References

1. Van Ginkel, H.J., Rodrigues Dias, M.A.: Institutional and political challenges of accreditation at the international level, Report: Higher Education in the World 2007: Accreditation for Quality Assurance: What is at Stake? (2007)
2. Ahmad, et al.: Service quality in higher education: Management student's perspective. Management department, Faculty of management and human resource development, University of Technology Malaysia (2004)
3. Abdullah, F.: The development of HEDPERF: a new measuring instrument of service quality for the higher education sector. Int. J. Cons. Stud. **30**(6), 569–581 (2006)
4. Balmer, J.M., van Riel, C.B., Jo Hatch, M., Schultz, M.: Relations between organizational culture, identity and image. Eur. J. Market. **31**(5/6), 356–365 (1997)
5. Alessandri, S.W., Yang, S.-U., Kinsey, D.F.: An integrative approach to university visual identity and reputation. Corp. Reput. Rev. **9**(4), 258–270 (2006)
6. Nguyen, N., LeBlanc, G.: Image and reputation of higher education institutions in students' retention decisions. Int. J. Educ. Manag. **15**(6), 303–311 (2001)
7. Dehghan, A., Dugger, J., Dobrzykowski, D., Balazs, A.: The antecedents of student loyalty in online programs. Int. J. Educ. Manag. **28**(1), 15–35 (2014)
8. Wang, S., Fu, Y.Y.: Applications of planned behavior and place image to visit intentions: a casino gaming context. In: Advances in Hospitality and Leisure, pp. 67–87 (2016)

9. Walsh, G., Mitchell, V.W., Jackson, P.R., Beatty, S.E.: Examining the antecedents and consequences of corporate reputation: a customer perspective. British J. Manag. **20**(2), 187–203 (2009)
10. Pratama, A.: Factors affecting students' learning interest in an accounting study programme: a study in Bandung city, West Java, Indonesia. Rev. Integr. Bus. Econ. Res. **6**(2), 295–311 (2017)
11. Sutter, N., Paulson, S.: Predicting college students' intention to graduate: a test of the theory of planned behavior. Coll. Stud. J. **50**(3), 409–421 (2017)
12. Khoi, B.H., Van Tuan, N.: Using smartPLS 3.0 to analyse internet service quality in vietnam. In: Anh, L.H., Dong, L.S., Kreinovich, V., Thach, N.N. (eds.) Econometrics for Financial Applications, Studies in Computational Intelligence, vol. 760, pp. 430–439. Springer, Cham (2018)
13. Latan, H., Noonan, R.: Partial Least Squares Path Modeling: Basic Concepts. Methodological Issues and Applications. Springer, Cham (2017)
14. Wong, K.K.-K.: Partial least squares structural equation modeling (PLS-SEM) techniques using smartPLS. Market. Bull. **24**(1), 1–32 (2013)
15. Henseler, J., Hubona, G., Ray, P.A.: Using PLS path modeling in new technology research: updated guidelines. Industr. Manag. Data Syst. **116**(1), 2–20 (2016)

Impact of Leverage on Firm Investment: Evidence from GMM Approach

Duong Quynh Nga[1(✉)], Pham Minh Dien[2], Nguyen Tran Cam Linh[1], and Nguyen Thi Hong Tuoi[1]

[1] Ho Chi Minh Open University, Ho Chi Minh City, Vietnam
{nga.dq,linh.ntc}@ou.edu.vn,
love_foreverl1190@yahoo.co.uk
[2] Central Committee of the Communist Party of Vietnam, Hanoi, Vietnam
minhdienbkttw@gmail.com

Abstract. By applying the Differential Generalized Method of Moment (DGMM) to the 2009–2014 data about 107 companies listed on Ho Chi Minh city Stock Exchange (HOSE), we show that financial leverage has a positive effect on the level of investment, and this effect is stronger in high growth companies.

Keywords: DGMM · Leverage · Investment decision

1 Introduction

In a company, decision making is very important. Financial decisions made by companies are most discussed in the existing literature of financial management and corporate finance. Corporate finance is concerned with investing, procuring, financing and managing corporate assets to achieve their goals [35]. From that, it is found that corporate finance relates to the main types of decisions: investment decisions, capital decisions, asset management decisions and dividends decisions. In particular, investment decisions are considered the most important decision among corporate finance decisions because it creates value for the business. However, in order to get an investment decision, the initial thing to do is to decide what kind of capital to provide for the purchase of the asset, using short-term or long-term capital, equity or loans. Therefore, there is a close relationship between investment decisions and capital decisions.

Under the first clause of the M&M theory Modigliani and Miller (1958), under no-tax conditions, the value of the levered firm[1] is equal to the value of the non-levered firm[2]. In other words, leverage does not affect investment decisions, but in fact, the market is imperfect and there are no assumptions as in M&M theory. So, leverage and investment are intertwined, and this relationship is endogenous, when the two elements interact.

---

[1] A company uses debt in its capital structure. They have a large percentage of debt relative to equity when compared against peers in the same industry. The value of a levered firm is equal to the value of an unlevered firm plus tax shield.

[2] A company without debt.

V. Kreinovich et al. (Eds.): ECONVN 2019, SCI 809, pp. 282–295, 2019.
https://doi.org/10.1007/978-3-030-04200-4_22

Vietnam is a developing country, most of the enterprises in Vietnam are small and medium size, therefore, for more capital to invest, businesses will usually choose to loan. However, actually, according to a report from the Ministry of Finance, enterprises using financial leverage on investment are currently in a state of difficulty when using too much debt, which will make these enterprises stay in the vicious cycle: bad debt, big interest, weak business and no new cash flow to reinvest; so the investment decision will be difficult. However, when the enterprise uses too little debt compared to the capital structure, the investment decision is not necessarily favorable. In addition, the growth opportunities of each enterprise also greatly affect this issue.

A business that has the right level of investment will create more value for the business; in particular, it will increase the value of the business, thereby increasing the property for the owner. The level of investment can be influenced by a variety of factors such as leverage[3], cash flow, return on asset (ROA), Tobin Q, size, etc. Among the factors that affect the level of investment, the study focuses only on financial leverage in the business.

This study will investigate the impact of financial leverage on the level of investment of enterprises on the HOSE; in particular, to find the effect is inverse or covariant, then the level of impact of financial leverage on the level of investment, and finally the difference in the impact of leverage on the level of investment between the companies have high and low growth opportunities, thereby finding solutions to improve the efficiency of the first private business.

2 Theoretical Background

According to Hillier et al. (2010), because of the conflict between shareholders, managers, and creditors, the relationship between leverage and investment is largely inverted, but on the growth opportunities of different companies, this mechanism is different. For low growth companies, on one hand, conflicts between shareholders and creditors make the company lack good investment opportunities; on the other hand, conflicts between shareholders and managers can cause excessive investment.

The over-investment will occur when the company has a low growth opportunity, which is when the company has no positive NPV projects. But managers want to increase the size of the company and increase free cash flow to carry out activities that can achieve their best interests while the interests of the company are ignored [5]; Van-Horne and Wachowicz (2001). Therefore, they continue to invest, even in negative NPV projects. This leads to a positive relationship between leverage and investment. But managers cannot continue to increase the level of debt. Debt can also serve as a safeguard against over-investment such as cash being paid to bondholders in order to limit the conduct of wasteful activities and possible bondholders to evaluate the executives [5]; Vo (2015); [26]. This leads to a negative relationship between leverage and investment, because managers do not want to pay interest and principal upon

[3] Leverage (Investopedia) results from using borrowed capital as a funding source. Leverage can also refer to the amount of debt a firm uses to finance assets.

demand that increases the risk of default. Sub-investment takes place in the condition that the company has high growth opportunities. In addition, managers are reluctant to pay external capital costs (whether or not they are affected by asymmetric information), which causes the increase of the risk of default [5]. This leads to a negative relationship between leverage and investment because of the limited amount of debt financing due to the cost of capital and the increased risk of default. Based on the above theories, the study proposes the first hypothesis as follows

H1: Financial leverage and investment are inversely related.

In addition, according to Jensen (1986) and Stulz (1990), a higher debt ratio in the low growth company would lead to a reduction in investment. With high growth companies, thanks to more investment opportunities, sometimes high risks of projects or asymmetric information, debt can undermine excessive investment caused by shareholders, managers, and reducing investment. Therefore, leverage in high growth companies impacts on investment less than low growth companies. Based on recent empirical theories and results, the second hypothesis is as follows:

H2: For companies with high growth opportunities, the negative relationship of leverage to investment is weaker than those with low growth opportunities.

3 Research Model

Research applied regression method, using DGMM model to find the impact of leverage on the level of investment of enterprises.

The research model is based on Li et al.'s model (2010). The reason for choosing this model is due to the similarity between the Chinese and Vietnamese markets when they are emerging economies, depending on import and export. In addition, according to Nguyet et al. (2014), the Chinese stock market reflects information on poor performance of the company compared to developed economies because the stock market tends to reflect market-level information rather than the company level, and the Vietnamese stock market is similar when it does not fully reflect the basic economic information of the business. The model that determines how leverage affects investment, the first independent variable is financial leverage, and investment is also influenced by other variables such as ROA, cash flow, efficient use of fixed assets, Tobin Q, company size. However, these variables are not necessarily sufficient so the equation will add the variables λ_t and μ_i, which represent the remaining variables, and finally add $\varepsilon_{i,t}$ is the error variable.

$$\begin{aligned} \frac{I_{i,t}}{K_{i,t-1}} &= \alpha + \beta_1 LEV_{i,t-1} + \beta_2 \frac{I_{i,t-1}}{K_{i,t-2}} + \beta_3 ROA_{i,t} \\ &+ \beta_4 \frac{CF_{i,t}}{K_{i,t-1}} + \beta_5 \frac{S_{i,t}}{K_{i,t-1}} + \beta_6 TQ_{i,t-1} + \beta_7 Size_{i,t} + \lambda_t + \mu_i + \varepsilon_{i,t} \end{aligned} \tag{1}$$

Therein:

i = 1, 2, 3, 4... 107 (107 Companies)
t = 2009, 2010, 2011 ... 2014 (6 periods)
$I_{i,t}$: is the long-term investment in the period t
$K_{i,t-1}$: capital accumulation, is the total assets of the previous period (the period t–1) and that is also the total assets at the beginning of the year
$I_{i,t}/K_{i,t-1}$ indicates the level of investment in the period t
$LEV_{i,t-1}$ is the financial leverage of the period t–1
$(I_{i,t-1}/K_{i,t-2})$ is the investment level at the period t–1
$ROA_{i,t}$ is the net asset income
$CF_{i,t}/K_{i,t-1}$ is the cash flow from the business
$S_{i,t}/K_{i,t-1}$ is the performance of fixed assets.
$TQ_{i,t-1}$ is Tobin Q at the period t–1,
$Size_{i,t}$ is the firm size, in log of total assets,
λ_t is the effect of time (the change coefficient varies with time).
μ_i is the effect of each particular unit unobserved;
$\varepsilon_{i,t}$ t is the error variable of the model.

According to previous research theories, there is a negative relationship between investment and leverage, but this relationship varies between high growth companies and low growth companies. To test this difference, the sample was divided into high growth companies - model (2) and low - model (3) based on the Tobin Q index representing the growth opportunities of the company.

With Tobin Q > 1, it represents companies with high growth opportunities, and companies with low growth opportunities for the opposite. This sampling was also carried out in previous studies by Lang (1996), Varouj et al. (2005).

Dependent variable:

- ***Level of investment*** $(I_{i,t}/K_{i,t-1})$:

This study uses the level of investment as a dependent variable. The level of investment is calculated by the ratio of capital expenditure $(I_{i,t}/K_{i,t-1})$. This is a measure of the company's investment, which eliminates the impact of enterprise size on investment. Therein, $I_{i,t}$: is the long-term investment in the period t. Capital Accumulation $K_{i,t-1}$: is the total assets of the previous period (the period t-1) and that is also the total assets at the beginning of the year.

Independent variables:

- *Financial leverage* ($LEV_{i,t-1}$):

Financial leverage is the ratio of total liabilities in year t over total assets in the period t–1. Total assets in the period t–1 are higher than the period t, because the distribution of interests between shareholders and creditors is often based on the initial financial structure.

If managers get too much debt, they will abandon projects that bring positive net present value. Moreover, it also supports both the theory of sub-investment and the theory of over-investment. Although the research focuses on the impact of financial leverage on investment levels, there are other factors that influence the level of investment according to the company investment theory. As a result, Consequently, the study adds elements such as: cash flow ($CF_{i,t}/K_{i,t-1}$), growth opportunities ($TQ_{i,t-1}$), efficient use of fixed assets $\left(S_{i,t}/K_{i,t-1}\right)$, investment level in the period t–1 $\left(I_{i,t-1}/K_{i,t-2}\right)$, net asset income ($ROA_{i,t}$), firm size ($Size_{i,t}$), time effect ($\lambda_t$) and unobserved specific unit effect (μ_i).

- *Cash flow ($CF_{i,t}/K_{i,t-1}$):*

According to Franklin and Muthusamy (2011), cash flow is measured by the gross profit before extraordinary items and depreciation, which is an important factor for growth opportunities

- *Growth opportunities ($TQ_{i,t-1}$):*

According to Phan Dinh Nguyen (2013), Tobin Q is used as a representation of the growth opportunities for businesses. The measurement of Tobin Q is the ratio of the market value of total assets and book value of total assets. Based on the research by Li et al. (2010), Tobin Q is calculated using the following formula:

$$\text{Tobin Q} = \frac{\text{Debt } + \text{share price x number of issued shares}}{\text{Book value of assets}}$$

Therein: Book value of assets = Total assets – Intangible fixed assets – Liabilities

Information of this variable is taken from the balance sheets and annual reports of the business. It can be said that investment opportunities affect the level of investment, the higher growth opportunities will make the level of investment more effective when businesses try to maximize the value of the company through the project has a positive net present value.

The study uses $TQ_{i,\ t-1}$ because it has a higher level of interpretation than t–1, when the distribution of interests between shareholders and creditors is often based on the initial financial structure

- ***Efficient use of fixed assets $\left(S_{i,t}/K_{i,t-1}\right)$:***

This variable is measured by the annual revenue divided by the fixed assets in the period t-1. A high efficient use of fixed assets ratio reflects the level of enterprise asset utilization, and vice versa, a low rate that reflects a low level of asset utilization. The latency of efficient use of fixed assets variables is explained by the fact that technology and projects often take a long time to get into operation, so the latency of this variable is used.

- ***Net asset income ($ROA_{i,t}$):***

According to Franklin and Muthusamy (2011), profitability is measured by the value of net profit and assets. It is calculated by the formula

$$\text{ROA} = \frac{\textit{Profit after tax}}{\textit{Total assets}}$$

- ***Firm size ($Size_{i,t}$):***

The study uses log of total assets, information of this variable is taken from the balance sheet. Data information is derived from secondary data sources, in particular, financial reports, annual reports and prospectuses of 107 non-financial companies obtained from HOSE from 2009 to 2014, including 642 observations. The study excludes observations that are financial institutions such as banks and finance companies, investment funds, insurance companies, and securities companies because of their different capital structure and structure for other business organizations.

Data collected for 6 years from 2009 to 2014, there is a total of 642 observations of enterprises with a full database. However, variables such as the level of investment in the sample are fixed assets in year t-1 and t-2, so the study will collect more data in 2007 and 2008 (Tables 1, 2, 3 and 7).

Table 1. Defining variables

No.	Variables	Description	Empirical studies	Expected mark
1	**Dependent variable**			
	Level of investment $\left(\frac{I_{i,t}}{K_{i,t-1}}\right)$	[Fixed asset in year t–1 fixed assets + Depreciation]/fixed assets in year t–1	Robert and Alessandra (2003); Catherine and Philip (2004); Frederiek and Cynthia (2008); Maturah and Abdul (2011); Yuan and Motohashi (2008, 2012); Varouj et al. (2005); Franklin and Muthusamy (2011); Ngoc Trang and Quyen (2013); Li et al. (2010)	
2	**Independent variables**			
	Leverage ($LEV_{i,t-1}$)	Total debt in year t/Total assets in year t–1	Maturah and Abdul (2011); Yuan and Motohashi (2008, 2012); Varouj et al. (2005); Franklin and Muthusamy (2011); Ngoc Trang and Quyen (2013); Phan Thi Bich Nguyet et al. (2014); Li et al. (2010)	–
	Level of investment in year t–1 $\left(\frac{I_{i,t-1}}{K_{i,t-2}}\right)$	[Fixed asset in year t–1 – Fixed asset in year t-2 + Depreciation]/Fixed asset in year t–2	Robert and Alessandra (2003); Catherine and Philip (2004); Li et al. (2010)	+

(*continued*)

Table 1. *(continued)*

No.	Variables	Description	Empirical studies	Expected mark
	Ratio of return on total assets ($ROA_{i,t}$)	Net income after tax/Total assets	Li et al. (2010); Ngoc Trang and Quyen (2013).	+
	Cash flow $\left(\frac{CF_{i,t}}{K_{i,t-1}}\right)$	(EBITDA – interest rate – tax) year t/fixed assets year t–1	Robert and Alessandra (2003); Frederiek and Cynthia (2008); Maturah and Abdul (2011); Yuan and Motohashi (2008, 2012); Varouj et al. (2005); Franklin and Muthusamy (2011); Ngoc Trang and Quyen (2013); Li et al. (2010); Lang et al. (1996)	+
	Efficient use of fixed assets $\left(\frac{S_{i,t}}{K_{i,t-1}}\right)$	Turnover in year t/Fixed assets in year t–1	Varouj et al. (2005); Li et al. (2010)	+
	Growth Opportunities–Tobin Q ($TQ_{i,t-1}$)	(Debt + share price x number of issued shares)/ Book value of assets Inside: Book value of assets = Total assets – Intangible fixed assets – Liabilities	Robert and Alessandra (2003); Maturah and Abdul (2011); Nguyen et al. (2008, 2012); Franklin and Muthusamy (2011); Varouj et al. (2005); Ngoc Trang and Quyen (2013); Nguyet et al. (2014); Li et al. (2010)	+
	Firm size ($Size_{i,t}$)	Log total assets in year t	Frederiek and Cynthia (2008); Nguyet et al. (2014); Li et al. (2010); Yuan and Motohashi (2012)	+

Table 2. Statistics table describing the observed variables

Observed variables	Full sample				High growth company (> 1)				Low growth company (< 1)			
	Medium	Std dev	Smallest	Largest	Medium	Std dev	Smallest	Largest	Medium	Std Dev	Smallest	Largest
$I_{i,t}/K_{i,t-1}$	0.366	1.117	–1.974	14.488	0.383	1.249	–1.368	11.990	0.351	0.984	–1.974	14.488
$LEV_{i,t-1}$	0.518	0.271	0.033	1.723	0.702	0.210	0.041	1.635	0.353	0.205	0.033	1.723
$ROA_{i,t}$	0.079	0.084	–0.169	0.562	0.042	0.056	–0.169	0.562	0.112	0.091	–0.158	0.428
$CF_{i,t}/K_{i,t-1}$	0.880	1.665	–3.978	28.219	0.698	0.907	–2.545	8.092	1.044	2.116	–3.978	28.219
$S_{i,t}/K_{i,t-1}$	9.477	11.649	0.216	75.117	10.519	12.783	0.216	75.117	8.539	10.455	0.223	64.019
$TQ_{i,t-1}$	1.247	1.168	0.032	6.703	2.141	1.138	1.000	6.703	0.443	0.252	0.032	0.997
$Size_{i,t}$	13.924	1.209	11.738	17.409	14.212	1.206	11.851	17.409	13.665	1.154	11.738	17.065

Source: Author's calculations, based on 642 observations of 107 companies obtained from the HOSE during the period 2009–2014.

Table 3. Hausman test for 3 case estimates

No.	Case estimates	Chi2	Prob(chi2)	Options
1	Full sample	77.46	0.000	Fixed effect
2	High growth company (> 1)	118.69	0.000	Fixed effect
3	Low growth company (< 1)	124.42	0.000	Fixed effect

Source: Author's calculations

4 Results

Looking at the statistics table, the average $I_{i,t}/K_{i,t-1}$ of the study was 0.366, while Lang's study (1996) was 0.122, Li Jiming was 0.0371, Varouj et al. (2005) was 0.17, Nguyet et al. (2014) was 0.0545, Jahanzeb and Naeemullah (2015) was 0.225.

The average $LEV_{i,t-1}$ of the whole sample size is 0.518, which is roughly equivalent to previous studies by Lang (1996) was 0.323, Li (2010) was 0.582, Phan Thi Bich Nguyet was 0.1062, Aivazian (2005) was 0.48, Jahanzeb and Naeemullah (2015) was 0.62.

The average Tobin Q of the whole sample is 1.247, compared with the previous studies, which is quite reasonable, with Lang (1996) was 0.961, Aivazian (2005) was 1.75, Li (2010) was 2.287, Nguyet (2014) was 1.1482, Jahanzeb and Naeemullah (2015) was 0.622, with the largest value of this study being 6,703, while Vo (2015) research on HOSE was 3.5555.

4.1 Regression Results

According to the analysis results, the coefficients Prob (chi2) are less than 0.05, so the H0 hypothesis is rejected; the conclusion is that using Fixed Effect will be more compatible

Check for Model Defects

Table 4 shows the matrix of correlations between the independent variables, and also the Variance Inflation Factor (VIF), an important indicator for recognizing multicollinearity in the model. According to Gujarati (2004), this index > 5 is a sign of high multi-collinearity, if the index of approximately 10 indicates a serious multicollinearity.

Between variable pairs, the correlation coefficient is less than 0.8, considering that the VIF of all variables to be less than 2. So there are no multilayers in the model.

Next, Table 5 includes the table A of the Wald Verification and Table B of the Wooldridge Verification to examine the variance and self-correlation of the model.

Tables 4 and 5 show the defect of the model; therefore, the study will use appropriate regression to address the aforementioned defect.

Table 6 presents regression results using the DGMM method, also known as GMM Arellano Bond (1991). So GMM is the regression method when there are endogenous phenomena and T-time series of small table data in the model; according to previous studies by Lang (1996), Varouj et al. (2005), etc., leverage and investment are

Table 4. Correlation matrix of independent variables

Full sample								
	$LEV_{i,t-1}$	$I_{i,t-1}/K_{i,t-2}$	$ROA_{i,t}$	$CF_{i,t}/K_{i,t-1}$	$S_{i,t}/K_{i,t-1}$	$TQ_{i,t-1}$	$Size_{i,t}$	VIF
$LEV_{i,t-1}$	1							1.93
$I_{i,t-1}/K_{i,t-2}$	0.0756	1						1.02
$ROA_{i,t}$	−0.3401*	0.0006	1					1.42
$CF_{i,t}/K_{i,t-1}$	0.0647	−0.0059	0.3435*	1				1.49
$S_{i,t}/K_{i,t-1}$	0.2505*	−0.0671	0.0441	0.4557*	1			1.4
$TQ_{i,t-1}$	0.6372*	0.1008*	−0.4062*	−0.0787*	0.1147*	1		1.84
$Size_{i,t}$	0.2775*	0.0771	0.0044	0.0836*	−0.0487	0.2227*	1	1.14
Mean VIF								1.46
High growth company (TQ > 1)								
	$LEV_{i,t-1}$	$I_{i,t-1}/K_{i,t-2}$	$ROA_{i,t}$	$CF_{i,t}/K_{i,t-1}$	$S_{i,t}/K_{i,t-1}$	$TQ_{i,t-1}$	$Size_{i,t}$	VIF
$LEV_{i,t-1}$	1							1.33
$I_{i,t-1}/K_{i,t-2}$	0.0528	1						1.05
$ROA_{i,t}$	−0.0261	0.0535	1					1.32
$CF_{i,t}/K_{i,t-1}$	0.2451*	−0.0876	0.3938*	1				1.62
$S_{i,t}/K_{i,t-1}$	0.3140*	−0.1118	0.0498	0.4730*	1			1.43
$TQ_{i,t-1}$	0.3393*	0.0969	−0.2317*	0.0092	0.0994	1		1.22
$Size_{i,t}$	0.2191*	0.0876	0.0889	0.0608	−0.0679	0.1179*	1	1.1
Mean VIF								1.3
Low growth company (TQ < 1)								
	$LEV_{i,t-1}$	$I_{i,t-1}/K_{i,t-2}$	$ROA_{i,t}$	$CF_{i,t}/K_{i,t-1}$	$S_{i,t}/K_{i,t-1}$	$TQ_{i,t-1}$	$Size_{i,t}$	VIF
$LEV_{i,t-1}$	1							1.6
$I_{i,t-1}/K_{i,t-2}$	0.0417	1						1.01
$ROA_{i,t}$	−0.151*	0.014	1					1.22
$CF_{i,t}/K_{i,t-1}$	0.1636*	0.0473	0.3216*	1				1.68
$S_{i,t}/K_{i,t-1}$	0.1951*	−0.014	0.1219*	0.5518*	1			1.53
$TQ_{i,t-1}$	0.5616*	0.0516	−0.2609*	−0.0386	0.0278	1		1.55
$Size_{i,t}$	0.1364*	0.0373	0.1303*	0.1435*	−0.0729	0.0407	1	1.09
Mean VIF								1.38

*: statistically significant at 5%
Source: Test results from Stata software

interrelated, leading to being endogenous in the model. In addition, according to Richard et al. (1992), TQ variables are also endogenous with investment. Regression Models for 7 Variables (Level of Investment, Leverage, ROA, Cash Flow, Efficient use of fixed assets, Tobin Q, Firm Size), and lag 1 of Investment Level.

The regression results from the model (1), (2) and (3) will lead to the conclusion of accepting or rejecting the hypothesis given in Chapter 3.

Table 5. Variance and self-correlation checklist

Table A: Wald verification

No.	Cases	Chi2	Prob (chi2)	Verification results	Conclusion
1	Full sample	8.5E+05	0.000	H0 is rejected	There is variance
2	High growth company TQ (> 1)	2.1E+33	0.000	H0 is rejected	There is variance
3	Low growth company TQ (< 1)	1.5E+36	0.000	H0 is rejected	There is variance

Table B: Wooldridge verification

No.	Cases	F	Prob (F)	Verification results	Conclusion
1	Full sample	57.429	0.000	H0 is rejected	There is correlation
2	High growth company TQ (> 1)	29.950	0.000	H0 is rejected	There is correlation
3	Low growth company TQ (< 1)	10.360	0.002	H0 is rejected	There is correlation

Source: Test results from Stata software

Estimated results by DGMM method showed that:

- Variables are endogenous in estimation: Leverage and Tobin Q (implemented in GMM content), the remaining variables are exogenous: lag 1 of Investment Level, ROA, Cash Flow, Efficient use of fixed assets, Company size (expressed in the iv_instrument variable) when carrying out the empirical modeling.
- For the self-correlation of the model, the Arellano-Bond level 2 test, AR (2) shows that the variables have no correlation in the model.
- On verifying endogenous limits in the model, Sargan's test confirms that instrument variables are exogenous, i.e. not correlated with the residuals.

Observing the regression model we see:

- The $LEV_{i,t-1}$ is significant in all three cases and all have the same effect on $I_{i,t}/K_{i,t-1}$.
- The $ROA_{i,t}$ is significant in cases 1 and 3 and is inversely related to $I_{i,t}/K_{i,t-1}$.
- The $CF_{i,t}/K_{i,t-1}$ are significant in all three models, having a similar relationship with $I_{i,t}/K_{i,t-1}$ in models 1 and 3, while the second model is inverted.
- The $S_{i,t}/K_{i,t-1}$ are significant in both cases 1 and 2 and all have the same effect on $I_{i,t}/K_{i,t-1}$.
- The $TQ_{i,t-1}$ is significant in model 2, having a relationship with $I_{i,t}/K_{i,t-1}$.
- The $Size_{i,}$ is significant in models 1 and 3, showing inverse effects with $I_{i,t}/K_{i,t-1}$.

The empirical results show that financial leverage is positively correlated with the level of investment, and this relationship is stronger in high growth companies.

Table 6. Regression results

Observed variables	$I_{i,t}/K_{i,t-1}$		
	Full sample	High growth company TQ (> 1)	Low growth company TQ (< 1)
	(1)	(2)	(3)
$I_{i,t-1}/K_{i,t-2}$	**–0.20761*****	**–0.34765*****	**–0.09533****
	(0.000)	(0.006)	(0.040)
$LEV_{i,t-1}$	**2.97810****	**4.95768*****	**2.23567*****
	(0.047)	(0.004)	(0.002)
$ROA_{i,t}$	**–3.95245****	**–4.48749**	**–2.87445*****
	(0.020)	(0.357)	(0.010)
$CF_{i,t}/K_{i,t-1}$	**0.31868*****	**–1.12392***	**0.28351****
	(0.006)	(0.10)	(0.018)
$S_{i,t}/K_{i,t-1}$	**0.06949*****	**0.16610*****	**0.00414**
	(0.001)	(0.000)	(0.765)
$TQ_{i,t-1}$	**0.20673**	**0.76265****	**–1.05025**
	(0.486)	(0.038)	(0.294)
$Size_{i,t}$	**–1.23794***	**–2.63434**	**–0.75111***
	(0.059)	(0.233)	(0.058)
Obs	321	119	192
AR (2)	0.144	0.285	0.783
Sargan test	0.707	0.600	0.953

Note: * $p < 0.1$, ** $p < 0.05$, *** $p < 0.01$
Source: Test results from Stata software

Table 7. Regression models are rewritten

No.	Cases	The regression model is rewritten
1	Full sample	$I_{i,t}/K_{i,t-1} = -0.20761\ I_{i,t-1}/K_{i,t-2} + 2.97810\ LEV_{i,t-1-3}.95245\ ROA_{i,t} + 0.31868\ CF_{i,t}/K_{i,t-1} + 0.06949\ S_{i,t}/K_{i,t-1-1}.23794\ Size_{i,t}$
2	High growth company TQ (> 1)	$I_{i,t}/K_{i,t-1} = -0.34765\ I_{i,t-1}/K_{i,t-2} + 4.95768\ LEV_{i,t-1-1}.12392\ CF_{i,t}/K_{i,t-1} + 0.1661\ S_{i,t}/K_{i,t-1} + 0.76265\ TQ_{i,t-1}$
3	Low growth company TQ (< 1)	$I_{i,t}/K_{i,t-1} = -0.09533\ I_{i,t-1}/K_{i,t-2} + 2.23567\ LEV_{i,t-1-2}.87445\ ROA_{i,t} + 0.28351\ CF_{i,t}/K_{i,t-1} - 0.75111\ Size_{i,t}$

In experimental terms, these results are not consistent with the initial expectation; the following is an analysis of the impact of leverage on the level of investment.

Financial Leverage
The impact of financial leverage on the level of investment is contrary to the initial expectation of the regression across the sample. The effect was quite strong, with other factors remaining unchanged, when financial leverage increased by one unit, the level

of investment increased 2.98 units. When leverage increases, it increases investment, in other words, the more debt the company makes, the higher the investment in fixed assets is. The impact remains unchanged when it comes to companies with low and high growth opportunities, especially in high growth companies, leverage that has a stronger impact on investment, as expected and as mentioned in previous research by Ross (1977), Jensen (1986), Ngoc Trang and Quyen (2013). This shows that companies with high growth opportunities can easily access loans through their relationships, and invest as soon as they have a good chance.

The Ratio of Return on Total Assets

On the whole sample, given that other factors remained unchanged, when the return on total assets increased by one unit, the investment was reduced by 3.95 units. The relationship between ROA and level of investment found in this study is the inverse relationship for cases 1 and 3. This is in contrast to previous studies by Ngoc Trang and Quyen (2013), Li et al. (2010), found a positive correlation between ROA and investment. Since these companies can look for loans through their relationship without having to rely on financial ratios to prove the financial condition of the company.

Cash Flow

In the whole sample, given that other factors remained unchanged, when the cash flow increased one unit, the investment level increased by 0.31 units. Cash flow has the same impact on the return on investment in the sample and in the low growth companies. This is consistent with previous studies by Varouj et al. (2005), Li et al. (2010), Lang et al. (1996). The investment of the company in the whole sample depends on internal cash flow, as more cash flow can be used in investment activities. While the company has high growth opportunities, the cash flow is inversely related to investment, which indicates that high growth companies are not dependent on internal cash flow. You can use the relationship to find an easy loan.

Efficient Use of Fixed Assets

In the whole sample, with other factors remaining unchanged, when the efficient use of fixed assets increased by one unit, the investment increased by 0.32 units. Research indicates that sales have a positive relationship with investment levels in cases 1 and 2, agreed with Varouj et al. (2005), Li et al. (2010), Lang et al. (1996), Ngoc Trang and Quyen (2013), as the company has the higher sales from the efficient use of fixed assets leading to increase the production of the company, to meet that demand, the company will strengthen invest by expanding the production base, increasing investment for the company.

Tobin Q

The regression is carried out across the sample and in the low growth companies, the results show that the relationship between Tobin Q's and the level of business investment was not found. However, when the regression is under case 2 with high growth opportunities, this effect is similar (see Varouj et al. (2005), Li et al. (2010), Lang et al. (1996), Nguyet et al. (2014)). Explaining this impact, companies with high growth opportunities will make investment opportunities more efficient; therefore there will be more investment. With a full sample, Tobin Q has no effect. With the empirical results of Abel (1979) and Hyashi (1982), Tobin Q is consistent with the neoclassical model given the perfect market conditions, the production function and adjustment cost. To meet certain conditions, such as perfect competition, profitable return on a scale of production technology, the company can control the capital flow and predefined equity investments. And with data from experimental results by Goergen and Renneboog (2001) and Richardson (2006), they argue that Tobin's Q is not an explanatory variable for ideal investment because it only includes opportunities growth in the past.

Company Size
In the whole sample, with other factors remaining unchanged, when the size of the company increased one unit, the investment level decreased by 1.24 units. The size of the company has a inverse impact on the level of investment in the regression across the sample and in companies with low growth opportunities. This indicates that as the company has more assets, the more difficult it is for the company to control, the less likely it is to invest [according to Ninh et al. (2007)]. While in companies with high growth opportunities, this relationship was not found in the study.

5 Conclusion

With the number of 107 companies obtained from the HOSE, including 642 observations during the period 2009–2014, the analysis results show that:

- Financial leverage has a positive impact on the company's investment, which is consistent with previous studies by Ross (1977), Jensen (1986), Nguyen Thi Ngoc Trang and Trang Thuy Quyen (2010).
- The level of impact of financial leverage is quite high: under the condition that other variables are constant, when the leverage is increased by 1 unit, the investment level increases by 2,978 units.
- There is a difference in the impact of financial leverage on the level of investment between companies that have high and low growth opportunities. Specifically, the company has a high growth opportunity, a strong correlation of 2.72201 units compared to its low growth.

References

Franklin, J.S., Muthusamy, K.: Impact of leverage on firms investment decision. Int. J. Sci. Eng. Res. **2**(4), 1–16 (2011)

Goergen, M., Renneboog, L.: Investment policy, internal financing and ownership concentration in the UK. J. Corp. Finance **7**, 257–284 (2001)

Hillier, D., Jaffe, J., Jordan, B., Ross, S., Westerfield, R.: Corporate Finance. First European Edition, McGraw-Hill Education (2010)

Jahanzeb, K., Naeemullah, K.: The impact of leverage on firm's investment. Res. J. Recent Sci. **4**(5), 67–70 (2015)

Jensen, M.C.: Agency costs of free cash flow, corporate finance and takeovers. Am. Econ. Rev. **76**(2), 323–329 (1986)

Modigliani, F., Miller, M.H.: The cost of capital, corporation finance and the theory of investment. Am. Econ. Rev. **48**(3), 261–297 (1958)

Myers, S.C.: Capital structure. J. Econ. Perspect. **15**(2), 81–102 (2001)

Myers, S.C.: Determinants of corporate borrowing. J. Finan. Econ. **5**, 147–175 (1977)

Myers, S.C., Majluf, N.S.: Corporate financing and investment decisions when firms have information that investors do not have. J. Finan. Econ. **13**(2), 187–221 (1984)

Kiều, N.M.: Tài chính doanh nghiệp căn bản. Nhà xuất bản lao động xã hội (2013)

Ngọc Trang, N.T., Quyên, T.T.: Mối quan hệ giữa sử dụng đòn bẩy tài chính và quyết định đầu tư. Phát triển & Hội nhập **9**(19), 10–15 (2013)

Pawlina, G., Renneboog, L.: Is investment-cash flow sensitivity caused by agency costs or asymmetric information? Evidence from the UK. Eur. Finan. Manag. **11**(4), 483–513 (2005)

Nguyen, P.D., Dong, P.T.A.: Determinants of corporate investment decisions: the case of Vietnam. J. Econ. Dev **15**, 32–48 (2013)

Nguyệt, P.T.B., Nam, P.D., Thảo, H.T.P.: Đòn bẩy và hoạt động đầu tư: Vai trò của tăng trưởng và sở hữu nhà nước. Phát triển & Hội nhập **16**(26), 33–40 (2014)

Richard, B., Stephen, B., Michael, D., Fabio, S.: Investment and Tobin's Q. evidence from company panel data. J. Econ. **51**, 233–257 (1992)

Richardson, S.: Over-investment of free cash flow. Rev. Account. Stud. **11**(2), 159–189 (2006)

Robert, E.C., Alessandra, G.: Cash flow, investment, and investment opportunities: new tests using UK panel data. Discussion Papers in Economics, No. 03/24, ISSN 1360-2438, University of Nottingham (2003)

Ross, G.: The determinants of financial structure: the incentive signaling approach. Bell J. Econ. **8**, 23–44 (1977)

Stiglitz, J., Weiss, A.: Credit rationing in markets with imperfect information. Am. Econ. Rev. **71**, 393–410 (1981)

Stulz, R.M.: Managerial discretion and optimal financing policies. J. Finan. Econ. **26**, 3–27 (1990)

Van-Horne, J.-C., Wachowicz, J.M.: Fundamentals of Financial Management. Prentice Hall, Upper Saddle River (2001)

Varouj, A., Ying, A., Qiu, J.: The impact of leverage on firm investment: Canadian evidence. J. Corp. Finan. **11**, 277–291 (2005)

Vo, X.V.: The role of corporate governance in a transitional economy. Int. Finan. Rev. **16**, 149–165 (2015)

Yuan, Y., Motohashi, K.: Impact of Leverage on Investment by Major Shareholders: Evidence from Listed Firms in China. WIAS Discussion Paper No. 2012-006 (2012)

Zhang, Y.: Are debt and incentive compensation substitutes in controlling the free cash flow agency problem? J. Finan. Manag. **38**(3), 507–541 (2009)

Oligopoly Model and Its Applications in International Trade

Luu Xuan Khoi[1(✉)], Nguyen Duc Trung[2], and Luu Xuan Van[3]

[1] Forecasting and Statistic Department, State Bank of Vietnam, Hanoi, Vietnam
khoi.luuxuan@sbv.gov.vn
[2] Banking University of Ho Chi Minh City, Ho Chi Minh City, Vietnam
trungnd@buh.edu.vn
[3] Faculty of Information Technology and Security, People's Security Academy, Hanoi, Vietnam
vanlx.hvan@gmail.com

Abstract. Each firm in the oligopoly plays off of each other in order to receive the greatest utility, expressed in the largest profits, for their firm. When analyzing the market, decision makers develop sets of strategies to respond the possible actions of competitive firms. In international stage, firms are competitive and they have different business strategies, their interaction becomes essential because the number of competitors is increased. This paper will provide an examination in international trade balance and public policy under Cournot's framework. The model shows how the oligopolistic firm can decide the business strategy to maximize its profit given others' choice, and how the public maker can find out the optimal tariff policy to maximize its social welfare. The discussion in this paper can be significant for both producers in deciding their quantities needed to be sold in not only domestic market but also international stage in order to maximize their profits and governments in deciding the tariff rate on imported goods to maximize their social welfare.

Keywords: Cournot model · International trade · Public policy Oligopoly

1 Introduction

It may be unusual that countries simultaneously import and export same type of goods or services with their international partners (intra-industry trade). However, in general, there are a range of benefits of intra-industry trade offering businesses and countries engaging in it. The benefits of intra-industry trade have been obvious because it reduce the production cost that can be beneficent to consumers. It also gives opportunity for businesses to benefit from the economies of scale, as well as use their comparative advantages and stimulates innovation in industry. Beside to benefits from intra-industry trade, the role of government is also important by using its power to protect domestic industry from dumping.

V. Kreinovich et al. (Eds.): ECONVN 2019, SCI 809, pp. 296–310, 2019.
https://doi.org/10.1007/978-3-030-04200-4_23

Government can apply tariff barrier on imported goods to foreign manufacturers with the aim of increasing the price of imported goods and making them more expensive to consumers. In this international background, managers need to decide the quantity sold in not only domestic market but also other markets under tariff barrier from foreign countries. We consider a game in which the players are firms, nations and strategies are choices of outputs and tariffs. The appropriate game-theoretic model for international trade is the non-cooperate game. The main methods to analyze the strategies of players in this model are developed by the theoretical model: "Cournot Duopoly" - the subject of increased interest in recent years. The target of this paper is to examine the application of Cournot oligopoly analysis to non-collusive firms' behavior in international stage and suggest to decision makers the necessary outcome to maximize their profits as well as the best policy in tariff rate applied by the government. We develop the quantity-setting model under classical Cournot competition in trade theory to find out the equilibrium production between countries in the case that tariffs are imposed by countries to protect its domestic industry and prevent dumping from foreign firms. Section 2 recalls the Cournot oligopoly model in background. Section 3 develops the 2-market models with 2 firms competing in the presence of tariff under Cournot behaviors and examines the decision of Governments on tariff rate in considering to its social welfare. In Sect. 3, we can realize the impact of tariff difference on equilibrium price and the quantity of production between 2 countries. Moreover, both governments tend to decide the same tariff rate for importing goods with the aim of maximizing its welfare benefits. Section 4 analyzes the model, in general, with n monopolist firms competing in the international trade stage. When n become larger, the difference between equilibrium prices will be equal to the difference between tariff rates as country which imposes the higher tariff rate will have the higher equilibrium price in its domestic market. In addition to that, there will be no difference between the total quantities each firm should produce to maximize its profits when the number of trading countries (or firms) becomes larger. Section 4 also considers to welfare benefits of countries and the decision of governments on tariff rates to maximize its domestic welfare. In this section, we also find out that if there is any agreement between countries to reduce its tariff on imported goods, the social welfare in all country could be higher. Section 5 contains concluding remarks.

2 Review of Cournot Oligopoly Model

Cournot Oligopoly Model is a simultaneous-move quantity-setting strategic game of imperfect quantity competition in which firms (main players), assumed to be perfect substitutes with identical cost functions compete with homogeneous products by choosing its outputs strategically in the set of possible outputs with any nonnegative amount, and the market determines the price at which it is sold. In Cournot oligopoly model, firms recognize that they should account for the output decisions of their rivals, yet when making their own decision, they view their rivals' output as fixed. Each firm views itself as a monopolist on the residual demand curve – the demand left over after subtracting the output of its rivals. The payoff of each firm is its profit and their utility functions are increasing with their profits. Denote cost to firm i of producing q_i units: $C_i(q_i)$, where $C_i(q_i)$ is convex, nonnegative and increasing, given the overall produced amount ($Q = \sum_i q_i$), the price of the product is $p\,(Q)$ and $p\,(Q)$ is non-increasing with Q. Each firm chooses its own output q_i, taking the output of all its rivals q_{-i} as given, to maximize its profits: $\pi_i = p(Q)q_i - C_i(q_i)$.

The output vector $(q_1, q_2, ..., q_n)$ is a Cournot Nash Equilibrium if and only if (given q_{-i}):

$$\pi_i\,(q_i, q_{-i}) \geq \pi_i\,(q_i', q_{-i}) \text{ for all } i.$$

The first order condition (FOC) for firm i is given by:

$$\frac{\partial \pi_i}{\partial q_i} = p'(Q)q_i + p(Q) - C_i'(q_i).$$

To maximize the firm's profit, the FOC should be 0:

$$\frac{\partial \pi_i}{\partial q_i} = 0 \Leftrightarrow p'(Q)q_i + p(Q) - C_i'(q_i) = 0$$

The Cournot-Nash equilibrium is found by simultaneously solving the first order conditions for all n firms.

Cournot's work to economic theory "ranges from the formulation of the concept of demand function to the analysis of price determination in different market structures, from monopoly to perfect competition" (Vives 1989). The Cournot model of oligopolistic interaction among firms produces logical results, with prices and quantities that are between monopolistic (i.e. low output, high price) and competitive (high output, low price) levels. It has been successful to help understanding international trade under more realistic assumptions and recognized as the cornerstone for the analysis of firms' strategic behaviour. It also yields a stable Nash equilibrium, which is defined as an outcome from which neither player would like to change his/her decision unilaterally.

3 The Basic 2-Markets Model Under Tariff

3.1 Trade Balance Under Tariff of the Basic 2-Factors Model

This section will develop a model in which 2 export-oriented monopolist firms in 2 countries. One firm in each country (no entry) produces one homogeneous good. In the home market, $Q_d \equiv x_d + y_d$, where x_d denotes the home firm's quantity sold in the home market and y_d denotes the foreign firm's quantity sold in the home market. Similarly, in the foreign market, $Q_f \equiv x_f + y_f$, where x_f denotes home firm's quantity sold abroad and y_f denotes foreign firm's quantity in its market. Domestic demand $p_d(Q_d)$ and foreign demand $p_f(Q_f)$ imply segmented markets. Firms choose quantities for each market, given quantities chosen by the other firm. The main idea is that each firm regards each country as a separate market and therefore chooses the profit-maximizing quantity for each country separately. In the detection of dumping, each government applied a tariff fee in exporting goods from one country to the other, let t_d be the tariff imposed by Home government to Foreign firm and t_f be the tariff imposed by Foreign government to Home firm to prevent this kind of action and protect its domestic industry (mutual retaliation). Home and Foreign firms' profits can be written as the surplus remaining after total costs and tariff cost are deducted from its total revenue:

$$\pi_d = x_d p_d(Q_d) + x_f p_f(Q_f) - C_d(x_d, x_f) - t_f x_f$$

$$\pi_f = y_d p_d(Q_d) + y_f p_f(Q_f) - C_f(y_d, y_f) - t_d y_d$$

We assume that firms in 2 countries exhibit a Cournot-Nash type behavior in 2 markets. Each firm maximizes its profit with respect to own output, which yields the zero first-order conditions and negative second-order conditions. To simplify, we suppose that the demand function is linear with quantity sold in both markets and the slope of both function is -1. Home firm and Foreign firm have fixed costs f and f_1, respectively, and total costs of each firm are quadratic functions with quantities produced:

$$\begin{aligned} p_d(Q_d) &= a - (x_d + y_d) \\ p_f(Q_f) &= a - (x_f + y_f) \\ C_d(x_d, x_f) &= f + \frac{1}{2}k(x_d + x_f)^2 \\ C_f(y_d, y_f) &= f_1 + \frac{1}{2}k(y_d + y_f)^2 \end{aligned}$$

Where: $a > 0$ is the total demand in the Home market as well as in the Foreign market when the price is zero. Assume that a can be large enough to satisfy the positive value of price and optimal outputs of firms.

$k > 0$ is the slope of the marginal cost function with quantity produced.

From the above equation system, we can reach the first-order and second-order conditions:

$$\begin{cases} \dfrac{d\pi_d}{dx_d} = a - (2x_d + y_d) - k(x_d + x_f) & = 0 \\ \dfrac{d\pi_d}{dx_f} = a - (2x_f + y_f) - k(x_d + x_f) - t_f & = 0 \\ \dfrac{d\pi_f}{dy_d} = a - (x_d + 2y_d) - k(y_d + y_f) - t_d & = 0 \\ \dfrac{d\pi_f}{dy_f} = a - (x_f + 2y_f) - k(y_d + y_f) & = 0 \\ \dfrac{d^2\pi_d}{d^2x_d} = \dfrac{d^2\pi_d}{d^2x_f} = \dfrac{d^2\pi_f}{d^2y_d} = \dfrac{d^2\pi_f}{d^2y_f} = -(k+2) < 0 \end{cases}$$

$$\Leftrightarrow \begin{cases} x_d + x_f = \dfrac{2a - t_f}{2k+2} - \dfrac{y_d + y_f}{2k+2} \\ y_d + y_f = \dfrac{2a - t_d}{2k+2} - \dfrac{x_d + x_f}{2k+2} \end{cases} \qquad (1)$$

Because the second-order conditions of π_d with respect to x_d, x_f and π_f with respect to y_d, y_f are both negative, then Eq. (1) shows the reaction functions (best-response functions) for both firms. For any given output level chosen by foreign firm $(y_d + y_f)$ and given tariff rate t_f, the best-response function shows the profit-maximizing output level for home firm $(x_d + x_f)$ and vice versa.

Next, we will derive the Nash equilibrium in this model $(x_d^*, y_d^*, x_f^*, y_f^*)$ by solving the above equation system:

$$\begin{bmatrix} 0 & k & 1 & k+2 \\ k & 0 & k+2 & 1 \\ 1 & k+2 & 0 & k \\ k+2 & 1 & k & 0 \end{bmatrix} \begin{bmatrix} x_d \\ y_d \\ x_f \\ y_f \end{bmatrix} = \begin{bmatrix} a \\ a - t_f \\ a - t_d \\ a \end{bmatrix} \text{ or } A.u = b.$$

We can use the Crammer's rule to solve for the elements of u by replacing the i-th column of A by vector b to form the matrix A_i; then $u_i = |A_i|/|A|$. We have:

$$x_d^* = \frac{\begin{vmatrix} a & k & 1 & k+2 \\ a - t_f & 0 & k+2 & 1 \\ a - t_d & k+2 & 0 & k \\ a & 1 & k & 0 \end{vmatrix}}{|A|} = \frac{a}{2k+3} + t_d \frac{2k^2 + 4k + 3}{3(2k+1)(2k+3)} + t_f \frac{k(4k+5)}{3(2k+1)(2k+3)}$$

$$y_d^* = \frac{\begin{vmatrix} 0 & a & 1 & k+2 \\ k & a - t_f & k+2 & 1 \\ 1 & a - t_d & 0 & k \\ k+2 & a & k & 0 \end{vmatrix}}{|A|} = \frac{a}{2k+3} - t_d \frac{(4k+3)(k+2)}{3(2k+1)(2k+3)} - t_f \frac{2k(k+2)}{3(2k+1)(2k+3)}$$

$$x_f^* = \frac{\begin{vmatrix} 0 & k & a & k+2 \\ k & 0 & a - t_f & 1 \\ 1 & k+2 & a - t_d & k \\ k+2 & 1 & a & 0 \end{vmatrix}}{|A|} = \frac{a}{2k+3} - t_d \frac{2k(k+2)}{3(2k+1)(2k+3)} - t_f \frac{(4k+3)(k+2)}{3(2k+1)(2k+3)}$$

$$y_f^* = \frac{\begin{vmatrix} 0 & k & 1 & a \\ k & 0 & k+2 & a - t_f \\ 1 & k+2 & 0 & a - t_d \\ k+2 & 1 & k & a \end{vmatrix}}{|A|} = \frac{a}{2k+3} + t_d \frac{k(4k+5)}{3(2k+1)(2k+3)} + t_f \frac{2k^2 + 4k + 3}{3(2k+1)(2k+3)}$$

At this point, Home firm is producing an output of x_d^* in Home' market and x_f^* in Foreign's market, Foreign firm is producing an output of y_d^* in Home's market and y_f^* in Foreign's market. If Home firm produces x_d^* in Home' market and x_f^* in Foreign's market, then the best response for foreign firm is to produce y_d^* in Home' market and y_f^* in Foreign's market. Therefore, $(x_d^*, y_d^*, x_f^*, y_f^*)$ is the best response of firms to each other and neither firm has an incentive to derive its choice or the market will be in equilibrium. The equilibrium price in each market will be:

$$p_d^*(Q_d) = a - (x_d^* + y_d^*) = a\frac{2k+1}{2k+3} + t_d\frac{k+3}{3(2k+3)} - t_f\frac{k}{3\,(2k+3)} \tag{2}$$

$$p_f^*(Q_f) = a - (x_f^* + y_f^*) = a\frac{2k+1}{2k+3} - t_d\frac{k}{3(2k+3)} + t_f\frac{k+3}{3\,(2k+3)} \tag{3}$$

Moreover, the first-order-conditions and second-order-conditions of $p_d^*(Q_d)$ and $p_f^*(Q_f)$ with t_d and t_f are:

$$\begin{cases} \dfrac{dp_d^*(Q_d)}{dt_d} = \dfrac{k+3}{3(2k+3)} > 0, & \dfrac{d^2p_d^*(Q_d)}{d^2(t_d)} = -\dfrac{1}{(2k+3)^2} < 0 \\ \dfrac{dp_d^*(Q_d)}{dt_f} = -\dfrac{k}{3(2k+3)} < 0, & \dfrac{d^2p_d^*(Q_d)}{d^2(t_f)} = -\dfrac{1}{(2k+3)^2} < 0 \\ \dfrac{dp_f^*(Q_f)}{dt_d} = -\dfrac{k}{3(2k+3)} < 0, & \dfrac{d^2p_f^*(Q_f)}{d^2(t_d)} = -\dfrac{1}{(2k+3)^2} < 0 \\ \dfrac{dp_f^*(Q_f)}{dt_f} = \dfrac{k+3}{3(2k+3)} > 0, & \dfrac{d^2p_f^*(Q_f)}{d^2(t_f)} = -\dfrac{1}{(2k+3)^2} < 0 \end{cases}$$

Thus, in homogeneous condition of other factors, when the tariff tax imposed by Home country increases (t_d increases), the equilibrium price of this good in Home market increases ($p_d^*(Q_d)$ increases) and the equilibrium price in Foreign market decreases ($p_f^*(Q_f)$ decreases) with declining rates as its negative second-order-conditions of $p_d^*(Q_d)$ and $p_f^*(Q_f)$ with t_d and t_f. Similarly, in homogeneous condition of other factors, when the tariff tax imposed by Foreign country increases (t_f increases), the equilibrium price of this good in Home market decreases ($p_d^*(Q_d)$ decreases) and the equilibrium price in Foreign market increases ($p_f^*(Q_f)$ increases) with declining rates as its negative second – order – conditions $p_d^*(Q_d)$ and $p_f^*(Q_f)$ with t_d and t_f. These results can be explained by the fact that, increased tariff will affect to the price of imported goods equivalently and because of this, domestic industry benefits from a reduction of threat of competition, they are not forced to improve their productivity and reduce their prices, this will lead to higher price facing to domestic consumers. Besides that, from (2) and (3) we can verify the effects of the difference in tariff rate on prices:

$$p_d^*(Q_d) - p_f^*(Q_f) = \frac{1}{3}(t_d - t_f) \Leftrightarrow \frac{p_d^*(Q_d) - p_f^*(Q_f)}{t_d - t_f} = \frac{1}{3}.$$

Thus, the proportion between the difference in equilibrium prices of 2 countries and the difference between tariff rates is $\frac{1}{3}$ and if the tariff rate of a country is higher than the other, the equilibrium price in this country will be higher with homogeneous other conditions.

Finally, the equilibrium total productions of 2 firms, Z_d^* and Z_f^*, are:

In Home firm:

$$Z_d^* \equiv x_d^* + x_f^* = \frac{2a}{2k+3} + \frac{t_d}{(2k+1)(2k+3)} - \frac{2t_f\,(k+1)}{(2k+1)(2k+3)}$$

In Foreign firm:

$$Z_f^* \equiv y_d^* + y_f^* = \frac{2a}{2k+3} - \frac{2t_d(k+1)}{(2k+1)(2k+3)} + \frac{t_f}{(2k+1)(2k+3)}$$

So, $Z_d^* - Z_f^* = \dfrac{t_d - t_f}{2k+1}$.

That means the difference in tariff rates will affect to the difference between optimal quantities of productions in 2 firms in homogeneous condition of other factors. Firm in country which has the higher tariff rate will produce the larger output than the other within country which has the lower tariff rate although it has to face with higher equilibrium price in its market. With the same elasticity of demand in 2 markets, export – oriented firm which has to face the lower tariff from imported country will have more incentive to manufacture the larger amount of its goods due to the lower input cost for each unit of output produced.

3.2 Welfare Policy Within 2 Countries

Brander and Krugman (1983) examined welfare effects under free entry to consumers only. However, the government considered welfare to not only its citizens but domestic industry and its revenue. So, the government imposes a proper tariff on imported goods to prevent dumping from other countries with the aim of maximizing its welfare. The optimal tariff rate imposed to import goods is determined to maximize the welfare of one country (W). The total welfare of a country can be determined as the sum of consumer surplus, tax revenue and firm profit:

$$\text{W} = \text{Consumer Surplus } (CS) + \text{Tax Revenue } (TR) + \text{Firm Profit } (\pi)$$

$$\text{Consumer Surplus } (CS) = \int_0^Q p(Q)dQ - p(Q)Q = \frac{1}{2}Q^2.$$

$$\text{Tax Revenue } (TR) = \text{Tariff rate } \times \text{ Imported good amount.}$$

$$\text{Firm Profit } (\pi) = \text{Total Revenue } - \text{Total Cost } - \text{Tariff Cost.}$$

$$\text{In Home country: } W_d = CS_d + TR_d + \pi_d = \frac{1}{2}\left(Q_d^*\right)^2 + t_d y_d^* + \pi_d^*$$

$$\text{In Foreign country: } W_f = CS_f + TR_f + \pi_f = \frac{1}{2}\left(Q_f^*\right)^2 + t_f x_f^* + \pi_f^*$$

Using the first-order and second-order conditions in W_d and W_f to find out the best-response functions of t_d and t_f:

$$\begin{cases} \frac{dW_d}{dt_d} = 0 \\ \frac{dW_f}{dt_f} = 0 \end{cases} \Leftrightarrow \begin{matrix} Q_d^* \frac{dQ_d^*}{dt_d} + y_d^* + t_d \frac{dy_d^*}{dt_d} + \frac{d\pi_d^*}{dt_d} = 0 \\ Q_f^* \frac{dQ_f^*}{dt_f} + x_f^* + t_f \frac{dx_f^*}{dt_f} + \frac{d\pi_f^*}{dt_f} = 0 \end{matrix} \Leftrightarrow \begin{cases} t_d = -\frac{Q_d^* dQ_d^*/dt_d + y_d^* + d\pi_d^*/dt_d}{dy_d^*/dt_d} \\ t_f = -\frac{Q_f^* dQ_f^*/dt_f + x_f^* + d\pi_f^*/dt_f}{dx_f^*/dt_f} \end{cases}$$

We can express the reaction functions for both governments in deciding their tariff rate as $t_d = f(t_f)$ and $t_f = f(t_d)$. When the foreign country increases the tariff rate, the home country needs to decrease the tariff rate in order to maximize its welfare and vice versa.

Proposition 1: In Nash equilibrium, both countries will impose the same tariff rates to imported goods: $t_d^* = t_f^* = t^*$ (Proof: See Appendix A).

However, whether this Nash equilibrium is Pareto efficiency or not, we need to examine the case that 2 nations collude to reduce its tariff barrier on imported goods.

When governments collude to impose the same tariff rate on imported goods: $t_d^* = t_f^* = t$, the social welfare in each country is:

In Home country:

$$W_d = \frac{1}{2}{Q_d}^2 + ty_d + \pi_d = \frac{(2a-t)[2a(k+2)+t(k+1)]}{2(2k+3)^2} - f_d.$$

FOCs with t: $\frac{dW_d}{dt} = \frac{-2a - 2t(k+1)}{2(2k+3)^2} < 0$ with $t > 0$.

In Foreign country:

$$W_f = \frac{1}{2}{Q_f}^2 + tx_f + \pi_f = \frac{(2a-t)[2a(k+2)+t(k+1)]}{2(2k+3)^2} - f_f.$$

FOCs with t: $\frac{dW_f}{dt} = \frac{-2a - 2t(k+1)}{2(2k+3)^2} < 0$ with $t > 0$.

Thus, W_d and W_f increase when t decreases, or social welfares in both countries can increase if they agree to reduce its tariff rate. That means Pareto efficiency is not satisfies if two nations complete with each other to protect its domestic industry and maximize its social welfare.

In the Nash equilibrium, two countries impose the same tariff rate on the export country and firms do not have an incentive to unilaterally deviate by altering their output levels: the chosen quantity maximizes the profits of each firm given the quantities chosen by the other firm. Two firms decide the same total amount of its product quantity and the difference in firms' profit depends only on the difference in fixed costs. However, because this Nash equilibrium is not Pareto efficiency, if both countries collude to decide the lower tariff rate on imported goods; the social welfare will be increased.

4 Model Expansion

4.1 Trade Balance Under Tariff of the N-Factors Model

In this section, we expand the model with n monopolist firms assuming oligopolistic competition in the international trade stage and reconsider the choice of optimal policy instruments. Let consider that there are n identical countries, and that each country has one firm i producing commodity Z_i. There is tariff fees incurred in exporting goods from one country to the others. Each firm regards each country as a separate market and therefore chooses the profit-maximizing quantity for each country separately. Each firm has a Cournot perception: it assumes the other firm will hold output fixed in each country. Firm i $(i = 1, 2, 3, \ldots, n)$ produces output x_{ii} for domestic consumption and output x_{ij} in country j. Firm i has fixed costs f_i and total costs (C_i) of each firm are quadratic functions with quantities produced. The tariff rate imposed by country i in imported goods is t_i where $t_i \geq 0$. Using $p_i(Q_i)$ to denote the demand function in country i and $Q_i = \sum x_{ji}$ be the total quantity sold in market i. Then, profits of firm i can be written, respectively, as:

$$\pi_i = \sum_{j=1}^{n} x_{ij} p_j(Q_j) - C_i(Z_i) - \sum_{j=1,n,j \neq i} t_j x_{ij}$$

Firm i maximize its profit by producing quantity $Z_i = \sum_{j=1}^{n} x_{ij}$ such that it satisfies the FOCs:

$$\frac{d\pi_i}{dx_{ij}} = p_j(Q_j) + x_{ij}\frac{dp_j}{dx_{ij}} - \frac{dC_i(Z_i)}{dx_{ij}} - t_j = 0 \ (j = \overline{1,n}, j \neq i)$$

$$\frac{d\pi_i}{dx_{ii}} = p_i(Q_j) + x_{ii}\frac{dp_i}{dx_{ii}} - \frac{dC_i(Z_i)}{dx_{ii}} = 0$$

Suppose that: $p_j(Q_j) = a - Q_j \ \forall j \in 1, 2, 3, \ldots, n$.

$$C_i(Z_i) = f_i + \frac{1}{2}k{Z_i}^2$$

We have n^2 equation and n^2 variables, the reaction function of firm i (Z_i) is a function of $(n-1)$ other firms. By solving n^2 equation, we can find out the equilibrium point of firms. From some calculations (See Appendix B), we have the following results:

The total quantity Z_i each firm i should produce to maximize its profits is:

$$Z_i^* = \sum_{j=1}^{n} x_{ij}^* = \frac{na}{nk+n+1} + \frac{t_i}{nk+1} - (nk+2)\frac{\sum_{j=1}^{n} t_j}{(nk+1)(nk+n+1)} (i = 1, 2, \ldots, n)$$

The optimal quantity Q_i^* sold in country i is:

$$Q_i^* = \sum_{j=1}^{n} x_{ji}^* = \frac{na}{nk+n+1} - \frac{n-1}{n+1}t_i + \frac{k(n-1)}{(n+1)(nk+n+1)}\sum_{j=1}^{n} t_j$$

The equilibrium price in country i is:

$$p_i^* = a - Q_i^* = \frac{a(nk+1)}{nk+n+1} + \frac{n-1}{n+1}t_i - \frac{k(n-1)}{(n+1)(nk+n+1)}\sum_{j=1}^{n} t_j$$

The optimal quantity each firm i produce in domestic market i is:

$$x_{ii}^* = p_i^* - kZ_i^*$$

The optimal quantity each firm i produce in foreign market j is:

$$x_{ij}^* = p_j^* - kZ_i^* - t_j$$

From these above results, we have the difference in equilibrium prices as well as quantity produced between 2 countries x and y:

$$p_x^* - p_y^* = \frac{n-1}{n+1}(t_x - t_y)$$

When $n \rightarrow \infty$, $\frac{n-1}{n+1} \rightarrow 1$, the difference between equilibrium prices will be equal to the difference between tariff rates as country which imposes the higher tariff rate will have the higher equilibrium price in its domestic market. Moreover, the difference in total quantity produced by 2 firms, x and y, that maximize its profits:

$$Z_x^* - Z_y^* = \frac{1}{nk+1}(t_x - t_y)$$

Thus, when $n \rightarrow \infty$, $Z_x \rightarrow Z_y$ means the difference between the total quantities each firm should produce to maximize its profits is insignificant when the number of country (or firm) becomes larger.

4.2 Welfare Policy Within n Countries

The optimal tariff rate imposed to import goods is determined to maximize the welfare in this country (W_i). The total welfare of a country can be determined as the sum of consumer surplus, tax revenue and firm profit:

$$W_i = Consumer\ Surplus\ (CS_i) + Tax\ Revenue\ (TR_i) + Firm\ Profit\ (\pi_i)$$

$$\text{Consumer Surplus of country } i\ (CS_i) = \int_0^{Q_i} p(Q_i)dQ_i - p(Q_i).Q_i = \frac{1}{2}{Q_i}^2$$

$$\text{Tax Revenue } (TR) = t_i \sum_{j=1, j \neq i}^{n} x_{ji} = t_i(Q_i - x_{ii}) = t_i(2Q_i + kZ_i - a)$$

$$\text{Firm Profit } (\pi_i) = \sum_{j=1}^{n} x_{ij} p_j(Q_j) - C_i(Z_i) - \sum_{j=1, j \neq i}^{n} t_j x_{ij}$$

Thus, the total welfare of country i will be determined as:

$$W_i = \frac{1}{2}Q_i^2 + t_i(2Q_i + kZ_i - a) + \pi_i$$

The equilibrium tariff rates in country i $(i = 1, 2, 3 \ldots, n)$ satisfy the FOCs on W_i:

$$\frac{dW_i}{dt_i} = Q_i\frac{dQ_i}{dt_i} + (2Q_i + kZ_i - a) + t_i\left(2\frac{dQ_i}{dt_i} + k\frac{dZ_i}{dt_i}\right) + \frac{d\pi_i}{dt_i} = 0$$

$$\Leftrightarrow t_i = -\frac{Q_i\frac{dQ_i}{dt_i} + (2Q_i + kZ_i - a) + \frac{d\pi_i}{dt_i}}{2\frac{dQ_i}{dt_i} + k\frac{dZ_i}{dt_i}} (i = 1, 2, 3, \ldots, n)$$

(The best-response functions of t_i).

Proposition 2: In Nash equilibrium, both countries will impose the same tariff rates to imported goods: $t_1 = t_2 = t_3 = \ldots. = t_n = t^*$. (Proof: See Appendix C).

However, whether this Nash equilibrium is Pareto efficiency or not, we need to examine the case that nations collude to reduce its tariff barrier on imported goods.

When governments collude to impose the same tariff rate on imported goods: $t_i = t \; \forall i = 1, n$, the social welfare in country i is:

$$\begin{aligned} W_i &= \frac{1}{2}{Q_i}^2 + t_i(2Q_i + kZ_i - a) + \pi_i \\ &= \frac{[na - (n-1)t][a(nk + n + 2) + (n-1)(k+1)t]}{2(nk + n + 1)^2} - f_i \end{aligned}$$

$$\text{FOCs: } \frac{dW_i}{dt} = \frac{-2(n-1)a - 2t(k+1)(n-1)^2}{2(nk + n + 1)^2} < 0 \text{ with } t > 0.$$

Thus, W_i increases when t decreases, or social welfares in both countries can increase if they agree to reduce its tariff rate. That means Pareto efficiency is not satisfies if nations complete with each other to protect its domestic industry and maximize its social welfare. In the Nash equilibrium, the tariff rates imposed by countries tend to converge to t^* and no country is willing to change its tariff rate unilaterally because the chosen quantity maximizes the profits of each firm given the quantities chosen by the other firms. Two firms decide the same total amount of its product quantity, that will lead to the unique price among markets and the difference in firms' profit depends only on the difference in fixed costs. However, because this Nash equilibrium tariff rate in a Cournot oligopoly is not Pareto efficient, if some degree of cooperation can be achieved, both countries can be made better off by simultaneous tariff reductions[1]. Although the government budget may be reduced by lowering the tariff rate on imported goods, the consumer surplus as consumers can enjoy the lower equilibrium price and producer surplus as producers get more profits will compensate this loss.

5 Conclusion

We can find that firms will decide the optimal quantity produced and sold in each international market to maximize its profits in the presence of tariff rate imposed by imported countries. The difference in tariff rates between countries will be proportional to the difference in strategic equilibrium prices as countries that impose higher tariff rate will have higher equilibrium prices. Moreover, although the difference in tariff rates imposed by countries has a significant influence to the difference in total quantity produced by firms in the case of a small number of countries participating to international trade, the difference between the total quantities each firm should produce to maximize its profits is insignificant when the number of country (or firm) becomes larger. Moreover, governments should impose the same tariff rate with imported goods in equilibrium to maximize its national economic welfare. In that case, firms decide the same total quantity produced for maximizing its profit and the equilibrium prices will be identical in every market. Finally, if there is any agreement between countries to reduce tariff on imported goods, the social welfare in all country could be higher as people can enjoy the lower price and firms can get higher profits that will compensate to the reducing government budget.

A Appendix (Proof of Proposition 1)

From the equation system, we have the FOCs of the welfare functions equal to zero:

$$\begin{cases} \dfrac{dW_d}{dt_d} = 0 \\ \dfrac{dW_f}{dt_f} = 0 \end{cases} \Leftrightarrow \begin{cases} Q_d^* \dfrac{dQ_d^*}{dt_d} + y_d^* + t_d \dfrac{dy_d^*}{dt_d} + \dfrac{d\pi_d^*}{dt_d} = 0 \\ Q_f^* \dfrac{dQ_f^*}{dt_d} + x_f^* + t_f \dfrac{dx_f^*}{dt_f} + \dfrac{d\pi_f^*}{dt_f} = 0 \end{cases}$$

From the optimal quantity produced by firms, equilibrium prices and quantity sold in markets, we can derive:

$$Q_d^* + \frac{1}{3} t_d = Q_f^* + \frac{1}{3} t_f$$

$$Z_d^* - \frac{1}{2k+1} t_d = Z_f^* - \frac{1}{2k+1} t_f \text{ and } p_d^* - \frac{1}{3} t_d = p_f^* - \frac{1}{3} t_f$$

[1] This result is equivalent to the conclusion in "Game Theory in International Economics", McMillan (2008).

Thus, there is a unique solution of this equation system: $t_d^* = t_f^* = t^*$. The second-order conditions of the welfare functions:

$$\frac{d^2W_d}{dt_d^2} = \frac{d^2W_f}{dt_f^2} = \left(\frac{dQ_d}{dt_d}\right)^2 + 2\frac{dy_d}{dt_d} + \frac{d^2\pi_d}{dt_d^2}$$
$$= \left(-\frac{k+3}{3(2k+3)}\right)^2 - \frac{2(4k+3)(k+2)}{3(2k+1)(2k+3)} + \frac{2}{9} - \frac{k}{(2k+1)^2(2k+3)^2} < 0$$

Therefore, in Nash equilibrium, both countries will impose the same tariff rates to imported goods: $t_d^* = t_f^* = t^*$.

B Appendix

The first-order conditions to maximize firms' profits show:

$$\begin{cases} \dfrac{d\pi_i}{dx_{ij}} = p_j(Q_j) + x_{ij}\dfrac{dp_j}{dx_{ij}} - \dfrac{dC_i(Z_i)}{dx_{ij}} - t_j = 0 (j = \overline{1,n}, j \neq i) \\ \dfrac{d\pi_i}{dx_{ii}} = p_i(Q_j) + x_{ii}\dfrac{dp_i}{dx_{ii}} - \dfrac{dC_i(Z_i)}{dx_{ii}} = 0 \end{cases}$$

Suppose that: $p_j(Q_j) = a - Q_j \quad \forall j \in 1, 2, 3, \ldots, n$ and $C_i(Z_i) = f_i + \frac{1}{2}kZ_i^2$
We have n^2 equation and n^2 variables:

$$\begin{cases} \dfrac{d\pi_i}{dx_{ij}} = p_j(Q_j) + x_{ij}\dfrac{dp_j}{dx_{ij}} - \dfrac{dC_i(Z_i)}{dx_{ij}} - t_j = 0 \left(j = \overline{1,n}, j \neq i\right) \\ \dfrac{d\pi_i}{dx_{ii}} = p_j(Q_j) + x_{ii}\dfrac{dp_i}{dx_{ii}} - \dfrac{dC_i(Z_i)}{dx_{ii}} = 0 \end{cases}$$

$$\Leftrightarrow \begin{cases} a - Q_j - x_{ij} - kZ_i - t_j & = 0 \\ a - Q_i - x_i i - kZ_i & = 0 \end{cases} \tag{4}$$

From (4), we have: $\forall i = \overline{1, n}$

$$Z_i^* = \sum_{j=1}^{n} x_{ij}^* = \frac{na}{nk+n+1} + \frac{t_i}{nk+1} - \frac{(nk+2)}{(nk+1)(nk+n+1)}\sum_{j=1}^{n} t_j$$
$$Q_i^* = \sum_{j=1}^{n} x_{ji}^* = \frac{na}{nk+n+1} - \frac{n-1}{n+1}t_i + \frac{k(n-1)}{(n+1)(nk+n+1)}\sum_{j=1}^{n} t_j$$

The equilibrium price in country i will be:

$$p_i^* = a - Q_i^*$$
$$= \frac{a(nk+1)}{nk+n+1} + \frac{n-1}{n+1}t_i - \frac{k(n-1)}{(n+1)(nk+n+1)}\sum_{j=1}^{n} t_j$$

From (4), the optimal quantity each firm i produce in domestic market i will be:

$$x_{ii}^* = p_i^* - kZ_i^*.$$

The optimal quantity each firm i produce in foreign market j is:

$$x_{ij}^* = p_j^* - kZ_i^* - t_j.$$

C Appendix (Proof of Proposition 2)

The optimal quantity each firm i produce in domestic market i will be:

$$x_{ii}^* = p_i^* - kZ_i^*$$

The optimal quantity each firm i produce in foreign market j will be:

$$x_{ij}^* = p_j^* - kZ_i^* - t_j$$

We have, the total welfare of country i will be determined as:

$$W_i = \frac{1}{2}{Q_i}^2 + t_i(2Q_i + kZ_i - a) + \pi_i$$

Where:

$$\pi_i = \sum_{j=1}^{n} x_{ij} p_j(Q_j) - C_i(Z_i) - \sum_{j=1, j\neq i}^{n} t_j x_{ij} \tag{5}$$

Using the first-order-conditions of W_i:

$$\frac{dW_i}{dt_i} = 0 \Leftrightarrow Q_i \frac{dQ_i}{dt_i} + (2Q_i + kZ_i - a) + t_i(2\frac{dQ_i}{dt_i} + k\frac{dZ_i}{dt_i}) + \frac{d\pi_i}{dt_i} = 0$$

From (5):

$$\begin{aligned}\frac{d\pi_i}{dt_i} &= \sum_{j=1}^{n}\left(\frac{dx_{ij}}{dt_i}p_j(Q_j) + x_{ij}\frac{dp_j(Q_j)}{dt_i}\right) - \frac{dC_i(Z_i)}{dt_i} - \sum_{j=1, j\neq i}^{n} t_j \frac{dx_{ij}}{dt_i} \\ &= 2\sum_{j=1}^{n}\frac{dp_j}{dt_i}x_{ij} - kZ_i\frac{dZ_i}{dt_i} (\text{By using } x_{ii} = p_i - kZ_i \text{ and } x_{ij} = p_j - kZ_i - t_j)\end{aligned}$$

Thus, the first-order-conditions of W_i $(i = 1, 2, 3, \ldots, n)$ can be expressed as:

$$\begin{aligned}&\frac{dW_i}{dt_i} = Q_i\frac{dQ_i}{dt_i} + (2Q_i + kZ_i - a) + t_i\left(2\frac{dQ_i}{dt_i} + k\frac{dZ_i}{dt_i}\right) + \frac{d\pi_i}{dt_i} = 0 \\ &\Leftrightarrow Q_i\frac{dQ_i}{dt_i} + (2Q_i + kZ_i - a) + t_i\left(2\frac{dQ_i}{dt_i} + k\frac{dZ_i}{dt_i}\right) + 2\sum_{j=1}^{n}\frac{dp_j}{dt_i}x_{ij} - kZ_i\frac{dZ_i}{dt_i} = 0\end{aligned}$$

From these above conditions, we have the best-response functions of t_i:

$$t_i = -\frac{Q_i \frac{dQ_i}{dt_i} + (2Q_i + kZ_i - a) + 2\sum_{j=1}^{n} \frac{dp_j}{dt_i} x_{ij} - kZ_i \frac{dZ_i}{dt_i}}{2\frac{dQ_i}{dt_i} + k\frac{dZ_i}{dt_i}} \qquad (i = 1, 2, 3, \ldots, n)$$

In order to find out the Cournot-Nash equilibrium, we need to solve the equation system of n variables $(t_1, t_2, t_3, \ldots, t_n)$. From the optimal quantity produced by firms, equilibrium prices and quantity sold in markets in Appendix B, we can derive:

$Q_i + \frac{n-1}{n+1} t_i$, $Z_i - \frac{1}{nk+1} t_i$ and $p_i - \frac{n-1}{n+1} t_i$ are constant with $i = 1, 2, 3, \ldots, n$.

Thus, there is a unique Nash equilibrium: $t_1 = t_2 = t_3 = \ldots. = t_n = t^*$.

References

Schotter, A.: Microeconomics: A Modern Approach. Cengage Learning (2008)

McMillan, J.: Game Theory in International Economics. Harwood Academic Publishers (2008)

Brander, J., Krugman, P.: A 'reciprocal dumping' model of international trade. J. Int. Econ. **15**(3/4), 313–321 (1983)

Toshimitsu, T.: The choice of optimal protection under oligopoly: import tariff vs production subsidy. Jpn. Econ. Rev. **53**(3), 301–341 (2002)

McKeown, T.J.: Firms and tariff regime change: explaining the demand for protection. World Polit. **36**(02), 215–233 (1984)

Vives, X.: Cournot and the oligopoly problem. Eur. Econ. Rev. **33**(2–3), 503–514 (1989)

Varian, H.R.: Intermediate Microeconomics. W.W. Norton & Company, New York (1999)

Energy Consumption and Economic Growth Nexus in Vietnam: An ARDL Approach

Bui Hoang Ngoc(✉)

Graduate School, Ho Chi Minh City Open University,
Ho Chi Minh City, Vietnam
ngocbh.16ae@ou.edu.vn

Abstract. This paper explores the relationship between electricity consumption, petroleum consumption and economic growth in Vietnam. The Cobb-Douglas production is used over the period of 1980–2014. We have applied the ARDL bounds testing approach and found that cointegration exists among the included variables. Further, the Granger causality test affirms that there is one-way causality between electricity consumption, petroleum consumption and economic growth. The empirical results provide a strong statistical evidence that electricity consumption has a positive impact on the economic growth of Vietnam both in the short-term and long-term. This study suggests that policy-makers should establish concrete plans to increase investment in the electricity sector, and explore new sources of energy to achieve sustainable economic development for the long-run.

Keywords: ARDL · Energy consumption · Economic growth
Vietnam

1 Introduction

Rostow (1990) presented five steps through which all countries must pass to become developed: the traditional society, the pre-conditions for take off into self-sustaining growth, the drive to maturity, and the age of high mass consumption. The prerequisite for successful completion of the pre-conditions for take off includes: *(i)* The productive investment rises from 5% to 10% or more of national income or net national product, *(ii)* One or more of the substantial manufacturing sectors witnesses high growth rate, *(iii)* The political, social and institutional framework improves to meet the requirement of expansion in the modern sectors. Under these three conditions, developing economies are subject to certain constraints, in particular restrictions on the sources of funds and the availability of natural resources.

V. Kreinovich et al. (Eds.): ECONVN 2019, SCI 809, pp. 311–322, 2019.
https://doi.org/10.1007/978-3-030-04200-4_24

Nowadays, energy is becoming a central component of an economy for aggregate supply and aggregate demand. In aggregate demand, electricity is an essential product for consumers to maximize the benefits of daily life, and even electricity demonstrates the civilization of the country/region relative to other countries/regions (Aytac and Guran 2011). According to the International Energy Agency (IEA), the world's primary energy demand will continue to increase (about 1.4% annually until 2035), this is especially true in fast-growing countries like China, Brazil and India.

As in many other developing countries, Vietnam now must deal with the problem of economic growth with great pressure as the demand for energy increases to meet consumer demand and the enlargement of the economy. Since the economic reforms of 1986, while Gross Domestic Product (GDP) increased from 44.06 US Dollar in 1980 to 2,012.05 US Dollar in 2014, the total electricity consumption of Vietnam also increased rapidly from 3.3 billion kWh in 1980 to 140.72 billion kWh in 2015. In 2014, the electricity consumption of the industrial sector was the largest, accounting for 53.9%, the residential sector accounted for 35.6%, the service sector accounted for 4.8%, the rest was for the regions other.

The relationship between energy consumption (EC) and gross domestic product (GDP) is stated in four hypotheses. The study of Kraft and Kraft (1978) supported the hypothesis that the Conversation (GDP–>EC) is a foundational work, finding a one-way causal relationship from economic growth to electricity consumption in the US economy in the period 1947–1974. However, Tang (2009) advocated the Feedback (GDP<–>EC) hypothesis in the Malaysian economy from 1970 to 2005. Besides, the study of Abdullah (2013) for India in the period 1975–2008 supported the Growth (EC–>GDP). Although, the number of studies that did not find the relationship between these two variables was less, the study of Akpan and Akpan (2012) in the case of Nigeria supported the neutrality hypothesis (GDP $\neq$ EC, EC $\neq$ GDP). Therefore, the aim of this paper is to test the causal relationship between energy consumption and economic growth to provide empirical evidence to help the government to make policy decisions, to ensure energy security, and to promote economic development for Vietnam.

The remainder of the paper is as follows: Sect. 2 presents theoretical background and reviews the relevant literature, Sect. 3 shows model construction, data collection and the econometric method, Sect. 4 presents results interpretations and Sect. 5 concludes and limits the results and points out some policy implications.

2 Theoretical Background and Literature Reviews

The exogenous growth theory of Solow (1956) agree that output is determined by two factors: capital and labor. The general form of production is given follow: Y = f (K, L) or Y = A. K^{α}. L^{β}. Where, Y is real gross domestic product, and K and L indicate real capital and labor respectively. A represents technology. The output elasticity with respect to capital and labor is α and β respectively. If we are based on the theory of exogenous growth, we will not find any relationship between energy consumption and economic growth.

However, the boom of the industrial revolution, especially since the personal computer and the internet appeared, science and technology has gradually become the *"production force"*. Arrow (1962) proposed learning-by-doing growth theory, Romer (1990) gave out the theory of endogenous growth. Both Arrow and Romer arguing that technological progress must be endogenous, that is, it directly impacts on economic growth. Romar performed the production function in the form of: Y = f (K, L, T) or $Y = A.\ K^{\alpha}.\ L^{\beta}.\ T^{\lambda}$. T is the technological progress of the country/enterprise at time t. We find the relationship between technology and energy consumption, because technology is considered to be an external factor that may be related to energy. Technologies only operate when the availability of useful energy provides sufficiently. The technology referred to be plant, machinery or the process of converting inputs into output products. If there is not enough power supply (in this case is electricity or petroleum), these technologies will be useless. Therefore, energy in general, is essential to ensure that technology is used and that it becomes an essential input for economic growth.

Energy is considered a key industry in many countries, so the interrelationship between energy consumption (EC) and economic growth (GDP) has been studied quite early. Kraft and Kraft (1978) considered to be the founding of a one-way causal relationship about the economic growth affected the consumption of electricity in the United State economy during 1947–1974. Follow-up studies in other countries/regions are also aimed at testing and confirming this relationship under specific conditions. If the EC and GDP have a two-way causal relationship (EC<–>GDP), this suggests that an additional relationship, an increase in energy consumption, would have a positive impact on economic growth and vice versa. On the one hand, if only one-way GDP affects the EC (GDP–>EC), it reflects that country/region is less dependent on energy. On the other hand, the EC affects GDP (EC–>GDP), the role of energy needs to be considered in national energy policy, since the initial investment cost for power plants is very high. There are several studies that do not find a relationship between these two variables, the explanation must be put in the context of specific research because energy consumption is highly dependent on scientific and technical level, the living standard of the people, the geographical location, the weather as well as the consumption habits of the people, enterprises or national energy policies, etc. A summary of the results of the study on the relationship between EC and GDP is presented in Table 1. The results in Table 1 show that the relationship between energy consumption (EC) and GDP in each country/region is not uniform. This is a proof, for the need to test this causal relationship with Vietnam.

Table 1. Summary of existing empirical studies

Author(s)	Countries	Methodology	Conclusion
Tang (2009)	Malaysia	ARDL, Granger	EC<->GDP
Esso (2010)	7 countries	Cointegration, Granger	EC<->GDP
Aslan et al. (2014)	United State	ARDL, Granger	EC<->GDP
Kyophilavong et al. (2015)	Thailand	VECM, Granger	EC<->GDP
Ciarreta and Zarraga (2007)	Spain	Granger	GDP->EC
Canh (2011)	Vietnam	Cointegration, Granger	GDP->EC
Hwang and Yoo (2014)	Indonesia	ECM & Granger causality	GDP->EC
Abdullah (2013)	India	VECM - Granger	EC->GDP
Wolde-Rufael (2006)	17 countries	ARDL & Granger causality	No relationship
Acaravci and Ozturk (2012)	Turkey	ARDL & Granger causality	No relationship
Kum et al. (2012)	G7 countries	Panel - VECM	PC->GDP
Shahbaz et al. (2013)	Pakistan	ARDL & Granger causality	PC->GDP
Shahiduzzaman and Alam (2012)	Australia	Cointegration, Granger	PC->GDP
Yoo (2005)	Korea	Cointegration, ECM	EC->GDP
Sami (2011)	Japan	ARDL, VECM, Granger	GDP->EC
Jumbe (2004)	Malawi	Cointegration, ECM	EC<->GDP
Long et al. (2018)	Vietnam	ARDL, Toda & Yamamoto	EC->GNI

3 Research Models

The main objective of the present paper is to investigate the relationship between electricity consumption and economic growth using the data of Vietnam over the period of 1980–2014. We use the Cobb-Douglas production function. The general form of production is given follow: $Y = A.\ K^{\alpha}.\ L^{\beta}$. (1). Where, Y is real gross domestic product, and K and L indicate real capital and labor respectively. A represents technology. The output elasticity with respect to capital and labor is α and β respectively. When Cobb–Douglas technology is constrained to $(\alpha + \beta = 1)$, we get constant returns to scale. We augment the Cobb–Douglas production function by assuming that technology can be determined by the level of energy consumption.

Because capital is not considered in this study. Thus, the model is constructed as following: $A_t = \varphi.EC_t^{\sigma}$. Where φ is time-invariant constant. Then (1) is rewritten as: $Y = \varphi.EC^{\sigma}.K^{\alpha}.L^{\beta}$. Following Shahbaz and Feridun (2012), Tang (2009), Abdullah (2013), Ibrahiem (2015) we divide both sides by population and get each series in per capita terms; but leave the impact of labor constant. By taking the log, the linearized Cobb–Douglas function is modeled as follows:

$$LnGDP_t = \beta_0 + \beta_1 LnEC_t + \beta_2 LnPC_t + u_t$$

Where: u_t denotes error, data is collected from 1980 to 2014, sources and detailed illustrations of variables are shown in Table 2.

Table 2. Sources and measurement method of variables in the model

Variable	Description	Unit	Source
LnGDP	is logarithms of the Gross Domestic Product per capita (in constant 2010 US Dollar)	US Dollar	UNCTAD
LnEC	is logarithms of total electricity consumption	Billion kWh	IEA
LnPC	is logarithms of total petroleum consumption	Thousand tonnes	IEA

The study uses the ARDL, that is introduced by Pesaran et al. (2001) have some of the following advantages: *(i)* the variables in the model just ensure maximum stationary at order one, they can stationary at the same order (integrated of order zero I(0) or integrated of order one I(1)), *(ii)* It is possible to avoid endogenous and more reliable problems for small observations by the addition lag variable of the dependent variable to the independent variable, *(iii)* Short-term and long-term impact coefficients can be estimated at the same time, the correction error model can integrate short-term and long-term equilibrium without missing information in the long run, *(iv)* Model is self-selectable optimal lag, accepting the optimal lag of the variables can be different, thus significantly improving the fit of the model (Davoud et al. 2013 and Nkoro and Uko 2016).

Then, the research model can be expressed as an ARDL model as follows:

$$\begin{aligned}\Delta LnGDP_t &= \beta_0 + \beta_1 LnGDP_{t-1} + \beta_2 LnEC_{t-1} + \beta_3 LnPC_{t-1} \\ &+ \sum_{i=0}^{m} \beta_{4i} \Delta LnGDP_{t-i} + \sum_{i=0}^{m} \beta_{5i} \Delta LnEC_{t-i} + \sum_{i=0}^{m} \beta_{6i} LnPC_{t-i} + \mu_t\end{aligned} \quad (1)$$

Where, Δ: is the first differenced. β_1, β_2, β_3: long-term coefficients. m is optimum lag. μ_t: error term.

The steps of testing include: (1) testing stationary of variables in the model, (2) Estimate model 1 by the ordinary least squares method (OLS), (3) Calculate the statistical value F to determine if there exists a long-term relationship between the variables. If there is a long-term co-integration relationship, the Error Correction Model (ECM) is estimated based on the following equation:

$$\begin{aligned}LnGDP_t &= \lambda_0 + \alpha.ECM_{t-1} + \sum_{i=0}^{p} \lambda_{1i} \Delta LnGDP_{t-i} + \sum_{i=0}^{q} \lambda_{2i} \Delta LnEC_{t-i} \\ &+ \sum_{i=0}^{s} \lambda_{3i} \Delta LnPC_{t-i} + \tau_t\end{aligned} \quad (2)$$

To select the lag value p, q, s in Eq. 2 model selection criteria such as AIC, SC, HQ information criteria, Adjusted R-squared are used. The best estimated model is the model which has the minimum information criteria or the maximum R-squared value. And if $\alpha \neq 0$ and statistically significant then the coefficient of α will show the rate of adjustment of the GDP per capita back to equilibrium after a short-term shock, (4) In addition to the research results are reliable, the author will test the additional diagnostics include: test of residual serial correlation, Normality test and heteroscedasticity test, the CUSUM (Cumulative Sum of Recursive Residuals) and CUSUMSQ (Cumulative Sum of Square Recursive Residuals) to check the stability of the long run and short run coefficients.

4 Research Results and Discussion

4.1 Descriptive Statistics

After the opening of the economy in 1986, the Vietnamese economy has made many positive changes. Vietnam's total electricity consumption also increased rapidly from 3.3 billion kWh in 1980 to 125 billion kWh in 2014. Total petroleum consumption also increased from 53,808 thousand tonnes in 1980 to 825,054 thousand tonnes in 2014. Descriptive statistics of variables are presented in Table 3.

Table 3. Descriptive statistics of the variables

Variables	Mean	Std. Deviation	Min	Max
LnGDP	5.63	1.22	3.52	7.61
LnEC	2.80	1.21	1.19	4.81
LnPC	12.38	0.99	10.89	13.78

4.2 Empirical Results

Unit Root Analysis

First, a test for stationarity is used to ensure that no variable is stationary at I(2) (a condition for using the ARDL model). Augmented Dickey-Fuller Test (ADF) (Dickey and Fuller 1981) is a popular method for studying time series data. We use the KPSS (Kwiatkowski-Phillips-Schmidt-Shin) and Phillips and Perron (1988) tests to ensure accuracy of the results obtained. The results of these tests shown in Table 4 suggest that with ADF, PP and KPSS tests, variables are stationary at I(1). Therefore, the application of the ARDL into the model is reasonable.

Table 4. Unit root test

Variable	ADF test	Phillips-Perron test	KPSS test
LnGDP	–4.001**	–2.927	0.047
ΔLnGDP	–4.369***	–5.035***	0.221***
LnEC	–0.537	–3.140	0.173**
ΔLnEC	–2.757*	–2.703*	0.189**
LnPC	–0.496	–0.977	0.145*
ΔLnPC	–5.028***	–5.046***	0.167**

Notes: ***, ** and * respectively showed for the significance level of 1%; 5% and 10%.

Cointegration Test
The Bounds testing approach was employed to determine the presence of cointegration among the series. The Bounds testing procedure is based on the joint F-statistics. The maximum lag value was selected to be m = 3 in Eq. 1.

Table 5. Optimum lag

Lag	AIC	SC	HQ
0	1.627240	1.764652	1.672788
1	–8.054310	–7.504659	–7.872116
2	–7.907131	–6.945242	–7.588292
3	–7.522145	–6.148018	–7.066661

In Table 5, AIC, SC values and F-statistics for the null hypothesis: $\beta_1 = \beta_2 = \beta_3 = 0$ are given. The optimum lag is selected relying on the minimizing the AIC and SC. Equation 1, the minimum AIC and SC values were obtained when the lag value m was equal to m = 1. Since F-statistics for this model is higher than upper critical values by Pesaran et al. (2001) in all cases, it was concluded that there is a cointegration which means a long-run relationship among the series. According to AIC, SC and Hannan-Quinn information criteria, the best model for Eq. 1 is ARDL(2, 0, 0) model which means p = 2, q = s = 0, selecting the maximum lag values p = q = s = 4. The F-statistics = 10.62 is more than the upper critical value = 5.00 at 0.1 level of significant, so the null hypothesis of no cointegrating relationship is rejected. It is concluded that there is a cointegrating relationship between the variables in long term. The results of Bounds test are shown in Table 6.

Granger Causality Test
To confirm the relationship between the variables, paper proceed to the Granger causal analysis (Engle and Granger 1987) with the null hypothesis is not causal.

According to the test results shown in Table 7, the LnEC has a causal relationship Granger with the LnGDP variable, LnPC and LnGDP, LnPC and LnEC. To illustrate the causal relationship between the three variables LnGDP, LnEC and LnPC are shown in Fig. 1 and Table 7.

Table 6. Results of Bounds test

F-Bounds test		Null hypothesis: No levels relationship		
Test statistic	Value	Signif.	I(0)	I(1)
			Asymptotic: n = 1000	
F-statistic	10.62459	10%	2.63	3.35
k	2	5%	3.1	3.87
		2.5%	3.55	4.38
		1%	4.13	5

Table 7. The Granger causality test

Null Hypothesis:	Obs	F-Statistic	Prob.
LnEC does not Granger Cause LnGDP	33	7.28637	0.0028
LnGDP does not Granger Cause LnEC		1.98982	0.1556
LnPC does not Granger Cause LnGDP	33	6.86125	0.0038
LnGDP does not Granger Cause LnPC		0.34172	0.7135
LnPC does not Granger Cause LnEC	33	5.53661	0.0094
LnEC does not Granger Cause LnPC		1.83268	0.1787

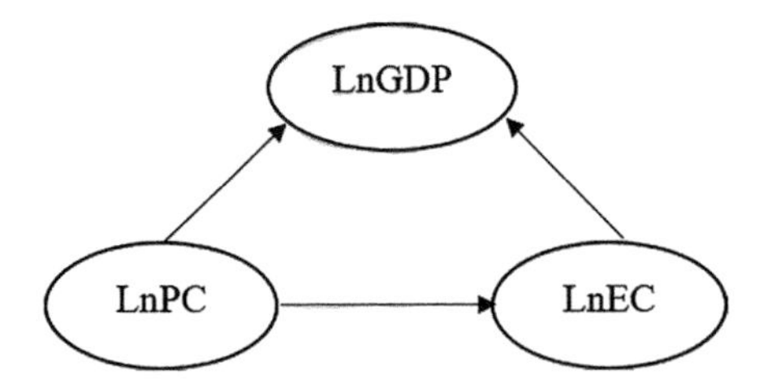

Fig. 1. Plot of the Granger causality test

The Short-Run Estimation

There is a cointegration relationship between the variables of the model in long-term, the paper continue to estimate the correction error model to determine the

Table 8. The short-run estimation

Variables	Coefficient	Std. Dev	t-statistic	Prob
ECM(-1)	−0.365629	0.053303	−6.859429	0.0000
ΔLnGDP(-1)	0.475094	0.085079	5.584173	0.0000
LnEC	0.244107	0.082847	2.946473	0.0064
LnPC	0.123986	0.087742	1.413086	0.1687
Intercept	−0.125174	0.816773	−0.153254	0.8793

coefficient of error correction term. The estimating ARDL(2, 0, 0) model results are presented in Table 8.

Estimated results show that the coefficient of $\alpha = -0.365$ is statistically significant at 1%. The coefficient of the error correction term is negative and significant as expected. When GDP per capita are far away from their equilibrium level, it adjusts by almost 36.5% within the first period (year). The full convergence to equilibrium level takes about 3 period (year). In the case any of shock to the GDP per capita, the speed of reaching equilibrium level is fast and significant. Electricity consumption is positive and significant, but petroleum consumption is positive and no significant.

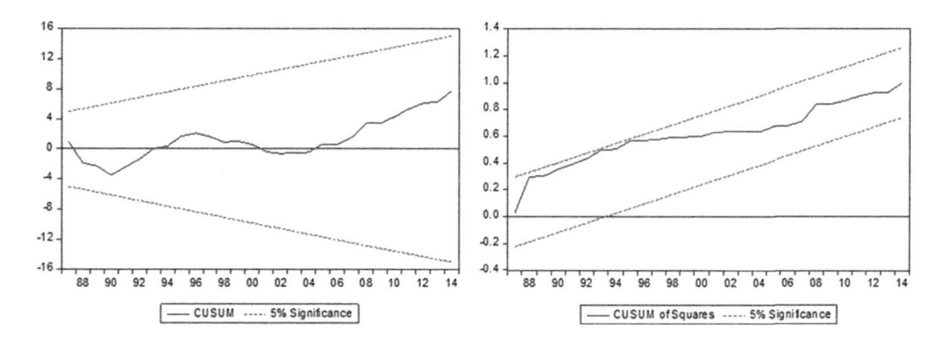

Fig. 2. Plot of the CUSUM and CUSUMSQ

The Long-Run Estimation

Next, paper estimate the long-term results of the effects of energy consumption on Vietnam's per capita income over the period 1980–2014. The long-run estimation results are shown in Table 9. Both coefficients have the expected signs. Electricity consumption is positive and significant, but petroleum consumption is positive and no significant. Accordingly, with other conditions unchanged, a 1% increase in electricity consumption will increase the GDP per capita by 0.667%.

In this model, all diagnostics are well. Lagrange multiplier test for serial correlation, in addition to the normality tests and the test for heteroscedasticity were performed. Serial correlation: $\chi^2 = 0.02$ (Prob = 0.975), Normality: $\chi^2 = 6.03$ (Prob = 0.058), Heteroscedasticity: $\chi^2 = 16.98$ (Prob = 0.072). Finally, the stability of the parameters was tested. For this purpose, it was drawn the CUSUM and CUSUMSQ graphs in Fig. 2. From this figure, statistic are between the critical bounds which imply the stability of the coefficients.

4.3 Discussions and Policy Implications

The experimental results of the study were consistent with Walt Rostow's take-off phase, similar to other conclusions of other studies for countries/regions with the same starting points and conditions to Vietnam, as Tang (2009) studied for the Malaysian economy from 1970 to 2005, Abdullah (2013) studied for the Indian economy from 1975–2008, Odhiambo (2009) studied for the Tanzania economy 1971–2006 period or Ibrahiem (2015) discussed for the Egyptian economy ... This is reasonable, according to Shahbaz et al. (2013) concluded that energy is an indispensable resource/input for all economic activity. Energy efficiency does not only imply cost savings but also improves profitability through increased labor productivity. Shahiduzzaman and Alam (2012) also states that *"even if we can not conclude that energy is finite, more efficient use of existing energy also increases the wealth of the nation"*.

The interesting insights drawn from this study leads us suggest a few notes when applying this result into practice as follows:

Firstly, Vietnam should strive to develop the electricity industry. The coefficient β of the LnEC variable is 0.667 and is statistically significant. This result supports the Growth (EC–>GDP) hypothesis, which implies that Vietnam's economic growth depends on electricity consumption. Thus, in the national electricity policy, it is necessary to calculate the speed of electricity development in line with the speed of economic development.

Secondly, energy consumption helps economic growth for Vietnam, this does not mean that Vietnam must build a lot of power plants. Efficient use of electricity, switching off unnecessary equipment, reducing the loss of power transmission... It is also a way for Vietnam to increase its electricity output.

Thirdly, with favorable geographical position, Vietnam has great potential to develop alternative energy sources substitute for electricity such as: Solar energy, wind energy, biofuels, geothermal ... these are more environmentally friendly

Table 9. The long-run estimation

Variable	Coefficient	Std. Error	t-Statistic	Prob.
LnEC	0.667637	0.174767	3.820149	0.0007
LnPC	0.339105	0.217078	1.562131	0.1295
Intercept	−0.342352	2.220084	−0.154207	0.8786
EC = LnGDP − (0.6676 * LnEC + 0.3391 * LnPC − 0.3424)				

energies. Exploit and convert to these sources of energy. This is of great importance in terms of socio-economic, energy security and sustainable development.

5 Conclusion

In the process of development, the need for capital to invest in infrastructure, social security, education, health care, defense, etc. ... is always great. The pressure to maintain a positive growth rate and improve the spiritual life of the people requires the Government to develop a comprehensive and synchronization, with data from 1980–2014, by using the ARDL approach and Granger causality test. Paper conclude that energy consumption has a positive impact on Vietnam's economic growth in both short and long term. In addition, we also found a one-way causal relationship Granger from energy consumption to economic growth (EC–>GDP), support for the Growth hypothesis.

Although the number of observations and test results are satisfactory, it must be noted that the data of the study is not long enough, the climate of Vietnam (winter is rather cold, summer is relatively hot) is also a cause for high energy consumption. Besides, the study did not analyze in detail the impact of power consumption by industrial sector, population sector to economic growth. This is the direction for further research.

References

Rostow, W.W.: The Stages of Economic Growth: A Non-communist Manifesto, 3rd edn. Cambridge University Press, Cambridge (1990)

Aytac, D., Guran, M.C.: The relationship between electricity consumption, electricity price and economic growth in Turkey: 1984–2007. Argum. Oecon. **2**(27), 101–123 (2011)

Kraft, J., Kraft, A.: On the relationship between energy and GNP. J. Energy Dev. **3**(2), 401–403 (1978)

Tang, C.F.: Electricity consumption, income, foreign direct investment, and population in Malaysia: new evidence from multivariate framework analysis. J. Econ. Stud. **36**(4), 371–382 (2009)

Abdullah, A.: Electricity power consumption, foreign direct investment and economic growth. World J. Sci. Technol. Subst. Dev. **10**(1), 55–65 (2013)

Akpan, U.F., Akpan, G.E.: The contribution of energy consumption to climate change: a feasible policy direction. J. Energy Econ. Policy **2**(1), 21–33 (2012)

Solow, R.M.: A contribution to the theory of economic growth. Q. J. Econ. **70**(1), 65–94 (1956)

Arrow, K.: The economic implication of learning-by-doing. Rev. Econ. Stud. **29**(1), 155–173 (1962)

Romer, P.M.: Endogenous technological change. J. Polit. Econ. **98**(5, Part 2), 71–102 (1990)

Esso, L.J.: Threshold cointegration and causality relationship between energy use and growth in seven African countries. Energy Econ. **32**(6), 1383–1391 (2010)

Aslan, A., Apergis, N., Yildirim, S.: Causality between energy consumption and GDP in the US: evidence from wavelet analysis. Front. Energy **8**(1), 1–8 (2014)

Kyophilavong, P., Shahbaz, M., Anwar, S., Masood, S.: The energy-growth nexus in Thailand: does trade openness boost up energy consumption? Renew. Sustainable Energy Rev. **46**, 265–274 (2015)
Ciarreta, A., Zarraga, A.: Electricity consumption and economic growth: evidence from Spain. Biltoki 2007.01, Universidad del Pais Vasco, pp. 1–20 (2007)
Canh, L.Q.: Electricity consumption and economic growth in VietNam: a cointegration and causality analysis. J. Econ. Dev. **13**(3), 24–36 (2011)
Hwang, J.H., Yoo, S.H.: Energy consumption, CO2 emissions, and economic growth: evidence from Indonesia. Qual. Quant. **48**(1), 63–73 (2014)
Wolde-Rufael, Y.: Electricity consumption and economic growth: a time series experience for 17 African countries. Energy Policy **34**(10), 1106–1114 (2006)
Acaravci, A., Ozturk, I.: Electricity consumption and economic growth nexus: a multivariate analysis for Turkey. Amfiteatru Econ. J. **14**(31), 246–257 (2012)
Kum, H., Ocal, O., Aslan, A.: The relationship among natural gas energy consumption, capital and economic growth: bootstrap-corrected causality tests from G7 countries. Renew. Sustain. Energy Rev. **16**, 2361–2365 (2012)
Shahbaz, M., Lean, H.H., Farooq, A.: Natural gas consumption and economic growth in Pakistan. Renew. Sustain. Energy Rev. **18**, 87–94 (2013)
Shahiduzzaman, M., Alam, K.: Cointegration and causal relationships between energy consumption and output: assessing the evidence from Australia. Energy Econ. **34**, 2182–2188 (2012)
Ibrahiem, D.M.: Renewable electricity consumption, foreign direct investment and economic growth in Egypt: an ARDL approach. Procedia Econ. Financ. **30**(2015), 313–323 (2015)
Pesaran, M.H., Shin, Y., Smith, R.J.: Bounds testing approaches to the analysis of level relationships. J. Appl. Econom. **16**(3), 289–326 (2001)
Davoud, M., Behrouz, S.A., Farshid, P., Somayeh, J.: Oil products consumption, electricity consumption-economic growth nexus in the economy of Iran: a bounds test co-integration approach. Int. J. Acad. Res. Bus. Soc. Sci. **3**(1), 353–367 (2013)
Nkoro, E., Uko, A.K.: Autoregressive Distributed Lag (ARDL) cointegration technique: application and interpretation. J. Stat. Econom. Methods **5**(4), 63–91 (2016)
Engle, R., Granger, C.: Cointegration and error correction representation: estimation and testing. Econometrica **55**, 251–276 (1987)
Dickey, D.A., Fuller, W.A.: Likelihood ratio statistics for autoregressive time series with a unit root. Econometrica **49**, 1057–1072 (1981)
Phillips, P.C.B., Perron, P.: Testing for a unit root in time series regression. Biomtrika **75**(2), 335–346 (1988)
Odhiambo, N.M.: Energy consumption and economic growth nexus in Tanzania: an ARDL bounds testing approach. Energy Policy **37**(2), 617–622 (2009)
Jumbe, C.B.L.: Cointegration and causality between electricity consumption and GDP: empirical evidence from Malawi. Energy Econ. **26**, 61–68 (2004)
Sami, J.: Multivariate cointegration and causality between exports, electricity consumption and real income per capita: recent evidence from Japan. Int. J. Energy Econ. Policy **1**(3), 59–68 (2011)
Yoo, S.H.: Electricity consumption and economic growth: evidence from Korea. Energy Policy **33**, 1627–1632 (2005)
Long, P.D., Ngoc, B.H., My, D.T.H.: The relationship between foreign direct investment, electricity consumption and economic growth in Vietnam. Int. J. Energy Econ. Policy **8**(3), 267–274 (2018)
Shahbaz, M., Feridun, M.: Electricity consumption and economic growth empirical evidence from Pakistan. Qual. Quant. **46**(5), 1583–1599 (2012)

The Impact of Anchor Exchange Rate Mechanism in USD for Vietnam Macroeconomic Factors

Le Phan Thi Dieu Thao[1], Le Thi Thuy Hang[2], and Nguyen Xuan Dung[2(✉)]

[1] Faculty of Finance, Banking University of Ho Chi Minh City, Ho Chi Minh City, Vietnam
dieuthaodhnh@gmail.com

[2] Faculty of Finance and Banking, University of Finance – Marketing, Ho Chi Minh City, Vietnam
hangleufm@gmail.com, max.nxd@gmail.com

Abstract. In this study, the author assessed the effects and impacts of the anchor exchange rate mechanism in USD for the macroeconomic factors of Vietnam by using the VAR autoregressive vector model and analytics of impulse reaction function, covariance decomposition. The study focused on three specific variables in the country: real output, price level of goods and services; and money supply. The results show that the change in the USD/VND exchange rate may have a significant impact on the macroeconomic variables of Vietnam. More specifically, the devaluation of the VND against the USD led to a decline in gross domestic product (GDP) and as a result tightening monetary policy. These results are quite robustly analyzed through the verification of econometric models for time series.

Keywords: Exchange rate USD/VND · Anchor in USD
Macroeconomic factors · Vietnam · VAR

1 Introduction

The size of Vietnam's GDP is too small compared to the size of GDP in Asia in particular and the world in general. Vietnam, with its modest economic potential, is required to maintain a large trade opening to attract foreign investment. However, the level of commercial diversification of Vietnam is not high, the United States remains a strategic partner and the USD remains the key currency used by Vietnam in international payments. On the other hand, the exchange rate mechanism of Vietnam in the direction of anchoring the exchange rate in USD, the fluctuation of exchange rates between other strong currencies to VND is calculated based on the fluctuation of the exchange rate between USD and VND. The anchor exchange rate mechanism in USD has led Vietnam's economy too dependent on USD for its payment and credit activities. Shocks of USD/VND exchange rate with abnormal fluctuations after Vietnam's integration to the WTO have greatly affected the business activities of enterprises and economic activities.

V. Kreinovich et al. (Eds.): ECONVN 2019, SCI 809, pp. 323–351, 2019.
https://doi.org/10.1007/978-3-030-04200-4_25

Kinnon's (2000–2001) study showed that all East Asian countries except Japan, which originated in the Asian economic crisis of 1997–1998 had fixed exchange rates regime or anchor in USD and was also called as "East Asian Dollar Standard". Fixing the exchange rate and anchoring exchange rates in a single currency, the US dollar, has made countries face the shocks of international economic crises caused to the domestic economy, especially the exchange rate shocks. Over-concentration on trade proportion in some countries and not using other strong currencies except USD to pay for international business transaction will create risks associated with exchange rate fluctuations and that is a great obstacle to the process of national integration and development, causing the vulnerability of the domestic economy to the exchange rate shocks.

Thus, proceed from the study and the actual situation has shown the relation between the exchange rate anchor mechanism in USD and the economic situation of the country. How has the growth of a nation's economy been affected by the exchange rate shock of that country's domestic currency against USD has drawn the attention of investors, policy planners and researchers for decades. This study will provide an overview of the USD/VND exchange rate shock affecting macroeconomic factors in Vietnam, showing the importance of the exchange rate policy in general for economic variables. The USD/VND exchange rate is a variable that influences the behavior of some other relevant variables such as: consumer price index, money supply, interest rates and economic growth rates.

The rest of the paper is structured as follows. In the next section, we present basic information to promote our research, briefly describe Vietnam's exchange rate mechanism, and highlight the relevant experimental documents. Section 2 outlines our experimental approach. Specifically, the study uses the automated vector model (VAR) to assess the impact of exchange rate fluctuation between USD and VND on Malaysia's economic efficiency. We rely on the analysis of variance and impulse reaction functions to capture the experimental information in the data. Section 3 presents and preliminary describes the sequence of data. Then the estimated results are presented and discussed in Sect. 4. Finally, Sect. 5 concludes with a summary of the main results and some concluding remarks. At the same time, the study will also contribute to suggestion for the selection of appropriate exchange rate management policy for Vietnam.

2 Exchange Rate Management Mechanism of Vietnam and Some Experimental Researches

Exchange Rate Management Mechanism of Vietnam

The official exchange rate of USD/VND is announced daily by the State Bank and is determined on the basis of the actual average exchange rate on the interbank exchange market on the previous day. The establishment of this new exchange rate mechanism is to change the fixed exchange rate mechanism with wide amplitude applied in the previous period, in which the new USD/VND exchange rate was determined based on the interbank average exchange rate and amplitude +/(−)%, which is the basis for commercial banks to determine the daily USD/VND exchange rate. The State Bank will adjust the supply or demand for foreign currency by buying or selling foreign currencies on the interbank market in order to adjust and stabilize exchange rates. This exchange rate policy is appropriate for the country always in deficit status and balance of payment often in deficit status, foreign currency reserves are not large and inflation is not really well controlled. In general, Vietnam has applied a fixed anchor exchange rate mechanism, the interbank average exchange rate announced by the State Bank is kept constant. Although USD fluctuates in the world market, but in the long period, the exchange rate in Vietnam is stable at about 1–3% per annum. That stability shades the exchange rate risk, even if USD is the currency that accounts for a large proportion of the payment. However, when impacted by the financial crisis in East Asia, Vietnam was forced to devaluate VND to limit the negative impacts of the crisis on the Vietnamese economy. At the same time, the sudden exchange rate adjustment has increased the burden of foreign debt, causing great difficulties for foreign-owned enterprises, even pushing more businesses into losses. This is the price to pay when maintaining the fixed exchange rate policy by stabilizing the anchor exchange rate in USD for too long. And the longer the fixed persistence time, the greater the commutation for policy planners. Since 2001, the adjusted anchor exchange rate mechanism has been applied. The Government has continuously adjusted the exchange surrender rate for economic organizations with foreign currency revenue in a gradually descending manner, namely: the exchange surrender rate was 50% in 1999; the exchange surrender rate decreased to 40% in 2001; the exchange surrender rate decreased to 30% in 2002. In 2005, Vietnam declared the liberalization of frequent transactions through the publication of the Foreign Exchange Ordinance. The exchange rate mechanism has been gradually floated since at the end of 2005 the International Monetary Fund (IMF) officially recognized that Vietnam fully implemented the liberalization of frequent transactions. Since 2006, the foreign exchange market of Vietnam has begun to bear real pressure of international economic integration. The amount of foreign currency poured into Vietnam began to increase strongly. The World Bank (WB) and the International Monetary Fund (IMF) have also warned that the State Bank of Vietnam should increase the flexibility of the exchange rate in the context of increasing capital pour into Vietnam. The timely exchange rate intervention will contribute to reducing the pressure on the monetary management of the State Bank. A series of changes by the State Bank of Vietnam aimed at helping the exchange rate management mechanism in line with current conditions in Vietnam, especially in terms of heightening marketability, flexibility and is more active with the market fluctuations, especially the emergence of external factors is clear in recent times, when the exchange rate floating destination can not be achieved immediately.

Vietnam Exchange Rate Management Policy Remarks: Firstly, the size of Vietnam's GDP is too small compared to the size of GDP in Asia as well as the world, so the trade opening of Vietnam can not be more narrowed, the difference of Vietnam's inflation compared with countries with very high trading relationships, it is impossible to implement the floating exchange rate mechanism right away. Secondly, the anchoring of the VND exchange rate in USD, while the position of USD has decreased, Vietnam's trade relations with other countries increased significantly, leading to the anchoring of the exchange rate according to USD has affected trade and investment activities with partners. Thirdly, the central exchange rate announced daily by the State Bank does not always reflect the real supply and demand of the market, especially when the excess or tension of foreign currency occurs. Fourthly, the process of trade liberalization is more and more widespread, the free-capital balance and the exchange rate management mechanism should avoid the condition of less flexibility, rigidity and non-market status which will greatly affect to the economic.

Impact Experimental Studies of Exchange Rate Management Mechanism on Macroeconomic Factors
The choice of exchange rate mechanism was more greatly noticed in international finance after the collapse of the Bretton Wood system in the early 1970s (Kato and Uctum 2007). Moreover, exchange rate mechanism is classified according to the following rules concerning the level of foreign exchange market intervention by monetary authorities (Frenkel and Rapetti 2012). Traditionally, the exchange rate regime is divided into two types: Fixed and floating exchange rate mechanism. A fixed exchange rate mechanism is often defined as the commitment of monetary authorities to intervene in the foreign exchange market to maintain a certain fixed rate for the national currency against another currency or a basket of currencies. The floating exchange rate regime is often defined as the monetary authority's commitment to determine the exchange rate established by market forces through the supply and demand of the market.

Moreover, between fixed and floating exchange rate mechanisms, there exists an alternative system to maintain certain flexibility. They are known as intermediate or soft mode. These include anchor under many basket of foreign currencies, adjustable anchor and mixed exchange rate mechanism, detailed study of intermediate mechanisms provided in Frankel (2003), Reinhart and Rogoff (2004), and Donald (2007). Trading between two different countries will occur based on a specific currency fixed by both countries for commercial purposes and determine the value of the currency of the country against the currencies of other countries based on the above currency are referred to as currency price anchor (Mavlonov 2005). The choice of USD as an anchor monetary has been based primarily on the dominance of the accounts of this currency in international trade. Continued with the USD which was selected for a number of reasons, most of which is export stability and financial revenue (when revenue is a major component of the state budget), the reliability of monetary policy when the anchor exchange rate in USD will increase and to protect the values of major financial assets in USD prevailing from exchange rate fluctuations. Anchoring exchange rate in USD has met the expectations of the economy in a considerable time. Anchoring exchange rate in USD has helped to eliminate or at least mitigate exchange rate risk and to stabilize the fluctuation of major USD financial assets of countries. It also reduces the cost of commercial transactions, financing and investment incentives. Internally, exchange rate stabilization has helped countries avoid nominal shocks and help maintain international competitiveness of economies (Kumah 2009; Khan 2009).

However, there is no unification in the optimal exchange rate mechanism or through factors that make a country choose a particular exchange rate mechanism (Kato and Uctum 2007). According to Frankel (1999, 2003), no single exchange rate regime is right for all countries, or at all times. The choice of a proper exchange rate regime depends primarily on the circumstances of the country as well as in terms of time.

Based on traditional theoretical documents, the most common criteria for determining the optimal exchange rate regime are the macroeconomic and financial stability in the face of nominal or real shocks (Mundell 1963). In the context of studies on the exchange rate regime affecting the economy of each country, this study aims to examine the appropriateness of the fixed exchange rate system anchore in available USD of Vietnam.

3 Research Method and Data

VAR Regression Model

The VAR model is a autoregressive vector model combining two uinvariate autoregression (AR) and simultaneous equations - Ses. VAR is a system of dynamic linear equations, all variables in the system are considered as endogenous variables, each equation (of each endogenous variable) in the system is explained by its delay variables and other variables in the system. In terms of the nature of the VAR model, it is commonly used to estimate the relationship between macroeconomic variables in terms of stop time series and this impact is time-delayed because the VAR method pay no attention to the endogenous nature of the economic variables in the model, it is common for macroeconomic variables to be endogenous meaning the interactions with each other, which will affects the degree of reliability of the regression results for the one-single dimensional equation regression research method.
The VAR model has two time series: y_{1t}, y_{2t} with the latency is 1

$$\begin{cases} y_{1t} = \alpha_{10} + \alpha_{11} y_{1,t-1} + \alpha_{12} y_{2,t-1} + u_{10} \\ y_{2t} = \alpha_{20} + \alpha_{21} y_{1,t-1} + \alpha_{22} y_{2,t-1} + u_{10} \end{cases}$$

$$\begin{bmatrix} y_{1t} \\ y_{2t} \end{bmatrix} = \begin{bmatrix} \alpha_{10} \\ \alpha_{20} \end{bmatrix} + \begin{bmatrix} \alpha_{11} & \alpha_{12} \\ \alpha_{21} & \alpha_{22} \end{bmatrix} \begin{bmatrix} y_{1,t-1} \\ y_{2,t-1} \end{bmatrix} + \begin{bmatrix} u_{10} \\ u_{10} \end{bmatrix}$$

$$y_t = A_0 + A_1 y_{t-1} + u_t$$

General formula for multiple-variable VAR models:

$$y_t = Dd_t + A_1 y_{t-1} + \ldots + A_p y_{t-1} + u_t$$

In which, $y_t = (y_{1t}, y_{2t}, \ldots y_{nt})$ is the endogenous vector series ($n \times 1$) according to time series t, D is the matrix of the intercept coefficient d_t, A_i coefficient matrix ($k \times k$) for $i = 1, \ldots, p$ of endogenous variables with the lag y_{tp}. u_t is the white noise error of the equations in the system whose covariance matrix is the unit matrix $E(u_t, u_t') = 1$.

The VAR model is a basic tool in econometric analysis with many applications. Among them, a VAR model with random fluctuations, proposed by Primiceri (2005), is widely used, especially in the analysis of macroeconomic issues due to its many outstanding advantages. Firstly, the VAR model does not distinguish endogenous and exogenous variables during regressive process and all variables are considered endogenous variables, variables in the endogenous model do not affect the level of reliability of the model. Second, the VAR model is executed when the value of a variable is expressed as a linear function of the past or delay values of that variable and all other variables in the model, so that it can be estimated by the OLS method without using any other complex system method such as least squares of the two stages (2SLS) or unrelated regression (SURE). Thirdly, the VAR built-in convenient measurement tools such as the push reaction function and the variance disintegrate analysis... which helps clarify how the dependent variable responds to a shock in one or many equations of the system. In addition, the VAR model does not require sequences of data for in a too long time, so it can be used in developing economies. From the advantages of the VAR model, the author proceeds step by step. These steps include: (1) unit and co-linkage tests, (2) VAR test and estimation and (3) variance disintegrate analysis and pulse reaction functions. In addition to providing information on the time characteristics of variables, step (1) requires a preliminary analysis of the data series to determine the proper characteristics of the VAR in step (2). Meanwhile, step (3) evaluates the estimated VAR results.

Describing the Variables of the Model

There are four variables according to the study, namely GDP, CPI, M2 and USD/VND exchange rate will be explained below:

The nominal exchange rate (NER) between two currencies is defined as the price of a currency expressed in the number of other currencies. Specifically, the NER only indicates the swap value between currency pairs without showing the Purchasing Power of that foreign currency in the domestic market. Thus, *the real exchange rate (RER)*, which is usually defined as the adjusted nominal exchange rate for the differences in the price of the traded and non-traded goods, is used.

Gross Domestic Product (GDP) is the value of all final goods and services produced nationally in a given period of time.

The Consumer Price Index (CPI) is an indicator to reflect the relative change in consumer prices over time. Because the index is based only on a basket of goods that represents the entire consumer goods.

Money supply refers to the supply of money in the economy to meet the demand for purchasing of goods, services, assets, etc. of individuals (households) and enterprises (excluding financial organizations).

Money in circulation is divided into parts:

M1 (narrow money) is called transaction money, that is the actual amounts used for trading goods, including: precious metals and paper money issued by the State Bank; demand deposits or payment deposits; traveller's cheques.

M2 (broad money) is the currency that can be easily converted into cash for a period of time including: M1; term deposits; saving money; short-term debt papers; short-term money market deposits.

M3 consists of M2; term deposits; long-term debts, long-term money market deposits.

In fact, there may be more variables that are considered to be suitable for the current analysis. However, the model that the author uses requires sufficient number of observations. With the latency length of the data series, the addition of a variable in the system can quickly make the regression process ineffective. The model is considered to have only three variables in the country but they are sufficient variables to express the conditions in the commodity market (GDP, CPI) and monetary (M2). The variables of the model are taken a logarithm apart from the GDP variable (%), calculated as follows (Tables 1 and 2):

Table 1. Sources of the variables used in the model

Variables	Symbols	Variable calculation	Sources
Vietnamese domestic products	GDP	GDP (%)	ADB
Consumer price	LNCPI00	The CPI is calculated by CPI of each year with base year (1st quarter 2000), then logarithmize	IFS
Money supply	LNM2	Total payments in the economy, the logarithmize	IFS
USD/VND real exchange rate	LNRUSDVND00	The RER is calculated by exchange rate of each year with base year (1st quarter 2000), then logarithmize	IFS
USD/VND nominal exchange rate	LNUSDVND00	The average interbank rate is calculated by exchange rate of each year with base year (1st quarter 2000), then logarithmize	IFS

Source: General author's summary

Table 2. Statistics describes the variables used in the model

Variables	Sign	Average	Median	Standard deviation	Smallest value	Biggest value	Number of observations
Vietnam output	GDP	6.71	6.12	1.34	3.12	9.50	69
Consumer price	LNCPI00	5.15	4.83	0.43	4.58	5.75	69
Money supply	LNM2	21.01	20.35	1.15	19.10	22.70	69
USD/VND exchange rate	LNRUSDVND00	4.49	4.39	0.18	4.26	4.74	69

Source: General author and calculation

Research Data

The data used in the quarterly analysis includes the period 2000.Q1–2017.Q1. The national output of Vietnam (GDP) is taken in percentage from ADB's international financial statistics. The variable that represents inflation used commonly is the consumer price index (CPI), the variable that represents currency is the large money supply (M2) and the USD/VND exchange rate variable is taken from the IMF financial statistics (IFS).

4 Research Results and Discussion

The Test of the Model

Testing the stationarity of data series, the unit root test result of testing showed that with the significance level $\alpha = 0.05\%$ the Ho hypothesis was accepted about the existence of unit root so the LNRUSDVND00, GDP, LNM2 and LNCPIVN00 series did not stop at the difference d = 0. Continuously, the test was conducted at a higher difference level. The unit root test result showed that with the significance level $\alpha = 0.05\%$, the Ho hypothesis was rejected of the existence of the unit root, so the LNRUSDVND00, GDP, LNM2, and LNCPI series at the difference levels of 1 and 2 as follows: LNRUSDVND00$\approx$I (1); GDP$\approx$I (1); LNM2$\approx$I (2); LNCPI00$\approx$I (1). Thus, the data series did not stop at the same level of difference (Table 3).

Table 3. Augmented Dickey-Fuller test statitic

Null hypothesis	t-Statistic	Prob.*
LNRUSDVND00 has a unit root (d = 1)	−4.852368	0.0002
GDP has a unit root (d = 1)	−8.584998	0.0000
LNCPI00 has a unit root (d = 1)	−4.808421	0.0002
LNM2 has a unit root (d = 2)	−6.570107	0.0000

Source: General author and calculation

Testing optimal selection of latency for the model: Using the LogL, AIC and SC criteria to determine optimal latency for the model. In this case the FPE, AIC, SC and HQ criteria should be used and the optimum latency selection result was p = 3 (Table 4).

Table 4. VAR lag order selection criteria

Endogenous variables: D(LNRUSDVND00) D(GDP) D(LNCPI00) D(LNM2,2)						
Lag	LogL	LR	FPE	AIC	SC	HQ
0	359.9482	NA	1.45e−10	−11.29994	−11.16387	−11.24643
1	394.5215	63.65875	8.07e−11	−11.88957	−11.20921*	−11.62198
2	419.9293	43.55613	6.03e−11	−12.18823	−10.96358	−11.70657
3	449.1182	46.33173*	4.03e−11*	−12.60693*	−10.83799	−11.91120*
4	458.8852	14.26281	5.07e−11	−12.40905	−10.09583	−11.49925

Source: General author and calculation

Causality test. Granger's Wald Tests testing assisted in determining variables included in the model were endogenous or exogenous variables that were necessary for inclusion in the model or not. The result showed that at the significance level $\alpha = 0.1$, LNCPIVN and LNM2 had an effect on LNRUSDVND00 (10%); At the significance level of α = 0.05, LM2 affected LRUSDVND (5%); At a significance level of α = 0.2, GDP had an impact on LNRUSDVND00 (20%). Thus, the variables introduced into the model were endogenous variables and necessary for the model (Table 5).

Table 5. VAR granger causality/block exogeneity wald tests

	Dependent variable: D(LNRUSDVND00)		
	Excluded	Chi-sq df	Prob.
D(GDP___)	3.674855	2	0.1592
D(LN_CPI_VN	5.591615	2	0.0611
D(LNM2,2)	4.826585	2	0.0895
	Dependent variable: D(LM2)		
	Excluded	Chi-sq df	Prob.
D(LNRUSDVND00)	3.674855	2	0.1592
D(LN_CPI_VN	5.591615	2	0.0611
D(LNM2,2)	4.826585	2	0.0895

Source: General author and calculation

Testing the white noise of the residue. The residue of the VAR model must be white noise, the new VAR model can be used for forecasting. The result showed that the p-value < α (α = 0.05) was from the 4th latency. There should be a self-correlation from the 4th latency. So the appropriate latency of the p = 3 model, then the residue of the model was white noise. The VAR model is appropriate for regression (Table 6).

Table 6. VAR residual portmanteau tests for autocorrelations

Lags	Q-Stat	Prob.	Adj Q-Stat	Prob.	Df
1	3.061755	NA*	3.110355	NA*	NA*
2	22.01334	NA*	22.67328	NA*	NA*
3	33.32862	NA*	34.54505	NA*	NA*
4	50.54173	0.0000	52.90570	0.0000	16
5	59.58451	0.0022	62.71482	0.0009	32
6	77.94157	0.0040	82.97088	0.0013	48
7	88.40769	0.0234	94.72232	0.0076	64
8	107.7682	0.0210	116.8487	0.0045	80
9	127.3510	0.0178	139.6358	0.0024	96
10	140.0949	0.0373	154.7398	0.0047	112
11	153.3520	0.0628	170.7483	0.0069	128
12	176.8945	0.0324	199.7237	0.0015	144

Source: General author and calculation

Testing the stability of the model. To test the stability of the VAR model, using the AR Root Test to consider roots or individual values less than 1 or both within a unit circle, the VAR model achieves stability. The results showed that the roots (with k * p = 4 * 3 = 12 roots) were smaller than 1 or both within a unit circle, so the VAR model is stable (Table 7).

Table 7. Testing the stability of the model

Root	Modulus
0.055713 − 0.881729i	0.883487
0.055713 + 0.881729i	0.883487
−0.786090	0.786090
−0.005371 − 0.783087i	0.783106
−0.005371 + 0.783087i	0.783106
0.628469 − 0.148206i	0.645708
0.628469 + 0.148206i	0.645708
−0.475907	0.475907
−0.203825 − 0.348864i	0.404043
−0.203825 + 0.348864i	0.404043
−0.002334 − 0.287802i	0.287811
−0.002334 + 0.287802i	0.287811

Source: General author and calculation

The Result of the VAR Model Analysis

According to Kinnon (2002), in China, Hong Kong and Malaysia appeared a pegged exchange rate with fixed dollar. Other East Asian countries (except Japan) pursued the looser fixing, but with the dollar was tight. Because USD was the dominant currency for all trade and international capital flows, and smaller East Asian economies pegged in USD to minimize settlement risk and fix their domestic prices. But this made them vulnerable to shocks.

From the VAR model, variance resolutions and impulse response functions will be performed and used as tools to evaluate the dynamic interaction and the strength of causal relationships between variables in the system. Moreover, the pulse response functions monitor the directional response of a variable with one standard deviation shock in the other variables. These functions capture both the direct and indirect effects of innovation on a variable of interest, thus allowing us to fully appreciate their dynamic linkage. The author used the Cholesky coefficient as suggested by Sims (1980) to identify shocks in the system. However, this method may be sensitive to the sequence of variables introduced into the model. In the case of the subject, the author put the variables in the following way: LNRUSDVND00, GDP, LNCPIVN00, LNM2. The order reflects the heterogeneity or relative diversity of these variables. The exchange rate will be exogenous with other variables, the exchange rate is then followed by the variables from the commodity market and finally a currency change. Real GDP and actual prices are very slow to adjust, so it should be considered to be exogenous more than money supply.

Impulse Response Functions

As seen from the figure, the direction of the GDP reaction to change shocks in other variables it is theoretically reasonable. Although GDP does not seem to respond significantly to the innovation of LNCPIVN00, GDP responds positively and resonates with a standard deviation in LNM2 at short sight. However, the impact of expanding money supply on real output will be negligible in longer terms. Thus, the standard view that the expansion of the money supply has a real short-term impact that is often affirmed in the author's analysis (Fig. 1).

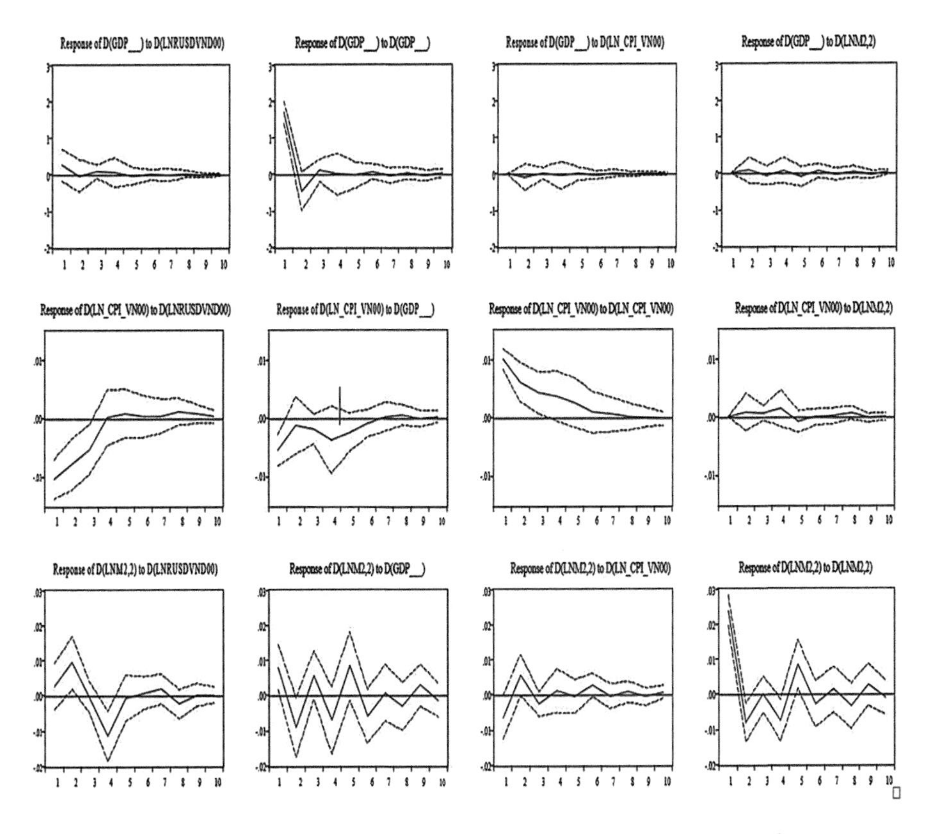

Fig. 1. Impulse response functions *Source: General author and calculation*

In the case of LNRUSD/VND00, devalued shocks of VND lead to an initial negative reaction to real GDP, meaning from the 1st - 2nd period. After that, GDP reverses strong reaction from the 3rd - 5th period. However, in the long term, the reaction of GDP fluctuates insignificantly; Therefore, it seems that shocks in the VND devaluation do not seem to have a severe and permanent impact on real output.

The author also notes the positive response of the LNCPIVN00 price to the change of real output and the fluctuation of LNM2, which should be expected. LNM2 money supply seems to react positively to changes in the real output value, it is not affected by sudden shocks. The devaluation shocks of VND as well as expansion of money supply has a strong impact on the price of LNCPIVN00 and the level of change is maintained longer. On the other hand, the money supply of LNM2 starts to change after VND devalued and increased strongly in the first period, then reversed and fluctuated much later, reflecting the monetary policy response to the monetary depreciation of the exchange rate.

Going back to the main objective of the topic, the result of the analysis is suitable to the view that the fluctuation of the USD/VND exchange rate is significant for a country with a large US dollar density and pegging exchange rate into the big US Dollar in the exchange rate policy like Vietnam presented at the beginning of the chapter. In addition to its influence on actual output value, the depreciation of VND seems to exert stronger pressure on CPI and M2 money supply, especially in longer periods. At the same time, in the event that currency change reacts to an exchange rate shock, the decline in money supply appears to be longer.

Variance Decompositions

The disintegration of variance of the error when predicting variables in the VAR model is the separation of the contribution of other time series as well as of the time series itself in the variance of the forecast error (Table 8).

Table 8. Variance decomposition

Variance decomposition to D(LNRUSD/VND00)			
Period	D(GDP)	D(LNCPIVN00)	D(LNM2)
1	2.302213	44.85235	1.063606
2	2.167654	49.60151	9.982473
3	2.390899	50.26070	9.623628
4	2.506443	46.70575	18.53786
5	2.527105	45.41120	16.61573
6	2.518650	45.25015	16.06629
7	2.524861	45.22999	16.24070
8	2.533009	45.31045	16.32126
9	2.540961	45.38759	16.14722
10	2.539904	45.39267	16.10966

Source: General author and calculation

The results of the disintegration of variance are suitable to the above findings and more importantly, it should be determined the relative importance of the LNRUSD/VND00 exchange rate for the actual output value in the country, price and money supply. Although the forecast error in GDP due to the fluctuation of LNRUSD/VND00 is about 2.5%. A similar model can also be recorded for other variables. However, the fluctuation of the LNRUSD/VND00 exchange rate accounts for about 45% of changes in LNCPIVN00. Meanwhile, the LNRUSD/VND00 variants explain more than 16% of the LNM2 forecast error from the fourth period onwards. This shows the significant impact of LNRUSD/VND00 exchange rate fluctuation for the price LNCPIVN00 and LNM2 money supply.

5 Conclusion

Vietnam has maintained a stable exchange rate system for a long time. In recent difficulties when Vietnam has joined the WTO, the flows of capital have rushed in and impacted and created great exchange rate shocks to the economy, Vietnam has really fixed VND to USD by operating under two central USD/VND exchange rate tools and the amplitude of oscillation in the current exchange rate policy. While ensuring the stability of the USD/VND, the pegging of exchange rate to the US dollar may increase the vulnerability of Vietnamese macro factors in practice.

The results of the study in Sect. 4 show that the fluctuation of the USD/VND exchange rate has impacted on the macroeconomic factors of Vietnam. And this level is significant for a country with a large USD density and pegging exchange rate into the big US Dollar in the exchange rate policy like Vietnam. In addition to its influence on actual output value, the depreciation of VND seems to exert stronger pressure on CPI and M2 money supply. Although the contribution in fluctuation of GDP due to the fluctuation of USD/VND exchange rate is only about 2.5% but the fluctuation of the USD/VND exchange rate accounts for about 45% of the fluctuation of CPI. Meanwhile, USD/VND exchange rate explains more than 16% of the M2 fluctuation from the fourth period onwards. That shows the significant impact of the USD/VND exchange rate fluctuation for the CPI price and M2 money supply.

The results have contributed to the debate about the choice of the way for arranging exchange rates between the flexible exchange rate regime and the fixed exchange rate one. The author believes that for small countries that depend much on international trade and foreign investment and have attempted to liberalize the financial market like Vietnam, the exchange rate stability is extremely important. In the context of Vietnam, the author suggests that the floating exchange rate system may not be appropriate. The inherent high exchange rate fluctuation in free floating regime may not only hinder international trade but also make the economy face the risk of excessive exchange rate fluctuation. With relatively underdeveloped financial markets, the cost of exchange rate fluctuation and risks can be significant.

Appendix 1: Latency Test of Time Series

Stationarity Test of the LNRUSDVND00 Series

Augmented Dickey-Fuller Unit Root Test on LNRUSDVND

Null Hypothesis: LNRUSDVND has a unit root
Exogenous: Constant
Lag Length: 1 (Automatic - based on SIC, maxlag=10)

		t-Statistic	Prob.*
Augmented Dickey-Fuller test statistic		-0.695152	0.8405
Test critical values:	1% level	-3.531592	
	5% level	-2.905519	
	10% level	-2.590262	

*MacKinnon (1996) one-sided p-values.

Augmented Dickey-Fuller Test Equation
Dependent Variable: D(LNRUSDVND)
Method: Least Squares
Date: 08/15/17 Time: 14:44
Sample (adjusted): 2000Q3 2017Q1
Included observations: 67 after adjustments

Variable	Coefficient	Std. Error	t-Statistic	Prob.
LNRUSDVND(-1)	-0.007807	0.011231	-0.695152	0.4895
D(LNRUSDVND(-1))	0.473828	0.112470	4.212915	0.0001
C	0.074773	0.111376	0.671354	0.5044

R-squared	0.217169	Mean dependent var	-0.004952
Adjusted R-squared	0.192705	S.D. dependent var	0.017966
S.E. of regression	0.016142	Akaike info criterion	-5.371037
Sum squared resid	0.016676	Schwarz criterion	-5.272319
Log likelihood	182.9297	Hannan-Quinn criter.	-5.331974
F-statistic	8.877259	Durbin-Watson stat	2.037618
Prob(F-statistic)	0.000396		

Augmented Dickey-Fuller Unit Root Test on D(LNRUSDVND00)

Null Hypothesis: D(LNRUSDVND00) has a unit root
Exogenous: Constant
Lag Length: 0 (Automatic - based on SIC, maxlag=10)

		t-Statistic	Prob.*
Augmented Dickey-Fuller test statistic		-4.852368	0.0002
Test critical values:	1% level	-3.531592	
	5% level	-2.905519	
	10% level	-2.590262	

*MacKinnon (1996) one-sided p-values.

Augmented Dickey-Fuller Test Equation
Dependent Variable: D(LNRUSDVND00,2)
Method: Least Squares
Date: 08/15/17 Time: 14:45
Sample (adjusted): 2000Q3 2017Q1
Included observations: 67 after adjustments

Variable	Coefficient	Std. Error	t-Statistic	Prob.
D(LNRUSDVND00(-1))	-0.537667	0.110805	-4.852368	0.0000
C	-0.002637	0.002041	-1.292206	0.2009

R-squared	0.265914	Mean dependent var	5.40E-05
Adjusted R-squared	0.254620	S.D. dependent var	0.018622
S.E. of regression	0.016078	Akaike info criterion	-5.393365
Sum squared resid	0.016802	Schwarz criterion	-5.327554
Log likelihood	182.6777	Hannan-Quinn criter.	-5.367324
F-statistic	23.54548	Durbin-Watson stat	2.014020
Prob(F-statistic)	0.000008		

Stationarity Test of the GDP Series

Augmented Dickey-Fuller Unit Root Test on GDP___

Null Hypothesis: GDP___ has a unit root
Exogenous: Constant
Lag Length: 2 (Automatic - based on SIC, maxlag=10)

		t-Statistic	Prob.*
Augmented Dickey-Fuller test statistic		-2.533289	0.1124
Test critical values:	1% level	-3.533204	
	5% level	-2.906210	
	10% level	-2.590628	

*MacKinnon (1996) one-sided p-values.

Augmented Dickey-Fuller Test Equation
Dependent Variable: D(GDP___)
Method: Least Squares
Date: 08/15/17 Time: 14:32
Sample (adjusted): 2000Q4 2017Q1
Included observations: 66 after adjustments

Variable	Coefficient	Std. Error	t-Statistic	Prob.
GDP___(-1)	-0.371004	0.146452	-2.533289	0.0138
D(GDP___(-1))	-0.184671	0.136200	-1.355884	0.1801
D(GDP___(-2))	-0.381196	0.118082	-3.228228	0.0020
C	2.464461	0.994385	2.478376	0.0159

R-squared	0.390524	Mean dependent var	-0.027136
Adjusted R-squared	0.361033	S.D. dependent var	1.671454
S.E. of regression	1.336083	Akaike info criterion	3.476054
Sum squared resid	110.6773	Schwarz criterion	3.608760
Log likelihood	-110.7098	Hannan-Quinn criter.	3.528492
F-statistic	13.24223	Durbin-Watson stat	2.129064
Prob(F-statistic)	0.000001		

Augmented Dickey-Fuller Unit Root Test on D(GDP___)

Null Hypothesis: D(GDP___) has a unit root
Exogenous: Constant
Lag Length: 2 (Automatic - based on SIC, maxlag=10)

		t-Statistic	Prob.*
Augmented Dickey-Fuller test statistic		-8.584998	0.0000
Test critical values:	1% level	-3.534868	
	5% level	-2.906923	
	10% level	-2.591006	

*MacKinnon (1996) one-sided p-values.

Augmented Dickey-Fuller Test Equation
Dependent Variable: D(GDP___,2)
Method: Least Squares
Date: 08/15/17 Time: 14:32
Sample (adjusted): 2001Q1 2017Q1
Included observations: 65 after adjustments

Variable	Coefficient	Std. Error	t-Statistic	Prob.
D(GDP___(-1))	-2.482507	0.289168	-8.584998	0.0000
D(GDP___(-1),2)	0.924875	0.201544	4.588937	0.0000
D(GDP___(-2),2)	0.276490	0.122439	2.258185	0.0275
C	-0.040440	0.167361	-0.241636	0.8099

R-squared	0.756951	Mean dependent var	-0.033892
Adjusted R-squared	0.744998	S.D. dependent var	2.672001
S.E. of regression	1.349301	Akaike info criterion	3.496614
Sum squared resid	111.0574	Schwarz criterion	3.630423
Log likelihood	-109.6400	Hannan-Quinn criter.	3.549410
F-statistic	63.32599	Durbin-Watson stat	2.066937
Prob(F-statistic)	0.000000		

Stationarity Test of the LNCPI00 Series

Augmented Dickey-Fuller Unit Root Test on LN_CPI_VN00

Null Hypothesis: LN_CPI_VN00 has a unit root
Exogenous: Constant
Lag Length: 2 (Automatic - based on SIC, maxlag=10)

		t-Statistic	Prob.*
Augmented Dickey-Fuller test statistic		-0.358024	0.9096
Test critical values:	1% level	-3.533204	
	5% level	-2.906210	
	10% level	-2.590628	

*MacKinnon (1996) one-sided p-values.

Augmented Dickey-Fuller Test Equation
Dependent Variable: D(LN_CPI_VN00)
Method: Least Squares
Date: 08/15/17 Time: 14:39
Sample (adjusted): 2000Q4 2017Q1
Included observations: 66 after adjustments

Variable	Coefficient	Std. Error	t-Statistic	Prob.
LN_CPI_VN00(-1)	-0.001607	0.004490	-0.358024	0.7215
D(LN_CPI_VN00(-1))	0.728427	0.122651	5.939007	0.0000
D(LN_CPI_VN00(-2))	-0.240407	0.120731	-1.991266	0.0509
C	0.017442	0.023102	0.754973	0.4531

R-squared	0.387406	Mean dependent var	0.017801
Adjusted R-squared	0.357765	S.D. dependent var	0.018929
S.E. of regression	0.015170	Akaike info criterion	-5.480326
Sum squared resid	0.014268	Schwarz criterion	-5.347620
Log likelihood	184.8508	Hannan-Quinn criter.	-5.427888
F-statistic	13.06968	Durbin-Watson stat	1.915090
Prob(F-statistic)	0.000001		

Augmented Dickey-Fuller Unit Root Test on D(LN_CPI_VN00)

Null Hypothesis: D(LN_CPI_VN00) has a unit root
Exogenous: Constant
Lag Length: 1 (Automatic - based on SIC, maxlag=10)

		t-Statistic	Prob.*
Augmented Dickey-Fuller test statistic		-4.808421	0.0002
Test critical values:	1% level	-3.533204	
	5% level	-2.906210	
	10% level	-2.590628	

*MacKinnon (1996) one-sided p-values.

Augmented Dickey-Fuller Test Equation
Dependent Variable: D(LN_CPI_VN00,2)
Method: Least Squares
Date: 08/15/17 Time: 14:39
Sample (adjusted): 2000Q4 2017Q1
Included observations: 66 after adjustments

Variable	Coefficient	Std. Error	t-Statistic	Prob.
D(LN_CPI_VN00(-1))	-0.516129	0.107339	-4.808421	0.0000
D(LN_CPI_VN00(-1),2)	0.245142	0.119171	2.057061	0.0438
C	0.009225	0.002621	3.518937	0.0008

R-squared	0.268471	Mean dependent var	0.000319
Adjusted R-squared	0.245248	S.D. dependent var	0.017340
S.E. of regression	0.015064	Akaike info criterion	-5.508564
Sum squared resid	0.014297	Schwarz criterion	-5.409034
Log likelihood	184.7826	Hannan-Quinn criter.	-5.469235
F-statistic	11.56052	Durbin-Watson stat	1.913959
Prob(F-statistic)	0.000053		

Stationarity Test of the LNM2 Series

Augmented Dickey-Fuller Unit Root Test on LNM2

Null Hypothesis: LNM2 has a unit root
Exogenous: Constant
Lag Length: 0 (Automatic - based on SIC, maxlag=10)

		t-Statistic	Prob.*
Augmented Dickey-Fuller test statistic		-2.520526	0.1151
Test critical values:	1% level	-3.530030	
	5% level	-2.904848	
	10% level	-2.589907	

*MacKinnon (1996) one-sided p-values.

Augmented Dickey-Fuller Test Equation
Dependent Variable: D(LNM2)
Method: Least Squares
Date: 08/15/17 Time: 14:42
Sample (adjusted): 2000Q2 2017Q1
Included observations: 68 after adjustments

Variable	Coefficient	Std. Error	t-Statistic	Prob.
LNM2(-1)	-0.007158	0.002840	-2.520526	0.0141
C	0.204764	0.059678	3.431126	0.0010

R-squared	0.087806	Mean dependent var	0.054561
Adjusted R-squared	0.073985	S.D. dependent var	0.027481
S.E. of regression	0.026445	Akaike info criterion	-4.398565
Sum squared resid	0.046155	Schwarz criterion	-4.333285
Log likelihood	151.5512	Hannan-Quinn criter.	-4.372699
F-statistic	6.353049	Durbin-Watson stat	1.696912
Prob(F-statistic)	0.014143		

Augmented Dickey-Fuller Unit Root Test on D(LNM2)

Null Hypothesis: D(LNM2) has a unit root
Exogenous: Constant
Lag Length: 3 (Automatic - based on SIC, maxlag=10)

		t-Statistic	Prob.*
Augmented Dickey-Fuller test statistic		-2.495658	0.1213
Test critical values:	1% level	-3.536587	
	5% level	-2.907660	
	10% level	-2.591396	

*MacKinnon (1996) one-sided p-values.

Augmented Dickey-Fuller Test Equation
Dependent Variable: D(LNM2,2)
Method: Least Squares
Date: 08/15/17 Time: 14:42
Sample (adjusted): 2001Q2 2017Q1
Included observations: 64 after adjustments

Variable	Coefficient	Std. Error	t-Statistic	Prob.
D(LNM2(-1))	-0.499503	0.200149	-2.495658	0.0154
D(LNM2(-1),2)	-0.250499	0.175846	-1.424537	0.1596
D(LNM2(-2),2)	-0.279503	0.148116	-1.887055	0.0641
D(LNM2(-3),2)	-0.397127	0.116709	-3.402713	0.0012
C	0.025994	0.011434	2.273386	0.0267

R-squared	0.489874	Mean dependent var	-0.000194
Adjusted R-squared	0.455289	S.D. dependent var	0.033700
S.E. of regression	0.024872	Akaike info criterion	-4.475219
Sum squared resid	0.036499	Schwarz criterion	-4.306556
Log likelihood	148.2070	Hannan-Quinn criter.	-4.408774
F-statistic	14.16444	Durbin-Watson stat	1.846672
Prob(F-statistic)	0.000000		

Augmented Dickey-Fuller Unit Root Test on D(LNM2,2)

Null Hypothesis: D(LNM2,2) has a unit root
Exogenous: Constant
Lag Length: 4 (Automatic - based on SIC, maxlag=10)

		t-Statistic	Prob.*
Augmented Dickey-Fuller test statistic		-6.570107	0.0000
Test critical values:	1% level	-3.540198	
	5% level	-2.909206	
	10% level	-2.592215	

*MacKinnon (1996) one-sided p-values.

Augmented Dickey-Fuller Test Equation
Dependent Variable: D(LNM2,3)
Method: Least Squares
Date: 08/15/17 Time: 14:42
Sample (adjusted): 2001Q4 2017Q1
Included observations: 62 after adjustments

Variable	Coefficient	Std. Error	t-Statistic	Prob.
D(LNM2(-1),2)	-3.382292	0.514800	-6.570107	0.0000
D(LNM2(-1),3)	1.843091	0.452682	4.071493	0.0001
D(LNM2(-2),3)	1.181569	0.339304	3.482336	0.0010
D(LNM2(-3),3)	0.498666	0.229630	2.171604	0.0341
D(LNM2(-4),3)	0.356697	0.123708	2.883383	0.0056
C	-0.001480	0.003162	-0.468034	0.6416

R-squared	0.819239	Mean dependent var	0.000606
Adjusted R-squared	0.803100	S.D. dependent var	0.055894
S.E. of regression	0.024802	Akaike info criterion	-4.463996
Sum squared resid	0.034449	Schwarz criterion	-4.258145
Log likelihood	144.3839	Hannan-Quinn criter.	-4.383174
F-statistic	50.76036	Durbin-Watson stat	1.964479
Prob(F-statistic)	0.000000		

Appendix 2: Optimal Lag Test of the Model

VAR Lag Order Selection Criteria
Endogenous variables: D(LNRUSDVND00) D(GDP___) D(LN_CPI_VN00) D(LNM2,2)
Exogenous variables: C
Date: 08/15/17 Time: 10:17
Sample: 2000Q1 2017Q1
Included observations: 63

Lag	LogL	LR	FPE	AIC	SC	HQ
0	359.9482	NA	1.45e-10	-11.29994	-11.16387	-11.24643
1	394.5215	63.65875	8.07e-11	-11.88957	-11.20921*	-11.62198
2	419.9293	43.55613	6.03e-11	-12.18823	-10.96358	-11.70657
3	449.1182	46.33173*	4.03e-11*	-12.60693*	-10.83799	-11.91120*
4	458.8852	14.26281	5.07e-11	-12.40905	-10.09583	-11.49925

* indicates lag order selected by the criterion
LR: sequential modified LR test statistic (each test at 5% level)
FPE: Final prediction error
AIC: Akaike information criterion
SC: Schwarz information criterion
HQ: Hannan-Quinn information criterion

Appendix 3: Granger Causality Test

VAR Granger Causality/Block Exogeneity Wald Tests
Date: 08/15/17 Time: 10:24
Sample: 2000Q1 2017Q1 Included observations: 64

Dependent variable: D(LNRUSDVND00)

Excluded	Chi-sq	df	Prob.
D(GDP___)	3.674855	2	0.1592
D(LN_CPI_VN	5.591615	2	0.0611
D(LNM2,2)	4.826585	2	0.0895
All	12.04440	6	0.0610

Dependent variable: D(GDP___)

Excluded	Chi-sq	df	Prob.
D(LNRUSDVN	0.063974	2	0.9685
D(LN_CPI_VN	0.147563	2	0.9289
D(LNM2,2)	0.363190	2	0.8339
All	0.875545	6	0.9899

Dependent variable: D(LN_CPI_VN00)

Excluded	Chi-sq	df	Prob.
D(LNRUSDVN	3.874508	2	0.1441
D(GDP___)	2.593576	2	0.2734
D(LNM2,2)	0.902341	2	0.6369
All	8.224893	6	0.2221

Dependent variable: D(LNM2,2)

Excluded	Chi-sq	df	Prob.
D(LNRUSDVN	15.68422	2	0.0004
D(GDP___)	1.281235	2	0.5270
D(LN_CPI_VN	1.464528	2	0.4808
All	24.54281	6	0.0004

Appendix 4: White Noise Error Test of Residuals

VAR Residual Portmanteau Tests for Autocorrelations
Null Hypothesis: no residual autocorrelations up to lag h
Date: 10/19/17 Time: 07:50
Sample: 2000Q1 2017Q1
Included observations: 64

Lags	Q-Stat	Prob.	Adj Q-Stat	Prob.	Df
1	3.061755	NA*	3.110355	NA*	NA*
2	22.01334	NA*	22.67328	NA*	NA*
3	33.32862	NA*	34.54505	NA*	NA*
4	50.54173	0.0000	52.90570	0.0000	16
5	59.58451	0.0022	62.71482	0.0009	32
6	77.94157	0.0040	82.97088	0.0013	48
7	88.40769	0.0234	94.72232	0.0076	64
8	107.7682	0.0210	116.8487	0.0045	80
9	127.3510	0.0178	139.6358	0.0024	96
10	140.0949	0.0373	154.7398	0.0047	112
11	153.3520	0.0628	170.7483	0.0069	128
12	176.8945	0.0324	199.7237	0.0015	144

*The test is valid only for lags larger than the VAR lag order.

df is degrees of freedom for (approximate) chi-square distribution

Appendix 5: Stability Test of the Model

VAR Stability Condition Check

Roots of Characteristic Polynomial
Endogenous variables: D(LNRUSDVND00) D(GDP__
Exogenous variables: C
Lag specification: 1 3
Date: 08/24/17 Time: 15:54

Root	Modulus
0.055713 - 0.881729i	0.883487
0.055713 + 0.881729i	0.883487
-0.786090	0.786090
-0.005371 - 0.783087i	0.783106
-0.005371 + 0.783087i	0.783106
0.628469 - 0.148206i	0.645708
0.628469 + 0.148206i	0.645708
-0.475907	0.475907
-0.203825 - 0.348864i	0.404043
-0.203825 + 0.348864i	0.404043
-0.002334 - 0.287802i	0.287811
-0.002334 + 0.287802i	0.287811

No root lies outside the unit circle.
VAR satisfies the stability condition.

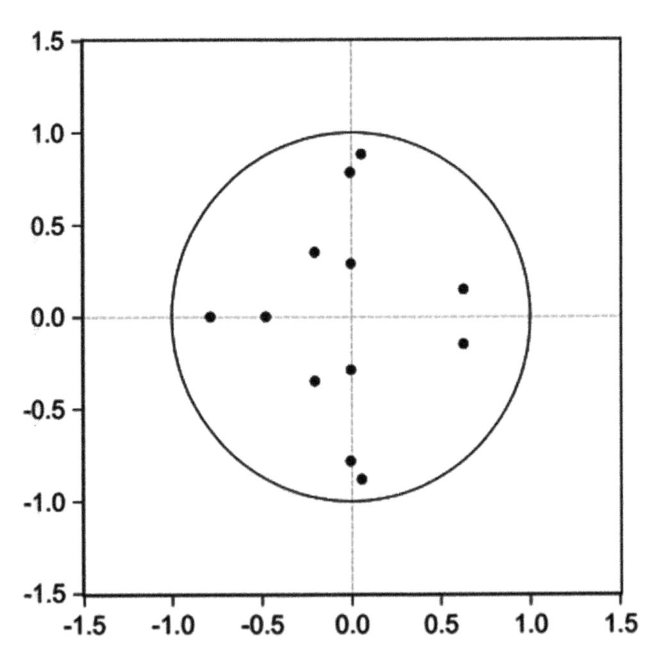

Appendix 6: Impulse Response of the Model

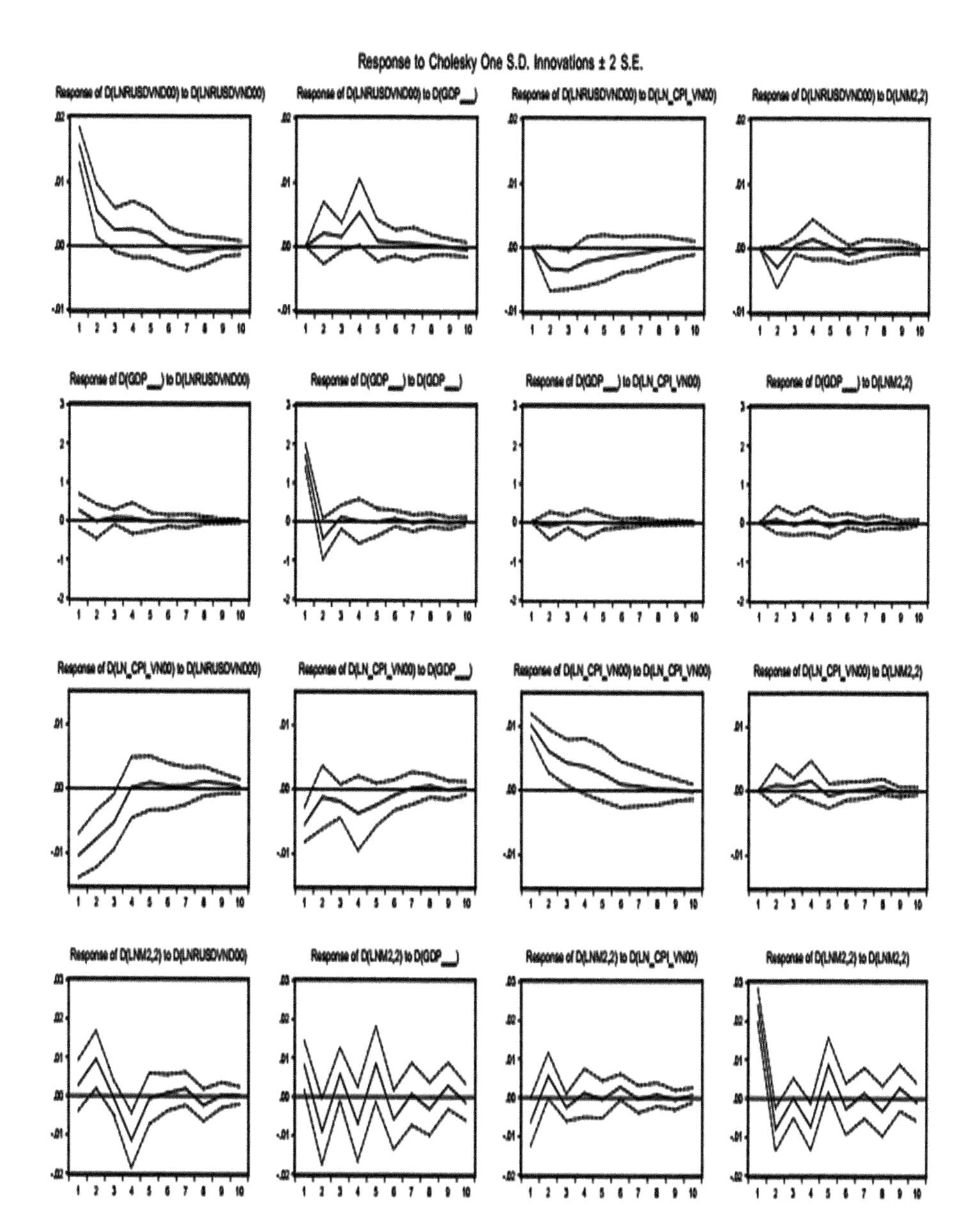

Appendix 7: Variance Decomposition of the Model

Variance Decomposition of D(LNRUSDVND00):

Period	S.E.	D(LNRUSDV	D(GDP___)	D(LN_CPI_V	D(LNM2,2)
1	0.015618	100.0000	0.000000	0.000000	0.000000
2	0.017255	91.95687	1.530619	3.638528	2.873983
3	0.017855	87.85880	2.187791	7.224460	2.728945
4	0.019005	79.46887	9.937881	7.669485	2.923765
5	0.019194	78.92539	9.973457	8.211390	2.889767
6	0.019259	78.39822	10.01194	8.485401	3.104440
7	0.019305	78.28042	10.02105	8.607173	3.091362
8	0.019322	78.27990	10.02943	8.604155	3.086518
9	0.019324	78.27396	10.02779	8.603456	3.094794
10	0.019333	78.22262	10.06725	8.597135	3.112999

Variance Decomposition of D(GDP___):

Period	S.E.	D(LNRUSDV	D(GDP___)	D(LN_CPI_V	D(LNM2,2)
1	1.734062	2.302213	97.69779	0.000000	0.000000
2	1.798063	2.167654	97.30540	0.284768	0.242177
3	1.804578	2.390899	96.98189	0.288063	0.339144
4	1.807590	2.506443	96.65930	0.337351	0.496906
5	1.810514	2.527105	96.36550	0.336424	0.770975
6	1.813562	2.518650	96.20009	0.370651	0.910606
7	1.814533	2.524861	96.13982	0.376686	0.958628
8	1.815408	2.533009	96.08246	0.382024	1.002506
9	1.816423	2.540961	96.02366	0.384548	1.050833
10	1.816853	2.539904	96.00118	0.386930	1.071991

Variance Decomposition of D(LN_CPI_VN00):

Period	S.E.	D(LNRUSDV	D(GDP___)	D(LN_CPI_V	D(LNM2,2)
1	0.015495	44.85235	12.43511	42.71254	0.000000
2	0.018472	49.60151	9.202773	40.95943	0.236292
3	0.019761	50.26070	8.918763	40.49315	0.327382
4	0.020501	46.70575	11.52800	40.92422	0.842035
5	0.020832	45.41120	12.47947	41.15871	0.950620
6	0.020876	45.25015	12.60773	41.19544	0.946674
7	0.020892	45.22999	12.59886	41.21827	0.952878
8	0.020945	45.31045	12.60754	41.02094	1.061065
9	0.020961	45.38759	12.58796	40.95768	1.066770
10	0.020967	45.39267	12.60049	40.94027	1.066565

Variance Decomposition of D(LNM2,2):

Period	S.E.	D(LNRUSDV	D(GDP___)	D(LN_CPI_V	D(LNM2,2)
1	0.026358	1.063606	9.421715	5.803830	83.71085
2	0.030997	9.982473	15.36009	7.474443	67.18300
3	0.031640	9.623628	18.06035	7.834814	64.48121
4	0.035229	18.53786	18.57076	6.439528	56.45185
5	0.037252	16.61573	21.75409	5.776495	55.85369
6	0.037931	16.06629	23.41227	6.120082	54.40136
7	0.038009	16.24070	23.35379	6.105812	54.29970
8	0.038360	16.32126	23.58148	6.042720	54.05454
9	0.038570	16.14722	23.88205	5.994738	53.97599
10	0.038617	16.10966	23.95569	6.019428	53.91522

References

Frankel, J.: Experience of and lessons from exchange rate regimes in emerging economies. Johan F. Kennedy School of Government, Harvard University (2003)

Frenkel, R., Rapetti, M.: External fragility or deindustrialization: what is the main threat to Latin American countries in the 2010s? World Econ. Rev. **1**(1), 37–56 (2012)

MacDonald, R.: Solution-Focused Therapy: Theory, Research and Practice, p. 218. Sage, London (2007)

Mavlonov, I.: Key Economic Developments of the Republic of Uzbekistan. Finance India (2005)

Mundell, R.: Capital mobility and stabilization policy under fixed and flexible exchange rates. Can. J. Econ. Polit. Sci. **29**, 421–431 (1963)

Reinhart, C., Rogoff, K.: The modern history of exchange rate arrangements: a reinterpretation. Q. J. Econ. **CXIX**(1), 1–48 (2004)

Kato, I., Uctum, M.: Choice of exchange rate regime and currency zones. Int. Rev. Econ. Finan. **17**(3), 436–456 (2007)

Khan, M.: The GCC monetary union: choice of exchange rate regime. Peterson Institute International Economics, Washington, Working Paper No. 09-1 (2009)

Kumah, F.: Real exchange rate assessment in the GCC countries-a trade elasticities approach. Appl. Econ. **43**, 1–18 (2009)

The Impact of Foreign Direct Investment on Structural Economic in Vietnam

Bui Hoang Ngoc(✉) and Dang Bac Hai

Graduate School, Ho Chi Minh Open University, Ho Chi Minh city, Vietnam
ngocbh.16ae@ou.edu.vn, haidb.16ae@ou.edu.vn

Abstract. This study examines the impact of FDI inflows on the sectoral economic structure of Vietnam. With data from the first quarter of 1999 to the fourth quarter of 2017 and the application of the vecto autoregression model (VAR), the econometric analysis provides second key results. First, there is a strong statistical evidence that foreign direct investment has a direct impact on Vietnam's sectoral economic structure. Accordingly, this impact makes the proportion of agriculture and industry tends to decrease, the proportion of the service sector tends to increase. Second, industry support active FDI attraction to Vietnam. This result is an important suggestion for policy-maker in planning directions for development investment and structural transformation in Vietnam.

Keywords: FDI · Economic structure · Vietnam

1 Introduction

Development is essential for Vietnam as it leads to an increase in resources. However, economic development should be understood not only as an increase in the scale of the economy but also as a positive change in the economic structure. Indeed, structural transformation is the reorientation of economic activity from less productive sectors to more productive ones (Herrendorf et al. 2011), and can be assessed from three ways: (i) First, structural transformation happens in a country, when the share of its manufacturing value added in GDP increases. (ii) Second, structural transformation of an economy occurs when labor gradually shifts from primary sector to secondary sector and from secondary sector to tertiary sector. In other words, it is the displacement of labor from sectors with low productivity to sector with high-productivity, both in urban than rural areas. (iii) Finally, structural transformation takes place when total factor of productivity (TFP) increases. Although it is difficult to determine the factors explaining a higher increase in TFP, there is an agreement on the fact that there is a positive correlation between institutions, policies and productivity growth.

The economic restructuring reflects the level of development of the productive forces, manifested mainly on two sides: (i) The more productive the production force facilitates the process of division of social labor becomes profound (ii) the

V. Kreinovich et al. (Eds.): ECONVN 2019, SCI 809, pp. 352–362, 2019.
https://doi.org/10.1007/978-3-030-04200-4_26

development of social labor division has made the market economy stronger, economic resources are allocated more effectively. The change in both quantity and quality of structural transformation, especially the sectoral economic structure will shift from a broader economic growth model to an in-depth economic growth model. A country has reasonable economic structure. It will promote a harmonious and sustainable development of the economy and vice versa.

2 Literature Reviews

Structural change is the efficient re-allocation of resources across sectors in an economy that is a prominent feature of economic growth. Structural change plays an important role in driving economic growth and improving labor productivity. This has been proven by many influential studies, such as Lewis (1954), Clark (1957), Kuznets (1966), Denison (1967), Syrquin (1988), Lin (2009). The natural expectation of structural change dynamics is the continual shift of inputs from low-productivity industries to high-productivity industries that continuously increase the productivity of the whole economy. The factors that affect the economic transformation of a nation or a locality such as science, technology, labor, institutional environment and policy, resources and comparative advantage of the nation or the local, level of integration of the economy ... In addition, the need for investment capital is also an indispensable factor, especially foreign capital. The relationship between foreign direct investment (FDI) and the economic transformation process is found in both academic and practical fields.

Academic Field: The theory of competitive advantage to explain the phenomenon of trade between countries and later applied to explain international investment. According to the content of this theory, all countries have comparative advantages in terms of investment factors (capital, labor, technology), especially between developed and developing countries, FDI will bring benefits to both parties. Even if one of the two countries can produce all goods cheaper than the other. Although each country may have higher or lower productivity than other countries, each country still has a certain advantage in terms of other production conditions. This theory of FDI will create conditions for countries to specialize and allocate labor more effectively than simply based on domestic production. For example, multinational companies (MNCs) from industrialized countries are scrutinizing the potential and strengths of each developing country to take part in a production line in a suitable developing country.

This assignment is often appropriate for many production sectors, which require different levels of engineering (automotive, motorcycle, electronics). Under the control of parent companies, these products will be imported or exported within the MNCs or gathered in a particular country to assemble complete products for export or consumption. Thus, through the form of direct investment MNCs companies have participated in adjusting the economic structure in the developing country. The structural theory that Hymer (1960) and Hirschman (1958) have analyzed and explained clearly the role of FDI in the process of economic structural change, especially the structure of industries in the developing countries. FDI is considered as an important channel for capital mobility, technology transfer, and distribution network development...for the developing countries. This will not only give the opportunity to receive capital, technology and management experience for the process of industrialization and modernization, but also help the developing countries to take advantage of and take over the impact of economic restructuring. developed countries and participate in the new international division of labor.

This is an important factor in increasing the proportion of industry and reducing the proportion of traditional industries (agriculture, mining). The theory of "flying saucers" was introduced by Akamatsu (1962). This theory points to the importance of the factors of production in the product development stages that have resulted in the rule of the shift of advantages. Developed countries always have the need to shift their old-fashioned industries, out-of-date technologies, aging products so that they can concentrate on developing new industries and techniques and prolonging their technology and products. Similarly, less developed industrialized countries (NICs) also have the need to shift their investment in technologies and products that have lost a comparative advantage to less developed countries. Often, the technology transfer process in the world takes the form of "flying saucers", which means that developed countries transfer technology, equipment to developed countries or NICs. In turn, these countries will shift their investments to developing countries or less developing countries. In addition, the relationship between FDI in the growth of individual economic sectors, economic regions and economic sectors also affect the economic shift in width and depth. This relationship is reflected through the Harrod-Domar model, which is evident in the ICOR coefficient. The ICOR coefficient of the model reflects the efficiency of the use of investment capital, including FDI and mobilized capital for investment in GDP growth of economic sectors, economic regions and economic sectors. The smaller the ICOR coefficient, the greater the efficiency of capital use for economic growth and vice versa. Therefore, in order to transform the national and local economies, FDI plays a very important role.

Practical Field: According to Prasad et al. (2003) with the attraction of long-term investment and capital controls, foreign-invested enterprises can facilitate the transfer of capacity. (technology and management) and provide a participatory approach to the regional and global value chain. Thus, FDI can generate productivity gains not only for the company but also for the industry. FDI is increasing the competitiveness within the ministry, foreign investment forces domestic firms to improve efficiency and promote ineffective businesses. So it will improve overall productivity within the sector. In addition, the technology and methodologies of foreign firms can be transferred to domestic firms in the same industry (horizontal spillover) or along the supply chain (vertical diffusion) through moving labor and goods. In turn, these countries will shift their investments to developing countries or less developing countries. In addition, the relationship between FDI in the growth of individual economic sectors, economic regions and economic sectors also affect the economic shift in width and depth.

In addition, the technology and methodologies of foreign firms can be transferred to domestic firms in the same industry (horizontal spillover) or along the supply chain (vertical diffusion) through moving labor and goods. As a result, increased labor productivity creates more suitable jobs and shifts towards higher value-added activities (Orcan and Nirvikar 2011). In the commodity development phase, African countries are struggling due to low labor productivity and outdated manufacturing, foreign investment can catalyze the structural shift needed to boost growth (Sutton et al. 2010). Investment-based strategies that encourage adoption and imitation rather than creativity are particularly important for policy-maker in countries in the early stages of development (Acemoglu et al. 2006).

The experience of East Asian nations during the past three decades has made it clear that, in the globalization phase, foreign capital may help to upgrade or diversify the structure of industries in those capital attraction countries (Chen et al. 2014). According to Hiep (2012) pointed out: the process of economic restructuring in the direction of industrialization and modernization in Vietnam needs capital and technology strengths of multinational companies. In fact, over the past 20 years, direct investment from multinational companies has contributed positively to the economic transition. Hung (2010) analyzed the impact of FDI on the growth of Vietnam's economy during 1996–2001 and concluded:

+ The proportion of FDI in GDP of an economic sector increased by 1%, the GDP of that sector will increase to 0.041%. This includes expired FDI projects and annual dissolutions.
+ The proportion of FDI in the GDP of an economic sector increased by 1%, the GDP of that sector will increase to 0.053%. This result is more accurately reflected by the elimination of expired and dissolution FDI projects, which will not take part in production and FDI sectors that have a stronger impact on the economy.
+ If FDI in the GDP of a sector decreases by 1%, it will directly reduce the GDP of the economy by 0.183%.

From the results of this analysis, FDI has shown no significant impact on economic growth. This impact can cause the proportion of sectors in the economic structure to increase or decrease in different proportions, resulting in a shift in the economic structure. Therefore, to attract FDI to increase the proportion of GDP in general and the share of FDI in GDP of the economic sector, thereby creating growth for each economic sector to contribute to the economic restructuring.

3 Research Models

The purpose of this study is to examine the impact of FDI on the sectoral economic structure of Vietnam, with three basic sectors: (i) agriculture, forestry and fisheries, (ii) industry and construction, (iii) service sector, so the research model is divided into three models:

$$Agr\_rate_t = \beta_0 + \beta_1 LnFDI_t + u_t \quad (1)$$

$$Ind\_rate_t = \beta_0 + \beta_1 LnFDI_t + u_t \quad (2)$$

$$Ser\_rate_t = \beta_0 + \beta_1 LnFDI_t + u_t \quad (3)$$

Where: u is the error of the model, t is the study time from the first quarter of 1999 to the fourth quarter of 2017. The source and other variables are illustrated in Table 1.

Table 1. Sources and measurement method of variables in the model

Variable	Description	Unit	Source
Agr_rate	is share of GDP of agriculture, forestry and fisheries compare with total GDP	%	GSO & CEIC
Ind_rate	is share of GDP of industry and construction compare with total GDP	%	GSO & CEIC
Ser_rate	is share of GDP of service sector compare with total GDP	%	GSO & CEIC
LnFDI	is logarithm of total FDI net inflows	Million US Dollar	UNCTAD

https://www.ceicdata.com/en/country/vietnam, *GSO is Vietnam Government Statistics Organization*

4 Research Results and Discussion

4.1 Descriptive Statistics

After 1986, the Vietnamese economy has made many positive changes. Income per capital increased from USD 80.98 in 1986 to USD 2,170.65 in 2016 (at constant 2010 prices). The capital and number of FDI projects poured into Vietnam also increased rapidly, as of March 2018, 126 countries and territories have investment projects still valid in Vietnam. It can be said that FDI is an important factor contributing significantly to the industrial restructuring in the direction of industrialization in Vietnam and the proportion of industry to GDP increase due to significant FDI sector. In general, FDI has appeared in all sectors, but FDI is still most attracted to the industry, in which the processing and manipulation industries are also the large contributions of FDI attraction.

In the early stages of attracting foreign direct investment, FDI inflows were directed towards the mining and import-substituting industries. However, this trend has changed since 2000. Accordingly, FDI projects in the processing and export industries have increased rapidly. These are contributing to the increase in total export turnover and the shift of export structure of Vietnam. Over time, the orientation for attracting foreign direct investment in the field of industry and construction has changed in terms of specific fields and products, it is still oriented towards encouraging the production of new materials, hi-tech products, information technology, mechanical engineering, precision mechanical equipment, electronic products and components... This is also a project that has the potential to create high value-added and Vietnam has a comparative advantage when attracting FDI. Data on foreign direct investment in Vietnam by economic sector in 2017 are shown in Table 2.

Table 2. 10 sectors to attract more foreign direct investment in Vietnam

No.	Sectors	Number of projects	Total registered capital
1	Processing industry, manufacturing	12,456	186,127
2	Real estate business activities	635	53,164
3	Production, distribution of electricity, gas, water	115	20,820
4	Accommodation and catering	639	12,008
5	Construction	1,478	10,729
6	Wholesale and retail	2,790	6,186
7	Mining	104	4,914
8	Warehouse and Transport	665	4,625
9	Agriculture, forestry and fisheries	511	3,518
10	Information and communication	1,648	3,334

Source: Foreign investment agency, Ministry of Planning and Investment, Vietnam. Unit: million US Dollar

It is worth mentioning that the appearance of FDI and development of this sector has contributed directly to the economic restructuring of Vietnam. Agricultural sector ranges from 11.2% to 25.8%, while the industrial sector ranges from 32.4% to 44.7% and the service sector accounts for a high proportion, ranging from 37.3% to 46.8%. Statistics describing changes in economic structure in three main categories of Vietnam from the first quarter of 1999 to the fourth quarter of 2017 are illustrated in Table 3.

Table 3. Descriptive statistics of the variables

Variables	Mean	Std. deviation	Min	Max
Agr_rate	0.192	0.037	0.112	0.258
Ind_rate	0.388	0.322	0.325	0.447
Ser_rate	0.403	0.024	0.373	0.468
LnFDI	6.941	0.952	5.011	8.44

4.2 Unit Root Test

In time series data analysis, the unit root test must be taken first on order to identify the stationary properties of the relevant variables, and to avoid the spurious regression results. The three possible forms of the ADF test (Dickey and Fuller, 1981) are given by the following equations:

$$\Delta Y_t = \beta.Y_{t-1} + \sum_{i=1}^{k} \rho_i.\Delta Y_{t-i} + \varepsilon_t$$

$$\Delta Y_t = \alpha_0 + \beta.Y_{t-1} + \sum_{i=1}^{k} \rho_i.\Delta Y_{t-i} + \varepsilon_t$$

$$\Delta Y_t = \alpha_0 + \beta.Y_{t-1} + \alpha_2.T + \sum_{i=1}^{k} \rho_i.\Delta Y_{t-i} + \varepsilon_t$$

Where: Δ is the first difference, ε_t is error. Phillips and Perron (1988) developed a generalization of the ADF test procedure that allows for fairly mild assumptions concerning the distribution of error. The test regression for the Phillips and Perron (PP) test is the AR(1) process:

$$\Delta Y_{t-1} = \alpha_0 + \beta.Y_{t-1} + \varepsilon_t$$

Test stationary of variables by methods of ADF, PP are shown in Table 4.

Table 4 shows that only the Ser_rate variable is stationary at I(0) and all variables stationary at I(1), so regression analysis must use differential variables.

4.3 Optimal Selection Lag

In time series data analysis, determining optimizing lag is especially important. If the lag is too long, the estimation will be ineffective; otherwise, if the lag is too short, the residuals of the estimate do not satisfy the white noise which makes the deviation of the analysis result. The basis for choosing the optimal lag are standards such as: the Akaike Information Criterion, the Schwart Bayesian Criterion, and the Hannan Quinn Information Criterion. According to AIC, SC, and HQ, the optimal lag has the smallest index. The results for the optimal lag of Eqs. 1, 2 and 3 are shown in Table 5.

Results show that all three AIC, SC and HQ criteria indicate the optimal lag of the Eqs. 1, 2 and 3 used in the regression analysis is lag = 5.

Table 4. Unit root test

Variable	Level		First difference	
	ADF	PP	ADF	PP
Agr_rate	−0.913	−7.225***	−3.191**	−38.64***
Ind_rate	−1.054	−4.033***	−2.089	−17.82***
Ser_rate	−2.953**	−6.268***	−3.547***	−26.81***
LnFDI	−0.406	−1.512	−9.312***	−27.98**

*Notes: ***, ** & *indicate 1%; 5% and 10% level of significance.*

Table 5. Results of optimal selection lag for Eqs. 1, 2 and 3

Equation	Lag	AIC	SC	HQ
1	5	−6.266289*	−5.553965*	−5.983687*
2	5	−5.545012*	−4.832688*	−5.262409*
3	5	−5.437267*	−4.724943*	−5.154664*

4.4 Empirical Results and Discussions

Since the variables are stationary at I(1), the optimal lag of the model is 5, and between the non-cointegration variables, the article applies the vecto autoregressive model (VAR) to examine the effect of FDI to the economic structure of Vietnam in the period 1999–2017. Estimated results using the VAR model with a lag = 5 are shown in Table 6. The empirical results provide a multidimensional view of the relationship between foreign direct investment and the three groups of the sectoral economic structure of Viet Nam, as follows:

a. The relationship between FDI and agriculture, forestry and fisheries

For the agricultural sector, the regression results show the opposite effect for FDI and statistically significant. That means increased foreign direct investment will reduce the proportion of this sector in GDP. The results also show that the agricultural sector is not attractive to foreign direct investors. When the share of the agricultural sector increases, attracting FDI tends to decrease. The change in share of agricultural sector in the previous period did not affect the share of agricultural sector in the future. This result is also consistent with the conclusions of Grazia (2018), Sriwichailamphan et al. (2008), Slimane et al (2016). According to Grazia (2018), FDI in land by developing-country investors negatively influence food security by decreasing cropland due to home institutional pressure to align to national interests and government policy objectives, in addition to negative spillovers.

Table 6. Empirical results by VAR model

Equation	Variables	Coefficient		Coefficient	
1	**Dependent variables**	**Agr_rate**	**Prob**	**LnFDI**	**Prob**
1	Agr_rate	−0.0743	0.492	−6.086	0.000
	LnFDI	−0.0189	0.000	0.799	0.000
	Intercept	0.3331	0.000	2.723	0.000
2	**Dependent variables**	**Ind_rate**	**Prob**	**LnFDI**	**Prob**
2	Ind_rate	0.574	0.000	5.009	0.007
	FDI	−0.010	0.001	0.895	0.000
	Intercept	0.236	0.000	−1.093	0.211
3	**Dependent variables**	**Ser_rate**	**Prob**	**LnFDI**	**Prob**
3	Ser_rate	−0.047	0.675	3.025	0.198
	LnFDI	0.011	0.000	0.864	0.000
	Intercept	0.349	0.000	−0.129	0.895

b. The relationship between FDI and industry, construction

The industrial sector, particularly the manufacturing industry, is always attractive to foreign direct investors. With the advantage of advanced economies, multinational corporations invest heavily in the industrial sector and for innovative research. This is a sector that is less labor intensive, can be produced on a large scale, has a stable profit margin and is less dependent on weather conditions such as agriculture. The regression results in Table 6 show that FDI reduces the share of industry and construction in contributing to the GDP of the Vietnamese economy. This is perfectly reasonable, because businesses have invested in factories and machinery...They have to take into account the volatility of the market and not simply convert these assets into cash. Interestingly, both the FDI attraction to the industrial sector and the proportion of the previous industry all encourage FDI attraction at the moment.

c. The relationship between FDI and service sector

Attracting FDI increases the share of the service sector. Although pointing out the optimal proportions for an economy are many different views, the authors suggest that increasing the proportion of FDI in the service sector to the Vietnamese economy is a good sign because: (i) The service sector uses less natural resources and therefore does not cause resource depletion and it causes less pollution than the industrial sector, (ii) The labor-intensive sector should reduce the employment pressure for state management agencies, (iii) The service sector is involved in both the previous and next stage of the agricultural and industrial sectors, (iv) The service sector is involved in both the previous and next stage of the agricultural and industrial sectors. Therefore, the development of the service sector is also indirectly supporting the development of the remaining sectors in the economy.

5 Conclusions and Implication Policy

Since the economic reform in 1986, the Vietnam economy has made many positive and profound changes in many fields of socio-economic life. The orientation and maintenance of an optimal economic structure will help Vietnam not only exploiting the comparative advantage, but also harmonious and sustainable development. With data from the first quarter of 1999 to the fourth quarter of 2017 and the application of the vecto autoregressive model (VAR), the article finds statistical evidence that foreign direct investment has a direct impact on Vietnam's sectoral economic structure. The authors also note some points when applying the results of this study to the practice as follows:

Firstly: The conclusion of the study is that FDI has changed the proportion of economic structure by sector of Vietnam. Accordingly, this impact makes the proportion of agriculture and industry tends to decrease, the proportion of the service sector tends to increase. This result does not imply that the sector is the most important, as sectors in the economy both support each other and oppose each other in a unified whole.

Secondly: The optimal share of each sector was not solved in this study. Therefore, in each period, the proportion of sectros depends on the weather, natural disasters and the orientation of the Government. Attracting foreign direct investment is only one way to influence the economic structure.

References

Lewis, W.A.: Economic development with unlimited supplies of labour. Econ. Soc. Stud. Manch. Sch. **22**, 139–191 (1954)

Clark, C.: The Conditions of Economic Progress, 3rd edn. Macmillan, London (1957)

Kuznets, S.: Modern Economic Growth: Rate Structure and Spread. Yale University Press, London (1966)

Denison, E.F.: Why Growth Rates Differ. Brookings, Washington DC (1967)

Syrquin, M.: Patterns of structural change. In: Chenery, H., Srinavasan, T.N. (eds.) Handbook of Development Economics. North Holland, Amsterdam (1988)

Lin, J.Y.: Economic Development and Transition. Cambridge University Press, Cambridge (2009)

Hymer, S.H.: The International Operations of National Firms: A Study of Direct Foreign Investment. The MIT Press, Cambridge (1960)

Hirschman, A.O.: The Strategy of Economic Development. Yale University Press, New Haven (1958)

Akamatsu, K.: Historical pattern of economic growth in developing countries. Dev. Econ. **1**, 3–25 (1962)

Prasad, M., Bajpai, R., Shashidhara, L.S.: Regulation of Wingless and Vestigial expression in wing and haltere discs of Drosophila. Development **130**(8), 1537–1547 (2003)

Orcan, C., Nirvikar, S.: Structural change and growth in India. Econ. Lett. **110**, 178–181 (2011)

Sutton, J., Kellow, N.: An Enterprise Map of Ethiopia. Internation Cente Growth, London (2010)

Acemoglu, D., Aghion, P., Zilibotti, F.: Distance to frontier, selection, and economic growth. J. Eur. Econ. Assoc. **4**, 37–74 (2006)

Chen, Y.-H., Naud, C., Rangwala, I., Landry, C.C., Miller, J.R.: Comparison of the sensitivity of surface downward longwave radiation to changes in water vapor at two high elevation sites. Environ. Res. Lett **9**(11), 127–132 (2014)
Herrendorf, B., Rogerson, R., Valentinyi, A.: Two perspectives on preferences and structural transformation. Institute of Economics, Centre for Economic and Regional Studies, Hungarian Academy of Sciences, IEHAS Discussion Papers, 1134 (2011)
Hiep, D.V.: The impact of FDI on structural economic in Vietnam. J. Econ. Stud. **404**, 23–30 (2012)
Hung, P.V.: Investment policy and impact of investment policy on economic structure adjustment: the facts and recommendations. Trade Sci. Rev. **35**, 3–7 (2010)
Dickey, D.A., Fuller, W.A.: Likelihood ratio statistics for autoregressive time series with a unit root. Econometrica **49**, 1057–1072 (1981)
Phillips, P.C.B., Perron, P.: Testing for a unit root in time series regression. Biomètrika **75**(2), 335–346 (1988)
Slimane, M.B., Bourdon, M.H., Zitouna, H.: The role of sectoral FDI in promoting agricultural production and improving food security. Int. Econ. **145**, 50–65 (2016)
Grazia, D.S.: The impact of FDI in land in agriculture in developing countries on host country food security. J. World Bus. **53**(1), 75–84 (2018)
Sriwichailamphan, T., Sriboonchitta, S., Wiboonpongse, A., Chaovanapoonphol, Y.: Factors affecting good agricultural practice in pineapple farming in Thailand. Int. Soc. Hortic. Sci. **794**, 325–334 (2008)

A Nonlinear Autoregressive Distributed Lag (NARDL) Analysis on the Determinants of Vietnam's Stock Market

Le Hoang Phong[1,2](✉), Dang Thi Bach Van[1], and Ho Hoang Gia Bao[2]

[1] School of Public Finance, University of Economics Ho Chi Minh City, 59C Nguyen Dinh Chieu, District 3, Ho Chi Minh City, Vietnam
lhphong@hcmulaw.edu.vn, bachvan@ueh.edu.vn

[2] Department of Finance and Accounting Management, Faculty of Management, Ho Chi Minh City University of Law, 02 Nguyen Tat Thanh, District 4, Ho Chi Minh City, Vietnam
hhgbao@hcmulaw.edu.vn

Abstract. This study examines the impacts of some macroeconomic factors, including exchange rate, interest rate, money supply and inflation, on a major stock index of Vietnam (VNIndex) by utilizing monthly data from April, 2001 to October, 2017 and employing Nonlinear Autoregressive Distributed Lag (NARDL) approach introduced by Shin et al. [33] to investigate the asymmetric effects of the aforementioned variables. The bound test verifies asymmetric cointegration among the variables, thus the long-run asymmetric influences of the aforesaid macroeconomic factors on VNIndex can be estimated. Besides, we apply Error Correction Model (ECM) based on NARDL to evaluate the short-run asymmetric effects. The findings indicate that money supply improves VNIndex in both short-run and long-run, but the magnitude of the negative cumulative sum of changes is higher than the positive one. Moreover, the positive (negative) cumulative sum of changes of interest rate has negative (positive) impact on VNIndex in both short-run and long-run, but the former's magnitude exceeds the latter's. Furthermore, exchange rate demonstrates insignificant effects on VNIndex. Also, inflation hampers VNIndex almost linearly. This result provides essential implications for policy makers in Vietnam in order to successfully manage and sustainably develop the stock market.

Keywords: Macroeconomic factors · Stock market
Nonlinear ARDL · Asymmetric · Bound test

1 Introduction

Vietnam's stock market was established on 20 July, 2000 when Ho Chi Minh City Securities Trading Center (HOSTC) was officially opened. For nearly two decades, Vietnam's stock market has grown significantly when the current market capitalization occupies 70% GDP, compared to 0.28% in the year 2000 with only 2 listed companies.

V. Kreinovich et al. (Eds.): ECONVN 2019, SCI 809, pp. 363–376, 2019.
https://doi.org/10.1007/978-3-030-04200-4_27

It is obvious that the growth of stock market has become an important source of capital and played an essential role in contributing to the sustainable economic development. Accordingly, policy makers must pay attention to the stable development of stock market, and one crucial aspect to be considered is the examination of the stock market's determinants, especially macroeconomic factors.

We conduct this consequential study to evaluate the impacts of macroeconomic factors on a major stock index of Vietnam (VNIndex) by NARDL approach. The main content of this study complies with a standard structure in which literature review is presented first, followed by estimation methodology and empirical results. Crucial tests and analyses including unit root test, bound test, NARDL model specification, diagnostic tests and estimations of short-run and long-run impacts are also demonstrated.

2 Literature Review

Stock index represents the prices of virtually all stocks on the market. As stock price of each company is affected by economic circumstances, stock index is also impacted by micro- and macroeconomic factors.

There are many theories that can explain the relationship between stock index and macroeconomic factors, and among them, Arbitrage Pricing Theory (APT) has been extensively used in studies scrutinizing the relationship between stock market and macroeconomic factors. Nonetheless, the APT model has a drawback as it assumes the constant term to be a risk-free rate of return [3]. Other models, however, presume the stock price as the current value of all expected future dividends [5], and it is calculated as follows:

$$P_t = \sum_{i=1}^{\infty} \frac{1}{(1+\rho)^i} \cdot E(d_{t+i}|h_t). \tag{1}$$

where P_t is the stock price at time t; ρ is the discount rate; d_t is the dividend at time t; h_t is the collection of all available information at time t. Equation (1) consists of 3 main elements: the growth of stock in the future, the risk-free discount rate and the risk premium contained in ρ; see, e.g., [2].

Stock price reacts in the opposite direction with a change in interest rate. An increase in interest rate implies that investors have higher profit expectation, and thus, the discount rate accrues and stock price declines. Besides, the relationship between interest rate and investment in production can be considerable because high interest rate discourages investment, which in turn lowers stock price. Consequently, interest rate can influence stock price directly through discount rate and indirectly through investment in production. Both the aforementioned direct and indirect impacts make stock price negatively correlate with interest rate.

Regarding the impact of inflation, stock market is less attractive to investors when inflation increases because their incomes deteriorate due to the decreasing value of money. Meanwhile, higher interest rate (in order to deal with inflation) brings higher costs to investors who use leverage or limits capital flow into the stock market or diverts the capital to other safer or more profitable investment types. Furthermore, the fact that revenues of companies are worsened by inflation, together with escalating costs (capital costs, input costs resulting from demand-pull inflation), aggravates the expected profits, which negatively affects their stock prices. Hence, inflation has unfavorable impact on stock market.

Among macroeconomic factors, money supply is often viewed as an encouragement for the growth of stock market. With expansionary monetary policy, interest rate is lowered, companies and investors can easily access capital, which fosters stock market. In contrast, with contractionary monetary policy, stock market is hindered.

Export and import play an important role in many economies including Vietnam, and exchange rate is of the essence. When exchange rate increases (local currency depreciates against foreign currency), domestically produced goods become cheaper, and thus, export is enhanced and exporting companies' performances are improved while the import side faces difficulty, which in turn influences stock market. Also, incremental exchange rate attracts capital flow from foreign investors into stock market. The effect of exchange rate, nevertheless, can vary and be subject to specific situations of listed companies on the stock market as well as the economy.

Empirical researches find that stock index is influenced by macroeconomic factors such as interest rate, inflation, money supply, exchange rate, oil price, industrial output, etc. Concerning the link between interest rate and stock index, many studies conclude the negative relationship. Rapach et al. [29] show that interest rate is one of the consistent and reliable predictive elements for stock profits in some European countries. Humpe and Macmillan [12] observe negative impact of long-term interest rate on American stock market. Peiró [21] detects negative impact of interest rate and positive impact of industrial output on stock markets in France, Germany and UK, which is similar to the subsequent repetitive study of Peiró [22] in the same countries. Jareño and Navarro [14] confirm the negative association between interest rate and stock index in Spain. Wongbangpo and Sharma [32] find negative connection between inflation and stock indices of 5 ASEAN countries (Indonesia, Malaysia, Philippines, Singapore and Thailand); in the meantime, interest rate has negative linkage with stock indices of Singapore, Thailand and Philippines.

Hsing [11] indicates that budget deficit, interest rate, inflation and exchange rate have negative relationship with stock index in Bulgaria over the 2000–2010 period. Naik [18] employs VECM model on quarterly data from 1994Q4 to 2011Q4, finds that money supply and industrial production index improve the stock index of India, while inflation exacerbates it, and the roles of interest rate and exchange rate are statistically insignificant. Vejzagic and Zarafat [31] conclude that money supply fosters the stock market of Malaysia, while inflation and exchange rate hamper it. Gul and Khan [9] explores that exchange rate has positive impact on KSE 100 (the stock index of Pakistan) while that of money supply is negative. Ibrahim and Musah [13] examine Ghana's stock market from

October 2000 to October 2010 by using VECM model and denote enhancing causation of inflation and money supply, while interest rate, exchange rate and industrial production index bring discouraging causality. Mutuku and Ng'eny [17] use VAR method on quarterly data from 1997Q1 to 2010Q4 and find that inflation has negative effect on Kenya's stock market while other factors such as GDP, exchange rate and bond interest have positive impacts. In Vietnam, Nguyet and Thao [19] explored that money supply, inflation, industrial output and world oil price can facilitate stock market while interest rate and exchange rate hinder it during July 2000 and September 2011.

From the above literature review, we include 4 factors (inflation, interest rate, money supply and exchange rate) in the model to explain the change of VNIndex.

3 Estimation Methodology

3.1 Unit Root Test

Stationarity is of the essence in scrutinizing time series data. A time series is stationary if its mean and variance do not change over time. Stationarity can be tested by several methods: ADF (Augmented Dickey-Fuller) [7], Phillips-Perron [26], and KPSS [16]. In several papers, the ADF test is often exploited in unit root test.

The simplest case of unit root testing considers an AR(1) process:

$$Y_t = m \cdot Y_{t-1} + \varepsilon_t. \tag{2}$$

where Y_t denotes the time series; Y_{t-1} indicates the one-period-lagged value of Y_t; m is the coefficient; and ε_t is the error term. If $m < 1$, the series is stationary (i.e. no unit root). If $m = 1$, the series is non-stationary (i.e. unit root exists) The aforesaid verification for unit root is normally known as Dickey–Fuller test, which can be alternatively expressed as follows by subtracting Y_{t-1} in each side of the AR(1) process:

$$\Delta Y_t = (m-1) \cdot Y_{t-1} + \varepsilon_t. \tag{3}$$

Let $\gamma = m - 1$, the model then becomes:

$$\Delta Y_t = \gamma \cdot Y_{t-1} + \varepsilon_t. \tag{4}$$

Now, the conditions for stationarity and non-stationarity are respectively $\gamma < 0$ and $\gamma = 0$. Nonetheless, the Dickey–Fuller test is only valid in case of AR(1) process. If AR(p) process is necessitated, the Augmented Dickey-Fuller (ADF) test must be employed because it permits p lagged values of Y_t as well as the inclusion of a constant and a linear time trend, which is written as follows:

$$\Delta Y_t = \alpha + \beta \cdot t + \gamma \cdot Y_{t-1} + \sum_{j=1}^{p} (\phi_j \cdot \Delta Y_{t-j}) + \varepsilon_t. \tag{5}$$

In Eq. (5), α, β, and p are respectively the constant number, linear time trend coefficient and autoregressive order of lag. When $\alpha = 0$ and $\beta = 0$, the series is a random walk without drift, and in case only $\beta = 0$, the series is a random walk. The null hypothesis of ADF test states that Y_t has unit root and there is no stationarity. The alternative hypothesis states that Y_t has no unit root and the series is stationary. In order to test for unit root. ADF test statistic is compared with a corresponding critical value: if the absolute value of the test statistic is smaller than that of the critical value, the null hypothesis cannot be rejected. In case the series is non-stationary, its difference is used. If the time series is stationary at level, it is called I(0). If the time series is non-stationary at level but the stationarity is achieved at the first difference, it is called I(1).

3.2 Cointegration and NARDL Model

Variables are deemed to be cointegrated if there exists a stationary linear combination or long-term relationship among them. For testing cointegration, traditional methods such as Engle-Granger [8] or Johansen [15] are frequently employed.

Nevertheless, when variables are integrated at I(0) or I(1), the 2-period-residual-based Engle-Granger and the maximum-likelihood-based Johansen methods may produce biased results regarding long-run interactions among variables [8,15]. Relating to this issue, Autoregressive Distributed Lag (ARDL) method proposed by Pesaran and Shin [24] give unbiased estimations regardless of whether I(0) and I(1) variables exist in the model.

ARDL model in analyzing time series data has 2 components: "DL" (Distributed Lag)-independent variables with lags can affect dependent variable and "AR" (Autoregressive)-lagged values of the dependent variable can also impact its current value. Going into detail, the simple case ARDL(1,1) is displayed as:

$$Y_t = \alpha_0 + \alpha_1 \cdot Y_{t-1} + \beta_0 \cdot X_t + \beta_1 \cdot X_{t-1} + \varepsilon_t. \tag{6}$$

ARDL(1,1) model shows that both independent and dependent variables have the lag order of 1. In such case, the regression coefficient of X in the long-run equation is as follows:

$$k = \frac{\beta_0 + \beta_1}{1 - \alpha_1}. \tag{7}$$

ECM model based on ARDL(1,1) can be shown as:

$$\Delta Y_t = \alpha_0 + (\alpha_1 - 1) \cdot (Y_{t-1} - k \cdot X_{t-1}) + \beta_0 \cdot \Delta X_{t-1} + \varepsilon_t. \tag{8}$$

The general ARDL model for one dependent variable Y and a set of independent variables X_1, X_2, X_3,..., X_n is denoted as $ARDL(p_0, p_1, p_2, p_3, ..., p_n)$, in which p_0 is the lag order of Y and the rest are respectively the lag orders of

X_1, X_2, X_3,..., X_n. $ARDL(p_0, p_1, p_2, p_3, ..., p_n)$ is written as follows:

$$Y_t = \alpha + \sum_{i=1}^{p_0} (\beta_{0,i} \cdot Y_{t-i}) + \sum_{j=0}^{p_1} (\beta_{1,j} \cdot X_{1,t-j}) + \sum_{k=0}^{p_2} (\beta_{2,k} \cdot X_{2,t-k}) + \sum_{l=0}^{p_3} (\beta_{3,l} \cdot X_{3,t-l}) + ... + \sum_{m=0}^{p_n} (\beta_{n,m} \cdot X_{n,t-m}) + \varepsilon_t. \quad (9)$$

ARDL methods begins with bound test procedure to identify the cointegration among the variables – in other words the long-run relationship among the variables [23]. The Unrestricted Error Correction Model (UECM) form of ARDL is shown as:

$$\Delta Y_t = \alpha + \sum_{i=1}^{p_0} (\beta_{0,i} \cdot \Delta Y_{t-i}) + \sum_{j=0}^{p_1} (\beta_{1,j} \cdot \Delta X_{1,t-j}) + \sum_{k=0}^{p_2} (\beta_{2,k} \cdot \Delta X_{2,t-k}) + \sum_{l=0}^{p_3} (\beta_{3,l} \cdot \Delta X_{3,t-l}) + ... + \sum_{m=0}^{p_n} (\beta_{n,m} \cdot \Delta X_{n,t-m}) \quad (10)$$
$$+ \lambda_0 \cdot Y_{t-1} + \lambda_1 \cdot X_{1,t-1} + \lambda_2 \cdot X_{2,t-1} + \lambda_3 \cdot X_{3,t-1} + ... + \lambda_n \cdot X_{n,t-1} + \varepsilon_t.$$

We test these hypotheses to find the cointegration among variables: the null hypothesis H0: $\lambda_0 = \lambda_1 = \lambda_2 = \lambda_3 = ... = \lambda_n = 0$: (no cointegration) against the alternative hypothesis H1: $\lambda_0 \neq \lambda_1 \neq \lambda_2 \neq \lambda_3 \neq ... \neq \lambda_n \neq 0$. (there exists cointegration among variables). The null hypothesis is rejected if the F statistic is greater than the upper bound critical value at standard significance level. If the F statistic is smaller than the lower bound critical value, H0 cannot be rejected. In case the F statistic lies between the 2 critical values, there is no conclusion about H0.

After the cointegration among variables is identified, we need to make sure that ARDL model is stable and trustworthy by conducting relevant tests: Wald test, Ramsey's RESET test using the square of the fitted values, Larange multiplier (LM) test, CUSUM (Cumulative Sum of Recursive Residuals) and CUSUMSQ (Cumulative Sum of Square of Recursive Residuals), which allows some important examination such as serial correlation, heteroscedasticity and the stability of residuals. After the ARDL model's stability and reliability are confirmed, short-run and long-run estimations can be implemented. Besides the flexibility of allowing both I(0) and I(1) in the model, ARDL approach to cointegration provides several more advantages over other methods [27,28]. Firstly, ARDL can generate statistically significant result even with small sample size, while Johansen cointegration method requires a larger sample size to attain significance [25]. Secondly, while other cointegration techniques require the same lag orders of variables, ARDL allows various ones. Thirdly, ARDL technique estimates only one equation by OLS method rather than a set of equations like other techniques [30]. Finally, ARDL approach outputs unbiased long-run estimations, provided that some of the variables in the model are endogenous [10,23].

Based on the benefits of ARDL model, in order to evaluate the asymmetric impacts of independent variables (i.e. exchange rate, interest rate, money supply and inflation) on VNIndex, we employ NARDL (Non-linear Autoregressive

Distributed Lag) model proposed by Shin et al. [33] under the conditional error correction version displayed as follows:

$$\begin{aligned}
\Delta LVNI_t = \alpha &+ \sum_{i=1}^{p_0}(\beta_{0,i}\cdot \Delta LVNI_{t-i}) + \sum_{j=0}^{p_1^+}(\beta_{1,j}^+\cdot \Delta LEX_{t-j}^+) + \sum_{j=0}^{p_1^-}(\beta_{1,j}^-\cdot \Delta LEX_{t-j}^-)\\
&+ \sum_{k=0}^{p_2^+}(\beta_{2,k}^+\cdot \Delta LMS_{t-k}^+) + \sum_{j=0}^{p_2^-}(\beta_{2,k}^-\cdot \Delta LMS_{t-k}^-) + \sum_{l=0}^{p_3^+}(\beta_{3,l}^+\cdot \Delta LDR_{t-l}^+)\\
&+ \sum_{l=0}^{p_3^-}(\beta_{3,l}^-\cdot \Delta LDR_{t-l}^-) + \sum_{m=0}^{p_4^+}(\beta_{4,m}^+\cdot \Delta CPI_{t-m}^+) + \sum_{m=0}^{p_4^-}(\beta_{4,m}^-\cdot \Delta CPI_{t-m}^-) \qquad (11)\\
&+\lambda_0\cdot LVNI_{t-1} + \lambda_1^+\cdot LEX_{t-1}^+ + \lambda_1^-\cdot LEX_{t-1}^- + \lambda_2^+\cdot LMS_{t-1}^+ + \lambda_2^-\cdot LMS_{t-1}^-\\
&+\lambda_3^+\cdot LDR_{t-1}^+ + \lambda_3^-\cdot LDR_{t-1}^- + \lambda_4^+\cdot LCPI_{t-1}^+ + \lambda_4^-\cdot LCPI_{t-1}^- + \varepsilon_t.
\end{aligned}$$

In equation (11), $LVNI$ is the natural logarithm of VNIndex; LEX is the natural logarithm of exchange rate; LMS is the natural logarithm of money supply (M2); LDR is the natural logarithm of deposit interest rate (% per annum); CPI is the natural logarithm of the index that represents inflation. The "+" and"−" notations of the independent variables respectively denote the partial sum of positive and negative changes; specifically:

$$\begin{aligned}
LEX_t^+ &= \sum_{i=1}^{t}\Delta LEX_i^+ = \sum_{i=1}^{t}\max(\Delta LEX_i,0)\\
LEX_t^- &= \sum_{i=1}^{t}\Delta LEX_i^- = \sum_{i=1}^{t}\min(\Delta LEX_i,0)\\
LMS_t^+ &= \sum_{i=1}^{t}\Delta LMS_i^+ = \sum_{i=1}^{t}\max(\Delta LMS_i,0)\\
LMS_t^- &= \sum_{i=1}^{t}\Delta LMS_i^- = \sum_{i=1}^{t}\min(\Delta LMS_i,0)\\
LDR_t^+ &= \sum_{i=1}^{t}\Delta LDR_i^+ = \sum_{i=1}^{t}\max(\Delta LDR_i,0)\\
LDR_t^- &= \sum_{i=1}^{t}\Delta LDR_i^- = \sum_{i=1}^{t}\min(\Delta LDR_i,0)\\
LCPI_t^+ &= \sum_{i=1}^{t}\Delta LCPI_i^+ = \sum_{i=1}^{t}\max(\Delta LCPI_i,0)\\
LCPI_t^- &= \sum_{i=1}^{t}\Delta LCPI_i^- = \sum_{i=1}^{t}\min(\Delta LCPI_i,0)\,. \qquad (12)
\end{aligned}$$

Similar to the linear ARDL method, Shin et al. [33] introduces the bound test for identifying asymmetrical cointegration in the long-run. The null hypothesis states that the effect is symmetrical in the long-run (H0: $\lambda_0 = \lambda_1^+ = \lambda_1^- = \lambda_2^+ = \lambda_2^- = \lambda_3^+ = \lambda_3^- = \lambda_4^+ = \lambda_4^- = 0$). On the contrary, the alternative hypothesis states that the effect is asymmetrical in the long-run (H1: $\lambda_0 \neq \lambda_1^+ \neq \lambda_1^- \neq \lambda_2^+ \neq$

$\lambda_2^- \neq \lambda_3^+ \neq \lambda_3^- \neq \lambda_4^+ \neq \lambda_4^- \neq 0$). The F statistic and critical values are also used to give conclusion about H0. If H0 is rejected, there exists asymmetrical effect.

When cointegration is identified, the calculation procedure of NARDL is similar to that of the traditional ARDL. Also, Wald test, functional form, Larange multiplier (LM) test, CUSUM (Cumulative Sum of Recursive Residuals) and CUSUMSQ (Cumulative Sum of Square of Recursive Residuals) are necessary to ensure the trustworthiness and stability of NARDL model.

4 Estimation Sample and Data

We use monthly data from April, 2001 to October, 2017. The variables are described in Table 1.

Table 1. Descriptive statistics.

Variable	Obs	Mean	Std. Dev.	Max	Min
LVNI	199	6.03841	0.494204	7.036755	4.914198
LEX	199	9.803174	0.146436	10.01971	9.553859
LMS	199	14.20515	1.099867	15.83021	12.28905
LDR	199	1.987935	0.333566	2.842581	1.543298
LCPI	199	2.368312	0.934708	4.036674	-1.04759

Source: Authors' collection and calculation

LVNI is the natural logarithm of VNIndex which is retrieved from Ho Chi Minh City Stock Exchange (http://www.hsx.vn). *LEX* is the natural logarithm of exchange rate. *LMS* is the natural logarithm of money supply (M2). *LDR* is the natural logarithm of deposit interest rate (% per annum). *LCPI* is the natural logarithm of the index that represents inflation. In this study, we apply the inverse hyperbolic sine transformation formula mentioned in Burbidge et al. [4] to deal with negative value of inflation (see also e.g., [1,6]). The macroeconomic data is collected from IMF's International Financial Statistics.

5 The Empirical Results

Whereas unit root test is not compulsory for ARDL approach, we utilize Augmented Dickey-Fuller (ADF) test and Phillips-Perron (PP) test to confirm that the variables are not integrated at second level difference so that F-test is trustworthy [20,28].

Table 2. ADF and PP tests results for non-stationarity of variables.

	ADF test statistic		PP test statistic	
Variable	Intercept	Intercept and trend	Intercept	Intercept and trend
$LVNI_t$	–1.686	–2.960	–1.420	–2.324
$\Delta LVNI_t$	–10.107***	–10.113***	–10.107***	–10.157***
LEX_t	–0.391	–1.449	–0.406	–1.5108
ΔLEX_t	–15.770***	–15.730***	–15.792***	–15.751***
LMS_t	–2.298	0.396	–1.957	0.047
ΔLMS_t	–11.914***	–12.207***	–12.138***	–12.305***
LDR_t	–2.336	–2.478	–1.833	–1.907
ΔLDR_t	–8.359***	–8.452***	–8.5108***	–8.598***
$LCPI_t$	–3.489***	–3.261**	–3.722***	–3.682**

Note: ***, ** and * are respectively the 1%, 5% and 10% significance level.
Source: Authors' collection and calculation

The result of ADF test and PP test (displayed in Table 2) denotes that $LCPI$ is stationary at level while $LVNI$, LEX, LMS, and LDR are stationary at first level difference, which means that the variables are not integrated at second level difference. Thus, the F statistic shown in Table 3 is valid for cointegration test among variables.

Table 3. The result of bound tests for cointegration test

	90%		95%		97.5%		99%	
F statistic	I(0)	I(1)	I(0)	I(1)	I(0)	I(1)	I(0)	I(1)
4.397**	2.711	3.800	3.219	4.378	3.727	4.898	4.385	5.615

Note: The asterisks ***, ** and * are respectively the 1%, 5% and 10% significance level.
Source: Authors' collection and calculation

From Table 3, the F statistic (4.397) is larger than the upper bound critical value (4.378) at 5% significance level, which indicates the occurrence of cointegration (or long-run relationship) between VNIndex and its determinants. Next, according to Schwartz Bayesian Criterion (SBC), the maximum lag order equals 6 to save the degree of freedom. Also, based on SBC, we can apply NARDL (2, 0, 0, 0, 0, 1, 0, 0, 0) demonstrated in Table 4.

Table 4. Results of asymmetric ARDL model estimation.

Dependent variable: $LVNI$		
Variable	Coefficient	t-statistic
$LVNI_{t-1}$	1.1102***	15.5749
$LVNI_{t-2}$	-0.30426***	-4.7124
LEX_t^+	0.12941	0.45883
LEX_t^-	-1.4460	-1.3281
LMS_t^+	0.30997***	4.2145
LMS_t^-	2.3502***	2.5959
LDR_t^+	-0.58472***	-3.2742
LDR_{t-1}^+	0.45951**	2.4435
LDR_t^-	0.13895***	2.6369
$LCPI_t^+$	-0.034060**	-2.3244
$LCPI_t^-$	-0.030785**	-1.9928
$Constant$	1.0226***	4.4333
$Adj - R^2 = 0.97200$		
$DW - statistics = 1.8865$		
$SE\ of\ Regression = 0.083234$		
Diagnostic tests	A: Serial Correlation	$ChiSQ(12) = 0.0214$ [0.884]
	B: Functional Form	$ChiSQ(1) = 1.4231$ [0.233]
	C: Normality	$ChiSQ(2) = 0.109$ [0.947]
	D: Heteroscedasticity	$ChiSQ(1) = 0.2514$ [0.616]

Note: ***, ** and * are respectively the 1%, 5% and 10% significance level.
A: Lagrange multiplier test of residual serial correlation
B: Ramsey's RESET test using the square of the fitted values
C: Based on a test of skewness and kurtosis of residuals
D: Based on the regression of squared residuals on squared fitted values
Source: Authors' collection and calculation

Table 4 denotes that the overall goodness of fits of the estimated equations is very high (approximately 0.972), which means 97.2% of the fluctuation in VNIndex can be explained by exchange rate, interest rate, money supply and inflation. The diagnostic tests show no issue with our model.

Figures 1 and 2 illustrate CUSUM and CUSUMSQ tests. As cumulative sum of recursive residuals and cumulative sum of square of recursive residuals both are within the critical bounds at 5% significance level, our model is stable and trustworthy to estimate short-run and long-run coefficients.

The estimation result of asymmetrical short-run and long-run coefficients of our NARDL model is listed in Table 5.

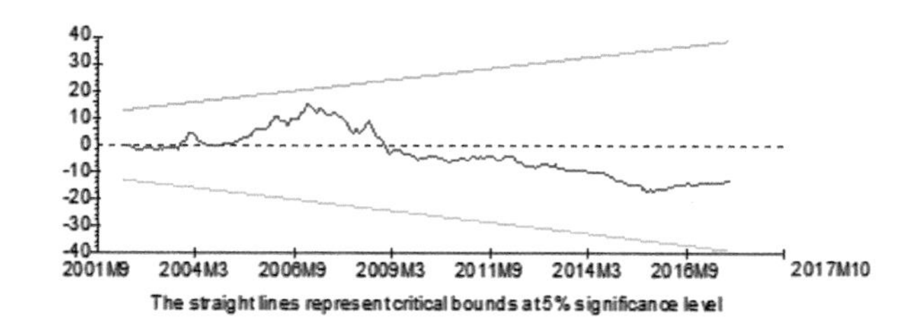

Fig. 1. Plot of cumulative sum of recursive residuals (CUSUM)

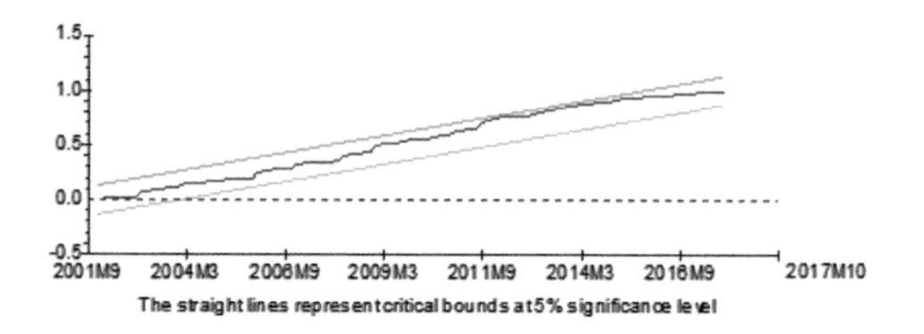

Fig. 2. Plot of cumulative sum of squares of recursive residuals (CUSUMSQ)

The error correction term EC_{t-1} is negative and statistically significant at 1% level, and thus, it once again shows the evidence of cointegration among variables in our model and indicates the speed of adjustment from short-run towards long-run [28].

6 Conclusion

This study analyzes the impacts of some macroeconomic factors on Vietnam's stock market. The result of Non-linear ARDL approach indicates statistically significant asymmetrical effects of money supply, interest rate and inflation on VNIndex.

Specifically, money supply increases VNIndex in both short-run and long-run, and there is considerable difference between the negative cumulative sum of changes and the positive one where the magnitude of the former is much more than that of the latter.

The positive cumulative sum of changes of interest rate worsens VNIndex, whereas the negative analogue improves VNIndex. Besides, in the short-run, the effect of the positive component is substantially higher than the negative counterpart, yet the reversal is witnessed in the long-run.

Both the positive and negative cumulative sum of changes of inflation exacerbate VNIndex. Nonetheless, the asymmetry between them is relatively weak, thus akin to the negative linear connection between inflation and VNIndex reported by existing empirical studies in Vietnam. Consequently, inflation is normally deemed as "the enemy of stock market", and it necessitates effective policies so that the macroeconomy can develop sustainably, which in turn fosters

Table 5. Result of asymmetric short-run and long-run coefficients.

Asymmetric long-run coefficients (dependent variable: $LVNI_t$)		
Variable	Coefficient	t-statistic
LEX_t^+	0.66680	0.46230
LEX_t^-	-7.4509	-1.2003
LMS_t^+	1.5972***	8.9727
LMS_t^-	12.1097***	2.8762
LDR_t^+	-0.64513***	-2.7839
LDR_t^-	0.71594***	2.9806
$LCPI_t^+$	-0.17550***	-2.5974
$LCPI_t^-$	-0.15862**	-1.9998
$Constant$	5.2689***	14.7685
Asymmetric short-run coefficients (dependent variable: $\Delta LVNI_t$)		
Variable	Coefficient	t-statistic
$\Delta LVNI_{t-1}$	0.30426***	4.7124
ΔLEX_t^+	0.12941	0.45883
ΔLEX_t^-	-1.4460	-1.3281
ΔLMS_t^+	0.30997***	4.2145
ΔLMS_t^-	2.3502***	2.5959
ΔLDR_t^+	-0.58472***	-3.2742
ΔLDR_t^-	0.13895***	2.6369
$\Delta LCPI_t^+$	-0.034060**	-2.3244
$\Delta LCPI_t^-$	-0.030785**	-1.9928
$Constant$	1.0226***	4.4333
EC_{t-1}	-0.19408***	-5.42145

Note: The asterisks ***, ** and * are respectively the 1%, 5% and 10% significance level.
Source: Authors' collection and calculation

the stable growth of stock market, attracts capital from foreign and domestic investors and increases their confidence. Also, the State Bank of Vietnam needs flexible approaches to manage money supply and interest rate based on market mechanism; specifically, monetary policy should be established in accordance with the overall growth strategy for each period and continuously monitored so as to avoid instant shocks that aggravate the economy as well as stock market investors.

Finally, the findings recommend stock market investors to notice the changes in macroeconomic factors as they have considerable effects on, and can be employed as indicators of, the stock market.

Acknowledgments. This study has received funding from the European Union's Horizon 2020 research and innovation programme under the Marie Sklodowska-Curie grant agreement No 734712.

References

1. Arcand, J.L., Berkes, E., Panizza, U.: Too much finance?, IMF Working Paper, WP/12/161 (2012)
2. Boyd, J.H., Hu, J., Jagannathan, R.: The stock market's reaction to unemployment news: why bad news is usually good for stocks? J. Finan. **60**(2), 649–672 (2005)
3. Brahmasrene, T., Komain, J.: Cointegration and causality between stock index and macroeconomic variables in an emerging market. Acad. Account. Finan. Stud. J. **11**, 17–30 (2007)
4. Burbidge, J.B., Magee, L., Robb, A.L.: Alternative transformations to handle extreme values of the dependent variable. J. Am. Stat. Assoc. **83**(401), 123–127 (1988)
5. Cochrane, J.H.: Production-based asset pricing and the link between stock returns and economic fluctuations. J. Finan. **46**(1), 209–237 (1991)
6. Creel, J., Hubert, P., Labondance, F.: Financial stability and economic performance. Econ. Model. **48**, 25–40 (2015)
7. Dickey, D.A., Fuller, W.A.: Distribution of the estimators for autoregressive time series with a unit root. J. Am. Stat. Assoc. **74**(366), 427–431 (1979)
8. Engle, R.F., Granger, C.W.J.: Co-integration and error correction: representation, estimation, and testing. Econometrica **55**(2), 251–276 (1987)
9. Gul, A., Khan, N.: An application of arbitrage pricing theory on KSE-100 index; a study from Pakistan (2000–2005). IOSR J. Bus. Manag. **7**(6), 78–84 (2013)
10. Harris, R., Sollis, R.: Applied Time Series Modelling and Forecasting. Wiley, West Sussex (2003)
11. Hsing, Y.: Impacts of macroeconomic variables on the stock market in Bulgaria and policy implications. J. Econ. Bus. **14**(2), 41–53 (2011)
12. Humpe, A., Macmillan, P.: Can macroeconomic variables explain long-term stock market movements? a comparison of the US and Japan. Appl. Finan. Econ. **19**(2), 111–119 (2009)
13. Ibrahim, M., Musah, A.: An econometric analysis of the impact of macroeconomic fundamentals on stock market returns in Ghana. Res. Appl. Econ. **6**(2), 47–72 (2014)
14. Jareño, F., Navarro, E.: Stock interest rate risk and inflation shocks. Eur. J. Oper. Res. **201**(2), 337–348 (2010)
15. Johansen, S.: Statistical analysis of cointegration vectors. J. Econ. Dyn. Control **12**(2–3), 231–254 (1988)
16. Kwiatkowski, D., Phillips, P.C.B., Schmidt, P., Shin, Y.: Testing the null hypothesis of stationarity against the alternative of a unit root: how sure are we that economic time series have a unit root? J. Econ. **54**(1–3), 159–178 (1992)
17. Mutuku, C., Ng'eny, K.L.: Macroeconomic variables and the Kenyan equity market: a time series analysis. Bus. Econ. Res. **5**(1), 1–10 (2015)
18. Naik, P.K.: Does stock market respond to economic fundamentals? time series analysis from Indian data. J. Appl. Econ. Bus. Res. **3**(1), 34–50 (2013)
19. Nguyet, P.T.B., Thao, P.D.P.: Analyzing the impact of macroeconomic factors on Vietnam's stock market. J. Dev. Integr. **8**(18), 34–41 (2013)

20. Ouattara, B.: Modelling the long run determinants of private investment in Senegal, The School of Economics Discussion Paper Series 0413, The University of Manchester (2004)
21. Peiró, A.: Stock prices, production and interest rates: comparison of three European countries with the USA. Empirical Econ. **21**(2), 221–234 (1996)
22. Peiró, A.: Stock prices and macroeconomic factors: some European evidence. Int. Rev. Econ. Finan. **41**, 287–294 (2016)
23. Pesaran, M.H., Pesaran, B.: Microfit 4.0 Window Version. Oxford University Press, Oxford (1997)
24. Pesaran, M.H., Shin, Y.: An autoregressive distributed lag modeling approach to cointegration analysis. In: Strom, S. (ed.) Econometrics and Economic Theory: The Ragnar Frisch Centennial Symposium, pp. 371–413. Cambridge University Press, Cambridge (1998)
25. Pesaran, M.H., Shin, Y., Smith, R.J.: Bounds testing approaches to the analysis of level relationships. J. Appl. Econ. **16**(3), 289–326 (2001)
26. Phillips, P.C.B., Perron, P.: Testing for a unit root in time series regression. Biometrika **75**(2), 335–346 (1988)
27. Phong, L.H., Bao, H.H.G., Van, D.T.B.: The impact of real exchange rate and some macroeconomic factors on Vietnam's trade balance: an ARDL approach. In: Proceedings International Conference for Young Researchers in Economics and Business, pp. 410–417 (2017)
28. Phong, L.H., Bao, H.H.G., Van, D.T.B.: Testing J–curve phenomenon in vietnam: an autoregressive distributed lag (ARDL) approach. In: Anh, L., Dong, L., Kreinovich, V., Thach, N. (eds.) ECONVN 2018. Studies in Computational Intelligence, vol. 760, pp. 491–503. Springer, Cham (2018)
29. Rapach, D.E., Wohar, M.E., Rangvid, J.: Macro variables and international stock return predictability. Int. J. Forecast. **21**(1), 137–166 (2005)
30. Srinivasana, P., Kalaivanib, M.: Exchange rate volatility and export growth in India: an ARDL bounds testing approach. Decis. Sci. Lett. **2**(3), 192–202 (2013)
31. Vejzagic, M., Zarafat, H.: Relationship between macroeconomic variables and stock market index: co-integration evidence from FTSE Bursa Malaysia Hijrah Shariah Index. Asian J. Manag. Sci. Educ. **2**(4), 94–108 (2013)
32. Wongbangpo, P., Sharma, S.C.: Stock market and macroeconomic fundamental dynamic interactions: ASEAN-5 countries. J. Asian Econ. **13**(1), 27–51 (2002)
33. Shin, Y., Yu, B., Greenwood-Nimmo, M.: Modeling asymmetric cointegration and dynamic multipliers in a nonlinear ARDL framework. In: Horrace, W.C., Sickles, R.C. (eds.) Festschrift in Honor of Peter Schmidt: Econometric Methods and Applications, pp. 281–314. Springer Science & Business Media, New York (2014)

Explaining and Anticipating Customer Attitude Towards Brand Communication and Customer Loyalty: An Empirical Study in Vietnam's ATM Banking Service Context

Dung Phuong Hoang(✉)

Faculty of International Business, Banking Academy, Hanoi, Vietnam
dunghp@hvnh.edu.vn

Abstract. Purpose: This research investigates the impacts of perceived value, customer satisfaction and brand trust that are formed by customers' experience with the ATM banking service on brand communication, also known as customer attitude towards their banks' marketing communication efforts, and loyalty. In addition, the mediating roles of brand communication and trust in such relationships are also examined.

Design/methodology: The conceptual framework is developed from the literature. A structural equation model linking brand communication to customer satisfaction, trust, perceived value and loyalty is tested using data collected from a survey with 389 Vietnamese customers of the ATM banking service. SPSS 20 and AMOS 22 were used to analyze the data.

Findings: The results indicate that customers' perceived value and brand trust resulted from their usage of ATM banking service directly influence their attitudes toward the banks' follow-up marketing communication which, in turn, have an independent impact on bank loyalty. More specifically, how ATM service users react to their banks' controlled marketing communication efforts mediates the impacts of bank trust and perceived costs that were formed by customers' experience with the ATM service on customer loyalty. In addition, brand trust is found to have mediating effect in the relationship between either customer satisfaction or perceived value and customer loyalty.

Originality/value: The study treats brand communication as an dependent variable to identify factors that help either explain or anticipate how a customer reacts to their banks' marketing communication campaigns and to what extent they are loyal.

Keywords: Brand communication · Customer satisfaction · Brand trust
Perceived value · Customer loyalty · Vietnam

Paper type: Research paper.

V. Kreinovich et al. (Eds.): ECONVN 2019, SCI 809, pp. 377–401, 2019.
https://doi.org/10.1007/978-3-030-04200-4_28

1 Introduction

The ATM is usually regarded as a distinct area of banking services, one that rarely changes and operates separately from mobile or Internet banking. Since ATM service is relatively simple so that every customer with even little amount of money can use, it is often offered to first-use bank customers and helps banks easily initiate customer relationships for further sales effort. In other words, while having customers use ATM service, banks may aim at two purposes which are persuading customers to use other banking services through follow-up marketing communication efforts and enhancing customer loyalty.

Having more response rate over advertising and sales promotion is always the ultimate goal of advertisers and marketing managers. Therefore, the relationship between brand communication and other marketing variables has been the focus of many previous researches. The literature reveals two perspectives in defining brand communication. In the first perspective, brand communication is defined as an exogenous variable which reflects what and how the companies communicate to their customers (Keller and Lehmann 2006; Runyan and Droge 2008; Sahin et al. 2011). On the other hand, brand communication is regarded as consumers' attitudes or feelings towards the controlled communications (Grace and O'Cass 2005) or also called "customer dialogue" which is measured by customers' readiness to engage in the dialogue with the company (Grigoroudis and Siskos 2009). In this study, we argue that measuring and anticipating brand communication as customers' attitudes is more important than merely describing what and how a firm communicates with its customers. We, therefore, take customer attitude approach in relation to brand communication definition.

Although the direct effect of brand communication on customer loyalty in which brand communication is treated as an exogenous variable has been affirmed in many previous studies (Bansal and Taylor 1999; Grace and O'Cass 2005; Jones et al. 2000; Keller and Lehmann 2006; Ranaweera and Prabhu 2003; Runyan and Droge 2008; Sahin et al. 2011), there are very few research which investigate the determinants of customer attitude towards a brand's controlled communication. According to Grigoroudis and Siskos (2009), how a customer reacts and perceives to the supplier's communication is influenced by their satisfaction formed by previous transactions. In expanding the model suggested by Grigoroudis and Siskos (2009), this study, upon Vietnam banking sector, adds perceived value and brand trust which are also formed by customers' previous experience with the ATM service as determinants of customers' attitudes towards their banks' further marketing communication efforts and further tests the mediating roles of brand communication in the effects that customer satisfaction, perceived value and brand trust may have on bank loyalty.

The main purpose of the current research is, therefore, to investigate the role of brand communication in its relationship with perceived value, customer satisfaction and brand trust in influencing customer loyalty. While each of these variables may independently affect customer loyalty, some of them may have mediating effects on others' influences on customer loyalty. Specifically, this study will follow the definition of brand communication as consumers' attitudes towards brand communication to test two ways that brand communication can influence customer loyalty:

(1) its direct positive effect on customer loyalty; and
(2) its moderating role on the effects of brand trust, customer satisfaction and perceived value on customer loyalty

This study also gives an insight into relationships concerning the linkages among perceived value, customer satisfaction, brand trust and customer loyalty that have already been empirically studied in several other contexts. This becomes significant because of the particular nature of the context studied. ATM banking service is featured by low personal contact, high technology involved and continuous transaction. In such a competitive ATM banking industry where a person can hold several ATM cards in Vietnam, customers' attitudes towards service providers and service value may have special characteristics that, in turn, alter the way customer satisfaction, perceived value and brand trust are interrelated and their influences on customer loyalty in comparison to other previous studies. Analyzing the interrelationships between these variables in one single model, this research aims at investigating in depth their direct effects and mediating effects on customer loyalty especially in the special context of Vietnam banking sector.

2 Theoretical Framework and Hypotheses Development

Conceptual Framework

The conceptual framework in this study is developed from the SWISS Consumer Satisfaction Index Model proposed by Grigoroudis and Siskos (2009). According to this model, customer dialogue is measured by three dimensions including the customers' readiness to engage in the dialogue with the company, whether the customers consider getting in touch with their suppliers easy or difficult, and customer satisfaction in communicating with the suppliers. Customer dialogue, therefore, reflects partly customers' attitudes towards brand communication. Furthermore, the model points out that customer satisfaction which is formed by customers' experience and brand attitudes through previous brand contacts has a direct effect on customer dialogue. In other words, customer satisfaction affects significantly their attitudes towards brand communication which, in turn, positively enhance customer loyalty. Similarly, Angelova and Zekiri (2011) have affirmed that satisfied customers are more open to the dialogue with their suppliers in the long term, and the loyalty eventually increases or in other words, how customers' reaction to brand communication has a mediating effect on the relationship between customer satisfaction and loyalty.

Thus, in our model, customer satisfaction is posited as driving customer loyalty while attitudes toward brand communication, shortly called brand communication mediate such relationship. Since other variables such as brand trust and perceived value are also formed through the framework of the existing business relations like customer satisfaction is and were proven to have significant effects on customer loyalty in previous studies, this study expands the SWISS Customer Satisfaction Index's model to include brand trust and perceived value as proposed in Fig. 1.

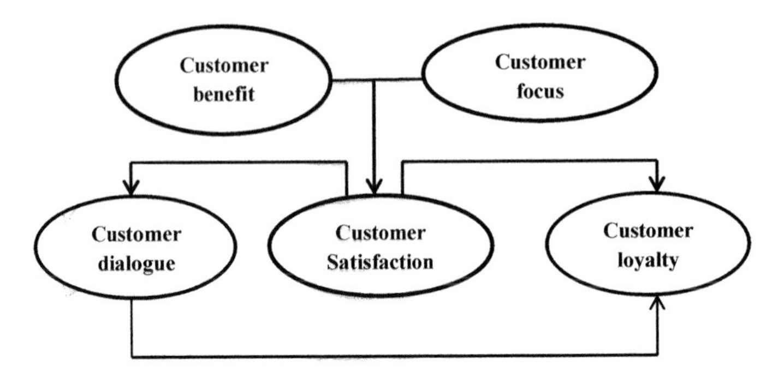

Fig. 1. SWISS consumer satisfaction index model (Grigoroudis and Siskos 2009).

The following part will clarify the definitions and measurement scales of the key constructs, followed by the theoretical background and empirical evidence supporting the hypothesis indicated in the proposed conceptual framework. Since customers' attitudes towards brand communication and its relationship with other variables are the primary focus of this study, the literature review about brand communication will be placed first.

Brand Communication

In service marketing, since services lack the inherent physical presence such as packaging, labeling, and display, company brand becomes paramount. Brand communication is when brand ideas or images are marketed so that target customers can perceive and recognize the distinctiveness or unique selling points of a service company's brand. Due to the rapid development of advanced information technology, today brand communication can be conducted via either in-person with service personnel or various media such as TV, print media, radio, direct mail, web site interactions, social media, and e-mail before, during, and after service transactions. According to Grace and O'Cass (2005), service brand communication can be either controlled or uncontrolled. Controlled communications consist of advertising and promotional activities which aim to convey brand messages to consumers, therefore, consumers' attitudes or feelings towards the controlled communication will affect directly customers' attitudes or intentions to use the brand. Uncontrolled communications includes WOM and non-paid publicity in which positive WOM and publicity help enhance brand attitudes (Bansal and Voyer 2000) while negative ones may diminish customers' attitudes toward the brand (Ennew et al. 2000). In addition, brand communication can be regarded as one-way or indirect communication and two-way or direct communication depending on how the brand interacts with the customers and whether brand communication can create dialogue with customers (Sahin et al. 2011). In the case of two-way communication, brand communication is also regarded as customer dialogue, an endogenous variable that is explained by customer satisfaction (Bruhn and Grund 2000). This study focuses on controlled brand communication including advertising and promotional campaigns which are either communicated indirectly through TV, radio, Internet or create two-way interactions such as advertising and promotional initiatives which are conducted on social media, telephone or through presentation and small talk by salespersons.

Although brand communication is an important metric of relationship marketing, there have been still controversies about what brand communication is about and how to measure it. According to Ndubisi and Chan (2005); Ball et al. (2004) and Ndubisi (2007), brand communication refers to the company's ability to keep in touch with customers, provide timely and trustworthy information, and communicate proactively, especially in case of a service problem. However, according to Grace and O'Cass (2005), brand communication is defined as consumers' attitudes or feelings towards the brand's controlled communications. In other words, brand communication may be measured as either how well the firm does for marketing the brand or how customers react and feel about the advertising and promotional activities of the brand. In this study, brand communication is measured as customers' attitudes towards advertising and promotional activities of a brand

Satisfaction, Trust, Perceived Value and Customer Loyalty

Satisfaction

Customer satisfaction is a popular customer-oriented metric for managers in quality control and marketing effectiveness evaluation across different types of products and services. Customer satisfaction can be defined as an effective response or estate resulting from a customer's evaluation of their overall product consumption or service experience upon the comparison between the perceived product or service performance and pre-purchase expectations (Fornell 1992; Halstead et al. 1994; Cronin et al. 2000). Specifically, according to Berry and Parasuraman (1991), in service marketing, each consumer forms two levels of service expectations: a desired level and an adequate level. The area between two these levels is called a zone of tolerance, also defined as a range of service performance within which customer satisfaction is achieved. Thereby, if perceived service performance exceeds the desired level, customers are pleasantly surprised and their loyalty is better strengthened. The literature reveals two primary methods to measure customer satisfaction including transaction specific measure which covers customers' specific satisfaction towards each transaction with the service provider (Boulding et al. 1993; Andreassen 2000) and cumulative measure of satisfaction which refers to overall customer scoring based on all brand contacts and experiences overtime (Johnson and Fornell 1991; Anderson et al. 1994; Fornell et al. 1996; Johnson et al. 2001; Krepapa et al. 2003). According to Rust and Oliver (1994), the cumulative satisfaction perspective is more fundamental and useful than the transaction-specific one in anticipating consumer behavior. Besides, the cumulative satisfaction has been adopted more popularly in many studies (Gupta and Zeithaml 2006). This study, therefore, will measure customer satisfaction under the cumulative perspective.

Customer Trust
Trust is logically and experientially one of the critical determinants of customer loyalty (Garbarino and Johnson 1999; Chaudhuri and Holbrook 2001; Sirdeshmukh et al. 2002). According to Sekhon et al. (2014), while trustworthiness refers to a characteristic of a brand, a product or service or an organization to be trusted; trust is the customers' willingness to depend on or cooperate with the trustee upon either cognitive base (i.e. reasoning assessment of trustworthiness) or affective base (i.e. resulted from care, concern, empathy, etc.). Trust is driven by two main components including performance or creditability which refers to the expectancy that what the firm say or offer can be relied on and its promises will be kept (Ganesan 1994; Doney and Cannon 1997; Garbarino and Johnson 1999; Chaudhuri and Holbroook 2001) and benevolence which is the extent that the firm cares and works for the customer's welfare (Ganesan 1994; Doney and Cannon 1997; Singh and Sirdeshmukh 2000; Sirdeshmukh et al. 2002).

Perceived Value
Perceived value, also known as customer perceived value, is an essential metric in relationship marketing since it is the key determinant of customer loyalty (Bolton and Drew 1991; Sirdeshmukh et al. 2002). The literature reveals different definitions about customer perceived value. According to Zeithaml (1988), perceived value reflects customers' cognitive and utilitarian perception in which "perceived value is the customer's overall assessment of the utility of a product based on perceptions of what is received and what is given". In other words, perceived value represents trade-off between what customers get (i.e. benefits) and what they pay (i.e. price or costs). Another definition of perceived value is proposed by Woodruff (1997) in which perceived value is defined as "a customer' s perceived preference for, and evaluation of, those product attributes, attribute performances, and consequences arising from use that facilitates achieving the customer's goals and purposes in use situations". However, this definition is too complicated since it combines both pre- and post-purchase context, both preference and evaluation as cognitive perceptions and multiple criteria (i.e. product attributes, usage consequences, and customer goals) that make it difficult to be measured and conceptualized (Parasuraman 1997). Therefore, this study adopts the clearest and most popular definition of perceived value which is proposed by Zeithaml (1988). The literature reveals two key dimensions of customer perceived value which are post-purchase functional and affective values (Sweeney et al. 1996; Sweeney and Soutar 2001; Moliner et al. 2005) both of which are valuated upon the comparison between the cognitive benefits and costs (Grewal et al. 1998; Cronin et al. 2000). Specifically, post-purchase perceived functional values are measured upon five indicators including installations, service quality, professionalism of staff, economic costs and non-economic costs (Sweeney et al. 1996; Sweeney and Soutar 2001; Moliner et al. 2000; Singh and Sirdeshmukh 2000). Meanwhile, the affective component of perceived value refers to how customers feel when they consume the product or experience service and how others see and evaluate them when they are customers of a

specific provider (Mattson 1991; De Ruyter et al. 1997). Depending on different contexts and product or service characteristic, some studies many only focus on the functional value while others concentrate on the affective value or both of them. In this study, the primary benefit that ATM banking service provides to customers is functional value, therefore, customer perceived value of ATM banking service is measured upon the measurement items for the functional value proposed by Singh and Sirdeshmukh (2000). There is a great equivalence between the measurement model by Singh and Sirdeshmukh (2000) and the definition of perceived value by Zeithaml (1988). The installations, service quality and professionalism of staff can be considered as "perceived benefits" that customers receive while economic costs and non-economic costs can be regarded as "perceived costs" that customers must sacrifice.

Customer Loyalty

Due to the increasing importance of relationship marketing in recent years, there has been rich literature on customer loyalty as a key component of relationship quality and business performance (Berry and Parasuraman 1991; Sheth and Parvatiyar 1995). The literature defines customer loyalty differently. From a behavioral perspective, customer loyalty is defined as biased behavioral response reflected by repeat purchasing frequency (Oliver 1999). However, further studies have pointed out that commitment to rebuy should be the essential feature of customer loyalty, instead of simply purchasing repetition since purchasing frequency may be resulted from convenience purposes or happenstance buying while multi-brand loyal customers may be not detected due to infrequent purchasing (Jacoby and Kyner 1973; Jacoby and Chestnut 1978). Upon behavioral and psychological components of loyalty, Solomon (1992) and Dick and Basu (1994) distinguish two levels of customer loyalty which are loyalty based on inertia resulted from habits, convenience or hesitance to switch brands and true brand loyalty resulted from conscious decision of purchasing repetition and motivated by positive brand attitudes and highly brand commitment. Obviously, true brand loyalty is what companies want to achieve the most. Recent literature about measuring true brand loyalty reveals different measurement items of customer loyalty, but most of them can be categorized into two dimensions: behavioral and attitudinal brand loyalty (Maxham 2001; Beerli 2002; Teo et al. 2003; Algesheimer et al. 2005; Morrison and Crane 2007). Specifically, behavioral loyalty refers to in-depth commitment to rebuy or consistently favor a particular brand, product or service in the future in spite of influences and marketing efforts that may encourage brand switching. Meanwhile, attitudinal loyalty is driven by the intention to repurchase, the willingness to pay a premium price for the brand, and the tendency to endorse the favorite brand with positive WOM. In this study, true brand loyalty is measured upon both behavioral and attitudinal components using the constructs proposed by Beerli (2002).

The Relationships Linking Brand Communication and Satisfaction, Trust, Perceived Value

Previous studies found that customer satisfaction based on their brand experiences has a significant impact on their satisfaction in communicating with the brands (Grigoroudis and Siskos 2009). Similarly, Angelova and Zekiri (2011) affirmed that customer satisfaction positively affects their readiness and openness to brand communication. In addition, according to Berry and Parasuraman (1991), customers' experience-based beliefs and perceptions about service concept, quality and perceived value towards a brand are so powerful that they can diminish the effects of company-controlled communications that conflict with actual customer experience. In other words, favorable attitudes towards a brand's communication campaigns cannot be achieved without positive evaluation of service that the customers have experienced. Besides, strong brand communication can draw new customers but cannot compensate for a weak service. Moreover, service reliability which is a component of trust in terms of performance or credibility is found to surpass quality of advertising and promotional inducements in affecting customers' attitudes towards brand communication and the brand itself (Berry and Parasuraman 1991).

Since this study focuses on brand communication to current customers who have already experienced the services offered by the brand, it is crucial to view attitudes towards brand communication as an endogenous variable which is influenced by the customers' brand experiences and evaluation such as customer satisfaction, brand trust and perceived value.

Based on the existing literature and the above discussions, the following hypotheses are proposed:

H1: Customer satisfaction has a positive effect on brand communication
H2: Brand trust has a positive effect on brand communication
H3a: Perceived benefit has a positive effect on brand communication
H3b: Perceived cost has a positive effect on brand communication

The Relationship Between Brand Communication and Customer Loyalty
According to Grace and O'Cass (2005), the more favorable feelings and attitudes a consumer forms towards the controlled communications of a brand are, the more effectively the brand messages are transferred. As a result, the favorable consumers' attitudes towards the controlled communications will enhance customers' intention to purchase or repurchase the brand. The direct positive impact of brand communication on customer loyalty has been confirmed in many previous studies (Bansal and Taylor 1999; Jones et al. 2000; Ranaweera and Prabhu 2003; Grace and O'Cass 2005). In line with the existing research, this study hypothesizes that:

H4: Brand communication has a positive effect on customer loyalty

Mediating Role of Customers' Attitude Towards Brand Communications
According to the SWISS Consumer Satisfaction Index Model, two dimensions of customer dialogue including the customers' readiness to engage in the brand's communication initiatives and their satisfaction in communicating with the brand mediate the relationship between customer satisfaction and customer loyalty (Grigoroudis and Siskos 2009). Moreover, Angelova and Zekiri (2011) also point out that customer satisfaction positively affects customer readiness and openness to brand communication in the long term, and how customers react to brand communication will mediate the relationship between customer satisfaction and customer loyalty. To date, there is hardly study which has tested the mediating role of customers' attitudes towards brand communication in the relationship between either brand trust and customer loyalty or perceived value and customer loyalty.

Regarding the mediating role of brand communication, the following hypotheses are proposed:

H5a: Brand communication mediates partially or totally the relationship between brand trust and customer loyalty, in such a way that the greater the brand trust, the greater the customer loyalty
H5b: Brand communication mediates partially or totally the relationship between customer satisfaction and customer loyalty, in such a way that the greater the customer satisfaction, the greater the customer loyalty
H5c: Brand communication mediates partially or totally the relationship between perceived benefit and customer loyalty, in such a way that the greater the perceived value, the greater the customer loyalty
H5d: Brand communication mediates partially or totally the relationship between perceived cost and customer loyalty, in such a way that the greater the perceived value, the greater the customer loyalty

The Relationships Linking Customer Satisfaction, Brand Trust, Perceived Value and Customer Loyalty
In this study, the relationships among customer satisfaction, brand trust, perceived value and customer loyalty in the presence of brand communication are investigated as a part of the proposed model.

Since loyalty is the key metric in relationship marketing, previous studies confirmed various determinants of customer loyalty including customer satisfaction, brand trust and perceived value. Specifically, brand trust is affirmed as an important antecedent to customer loyalty upon various industries (Chaudhuri and Holbrook 2001; Delgado et al. 2003; Agustin and Singh 2005; Bart et al. 2005; Chiou and Droge 2006 and Chinomona 2016). Besides, customer satisfaction is found to positively affect customer loyalty in many studies (Hallowell 1996; Dubrovski 2001; Lam and Burton 2006; Kaura 2013; Saleem et al. 2016). However, according to Andre and Saraviva (2000) and Ganesh et al. (2000), both satisfied and dissatisfied customers have tendency to switch their providers, especially in case of small product differentiation and low customer involvement (Price et al. 1995). On the contrary, all studies about perceived value have confirmed that customers' decision of whether or not to continue the relationship with their providers is made based on evaluation of perceived value or in other words, perceived value has a significant positive impact on customer loyalty (Bolton and Drew 1991; Chang and Wildt 1994; Holbrook 1994; Sirdeshmukh et al. 2002).

In addition, the literature also reveals the relationships among customer satisfaction, perceived value and brand trust. Few studies have shown that perceived value positively affects brand trust (Jirawat and Panisa 2009) and also directly influence customer satisfaction (Bolton and Drew 1991; Jirawat and Panisa 2009). Moreover, the impact of perceived value on customer loyalty is totally mediated via customer satisfaction (Patterson and Spreng 1997). Furthermore, the mediating role of trust on the relationship between customer satisfaction and customer loyalty has also been confirmed (Bee et al. 2012). Based on the above literature review and discussion, the following hypotheses are proposed:

H6: Brand trust positively affects customer loyalty
H7: Customer satisfaction positively affects customer loyalty
H8a: Perceived benefit positively affects customer loyalty
H8b: Perceived cost positively affects customer loyalty
H9: Customer satisfaction positively affects brand trust
H10a: Perceived benefit positively affects brand trust
H10b: Perceived cost positively affects brand trust
H11a: Perceived benefit positively affects customer satisfaction
H11b: Perceived cost positively affects customer satisfaction
H12a: Brand trust mediates partially or totally the relationship between customer satisfaction and customer loyalty, in such a way that the greater the customer satisfaction, the greater the customer loyalty
H12b: Brand trust mediates partially or totally the relationship between perceived benefit and customer loyalty, in such a way that the greater the perceived benefit, the greater the customer loyalty
H12c: Brand trust mediates partially or totally the relationship between perceived cost and customer loyalty, in such a way that the greater the perceived cost, the greater the customer loyalty
H13a: Customer satisfaction mediates partially or totally the relationship between perceived benefit and customer loyalty, in such a way that the greater the perceived benefit, the greater the customer loyalty
H13b: Customer satisfaction mediates partially or totally the relationship between perceived cost and customer loyalty, in such a way that the greater the perceived cost, the greater the customer loyalty

The Mediating Role of Trust in the Relationship Between Each of Perceived Value and Customer Satisfaction and Attitudes Towards Brand Communication
To date, there is hardly study which tested the mediating role of brand trust in the relationship between either customer satisfaction and brand communication or perceived value and brand communication. This study will test the following hypotheses:

H14a: Brand trust mediates partially or totally the relationship between perceived benefit and brand communication, in such a way that the greater the perceived benefit, the greater the brand communication
H14b: Brand trust mediates partially or totally the relationship between perceived cost and brand communication, in such a way that the greater the perceived cost, the greater the brand communication
H14c: Brand trust mediates partially or totally the relationship between customer satisfaction and brand communication, in such a way that the greater the customer satisfaction, the greater the brand communication.

The conceptual model is proposed as shown in Fig. 1 below:

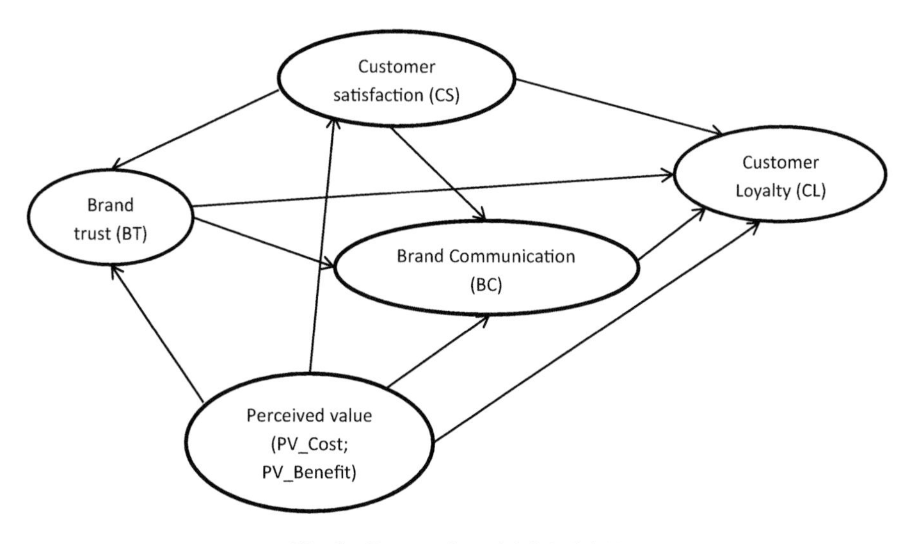

Fig. 2. Proposed model (Model 1)

Model 1's equations are as follows:

$$\begin{cases} CS = \beta_1 PV\_Cost + \beta_2 PV\_Benefit + \varepsilon_{CS} \\ BT = \gamma_1 CS + \gamma_2 PV\_Cost + \gamma_3 PV\_Benefit + \varepsilon_{BT} \\ BC = \phi_1 CS + \phi_2 PV\_Cost + \phi_3 PV\_Benefit + \phi_4 BT + \varepsilon_{BC} \\ CL = \lambda_1 CS + \lambda_2 PV\_Cost + \lambda_3 PV\_Benefit + \lambda_4 BT + \lambda_5 BC + \varepsilon_{CL} \end{cases}$$

3 Research Methodology

In order to test the proposed research model, a quantitative survey was designed. Measurement scales were selected from previous studies in the service industry. Customer attitude towards the controlled communications was measured with six items adapted from Zehir et al. (2011) covering the cognitive (e.g. "The advertising and promotions of this bank are good" and "The advertising and promotions of this bank do good job"); affective (e.g. "I feel positive towards the advertising and promotions of this bank"; "I am happy with the advertising and promotions of this bank" and "I like the advertising and promotions of this bank") and behavioral (e.g. "I react favorably to the advertising and promotions of this bank") aspects of an attitude. Consistent with the conceptualization discussed above, brand trust was scored through three items adapted from Ball (2004) for banking sector which represents overall trust (e.g. "Overall, I have complete trust in my bank") and both of two components of trust including performance or creditability (e.g. "The bank treats me in an honest way in every transaction") and benevolence (e.g. "When the bank suggests that I buy a new product it is because it is best for my situation"). Perceived value was tapped through eleven items proposed by Singh and Sirdeshmukh (2000) and once adapted by Moliner (2009). However, this study categorizes the eleven items into two dimensions of perceived value which are perceived benefit and perceived cost as defined by Zeithaml (1988). As a result, the paths to and from the perceived cost and perceived benefit are tested separately in the proposed model. Customer satisfaction was measured upon the cumulative perspective in which overall customer satisfaction was scored using a five-point Likert-scale from 'Highly Dissatisfied (1)' to 'Highly Satisfied (5)'. Finally, customer loyalty was measured with three items representing both behavioral and attitudinal components as proposed by Beerli (2002) adapted in banking sector.

The questionnaire was translated into Vietnamese and pretested with twenty Vietnamese bank customers so as to make sure its comprehension; easy-to-understand language and phraseology; ease of answering; practicality and length of the survey (Hague et al. 2004). The survey was conducted in Hanoi where is home to majority of both national and foreign banks in Vietnam. Data collection was conducted during March of 2018 through face-to-face with bank customers of at 52 ATM points which were randomly selected from the lists of all ATM addresses disclosed by 25 major banks in Hanoi city. The survey finally yielded 389 usable questionnaires in which 63 percent are filled by female respondents and the rest by male respondents. 82 percent of respondents were aged between 20 and 39 while only 4 percent were from 55 and above. These figures reflect the dominance of the young customer segment in the Vietnam ATM banking market.

4 Results

The guidance on the use of structural equation modeling in practice suggested by Anderson and Gerbing (1988) was adopted to assess the measurement model of each construct before testing the hypothesis. Firstly, exploratory factor analysis (EFA) on SPSS and confirmatory factor analysis (CFA) on AMOS 22 were conducted for testing the convergent validity of measurement items used for each latent variable. Based on statistical results and theoretical backgrounds, some measurement items were dropped from the initial pool of items and only the final selected items were subjected to the further EFA and hypothesis testing.

According to CFA results, items which loaded less than 0.5 should be deleted. Upon this guidance, four items from perceived value's scale were removed from the original set of items. It was verified that the removal of these items did not harm or alter the intention and meaning of the constructs.

After the valid collection of items for perceived value, brand trust, brand communication and customer loyalty was finalized, an exploratory factor analysis was conducted in which five principal factors emerged upon the extraction method followed by varimax rotation. These five factors fitted the initial intended meaning of all constructs in which perceived value items were convergent to two factors representing perceived benefit and perceived cost. The results confirmed the construct validity and demonstrated the unidimensionality for the measurement of constructs (Straub 1989). Table 1 shows the mean, standard deviation (SD), reliability coefficients, and inter-construct correlations for each variable. Since customer satisfaction is measured with only one item, it is treated as an observed variable and there is no reliability coefficient value for it.

Table 1. Mean, SD, reliability and correlation of constructs

	PV_Cost	PV_Benefit	BT	BC	CL	CS	Mean	SD	Reliability
PV_Cost	1						3.11	0.635	0.762
PV_Benefit	0.619	1					3.24	0.676	0.659
BT	0.650	0.550	1				3.15	0.570	0.695
BC	0.518	0.509	0.555	1			3.51	0.495	0.829
CL	0.349	0.290	0.532	0.466	1		3.24	0.690	0.797
CS	0.423	0.314	0.480	0.307	0.571	1	3.48	0.676	___

Table 2. Confirmatory factor analysis results

Construct scale items	Factor loading	t-value
PV_Cost (strongly agree-strongly disagree)		
The money spent is well worth it	0.730	9.193
The service is good for what I pay every month	0.788	9.458
The economic cost is not high	0.632	8.547
The waiting lists are reasonable	0.521	___
PV_Benefit (strongly agree-strongly disagree)		
The installations are spacious, modern and clean	0.674	8.573
It is easy to find and to access	0.598	8.140
The quality was maintained throughout the contact	0.608	___
BC (strongly agree-strongly disagree)		
I react favourably to the advertising and promotions of this bank	0.587	9.066
I feel positive towards the advertising and promotions of this bank	0.729	10.452
The advertising and promotions of this bank are good	0.750	10.625
The advertising and promotions of this bank do good job	0.657	9.791
I am happy with the advertising and promotions of this bank	0.718	10.355
I like the advertising and promotions of this bank	0.576	___
BT (strongly agree-strongly disagree)		
Overall, I have complete trust in my bank	0.710	10.228
When the bank suggests that I buy a new product it is because it is best for my situation	0.601	9.607
The bank treats me in an honest way in every transaction	0.654	___
CL (strongly agree-strongly disagree)		
I do not like to change to another bank because I value the selected bank	0.773	___
I am a customer loyal to my bank	0.779	13.731
I would always recommend my bank to someone who seeks my advice	0.715	12.890

Notes: Measurement model fit details: CMIN/df = 1.911; p = .000; RMR = 0.026; GFI = 0.930; CFI = 0.944; AGFI = 0.906; RMSEA = 0.048; PCLOSE = 0.609; "___" denotes loading fixed to 1

Upon these findings, a CFA was conducted on this six-factor model. The results from AMOS 22 revealed a good model fit (CMIN/df = 1.911; p = .000; RMR = 0.026; GFI = 0.930; CFI = 0.944; AGFI = 0.906; RMSEA = 0.048; PCLOSE = 0.609). The factor loadings and t -values resulted from the CFA are presented in Table 2. The table demonstrates confirmation of convergent validity for the measurement constructs since all factor loadings were statistically significant and higher than the cut-off value of 0.4 suggested by Nunnally and Bernstein (1994).

Among six factors, two factors which are perceived cost and brand communication had Average Variance Extracted (AVE) value slightly lower than the recommended level of 0.5 indicating low convergent validity. However, all of AVE values are greater than the square of correlations between each two constructs. Therefore, the discriminant validity of the constructs was still confirmed.

Overall, the EFA confirmed the unidimensionality of the constructs and the CFA indicated their significant convergent and discriminant validity. Therefore, this study retains the constructs with its measurement items as shown in Table 2 to conduct the hypothesis testing (Table 3).

Table 3. Average variance extracted and discriminant validity test

	PV_Cost	PV_Benefit	BC	BT	CL
PV_Cost	0.497				
PV_Benefit	0.383	0.530			
BC	0.268	0.259	0.488		
BT	0.422	0.302	0.308	0.647	
CL	0.121	0.084	0.217	0.283	0.503

Figure 2 shows the proposed model of hypothesized relationships which were tested through a path analysis procedure conducted in AMOS 22. This analysis method is recommended by (Oh 1999) to allow both direct and indirect relationships indicated in the model are simultaneously estimated and thereby, the significance and magnitude of all hypothesized interrelationships among all variables presented in one framework can be tested. The model fit indicators suggested by AMOS 22 shows that the proposed model reflects a reasonably good fit to the data.

Table 4 exhibits the path coefficients in the original proposed model and modified models. Since the interrelationships of attitude towards brand communication with other variables and their impacts on customer loyalty are the primary focuses of this research, the coefficients of paths to and from brand communication and paths to customer loyalty are placed first.

Table 4. Path coefficients

Construct path	Coefficients	Model 1 (original)	Model 2 (without BC)	Model 3 (without BT)	Model 4 (without CS)	Model 5 (without BC, BT and CS)
PV_Cost to BC	ϕ_2	0.158		0.292*	0.158	
PV_Benefit to BC	ϕ_3	0.167*		0.216*	0.166*	
BT to BC	ϕ_4	0.244*			0.254*	
CS to BC	ϕ_1	0.008		0.052		
BC to CL	λ_5	0.417**		0.525**	0.430**	
PV_Cost to CL	λ_2	−0.177	−0.113	−0.021	−0.056	0.421*
PV_Benefit to CL	λ_3	−0.077	−0.006	−0.026	−0.081	0.141
BT to CL	λ_4	0.359*	0.458**		0.540**	
CS to CL	λ_1	0.384*	0.387*	0.444**		
PV_Cost to CS	β_1	0.603**	0.599**	0.615**		
PV_Benefit to CS	β_2	0.104	0.107	0.108		
PV_Cost to BT	γ_2	0.513**	0.527*		0.608*	
PV_Benefit to BT	γ_3	0.207*	0.201*		0.226*	
CS to BT	γ_1	0.179*	0.186**			
Fit indices						
CMIN/df		1.911	1.967	1.993	1.946	2.223
CFI		0.944	0.959	0.949	0.943	0.963
GFI		0.930	0.954	0.939	0.931	0.966
AGFI		0,906	0.929	0.916	0.908	0.941
RMR		0.026	0.028	0.026	0.027	0.03
RMSEA		0.048	0.05	0.051	0.049	0.056
PCLOSE		0.609	0.487	0.447	0.534	0.264

Notes: *$p < 0.05$ and **$p < 0.001$

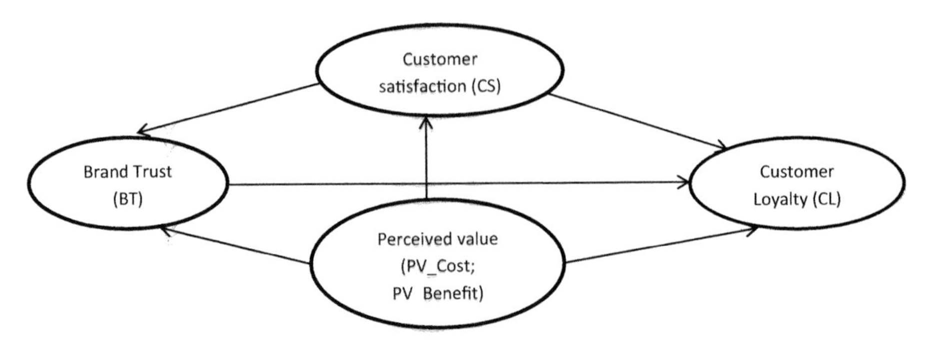

Fig. 3. Model 2

Model 2's equations are as follow:

$$\begin{cases} CS = \beta_1 PV\_Cost + \beta_2 PV\_Benefit + \varepsilon_{CS} \\ BT = \gamma_1 CS + \gamma_2 PV\_Cost + \gamma_3 PV\_Benefit + \varepsilon_{BT} \\ CL = \lambda_1 CS + \lambda_2 PV\_Cost + \lambda_3 PV\_Benefit + \lambda_4 BT + \varepsilon_{CL} \end{cases}$$

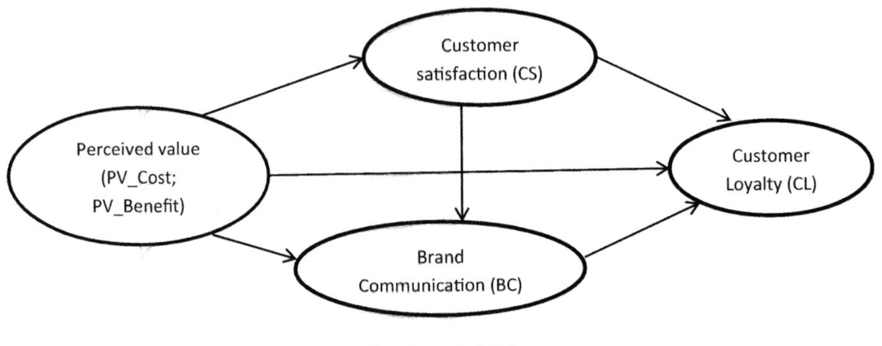

Fig. 4. Model 3

Model 3's equations are as follow:

$$\begin{cases} CS = \beta_1 PV\_Cost + \beta_2 PV\_Benefit + \varepsilon_{CS} \\ BC = \phi_1 CS + \phi_2 PV\_Cost + \phi_3 PV\_Benefit + \varepsilon_{BC} \\ CL = \lambda_1 CS + \lambda_2 PV\_Cost + \lambda_3 PV\_Benefit + \lambda_5 BC + \varepsilon_{CL} \end{cases}$$

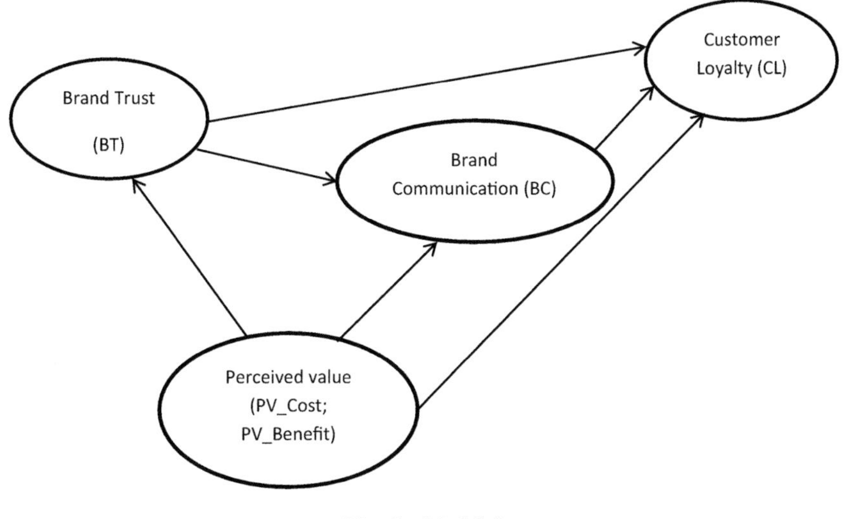

Fig. 5. Model 4

Model 4's equations are as follow:

$$\begin{cases} BT = \gamma_2 PV\_Cost + \gamma_3 PV\_Benefit + \varepsilon_{BT} \\ BC = \phi_2 PV\_Cost + \phi_3 PV\_Benefit + \phi_4 BT + \varepsilon_{BC} \\ CL = \lambda_2 PV\_Cost + \lambda_3 PV\_Benefit + \lambda_4 BT + \lambda_5 BC + \varepsilon_{CL} \end{cases}$$

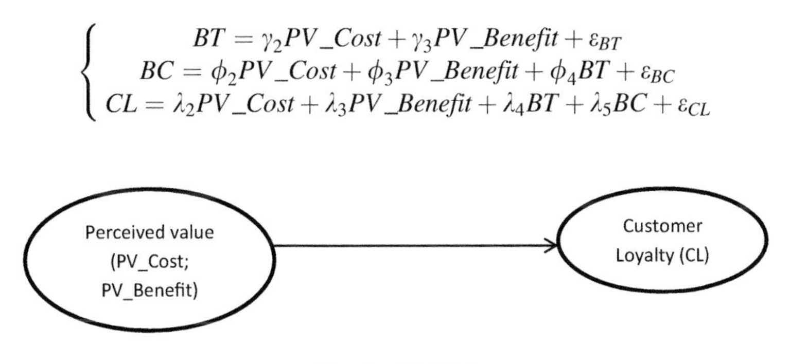

Fig. 6. Model 5

Model 5's equation is as follow:

$$CL = \lambda_2 PV\_Cost + \lambda_3 PV\_Benefit + \varepsilon_{CL}$$

Among the paths to brand communication, it is found that each of perceived benefit and brand trust has a positive effect on brand communication (support H2 and H3a) whereas the effects of perceived cost and customer satisfaction on brand communication were both not significant (reject H1, H3b, H14c). Brand communication, in turn, has a positive effect on customer loyalty (support H4). Similarly, customer satisfaction and brand trust also have direct significant positive effects on customer loyalty (support H6 and H7). In accordance to other studies' findings, the results also revealed that customer satisfaction has a significant positive impact on brand trust (support H9).

With regards to the relationships between perceived value and brand trust or customer satisfaction which have been tested in many previous researches, the findings demonstrated a closer look on the effect of two principal factors of perceived value, perceived cost and perceived benefit on brand trust and customer satisfaction. Specifically, perceived cost has a significant direct effect on customer satisfaction and brand trust (support H10b and H11b). The same direct effect has not seen in the case of perceived benefit (reject H10a and H11a).

In the original proposed model, there are three hypothesized mediators to be tested including brand communication, brand trust and customer satisfaction. In order to test the mediating roles of these variables, different models (Model 2, Model 3, Model 4 and Model 5) shown Figs. 3, 4, 5 and 6 were tested so that the strength of relationships among variables were compared with those in the original full Model 1.

Specifically, Model 2 which excludes brand communication is compared with Model 1 (the original model) to test the mediating role of brand communication. Similarly, Model 3, Model 4 and Model 5 present the removal of brand trust or customer satisfaction or all of brand communication, brand trust and customer satisfaction accordingly so that they are compared with Model 1 to test the mediating roles of brand trust, customer satisfaction or all of brand communication, brand trust, and customer satisfaction together. Table 4 presents the comparison of coefficients resulted from each model.

Comparing data of Model 1 and those of Model 2, it is found that:

- Both customer satisfaction and brand trust have significant positive effects on customer loyalty in Model 1 and Model 2
- In the absence of brand communication, the effect brand trust has on customer loyalty is greater than that in the presence of brand communication
- Customer satisfaction has no significant effect on brand communication and whether brand communication is included in the model or not, the effect that customer satisfaction has on customer loyalty is nearly unchanged

Based on the above findings and the mediating conditions suggested by Baron and Kenny (1986), it is concluded that the relationship between brand trust and customer loyalty is partially mediated by brand communication, and therefore supports H5a in such a way that the greater the trust, the greater the loyalty. However, brand communication is not the mediator in the relationship between customer satisfaction and customer loyalty (reject H5b)

In comparison of data from Model 1 and those of Model 3, it is found that:

- Customer satisfaction has a positive significant effect on customer loyalty in both Model 1 and Model 3. In the absence of brand trust, the effect customer satisfaction has on customer loyalty is greater than that in the presence of brand trust
- Perceived benefit has a positive significant effect on brand communication in both Model 1 and Model 3. In the absence of brand trust, the effect perceived benefit has on brand communication is greater than that in the presence of brand trust
- In the full Model 1, perceived cost has no significant effect on brand communication but when brand trust is removed or in Model 3, perceived cost has proven to have significant positive effect on brand communication

Based on the above results and the mediating conditions suggested by Baron and Kenny (1986), it is concluded that:

- The relationship between customer satisfaction and customer loyalty is partially mediated by brand trust in such a way that the greater the customer satisfaction, the greater the customer loyalty (support H12a)
- The relationship between perceived benefit and brand communication is partially mediated by brand trust and the relationship between perceived cost and brand communication is totally mediated by brand trust in such a way that the greater the perceived cost, the greater the brand communication (support H14a and H14b)

In comparison of data from Model 1, Model 2, Model 3, Model 4 and Model 5, it is found that both perceived cost and perceived benefit have no significant effect on customer loyalty when each of brand communication, brand trust or customer satisfaction is absent. Only when all of brand communication, brand trust and customer satisfaction are removed from the original full model, perceived cost is proven to have a significant positive effect on customer loyalty whereas the same relationship between perceived benefit and customer loyalty was not seen. Actually, we even tested the relationships between each of perceived cost and perceived benefit and customer loyalty in three more models when each pair of brand trust and customer satisfaction, brand communication and customer satisfaction and brand trust and brand communication are absent but no significant effect was found. Based on this finding, we concluded that only perceived cost has a significant positive effect on customer loyalty (support a part of H8b). In addition, the relationship perceived cost and customer loyalty is totally mediated by three variables which are brand trust, customer satisfaction and brand communication (support H5d, H12c and H13b). However, perceived benefit has no effect on customer loyalty (reject H8a, H5c, H12b and H13a)

5 Discussion and Managerial Implication

This research provides insights into the relationships among perceived value, brand trust, customer satisfaction, customer loyalty and attitude towards brand communication. In contrast with previous studies in which brand communication is regarded as an exogenous variable whose direct effect on customer satisfaction, customer loyalty and brand trust were analyzed separately, this study was based on the conceptual framework drawn from the Swiss Consumer Satisfaction model to view attitude towards brand communication as an endogenous variable which may be affected by customer satisfaction, perceived value or customer trust resulted from customer experience with the brand. Specifically, this study examined the combined impacts of customer satisfaction, perceived value or customer trust on brand communication and the mediating role of brand communication in the relationships between such variables and customer loyalty. Moreover, it also took closer to the interrelationships among perceived value, brand trust, customer satisfaction and customer loyalty in which two principal factors of perceived value, perceived costs and benefits, are treated as two separate variables and test the mediating effects of perceived benefit, perceived cost and customer satisfaction to customer loyalty, all in one single model.

The results reveal that attitude towards brand communication is significantly influenced by brand trust and perceived value in terms of both perceived cost and perceived benefit in which brand trust has a mediating effect on the relationship between perceived value and brand communication. In addition, attitude towards brand communication has both an independent effect as well as a mediating effect on customer loyalty through customer trust and perceived cost. The indirect effect of perceived cost on customer loyalty through attitude towards brand communication may be more due to calculative commitment, whereas indirect effect of trust on customer loyalty though attitudes towards brand communication as well as the direct effect of attitudes towards brand communication on customer loyalty may be more from affective commitment (Bansal et al. 2004). This finding extends previous studies on brand communication treating it as a factor aiding customer loyalty independent of existing brand attitudes and perceived value. Contrary to expectation and the suggestion of the Swiss Customer Satisfaction Index, the direct relationship between customer satisfaction and attitude toward brand communication was not found significant. This may be because of the particular context in which this relationship was tested upon Vietnamese customers in the Vietnam ATM service industry. This finding implies that the banks still have opportunities for service recovery and gain back customer loyalty since it is likely that even disappointed customers are still open to brand communication and expect something better from their banks.

This study also supports and expands some other important relationships that have already been empirically studied in several other contexts. These relationships concern the linkages among perceived value, brand trust, customer satisfaction and customer loyalty. Brand trust was found to play the key role in the nature of the relationship between either customer satisfaction or perceived value and customer loyalty since it not only has a direct impact on customer loyalty but also mediates totally the effect of perceived value and customer loyalty as well as mediates partially the relationship between customer satisfaction and customer loyalty. However, this study provides a further understanding about the role of perceived value with two separate principal factors including perceived benefit and perceived cost in which only perceived cost has a direct effect on customer satisfaction, brand trust and customer loyalty in this particular Vietnam ATM banking service context while such effects of perceived benefit were not found.

The findings of this study are significant from the point of view of both academic researchers and the marketing practitioners, especially advertisers as they describe the impacts of controllable variables on attitude vis-à-vis brand communication and customer loyalty in the banking industry. The study points out the multiple paths to customer loyalty from customer satisfaction and perceived value through brand trust and how customers react to marketing communication activities of banks. Overall, the findings suggest that the banks may benefit from pursuing a combined strategy of increasing brand trust and encouraging positive attitudes towards brand communication both independently and in tandem. The attitude vis-à-vis brand communication should be managed like perceived value and customer satisfaction in anticipating and enhancing customer loyalty. In addition, by achieving high brand trust through higher satisfaction and better value provisions for ATM service, the banks can trigger more positive attitudes and favorable reactions towards their marketing communication efforts for other banking services, thereby, further aiding customer loyalty. This has an important management implication, especially in Vietnam banking service market where customers are bombarded by promotional offers from many market players which aim at capturing existing customers of other service providers and even satisfied customers consider switching to the new provider. Moreover, since perceived value is formed by two principal factors including perceived costs and perceived benefits, it is crucial to separate them when analyzing the impact of perceived value on other variables since their effects may be totally different. In this particular ATM service in Vietnam where the banks provides similar benefits to customers, only perceived costs determine customers' satisfaction, brand trust and customer loyalty. With the knowledge of various paths to customer loyalty and determinants of attitude towards brand communication, the banks are able to design alternative strategies to improve its marketing communication effectiveness aimed at strengthening customer loyalty.

Limitations and Future Research

This study faces some limitations. First, the data are collected from only business to customer market of a single ATM service industry while perceived value, trust, customer satisfaction and especially attitude towards brand communication in various contexts may be different. Second, regarding sample size, although suitable sampling methods with adequate sample representation were used, a larger sample size with wider age range may be more helpful and effective for the path analysis and managerial implication. Third, this study adopted only a limited set of measurement items due to concerns about model parsimony and data collection efficiency. For example, customer satisfaction may be measured as a latent variable with multiple dimensions; this research considered it as an observed variable. Besides, perceived value can be measured upon even 5 factors, this study focused only on some selected measures based mainly on their relevance to the context studied.

Further studies could also look at the perceived value in the relationships concerned with attitude towards brand communication, customer loyalty, customer satisfaction or brand trust with the full six dimensions of perceived value suggested by the GLOVAL scale (Sanchez et al. 2006) including functional value of the establishment (installations), functional value of the contact personnel (professionalism), functional value of the service purchased (quality) and functional value price. Besides, future studies which separate different types of promotional tools in analyzing the relationship between attitude towards brand communication and other variables may draw more helpful implication for advertisers and business managers. Moreover, future research could also investigate these relationships in different product or market contexts where the nature of customer loyalty may be different.

References

Agustin, C., Singh, J.: Curvilinear effects of consumer loyalty determinants in relational exchanges. J. Mark. Res. **8**, 96–108 (2005)

Algesheimer, R., Dholakia, U.M., Herrmann, A.: The social influence of brand community; evidence from European car clubs. J. Mark. **69**, 19–34 (2005)

Anderson, J.C., Gerbing, D.W.: Structural equation modeling in practice: a review and recommended two-step approach. Psychol. Bull. **103**, 411–423 (1988)

Anderson, E.W., Fornell, C., Lehmann, R.R.: Customer satisfaction, market share, and profitability: findings from Sweden. J. Mark. **58**, 53–66 (1994)

Andre, M.M., Saraviva, P.M.: Approaches of Portuguese companies for relating customer satisfaction with business results. Total Qual. Manag. **11**(7), 929–939 (2000)

Andreassen, T.W.: Antecedents to satisfaction with service recovery. Eur. J. Mark. **34**, 156–175 (2000)

Angelova, B., Zekiri, J.: Measuring customer satisfaction with service quality using American Customer Satisfaction Model (ACSI Model). Int. J. Acad. Res. Bus. Soc. Sci. **1**(3), 232–258 (2011)

Beerli, A., Martın, J.D., Quintana, A.: A model of customer loyalty in the retail banking market. Las Palmas de Gran Canaria (2002)

Bansal, H.S., Taylor, S.F.: The service provider switching model (SPSM): a model of consumer switching behaviour in the service industry. J. Serv. Res. **2**(2), 200–218 (1999)

Bansal, H., Voyer, P.: Word-of-mouth processes within a service purchase decision context. J. Serv. Res. **3**(2), 166–177 (2000)

Bansal, H.P., Irving, G., Taylor, S.F.: A three component model of customer commitment to service providers. J. Acad. Mark. Sci. **32**, 234–250 (2004)

Baron, R.M., Kenny, D.A.: The moderator – mediator variable distinction in social psychological research: conceptual, strategic, and statistical considerations. J. Pers. Soc. Psychol. **51**(6), 1173–1182 (1986)

Bart, Y., Shankar, A., Sultan, F., Urban, G.L.: Are the driandrs and role of online trust the same for all web sites and consumers? A large-scale exploratory empirical study. J. Mark. **69**, 133–152 (2005)

Bee, W.Y., Ramayah, T., Wan, N., Wan, S.: Satisfaction and trust on customer loyalty: a PLS approach. Bus. Strategy Ser. **13**(4), 154–167 (2012)

Berry, L.L., Parasuraman, A.: Marketing Services: Competing Through Quality. The Free Press, New York (1991)

Bolton, R.N., Drew, J.H.: A multistage model of customers' assessment of service quality and value. J. Consum. Res. **17**, 375–384 (1991)

Boulding, W., Kalra, A., Staelin, R., Zeithaml, V.A.: A dynamic process model of service quality: from expectations to behavioral intentions. J. Mark. Res. **30**, 7–27 (1993)

Bruhn, M., Grund, M.: Theory, development and implementation of national customer satisfaction indices: the Swiss Index of Customer Satisfaction (SWICS). Total Qual. Manag. **11**(7), 1017–1028 (2000)

Chang, T.Z., Wildt, A.R.: Price, product information, and purchase intention: an empirical study. J. Acad. Mark. Sci. **22**, 16–27 (1994)

Chaudhuri, A., Holbrook, B.M.: The chain of effects from brand trust and brand affects to brand performance: the role of brand loyalty. J. Mark. **65**, 81–93 (2001)

Chiou, J.S., Droge, C.: Service quality, trust, specific asset investment, and expertise: direct and indirect effects in a satisfaction-loyalty framework. J. Acad. Mark. Sci. **34**(4), 613–627 (2006)

Chinomona, R.: Brand communication, brand image and brand trust as antecedents of brand loyalty in Gauteng Province of South Africa. Afr. J. Econ. Manag. Stud. **7**(1), 124–139 (2016)

Cronin, J.J., Brady, M.K., Hult, G.T.M.: Assessing the effects of quality, value, and customer satisfaction on consumer behavioral intentions in service environments. J. Retail. **76**(2), 193–218 (2000)

De Ruyter, K., Wetzels, M., Lemmink, J., Mattson, J.: The dynamics of the service delivery process: a value-based approach. Int. J. Res. Mark. **14**(3), 231–243 (1997)

Delgado, E., Munuera, J.L., Yagüe, M.J.: Development and validation of a brand trust scale. Int. J. Mark. Res. **45**(1), 35–54 (2003)

Dick, A.S., Basu, K.: Customer loyalty towards an integrated framework. J. Acad. Mark. Sci. **22** (2), 99–113 (1994)

Doney, P.M., Cannon, J.P.: An examination of the nature of trust in buyer-seller relationships. J. Mark. **61**, 35–51 (1997)

Dubrovski, D.: The role of customer satisfaction in achieving business excellence. Total Qual. Manag. Bus. Excel. **12**(7–8), 920–925 (2001)

Ball, D., Coelho, P.S., Machás, A.: The role of communication and trust in explaining customer loyalty: an extension to the ECSI model. Eur. J. Mark. **38**(9/10), 1272–1293 (2004)

Ennew, C., Banerjee, A.K., Li, D.: Managing word of mouth communication: empirical evidence from India. Int. J. Bank Mark. **18**(2), 75–83 (2000)

Fornell, C.: A national customer satisfaction barometer: the Swedish experience. J. Mark. **56**(1), 6–21 (1992)

Fornell, C., Johnson, M.D., Anderson, E.W., Cha, J., Everitt Bryant, B.: Growing the trust relationship. J. Mark. **60**(4), 7–18 (1996)

Ganesan, S.: Determinants of long-term orientation in buyer-seller relationships. J. Mark. **58**(2), 1–19 (1994)

Ganesh, J., Arnold, M.J., Reynolds, K.E.: Understanding the customer base of service providers: an examination of the differences between switchers and stayers. J. Mark. **64**, 65–87 (2000)

Garbarino, E., Johnson, M.K.: The different roles of satisfaction, trust and commitment in customer relationships. J. Mark. **63**, 70–87 (1999)

Grace, D., O'Cass, A.: Examining the effects of service brand communications on brand evaluation. J. Prod. Brand Manag. **14**(2), 106–116 (2005)

Grewal, D., Parasuraman, A., Voss, G.: The roles of price, performance and expectations in determining satisfaction in service exchanges. J. Mark. **62**(4), 46–61 (1998)

Grigoroudis, E., Siskos, Y.: Customer Satisfaction Evaluation: Methods for Measuring and Implementing Service Quality. Springer Science & Business Media (2009)

Gupta, S., Zeithaml, V.: Customer metrics and their impact on financial performance. Mark. Sci. **25**(6), 718–739 (2006)

Hallowell, R.: The relationship of customer satisfaction, customer loyalty, and profitability: an empirical study. Int. J. Serv. Ind. Manag. **7**(4), 27–42 (1996)

Halstead, D., Hartman, D., Schmidt, S.L.: Multisource effects on the satisfaction formation process. J. Acad. Mark. Sci. **22**(2), 114–129 (1994)

Hague, P.N., Hague, N., Morgan, C.: Market Research in Practice: A Guide to the Basics. Kogan Page Publishers, London (2004)

Holbrook, M.B.: The nature of customer value. In: Rust, R.T., Oliver, R.L. (eds.) Service Quality: New Directions in Theory and Practice, pp. 21–71. Sage Publications, London (1994)

Jacoby, J., Kyner, R.: Brand Loyalty: Measurement and Management. John Wiley & Sons, New York (1973)

Jacoby, J., Chestnut, R.W.: Brand Loyalty: Measurement and Management. Wiley & Sons, New York, NY (1978)

Jirawat, A., Panisa, M.: The impact of perceived value on spa loyalty and its moderating effect of destination equity. J. Bus. Econ. Res. **7**(12), 73–90 (2009)

Jones, M.A., Mothersbaugh, D.L., Beatty, S.E.: Switching barriers and repurchase intentions in services. J. Retail. **76**(2), 259–274 (2000)

Johnson, M.D., Fornell, C.: A framework for comparing customer satisfaction across individuals and product categories. J. Econ. Psychol. **12**, 267–286 (1991)

Johnson, M.D., Gustafsson, A., Andreason, T.W., Lervik, L., Cha, G.: The evolution and future of national customer satisfaction index models. J. Econ. Psychol. **22**, 217–245 (2001)
Kaura, V.: Antecedents of customer satisfaction: a study of Indian public and private sector banks. Int. J. Bank Mark. **31**(3), 167–186 (2013)
Keller, K.L., Lehmann, D.R.: Brands and branding: research findings and future priorities. Mark. Sci. **25**(6), 740–759 (2006)
Krepapa, A., Berthon, P., Webb, D., Pitt, L.: Mind the gap: an analysis of service provider versus customer perception of market orientation and impact on satisfaction. Eur. J. Mark. **37**, 197–218 (2003)
Lam, R., Burton, S.: SME banking loyalty (and disloyalty): a qualitative study in Hong Kong. Int. J. Bank Mark. **24**(1), 37–52 (2006)
Mattson, J.: Better Business by the ABC of Values. Studentliteratur, Lund (1991)
Maxham, J.G.I.: Service recovery's influence on consumer satisfaction, word-of-mouth, and purchase intentions. J. Bus. Res. **54**, 11–24 (2001)
Moliner, M.A.: Loyalty, perceived value and relationship quality in healthcare services. J. Serv. Manag. **20**(1), 76–97 (2009)
Moliner, M.A., Sa´nchez, J., Rodrı´guez, R.M., Callarisa, L.: Dimensionalidad del Valor Percibido Global de una Compra. Revista Espan˜ ola de Investigacio´ n de Marketing Esic **16**, 135–158 (2005)
Morrison, S., Crane, F.: Building the service brand by creating and managing an emotional brand experience. J. Brand Manag. **14**(5), 410–421 (2007)
Ndubisi, N.O., Chan, K.W.: Factorial and discriminant analyses of the underpinnings of relationship marketing and customer satisfaction. Int. J. Bank Mark. **23**(3), 542–557 (2005)
Ndubisi, N.O.: A structural equation modelling of the antecedents of relationship quality in the Malaysia banking sector. J. Financ. Serv. Mark. **11**, 131–141 (2006)
Nunnally, J.C., Bernstein, I.H.: Psychometric Theory, 3rd edn. McGraw-Hill, New York (1994)
Oh, H.: Service quality, customer satisfaction, and customer value: a holistic perspective. Int. J. Hosp. Manag. **18**(1), 67–82 (1999)
Oliver, R.L.: Whence consumer loyalty? J. Mark. **63**(4), 33–44 (1999)
Parasuraman, A.: Reflections on gaining competitive advantage through customer value. J. Acad. Mark. Sci. **25**(2), 154–161 (1997)
Patterson, P.G., Spreng, R.W.: Modelling the relationship between perceived value, satisfaction, and repurchase intentions in business-to-business, services context: an empirical examination. J. Serv. Manag. **8**(5), 414–434 (1997)
Phan, N., Ghantous, N.: Managing brand associations to drive customers' trust and loyalty in Vietnamese banking. Int. J. Bank Mark. **31**(6), 456–480 (2012)
Price, L., Arnould, E., Tierney, P.: Going to extremes: managing service encounters and assessing provider performance. J. Mark. **59**(2), 83–97 (1995)
Ranaweera, C., Prabhu, J.: The influence of satisfaction, trust and switching barriers on customer retention in a continuous purchase setting. Int. J. Serv. Ind. Manag. **14**(4), 374–395 (2003)
Runyan, R.C., Droge, C.: Small store research streams: what does it portend for the future? J. Retail. **84**(1), 77–94 (2008)
Rust, R.T., Oliver, R.L.: Service quality: insights and managerial implication from the frontier. In: Rust, R., Oliver, R.L. (eds.) Service Quality: New Directions in Theory and Practice, pp. 1–19. Sage, Thousand Oaks (1994)
Saleem, M.A., Zahra, S., Ahmad, R., Ismail, H.: Predictors of customer loyalty in the Pakistani banking industry: a moderated-mediation study. Int. J. Bank Mark. **34**(3), 411–430 (2016)
Sanchez, J., Callarisa, L.L.J., Rodrı´guez, R.M., Moliner, M.A.: Perceived value of the purchase of a tourism product. Tour. Manag. **27**(4), 394–409 (2006)

Sahin, A., Zehir, C., Kitapçi, H.: The effects of brand experiences, trust and satisfaction on building brand loyalty; an empirical research on global brands. In: The 7th International Strategic Management Conference, Paris (2011)
Sekhon, H., Ennew, C., Kharouf, H., Devlin, J.: Trustworthiness and trust: influences and implications. J. Mark. Manag. **30**(3–4), 409–430 (2014)
Sheth, J.N., Parvatiyar, A.: Relationship marketing in consumer markets: antecedents and consequences. J. Acad. Mark. Sci. **23**(4), 255–271 (1995)
Singh, J., Sirdeshmukh, D.: Agency and trust mechanisms in customer satisfaction and loyalty judgements. J. Acad. Mark. Sci. **28**(1), 150–167 (2000)
Sirdeshmukh, D., Singh, J., Sabol, B.: Consumer trust, value, and loyalty in relational exchanges. J. Mark. **66**, 15–37 (2002)
Solomon, M.R.: Consumer Behavior. Allyn & Bacon, Boston (1992)
Straub, D.: Validating instruments in MIS research. MIS Q. **13**(2), 147–169 (1989)
Sweeney, J.C., Soutar, G.N., Johnson, L.W.: Are satisfaction and dissonance the same construct? A preliminary analysis. J. Consum. Satisf. Dissatisf. Complain. Behav. **9**, 138–143 (1996)
Sweeney, J., Soutar, G.N.: Consumer perceived value: the development of a multiple item scale. J. Retail. **77**(2), 203–220 (2001)
Teo, H.H., Wei, K.K., Benbasat, I.: Predicting intention to adopt interorganizational linkages: an institutional perspective. MIS Q. **27**(1), 19–49 (2003)
Woodruff, R.: Customer value: the next source for competitive advantage. J. Acad. Mark. Sci. **25** (2), 139–153 (1997)
Zehir, C., Sahn, A., Kitapci, H., Ozsahin, M.: The effects of brand communication and service quality in building brand loyalty through brand trust; the empirical research on global brands. In: The 7th International Strategic Management Conference, Paris (2011)
Zeithaml, V.A.: Consumer perceptions of price, quality, and value: a means-end model and synthesis of evidence. J. Mark. **52**, 2–22 (1988)

Measuring Misalignment Between East Asian and the United States Through Purchasing Power Parity

Cuong K. Q. Tran[1(✉)], An H. Pham[1], and Loan K. T. Vo[2]

[1] Faculty of Economics, Van Hien University, Ho Chi Minh City, Vietnam
cuong.tqk@vnp.edu.vn , hoangan.tcnh@gmail.com
[2] HCM City Open University, Ho Chi Minh City, Vietnam
loan.vtk@ou.edu.vn

Abstract. The aim of this research is to measure the misalignment between East Asian countries and the United States using Dynamic Ordinary Least Square through Purchasing Power Parity (PPP) approach. Unit root test, Johansen Co-integraion test, Vector Error Correction Model are employed to investigate the relationship of PPP between these countries. The results indicate that only four countries namely, Vietnam, Indonesia, Malaysia and Singapore, have the existence of purchasing power parity with the United States. The exchange rate residual implies that the fluctuation of misalignment depends on the exchange rate regime such as in Singapore. In addition, it indicates that all domestic currencies experience a downward trend and are overvalued before the financial crisis. After this period, all currencies fluctuate. Currently, only Indonesian currency is undervalued in comparison to USD.

Keywords: PPP · Real exchange rate · VECM
Johansen cointegration test · Misalignment · DOLS

1 Introduction

Purchasing Power Parity (PPP) is one of the most interesting issues in international finance and it has crucial influence on economies. Firstly, using PPP enables economists to forecast the exchange rate in long-term and short-term course because exchange rate tends to move in the same direction of PPP. The valuation of real exchange rate is very important for developing countries like Vietnam. Kaminsky et al. (1998) and Chinn (2000) state that the appreciation of the exchange rate can lead to the crisis of emerging economies. It also affects not only on international commodity market but also international finance. Therefore, policy makers and managers of enterprises should have suitable plans and strategies to deal with the situation of exchange rate volatility. Secondly, exchange rate is very important to trade balance or balance of payment of a country. Finally, PPP helps to change economies ranking via adjusting

V. Kreinovich et al. (Eds.): ECONVN 2019, SCI 809, pp. 402–416, 2019.
https://doi.org/10.1007/978-3-030-04200-4_29

Gross Domestic Product per Capita. As a consequence, the existence of PPP has become one of the most controversial issues in the world. In short, PPP is a good indicator for policy makers, multinational enterprises and exchange rate market participants to have suitable strategies to develop.

However, the existence of PPP is still questionable. Coe and Serletis (2002), Tastan (2005) and Kavkler et al. (2016) find that the PPP does not exist. Nevertheless, Baharumshah et al. (2010), Dilem (2017) claim the relationship between Turkey and his main trading partners. It is obvious that the results of PPP depend on countries; currencies and methodologies which are used to conduct research

In this paper, the authors aim to find out the existence of PPP between East Asian countries and the United States. After that, they will measure the misalignment between these countries and United States.

This paper includes four sections: Sect. 1 presents the introduction, Sect. 2 reviews the literature for PPP approach; Sect. 3 describes the methodology and data collecting procedure; and Sect. 4 provides results and discussion.

2 Literature Review

Salamanca School in Spain was the first school to introduce the PPP in the 16th century. At that time, the meaning of PPP was basically about the price level of every country that should be the same when the common currency was changed (Rogoff 1996). PPP was then introduced by Cassel in 1918. After that, PPP became the benchmark for a central bank in building up the exchange rates and the resources for studying about exchange rate determinants. Balassa and Samuelson then were inspired by Cassel's PPP model when setting up their models in 1964. They worked independently and provided the final explanation of the establishment of the exchange rate theory based on the absolute PPP (Asea and Corden 1994). It can be explained that when any amount of money is exchanged into the same currency, the relative price of each good in different countries should be the same. There are two versions of PPP, namely the absolute and relative PPP (Balassa 1964). According to the first version, Krugman et al. (2012) define the absolute PPP as the exchange rate of pair countries equal to the ratio of the price level of those countries, meaning as follows:

$$s_t = \frac{p_t}{p_t^*} \tag{1}$$

On the other hand, Shapiro (1983) states that the relative PPP can be defined as the ratio of domestic to foreign prices equal to the ratio change in the equilibrium exchange rate. There is a constant k modifying the relationship between the equilibrium exchange rate and price levels, as presented below:

$$s_t = k * \frac{p_t}{p_t^*}$$

In the empirical studies, checking the validity of PPP by unit root test was popular in 1980s based on Dickey and Fuller approach, nevertheless, this approach has the low power (Ender & Granger 1998).

After that, Johansen (1988) developed a method of conducting VECM, which has become the benchmark model for many authors to test PPP approach.

The studies of PPP approach have linear and nonlinear models. With the linear model, it can be seen that almost papers use the cointegration test, the Vector Error Correction Model (VECM), or unit root test to check whether or not all variables move together or their means are reverted. With the latter, most studies apply the STAR-family model (Smooth Transition Auto Regressive) and then use the nonlinear unit root test for the real exchange rate in the nonlinear model framework.

2.1 Linear Model for PPP Approach

The stationary of real exchange rate by using unit root test was tested by Tastan (2005) and Narayan in 2005. At the same time, there was an attempt from Tastam to search for the stationary of real exchange rate between Turkey and four other partners: the US, England, Germany, and Italy. From 1982 to 2003, the empirical result stated non-stationary in the long run between Turkey and the US, Turkey and England as well. While this author just used single country, Narayan examined 17 OECD countries in which his results were different If he uses currencies based on the US dollar, the three countries, France, Portugal and Denmark, will be satisfied. If the usage of currency is German based, Deutschmark, seven countries will be satisfied. In addition, univariate techniques were applied to find out the equilibrium of the real exchange rate. However, Kremers et al. (1992) argued that technique might suffer low power against multivariate approach because the deception of improper common factor could be limited in the ADF test.

After Johansen's development of a method of conducting VECM in 1988, there has been various papers applied it to test PPP. Therefore, Chinn (2000) estimated whether the East Asian currencies were overvalued or undervalued with VECM. The results showed that the currencies of Hong Kong, Indonesia, Thailand, Malaysia, the Philippines and Singapore were overvalued. Duy et al. (2017) indicated the PPP exist between Vietnam and United States and VND is fluctuated in comparison to USD.

Besides Chinn, there are many authors using the technique VECM to conduct tests of the PPP theory. There are some papers that have the validity in empirical studies such as Yazgan (2003), Doğanlar et al. (2009), Kim (2011), Kim and Jei (2013), Jovita (2016), Bergin et al. (2017) and some papers does not have the validity such as Basher et al. (2004), Doğanlar (2006).

2.2 Nonlinear Model for PPP Approach

Baharumshah et al. (2010), Ahmad and Glosser (2011) have applied the nonlinear regression model in recent years. However, Sarno (1999) stated that when he used the STAR model, the presumption of real exchange rate could lead to wrong conclusions.

The KSS test was developed by Kapetanios et al. (2003) to test unit root for 11 OECD countries, and applied the nonlinear Smooth Transition Auto Regressive model. They used monthly data during 41 years from 1957 to 1998 and the US dollar as a numeraire currency. While the KSS test did not accept unit root in some cases, the ADF test provided reverse results, implying that the KSS is superior to ADF test. Furthermore, Liew et al. (2003) used KSS test to check whether RER is stationary in the context of Asia. In his research, the data was collected in 11 Asian countries with quarterly bilateral exchange rate from 1968 to 2001 and US dollar and Japanese Yen represented as the Japanese currencies. The results showed that the KSS test and ADF test conflicted to each other when it comes to the unit root. Particularly, the ADF test can be applied in all cases, whereas the KSS test was not accepted in eight countries with US dollar numeraire and six countries where YEN was considered as a numeraire. The other kinds of unit root test for nonlinear model were applied by Saikkonen and Lutkepol (2002) and Lanne et al. (2002), then used by Assaf (2008) to test the stability of the real exchange rate (RER) in eight EU countries. They came to the conclusion that there was no stationary of the RER in the structural breaks after the appearance of the Bretton Woods era, which can be explained that the authorities may interfere with the exchange market to decide its value.

Besides, Baharumshah et al. (2010) attempted to test the nonlinear mean reverting of six Asian countries based on nonlinear unit root test and the STAR model. The authors used quarterly the data from 1965 to 2004 and US dollar as a numeraire currency. This was a new approach to test the unit root of the exchange rate for some reasons. First, real exchange rate was proved to be nonlinear, then the unit root of real exchange rate was tested in nonlinear model. The evidence indicated that RER of these countries were nonlinear, which mean reverting and the misalignment of these currencies should be calculated with US dollar as a numeraire. This evidence may lead to different results with the ADF test for unit root.

In this paper, the authors apply Augmented Dickey Fuller (ADF) test, the Phillips-Perron (PP) test, and the Kwiatkowski, Phillips, Schmidt, and Shin (KPSS) test to explore the time series data whether it is stationary or not. The three test are the most popular tests which are used for the linearity unit root test, such as Kadir and Bahadr (2015), Arize et al. (2015). And this is similar to the paper of Huizhen et al. (2013), Bahmani-Oskooeea (2016) for estimating the univariate time series unit root test.

3 Methodology and Data

3.1 Methodology

Taking the log from the Eq. (1) we have:

$$\log(s_t) = \log(p_t) - \log(p_t^*)$$

So when we run regression, the formula is:

$$s_t = c + \alpha_1 p_t + \alpha_2 p_t^* + \varepsilon_t$$

where: s: is the natural log exchange rate in countries i [1]

p_t : is domestic price of countries i and measured by the natural log CPI of countries

p^* : is domestic price of United States and measured by the natural log CPI of the US.

Because of time series data, the most important issue is that s, p, and p^* stationary or nonstationary. If the variable is nonstationary, there will be spurious when we run the model.

Step 1: Testing s, p, and p^* stationary or nonstationary

Augmented Dickey Fuller Test

A time series is an Augmented Dickey Fuller test based on the equation below:

$$\Delta Y_t = \beta_1 + \beta_2 t + \beta_3 Y_{t-1} + \sum_{i=1}^{n} \alpha_i \Delta Y_{t-1} + \varepsilon_t$$

where: ε_t is a pure white noise error term and n the maximum length of lagged dependent variables.

$$H_0 : \beta_3 = 0 \tag{2}$$

$$H_1 : \beta_3 \neq 0 \tag{3}$$

If the absolute value t* exceeds ADF critical value, the null hypothesis could not be rejected, and this result implies that the variable is nonstationary.

If the ADF critical value is greater than the absolute value t^*, the null hypothesis will fail to reject, and this result suggests the stationary of the variables.

The Phillips-Perron (PP) Test

Phillips and Perron (1998) suggest another (nonparametric) method of controlling for serial correlation when checking for a unit root. The PP method computes the non-augmented DF test Eq. (2) and modifies the -ratio of the coefficient therefore serial correlation does not affect the asymptotic distribution of the test statistic. The PP test is conducted on the statistic:

$$\tilde{t}_\alpha = t_\alpha \left(\frac{\gamma_0}{f_0}\right)^{1/2} - \frac{T(f_0 - \gamma_0)(se(\widehat{\alpha}))}{2 f_0^{1/2} s} \tag{4}$$

where $\widehat{\alpha}$ is the estimate, and t_α the -ratio of α, $se(\widehat{\alpha})$ is coefficient standard error, and s is the standard error of the test regression. In addition, γ_0 is a consistent estimate of the error variance.

[1] i represents for the countries: Vietnam, Thailand, Singapore, Philippine, Malaysia, Korea, Indonesia and Hongkong.

The remaining term, f_0, is an estimator of the residual spectrum at frequency zero.

The conclusion for times series data whether stationary or not is the same as ADF test.

The Kwiatkowski, Phillips, Schmidt, and Shin (KPSS) Test
In the contrast of the other unit root tests in time series, the KPSS (1992) test is assumed to be (trend-) stationary under the null. The KPSS statistic is based on the error term of the OLS regression of on the exogenous variables:

$$y_t = x_t'\delta + u_t$$

The LM statistic is be defined as:

$$LM = \sum_t S(t)^2/(T^2 f_0)$$

where f_0, is an estimator of the residual spectrum at frequency zero and $S(t)$ is a cumulative residual function:

$$S(t) = \sum_{r=1}^{t} \hat{u}_r$$

The H_0 is that the variable is stationary.

The H_A is that the variable is nonstationary.

If the LM statistic is larger than the critical value, then the null hypothesis is rejected; as a result, the variable is nonstationary.

Step 2: Test of cointegration.

Johansen (1988) used the following VAR system to analyze the relationship among variables.

$$\Delta X_t = \Gamma_1 \Delta X_{t-1} + \cdot + \Gamma_{k-1} \Delta X_{t-(k-1)} + \Pi X_{t-k} + \mu + \varepsilon_t$$

where $X(q, 1)$ is the vector of observation of q variables at time t,

μ: the (q, 1) vector of constant terms in each equation

ε_t: $(q, 1)$ vector of error terms. $\Gamma i(q, q), \Gamma(q, q)$ are matrices of coefficients.

There were two tests in the Johansen (1988) procedure, which are Trace test and Maximum Eigenvalue to check the vectors cointegration. Trace test can be calculated by the formula as follows:

$$LRtr(r/k) = -T \sum_{i=r+1}^{k} \log(1 - \lambda i)$$

where r is the number of cointegrated equation $r = 0, 1, \ldots k - 1$ and k is the number of endogenous variables.

H_0: r is the number of cointegrated equations.

H_1: k is the number cointegrated equations.

We can also calculate the maximum Eigenvalue test by the formula below:

$$\mathrm{LR}\max(r/k+1) = -T\log(1-\lambda)$$

Null hypothesis: r is the number cointegrated equations

Alternative hypothesis: $r+1$ is the number cointegrated equations

After using Johansen (1988) procedure, all the variables will be evaluated to see whether they are cointegration or not. If yes, it can be concluded that the three variables have a long run relationship or one or three variables will come back to the mean.

Step 3: Vector Error Correction Model (VECM)

If there is the cointegrated among the series, the long-term relationship happen; therefore VECM can be applied. The regression of VECM has the form as follow:

$$\Delta e_t = \delta + \pi e_{t-1} + \sum_{i=1}^{\rho-1} \Gamma_i \Delta e_{t-1} + \varepsilon_t$$

where $e_t : n \times 1$ the exchange rates matrix, $\pi = \alpha\beta : \alpha$ is $n \times r$ and β is $r \times n$ matrices of the error correction term, $\Gamma_i : n \times n$ the short-term coefficient matrix, and $\varepsilon_t : n \times 1$ vector of iid errors

If Error Correction Term is negative and significant in sign, there will be a steady long term relative among variables.

Step 4: Measuring misalignment

Using the simple approach that was provided by Stock and Watson (1993), Dynamic Ordinary Least Square (DOLS), to measure the misalignment between countries i and the United States. Stock-Watson DOLS model is specified as follows:

$$Y_t = \beta_0 + \vec{\beta} X + \Sigma_{j=-q}^{p} \vec{d_j} \Delta X_{t-1} + u_t$$

where Y_t : Dependent variable

X : Matrix of explanatory variables

β : Cointegrating vector; i.e., represent the long-run cumulative multipliers or, alternatively, the long-run effect of a change in X on Y

p : lag length

q : lead length

3.2 Data

As being mentioned above, this paper aims to find out the validity of PPP in East Asian countries with United States. For that reason, nominal exchange rate (defined at domestic currency per US dollar, the consumer price index (CPI) of country i and the U.S are in logarithm form. All data span monthly from 1997:1 to 2018:4, except Malaysia data covers from 1997:1 to 2018:3 and data of Vietnam begins from 1997:1 to 2018:2. All data were collected from IFS (International Financial Statistic).

4 Results and Discussion

4.1 Unit Root Test

We applied the ADF, PP and KPSS test to examine the stationary of consumer price index and nominal exchange rate of countries i and U.S. All variables have log form.

Table 1. Unit root test for the CPI

Countries	ADF		KPSS		Phillips - Perron	
	Level	1st difference	Level	1st difference	Level	1st difference
Vietnam	−0.068	−3.120**	2.035	0.296*	0.201	−9.563**
United States	−0.973	−10.408***	2.058	0.128*	−1.060	−8.289**
Thailand	−1.800	−10.864***	2.065	0.288*	−1.983	−10.802**
Singapore	−0.115	−6.458***	1.970	0.297*	0.006	−18.348**
Philippines	−2.341	−7.530***	2.068	0.536***	−2.673	−11.596**
Malaysia	−0.313	−11.767***	2.066	0.046*	−0.311	−11.730**
Korea	−2.766	−10.954***	2.067	0.549***	−2.865	−10.462**
Indonesia	−5.632	−5.613***	0.347	0.077**	−3.191	−7.814**
Hong Kong	1.4000	−5.326	1.395	1.022	1.491	−15.567**

Note: *, **, *** indicate significant at 10%, 5% and 1% levels respectively.

Table 1 shows the results of unit root test in time series of the CPI of countries i and U.S. At level, all variables have their t-statistic greater than the critical value. As a result, they have unit root or nonstationary at level or $I(0)$.

On the contrary, at the first difference, almost the variables have the smaller t-statistic than the critical value except Philippine and Korea at 1% and Hong Kong in KPSS test. For this reason, PPP does not hold between Philippine, Korea, Hong Kong. As a consequence, Philippine, Korea, Hong Kong will be ignored when conducting VECM. In short, the CPI of all other countries have stationary or they are cointegrated at $I(1)$[2].

The Table 2 shows the unit root test for nominal exchange rate for the rest 6 countries. Although KPSS and PP test prove Thailand cointegrated at $I(1)$, the ADF test point out stationary at level. Under the circumstances, PPP does not exist between Thailand and United States. To sum up, the unit root test does not support PPP for Philippine, Korea, Hong Kong and Thailand with United States.

As being analyzed above, the variables are nonstationary at level and stationary at first difference; therefore, they cointegrated at $I(1)$ or at the same order. As a result, Johansen (1988) procedure was examined to investigate the cointegration among these time series.

[2] All variables are conducted with intercept except Indonesia in ADF test.

Table 2. Unit root test for the nominal exchange rate

Countries	ADF		KPSS		Phillips - Perron	
	Level	1st difference	Level	1st difference	Level	1st difference
Vietnam	−0.068	−3.120**	2.035	0.296*	0.201	−9.563**
United States	−0.973	−10.408***	2.058	0.128*	−1.060	−8.289**
Thailand	−1.800	−10.864***	2.065	0.288*	−1.983	−10.802**
Singapore	−0.115	−6.458***	1.970	0.297*	0.006	−18.348**
Malaysia	−0.313	−11.767***	2.066	0.046*	−0.311	−11.730**
Indonesia	−5.632	−5.613***	0.347	0.077**	−3.191	−7.814**

Note: *, **, *** indicate significant at 10%, 5% and 1% levels respectively.

4.2 Optimal Lag

We have to choose optimal lag before conducting Johansen (1988) procedure. In view package, five lags length criteria have the same power. Therefore, if one lag is dominated by many criterions, this lag will be selected or else every lag is used for every case in VECM.

Table 3. Lag criteria

Criterion	LR	FPE	AIC	SC	HQ
Vietnam	**3**	**3**	**3**	2	3
Singapore	**6**	**6**	**6**	2	4
Malaysia	6	**3**	**3**	**2**	**2**
Indonesia	**6**	**6**	**6**	2	3

LR: sequential modified LR test statistic (each test at 5% level)
FPE: Final prediction error
AIC: Akaike information criterion
SC: Schwarz information criterion
HQ: Hannan-Quinn information criterion

Table 3 illustrates the lag-length criteria that was choosen for the rest of 4 countries when conducting Johansen (1988). Singapore and Indonesia are dominated by lag 6. Lag 3 is used for Vietnam. However, Malaysia has two lags, 2 and 3. In other words, 3-lag and 2-lag were chosen for conducting Johansen (1988) procedure or testing cointegration of Malaysia.

4.3 Johansen (1988) Procedure for Cointegration Test

For the reasons, all the variables are cointegrated at the first order I(1), Johansen (1988) cointegration was conducted to test the long run relationship among variables.

Table 4. Johansen (1988) cointegration test

Variable	Vietnam	Singapore	Malaysia		Indonesia
Lags	3	6	3	2	6
Cointegration equation	1**	2**	1*	1*	1**

Note: *, ** indicate significant at 10% and 5% levels respectively.

Table 4 presents the Johansen (1988) cointegration test. The results indicate that Trace test and/or Eigenvalue test were statistically significant at 5% for Vietnam, Singapore and Indonesia and 10% for Malaysia both 3-lag and 2-lag. Hence, the null hypothesis of $r = 0$ is rejected. $R = 0$ implies one (Vietnam, Malaysia and Indonesia) and two (Singapore) cointegration equation in the long run, so the VECM can be used for further investigation of variables.

4.4 Vector Error Correction Model

The Table 5 suggests the long run relationship of PPP between 4 countries and United States. C(1) has negative in value and significant in sign (Prob less than 5%), is error correction term. This implies that the variables move along together or have mean reverting. As a result, PPP exists between Vietnam, Singapore, Malaysia and Indonesia with the U.S.

In conclusion, ADF, KPSS, PP test, Johansen Cointegration and Vector Error Correction Model prove that PPP hold between these countries and the U.S. This is a good indicator for policy makers, multinational firms and exchange rate market members to set their plans for future activities.

4.5 Measuring the Misalignment Between 4 Countries and the United States Dollar

Because of the existence of PPP between four countries and the United States, DOLS approach is used to calculate the exchange rate misalignment between these countries.

Table 5. The speed of adjustment coefficient of long run

Countries		Coefficient	Std. Error	t-Statistic	Prob.
Vietnam	C(1)	−0.0111	0.0349	−3.183	0.0017
Singapore		−0.0421	0.0188	−2.2397	0.0261
Malaysia (lag 2)		−0.0599	0.01397	−4.2854	0
Malaysia (lag 3)		−0.0643	0.01471	−4.3751	0
Indonesia		−0.0185	0.00236	−7.8428	0

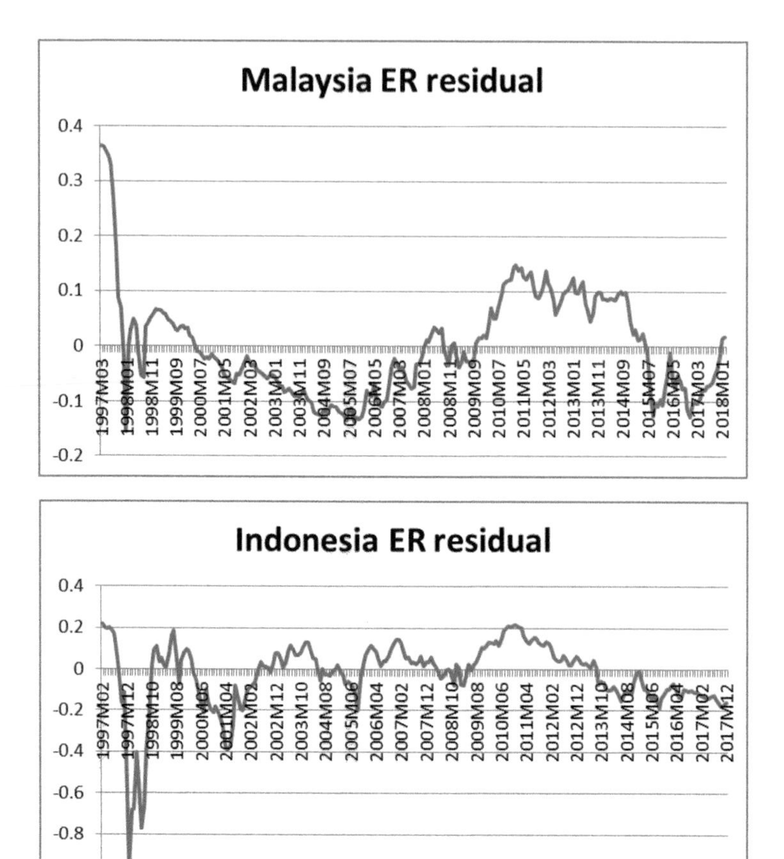
Malaysia ER residual
0.4
0.3
0.2
0.1
0
-0.1
-0.2
1997M03
1998M01
1998M11
1999M09
2000M07
2001M05
2002M03
2003M01
2003M11
2004M09
2005M07
2006M05
2007M03
2008M01
2008M11
2009M09
2010M07
2011M05
2012M03
2013M01
2013M11
2014M09
2015M07
2016M05
2017M03
2018M01
Indonesia ER residual
0.4
0.2
0
-0.2
-0.4
-0.6
-0.8
-1
-1.2
1997M02
1997M12
1998M10
1999M08
2000M06
2001M04
2002M02
2002M12
2003M10
2004M08
2005M06
2006M04
2007M02
2007M12
2008M10
2009M08
2010M06
2011M04
2012M02
2012M12
2013M10
2014M08
2015M06
2016M04
2017M02
2017M12

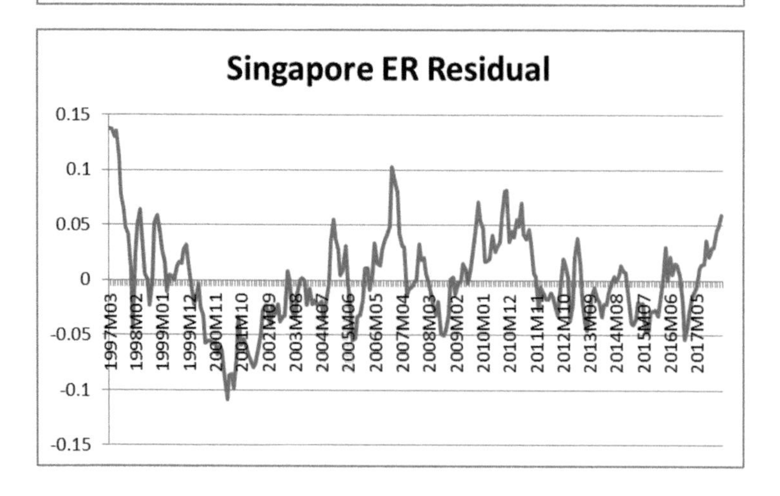
Singapore ER Residual
0.15
0.1
0.05
0
-0.05
-0.1
-0.15
1997M03
1998M02
1999M01
1999M12
2000M11
2001M10
2002M09
2003M08
2004M07
2005M06
2006M05
2007M04
2008M03
2009M02
2010M01
2010M12
2011M11
2012M10
2013M09
2014M08
2015M07
2016M06
2017M05

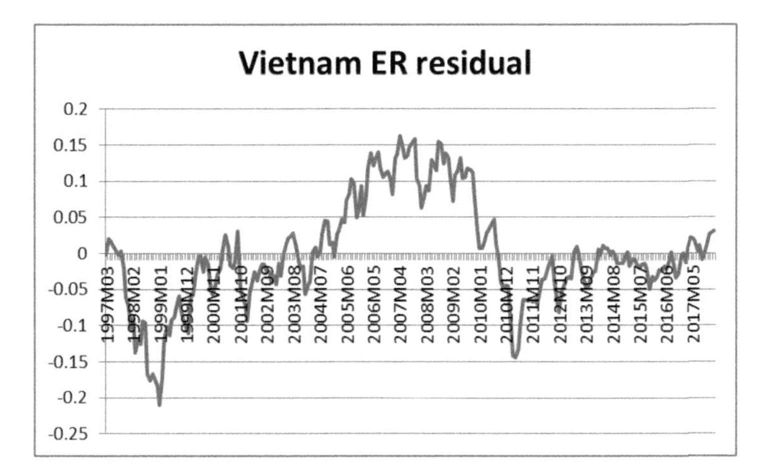

As can be seen from the graphs, the ER residual (the misalignment) of these countries had downward trend during the 1997 financial crisis and widely fluctuated during the whole period. After the crisis, in the 2000s, Malaysia with the fix exchange rate regime made the currency undervalued and this caused the surplus of the current account. To deal with the current account surplus, Malaysia shifted exchange rate to managed floating regime. The new exchange rate regime explained the exchange rate which had the upward trend after that. From 2009, to deal with short-term money inflow, the government used the high "soft" capital controls (Mei-Ching et al. 2017) which caused it to be overvalued of rigid during this period. Afterwards, rigid undervalued and fluctuated. Recently, the rigid has a little bit been overvalued.

Indonesia has been pursuing the floating exchange rate regime and free capital flows since Asia financial crisis. The misalignment of Indonesia's rupiah currency is not stable. The deviation is larger (from −0.4 to 0.2) compared to others countries after finishing the crisis. From the middle year 2002 to the beginning of 2009, the Indonesia's rupiah currency was overvalued except the period 2004:5 to 2005:10. Being similar to Malaysia, facing hot money inflows from 2009 (Mei-Ching et al. 2017), Indonesia feared the domestic currency could not be competitive to other currencies. As a result, Indonesia was one of the highest "soft" capital controls. Besides, Bank Indonesia Regulation No. 16/16/PBI/2014 in 2014 has made Indonesia's rupiah currency undervalued until now.

Since 1980s, Singapore's monetary policy has focused on the exchange rate than interest rate compared to other countries. The exchange rate system is taken the basket, band and crawl (BBC) by the Monetary Authority of Singapore (MAS). As can be seen from the graph, Singapore ER residual is very stable when comparing to the other countries. (from −0.1 to 0.1). Because the MAS pursuits Singapore dollar against a basket of currencies of its main trading partners. In contrast of Indonesia and Malaysia, facing the shot-term money, Singapore did not fear the competitive level of domestic currency therefore Singapore has the lowest "soft" control capital

In this paper, the result of misalignment of VND compared to USD is quite similar to the papers of Duy et al. (2017). They all share their agreement that VND was overvalued from 2004:4 to 2010:8. The main difference of the two papers goes for research result. While the authors claim that VND was undervalued from 1997:8 to 2004:3, Duy et al. (2017) show that it was overvalued from 1999 to 2003. The financial crisis happened and lead to the depreciation of all currencies. Therefore, our paper has more consistent evidence.

This paper examines the relationship of Purchasing Power Parity (PPP) between East Asian countries and the United States in Johansen cointegration and VECM frameworks. Using monthly data from 1997:1 to 2018:4, the econometrics tests proved that the PPP theory hold between Vietnam, Singapore, Malaysia and Indonesia with the U.S while it does not support for PPP between Thailand, Philippines, Korea and Hong Kong. After that, DOLS was applied to measure misalignment between VND, SGD, Rigid, Indonesia's rupiah to USD. The authors found out the misalignment had downward trend and fluctuated after Asian financial crisis. Recently, VND, SGD and Rigid are overvalued while Indonesia's rupiah is still undervalued.

References

Ahmad, Y., Glosser, S.: Searching for nonlinearities in real exchange rates. Appl. Econ. **43**(15), 1829–1845 (2011)

Kavkler, A., Bori, D., Bek, J.: Is the PPP valid for the EA-11 countries? New evidence from nonlinear unit root tests. Econ. Res.-Ekonomska Istraivanja **29**(1), 612–622 (2016). https://doi.org/10.1080/1331677X.2016.1189842

Asea, P.K., Corden, W.M.: The Balassa-Samuelson model: an overview. Rev. Int. Econ. **2**(3), 191–200 (1994)

Assaf, A.: Nonstationarity in real exchange rates using unit root tests with a level shift at unknown time. Int. Rev. Econ. Financ. **17**(2), 269–278 (2008)

Baharumshah, Z.A., Liew, K.V., Chowdhury, I.: Asymmetry dynamics in real exchange rates: new results on East Asian currencies. Int. Rev. Econ. Financ. **19**(4), 648–661 (2010)

Bahmani-Oskooeea, T.C., Kuei-Chiu, L.: Purchasing power parity in emerging markets: a panel stationary test with both sharp and smooth breaks. Econ. Syst. **40**, 453–460 (2016)

Balassa, B.: The purchasing-power parity doctrine: a reappraisal. J. Polit. Econ. **72**(6), 584–596 (1964)

Basher, S.A., Mohsin, M.: PPP tests in cointegrated panels: evidence from Asian developing countries. Appl. Econ. Lett. **11**(3), 163–166 (2004)

Chinn, D.M.: Before the fall: were East Asian currencies overvalued? Emerg. Mark. Rev. **1**(2), 101–126 (2000)

Coe, P., Serletis, A.: Bounds tests of the theory of purchasing power parity. J. Bank. Financ. **26**, 179–199 (2002)

Dilem, Y.: Empirical investigation of purchasing power parity for Turkey: evidence from recent nonlinear unit root tests. Cent. Bank Rev. **17**(2017), 39–45 (2017)

Doğanlar, M.: Long-run validity of Purchasing Power Parity and cointegration analysis for Central Asian countries. Appl. Econ. Lett. **13**(7), 457–461 (2006)

Doğanlar, M., Bal, H., Ozmen, M.: Testing long-run validity of purchasing power parity for selected emerging market economies. Appl. Econ. Lett. **16**(14), 1443–1448 (2009)
Duy, H.B., Anthony, J.M., Shyama, R.: Is Vietnam's exchange rate overvalued? J. Asia Pac. Econ. **22**(3), 357–371 (2017). https://doi.org/10.1080/13547860.2016.1270041
Johansen, S.: Statistical analysis of cointegrated vectors. J. Econ. Dyn. Control **12**(2–3), 231–254 (1988)
Jovita, G.: Modelling and forecasting exchange rate. Lith. J. Stat. **55**(1), 19–30 (2016)
Huizhen, H., Omid, R., Tsangyao, C.: Purchasing power parity in transition countries: old wine with new bottle. Japan World Econ. **28**(2013), 24–32 (2013)
Kadir, K., Bahadr, S.T.: Testing the validity of PPP theory for Turkey: nonlinear unit root testing. Procedia Econ. Financ. **38**(2016), 458–467 (2015)
Kapetaniosa, G., Shinb, Y., Snell, A.: Testing for a unit root in the nonlinear STAR framework. J. Econom. **112**(2), 359–379 (2003)
Kaminsky, G., Lizondo, S., Reinhart, C.M.: Leading indicators of currency crises. IMF Staff Papers **45**(1), 1–48 (1998). http://www.jstor.org/stable/3867328
Kim, H.-G.: VECM estimations of the PPP reversion rate revisited. J. Macroecon. **34**, 223–238 (2011). https://doi.org/10.1016/j.jmacro.2011.10.004
Kim, H.-G., Jei, S.Y.: Empirical test for purchasing power parity using a time-varying parameter model: Japan and Korea cases. Appl. Econ. Lett. **20**(6), 525–529 (2013)
Kremers, M.J.J., Ericsson, R.J.J.M., Dolado, J.J.: The power of cointegration tests. Oxford Bull. Econ. Stat. **54**(3), 325–348 (1992). https://doi.org/10.1111/j.1468-0084.1992.tb00005.x
Krugman, R.P., Obstfeld, M., Melitz, J.M.: Price levels and the exchange rate in the long run. In: Yagan, S. (ed.) International Economics Theory and Policy, pp. 385–386. Pearson Education (2012)
Kwiatkowski, D., Phillips, P., Schmidt, P., Shih, Y.: Testing the null hypothesis of stationarity against the alternative of a unit root: how sure are we that economic time series have a unit root? J. Econom. **54**(1992), 159–178 (1992)
Lanne, M., Ltkepohl, H., Saikkonen, P.: Comparison of unit root tests for time series with level shifts. J. Time Ser. Anal. **23**(6), 667–685 (2002). https://doi.org/10.1111/1467-9892.00285
Mei-Ching, C., Sandy, S., Yuanchen, C.: Foreign exchange intervention in Asian countries: what determine the odds of success during the credit crisis? Int. Rev. Econ. Financ. **51**(2017), 370–390 (2017)
Narayan, P.K.: New evidence on purchasing power parity from 17 OECD countries. Appl. Econ. **37**(9), 1063–1071 (2005)
Bergin, P.R., Glick, R., Jyh-Lin, W.: "Conditional PPP" and real exchange rate convergence in the euro area. J. Int. Money Financ. **73**(2017), 78–92 (2017)
Rogoff, K.: The purchasing parity puzzle. J. Econ. Lit. **34**, 647–668 (1996). http://scholar.harvard.edu/rogoff/publications/purchasing-power-parity-puzzle
Saikkonen, P., Ltkepohl, H.: Testing for a unit root in a time series with a level shift at unknown time. Econom. Theory **18**(2), 313–348 (2002)
Sarno, L.: Real exchange rate behavior in the Middle East: a re-examination. Econ. Lett. **66**(2), 127–136 (1999)
Shapiro, C.A.: What does purchasing power parity mean? J. Int. Money Financ. **2**(3), 295–318 (1983)
Stock, J., Watson, M.: A simple estimator of cointegrating vectors in higher order integrated systems. Econometrica **61**(4), 783–820 (1993)
Tastan, H.: Do real exchange rates contain a unit root? Evidence from Turkish data. Appl. Econ. **37**(17), 2037–2053 (2005)

Yazgan, E.: The purchasing power parity hypothesis for a high inflation country: a re-examination of the case of Turkey. Appl. Econ. Lett. **10**(3), 143–147 (2003)

Arize, A.C., Malindretos, J., Ghosh, D.: Purchasing power parity-symmetry and proportionality: evidence from 116 countries. Int. Rev. Econ. Financ. **37**, 69–85 (2015)

Enders, W., Granger, C.W.J.: Unit-Root Tests and Asymmetric Adjustment With an Example Using the Term Structure of Interest Rates. J. Bus. Econ. Stat. **16**(3), 304–311 (1998)

Phillips, P.C.B., Perron, P.: Testing for a Unit Root in Time Series Regression. Biometrika **75**(2), 335–346 (1998)

Determinants of Net Interest Margins in Vietnam Banking Industry

An H. Pham[1(✉)], Cuong K. Q. Tran[1], and Loan K. T. Vo[2]

[1] Faculty of Economics, Van Hien University, Ho Chi Minh City, Vietnam
hoangan.tcnh@gmail.com, cuong.tqk@vnp.edu.vn
[2] HCM City Open University, Ho Chi Minh City, Vietnam
loan.vtk@ou.edu.vn

Abstract. This study analyses determinants of net interest margins (NIM) in Vietnam banking industry. The paper uses the secondary data of 26 banks with 260 observations for the period 2008–2017 and applies the panel data regression method. The empirical results indicate that lending scale, capitalization, inflation rate have positive impacts on net interest margin. In contrast, Managerial efficiency has a negative impact on net interest margin. However, bank size, credit risk, and loan to deposit ratio are statistically insignificant to net interest margin.

Keywords: Net interest margin · NIM · Commercial banks
Panel data · Vietnam

1 Introduction

The efficiency of banking operations has always been an issue that gets great concern for bank managers, as it is the key factor of sustainable profit, which enables the bank to develop and become competitive in the international environment.

A competitive banking system will create a higher efficiency and a lower NIM (Sensarma and Ghost 2004). High profit return ratio causes significant obstacles to intermediaries, for example more savings encouraged by lower borrowing interest rate and reduced investment opportunities of the banks as a result of higher lending rate (Fungáčová and Poghosyan 2011). Therefore, banks expect to run their intermediate functionality with the lowest cost possible which is possible to promote economic growth.

NIM ratio is both a measure for the effectiveness and profitability, and a core indicator because it often accounts for about 70–85% the total income of a bank. As a consequence, the higher this ratio is, the higher the bank's income will be. It indicates the ability of the Board of Directors and employees in maintaining the growth of incomes (mainly from loans, investments and service fees) compared with the increase in cost (mainly from interest cost for deposits, monetary market's debts) (Rose 1999).

V. Kreinovich et al. (Eds.): ECONVN 2019, SCI 809, pp. 417–426, 2019.
https://doi.org/10.1007/978-3-030-04200-4_30

Therefore, research on determinants of net interest margins in Vietnam banking industry is necessary. The result of this study can serve as a scientific basis for bank managers to make suitable decisions, bring good efficiency and increase the attractiveness of their stocks.

2 Literature Review and Previous Studies

2.1 Net Interest Margin

To calculate the operating effectiveness of any banks, we often analyse Return on Equity (ROE), Return on Asset (ROA), Net interest margin (NIM) and interest spread (Rose 1999). Hemple et al. (1986) claim that NIM is helpful in measuring changes in interest spread and comparing profit between banks.

Net interest margin ratio is one of the most important measurements to quantify financial effectiveness in an intermediary institution (Golin 2001). Net interest margin is defined by net interest income over total earning asset.

$$\text{NIM} = \frac{\begin{pmatrix} \text{Interest income} & & \text{Interest expense on} \\ \text{from loans} & - & \text{deposits and other} \\ \text{and investments} & & \text{borrowed funds} \end{pmatrix}}{\text{Total earning asset}}$$

2.2 Factors Influencing Net Interest Margin

Based on previous researches in Russia, Turkey, China, Lebanon and Fiji, the authors identify similarities between Vietnam and these nations, and thereby suggests some factors which have impacts on net interest margin, including:

Size

Studies of Maudos and Guevara (2004), Ugur and Erkus (2010) find positive relation between lending scale and bank's net interest margin, where large average operating scale leads to higher market risk and credit risk, increasing the possibility of losses. Meanwhile, researches of Fungáčová and Poghosyan (2011), Hamadi and Awdeh (2012) show the negative effect of bank size on NIM, where large banks with high credit ratings earn their profit from economy of scale and have low NIMs. In Vietnam, banks with large size have advantages because they can utilize to raise capital at low cost, such as: large network of operation with many branches, wide variety of products and service, etc. to make higher profit.

Lending Scale (LAR)

Maudos and Guevara (2004), Maudos and Solís (2009), Hamadi and Awdeh (2012), Pham et al. (2018) find positive relation between lending scale and NIM. Where market risk and credit risk occur, larger lending scale leads to bigger losses for the bank. In contrast, researches of Hawtrey and Liang (2008), Zhou and Wong (2008), Kasman et al. (2010) indicate negative relation between LAR and NIM. Large banks can offer bigger loans with lower interest rate than small ones, leading to lower interest income. In Vietnam, lending is the most significant operation which brings income to banks, so ones with big loan size will have higher NIM.

Credit Risk (CR)
Credit risk is the risk that customers aren't able to repay the debt at its maturity. The research of Angbazo (1997) states that credit risk impacts banks' interest income in a positive relation. Banks which are lending out more money face higher credit risk and thus have to maintain more reserve; this forces them to charge more interest on their loans in order to make up for the expected losses, causing a positive relation (Garza-García 2010). More studies have found the positive relation between credit risk and net interest margin, namely Maudos and Guevara (2004), Doliente (2005), Maudos and Solís (2009), Kasman et al. (2010), Gounder and Sharma (2012), Tarus et al. (2012).

Equity Capital (CAP)
According to the IMF (2006), the ratio of equity over total asset is used as one of the recommended indicators to assess the financial health of a commercial bank. Most studies have found positive correlation between CAP and NIM (Brock and Suarez (2000), Saunders and Schumacher (2000); Maudos and Guevara (2004); Doliente (2005); Hawtrey and Liang (2008); Maudos and Solís (2009); Garza-García (2010); Ugur and Erkus (2010); Kasman et al. (2010); Fungáčová and Poghosyan (2011)), Pham et al. (2018)). Raising the capital will increase the mediate cost of keeping equity more than loans due to taxes and diluting shareholders' rights. The increase in mediate cost is often recovered through an increase in the interest rate spread. Whenever capital is too high, the manager is pressured to increase profit margin.

Loan/Deposit Ratio (LDR)
An increase in LDR indicates that a bank has less "onlay" to finance for its growth and protect itself from unexpected withdrawals, especially for banks which depend much on deposits for their growth. When LDR is at a relatively high level, bank's managers rarely want to give out loans and investments. In addition, they will be more cautious when LDR increases and demand a tightened credit line, therefore, interest rate tends to increase (Rose 1999). Most experimental researches show that LDR shares a positive correlation with NIM (Ahmad et al. (2011); Hamadi and Awdeh (2012)).

Management Efficiency (CTI)
High management efficiency helps banks maximize profit and minimize cost, allowing them to reduce the expenses for each dollar of income (Ugur and Erkus (2010)). High management efficiency also enhances managing responsibility to lessen cost and invest in more earning assets (Angbazo (1997); Maudos and Guevara (2004)). Consequently, the higher management effect is, the lower CTI and higher NIM get. Studies of Zhou and Wong (2008), Maudos and Solís (2009), Garza-García (2010), Kasman et al. (2010), Gounder and Sharma (2012), Hamadi and Awdeh (2012) also used this ratio to measure management

efficiency and came to the same conclusion of the negative correlation between management efficiency and bank's NIM.

Inflation Rate (INFL)
The increasing of inflation rate will lead to the soaring of net interest margin in banks and in vise versa. Specifically, when inflation rate rises up, it will drive loan interest to up, causing the hikes of NIM.

Even if banks do not anticipate inflation correctly, in the long term, interest rates would be adjusted to reflect the inflation premium, which would also increase the interest margin (Tarus et al. 2012). Most studies have found positive correlation between INFL and NIM (Kasman et al. 2010; Ugur and Erkus 2010 and Hamadi and Awdeh 2012).

The authors' experimental study mainly focuses on banks in one area, a group of countries such as Southeast Asia, OECD, EU, Europe, the United States, or a particular country such as America, Lebanon, Turkey, China, ect. There are few researches on factors affecting the net interest margin of banks in Vietnam. The data of the researches mentioned above is mainly from the period before 2008, plus only a few studies with research data of 2010. There is no study with data from the period 2008–2017.

3 Methodology and Data

3.1 The Model

Based on research models of Fungáčová and Poghosyan (2011), Gounder and Sharma (2012), Hamadi and Awdeh (2012), Pham et al. (2018) this study applies the following model:

$$\begin{aligned} NIM = &\beta_0 + \beta_1 \cdot SiZE + \beta_2 \cdot LAR + \beta_3 \cdot CR + \beta_4 \cdot CAP \\ &+ \beta_5 \cdot LDR + \beta_6 \cdot CTI + \beta_7 \cdot INFL + u_{it} \end{aligned}$$

Where:

NIM: Net interest margin; SIZE: Bank size; LAR: Lending size; CR: Credit risk; CAP: Equity Capital; LDR: Loan over Deposit ratio; CTI: Management Efficiency; INFL: Inflation rate.

3.2 Variable Measurements

The description of how to calculate variables and the expected signs are detailed in Table 1.

3.3 The Data

Data in this study was taken from the audited financial statements of Vietnamese Banks and the index report of International Monetary Fund in the period 2008–2017. Up to December 31, 2017, Vietnam has had a total of 35 Commercial

Table 1. Describing table for variables and expected signs

Variable	Description	Measurement	Expected sign	Previous studies
Dependent				
NIM	Net interest margin	(Interest income – Interest Expense)/Total earning asset		
Independent				
SIZE	Bank size	Logarithm of total asset	+	Maudos and Guevara (2004), Ugur and Erkus (2010)
LAR	Lending size	Loan Outstanding/Total asset	+	Hamadi and Awdeh (2012), Maudos and Guevara (2004), Maudos and Solís (2009), Pham et al. (2018)
CR	Credit Risk	Credit Provision/Total Loan Outstanding	+	Doliente (2005), Garza-García (2010), Gounder and Sharma (2012), Kasman et al. (2010), Maudos and Guevara (2004), Maudos and Solís (2009), Tarus et al. (2012)
CAP	Equity Capital	Equity/Total Asset	+	Doliente (2005), Fungáčová and Poghosyan (2011), Garza-García (2010), Hawtrey and Liang (2008), Kasman et al. (2010), Maudos and Guevara (2004), Maudos and Solís (2009), Saunders and Schumacher (2000), Ugur and Erkus (2010), Pham et al. (2018)
LDR	Loan to Deposit ratio	Total Loan/Total Deposit	+	Hamadi and Awdeh (2012), Ahmad et al. (2011)
CTI	Management Efficiency	Operating cost/Total income	–	Garza-García (2010), Gounder and Sharma (2012), Hamadi and Awdeh (2012), Kasman et al. (2010), Maudos and Solís (2009), Ugur and Erkus (2010), Zhou and Wong (2008).
INFL	Inflation rate	Annual rate of inflation	+	Kasman et al. (2010), Ugur and Erkus (2010), Hamadi and Awdeh (2012)

Banks. Data is collected after eliminating banks with lacking or unclear information. The result is a random balance panel data involving 26 banks and 260 observations, which accounts for about 70.3% of the Vietnamese banking system. Hence, one can say that those selected bank has right to represent commercial banks in Vietnam. Table 2 describes mean, standard deviation, min and max of variables.

This research employs 3 methods: Pooled OLS Regression, Fixed Effects Model and Random Effects Model. In addition, it uses Hausman Test (1978) to select the suitable model. After choosing one, the variance of the constant

Table 2. Describing observed variables

Variable	Mean	Standard deviation	Min	Max
NIM	0.0261	0.0115	−0.0063	0.0742
SIZE	18.0814	1.2430	14.7945	20.9075
LAR	0.5273	0.1307	0.1737	0.8517
CR	0.0128	0.0055	0.0021	0.037
CAP	0.1076	0.0608	0.035	0.4624
LDR	0.8756	0.2292	0.3719	2.0911
CTI	0.5287	0.1542	0.2251	1.1152
INFL	0.0843	0.0683	0.006	0.231

error and its autocorrelation are tested to determine the appropriate regression model. As the last step, the variables are sorted based on the statistics.

4 Empirical Result and Discussion

4.1 Empirical Result

The study examines the possibility of multicollinearity between the variables by setting up a correlation matrix of the variables and calculating VIF indicators, as being presented in Table 3.

Results show that none of the correlation coefficient between pairs of variables exceeds 0,8. The largest VIF index of the independent variables in this study is 2.64, less than 5 (Gujarati 2004). Therefore, the multicollinearity phenomenon in this research's models is negligible.

Table 3. Matrix of correlation between the variables

	SIZE	LAR	CR	CAP	LDR	CTI	INFL	VIF
SIZE	1							**2.64**
LAR	0.1291	1						1.71
CR	0.3549	−0.0876	1					1.19
CAP	−0.7288	0.0027	−0.2532	1				2.44
LDR	−0.2625	0.5206	−0.2309	0.3909	1			1.97
CTI	−0.0788	−0.0556	0.0117	−0.1055	−0.2121	1		1.19
INFL	−0.3493	−0.1991	−0.0495	0.3462	0.2397	−0.2505	1	1.41

The results of the regression model are shown in Table 4, this study conducted the Hausman test to select the appropriate model. The Hausman test result gives the statistical value Chi-Square of 33.39 with Prob.Chi-Square of 0.0000.

As can be seen, Prob is less than 5%, which allows to reject hypothesis H0, and accept hypothesis H1 in Hausman test - no correlation exists between the random element of the bank and independent variables. Thus, the study will select the fixed effects regression model (FEM) to analyze the results. Next, the researchers conducted testing on the variance of the constant error as well as its autocorrelation.

The test results of the variance of the constant error (White test), Prob.Chi - Square = 0.000, less than 5% and the result of error autocorrelation (Breusch-Godfrey test), Prob.F (1,25) = 0.000, also less than 5%. These results show that the model has both the phenomenons of changing variance and autocorrelation of errors. According to Wooldridge (2002), the solution to changing variance errors & error autocorrelation is applying the regression model with the general least squares method (General Least Square -GLS). Table 4 presents the regression results of using GLS method to estimate the regression coefficients.

Table 4. Regression result

Variables	Model			
	Pooled (p-value)	FEM (p-value)	REM (p-value)	GLS (p-value)
Constant	0.0090 (0.540)	−0.0694 (0.001)	−0.0328(0.058)	−0.0011(0.932)
SIZE	0.0005(0.493)	0.0044***(0.000)	0.0025***(0.005)	0.0007(0.276)
LAR	0.0271***(0.000)	0.0244***(0.000)	0.0233***(0.000)	0.0214***(0.000)
CR	0.1267(0.247)	0.2946***(0.004)	0.2641***(0.008)	0.0931(0.282)
CAP	0.0796***(0.000)	0.0862***(0.000)	0.0814***(0.000)	0.0735***(0.000)
LDR	−0.0018(0.587)	0.0031(0.310)	0.0027(0.376)	0.0008(0.713)
CTI	−0.0285***(0.000)	−0.0290***(0.000)	−0.0266***(0.000)	−0.0195***(0.000)
INFL	0.0026(0.786)	0.0209**(0.020)	0.0112(0.178)	0.0152***(0.005)
Adjusted R^2	0.3942	0.1808	0.2981	-
F-statistic/Wald.Chi2	25.08(0.000)	25.20(0.000)	178.42(0.000)	197.81(0.000)
Hausman test	33.39***(0.0000)			

*Note: *, ** and *** have Statistical significance at 10%, 5% and 1% respectively.*

4.2 Discussion

In this section, the research focuses on the results of the regression model using GLS method.

The first variable, Lending scale (LAR), shares a positive correlation with NIM. The more Vietnam commercial banks enlarge their lending scale, the higher NIM is. These results are consistent with previous findings of Maudos and Guevara (2004) in Europe, Maudos and Solís (2009) in Mexico, Hamadi and Awdeh (2012) in Lebanon and Pham et al. (2018) in Vietnam. In Vietnam, lending makes up the most traditional and major activities of banks (about 70–80% of bank operations). Therefore, most banks tend to focus on lending activities, their main channel of profits.

Equity Capital (CAP) has a positive correlation with NIM of Vietnam commercial banks, demonstrating the importance of scale of equity in improving the banks' NIM. This study shows that better - capitalized banks face lower risk of default. Moreover, a strong capital structure is essential for banks to operate in developing economies, as it provides more power for banks to survive during times of financial crisis and increase the level of security provided to depositors when facing with the conditions of macroeconomic instability. These results are consistent with previous research findings: Brock and Suarez (2000), Saunders and Schumacher (2000), Maudos and Guevara (2004), Doliente (2005), Hawtrey and Liang (2008), Maudos and Solís (2009), Garza-García (2010), Kasman et al. (2010), Ugur and Erkus (2010), Fungáčová and Poghosyan (2011), Pham et al. (2018).

Whether management efficiency is good or do not depends on the ratio of operating cost over total income (CTI), the result pointed out that CTI has negative correlation with NIM. The study results are consistent with previous research findings: Angbazo (1997), Maudos and Guevara (2004), Zhou and Wong (2008), Maudos and Solís (2009), Garza-García (2010), Kasman et al. (2010), Ugur and Erkus (2010), Gounder and Sharma (2012), Hamadi and Awdeh (2012). In the period of 2008–2017, the Vietnamese economy has been facing many difficulties, banks had to go through large-scale reorganizations of their administration and operating systems in order to improve management efficiency as well as to clearly define the responsibilities and authorities of departments at different levels. Up to now, Vietnamese banks' administration and management efficiency has become more professional, with access to management knowledge from technology transfers and strategic cooperations.

Finally, the research results illustrate the positive correlation between inflation rate (INFL) and NIM, which represents the situation of Vietnamese banking system where the increase of inflation hikes interest rate of loans rise. The study results are consistent with studies by Kasman et al. (2010), Ugur and Erkus (2010) and Hamadi and Awdeh (2012).

5 Conclusions and Implications

The paper examined 7 factors that affect the net interest margins in Vietnam banking industry from 2008 to 2017. The chosen data is panel data. After analyzing and testing hypotheses violations, the study has applied regression models with GLS method. Research results indicate that in Vietnam, lending scale (LAR), scale of equity (CAP), and inflation rate (INFL) may positively impact the NIM of banks; while management efficiency (CTI) of Vietnam commercial banks has negative impact on it. Bank size (SIZE), credit risk (CR) and ratio of loans on deposits (LDR) are statistically insignificant to the NIMs of Vietnamese commercial banks.

From the result in Table 4, the authors suggest some solutions to enhance Net Interest Margin of Vietnam commercial bank, as below:

Widening Lending Scale

Lending scale has positive effect on NIM. Increasing bank loan means increasing NIM of banks. However, if banks widen lending scale without tight control, the consequence is of great concerns, for example, it may lead to imbalanced safety or increased inflation. To sum up, along with expanding their lending scale, banks need to ensure credit security is in accordance with the State Bank's regulations.

Increasing Equity

Owned equity scale impacts NIM of commercial banks in the same direction. As the scale of owned equity of one bank grows, NIM also grows. There are many ways to increase equity capital such as issuing additional shares in the market; selling shares to strategic partners which are local banks, foreign banks, domestic corporations, foreign investors; implementing dividends by shares; using the equity surplus of the last year to raise funds for this year, setting up the fund from profits of previous years. Depending on the strength and the specific situation in each period, banks will have different methods to raise capital that assure sustainable fund as well as the benefit of shareholders in the bank.

Improving the Efficiency of the Management of Commercial Banks

Management Effectiveness has opposite impact on NIM. Increasing management efficiency leads to decreased NIM, because when control is too tight, lending size will be narrowed as a result. To ensure effective management, commercial banks need to restructure and rearrange each of their business functions, governance and administration; they need to logically sort out and arrange the development of personnel staff and business managers who are highly-qualified, with sense of responsibility and good ethics. They also needs to modernize the IT system and develop the risk management system in accordance with the principles of the Basel Committee's standards.

Inflation Issue:

Inflation rate has positive effect on NIM of commercial banks in Vietnam. Therefore, the policy makers in Vietnam need to have a suitable plan to control inflation rate so as to keep NIM at low level. The Vietnamese government has set the priority on controlling inflation rate and stabilize the economy growth in macro level since 2012. The strategy has achieved some certain success such as reducing inflation rate from 13%, in the period of 2008 to 2012, to 3,53% in 2017.

There are some limitations in this study. Firstly, the authors just focus on commercial banks only, lack of information for foreign and joint-banks, thus, the study cannot provide the whole assessment of banking development in Vietnam as well as give comparison among banks. Secondly, the study does not investigate the difference between periods of before and after the financial crisis of Vietnamese commercial banks system. These limitations implicit implications for future research that aims to explore net interest margin (NIM) of others apart from commercial banks and provide comparison of banks in Vietnam and the ones in ASEAN.

References

Ahmad, R., Shahruddin, S.S., Tin, L.M.: Determinants of bank profits and net interest margins in East Asia and Latin America, Working paper series (2011). http://papers.ssrn.com/sol3/papers.cfm?abstract_id=1912319. Accessed 10 June 2018

Angbazo, L.: Commercial bank net interest margins, default risk, interest rate risk and off-balance sheet banking. J. Bank. Financ. **21**(1), 55–87 (1997)

Brock, P.L., Suarez, L.R.: Understanding the behavior of bank spreads in Latin America. J. Dev. Econ. **63**(1), 113–134 (2000)

Doliente, J.S.: Determinants of bank net interest margins in Southeast Asia. Appl. Financ. Econ. Lett. **1**(1), 53–57 (2005)

Fungáčová, Z., Poghosyan, T.: Determinants of bank interest margins in Russia: does bank ownership matter? Econ. Syst. **35**(4), 481–495 (2011)

Garza-García, J.G.: What influences net interest rate margins? Developed versus developing countries. Banks Bank Syst. **5**(4), 32–41 (2010)

Golin, J.: The Bank Credit Analysis Handbook: A Guide for Analysts, Bankers and Investors. Wiley, Singapore (2001)

Gounder, N., Sharma, P.: Determinants of bank net interest margins in Fiji, a small island developing state. Appl. Financ. Econ. **22**(19), 1647–1654 (2012)

Gujarati, D.: Basic Econometrics, 4th edn. Tata McGraw Hill, New Delhi (2004)

Hamadi, H., Awdeh, A.: The determinants of bank net interest margin: evidence from the lebanese banking sector. J. Money Invest. Bank. **23**(3), 85–98 (2012)

Hawtrey, K., Liang, H.: Bank interest margins in OECD countries. N. Am. J. Econ. Financ. **19**(3), 249–260 (2008)

Hempel, G., Coleman, A., Simonson, D.: Bank Management: Text and Cases, 2nd edn. Wiley, New York (1986)

IMF. Financial Soundness Indicators Compilation Guide (2006). http://www.imf.org/external/pubs/ft/fsi/guide/2006/index.htm. Accessed 15 June 2018

Kasman, A., Tunc, G., Vardar, G., Okan, B.: Consolidation and commercial bank net interest margins: evidence from the old and new European Union members and candidate countries. Econ. Model. **27**(3), 648–655 (2010)

Maudos, J., Guevara, J.F.D.: Factors explaining the interest margin in the banking sectors of the European Union. J. Bank. Financ. **28**(9), 2259–2281 (2004)

Maudos, J., Solís, L.: The determinants of net interest income in the Mexican banking system: an integrated model. J. Bank. Financ. **33**(10), 1920–1931 (2009)

Pham, A.H., Vo, L.K.T., Tran, C.K.Q.: The impact of ownership on net interest margin of commercial bank in Vietnam. In: Anh, L., Dong, L., Kreinovich, V., Thach, N. (eds.) Econometrics for Financial Applications, ECONVN 2018. Studies in Computational Intelligence, vol. 760, pp. 744–751. Springer, Cham (2018)

Rose, P.S.: Commercial Bank Management. Irwin/McGraw-Hil, Boston (1999)

Sensarma, R., Ghosh, S.: Net interest margin: does ownership matter? VIKALPA J. Decis. Makers **29**(1), 41–47 (2004)

Saunders, A., Schumacher, L.: The determinants of bank interest margins: an international study. J. Int. Money Financ. **19**(6), 813–832 (2000)

Tarus, D.K., Chekol, Y.B., Mutwol, M.: Determinants of net interest margins of commercial banks in Kenya: a panel study. Procedia Econ. Financ. **2**, 199–208 (2012)

Ugur, A., Erkus, H.: Determinants of the net interest margins of banks in Turkey. J. Econ. Soc. Res. **12**(2), 101–118 (2010)

Zhou, K., Wong, M.C.S.: The determinants of net interest margins of commercial banks in Mainland China. Emerg. Mark. Financ. Trade **44**(5), 41–53 (2008)

Wooldridge, J.: Econometric Analysis of Cross Section and Panel Data. MIT Press, Cambridge (2002)

Economic Integration and Environmental Pollution Nexus in Asean: A PMG Approach

Pham Ngoc Thanh[1], Nguyen Duy Phuong[1(✉)], and Bui Hoang Ngoc[1,2,3]

[1] University of Labour and Social Affairs, Hanoi Campus, Hanoi City, Vietnam
phamngocthanhulsa@gmail.com, nguyenduyphuong@ulsa.edu.vn
[2] University of Labour and Social Affairs,
Ho Chi Minh Campus, Ho Chi Minh City, Vietnam
ngocbh@ldxh.edu.vn
[3] Graduate School - Ho Chi Minh City Open University,
Ho Chi Minh City, Vietnam

Abstract. The nexus between economic integration and environmental pollution has been intensively analyzed by a number of studies, but the empirical evidence more often than not remains controversial and ambiguous. This research applies the estimation technique Pooled Mean Group (PMG) introduced by Pesaran et al. (1999) and the cointegration test of Fisher-Johansen to examine the impacts of economic integration on environmental quality (measured by the CO_2 emissions per capita) in Asean 8 countries during the 1986–2014 period. The empirical results provide a strong statistical evidence that economic integration increases environmental pollution in Asean countries, yet there exists an inverted U-shape of ecological Kuznets curve. The time required to return to equilibrium is 4 years, and the turning point's GDP per capita is about 9,400 US Dollar/year (at constant 2010 prices). This research suggests that policy-makers should control the environmental standards in the projects to improve environmental pollution, to achieve sustainable economic development in the long-run.

Keywords: Economic integration · CO_2 emissions
Environmental pollution · Asean

1 Introduction

Environmental pollution is obviously harmful to the development of the nature. More importantly, it can threaten the wellbeing and lives of people. According to the International Energy Agency, 2 vital factors that lead to environmental pollution are: energy consumption and economic growth. As most Asean countries are developing countries, energy demanded for economic growth always creates great pressure. Economic activities mainly use these 3 types of fuel: coal,

V. Kreinovich et al. (Eds.): ECONVN 2019, SCI 809, pp. 427–439, 2019.
https://doi.org/10.1007/978-3-030-04200-4_31

petroleum and gas. They are contributors to the majority of CO_2 emissions into the environment.

Obviously, the issue of environmental management will be more complicated once the economic integration of each country becomes more intensive and extensive. With the dizzying-speed development of global economic integration and trade freedom, followed by the growth of global economy, people attach greater weight to how such trends will influence the environment. Using panel data for 83 countries in the 1985–2013 period, Wanhai and Zhike (2018) discover the spillover effect of economic intergration on CO_2 emissions; Hoffman et al. (2005), Nadia and Merih (2016) notice negative impacts of economic integration on environmental quality of low- and middle-income countries. Bo et al. (2017) identify the spillover effect of CO_2 emissions' rising rate in 7 regions of China.

Integration and economic integration are broad and abstract concepts. According to Machlup (1977), integration is "the process of combining separate economies into a larger economic region". According to Balassa (1961) economic integration is defined as "the abolition of discrimination within an area". Nowadays, in bilateral and multilateral relationships, a country can promote integration of many dimensions: political integration, socio-cultural integration, defense-security integration, yet economic integration is still fundamental. Then, how does economic integration affect environmental pollution? The answer lies in the methods by which economic integration is measured. There are now three main measures of economic integration including: (i) general integration, (ii) financial integration and (iii) trade openness. Edison et al. (2002) propose the formula for financial integration measurement based on two types of measures: First, measure of just FDI inflows and second, measure of FDI inflows plus outflows. There have been a number of empirical research on the impacts of FDI on environmental quality such as research of Pao and Tsai (2011), Omri et al. (2014), Boluk and Mert (2015), Zhu et al. (2016), Baek (2015), whose conclusions are discrepant. A typical example is the research of Hoffman et al. (2005) on the relationship between FDI and environmental pollution for 3 types of countries. The authors find statistical evidence to conclude that in underdeveloped countries, increases in CO_2 emissions will attract more FDI. In developing countries, more FDI attraction will compound the problem of environmental pollution. However, in developed countries, no relationship is found between these 2 elements.

Some studies tend to measure the incidence of economic integration using the degree of trade openness. Antweiler et al. (2001) find that trade openness is associated with reduced pollution as proxied by SO_2 concentrations. A more recent study by Zhang et al. (2017) also reports that trade openness negatively and significantly affects emissions in 10 newly industrialized countries. However, some studies find that the impact of trade on environnmental quality is varied by the level of income. Le et al. (2016) demonstrate that trade openness has a begin effect on the environment in high-income, but a harmful effect in low and middle income countries.

This research aims at clarifying the impacts of economic integration on environmental pollution in Asean 8 countries including: Indonesia, Laos, Myanmar, Malaysia, Philippines, Singapore, Thailand and Vietnam in the 1986–2014 period, with an approach different from previous research as follows:

Firstly, previous research tend to use FDI, trade openness variable to represent economic integration, which affects environmental pollution. This research uses overall index of globalization (KOF index), which is created by Dreher (2006) with three main dimensions: economic globalization index (36%), social globalization index (37%) and political globalization index (27%).

Secondly, beside analyzing the impacts of economic integration on environmental pollution, the paper will examine the inverted U-shape of environmental Kuznets curve in the relationship of economic growth and CO_2 emissions in Asean 8 countries during the 1986–2014 period. If the inverted U-shape does exist, the research will calculate the value of its turning point.

2 Theoretical Background and Literature Review

Kuznets (1955) proposes a hypothesis of an inverted U-shape that shows the relationship between economic growth and environmental quality, implying that environmental degradation increases with output during the early stages of economic growth, but then declines with output after arriving at a threshold (later called environmental Kuznets curve - EKC). The EKC hypothesis implies that the environment changes from an inferior good at lower income levels to a normal good at some point by policies that both protect the environment and promote economic development.

Even thought there has been many published articles on the effects of economic integration on environment in recent year, there are still many aspects of this concept that need further scrutiny. Wanhai and Zhike (2018) discover indirect effects of globalization on the amounts of gas emissions in 83 countries over the world in the 1985–2013 period. Accordingly, if a country is neighbored by countries with better environmental quality, it is obliged to positively modify its environmental criteria. This conclusion is agreed by research of Burnett et al. (2013) and Jorgenson et al. (2014). Bo et al. (2017) use input-output panel data to analyze spillover effect of economic activities and investments on CO_2 emissions in 7 economic regions of China in the 2007–2010 period. Then they conclude that if economic activities are changed for consumption or export, CO_2 emissions will decline, while their change for production technology investment has ambiguous results: increase in CO_2 emissions in some regions and decrease in others. The authors also suggest modifying negative environmental behaviors of an enterprise with an entire supply chain. Direct and continuous interaction with customers and suppliers will have enterprises promote modification of environmental criteria.

This paper summarizes results of other empirical research on the impacts of economic integration and financial integration on CO_2 emissions as in Table 1. The differences in prior research and the importance of environmental preservation highlight the necessity of further research on this issue.

Table 1. Sumary of empirical results

Author(s)	Countries	Methods	Conclusions
Pao and Tsai (2011)	BRIC	VECM	FDI <-> CO_2
Soytas and Sari (2009)	Turkey	ARDL, Toda & Yamamoto	CO_2 <-> GDP
Baek (2015)	Asean 5	Pool Mean Group	FDI -> CO_2
Dijkagraff et al. (2005)	OECD	FEM	No relationship
Dinh et al. (2014)	Vietnam	ECM & Granger causality	FDI $\neq$ CO_2
Ang (2008)	Malaysia	ECM & Granger causality	CO_2 -> GDP
Boluk and Mert (2015)	Turkey	ARDL & Granger causality	GDP -> CO_2
Lee (2013)	Asean	Panel cointegration & ECM	FDI -> GDP, CO_2 <-> GDP, FDI $\neq$ CO_2
Liu et al. (2018)	China	Spatial regression	FDI -> SO_2; GDP -> SO_2, CO_2
Zhang and Zhou (2016)	China	Spatial regression	FDI -> CO_2

3 Research Model

This research aims at examining the impacts of economic integration on CO_2 emissions in Asean countries in the 1986–2014 period. On the basis of previous research of Halicioglu (2009), Boluk and Mert (2015) and Dinh et al. (2014), this paper proposes a model as follows:

$$CO_{2(it)} = (\beta_0 + v_i) + \beta_1 \mathrm{KOF}_{it} + \beta_2 GDP_{it} + \beta_3 GDP^2_{it} + u_{it} \quad (1)$$

Note: i = 1, 2, ..., 8 corresponding to Indonesia, Laos, Myanmar, Malaysia, Philippines, Singapore, Thailand and Vietnam.

t: time studied during the 1986–2014 period. u: denotes error, v is distinct feature of each country. Data is collected from 1986 to 2014, sources and detailed illustrations of variables are shown in Table 2.

Presently, the Asean consists of 10 countries, yet KOF data of Brunei and Cambodia cannot be found, which is unfortunate and unintended. The KOF index was created and introduced by Dreher in 2006. KOF index is a composite index calculated from 3 indexes: economic globalization index (36%), social globalization index (37%) and political globalization index (27%). KOF index is published annually by Swiss Economic Institution with a scale from zero to 100 point, higher values present higher degree of globalization.

Table 2. Sources and measurement method of variables

Variable	Decription & Measurement	Unit	Source
CO_2	is CO_2 emissions per capita	Metric ton	IEA
KOF	is Overall index of globalization	Point	Swiss Economic Institution
GDP	is the Gross Domestic Product per capita (in constant 2010 US Dollar)	US Dollar	UNCTAD
GDP^2	is GDP square	US Dollar	UNCTAD

Under the EKC hypothesis, in Eq. 1 β_2 is expected positive sign and the β_3 is expected negative sign. That the β_2 is positive means that the greater economic growth the greater carbon emissions. At the same time, that the β_3 is significant and negative means that there is a turning down point on the curve. At this point, increasing economic growth begins to make carbon emissions reduction. In this situation, the peak point of GDP is calculated to be $\beta_2/\left|2.\beta_3\right|$. However, when the β_3 is insignificant, carbon emissions increase monoto- nously. Moreover, the empirical evidence more often than not remains controversial and ambiguous.

4 Research Results and Discussion

4.1 Decriptive Statistics

According to the United Nations Conference on Trade and Development (UNCTAD) and the World Bank, Asean is an active economic region, in which integration is increasingly intensive and extensive. Annually, Asean countries attracts more than 100 billion USD of FDI and enters into a number of negotiation as well as signature of trade and investment agreements with many other regions over the world (UNCTAD, 2016). While the economic growth rate reaches relatively high and stable levels, the economic growth is accompanied by intensified environmental pollution. The International Energy Agency provides data showing that the average CO_2 emissions of Singapore and Malaysia are twice as high as the average global emissions. Although the pollution levels of Indonesia, Philippines, Thailand and Vietnam are still below the global average level, it does not mean that these countries are not negatively affected by environmental pollution, considering its complex continuation in large cities and industrial estates, etc., where average statistics are unable to reflect the situation precisely. The descriptive statistics of variables are shown in Table 3.

Table 3. Decriptive statistics

Variables	Mean	Std.Deviation	Min	Max
CO_2	1.776	2.040	0.047	8.033
KOF	52.54	17.94	21.84	83.15
GDP	6,324	11,205	160.3	52,068

4.2 Stationarity Test

Nelson and Plosser (1982) claim that most economic variables have a trend of increasing over time, thus time series are usually non-stationary at level. Thus, to avoid spurious regression, this paper will examine the stationarity of variables in the model. The ADF test of Dickey and Fuller (1981) for time series data and tests of Levin-Lin-Chu (2002); Breitung (2000) for panel data are illustrated as follows:

a. ADF test: $\Delta Y_t = \alpha_0 + \beta.Y_{t-1} + \sum_{i=1}^{k} \rho_i.\Delta Y_{t-i} + \gamma.T + \varepsilon_t$

b. Breitung test: $\Delta Y_{it} = \rho_i Y_{it-1} + X_{it}\delta_i + \varepsilon_{it}$

Note: Δ is first difference, ε is residual, T is trend, i is cross-section unit, t is time period of the observation in panel data. In Breitung test, X_{it} is exogenous variable fixed for each cross-section unit. If $|\rho_i| < 1$, Y_i is non-stationary. If $|\rho_i| = 1$, Y_i is stationary. The results of variable stationarity test using the method of Levin-Lin-Chu, Breitung, Im-Pesara-Shin (2003), Augmented Dickey & Fuller, Phillips and Perron (1988) are shown in Table 4. Data at level is labeled as I(0), data at first difference as I(1).

Table 4. Unit root test

	LLC		Breitung		IPS		ADF		PP	
Variable	I(0)	I(1)	I(0)	I(1)	I(0)	I(1)	I(0)	I(1)	I(0)	I(1)
CO_2	0.15	−6.37***	0.81	−0.667***	1.81	9.95***	8.11	110***	15.8	133***
KOF	−3.06***	−5.68***	1.76	−4.18***	0.20	−6.58***	16.76	73.1***	14.07	114.8***
GDP	−0.02	−5.86***	0.37	−3.94***	3.36	−4.83***	6.36	51.9***	8.71	81.6***
GDP^2	6.42	−3.82***	9.81	0.57	9.90	−2.66***	0.63	36.3***	0.72	71.3***

*Notes: ***, ** & * indicate 1%; 5% and 10% level of significance*

Table 4 shows that all variables are stationary at first difference, thus the regression analysis needs to use I(1) variables, which satisfies the conditions to apply the Pooled Mean Group (PMG) method by Pesaran et al. (1999).

4.3 Panel Cointegration Test

Having established that all of our variables are I(1), we proceed to test the null of no cointegration. First, we report Pedroni (1999) ADF-based and PP-based cointegration tests as well as Kao (1999) ADF-based tests. Panel cointegration test results in Fig. 1 show that in Pedroni test, 5/7 tests suggest the rejection of the no cointegration, which means long-run cointegration does exist between variables of Eq. 1. Despite the fact that Pedroni's and Kao's cointegration tests are applied to the demeaned data, a procedure suggested in case of suspected cross sectional dependence, strictly speaking these tests do not account for this kind of dependence. Second, to check the robustness of our results we also apply the error-correction-based panel cointegration tests proposed by Johansen (1996). Result of Johansen's test are presented in Fig. 1.

Pedroni Test	Statistic	Prob
Panel v-Statistic	-2.353234	0.9907
Panel rho-Statistic	-2.951320	0.0016
Panel PP-Statistic	-5.231060	0.0000
Panel ADF-Statistic	-1.307242	0.0073
Group rho-Statistic	0.159603	0.5634
Group PP-Statistic	-2.774427	0.0028
Group ADF-Statistic	-4.383895	0.0000
Kao Test		
ADF	-1.569887	0.0582

Cointegration test results using the method of Fisher - Johansen

No. of CE(s)	Fisher Stat. (from trace test)	Prob	Fisher Stat. (from eigenvalue test)	Prob
r = 0	127.8	**0.0000**	85.68	**0.0000**
r = 1	58.42	**0.0000**	46.78	**0.0001**
r = 2	26.26	**0.0504**	25.99	**0.0542**

Fig. 1. Panel cointegration test results

According to Johansen's test, the number of cointegrations is 1, which mean there is panel evidence of a long-run relationship between emissions, income per capita and economic integration across the 8 countries under study. The existence of cointegrations means that estimation results of ordinary least squares (OLS) will be biased. Recently, Pesaran and Smith (1995), Pesaran et al. (1999) introduce a new method of cointegrating estimation called Mean Group (MG) and Pooled Mean Group (PMG) for panel data.

Both of these estimators are based on the maximum likelihood procedure and the autoregressive distributed lags (ARDL), considering the long-run equilibrium as well as accounting for dynamic heterogeneity of the adjustment process. Specifically, the PMG imposes a restriction on the long-run parameters to be similar across panel members, but allows the short-run parameter (together with the speed of adjustment), intercepts, and error variances to be different across the panel (Kim et al., 2010). Although the MG estimates are consistent, Pesaran and Smith (1995) caution that if the long-run homogeneity restrictions are correct, the PMG becomes more appropriate because the MG estimates will be inefficient, which may yield misleading results.

4.4 Empirical Result

Table 4 shows that all variables are stationary at I(1), which satisfies the conditions to use Pooled Mean Group method. Empirical results are presented in Table 5.

Table 5. Empirical result

Variable	Coefficient	Std. Error	t-Statistic	Prob.*
Long Run Equation				
ECT(-1)	-0.276693	0.300879	-0.919617	0.0602
KOF	0.038503	0.003295	11.68555	0.0000
GDP	0.000964	0.000101	9.539411	0.0000
GDP^2	$-5.14E-08$	$8.49E-09$	-6.059984	0.0000

Empirical results in Table 5 present that KOF is positive with a significance level of 1%. This confirms that economic integration increases environmental pollution in Asean countries. Error correction term ECT(-1) is negative with a significance level of 10%, inferring that environmental quality is capable of returning to equilibrium after each short-run "shock" in economic integration and economic growth. Correction time is relatively long, the full convergence to equilibrium level takes about 4 years ($= -1/ECT(-1)$).

GDP is positive, GDP^2 is negative with both significance levels of 1%. This is similar to expected results, which means the inverted U-shaped of Kuznets's hypothesis does exist in the relationship between economic growth and environmental quality in the Asean. The turning point is determined to be at $\beta_2/\,|2.\beta_3| \cong$ 9,377 US Dollar/year. Among 8 countries studied, only Singapore and Malaysia's income per capita exceed this level. However, these are the 2 countries with the highest average CO_2 emissions per capita in the region. While the CO_2 emissions of Singapore is about to decrease, the CO_2 emissions of other countries are expecting an increasing trend. The gaps between present incomes to the turning point are quite large for other 6 countries. According to the UNCTAD (2016) income per capita (at constant 2010 prices) of Indonesia is 3,974 USD; Laos 1,683 USD; Malaysia 11,031 USD; Myanmar 1,175 USD; Philippines 2,753 USD; Singapore 52,458 USD; Thailand 5,962 USD; Vietnam 1,735 USD.

4.5 Granger Causality Test

Lastly, this paper examines the causal relationships between variables. Dumitrescu et al. (2012) develops Engle and Granger (1987) testing techniques for panel data. Causal relationship between 2 variables X and Y for panel data is illustrated as follows:

$$X_t = \alpha_{0,i} + \sum_{j=1}^{m} \alpha_{1,i} X_{i,t-j} + \sum_{j=1}^{m} \beta_{1,i} Y_{i,t-j} + \varepsilon_{i,t}$$

$$Y_t = \alpha_{0,i} + \sum_{j=1}^{m} \alpha_{1,i} Y_{i,t-j} + \sum_{j=1}^{m} \beta_{1,i} X_{i,t-j} + \mu_{i,t}$$

Note: t is time period of panel data, i is cross-section unit in panel data, m is optimal lag length. If there exist $\alpha_{1,i}$ and $\beta_{1,i} \neq 0$ for all i, j. It can be concluded that causality does exist between X and Y. (X Y) pairs in this research are (CO_2 KOF), (CO_2 GDP) and (KOF GDP). Results of Granger causality test of variables in the model using Dumitrescu et al. (2012) method are shown in Table 6 and Fig. 2.

The test results show that there are two-way causality between GDP and CO_2 emissions and one-way causality between KOF and CO_2 emissions. This supports the conclusion that economic integration increases environmental pollution in Asean countries.

Table 6. Result of Granger causality test

Null hypothesis:	W-Stat.	Zbar-Stat.	Prob.
KOF does not homogeneously cause CO_2	6.90377	5.47007	0.0000
CO_2 does not homogeneously cause KOF	3.02583	0.96037	0.3369
GDP does not homogeneously cause CO_2	9.07477	7.99476	0.0000
CO_2 does not homogeneously cause GDP	5.40707	3.72954	0.0002
GDP does not homogeneously cause KOF	2.39675	0.22880	0.8190
KOF does not homogeneously cause GDP	2.32302	0.14306	0.8862

Narayan et al. (2010) and Baek (2015) claim that except from Singapore, Asean countries are all developing countries, which means pressure of economic growth will exceed pressure of environmental pollution. With 36% of KOF index calculated from financial integration index, this conclusion is in agreement with conclusions of Pao and Tsai (2011); Lee (2013); Liu et al. (2018) and Baek

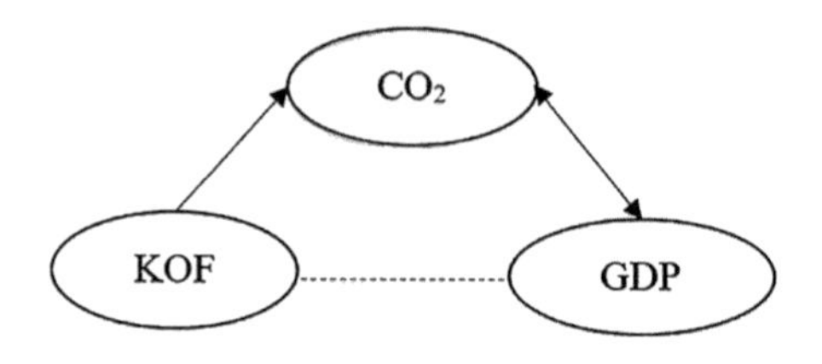

Fig. 2. Causal relationships between variables

(2015). In Vietnam, the relationships between economic integration, FDI and environmental pollution has been rarely researched with time series data. Dinh et al. (2014) do not find statistical evidence to affirm that FDI has impacts on environmental pollution or relaxation of environmental regulations will help Vietnam attract more FDI.

5 Conclusion and Policy Implications

The pursuit of a green and clean living environment is natural and rightful to every society member. With the data of the 1986–2014 period, this research applies Pooled Mean Group estimation method proposed by Pesaran et al. (1999) and Granger causality test using Dumitrescu et al. (2012) method to affirm these 2 following points:

(i): This study adds another empirical research on the impacts of economic integration on environmental pollution in the Asean 8. Evidence shows that Kuznets's hypothesis of inverted U-shaped does exist in the relationship between economic growth and environmental quality in the Asean. The turning point is at approximately 9.400 US Dollar/year (at constant 2010 prices).

(ii): There exist two-way Granger causality between economic growth and CO_2 emissions, one-way causality between economic integration and CO_2 emissions of Asean countries. Its inference is that both economic growth and economic integration will have causal effects on the environment without effective control of the government and its residents.

Based on research results, this paper suggests some considerations when applying these results in practice as follows:

Firstly, the Governments of Asean countries should not wait until the turning point to improve environmental behaviors. Increase in CO_2 emissions will directly threaten the well-being of their residents. As the environmental pollution has been a long-term problem, environmental conservation is also a long-term process. Thus, a little effort of an individual or an organization will contribute to the heightened awareness of environmental preservation and conservation.

Secondly, it is evident that economic integration increases environmental pollution. However, economic integration encourages FDI attraction. Given that the slightest manifestation of environmental pollution is discouraged, it is unrealistic to prohibit FDI projects from producing pollutant. Hence, Government administrative agencies are suggested to improve the verification and validation of FDI projects before, during and after investment processes so that CO_2 emissions can be kept under acceptable level.

The Asean is experiencing more serious environmental pollution, and there still exist differences between this paper's conclusions with others'. Future research are encouraged to add variables of energy consumption, or to measure other types of environmental pollution such as water pollution, litter pollution, noise pollution, etc. as according to Duong and Trinh (2017), environmental pollution in Vietnam is mainly caused by the usage of coal, petroleum and gas. Furthermore, other estimation methods such as spatial regression can be applied. Nearby regions and nations tend to have strong economic interactions due to many factors such as investment flows' direction, labour force and import-export turnover. Similarities in geography, climate and natural resources result in replication of effective policies, which creates the spillover effect of economic policies, including FDI attraction and environmental management policies, on neighboring regions and nations.

References

Ang, J.: Economic development, pollutant emissions and energy consumption in Malaysia. J. Policy Model. **30**(2), 271–278 (2008)

Antweiler, W., Copeland, B., Taylor, S.: Is free trade good for the environment? Am. Econ. Rev. **91**(4), 877–908 (2001)

Baek, J.: The new look at the FDI-Income-Energy-Environment nexus: dynamic panel data analysis of Asean. J. Energy Policy **91**, 22–27 (2015)

Balassa, B.: The Theory of Economic Integration. Richard D. Irwin, Homewood (1961)

Bo, M., Jianguo, W., Robbie, A., Hao, X., Jinjun, X., Glen, P.: Spatial spillover effects in determining Chinas Regional CO_2 emissions growth: 2007–2010. Energy Econ. **63**, 161–173 (2017)

Boluk, G., Mert, M.: The renewable energy, growth and environmental Kuznets curve in Turkey: an ARDL approach. Renew. Sustain. Energy Rev. **52**, 587–595 (2015)

Breitung, J.: The local power of some unit root tests for panel data. Adv. Econ. **15**, 61–177 (2000)

Burnett, J.W., Bergstrom, J.C., Dorfman, J.H.: A spatial panel data approach to estimating US state-level energy emissions. Energy Econ. **40**, 396–404 (2013)

Dickey, D.A., Fuller, W.A.: Likelihood ratio statistics for autoregressive time series with a unit root. Econometrica **49**, 1057–1072 (1981)

Dijkgraaf, E., Herman, R.J.V.: A test for parameter heterogeneity in CO_2 panel EKC estimations. Environ. Resour. Econ. **32**(2), 229–239 (2005)

Dinh, H.L., Lin, S.M.: CO_2 emissions, energy consumption, economic growth and FDI in Vietnam. Manag. Glob. Trans. **12**(3), 219–232 (2014)

Dreher, A.: Does globalization affect growth? Evidence from a new index of globalization. Appl. Econ. **38**(10), 1091–1110 (2006)

Dumitrescu, E.-I., Hurlin, C.: Testing for granger non-causality in heterogeneous panel. Econ. Model. **29**, 1450–1460 (2012)
Duong, M.H., Trinh, N.H.A.: Two scenarios for carbon capture and storage in Vietnam. Energy Policy **110**, 559–569 (2017)
Edison, H.J., Levine, R., Ricci, L., Slot, T.: International financial integration and economic growth. J. Int. Money Financ. **21**(6), 749–776 (2002)
Engle, R.F., Granger, C.W.J.: Co-integration and error correction: representation, estimation, and testing. Econometrica **55**, 251–276 (1987)
Halicioglu, F.: An econometric study of CO_2 emissions, energy consumption, income and foreign trade in Turkey. Energy Policy **37**, 1156–64 (2009)
Hoffmann, R., Lee, C.G., Ramasamy, B., Yeung, M.: FDI and pollution: a granger causality test using panel data. J. Int. Dev. **17**(3), 1–13 (2005)
Johansen, S.: Likelihood-Based Inference in Cointegrated Vecto Auto-Regressive Models, 2nd edn. Oxford University Press, Oxford (1996)
Jorgenson, A.K., Givens, J.E.: Economic globalization and environmental concern: a multilevel analysis of individuals within 37 nations. Environ. Behav. **46**(7), 848–871 (2014)
Kao, C.: Spurious regression and residual-based tests for cointegration in panel data. J. Econ. **90**, 1–44 (1999)
Kuznets, S.: Economic growth and income inequality. Am. Econ. Rev. **45**, 1–28 (1955)
Kim, D.H., Lin, S.C., Suen, Y.B.: Dynamic effects of trade openness on financial development. Econ. Model. **27**(1), 254–261 (2010)
Le, T.H., Chang, Y., Park, D.: Trade openness and environmental quality: international evidence. Energy Policy **92**, 45–55 (2016)
Lee, W.J.: The contribution of foreign direct investment to clean energy use, carbon emissions and economic growth. Energy Policy **55**, 483–489 (2013)
Levine, A., Lin, C.F., Chu, C.S.: Unit root tests in panel data: asymptotic and finite-sample properties. J. Econ. **108**, 1–24 (2002)
Liu, Q., Wang, S., Zhang, W., Zhan, S., Li, J.: Does foreign direct investment affect environmental pollution in China's cities? a spatial econometric perspective. Sci. Total. Environ. **613**, 521–529 (2018)
Machlup, F.: A History of Thought on Economic Integration. Columbia University Press, New York (1977)
Nadia, D., Merih, U.: Globalization and the environmental impact of sectoral FDI. Econ. Syst. **40**(4), 582–594 (2016)
Narayan, P.K., Narayan, S.: Carbon dioxide emissions and economic growth: panel data evidence from developing countries. Energy Policy **38**, 661–666 (2010)
Nelson, C., Plosser, C.: Trends and random walks in macroeconmic time series: some evidence and implications. J. Monet. Econ. **10**(2), 139–162 (1982)
Omri, A., Khuong, N.D., Rault, C.: Causal interactions between CO_2 emissions, FDI, and economic growth: evidence from dynamic simultaneous-equation models. Econ. Model. **42**, 382–389 (2014)
Pao, H.T., Tsai, C.M.: Multivariate Granger causality between CO_2 emissions, energy consumption, FDI and GDP: evidence from a panel of BRIC Countries. Energy Econ. **36**, 685–693 (2011)
Pedroni, P.: Critical values for cointegration tests in heterogeneous panels with multiple regressors. Oxf. Bull. Econ. Stat. **61**, 653–670 (1999)
Pesaran, M.H., Smith, R.J.: Estimating long-run relationships from dynamic heterogeneous panels. J. Econ. **68**, 79–113 (1995)
Pesaran, M.H., Shin, Y., Smith, R.J.: Pooled mean group estimation of dynamic heterogeneous panels. J. Am. Stat. Assoc. **94**(446), 621–634 (1999)

Phillips, P.C., Perron, P.: Testing for a unit root in time series regression. Biometrika **75**, 335–346 (1988)

Soytas, U., Sari, R.: Energy consumption, economic growth, and carbon emissions: challenges faced by an EU candidate member. Ecol. Econ. **68**(6), 1667–1675 (2009)

You, W., Lv, Z.: Spillover effects of economic globalization in CO_2 emissions: a spatial panel approach. Energy Econ. **73**, 248–257 (2018)

Zhang, C., Zhou, X.: Does foreign direct investment lead to lower CO_2 emissions? Evidence from a regional analysis in China. Renew. Sustain. Energy Rev. **58**, 943–951 (2016)

Zhang, Y.J.: The impact of financial growth on carbon emissions: an empirical analysis in China. Energy Policy **39**, 2197–2203 (2011)

Zhang, S., Liu, X., Bae, J.: Does trade openness affect CO_2 emissions: evidence from ten newly industrialized countries? Environ. Sci. Pollut. Res. **24**(21), 17616–17625 (2017)

Zhu, H., Duan, L., Guo, Y., Yu, K.: The effect of FDI, economic growth and energy consumption on carbon emissions in ASEAN-5: evidence from panel quantile regression. Econ. Model. **58**, 237–248 (2016)

The Threshold Effect of Government's External Debt on Economic Growth in Emerging Countries

Yen H. Vu(✉), Nhan T. Nguyen, Trang T. T. Nguyen, and Anh T. L. Pham

Banking Faculty, Banking Academy of Vietnam, 12 Chua Boc Street, Dong Da District Hanoi, Vietnam
{yenvh, nhantn, trangntt, lamanh}@hvnh.edu.vn

Abstract. This paper aims to examine the threshold effect of Government's external debt on economic growth in a group of 10 emerging countries. By employing panel data of 10 countries for the period from 2005 to 2015, our empirical results indicate that the threshold of Government's external debt to domestic product (GDP) ratio is 33.17%. We estimate that 1% rise in government external debt ratio corresponds with 0.056% rise in GDP at the level lower than 33.17% of GDP, showing a positive correlation between economic growth and the explanatory variables namely external debt. However, every additional 1% rise in debt-to-GDP ratio beyond the debt threshold costs 0.02% of annual average GDP growth.

Keywords: Threshold · External debt of government · Panel data

JEL Classification Numbers: F34 · C23 · C24

1 Introduction

In recent years, Vietnam is considered as one of the most dynamic emerging countries in East Asia region with significant but unsustainable economic growth. This is due to the fact that Vietnam has benefited from a program of internal restructuring, a transition from the agricultural base toward manufacturing and services, and a demographic dividend powered by a youthful population, especially since the accession of the country in the World Trade Organization in 2007, normalizing trade relations with the United States and ensuring that the economy is consistently ranked as one of Asia's most attractive destinations for foreign investors, becoming one of Asia's most attractive destinations for foreign investors. This had led to the fact that public debt and Government's external debt ratio have increased at an accelerating rate. Even though these ratios are considered "safe" and under control, there would be numerous problems in terms of ineffective public investment and public sectors in association with persistent budget deficit that the government has to solve in order to reach external debt sustainability. If the policymakers do not impose strict measures in the process of fiscal reform and foreign debt control, Vietnam will likely face debt crisis as the case of

V. Kreinovich et al. (Eds.): ECONVN 2019, SCI 809, pp. 440–451, 2019.
https://doi.org/10.1007/978-3-030-04200-4_32

Greece and the other crisis countries in the Eurozone. Therefore, this study focuses on investigating the government's foreign debt threshold in emerging countries which have similar macro-economic conditions as but enjoy higher level of development than Vietnam. Based on our empirical findings, some policy implications will be provided for Vietnam in controlling and managing external debt of the government.

Determining the external debt threshold means seeking the optimal debt-GDP ratio that the economy are able to maintain without exerting a negative impact on economic growth. Beyond the threshold, however, an increase in the debt ratio will have a significant adverse influence on economic growth and erode the economy as a whole. Adopting the Hansen threshold effect model applied on a sample of 10 emerging countries in 2005–2015, the study shows that estimated results show that government external debt threshold is 33.17%. Specifically, GDP growth rate increases by 0.056% on average before debt ratio reaches the threshold and decreases by 0.02% when the debt ratio exceeds 33.17%.

The paper is organized as follows. Section 2 reviews the literature on effect of external debt on economic growth. Section 3 presents our panel threshold model and presents the empirical findings on debt-threshold effects and the impact of debt accumulation on economic growth. Some concluding remarks are provided in Sect. 4.

2 Literature Review

There has been much the previous empirical literature on the relationship between debt and growth and until recently a lot of research has focused on the role of external debt in both developed and developing countries. Some concluded a significant positive relationship between the two variables because it is found that external debts can stimulate the economic growth by meeting the domestic investment need for national projects and encouraging domestic production. External debt at a reasonable level has increased the aggregate demand, return on investment, and thereby boosting investment despite rising interest rates (Eisner et al 1984). Besides, Sachs (1998) argued countries using external debts would be able to boost their economies. However, some studies provided evidence of adverse effect of external debts on economic growth. Modigliani (1961) stated that governments would raise taxes to repay the external debts, thereby reducing the net income of households, which would lead to an reduction in expected returns and barely stimulate investment for economic growth. In addition, as external debts put a pressure on interest rates, thus reducing private investment and causing economic slowdowns (Friedman 1988). Safia et al. (2009) conducted a study on 24 developing countries during the period 1976–2003 and confirmed that the high level of debt had negative effects on economic growth due to the debt overhang problem. Therefore, these contrasting empirical findings suggest that external debts can have either a positive or negative impact on economic growth. Reasonable levels of external debt that help finance productive investment may be expected to enhance economic growth, but beyond certain levels additional indebtedness may impede growth in a way that it can slow down the growth process or impact growth negatively. In another study by Clements et al. (2003) with the employment of the Generalized Method of Moments approach and fixed-effect model for 55 low income countries in the period from 1970 to 1999, the empirical results show that the threshold level of the external debt is in the range of 30 and 37%. Frimpong et al. (2006) studied on a lager sample size with 93 developing countries during the period from 1969 to 1998 and proved that there is a significant negative effect of external debt on GDP growth when the debt-GDP ratio varies from 35 to 40%. Aside from external debt, this study also included other independent variables such as domestic investment, economic openness, and foreign direct investment. According to Tokunbo et al. (2007), the threshold level of external debt of Nigeria from 1970 to 2003 is 60% of GDP. Other empirical research share the similar threshold effect of external debt on the growth of the economy such as studies by Savvides (1992), Pattillo et al. (2002), Moss et al. (2003), and Safia (2009).

In Vietnam, Nguyen H. T. (2012) proved a non-linear relationship of external debt and economic growth during the period from 1986 to 2009 and found the threshold level of external debt for Vietnam is approximate 65%. A one percent increase in the external debt-real GDP ratio led to a growth in the real GDP by USD 15.76987 million per year. If the external debt–real GDP ratio exceeds 65%, the real GDP would decrease by USD 22.9528 million per annum. In addition to external debt, economic openness (as measured by the sum of exports and imports) has also slightly contributed to enhancing economic growth. In addition, Nguyen and Pham (2013) also identified the external debt threshold for Vietnam during 1985-2011. The research plots the discrete points of GDP and external debt-GDP ratios on a second-order curve and determining the maximum point. The authors also argued that the threshold level of external debt for Vietnam during the research period is 65% of GDP. If the external debt ratio is less than 65% of GDP, an increase in external debt will lead to a corresponding increase in economic growth, and a ratio that exceeds 65% will create a heavy burden on Vietnam's economic growth. The authors used the ratio of external debt to GDP and the openness of the economy as the independent variables to explain economic growth.

While many studies have analyzed the effect of external debt on economic growth in Vietnam and emerging markets, limited attention has been focused on identifying the threshold effect of the government external debt. This paper seeks to fill this literature gap by investigating whether the debt-growth relation varies with the level of indebtedness or in other words, determining the optimal government external debt ratio of a group of 10 emerging countries that these economies can maintain without having a negative impact on economic growth.

3 Methodology

3.1 Model Selection

To assess the implication of foreign debt and estimate the threshold of foreign debt to economic growth, following Pham and Nguyen (2015), the study employs panel threshold model proposed by Hansen (2000)

$$\begin{aligned} y_{it} =& \mu_i + \beta_1' x_{it} I(q_{it} < \gamma_1) + \beta_2' x_{it} I(\gamma_1 < q_{it} < \gamma_2) + \beta_3' x_{it} I(\gamma_2 < q_{it} < \gamma_3) \\ &+ \beta_4' x_{it} I(q_{it} > \gamma_3) + \theta Z_{it} + e_{it} + \mu_i \end{aligned} \tag{1}$$

where, y_{it} denotes the growth of real GDP, q_{it} is logarithm of government's foreign debt/GDP ratio, γ_1 is threshold need to estimate; *I(,)* is the indicator function. With *I* = *1* if the condition inside in blankets is satisfied, otherwise is equal), μ_i and e_{it} are illustrated for the fix effect over the time and space, which are not observed.

Unlike Hansen (2000), in this study, x_{it} is only the variable changing over threshold, in particular, that is variable Ln (government foreign debt/nominal GDP). Z_{it} is vector controlling the impact of other macroeconomic variables on the economic growth at the points of threshold.

In the function (1), in the case of existing one threshold, β_1' will be estimated when the foreign debt is under the threshold point. When this debt is greater than threshold, β_2' will be estimated. If threshold exists, β_1' is expected to be positive while β_2' is expected to be negative. If the model has more than one threshold, β_2' is still expected to be positive while at least one of three slope namely $\beta_2', \beta_3', \beta_4'$ is expected to be negative.

The function (1) is also used to evaluate the effect of other macroeconomic variables on GDP at the threshold point through vector Z_{it}. Z_{it} includes 6 variables, namely: economic openness, inflation, aggregate investment, government spending, real exchange rate and lending rate.

To estimate the threshold for panel data set, besides threshold-estimating model, the estimating method is also very important. Normally, three method including Pool OLS, Fixed Effect (FE) and Random Effect (RE) are under consideration. However, in this research, the result of the test such as F, Wald, Pesaran, Breusch and Pagan Lagrangian multiplier and Hausman indicate that FE is the most appropriate method to estimate the threshold model (Table 1).

Table 1. The test for choosing the most appropriate method

Type of test	Goal	Test statistic	Conclusion
F test	Test the existence of threshold model of three method POLS, FE, RE	– POLS: Pro > F = 0.000 – FE: Pro > F = 0.000 – RE: Pro > chi2 = 0.000	Threshold model all exists with these methods
Wald Test	Test the existence of heteroskedasticity u_i for the FE	Prob > chi2 = 0.0000	u_i exist then FE is more appropriate than POLS
Pesaran Test	Test whether RE is suitable or not	Pr = 0.0000	RE is likely to be an appropriate method
Breusch and Pagan Lagrangian multiplier Test	Select between OLS and RE	Prob > Chibar2 = 0.0003 < 0.05	RE is more suitable to estimate threshold
Hausman	Select between RE and FE	Prob > chi2 = 0.000 < 0.05	FE is more appropriate than RE

(Source: The authors)

3.2 Data and Statistics

Among the emerging nations, the study selected 10 countries including: Thailand, Brazil, Chile, Indonesia, Malaysia, Mexico, Russia, Peru, South Africa and Poland to estimate the threshold effect model. Besides using the data on the external debt, normal GDP, the growth rate of real GDP, this research also collects statistics concerning inflation, government spending, real exchange rate, openness (import+export) and lending rate between the period 2005 Quarter 1st and 2015 quarter 4th in these countries. This is because according to Ramzan and Ahmad (2014), the effect of external debt on economic growth is also influenced by the macroeconomic policies such as monetary policy, fiscal policy and trade policy.

Statistics for such variables in this model were retrieved from IMF, FED and these countries' central bank. The model variables include: the growth rate of real GDP (rgdp-%), normal GDP measured in domestic currency (ngdp), government spending-billion domestic currency (spending), government investment-billion domestic currency (inv), real exchange rate (er), total trade-billion domestic currency (opness), external debt-billion domestic currency (exdebt), lending rate (Lendrate) and inflation. Based on the data of these variables, the study coded such variables as listed at the Table 2.

Table 2. The form of variables

Variables	Types	Note
GDPG	Dependent Variable	The growth rate of GDP
EXDEBT	Threshold Variable	Ln(external debt/GDP)
INVEST	Explanatory variable	Ln(investment/nominal GDP),
SPENDING	Explanatory variable	Ln(government spending/nominal GDP)
INF	Explanatory variable	inflation
EX	Explanatory variable	Ln(real exchange rate)
LENDRATE	Explanatory variable	Ln(lending rate)
OPNESS	Explanatory variable	Ln(total trade/nominal GDP)

For the threshold variable- EXDEBT, the impact of this variable on the economic growth is both negative and positive. While Mohamed (2013) and Daud and Podivinsky (2012) indicate that external debt has a negative effect on economic growth in Tunisia and other 31 developing countries, Butts et al. (2012) find out that for the period from 1970 to 2003, in short-term, the relationship between external debt and economic growth is negative. INVEST, SPENDING and OPNESS are expected to have a positive impact on GDP growth as these variables are deemed as the elements of aggregate demand. While INF is expected to have a negative impact on GDP growth, ER is expected to be positively correlated with economic growth, as the increase in exchange rate is likely to encourage export. The impact of lending rate on GDP growth is negative as the upward trajectory in the lending rate is likely to have negative effect on investment and consumption.

Based on the results displayed in Table 3, it could be seen that GDPG fluctuates between −0.11 and 0.18, which is equivalent to the GDP growth being from −11% to 18%/quarterly. The mean of GDPG is 0.0388 equivalent to 3.88%. The maximum of EXDEBT is 0.2366 and minimum is −4.2953. This means the rate of external debt over GDP is from 0.0136 to 1.267 times. The value of SPENDING is from −2.8 to −1.467, which means government spending accounts from 6% to 20% of GDP. This number for INVEST is from 12.19% to 35.43%. For INFLATION, the statistic of this variable points out that the quarterly inflation of these nations fluctuates from −3% to 17%. This number of ER is from 4.14 to 4.76. The value of LENDRATE fluctuates from 1.194 to 4.026 and OPNESS between −3.392 and 4.642.

Table 3. The summarization of statistic variables

Variable	Observation	Mean	Std. Dev.	Min	Max
GDPG	440	0.038848	0.033128	−0.11151	0.185709
EXDEBT	440	−1.770112	1.34871	−4.29532	0.236682
SPENDING	440	−1.96697	0.298377	−2.8013	−1.4647
INVEST	440	−1.50924	0.200026	−2.1043	−1.0375
INFLATION	440	0.046611	0.032259	−0.03027	0.17793
ER	440	4.554132	0.091345	4.141435	4.763524
LENDRATE	440	2.345373	0.636407	1.194932	4.02416
OPNESS	440	.0304871	1.931224	-3.392434	4.624131

(Source: The authors)

To ensure the accuracy of the estimating model, the test for multicollinearity among variable is conducted through variance inflation factors (VIF) indicator (Table 4).

Table 4. The result of multicollinearity test

Variable	VIF	1/VIF
OPNESS	2.09	0.479359
EXDEBT	1.91	0.522872
SPENDING	1.83	0.547594
LENDRATE	1.63	0.613972
ER	1.48	0.674416
INFLATION	1.40	0.711775
INVEST	1.36	0.736290
Mean VIF	1.67	

(Source: The authors)

All VIF indicator is less than 10, this implies that multicollinearity does not exist.

The stationary test of all variables is also conducted through Levin- Lin- Chu test. Table 5 summarizes the result of stationary test.

Table 5. The result of stationary test

Variable	Statistic	P-value	Conclusion
GDPG	0.5768	0.000	Stationary at the level of 5%
EXDEBT	−5.6066	0.000	Stationary at the level of 5%
INVEST	−6.8741	0.000	Stationary at the level of 5%
SPENDING	−10.8545	0.000	Stationary at the level of 5%
INF	−4.0703	0.000	Stationary at the level of 5%
ER	−1.9111	0.028	Stationary at the level of 5%
LENDRATE	2.6435	0.041	Stationary at the level of 5%
OPNESS	0.1970	0.000	Stationary at the level of 5%

(Source: The authors)

The test result indicate that all variables are stationary at the level of 5%. Thus, the estimating results are reliable (Table 6).

Table 6. Empirical and threshold testing results

Number of threshold	Threshold value	Low value	High value	P-value threshold testing with 5% significant	Fstat	Crit5	Crit 1
0–1	−1.1034	−1.1187	−0.9865	0.0267	23.28	20.325	28.689
1–2	−2.5419	−2.6412	−2.5293	0.64	6.71	21.682	29.219
2–3	−0.2818	−0.3172	−0.2782	0.6733	7.23	21.449	26.175

(Source: The authors)

3.3 Empirical Results

The results show that there are three threshold levels at –1.1034; −2.5419; −0.2818 and −1.8398, which are equivalent to 33.17%; 7.87%; 75.44%; and 15.88% respectively in the external debt- GDP ratios. Following Hansen (2000) and Wang (2015), this study applies bootstrap technique and p-value and F- test to test the existence of these the threshold levels. The result of these test suggests that the model only has the unique threshold at −1.1034 with P- value = 0.0267 < 0.05 and F-stat = 23.28 > F-crit 5% = 20.325. The threshold at −1.1034 is equivalent to 33.17%external debt- GDP ratio. With the 95% confident, the threshold level varies from −1.1187 to 0.9865, equivalent to external debt- GDP ratio, varying in the range of 32.68% and 37.28%.

Along with the identified threshold level, the results of threshold effect model for Government's external debt on economic growth in a group of ten emerging countries a follows (Table 7):

Table 7. Results of threshold effect model

	GDPG	Coefficient	Std. Err.
	SPENDING	1.713103*** [0.000]	1.131343
	INVEST	3.208597** [0.018]	0.9066967
	INFLATION	−2.728852 [0.688]	6.793504
	ER	3.544812* [0.061]	1.885803
	LENDRATE	−3.184615*** [0.000]	0.7037365
	OPNESS	4.66771*** [0.000]	0.8699659
	EXDEBT		
0		5.58934*** [0.009]	0.6023942
1		−2.22041*** [0.001]	0.6645716
	Cons	−17.72781* [0.057]	9.293655

Note:
p-values in brackets; ***, **, * indicate significance at 1%, 5% and 10%, respectively.
(Source: The authors)

Based on results of estimation, the threshold effect model for Government's external debt on economic growth in a group of 10 emerging countries is featured below:

$$y_{it} = \mu_{it} + 5.58934\,EXDEBT * I(EXDEBT < -1.1034) + (-2.22041)EXDEBT * I(EXDEBT > -1.1034) + 1.7131\,SPENDING + 3.208597\,INVEST + (-2.728852)INFLATION + 3.544812\,ER + (-3.184641)LENDRATE + 4.66771\,OPNESS + e_i \quad (2)$$

The estimation shows that with 5% significant and the threshold level of 33.17%, the impact of government external debt on economic growth is 5.58934. It means that at the level lower than 33.17% of GDP, every percent increase in Government's external debt ratio is associated with 0.056% rise in GDP. However, as the debt ratio moves beyond 33.17%, the impact of Government's external debt on economic growth is −2.2204. It means that as the debt ratio moves beyond 33.17%, the effect on economic growth shifts from positive to negative and GDP growth rate will decrease by 0.02% for 1% rise in debt ratio.

4 Conclusion

This paper targets on determining the threshold effect of government external debt-to-GDP ratios on economic growth. By adopting Hansen threshold estimation approach (1996, 2000), our findings indicate that emerging markets have the threshold of Government's foreign debt ratio of 33.17%, relative lower than that in developed countries. The estimation results show that Government foreign debt contributes positively to output growth when this ratio is below 33.17% but acts as a deterrent for economic growth when it moves beyond 33.17% of GDP. However, before reaching the optimal debt ratio, the negative impacts on GDP have been recorded. Research of Checheria and Rother (2010) conducted in European countries has indicated that the threshold of public debt ratios is in range of 90–100% of GDP. The confidence interval for debt turning point suggest that the negative growth effect of high debt may be existent from the level of around 70–80% of GDP, therefore, the government needs more prudent indebtedness policies before adjusting the threshold.

Estimation results also bring policy implications for the case of Vietnam, since our sample includes ten emerging markets which have similar background as Vietnam. These countries have a threshold Government's external debt ratio of 33.17% which means Vietnam which has a relatively poorer economic performance will likely have a lower level of threshold. Currently, Vietnamese sovereign debt in foreign currencies and in general is about 25% and 65% of GDP respectively. These ratios are relatively high and considered alarming in the text of existing deficit budget and poor public investment. Therefore, policy-makers in Vietnam should reduce debt ratio consisting of foreign debt and public debt by taking measures such as implementing fiscal reform, reducing administration spending, restructuring State owned enterprises sector in order to maintain fiscal sustainability and macroeconomic stability.

Appendix 1.

The result of Wald test in FE method

Modified Wald test for groupwise heteroskedasticity in FE regression model
H0: sigma(i)^2 = sigma^2 for all i
Chi2 (10) = 438.14
Prob>chi2 = 0.0000

Appendix 2.

The result of Pesaran test

Pesaran's test of cross sectional independence = 12.739; Pr = 0.0000
Average absolute value of the off-diagonal elements = 0.363

Appendix 3.

The result of Breusch and Pagan Lagrangian test

	Varsd = sqrt (Var)	
GDP	0.0010975	0.0331282
e	0.0008235	0.0286967
u	0.0000931	0.0096463

Test: Var (u) = 0
chibar2 (01) = 11.77
Prob > chibar2 = 0.0003

Appendix 4.

The result of Hausman test

	FE	RE	Difference	S.E.
EXDEBT	0.0096841	0.0047847	0.0048993	0.0034697
SPENDING	−0.0514676	−0.0419642	−0.0095034	0.008002
INVEST	−0.0093191	−0.0137731	0.004454	0.0038296
INFLATION	−0.0949191	−0.0476258	−0.0472933	0.0368124
ER	0.1008677	0.072941	0.0279267	0.0075403
LENDRATE	−0.0174574	0.0016936	−0.0191509	0.0058338
OPNESS	0.0593208	0.002825	0.0564958	0.0112239
Cons	−0.4623925	−0.3836922	−0.0787003	0.0317918

Test: H0: difference in coefficients not systematic
Chi2 = 44.42
Pro > Chi2 = 0.0000

Appendix 5.

Debt threshold estimator

Model	Threshold	Lower	Upper
Th-1	−1.1034	−1.1187	−0.9865
Th-21	−2.5419	−2.6412	−2.5293
Th-22	−0.2818	−0.3172	−0.2782
Th-3	−1.8398	−1.8403	−1.8372

Appendix 6.

Debt threshold effect test

Threshold	RSS	MSE	Fstat	Prob	Crit10	Crit5	Crit1
Single	0.3010	0.0008	23.28	0.0267	17.2186	20.3255	28.6897
Double	0.2960	0.0007	6.71	0.6400	17.2783	21.6821	29.2193
Triple	0.2907	0.0007	7.23	0.6733	17.7834	21.4489	26.1754

References

Checherita, C., Rother, P.: The impact of high and growing government debt on economic growth–an empirical investigation for the Euro Area, Working Paper, No 1237/August 2010 (2010)

Clements, B., Bhattacharya, R., Nguyen, T.Q.: External Debt, public investment and growth in low-income countries, IMF working paper, No 03/249 (2003)

Eisner, R., Pieper, P.J.: A new view of the federal debt and budget deficits. Am. Econ. Rev. **74**, 11–29 (1984)

Friedman, B.M.: Day of Reckoning: The Consequences of American Economic Policy Under Reagan and After. Random House, New York (1988)

Frimphong, J.M., Oteng-Abayie, E.F.: The impact of external debt on economic growth in ghana: a cointegration analysis. J. Sci. Technol. **26**(3), 121–130 (2006)

Hansen, B.E.: Sample splitting and threshold estimation. Econometrica **68**(3), 575–603) (2000)

Modigliani, F.: Long run implications of alternative fiscal policies and the burden of the national debt. Econ. J. **71**, 730–755 (1961)

Moss, T.J., Chiang, H.S.: The other costs of high debt in poor countries: growth, policy dynamics, and institutions, Debt sustainability issue paper, World Bank, no. 3 (2003)

Nguyen, H.T.: The relationship between external debt and economic growth in Vietnam. Dev. Integr. J. **4**(14) (2012)

Nguyen, X.T., Pham, T.K.V.: Determine the debt threshold to 2020 for Vietnam by Laffer model. Bank. J. **18**, 13–16 (2013)

Pattillo, C., Poirson, H., Ricci, L.: What are the channels through which external debt affects growth? IMF Working Paper, WP/04/15 (2004)
Pham, T.A., Nguyen, H.N.: Effect of public debt threshold to economic growth- the implication for Vietnam. Econ. Dev. J. **216**(II) (2015)
Safia, S. (2009), Does External Debt Affect Economic Growth: Evidence from Developing Countries
Savvides, A.: Investment slowdown in developing countries during the 1980s: debt overhang or foreign captial inflow? Int. Rev. Soc. Sci. **45**(3), 363–378 (1992)
Osinubi, T.S., et al.: Budget deficits, external debt and economic growth in Nigeria (2007)
Wang, Q.: Fixed-effect panel threshold model using stata. Stata J. **15**(1), 121–134 (2015)

Value at Risk of the Stock Market in ASEAN-5

Petchaluck Boonyakunakorn[1(✉)], Pathairat Pastpipatkul[1], and Songsak Sriboonchitta[2]

[1] Faculty of Economics, Chiang Mai University, Chiang Mai, Thailand
petchaluckecon@gmail.com, ppthairat@hotmail.com

[2] Puay Ungphakorn Center of Excellence in Econometrics, Faculty of Economics, Chiang Mai University, Chiang Mai, Thailand
songsakecon@gmail.com

Abstract. This paper analyzes the Value at Risk (VaR) of ASEAN-5 stock market indexes by employing Bayesian MSGARCH models. The estimated MSGARCH models with two-regime results show that the two regimes have different unconditional volatility levels and volatility persistence for all ASEAN-5 stock return. This different parameter estimate shows that the volatility process evolution is heterogeneous across the two regimes. Therefore, MSGARCH with two-regime should provide a better result than the standard GRACH model since Markov-switching model can capture characterize the time series behaviors in different regimes. For the estimated VaR results, we found that Philippines stock return presents the highest risk, whereas it provides the highest average yield among ASEAN-5 which is attractive for risk-lover investors. Malaysia is the preferred one for the risk-averse investors since it presents the lowest VaR, but provides a high return. Thailand stock return offers the median risk and median returns among ASEAN-5. Singapore stock return presents a high VaR estimate, but provides the lowest yield, being the most not attractive for investors.

Keywords: Value-at-Risk · ASEAN-5 · Stock market
Markov-switching

1 Introduction

With the rapidly globalized financial market, the number of foreign investments has significantly increased in the Association of Southeast Asian Nations (ASEAN). This is the consequence of the rapid pace of ASEAN during the twentieth century. The financial markets of ASEAN also improved their policies to facilitate foreign investment (Wang and Liu [12]). Therefore, these above reasons have attracted the international investors, who attempt to seek opportunities to diversify their portfolios by exploring higher returns from any investment. However, the high return is not the only determinant factor that the investors consider. They also take account of the risk since the higher return carries typically

V. Kreinovich et al. (Eds.): ECONVN 2019, SCI 809, pp. 452–462, 2019.
https://doi.org/10.1007/978-3-030-04200-4_33

with higher risk. As a result, we would like to investigate the risk in ASEAN-5 stock market indexes consisting of the Jakarta Stock Exchange (JSK) of Indonesia, the Kuala Lumpur Stock Exchange (KLSE) of Malaysia, the Philippines Stock Exchange (PSE), the Stock Exchange of Thailand (SET), and the Singapore Strait's Time Index (STI).

Only a few papers investigated the ASEAN-5 stock index returns, such as the studies of Guidi and Gupta [7] and Kiwiriyakun [8]. The former showed evidence that among ASEAN-5, the Indonesian stock return has the most significant volatility response to a negative shock. Meanwhile, the latter proposed that Thailand stock market (SET) has the highest risk premium followed by Singapore, Malaysia and Philippine. Philippines stock return also has the negative risk premium. Our study objective is to examine the risk of stock return indexes in ASEAN-5, and investigate the risk-return performance by using the average return and Value at Risk (VaR), helping investors to seek the opportunities to diversify their portfolio in which countries can make gain more profits based on the risks they are facing.

For measuring the market risk, one of the most widely used indicators is Value at Risk (VaR) proposed by J.P. Morgan in the 80s. The concept is linked with the losses. VaR has been an increasingly important measure of risk since the Basel Committee required that banks should cover losses on their trade portfolio over a ten days horizon with 99% of the time. To estimate the VaR for five stock ASEAN index, we employ GARCH models. In the beginning, the stock returns are assumed to be normal distributed. Then, many researchers found that the residuals of financial series frequently are non-normally distributed with exhibiting skewness, and excess kurtosis. According to Aumeboonsuke [2], it was found that the beta and returns are asymmetric in both up and down periods in Thailand, Singapore and Malaysia market return by using the conditional capital asset pricing model. The misspecification of GARCH model would lead to underestimate or overestimate of the VaR. After that, some researchers improved the VaR estimation by taking asymmetric of the distribution into consideration. For example, Guidi and Gupta [7] considered the asymmetry in forecasting volatility of ASEAN-5 stock return by using Asymmetric-Threshold-GARCH (TARCH) model with student-t and GED distribution. Thus, this study we consider asymmetric by using both asymmetric and symmetric GARCH type models with different error distributions. However, these models do not take account of nonlinear structure in the variance structure, which might lead to biased results. This is because GARCH models have the persistent behavior of shocks to conditional variance.

Many financial series often show evidence of the existence of regime change between normal volatility and a high volatility state. When the market goes back to normal state, the conditional variance will be overestimated. When the market changes from volatility state to high volatility state, the conditional variance will be underestimated as a consequence. To cope with this issue, we apply Markov-Switching (MS) model with GARCH model, since MS can capture characterize the time series behaviors in different regimes. According to Billio and Pelizzon [3], it was found that switching regimes provide more accurate result in estimating the VaR for stock series than the standard one. Marcucci [9] and Sajjad et al. [10] confirmed that MS-GARCH model has better performance in VaR estimation for stock volatility in both the US and the UK compared with standard GARCH models, respectively. The MSGARCH-type framework thus takes account of regime changes and the asymmetry in both conditional variance and distribution of error terms of in stock return data in VaR estimation.

However, the estimation MS-GARCH models by using maximum likelihood technique are cumbersome. Bayesian approach is an alternative to estimate MS-GARCH model, since the Bayes factors allow to determine the two specifications for the transition probabilities' dynamics. This paper therefore employs Bayesian MSGARCH models to analyze the risk of ASEAN-5 stock market indexes. To our knowledge, this is the first empirical result with accesses the risk of ASEAN-5 by using Bayesian MSGARCH models. The advantages of this paper consist of many aspects. Firstly, we concentrate on the stock return of ASEAN which belong to a group of emerging countries. Secondly, we consider the asymmetric effects by using various error distributions and obtaining skewness into any unimodal and symmetric distribution. Finally, we use MS to allow the regime changes that volatility process evolution is different between the two regimes.

The remainder of this paper is organized as follows. Section 2 presents the model specification consisting of Markov-Switching GARCH (MS-GARCH) Model, Bayesian Inference, and Value at Risk (VaR) and Expected-Shortfall (ES). Section 3 describes the summary statistics as well as the unit root test results, and the empirical results are in Sect. 4. Section 5 provides some brief concluding remarks.

2 Model Specification

2.1 Markov-Switching GARCH (MS-GARCH) Model

Let S_t be an ergodic Markov chain on a finite set $S = \{1, .., K\}$ with transition probabilities matrix

$$\mathbf{P} = \begin{bmatrix} p_{11} \; p_{21} \\ p_{12} \; p_{22} \end{bmatrix} = \begin{bmatrix} p & (1-q) \\ (1-p) & q \end{bmatrix}, \tag{1}$$

where $p_{ij} = \Pr(S_t = i \,|S_{t-1} = j\,)$ In this study, the state variable (S_t) takes value 0 or 1 referring to a two-state.

This study considers the lag order (1,1) model, as it is sufficient to capture the volatility clustering in financial series data (Brooks [4]). The lag (1,1) refers to one lag of ARCH effect and one lag of the moving average. Many applications of financial series also use the basic GARCH (1,1) model and find that it fits the changes in conditional variance.

2.1.1 MS-GARCH (1,1) Model

$$\sigma_t^2 = \gamma_{S_t} + \alpha_{1,S_t} y_{t-1}^2 + \beta_{S_t} \sigma_{t-1}^2, \tag{2}$$

where σ_t^2 is the conditional variance; α_{1,S_t} and β_{S_t} are the coefficient of the GARCH process. We have $\theta_{S_t} = (\gamma_{S_t}, \alpha_{1,S_t}, \beta_{S_t})^T$. To guarantee the positivity of the conditional variance, we impose the restrictions $\gamma_{S_t} > 0$, $\alpha_{1,S_t} \geq 0$, $\beta_{S_t} \geq 0$.

2.1.2 MS-EGARCH (1,1) Model

$$\ln\left(\sigma_t^2\right) = \gamma_{S_t} + \alpha_{1,S_t}\left(|\eta_{S_t,t-1}| - E\left(|\eta_{S_t,t-1}|\right)\right) + \alpha_{2,S_t} y_{t-1} + \beta_{S_t} \ln\left(\sigma_{t-1}^2\right), \tag{3}$$

where the expectation $E\left(|\eta_{S_t,t-1}|\right)$ is taken with respect to the distribution conditional on state S_t. We have $\theta_{S_t} = (\gamma_{S_t}, \alpha_{1,S_t}, \alpha_{2,S_t}, \beta_{S_t})^T$. This specification takes into account of leverage effect referring that the positive value has less impact on the conditional volatility compared with the past negative value. This model requires $\beta_{S_t} < 1$ for covariance-stationarity in each state.

2.1.3 MS-GJR-GARCH (1,1) Model

$$\sigma_t^2 = \gamma_{S_t} + \left(\alpha_{1,S_t} + \alpha_{2,S_t} I\left\{y_{t-1} < 0\right\}\right) y_{t-1}^2 + \beta_{S_t} \sigma_{t-1}^2, \tag{4}$$

where the indicator function (I) is defined to be 1 if the condition holds and 0 otherwise. We have $\theta_{S_t} = (\gamma_{S_t}, \alpha_{1,S_t}, \alpha_{2,S_t}, \beta_{S_t})^T$. To ensure the positivity of the conditional variance we impose the restrictions $\gamma_{S_t} > 0$, $\alpha_{1,S_t} \geq 0$, $\alpha_{2,S_t} \geq 0$, and $\beta_{S_t} \geq 0$. The degrees of asymmetry in the conditional volatility is governed by the parameter α_{2,S_t}.

2.1.4 MS-TGARCH(1,1) Model

$$\sigma_t = \gamma_{S_t} + \left(\alpha_{1,S_t} I\left\{y_{t-1} \geq 0\right\} - \alpha_{2,S_t} I\left\{y_{t-1} < 0\right\}\right) y_{t-1} + \beta_{S_t} \sigma_{t-1} \tag{5}$$

We have $\theta_{S_t} = (\gamma_{S_t}, \alpha_{1,S_t}, \alpha_{2,S_t}, \beta_{S_t})^T$. We impose the restrictions to ensure the positivity of the conditional variance $\gamma_{S_t} > 0$, $\alpha_{1,S_t} \geq 0$, $\alpha_{2,S_t} \geq 0$, and $\beta_{S_t} \geq 0$.

2.2 Distribution of GARCH Model

In general, the Normal distribution (norm) is used for conditional distribution. However, the financial time series are often found the leptokurtosis of the empirical distribution. To capture the fat-tail distribution in stock return, we apply student-t (std), and Generealized Error Distribution (ged). Despite being considered fat-tail, we also consider the skewness of conditional distribution for normal, student-t, and GED distribution, which are identified as "snorm" , "sstd" and "sged", respectively.

2.2.1 Normal Distribution

The probability density function (PDF) of the standard normal distribution can be expressed as

$$f(\eta) = \frac{1}{\sqrt{2\pi}} e^{-\frac{1}{2}\eta^2}, \ \eta \in \mathbf{R} \tag{6}$$

which may be maximized with respect to $(\beta, \gamma, \sigma, \delta)$.

2.2.2 Student-t Distribution

The PDF of the standardized student-t distribution can be expressed as

$$f(\eta, \nu) = \frac{\Gamma\left(\frac{\nu+1}{2}\right)}{\sqrt{(\nu-2)\,\pi}\,\Gamma\left(\frac{\nu}{2}\right)} \left(1 + \frac{\eta^2}{(\nu-2)}\right)^{-\frac{\nu+1}{2}}, \ \eta \in \mathbf{R}, \tag{7}$$

where $\Gamma()$ is the Gamma distribution. To guarantee the second order moment exists, the constraint of v has to be higher than two. The kurtosis of the distribution is higher for lower v.

2.2.3 GED Distribution

The PDF of the standardized generalized error distribution (GED) can be expressed as

$$f(\eta; \nu) = \frac{ve^{-1/2|\eta/\lambda|^{\nu}}}{\lambda 2^{(1+1/\nu)}\Gamma(1/\nu)}, \ \lambda = \left(\frac{\Gamma(1/\nu)}{4^{1/\nu}\Gamma(3/\nu)}\right)^{1/2}, \ \eta \in \mathbf{R}, \tag{8}$$

where v is the shape parameter which has to be greater than zero.

Despite of the evidence of heavy tail in financial time series, many empirical distribution are also found to be skewed. Fernández and Steel [5] proposed how to obtain skewness into any unimodal and symmetric univariate distribution through the added parameter ξ. Giot and Laurent [6] applied a skewed student distribution in estimating VaR for stock index, and found that it has a better performance than the standard symmetric distribution.

2.3 Bayesian Inference

The kernel of posterior density $f(\Psi | I_T)$ is obtained from the combination of the likelihood function $L(\Psi | I_T)$ and a prior $f(\Psi)$. For the prior density $f(\Psi)$ in this study, we follow the study of Ardia et al. [1], in which their prior is built from independent diffuse priors as follows:

$$
\begin{aligned}
& f(\Psi)=f(\theta_1,\xi_1)\ldots f(\theta_K,\xi_K)\,f(P)\\
& f(\theta_K,\xi_K)\propto f(\theta_k)\,f(\xi_k)\,I\{(\theta_K,\xi_K)\in CSC_k\}\ (S_t=1,\ldots,K)\\
& f(\theta_K)\propto f_N\left(\theta_k;\mu_{\theta_k},diag\left(\sigma^2_{\theta_k}\right)\right)I\{\theta_k\in PC_k\}\ (S_t=1,\ldots,K)\\
& f(\xi_K)\propto f_N\left(\xi_k;\mu_{\xi_k},diag\left(\sigma^2_{\xi_k}\right)\right)I\{\xi_{k,1}>0,\xi_{k,2}>2\}\ (S_t=1,\ldots,K)\\
& f(P)\propto\prod_{i=1}^{K}\left(\prod_{j=1}^{K}p_{i,j}\right)I\{0<p_{i,j}<1\},
\end{aligned}
\tag{9}
$$

where $\Psi = (\theta_1, \xi_1, ..., \theta_K, \xi_K, P)$ is the vector of model parameters. CSC_{S_t} is the covariance-stationarity condition in a state S_t, PC_{S_t} defines the positive condition in the state S_t, $\xi_{S_t,1}$ denotes the asymmetry parameter, $\xi_{S_t,2}$ denotes the tail parameter of the skew-Student t distribution in state S_t, $f_N(.;\mu,\Sigma)$ defines the multivariate Normal density with mean μ and variance Σ.

The likelihood function $L(\Psi\,|I_T)$ is $L(y_t\,|\Psi, I_{t-1}) = \prod_{t=1}^{T} f(y_t\,|\Psi, I_{t-1})$, where $f(y_t\,|\Psi, I_{t-1})$ is the density of y_t given by its past observations (I_{t-1}), and model parameters. The conditional density of y_t for the MSGARCH model is expressed as

$$f(y_t\,|\Psi, I_{t-1}) = \sum_{i=1}^{K}\sum_{j=1}^{K} p_{i,j} z_{i,t-1} f_D(y_t\,|s_t = j, \Psi, I_{t-1}), \tag{10}$$

where $z_{i,t-1} = P[s_{t-1} = i\,|\Psi, I_{t-1}]$ is the filtered probability of state i at time $t-1$. The condition density of in state y_t given by Ψ and I_{t-1} is $f_D(y_t\,|s_t = k, \Psi, I_{t-1})$. After we obtain the posterior density function, we employ Markov Chain Monte (MCMC) for numerical integration. The marginal posterior density function and the state variables are obtained by integrating the posterior density function. We follow Vihola [11] that samples are produced from the posterior distribution with adaptive MCMC algorithm. The benefit is that converge of Markov chain is faster as when it is coercing the acceptance rate, it also learns the shape of the target distribution. This algorithm also guarantees a positive variance and covariance-stationarity of the conditional variance.

2.4 Value at Risk (VaR) and Expected-Shortfall (ES)

The VaR measures the threshold value such that the probability of observing a loss more massive or equal to it in a given time horizon is equal to α. The ES estimates the expected loss below the VaR level. The VaR estimation in $T+1$ at risk level α can be expressed as

$$VaR^{\alpha}_{T+1} = \inf\{y_{T+1} \in |F(y_{T+1}\,|I_T) = \alpha\}, \tag{11}$$

where $F(y_{T+1}\,|I_T)$ is the 1-step ahead CDF evaluated in y. The ES is defined as

$$ES^{\alpha}_{T+1} = E\left[y_{T+1}\,\middle|y_{T+1} \le VaR^{\alpha}_{T+1}, I_T\right] \tag{12}$$

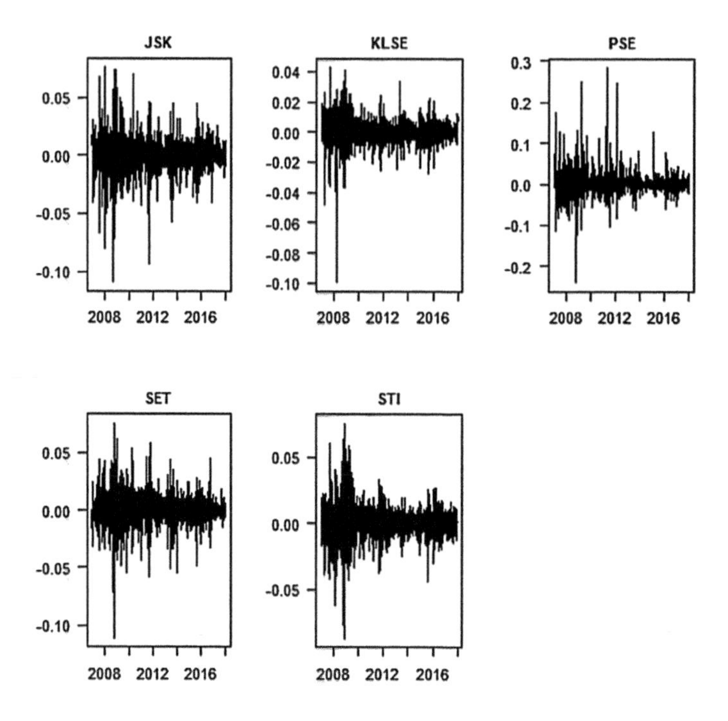

Fig. 1. Stock return of ASEAN-5

Table 1. Descriptive of the ASEAN-5 stock return

Country	INDONESIA	MALAYSIA	PHILIPPINES	THAILAND	SINGAPORE
Min	−0.1095	−0.0998	−0.2412	−0.1109	−0.087
Mean	0.0004	0.0002	0.0005	0.0003	0.0001
Max	0.0762	0.0426	0.2845	0.0755	0.0753
Skewness	−0.6597	−1.178	1.9247	−0.6665	−0.1814
Kurtosis	11.8604	19.3005	32.3404	11.5715	10.0314
St.Dev	0.0131	0.0073	0.0227	0.0119	0.0112

3 Empirical Results

3.1 Data and Descriptive Statistics

For analysis value at risk of five ASEAN stock market indexes, we choose the Jakarta Stock Exchange (JSK) of Indonesia, the Kuala Lumpur Stock Exchange (KLSE) of Malaysia, the Philippines Stock Exchange (PSE), the Stock Exchange of Thailand (SET), and the Singapore Strait's Time Index (STI). The sample data is retrieved from DATASTREAM. The study period covers from January 1, 2007 to December 30, 2017. The return of the stock index is constructed from the first difference of logarithmic stock price index for each country.

Table 1 presents the summary descriptive statistics and the result of unit root test for the stock return in the ASEAN-5 during the period of interest.

The average of the stock return in ASEAN-5 varies between 0.0001 and 0.0005. Singapore yields the lowest stock return, whereas the Philippines yields the highest stock return. However, Philippines presents the highest standard deviation referred to the risk of the stock market. Malaysia provides the lowest risk according to the lowest standard deviation. Only Philippines stock return has the positive skewness indicating that the right-handed tail of Philippines stock return is larger than the left-handed tail. For the other stocks in ASEAN-5, their distributions are skewed to the left.

4 Empirical Results

Figure 1 depicts the volatility in the ASEAN-5 stock returns covering the period 2007–2017. We can see that the high stock return volatility of ASEAN-5 follows both global and financial crisis in 2008 and 2012, respectively. During the studied period, the largest driver of volatility is the global crisis. Due to the financial crisis, Philippines stock return was the most affected. Overall, Philippines presents the highest volatility among the ASEAN-5 that its return changes between −0.3 and 0.3.

We apply package MSGARCH in R to estimate the Bayesian MSGARCH (1,1)-type models with different error distributions. For MCMC algorithm, we

Table 2. DIC criterion for each stock

GARCH-model	Distribution	INDONESIA	MALAYSIA	PHILIPPINES	THAILAND	SINGAPORE
GARCH	norm	−18015.298	−21293.306	−16156.499	−18517.931	−19119.054
	snorm	−18060.744	−21312.564	−16112.838	−18530.516	−19114.86
	std	−18576.516[a]	−21360.938	−20152.417	−19130.999[a]	−19141.945
	sstd	−18118.285	−21378.304	−20374.321	−18576.036	−19168.108
	ged	−18138.535	−21430.119	−19664.31	−18664.726	−19157.439
	sged	−18147.074	−21411.622	−17105.161	−18592.618	−19168.902
GJR-GARCH	norm	−18045.618	−21315.296	−16164.698	−18544.555	−19195.156
	snorm	−18054.663	−21294.493	−16182.613	−18532.046	−19169.443
	std	−18102.435	−20894.945	−20413.452[a]	−18596.538	−19190.363
	sstd	−18167.725	−21392.903	−19287.637	−18626.052	−19199.782
	ged	−18231.486	−21494.532[a]	−19834.958	−17477.987	−19212.73
	sged	−18173.606	−21409.897	−16924.059	−18629.949	−19209.515
TGARCH	norm	−18038.552	−21345.631	−16120.001	−18555.969	−19184.42
	snorm	−18100.905	−21309.698	−16142.756	−18566.102	−19161.64
	std	−18133.844	−21175.549	−18254.61	−18610.936	−19183.707
	sstd	−18160.689	−21384.71	−16157.674	−18632.966	−19193.79
	ged	−18145.712	−21273.806	−17040.532	−18641.212	−15245.987
	sged	−18132.359	−21383.772	−16805.029	−18606.362	−19196.139
EGARCH	norm	−18013.957	−21258.902	−15245.545	−18488.826	−19132.652
	snorm	−18068.935	−21171.537	−16137.798	−18538.321	−19147.67
	std	−18002.637	−21269.066	−16449.9	−18549.379	−19143.829
	sstd	−17982.832	−21364.848	−16221.968	−18582.058	−19151.982
	ged	−18120.707	−21428.238	−18945.733	−18628.666	−19192.892
	sged	−17770.002	−21337.772	−16174.276	−18632.167	−19201.518[a]

Note: [a] indicates the minimum DIC

Table 3. Bayesian Parametric estimation. Posterior Standard deviation is given in parenthesis

	Parameters	INDONESIA Posterior Mean		MALAYSIA Posterior Mean		PHILIPPINES Posterior Mean		THAILAND Posterior Mean		SINGAPORE Posterior Mean	
Regime 1	$\alpha_{0,1}$	0.0000 (0.0000)		0.0000 (0.0000)		0.0000 (0.0000)		0.0000 (0.0000)		-0.1277 (0.0000)	
	$\alpha_{1,1}$	0.0000 (0.0000)		0.0016 (-0.0057)		0.0000 (0.0000)		0.0077 (-0.0181)		0.0867 (-0.0213)	
	$\alpha_{2,1}$	-		0.5763 (-0.2192)		0.0001 (0.0000)		-		-0.0773 (-0.0115)	
	β_1	0.0000 (0.0000)		0.5547 (-0.1453)		0.0001 (0.0000)		0.1617 (-0.335)		0.9871 (-0.001)	
	u_1	2.1018 (-0.0007)		0.8153 (-0.3734)		2.1025 (-0.0009)		2.5938 (-1.5668)		1.5164 (-0.0715)	
	ϕ_1	-		-		-		-		0.9382 (-0.0317)	
Regime 2	$\alpha_{0,2}$	0.0000 (0.0000)		0.0000 (0.0000)		0.0022 (-0.0003)		0.0000 (0.0000)		-0.1094 (-0.0141)	
	$\alpha_{1,2}$	0.1117 (-0.0164)		0.0235 (-0.0095)		0.0009 (-0.0013)		0.1103 (-0.0193)		0.1624 (-0.0396)	
	$\alpha_{2,2}$	-		0.0998 (-0.2125)		1.5967 (-0.1251)		-		-0.0849 (-0.0312)	
	β_2	0.8797 (-0.0171)		0.9212 (-0.1145)		0.1772 (-0.0603)		0.8823 (-0.0272)		0.9871 (-0.0016)	
	u_2	6.2181 (-0.6574)		1.9688 (-0.4097)		2.1093 (-0.0203)		9.0436 (-3.1004)		1.5696 (-0.0766)	
	ϕ_2	1.0000 (2.0000)		-		-		-		0.9705 (-0.042)	
Posterior mean	Transition matrix	1	2	1	2	1	2	1	2	1	2
	1	0.37	0.631	0.441	0.559	0.355	0.645	0.24	0.76	0.999	0.0004
	2	0.053	0.947	0.382	0.619	0.221	0.779	0.09	0.911	0.0004	0.999
	State	1	2	1	2	1	2	1	2	1	2
	State probablilities	0.077	0.923	0.406	0.594	0.255	0.745	0.105	0.895	0.517	0.483

Note: The numbers in the parentheses are the posterior standard deviation.

use 5,000 burn-in draws and build the posterior with the 12,500 iterations. We thinned at very fifth to diminish the autocorrelation in the posterior draws. Then, we select the best-fitted two-regime MSGARCH-type model based on the minimum deviance information criterion (DIC) as shown in Table 2 for each stock return index. The best-fitted model for Indonesia stock return is two-regime MSGARCH with student-t distribution, for Malaysia stock return is two-regime GJR-GARCH with GED distribution, for Philippines stock return is two-regime MS-GJR-GARCH with student-t distribution, for Thailand is two-regime MSGARCH with student-t distribution, and for Singapore return is MS-EGARCH with skewed GED distribution.

The stock return volatility is separated into two regimes, which are high volatility and low volatility. The high volatility regime is related to high stock market return deviations, whereas the low volatility regime is related to small stock market return volatility. Estimated parameters in Table 3 show that the two regimes have different unconditional volatility levels and volatility persistence for all ASEAN-5 stock return. Therefore, it confirms that the MSGARCH with two-regime should provide a better result than the standard one. This model also provides the posterior mean stable probabilities of being both in the first and the second regime. Overall, Singapore provides the highest possibility of being in the first regime which is 51.7%, whereas Indonesia has the highest possibility of being in the second regime, which is possibility 92.3%.

Table 4. Estimated VaR and ES results

Stocks	VaR(1%)	Rank (VaR)	State	VaR(1%)	Rank (Return)	Rank (SD)	ES(1%)
INDONESIA	−0.0199	2	1	−0.001	2	4	−0.0252
			2	−0.02			
MALAYSIA	−0.0101	1	1	−0.011	4	1	−0.0121
			2	−0.01			
PHILIPPINES	−0.0655	5	1	−0.001	1	5	−0.1044
			2	−0.074			
THAILAND	−0.0132	3	1	−0.012	3	3	−0.017
			2	−0.018			
SINGAPORE	−0.0137	4	1	−0.014	5	2	−.0164
			2	−0.015			

Table 4 shows the estimated results of VaR and ES. We found that Philippines stock return shows the highest risk in both standard deviation and the VaR, whereas it offers the highest stock return among ASEAN-5, which is attractive for risk-lover investors. The ES of Philippines stock return provides the highest expected loss, which is 10.44%. Malaysia is appropriate for the risk-averse investor, since it presents the lowest VaR, but high returns. For Thailand stock return, it provides the average both risk and return. Singapore offers a high VaR estimate, but provides the lowest yield, being the most not attractive for investors.

5 Conclusion

We aim to analyze Value at Risk of ASEAN-5 stock market returns by employing Bayesian MSGARCH-type models. We found that Philippines stock return presents the highest risk investigating both standard deviation as well as the VaR, whereas it provides the highest stock return among ASEAN-5, which is attractive for risk-lover investors. Malaysia is the preferred one for the risk-averse investor, since it presents the lowest VaR, but high return. Indonesia also offers the small VaR and high return which is also attractive for risk-averse investors. Thailand stock return provides the median both risk and returns. Singapore presents a high VaR estimate, but provides the lowest return, being the most not attractive for investors. The stock return volatility in this study, we consider two regimes. The first regime refers to high volatility related to high stock market return deviations. Meanwhile, the second regime refers to low volatility related to small stock market return volatility. According to the posterior mean stable probabilities of being in each regime, Singapore provides the highest possibility of being in the first regime of 51.7%, whereas Indonesia has the highest possibility of being in the second regime of 92.3%.

References

1. Ardia, D., Bluteau, K., Boudt, K., Peterson, B., Trottier, D.A.: MSGARCH: Markov-switching GARCH models in R. R package version 0.17, 7 (2016)
2. Aumeboonsuke, V.: The vitality of beta in the ASEAN stock markets. Invest. Manag. Financ. Innov. **11**(3), 81–86 (2014)
3. Billio, M., Pelizzon, L.: Value-at-risk: a multivariate switching regime approach. J. Empir. Financ. **7**(5), 531–554 (2000)
4. Brooks, C.: RATS Handbook to Accompany Introductory Econometrics for Finance. Cambridge Books (2008)
5. Fernández, C., Steel, M.F.: On Bayesian modeling of fat tails and skewness. J. Am. Stat. Assoc. **93**(441), 359–371 (1998)
6. Giot, P., Laurent, S.: Value-at-risk for long and short trading positions. J. Appl. Econ. **18**(6), 641–663 (2003)
7. Guidi, F., Gupta, R.: Forecasting volatility of the ASEAN-5 stock markets: a nonlinear approach with non-normal errors. Discussion Papers Finance (14) (2012)
8. Kiwiriyakun, M.: The Risk-return Relationship in ASEAN-5 Stock Markets: An Empirical Study Using Capital Asset Pricing Model (Doctoral dissertation, Faculty of Economics, Thammasat University) (2013)
9. Marcucci, J.: Forecasting stock market volatility with regime-switching GARCH models. Stud. Nonlinear Dyn. Econ. **9**(4), 1–53 (2005)
10. Sajjad, R., Coakley, J., Nankervis, J.C.: Markov-switching GARCH modelling of value-at-risk. Stud. Nonlinear Dyn. Econ. **12**(3), 1–31 (2008)
11. Vihola, M.: Robust adaptive Metropolis algorithm with coerced acceptance rate. Stat. Comput. **22**(5), 997–1008 (2012)
12. Wang, Y., Liu, L.: Spillover effect in Asian financial markets: a VAR-structural GARCH analysis. China Financ. Rev. Int. **6**(2), 150–176 (2016)

Impacts of Monetary Policy on Inequality: The Case of Vietnam

Nhan Thanh Nguyen(✉), Huong Ngoc Vu, and Thu Ha Le

Banking Faculty, Banking Academy of Vietnam, Hanoi, Vietnam
{nhannt,huongvn,thulh}@hvnh.edu.vn

Abstract. This paper mainly concentrates on examining the impact of monetary policy on income inequality in Vietnam from 2001 to 2014. In our study, monetary policy shock is represented by the difference between the real and targeted growth rates of money supply of the State Bank of Vietnam (SBV), while income inequality is measured by Gini coefficients. The results of VAR model show that monetary policy has a small and lagged effect on income inequality. Besides monetary policy, inflation, education and unemployment are also found to have significant impact on income inequality, while economic growth has insignificant effect on this variable. Based on these findings, we suggest that the SBV should pay more attention at the inequality consequences caused by its monetary policy.

Keywords: Monetary policy · Income inequality
Monetary policy shocks

1 Introduction

1.1 The Trend of Inequality in Vietnam

Vietnam has experienced rapid economic growth in the last 30 years, characterized by rising average incomes and a significant fall in the number of people living in poverty. However, there is now a growing gap between rich and poor in Vietnam. According to the Standardized World Income Inequality Database (SWIID) [20], the Gini coefficient[1] increased from 40.1 to 42.2 in the 22-year period from 1992 to 2014, indicating that income inequality rose in that period.

[1] Gini coefficient is the ratio of the area between the actual income distribution curve and the line of perfect income equality over the total area under the line of perfect income equality. Formally, let x_i be a point on the x-axis, and y_i a point on the y-axis, then $Gini = 1 - \sum_{i=1}^{N}(x_i - x_{i-1})(y_i + y_{i-1})$. When there are N equal intervals on the x-axis, then $Gini = 1 - \frac{1}{N}\sum_{i=1}^{N}(y_i + y_{i-1})$.

V. Kreinovich et al. (Eds.): ECONVN 2019, SCI 809, pp. 463–476, 2019.
https://doi.org/10.1007/978-3-030-04200-4_34

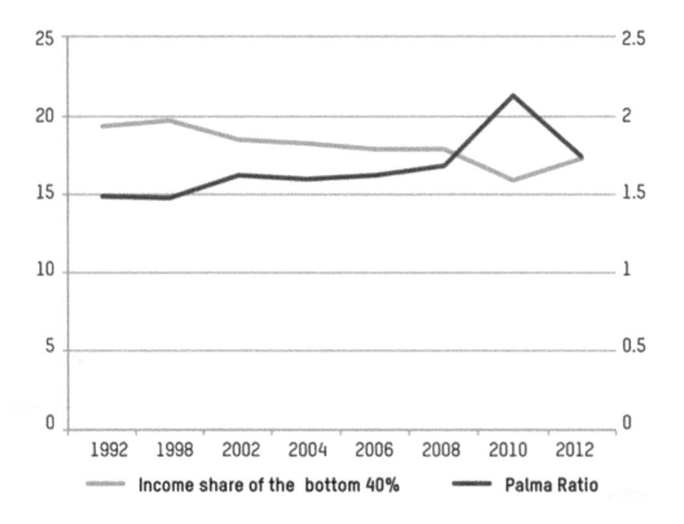

Fig. 1. Changes in income inequality in Vietnam, 1992–2012 [17]

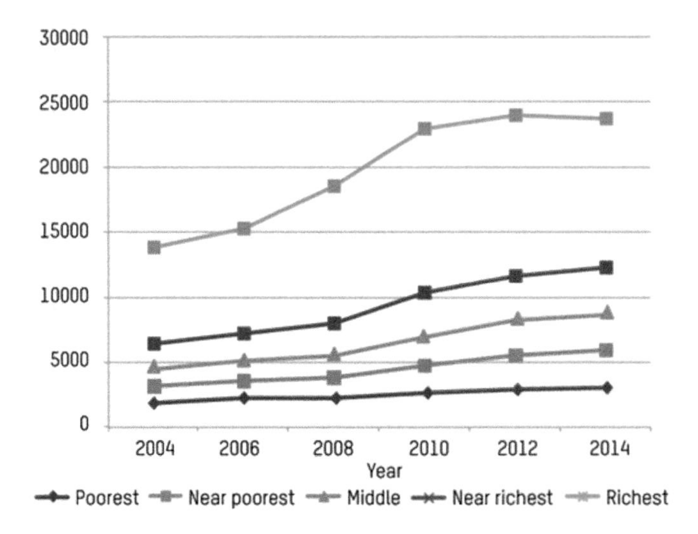

Fig. 2. Per capita income, by income quintiles, 2004–2014 [17]

Moreover, as Oxfam (2017) [17] reported, between 1992 and 2012, the Palma ratio (which measures the ratio between the income share of the top 10% to the bottom 40%) increased by 17%, mostly driven by a decline in the income share of the bottom 40% of the population (Fig. 1). This indicated that the poorest sections of the population have not benefited as much as the rest. Furthermore, the distribution of the benefits of growth has become more unequal in recent years. In other words, income distribution has been increasingly polarizing over time. While there are small income differences between the first four quintiles of the distribution (the bottom 80%), there is a large gap between these and the richest quintile (the top 20%), and this gap has been widening since 2004 (Fig. 2).

1.2 Monetary Policy and Inequality in Vietnam

Monetary policy involves the use of monetary instruments to regulate or control the volume, the cost, the availability and the direction of money and credit in an economy to achieve some macroeconomic objectives such as price stability, full employment and sustainable economic growth (Mishkin (2013) [12]). In Vietnam, monetary policy is implemented by the State Bank of Vietnam (SBV). According to the 2010 Law on the SBV, its monetary policy aims at "currency value stability which is denoted by the inflation rate and decisions on the use of tools and measures to obtain the set objective". In other words, the main objective of the SBV's monetary policy is to stabilize currency's value and control inflation rate. Furthermore, the SBV announces annual targets for total liquidity (M2) and credit to the economy and uses monetary instruments including direct instruments (i.e. setting credit growth limitation, applying ceiling interest rates, and stipulating lending rates in prioritized areas) and indirect instruments (i.e. reserve requirements, refinancing policy and open market operations) to achieve this target.

Based on the Law, it can be seen that addressing inequality is not a direct objective of the SBV's monetary policy. However, in pursuing its macroeconomic objectives, the instruments used by the SBV might potentially affect inequality. According to Furceri et al. (2016) [8], the effect of monetary policy on inequality is ambiguous as the quantitative importance of different transmission channels can result in its increase or decrease. For example, expansionary monetary policy can increase inequality by boosting inflation as lower-income households tend to hold more liquid assets and thus tend to be influenced more by inflation. On the other hand, expansionary monetary policy lowers interest rates which will benefit borrowers - generally those less wealthy, therefore it can reduce inequality.

To summarize, inequality in Vietnam has been increasing for the last two decades, while the SBV's monetary policy is expected to potentially impact inequality. Therefore, it is important to analyze the relationship between the two variables. In this paper, we first review some research on the link between monetary policy and inequality in different countries. After that, we describe our data, model and results on this issue. We then discuss these results and finally come up with some implications for Vietnam.

2 Literature Review

2.1 The Relationship Between Monetary Policy and Inequality

Monetary policy usually refers to central banks' actions to achieve specified targets, for example maximum employment, stable prices, and moderate long-term interest rates. A number of theoretical channels have been proposed by which monetary policy might influence inequality. However, none of them provides a clear answer of the relationship because each depends on the distribution of population characteristics and the association with different types of income as well as assets and liabilities.

Nakajima (2015) [13] analyzed the relationship between conventional monetary policy and inequality in theory and suggested five channels through which monetary policy might affect inequality. These channels could be described as following:

(i) Inflation tax channel: Expected inflation acts as a regressive consumption tax which disproportionately erodes the purchasing power of lower-income households who hold a larger fraction of their assets in cash, thereby increasing inequality.
(ii) Savings redistribution channel: Increases in unexpected inflation reduce the real value of nominal assets and liabilities, making borrowers better off at the expense of lenders, because the real value of nominal debts declines. Therefore, the effect of monetary policy on inequality depends on how these assets are distributed across the population.
(iii) Interest rate exposure channel: Declines in real interest rates increases financial asset prices because the interest rate used to discount future cash flows reduces. Net savers whose wealth is concentrated in short-duration assets (like CDs or T-bills) and net borrowers whose liabilities are of relatively long-duration (like fixed-rate mortgages) benefit from expansionary monetary policy, since it decreases real interest rates. On the contrary, net savers whose wealth are concentrated in long-duration assets (like Treasure bonds) and of net borrowers whose liabilities are of relatively short-duration (like adjustable-rate mortgages) lose as real interest rates reduce. However, the effect of monetary policy on inequality also depends on the distribution of these assets and liabilities across the population.
(iv) Earnings heterogeneity channel: Changes in monetary policy might differently affect labour earnings, depending on where a household is in the earnings distribution. While earnings at the bottom of the distribution are mainly influenced by changes in working hours and unemployment rate, earnings at the top are mainly influenced by changes in hourly wages. Therefore, monetary policy which affects these variables differently might produce redistributive income effects.
(v) Income composition channel: Households' incomes are contributed by different sources, e.g. business and capital income, labour income and transfer income (like unemployment benefits). Each of these sources might respond differently to changes in monetary policy. Therefore, monetary policy might create different impacts on different class of population, or inequality.

2.2 Empirical Evidence

Since there is no clear implication on the effects of monetary policy in theory, empirical evidence on these effects is still limited and inconclusive.

Carpenter and Rodgers (2004) [3] and Doepke and Schneider (2006) [6] did not focus on the link between monetary policy and inequality, but provided evidence that monetary policy might considerably impact income distribution in the United States. While the former indicated that monetary policy has a disproportionate effect on the unemployment rate of different population groups, the latter suggested even moderate inflation may lead to significant redistribution of wealth in the economy. With the view that inflation is always and everywhere a monetary phenomenon, both studies imply a link between monetary policy and inequality through the distribution of income. This issue is developed further by Guerello (2016) [9] who found that conventional monetary policy has a small effect on income distribution in the Euro area.

Focusing on the impact of monetary policy on inequality, research findings are divided into three groups:

Firstly, monetary policy does not have significant impact on inequality. This view is supported by O'Farrell et al. (2016) [16] and Inui et al. (2017) [10]. While the former studied the effects of monetary policy on inequality through its impacts on returns on assets, the cost of debt servicing and asset prices in selected advanced economies and found that expansionary monetary policy has a priori ambiguous and small effects on inequality, the latter studied effects of monetary policy shocks on inequality in Japan by using the micro-level data of Japanese households from 1981 to 2008 and found that expansionary monetary policy shock increased income inequality in the period before the 2000 s, but the effect became insignificant when earnings inequality across all households was considered.

Secondly, contractionary monetary policy increases inequality. Coinbion et al. (2016) [4] studied the effects and historical contribution of monetary policy shocks to consumption and income inequality in the United States since 1980. In this paper, they used the method developed by Romer and Romer (2004) [18], which measures monetary policy shocks by changes in the target Federal Funds rate at each FOMC meeting from 1969 to 1996, and extended the dataset until 2008. To measure inequality, the authors used Gini coefficients of levels, cross-sectional standard deviations of log levels, and differences between individual percentiles of the cross-sectional distribution of log levels. They found that monetary shocks might significantly affect cyclical variation in income and consumption inequality. Moreover, contractionary monetary policy systematically increases inequality in labour earnings, total income, consumption and total expenditures. This point of view is similar to Furceri et al. (2016) [8]. They also used Gini coefficients as the measure of income inequality, but followed Auerbach and Gorodnichenko (2013) [1] to measure monetary policy shocks. In particular, they computed the forecast error of the policy rates (i.e. the difference between the actual policy rate and the rate expected by analysts of the same year), and then regressed for each country the forecast errors of the policy rates on similarly computed forecast errors of inflation and output growth to get the residual which captures exogenous monetary policy shocks. By using the dataset of 32 advanced and emerging market countries over the period 1990–2013, the authors also found that contractionary monetary policy increases income inequality. However, their new finding is the effect depends on the type of shocks, the state of the business cycle, the share of labour income and redistribution policies. In particular, the effect is larger for positive monetary policy shocks, especially during expansions and for countries with higher labour share of income and smaller redistribution policies. Furthermore, the authors contributed to the literature by suggesting that unexpected increases in policy rates increase inequality, while the opposite is true for changes in policy rates driven by economic growth. Other research conducted by Bivens (2015) [2] argued that the Fed's very low interest rates and large-scale asset purchases attempt to push the economy closer to full employment, and thus reduce inequality. In other word, the Fed's expansionary monetary policy can lower inequality by moving the economy to potential output.

Thirdly, expansionary monetary policy increases inequality. Domanski et al. (2016) [7] analyzed the potential effect of monetary policy on wealth inequality through its impact on interest rates and asset prices. By exploring the recent evolution of household wealth inequality in advanced economies, particularly valuation effects on household assets and liabilities, the authors found that rising equity prices are the key driver of wealth inequality, while low interest rates and rising bond prices have a negligible impact on this variable. Therefore, expansionary monetary policy which boosted equity prices is suggested to increase wealth inequality. By focusing on the long run relation between monetary policy and income inequality in the United States, Davtyan (2016) [5] had a similar finding that contractionary monetary policy decreased income inequality. Another study by Saiki and Frost (2014) [19] analyzed the distributional impact of unconventional monetary policy in Japan and found that unconventional monetary policy widened income inequality, especially after 2008.

In Vietnam, the link between economic growth and inequality is well researched, for example by Nguyen (2014) [14], Nguyen (2015) [15] or Le and Nguyen (2016) [11]. By using Gini coefficients to represent income inequality, these authors analyzed the positive relationship between economic growth and inequality in Vietnam in recent periods. However, monetary policy and inequality is a relatively new topic and there has been no study dealing with this relation. To summarize, the literature suggests that there is a relationship between monetary policy and inequality. However, this area is still under-researched as the direction and magnitude of the effect is inconclusive, and papers mostly focus on advanced economies, particularly the United States, the Euro area and Japan. Therefore, to contribute to the literature, we conduct a research on the effect of monetary policy on income inequality in Vietnam from 2001 to 2014.

3 Empirical Evidence of Vietnam

3.1 Data

For the measurement of inequality, this paper uses Gini coefficient, following the previous studies including Coibion et al. (2016) [4] and others. Similarly to the current work of Coinbion et al. (2016) [4], our study also employs monetary policy shocks as a measurement of monetary policy. Following the method developed by Romer and Romer (2004) [18], in our study, the monetary policy shock is measured as the difference between the real and targeted money growth. Moreover, other relevant macroeconomic variables including real GDP, inflation rate and unemployment rate are also employed. Since the measurement of Vietnam's unemployment rate is sometimes ambiguous, another social indicator - Education Index (EDU) - is added into the model.

The time-series data of Vietnam's real GDP growth (GDP) and inflation rate (INF) is collected from Vietnam's General Statistics Office - GSO, while the International Financial Statistics and the UNDP database provide the data of Vietnam's unemployment rate (UNEMP) and education index (EDU), respectively. The SHOCK variable capturing the difference between real and targeted money growth rate is collected and calculated from the data source of State Bank of Vietnam - SBV. Besides, Gini coefficient (GINI) measures the inequality in equivalized household disposable income; and this data is collected from the Standardized World Income Inequality Database (SWIID) [20]. The investigated period is from 2001 to 2014, in which the SBV have started to use full package of monetary instruments and have a more precise calculation of money supply.

For the purpose that all input variables in VAR model are stationary, the growth rates of GINI, SHOCK, UNEMP variables are generated and proved to be stationary at their own levels through unit root tests. Apart from those variable, three other variables INF, EDU, and GDP are proved to be stationary at their own levels. Therefore, six variables employed into the VAR model are GSHOCK, INF, GDP, GUNEMP, EDU, and GGINI, whose statistic summary is shown in Fig. 3.

Lag	LogL	LR	FPE	AIC	SC	HQ
0	440.2556	NA	1.62e-15	-17.02963	-16.80236	-16.94278
1	761.1215	553.6510	2.30e-20	-28.20084	-26.60993	-27.59291
2	914.9742	229.2707	2.40e-22	-32.82252	-29.86796	-31.69349
3	1008.408	117.2503	2.95e-23	-35.07483	-30.75663*	-33.42471
4	1076.107	69.02662*	1.18e-23	-36.31793	-30.63609	-34.14673
5	1134.045	45.44169	9.28e-24*	-37.17825*	-30.13277	-34.48596*

*** indicates lag order selected by the criterion**

Fig. 3. Statistic summary of variables

3.2 Model

To assess the effect of monetary policy on inequality in Vietnam, this paper applies the Vector Autoregression (VAR) model. The regressed variables include the growth rate of monetary policy shock (GSHOCK), inflation rate (INF), real GDP growth rate (GDP), the growth rate of unemployment rate (GUNEMP), education index (EDU), and the growth rate of Gini coefficient (GGINI), as they are proved to be stationary. The Cholesky ordering in VAR model is GSHOCK, INF, GDP, GUNEMP, EDU, and GGINI, as the impact of monetary policy on income inequality can be affected by changes in the Vietnamese macro economy. The statistic summary of variables is given in Fig. 3.

The lag of three periods is chosen, as recommended by the Schwarz Information Criterion (SC), according to Fig. 4.

Lag	LogL	LR	FPE	AIC	SC	HQ
0	440.2556	NA	1.62e-15	-17.02963	-16.80236	-16.94278
1	761.1215	553.6510	2.30e-20	-28.20084	-26.60993	-27.59291
2	914.9742	229.2707	2.40e-22	-32.82252	-29.86796	-31.69349
3	1008.408	117.2503	2.95e-23	-35.07483	-30.75663*	-33.42471
4	1076.107	69.02662*	1.18e-23	-36.31793	-30.63609	-34.14673
5	1134.045	45.44169	9.28e-24*	-37.17825*	-30.13277	-34.48596*

*** indicates lag order selected by the criterion**

Fig. 4. Lag specification

Therefore, the regressed equations of VAR model are expressed as follow:
$Y_t = c + \Phi_1.Y_{t-1} + \Phi_2.Y_{t-2} + \Phi_3.Y_{t-3} + \varepsilon_t$
where Y_t is:

$$(GSHOCK_t\ INF_t\ GDP_t\ GUNEMP_t\ EDU_t\ GGINI_t)$$

Φ_1, Φ_2 and Φ_3 are (6×6) matrixes of coefficients for Y_{t-1}, Y_{t-2} and Y_{t-3}, respectively;

c and ε_t are (6×1) vectors of constants and error terms.

Moreover, all inverse roots of Autoregressive (AR) characteristic polynomial are less than 1 (see Figs. 5 and 6), proving that the VAR model is stationary and the estimated output is considered to be reliable. The VAR model is also proved to have no cross terms, or autocorrelations, through the White Heteroskedasticity Test (see Fig. 7).

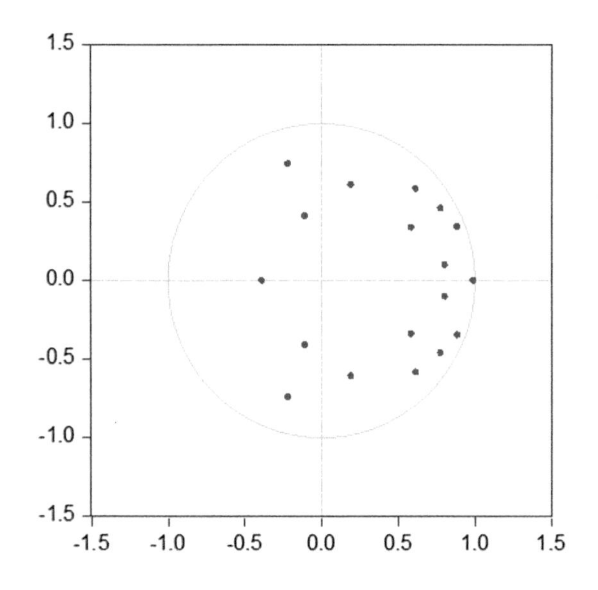

Fig. 5. Inverse roots of autoregressive characteristic polynomial

3.3 Limitations

Although the necessary tests are conducted and prove the validity of our VAR model, this empirical model has a limitation due to data availability. Foremost, for VAR model, the applied time-series data is required to be long enough and contains many observations. However, due to the Vietnam's annual data availability of monetary policy shock, unemployment rate, education index, and Gini coefficient from 2001 to 2014, the total number of observations is 14, which is not enough to assure the validity of the VAR model. Therefore, to handle with these difficulties, the annual data of these variables is interpolated into quarterly data by some popular interpolation methods including cubic spline and cardinal spline. Despite of this limitation, the validity of this VAR model has been proved through various tests and the results which are produced by this VAR model is considered to be reliable and ready to be used for further discussions.

3.4 Discussion

First of all, the impact of monetary policy on inequality in Vietnam is found to be small. Specifically, the impulse response function of GGINI to GSHOCK is always above the zero-line and GSHOCK is responsible for only about 5% of the change in GGINI, according to Fig. 9. Both Variance Decomposition Computation and Impulse Response which are produced from the VAR model show the lagged effect of monetary policy on inequality, as the proportion of the GGINI's fluctuation which is due to the monetary policy shock is greater for further periods. Moreover, this positive impact of money supply on inequality can be explained by the inflation tax channel. In particular, as the intermediate target

Root	Modulus
0.990462	0.990462
0.886939 - 0.345623i	0.951901
0.886939 + 0.345623i	0.951901
0.778446 + 0.461544i	0.904987
0.778446 - 0.461544i	0.904987
0.615718 - 0.582632i	0.847685
0.615718 + 0.582632i	0.847685
0.806000 - 0.101574i	0.812375
0.806000 + 0.101574i	0.812375
-0.219291 - 0.740684i	0.772464
-0.219291 + 0.740684i	0.772464
0.585929 - 0.340281i	0.677572
0.585929 + 0.340281i	0.677572
0.192495 - 0.607162i	0.636946
0.192495 + 0.607162i	0.636946
-0.108408 - 0.409569i	0.423673
-0.108408 + 0.409569i	0.423673
-0.388288	0.388288

No root lies outside the unit circle.

VAR satisfies the stability condition.

Fig. 6. VAR stability condition check

Joint test:

Chi-sq	df	Prob.
790.1624	756	0.1887

Fig. 7. VAR residual heteroskedasticity tests

of the SBV's monetary policy is money supply, the positive difference between real and targeted money growth generally enhances expected inflation in the domestic economy. Therefore, income inequality is increased, as the negative effect of expected inflation is relatively stronger on poor people. Therefore, the impact of monetary policy on inequality is shown significantly through inflation (Fig. 8).

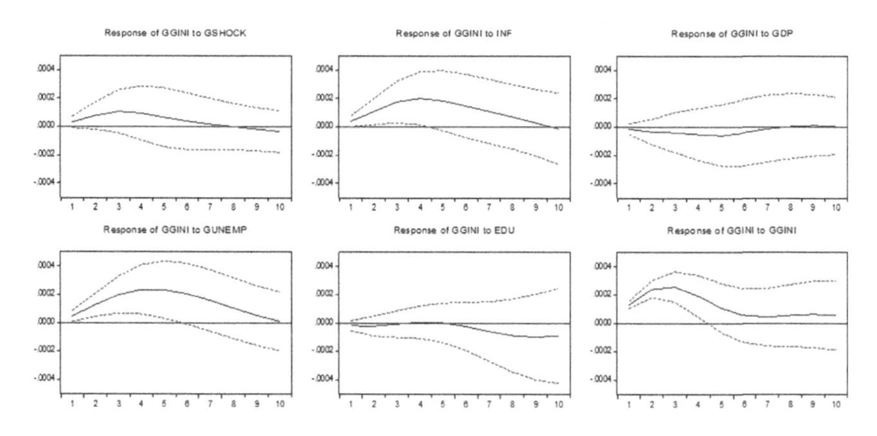

Fig. 8. Responses of GGINI to the shocks of other variables

Secondly, the impact of inflation on inequality in Vietnam is stable and significant. According to Fig. 9, the inflation rate of Vietnam's economy explains around 20% of the change in income inequality of Vietnamese people. According to the Impulse Response of GGINI to GINF, the increase in inflation rate promotes the increase in income inequality of Vietnam which is presented by the Gini coefficient, as the impulse response graph is always above the zero-line. Practically, inflation decreases the real incomes of all individuals in the economy through increasing the prices of consumption goods and services. However, it can also be seen that the poor people usually hold a relatively larger fraction of

Period	S.E.	GSHOCK	INF	GDP	GUNEMP	EDU	GGINI
1	0.000150	4.636243	7.216750	0.623642	9.602256	1.519316	76.40179
2	0.000342	6.080287	11.58796	0.987547	15.38131	0.689542	65.27336
3	0.000517	6.991502	16.67204	0.943486	21.50408	0.321131	53.56776
4	0.000642	6.735540	20.71611	1.235128	27.42458	0.216728	43.67192
5	0.000722	6.176132	22.95062	1.619775	32.28431	0.171841	36.79731
6	0.000769	5.695888	23.91779	1.684020	35.56134	0.244591	32.89638
7	0.000797	5.361163	24.22535	1.591732	37.12199	0.759351	30.94042
8	0.000814	5.144202	24.04198	1.530766	37.28279	1.867220	30.13305
9	0.000824	5.050812	23.55580	1.510477	36.72673	3.220957	29.93523
10	0.000832	5.121065	23.16607	1.491354	36.10994	4.262786	29.84879

Fig. 9. Variance decomposition of GGINI

their assets in cash than the richer people do. Therefore, the effect of inflation is usually stronger on the poorer people, which increases income inequality in society. This result confirms the importance of inflation tax channel in the impact of monetary policy on inequality in Vietnam.

Thirdly, we do not find strong evidence on the effect of economic growth on inequality. Specifically, according to Variance Decomposition analysis, GGDP is responsible for only approximately 1% of the change in income inequality in Vietnam. On one hand, the growth of global economy throughout these years is along with the development of technology, in which the industrial innovation 4.0 is the most noticed. The rapid development of the high technology increases the inequality in the society, as the people who are able to access these technology and also are the middle income class or higher will get more benefit compared to the low income class. Therefore, in general, there should be a positive relationship between economic growth and inequality. However, the rapid economic growth that Vietnam's economy has been experiencing throughout these years is mainly based on the growth of invested capital, especially foreign direct investment (FDI) which contributes to more than 20% of the Vietnam's economic growth. This factor promotes the employment of low income people and thus reduces income inequality. Therefore, the economic growth in Vietnam is found not to have strong effect on the income inequality.

Fourthly, while EDU can explain about 1% change of GGINI, GUNEMP is responsible for about 20%–30% of the change in GGINI. Specifically, education has a negative and lagged impact on inequality in Vietnam, which suggests education promotion policy of the Vietnamese government might improve inequality in the country. Meanwhile, the VAR model's result shows the significant impact of unemployment rate on the income inequality in Vietnam. Indeed, when the economy experiences an increased unemployment rate, the less skilled-labor who are also the poorer people in society will be unemployed first and suffer more than the other population group. This makes the gap between the rich and poor people in society greater. Therefore, both practical fact and empirical evidence show the importance of unemployment in causing inequality. Indeed, according to the VAR model's results, the explanatory power of GUNEMP over GGINI is only less than the impact of GGINI on itself, which is estimated to be roughly 70% in the first period and gradually decreased in the following periods.

4 Conclusion and Implications

This paper has revealed the relationship between monetary policy and income inequality in Vietnam from 2001 to 2014 and found that monetary policy has a small and lagged effect on income inequality. This finding is consistent with the majority of previous studies including Coinbion et al. (2016) [4] and Furceri et al. (2016) [8]. Besides monetary policy, inflation, education and unemployment are also found to have a significant impact on income inequality, while economic growth has insignificant effect on this variable.

Moreover, since monetary policy is found to have a potential effect on income inequality, the SBV should pay more attention at the inequality consequences caused by its monetary policy. In practice, decisions regarding the redistribution of income or income inequality are usually considered to be the province of fiscal policy. However, it might be impossible to avoid these consequences of monetary policy. If these effects are relatively small compared with the ways in which monetary policy affects all segments of the population equally, these consequences might be less of a concern. Nevertheless, monetary policymakers should consider these effects carefully so that their policy will not exacerbate income inequality further.

References

1. Auerbach, A., Gorodnichenko, Y.: Fiscal multipliers in recession and expansion. In: Alesina, A., Giavazzi, F. (eds.) Fiscal Policy After the Financial Crisis. NBER Books, National Bureau of Economic Research Inc., Cambridge (2013)
2. Bivens, J.: Gauging the impact of the fed on inequality during the great recession. Hutchins Center on Fiscal and Monetary Policy at Brookings, WP12 (2015)
3. Carpenter, S., Rodgers III, W.: The disparate labor market impacts of monetary policy. J. Policy Anal. Manag. **23**(4), 813–830 (2004)

4. Coibion, O., Gorodnichenko, Y., Kueng, L., Silvia, J.: Innocent Bystanders? Monetary Policy and Inequality in the U.S., Unpublished manuscript, University of Texas-Austin (2016)
5. Davtyan, K.: Income inequality and monetary policy: an analysis on the long run relation. Research Institute of Applied Economics, Working Paper 2016/04 (2016)
6. Doepke, M., Schneider, M.: Inflation and the redistribution of wealth. J. Polit. Econ. **114**(6), 1069–1097 (2006)
7. Domanski, D., Scatigna, M., Zabai, A.: Wealth inequality and monetary policy. BIS Q. Rev. March, 45–64 (2016)
8. Furceri, D., Loungani, P., Zdzienicka, A.: The Effects of Monetary Policy Shocks on Inequality, IMF Working Paper 16/245 (2016)
9. Guerello, C.: Conventional and unconventional monetary policy vs. households income distribution: an empirical analysis for the euro area. CEPWEB (2016). http://www.cepweb.org/wp-content/uploads/Guerello.pdf
10. Inui, M., Sudo, N., Yamada, T.: Effects of monetary policy shocks on inequality in Japan. Bank of Japan Working Paper Series, no. 17-E-3 (2017)
11. Le, H.P.L., Nguyen, N.A.T.: Impact of inequality on economic growth in Vietnam during the 2002–2012 period. J. Sci. **3**(48), 33–44 (2016). Ho Chi Minh City Open University
12. Mishkin, F.S.: The Economics of Money, Banking, and Financial Markets, 10th edn. Pearson Education, New York (2013)
13. Nakajima, M.: The Redistributive Consequences of Monetary Policy, Business Review, Second Quarter 2015, Federal Reserve Bank of Philadelphia Research Department (2015)
14. Nguyen, H.B.: Inequality and pro-poor economic growth in Vietnam. Econ. Dev. Rev. **289**, 2–22 (2014)
15. Nguyen, T.H.: Impact of income inequality on economic growth in Vietnam. J. Econ. Dev. **216**, 18–25 (2015)
16. O'Farrell, R., Rawdanowicz, L., Inaba, K.: Monetary Policy and Inequality, OECD Working Papers No. 1281, February 2016 (2016)
17. Oxfam: Even It Up: How to Tackle Inequality in Vietnam, Oxfam briefing paper, January 2017 (2017)
18. Romer, C.D., Romer, D.H.: A new measure of monetary shocks: derivation and implications. Am. Econ. Rev. **94**, 1055 (2004)
19. Saiki, A., Frost, J.: How does unconventional monetary policy affect inequality? Evidence from Japan, DNB Working Paper No. 423, May 2014 (2014)
20. Solt, F.: The Standardized World Income Inequality Database, Social Science Quarterly 97, SWIID Version 6.0, July 2017 (2016)

Earnings Quality: Does State Ownership Matter? Evidence from Vietnam

Tran Minh Tam[1,2(✉)], Le Quang Minh[1,2], Le Thi Khuyen[1,2], and Ngo Phu Thanh[1,2]

[1] Banking University, HCMC, Ho Chi Minh City, Vietnam
{tamtm,khuyenlt}@buh.edu.vn
[2] University of Economics and Law, Ho Chi Minh City, Vietnam
{minhlq,thanhnp}@uel.edu.vn

Abstract. In regard to profitability and efficiency, although there are mixed empirical evidence, private ownership is predominantly proven to be superior to state ownership. Motivated by this phenomenon, we are curious whether the earnings numbers released by privately owned enterprises are better than the ones published by state-owned companies in term of earnings quality. The results in this research reveal an interesting picture of earnings quality in the Vietnamese financial market. Specifically, using a matched sample of state-owned companies and privately owned companies, we show that private firms are more likely to "manipulate" their reported earnings number than state-owned firms in the Vietnamese financial market. Based on the result, we recommend that when assessing the quality of earnings, analysts and investors should pay more attention to those released by privately owned enterprises.

Keywords: State ownership · Private ownership · Earnings quality Matched sample

1 Introduction

The story of state-owned corporations in Vietnam always attracts public attention. Unfortunately, state-owned enterprises (SOEs) are often known because of their inefficiency in performance. In Vietnam, SOEs appear in key industries of the economy, including electricity, oil and gas, mining and quarrying, water supply, as well as investment in industries and areas where other types of businesses are not allowed or do not want to invest.

According to the Vietnam Development Report (2012) [38], the OECD report (2013) [33] and the study on SOEs in Vietnam (Nguyen and Freeman 2009) [27], the development of SOEs in Vietnam can be divided into two stages: (1) after the Doi Moi reform to 2009, and (2) from 2010 to present. The first phase is the 20 years after the Doi Moi reform and the opening of the integration of the Vietnamese economy, the number of SOEs have been equitized, but the efficiency was not high. Later, Vietnam prepared to join the World Trade Organization (WTO)

V. Kreinovich et al. (Eds.): ECONVN 2019, SCI 809, pp. 477–496, 2019.
https://doi.org/10.1007/978-3-030-04200-4_35

in 2006, the Vietnamese government recognized that most domestic companies, including the General Corporations (GCs), were too weak to withstand competition from foreign companies. Therefore, the government decided to establish the State Economic Groups (SEGs) and provided them with privileges as well as autonomy to help them compete with foreign companies. This move was reinforced by a forecast that with a revised industrial policy, in which the SEGs would play a catalytic role, Vietnam could transform itself into a modern and prosperous country (Vietnam Development Report, 2012) [38]. In addition, the government also issued Decree 101/2009/ND-CP in 2009 to set up several development targets for SEGs such as facilitating the improvement of other industries and introducing advanced and latest technologies in Vietnam.

However, the performance of SEGs did not meet the expectation and the bankrupt-cy of Vinashin (Shipbuilding Industry Corporation) in 2010 was a warning signal, emergency call for the reform. The OECD report (2013) [33] indicated that the Vietnamese government tried to promote the restructuring and equitization process of SOEs by establishing a Steering Committee for restructuring SOEs, headed by the Finance Minister (2011) and completely modifying the legal framework for equitization (The Vietnamese Government, 2011) [37] as well as separating the state regulation function from the state ownership rights. Furthermore, to comply with Decision No. 929/QD-TTg, nearly 900 SOEs would be restructured and equitized in the period 2011–2015 (OECD, 2013) [33]. According to the General Statistics Office of Vietnam (GSO), Vietnam had approximately 518,000 firms as of the beginning of 2017, about 1.5 times as in 2012. Meanwhile, the number of SOEs in Vietnam was more than 2,700 in 2017, 18.3% less compared to 2012 (equivalent to 607 firms), thanks to the government's efforts to privatize SOEs.

Regarding profitability and efficiency, as what we briefly described above, the inferiority of SOEs in Vietnam is unquestionable. However, in the financial market, in addition to profitability and efficiency indicators, there are other important indicators that analysts and investors always look for. One of them is the firms' net income.

Theoretical research argues that earnings announcements are one of the important signals used by managers to transmit information to the public about a firm's future prospects (Aharony and Swary 1980) [1]. Furthermore, the content of corporate earnings announcements is obviously important for investors. It is assumed that such information will be significant for investors and reflected in stock price movements as soon as the information is publicly released to the market (Hussin et al. 2010) [18]. Therefore, managers generally try to manage earnings figures in the direction that they get the most benefits. The actions of managers that use judgment in earnings data are known as "earnings management" (Healy and Wahlen 1999, p. 368) [17]. Earnings management, in turn, reduces "earnings quality" (Dechow and Schrand 2004, p. 5) [10]. The terms "earnings management" and "earnings quality" will be discussed clearly in Sect. 2.

Given the inferiority of the Vietnamese SOEs in performance and the importance of reported earnings number, we are curious whether there is the same problem occurs in earnings quality? Particularly, we want to answer whether earnings numbers released by state-owned firms are less reliable than those published by privately owned firms. We believe the mentioned issue is extremely important, not only for analysts and investors but also for policymakers as well since the quality of reported earnings number is critical for evaluating current operating performance, projecting future operating performance and valuing the intrinsic value of companies (Dechow and Schrand 2004) [10]. Although this is an interesting question for the Vietnamese financial market, to our best knowledge, there is no existing research on this issue in Vietnam. Therefore, our main contribution to the literature of the field is to provide empirical evidence to see whether there is a link between state-owned firms, privately owned firms, and earnings quality and in what direction that relationship occurs using data in the Vietnamese financial market. Based on what we find, we also point out the implication that is valuable to practitioners, especially analysts and investors.

Our research is organized as follows: Sect. 2 is going to review literature in the field. Section 3 then presents our research design and data. After that, we move to Sect. 4 to discuss empirical results while Sect. 5 figures out implication and concludes the paper.

2 Literature Review

2.1 Earnings Quality Definition

Although scholars have focused on the quality of an earnings number very early, however, until now, it would be still challenging to find a consensus definition on the term. Among the thousands of articles in the field, there are the most used two terms: the first one is "earnings management" and the second one is"earnings quality". In this section, we review these two terms in turn. Based on the review, we then make a summary to choose the definition that we are going to use throughout this paper.

In their frequently cited work, Healy and Wahlen (1999) [17] define the term "earnings management" that is appropriate from the standard setters' point of view:

> Earnings management occurs when managers use judgment in financial reporting and in structuring transactions to alter financial reports to either mislead some stakeholders about the underlying economic performance of the company or to influence contractual outcomes that depend on reported accounting numbers (Healy and Wahlen 1999, p. 368) [17].

According to this definition, "earnings management" would involve two aspects: (1) management discretion over financial reporting process and (2) that management discretion is intentional to conceal the company's true financial performance.

The second term that is often used in the field is "earnings quality". Dechow and Schrand (2004) [10] stand on the viewpoint of analysts and define "earnings quality" as follows:

> From this perspective, a high-quality earnings number is one that accurately reflects the company's operating performance, is a good indicator of future operating performance, and is a useful summary measure for assessing firm value (Dechow and Schrand 2004, p. 5) [10].

Besides, they emphasize further that: "... earnings to be of high quality when the earnings number accurately annuitizes the intrinsic value of the firm" (Dechow and Schrand 2004, p. 5) [10]. Based on the definition, to be evaluated as high quality, an earnings number has to consist of the following characteristics: (1) persistence, (2) predictability and (3) the ability to capture the intrinsic value of a company (Dechow and Schrand 2004) [10]. In our opinion, we prefer the definition of earnings quality by Dechow and Schrand (2004) [10] to the definition of earnings management by Healy and Wahlen (1999) [17] because the definition of earnings quality has already been related to the definition of earnings management. To be classified as a high-quality earnings number, that number must be reported with an absence of earnings management. Therefore, throughout this research, we refer the term earnings quality to the one defined by Dechow and Schrand (2004) [10].

In addition, Dechow et al. (2010) [11] make an excellent review and complement that the use of the earnings quality definition is contextual. They, therefore, suggest that researchers should firstly describe the specific decision context clearly. Based on that, researchers then propose the proxies of earnings quality that are relevant to the defined context. We agree with this recommendation but we postpone the discussion about the decision context and our chosen proxies to measure earnings quality until we describe our research design in Sect. 3.

2.2 The Determinants of Earnings Quality

Many researchers have found various factors that can affect firms' earnings quality[1]. In this section, we only review the factors that are relevant to the model we use in this paper.

Firm size

Jensen and Meckling (1978) [19] and Watts and Zimmerman (1978) [39] argue that under public pressure, large firms are often subject to social responsibilities, tighten regulations, or higher tax rates. Large firms, therefore, find themselves a strong incentive to run away from these issues. There are many techniques to do so and one of them is to decrease reported earnings number. Hence, according to this hypothesis, large firms are in favor of accounting methods that can decrease net income. Motivated by this theory, Hagerman and Zmijewski (1979) [16] provide empirical evidence that large firms tend to avoid public scrutiny by choosing income-deflating accounting methods such as depreciation methods and the investment tax credit. However, Hagerman and Zmijewski (1979) [16] also state that the results are mixed because when they test with another accounting

[1] See Dechow et al. (2010) [11] for an excellent review of the determinants of earnings quality.

methods like inventory and pension costs amortization, firm size does not have any effects.

Nevertheless, in the literature, political pressure is not the only reason that can explain the relationship between firm size and earnings quality. Kinney and McDaniel (1989) [23] document that there is a relationship between firm size and the quality of internal control. The internal control quality, in turn, can affect the probability that errors occur in reporting quarterly earnings. Kinney and McDaniel (1989) [23] show that smaller firms are in a higher probability of having internal control inferiority and hence, more often to correct previously reported quarterly earnings.

Leverage
There is a hypothesis proposed by Watts and Zimmerman (1986) [40] regarding the debt covenant. According to Watts and Zimmerman (1986) [40], firms closer to debt agreement constraints are more likely to be in favor of accounting methods that boost earnings up. They do that to avoid violating debt covenants. This hypothesis is strongly supported by empirical evidence. Bowen et al. (1981) [5] examine potential factors that can affect firms' decisions to capitalize interest or expense. They find out that firms closer to debt covenant violation are more likely to capitalize interest, compared to the control group. Similarly, Daley and Vigeland (1983) [9] show that companies with high leverage are more willing to capitalize research and development costs.

Profitability
Prior studies suggest that firms with poorer financial performance have more incentives to engage in earnings management (Dechow et al. 2010) [11]. Motivated by this hypothesis, Balsam et al. (1995) [2] find that firms with lower changes in ROA are more likely to adopt income-accelerating accounting methods. In addition, Keating and Zimmerman (1999) [22] document that companies with weaker performance are more willing to adopt accounting methods to boost earnings up for all assets.

Other determinants
In addition to the above factors, prior studies also find other determinants are likely to have effects on earnings quality. The results found by Katz (2009) [21] show that quick ratio, the percentage of cash to total assets and whether a firm report a loss in a given year can have statistically significant relationships with earnings quality.

2.3 State Ownership, Private Ownership and Earnings Quality

To our knowledge, until now, research focusing on the relationship between state ownership and earnings quality is still limited. This might be due to the reason that this type of ownership no longer plays significant roles in developed countries. Because of the shortage of the literature in this relationship, our strategy in this section is that we firstly review the relationship between state ownership and firm performance. We then link that relationship to earnings quality to see whether the same relationship exists between state ownership and earnings quality.

In term of performance, there is a common belief that state-owned firms are less efficient and perform poorer than privately owned firms. In reality, the empirical evidence is mixed. On the one hand, Boardman and Vining (1989) [4] document that privately owned enterprises are substantially more profitable than state-owned enterprises. On the other hand, however, Caves and Christensen (1980) [6] provide empirical evidence to show that government-owned firms do not necessarily perform worse than private enterprises in the railroad industry in Canada. Similarly, Martin and Parker (1995) [26] find no evidence to support the claim that private ownership is more profitable than public ownership in the UK. Motivated by mixed empirical evidence, Dewenter and Malatesta (2001) [14] enlarge samples, lengthen periods and control more additional factors to show the robust result that private firms are not only statistically profitable than state-owned firms but the difference between the two groups is significantly large. To our knowledge, although the empirical evidence is mixed in the field, however, the evidence supporting the superior of privately owned firms seem to dominate. The difference in performance between state-owned firms and private firms might be due to the differences in managerial style, political factors or incentives to name a few. This phenomenon raises us an interesting question: given various different factors among state-owned companies and private companies, would they differ in the quality of reported earnings number?

Currently, there is no extant theory addressing the relationship between state ownership, private ownership, and earnings quality. The existing theories mainly focus on the difference in performance, efficiency or innovation between state ownership and private ownership. From the contracting perspective, Shleifer (1998) [29] prefers private ownership to state ownership and argues that private ownership has stronger incentives to innovate and use resources efficiently. The other well-known theory that is relevant to the issue of ownership type is the principal-agent problem theory. Although the agency problems exist in both types of ownership, privately owned firms are, however, expected to face less serious agency problems than state-owned firms, and hence private companies are likely to operate more efficiently than government-owned companies. There is an interesting point inferred from this argument, that is because of less severe agency problems, privately owned enterprises have weaker incentives to manipulate their earnings number (Ding et al. 2007) [15]. Nevertheless, the empirical evidence in China found by Ding et al. (2007) [15] reveals the opposite picture. Ding et al. (2007) [15] analyze 273 government-owned firms and private firms in China in 2002 and the results show that private companies are more likely to manipulate their earnings number than government-owned companies. The authors explain that because in the Chinese financial market, private firms are still in a weaker position than state-owned firms, therefore, private firms try to make management discretion over their reported earnings number to upgrade their position. We are uncertain whether this result is true in the Vietnamese financial market. And as we mentioned in Sect. 1, because the quality of earnings number is definitely critical to all stakeholders in the Vietnamese financial market and given the fact that the existence of state-owned firms in the Vietnamese financial market is still significant, therefore, in this paper, we are eager to answer the question: do state-owned firms and privately owned firms differ in earnings quality and in what direction that relationship occurs?

3 Research Design and Data

3.1 Research Design

To find the answer for the central question of this paper is whether there is a difference in earnings quality between state ownership and private ownership, we utilize the matching approach. The matching procedure is a widely-used method among researchers in the field to mitigate the problem of omitted variables (Barth et al. 2008) [3], (DeFond and Jiambalvo 1994) [13], (Ding et al. 2007) [15], (Lang et al. 2003) [24], (Lang et al. 2006) [25], (Sweeney 1994) [31]. To implement this strategy, previous studies match treated firms with control firms based on industry, time and one attribute that can potentially impact on earnings quality. The reason to match firms within industry and in the same period because those firms are subject to the same financial accounting process and economic events (Sweeney 1994) [31]. In addition, the selection of one attribute to form matched sample varies among researchers. Firms can be matched on total assets, equity market value or revenues growth rate.

Adopting this strategy in our context, rather than matching in industry, we match state-owned firms (the treated group) with private firms (the control group) in the same sector and in the same period. We come to this decision for several reasons. Firstly, in our dataset, the number of firms in a specific industry is not large and a good matching procedure requires a rich source of data, therefore matching in industry would be inappropriate in our circumstance. Secondly, although we match firms at the sector level, we believe that within sector, firms also use similar accounting processes and also have to face with similar economics event. The last point in forming our matched sample is to figure out one attribute to match firms within each sector. After careful consideration, we decide to match state-owned firms with their counterparts based on the average annual growth rate of revenues in the concerned period. There are some factors that have driven our decision. Firstly, the growth rate of revenue is expected to have an impact on earnings quality of a firm (Lang et al. 2006) [25]. Secondly, compared to measures of size such as total assets or equity market value, revenues are more relevant to earnings number and earnings quality. We, therefore, use the average annual growth rate of revenues in the studied period to match the treated group and the control group. The formula to calculating this growth rate is as follows:

$$\text{Growth rate} = \sqrt[n-1]{\frac{\text{Revenues}_{\text{the end year of the studied period}}}{\text{Revenues}_{\text{the beginning year of the studied period}}}} - 1 \quad (1)$$

In formula (1), n is the number of years in the studied period. After forming matched sample, we then estimate a panel data regression using the following model:

$$EQ_{it} = \alpha_0 + \alpha_1 * STATE_{it} + \alpha_2 * SIZE_{it} + \alpha_3 * LEVERAGE_{it} + \alpha_4 * PROFIT_{it} + \alpha_5 * LIQUIDITY_{it} + \alpha_6 * CASH_{it} + \alpha_7 * LOSS_{it} + \epsilon_{it} \quad (2)$$

The reason to involving the variables in model (2) is based on the review of the determinants of earnings quality that we discussed in Sect. 2 and with a reference to previous research (Katz 2009) [21], (Lang et al. 2003) [24], (Lang et al. 2006) [25]. In model (2), the subscript i denotes firm i while the subscript t denotes year t. The calculation of the variables in model (2) desires for further explanation.

3.1.1 Earnings Quality Measures

The variable *EQ* in model (2) is earnings quality. There are various measures in the field have been developed to evaluate how good an earnings number of a firm is. Dechow et al. (2010) [11] point out that all measures of earnings quality have both advantages and disadvantages and the use of one measure should be in respect of a specific decision context. In alignment with this recommendation, we firstly set out the decision context using in this paper and from that point, we decide to select appropriate measures of earnings quality.

In this paper, we stand on the perspectives of analysts and investors. When analysts and investors look for a potential investment opportunity, they rely on firms' earnings number to evaluate firms' operating performance, forecast firms' future operating performance and value firms' intrinsic value (Dechow and Schrand 2004) [10]. Based on these analyses, they then make investment decision. We use this decision context throughout this paper and the definition of the term earnings quality we use is referred to the one defined by Dechow and Schrand (2004) [10].

Since there is no perfect single measure of earnings quality and to be consistent with our decision context, in this study, we propose to combine the following three measures to be proxies for earnings quality. These measures are relevant to our decision context.

The primary measure of earnings quality in our research is based on the Jones model which is developed by Jones (1991) [20]. Model (3) below represents the idea of the Jones model.

$$Acc_{it} = \alpha_i + \beta_{1i} * \Delta REV_{it} + \beta_{2i} * PPE_{it} + \varepsilon_{it} \quad (3)$$

In model (3), the variables are calculated as follows:

- Acc_{it}: Total accruals for firm i in year t.
  More specifically, total accruals can be calculated as: Acc_{it} = ΔCurrent assets$_t$ − ΔCurrent liabilities$_t$ − ΔCash$_t$ + ΔShort term debt included in current liabilities$_t$ − Depreciation and armotization expense$_t$.
- ΔREV_{it}: change in revenues of firm i in in year t and is equal to revenues of that company in year t minus revenues in year $t-1$.
- PPE_{it}: Gross property, plant and equipment in year t of company i.

All variables in model (3) are scaled by total assets of firm i in year t−1. However, in this paper, we divide them by average total assets of firm i in year t−1 because average total assets represent better the size of company i in year t−1 than total assets of that company at the end of year t−1. The residual in model (3) is the measure of earnings quality. The intuition behind this model is that after regressing total accruals on their contributing economic factors, revenues and gross, property plant and equipment, the residuals from the model will capture abnormal accruals which is an appropriate measure of management discretion (Dechow et al. 2010) [11]. This means the higher the residuals are, the lower the earnings quality of a firm is. In addition, Cohen et al. (2008) [8] point out that model (3) can be estimated within industry so that economic conditions in each industry that have potential effects on total accruals can be controlled for. We adopt the same strategy in this paper, nevertheless, instead of industry, we estimate model (3) within sector.

The second measure of earnings quality we use in this research is derived from the modified Jones model. Dechow et al. (1995) [12] argue that because firms are more tempted to make management discretion on credit sales than cash sales, therefore, they introduce a modified version of the Jones model to account for the change in credit sales. Particularly, the model they use is:

$$Acc_{it} = \alpha_i + \beta_{1i} * (\Delta REV - \Delta REC)_{it} + \beta_{2i} * PPE_{it} + \varepsilon_{it} \tag{4}$$

All variables in model (4) are exactly the same as in model (3) except for $\Delta \mathrm{REC}_{\mathrm{it}}$ which is equal to net receivables of firm i in year t minus net receivables of that firm in year t−1. And like the Jones model, all variables in model (4) are deflated by lagged total assets. In this research, we divide them by lagged average total assets. And like model (3), we estimate model (4) within sector.

The last metric to gauge earnings quality we employ in this study is just simply total accruals itself. In alignment with previous studies (Dechow et al. 2010) [11], (Richardson et al. 2005) [28] and with a reference to CFA program curriculum (2012) [7], we use the following formula to calculate total accruals:

$$\text{Total Accruals} = \text{Net Operating Assets}_t - \text{Net Operating Assets}_{t-1} \tag{5}$$

In formula (5), the subscript t denotes year t and net operating assets are equal to operating assets minus operating liabilities. Operating assets are equal to total assets minus cash and cash equivalents while operating liabilities are equal to total liabilities minus total debts. After calculating total accruals using formula (5), we then divide it by average net operating assets and the result is our third measure of earnings quality. The intuition behind this metric is that in an earnings number of a firm, there are two components: the first one is cash component and the second one is accruals component. The accruals component is less persistent than the cash component (Sloan 1996) [30] and therefore, if an earnings number has more accruals component, it is likely to be less persistent in future and hence, its quality is lower. This means an increase in total accruals is associated with a decrease in earnings quality.

3.1.2 State Ownership

In model (2), the variable $STATE$ is a dichotomous variable, equals 1 if the firm in concern is state-owned firm and 0 otherwise. According to the Vietnamese Securities Law 2006 [35], a person or an entity that holds at least 5% of a firm's voting shares is defined as "leading shareholder". We use this 5% threshold to separate state-owned firms from private firms. Specifically, if in a firm's ownership structure, the share of state ownership is larger than or equal to 5%, then the variable $STATE$ is assigned 1 and 0 otherwise. In our opinion, the 5% threshold to categorize whether a firm is a state-owned firm or a private firm is reasonable. In our database, if a firm has a presence of state ownership in its ownership structure, then that share of state ownership is always larger than 5%. In addition, the presence of state ownership often has a considerable voice in a firm's decisions. This means even the presence of state ownership can be theoretically as small as 5% of a firm's voting shares, that existence, however, is still significant to any firm's critical decisions.

3.1.3 Control Variables

Putting the variable EQ and the variable $STATE$ aside, all the rest variables in model (2) are control variables. According to previous research that we have mentioned in Sect. 2, these variables are likely to have effects on firm's earnings quality, hence, it is necessary to control for.

The calculation of these variables is based on the research conducted by Katz (2009) [21]. In model (2), $SIZE$ is the natural logarithm of total assets, $LEVERAGE$ is equal to total debts divided by total assets. $PROFIT$ stands for profitability and we use return on assets (ROA) to be proxy for the variable $PROFIT$. $PROFIT$ is calculated as net income plus interest expenses after tax deduction, the result is then divided by average total assets. $LIQUIDITY$ is equal to cash and cash equivalents plus short-term investment and short-term account receivables, and then divided by current liabilities. $CASH$ is calculated by aggregating cash and cash equivalents and short-term investments then divided by total assets. Finally, $LOSS$ is a dummy variable, which is equal to 1 if the earnings number of a firm in a given year is negative, and 0 otherwise.

3.2 Data

In our database, we include firms in both the Hochiminh Stock Exchange (HOSE) and the Hanoi Stock Exchange (HNX). In alignment with Sloan (1996) [30], we exclude firms in banking, insurance, financial services and real estate industries because they are subject to different financial accounting process. To calculate variables in model (2), we use firms' financial statements which are obtained from Fiinpro Platform provided by Stoxplus while data of state ownership is provided by Vietstock[2].

[2] Stoxplus and Vietstock are the leading financial information services companies in Vietnam.

To implement our research design, we first match each firm in the state-owned group with its twin in the privately owned group that is in the same sector, in the same studied period and has the nearest value of revenues growth rate in the studied period. We obtain the sector classification of firms from Thomson Reuters who use the Global Industry Classification Standard (GICS). In regard to the studied period, we decide to use the period 2010–2015. The period 2010–2015 is chosen because in our database, this period offers us the most complete financial data and the most reliable data of state ownership. These conditions are the most important ones to form a good matched sample. Before matching, we have 601 firms in our database, unequally distributing among 8 sectors, including: consumer discretionary, consumer staples, energy, health care, industrials, information technology, materials and utilities. After matching, we have 224 firms remaining in our sample, in which 112 firms are state-owned firms and 112 are privately owned firms. With the period from 2010 to 2015, our panel data finally consists of 1344 firm-year observations. The 224 firms are now unequally distributed among only 7 sectors because we are unable to identify matched pairs in the utilities sector. Table 1 below presents the distribution of firms among sectors.

Table 1. Matched firms breakdown by sectors[a]

Sector	Number of state-owned firms	Percentage of state-owned firms (%)	Number of privately owned firms	Percentage of privately owned firms (%)
Consumer Discretionary	12	10.71	12	10.71
Consumer Staples	13	11.61	13	11.61
Energy	2	1.79	2	1.79
Health Care	4	3.57	4	3.57
Industrials	48	42.86	48	42.86
Information Technology	6	5.36	6	5.36
Materials	27	24.11	27	24.11
Total	112	100.00	112	100.00

Source: Author calculations

[a]Because each firm of the state-owned group is matched exactly with one firm in the privately owned group, the number of state-owned firms are exactly the same as the number of privately owned firms in our matched sample.

Table 1 shows that most firms concentrate on industrials and materials where the number of state-owned firms in industrials and materials are 48, corresponding to 42.86% and 27, corresponding to 24.11% of total state-owned firms, respectively. In addition, there are 12 state-owned firms in consumer discretionary sector and 13 others in consumer staples. These two sectors constitute 22.32% of the group. In our matched sample, energy sector contributes the least, only 2 state-owned firms.

After matching on the average annual growth rate of the period 2010–2015, it is worth to check the balance on this attribute again between the state-owned group and the privately owned group. Table 2 exhibits the result of this balance test.

Table 2. Balance check on the average annual growth rate of the period 2010–2015 between state-owned firms and privately owned firms

Sector	The average annual growth rate of the period 2010–2015 (State-owned firms)	The average annual growth rate of the period 2010–2015 (Privately-owned firms)	Difference[a]
Consumer Discretionary	0.1135	0.0729	−0.0406 (0.0420)
Consumer Staples	0.0849	0.0517	−0.0332 (0.0339)
Energy	−0.0577	0.0701	0.1278 (0.2043)
Health Care	0.0418	0.0819	0.0401 (0.0514)
Industrials	0.0921	0.0857	−0.0063 (0.0278)
Information Technology	0.0993	0.0386	−0.0607 (0.1522)
Materials	0.0839	0.0622	−0.0217 (0.0289)
Total	0.0875	0.0718	−0.0157 (0.0171)

Source: Author calculations

[a]Standard errors in parentheses.

Table 2 shows that in any sector, there is no statistically significant difference in the average annual growth rate of the period 2010–2015 between state-owned firms and private firms. This result affirms out matching procedure is successful.

4 Descriptive Statistics and Empirical Results

4.1 Descriptive Statistics

Table 3 presents the descriptive statistics of all variables using in this research. EQ.1 is our primary measure of earnings quality which is calculated from the Jones model while EQ.2 is the second measure, obtaining from the modified Jones model and EQ.3 is the third one, that is total accruals. Because the only difference between the Jones model and the modified Jones model is the later one adjusts for the change in credit sales, therefore values of EQ.1 seem to be nearly the same as values of EQ.2 in Table 3. The mean of EQ.1 is $-1.85 * 10^{-10}$ and the mean of EQ.2 is $-1.11 * 10^{-10}$. While EQ.1 and EQ.2 capture abnormal accruals, EQ.3 measure total accruals, including normal and abnormal accruals, hence values of EQ.3 are quite different from those of EQ.1 and EQ.2. The mean of EQ.3 is 0.0998.

In our matched sample, the number of state-owned firms and the number of private firms are equal, therefore each group contributes 50% of the sample. In regard to firms' financial structure, there are some firms do not use debts at all, on the other hand, firms can finance their total assets by debts up to 75.81% in a given year. On average, in a typical firms' financial structure in a particular year, 23.33% of total assets are financed from debt. Profitability varies widely among firms. Average ROA of a firm in a year is about 6.14%. The most profitable firm can use every 100 dollars efficiently to create 67.90 dollars for both equity owners and debt holders while the worst one has ROA of −45.71%. Concerning liquidity, in some firms, their ability to meet short term obligations can be problematic since their quick ratios can be as low as 0.0269 while the other firms seem to be very safe with their liquidity assets are as more than 16 times as their current liabilities. On average, firms' liquidity is guaranteed. The amount of cash and cash equivalents and short-term investments holding by companies are also discrepant. Some firms seem not to reserve cash at all, the other firms, however, can hold cash and cash equivalents and short-term investment up to more than 90% of their total assets.

Table 3. Descriptive statistics

Variables	Observations	Mean	Standard deviation	Min	Max
Dependent variables					
EQ1	1,344	$-1.85*10^{-10}$	0.1885	−1.0621	2.0553
EQ2	1,344	$-1.11*10^{-10}$	0.1919	−1.0657	2.0530
EQ3	1,344	0.0998	0.3549	−3.1050	2.6545
Independent variable					
STATE	1,344	0.5000	0.5002	0.0000	1.0000
Control variables					
SIZE	1,344	26.8085	1.3718	23.4499	31.5191
LEVERAGE	1,344	0.2333	0.1940	0.0000	0.7581
PROFIT	1,344	0.0614	0.0677	−0.4571	0.6790
LIQUIDITY	1,344	1.1079	1.2526	0.0269	16.3197
CASH	1,344	0.1269	0.1357	0.0001	0.9437
LOSS	1,344	0.0640	0.2448	0.0000	1.0000

Source: Author calculations

Table 4 presents correlation coefficients among variables. In Table 4, the correlation coefficient between STATE and EQ.1 is −0.0702, signaling that state-owned firms might have their earnings quality better than their counterpart's. All coefficients in Table 4 are small. In addition, the Variance Inflation Factor (VIF) of all variables in Table 4 are smaller than 10, indicating the absence of severe multicollinearity problem in the model.

4.2 Empirical Results

Table 5 exhibits the regression results with our primary measure of earnings quality. Column (1) shows the results of pooled OLS while column (2) presents the results when we use fixed effect for sectors. In both regressions, we use robust

Table 4. Correlation matrix of all variables

	EQ1	STATE	SIZE	LEVERAGE	PROFIT	LIQUIDITY	CASH	LOSS
EQ1	1.0000							
STATE	−0.0702	1.0000						
SIZE	0.0551	0.0277	1.0000					
LEVERAGE	0.0947	−0.0141	0.4243	1.0000				
PROFIT	0.0053	0.2212	0.0138	−0.2325	1.0000			
LIQUIDITY	0.0449	0.0715	−0.2200	−0.4551	0.2785	1.0000		
CASH	−0.0911	0.0657	−0.1224	−0.3995	0.3825	0.5293	1.0000	
LOSS	−0.0844	−0.1216	−0.0084	0.1343	−0.4007	−0.1072	−0.1513	1.0000

Source: Author calculations

standard errors to control for potential heteroskedasticity and serial correlation. Table 5 shows that the results are robust regardless of methods. Our main concern, the estimated coefficient of the variable STATE is −0.0335 in pooled OLS ($p = .001$, 95% CI [−0.05, −0.01]) and is −0.0337 in fixed effects ($p = .001$, 95% CI [−0.05, −0.01]). This means that compared to their counterparts, state-owned firms have better earnings quality. This result is consistent with the findings by Ding et al. (2007) [15] in China. Ding et al. (2007) [15] claim that in the Chinese financial market, state-owned firms are still in a better position than privately owned firms, therefore the latter ones are induced to make more management discretion over their earnings to improve their position. In our opinion, this argument is plausible in the Vietnam circumstance. In the Vietnamese financial market, state-owned firms are often easier than their counterparts to access loans from banks. This might create a strong incentive for privately owned firms to boost their earnings up to make them become more profitable and therefore, get them easier to access loans from the banking system. Besides, because in private firms, managers are often holding a significant amount of shares of those companies and boosting earnings could make their companies become more attractive and their stock prices would increase, creating a capital gain and hence, an obvious financial benefit for managers to engage in management discretion. This is not the case for state-owned firms. In state-owned firms, the presence of state ownership often appears through a representative. That representative does not truly own the company, the true owner is “the state”. And the point is since he is just simply a legal representative for the state capital, he finds no financial incentive to adjust the firm’s earnings upward. And therefore, state-owned firms appear with better earnings quality than privately owned firms.

In addition to our main variable, the coefficients of control variables also reveal exciting pictures about earnings quality in the Vietnamese financial market. In Table 5, while size and profitability do not play any role in firms’ earnings quality, leverage, liquidity, cash and loss seem to do affect, however. The signs of the coefficients of the variables leverage, liquidity, cash and loss are consistent with the results found by Katz (2009) [21].

The coefficient of the variable leverage is 0.1178 and is positive ($p = .003$, 95% CI [0.04, 0.20]). There are empirical evidence proving that firms with high leverage have more incentives to make management discretion over earnings

Table 5. Regression results. Dependent variable: Earnings Quality (EQ1)

Variables	Dependent Variables: Earnings Quality (EQ1)	
	(1) Pooled OLS[a]	(2) Fixed Effects[a]
STATE	−0.0335*** (.01019)	−0.0337*** (0.0102)
SIZE	0.0034 (0.0038)	0.0032 (0.0039)
LEVERAGE	0.1178*** (0.0402)	0.1222*** (0.0428)
PROFIT	0.0506 (0.0973)	0.0613 (0.1031)
LIQUIDITY	0.0262*** (0.0075)	0.0263*** (0.0075)
CASH	−0.2074*** (0.0579)	−0.2085*** (0.0581)
LOSS	−0.0831*** (0.0178)	−0.0822*** (0.0180)
Constant	−0.1034 (0.1005)	−0.1003 (0.1029)
Observations	1,344	1,344
R-squared	0.0497	0.0502

Source: Author calculations
[a]Robust standard errors in parentheses. *** Significant at the 1% level. ** Significant at the 5% level. * Significant at the 10% level.

number because they want to avoid violating debt covenants (Dechow et al. 2010) [11]. Therefore, a high level of debt firms use to finance their assets would trigger managers' incentives to increase accruals component in earnings numbers and therefore, worsen the quality of those earnings numbers.

The coefficient of the variable liquidity, which is exactly the firm's quick ratio, is 0.0262 and is positive ($p < .001$, 95% CI [0.01, 0.04]). This is because the calculation of quick ratio involves account receivables that are fundamentally accruals component. Therefore, a higher quick ratio implies a higher accruals component of an earnings number and therefore a lower quality of that earnings number.

The coefficient of the variable cash is −0.2074 and is negative ($p < .001$, 95% CI [−0.32, −0.09]). This is because when cash increases, it is often associated with an increase in the cash component of an earnings number. And because the higher the cash component is, the lower the accruals component is and therefore, the better the quality of an earnings number is.

The coefficient of the variable loss is −0.0831 and is negative ($p < .001$, 95% CI [−0.12, −0.05]). Prior studies document a phenomenon called loss avoidance behavior (Dechow et al. 2010) [11]. That is managers tend to avoid record a loss in the income statement and therefore, try to boost their earnings number up. This implies that because managers do not want to report a loss, hence if a loss is reported, meaning that managers have no available management discretion other than presenting a loss, then in this situation a loss is likely to reflect the true performance of a company. Therefore, in term of earnings quality, a loss shows a better signal than a profit.

As we mentioned before, in addition to the primary measure of earnings quality obtained from the Jones model, we also use the modified Jones model and total accruals to be other proxies for earnings quality. Table 6 presents the regression result in which the dependent variable is our second metric of earnings quality that is calculated from the modified Jones model while Table 7 exhibits the regression results that we use the third measure of earnings quality, total accruals.

Table 6. Regression results. Dependent variable: Earnings Quality (EQ2)

Variables	Dependent Variables: Earnings Quality (EQ2)	
	(1) Pooled OLS[a]	(2) Fixed Effects[a]
STATE	−0.0370*** (0.0104)	−0.0373*** (0.0104)
SIZE	0.0025 (0.0038)	0.0023 (0.0040)
LEVERAGE	0.1216*** (0.0406)	0.1264*** (0.0431)
PROFIT	0.0760 (0.0973)	0.0889 (0.1032)
LIQUIDITY	0.0253*** (0.0074)	0.0254*** (0.0074)
CASH	−0.1999*** (0.0576)	−0.2007*** (0.0579)
LOSS	−0.0944*** (0.0179)	−0.0934*** (0.0181)
Constant	−0.0785 (0.1030)	−0.0767 (0.1054)
Observations	1,344	1,344
R-squared	0.0507	0.0512

Source: Author calculations
[a]Robust standard errors in parentheses. *** Significant at the 1% level. ** Significant at the 5% level. * Significant at the 10% level.

Because there are not too many discrepancies between our primary and second measure of earnings quality, therefore, the regression results in Table 6 do not change significantly from the results in Table 5. And in fact, the coefficients of all variables as well as the sign of these coefficients in Table 6 remain stable, compared to what is found in Table 5. This supports the robustness of the results found in Table 5.

Turning to Table 7 when we use total accruals to assess earnings quality, the results now change significantly from Tables 5 and 6. This is understandable because the intuition behind the first and second measures of earnings quality in this paper is only to evaluate the abnormal accruals while the third measure is calculating the total accruals. Therefore, it is expected that there should be a change in the coefficients of the variables in the model. However, the most important point is that the signs of all the coefficients remain the same as before. This means the direction in which these variables affect firms' earnings quality are confirmed.

Table 7. Regression results. Dependent variable: Earnings Quality (EQ3)

Variables	Dependent Variables: Earnings Quality (EQ3)	
	(1) Pooled OLS[a]	(2) Fixed Effects[a]
STATE	−0.0684*** (0.0201)	−0.0696*** (0.0203)
SIZE	0.0023 (0.0071)	0.0026 (0.0071)
LEVERAGE	0.3057*** (0.0731)	0.3209*** (0.0724)
PROFIT	0.6257*** (0.2228)	0.6690*** (0.2226)
LIQUIDITY	0.0327* (0.0182)	0.0320* (0.0182)
CASH	−0.2955** (0.1483)	−0.2901* (0.1498)
LOSS	−0.1551*** (0.0383)	−0.1544*** (0.0377)
Constant	−0.0266 (0.1807)	−0.0310 (0.1828)
Observations	1,344	1,344
R-squared	0.0591	0.0612

Source: Author calculations

[a]Robust standard errors in parentheses. *** Significant at the 1% level. ** Significant at the 5% level. * Significant at the 10% level.

5 Concluding Remarks

In this paper, we document that state-owned firms have better earnings quality than privately owned firms in the Vietnamese financial market. This is because private firms have financial benefits to make management discretion over earnings number. Boosting earnings number up could result in favorable conditions for an increase in stock price, which in turn, creates capital gain for managers who also hold a significant amount of shares in those companies. In addition, manipulating earnings number make private firms easier to access loans from banks. The finding in this paper implies a meaningful lesson for analysts and investors. When looking for a potential investment opportunity, analysts and investors should be very cautious on earnings number of privately owned corporations. This does not mean analysts and investor can ignore the same problem in state-owned firms, but rather, the evidence in this research encourages them to put more weight in private firms, compared to state-owned firms while making a scrutiny look over earnings number of all firms.

Although this paper has achieved some considerable results, it, however, still goes with some limitations. Firstly, although this paper uses the matching approach which is a popular method in the field, however, whether the matching procedure can completely eliminate the problem of omitted variables is not guaranteed. This has been mentioned by Barth et al. (2008) [3]. Secondly, because of limitation of data availability, this research only employs matched sample of 112 state-owned firms and 112 privately owned firm from 2010 to 2015. The number of firms is not large, the chosen period is short and the distribution of firms among sectors might not represent well for the whole market and so, the results achieved is still limited. Therefore, a more well-designed research is demanded for future research.

References

1. Aharony, J., Swary, I.: Quarterly dividend and earnings announcements and stockholders' returns: an empirical analysis. J. Financ. **35**(1), 1–12 (1980)
2. Balsam, S., Haw, I., Lilien, S.: Mandated accounting changes and managerial discretion. J. Account. Econ. **20**, 3–29 (1995)
3. Barth, M.E., Landsman, W.R., Lang, M.H.: International accounting standards and accounting quality. J. Account. Res. **46**(3), 467–498 (2008). https://onlinelibrary.wiley.com/doi/full/10.1111/j.1475-679X.2008.00287.x. Accessed 16 June 2018
4. Boardman, A.E., Vining, A.R.: Ownership and performance in competitive environments: a comparison of the performance of private, mixed, and state-owned enterprises. J. Law Econ. **32**(1), 1–33 (1989)
5. Bowen, R., Lacey, J., Noreen, E.: Determinants of the corporate decision to capitalize interest. J. Account. Econ. **3**, 151–179 (1981)
6. Caves, D.W., Christensen, L.R.: The relative efficiency of public and private firms in a competitive environment: the case of Canadian railroads. J. Polit. Econ. **88**(5), 958–976 (1980)
7. CFA Institute: Financial Reporting and Analysis, CFA Program Curriculum, Level 2, vol. 2, pp. 343–415 (2012)
8. Cohen, D., Dey, A., Lys, T.: Real and accrual-based earnings management in the pre- and post-Sarbanes-Oxley periods. Account. Rev. **83**, 757–787 (2008)
9. Daley, L., Vigeland, R.: The effects of debt covenants and political costs on the choice of accounting methods: the case of accounting for R&D costs. J. Account. Econ. **5**, 195–211 (1983)
10. Dechow, P.M., Schrand, C.M.: Earnings Quality. The Research Foundation of CFA Institute, USA (2004)
11. Dechow, P.M., Ge, W., Schrand, C.M.: Understanding earnings quality: a review of the proxies, their determinants and their consequences. J. Account. Econ. **50**(2–3), 344–401 (2010)
12. Dechow, P., Sloan, R., Sweeney, A.: Detecting earnings management. Account. Rev. **70**, 193–225 (1995)
13. DeFond, M., Jiambalvo, J.: Debt covenant violation and manipulation of accruals. J. Account. Econ. **17**, 145–176 (1994)
14. Dewenter, K.L., Malatesta, P.H.: State-owned and privately owned firms: an empirical analysis of profitability, leverage, and labor intensity. Am. Econ. Rev. **91**(1), 320–334 (2001)
15. Ding, Y., Zhang, H., Zhang, J.: Private vs state ownership and earnings management: evidence from Chinese listed companies. Corp. Gov. Int. Rev. **15**(2), 223–238 (2007). https://onlinelibrary.wiley.com/doi/full/10.1111/j.1467-8683.2007.00556.x. Accessed 12 June 2018
16. Hagerman, R., Zmijewski, M.: Some economic determinants of accounting policy choice. J. Account. Econ. **1**, 141–161 (1979)

17. Healy, P.M., Wahlen, J.M.: A review of the earnings management literature and its implications for standard setting. Account. Horiz. **13**(4), 365–383 (1999)
18. Hussin, B.M., Ahmed, A.D., Ying, T.C.: Semi-strong form efficiency: market reaction to dividend and earnings announcements in Malaysian stock exchange. IUP J. Appl. Finan. **16**, 36–60 (2010)
19. Jensen, M.C., Meckling, W.H.: Can the corporation survive? Financ. Anal. J. **34**(1), 31–37 (1978)
20. Jones, J.: Earnings management during import relief investigations. J. Account. Res. **29**, 193–228 (1991)
21. Katz, S.P.: Earnings quality and ownership structure: the role of private equity sponsors. Account. Rev. **84**(3), 623–658 (2009)
22. Keating, A., Zimmerman, J.: Depreciation-policy changes: tax, earnings management, and investment opportunity incentives. J. Account. Econ. **28**, 359–389 (1999)
23. Kinney, W., McDaniel, L.: Characteristics of firms correcting previously reported quarterly earnings. J. Account. Econ. **11**, 71–93 (1989)
24. Lang, M., Raedy, J., Yetman, M.: How representative are firms that are cross-listed in the United States? An analysis of accounting quality. J. Account. Res. **41**, 363–386 (2003)
25. Lang, M., Raedy, J., Wilson, W.: Earnings management and cross listing: are reconciled earnings comparable to US earnings? J. Account. Econ. **42**, 255–283 (2006)
26. Martin, S., Parker, D.: Privatization and economic performance throughout the UK business cycle. Manag. Decis. Econ. **16**(3), 225–237 (1995)
27. Nguyen, V.T., Freeman, N.J.: State-owned enterprises in Vietnam: are they 'crowding out' the private sector? Post Communist Econ. **21**, 227–247 (2009)
28. Richardson, S., Sloan, R., Soliman, M., Tuna, I.: Accrual reliability, earnings persistence and stock prices. J. Account. Econ. **39**, 437–485 (2005)
29. Shleifer, A.: State versus private ownership. J. Econ. Perspect. **12**, 133–150 (1998)
30. Sloan, R.: Do stock prices fully reflect information in accruals and cash flows about future earnings? Account. Rev. **71**, 289–315 (1996)
31. Sweeney, A.: Debt-covenant violations and managers' accounting responses. J. Account. Econ. **17**, 281–308 (1994)
32. The General Statistics Office of Vietnam, Press release on the preliminary results of the 2017 Economic Census (2018). http://www.gso.gov.vn/default.aspx?tabid=382&ItemID=18686. Accessed 20 June 2018
33. The OECD: Structural Policy Country Notes Vietnam, Structural Policy Challenges for Southeast Asian Countries, pp. 1–18 (2013)
34. The Prime Minister of The Vietnamese Government: Decision No. 929/QD-TTg (2012). http://vanban.chinhphu.vn/portal/page/portal/chinhphu/hethongvanban?class_id=2&_page=1&mode=detail&document_id=162394. Accessed 20 June 2018
35. The Vietnamese Congress: The Vietnamese Securities Law (2006). http://vanban.chinhphu.vn/portal/page/portal/chinhphu/hethongvanban?class_id=1&_page=3&mode=detail&document_id=80082. Accessed 16 June 2018
36. The Vietnamese Government: Decree 101/2009/ND-CP (2009). http://www.moj.gov.vn/vbpq/lists/vn
37. The Vietnamese Government: Decree 59/2011/ND-CP (2011). http://vanban.chinhphu.vn/portal/page/portal/chinhphu/hethongvanban?class_id=1&_page=1&mode=detail&document_id=101801. Accessed 20 June 2018

38. Vietnam Development Report: Market Economy for a Middle-Income Vietnam, Joint Donor Report to the Vietnam Consultative Group Meeting December 06, 2011 (2012)
39. Watts, R.L., Zimmerman, J.L.: Towards a positive theory of the determination of accounting standards. Account. Rev. **53**(1), 112–134 (1978)
40. Watts, R., Zimmerman, J.: Positive Accounting Theory. Prentice-Hall Inc., Englewood Cliffs (1986)

Does Female Representation on Board Improve Firm Performance? A Case Study of Non-financial Corporations in Vietnam

Anh D. Pham[1(✉)] and Anh T. P. Hoang[2]

[1] Research Institute for Banking, Banking Academy of Vietnam, 12 Chua Boc St., Dong Da Dist., Hanoi, Vietnam
anhpd@hvnh.edu.vn

[2] School of Finance, University of Economics Ho Chi Minh City, 196 Tran Quang Khai St., Dist. 1, Ho Chi Minh City, Vietnam
anhtcdn@ueh.edu.vn

Abstract. This paper evaluates the impact of board gender diversity on the performance of 170 non-financial corporations listed on the Vietnamese stock exchange over the period 2010–2015. Empirical results suggest that gender diversity measured by the proportion of female directors on board and the number of female directors on board positively affects firm performance. Such positive effect is primarily derived from women directors' executive power and management skills rather than their independence status. Besides, we found evidence that boards with at least three female members exert a stronger positive effect on firm performance than boards with two or fewer female members.

Keywords: Board gender diversity · Firm performance · Board chair Critical mass

1 Introduction

In East Asian cultures including Vietnam, there have been long held misconceptions about the role of women in society. People in these countries tend to have preconceived notions of gender prejudice in the the belief that the duty of women was limited in their homes, viz. taking care of their family and doing the housework. Nevertheless, in recent years, the position of women in families in particular and in society in general has been strengthened a great deal. Women have been engaging in as many professions as men have, particularly in doing business. According to a report by Vietnam Chamber of Commerce and Industry (VCCI), in 2014, one in every four businesses would have female directors on board. Alongside their ownership role in small and medium enterprises (SMEs), business women have assisted major corporations in, step by step, coping with difficulties, growing and striving for success, at both domestic and international level.

Although Vietnam has scored some notable successes in attaining the goal of gender equality, today's women still encounter countless difficulties in different areas, especially in political and economic aspects. A range of empirical studies reveal that,

V. Kreinovich et al. (Eds.): ECONVN 2019, SCI 809, pp. 497–509, 2019.
https://doi.org/10.1007/978-3-030-04200-4_36

unlike males, females supposedly fall far short of requisite qualities and talents to be successful as they tend to associate themselves with friendliness and sharing mind (best known as *social service-oriented model*), rather than rewardingness (best known as *performance-oriented model*), and unfortunately, the latter is believed to be a must-have quality of a genuine commander (Eagly and Johannesen-Schmidt 2001). In addition, Kanter (1977) argues that observers are inclined to distort the image of female executives by closely relating their image with femininity rather than the distinct qualities of a leader. Indeed, in this regard, the role women play in Vietnamese society has not received adequate attention of the government. According to the World Bank statistics, in 2014, there were approximately 23% of household businesses and 71% of SMEs headed by female, while the proportion of female to male employed stood at 88.7%. As shown from the World Bank survey, there still exist doubts and prejudices in Vietnam about whether the capacity and quality of women could contribute significantly to the development of the enterprise community in particular and the economy as a whole.

Unlike Norway, Italy or any other European nations where regulations governing the number of women on board of directors are stringently enforced, there has been a lack of government intervention on this aspect in Asian countries, particularly Vietnam. Hence, along with the increasing social recognition of the women's role, this paper seeks to clarify the impact of board gender diversity on firm performance in Vietnam during the period from 2010 to 2015.

2 Review of Related Literature

2.1 Theoretical Background

Agency Theory

Agency problems arise in businesses when the managers act not in the best interests of the shareholders. A solution to this issue is to extend the supervision by the board of directors. Fama and Jensen (1983) highlight that efficient guidance and monitoring from the board is the key to resolving such conflicts of interest. Gender diversity is expected to help enhance board oversight since hiring members with different backgrounds might help fortify diversity in multiple aspects of supervision, and as a result, a wide range of questions could be raised in the boardroom to illuminate the status quo. Since women tend to assume their responsibility on board in earnest, this might lead to more civilized behaviors, and thereby strengthening the soundness of corporate governance (Singh and Vinnicombe 2004). There is ample evidence that female directors appear more proactive in monitoring activities, for instance, Gul et al. (2008) indicate that boards with greater gender diversity would require a higher degree of control and management effort, thence firm performance could be improved.

Resource Dependence Theory

Pfeffer and Salancik's (1978) resource dependence theory acknowledges that businesses are contingent upon external resources to survive and this could pose a risk to them. In order to minimise such dependency and uncertainty, firms could establish relationship with external entities who possess these resources. According to Pfeffer and Salancik (1978), advice and counseling, legitimacy and communication channels are deemed the three most important benefits to corporate board linkages. As regards the issue of advice and counseling, existing literature suggests that gender-diverse boards have higher-quality board meetings on complex issues, some of which might be difficult for all-male boards (Huse and Solberg 2006; Kravitz 2003). Concerning legitimacy, business activities could be legitimated by accepting societal values and norms. "Value in diversity" assumption by Cox et al. (1991) points out that, as gender equality has become a growing tendency in society, businesses could acquire legitimacy when appointing women to the board of directors. Through communication channels, female leaders, with their practical experiences and perspectives, could perform their duties better in connecting their business to female clients, female workers and to society as a whole (Campbell and Minguez-Vera 2008).

Critical Mass Theory

According to Kramer et al. (2006), the critical mass theory refers to the fact that a subgroup must reach a certain size in order to affect the overall group. As indicated in Asch (1951, 1955) studies, the efficiency derived from a subgroup's pressure could be markedly improved when the group size equals three, yet the increase in group size might contribute only a small fraction to the overall effect. Accordingly, it is proposed in the majority of related literature that three be generally the starting point (critical mass level) that has an impact on group formation (Bond 2005; Nemeth and Kwan 1987). Based on previous arguments, recent studies on board gender diversity (Erkut et al. 2008; Konrad et al. 2008) suggest that if there are at least three female directors on board, the critical mass level for female members will be met. Based on in-depth inverviews and group discussion among 50 female directors, research findings reveal that boards with at least three women could alter the general working style, thus influencing the boardroom's dynamics. Under the circumstances, the women's voices and opinions may gain more weight and thus the dynamics of the board would improve significantly.

2.2 Empirical Evidence

Corporate governance is a subject of considerable debate across nations. The rationale behind these discussions, as indicated by Carter et al. (2010), is the tendency for gender diversity being disregarded in both management and board of directors of major corporations. Due to this, 16 countries are demanding a quota with higher number of women directors on board, and concurrently, many others set voluntary quotas in their corporate governance laws (Rhode and Packel 2014).

Empirical studies on the impact of board gender diversity on firm performance have yielded mixed results. An enormous number of scholars admitted the positive influence of gender diversity on firm's financial performance. For instance, Carter et al. (2003) found a positive association between proportion of female directors in the boardroom and firm value using Tobin's Q measurement on a sample of Fortune 1000 public companies. This finding is favoured by Erhard et al. (2003), who witnessed that gender diversity on board in U.S firms has helped enhance surveillance effectiveness and corporate performance as measured by ROA and ROI. Positive relationship between the percentage of female directors on board and Tobin's Q of Spanish enterprises was, once again, brought to light by Campell and Minquez-Vera (2008). Liu et al. (2014) documented the robust positive impact of female participation on board on the ROA and ROE of selected firms in China. Mahadeo et al. (2012) studied enterprises in Mauritius and pointed to a marked difference in corporate performance effect of between gender-diverse boards and all-male boards. Other studies also indicate a favourable relation between board gender diversity and business performance, for instance, studies for the case of France by Sabatier (2015), or Spain by Martín-Ugedo and Minquez-Vera (2014). Unlike the vast majority of empirical literature on listed companies, Martín-Ugedo and Minquez-Vera (2014) conducted their study on a sample of SMEs. Besides the positive relationship as aforementioned, some have shown that in case gender diversity in the boardroom is enhanced, there would be a gradual decline in firm performance (Adams and Ferreira 2009). This study acknowledges that female directors makes the monitoring process become closer, yet, as for countries with strong shareholder defense, higher degree of gender diversity on board might possibly lead to over-supervision which in turn has a negative impact on the business performance.

In addition to positive and negative results, a number of studies found no evidence of the impact of board gender diversity on firm performance. Using Tobin's Q measurement, Rose (2007) did not find any significant link between board gender diversity and corporate performance. This result is reinforced by Carter et al. (2010). Although female participation in the boardrooms is considered to deliver a stronger business performance, there still remain studies showing no evidence that gender diversity on board helps boost business value (Farrell and Hersch 2005). Francoeur et al. (2008) examined women participation in senior management and governance boards of Canadian enterprises and concluded that extra income derived from board gender diversity proves sufficient to catch up with ordinary stock returns, yet not superior to alternative board models. Furthermore, during the financial crisis period, board gender diversity was found to have no impact on corporate performance (Engelen et al. 2012).

3 Data and Methodology

3.1 Data

The study obtains data from 170 non-financial corporations listed on Hanoi Stock Exchange (HNX) and Ho Chi Minh Stock Exchange (HOSE) between 2010 and 2015. In this paper, we exclude firms in financial and public utility areas from our sample. The removal of public utility firms is due to the fact that these firms often receive subsidies from the government to bring welfare to society, hence their operations are deemed economically inefficient. Meanwhile, the motive for the exclusion of financial firms is that, the capital structure of these firms is completely different from that of ordinary businesses, furthermore, it fails to reflect accurately the objective of our research. Data on the characteristics and structure of the board as well as financial performance of these corporations are collected from their annual reports published on VietStock.vn.

3.2 Econometric Model

To gauge the effect of board gender diversity on firm performance, we follow the regression model developed by Liu et al. (2014). Our model is constructed as follows:

$$Firm\_Performance_{it} = \gamma \, Gender\_Diversity_{it} + \beta_m Board\_Char_{it,m} + \beta_n Firm\_Char_{it,n} + \alpha_i + \lambda_t + \varepsilon_{it} \quad (1)$$

where:

Firm_Performance: Two proxies chosen to measure firm performance in this paper include: (1) return on sales (ROS), calculated as the ratio of net income to sales; (2) return on assets (ROA), calculated as the ratio of net income to total assets.

Gender_Diversity: is a measure of board gender diversity. In a diverse range of studies, the percentage of female directors on board are employed to quantify board gender diversity (Adams and Ferreira 2009; Ahern and Dittmar 2012). Less common alternative measures might encompass the number of female executives on board and a dummy variable linked to the critical mass threshold, across which female directors involvement has a significant impact on firm performance (Simpson et al. 2010). In this study, we use both the percentage of female directors on board (%Women) and the number of female directors on board as measures of board gender diversity. %Women could be split into two groups: (i) the percentage of women independent directors (% IndependentWomen) and (ii) the percentage of women executive directors (%ExecutiveWomen). The number of female directors is defined under a set of three dummy variables as follows: D_1Woman equals 1 if the board has one female director and 0 otherwise; D_2Women equals 1 if the board has two female directors and 0 otherwise; D_3Women equals 1 if the board has greater than or equal three female directors and 0 otherwise.

Aside from that, **Board_Char** and **Firm_Char** are control variables representing the characteristics of the board and the characteristics of the firm:

- **Board_Char** (board characteristics) consists of the percentage of independent directors on board (%Independent), the natural log of the board size (Ln_BoardSize) and the dummy Duality (equals 1 if the CEO is also board chair and 0 otherwise).
- **Firm_Char** (firm characteristics) includes the dummy Woman_CEO (equals 1 if the CEO is female and 0 otherwise), the natural log of the number of employees (Ln_Employee), the debt ratio (Leverage) and the natural log of the number of years for which a firm is listed on exchange (Ln_FirmAge).

To estimate panel data, we may opt for either pooled OLS, fixed effect or random effect model. Nevertheless, the proposition by Hermalin and Weisbach (1998) that the board of directors is determined to be endogenous seems theoretically and empirically reasonable. Clearly, firm performance is not only the consequence of actions from the prior directors on board, but a key criterion for selecting board members in the future. These authors also prove that poor performance could possibly lead to higher degree of independence, which is measured by the number of independent directors on board. Therefore, to address endogeneity issue, we employs system GMM (Generalized Method of Moments) estimation on the recommendation of De Andres and Vallelado (2008).

4 Results

4.1 Summary Statistics

Section A of Table 1 presents the summary statistics on firm performace by ROS and ROA. The mean ROS and ROA are 8% and 7%, respectively.

Table 1. Summary statistics for variables in the models

Variable	Obs	Mean	Std. Dev.	Min	Max
Section A: Firm performance					
ROS (Net income/Sales)	1020	0.08	0.42	−11.04	1.90
ROA (Net income/Assets)	1020	0.07	0.09	−0.39	0.89
Section B: Women directors					
%Women	1020	0.14	0.15	0	0.60
%IndependentWomen	1020	0.04	0.09	0	0.50
%ExecutiveWomen	1020	0.10	0.13	0	0.60
D_1Woman	1020	0.37	0.48	0	1
D_2Women	1020	0.16	0.36	0	1
D_3Women	1020	0.04	0.20	0	1
Woman_Chair	1020	0.09	0.29	0	1
Section C: Control variables					
Board characteristics					
%Independent	1020	0.27	0.24	0	0.86
Ln_BoardSize	1020	1.69	0.23	0.69	2.40
Duality	1020	0.38	0.49	0	1
Firm characteristics					
Woman_CEO	1020	0.21	0.41	0	1
Ln_Employee	1020	6.35	1.27	2.20	10.09
Leverage	1017	0.48	0.22	0	1.11
Ln_FirmAge	1020	1.37	0.65	0	2.48

(Source: The authors)

Section B of Table 1 reports statistics on board gender diversity measures. As can be seen, 14% of all directors are female in the complete sample. Approximately 4% (10%) of all directors are female independent directors (female executive directors), suggesting that 28.6% of female directors are independent and the remainder hold executive or management positions. Out of 1,020 firm-year observations, 37%, 16% and 4% have one woman, two women and three or more women on their boards, and the remaining 43% have no women directors on board. Meanwhile, 9% of the board chair is female. Compared to China, the percentage of female directors on board of Vietnamese listed firms is 3.8% higher, i.e. Vietnam has achieved greater gender equality at the corporate level. Yet, the percentage of female directors on board in Vietnam seems still far lower than developed economies such as France, Finland (both at 30%) and Norway (39%). The underlying reason for these differences lies in the policy issues, i.e. these countries have been making every effort to attain the right balance between men and women on the board of directors by introducing a binding percentage of female directors on board upon listed company, meanwhile, Vietnam has not approved any specific policy in this regard.

According to the summary statistics for control variables (Section C of Table 1), there are 5 to 6 members in an average board, about 27% of board members are independent and 38% of board chairs are also chief executives of the same corporation. As regards firm characteristics, our statistics reveal that 21% of CEOs in the sample are women. An average listed firm has 572 employees with nearly 4 years of listing history and a financial leverage of 48%.

4.2 Impact of the Percentage of Women Directors on Board on Firm Performance

First, we seeks to clarify whether the percentage of female directors on board (% Women) has a significant impact on firm performance. Table 1 contains the results of the main regression model, where board gender diversity is measured by %Women and firm performance is measured by either ROS or ROA. It is evident from the statistical summary for the GMM estimates (at the bottom of Table 2) that Hansen's over-identification and AR(2) testing conditions are satisfactorily met. This implies estimation results from our models are reliable.

Our results show that female directors (%Women) have a positive influence on the firm performance measured by both ROS and ROA. This finding is consistent with the resource dependence theory, which claims that firms assemble benefits through three channels: advice and counseling, legitimacy and communication (Pfeffer and Salancik, 1978). The gender-diverse board could help reinforce the three channels. For instance, businesses may supplement female entrepreneurs to their board to sustain relationships with their female trade partners and consumers. Some firms regard their female leaders as fresh inspiration and connections with their female workers. Others, meanwhile, desire to incorporate female views in every key decisions of the board. Hence, gender diversity on board helps strengthen the board's reputation and the quality of their decisions, thereby benefiting businesses as a whole.

The percentage of independent directors (%Independent) has an inverse influence on the performance of the business. The reason for this might primarily be due to the fact that, a majority of Vietnamese listed companies fail to meet the required rate of 20% for independent directors on board, as stipulated in the Circular number 121/2012/TT-BTC of the Vietnam's Ministry of Finance. Later on, the Law on

Table 2. Impact of the percentage of women directors on board on firm performance

	ROS (Net income/Sales) (1)	ROA (Net income/Assets) (2)
%Women	0.061	0.064**
	[0.220]	[0.022]
%Independent	–0.049**	–0.059***
	[0.013]	[0.001]
Ln_BoardSize	0.193***	0.014
	[0.000]	[0.434]
Duality	0.025***	0.018*
	[0.000]	[0.051]
Woman_CEO	0.123	0.016***
	[0.600]	[0.002]
Ln_Employee	0.123***	0.007
	[0.000]	[0.143]
Leverage	–0.263***	–0.136***
	[0.000]	[0.000]
Ln_FirmAge	0.002	–0.021***
	[0.747]	[0.000]
Obs	847	847
AR(1)	0.001	0.006
AR(2)	0.948	0.239
Hansen test	0.385	0.512

Notes: p-values in brackets; ***, **, * indicate significance at 1%, 5% and 10%, respectively.
(Source: The authors)

Enterprises 2014 has redefined the criteria and conditions of independent board member under Article 5, Clause 2; nevertheless, in the current situation of Vietnam, entrepreneurs argue that it is not an easy task to hunt for members considered eligible for independence, academic qualifications, real-world experience and social status on duty. Since a majority of firms in Vietnam fail to meet the required number of independent members, the role of independent members seems negligible in the decision-making process of the board. Hence, it is understandable why independence is limited to the fulfilment of its role.

4.3 Impact of Independent Versus Executive Women Directors and Women Board Chairs on Firm Performance

Independent board members could affect firm performance via the monitoring channel due to their independence in operation, while the executive directors via the executive channel due to their executive power and management skills. Hence, this study seeks to investigate via which channel women directors on board could influence firm performance. First, we separate the women directors into two groups: independent directors and executive directors. Afterwards, we utilise the percentage of women independent directors (%IndependentWomen) and the percentage of women executive directors

(%ExecutiveWomen) for the regression model, instead of the percentage of female directors (%Women) as previously regarded. Statistical tests in Table 3 show that our model completely satisfies the Hansen and AR(2) test conditions. Thus, the estimation results are reliable.

Table 3. Impact of independent versus executive women directors on firm performance

	ROS (Net income/Sales) (1)	ROA (Net income/Assets) (2)
%IndependentWomen	0.464 [0.146]	–0.030 [0.329]
%ExecutiveWomen	0.481 [0.123]	0.144*** [0.000]
Woman_CEO	0.053*** [0.006]	0.010*** [0.007]
Obs	847	847
AR(1)	0.002	0.004
AR(2)	0.476	0.334
Hansen test	0.999	0.670

Notes: p-values in brackets; *** indicates significance at 1%.
(Source: The authors)

Table 3 reveals that.%IndependentWomen has no significant impact on both ROS and ROA, while %ExecutiveWomen has a positive influence on ROA. These findings provide strong evidence that, in Vietnamese corporations, female directors could only enhance firm performance via the executive channel. Aside from that, the study documents the positive impact of Woman_CEO on both ROS and ROA. Compared to a firm with no female executives, a firm with female executives is associated with a 0.05% higher ROS and a 0.01% higher ROA. This once again confirms the presence of female executive directors helps improve firm's financial performance.

Table 4. Impact of women board chair on firm performance

	ROS (Net income/Sales) (1)	ROA (Net income/Assets) (2)
Woman_Chair	0.263* [0.053]	0.093*** [0.000]
%Independent	–0.309* [0.063]	–0.068*** [0.000]
Woman_CEO	0.083*** [0.000]	0.017*** [0.000]
Obs	847	847
AR(1)	0.002	0.005
AR(2)	0.331	0.291
Hansen test	0.656	0.281

Notes: Woman_Chair = 1 if the board chair is female and 0 otherwise; p-values in brackets; ***,* indicate significance at 1% and 10%, respectively.
(Source: The authors)

To further investigate the executive effect, the paper judges whether a female board chair has any positive effect on firm performance, since a female board chair is often served by an executive director, rather than an independent director, in Vietnamese listed corporations. We re-estimate the model in Table 3 by substituting board gender diversity measure (%Women) in model (1) for a dummy namely Woman_Chair (Woman_Chair equals 1 if the board chair is female and 0 otherwise). Results in Table 4 show that financial performance of firms with a woman board chair appears higher than firms without a woman board chair. These findings on Woman_Chair have further stressed the role of female executives in boosting the firm's financial outcome.

4.4 Impact of the Number of Women on Board on Firm Performance

Liu et al. (2014) suggest that three female directors among a total of 15 members on board could exert a stronger impact than only one female member among five board members, although both cases have similar proportions of women. According to Kramer et al. (2006), the critical mass theory highlights that a subgroup must reach a threshold in order to affect the overall population. Thus, female directors must also attain a certain scale in order to have influence on the board, and hence rule over the firm performance.

Table 5. Impact of the number of women on board on firm performance

	ROS (Net income/Sales) (1)	ROA (Net income/Assets) (2)
D_1Woman	–0.155**	–0.007*
	[0.016]	[0.084]
D_2Women	0.161*	0.010**
	[0.064]	[0.036]
D_3Women	0.177*	0.056***
	[0.096]	[0.000]
Obs	847	847
AR(1)	0.001	0.005
AR(2)	0.562	0.265
Hansen test	0.975	0.574

Notes:
• D_1Woman = 1 if the board has 1 female director and 0 otherwise
• D_2Women = 1 if the board has 2 female directors and 0 otherwise
• D_3Women = 1 if the board has 3 or more female directors and 0 otherwise; p-values in brackets; ***, **, * indicate significance at 1%, 5% and 10%, respectively.
(Source: The authors)

There are an enormous number of arguments on how many female members should be in a boardroom (Burke and Mattis 2000; Carver and Oliver 2002; Huse and Solberg 2006; Singh et al. 2007), yet, in several countries, female directors on board appear to be a 'token' representation (Daily and Dalton 2003; Kanter 1977; Singh et al. 2004; Terjesen et al. 2008). Manifold research on female entrepreneurs has endeavoured to work out the threshold number of women members on board, beyond which the influence of women on firm value could genuinely be perceived, yet no specific conclusions is drawn for this matter. Hence, in this paper, we intend to unveil the critical mass theory using a set of three female director dummies (D_1Woman, D_2Women, D_3Women) as measures of board gender diversity in regression model (1). Specifically, the paper tests the relationship between different groups (one female, two females and at least three females) and the firm performance. The results are presented in Table 5. As can be seen, Hansen and AR(2) tests in both models come up with p-values in excess of 10% significance level. Hence, the reliability of our GMM estimation models is guaranteed.

We first focus on the impact of D_1Woman and D_2Women on financial performance measured by ROS and ROA. Table 5 reveals that a board with a lone female director (D_1Woman) has an inverse influence on firm performance. Nevertheless, as the number of females rises, empirical findings reveal that a board with two (D_2Women) or three (D_3Women) female members will have a positive impact on the performance of the business. These results strongly support the critical mass theory, in a way that *"one female on board is just like a 'token', two females only show their presence, and three females on board will have a say in the decision-making"* (Kristie 2011).

5 Conclusions

This study investigates board gender diversity, viz. the influence of female directors on the financial performance of 170 non-financial firms in Vietnam between 2010 and 2015. Empirical results reveal that gender diversity on board makes a positive contribution to business performance (as measured by ROS and ROA). On the other hand, women directors are found to have a positive influence on firm performance via executive channel rather than monitoring channel. Another striking finding is that boards with three or more female directors have stronger influence on the firm performance than boards with two or fewer female members.

As empirically proven, board gender diversity seems, to certain extent, useful in the case of Vietnamese publicly listed companies. Such judgement would likely open up several policy implications to government regulators, for instance, in terms of legislation, until the present time, since there has been no specific regulation on the number of women required to achieve in listed corporations, the paper could provide policy-makers with compelling arguments in setting minimum standards for the number of women directors on board in Vietnamese listed companies. In addition, it is universally accepted in existing literature that sound corporate governance could help reduce agency issues, thanks to which firm performance would be enhanced. Empirical evidence also indicates that gender-diverse board might probably improve corporate governance shortcomings (Gul et al. 2011). Typically, female directors could enhance corporate governance through improved monitoring and supervision in management activities (Adams and Ferreira 2009; Gul et al. 2008). Furthermore, as their involvement in board meetings might enhance the quality of discussion on complicated issues, the probability of error in making key decisions would be mitigated (Huse and Solberg

2006; Kravitz 2003). Based on this argument, it is concluded that board gender diversity seems to be a viable solution to poor governance of Vietnamese businesses for the time being.

This paper has provided a profound insight into the role of women in the context of economic development and restructuring in Vietnam. To foster this role, it is vital that the government, the business community and society as a whole create favorable conditions for the maximization of female entrepreneurs' potentials, which could, accordingly, contribute actively to the overall development achievements.

References

Adams, R.B., Ferreira, D.: Women in the boardroom and their impact on governance and performance. J. Financ. Econ. **94**(2), 291–309 (2009)

Ahern, K., Dittmar, A.: The changing of the boards: the impact on firm valuation of mandated female board representation. Q. J. Econ. **127**(1), 137–197 (2012)

Asch, S.E.: Effects of group pressure on the modification and distortion of judgments. In: Guetzkow, H. (ed.) Groups, Leadership and Men, pp. 177–190. Carnegie Press, Pittsburgh (1951)

Asch, S.E.: Opinions and social pressure. Sci. Am. **193**(5), 31–35 (1955)

Bond, R.: Group size and conformity. Group Process. Intergroup Relat. **8**(4), 331–354 (2005)

Burke, R., Mattis, M.: Women on corporate boards of directors: where do we go from here? In: Burke, R., Mattis, M. (eds.) Women on Corporate Boards of Directors, pp. 3–10. Kluwer Academic, Netherlands (2000)

Campbell, K., Minguez-Vera, A.: Gender diversity in the boardroom and firm financial performance. J. Bus. Ethics **83**(3), 435–451 (2008)

Carter, D.A., D'Souza, F., Simkins, B.J., Simpson, W.: The gender and ethnic diversity of US boards and board committees and firm financial performance. Corp. Gov. Int. Rev. **18**(5), 396–414 (2010)

Carter, D.A., Simkins, B.J., Simpson, W.G.: Corporate governance, board diversity, and firm value. Financ. Rev. **38**(1), 33–53 (2003)

Carver, J., Oliver, C.: Corporate Boards that Create Value: Governing Company Performance from the Boardroom. Wiley, San Francisco (2002)

Cox, T., Lobel, S., McLeod, P.: Effects of ethnic group cultural differences on cooperative and competitive behavior on a group task. Acad. Manage. J. **34**(4), 827–847 (1991)

Daily, C.M., Dalton, D.R.: Women in the boardroom: a business imperative. J. Bus. Strategy **24**(5) (2003)

De Andres, P., Vallelado, E.: Corporate governance in banking: the role of the board of directors. J. Bank. Finance **32**(12), 2570–2580 (2008)

Eagly, A.H., Johannesen-Schmidt, M.C.: The leadership styles of women and men. J. Soc. Issues **57**(4), 781–797 (2001)

Engelen, P.J., van den Berg, A., van der Laan, G.: Board diversity as a shield during the financial crisis. Corporate Governance, pp. 259–285. Springer, Heidelberg (2012)

Erhardt, N.L., Werbel, J.D., Shrader, C.B.: Board of director diversity and firm financial performance. Corp. Gov. Int. Rev. **11**(2), 102–111 (2003)

Erkut, S., Kramer, V.W., Konrad, A.M.: Critical mass: does the number of women on a corporate board make a difference, women on corporate boards of directors. Int. Res. Pract. 350–366 (2008)

Fama, E.F., Jensen, M.C.: Separation of ownership and control. J. Law Econ. **26**(2), 301–325 (1983)
Farrell, K.A., Hersch, P.L.: Additions to corporate boards: the effect of gender. J. Corp. Financ. **11**(1–2), 85–106 (2005)
Francoeur, C., Labelle, R., Sinclair-Desgagne, B.: Gender diversity in corporate governance and top management. J. Bus. Ethics **81**(1), 83–95 (2008)
Gul, F.A., Srinidhi, B., Ng, A.C.: Does board gender diversity improve the informativeness of stock prices? J. Account. Econ. **51**(3), 314–338 (2011)
Gul, F.A., Srinidhi, B., Tsui, J.S.L.: Board diversity and the demand for higher audit effort, SSRN (2008). https://ssrn.com/abstract=1359450. Accessed 20 Aug 2018
Hermalin, B.E., Weisbach, M.S.: Endogenously chosen boards of directors and their monitoring of the CEO. Am. Econ. Rev. **88**(1), 96–118 (1998)
Huse, M., Solberg, A.: How Scandinavian women make and can make contributions on corporate boards. Women Manage. Rev. **21**(2), 113–130 (2006)
Kanter, R.M.: Men and Women of the Corporation. Basic Books, New York (1977)
Konrad, A.M., Kramer, V., Erkut, S.: Critical mass: the impact of three or more women on corporate boards. Organ. Dyn. **37**(2), 145–164 (2008)
Kramer, V.W., Konrad, A.M., Erkut, S., Hooper, M.J.: Critical mass on corporate boards: Why three or more women enhance governance. Wellesley Centers for Women, Wellesley (2006)
Kravitz, D.A.: More women in the workplace: is there a payoff in firm performance? Acad. Manage. Perspect. **17**(3), 148–149 (2003)
Kristie, J.: The power of three. Director Boards **35**(5), 22–32 (2011)
Liu, Y., Wei, Z., Xie, F.: Do women directors improve firm performance in China? J. Corp. Finance **28**, 169–184 (2014)
Mahadeo, J.D., Soobaroyen, T., Hanuman, V.O.: Board composition and financial performance: uncovering the effects of diversity in an emerging economy. J. Bus. Ethics **105**(3), 375–388 (2012)
Martín-Ugedo, J.F., Minguez-Vera, A.: Firm performance and women on the board: Evidence from Spanish small and medium-sized enterprises. Feminist Econ. **20**(3), 136–162 (2014)
Nemeth, C.J., Kwan, J.L.: Minority influence, divergent thinking and detection of correct solutions. J. Appl. Soc. Psychol. **17**(9), 788–799 (1987)
Pfeffer, J., Salancik, G.R.: The External Control of Organizations: A Resource Dependence Perspective. Harper & Row, New York (1978)
Rhode, D.L., Packel, A.K.: Diversity on corporate boards: how much difference does difference make. Delaware J. Corp. Law **39**(2), 377–426 (2014)
Rose, C.: Does female board representation influence firm performance? The Danish evidence. Corp. Gov. Int. Rev. **15**(2), 404–413 (2007)
Sabatier, M.: A women's boom in the boardroom: effects on performance? Appl. Econ. **47**(26), 2717–2727 (2015)
Simpson, W., Carter, D., D'Souza, F.: What do we know about women on boards? J. Appl. Finance **20**(2), 27–39 (2010)
Singh, V., Vinnicombe, S.: Why so few women directors in top UK boardrooms? evidence and theoretical explanations. Corp. Gov. Int. Rev. **12**(4), 479–488 (2004)
Singh, V., Vinnicombe, S., Terjesen, S.: Women advancing onto the corporate board. In: Bilimoria, D., Piderit, S.K. (eds.) Handbook on Women in Business and Management, pp. 304–329. Edward Elgar, Cheltenham (2007)
Terjesen, S., Singh, V.: Female presence on corporate boards: a multi-country study of environmental context. J. Bus. Ethics **83**(1), 55–63 (2008)

Measuring Users' Satisfaction with University Library Services Quality: Structural Equation Modeling Approach

Pham Dinh Long[1(✉)], Le Nam Hai[2], and Duong Quynh Nga[1]

[1] Hochiminh City Open University, Ho Chi Minh City, Vietnam
long.pham@ou.edu.un
[2] Industrial University of Hochiminh City, Ho Chi Minh City, Vietnam

Abstract. The purpose of this research is to measure users' satisfaction of library services quality in university sector. The survey was conducted on 525 students who are economic majors, including business administration, banking and finance, accounting and auditing, commerce and tourism. We combined exploratory factor analysis with structural equation modeling (SEM) to develop the research model. The findings show five dimensions such as place, assurance, responsiveness, reliability and service information having positive impacts on users' satisfaction for library services quality. The results also indicate differences in satisfaction levels between different groups with regard to gender and study time.

Keywords: User satisfaction · Service quality · Library service
Structural equation modeling

1 Introduction

Libraries play an important role in university education. A good library service enables customers to use the library resources effectively. This benefit is only achieved when libraries know their customers' requirements and expectations (Awan and Mahmood [2]). Moreover, libraries support the innovation of teaching and learning methods and the self-study process of learners by proving access to new knowledge.

There have been several studies that dealing with library service quality. For example, Awan and Mahmood [2] set up a model for measurement of library service quality with 6 dimensions and 30 items. On the other hand, Wang and Shieh [18] suggest that library service quality is influenced by 5 dimensions that have significantly positive effects on users' satisfaction.

Our study measures the user perception for library services quality in the university context. The library services quality, in this paper, can be understood within 5 dimensions, including place, assurance, responsiveness, reliability, and service information. With the use of survey-based data, the sample of the study consists of 525 economic students selected by stratified sampling method. According to our findings, all 5 dimensions of a library service quality are correlated with users' satisfaction. Furthermore, the satisfaction levels vary depending on gender and study time of the users.

V. Kreinovich et al. (Eds.): ECONVN 2019, SCI 809, pp. 510–521, 2019.
https://doi.org/10.1007/978-3-030-04200-4_37

The current research is structured as follows. Section 2 represents the relevant theoretical backgrounds and hypothesis development in the research. Meanwhile, Sect. 3 describes our empirical methodology. The data collection procedure and descriptive statistics are also shown in this section. Section 4 discuss the results, and Sect. 5 concludes.

2 Theoretical Backgrounds and Hypothesis Development

2.1 Measuring Service Quality

In the literature, there are many directions to measure service quality. According to Lehtinen and Lehtinen [9], evaluating service quality bases on two issues, the process of service delivery and outcomes of service. On the other hand, Grooncross [6] emphasizes technical quality, functional quality and images when he defines the service quality. Parasuaraman et al. [14, 15] propose a 5-gap model as a service quality measurement scale, called SERVQUAL scale. Another major contribution in this regard comes from Cronin and Taylor [4], who introduce SERVPERF scale developed from the SERVQUAL scale. Besides, service quality is also described as an interactive process between customers and employees (Svensson [17]). Moreover, Liqual+ instrument was used in many studies of library servive quality such as Cook et al. [3], Miller [11].

2.2 Measuring Library Service Quality

Awan and Mahmood [2] use Confirmatory factor analysis to develop the scale of library service quality measurement with 6 dimensions and 30 items. They include reliability, access, responsiveness, assurance, communication and empathy. By employing the SERVPERF scale of Cronin and Taylor [4], Wang and Shieh [18] measure library service quality with 5 dimensions: tangibles, responsiveness, reliability, assurance and empathy. In the same line of research, however, Cook et al. [3] use a new measurement tool, LibQUAL, which is developed by the Association of Research Libraries (ARL) basing on SERVQUAL scale. The research provide 6 factors that affect the library services quality, including: place, reliability, self – reliance, access to information and comprehensive collections.

2.3 The Relationship Between Service Quality and Satisfaction

Service quality and customer satisfaction are two different concepts. Service quality focuses on addressing components or a specific aspect of the service provided, which are modified depending on the type of service. The satisfaction inclines emotions when customers use the service. According to Parasuraman et al. [13], the relationship between service quality and customer satisfaction is causality. Service quality is one of the factors to evaluate customer satisfaction (Zeithalm and Bitner [19]). It is shown in Fig. 1. Moreover, service quality is the premise of customer satisfaction and this clearly shows in Spreng and Mackoy [16].

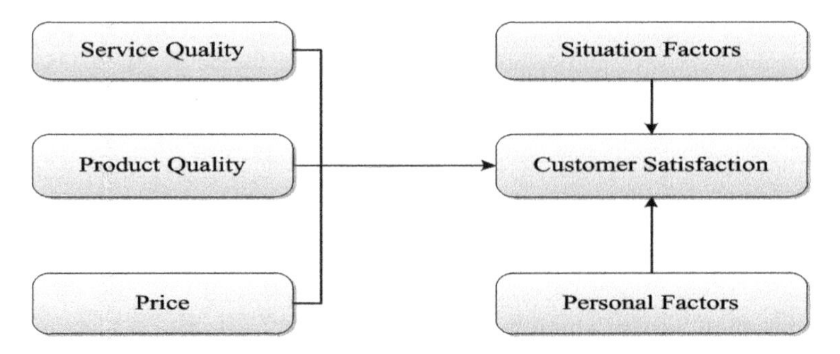

Fig. 1. The relationship between service quality and satisfaction *Resource: Zeithalm and Bitner* [19]

2.4 Research Model and Hypothesis

Based on the theoretical backgrounds, this study sets up a research model with 6 dimensions that affect student satisfaction. As indicated above, these factors have a positive effect on the satisfaction of library' users. In this study, the research hypothesis is laid out as follows:

Assurance: It is shown through professional qualifications staff, as well as serving attitude when providing services to readers. When using library services, customers (students) will contact staff in different ways. Therefore, library service quality is influenced by assurance. In addtion, several empirical studies confirm that a higher level of assurance is related to a higher level of customer satisfaction (Wang and Shieh [18]; Awan and Mahmood [2]). Thus, we propose the following hypothesis:

H_1: Assurance has a positive effect on customer satisfaction.

Responsiveness: In this study, the responsiveness is shown in terms of the availability of library materials and providing timely service. Awan and Mahmood [2] demonstrate that responsiveness is one of service quality dimensions in the model to measure library service quality. Thus, we have the following hypothesis:

H_2: Responsiveness has a positive effect on customer satisfaction.

Empathy: Empathy reflects the interest of the library staff, the timely grasp the needs of library users and the convenience of time when using services. On the other hand, Wang and Shieh [18] shown that empathy significantly increases the users' satisfaction. Similarly, we propose the following hypothesis:

H_3: Empathy has a positive effect on customer satisfaction.

Reliability: Reliability implies the commitments of service providers. Awan and Mahmood [2] indicate the importance of reliability on users' satisfaction. Following their study, this research proposes that:

H_4: Reliability has a positive effect on customer satisfaction.

Library as a Place: This factor refers the location, environment and space in which the service is provided. Library's users always would like to have a quiet and comfortable space, convenient location as well as a good environment. In this study, we test the following hypothesis:

H_5: Library as place has a positive effect on customer satisfaction.

Information Control: Information control reflects user guide information and the ability to access information. It is argued that when their users receive a clear guide information from the libraries, they can use the library service in an accurate way. Accordingly, the users' satisfaction increases. Thus, this study proposes that:

H_6: Information control has a positive effect on customer satisfaction.

3 Methodology

3.1 Measure Development

Indicators measuring the factors of assurance, responsiveness, empathy, reliability, library as place and information control are developed from Awan and Mahmood [2]; Cock et al. [3]; Miller [11]; Wang and Shieh [18]. Meanwhile, the variable of customer satisfaction is measured by 7 indicators which are developed from Cock et al. [3]. Specific information of indicators measuring each of the factors are shown in Table 1.

Table 1. Indicators of measuring factors

Factor	Item	Sources
Assurance	The staff is friendly and courteous. AS1 The staff is available to answer questions. AS2 Employees who have the knowledge to answer user questions. AS3 Employees who deal with users in caring fashion. AS4 Thorough understanding of the collection by the staff. AS5	Awan and Mahmood [2] Wang and Shieh [18] Miller [11]
Responsiveness	Willingness to help users. RES1 Having plenty of documents. RES2 The number of documents fully meet peak times. RES3 Borrowing and returning books without long waiting time. RES4 Using utility without long waiting time. RES5	Awan and Mahmood [2] Wang and Shieh [18]
Empathy	Employees are caring. EM1 Employees who understand the needs of their users. EM2 Employee who instill confidence in users. EM3 Serving time convenient. EM4	Awan and Mahmood [2] Wang and Shieh [18] Miller [11]

(*continued*)

Table 1. (*continued*)

Factor	Item	Sources
Reliability	Providing service matches quality commitment. RE1 Providing services are timely. RE2 Services match the program before. RE3 I believe that the library service improves efficiently my study. RE4 Giving correct answer to student's questions. RE5 The loan and return records are accurate. RE6 The library provides trustworthy information. RE7	Awan and Mahmood [2] Wang and Shieh [18]
Library as place	Quiet space for individual activities. LA1 Library space that are convenient for group activities. LA2 A comfortable and inviting location. LA3 Suitable location for studying and researching. LA4 Building and layout are good. LA5 The environment is clean. LA6	Cock et al. [3] Wang and Shieh [18]
Information control	Making electronic resources accessible form my home.IC1 A library website enabling me to locate information on my own. IC2 Modern equipment the lets me easily access needed information.IC3 Print and/or electronic collections I require for my study and research.IC4 Easy-to-use access tools that allow me to find things on my own.IC5 Keeping databases updated and running. IC6 Guide information is obvious.IC7 The collection is arranged systematically. IC8	Awan and Mahmood [2] Cock et al. [3] Miller [11]
Customer satisfaction	The library helps me stay abreast of developments in my study or research. SA1 The library enables me to be more efficient in my academic pursuits. SA2 The library provides me with information skills I need in my work or study.SA3 The library helps me distinguish between trustworthy and untrustworthy information. SA4 In general, I am satisfied with the way in which I am treated at the library. SA5 In general, I am satisfied with library support for my learning, research. SA6 In general, I am satisfied with library services.SA7	Cock et al. [3]

3.2 Data Collection Procedure

The survey was conducted in November 2013 with 525 respondents. They are economic students, except for not using library services.

The descriptive statistics of the sample are shown in Table 2. Of total 516 valid participants, 58.1% are males, 41.9% are females. The participants come from four sectors: business administration, banking and finance, accounting and auditing and commerce tourism. The respondents in accounting and auditing make up the highest share at 30.4%, while the commerce tourism is at the lowest rate of 18.2%.

Table 2. Descriptive statistics of respondent characteristics.

Variables		Count	%
Gender	Male	216	58.1%
	Female	300	41.9%
Sector	Business administration	127	24.6%
	Banking and finance	157	26.7%
	Accounting and auditing	138	30.4%
	Commerce tourism	94	18.2%
Period	Freshmen	63	12.2%
	Sophomore	159	30.8%
	Junior	179	34.7%
	Senior	115	22.3%
Training levels	University	274	53.1%
	College	154	29.8%
	Vocational College	88	17.1%

3.3 The Criteria for Fit Indices

This research applies Structural Equational Modeling (SEM) that combines the measurement model (Confirmatory factor analyze) and structural model (regression or path analysis) into a simultaneous statistical test (Garver and Mentzer [5]; Aaker and Bagozi [1]). Accordingly, we use indices such as: RMSEA (Root mean squared approximation of error); CFI (Comparative fit index); TLI Tucker – Lewis index) to evaluate overall model fit. These indices formula can be see from Table 3.

Table 3. The Criteria for fit indices

Indices	Formula	Criterion	Sources
RMSEA	$\frac{\sqrt{\mathcal{X}^2-df}}{\sqrt{df(N-1)}}$ Where: N is the sample size df is the degrees of freedom of the model $\mathcal{X}^2$ is The Chi - square test	Between 0.05 to 0.08	Medsker, Williams, Holahan [10]
CFI	$\frac{d(null\ model) - d(proposed\ model)}{d(null\ model)}$ Where: $d = \mathcal{X}^2 - df$ $\mathcal{X}^2$ is The Chi - square test df is the degrees of freedom of the model	This index ranges from 0 to 1	Medsker, Williams, Holahan [10]
TLI	$\frac{\frac{\mathcal{X}^2}{df(null\ model)} + \frac{\mathcal{X}^2}{df(Proposed\ model)}}{\frac{\mathcal{X}^2}{df1(null\ model)} - 1}$ Where: $\mathcal{X}^2$ is The Chi - square test df is the degrees of freedom of the model	This index is 0.9 or greater	Hulland, Chow, Lam [8]

4 Result

4.1 Exploratory Factor Analyze

Exploratory factor analysis (EFA) is used to test indicators. Some of them are deleted because the factor loading is less than 0.5 (Hair el al [7]). Before using EFA, Cronbach' Alpha coefficient is a precondition to remove inappropriate items that have total correlation less than 0.3 (Nunnally and Burnstein [12]).

The results are shown in Table 4. Accordingly, there are 7 independent factors that influence satisfaction, including: library as place, assurance, access electronic resources, reliability, empathy, responsiveness and service information.

4.2 Adjusted Model and Hypotheses

H_1: Library as place has a positive effect on customer satisfaction
H_2: Assurance has a positive effect on customer satisfaction
H_3: Reliability has a positive effect on customer satisfaction
H_4: Responsiveness has a positive effect on customer satisfaction
H_5: Service information has a positive effect on customer satisfaction
H_6: Empathy has a positive effect on customer satisfaction
H_7: Access electronic resources have a positive effect on customer satisfaction

Table 4. Exploratory factor analysis result

Factors	Items	Factor loading	KMO	
Library as place	LA1	**.750**	.868	
	LA3	**.687**	**Bartlett's Test of Sphericity**	
	LA4	**.648**	Approx. Chi-Square	3519.072
	LA1	**.622**	df	300
	LA5	**.588**	Sig.	.000
	LA6	**.518**		
Assurance	AS2	**.748**	**Initial Eigenvalues**	
	AS3	**.669**	1.038	
	AS1	**.628**	**Rotation Sums of Squared Loadings**	
	AS4	**.586**	58.560	
	AS5	**.565**		
Access electronic resources	IC3	**.710**		
	IC1	**.708**		
	IC5	**.680**		
	IC2	**.677**		
Reliability	RE2	**.740**		
	RE3	**.725**		
	RE1	**.699**		
Empathy	EM2	**.767**		
	EM1	**.754**		
	EM3	**.519**		
Responsiveness	RES3	**.830**		
	RES2	**.806**		
Service Information	IC8	**.757**		
	IC7	**.644**		
Customer satisfaction	SA7	**.797**		
	SA6	**.782**		
	SA5	**.708**		
	SA3	**.706**		
	SA2	**.681**		
	SA4	**.639**		

4.3 Confirmatory Factor Analyze

In next step, we use a confirmatory factor analyze to test items such as: standard loadings, reliability…All the item standardized regression weights are greater than 0.5 except for AS4 (deleted). Additional, we also use Cronbach's alpha and composite reliability to test the reliability of each factor. The analysis results show that Cronbach's alpha satisfies requirements (from 0.6 to 0.8) (Table 5).

Table 5. Confirmatory factor analysis result

Constructs	Items	Standard loadings	C.R	Cronbach Alpha
Library as place	LA6	.571	6.353	**0.7675**
	LA5	.534		
	LA4	.654		
	LA3	.605		
	LA2	.656		
	LA1	.548		
Assurance	AS5	.519	5.594	**0.7088**
	AS3	.556		
	AS2	.727		
	AS1	.639		
Access electronic resources	IC5	.605	6.770	**0.7487**
	IC3	.640		
	IC2	.725		
	IC1	.640		
Reliability	RE3	.618	7.028	**0.7431**
	RE2	.695		
	RE1	.784		
Empathy	EM3	.631	6.821	**0.6635**
	EM2	.604		
	EM1	.655		
Responsiveness	RES3	.659	5.921	**0.6507**
	RES2	.729		
Service Information	IC8	.638	6.567	**0.6213**
	IC7	.704		
Customer satisfaction	SA7	.751	9.464	**0.8164**
	SA6	.733		
	SA5	.660		
	SA4	.549		
	SA3	.610		
	SA2	.601		

4.4 The Structural Equation Modeling Analyze

We use AMOS 20.0 to set up SEM model and to test model fit. After deleting 2 constructs having p-value > 0.05, empathy and access electronic resources, the model contains 23 items and 6 constructs: library as place (LA), assurance (AS), reliability (RE), responsiveness (RES), service Information (SI) and customer satisfaction (SA). The results to examine the fitness of our research model are shown in Table 6. As we can see, all fit indices such as, CMIN/DF, RMSEA, GFI, CFI, IFI, AGFI, TLI, are acceptable. Therefore, it is suggested that our research model fits the empirical data at an acceptable level.

Table 6. Model fit

Fit indices	CMIN/DF	RMSEA	GFI	CFI	IFI	AGFI	TLI
Recommend value	<3	<0.08	>0.9	>0.9	>0.9	>0.9	>0.9
Value in this study	2.053	0.045	0.927	0.932	0.933	0.907	0.920

In addition, as can be seen from Fig. 2, the effects of assurance, reliability and service information on students' satisfaction are stronger than the other effects. Service information among them has the greatest influence, corresponding to the correlation coefficient 0.349, followed by reliability (0.312) and assurance (0.212). By contrast, responsiveness has the least effect on students' satisfaction with its coefficient of 0.093. Overall, it can be suggested that the library should improve its assurance, reliability and service information to increase its users' satisfaction (Fig. 3).

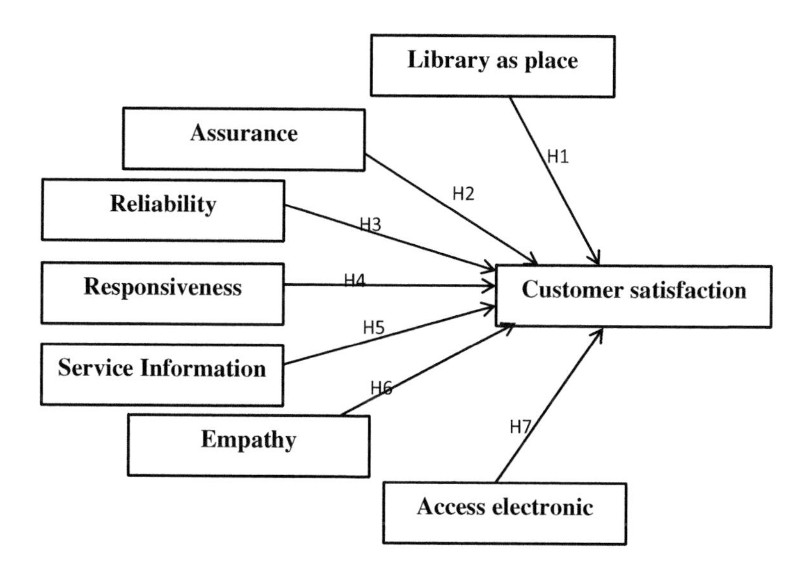

H_1: Library as place has a positive effect on customer satisfaction
H_2: Assurance has a positive effect on customer satisfaction
H_3: Reliability has a positive effect on customer satisfaction
H_4: Responsiveness has a positive effect on customer satisfaction
H_5: Service information has a positive effect on customer satisfaction
H_6: Empathy has a positive effect on customer satisfaction
H_7: Access electronic resources have a positive effect on customer satisfaction

Fig. 2. Adjusted research model

Besides, the initial hypothesis of the impact of each factor on students' satisfaction is acceptable. The findings are shown in Table 7.

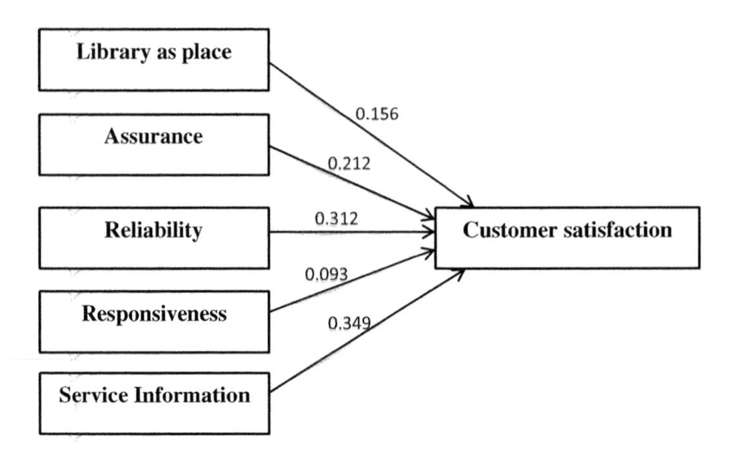

Fig. 3. Results of the structure model analysis

Table 7. Results of hypothesis

Hypothesis				Estimate	S.E.	P
H_1	SA	<—	LA	.156	.068	.021
H_2	SA	<—	AS	.212	.059	***
H_3	SA	<—	RE	.312	.076	***
H_4	SA	<—	RES	.093	.041	.025
H_5	SA	<—	SI	.349	.089	***

Moreover, our empirical findings indicate differences in satisfaction levels between different groups with respect to gender and study time. It is suggested that female students have higher level of satisfaction than male students (Table 8). In terms of study time, freshmen have the highest satisfaction level. Whereas, the satisfaction levels of third-year and fourth-year students are lowest.

Table 8. Results of independent Sample T-Test

	Mean	Mean difference	Levence's test		T-test for equality of means	
Male	3.1574	.26370	F	sig	df	Sig. (2-tailed)
Female	3.4211		6.867	.009	514	.000

5 Conclusion

The current paper focuses on measuring the students' satisfaction in relation to the library service offered. The results found that library as place, assurance, responsiveness, reliability and service information having positive impacts on users' satisfaction for library services quality. Knowing what factors have a strong impact on satisfaction to maintain and improve. Thus, this research provides helpful inputs into the policy design for quality improvement of university libraries. A further study will examine differences in user's satisfaction levels between different groups with respect to majors such as: technical majors.

References

1. Aaker, D.A., Bagozzi, R.P.: Unobservable variables in structural equation models with an application in industrial selling. J. Mark. Res. **16**(2), 147–158 (1979)
2. Awan, M.U., Mahmood, K.: Development of a service quality model for academic libraries. Qual. Quant. **47**(2), 1093–1103 (2013)
3. Cook, C., Heath, F., Thompson, B., Thompson, R.: LibQUAL+ : service quality assessment in research libraries. IFLA J. **27**(4), 264–268 (2001)
4. Cronin, J.J., Taylor, S.A.: Measuring service quality: a reexamination and extension. J. Market. **6**, 55–68 (1992)
5. Garver, M.S., Mentzer, J.T.: Logistics research methods: employing structural equation modelling to test for construct validity. J. Bus. Logistics **20**, 33–57 (1999)
6. Gronroos, C.: A service quality model and its marketing implications. Eur. J. Mark. **18**(4), 36–44 (1984)
7. Hair, J.F., Black, W.C., Babin, B.J., Anderson, R.E., Tatham, R.L.: Multivariate Data Analysis, 5 th edn. Prentice–Hall, Upper Saddle River (1998)
8. Hulland, J., Chow, Y.H., Lam, S.: Use of causal models in marketing research: a review. Int. J. Res. Mark. **13**(2), 181–197 (1996)
9. Lehtinen, U., Lehtinen, J.R.: Service Quality: A Study of quality Dimensions, Working Paper, Service Management Institute, Helsinki, Finland (1982)
10. Medsker, G.J., Williams, L.J., Holahan, P.J.: A review of current practices for evaluating causal models in organizational behavior and human resources management research. J. Manage. **20**(2), 439–464 (1994)
11. Miller, K.F.: Service Quality In Academic Libraries: An Analysis Of LibQUAL+^{TM} Score And Institutional Characteristics, The University of Central Florida Orlando, Florida (2008)
12. Nunally, J.C., Berntein, I.H.: Psychometric Theory, 3rd edn. McGraw-Hill, NewYork (1994)
13. Parasuraman, A., Berry, L.L., Zeithaml, V.A.: More on improving service quality measurement. J. Retail. **69**(1), 140–147 (1993)
14. Parasuraman, A., Zeithaml, V.A., Berry, L.L.: A conceptual model of service quality and its implications for future research. J. Market. **49**(3), 41–50 (1985)
15. Parasuraman, A., Zeithaml, V.A., Berry, L.L.: SERVQUAL: a multiple item scale for measuring consumer perception of service quality. J. Retail. **64**(1), 12–37 (1988)
16. Spreng, R.A., Mackoy, R.D.: An empirical examination of the antecedents of perceived service quality and satisfaction. J. Retail. **72**(2), 201–214 (1996)
17. Svensson, G.: A triadic network approach to service quality. J. Serv. Mark. 158 (2002)
18. Wang, I.M., Shieh, C.J.: The relationship between service quality and customer satisfaction: the example of CJCU library. J. Inf. Optim. Sci. **27**(1), 193–209 (2006)
19. Zeithaml, V.A., Bitner, M.J.: Services Marketing: Integrating Customer Focus Across the Firm, 2nd edn. Irwin/McGraw-Hill, Boston (2000)

Analysis of the Factors Affecting Credit Risk of Commercial Banks in Vietnam

Hoang Thi Thanh Hang(✉), Vo Kieu Trinh(✉), and Ha Nguyen Tuong Vy(✉)

Banking University of Ho Chi Minh City, 36 Ton That Dam str, Dist 1, Ho Chi Minh City, Vietnam
hanghtt@buh.edu.vn, trinhvo29797@gmail.com, vy.hanguyen273@gmail.com

Abstract. The research applied FEM-REM model using panel data in order to define the internal and macro determinants affecting the credit risk of 20 commercial banks for the period of 2006–2017. This paper concentrated on the credit risk measured by non-performing loans ratio. The macroeconomic variables are comprised the GDP growth, real interest rate, unemployment rate, and the bank-specific variables are determined by the return on assets with the 1-year time lag, the loan growth, loan loss provision. The results indicate that the the unemployment rate, real interest rate affect negatively the credit risk. Additionally, the credit risk is affected positively by loan loss provision.

Keywords: Internal factors · Macro factors · Credit risk · Commercial banks

1 Introduction

Credit risk has attracted considerable topic in Viet Nam's economy in recent years. In order to solve bad debts, resolution 42/2017/QH14 was issued with the aim of restructuring the bad debts and enhance the quality of the commercial banks' assets. According to National Financial Supervisory Commission (NFSC), thanks to the aggressive bad loan resolution from the commercial banks, the bad loan ratio by the end of 2017 decreased sharply significantly from 11,9% to 9.5% compared to the same period of 2016. However, in the first quarter of 2018, the bad debt ratio rose again in almost commercial banks. Until now, there has been many studies about credit risk in Vietnam, but almost they applied the qualitative method. Therefore, quantitative approach measures the impact of macro and internal factors on credit risk is significant in the current context, to help the Vietnamese policymakers supervise and sustain the stability of the banking system in particular and the economy in general.

V. Kreinovich et al. (Eds.): ECONVN 2019, SCI 809, pp. 522–532, 2019.
https://doi.org/10.1007/978-3-030-04200-4_38

2 Theory Basis and Empirical Studies

2.1 Theory Basis

In term of the macro-economy, according to Financial Accelerator Effect (Bernanke and Gertler 1995), changes in the credit market cause changes in the economy; economic disturbances can be amplified and affect credit market, in return. When central bank implements tight monetary policy, base interest rate increases leading to the increase of market and interbank interest rate. When market interest rate increases, bond prices and other financial derivatives that bank's holding decrease that might lead to the decrease in bank capital, so that lending supply decreases (Van den Heuvel 2009). Besides, increased market interest rate causes the decrease in asset's value, leading to the decrease in economic agent's net worth; so these borrowers have to pay the higher cost of debt. At that point, they will consider carefully when taking on a loan, which causes the decrease in investment and consumption. Overall, the effect of interest rate will cause the decline in lending supply demand. Thus, GDP decreases because of decreased investment and consumption, leading to economic depression, cash flow decreases which affects repayment capacity and arises credit risk.

To explain the relationship between credit growth and non-performing loan, Keeton (1999) described changes in three factors: expected rate of return, creditworthiness and total loan amount. When commercial banks are willing to increase total loan amount, they accept lower expected rate of return by lowering credit standards. Thus, loan quality decreases because high-risk customers are easily disbursed. So that, there is the positive relationship between credit growth and non-performing loan due to increasing total loan amount. Increased credit growth may also be from economic agents' demand for credit, commercial banks might tighten credit standards to improve the quality of loans that will lower credit risk. The inverse relationship between credit growth and credit risk is also found out when labor productivity increases. It is because the higher labor productivity is, the higher repayment capacity is. Hence, agents can increase future cash flow when investing in large projects which thanks to the increased labor productivity. Overall, increased labor productivity partly contributes to increasing total loan amount and the expected rate of return, resulting in lower credit risk.

According to Bad management theory (Berger and Deyoung 1997), low cost efficiency relates to bad management. In the research of Peristiani (1996), there is the positive relationship between cost efficiency and bank's quality management ranking. Thus, bad management in the bank might lead to credit risk.

According to Berger and Deyoung (1997), "moral hazard" is a classic problem of taking a risk. When a party to a transaction has an incentive to take the risk in an attempt to earn the profit, but if the failure occurs, it will affect the other party. Moral hazard may be from banks. According to Keeton and Morris (1987), low-capital banks are willing to increase the level of risk in their loan portfolio. In pursuit of profit, banks tend to loosen credit policy, fewer constraints on their clients as well as irresponsibility of bank staffs in controlling and appraising loan applications. This leads to higher non-performing loans in the future. Moral hazard may also be from customers, asymmetric information between customer and bank, customers use loan wrongly, invest in high-risk aspects. These also lead to credit risk.

2.2 Empirical Studies

The research of Abid, Ouertani and Zouari (2014) uses panel data of 16 banks in Tunisia and GMM estimation method to analyze economic indicators, such as GDP growth with the 1-year lag, lending interest rate. The research points out the positive relationship between lending interest rate and non-performing loan. The research of Zribi and Boujelbène (2011) studied the macroeconomic impact on the credit risk of 10 commercial banks in Tunisia in the period from 1995 to 2008. Using FEM-REM estimator shows that GDP growth, interest rate have a large impact on credit risk. In Vietnam, Vo Thi Quy and Bui Ngoc Toan (2014) use panel data and GMM estimation method to study the factors affect the credit risk of Vietnam commercial bank system during the period from 1999 to 2012. The study shows that GDP growth with one-year lag has the opposite effect on credit risk. Besides, in the period from 2005 to 2013, Nguyen Quoc Anh and Nguyen Huu Thach (2015) using data of 26 commercial banks in Vietnam found out the inverse relationship between GDP growth and non-performing loan rate.

Mondal (2016) examined the relationship between unemployment rate and non-performing loan in 22 commercial banks in Bangladesh in the period from 2005 to 2014. The author using Granger methodology points out the positive relationship between non-performing loan and unemployment rate. Louzis et al. (2010) examined the relationship between unemployment rate and non-performing loan in 9 largest commercial banks in Greece during the period from 2003–2009. The author considered separately on three types of loans (mortgage loans, consumer loans, and business loans). By using GMM estimation method, regression result shows that the unemployment rate with one-year lag positively affects each type of loan, especially business loans.

The research of Foos et al. (2010) studied the impact of credit growth on credit risk in 16.000 commercial banks of 16 developed countries from 1997–2007. By using GMM and OLS estimation method, the authors point out that credit growth causes increased non-performing loan in the next 2 or 3 years. Credit risk is measured by loan loss reservation over the total loan the last year. The result converges with the research of Nguyen Thuy Duong and Tran Thi Thu Huong (2007). In the research, the authors use GMM estimation method and data of 20 commercial banks in Vietnam in the period from 2006 to 2014, credit risk is measured by non-performing loan over total loan. In a developing country like Vietnam, credit growth with one-year lag has a huge impact on credit risk. However, in the research of Vo Thi Quy and Bui Ngoc Toan (2014) studied on 24 commercial banks in Vietnam, credit risk is measured by loan loss reservation over total loan the last year, the result shows the inverse relationship between credit growth and credit risk.

Hasna Chabi (2016) indicated the negative relationship between the credit risk and return on assets (ROA) of 10 listed commercial banks in Tunisian, The authors estimated the credit risk by non-performing loans ratio and the quality of loans. With the credit risk measured by the bad debt ratio, the authors used the OLS method. For the credit risk measured by the quality of loans, which was based on classifying the repayment capacity in the range from 1 to 4, the authors applied the probit analysis model. Both results showed the profitability's bank is associated negatively with the non-performing loans ratio in the future. However, by using GMM estimation method for 129 commercial banks in Spain, the research of Garciya-Marco and Robles-Fernandez (2008) found the empirical evidence proving that higher return on assets will increase the credit risk. Specifically, when considering the two types of banks (commercial banks and savings banks), this result was explored in the commercial banks. However, the authors did not find statistical significance in the case of saving banks.

Hasan and Wall (2014) showed the positive association between the bad debt ratio and loan losses reserves of 21 countries throughout the world. This implies that provision is a method to control the risk in the commercial banks. The manager based on the experience in the past, if they assume that the impaired loans rose, they will increase the allowance for bad debts. Besides, this research also concluded that the commercial banks use the provision as the method to adjust the profit. However, Boudgria et al. (2009) used FEM-REM estimation method found the negative association between the lagged loan losses reserves and bad debts ratio. With the sample of 59 commercial banks for the period 2002–2006, the lower loan losses reserves ratio will increase the non-performing loans next year. The authors assumed that the banks have low credit risk if they have a high provision for bad debts. Making the provision reflects the attitude of each country in the banking sector. In countries where risk management is important in the banking strategy, provision is used largely to reduce the risk.

3 Methodology and Data

3.1 Data

In order to examine the impact of factors on credit risk of commercial banks in Vietnam, the article used the data collected from financial statements, annual reports of banks in the period 2006–2017 and macroeconomic data from World Bank. The merged banks in this period were excluded from the research sample because of inconsecutive data. Therefore, the research sample of this study includes 20 commercial banks operating in Vietnam during the period 2006–2017.

3.2 Methodology

The research uses quantitative method. Specifically, using Eviews software to calculate descriptive statistics, analyze coefficient matrix, check multicollinearity phenomenon. Then, research performs Fixed Effect Model (FEM) and Random Effect Model (REM) in turn. The difference between these two econometric models is that FEM supposes cross section units are different at fixed intercepts, REM supposes cross section units are different at errors. Thus, Hausman test is used to choose which model is suitable. There are two hypotheses in Hausman test:

$$H_0 : \mathrm{Cove}(\varepsilon_i; X_{i,t}) = 0 (\text{choose REM})$$

$$H_1 : \mathrm{Cove}(\varepsilon_i, X_{i,t}) \neq 0 (\text{choose FEM})$$

If significant value $\alpha > $ P-value, hypothesis H_0 is excluded, FEM is suitable. In contrast, REM is suitable.

4 Model and Hypotheses

The study examines the impact of factors on credit risk based on the approach of Messai and Jouini (2013). The relationship is presented as the following equation:

$$\frac{\mathbf{NPL}}{\mathbf{TL}}_{i,t} = \beta_0 + \beta_1 \Delta \mathbf{GDP}_{t-1} + \beta_2 \mathbf{RIR}_t + \beta_3 \mathbf{UN}_t + \beta_4 \mathbf{ROA}_{i,t-1} + \beta_5 \left(\frac{\mathbf{LLR}}{\mathbf{TL}}\right)_{i,t} + \beta_6 \Delta \mathbf{Loan}_{i,t} + \varepsilon_{i,t}$$

The definition of credit risk is described by Basel Committee on Bank Supervision is: "Credit risk is the potential that a bank borrower or counterparty will fail to meet its obligations in accordance with agreed terms" (BIS 2000, p. 5)

Credit risk can be measured in two ways: non-performing loan over total loan or loan loss provision over the total loan. According to Messai and Jouini (2013), credit risk is measured by non-performing loan over the total loan. Regarding Circular No.02/2013/TT-NHNN of State Bank of Vietnam about loan classification: Loans are classified into 5 groups: Certain, special mentioned, sub-standard, doubtful, bad. Non-performing loans are sub-standard, doubtful and bad loans. This classification is based on 90 days or more past due and incapacity to repay the principal and interest of borrowers

ΔGDPt-1 is changed in GDP growth with the one-year lag. Based on the analysis presented in Chapter 2, the authors expect there is the inverse relationship between GDP growth with one-year lag and credit risk.

Hypothesis 1: Change in GDP growth with one-year lag has the inverse relationship with credit risk

RIRt is the real interest rate, measured by subtracting inflation from the lending rate. Based on the analysis presented in Chapter 2, the authors expect there is the relationship in the same direction between real interest rate and credit risk

Hypothesis 2: Real interest rate has the relationship in the same direction with credit risk

UNt is unemployment rate. Based on the analysis presented in Chapter 2, the authors expect there is the relationship in the same direction between the unemployment rate and credit risk

Hypothesis 3: Unemployment rate has the relationship in the same direction with credit risk

ROAi,t-1 is return on assets with one-year lag, measured by the percentage of net profit over an average total asset. Based on the analysis presented in Chapter 2, the authors expect there is the inverse relationship between ROA with one-year lag and credit risk

Hypothesis 4: ROA with one-year lag has the inverse relationship with credit risk

LLR/TLi,t is loan loss reservation, measured by loan loss reservation over the total loan. Based on the analysis presented in Chapter 2, the authors expect there is the relationship in the same direction between loan loss reservation and total loan.

Hypothesis 5: Loan loss reservation has the relationship in the same direction with total loan

ΔLoani,t is the growth rate of loan, measured by the percentages of change in the loan in year t and year t-1. Based on the analysis presented in Chapter 2, the authors expect there is the relationship in the same direction between the growth rate of loan and credit risk (Table 1).

Table 1. Detailed descriptions and expectations of variables in the model

Variables	Description	Formula	Expectation	Previous studies
Dependent variable: non-performing loan: $NPL_{i,t} = \frac{\text{Non-performing loans}}{\text{Total loan}}(\%)$				
Independent variables				
ΔGDP_{t-1}	Change in GDP growth with one-year lag	$\Delta GDP_{t-1} = \frac{GDP_{t-1} - GDP_{t-2}}{GDP_{t-2}}(\%)$	–	Luc Laeven and Giovanni Majnoni (2002), Abhiman Das and Saibal Ghosh (2007), Ner Klein (20013)
RIR_t	Real interest rate	RIR = Nominal interest rate - Inflation	+	Gonzalez-Hermosillo (1997), Berge and Boye (2007), Aver (2008), Nkusu (2011), Louzis and ctg (2012)
UN_t	Unemployment rate	$UN_t = \frac{\text{Unemployed people year t}}{\text{Labor force year t}}(\%)$	+	Berge and Boye (2007), Nkusu (2011), Vogiazas and Nikolaidou (2011), Võ Văn Tình (2017)
$ROA_{i,t-1}$	Return on assets	$ROA_{i,t-1} = \frac{\text{Net income year t-1}}{\text{Average total assets year t-1}}(\%)$	–	Messai & Jouini (2013), Godlewski (2004), Trương Đông Lộc and Nguyễn Văn Thép (2015)
$\frac{LLR}{TL}_{i,t}$	Loan loss reservation over total loan	$\frac{LLR}{TL}_{i,t} = \frac{\text{Loan loss reserve}}{\text{Total loan}}(\%)$	+	Messai and Jouini (2013), Võ Văn Tình (2017), Nguyễn Thị Hồng Vinh (2017)
$\Delta Loan_{i,t}$	Growth rate of loan	$\Delta Loan_{i,t} = \frac{\text{Total loan year t} - \text{Total loan year t-1}}{\text{Total loan year t-1}}(\%)$	+	Nguyễn Thị Ngọc Điệp and Nguyễn Minh Kiều (2015), Nguyen Thuy Duong and Tran Thi Thu Huong (2016), Nguyen Thi Hong Vinh (2017)

Source: Compiled by the authors

Hypothesis 6: the Growth rate of loan has the relationship in the same direction with credit risk.

5 Result

5.1 Descriptive Statistics

Before the authors conducted the model estimation to examine factors affect credit risk of commercial banks in Vietnam, the authors generated the descriptive statistics of variables in the article to generalize research variables through the mean, standard deviation, minimum value, median and maximum value. Table 2 presents descriptive statistics of variables in the article. The non-performing loan indicator has the average value of 2.09 with the standard deviation of 1.64.

Table 2. Descriptive statistics of variables in the model

Variables	Means	Standard deviation	Minimum value	Median	Maximum value	Observations
NPL	2.094	1.639	0.000	1.950	20.000	235
Δ GDP	6.323	0.795	5.200	6.200	7.547	235
RIR	2.562	3.736	–5.616	3.628	7.614	235
UN	2.270	0.224	1.800	2.300	2.600	235
LLR	1.189	0.631	0.060	1.100	3.702	235
ROA	1.191	0.953	–1.500	1.089	5.950	235
LOAN	48.339	112.229	–31.294	25.020	1131.726	235

5.2 Correlation Analysis

The correlation coefficients between the different variables are low. These coefficients are totally less than 1. Therefore, the study concludes that the model does not have multicollinearity phenomenon. In particular, these variables such as UN, RIR, ΔLoans, ROAt-1 have a negative influence on the dependent variables NPL; there is the positive relationship between ΔGDP, LLR/TL, and NPL (Table 3).

Table 3. Matrix of the correlation between the variables in the model and bank profitability

	NPL	$\Delta\ GDP_{t-1}$	RIR_t	*UN*	$LLR/TL_{i,t}$	$ROA_{i,t-1}$	$\Delta\ Loans_{i,t}$
NPL	1.000000						
Δ GDP$_{t-1}$	–0.199170	1.000000					
RIR$_t$	–0.119198	–0.159014	1.000000				
UN	–0.274085	0.255136	0.074350	1.000000			
LLR/TL$_{i,t}$	0.476283	–0.311189	0.103880	–0.260726	1.000000		
ROA$_{i,t-1}$	–0.102484	0.285970	–0.405009	0.198011	–0.199321	1.000000	
Δ Loans$_{i,t}$	–0.208233	0.298791	–0.016480	0.250197	–0.274334	0.356224	1.000000

5.3 Randem Effects Testing

Redundant Fixed Effects Tests

Equation: Untitled

Test cross-section fixed effects

Effects Test	Statistic	d.f.	Prob.
Cross-section F	1.562359	(19,209)	0.0681
Cross-section Chi-square	31.210279	19	0.0383

The assumption of homogeneity of cross-unit is rejected at a level of 10%. The use of FEM-REM model is more appropriate than Pooled OLS-model

5.4 Hausman Test

Hausman test is used to select between FEM and REM in the model

Correlated Random Effects - Hausman Test

Equation: Untitled

Test cross-section random effects

Test Summary	Chi-Sq. Statistic	Chi-Sq. d.f.	Prob.
Cross-section random	4.647214	6	0.5898

The assumption of the correlation between the independents variable and the omitted variables is approved due to Prob = 0.5898 > α = 10%. As a result, REM model is more appropriate than FEM model.

5.5 Autocorrelation Phenomenon

Durbin-Watson coefficient in the REM model is 2.034, which is in the range from 1 to 3. Therefore the autocorrelation phenomenon does not exsit in the model.

5.6 The Estimation Result of REM Method

Dependent Variable: NPL

Method: Panel EGLS (Cross-section random effects)

Date: 06/27/18 Time: 12:55

Sample: 2006- 2017

Periods included: 12

Cross-sections included: 20

Total panel (unbalanced) observations: 235

Swamy and Arora estimator of component variances

Variable	Coefficient	Std. Error	t-Statistic	Prob.
C	3.653688	1.245294	2.933995	0.0037
GDP_T_2	-0.089270	0.127464	-0.700356	0.4844
REALIR	-0.075137	0.027541	-2.728221	0.0069
UN	-0.899078	0.439571	-2.045353	0.0420
LLR_TL	1.110668	0.168840	6.578219	0.0000
ROA_T_1	-0.045070	0.121017	-0.372428	0.7099
DELTA_LOAN	-0.000695	0.000942	-0.738090	0.4612

The result shows that there are three variables which are statistically insignificant: ΔGDPt-1, ΔLoans, ROAt-1. This means that the study does not find the relationship between GDP growth with one-year lag (ΔGDPt-1), the loan growth (ΔLoans), the return on assets with one-year lag (ROAt-1) and the credit risk.

Regarding the real interest rate, it was statistically significant at the level of 1%. We concluded that higher real interest rate is associated with high level of credit risk. This finding is completely opposite to the expectation. However, in the context of high inflation such as Vietnam, this result is consistent. In the period of 2007–2008, due to the impact of the world economic crisis, the bad debt ratio increased, however, as the inflation rate of this period was higher than that of the lending interest rate, leading to real interest rates decreased. For the period of 2009–2011, the bad debt ratio increased considerably because of the rapid credit growth. However, during this interval, the real interest rates plummeted. From 2013 to 2017, the interest rate increased slightly but the bad debt ratio decreased due to the establishment of the Vietnam Asset Management Company (VAMC), which partly helped to handle bad debts of commercial banks.

The empirical result also revealed there is a negative effect between the unemployment rate and the credit risk at the level of 5%. This finding is contrary to the expectation of authors. However, this is appropriate to Vietnam's economy in the period of the period 2006–2008 and the period 2010–2012, the unemployment rate decreased while the non-performing loans rose. In the years 2008–2010 and 2012–2015, the unemployment rate increased while the impaired loans declined.

Besides, the authors explored the significant positive association between loan losses reserves and the credit risk at the level of 5%. This finding is suitable for the authors' expectation and diverges from the results revealed by Boudgria et al. (2009), Nguyen Quoc Anh and Nguyen Huu Thach (2015). This is explained that higher provision for bad debts, the lower standard loans they get. And this also means repayment capacity of debtors is low, so the credit risk increases

6 Conclusion and Recommendation

6.1 Conclusion

Credit risk is the most substantial challenges that the commercial banks have to face. Additionally, this issue has been also attracting the attention of the policy makers in the banking sector. The research examined the factors impacting on the credit risk in the commercial banks during the period from 2006 to 2017. By using REM method for the sample of 20 commercial banks, the results indicate the significant and negative relationship between unemployment rate, real interest rate and the credit risk at the level of 5%. Besides, the authors also explored the significant positive association between loan losses reserves and the credit risk at the level of 5%. This means that the more provisions the banks make for credit grants, the higher impaired loans they grant. However, regarding the influence of the lagged GDP growth, the return on assets (ROA) and the loan growth on the credit risk, it is not significant.

Concerning the level of influence, the regression results illustrated the credit risk is most affected by the loan loss provision ratio. This implies that when the loan loss provision increases, the banks believe that the probability of capital losses will rises as well as the performance is low. As a result, the bad debts will increase in the future.

Meanwhile, the results also showed a negative correlation between unemployment rate, real interest rate, and the credit risk. This empirical evidence is different from previous studies. This is explained by the implementing in the State bank's policy that commercial banks focused on restructuring bad debts through debt trading, which helps to reduce bad debt rates while interest rates tend to increase

6.2 Recommendations

The result showes that the loan loss provision ratio influenced positively on the credit risk. Therefore, the State Bank should supervise regularly and monitor rigorously the commercial banks to guarantee that the banks comply with regulations. Specifically, according to Circular 02/2013/TT-NHNN, commercial banks need to restructure, classify impaired loans and make sufficient provision for each group of uncollected loans. When commercial banks make provision for non-performing loans, the government can base on this indicator to control the credit risk in the future.

Resolution no. 42 on handling bad debt in credit institutions is still many inadequacies. Specifically, in Resolution 42 there is a simplified procedure for liquidating the collaterals. But if the debtor does not cooperate, the collateral liquidation will not be solved. In addition, the resolution does not specify the ownership of commercial banks for the selling the securities. Currently, commercial banks only have the right to possess property if the borrower's default. However, the bank wants to liquidate or lease the collateral, they must have the right of disposal. Therefore, bad debt cannot be resolved definitely. As a result, the SBV should have obvious legal documents guiding the implementation of collateral liquidation in order to accelerate bad debt resolution.

References

Berger, A., DeYoung, R.: Problem loans and cost efficiency in commercial banks. J. Bank. Finance **21**, 849–870 (1997)

Bernanke, B., Gertler, M.: Inside the black box: the credit channel of monetary policy transmission. J. Econ. Perspect. **9**(4), 27–48 (1995)

BIS Principles for the Management of Credit Risk, p. 5 (2000)

Boudriga, A., Boulila Taktak, N., Jellouli, S.: Banking supervision and nonperforming loans: a cross-country analysis. J. Financ. Econ. Policy **1**(4), 286–318 (2009)

Chaibi, H., Ftiti, Z.: Credit risk determinants: Evidence from a cross-country study. Res. Int. Bus. Financ. **33**, 1–16 (2015)

Foos, D., Norden, L., Weber, M.: Loan growth and riskiness of banks. J. Bank. Finance **34**, 217–228 (2010)

García-Marco, T., Robles-Fernández, M.: Risk-taking behaviour and ownership in the banking industry: the Spanish evidence. J. Econ. Bus. **60**(4), 332–354 (2008)

Hasan, I., Wall, L.: Determinants of the loan loss allowance: some cross-country comparisons. Financ. Rev. **39**(1), 129–152 (2004)

Keeton, W.R.: Does faster loan growth lead to higher loan losses? Econ. Rev. Fed. Reserve Bank Kansas City **84**(2), 57 (1999)

Keeton, W.R., Morris, C.S.: Why do banks' loan losses differ?. Econ. Rev. Fed. Reserve Bank Kansas City, 3–21 (1987)

Louzis, D.P., Vouldis, A.T., Metaxas, V.L.: Macroeconomic and bank-specific determinants of non-performing loans in Greece: a comparative study of mortgage, business and consumer loan portfolios. J. Bank. Financ. **36**(4), 1012–1027 (2012)

Mondal, T.: Sensitivity of non-performing loan to macroeconomic variables: empirical evidence from banking industry of Bangladesh. Glob. J. Mange. Bus. Res. **16**(4), 21–28 (2016)

Duong, N.T., Huong, T.T.T.: The Analysis of major credit risk factors – the case of the Vietnamese commercial banks. Int. J. Financ. Res. **1**(8), 33–42 (2016)

Anh, N.Q., Thach, N.H.: Factors affecting credit risk- practical evidences at Vietnamese commercial banks. Sci. J. An Giang Univ. **1**(1), 27–39 (2015)

Van den Heuvel, J.S.: The Bank Capital Channel of Monetary Policy. Working paper. University of Pennsylvania (2009)

Quy, Q.T., Toan, B.N.: Factors affecting credit risk of Vietnamese commercial banks. J. Open Univ. HCMC **3**(36), 16–25 (2014)

Analysis of Monetary Policy Shocks in the New Keynesian Model for Viet Nams Economy: Rational Expectations Approach

Nguyen Duc Trung[1], Le Dinh Hac[1], and Nguyen Hoang Chung[2](✉)

[1] Banking University, 36 Ton That Dam Street, District 1, Ho Chi Minh City, Vietnam
ndtrunghvnh@gmail.com, hacld@buh.edu.vn

[2] Thu Dau Mot University, 06 Tran Van On Street, Thu Dau Mot City, Binh Duong Province, Vietnam
chung.nguyenhoang68@gmail.com

Abstract. The study estimates the SVAR model for Vietnam's economy based on the New Keynesian model for the small open economy, taking into account the forward-looking behavior exhibited by economic agents. Deep structural parameters are identified by placing exclusion restrictions on the VAR innovations and the covariance matrix. The study affirms the important role of interest rates through monetary policy shocks in controlling inflation and macroeconomic stability. In addition, the exchange rate shock reflects Viet Nam's exchange rate management mechanism, which is in line with the SBV's management objective in an opened economy but with a controlled flow of capital. In addition, the aggregate demand shock shows the impact of interest rates and inflation on output volatility of the economy in the way of reducing interest rates, stabilizing inflation but increasing output, this is also the SBV's final goal. In contrast, this study shows that the approach method is not general, has not set the rational expectations channel in combination with other transmission channels for analysis; the transparency, public credibility with the SBV policy or the control of capital flows in Vietnam are also a barrier for this channel to maximize its effectiveness. Finally, the study reaffirms the crucial role of monetary policy in managing monetary policy effectively, recommending the SBV to improve its planning, analyzing and forecasting policy towards a sustainable and stable growth.

Keywords: Monetary policy · New Keynesian · Structural VAR
Rational expectations

1 Introduction

The study builds a New Keynesian small open economy model with the SVAR approach. This model is a combination of structured shocks and rational expectations of agents in the economy to evaluate how macro variables such as output,

V. Kreinovich et al. (Eds.): ECONVN 2019, SCI 809, pp. 533–566, 2019.
https://doi.org/10.1007/978-3-030-04200-4_39

inflation, exchange rates and interest rates respond to impacts from their own shocks. While most studies have estimated New Keynesian model using a variety of methods, this study still uses the Keynesian model with the small sample size of the structural parameter of Cho and Moreno [17]. The study also reinforces the Government's policies as well as the empirical evidence of previous studies on monetary policy. Especially, this study underlines the crucial role of policy rate and exchange rate manipulation which are important conditions for Vietnam's broader integration into the global economy with a stable inflation towards macroeconomic stability.

2 Background Theories and Overview of Monetary Policy Transmission Studies

2.1 Econometric Theory of the New Keynesian Model

2.1.1 Overview of the New Keynesian Theory

In recent years, a series of papers have been published to discuss gap in the standard AS/AD model analysis such as Colander [23], to provide an alternative macroeconomic framework to the IS/LM-AS/AD[1] model. Among them, the most influential models are IS - MP (monetary policy) - IA (inflation adjustment) model by Romer [67], Taylor [70] và Romer [67] pointed out that modern central banks no longer targeted at money supply but followed a real interest rate rule and replaced price level with inflation in the AS/AD model. The evidence has been a study of the Keynesian model without the LM curve [67] and replaced by the interest rate - a Taylor rule [70] for the IS-LM - AS framework. The Walsh model [75] emphasized that central banks did not set exogenous interest rates but substitute inflation targets; the AD (aggregate demand) - PA (price adjustment) model of Weerapana [77]; the IS - PC (Phillips) - MR (monetary policy rule) model of Carlin and Soskice [12] and the BMW model of Bofinger, Mayer và Wollmershäuser [11] presented rich graphs to analyze Keynesian models which describe the interaction between inflation, output gap and interest rates. This has been considered one of the empirical models allowing easy comparison of optimized and simplistic monetary policy rules provides more significant advantages over the IS/LM model. This form can analyze important concepts such as target inflation and the monetary policy rule, compared to other recent methods of Romer [67]; Taylor [71] and Walsh [75], the BMW explicitly refers to counterparts the central bank's response to the demand shocks and focusing on the central bank's credibility, plays a key role in curbing inflation.

This new Keynesian model also describes the general interaction of three sets of macroeconomic objects. The IS equation describes the behavior of the household, the AS (the aggregate supply) introduces the entrepreneur's price decision and the MR (monetary rule) specification of MP execution by the central

[1] IS/LM - AS/AD: Investment Saving/Liquidity Preference Money Supply - Aggregate Supply/Aggregate Demand [49].

bank. It can be understood that this model structure focuses on the construction of monetary policy in central banks, derived from the study of Gali and Monacelli [29], the approach of Lubik and Schorfheide [44]. The open economy can engage in inconsistency time as well as inconsistency transactions in order to meet consumer demand and escape constraints of the closed economy.

The model is similar to the New Keynes models contributed by Clarida, Gali and Gertler [20–22], the main building blocks in all models include: (i) An IS equation illustrates the relationship between the output gap and real interest rates. (ii) A Phillips curve illustrates the relationship between inflation and the output gap. (iii) And the monetary policy rules are evaluated according to or derived from the loss function. While the IS equation existed in the New Keynesian model (though it is now rooted in solid micro-economy), significant innovations to the IS/LM-AS/AD model are: Monetary policy is described by the interest rate rule (instead of the money supply rule), the inflation introduced into the model (instead of the price level), and the supply side of the economy is synthesized by the Phillips curve (instead of the AS curve which is no longer suitable). In addition, in order to expand the new Keynesian model for the open economy, consideration should be given to the influence of international commodity markets and international financial markets on the domestic economy.

The development of New Keynesian models has also expanded to include additional market friction or market failures in the labor market or financial market (this extension is particularly significant after the global financial crisis of 2008). For central banks, the application of this model is also in a more widespread trend, gradually replacing classical econometric models, especially in central banks pursuing targeting inflation[2]. Then, foreign shocks, such as trade prices, can change the fluctuations of the domestic business cycle.

2.1.2 New Keynesian Model Structure with 3 Equations

According to Nguyen [57], the standard New Keynesian model includes: A New Keynesian Phillips, an IS curve, and a Taylor rule [70]. Specifically in the empirical modeling for Vietnam, the new Keynesian model is rewritten as follows:

$$\begin{aligned} x_t &= a_1 x_{t-1} + a_2 x_{t-2} + a_3(i_t - \pi_t) + u_t, \\ \pi_t &= b_1 \pi_{t-1} + b_2 \pi_{t-2} + b_3 x_t + e_t, \\ i_t &= c_1 i_{t-1} + c_2 i_{t-2} + c_3 \pi_t + c_4 x_t + v_t. \end{aligned} \tag{1}$$

In (1), where x_t, π_t & i_t are the output gap[3], the difference in inflation (compared to target inflation) and the interest rate difference (relative to the natural rate), u_t, e_t and v_t are the shock of the IS equation, the New Keynesian Phillips (NKP) equation and the monetary policy equation.

---

[2] There are approximately 20 central banks in 2012 according to Hamilton [31].

[3] The output gap is defined as the difference between the actual output of the economy and its potential output. (Bank of Canada, 2012). This text can be found at: bankofcanada.ca - search for "backgrounders".

2.1.3 New Keynesian Model Structure with 4 Equations

In addition, for simplification purpose, the reduced-form Keynesian model used for this study as a prior information source is a variant of the model proposed by Lubik and Schorfheide [44], a simple form of Galí and Monacelli [29]. The model of Lubik and Schorfheide [44], a simple form of Galí and Monacelli [29]. The model of Lubik and Schorfheide [44] has been used for the study in New Zealand by Lees et al. [42], or was previously mentioned [16]; inherited from the study by Del Negro and Schorfheide [53] and further developed by Hodge et al. [32]; Used to estimate the Australian economy and this model is also seen as a New Keynesian model that can represent the small and open economy as Vietnam. One thing to note is that this model ignores many of the features of the traditional model to ensure data compatibility such as consumer habits or consumer price index (CPI). Although this model is also based on micro foundation, that is not what the study is intended to do, the paper focuses only on providing a brief summary of important linear logarithm equations in the model [29,30,43,44] and described briefly by 4 equations: (i) the IS curve equation; (ii) the Phillips curve equation; (iii) the monetary policy equation, assuming the nominal interest rate is adjusted in part $(1 - \rho R)$ over the Talor rule and Clarida, Galí and Gertler [22]; (iv) The change in term of trade equation included in the Phillips curve reflects consumption goods being imported and assumes Purchasing Power Parity (PPP) [58].

2.1.4 The Model Structure Applied in Vietnam (Developed by IMF)

In Viet Nam, Dizioli and Schmittmann [25], the development of the Dynamic Stochastic General Equilibrium (DSGE) model for Vietnam is a new, small-scale Keynesian model inherited from the model [6–8] includes: (i) an aggregate demand equation (output gap estimated by HP filter)[4], (ii) two Phillips curve equations (food and fuel + inflation (iii) an uncovered interest rate (UIP) equation and (iv) an equation for the monetary policy rule (Taylor rule and exchange rate target)[5]. The estimation method used is Bayesian with the major parameter values in the model based on the theoretical basis of the current forecasting and policy analysis system FPAS (Forecasting and Policy Analysis System). The pre-selected average value reflects an overview of the structural characteristics of the Vietnamese economy. The four equations are the core of the model: (1) the aggregate demand curve describing the expected and the past output, the real interest rate and real exchange rate; (2) two Phillips curves relative to the composition of past inflation, expected inflation and exchange rates; (3) UIP and (4) the monetary policy rules, which are functions of output gap, exchange rate and deviations of expected inflation from target inflation.

[4] According to Leu [43].

[5] Equations from (i) - (iv) are derived from the IMF's study [25].

2.1.5 A New Keynesian SVAR Model

This study introduces the New Keynesian SVAR model of Vietnam's economy, where the identification is based on a small open economy with short-term fluctuations. The economic agents including central banks are assumed to have the same information that shapes rational expectations. In order to determine the SVAR model for rational expectations, the study estimates the deep-structure parameters using the method of Keating [37]. Deep structural parameters from utility functions and technological constraints of economic agents in the economy. Their identification is desirable because they do not change for policy and consequently the result is not the object of Lucas critique [43]. The deep structural parameters in this model are determined by imposing constraints on residuals of the VAR and the covariance matrix while ignoring the unconstrained dynamic lags. Dhrymes and Thomakos [24] use the same method to impose exclusionary constraints on an opened - small economic model. Their research differs from this article in three respects: the open economic structure model is not based on time optimization; the method to solve rational expectations are different; and no exclusion of exogenous variables. This study describes the new Keynesian model that is consistent with the underlying behavior of intertemporally optimizing economic agents. The aggregation relationships that formed the contemporaneous part of the SVAR model are derived from dynamic general equilibrium setting, in which agents are assumed to be rational and forward-looking.

2.1.5.1. IS Equation

Output gap is described by the opened - economy equation derived from the representative household maximization in the context of a small open economy [46,47].

$$x_t = \alpha_0 + E_t x_{t+1} - \alpha_1 (i_t - E_t \pi_{t+1}) + \alpha_2 (s_t + p_t^* - p_t) + \varepsilon_t^x \tag{2}$$

x_t is the output gap, i_t is the short-term nominal interest rate, π_t is the inflation rate, s_t is the exchange rate denoting the corresponding domestic currency of a unit of foreign currency, p_t is the price, denote (*) indicates foreign variables, and E_t is the expected condition based on the information available to the economic agents at time t.

Equation (2) represents the aggregate demand of a small open economy that is interpreted as the aggregate demand shock. Its main difference from the traditional IS curve is the dependence of current output on the level of future expected output ($E_t x_{t+1}$) through the effects of the expected future marginal utility of consumption balancing current marginal utility. Current output increases when future output is expected to be higher because individuals desire to achieve a balanced consumption portfolio. Individuals expect higher consumption in the next periods due to higher expected output, which induces consumers to spend more at present to increase consumption. The real interest effect, $(i_t - E_t \pi_{t+1})$, on the present output, is negative, reflecting the intertemporal substitution of consumption. The increase in value of the real exchange rate $(s_t + p_t^* - p_t)$, which is the real depreciation of the domestic currency, is expected to increase current output through an expenditure-switching effect.

2.1.5.2. AS Equation

The key theme in the New Keynesian model related to price adjustment is the combination of nominal rigidity and optimizing behavior of firms that generate dynamic inflation expectations. The new Phillips Keynesian curve (NKPC) derives from Calvo's [13] nominal pricing model, compared to an external shock, ε_t^{π}, which is given by:

$$\pi_t = \beta_0 + \beta_1 E_t \pi_{t+1} + \beta_2 x_t + \varepsilon_t^{\pi} \tag{3}$$

In each period, a fraction of the monopolistically competitive firms is allowed to adjust their prices (but they may choose not to do so) according to a fixed probability $(1-\omega)$. Therefore, the estimated interval between price adjustments is $1/(1-\omega)$ and probability ω measures the degree of prices in the economy. Companies maximize the expected discount value of current and future profits to obtain the optimal pricing decisions. Firms that can reset (and select) prices in the current period know that the new price will remain unchanged during $1/(1-\omega)$. Therefore, companies are concerned with future inflation because it may not be able to adjust prices for some period in the future. Thus, the current inflation depends on expected future inflation and $\beta_1 < 1$ is the subjective discount factor of the firm. This is a significant difference with the standard Phillips curve where expected current inflation, $E_{t-1}\pi_t$, enters the equation instead.

2.1.5.3. Uncovered Interest Parity (UIP)

A standard feature in most small open economic models is the inclusion of uncovered interest rate balances to describe the nominal exchange rate:

$$s_t = E_t s_{t+1} - (i_t - i_t^*) + \varepsilon_t^s \tag{4}$$

where time-varying risk-premium shocks ε_t^s reflect temporary change from interest rates parity (UIP) condition.

2.1.5.4. A Forward-Looking Monetary Policy Rule

$$i_t = \gamma_0 + \pi_t + \gamma_1(E_t \pi_{t+1} - \pi^T) + \gamma_2 x_t + \varepsilon_t^i \tag{5}$$

where π^T is the target inflation rate and ε_t^i represents the monetary policy shock. The specification follows closely follows the forward-looking rules discussed by Clarida et al. [20–22]. The central bank is assumed to react to the expected future inflation deviations from its target value and the current output gap deviations from its trend value. With positive inflation and/or output deviations from the corresponding target values, Eq. (29) calls upon the central bank to adjust the nominal interest rate more than once so that the real interest rate $(i_t - \pi_t)$ increases enough to narrow the positive gap. Thus, a priori $\gamma_1, \gamma_2 > 0$ if the central bank pursues a stable policy for movements in inflation and output gap.

2.2 Rational Expectations Econometrics

The Lucas critique [45] underestimates value and reduce reliability of traditional econometric policy analysis. According to this point of view, expectations play an important role in macroeconomics, altering traditional methods which do not allow the expectation factor to adapt to policy changes. To understand Lucas's argument, the study will analyze econometric policy through the econometric model. The model includes equations that describe the relationships of hundreds of variables. These relationships exist for constants and use of data in the past. Lucas's challenge to this policy evaluation is based on the simple principle of rational expectations theory. The way in which expectations are formed (the relationship of expectations for information in the past) changes as the behavior of the forecast variable changes. So when policy changes, the relationship between expectations and information in the past will change due to economic behavior is affected by expectations, the relationship in the econometric model will change. Econometric models are built from past data which are no longer the appropriate model for evaluating policy responses or even misleading[6]. The Lucas critique points out not only that conventional econometric models cannot be used for policy evaluation, but also that the public's expectations about a policy will influence the response to that policy [49] (Table 1).

An important aspect of the Friedman-Phelps hypothesis is that a natural inflation rates can be confusing to entrepreneurs and households but long - term sustainable inflation will not promote employment. This can be seen that expected inflation will adjust to any continuous price fluctuation in the economy. In the early 1970s, Lucas [45] made further progress and demonstrated that public and market expectations for policy actions have an important impact on all aspects of the economy[7]. Rational expectations theory refers to the optimization behavior of economic agents, and therefore their expectations for future variables are optimized to use all the available information. The basic insight of a reasonable expectation revolution is that expectations of a looking - forward monetary policy have a significant impact on the development of economic activity. As a result, reasonable expectations become a systematic component of the actions of policy - makers, or key actors in the implementation of monetary policy. In fact, the consideration of policy expectations in the future has become the central element of monetary theory that highlighted in recent research by Woodford [78]. And this result has a crucial implications, such as issues related to the systematic behavior of policymakers likely to be conducive to macroeconomic stability and GDP

[6] According to Mishkin [49], when the Fed wants to evaluate the changes in the economy by raising short-term interest rates from 5% to 8%. The term structure equation used with past data indicates that there is a slight variation in long-term interest rates. However, if the public finds that short-term interest rates are rising steadily, the rational expectation theory suggests that the economic agents will no longer expect short-term interest rates to increase at present. Instead, when they see interest rates increase by 8%, they will expect the average short-term interest rate to rise significantly. Then the long-term interest rates will increase sharply, but not in the interest rates's term structure. Therefore, it can be seen that evaluating the results of the Fed's policy change with econometric models can lead to misleading.

[7] The rational expectations theory was first introduced in 1961s by Muth but the public has just really focused on this theory when Lucas introduced it at the Carnegie-Rochester Conference (1973–1976).

Table 1. The comparison of three models

Model	Response to unanticipated expansionary policy	Response to anticipated expansionary policy	The benefit from activity policy	Response to unanticipate control - inflation policy	Response to anticipated control - inflation policy	The credibility's agents important to efficient control - inflation policy
Traditional model	$Y\uparrow, P\uparrow$	$Y\uparrow, P\uparrow$ same amount as when policy is unanticipated	Yes	$Y\downarrow, \pi\downarrow$	$Y\downarrow, \pi\downarrow$ same amount as when policy is unanticipated	No
New classical model[a]	$Y\uparrow, P\uparrow$	Y unchanged, $P\uparrow$ by more than when policy is unanticipated	No	$Y\downarrow, \pi\downarrow$	Y unchanged, $\pi\downarrow$ by more than when policy is unanticipated	Yes
New Keynesian model[b]	$Y\uparrow, P\uparrow$	$Y\uparrow$ by less than, $P\uparrow$ by more than when policy is unanticipated	Yes, but designing a beneficial policy is difficult	$Y\downarrow, \pi\downarrow$	$Y\downarrow$ by less, $\pi\downarrow$ by more than when policy is unanticipated	Yes

Source: Mishkin [49]

[a]The assumption of rational expectations of wages and price flexibility shows an upward - expected price that has the effect of increasing prices and wages. It leads to an inefficient policy recommendations because the expected policies have no effect on output; Empirical studies show that only unanticipated policies have an impact on output.

[b]The model assumes rational expectation but it is characterized by rigid prices and wages. The model also distinguishes between the expected and unexpected policy effects. However, the expected policy has a smaller impact than the unexpected policy on output that indicates the role of the expected policy has an effect on the volatility of output.

growth., i.e. when the government announces change policy, the economic agents will change their expectations. One response is to estimate a structural rational expectations model that overcome these arguments. Private agents' optimising behaviour is combined with the complete knowledge of the structural parameters of the economy and the underlying stochastic forcing processes. The solutions to the dynamic rational expectations models yield restrictions across equations from rational expectations and structural embedded in the optimisation problem assumptions to determine the structural parameters in the model. In addition, the SVAR model typically includes a contemporaneous system of behavioral relationships with unrestricted short-run lag dynamics. Instead of using lag restrictions to identify structural parameters, this study uses the Keating's [37] rational expectations scheme for the new Keynesian open economy.

The contemporaneous structural model described by Eqs. (2)–(5) is converted into a corresponding representation that is composed of structural disturbances and the reduced-form innovations. Economic agents forms future expectations using observable innovations that result from the dynamic structure of the economy. Therefore, the identification of deep parameters comes from the VAR residuals and restrictions constraints on the covariance matrix of the structural disturbances [43].

2.3 Foreign Studies

Raghavan and Silvapulle [65] - Monetary Policy Study of Malaysia, a small and open economy. Malaysia was a country in the financial crisis of 1997–1998, so the paper explored the transmission mechanism of monetary policy in the post-crisis period through the reaction of macro variables to shocks different. The results show that in the post-crisis period, domestic monetary policy is susceptible to price shocks and world output. This is also the result of research by Pagan, Catão and Laxton [62] in Brazil.

Pagan, Catão and Laxton [62] - The study uses the SVAR model in Brazil. The results show that the monetary policy transmission mechanism has a shorter lag than some developed countries, the change in interest rates has an impact on output and inflation in the short term. Since inflation targeting has been established in this country, the response of inflation to the monetary policy shock has become clearer and that improves the balance in the relationship between output and inflation in this country.

Ncube and Ndou [51] - The study examines the impact of monetary policy on macro variables via the policy transmission channel to economic agent as the household through consumption behavior. The study shows that the channel of asset and credit costs influencing real spending is statistically significant. With policy interest rates rising, consumption will decrease, thereby lowering inflationary pressures that could be attributed to excessive spending on asset and credit channels.

Acosta-Ormaechea and Coble [2] - The study was conducted in dollarized economies (Peru and Uruguay) and no dollarized economies (Chile and New Zealand). The results show that the interest rate channel is meaningful for countries that are not dollarized, but the exchange rate plays an important role in controlling inflation in dollarized countries. In addition, reforms in countries such as Peru and Uruguay, which reduced their dependence on the dollar, have changed the role of the monetary policy transmission channel from the exchange rate to the interest rate channel of the central bank. Increasing interest rates or tightening of the monetary policy curb as a result of rising exchange rates are also the result of Aastveit, Bjorland and Thorsrud [1] study in the Norway economy.

Aastveit, Bjorland and Thorsrud [1] - The paper deals with the mechanism of monetary transmission in the Norwegian economy, through the VAR model for small and open economies. The results of the study show a close relationship between exchange rate and monetary policy. As a result, the exchange rate rises immediately as the monetary policy shock tightens, then the amplitude decreases. For other macro variables, the response of the output is delayed for the monetary policy shock, and the shock also has a negative impact on consumer prices.

Leu [43] - A study of the New Keynesian SVAR model for Australia (1984–2009), assessing the impact of monetary policy on macro variables in the economy. The results reflect the impact of the four shocks on the Australian economy; aggregate demand shocks, aggregate supply shocks, monetary policy shocks (through interest rates) and exchange rate shocks. In particular, the positive monetary policy shock has the effect of narrowing output and inflation but high interest rates add to the exchange rate. A positive shock to the local currency (the appreciation of the currency) or a fall in exchange rates has been confirmed by the large exchange rate fluctuations that may have a negative impact on real output. The shock of aggregate demand, inflation increased with the increase of output forced the central bank to take measures to tighten control. Aggregate supply shock shows that the economy, when falling into inflation, central banks are lagging behind the policy response by raising nominal interest rates to curb inflation but negative real interest rates to reduce output spreads.

Mohanty [50] - This is also a study of monetary policy transmission made in India. The results of interest rate research and output growth have been negatively correlated with a two-quarter effect lag, a 3-quarter lag for inflation mitigation, and then fluctuate around 8–10 quarters. In spite of locating in the same South Asian region, the monetary policy shock in Bangladesh has a 6-month lag for output and a one-year for inflation. Meanwhile, there is an immediate impact on the exchange rate and liquidity [9].

Vinayagathasan [74] - The study explains the mechanism of monetary policy transmission of Sri Lanka. The study builds the SVAR model to assess the impact of macro variables through foreign exchange shock and the oil price shocks. In particular, the study uses a covariance function to explain the variability of macro variables. It shows the best interest rate shocks, followed by money supply shocks and exchange rate shocks to volatility. In addition, externalities and oil prices have no significant impact on the economy.

Kilinc and Tunc [40] - With the SVAR approach to study the impact of shocks in the monetary policy in Turkey, a small and open economy. The paper analyzes effects of four shocks; with two endogenous shocks are interest rate and risk premium, besides exogenous shocks are prices of imported goods and world output. A positive interest rate shock causes the exchange rate to fall (the value of the currency increases) and reduces inflation, while the risk premium causes the exchange rate to rise (negative value of the domestic currency) and negative impact to inflation. With the open economy in particular, the shock of world output has had an impact on Turkey's inflation.

2.4 Domestic Studies

Pham ([63] - Study on the impact of monetary policy on inflation, output and other macroeconomic variables. Using the four-variable SVAR model (industrial output, consumer price index, money supply and interest rates). The results show that when interest rates increase output and consumer price index (CPI), the lagged response of the consumer price index is relatively slower than the output variable. Rising interest rates also drag down M2 money supply. In addition, the response of output growth and inflation to its own shocks has a lag about one year. The M2 money supply shock and interest rates have a negligible effect on the dynamics of growth and inflation.

Hung and Pfau [34] - The paper uses the VAR model to examine the relation between monetary policy, real output, price, real interest rate, real exchange rate and other macro variables. The money supply is used as a direct tool to implement the monetary policy and the interest rate instrument is mediated from money supply to other macro variables. Research shows that monetary policy has a statistically significant effect on yield and price. However, the relationship between monetary policy and inflation is negligible. In addition, interest rate transmission has less impact on credit and foreign exchange. The impact of the monetary policy shock on output lasted from quarter 1 to quarter 2, suggesting that the change in output is highly dependent on the shock, but the impact of the shock on prices is sustained from 3 to 8 quarters.

Nguyen and Tran [61] - Study on Vietnam's inflation under the SVAR approach. The study found that inflation in Vietnam was mainly originated from the domestic sector, driven by cost push factor rather than demand pull one. Monetary policy in the country has a significant impact on inflation, especially M2 money supply. Exchange rates have an impact on domestic inflation, but the magnitude of impact is not significant and is mainly expressed as a transmission channel.

Tran and Nguyen [73] - The mechanism of monetary policy transmission in Vietnam is analyzed through the SVAR model. In particular, the article uses oil prices, the basic interest rate of the United States is the representatives of foreign variables; industrial output, inflation, money demand, interest rates, nominal multilateral exchange rates represent domestic variables. The paper discusses the breakdown of the structure and divides the research data into two phases before and after Vietnam's accession to WTO. The results show that Vietnam's inflation is sensitive to oil prices and exchange rates. Exogenous shocks and exchange rates have a significant impact on the change in interest rates. The multifaceted nominal exchange rate is subject to significant impacts of inflation and interest rate changes of the FED. Also in that year, research by Nguyen [59] on monetary policy transmission (2001–2012) showed that tightening monetary policy (interest rate increase) would reduce output but increase inflation (the effect In addition to widening the money supply and raising the exchange rate to increase output and lower inflation, the significant impact of M2 money supply on inflation [61].

Nguyen [59] - Using the VECM method to analyze how monetary policy channels in Vietnam affect economic activity and price, survey data from January 2001 to July 2012. The results of the study show evidence that tightening monetary policy by using interest rate tools would reduce output but increase the price puzzle, while expanding the money supply and increasing exchange rates resulted in increased output and reduced inflation. In addition, there is evidence of shock from stock prices that impact very little on output and price.

Dinh and Phan [26] - Analysis of the mechanism of transmission of interest rates from policy interest rates to market rates to retail rates in Vietnam and some other emerging economies in Asia through symmetry and asymmetry in interest rate transmission using the ECM model. Examining the impact of interest rate instability, rigidity during adjustment and leverage effects on transmission using ECM EGARCH (1, 1) - M model based on the study by Wang and Lee [76]. The empirical results of the paper show that the level of transmission from market interest rates to retail interest rates is not complete, with symmetric or asymmetric transmission mechanisms. In some cases, interest rate fluctuations increase the transmission range, but there are also some cases that show the opposite.

Nguyen [60] - Study on monetary policy and stock market in the period 2000 to 2013 to explore monetary policy channel through asset channels. The study was conducted using the SVAR model, which concluded that monetary policy had a strong transfer through the stock market through money supply while interest rates did not have a significant impact on the stock market at both VN-Index, HNX-Index and not change the price.

Tram, Vo and Nguyen [72] - The authors use the VAR model to study the impact of Vietnam's monetary policy through the channel of interest rate transmission in the period of 1/2000–7/2013 divided into two stages before and after the crisis in 2008. Research results show that the interest rate before 2008, but in the crisis there is a channel cost when the interest rates of the SBV increase leading to inflation (the prize puzzle). In addition, at this stage, Nguyen [60] found evidence of monetary policy transmission through the securities market by money supply.

Huynh, Le, Le and Hoang [35] - Analyze the impact of macro variables including money supply (MS), loan interest rate (LTR), consumer price index (CPI), exchange rate (EXR) and industrial output value (IP) to the Vietnam stock market (VN-Index) in the period from 2001 to 2013. Since the existing variables co-researchers should use the ECM model to determine the relationship between short-term variables and VECM to examine long-term equilibrium relationships. The results show that in the long term, MS and IP are in the same direction with VN-Index, LTR and CPI have opposite relationship with VN-Index. In case of shocks of macro variables, the VN-Index adjustment to equilibrium is quite slow.

Pham and Dinh [64] - This study has two main objectives. Firstly, the paper focuses on the exchange rate mechanism and its effects on the trade balance, inflation and macro environment in Vietnam over the past two decades. Next, based on its strengths and weaknesses, the paper attempts to argue for a more flexible exchange rate regime, rather than a rigidly anchored exchange rate regime. In addition, the study provides specific guidance on how to calculate NEER (Nominal Effective Exchange Rate).

Cao and Le [15] - The SVAR model used in the period 1995–2014 to estimate 11 macro variables divided into two regions: international and regional. Research results show that when money supply changes, it will affect interest rates, credit and exchange rates that affect the two variables are output and inflation. However, the money supply only affects the transmission channel in a short time. Money supply has a positive impact on output and has a negative impact on the price index but diminishes after 6 quarters. At the same time, research shows that interest rate, exchange rate and bank lending channels all have an impact on output and inflation in the short run.

3 Research Methods

3.1 Data and Sampling

Data for variables is collected quarterly from Q2/2000 to Q3/2017. The data series used for the study include: The policy interest rate (r): Prior to 2010, the IMF-IFS data is the treasury bill rate; Since 2010, this data has been standardized by IMF - IFS as policy rate. According to Nguyen [52], interest rates in the money market are divided into two groups: interest rate groups affected by the supply - demand relationship and the interest rate group announced by the Central Bank to run the monetary policy. According to Leu [43], the output gap (dy_hp) is calculated as the difference between the logarithm of the actual output and the potential output, with data from IMF - IFS (HP filter)[8]. The exchange rate (lne) derives from exchange rate of the USD/VND exchange rate of the IMF - IFS which take logarithm nepe (lne) is used to compute the data. The domestic inflation (dpi) is represented by the domestic consumer price index quarterly, taking data according to IMF - IFS. For foreign variables including the inflation (cpi_us) and the US Federal Funds rate (int_us) which get data from the IMF - IFS and the Board of Governors of the Federal Reserve (FED).

3.2 Order of Arranging Variables in the VAR Model

Study presented in the form of SVAR models with variable order: output gap, inflation, exchange rate, policy interest which based on Leu's [43]. The research order reflects the impact of the transmission from output to the policy interest rate. This is also the sequential order of the new Keynesian model for open economies, with low income based on empirical studies for African countries such as Uganda, Kenya, Tanzania [7,8], Andre et al. [5] and summarized in the study on the monetary transmission mechanism of low income countries [48].

[8] The HP filter is a univariate detrending algorithm that extracts the potential output component by minimising a particular loss function. The smoothness of the HP stochastic trend depends on the input value of an ad - hoc smoothness parameter. For quarterly data [33] recommended setting the smoothness parameter to 1600.

3.3 SVAR Identification Incorporating Rational Expectations

3.3.1 A Closed Economy New Keynesian Model

The study identified the SVAR model under the rational expectation [43] with a closed economy version of the New Keynesian model [21].

$$\begin{aligned} x_t &= \alpha_0 + E_t x_{t+1} - \alpha_1(i_t - E_t\pi_{t+1}) + \varepsilon_t^x \\ \pi_t &= \beta_0 + \beta_1 E_t\pi_{t+1} + \beta_2 x_t + \varepsilon_t^\pi \\ i_t &= \gamma_0 + \pi_t + \gamma_1(E_t\pi_{t+1} - \pi^T) + \gamma_2 x_t + \varepsilon_t^i \end{aligned} \tag{6}$$

The identification system is based on converting the contemporaneous structure system into an innovations representation system of equations including structural disturbance and VAR residuels, by subtracting each variable with the expected value at time $t-1$ of the variable carries the past information of that variable.

$$\varepsilon_t^i = e_t^i - e_t^\pi - \gamma_1(E_t\pi_{t+1} - E_{t-1}\pi_{t+1}) - \gamma_2 e_t^x$$

$$\begin{aligned} \varepsilon_t^\pi &= (\pi_t - E_{t-1}\pi_t) - \beta_1(E_t\pi_{t+1} - E_{t-1}\pi_{t+1}) + \beta_2(x_t - E_{t-1}x_t) \\ &= e_t^\pi - \beta_1(E_t\pi_{t+1} - E_{t-1}\pi_{t+1}) - \beta_2 e_t^x \end{aligned} \tag{7}$$

$$\begin{aligned} \varepsilon_t^i &= (i_t - E_{t-1}i_t) - (\pi_t - E_{t-1}\pi_t) + \gamma_1(E_t\pi_{t+1} - E_{t-1}\pi_{t+1}) - \gamma_2(x_t - E_{t-1}x_t) \\ &= e_t^i - e_t^\pi - \gamma_1(E_t\pi_{t+1} - E_{t-1}\pi_{t+1}) - \gamma_2 e_t^x \end{aligned}$$

Where $(x_t - E_{t-1}x_t)$, $(\pi_t - E_{t-1}\pi_t)$ & $(i_t - E_{t-1}i_t)$, are the current value of the output gap, inflation and interest rate VAR innovations respectively.

In (7), each structural disturbance is additionally related to one or both of the expectations adjustment processes of output gap and inflation, i.e. $(E_t x_{t+1} - E_{t-1}x_{t+1})$, $(E_t\pi_{t+1} - E_{t-1}\pi_{t+1})$. To calculate these two terms, the VAR form should be rewritten in the form of:

$$Y_t = AY_{t-1} + Qe_t \tag{8}$$

Or equivalently:

$$\begin{bmatrix} y_t \\ y_{t-1} \\ y_{t-2} \\ \vdots \\ y_{t-q+1} \end{bmatrix} = \begin{bmatrix} A_1 & A_2 & \ldots\ldots & A_q \\ I_n & 0_n & \ldots\ldots & 0_n \\ 0_n & I_n & 0_n \ldots & 0_n \\ \vdots & \vdots & \vdots \ddots & \vdots \\ 0_n & \ldots & 0_n \; I_n & 0_n \end{bmatrix} * \begin{bmatrix} y_{t-1} \\ y_{t-2} \\ y_{t-3} \\ \vdots \\ y_{t-q} \end{bmatrix} + \begin{bmatrix} I_n \\ 0_n \\ 0_n \\ \vdots \\ 0_n \end{bmatrix} * e_t \tag{9}$$

With q denotes the lag order of endogenous variables, I_n and 0_n are $(n \times n)$ identity and zero matrices respectively, and $n = 4$ are the endogenous variables. The j - *step* conditional expectation of (8) is:

$$E_t Y_{t+j} = (A)^j Y_t \tag{10}$$

Two vectors of length nq are generated to locate the forecasted variables:

$$r_x^{'} = (1, 0, 0, ..., 0) \text{ for the output gap} \tag{11}$$

$$r_\pi^{'} = (0, 1, 0, ..., 0) \text{ for inflation} \tag{12}$$

The expected future value of the output gap and the inflation of the next period, $j = 1$, are derived by premultiplying Eq. (10) by the vectors defined in (11 & 12).

$$E_t x_{t+1} = r_x^{'} A Y_t \tag{13}$$

$$E_t \pi_{t+1} = r_\pi^{'} A Y_t \tag{14}$$

Hence the expectation adjustment processes are the differences between the Eqs. (13, 14) and its own expected value at time $t-1$, which by using Eq. (8), are:

$$E_t x_{t+1} - E_{t-1} x_{t+1} = r_x^{'} A (Y_t - E_{t-1} Y_t) = r_x^{'} A Q e_t \tag{15}$$

$$E_t \pi_{t+1} - E_{t-1} \pi_{t+1} = r_\pi^{'} A (Y_t - E_{t-1} Y_t) = r_\pi^{'} A Q e_t \tag{16}$$

Inserting (15, 16) into the system of VAR innovations described by Eq. (7) as follows:

$$\begin{aligned} \varepsilon_t^x &= e_t^x - r_x^{'} A Q e_t + \alpha_1 (e_t^i - r_n^{'} A Q e_t) \\ \varepsilon_t^\pi &= e_t^\pi - \beta_1 r_\pi^{'} A Q e_t - \beta_2 e_t^x \\ \varepsilon_t^i &= e_t^i - e_t^\pi - \gamma_1 r_\pi^{'} A Q e_t - \gamma_2 e_t^x \end{aligned} \tag{17}$$

The forward-looking expectations expressed by Eq. (17) suggests nonlinear restrictions across the coefficients of each contemporaneous structural equation. Economic agents such as consumers, firms and the SBV are assumed to combine all relevant innovations in forecasting future value of endogenous variables.

3.4 An Open Economy New Keynesian Model

Instead of using lagged restriction constraints to define structural parameters, the paper uses Keating's (1990) [37] rational expectations determination system. Accordingly, the paper examines the convergence of simultaneous New Keynesian open-system equations from (2) through (5) in terms of structural disturbances and VAR innovations. In the process of conversion, the excess of exogenous variables (i.e. foreign price and the foreign interest rate) become internal factors within the system of VAR innovations to the four endogenous variables. Therefore, the residues now include only endogenous variable innovations:

$$\begin{aligned} \varepsilon_t^x &= e_t^x - (E_t x_{t+1} - E_{t-1} x_{t+1}) + \alpha_1 e_t^i - \alpha_1 (E_t \pi_{t+1} - E_{t-1} \pi_{t+1}) - \alpha_2 (e_t^s - \tfrac{e_t^\pi}{400}) \\ \varepsilon_t^\pi &= e_t^\pi - \beta_1 (E_t \pi_{t+1} - E_{t-1} \pi_{t+1}) - \beta_2 e_t^x \\ \varepsilon_t^s &= e_t^s - (E_t s_{t+1} - E_{t-1} s_{t+1}) + e_t^i \\ \varepsilon_t^i &= e_t^i - e_t^\pi - \gamma_1 (E_t \pi_{t+1} - E_{t-1} \pi_{t+1}) - \gamma_2 e_t^x \end{aligned} \tag{18}$$

where the domestic price innovation is equal to domestic inflation innovation over 400 [43].

Economic agents are required to update their future expectations on output gap, inflation and exchange rates, i.e. $(E_t x_{t+1} - E_{t-1} x_{t+1})$, $(E_t \pi_{t+1} - E_{t-1} \pi_{t+1})$, $(E_t s_{t+1} - E_{t-1} s_{t+1})$. In the closed economy model, all observable innovations contribute to the revision of future expectations. Because the endogenous residues are included in residues represented by Eq. (18), the article computes the expected revision processes through Eq. (8), $Y_t = AY_{t-1} + Qe_t$. With $n = 4$, the vectors of length nq are as follows:

$$\begin{aligned} r_x^{'} &= (1, 0, 0, ..., 0) \\ r_\pi^{'} &= (0, 1, 0, ..., 0) \\ r_s^{'} &= (0, 0, 1, ..., 0) \end{aligned} \tag{19}$$

The expectations revision processes are defined as follows:

$$\begin{aligned} E_t x_{t+1} - E_{t-1} x_{t+1} &= r_x^{'} AQe_t \\ E_t \pi_{t+1} - E_{t-1} \pi_{t+1} &= r_\pi^{'} AQe_t \\ E_t s_{t+1} - E_{t-1} s_{t+1} &= r_s^{'} AQe_t \end{aligned} \tag{20}$$

Apply Eq. (20) into the innovations system, obtain the open economy model below:

$$\begin{aligned} \varepsilon_t^x &= e_t^x - r_x^{'} AQe_t + \alpha_1 (e_t^i - r_\pi^{'} AQe_t) - \alpha_2 (e_t^s - \tfrac{e_t^\pi}{400}) \\ \varepsilon_t^\pi &= e_t^\pi - \beta_1 r_\pi^{'} AQe_t - \beta_2 e_t^x \\ \varepsilon_t^s &= e_t^s - r_s^{'} AQe_t + e_t^i \\ \varepsilon_t^i &= e_t^i - e_t^\pi - \gamma_1 r_\pi^{'} AQe_t - \gamma_2 e_t^x \end{aligned} \tag{21}$$

The system of Eq. (21) is the restriction constraint used in the model estimation to calculate the value of the deep structural parameters.

3.5 Estimation Method for SVAR Model

The model of open economy structural in matrix form:

$$\Gamma_0 y_t = \Gamma_1 y_{t-1} + ... + \Gamma_q y_{t-q} + \Lambda_0 z_t + \Lambda_1 z_{t-1} + ... + \Lambda_k z_{t-k} + \varepsilon_t,\ \varepsilon_t \tilde{} (0, D) \tag{22}$$

With $y_t = (x_t, \pi_t, s_t, i_t)^{'}$ is the vector of endogenous variables and $z_t = (p_t^*, i_t^*)'$ is the exogenous vectors, Γ_i and Λ_j are coefficient matrices for variables The vector $\varepsilon_t = (\varepsilon_t^x, \varepsilon_t^\pi, \varepsilon_t^s, \varepsilon_t^i)'$ contains the structural disturbances, 0 is a (4×1) vector of zeros, and **D** is a cross-level (4×4) diagonal variance - covariance matrix. Multiplying Eq. (22) with reduced - form VAR:

$$y_t = A_1 y_{t-1} + ... + A_q y_{t-q} + B_0 z_t + B_1 z_{t-1} + ... + B_k z_{t-k} + e_t,\ e_t \tilde{} (0, \Omega) \tag{23}$$

Where:

$$A_i = \Gamma_0^{-1}\Gamma_i,\ i = 1, ..., q,\ B_j = \Gamma_0^{-1}\Lambda_j,\ j = 0, 1, ..., k,\ e_t = \Gamma_0^{-1}\varepsilon_t \text{ and } \Omega = \Gamma_0^{-1}D\Gamma_0^{-1'} \quad (24)$$

The estimation proceeds consists of two steps. Step 1 involves estimating the reduced - form VAR specified by (23). Estimated parameters in **A** and the rational expectations restrictions conditions are expressed by the Eq. (21) given in Γ_0; In addition, limit conditions are placed on the contemporaneous exogenous coefficient matrix Λ_0. Step 2, the lagged conditions do not include constraint conditions and the study estimates the structural system (22) using the FIML method [37,43] with the assumption that distributed structural disturbances standard. The structural parameters are achieved by maximising the following log - likelihood function:

$$L = \sum_{t=1}^{T}\left[-\frac{n}{2}\ln(2\pi) - \frac{1}{2}\ln\left|\Gamma_0^{-1}D\Gamma_0^{-1'}\right| - \frac{1}{2}\varepsilon_t^{'}D^{-1}\varepsilon_t\right] \quad (25)$$

and the asymptotic standard errors are obtained as the inverse of the Hessian matrix.

4 Research Results

4.1 Stationary Test (unit Root Test) for Data Series

The condition for data series to be included in the VAR model estimation is the stationary. To test stationary, the author uses unit root tests for series of data: output gap (*dy_hp*), domestic inflation (*dpi*), natural logarithm nominal exchange rate (*lne*) and the refinancing rate (r), the United State (US) consumer price index (*cpi_us*) and the US federal interest rate (*int_us*).

This test assumes H_0: The data series has a unit root (i.e., unstationary). Based on the P-value of the test to accept or reject the null hypothesis (H_0). With the data series *pi*, $P - value = 0.1071(0.1827, 0.4425) > 0.1$; then, with a 10% significance level, the hypothesis H_0 is accepted. Or the data series *pi* is unstationary. Therefore, the author performs a first order difference for the pi string and continues to check whether the string has a stationary. The results show that for $P - value = 0.0000 < 0.01$, it can be concluded that *pi* was stopped when taking the first order difference (*dpi*) (Table 2).

4.2 Selection of Optimal Lag and Stability of the Model

For selection of the lag for endogenous variables in the model, the author will rely on a number of standard values (Table 3).

According to the LR, FPE, AIC and HQ standards, a four - lag for endogenous variables should be selected, while the SC and HQ standard should select

Table 2. Summary of unit root test results for data series

	Level			1st difference		
	Intercept	Trend and intercept	None	Intercept	Trend and intercept	None
dy_hp	0.0470	0.1665	0.0037	0.0103	0.0483	0.0006
pi	0.1071	0.1827	0.4425	0.0000	0.0001	0.0000
lne	0.0000	0.0000	0.0936	0.0000	0.0000	0.0000
r	0.0369	0.1087	0.4238	0.0000	0.0000	0.0000
cpi_us	0.0678	0.0698	0.2075	0.0017	0.0102	0.0001
int_us	0.0364	0.0313	0.0033	0.0107	0.0284	0.0008

Source: Author's collection and estimation of data by Eview 8

Table 3. Synthesis of selection standards for endogenous variables in the model

VAR lag order selection criterion[a] for endogenous variables in VAR model						
Lag	LogL	LR	FPE	AIC	SC	HQ
0	−684.4465	NA	18610.29	21.18297	21.31678	21.23576
1	−583.5028	186.3575	1365.161	18.56932	19.23836*	18.83330*
2	−562.0990	36.88039	1162.933	18.40305	19.60732	18.87821
3	−540.0516	35.27595	979.7753	18.21697	19.95648	18.90332
4	−518.0736	32.45979*	838.2281*	18.03303*	20.30778	18.93057

Source: Author's collection and estimation of data by Eview 8

[a]LR: LR test (level of significance 5%). FPE: Final priction error. AIC: Akaike information criterion. SC: Schawrz information criterion. HQ: Hannan-Quinn information criterion.

*Indicates lag order selected by the criterion.

a lag of 1 for endogenous variables. The study selects a lag of 4 to estimate the parameters in the model which is the same with Leu's [43] estimation result. Lag of exogenous variables is from 1 to 4 and in the VAR model (4, 1); VAR (4, 2); VAR (4, 3) and VAR (4, 4); The residuals of these equations are not autocorrelation (based on LM Test results) but VAR (4, 1) and VAR (4, 2). From the statistical lag of the above standards, it can be seen that 4 is the most appropriate lag for exogenous variables (Table 4).

4.3 Diagnostics of the Reduced-Form VAR

Estimation of SVAR models is done first by performing a series of statistical completeness tests of the reduced model [69]. The lags of endogenous and exogenous variables are allowed to be different. Dynamic structures are used to estimate

Table 4. Synthesis of external variable selection standards

Criteria	VAR (3, 4)	VAR (3, 3)	VAR (3, 2)	VAR (3, 1)
Akaike info criterion	17.74230*	17.94789*	18.32719*	18.22441*
Schwarz criterion	21.75656	21.69453	21.80621	21.43581
Hannan-Quinn Criterion	19.32618*	19.42618*	19.69989	19.49152
Log likelihood	−456.6248	−471.3065	−491.6337	−496.2934

Source: Author's collection and estimation of data by Eview 8
*Indicates lag order selected by the criterion.

the four lags used for endogenous and exogenous variables[9]. In addition, the study used the three seasonal dummies in each endogenous equation to eliminate the seasonal factor affecting the data series ($q_1 = 1\, if\, q_1$, $q_1 = 0$ if it is another quarter, similar to q_2, q_3, & q_4). After the regression of quadratic Eq. 4 and obtained the reduce - form VAR. Based on the results of the selected reduce - form VAR tests, the study conducts the remaining tests for a serial correlation, a standard deviation, and an ARCH at the appropriate significance level. At the same time, the stability of the model is tested. When these test results are evaluated together, there is a general support for the statistical completeness of the model.

4.4 Test of Residuals of Equations in the VAR Model

4.4.1 Autocorrelation Test

The study uses the LM test, with the hypothesis H_0: There is no autocorrelation. Consider the residuals of the system of equations, as Table 5 may accept the hypothesis H_0, or no autocorrelation exists.

4.4.2 Normal Distribution Test

The paper is based on the Jarque-Bera test, with the hypothesis H_0: There is a standard distribution. Consider the residual of the VAR model, $Pro. = 0.88 > 0.1$; Therefore, with the significance level of 10%, it is possible to accept the hypothesis H_0, or the standard excess, satisfying the necessary conditions to estimate VAR quantitatively.

[9] This lag is consistent with Leu's [43] study which is commonly used to obtain a result with time series data. Given the quarterly data and its relatively small sample size, the upper bound was set at 4 lags for both endogenous variables and 1 lag to 4 lags for exogenous variables. However, the system of equations in VAR (4, 1) and VAR (4, 2) failed to reject the null of no serial correlation. Therefore, the study examined the lag of exogenous variables to seek a more parsimonious specification. Using the likelihood ratio test in which VAR (4, 4) and VAR (4, 3) fit, there is no autocorrelation problem.

Table 5. Results of LM Test for residuals of equations in the VAR model

Lags	LM-Stat	Prob
1	32.47445	0.0087
2	11.08399	0.8043
3	19.85929	0.2266
4	14.02418	0.5969

Probs from chi-square with 16 df.

Source: Author's collection and estimation of data by Eview 8

4.4.3 ARCH Effect Test

The paper uses the Heteroskedasticity Test, with the hypothesis H_0: No effect on ARCH.

Consider the residual of the VAR model, $Prob = 0.4231 > 0.1$; It is therefore possible to accept the hypothesis H_0, or to conclude that no effect of ARCH exists.

4.4.4 VAR Model Stability Test

The results show that the values are in the range $[-1, 1]$, so the VAR model is stable (Fig. 1).

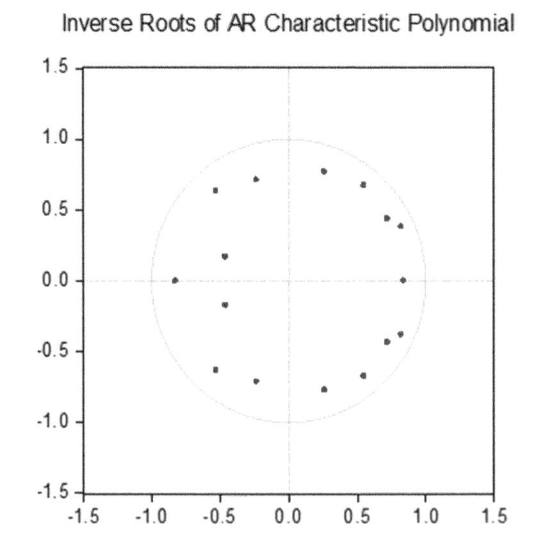

Fig. 1. Results of VAR model stability test *Source: Author's collection and estimation of data by Eview 8*

4.5 Determination of Limit Conditions in SVAR Estimation

Based on the matrix A obtained from the estimation parameters for the VAR reduction form, the paper identifies the following limitation conditions:

$$u_1 = e_1 - (0.63098 * e_1 - 0.55809 * e_2 - 0.5240 * e_3 + 0.63008 * e_4) + \alpha_1 * (e_4 - (-0.04468 * e_1 + 0.6812 * e_2 - 0.07787 * e_3 - 0.799624 * e_4)) - \alpha_2 * (e_3 - 0.0025 * e_2)$$

$$u_2 = e_2 - \beta_1 * (-0.04468 * e_1 + 0.6812 * e_2 - 0.07787 * e_3 - 0.7996 * e_4) - \beta_2 * e_1$$

$$u_3 = e_3 - (-0.048195 * e_1 - 0.13709 * e_2 - 0.1036 * e_3 + 0.04268 * e_4) + e_4$$

$$u_4 = e_4 - e_2 - \gamma_1 * (-0.04468 * e_1 + 0.6812 * e_2 - 0.07787 * e_3 - 0.7996 * e_4) - \gamma_2 * e_1$$

With e_1, e_2, e_3, e_4 respectively as residuals of the equations *dy_hp*, *dpi*, *lne*, *r* in VAR and u_1, u_2, u_3, u_4 respectively as structural disturbances of the equations *dy_hp*, *dpi*, *lne*, *r* in the SVAR model.

4.6 Contemporaneous Structural Parameter Estimation

Estimated structural parameters in the Eq. (21) are presented in Table 6

Table 6. Contemporaneous structural estimates

α_1	α_2	β_1	β_2	γ_1	γ_2
0.2863	0.3365	0.8921	0.0277	−0.6916	−0.0532
(0.2349)	(0.0074)***	(0.0000)***	(0.8123)	(0.0029)***	(0.7843)

Source: Author's collection and estimation of data by Eview 8
Note: P-values are in parenthese.
*, ** & ***indicate significance at 1%, 5% and 10% respectively.

Apart from $\alpha_1, \beta_2, \gamma_2$, the remaining contemporaneous parameter estimates are stable and statistically significant at 1%. From the IS equation, α_2 indicates that the real exchange rate rises or the domestic currency depreciate, domestic goods become cheaper than foreign goods, the increased commodity competitiveness augments export activity, promotes production in the country, thereby increasing output gap and aggregate demand. However, the coefficient α_1 is not statistically significant, suggesting that the effect from the interest rate transmission channel on the output is not as expected, largely due to the fact that the transparency and credibility of the public to the central bank's decisions are not big enough, capital inflows from credit growth have not yet focused on production and business activities, but dominated by securities and real estate channels.

For the AS equation, β_1 denotes the fluctuation of inflation influenced by public expectations (business activity of enterprises and consumption of individuals and households) in a rigid price environment [13]. Statistical significance estimation of β_1 implies that inflation expectations in Vietnam are mainly due to future inflation expectations [30]. Fuhrer and Moore [27] have found the same conclusion that inflation expectations depend on current inflation with empirical evidence of the post-war period. Indeed, the economic agents in Viet Nam are subjected to psychological pressure from the public. When consumers predict that inflation will increase, they demand a pay rise. This increases the cost of the enterprises, resulting in the rising cost of production and a rise in commodity prices, which in turn lead to a rise in the Consumer Price Index, bringing about higher inflation. In addition, the existence of the lag variables according to Roberts [66] is necessary for the agents in the inflation expectations [43] or the inflation's lag.

In addition, in the test study on Phillips Curve toward the future, the output gap is an important variable to the fluctuation of inflation [36] or the positive correlation between growth and inflation. However, in this study, the coefficient β_2 is not statistically significant, indicating that in the past time, the State Bank of Viet Nam (SBV) has performed relatively well the role of controlling inflation but has maintained GDP growth, trying to limit the growth with risk of high inflation which will destabilize the economy. Estimation of the structural parameter from the interest rate equation shows that the SBV policy interest rates have implemented policy interest rates to stabilize inflation and aggregate demand of the economy. The regression coefficient γ_1 shows that the short - term nominal interest rate adjusts for inflationary fluctuations to make a real interest rate change under the Taylor rule [70].

4.7 Impulse Response Function (IRF)

The IRF with a 95% confidence interval for four structural shocks: monetary policy shock, exchange rate shock, aggregate supply shock, and aggregate shocks are reported in Figs. 2, 3, 4 and 5[10]. Each structural shock has one standard deviation. In many cases, the confidence intervals show that the dynamic responses are statistically significant in the short run (in the first year of the initial shock). All reactions show mean - reversion which refects the stationary properties of the structural model.

4.7.1 Monetary Policy Shock

Prior to the monetary policy shock, the interest rate increased, but this did not seem to affect output gap when the volatility of output gap also increased accordingly in the first five periods. Thus, in Vietnam, the phenomenon of output puzzle (increased output when tightening monetary policy) exists. However, since the 6th period the output gap decreased, the monetary policy tightening has begun to take effect despite the lag. But when the interest rate increases by more than 3%, it has the potential to significantly reduce output (9th, 10th

[10] The calculation's method of confidence interval has based on bootstrapping technique with 5000 times simulation [68].

periods) of the economy, the central bank will reduce interest rates and maintain a growth rate of less than 2% to recover the output of the economy. Second, increased interest rate has the effect of lowering inflation. This is in line with the current policy of the central bank when interest rate is a tool implemented to curb inflation. However, in the 12th or from 24th to 28th periods, although interest rates continue to increase, it seems that the impact of lowering (or even decreasing) inflation is consistent with the positive correlation between interest rate and inflation. Especially in 2008, although the SBV raised very high interest rate (the policy interest rate increased from 8.25%year from January 2008 to 14%year in October 2008) but inflation was not reduced. Thus, it demonstrates that in this period in Vietnam there is a price puzzle phenomenon - price, inflation increase when tightening monetary policy - [73]. Exchange rate movements follow the theory, the raised interest rate shock makes the price of domestic currency increase or the exchange rate decrease, especially when the interest rate rises more than 3%, the exchange rate falls more than 1% and only when interest rate falls back less than 2% to stimulate exports, the exchange rate starts to increase again (11th–40th periods) in accordance with the theory of Uncovered Interest Parity (UIP). Thus, Vietnam does not have the phenomenon of exchange rate puzzle in long term period, which is consistent with results of the empirical study by Tran and Nguyen [73] and Leu [43].

In conclusion, the impossible trinity theory (Mundell - Flemming model) implies that effectiveness of the monetary policy depends on the exchange rate regime and the level of capital flow control in each country, particularly in the fixed exchange rate regime and the open capital flows, the monetary policy is independent and less important in the economy. However, the empirical results show that the monetary policy shock has an impact on the output gap, inflation and exchange rate or the crucial - positive role of the SBV in macro management to guide the market in order to maintain the stability through announcement of the central exchange rate[11], and the cross rate of VND versus other foreign currencies (Decision No. 2730/QD - 12/2015). Thus, the exchange rate is almost stable and the monetary policy is implemented as the Government imposes controls on foreign capital flows in the Vietnam‘s financial market.

[11] The central exchange rate is determined on the basis of reference to the weighted average exchange rate movement in the inter-bank foreign exchange market, the exchange rate movement in the international market of foreign currencies of several countries having trade, lending - borrowing relations or large investment with Vietnam, the macro - economic and national monetary balances in line with the SBV monetary policy targets. The new method of managing exchange rate allows exchange rate system to be determined more flexibly in line with the domestic supply and demand of foreign currencies, the fluctuations of exchange rate in the international market, while ensuring the role of the SBV in managing the monetary policy [19].

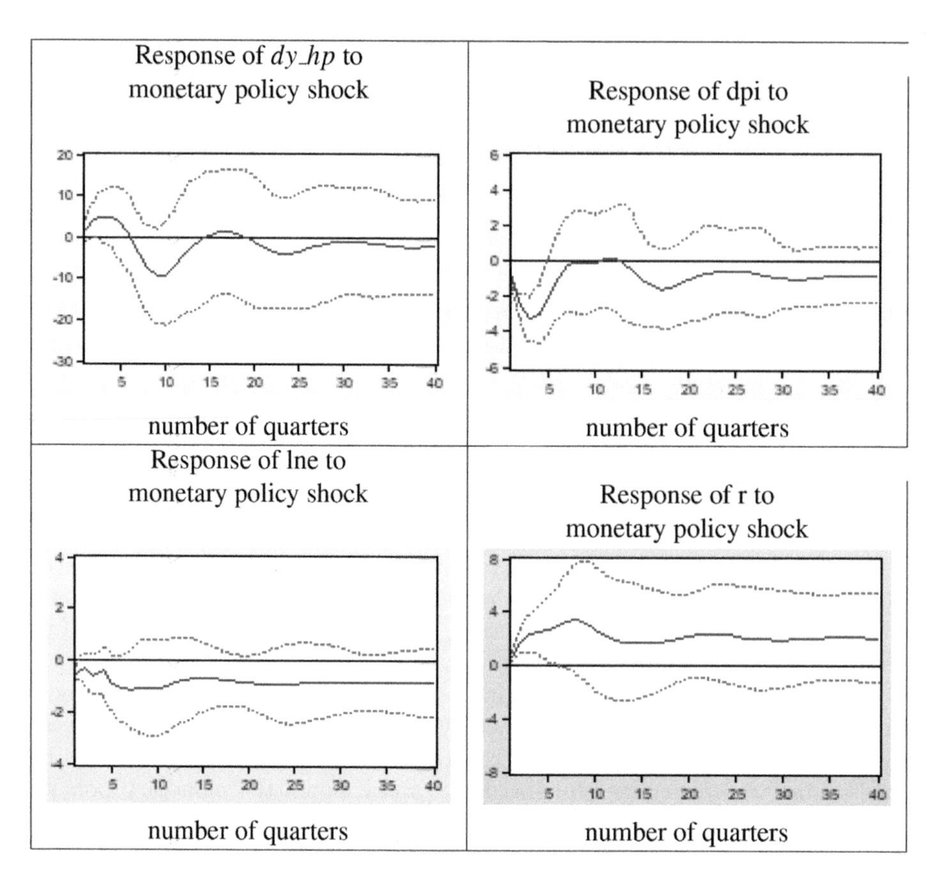

Fig. 2. Response of macroeconomic variables to monetary policy shock *Source: Author's collection and estimation of data by Eview 8*

4.7.2 Foreign Exchange Rate Shock

The exchange rate shock reduces the output gap and inflation, but increases interest rate. It can be seen that the new Keynesian open economy model, the exchange rate is not an explanatory variable in the interest rate rule [71]. However, Vietnam remains a country that tightly controls capital inflows and outflows and exchange rate fluctuations under the central bank's operation. As the exchange rate rises, commodity prices have increased in the short run, but due to price stickiness, it is not timely to adjust prices which cause profits of businesses to shrink, resulting in reduced production. In addition, foreign exchange reserves increase sharply to strengthen the SBV's commitment to stabilize the exchange rate[12] through the ability to intervene to sell foreign currencies to stabilize the forex market and exchange rate within the set band. So, the central

[12] According to Bloomberg (2008) on the level of currency stability of several currencies in Asia, VND is considered to be currency in the most stable group. As of 31/12/2017, the central exchange rate between VND and USD announced by the SBV was at 22,425 VND/USD, increasing by 1.2% compared to that in late 2016.

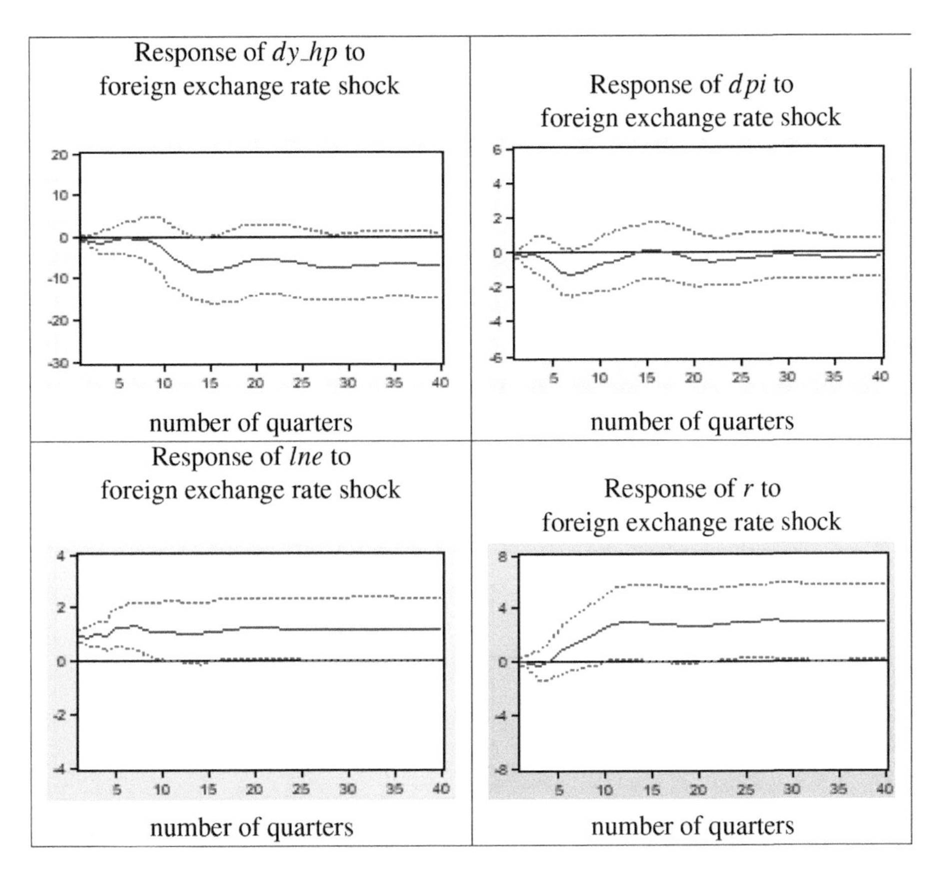

Fig. 3. Response of macroeconomic variables to foreign exchange rate shock *Source: Author's collection and estimation of data by Eview 8*

bank focused on attracting the amount of VND in circulation (reducing circulation to curtail inflation), in which credit tightening policy has led to shrink domestic capital inflows that reduce the output in the economy. However, the supply of foreign currencies can lead to a decrease in national foreign exchange reserves and slow down the exchange rate at the same time, the interest rate tool is used to adjust the increase to recover the value of the domestic currency simultaneously and to curb inflation. Thus, the smooth coordination of instruments in the currency market is contributed to stabilizing the exchange rate and output.

4.7.3 Aggregate Supply Shock

The aggregate supply shock or inflation shock increases the output gap by nearly 10% (in 9th–11th periods) while the exchange rate decreases (domestic currency appreciates). This proves the central bank's efforts to stabilize the exchange rate. In addition, in spite of rising inflation, the SBV has raised interest rate that has an impact on inflation reduction. The tightening of the monetary policy from the 5th to 11th periods, with interest rate increasing from 2% to 3%, has contributed to

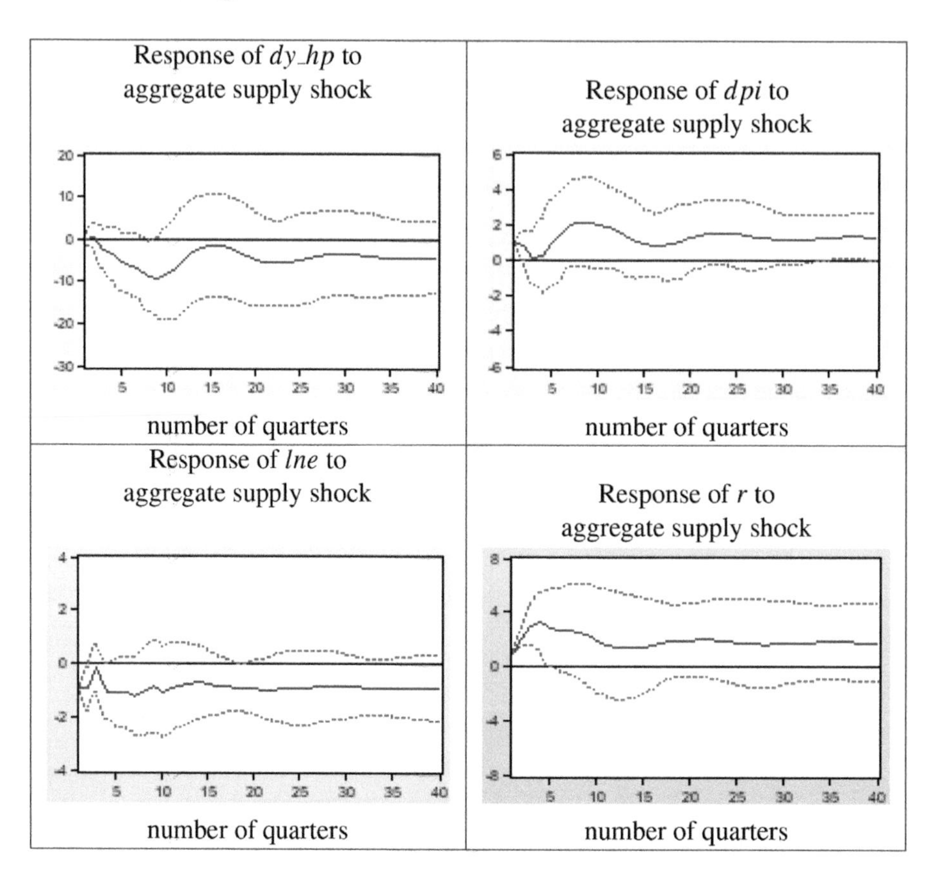

Fig. 4. Response of macroeconomic variables to aggregate supply shock *Source: Author's collection and estimation of data by Eview 8*

lowering domestic inflation (an increase from 2% to 1%). However, such tightening of the monetary policy has the effect of reducing output. Therefore, then the monetary policy adjusts to lower the interest rate (around 1%) to bring inflation increase again (up over 1%), narrowing the output gap. Additionally, the increased inflation shock causes interest rate to rise and theoretically the domestic currency or exchange rate falls but the margin does not exceed 2%. In the long term, the exchange rate has a sign of future recovery under the theory of UIP.

4.7.4 Aggregate Demand Shock

As shown in Fig. 5, the output gap and inflation increase as a positive aggregate demand shock impacts on the economy, but the inflation rate is negligible, then inflation tends to decrease on average 0.2% (40 periods). On the part of the SBV, before the aggregate demand shock, the monetary policy easing evidence showed that interest rate fell by an average of 1.2% in 40 periods. According to such developments, it can be seen that from 2012 onwards, the SBV has always

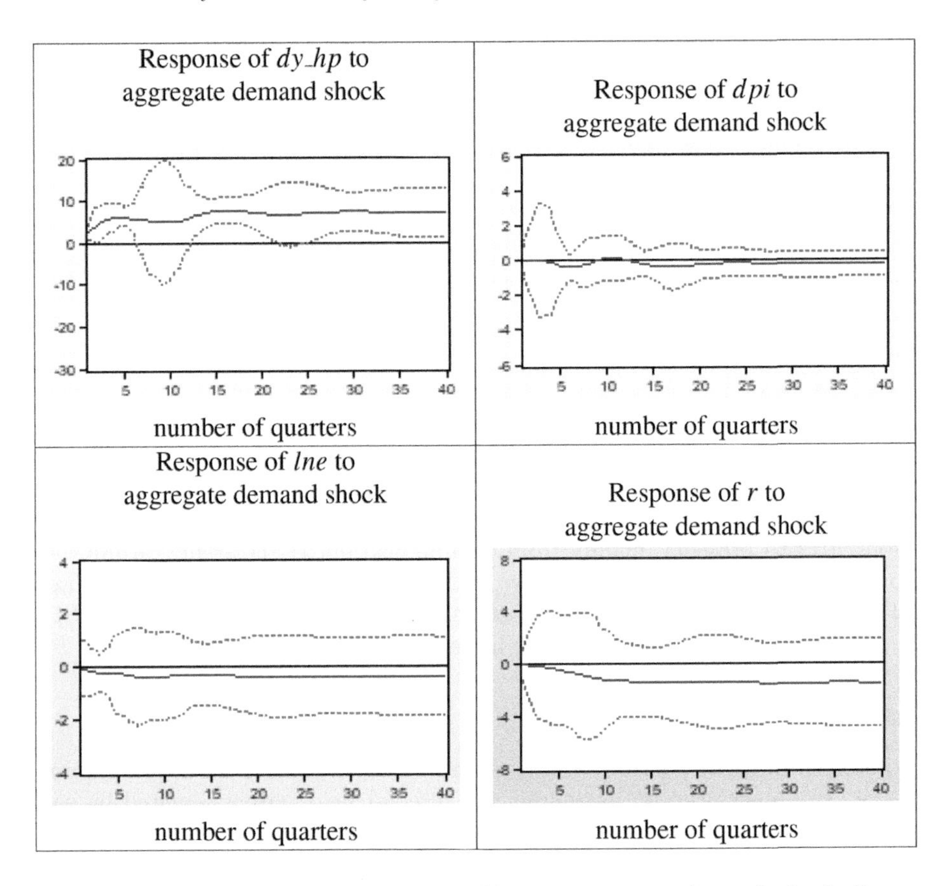

Fig. 5. Response of macroeconomic variables to aggregate demand shock *Source: Author's collection and estimation of data by Eview 8*

maintained the position of maintaining low inflation in short, medium and long term but still controlling not to let deflation happen. Having the ultimate goal in line with this macro context helps the SBV to use the right intermediary. Monetary supply is no longer a major factor in creating inflationary pressure. Instead, to ensure macroeconomic stability requires improvement and recovery from the troubled corporate system. Therefore, the interest rate decrease in the condition of output growth but not increase inflation is in line with the instruction 01 of the SBV[13]. Accordingly, the operating interest rates and short-term lending rates are reduced to coordinate with credit institutions to aim at supporting

[13] In Directive No. 01/CT - NHNN dated 10/1/2017 on implementing the monetary policy and ensuring safe and effective banking operations in 2017, the State Bank of Vietnam has guided" the management of interest rates in line with changes in macroeconomy, inflation and the monetary market to stabilize interest rates; On the basis of the ability to control inflation, stabilize the foreign exchange market, strive to reduce lending interest rates.

Table 7. Results of the variance decomposition of interest rate

Variance decomposition of r					
Period	Standard error	The 1st shock	The 2nd shock	The 3rd shock	The 4th shock
1	1.20	0.33	67.56	0.02	32.10
4	2.28	0.40	59.43	1.76	38.41
8	2.56	2.29	50.15	13.50	34.06
12	2.82	2.17	45.42	15.93	36.48
20	2.85	2.25	45.21	15.74	36.81
40	2.87	2.25	44.87	15.81	37.07

operating expenses for enterprises, contributing to economic growth under the policy of the Government but still ensure careful monitoring of controlled inflation. In conclusion, in order to achieve the intermediate and final objectives as analyzed above, there are many changes in the management of monetary policy (from 2012 to present), mainly from communication, transparency in information, mechanisms, policies and active and timely decisions and actions, closely following developments in the domestic and foreign financial markets of the SBV in particular and the banking sector in general.

4.8 Variance Decomposition

An assessment of the relative importance of four structural shocks at different limits can be achieved by examining the forecast error rate calculated by each shock[14], and is reported in Tables 6, 7, 8, 9 and 10.

First, Table 7 shows the forecast error rate in output corresponding to each shocks with the limit of 40. In addition to the role of interest rate itself, the inflation shock contributes primarily to explaining the changes in interest rate. This shock accounts for 67.56% of the quarter 1 fluctuation, down to 45.42% in the second quarter and further down to 44.87% in the 40th period. Inflation shock affects interest rate changes that gradually reduce in the long run. The rate shock is ranked third in terms of the impact on interest rate. In the meantime, the aggregate demand shock is ranked the last in terms of impact.

Second, Table 8 shows the contribution of shocks in explaining the volatility of exchange rate. The biggest impact factor is the inflation shock, followed by the exchange rate shock explaining the volatility itself. In addition to the interest rate factor, the role of interest rate also contributes to the explanation of exchange rate changes in addition to the impact of aggregate demand shock.

Third, Table 9 shows the impact of shocks on inflation volatility. Interest rates contributed primarily to the change in inflation, suggesting a clear role for the central bank in stabilizing inflation, contributing to fiscal stability. In addition, the impact of inflation explains itself in addition to the effects of exchange rate shock and output.

[14] The 1st shock: aggregate demand shock; the 2nd shock: inflation shock, the 3rd shock: exchange rate shock, the 4th shock: interest rate shock.

Table 8. Results of the variance decomposition of exchange rate

Variance decomposition of lne					
Period	Standard error	The 1st shock	The 2nd shock	The 3rd shock	The 4th shock
1	1.45	0.17	44.38	43.04	12.41
4	1.91	0.61	63.53	25.32	10.5
8	2.03	0.81	57.32	24.91	16.96
12	2.07	0.82	58.05	24.16	16.97
32	2.09	0.84	57.67	24.02	17.47
40	2.09	0.84	57.61	23.99	17.56

Table 9. Results of the variance decomposition of aggregate supply (inflation)

Variance decomposition of dpi					
Period	Standard error	The 1st shock	The 2nd shock	The 3rd shock	The 4th shock
1	1.52	0.14	37.35	0.70	61.81
4	2.35	0.45	26.81	1.03	71.70
8	3.13	0.91	25.45	4.29	69.35
12	3.18	1.29	25.55	5.53	67.63
32	3.37	1.56	24.99	6.26	67.19
40	3.41	1.61	25.13	6.42	66.84

Table 10. Results of the variance decomposition of aggregate demand (output gap)

Variance decomposition of dy_hp					
Period	Standard error	The 1st shock	The 2nd shock	The 3rd shock	The 4th shock
1	3.07	70.28	0.04	1.75	27.94
4	5.67	37.35	31.22	3.58	27.85
8	9.03	15.24	22.52	1.76	60.47
12	10.69	11.22	23.19	9.23	56.36
32	11.71	10.35	24.06	9.82	55.77
40	12.00	9.97	24.03	10.02	55.98

Finally, Table 10 shows the importance of each shock to variation in output. In particular, the shock itself explains in the short term with more than 70%, but from the eighth period onwards the interest rate shock and inflation have contributed to the change in output or economic growth. Third is the explanatory role of the exchange rate shock and finally in the 40th period is the contribution of aggregate demand shock.

5 Conclusion

By approaching SVAR in accordance with new Keynesian theory estimated with reasonable expectation, the study assessed the impact of macro variables in Viet Nam. The results of dynamic response simulation show that the positive policy interest rate shock has the effect of reducing output and inflation. The higher domestic currency interest rates cause the exchange rate to decrease, indicating that there is no sign of the existence of the exchange rate puzzle phenomenon in Vietnam during the study period, while tightening monetary policy aims at controlling inflation. However, the results of variance decomposition show that the positive aggregate shock increases over time showing that inflation is still rising when there is a monetary shock in the first periods, indicating that there is a price puzzle in Vietnam [73]. Next, the exchange rate shock reduces output, inflation but increases interest rate. Meanwhile, the supply-side inflation shock reduces output in the context of rising interest rate to curb inflation. Vietnam's exchange rate regime in this period operates under a theoretical mechanism. Finally, a surge in demand has impacted lower interest rate, which is seen as a move that shows a clear role for the central bank in managing macro stability with interest rate instruments, reducing dependence on currency supply. On the basis of lowering interest rate to support those in the economy stimulating production investment, but closely monitor the exchange rate movements to ensure growth and stable inflation [19]. In particular, in a small and open economy but with relatively tight capital control such as Vietnam, the impact of monetary policy on the exchange rate has shown that the Government's message is more flexible instead of anchoring to help Vietnam integrate into the world economy. Apart from some contributions, this study also has some limitations to be considered. First, the study only builds up a model of the structure using the expectation channel in the analysis of monetary policy transmission. However, the expectation channel cannot be used for independent analysis, but should be combined with other channels to increase the coverage of the study, so that this channel has the potential for public trust, and decisions of the SBV. The independence, transparency, credibility of the SBV in implementing the monetary policy and other policies will determine effectiveness of the expectation channel. Second, the study demonstrates the regulatory role of policy interest rate, central rate to macro variables such as output and inflation which the SBV implements to stabilize macroeconomy[15]. However, policy interest rate is

[15] Vietnam's economic growth and inflation rate since 2001 have shifted from "relatively high growth, low inflation" to "high growth, moderate inflation" (2004–2007), then moved to the position of "good growth, high inflation" (2008–2011), "low growth, low inflation" (2012–2014), and to this stage is" growth and inflation (2015–2017). Results of the last 3 years (2015–2017) show that the relative stability of the economy is a condition for the accumulation of factors necessary for the later high growth period (Chu, K. L, 2018) [19].

heavily influenced by administrative orders, which are only relevant to the primary financial market, leading in the long run to direct interest rate control that distorts the development of the financial market with potential risks causing macro instability. Therefore, the requirement for Vietnam is that there is a roadmap towards the scientific liberalization of interest rate to avoid negative effects from this roadmap. In addition, the stable exchange rate is a strong commitment of the regulator on the ability to intervene in the foreign exchange market, along with the transparency of information and publication of national foreign exchange reserves in a positive way which has contributed to reducing the drastically fluctuating exchange rate expectations, thereby eliminating the psychology of foreign currency speculation. In addition, the State Bank also closely coordinates with related functional units to inspect and handle violations of foreign exchange trading, bringing the USD mobilization rate to 0%, which is also the move to enhance effectiveness of exchange rate stabilizing policy. However, empirical evidence in the study has shown that the key contributing factor to this stability is the Vietnam's tight control of international capital flows, which creates unsustainable stability for the exchange rate, as in the trend of international integration, the openness of the economy is an important condition for increasing competitiveness and attracting investment of Viet Nam's economy.

References

1. Aastveit, K.A., BjØrnland, H.C., Thorsrud, L.A.: The world is not enough! Small open economies and regional dependence. Working paper 2011/16, Economics Department, Norges Bank (2011)
2. Acosta-Ormaechea, S., Coble, D.L.: Monetary transmission in dollarized and non-dollarized economies: the cases of Chile, New Zealand, Peru and Uruguay. IMF document de trabajo, WP/11/87 (2011)
3. An, S., Schorfheide, F.: Bayesian analysis of DSGE models. Econ. Rev. **26**(2–4), 113–172 (2006)
4. An, S., Schorfheide, F.: Bayesian analysis of DSGE models. Econ. Rev. **26**(2–4), 113–172 (2007). https://doi.org/10.1080/07474930701220071
5. Andrle, M.A., Berg, A., Morales, R., Portillo, R., Vleck, V.: Forecasting and monetary policy analysis in low income countries (1): food and non-food inflation in Kenya. Working Paper, International Monetary Fund, Washington, D.C (2013)
6. Berg, A., Karam, P., Laxton, D.: Practical model - based monetary policy analysis - a how to guide. IMF Working Paper, WP/06/81 (2006)
7. Berg, A., Portillo, R., Unsal, D.F.: On the optimal adherence to money targets in a new-Keynesian framework: an application to low-income countries. Working Paper, International Monetary Fund, Washington D.C (2010)
8. Berg, A., Charry, L., Portillo, R., Vlcek, J.: The monetary transmission mechanism in the tropics: a narrative approach. Working Paper no WP/13/197, International Monetary Fund, Washington D.C (2013)
9. Bhuiyan, R.: The effects of monetary policy shocks in Bangladesh: a Bayesian structural VAR approach. Int. Econ. J. **26**(2), 301–316 (2012)
10. Blanchard, O., Galí, J.: Labor markets and monetary policy: a new Keynesian model with unemployment. Am. Econ. J. Macroecon. **2**(2), 1–30 (2010)

11. Bofinger, P., Mayer, E., Wollmershauser, T.: The BMW model: a new framework for teaching monetary economics. J. Econ. Educ. 98– 117 (2005)
12. Carlin, W., Soskice, D.: The 3-equation new Keynesian model - a graphical exposition. Contrib. Macroecon. **5**(1), 13 (2005)
13. Calvo, G.A.: Staggered prices in a uility-maximizing framework. J. Monet. Econ. **12**(3), 383–398 (1983)
14. Calzolari, G., Panattoni, L., Weihs, C.: Computational efficiency of FIML estimation. J. Econ. **36**(3), 299–310 (1987)
15. Cao, T.Y.N., Le, T.G.: Using SVAR model for testing monetary transmission and suggesting for monetary policy in Viet Nam. J. Econ. Dev. **216**, 37–47 (2015)
16. Chen, S.S.: DSGE Models and Central Bank Policy Making: A Critical Review. Department of economics national Taiwan university, National Taiwan University (2010)
17. Cho, S., Moreno, A.: A structural Estimation and Interpretation of the New Keynesian Macro Model. Universidad de Navarra, Working Paper n0 14/03 (2003)
18. Christiano, L.J., Eichenbaum, M., Evans, C.: Nominal rigidities and the dynamic effects of a shock to monetary policy. J. Polit. Econ. **113**(1), 1–45 (2005)
19. Chu, K.L.: Monitoring monetary policy and the orientation in 2018. J. Financ. (2018). http://tapchitaichinh.vn/nghien-cuu-trao-doi/trao-doi-binh-luan/dieu-hanh-chinh-sach-tien-te-va-dinh-huong-trong-nam-2018-135403.html
20. Clarida, R., Galí, J., Gertler, M.: Monetary policy rules in practice: some international evidence. Eur. Econ. Rev. **42**, 1033–1067 (1998)
21. Clarida, R., Galí, J., Gertler, M.: The science of monetary policy: a new Keynesian perspective. J. Econ. Lit. **37**(4), 1661–1707 (1999)
22. Clarida, R., Galí, J., Gertler, M.: Monetary policy rules and macroeconomic stability: evidence and some theory. Q. J. Econ. **115**(1), 147–180 (2000)
23. Colander, D.: The stories we tell: a reconsideration of AS/AD analysis. J. Econ. Perspect. **9**(Summer), 169–88 (1995)
24. Dhrymes, P., Thomakos, D.: Structural VAR, MARMA, and open economy models. Int. J. Forecast. **14**, 187–198 (1998)
25. Dizioli, A., Schmittmann, J.M.: A Macro - model approach to monetary policy analysis and forecasting for Viet Nam. IMF Working Paper, WP/15/273 (2015)
26. Dinh, T.T.H., Phan, D.M.: The effectiveness of monetary policy through interest rate channel. J. Dev. Integr. **12**(22), 39–47 (2013)
27. Fuhrer, J., Moore, G.: Inflation persistence. Q. J. Econ. **110**, 127–159 (1995)
28. Galí, J.: Monetary Policy, Inflation, and the Business Cycle: An Introduction to the New Keynesian Framework. Princeton University Press, Princeton (2008)
29. Galí, J., Monacelli, T.: Monetary Policy and Exchange Rate Volatility in a Small Open Economy. Mimeo, Boston College (2002)
30. Gruen, D., Pagan, A., Thompson, C.: The phillips curve in Australia. J. Monet. Econ. **44**, 223–258 (1999)
31. Hamilton, J.D., Wu, J.C.: The effectiveness of alternative monetary policy tools in a zero lower bound environment. J. Money Credit. Bank. **44**, 3–46 (2012)
32. Hodge, A., Robinson, T., Stuart, R.: A small BVAR-DSGE model for forcasting the Australian Economy. RBA Research Discussion Paper 2008 - 04 (2008)
33. Hodrick, R., Prescott, E.C.: Postwar U.S. business cycles: an empirical investigation. J. Money Credit. Bank. **29**(1), 1–16 (1997)
34. Hung, L.V., Pfau, W.D.: VAR analysis of the monetary transmission mechanism in Vietnam. Appl. Econ. Int. Dev. **9**(1), 165–179 (2009)
35. Huynh, T.C.H., Le, T.L., Le, T.H.M., Hoang, T.P.A.: Testing macro - variables which effecting stock market in Viet Nam. J. Sci. **3**(2), 70–78 (2014)

36. Jondeau, E., Le Bihan, H.: Testing for the new Keynesian phillips curve: additional international evidence. Econ. Model. **22**, 521–550 (2005)
37. Keating, J.W.: Identifying VAR models under rational expectations. J. Monet. Econ. **25**(3), 453–476 (1990)
38. Keating, J.W.: Macroeconomic modeling with asymmetric vector autoregressions. J. Macroecon. **22**(1), 1–28 (2000)
39. Keynes, J.M.: The General Theory of Employment Interest and Money. Macmillan Cambridge University Press, New York (1936)
40. Kilinc, M., Tunc, C.: Identification of Monetary Policy Shocks in Turkey: A Structural VAR Approach, 1423 (2014)
41. Kydland, F.E., Prescott, E.C.: Time to build and aggregate fluctuations. Econometrica **50**(6), 1345–1370 (1982)
42. Lees, K., Matheson, T., Smith, C.: Open economy DSGE-VAR forecasting and policy analysis: head to head with the RBNZ published forecasts. CAMA Working Paper 5/2007 (2007)
43. Leu, S.C.Y.: A new Keynesian SVAR model of the Australian economy. Econ. Model. **28**(1), 157–168 (2011)
44. Lubik, T.A., Schorfheide, F.: Do central banks respond to exchange rate movements? A structural investigation. J. Monet. Econ. **54**(4), 1069–1087 (2007)
45. Lucas, R.E.: Econometric Policy Evaluations: A Critique. Carnegie - Rochester Conference Series on Public Policy. Elsevier Science Publishers B. V (North - Holland), 1983 (1976)
46. McCallum, B., Nelson, E.: Nominal incom e targeting in an open economy optimising model. J. Monet. Econ. **43**, 553–578 (1999)
47. McCallum, B., Nelson, E.: Monetary policy for an open economy: an alternative framework with optimising agents and sticky prices. Oxf. Rev. Econ. Policy **16**, 74–91 (2000)
48. Mishra, P., Montiel, P.J., Spilimbergo, A.: Monetary transmission in low-income countries: effectiveness and policy implications. IMF Econ. Rev. **60**(2), 270–302 (2012)
49. Mishkin, F.S.: The Economics of Money, Banking and Financial Markets. Pearson Education, London (2012)
50. Mohanty, D.: Evidence on interest rate channel of monetary policy transmission in India. In: Second International Research Conference at the Reserve Bank of India, pp. 1–2 (2012)
51. Ncube, M., Ndou, E.: Monetary policy transmission, house prices and consumer spending in South Africa: An SVAR approach. African Development Bank Group Working Paper, 133 (2011)
52. Nguyen, T.K.: Monitoring interest rate of the State Bank of Viet Nam. J. Bank. (2017)
53. del Negro, M., Schorfheide, F.: Priors from general equylibrium models for VARs. Int. Econ. Rev. **45**(2), 643–673 (2004)
54. del Negro, M., Schorfheide, F.: Monetary policy analysis with potentially misspecified models. IMF Working paper, 475 (2005)
55. del Negro, M., Schorfheide, F.: How good is what you've got? DSGE-VAR as a toolkit for evaluating DSGE models. Econ. Rev. **91**(2), 21–37 (2006)
56. del Negro, M., Schorfheide, F.: Inflation dynamics in a small openeconomy model under inflation targeting: Some evidence from Chile. Federal Reserve Bank of New York Staff Reports, No. 329 (2009)

57. Nguyen, D.T.: The application dynamic stochastic general equilibrium in analyzing aggregate demand in Viet Nam economy. Bank. Sci. Traning Rev. **167**, 17–19 (2016)
58. Nguyen, D.T., Nguyen, H.C.: A small - opened foracasting model for Viet Nam. J. Econ. Dev. **28**(10), 5–39 (2017). T10/2017
59. Nguyen, K.Q.B., et al.: The impact of monetary policy on Viet Nam economy. Scientific research works, University of Economics HCMc (2013)
60. Nguyen, P.C.: The transmission regime of monetary through the financial asset price channel: emperical evidence in Viet Nam. J. Dev. Integr. **19**(29), 11–18 (2014)
61. Nguyen, T.L.H., Tran, D.D.: The study of inflation in Viet Nam through SVAR method. J. Dev. Integr. **10**(20), 32–38 (2013)
62. Pagan, A.R., Catão, L., Laxton, D.: Monetary transmission in an emerging targeter: the case of Brazil. IMF Working Papers, pp. 1–42 (2008)
63. Pham, T.A.: Using SVAR model to identificate the effect of monetary policy and forecast inflation in Viet Nam, National Economics Universtiy (2008)
64. Pham, T.A., Dinh, T.M.: The exchange rate policy: what is the selection for Viet Nam? J. Econ. Dev., 210 (2014)
65. Raghavan, M., Silvapulle, P.: Structural VAR approach to Malaysian monetary policy framework: evidence from the pre-and post-Asian crisis periods. In: New Zealand Association of Economics, NZAE Conference, pp. 1–32 (2008)
66. Roberts, J.: How well does the new Keynesian sticky price model fit the data? Board of Governors of the Federal Reserve System, Finance and Economics Discussion Paper 2001-13 (2001)
67. Romer, D.: Keynesian macroeconomics without the LM curve. J. Econ. Perspect. **14**(2), 149–169 (2000)
68. Runkle, D.: Vector autoregressions and reality. J. Bus. Econ. Stat. **5**, 437–442 (1987)
69. Spanos, A.: The simultaneous equations model revisited: statistical adequacy and identification. J. Econ. **44**, 87–105 (1990)
70. Taylor, J.B.: Discretion versus policy rules in practice. In: Carnegie-Rochester Conference Series on Public Policy, vol. 39, pp. 195–214 (1993)
71. Taylor, J.B.: The role of the exchange rate in monetary-policy rules. Am. Econ. Rev. **91**(2), 263–267 (2001)
72. Tram, T.X.H., Vo, X.V., Nguyen, P.C.: Monetary transmission through interest rate channel in Viet Nam. J. Dev. **283**, 42–67 (2014)
73. Tran, N.T., Nguyen, H.T.: The regime of monetary transmission in Viet Nam, approaching through SVAR model. J. Dev. Integr. **10**(20), 8–16 (2013)
74. Vinayagathasan, T.: Monetary policy and real economy: a structural VAR approach for Sri Lanaka. National Graduate Institute for Policy Studies, 7–22 (2013)
75. Walsh, C.E.: Teaching inflation targeting: an analysis for intermediate macro. J. Econ. Educ. **33**, 333–346 (2000)
76. Wang, K.M., Lee, Y.M.: Market volatility and retail interest rate pass - through. Econ. Model. **26**, 1270–1282 (2009)
77. Weerapana, A.: Intermediate macroeconomics without the IS-LM model. J. Econ. Educ. **34**(3), 241–262 (2003)
78. Woodford, M.: Interest and Prices: Foundations of a Theory of Monetary Policy. Princeton University Press, Princeton (2003)

The Use of Fractionally Autoregressive Integrated Moving Average for the Rainfall Forecasting

H. P. T. N. Silva[1(✉)], G. S. Dissanayake[2], and T. S. G. Peiris[3]

[1] Department of Social Statistics, Faculty of Humanities and Social Science, University of Sri Jayewardenepura, Nugegoda, Sri Lanka
thanuja@sjp.ac.lk

[2] University of Sydney, Sydney, Australia

[3] Department of Mathematics, Faculty of Engineering, University of Moratuwa, Moratuwa, Sri Lanka

Abstract. A study of rainfall pattern and its variability in South Asian countries is vital as those regions are frequently vulnerable to climate change. Models for rainfall have been developed with different degrees of accuracy, since this key climatic variable is of importance at local and global level. This study investigates the rainfall behaviour using the long memory approach. Since the observed series consists of an unbounded spectral density at zero frequency, a fractionally integrated auto regressive model (ARFIMA) is fitted to explore the pattern and characteristics of the weekly rainfall in the city of Colombo. The maximum likelihood estimation (MLE) method was utilized to obtain estimates for model parameters. To evaluate the suitability of the method for parameter estimation, a Monte Carlo simulation was done with various fractionally differenced parameter values. Model selection was done based on the minimum of the mean absolute error and validated by the forecasting performance that was evaluated using an independent sample. The experimental result yielded a good prediction accuracy with a best fitted long range dependency model and a coverage probability of 95% in terms of prediction intervals that resulted in closer nominal coverage.

Keywords: Rainfall · Fractional differencing · Long-memory Maximum likelihood estimators · Forecasting

1 Introduction

Modelling rainfall is a challenging task for researchers due to the high degree of uncertainty in atmospheric behaviour. Observational evidence indicates that the climate change has significantly affected global community at a different level. Climate vulnerabilities are expected to be critical in Sri Lanka in the various sectors as agriculture, fisheries, water, health, urban development, human settlement, economic infrastructure, biodiversity and ecosystem in the country [22].

V. Kreinovich et al. (Eds.): ECONVN 2019, SCI 809, pp. 567–580, 2019.
https://doi.org/10.1007/978-3-030-04200-4_40

Information on key climatic variable predictions allow to various stakeholders to prompt themselves for action in order to reduce adverse impacts and enhance positive effects of climatic variation. Rainfall is the one of the most important climatic variable to tropical country like Sri Lanka and this is the variable which give erratic variation at any time in the country. Sri Lanka receives rainfall during the year, with a mean annual rainfall varying from 900 mm in the dry zone to over 5000 mm in the wet zone. Annual rainfall pattern in many parts of Sri Lanka are bimodal and predominantly governed by a seasonally varying monsoon system. Sri Lanka needs to address climate change adaptation to ensure the economic development by the careful investigating of the information on rainfall pattern and its variability which resulting from the predictions of the best fitted rainfall models in various regions. Rainfall analysis is not only important for agricultural areas but also for the urban areas since those areas engage with many activities such as construction, industrial planning, urban traffic, sewer systems, health, rainwater harvesting and climate monitoring. Rainfall is the main source of the hydrological cycle and provides practical benefits through its analysis. Thus, modelling rainfall is one of the key requirements in the country, some of the researchers made attempt to analyse weekly rainfall in Sri Lanka using percentile bootstrap approach to identify the extreme rainfall events [24]. Another study was carried out by the Silva and Peiris [25] to identify the most likelihood time period to form the extreme rainfall events during the South west moons time span by fitting best probability distribution for the weekly rainfall percentiles. Since the Sri Lanka is a developing country which hasn't high technology to sensitive to some important climatic information with related to rainfall is one of the reason cause to low prediction accuracy. However, researchers made effort to model rainfall of the country with increasing degree of accuracy using different techniques. Silva and Peiris [26] discussed problems faced in modelling rainfall which showed positive skewed distribution with longer tail to the right. Rainfall is one of the most difficult variables of the hydrological cycle to understand and model due to its high variability in both space and time [13]. However, several modelling strategies have been applied for the forecasting of rainfall in different areas all over the world. Box-Jenkins autoregressive integrated moving average (ARIMA) model has been widely used for rainfall modelling ([11,20,29,30]). Some of the researchers have made attempts to model rainfall using artificial neural networks ([10,18]). However, very few studies on rainfall in context of long memory can be identified in literature. Granger and Joyeux [15] and Hosking [17] initially proposed a long memory class of models, known as the fractionally integrated autoregressive moving average (ARFIMA) process for stochastic processes. The model defined as ARFIMA (p, d, q) allows the parameter "d" to take fractional values for differencing. There is a fundamental change in the correlation structure of the ARFIMA model, when compared with the correlation structure of the conventional ARIMA model ([6]). According to Granger and Joyeux [15], the slowly decaying autocorrelation exhibited in long range dependency or long memory models differ from stationary ARIMA models that decay exponentially. Many researchers proposed different methods to estimate the fractional differencing parameter. Porter-Hudak and Geweke [14] proposed a method for estimating the long memory differencing parameters based on a simple linear regression of the log periodogram. An approximate maximum likelihood method for parameter "d" was proposed by Fox and Taqqu [12]. An exact maximum likelihood estimation method for differencing parameter was introduced by Sowell [27]. Chen et al. [6] developed a regression type estimator of "d" using lag window spectral density estimators. Number of studies were carried out by comparing various properties of the ARFIMA model based on the estimation method used for the fractionally differencing parameter. (See [2,3,7,16,23]). Dissanayake [9] established a methodology to find an optimal lag order of a standard long memory ARFIMA series within a short process time duration and applied the theory to Nile river data.

Though short memory models have been developed for rainfall still there is a noticeable gap modeling persistent rainfall in view of long memory. The main goal of this study is to fit an ARFIMA model for a weekly rainfall data series in the city of Colombo by capturing the long range dependency features. The paper outline is shaped as follows. In Sect. 2, the long memory ARFIMA model is introduced and some properties of the model are discussed. The model parameter estimation procedure is also described within the section. The results of the Monte Carlo simulation which was used to evaluate the suitability and reliability of the parameter estimation procedure is presented in Sect. 3. Section 4 provides brief details on prediction intervals for forecasting values relevant to the utilized series. The results of weekly rainfall modelling are presented in Sect. 5. Final section, comprises of the conclusion and proposed suggestions.

2 ARFIMA Long Range Dependency Model

ARFIMA is a natural extension of the Box and Jenkins model with non-integer values assigned for d. The ARFIMA (p, d, q) model of a process $\{Y_t\}_{t\in Z}$ is given by the formula

$$\phi(B)\nabla^d(Y_t-\mu)=\psi(B)\varepsilon_t \tag{1}$$

Where μ is the mean of the process, $\{\epsilon_t\}_{t\in Z}$ is a white noise process with zero mean and variance σ_ϵ^2. B is the backward shift operator such that $y_{t-n}=B^n y_t$, $\phi(B)$ and $\theta(B)$ are autoregressive and moving average polynomials of order p and q respectively.

$$\phi(B)=\sum_{i=1}^{p}\phi_i B^i \quad 1\le i\le p \tag{2}$$

$$\psi(B)=\sum_{j=1}^{P}\psi_j B^j \quad 1\le j\le q \tag{3}$$

where d is called as the long memory parameter and differencing operator ∇^d is defined as,

$$\nabla^d=(1-B)^d=\sum_{k=0}^{\infty}\binom{d}{k}(-B)^k \tag{4}$$

Where $\binom{d}{k} = \frac{\Gamma(1+d)}{\Gamma(1+k)\Gamma(1+d-k)}$.

If $d > -0.5$ then the process is invertible and if $d < 0.5$ then the process is stationary. Therefore $d \in (-\frac{1}{2}, \frac{1}{2})$ shows that the process is stationary and invertible. The spectral density function of the $\{Y_t\}_{t \in Z}$ is $f(\omega)$ that can be written as

$$f(\omega) = (2sin\frac{\omega}{2})^{-2d} \quad 0 < \omega \leq \pi \tag{5}$$

$f(\omega) \approx \omega^{-2d} \quad \omega \to 0$

The spectral density function $f(\omega)$ is unbounded when the frequency is near zero. Also, the autocovariance function and correlation function of the process can be expressed as follows

$$\gamma_k = \frac{(-1)^k(-2d)!}{(k-d)!(-k-d)!} \tag{6}$$

$$\rho_k = \frac{d(1+d)...(k-1+d)}{(1-d)(2-d)(3-d)...(k-d)}(k = 1, 2, 3, 4...) \tag{7}$$

Hosking (1981) showed that the auto correlation function of the process satisfies the expression $\rho_k \approx k^{2d-1}$ when $0 < d < 1/2$. Thus, the autocorrelation of the ARFIMA process decays hyperbolically to zero as $k \to \infty$ and in contrast, the auto correlation function of the ARIMA process has a exponential decay. The process with $d = 0$ reduces to a short memory ARMA model.

Let Z denote a series of "n" observations with mean μ and variance σ_Z^2. If the decay parameter is considered as α, then the natural fractional differencing parameter "d" can be written as $d = (1 - \alpha)/2$.
The log likelihood function of the Exact Gaussian can be written as

$$l(\alpha, \sigma_y^2) = -\frac{1}{2}(log \quad det(\Gamma_n) + Z'\Gamma_n^{-1}Z') \tag{8}$$

The arfima package (See [28]) in R optimized the log likelihood function and obtained the exact maximum likelihood estimators. Two algorithms namely Durbin-Levinson and Trench algorithms were utilized to maximize the likelihood and obtain optimal simulation and forecasting results.

3 Result of the Monte Carlo Simulation

A number of Monte Carlo experiments were carried out to evaluate the performance of the maximum likelihood method used for parameter estimation. The simulation was done based on various fractional differencing parameter values with 1000 replications. The four different series lengths ($n = 100$, $n = 200$, $n = 500$ and $n = 1000$) were considered for the simulation. The simulation results provided fractionally differenced parameter estimates and corresponding standard and mean square errors. Monte Carlo experiment was conducted on a simulated ARFIMA(0,d,0) series with parameter values: $d = 0.1$, $d = 0.15$, $d = 0.3$ and $d = 0.45$.

The simulation was carried out using the R programming Language (Version 3.4.2) utilizing a HP11(8 GB, 64 bit) computer. The standard errors of the estimates $SD(\hat{d})$ and mean square error of the estimates $MSE(\hat{d})$ can be expressed as;

$$SD(\hat{d}) = \sqrt{\sum_{r=1}^{R}(\hat{d}_r - \hat{d})/R} \qquad MSE(\hat{d}) = \sum_{r=1}^{R}(\hat{d}_r - d)^2/R$$

Where $\hat{d}_r$ is the MLE of d for the rth replication. The value R denotes the number of replications ($R = 1000$ for all tabulated simulation results of this paper). Tables 1, 2, 3 and 4 present the average of the estimated d, corresponding standard error and MSE of the estimator.

According to the results in Tables 1, 2, 3 and 4, the performance of the maximum likelihood estimator is reasonably accurate. It can be clearly seen that the parameter bias has decreased with the increase in sample size. Furthermore,

Table 1. MLE of d for a generating process of ARFIMA(0,d,0) with d = 0.1. The results are based on 1000 Monte Carlo replications

n	$\hat{d}$	SD($\hat{d}$)	MSE($\hat{d}$)
100	0.0517	0.0912	0.0106
200	0.0748	0.0626	0.0045
500	0.0885	0.0367	0.0014
1000	0.0949	0.0254	0.0006

Table 2. MLE of d for a generating process of ARFIMA(0,d,0) with d = 0.15. The results are based on 1000 Monte Carlo replications

n	$\hat{d}$	SD($\hat{d}$)	MSE($\hat{d}$)
100	0.1048	0.0915	0.0104
200	0.1265	0.0593	0.0040
500	0.1408	0.0367	0.0014
1000	0.1456	0.0254	0.0006

Table 3. MLE of d for a generating process of ARFIMA(0,d,0) with d = 0.3. The results are based on 1000 Monte Carlo replications

n	$\hat{d}$	SD($\hat{d}$)	MSE($\hat{d}$)
100	0.2493	0.0877	0.0102
200	0.2726	0.0575	0.0040
500	0.2892	0.0362	0.0014
1000	0.2947	0.0251	0.0006

Table 4. MLE of d for a generating process of ARFIMA(0,d,0) with d = 0.45. The results are based on 1000 Monte Carlo replications

n	$\hat{d}$	SD($\hat{d}$)	MSE($\hat{d}$)
100	0.3774	0.0695	0.0101
200	0.4079	0.0477	0.0040
500	0.4310	0.0314	0.0013
1000	0.4435	0.0270	0.0007

the results provide evidence that the parameters become consistent with the increase in series length. As we expected the standard deviation and the MSE of the estimators have decreased with the increase in series length.

4 Forecast and Prediction Intervals

Forecasts are obtained based on the best fitted long memory model. However, predicting of future values along with their prediction intervals become more beneficial in long memory time series analysis. The lower (L) and upper (U) boundaries covering the forecast values with known probability are simply called prediction intervals of the form [L, U]. A detailed review of approaches in calculating interval forecast using time series was described in Chatfield [5]. Charles et al. [4] made an effort to make prediction intervals to forecast US core inflation values that provided a unique fractional model. Prediction intervals were utilized to forecast tourism demand by Chu [8]. Zhou et al. [31] suggested a prediction interval method to predict aggregates of future values derived from a long memory model. A new bootstrap method for autoregressive models was proposed by Hwang and Shin [19]. Ali et al. [1] suggested a Sieve bootstrap approach to construct intervals for a long memory model. Prediction interval approach was utilized to measure the uncertainty about long-run predictions by Muller and Watson [21].

5 Application

Sri Lanka is a tropical country in South Asian region located at the latitudes of 5° 55 N and 9° 51 N and the longitudes of 79° 41 E and 81° 53 E with an area of 65610 km^2 and the Colombo city is the commercial capital of Sri Lanka. Daily rainfall data of Colombo were collected from 1990 to 2015 from the Department of Meteorology, Sri Lanka for this analysis. The daily rainfall (mm) data has been converted into weekly rainfall by dividing a year into 52 weeks such that week 1 corresponds to 1–7 January, Week 2 corresponds to 8–14 January and so on. The data during the time span from 1990 to 2014 was used to build the model while the rest was used for model validation. To examine the temporal

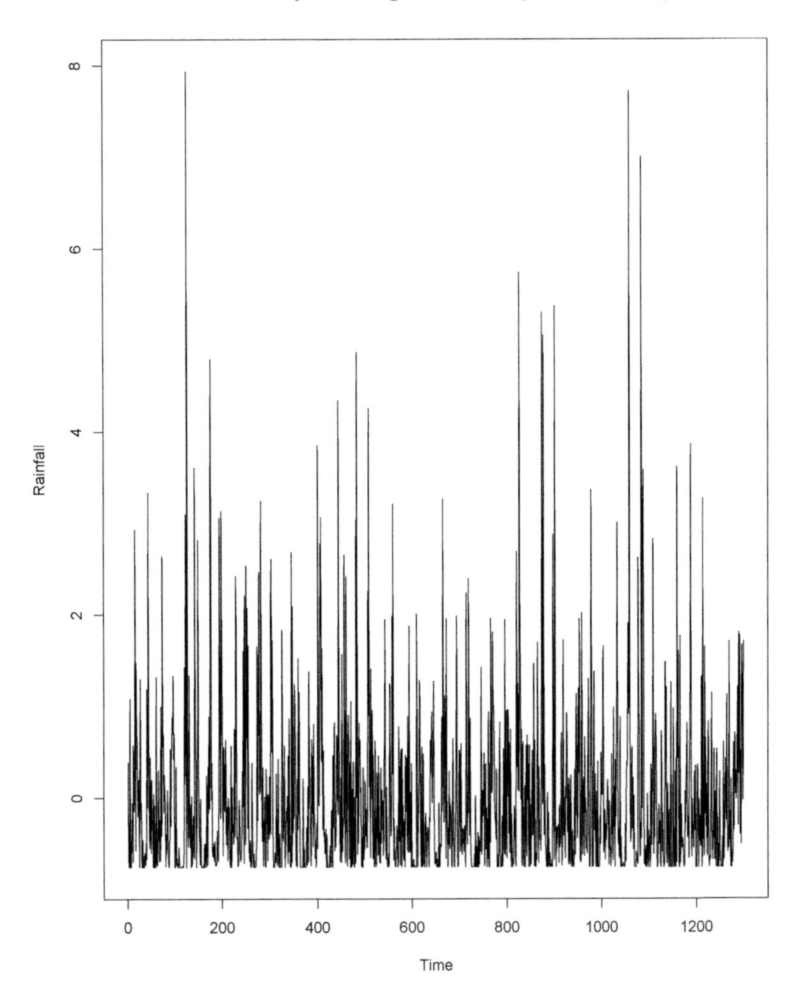

Fig. 1. Time series plot of weekly rainfall series from 1990 to 2014

variability of the rainfall series, time series plots were taken and presented in Fig. 1.

The time series plot explores the random behaviour of weekly rainfall during the considered time span from 1990 to 2014. In order to identify the correlation structure of the observed series, the autocorrelation and partial auto correlation plots were taken and those results are shown in Figs. 2 and 3 respectively.

In order to study the long memory features of the weekly rainfall series, the periodgram was obtained and presented in Fig. 4. The maximum spectrum density is 0.0385185 given at a frequency which is very close to zero. Based on those characteristics the series displays long memory. Thus, we conclude that the ARFIMA standard long memory model may be suitable for the observed weekly rainfall series. Long range correlation of observed data were considered

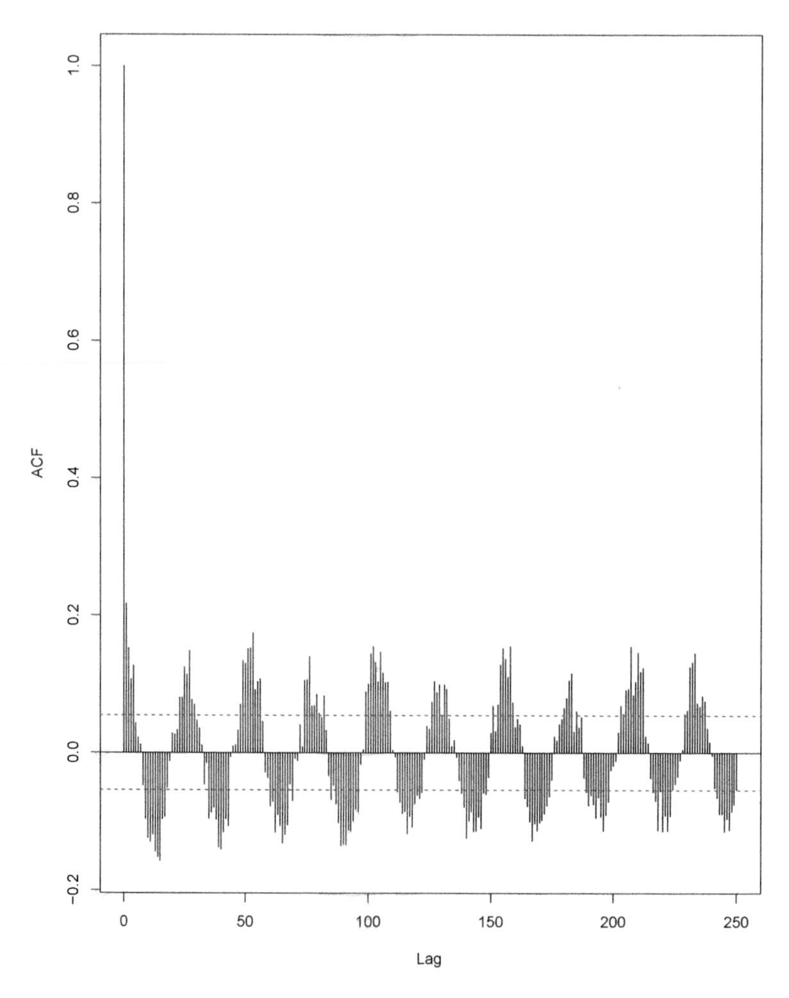

Fig. 2. Autocorrelation plot of the series from 1990 to 2014

in long memory modelling. Various ARFIMA models were fitted for the data set that vary from 1990 to 2014 (series length = 1300).

Those fitted models were employed to predict the weekly rainfall during the time span from 2014 to 2015 and best fitted model is selected with the minimum mean absolute error (MAE). The MAE can be written as,

$$MAE = \frac{1}{n}\sum_{i=1}^{n} |e_i|$$

Where e_i is the forecasting error and n is the length of the forecasting series.

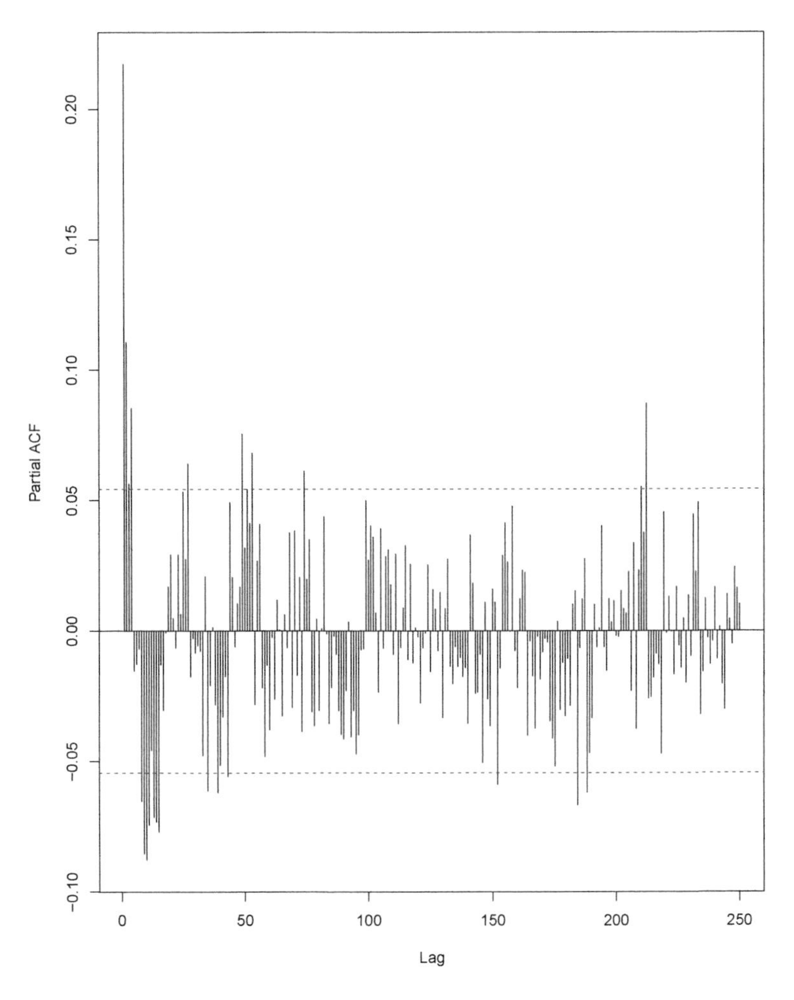

Fig. 3. Partial autocorrelation plot of the series from 1990 to 2014

The best fitted model and the corresponding parameter estimates are presented in Table 5. The ARFIMA (4,0,4) model was found to be the best fit for the weekly rainfall series returning the smallest MAE.

All model parameters except the constant are significant at the 0.05 level of significance. The residual analysis of the fitted model was performed and found the uncorrelated at a 5% level of significance. Furthermore, the model was tested for weekly rainfall data in 2015 and the result is presented in Table 6. Figure 5 illustrates the weekly rainfall over the year 2015 along with the predicted estimates.

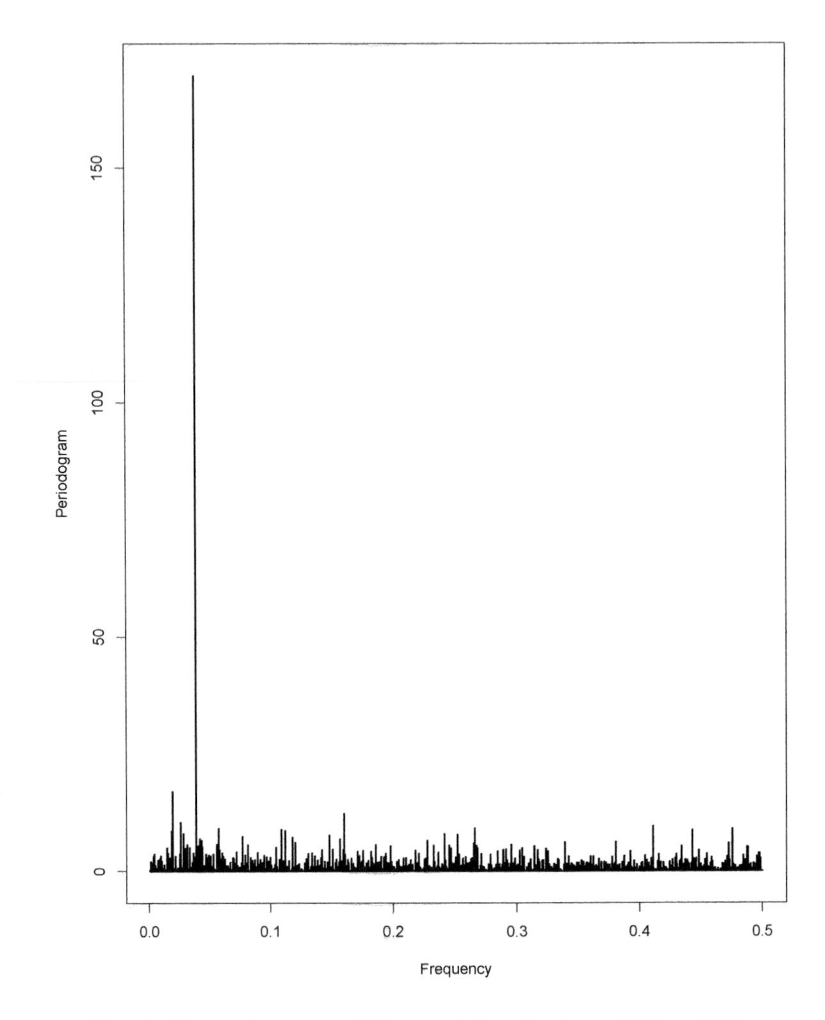

Fig. 4. The periodgram of the rainfall series from 1990 to 2014

Table 5. Fitted model for the weekly rainfall series ARFIMA (4,0,4) with p = 4, q = 4, d = 0.05792421

Coefficients	ϕ_1	ϕ_2	ϕ_3	ϕ_4	θ_1
Estimate	1.2059	−0.2493	0.5765	−0.6752	1.1243
Standard error	0.0242	0.0454	6.324e-07	6.324e-07	0.0231-Correct value (CV)
Z-value	4.9768e01	5.4903	9.1153e05	−1.0676e06	4.8638e01
Pr(>|Z|)	0.0000	0.0005	0.0000	0.0000	0.0000

Table 5. (*continued*)

Coefficients	θ_2	θ_3	θ_4	Constant	d
Estimate	−0.1131	0.5220	−0.6743	−0.0163	0.0579
Standard error	0.0365 (CV)	0.0354 (CV)	0.0215	0.0380	0.0276
Z-value	−3.0992	1.4735e01	−3.1363e01	−4.2907e-01	2.0950
Pr(>|Z|)	0.0019	0.0000	0.0000	0.6678	0.0361

Table 6. Absolute Forecast Error for independent sample (2015)

Absolute forecasting error in mm	ARFIMA number of weeks percentage
0–10	10(19.2)
11–15	6(11.5)
16–20	6(11.5)
21–25	4(7.7)
26–30	6(11.5)
31–35	1(1.9)
36–40	4(7.7)
41–45	1(1.9)
46–50	2(4.0)
More than 50	12(23.1)

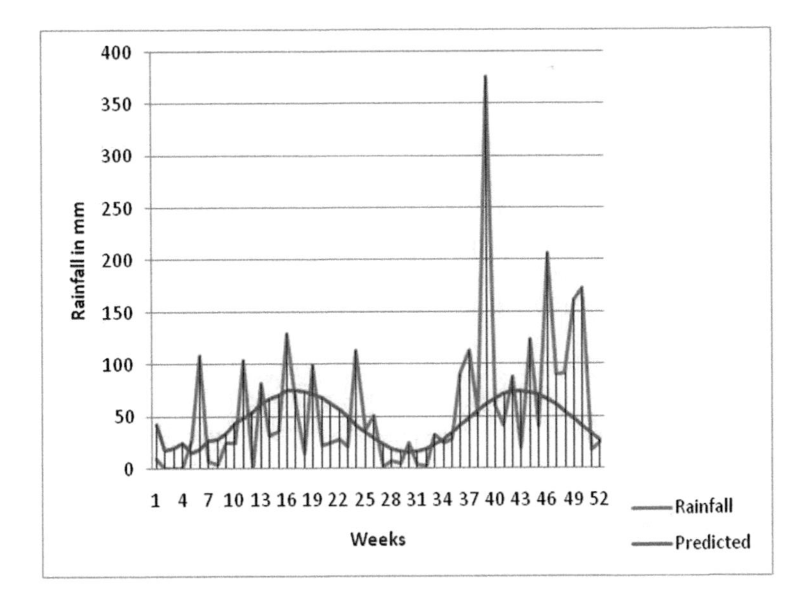

Fig. 5. Forecasted and actual weekly rainfall in 2015

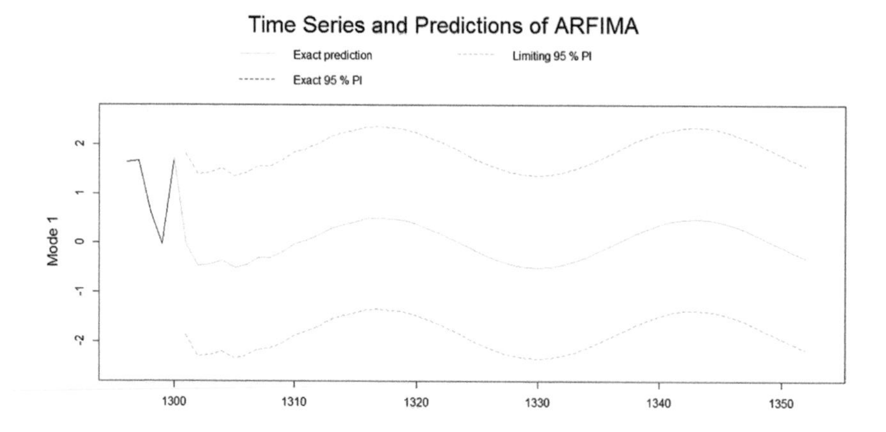

Fig. 6. Prediction intervals for forecasted rainfall values in 2015

According to Fig. 5, it can be seen that the predicted values are in considerable good agreement with the actual rainfall values. The result of the 95% prediction interval also provides encouraging prediction accuracy with a 93.23% coverage probability (Fig. 6).

6 Conclusion

Observed rainfall series illustrates long memory features with an unbounded spectral density. Therefore a standard long memory ARFIMA model was fitted to capture the rainfall pattern and its variability. The Monte Carlo simulation results prove the accuracy of the maximum likelihood method used to estimate the parameters of the model. Furthermore, it is noticed that the parameter bias has decreased and the parameters become consistent with the increase of the simulated series length. ARFIMA(4,0.0579,4) model was found to be the best fitted model that provided a minimum MAE. The out of sample prediction values give good conformity with the actual weekly rainfall in 2015. The 95% prediction intervals also give a promising result to capture the real dynamics of the persistent rainfall. For future work it is suggested that prediction intervals using the bootstrap re-sampling approach may forecast estimates with a higher degree of accuracy.

Acknowledgement. This study was partially funded by the University Research Grant, University of Sri Jayewardenepura, Sri Lanka under Grant (ASP/01/RE/HSS/2016/75).

References

1. Alamgir, A.A., Kalil, U., Khan, S.A., Khan, D.M.: A Sieve bootstrap approach to constructing prediction intervals for long memory time series. Res. J. Recent Sci. **4**(7), 93–99 (2015)

2. Beran, J., Feng, Y., Ghosh, S., Kulik, R.: Long memory Processes Probabilistic Properties and Statistic properties and Statistical Methods. Springer, Heidelberg (2013)
3. Chan, N.H., Palma, W.: Estimation of long-memory time series models. Adv. Econ. **20**, 89–121 (2006)
4. Charles, S.B., Franses, P.H., Ooms, M.: Inflation, forecast intervals and long memory regression models. Int. J. Forecast. **18**, 243–264 (2002)
5. Chatfield, C.: Calculating interval forecasts. J. Bus. Econ. Stat. **11**(2), 121–135 (1993)
6. Chen, G., Abraham, B., Peiris, S.: Lag window estimation of the degree of differencing in fractionally integrated time series models. J. Time Ser. Anal. **15**(5), 473–487 (1994)
7. Cheung, Y., Diebold, F.X.: On maximum likelihood estimation of the differencing parameter of fractionally integrated noise with unknown mean. J. Econ. **62**, 301–316 (1994)
8. Chu, F.L.: A fractionally integrated autoregressive moving average approach to forecast tourism demand. Tour. Manag. **29**, 79–88 (2008)
9. Dissanayake, G.S.: Rapid optimal lag order detection and parameter estimation of standard long memory time series. In: Causal Inference in Econometrics, pp. 17–28. Springer, Cham (2016)
10. Dubey, A.D.: Artificial neural network models for rainfall prediction in Pondicherry. Int. J. Comput. Appl. **120**(3), 30–35 (2015)
11. Eni, D., Adeyeye, F.J.: Seasonal ARIMA modelling and forecasting of rainfall in Warri Town, Nigeria. J. Geosci. Environ. Protection **3**, 91–98 (2015)
12. Fox, R., Taqqu, M.S.: Large sample properties of parameter estimates for strongly dependent stationary Gaussian time series. Ann. Stat. **14**(2), 517–532 (1986)
13. French, M.N., Krajewski, W.F., Cuykendall, R.R.: Rainfall forecasting in space and time using a neural network. J. Hydrol. **137**, 1–31 (1992)
14. Geweke, J., Hudak, S.P.: The estimation and application of long memory time series models. J. Time Ser. Anal. **4**, 221–238 (1983)
15. Granger, C.W.J., Joyeux, R.: An introduction to long-memory time series models and fractional differencing. J. Time Ser. Anal. **1**, 15–29 (1980)
16. Hauser, M.A.: Maximum likelihood estimators for ARMA and ARFIMA models: a Monte Carlo study. J. Stat. Plan. Inference **80**, 229–255 (1999)
17. Hosking, J.R.M.: Fractional differencing. Biometrika **68**, 165–176 (1981)
18. Hung, N.Q., Babel, M.S., Weesakul, S., Tripathi, N.K.: An artificial neural network model for rainfall forecasting in Bangkok, Thailand. Hydrol. Earth Syst. Sci. **13**, 1413–1425 (2009)
19. Hwang, E., Shin, D.W.: New bootstrap method for autoregressive models. Commun. Stat. Appl. Methods **20**(1), 85–96 (2013)
20. Momani, P.E.N.M.: Time series analysis model for rainfall data in Jordan: case study for using time series analysis. Am. J. Environ. Sci. **5**(5), 559–604 (2009)
21. Muller, U.K., Watson, M.W.: Measuring uncertainty about long-run predictions. Rev. Econ. Stud. **83**, 1711–1740 (2016)
22. National Climate Change Adaptation Strategy for Sri Lanka 2011 to 2016. Ministry of Environment (2010)
23. Palma, W.: Long Memory Time Series Theory and Methods. John Wiley and Sons, New Jersey (2007)
24. Silva, H.P.T.N., Peiris, T.S.G.: Analysis of weekly rainfall using percentile bootstrap approach. Int. J. Ecol. Dev. **32**(3), 97–106 (2017)

25. Silva, H.P.T.N., Peiris, T.S.G.: Statistical modeling of weekly rainfall: a case study in Colombo city in Sri Lanka. In: Proceedings of the 3rd Moratuwa Engineering Research Conference (MER Con), Sri Lanka, 29–31 May, pp. 241–246. IEEE (2017)
26. Silva, H.P.T.N., Peiris, T.S.G.: Accurate confidence intervals for Weibull percentiles using bootstrap calibration: a case study of weekly rainfall in Sri Lanka. Int. J. Ecol. Econ. Stat. **39**(3), 67–76 (2018)
27. Sowell, F.: Maximum likelihood estimation of stationary univariate fractionally integrated time series models. J. Econ. **53**, 165–188 (1992)
28. Veenstra, J.Q., McLeod, A.I.: Persistence and Anti-persistence: Theory and Software (Ph.D. thesis) (2013)
29. Wang, S., Feng, J., Liu, G.: Application of seasonal time series model in the precipitation forecast. Math. Comput. Model. **58**, 677–683 (2013)
30. Zakaria, S., Al-Ansari, N., Knutsson, S., Al-Badrany, T.: ARIMA models for weekly rainfall in the semi-arid Sinjar District at Iraq. J. Earth Sci. Geotech. Eng. **2**(3), 25–55 (2012)
31. Zhou, Z., Xu, Z., Wu, W.B.: Long-term prediction intervals of time series. IEEE Trans. Inf. Theory **56**(3), 1436–1446 (2010)

Detection of Structural Changes Without Using P Values

Chon Van Le(✉)

School of Business, International University - VNU HCMC,
Ho Chi Minh City, Vietnam
lvchon@hcmiu.edu.vn

Abstract. The econometrics of structural change has evolved a lot since the classical Chow [5] test. Several approaches have been proposed to find the unknown breakdate. But they could be invalid as it is claimed that the P value has been misused for the past one hundred years. This paper reviews other methods of detecting structural changes. Specifically, the Bayes factor can be used for a pairwise comparison of competing models. The Markov-switching model is an effective way of dealing with a number of discrete regimes. But if the regime is a continuous normal variable, the Kalman filter is a better resolution.

Keywords: Structural changes · Bayes factor
Markov-switching model · Kalman filter

1 Introduction[1]

The econometrics of structural change has received a lot of attention among scientists to search for standard methods to identify structural breaks. The first and classical test for structural change was proposed by Chow [5]. Under his testing procedure, a sample is divided into two subperiods, for each of which the parameters are estimated, and then the equality of the two sets of parameters is tested using an F statistic. However, the Chow test requires that the breakdate be known in advance. Otherwise, the test can be false because the candidate breakdate is endogenous (Hansen, [9]). Several options have been suggested to find the unknown breakdate. Quandt [17] recommended picking up the largest Chow statistic among all potential breakdates. Bai and Perron [2] introduced a sequential method for multiple structural changes. Chen [4] developed a two-stage local linear estimator for time-varying models with endogenous independent variables, and a Wald-type test for structural changes.

It seems that these efforts did not lead to a fully adequate solution of the problem as Goodman [6] and Nuzzo [16], among other critics, showed that the

[1] I am very grateful to Dr. Hung T. Nguyen for his valuable suggestions in this paper. Anonymous referees' comments are appreciated. All errors therein are mine.

V. Kreinovich et al. (Eds.): ECONVN 2019, SCI 809, pp. 581–595, 2019.
https://doi.org/10.1007/978-3-030-04200-4_41

P value has been misused for roughly the past one hundred years. It summarizes the data under a specific null hypothesis, that is, contains some empirical evidence, but not sufficient to make conclusive statements about the underlying population. In response, many researchers have recommended reducing the default P value threshold for statistical significance (or the alpha level) to more conservative values to decrease the risk of committing a Type I error. Initially, Melton [15] suggested using an alpha level of 0.01 instead of the typical 0.05 and recently Benjamin et al. [3] favored 0.005, which is argued to be able to avoid wrong conclusions in significance testing. Nevertheless, Trafimow et al. [19] demonstrated that such adjustments cannot mitigate the problems with significance tests. The journal Basic and Applied Social Psychology even published an Editorial [18], which officially bans the null hypothesis significance testing procedure (NHSTP). It asserts that the NHSTP is invalid and thus has to be removed from manuscripts prior to publication in this journal.

Since hypothesis testing is ultimately to secure an appropriate statistical model, there are several alternative methods to achieve this objective. If model selection is based on model probabilities, the Bayesian approach is more appropriate. Given a prior distribution of the model parameters and a sampling density structure, the Bayes factor allows us to make a pairwise comparison of competing models, including those that allow for structural changes.

In most cases breakdates are unknown or a change in regime cannot be considered as the outcome of a deterministic event. The change in regime should be instead a random variable (Hamilton, [8]). In addition, if the series has changed in the past, it can change again in the future, so forecasting should take this possibility into account. Therefore, a comprehensive model should describe the probability law that governs the shift from one regime to another. Such a model is the Markov-switching model, introduced by Hamilton [7], in which the latent state variable administering the regime shift follows a Markov chain. Another method is the Kalman filter, developed by Kalman [11], which updates knowledge of the state variable recursively on the availability of new data. This paper is to review methods applied to detect structural changes without using P values, namely the Bayes factor in Sect. 2, the Markov-switching model in Sect. 3, and the Kalman filter in Sect. 4. Conclusions follow in Sect. 5.

2 The Bayes Factor[2]

Suppose we wish to select a model from q candidate models $M_1, \ldots, M_q$. Assume that each model M_k is characterized by a probability density $f_k(\boldsymbol{x}|\boldsymbol{\theta}_k, M_k)$, where $\boldsymbol{\theta}_k$ is a $p_k \times 1$ vector of unknown parameters and $\boldsymbol{\theta}_k \in \Theta_k \subset R^{p_k}$, where Θ_k is the parameter space. Let $\pi_k(\boldsymbol{\theta}_k|M_k)$ denote the prior distribution of the parameter vector $\boldsymbol{\theta}_k$ under model M_k. From Bayes' theorem, the posterior probability of model M_k given a data set of n observations $\boldsymbol{D} = \{\boldsymbol{x}_1, \ldots, \boldsymbol{x}_n\}$ is

[2] This section is to a large extent based on Ando [1].

$$\Pr(M_k|\boldsymbol{D}) = \frac{\Pr(M_k)\Pr(\boldsymbol{D}|M_k)}{\sum_{j=1}^{q}\Pr(M_j)\Pr(\boldsymbol{D}|M_j)} = \frac{\Pr(M_k)\int f_k(\boldsymbol{D}|\boldsymbol{\theta}_k, M_k)\pi_k(\boldsymbol{\theta}_k|M_k)d\boldsymbol{\theta}_k}{\sum_{j=1}^{q}\Pr(M_j)\int f_j(\boldsymbol{D}|\boldsymbol{\theta}_j, M_j)\pi_j(\boldsymbol{\theta}_j|M_j)d\boldsymbol{\theta}_j}, \tag{1}$$

where $\Pr(M_k)$ is the prior probability of model M_k and $f_k(\boldsymbol{D}|\boldsymbol{\theta}_k, M_k)$ the likelihood function. Given an initial view of model uncertainty via the prior probabilities $\Pr(M_k)$ and $\pi_k(\boldsymbol{\theta}_k|M_k)$ for model M_k, we update our view of model uncertainty via the posterior model probability $\Pr(M_k|\boldsymbol{D})$ after having observed the data (Ando, [1]).

The fundamental data-dependent term $\Pr(\boldsymbol{D}|M_k)$ is the marginal likelihood of the data $\boldsymbol{D}$ under model M_k, representing the probability that the data are generated under the assumption of the model. The Bayes factor compares two models, for example M_k and M_l, based on their marginal likelihoods. It is defined as

$$\text{Bayes factor}(M_k, M_l) \equiv \frac{\Pr(\boldsymbol{D}|M_k)}{\Pr(\boldsymbol{D}|M_l)} = \frac{\int f_k(\boldsymbol{D}|\boldsymbol{\theta}_k, M_k)\pi_k(\boldsymbol{\theta}_k|M_k)d\boldsymbol{\theta}_k}{\int f_l(\boldsymbol{D}|\boldsymbol{\theta}_l, M_l)\pi_l(\boldsymbol{\theta}_l|M_l)d\boldsymbol{\theta}_l}. \tag{2}$$

The Bayes factor provides the evidence for model M_k against model M_l and chooses the model with largest marginal likelihood. Since

$$\frac{\Pr(M_k|\boldsymbol{D})}{\Pr(M_l|\boldsymbol{D})} = \frac{\Pr(\boldsymbol{D}|M_k)}{\Pr(\boldsymbol{D}|M_l)} \times \frac{\Pr(M_k)}{\Pr(M_l)},$$

or Posterior odds(M_k, M_l) = Bayes factor(M_k, M_l) × Prior odds(M_k, M_l), then the Bayes factor is the ratio of posterior odds and prior odds of the two models:

$$\text{Bayes factor}(M_k, M_l) = \frac{\text{Posterior odds}(M_k, M_l)}{\text{Prior odds}(M_k, M_l)}. \tag{3}$$

When the two models are equally probable a priori or $\Pr(M_k) = \Pr(M_l)$, the Bayes factor reduces to the ratio of posterior probabilities of M_k and M_l. Moreover, if we use the likelihood corresponding to the maximum likelihood estimates of the parameters for each model instead of the integral in Eq. (2), the Bayes factor becomes a classical likelihood-ratio test.

A value of the Bayes factor greater than 1 means that the data support model M_k more than M_l and vice versa. Jeffreys [10] and Kass and Raftery [12] provided tables for intepretation of the Bayes factor (Tables 1 and 2).

Regarding the location of the breakdates, it is common in the Bayesian econometric literature to adopt a diffuse prior such that equal weight is given to every possible breakdate. If there is a single breakdate, τ_1, the class of discrete uniform distributions simply sets

$$\pi(\tau_1) = \frac{1}{T-1}, \qquad \tau_1 = 1, \ldots, T-1. \tag{4}$$

Table 1. Jeffreys' scale for intepretation of the Bayes factor

Bayes factor	Evidence for M_k
$< 10^0$	Negative (supports M_l)
10^0 to $10^{1/2}$	Barely worth mentioning
$10^{1/2}$ to 10^1	Substantial
10^1 to $10^{3/2}$	Strong
$10^{3/2}$ to 10^2	Very strong
$> 10^2$	Decisive

Source: Jeffreys [10].

Table 2. Kass and Raftery's scale for intepretation of the Bayes factor

2ln(Bayes factor)	Bayes factor	Evidence for M_k
0 to 2	1 to 3	Not worth more than a bare mention
2 to 6	3 to 20	Positive
6 to 10	20 to 150	Strong
> 10	> 150	Very strong

Source: Kass and Raftery [12].

The prior is noninformative in the sense of favoring all candidate breakdates equally. However, Koop and Potter [13] indicated that this approach is no longer noninformative where there is more than one breakdate. In the case of, for example, two breakdates, Eq. (4) would be extended to represent the prior as

$$\pi(\tau_1, \tau_2) = \pi(\tau_1)\pi(\tau_2|\tau_1), \tag{5}$$

where

$$\pi(\tau_1) = \frac{1}{T-2}, \qquad \tau_1 = 1, \ldots, T-2, \tag{6}$$

$$\pi(\tau_2|\tau_1) = \frac{1}{T-\tau_1-1}, \qquad \tau_2 = \tau_1+1, \ldots, T-1. \tag{7}$$

The marginal prior for τ_2 calculated by summing the joint probability distribution over τ_1:

$$\pi(\tau_2 = k) = \frac{1}{T-2}\sum_{j=2}^{k}\frac{1}{T-j}, \qquad k = 2, \ldots, T-1,$$

gives more weight to breakdates near the end of the sample.

To solve this problem, Koop and Potter [13] suggested replacing Eq. (7) by

$$\pi(\tau_2|\tau_1) = \frac{1}{T-2}, \qquad \tau_2 = \tau_1 + 1, \ldots, T + \tau_1 - 2. \tag{8}$$

Then both $\pi(\tau_1)$ and $\pi(\tau_2|\tau_1)$ are uniform and have the same number of points of support. The prior probability does not pile up near the end of the sample. In addition, while the prior given by Eqs. (6) and (7) imposes strictly two breakdates in the sample, the prior given by (6) and (8) allows one breakdate to occur out of the sample. This property is highly desirable for the case where the number of breakdates is unknown.

3 Markov-Switching Model[3]

Since it is preferable to treat the number of breakdates as unknown and the change in regime as a random variable and sometimes temporary, Hamilton [8] argued that a probability law governing the change in regime should be included in the model. Let $\mathbf{y}_t$ denote an $(n \times 1)$ vector of observed endogenous variables, and $\mathbf{x}_t$ a $(k \times 1)$ vector of observed exogenous variables. Let $F_t = (\mathbf{y}_t', \mathbf{y}_{t-1}', \ldots, \mathbf{x}_t', \mathbf{x}_{t-1}', \ldots)'$ be a vector containing all information available up to date t. Let s_t denote the state or regime that the time series process was in at date t. It is assumed that s_t is a random variable that takes an integer value $\{1, 2, \ldots, N\}$, implying that there are N different regimes. If the process is in regime j at date t, the conditional density of $\mathbf{y}_t$ is

$$f(\mathbf{y}_t|s_t = j, \mathbf{x}_t, F_{t-1}; \boldsymbol{\theta}), \tag{9}$$

where $\boldsymbol{\theta}$ is a vector of parameters that characterizes the conditional density. Under N different regimes, there are N different densities which are collected in an $(N \times 1)$ vector $\boldsymbol{\delta}_t = [f(\mathbf{y}_t|s_t = 1, \mathbf{x}_t, F_{t-1}; \boldsymbol{\theta}), \ldots, f(\mathbf{y}_t|s_t = N, \mathbf{x}_t, F_{t-1}; \boldsymbol{\theta})]'$.

Assume that s_t evolves over time, following a Markov chain which does not depend on current or past $\mathbf{x}_t$ and past $\mathbf{y}_t$:

$$\Pr(s_t = j|s_{t-1} = i, s_{t-2} = l, \ldots, \mathbf{x}_t, F_{t-1}) = \Pr(s_t = j|s_{t-1} = i) = p_{ij}. \tag{10}$$

Equation (10) specifies that the probability that the process at date t is in regime j depends on only the regime at date $t-1$. Transition probabilities $\{p_{ij}\}_{i,j=1,\ldots,N}$ can be collected in an $(N \times N)$ transition matrix $\mathbf{P}$:

$$\mathbf{P} = \begin{bmatrix} p_{11} & p_{21} & \cdots & p_{N1} \\ p_{12} & p_{22} & \cdots & p_{N2} \\ \vdots & \vdots & \ddots & \vdots \\ p_{1N} & p_{2N} & \cdots & p_{NN} \end{bmatrix}, \quad \text{where} \sum_{j=1}^{N} p_{ij} = 1, \quad i = 1, \ldots, N. \tag{11}$$

[3] This section is to a large extent based on Hamilton [8].

A Markov chain can be represented by letting $\boldsymbol{\psi}_t$ be a random $(N \times 1)$ vector whose jth element equals unity if $s_t = j$ and zero otherwise. Hence,

$$\boldsymbol{\psi}_t = \begin{cases} (1, 0, 0, \ldots, 0)' & \text{if } s_t = 1, \\ \vdots & \vdots \\ (0, 0, 0, \ldots, 1)' & \text{if } s_t = N. \end{cases}$$

If $s_t = i$, the jth element of $\boldsymbol{\psi}_{t+1}$ is a random variable that equals unity with probability p_{ij} and zero with probability $1-p_{ij}$. Its expectation is p_{ij}. Therefore, the conditional expectation of $\boldsymbol{\psi}_{t+1}$ given $s_t = i$ is

$$E(\boldsymbol{\psi}_{t+1}|s_t = i) = [p_{i1}, \ldots, p_{iN}]' = \mathbf{P}\boldsymbol{\psi}_t,$$

which is the ith column of $\mathbf{P}$. The above equation can be rewritten as

$$E(\boldsymbol{\psi}_{t+1}|\boldsymbol{\psi}_t) = \mathbf{P}\boldsymbol{\psi}_t.$$

Moreover, the Markov property in Eq. (10) indicates that

$$E(\boldsymbol{\psi}_{t+1}|\boldsymbol{\psi}_t, \boldsymbol{\psi}_{t-1}, \ldots) = \mathbf{P}\boldsymbol{\psi}_t.$$

Then a Markov chain can be expressed in the form

$$\boldsymbol{\psi}_{t+1} = E(\boldsymbol{\psi}_{t+1}|\boldsymbol{\psi}_t, \boldsymbol{\psi}_{t-1}, \ldots) + \mathbf{v}_{t+1} = \mathbf{P}\boldsymbol{\psi}_t + \mathbf{v}_{t+1}, \tag{12}$$

where $\mathbf{v}_{t+1}$ denotes the innovation at date $t+1$ which is a martingale difference sequence. Equation (12) implies that

$$\boldsymbol{\psi}_{t+m} = \mathbf{v}_{t+m} + \mathbf{P}\mathbf{v}_{t+m-1} + \mathbf{P}^2\mathbf{v}_{t+m-2} + \cdots + \mathbf{P}^{m-1}\mathbf{v}_{t+1} + \mathbf{P}^m\boldsymbol{\psi}_t. \tag{13}$$

As we do not know which regime the process was in at date t, we can just make a probabilistic inference about it. Let $\Pr(s_t = j|F_t; \boldsymbol{\beta})$ denote our inference about the value of s_t based on information available up to date t and knowledge of the parameters $\boldsymbol{\theta}$ and transition probabilities p_{ij} that are gathered in a vector $\boldsymbol{\beta}$. The inference is a conditional probability that the tth observation was generated by regime j. These conditional probabilities $\Pr(s_t = j|F_t; \boldsymbol{\beta})_{j=1,\ldots,N}$ are stacked in an $(N \times 1)$ vector $\hat{\boldsymbol{\psi}}_{t|t} = [\Pr(s_t = 1|F_t; \boldsymbol{\beta}), \ldots, \Pr(s_t = N|F_t; \boldsymbol{\beta})]'$. Then $\hat{\boldsymbol{\psi}}_{t+1|t}$ contains forecasts of how likely the process is to be in regime j at date $t+1$ given information available up to date t.

We can find the optimal inference and forecast for date t by iterating on the equations:

$$\hat{\boldsymbol{\psi}}_{t|t} = \frac{\hat{\boldsymbol{\psi}}_{t|t-1} \odot \boldsymbol{\delta}_t}{\mathbf{1}'(\hat{\boldsymbol{\psi}}_{t|t-1} \odot \boldsymbol{\delta}_t)}, \tag{14}$$

$$\hat{\boldsymbol{\psi}}_{t+1|t} = \mathbf{P}\hat{\boldsymbol{\psi}}_{t|t}, \tag{15}$$

where $\mathbf{1}$ is an $(N \times 1)$ vector of 1s, and the symbol $\odot$ denotes element-by-element (element-wise) multiplication[4]. Given an initial value $\hat{\boldsymbol{\psi}}_{1|0}$ and an assumed value for the population parameter vector $\boldsymbol{\beta}$, we can iterate on Eqs. (14) and (15) to calculate $\hat{\boldsymbol{\psi}}_{t|t}$ and $\hat{\boldsymbol{\psi}}_{t+1|t}$ for $t = 1, \ldots, T$. There are several options for choosing the initial value. One approach is to set $\hat{\boldsymbol{\psi}}_{1|0}$ equal to an $(N \times 1)$ eigenvector $\boldsymbol{\pi}$ of the transition matrix $\mathbf{P}$[5]. Another approach is to set $\hat{\boldsymbol{\psi}}_{1|0} = \boldsymbol{\rho}$, where $\boldsymbol{\rho}$ is an $(N \times 1)$ vector of nonnegative constants that sum to unity. Or we can estimate $\boldsymbol{\rho}$ along with $\boldsymbol{\beta}$ by maximum likelihood subject to the constraint that $\mathbf{1}'\boldsymbol{\rho} = 1$ and $\rho_j \geq 0, j = 1, \ldots, N$.

When the iteration on Eqs. (14) and (15) is completed for all t with an assumed, fixed parameter vector $\boldsymbol{\beta}$, we obtain the log likelihood function $\mathbb{L}(\boldsymbol{\beta})$

[4] Recall that $\mathbf{x}_t$ is assumed to be exogenous, i.e., having no information about s_t beyond that contained in F_{t-1}. Thus, the jth element of $\hat{\boldsymbol{\psi}}_{t|t-1}$ can be rewritten as $\Pr(s_t = j|\mathbf{x}_t, F_{t-1}; \boldsymbol{\beta})$. The jth element of $\boldsymbol{\delta}_t$ is $f(\mathbf{y}_t|s_t = j, \mathbf{x}_t, F_{t-1}; \boldsymbol{\beta})$. The numerator in the right side of Eq. (14) is

$$\hat{\boldsymbol{\psi}}_{t|t-1} \odot \boldsymbol{\delta}_t = \begin{bmatrix} \Pr(s_t = 1|\mathbf{x}_t, F_{t-1}; \boldsymbol{\beta}) \\ \vdots \\ \Pr(s_t = N|\mathbf{x}_t, F_{t-1}; \boldsymbol{\beta}) \end{bmatrix} \odot \begin{bmatrix} f(\mathbf{y}_t|s_t = 1, \mathbf{x}_t, F_{t-1}; \boldsymbol{\beta}) \\ \vdots \\ f(\mathbf{y}_t|s_t = N, \mathbf{x}_t, F_{t-1}; \boldsymbol{\beta}) \end{bmatrix} = \begin{bmatrix} p(\mathbf{y}_t, s_t = 1|\mathbf{x}_t, F_{t-1}; \boldsymbol{\beta}) \\ \vdots \\ p(\mathbf{y}_t, s_t = N|\mathbf{x}_t, F_{t-1}; \boldsymbol{\beta}) \end{bmatrix}.$$

The denominator in the right side of Eq. (14) is

$$\mathbf{1}'(\hat{\boldsymbol{\psi}}_{t|t-1} \odot \boldsymbol{\delta}_t) = \sum_{j=1}^{N} p(\mathbf{y}_t, s_t = j|\mathbf{x}_t, F_{t-1}; \boldsymbol{\beta}) = f(\mathbf{y}_t|\mathbf{x}_t, F_{t-1}; \boldsymbol{\beta}).$$

Since

$$\frac{p(\mathbf{y}_t, s_t = j|\mathbf{x}_t, F_{t-1}; \boldsymbol{\beta})}{f(\mathbf{y}_t|\mathbf{x}_t, F_{t-1}; \boldsymbol{\beta})} = \Pr(s_t = j|F_t; \boldsymbol{\beta})$$

is the jth element of $\hat{\boldsymbol{\psi}}_{t|t}$, then Eq. (14) is proved.

Take expectations of Eq. (12) conditional on information available up to date t:

$$E(\boldsymbol{\psi}_{t+1}|F_t) = \mathbf{P}E(\boldsymbol{\psi}_t|F_t) + E(\mathbf{v}_{t+1}|F_t),$$

which is Eq. (15) because $\mathbf{v}_{t+1}$ is a martingale difference sequence with respect to F_t.

[5] The eigenvector $\boldsymbol{\pi}$ satisfies $\mathbf{P}\boldsymbol{\pi} = \boldsymbol{\pi}$. It is normalized so that its elements sum to unity, i.e., $\mathbf{1}'\boldsymbol{\pi} = 1$. Stack the two equations $(\mathbf{I} - \mathbf{P})\boldsymbol{\pi} = \mathbf{0}$ and $\mathbf{1}'\boldsymbol{\pi} = 1$:

$$\begin{bmatrix} \mathbf{I} - \mathbf{P} \\ \mathbf{1}' \end{bmatrix} \boldsymbol{\pi} = \begin{bmatrix} \mathbf{0} \\ 1 \end{bmatrix}$$

where $\mathbf{i}_{N+1}$ is the $(N+1)$th column of $\mathbf{I}_{N+1}$, and $\boldsymbol{\pi}$ is the $(N+1)$th column of $(\mathbf{A}'\mathbf{A})^{-1}\mathbf{A}'$.

Equation (11) implies that $\mathbf{P}'\mathbf{1} = \mathbf{1}$. Since a matrix and its transpose have the same eigenvalues, unity is an eigenvalue of the transition matrix $\mathbf{P}$. Suppose that all other eigenvalues of $\mathbf{P}$ have absolute values less than unity. Then the Markov chain is ergodic, and $\boldsymbol{\pi}$ is the vector of ergodic probabilities.

for the observed data F_T:

$$\mathbb{L}(\boldsymbol{\beta}) = \sum_{t=1}^{T} \log f(\mathbf{y}_t|\mathbf{x}_t, F_{t-1}; \boldsymbol{\beta}),$$

where $f(\mathbf{y}_t|\mathbf{x}_t, F_{t-1}; \boldsymbol{\beta})$ is the denominator in the right side of Eq. (14) (see footnote (4)). Then we can determine the value of $\boldsymbol{\beta}$ that maximizes the log likelihood.

The vector $\hat{\boldsymbol{\psi}}_{t|\tau} = [\Pr(s_t = 1|F_\tau; \boldsymbol{\beta}), \ldots, \Pr(s_t = N|F_\tau; \boldsymbol{\beta})]'$ represents a forecast about the regime for $t > \tau$, and the smoothed inference about the regime for $t < \tau$. Taking expectations of Eq. (13) conditional on information available up to date t, we obtain the m-period-ahead forecast of $\boldsymbol{\psi}_{t+m}$:

$$\begin{aligned} & E(\boldsymbol{\psi}_{t+m}|F_t) = \mathbf{P}^m E(\boldsymbol{\psi}_t|F_t) \\ \Leftrightarrow \quad & \hat{\boldsymbol{\psi}}_{t+m|t} = \mathbf{P}^m \hat{\boldsymbol{\psi}}_{t|t}. \end{aligned} \tag{16}$$

where $\hat{\boldsymbol{\psi}}_{t|t}$ is calculated from Eq. (14). We can use an algorithm[6] developed by Kim [14] to compute smoothed inferences:

$$\hat{\boldsymbol{\psi}}_{t|T} = \hat{\boldsymbol{\psi}}_{t|t} \odot \left[\mathbf{P}' \left(\hat{\boldsymbol{\psi}}_{t+1|T} (\div) \hat{\boldsymbol{\psi}}_{t+1|t}\right)\right], \tag{17}$$

where the symbol $(\div)$ denotes element-by-element (element-wise) division. The smoothed probabilities $\hat{\boldsymbol{\psi}}_{t|T}$ are determined by iterating on Eq. (17) backward for $t = T-1, T-2, \ldots, 1$. The iteration starts with $\hat{\boldsymbol{\psi}}_{T|T}$, obtained from Eq.

[6] Recall that the regime s_t is assumed to depend on past observations F_{t-1} through the value of s_{t-1}. Similarly, s_t depends on future observations through the value of s_{t+1}, that is,

$$\Pr(s_t = j|s_{t+1} = i, F_T; \boldsymbol{\beta}) = \Pr(s_t = j|s_{t+1} = i, F_t; \boldsymbol{\beta}).$$

Proof: We suppress the implicit dependence on $\boldsymbol{\beta}$ to simplify the notation. It must be the case that

$$\begin{aligned} \Pr(s_t = j|s_{t+1} = i, F_{t+1}) &= \Pr(s_t = j|s_{t+1} = i, \mathbf{y}_{t+1}, \mathbf{x}_{t+1}, F_t) = \frac{p(\mathbf{y}_{t+1}, s_t = j|s_{t+1} = i, \mathbf{x}_{t+1}, F_t)}{f(\mathbf{y}_{t+1}|s_{t+1} = i, \mathbf{x}_{t+1}, F_t)} \\ &= \frac{f(\mathbf{y}_{t+1}|s_t = j, s_{t+1} = i, \mathbf{x}_{t+1}, F_t)\Pr(s_t = j|s_{t+1} = i, \mathbf{x}_{t+1}, F_t)}{f(\mathbf{y}_{t+1}|s_{t+1} = i, \mathbf{x}_{t+1}, F_t)} \\ &= \Pr(s_t = j|s_{t+1} = i, \mathbf{x}_{t+1}, F_t) \quad (\mathbf{y}_{t+1} \text{ depends on only the current value } s_{t+1}) \\ &= \Pr(s_t = j|s_{t+1} = i, F_t) \quad (\mathbf{x}_{t+1} \text{ is strictly exogenous}). \end{aligned}$$

Similar reasoning indicates that

$$\begin{aligned} \Pr(s_t = j|s_{t+1} = i, F_{t+2}) &= \Pr(s_t = j|s_{t+1} = i, \mathbf{y}_{t+2}, \mathbf{x}_{t+2}, F_{t+1}) = \frac{p(\mathbf{y}_{t+2}, s_t = j|s_{t+1} = i, \mathbf{x}_{t+2}, F_{t+1})}{f(\mathbf{y}_{t+2}|s_{t+1} = i, \mathbf{x}_{t+2}, F_{t+1})} \\ &= \frac{f(\mathbf{y}_{t+2}|s_t = j, s_{t+1} = i, \mathbf{x}_{t+2}, F_{t+1})\Pr(s_t = j|s_{t+1} = i, \mathbf{x}_{t+2}, F_{t+1})}{f(\mathbf{y}_{t+2}|s_{t+1} = i, \mathbf{x}_{t+2}, F_{t+1})}. \end{aligned}$$

(14) for $t = T$. The algorithm is reliable only if the regime s_t follows a Markov chain in Eq. (10), the conditional density of $\mathbf{y}_t$ in Eq. (9) depends on only the

---

Since

$$f(\mathbf{y}_{t+2}|s_t = j, s_{t+1} = i, \mathbf{x}_{t+2}, F_{t+1}) = \sum_{k=1}^{N} p(\mathbf{y}_{t+2}, s_{t+2} = k|s_t = j, s_{t+1} = i, \mathbf{x}_{t+2}, F_{t+1})$$

$$= \sum_{k=1}^{N} f(\mathbf{y}_{t+2}|s_{t+2} = k, s_t = j, s_{t+1} = i, \mathbf{x}_{t+2}, F_{t+1})\Pr(s_{t+2} = k|s_t = j, s_{t+1} = i, \mathbf{x}_{t+2}, F_{t+1})$$

$$= \sum_{k=1}^{N} f(\mathbf{y}_{t+2}|s_{t+2} = k, s_{t+1} = i, \mathbf{x}_{t+2}, F_{t+1})\Pr(s_{t+2} = k|s_{t+1} = i, \mathbf{x}_{t+2}, F_{t+1})$$

$$= \sum_{k=1}^{N} p(\mathbf{y}_{t+2}, s_{t+2} = k|s_{t+1} = i, \mathbf{x}_{t+2}, F_{t+1}) = f(\mathbf{y}_{t+2}|s_{t+1} = i, \mathbf{x}_{t+2}, F_{t+1}),$$

then

$$\Pr(s_t = j|s_{t+1} = i, F_{t+2}) = \Pr(s_t = j|s_{t+1} = i, \mathbf{x}_{t+2}, F_{t+1}) = \Pr(s_t = j|s_{t+1} = i, F_{t+1})$$
$$= \Pr(s_t = j|s_{t+1} = i, F_t).$$

Generally, $\Pr(s_t = j|s_{t+1} = i, F_{t+m}) = \Pr(s_t = j|s_{t+1} = i, F_t), \quad m = 1, 2, \ldots$ ■

The smoothed inference for date t given information available up to date T is

$$\Pr(s_t = j|F_T) = \sum_{i=1}^{N} \Pr(s_t = j, s_{t+1} = i|F_T) = \sum_{i=1}^{N} \Pr(s_{t+1} = i|F_T)\Pr(s_t = j|s_{t+1} = i, F_T)$$
$$= \sum_{i=1}^{N} \Pr(s_{t+1} = i|F_T)\Pr(s_t = j|s_{t+1} = i, F_t).$$

Because

$$\Pr(s_t = j|s_{t+1} = i, F_t) = \frac{\Pr(s_t = j, s_{t+1} = i|F_t)}{\Pr(s_{t+1} = i|F_t)} = \frac{\Pr(s_t = j|F_t)\Pr(s_{t+1} = i|s_t = j, F_t)}{\Pr(s_{t+1} = i|F_t)}$$
$$= \frac{p_{ji}\Pr(s_t = j|F_t)}{\Pr(s_{t+1} = i|F_t)},$$

then

$$\Pr(s_t = j|F_T) = \sum_{i=1}^{N} \Pr(s_{t+1} = i|F_T)\frac{p_{ji}\Pr(s_t = j|F_t)}{\Pr(s_{t+1} = i|F_t)} = \Pr(s_t = j|F_t)\sum_{i=1}^{N} \frac{p_{ji}\Pr(s_{t+1} = i|F_T)}{\Pr(s_{t+1} = i|F_t)}$$
$$= \Pr(s_t = j|F_t)\left[p_{j1} \cdots p_{jN}\right]\left[\frac{\Pr(s_{t+1}=1|F_T)}{\Pr(s_{t+1}=1|F_t)} \cdots \frac{\Pr(s_{t+1}=N|F_T)}{\Pr(s_{t+1}=N|F_t)}\right]'$$
$$= \Pr(s_t = j|F_t)\mathbf{p}_j'\left[\hat{\boldsymbol{\psi}}_{t+1|T}(\div)\hat{\boldsymbol{\psi}}_{t+1|t}\right],$$

where $\mathbf{p}_j$ denotes the jth column of $\mathbf{P}$. Collect $\Pr(s_t = j|F_T)$ for $j = 1, \ldots, N$ in an $(N \times 1)$ vector:

$$\hat{\boldsymbol{\psi}}_{t|T} = \hat{\boldsymbol{\psi}}_{t|t} \odot \left[\mathbf{P}'\left(\hat{\boldsymbol{\psi}}_{t+1|T}(\div)\hat{\boldsymbol{\psi}}_{t+1|t}\right)\right].$$

current regime s_t, and $\mathbf{x}_t$, the vector of explanatory variables other than the lagged values of $\mathbf{y}_t$, is strictly exogenous.

The Markov-switching model is a useful way of dealing with a number of discrete regimes. Each regime is characterized by its own set of parameters. However, if the regime is a continuous normal variable, we cannot estimate countless sets of parameters. Moreover, the Markov-switching model produces conditional probabilities while the regime's distribution is now summarized by its mean and variance. In this case, the Kalman filter, which is discussed in the next section, is a better resolution.

4 Kalman Filter[7]

The unobserved regime can be examined explicitly using a separate equation in a state-space model. A general linear state-space model takes the form

$$\boldsymbol{s}_{t+1} = \boldsymbol{a}_t + \boldsymbol{G}_t\boldsymbol{s}_t + \boldsymbol{P}_t\boldsymbol{\zeta}_t, \tag{18}$$

$$\boldsymbol{y}_t = \boldsymbol{b}_t + \boldsymbol{Z}_t\boldsymbol{s}_t + \boldsymbol{\varepsilon}_t, \tag{19}$$

where $\boldsymbol{s}_t = (s_{1t}, \ldots, s_{mt})'$ is an $(m\times 1)$ state vector, $\boldsymbol{y}_t = (y_{1t}, \ldots, y_{kt})'$ is a $(k\times 1)$ observation vector, $\boldsymbol{a}_t$ and $\boldsymbol{b}_t$ are $(m \times 1)$ and $(k \times 1)$ deterministic vectors, $\boldsymbol{G}_t$ and $\boldsymbol{Z}_t$ are $(m \times m)$ and $(k \times m)$ coefficient matrices, $\boldsymbol{P}_t$ is an $m \times n$ matrix, and $\{\boldsymbol{\zeta}_t\}$ and $\{\boldsymbol{\varepsilon}_t\}$ are $(n \times 1)$ and $(k \times 1)$ Gaussian white noise series such that

$$\boldsymbol{\zeta}_t \sim N(\boldsymbol{0}, \boldsymbol{Q}_t), \qquad \boldsymbol{\varepsilon}_t \sim N(\boldsymbol{0}, \boldsymbol{H}_t),$$

where $\boldsymbol{Q}_t$ and $\boldsymbol{H}_t$ are $(n \times n)$ and $(k \times k)$ positive-definite matrices. The starting state $\boldsymbol{s}_1 \sim N(\boldsymbol{\mu}_{1|0}, \boldsymbol{\Sigma}_{1|0})$, where $\boldsymbol{\mu}_{1|0}$ and $\boldsymbol{\Sigma}_{1|0}$ are given, and $\boldsymbol{s}_1$ is independent of $\boldsymbol{\zeta}_t$ and $\boldsymbol{\varepsilon}_t$ for $t > 0$. The state Eq. (18) shows a first-order Markov Chain that regulates the state transition with innovation $\boldsymbol{\zeta}_t$. The observation Eq. (19) relates the observation vector $\boldsymbol{y}_t$ to the state vector $\boldsymbol{s}_t$ with the measurement error $\boldsymbol{\varepsilon}_t$.

The goal of the Kalman filter is to secure the conditional distribution of $\boldsymbol{s}_{t+1}$ given the information available up to date t, i.e., F_t, and the state-space model. Normality of the innovation $\boldsymbol{\zeta}_t$ translates into normal conditional distribution of $\boldsymbol{s}_{t+1}$ given F_t, that is, $\boldsymbol{s}_{t+1}|F_t \sim N(\boldsymbol{s}_{t+1|t}, \boldsymbol{\Sigma}_{t+1|t})$, where the conditional mean and covariance matrix are

$$\boldsymbol{s}_{t+1|t} = E(\boldsymbol{a}_t + \boldsymbol{G}_t\boldsymbol{s}_t + \boldsymbol{P}_t\boldsymbol{\zeta}_t|F_t) = \boldsymbol{a}_t + \boldsymbol{G}_t\boldsymbol{s}_{t|t}, \tag{20}$$

$$\boldsymbol{\Sigma}_{t+1|t} = \mathrm{Var}(\boldsymbol{s}_{t+1}|F_t) = \boldsymbol{G}_t\boldsymbol{\Sigma}_{t|t}\boldsymbol{G}_t' + \boldsymbol{P}_t\boldsymbol{Q}_t\boldsymbol{P}_t'. \tag{21}$$

From Eq. (19), the conditional mean of $\boldsymbol{y}_t$ given F_{t-1} is

$$\boldsymbol{y}_{t|t-1} = E(\boldsymbol{y}_t|F_{t-1}) = \boldsymbol{b}_t + \boldsymbol{Z}_t\boldsymbol{s}_{t|t-1}.$$

[7] This section is to a large extent based on Tsay [20].

Let $\boldsymbol{u}_t$ be the 1-step-ahead forecast error of $\boldsymbol{y}_t$ given F_{t-1}. Then

$$\boldsymbol{u}_t = \boldsymbol{y}_t - \boldsymbol{y}_{t|t-1} = \boldsymbol{Z}_t(\boldsymbol{s}_t - \boldsymbol{s}_{t|t-1}) + \boldsymbol{\varepsilon}_t. \tag{22}$$

Because $\boldsymbol{u}_t$ is a sequence of independent normal random vectors with zero conditional mean, i.e., $E(\boldsymbol{u}_t|F_{t-1}) = \boldsymbol{0}$, and is independent of F_{t-1}, its covariance is

$$\boldsymbol{V}_t = \text{Var}(\boldsymbol{u}_t|F_{t-1}) = \text{Var}(\boldsymbol{u}_t) = \boldsymbol{Z}_t\boldsymbol{\Sigma}_{t|t-1}\boldsymbol{Z}_t' + \boldsymbol{H}_t. \tag{23}$$

With $F_t = \{F_{t-1}, \boldsymbol{y}_t\} = \{F_{t-1}, \boldsymbol{u}_t\}$, we update[8]

$$\boldsymbol{s}_{t|t} = E(\boldsymbol{s}_t|F_{t-1}, \boldsymbol{u}_t) = \boldsymbol{s}_{t|t-1} + \text{Cov}(\boldsymbol{s}_t, \boldsymbol{u}_t|F_{t-1})\boldsymbol{V}_t^{-1}\boldsymbol{u}_t = \boldsymbol{s}_{t|t-1} + \boldsymbol{D}_t\boldsymbol{V}_t^{-1}\boldsymbol{u}_t, \tag{24}$$

where $\boldsymbol{D}_t = \text{Cov}(\boldsymbol{s}_t, \boldsymbol{u}_t|F_{t-1}) = \text{Cov}(\boldsymbol{s}_t, \boldsymbol{Z}_t(\boldsymbol{s}_t - \boldsymbol{s}_{t|t-1}) + \boldsymbol{\varepsilon}_t|F_{t-1}) = \boldsymbol{\Sigma}_{t|t-1}\boldsymbol{Z}_t'$. Substituting $\boldsymbol{s}_{t|t}$ in Eq. (24) into Eq. (20), we obtain

$$\boldsymbol{s}_{t+1|t} = \boldsymbol{a}_t + \boldsymbol{G}_t\boldsymbol{s}_{t|t-1} + \boldsymbol{G}_t\boldsymbol{D}_t\boldsymbol{V}_t^{-1}\boldsymbol{u}_t = \boldsymbol{a}_t + \boldsymbol{G}_t\boldsymbol{s}_{t|t-1} + \boldsymbol{K}_t\boldsymbol{u}_t, \tag{25}$$

where

$$\boldsymbol{K}_t = \boldsymbol{G}_t\boldsymbol{D}_t\boldsymbol{V}_t^{-1} = \boldsymbol{G}_t\boldsymbol{\Sigma}_{t|t-1}\boldsymbol{Z}_t'\boldsymbol{V}_t^{-1}, \tag{26}$$

which is the Kalman gain at date t.

We also update[9]

$$\begin{aligned}\boldsymbol{\Sigma}_{t|t} &= \text{Var}(\boldsymbol{s}_t|F_{t-1}, \boldsymbol{u}_t) = \text{Var}(\boldsymbol{s}_t|F_{t-1}) - \text{Cov}(\boldsymbol{s}_t, \boldsymbol{u}_t|F_{t-1})[\text{Var}(\boldsymbol{u}_t|F_{t-1})]^{-1}\text{Cov}(\boldsymbol{u}_t, \boldsymbol{s}_t|F_{t-1}) \\ &= \boldsymbol{\Sigma}_{t|t-1} - \boldsymbol{D}_t\boldsymbol{V}_t^{-1}\boldsymbol{D}_t' = \boldsymbol{\Sigma}_{t|t-1} - \boldsymbol{\Sigma}_{t|t-1}\boldsymbol{Z}_t'\boldsymbol{V}_t^{-1}\boldsymbol{Z}_t\boldsymbol{\Sigma}_{t|t-1}.\end{aligned} \tag{27}$$

Substituting $\boldsymbol{\Sigma}_{t|t}$ in Eq. (27) into Eq. (21) and using Eq. (26), we obtain

$$\begin{aligned}\boldsymbol{\Sigma}_{t+1|t} &= \boldsymbol{G}_t\boldsymbol{\Sigma}_{t|t-1}\boldsymbol{G}_t' - \boldsymbol{G}_t\boldsymbol{\Sigma}_{t|t-1}\boldsymbol{Z}_t'\boldsymbol{V}_t^{-1}\boldsymbol{D}_t'\boldsymbol{G}_t' + \boldsymbol{P}_t\boldsymbol{Q}_t\boldsymbol{P}_t' = \boldsymbol{G}_t\boldsymbol{\Sigma}_{t|t-1}(\boldsymbol{G}_t' - \boldsymbol{Z}_t'\boldsymbol{K}_t') + \boldsymbol{P}_t\boldsymbol{Q}_t\boldsymbol{P}_t' \\ &= \boldsymbol{G}_t\boldsymbol{\Sigma}_{t|t-1}\boldsymbol{L}_t' + \boldsymbol{P}_t\boldsymbol{Q}_t\boldsymbol{P}_t',\end{aligned} \tag{28}$$

where $\boldsymbol{L}_t = \boldsymbol{G}_t - \boldsymbol{K}_t\boldsymbol{Z}_t$.

Given the initial values $\boldsymbol{s}_{1|0}$ and $\boldsymbol{\Sigma}_{1|0}$, the Kalman filter algorithm for the state-space model is

$$\begin{aligned}\boldsymbol{u}_t &= \boldsymbol{y}_t - \boldsymbol{b}_t - \boldsymbol{Z}_t\boldsymbol{s}_{t|t-1}, \\ \boldsymbol{V}_t &= \boldsymbol{Z}_t\boldsymbol{\Sigma}_{t|t-1}\boldsymbol{Z}_t' + \boldsymbol{H}_t, \\ \boldsymbol{K}_t &= \boldsymbol{G}_t\boldsymbol{\Sigma}_{t|t-1}\boldsymbol{Z}_t'\boldsymbol{V}_t^{-1}, \\ \boldsymbol{L}_t &= \boldsymbol{G}_t - \boldsymbol{K}_t\boldsymbol{Z}_t, \\ \boldsymbol{s}_{t+1|t} &= \boldsymbol{a}_t + \boldsymbol{G}_t\boldsymbol{s}_{t|t-1} + \boldsymbol{K}_t\boldsymbol{u}_t, \\ \boldsymbol{\Sigma}_{t+1|t} &= \boldsymbol{G}_t\boldsymbol{\Sigma}_{t|t-1}\boldsymbol{L}_t' + \boldsymbol{P}_t\boldsymbol{Q}_t\boldsymbol{P}_t', \qquad t = 1, \ldots, T.\end{aligned} \tag{29}$$

[8] Note that $E(\boldsymbol{x}|\boldsymbol{y}, \boldsymbol{z}) = E(\boldsymbol{x}|\boldsymbol{y}) + \boldsymbol{\Sigma}_{\boldsymbol{xz}}\boldsymbol{\Sigma}_{\boldsymbol{zz}}^{-1}(\boldsymbol{z} - \boldsymbol{\mu}_{\boldsymbol{z}})$.

[9] Note that $\text{Var}(\boldsymbol{x}|\boldsymbol{y}, \boldsymbol{z}) = \text{Var}(\boldsymbol{x}|\boldsymbol{y}) - \boldsymbol{\Sigma}_{\boldsymbol{xz}}\boldsymbol{\Sigma}_{\boldsymbol{zz}}^{-1}\boldsymbol{\Sigma}_{\boldsymbol{zx}}$.

We can revise the Kalman filter to calculate the updated quantities $\boldsymbol{s}_{t|t}$ and $\boldsymbol{\Sigma}_{t|t}$ as follows

$$\begin{aligned}
\boldsymbol{u}_t &= \boldsymbol{y}_t - \boldsymbol{b}_t - \boldsymbol{Z}_t\boldsymbol{s}_{t|t-1},\\
\boldsymbol{D}_t &= \boldsymbol{\Sigma}_{t|t-1}\boldsymbol{Z}_t',\\
\boldsymbol{V}_t &= \boldsymbol{Z}_t\boldsymbol{\Sigma}_{t|t-1}\boldsymbol{Z}_t' + \boldsymbol{H}_t = \boldsymbol{Z}_t\boldsymbol{D}_t + \boldsymbol{H}_t,\\
\boldsymbol{s}_{t|t} &= \boldsymbol{s}_{t|t-1} + \boldsymbol{D}_t\boldsymbol{V}_t^{-1}\boldsymbol{u}_t,\\
\boldsymbol{\Sigma}_{t|t} &= \boldsymbol{\Sigma}_{t|t-1} - \boldsymbol{D}_t\boldsymbol{V}_t^{-1}\boldsymbol{D}_t',\\
\boldsymbol{s}_{t+1|t} &= \boldsymbol{a}_t + \boldsymbol{G}_t\boldsymbol{s}_{t|t},\\
\boldsymbol{\Sigma}_{t+1|t} &= \boldsymbol{G}_t\boldsymbol{\Sigma}_{t|t}\boldsymbol{G}_t' + \boldsymbol{P}_t\boldsymbol{Q}_t\boldsymbol{P}_t', \qquad\qquad t = 1, \ldots, T.
\end{aligned}$$

Smoothed State Vector and Its Covariance Matrix

Like the Markov-switching model, the Kalman filter can perform state space smoothing via the conditional distribution of $\boldsymbol{s}_t$ given F_T. Let $\boldsymbol{x}_t = \boldsymbol{s}_t - \boldsymbol{s}_{t|t-1}$ be the state prediction error. Then $\text{Var}(\boldsymbol{x}_t|F_{t-1}) = \text{Var}(\boldsymbol{s}_t|F_{t-1}) = \boldsymbol{\Sigma}_{t|t-1}$ and $\boldsymbol{x}_{t+1} = \boldsymbol{s}_{t+1} - \boldsymbol{s}_{t+1|t} = \boldsymbol{L}_t\boldsymbol{x}_t + \boldsymbol{P}_t\boldsymbol{\zeta}_t - \boldsymbol{K}_t\boldsymbol{\varepsilon}_t$. The 1-step-ahead forecast error in Eq. (22) can be rewritten as $\boldsymbol{u}_t = \boldsymbol{Z}_t\boldsymbol{x}_t + \boldsymbol{\varepsilon}_t$. Footnote (8) implies that

$$\begin{aligned}
\boldsymbol{s}_{t|T} &= E(\boldsymbol{s}_t|F_{t-1}, \boldsymbol{u}_t, \ldots, \boldsymbol{u}_T)\\
&= \boldsymbol{s}_{t|t-1} + \text{Cov}(\boldsymbol{s}_t, \boldsymbol{u}_t|F_{t-1})[\text{Var}(\boldsymbol{u}_t)]^{-1}\boldsymbol{u}_t + \ldots + \text{Cov}(\boldsymbol{s}_t, \boldsymbol{u}_T|F_{t-1})[\text{Var}(\boldsymbol{u}_T)]^{-1}\boldsymbol{u}_T\\
&= \boldsymbol{s}_{t|t-1} + \sum_{j=t}^{T} \text{Cov}(\boldsymbol{s}_t, \boldsymbol{u}_j|F_{t-1})\boldsymbol{V}_j^{-1}\boldsymbol{u}_j,
\end{aligned} \tag{30}$$

where

$$\text{Cov}(\boldsymbol{s}_t, \boldsymbol{u}_j|F_{t-1}) = E(\boldsymbol{x}_t\boldsymbol{u}_j'|F_{t-1}) = E[\boldsymbol{x}_t(\boldsymbol{Z}_j\boldsymbol{x}_j + \boldsymbol{\varepsilon}_j)'|F_{t-1}] = E(\boldsymbol{x}_t\boldsymbol{x}_j'|F_{t-1})\boldsymbol{Z}_j'. \tag{31}$$

Specifically,

$$\begin{aligned}
E(\boldsymbol{x}_t\boldsymbol{x}_t'|F_{t-1}) &= \boldsymbol{\Sigma}_{t|t-1},\\
E(\boldsymbol{x}_t\boldsymbol{x}_{t+1}'|F_{t-1}) &= E[\boldsymbol{x}_t(\boldsymbol{L}_t\boldsymbol{x}_t + \boldsymbol{P}_t\boldsymbol{\zeta}_t - \boldsymbol{K}_t\boldsymbol{\varepsilon}_t)'|F_{t-1}] = \boldsymbol{\Sigma}_{t|t-1}\boldsymbol{L}_t',\\
E(\boldsymbol{x}_t\boldsymbol{x}_{t+2}'|F_{t-1}) &= E[\boldsymbol{x}_t(\boldsymbol{L}_{t+1}\boldsymbol{x}_{t+1} + \boldsymbol{P}_{t+1}\boldsymbol{\zeta}_{t+1} - \boldsymbol{K}_{t+1}\boldsymbol{\varepsilon}_{t+1})'|F_{t-1}] = \boldsymbol{\Sigma}_{t|t-1}\boldsymbol{L}_t'\boldsymbol{L}_{t+1}',\\
&\;\;\vdots\\
E(\boldsymbol{x}_t\boldsymbol{x}_T'|F_{t-1}) &= \boldsymbol{\Sigma}_{t|t-1}\boldsymbol{L}_t'\boldsymbol{L}_{t+1}'\cdots\boldsymbol{L}_{T-1}'.
\end{aligned} \tag{32}$$

Substituting Eq. (32) into Eq. (31), then into Eq. (30), we obtain

$$\begin{aligned}
\boldsymbol{s}_{T|T} &= \boldsymbol{s}_{T|T-1} + \boldsymbol{\Sigma}_{T|T-1}\boldsymbol{Z}_T'\boldsymbol{V}_T^{-1}\boldsymbol{u}_T,\\
\boldsymbol{s}_{T-1|T} &= \boldsymbol{s}_{T-1|T-2} + \boldsymbol{\Sigma}_{T-1|T-2}\boldsymbol{Z}_{T-1}'\boldsymbol{V}_{T-1}^{-1}\boldsymbol{u}_{T-1} + \boldsymbol{\Sigma}_{T-1|T-2}\boldsymbol{L}_{T-1}'\boldsymbol{Z}_T'\boldsymbol{V}_T^{-1}\boldsymbol{u}_T,\\
\boldsymbol{s}_{t|T} &= \boldsymbol{s}_{t|t-1} + \boldsymbol{\Sigma}_{t|t-1}\boldsymbol{Z}_t'\boldsymbol{V}_t^{-1}\boldsymbol{u}_t + \boldsymbol{\Sigma}_{t|t-1}\boldsymbol{L}_t'\boldsymbol{Z}_{t+1}'\boldsymbol{V}_{t+1}^{-1}\boldsymbol{u}_{t+1}\\
&\quad + \cdots + \boldsymbol{\Sigma}_{t|t-1}\boldsymbol{L}_t'\boldsymbol{L}_{t+1}'\cdots\boldsymbol{L}_{T-1}'\boldsymbol{Z}_T'\boldsymbol{V}_T^{-1}\boldsymbol{u}_T,
\end{aligned}$$

for $t = T-2, T-3, \ldots, 1$, where $\boldsymbol{L}_t'\boldsymbol{L}_{t+1}' \cdots \boldsymbol{L}_{T-1}' = \boldsymbol{I}_m$ when $t = T$. The smoothed state vectors can be written as

$$\boldsymbol{s}_{t|T} = \boldsymbol{s}_{t|t-1} + \boldsymbol{\Sigma}_{t|t-1}\boldsymbol{k}_{t-1}, \tag{33}$$

where $\boldsymbol{k}_{T-1} = \boldsymbol{Z}_T'\boldsymbol{V}_T^{-1}\boldsymbol{u}_T$, $\boldsymbol{k}_{T-2} = \boldsymbol{Z}_{T-1}'\boldsymbol{V}_{T-1}^{-1}\boldsymbol{u}_{T-1} + \boldsymbol{L}_{T-1}'\boldsymbol{Z}_T'\boldsymbol{V}_T^{-1}\boldsymbol{u}_T$, and $\boldsymbol{k}_{t-1} = \boldsymbol{Z}_t'\boldsymbol{V}_t^{-1}\boldsymbol{u}_t + \boldsymbol{L}_t'\boldsymbol{Z}_{t+1}'\boldsymbol{V}_{t+1}^{-1}\boldsymbol{u}_{t+1} + \cdots + \boldsymbol{L}_t'\boldsymbol{L}_{t+1}' \cdots \boldsymbol{L}_{T-1}'\boldsymbol{Z}_T'\boldsymbol{V}_T^{-1}\boldsymbol{u}_T$, for $t = T-2, T-3, \ldots, 1$. The vector $\boldsymbol{k}_{t-1}$ is a weighted sum of the 1-step-ahead forecast errors $\boldsymbol{u}_j$ for $j > t-1$, and can be calculated recursively backward as

$$\boldsymbol{k}_{t-1} = \boldsymbol{Z}_t'\boldsymbol{V}_t^{-1}\boldsymbol{u}_t + \boldsymbol{L}_t'\boldsymbol{k}_t, \qquad t = T, T-1, \ldots, 1, \tag{34}$$

where $\boldsymbol{k}_T = \boldsymbol{0}$. Equations (33) and (34) constitute a backward recursion for the smoothed state vectors, where $\boldsymbol{s}_{t|t-1}$, $\boldsymbol{\Sigma}_{t|t-1}$, $\boldsymbol{L}_t$, and $\boldsymbol{V}_t$ are computed from the Kalman filter.

Regarding the covariance matrix of the smoothed state vector, footnote (9) indicates that

$$\begin{aligned}
\boldsymbol{\Sigma}_{t|T} &= \mathrm{Var}(\boldsymbol{s}_t|F_{t-1}, \boldsymbol{u}_t, \ldots, \boldsymbol{u}_T) \\
&= \boldsymbol{\Sigma}_{t|t-1} - \sum_{j=t}^{T} \mathrm{Cov}(\boldsymbol{s}_t, \boldsymbol{u}_j|F_{t-1})\boldsymbol{V}_j^{-1}[\mathrm{Cov}(\boldsymbol{s}_t, \boldsymbol{u}_j|F_{t-1})]' \\
&= \boldsymbol{\Sigma}_{t|t-1} - \boldsymbol{\Sigma}_{t|t-1}\boldsymbol{Z}_t'\boldsymbol{V}_t^{-1}\boldsymbol{Z}_t\boldsymbol{\Sigma}_{t|t-1} - \boldsymbol{\Sigma}_{t|t-1}\boldsymbol{L}_t'\boldsymbol{Z}_{t+1}'\boldsymbol{V}_{t+1}^{-1}\boldsymbol{Z}_{t+1}\boldsymbol{L}_t\boldsymbol{\Sigma}_{t|t-1} \\
&\quad - \cdots - \boldsymbol{\Sigma}_{t|t-1}\boldsymbol{L}_t'\boldsymbol{L}_{t+1}' \cdots \boldsymbol{L}_{T-1}'\boldsymbol{Z}_T'\boldsymbol{V}_T^{-1}\boldsymbol{Z}_T\boldsymbol{L}_{T-1} \cdots \boldsymbol{L}_{t+1}\boldsymbol{L}_t\boldsymbol{\Sigma}_{t|t-1} \\
&= \boldsymbol{\Sigma}_{t|t-1} - \boldsymbol{\Sigma}_{t|t-1}\boldsymbol{W}_{t-1}\boldsymbol{\Sigma}_{t|t-1},
\end{aligned} \tag{35}$$

where
$\boldsymbol{W}_{t-1} = \boldsymbol{Z}_t'\boldsymbol{V}_t^{-1}\boldsymbol{Z}_t + \boldsymbol{L}_t'\boldsymbol{Z}_{t+1}'\boldsymbol{V}_{t+1}^{-1}\boldsymbol{Z}_{t+1}\boldsymbol{L}_t + \cdots + \boldsymbol{L}_t' \cdots \boldsymbol{L}_{T-1}'\boldsymbol{Z}_T'\boldsymbol{V}_T^{-1}\boldsymbol{Z}_T\boldsymbol{L}_{T-1} \cdots \boldsymbol{L}_t$, and $\boldsymbol{L}_t'\boldsymbol{L}_{t+1}' \cdots \boldsymbol{L}_{T-1}' = \boldsymbol{I}_m$ when $t = T$. The matrix $\boldsymbol{W}_{t-1}$ can be rewritten as

$$\boldsymbol{W}_{t-1} = \boldsymbol{Z}_t'\boldsymbol{V}_t^{-1}\boldsymbol{Z}_t + \boldsymbol{L}_t'\boldsymbol{W}_t\boldsymbol{L}_t, \qquad t = T, T-1, \ldots, 1, \tag{36}$$

with the initial value $\boldsymbol{W}_T = \boldsymbol{0}$. Equations (35) and (36) constitute a backward recursion for the covariance matrices of the smoothed state vectors, where $\boldsymbol{\Sigma}_{t|t-1}$, $\boldsymbol{L}_t$, and $\boldsymbol{V}_t$ are computed from the Kalman filter.

Therefore, after using the Kalman filter in Eq. (29) to compute the quantities $\boldsymbol{u}_t$, $\boldsymbol{V}_t$, $\boldsymbol{K}_t$, $\boldsymbol{L}_t$, $\boldsymbol{s}_{t|t-1}$, $\boldsymbol{\Sigma}_{t|t-1}$ for $t = 1, \ldots, T$, we combine the two backward recursions

$$\begin{aligned}
\boldsymbol{k}_{t-1} &= \boldsymbol{Z}_t'\boldsymbol{V}_t^{-1}\boldsymbol{u}_t + \boldsymbol{L}_t'\boldsymbol{k}_t, \\
\boldsymbol{s}_{t|T} &= \boldsymbol{s}_{t|t-1} + \boldsymbol{\Sigma}_{t|t-1}\boldsymbol{k}_{t-1}, \\
\boldsymbol{W}_{t-1} &= \boldsymbol{Z}_t'\boldsymbol{V}_t^{-1}\boldsymbol{Z}_t + \boldsymbol{L}_t'\boldsymbol{W}_t\boldsymbol{L}_t, \\
\boldsymbol{\Sigma}_{t|T} &= \boldsymbol{\Sigma}_{t|t-1} - \boldsymbol{\Sigma}_{t|t-1}\boldsymbol{W}_{t-1}\boldsymbol{\Sigma}_{t|t-1},
\end{aligned} \tag{37}$$

with $\boldsymbol{k}_T = \boldsymbol{0}$ and $\boldsymbol{W}_T = \boldsymbol{0}$ to obtain $\boldsymbol{s}_{t|T}$ and $\boldsymbol{\Sigma}_{t|T}$ for $t = T, \ldots, 1$.

5 Conclusions

The econometrics of structural change has evolved a lot since the first and classical test introduced by Chow [5]. Because the breakdate in most cases is unknown a priori, several methods have been proposed by Quandt [17], Bai and Perron [2], Chen [4], among others. But these efforts could be invalid as it is claimed that the P value has been misused for roughly the past one hundred years. Then the Bayes factor can be employed for a pairwise comparison of competing models, including those that account for structural changes, based on a prior distribution of the model parameters and a sampling density structure.

If a change in regime is not considered as the outcome of a deterministic event, but instead a random variable, then a time series model should incorporate the probability law that governs the shift from one regime to another. The Markov-switching model, introduced by Hamilton [7], is an effective way of dealing with a number of discrete regimes. However, if the regime is a continuous normal variable, the Markov-switching model does not work and should be replaced by the Kalman filter. Both frameworks produce not only forecast about the regime but also smoothed inference about the regime given data obtained through some later date.

References

1. Ando, T.: Bayesian Model Selection and Statistical Modeling. Chapman and Hall/CRC, Boca Raton (2010)
2. Bai, J., Perron, P.: Estimating and testing linear models with multiple structural changes. Econometrica **66**(1), 47–78 (1998)
3. Benjamin, D.J., Berger, J.O., Johannesson, M., Nosek, B.A., Wagenmakers, E.-J., Berk, R., et al.: Redefine statistical significance. Nat. Hum. Behav. **2**, 6–10 (2018)
4. Chen, B.: Modeling and testing smooth structural changes with endogenous regressors. J. Econom. **185**(1), 196–215 (2015)
5. Chow, G.C.: Tests of equality between sets of coefficients in two linear regressions. Econometrica **28**(3), 591–605 (1960)
6. Goodman, S.: A dirty dozen: twelve *P*-value misconceptions. Semin. Hematol. **45**, 135–140 (2008)
7. Hamilton, J.D.: A new approach to the economic analysis of nonstationary time series and the business cycle. Econometrica **57**, 357–384 (1989)
8. Hamilton, J.D.: Time Series Analysis. Princeton University Press, Princeton (1994)
9. Hansen, B.E.: The new econometrics of structural change: dating breaks in U.S. labor productivity. J. Econ. Perspect. **15**(4), 117–128 (2001)
10. Jeffreys, H.: The Theory of Probability. Oxford University Press, Oxford (1961)
11. Kalman, R.E.: A new approach to linear filtering and prediction problems. J. Basic Eng. **82**(1), 35–45 (1960)
12. Kass, R.E., Raftery, A.E.: Bayes factors. J. Am. Stat. Assoc. **90**(430), 773–795 (1995)
13. Koop, G., Potter, S.: Prior elicitation in multiple change-point models. Int. Econ. Rev. **50**(3), 751–772 (2007)
14. Kim, C.J.: Dynamic linear models with markov-switching. J. Econom. **60**, 1–22 (1994)

15. Melton, A.: Editorial. J. Exp. Psychol. **64**, 553–557 (1962)
16. Nuzzo, R.: Statistical errors. Nature **506**, 150–52 (2014)
17. Quandt, R.: Tests of the hypothesis that a linear regression obeys two separate regimes. J. Am. Stat. Assoc. **55**, 324–330 (1960)
18. Trafimow, D., Marks, M.: Editorial. Basic Appl. Soc. Psychol. **37**, 1–2 (2015)
19. Trafimow, D., Amrhein, V., Areshenkoff, C.N., Barrera-Causil, C.J., Beh, E.J., Bilgiç, Y.K., et al.: Manipulating the alpha level cannot cure significance testing. Front. Psychol. **9**, 699 (2018)
20. Tsay, R.S.: Analysis of Financial Time Series. Wiley, Hoboken (2010)

Measuring Internal Factors Affecting the Competitiveness of Financial Companies: The Research Case in Vietnam

Doan Thanh Ha and Dang Truong Thanh Nhan(✉)

Banking University HCMC, Ho Chi Minh, Vietnam
nhan.dang.811@gmail.com

Abstract. Under the current trend of development and integration, financial companies are increasingly focusing on brand development, image enhancement and service quality improvement. They have also been encountering difficulties related to improving the competitiveness of interest rates and management capability. These factors have significant impacts on the competitiveness of financial companies in Vietnam. This study applies the Thompson - Strickland matrix model of internal factors to find and test variables significant to the competitiveness of financial firms. This empirical research on internal factors affecting the competitiveness of financial firms in Vietnam shows that there are eight internal factors that affect the competitiveness of those financial firms. All the eight internal factors have impact in the same direction on the competitiveness of those firms. Such means if managers of a financial company improve these factors positively, the competitiveness of the financial company will be reinforced. The improvement of financial companies' competitiveness will help to generate more financial growth, which would in return improve the internal factors.

Keywords: Competitiveness · Internal factors · Financial companies

1 Introduction

Vietnam has been in the process of deep integration with the world economy; finance and banking market has been increasingly developing in a complicated way. The role of credit institutions has been more important and should be promoted for the improvement of the effectiveness of the State Bank's monetary policy. Currently in Vietnam, financial companies are non-bank financial institutions which play a part of the system of credit institutions in Vietnam's financial market. Consequently, these financial companies are also influenced by the above trend. Especially, consumer credits have gone through a great development and become an attractive business area due to changes in consumption trends. In 2017, consumer credits had the growth rate of 65%, accounting for about 18% of total credit outstanding balance of the economy.

The study was conducted at 16 financial companies as of 12/2017 in Vietnam. A number of studies on the competitiveness of financial institutions have been conducted in Vietnam. However, there has been no specific study which identifies and measures the impact of internal factors on the competitiveness of Vietnamese financial

V. Kreinovich et al. (Eds.): ECONVN 2019, SCI 809, pp. 596–605, 2019.
https://doi.org/10.1007/978-3-030-04200-4_42

firms. Therefore, the identification, measurement and systematization of internal factors affecting the competitiveness of financial companies is essential for the enhancement of financial companies's competitiveness in specific and financial institutions' competitiveness in general.

2 Theoretical Background and Empirical Studies

There are many different views on competition in general and competitiveness in particular. In the economic science, the competitiveness refers to the way in which economic environment manages its competencies in order to achieve prosperity (Cişmaş and Stan 2010), proportionally generating more wealth than its competitors. A competitive system is based on the production systems that generate through their innovation capacity, the quality of products, adaptation to the market, competitive advantages, structures and specific resources capable to generate distinctive competencies (Dobrea and Gaman 2011).

There are many different views on competition in general and competitiveness in particular. According to Porter (1980), competition is firstly based on the ability to maintain low cost of production and then on product differentiation from competitors. The theory's focus is the proposition of the five-force model. According to Porter's theory, in any industry, there are five affecting factors which include the competition among existing companies, the threat of a new entrant entering the market, the risk of substitute products, the role of retail companies and ultimately the power of suppliers.

Thomson and Strickland (1998) proposed internal factors which affect the overall competitiveness of an enterprise based on factors such as image, credibility, technology, distribution network, product development, production costs, customer service, human resources, financial situation, advertising level, and the ability to manage changes. However, this research just identified the factors influencing enterprises' competitiveness and their importance as well as evaluated such factors based on the scoring method in order to compare the capacity among enterprises. This research did not identify the level of individual factor's impact on enterprises' competitiveness.

Kontoghiorghes (2003) identified key predictors of organizational competitiveness in a service firm in the health care service industry. This research took an integrated approach to organizational competitiveness and examined the critical variables for competitive performance in a service organization such as quality, technology, innovation practices and employee involvement and empowerment. The main limitation of this research is that its data was gathered from a single source and was conducted in a service organization in the health care industry. Replicating this study in other industries and environments would help to generalize the results to other settings.

Ambastha and Momaya (2004) pointed out that enterprises' competitiveness is influenced by factors such as: (1) resources (human resources, structure, culture, technology level, assets of enterprises); (2) process (strategy, management process, technological process, marketing process), (3) performance (cost, price, market share, new product development). However, this study only focused on competitiveness of enterprises in general without the differentiation in terms of scale, geography, operation area.

Givi et al. (2010) investigated the competitiveness of the Iranian banking system based on 27 comprehensive indicators of competitiveness. The study used the EFA exploratory factor analysis, the CFA confirmatory factor analysis, and the TOSIS technique to analyze and evaluate the competitiveness of Iranian banks. However, the research only concentrated on financial factors without the overview of other factors such as human resources, technology, marketing ...

Biekpe (2011) empirically investigated the degree of bank competitiveness and intermediation efficiency in Ghana. The study found several reasons accounting for the non-competitive behaviours of banks in Ghana which indirectly created barriers to entry or hamper competition among banks. The identified key factors included high overhead costs, economies of scale, persistently high demand for loans by government, periodic slippages in financial discipline and dominance of a few large banks. This study has some limitations such as the short sample period and the limited availability of both firm level and industry level data.

Sauka (2014) pointed out seven factors affecting competitiveness of firms at Latvia, including: (1) capability to access resources, (2) competences of employees; (3) financial resources, (4) business strategy, (5) environmental impact; (6) business capacity compared to competitors, (7) use of communication networks. However, this study only used statistical methods and did not mention the relationship between factors and competitiveness of enterprises.

Fonseka et al. (2014) investigated the impact of different sources of external financing and internal financial capabilities on competitiveness and sustainability. The study also examined the nature of their relationships related to regulations on external financing in Chinese capital market. The results show that firms' abilities to raise capital from existing shareholders, the public and easy access to bank financing are related positively to an advantage on firm's competitiveness within a industry. This research focused on sources of financial capability of Chinese listed firms' impact on competitiveness and sustainability. Its context was specifically a regulated market. Hence, it is necessary to replicate this study in other contexts.

In Vietnam, there have been some researches on the competitiveness of financial institutions and the operations of financial companies such as:

Mai (2014) with the research "*Impact of technology on the competitiveness of commercial banks*" used panel data from 2010–2015 of five commercial banks. The research results show that banks which increase the level of investment in technology have better competitiveness. However, the research only emphasized the measurement of technological factors and did not mention other factors such as personnel, finance, brand...

Hoang Thi Thanh Hang's research in Ho Chi Minh City (2012) developed a measurement scale of the competitiveness of financial leasing companies and identified the factors which affect the competitiveness of financial leasing companies. The limitation of this research is that the research objects were only leasing companies.

This study uses the Thomson - Strickland method based on the inheritance of 10 internal factors and measures the level of each factor's influence on the competitiveness of financial companies in Vietnam.

3 Methodology and Data

The authors apply the Thompson – Strickland method with the advantage which is that it is not necessary to gather all information about competitors. However, it is essential to have the overview of the market and clearly understand the companies selected as research objects. In this research, survey was conducted for the objects-departmental leaders and staff of 16 financial companies in Vietnam. The survey time period was from 01/2018 to 03/2018. From each company, the researchers randomly selected 20 staff for interviews. The sample size was determined based on the formula $n \geq m * 5$; m is the number of factors; m = 58. Therefore, sample size for survey was 350, higher than the minimum requirement at 290. Such sample size can ensure the reliability.

The authors conducted the survey with 350 samples by stratified sampling method. There were 320 valid samples that could be used as the basis for the research. The authors then inputed the survey data and processed the results using version-20 SPSS Statistics software.

Linear Regression Model

Based on the standard form of the linear regression equation, the competitiveness model for financial companies in Vietnam was constructed as below:

$$Y = \beta_1 X_1 + \beta_2 X_2 + \beta_3 X_3 + \beta_4 X_4 + \beta_5 X_5 + \beta_6 X_6 + \beta_7 X_7 + \beta_8 X_8 + \beta_9 X_9 + \beta_{10} X_{10}$$

Dependent variable Y = Competitiveness

$\beta_1 = > \beta_{10}$: slope coefficient of the relationship between independent variable X*i* and dependent variable Y

Independent variables: X_1, X_2, X_3, X_4, X_5, X_6, X_7, X_8, X_9, X_{10}:

X_1: Financial capacity; This factor includes the observed variables from FIN1 to FIN6

X_2: Management capacity; This factor includes the observed variables from MAN1 to MAN9

X_3: Human resource capacity; This factor includes the observed variables from HR1 to HR5

X_4: Product capacity; This factor includes the observed variables from PRO1 to PRO5

X_5: Marketing Capacity; This factor includes the observed variables from MAR1 to MAR10

X_6: Capacity of service quality; This factor includes the observed variables from SER1 to SER4

X_7: Capacity of interest rate competitiveness; This factor includes the observed variables from INT1 to INT4

X_8: Branding capacity; This factor includes the observed variables from REP1 to REP7

X_9: Technological capacity; This factor includes the observed variables from TEC1 to TEC4

X_{10}: Network development capacity; This factor consists of the observed variables from NET1 to NET3

Based on the proposed research model, the authors provide the following hypotheses for the research:

H1: The financial capacity of financial firms has a positive impact on the competitiveness of these financial companies.

H2: Management capacity of financial firms has a positive impact on the competitiveness of these financial firms.

H3: Human resource capacity of financial firms has a positive impact on the competitiveness of these financial firms.

H4: Product development capacity of financial companies has a positive impact on the competitiveness of these financial companies.

H5: Marketing capacity of financial companies has a positive impact on the competitiveness of these financial companies.

H6: Service quality capacity of financial companies has a positive impact on the competitiveness of these financial companies.

H7: Interest rate competitiveness of financial companies has a positive impact on the competitiveness of these financial companies.

H8: Branding capacity of financial companies has a positive impact on the competitiveness of these financial companies.

H9: Techonological capacity of financial companies has a positive impact on the competitiveness of these financial companies.

H10: Capacity of product/service distribution network development of financial companies has a positive impact on the competitiveness of these financial companies

4 Results and Discussion

Test of Scale Reliability Through Cronbach Alpha Coefficients

Internal factors	Signs	Cronbach Alpha	Comments
Financial capacity	FIN	0.714	Good scale
Management capacity	MAN	0.723	Good scale
Human resource capacity	HR	0.626	Usable scale
Product development capacity	PRO	0.719	Good scale
Marketing capacity	MAR	0.688	Usable scale
Service quality capacity	SER	0.625	Usable scale
Interest rate competitiveness capacity	INT	0.717	Good scale
Branding capacity	REP	0.759	Good scale
Technological capacity	TEC	0.674	Usable scale
Network development capacity	NET	0.801	Good scale

(Source: Results extracted from SPSS)

The results in the above table show that the Cronbach's Alpha coefficients of the 10 groups of factors affecting the competitiveness are different, ranging from 0.801 to 0.626; all the scales are usable. From the 58 observed variables, there were 6 excluded variables which were FIN1, MAN7, MAN3, MAR8, SER3, NET2, and all Corrected Items (Total Correlation) of the remaining 52 variables are greater than 0.3.

After analyzing Cronbach Alpha for independent variables, the researchers conducted Cronbach Alpha analysis for the competitiveness-dependent variable,; the competitiveness variable had the Cronbach Alpha coefficient = 0.798. The variable of product development capacity was excluded because of its correlation with the total variable (0.2713 < 0.3). After the exclusion of the product development capacity variable, the model has: one dependent variable and nine independent variables, following by 47 corresponding observed variables for exploratory factor analysis.

Exploratory Factor Analysis - EFA

The result of EFA for the dependent variable had the KMO = 0.759 ($0.5 \leq$ KMO ≤ 1). This test has statistical significance if Sig. <0.05. However, there were a dispersion of REP-branding capacity variable and the rotation matrix. The researchers excluded this variable from the model; the 8 remaining variables converged at 1 Component and there was no rotation matrix. Therefore, the 8 variables corresponding to the 8 remaining factors were suitable for EFA for dependent variables. The 8 remaining factors were: financial capacity, management capacity, human resource capacity, marketing capacity, service quality capacity, interest rate competitiveness, technological capacity, distribution network development capacity.

KMO and Bartlett's test		
Kaiser-Meyer-Olkin measure of sampling adequacy		.759
Bartlett's Test of Sphericity	Approx. Chi-Square	738.874
	df	36
	Sig.	.000

Source: Test results and analysis of the researchers

The result of EFA for independent variables shows that KMO = 0.762 ($0.5 \leq$ KMO ≤ 1), which was statistically significant when Sig. <0.05.

KMO and Bartlett's test		
Kaiser-Meyer-Olkin measure of sampling adequacy		.762
Bartlett's Test of Sphericity	Approx. Chi-Square	3375.835
	df	535
	Sig.	.000

Source: Test results and analysis of the researchers

The attached observed variables had the Factor loading coefficient >0.5, with the exception of the HR4 observed variable which had the Factor loading coefficient = 0.491 < 0.5. The researchers excluded this variable from the model; the model had 39 observed variables which were splitted and converged into 8 components corresponding to the 8 remaining independent variables in the model. The Factor loading coefficients of all the factors were all greater than 0.5. The 39 observed variables converged into 8 independent variables for correlation analysis and regression analysis.

Pearson Correlation Analysis
In this part of the study, as the correlation is the first condition for regression, Pearson correlation analysis was used to examine the linear correlation between the dependent variable and the independent variables. The suspicion sign was based on the sig value of the correlation among independent variables which was less than 0.05 and Pearson correlation value greater than 0.3. All the values of sig were less than 0.05; it can be concluded that the dependent variable and independent variables had linear correlation with each other; the variables in the model were suitable to be the premise for regression analysis.

Linear Regression Results

Signs	Independent variables	Bêta	Impact direction
NET	Network development capacity	0.362	+
MAN	Management capacity	0.314	+
MAR	Marketing capacity	0.298	+
HR	Human resource capacity	0.257	+
INT	Interest rate competitiveness capacity	0.213	+
FIN	Financial capacity	0.179	+
SER	Service quality capacity	0.152	+
TEC	Technological capacity	0.142	+

Source: Test results and analysis of the researchers

Regression model demonstrating the competitiveness of financial companies in Vietnam:

$$\begin{aligned} \text{COM} \; &= 0.362 * \text{NET} + 0.314 * \text{MAN} + 0.298 * \text{MAR} + 0.257 * \text{HR} \\ &+ 0.213 * \text{INT} + 0.179 * \text{FIN} + 0.152 * \text{SER} + 0.142\text{TEC}. \end{aligned}$$

Coefficients[a]

Model		Unstandardized coefficients		Standardized coefficients	t	Sig.	Collinearity statistics	
		B	Std. error	Beta			Tolerance	VIF
1	(Constant)	.309	.040		7.739	.000		
	FIN	.147	.007	.179	19.918	.000	.818	1.392
	MAN	.172	.008	.314	22.578	.000	.780	1.470
	HR	.165	.007	.257	20.653	.000	.652	1.330
	MAR	.169	.008	.298	22.270	.000	.612	1.405
	SER	.134	.007	.152	18.182	.000	.701	1.426
	INT	.159	.008	.213	20.056	.000	.787	1.456
	TEC	.121	.007	.142	17.544	.000	.674	1.293
	NET	.178	.007	.362	23.921	.000	.862	1.309

a. Dependent Variable: COM
(Source: Results extracted from SPSS)

The results show that the VIF coefficients of all independent variables are less than 10, so there is no multicollinearity phenomenon (Hoang and Chu 2015).

Heteroscedasticity Test

R2 = 0.758. This shows that 75.8% of the variation of the competitiveness of financial firms in Vietnam can be explained by eight independent factors.

Model Summary[b]

Model	R	R Square	Adjusted R Square	Std. Error of the Estimate	Durbin-Watson
1	.779[a]	.759	.758	.070	1.530

a. Predictors: (Constant) FIN, MAN, HR, MAR, SER, INT, TECH, NET

b. Dependent Variable: COM

The result of Durbin - Watson = 1.530 (1 < Durbin - Watson < 3) indicates that the Durbin – Watson coefficient nearly equals 2. This regression result is suitable.

Anova Test

ANOVA[a]

Model		Sum of squares	df	Mean square	F	Sig.
1	Regression	46.100	8	5.638	1169.683	.000[b]
	Residual	1.830	457	.005		
	Total	47.930	465			

a. Dependent Variable: COM

b. Predictors: (Constant), MAN, FIN, MAR, HR, INT, NET, TEC, REP

(Source: Results extracted from SPSS)

At the same time, the ANOVA test shows the Sig coefficient <0.05, indicating that the regression model was constructed in accordance with the collected data.

5 Implications and Conclusions

From the result of regression model: $COM = 0.362 * NET + 0.314 * MAN + 0.298 * MAR + 0.257 * HR + 0.213 * INT + 0.179 * FIN + 0.152 * SER + 0.142TEC$

The distribution network development capacity with the beta coefficient of 0.362 has the greatest impact in the same direction among the internal factors in the model. When the network development capacity of a financial company increases, their competitiveness also increases. Financial companies are using two main distribution channels: one refers to the stores selling products and services of retailers such as Thegioididong (Mobile World), FPT shop, Honda..; the second refers to the provision of products via online channels. Besides, the companies also develop services through mobile applications such as Momo, Epay, Payoo… However, in addition to expanding payment channels by association methods, financial firms need to strengthen their distribution network by identifying key areas with large market share and potential areas to develop appropriate branch networks.

Management capacity with the beta coefficient of 0.314 has impact in the same direction on competitiveness. Therefore, in order to improve management capacity, firstly leaders' competencies should be improved through courses of CFO, CEO, Sales Manager … or foreign professional managers can be hired.

Marketing capacity with the beta of 0.298 has impact in the same direction on competitiveness. Therefore, besides traditional marketing activities, the promotion of advertising on social networking sites such as facebook, zalo, viber … will reduce costs and cause high spreading effect for the companies.

Human resource capacity with the coefficient of 0.257 has impact in the same direction on competitiveness. In fact, human resources of financial companies are large and regularly variable, because a big part of such resources refer to collaborators instead of official employees. Therefore, financial companies need administration policies for collaborators to avoid turbulence which can influence the company operations. Additionally, financial companies need appropriate policies to attract talented employees and improve training programs to develop and enhance the human resource quality.

The interest rate competitiveness with the beta of 0.213 has impact in the same direction on competitiveness. In reality, the competitiveness of interest rates plays an important role and attracts many customers when they approach products and services of financial companies. However, the interest rates of consumer loans of financial companies are still high because those loans are mainly used in the unsecured-loan form which has simple procedures and quick disbursement and does not require customers to prove their income (high risk, high return). The development of attractive and transparent interest rate policies would help financial companies to have more competitive advantages.

The financial capacity with the beta of 0.179 has impact in the same direction on competitiveness. To improve this capacity, the most important thing is to increase the ability to raise capital by diversifying products and increasing charter capital. In addition, the 34/2013/TT-NHNN circular allows non-bank financial institutions, including financial companies, to issue valuable papers such as certificates of deposit for capital mobilization. This regulation gives financial companies more convenience and legal basis for capital mobilization, especially in terms of medium and long-term capital.

The service quality capacity has the beta of 0.152 has impact in the same direction on competitiveness. When the service quality capacity of a financial company increases, the competitiveness of such financial company also increases. The most importants attributes of consumer credits are to serve customers with the professionalism in quality and service, clear and transparent process, simple procedures, quick resolution and quick disbursement. Besides, financial companies also need to take care of customers dedicatedly, build a professional customer service system, timely support requirements and inquiries of customers to improve the service quality and increase competitiveness in the financial market.

Technological capacity with the beta coefficient of 0.142 has impact in the same direction on competitiveness. As a result, financial companies need to focus more on technology in product development and improve distribution channels through technology applications to increase accessibility, meet customer demand and enhance competitiveness.

In conclusion, the consumer lending market has been developing for more than 6 six years, but has grown dramatically from nearly 600,000 billion in 2016 to nearly one million billion in 2019. It indicates that financial companies need to put stronger efforts in competition. Based on the research results of 16 financial companies in Vietnam, the companies need to improve their network and management capacity… in order to improve their competitiveness.

References

Ambastha, A., Momaya, K.: Competitiveness of firms: review of theory, frameworks and models. Singap. Manag. Rev. **26**(1), 45–61 (2004). First half

Biekpe, N.: The competitiveness of commercial banks in Ghana. Afr. Dev. Rev. **23**(1), 75–87 (2011)

CGMA: Porter's Five Forces of Competitive Position Analysis. In: CGMA, 11 July 2013. https://www.cgma.org/resources/tools/essential-tools/porters-five-forces.html. Accessed July 2018

Cişmaş, L., Stan, L.-M.: Avantaj competitiv şi performanţă în contextul responsabilizării sociale a companiilor. Rom. Econ. J. Year XIII **35**(1), 149–173 (2010)

Dobrea, R., Găman, A.: Aspects of the correlation between corporate social responsibility and competitiveness of organization. Rev. Econ. Ser. Manag. **14**(1), 236 (2011)

Fonseka, M.M., Tian, G.L., Li, L.C.: Impact of financial capability on firms' competitiveness and sustainability: evidence from highly regulated Chinese market. Chin. Manag. Stud. **8**(4), 593–623 (2014)

Givi, H.E., Ebrahimi, A., Nasrabadi, M.B., Safari, H.: Providing competitiveness assessment model for state and private banks of Iran. Int. J. Appl. Econ. Financ. **4**(4), 202–219 (2010)

Hoang, T., Chu, N.M.N.: Analyzing Research Data with SPSS. Hong Duc Publisher, Ho Chi Minh City (2015)

Kontoghiorghes, C.: Identification of key predictors of organizational competitiveness in a service organization. Organ. Dev. J. **21**(2), 28 (2003)

Mai, B.D.: The impact of technology on the competitiveness of commercial banks. Vietnam Trade and Industry Review, 27 October 2014

National Financial Supervisor Commission: The overall report of financial market. In: NFSC (2017). http://www.nfsc.gov.vn/bao-cao-giam-sat/bao-cao-tong-quan-thi-truong-tai-chinh-2017. Accessed July 2018

Porter, M.E.: Competitive strategy: techniques for analyzing industries and competition. New York **300**, 28 (1980)

http://tapchicongthuong.vn/tac-dong-cua-cong-nghe-den-nang-luc-canh-tranh-cua-cac-ngan-hang-thuong-mai-20171026062467O3p0c488.htm

Sauka, A.: Measuring the competitiveness of Latvian companies. Balt. J. Econ. **14**(1–2), 140–158 (2014)

Thompson, A.A., Strickland, A.J.: Crafting and Implementing Strategy: Text and Readings. Richard D Irwin (1998)

Multi-dimensional Analysis of Perceived Risk on Credit Card Adoption

Trinh Hoang Nam[1(✉)] and Vuong Duc Hoang Quan[2]

[1] Hochiminh City Banking University, Ho Chi Minh City, Vietnam
namth@buh.edu.vn
[2] Hochiminh City Institute for Development Studies, Ho Chi Minh City, Vietnam
quanvuong.aca@gmail.com

Abstract. Credit card, a combination of non-cash payment and personal consumption credit, is a beneficial and convenient electronic service in the modern bank sector. However, this service has not been used widely in Vietnam. Even in the last few years, consumers tend to reduce using it. Thus, a comprehensive investigation of credit card usage becomes imperative for banks. This study applies an approach of perceived risk to explain consumer's intended use of credit card. Based on data collecting from structured self-administered questionnaires of 228 Vietnamese bank account payers, the analytical results illustrate that the intention to use credit cards is negatively influenced by risk perception, which is synthesized from psychological risk, financial risk, performance risk, security risk, privacy risk, social risk and time risk with decreasing contribution. Some recommendations are made to reduce consumer concerns in order to encourage them signing up and using credit card as a mean of payment for daily expenses.

Keywords: Credit card · Perceived risk · Intended use · SEM
Vietnam

1 Introduction

Facing increasingly fierce competition in the banking sector, banks have been deploying a lot of modern banking services, such as foreign exchange trading, factoring, financial derivatives, etc. Banks also have increasingly applied scientific and technological achievements, especially information technology, communication network as well as internet into banking activities, such as deploying electronic banking services, connecting electronic payment gateways or expanding nationwide network of merchants. Along with debit cards, credit cards are expected to become a popular electronic payment method. In recent years, credit card business is continuously growing both in volume and transaction value. However, this development is not really sustainable as consumers tend to reduce credit card payments [55].

V. Kreinovich et al. (Eds.): ECONVN 2019, SCI 809, pp. 606–620, 2019.
https://doi.org/10.1007/978-3-030-04200-4_43

As a mean of electronic payment, which is indispensable in electronic commerce, consumer adoption of credit cards is affected by their perceived of usefulness, perceived of ease of use, social influence as well as perceived risk, which is considered as a single construct like others in theoretical model [1,52,53]. However, perceived risk is not only a one-dimensional, but also multi-dimensional construct, which is synthesized from many facets, such as financial risk, performance risk, social risk, psychological risk, time risk, security risk and privacy risk [14,23]. This structure of risk perception has been mentioned in many studies on technology acceptance [13,20,30,58].

Therefore, this study aims to fill this gap by firstly proposing consumer's perceived risk as a multi-dimensional construct relevant to their adoption of credit card with data collected from nationwide survey. Based on the findings, some recommendations will be suggested to reduce consumer's risk perception on credit cards, and thereby incite them to sign up and use credit cards as a mean of electronic payment for daily expenses.

2 Literature Review

The concept of perceived risk was introduced in consumer behavior literature by Bauer [4]. Perceived risk is a compensation of potential loss of certain behavior and its importance [43]. In process of purchasing, consumers may confuse about product features as well as the consequences of using them [11]. Murray [37] suggested that perceived risk is perceived uncertainty about potential outcomes after actual behavior performed by consumers, who may not be satisfied by the ambiguity of incurred gains or losses. Stone and Gronhaug [48] defined perceived risk is an expected loss associated with particular buying behavior. Similarly, perceived risk refers primarily to consumer's subjective expectations for incident losses [14]. Chan and Lu [8] argued that consumers are affected by the risks which they are aware, whether these risks exist or not.

Cox and Rich [9] supposed that perceived risk is not a single component, it is overall perception about consumer's uncertainty of a specific buying's situation. Perceived risk is considered as a combination of financial risk. performance risk, physical risk, social risk, psychological risk and time risk [9,43]. With the introduction of residential Internet in early 1990s, privacy risk and security risk were used as two new components of perceived risk [14,23]. This approach of multi-dimension is used in many studies relevant to risk perception on e-commerce, e-payment and e-banking [20,21,26], which are closely linked to credit card [28].

In the context of modern society, there are many unforeseen hazards, that only can be limited but not eliminated completely and rational consumers are concerned about both benefits and uncertainties in making decisions [36]. Consumers are influenced by risks that they perceive [8]. Empirical evidences confirmed the determinant of multi-dimensional perceived risk on consumers' intention to use e-services [14], e-commerce [5,27], e-payment [58] and e-banking [30,44].

Credit card is referred as a mean of payment methods, which involves a concept of buying first and paying later [25]. For simplicity, credit card is best described as "buy now" and "pay later" option, where purchasing amount is limited in credit line. Using credit card as an alternative payment device in place of cash, checks and other forms of payment is an advantage in different situations [49]. Moreover, in the context that electronic commerce is developed increasingly and electronic payment becomes indispensable [35], credit cards, a product of electronic banking, play an important role in electronic commerce business as a favorite and favorable tool for electronic payment [29].

Credit card, however, is not always as good as expected and cardholders may face to potential losses when they do not anticipate all possible outcomes [51]. Some studies found that consumers are more likely to use credit card when they pay less attention to its uncertainties [40,52]. Disagreed with this conclusion, [49, 51,56] asserted that there is no evidence for the casual relationship between consumer's perceived risk and their intention to use credit card. Even when consumers are exposed to unforeseen hazards, they may still highly intent to use credit cards in banks' ongoing efforts to reduce these losses [53]. Despite some differences in results, an approach of single-dimensional perceived risk was applied in these studies. This approach referred perceived risk as a common perception, which is determined by some observed variables, and therefore it did not reflect consumer's valuation of different types of potential losses relevant to credit card use.

Based on the lack of studies on multi-dimensional perceived risk and its influences on consumer's adoption of credit card, this study proposes Credit card multifaceted perceived risk model, in which perceived risk on credit card is multi-dimensional construct, including financial risk, performance risk, psychological risk, social risk, time risk, security risk and privacy risk as described followed.

Perceived risk (PR) refers primarily to the consumers' subjective expectations for incident losses [4,11]. Frambach [17] assumed that a process of acquiring certain behavior is fast or slow depending on how individual perceives losses caused by this behavior. Mitchell [36] also stated that perceived uncertainty may lead consumers change their mind about adopting new technology. Consumer's risk perception about credit card is considered as potential losses of applying and using them [52,54]. Risk perception is found to have negative effect on consumer's adoption of electronic services [5,14,27,58]. Therefore, perceived risk may be considered as a determinant of intention to use credit card.

Hypothesis 1: Consumer's perceived risk on credit card negatively effects on their intention to use credit card.

Perceived financial risk (FIR) refers to potential losses on investment costs as well as maintenance cost of the product or using cost of service [18]. For credit card business, investment cost includes issuing fee, annual fees and other service fees required by credit card issuers [47]. Meanwhile, using cost of credit card is a cost of cash advance, interest at maturity and penalty for late payment [47]. According to Payam and Hamid [40], consumers are afraid that using cost of credit card is always higher than the cost of conventional consumer credit. They are not only anxious that using credit card may let them overspending [57], but also concerned that the payment is failed or the seller does not receive, even though the buyer has been charged in his account [28,44]. Finally, the financial risk on credit card is expanded to include the recurring potential for financial loss due to fraud when cardholders incur transactions that they do not perform [40]. Thus, perceived financial risk can be considered as an important component of overall perception risk on credit card with the following hypothesis:

Hypothesis 2: Financial risk perception increases overall perceived risk on credit card.

Perceived performance risk (PER) refers to potential losses when goods or services are unusable, or used incorrectly as designed or advertised, and therefore do not satisfy buyer's expectations [18]. Credit cardholders are granted a pre-credit by issuing bank to pay their bills and then be responsible for repaying in full and on time [31]. However, consumers are not always successful when making payments via credit card [16]. Credit card payments may be declined because merchants receive cash only [46]. Otherwise, credit card payments are electronic transactions, in which messages are exchanged through computers' network in limited period of time. Therefore, once the process of exchanging message is interrupted or beyond the time allowed, the transaction immediately ends in failure, or finishes successfully with unforeseen results [14,40]. The role of perceived performance risk for overall perceived risk on credit card is stated by the following hypothesis:

Hypothesis 3: Performance risk perception increases overall perceived risk on credit card.

Perceived social risk (SOR) refers to potential losses of status in one's social group as a result of adopting a product or service, looking foolish or untrendy because of unusual consumer behavior [48]. According to Davis et al. [10] consumers accept new technology because they appropriate it and they try to share this view to their relatives. Living in an environment where many acquaintances own and use credit card, consumers cannot stop observing and assessing their beneficial outcomes [22]. They feel inconvenient to not have a credit card and quickly try to own one to integrate into society [25]. The influence of perceived social risk on overall perceived risk relevant to credit card is illustrated in the following hypothesis:

Hypothesis 4: Social risk perception increases overall perceived risk on credit card.

Perceived psychological risk (PSR) is considered as potential losses relevant to consumers feel anxious and depressed during their purchase and use of product or service [36]. In many cases, consumers are disappointed, even dissatisfied when they cannot buy a product, or when they cannot use the purchased product as their expectation neither as suppliers' advertising [11,14]. Credit card provides many benefits for cardholders via electronic devices such as point of sale devices, automatic teller machines and electronic payment gateways [25]. The dependence on electronic devices makes consumers nervous and upset when waiting for licensing as well as dealing with a technical error or unlicensed error [27,58,59]. The contribution of psychological perceived risk to overall perceived risk on credit card is demonstrated as follows:

Hypothesis 5: Psychological risk perception increases overall perceived risk on credit card.

Perceived time risk (TIR) is supposed as the potential losses of time when consumers seek information, buy, use, and repair or replace when purchased product does not result as expected [48]. It may take time to seek and find out merchants that accept credit card payments [3]. In the context of high-tech crime tends to increase, learning how to use credit cards safely and effectively is becoming urgent, requiring serious investment in time, money and effort from consumers [40,58,59]. The hypothesis about the relationship between the perceived time risk and overall perceived risk on credit card is as follows:

Hypothesis 6: Time risk perception increases overall perceived risk on credit card.

Perceived security risk (SER) refers to potential losses when the online trading system is not secure, may be hacked or attacked by cyber criminals, lead to data or network resources are destroyed, disclosure or changed [59]. As a mean of electronic payment, security issues with credit cards are related to the safety of payment transactions which are conducted at electronic devices or the protection of information systems managed by organizations relevant to credit card business [26,33,39]. Once the information systems are not protected properly, crooks can access data, steal personal information, cause physical and mental damage to cardholders [42]. The relationship between perceived security risk and overall perceived risk on credit card is as follows:

Hypothesis 7: Security risk perception increases overall perceived risk on credit card.

Perceived privacy risk (PRR) refers to potential losses of control over personal information [23]. Electronic service providers ask consumers to provide personal information like name, phone numbers, consumption locations, etc. and are responsible for ensuring security and privacy of such information [14]. In credit card business, consumer's privacy information, such as personal information, credit card information, transactional information, is collected and stored by organizations involved in issuing and acquiring process. This information is used for investigating complaints or other purposes with its owner's consent. However, consumer's privacy information may be used for wrong purposes, shared with third parties or published on the Internet without their knowledge or permission [33,44,58]. The hypothesis about the relationship between privacy risk and overall perceived risk on credit cards is stated as follows:

Hypothesis 8: Privacy risk perception increases overall perceived risk on credit card.

Figure 1 illustrates the model for multi-dimensional perceived risk on intention to use credit card. This model consists of a structural equation and a measurement equation:

The structural equation: $\eta = \Gamma * \xi + \zeta$
The measurement equation: $y = \Lambda_y * \eta + \varepsilon$

The structural equation links the first-order factors representing seven facets of perceived risk and adoption of credit card (η) to the second-order factor (overall perceived risk, ξ), where Γ is the pattern coefficient (regression paths) and unique component ζ is the residual, irrespectively. The measurement equation links the observed variables (y) to their respectively hypothesized faceted risk perception as well as credit card adoption. The matrix Λ_y contains the loading of the observed variables on the first-order factors (η). The unique component ζ in the observed variables are represented by unexpected error term ε.

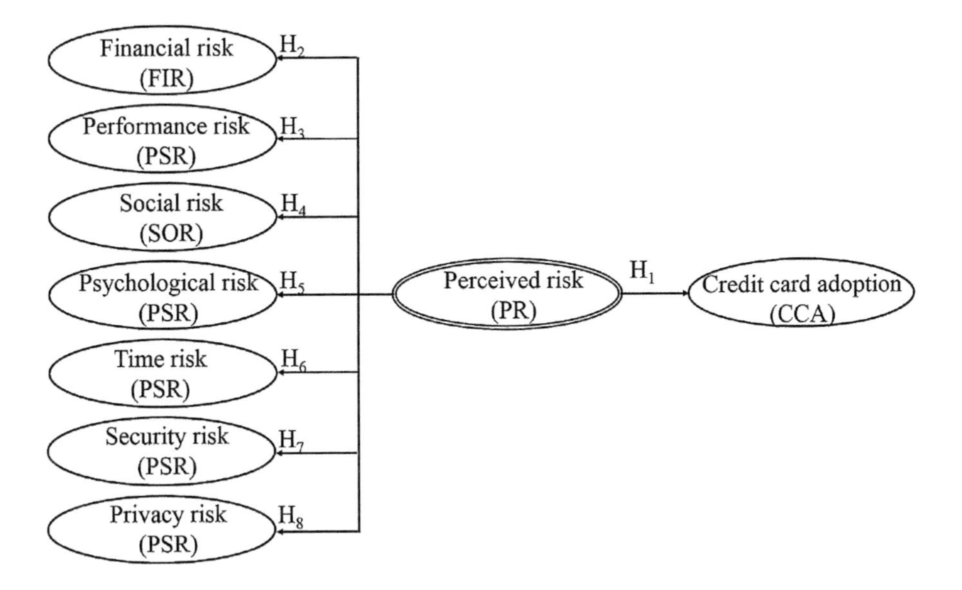

Fig. 1. Proposed research model

3 Methodology

This research conducted an online nationwide survey for bank account payers, who are highly potential for credit card application and use. Moreover, these consumers also have experienced in banking services, especially in using bankcards for paying their daily expenses, and may easily accept an invitation for new credit card or tend to use it in case of owning one.

The 27-item questionnaire focused on seven components of consumers' perceived risk and their intention to use credit card, which are corresponding to the proposed research model. They are the intention to use credit cards (4 items), financial risk (4 items), performance risk (4 items), social risk (3 items), psychological risk (3 items), time risk (3 items), security risk (3 items) and privacy risk (3 items), which are inherited from Amin [1], Payam and Hamid [40], Roy et al. [44] and Yang et al. [58]. This study used 5-point Likert scale from 1-completely disagrees to 5-completely agree.

The survey was conducted by using 246 respondents selected through convenient sampling, of which only 228 valid and usable samples were adopted, yielding a valid response rate of 92.68% among volunteered participants. The data from 228 respondents are compatible to 27 observed variables as recommended by Cattell [7]. Based on collected data, both exploratory factor analysis and confirmatory factor analysis are performed to select and arrange the significant variables to particular factors described in above measurement equation [6,12,19]. Finally, structural equation modeling is used for building the contribution of perceived risk and its effect on credit card adoption [2,6].

4 Findings

4.1 Profile of Respondents and Intention to Use Credit Cards

The data presented in Table 1 provides the demographic details on gender, marital status, occupation, age and highest level of academic qualification of the respondents. Of our samples, majority of the respondents are female (58.3%), married (55.3%) compared to male (41.7%) and single (44.7%). Survey participants are mostly young adulthood with 82.0% of them below the age of 45. The results also show that 52.7% of respondents have university education; 34.6% of them are postgraduates and only 12.7% remaining have college education. Regarding the respondents' occupation, their largest proportion belongs to financial services (46.0%), followed by public services (27.2%), trading services (22.4%) and industries (4.4%). However, the one-way ANOVA tests in comparing means of intention to use credit card insist that there is no significant difference between independent groups divided by these demographic variables, which is inconsistent to prior studies of Lydia et al. [32], Jusoh and Lin [24].

4.2 Factor Analysis

Applying exploratory factor analysis on data collected from survey questionnaires [6,12], 7 factors are extracted from 23 observed variables relevant to different concerns about potential losses on credit card use, which are available to the analysis because their loading factors are greater than 0.5 [19]. These extracted factors are suitable to the proposal model (see Table 2). The KMO coefficient is 0.802 with a statistical significance of 0.000, indicates that the EFA of the independent components is appropriate [19]. A total extracted variance of variables is 66.602%, greater than 50% as required by Anderson and Gerbing [2]. Four observed variables in intention to use credit cards (CCA) have high loading coefficients ($\geq$0.85) and its data variation is well explained ($\geq$78%). Therefore, the measurements were acceptable for Confirmatory Factor Analysis [6].

Table 1. Descriptive statistics and mean comparative analysis

Controlled variable	Frequency	Percent	Mean	ANOVA test
Gender				
Female	133	58.3	3.694	.580
Male	95	41.7	3.755	
Marital status				
Single	102	44.7	3.623	.131
Married	126	55.3	3.794	
Age				
Under 25	6	2.6	4.201	.530
From 25 to 35	105	46.1	3.657	
From 35 to 45	76	33.3	3.773	
From 45 to 55	35	15.4	3.686	
Above 55	6	2.6	3.833	
Regular income (monthly)				
Under 500 USD	37	16.2	3.905	.089
500 USD – 900 USD	93	40.8	3.581	
900 USD – 1,600 USD	64	28.1	3.734	
1,600 USD – 2,600 USD	32	14.0	3.813	
Above 2,600 USD	2	0.9	4.750	
Education				
College graduation	29	12.7	3.853	.392
University graduation	120	52.7	3.652	
Master graduation or higher	79	34.6	3.772	
Occupation				
Industrial	10	4.4	3.550	.498
Trading services	51	22.4	3.672	
Financial services	105	46.0	3.897	
Public services	62	27.2	3.637	

A Confirmatory factor analysis is applied for proposed model with 8 factors and 27 observed variables to examine the model-data fit. Absolut indices, such as Chi-square, Comparative Fit Index (CFI), Tucker and Lewis Index (TLI) and Root Mean Square Error Approximation (RMSEA) are used to evaluate measurement model. Empirical results are shown as follows: Chi-square/df = 1.587, CFI = 0.951, TLI = 0.942 and RMSEA = 0.051. These results indicate that the suggested model is appropriate [34]. Next, the validity of convergence is achievable because all factor loadings are greater than 0.5 (Table 2) and significant

Table 2. Factor analysis results

Constructs		EFA loading factors	CFA loading factors
Perceived Security Risk (SER): CR = 0.876; AVE = 0.703			
SER1	Payment technology is out of date	.837	.843
SER2	Payment devices are not secured properly	.916	.860
SER3	Payment systems may be attacked or hacked	.743	.812
Perceived Privacy Risk (PRR): CR = 0.910; AVE = 0.772			
PRR1	Others will know my personal details	.890	.897
PRR2	Others will misuse my personal details	.909	.891
PRR3	I will lose control of my personal details	.828	.846
Perceived Performance Risk (PER): CR = 0.851; AVE = 0.592			
PER1	Bill cannot be paid by credit cards	.751	.770
PER2	Credit cards do not solve my requirements	.816	.814
PER3	Using credit cards may lead to overspending	.782	.761
PER4	Credit cards are not good as advertisement	.729	.730
Perceived Financial Risk (FIR): CR = 0.874; AVE = 0.635			
FIR1	It will cost me money to use credit card	.759	.775
FIR2	I may be responsible for what I do not perform	.849	.824
FIR3	There is no compensation for lost money	.755	.794
FIR4	I am probably unable to repay on time	.814	.794
Perceived Time Risk (TIR): CR = 0.811; AVE = 0.598			
TIR1	It takes time to learn how to use	.843	.837
TIR2	It takes time to perform transactions	.812	.820
TIR3	It takes time to solve problems	.657	.649
Perceived Social Risk (SOR): CR = 0.874; AVE = 0.702			
SOR1	My relatives discourage me	.732	.782
SOR2	I am judged negatively by others	.860	.858
SOR3	I look foolish to others	.923	.871
Perceived Psychological Risk (PSR): CR = 0.745; AVE = 0.522			
PSR1	I feel anxious	.768	.704
PSR2	I feel frustrated	.764	.891
PSR3	I feel depressed	.636	.525
Credit card adoption (CCA): CR = 0.936; AVE = 0.787			
CCA1	I am desire to use	.944	.953
CCA2	I use it as soon as possible	.879	.861
CCA3	I will use it usually in the future	.875	.899
CCA4	I will encourage others to use	.853	.831

t-statistics [2]. Table 2 shows that the internal consistency reliability of measurements is accepted when overall reliability coefficient is greater than 0.6 and the values of corrected item-total correlation are greater than 0.3 for all observed variables [38,45]. Moreover, the AVE values of these constructs are between 0.522 and 0.787, which are greater than 0.5 as well as squares of their correlation coefficients, respectively, then each construct is a distinct construct and discriminant validity is acceptable [15]. Therefore, CFA results confirm that 27 observed variables are extracted into 8 constructs as well as the measurements are model-data fit, discriminant validity, unidimensionality, convergence validity and internal consistency reliability.

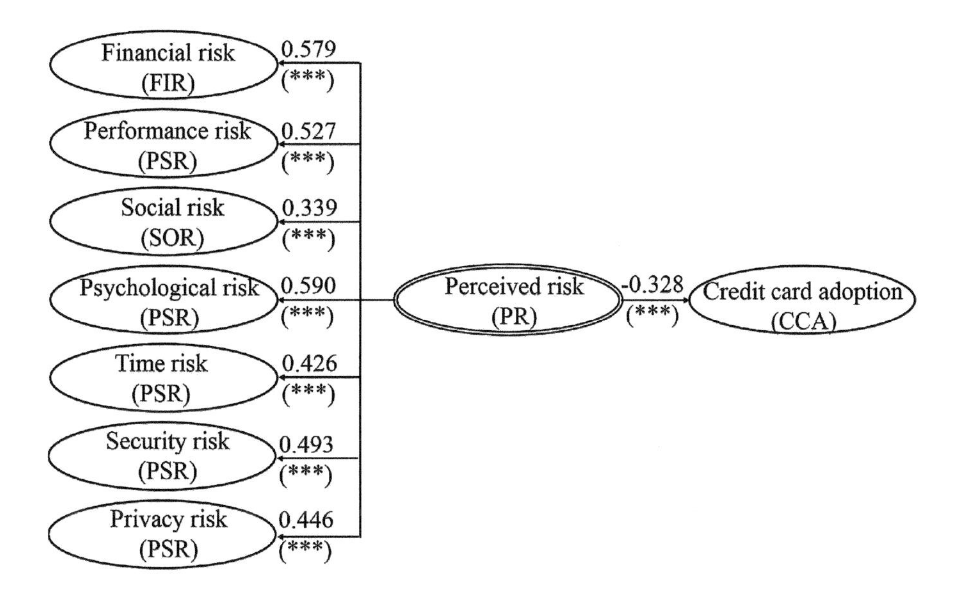

Fig. 2. Proposed research model and the result of SEM

4.3 Structural Equation Modeling

A Structural Equation Model (SEM) is conducted to test the structural model including consumer's multi facets of perceived risk and their intention to use credit cards, which are measured by 27 observed variables as mentioned in above CFA analysis. Consumer's overall perceived risk is synthesized from its dimensions and has direct influence on their adoption of credit cards. Figure 2 shows the whole SEM for the proposed model. All indicators (Chi-square/df = 1.625, CFI = 0.944, TLI = 0.938 and RMSEA = 0.052) show that the proposed model is appropriate for data collected from the market [34]. The result of SEM is described in Table 3. Whereby, consumer's perceived risk on credit cards is significantly contributed from financial risk, performance risk, social risk, psychological risk, time risk, security risk and privacy risk with coefficients of 0.579,

0.527, 0.339, 0.590, 0.426, 0.493 and 0.446, respectively. Next, consumers' perceived risk accounted 10.7% of the variance in their intention to use credit card with coefficient of −0.328. Therefore, all hypotheses are accepted.

Table 3. Results of the structural equation model

Hypothesis	Relationship			Estimate	S.E.	C.R.	P.	Result
H1	PR	→	CCA	−.328	.144	−3.486	***	Accepted
H2	FIR	←	PR	.579			***	Accepted
H3	PER	←	PR	.527	.181	4.507	***	Accepted
H4	SOR	←	PR	.339	.153	3.424	***	Accepted
H5	PSR	←	PR	.590	.199	4.449	***	Accepted
H6	TIR	←	PR	.426	.191	3.969	***	Accepted
H7	SER	←	PR	.493	.200	4.430	***	Accepted
H8	PRR	←	PR	.446	.203	4.253	***	Accepted

5 Discussion

Many unforeseen hazards are present in modern society, which can only be limited but not eliminated completely [36]. Consumers are influenced by the risks that they perceive [4,11], whether these risks are realistic or not [8]. The analytical results demonstrated that consumers' perceived risk significantly influences their intention to use credit cards ($\beta = -0.328$; $p = 0.000$). This result is consistent with prior worldwide studies on adoption of electronic commerce [21], electronic banking [30,59] as well as electronic payment [20,59].

The SEM results also proved that perceived risk is a multi-dimensional construct, which is synthesized from many facets, including financial risk, performance risk, social risk, psychological risk, time risk, security risk and privacy risk as proposed by Jarvenpaa and Todd [23], Featherman and Pavlou [14]. These perceived dimensions are found to significantly increase consumer's overall perceived risk on credit card, which are consistent with prior studies in studies on different electronic services [13,14,30,58,59].

The findings show that consumers are strongly concerned about psychological, financial and performance risks; conversely, they are less interested in time and social risks. However, security and privacy issues have been found to be considered moderately by consumers, although these dimensions are primarily concerned in studies on consumers' adoption of electronic financial services [26,27,39,49].

Thus, the research reaches its initial goal to determine the role of risk perception on intended use of credit card. Perceived risk, a potential loss of applying and using credit card, is synthesized from many types of losses. Finally, it is significant barrier of credit card adoption.

6 Conclusion

Expanding credit card business sustainably for banks can be a complex issue. The findings of this study indicate that Vietnamese consumers are seriously concerned about potential losses on credit card, which make them tend to close down on this mean of payment methods. Unlike to prior studies on credit card, this study proposed perceived risk as a second-order factor in a causal relationship with intention to use, which is shared with first-order factors representing different types of risks. Thus, this study, which contributes to the perceived risk theory with perceived risk as a multi-dimensional structure within the scope of modern banking research, is appropriate.

The study provides some managerial implications for bank managers. As the findings imply, banks should focus their resources on overcoming the risk aspects, which can help motivating potential consumers. Banks should advertise that credit card is not a risky service by providing positive reviews at point of sales or in mass media. The publicity of loss protection policies and service-level agreements may reduce potential losses of performance or finance. Additional effective risk-preventing policies may include money back guarantees, so that consumers feel more comfortable and safe with the system. Once consumers are fully aware of credit card, they are willing to accept and use it.

Although this study provides evidence for the multifaceted structure of consumer's perceived risk and its influence on their intention to use credit card, it still has several limitations. As a mean of electronic payment, consumer's intended behavior towards credit card may be explained by their perceived usefulness, perceived ease of use as well as social influences, which are primary constructs in consumer behavior literature. Otherwise, consumers are always rationale, they may accept to use credit card in case that they are aware of need to use or negative changes in work, earnings as well as spending in the future. Thereby, future studies should address these shortcomings.

References

1. Amin, H.: An analysis of mobile credit card usage intentions. Inf. Manag. Comput. Secur. **15**, 260–269 (2007)
2. Anderson, J.C., Gerbing, D.W.: Predicting the performance of measures in a confirmatory factor analysis with a pretest assessment of their substantive validities. J. Appl. Psychol. **76**, 732–740 (1991)
3. Arpita, K., Anshuman, K., Singh, S.: Factors affecting credit card use in India. Asia Pac. J. Mark. Logist. **24**, 236–256 (2012)
4. Bauer, R.A.: Consumer Behavior as Risk Taking, Dynamic Marketing for a Changing World, pp. 389–398. American Marketing Association, Chicago (1960)
5. Bianchi, C., Andrews, L.: Risk, trust, and consumer online purchasing behaviour: a Chilean perspective. Int. Mark. Rev. **29**, 253–276 (2012)
6. Byrne, B.M.: Structural Equation Modeling with Amos: Basic Concepts, Applications and Programming, 2nd edn. Taylor & Francis Group, New York (2010)
7. Cattell, R.B.: Matched determiners vs. factor invariance: a reply to Korth. Multivar. Behav. Res. **13**, 431–448 (1978)

8. Chan, S.C., Lu, M.T.: Understanding internet banking adoption and use behavior: a Hong Kong perspective. J. Glob. Inf. Manag. **12**, 21–43 (2004)
9. Cox, D.F., Rich, S.U.: Perceived risk and consumer decision making - the case of telephone shopping. J. Mark. Res. **1**, 32–39 (1964)
10. Davis, F.D., Bagozzi, R.P., Warshaw, P.R.: User acceptance of computer technology: a comparison of two theoretical models. Manag. Sci. **35**, 982–1003 (1989)
11. Derbaix, C.: Perceived risk and risk relievers: an empirical investigation. J. Econ. Psychol. **3**, 19–38 (1983)
12. Fabrigar, L.R., Wegener, D.T., MacCallum, R.C., Strahan, E.J.: Evaluating the use of exploratory factor analysis in psychological research. Psychol. Methods **4**, 272–299 (1999)
13. Farzianpour, F., Pishdar, M., Shakib, M.D., Toloun, S.H.: Consumers' perceived risk and its effect on adoption of online banking services. Am. J. Appl. Sci. **11**, 45–56 (2014)
14. Featherman, M.S., Pavlou, P.A.: Predicting e-services adoption: a perceived risk facets perspective. Int. J. Hum. Comput. Stud. **59**, 451–474 (2003)
15. Fornell, C., Larcker, D.F.: Evaluating structural equation models with unobservable variables and measurement error. J. Mark. Res. **18**, 39–50 (1981)
16. Forsythe, S.M., Shi, B.: Consumer patronage and risk perceptions in internet shopping. J. Bus. Res. **56**, 867–875 (2003)
17. Frambach, R.T.: An integrated model of organizational adoption and diffusion of innovations. Eur. J. Mark. **27**, 22–41 (1993)
18. Grewal, D., Gotlieb, J., Marmorstein, H.: The moderating effects of message framing and source credibility on the price-perceived risk relationship. J. Consum. Res. **21**, 145–153 (1994)
19. Hair, J., Black, W., Babin, B., Anderson, R., Tatham, R.: Multivariate Data Analysis, 7th edn. Pearson Education, Upper Saddle River (2014)
20. Hamid, N.R., Cheng, A.Y.: A risk perception analysis on the use of electronic payment systems by young adult. Inf. Sci. Appl. **10**, 26–35 (2013)
21. Hong, I.B.: Understanding the consumer's online merchant selection process: the roles of product involvement, perceived risk, and trust expectation. Int. J. Inf. Manag. **35**, 322–336 (2015)
22. Ismail, S., Amin, H., Shayeri, S.F., Hasim, N.: Determinants of attitude towards credit card usage. J. Pengur. **41**, 145–154 (2014)
23. Jarvenpaa, S.L., Todd, P.A.: Consumer reactions to electronic shopping on the world wide web. Int. J. Electron. Commer. **1**, 59–88 (1996)
24. Jusoh, Z.D., Lin, L.Y.: Personal financial knowledge and attitude towards credit card practices among working adults in Malaysia. Int. J. Bus. Soc. Sci. **3**, 176–185 (2012)
25. Kaynak, E., Harcar, T.: Consumer's attitudes and intentions towards credit card usage in an advanced developing country. J. Financ. Serv. Mark. **6**, 24–39 (2001)
26. Khalilzadeh, J., Ozturk, A.B., Bilgihan, A.: Security-related factors in extended UTAUT model for NFC based mobile payment in the restaurant industry. Comput. Hum. Behav. **70**, 460–474 (2017)
27. Kim, D.J., Ferrin, D.L., Rao, H.R.: A trust-based consumer decision-making model in electronic commerce: the role of trust, perceived risk, and their antecedents. Decis. Support. Syst. **44**, 544–564 (2008)
28. Kim, K.K., Prabhakar, B., Park, K.S.: Trust, perceived risk, and trusting behavior in internet banking. Asia Pac. J. Inf. Syst. **19**, 1–23 (2009)
29. Laudon, K.C., Traver, C.G.: E-commerce: Business, Technology, Society. Pearson Education, Upper Saddle River (2014)

30. Lee, M.C.: Factors influencing the adoption of internet banking: an integration of TAM and TPB with perceived risk and perceived benefit. Electron. Commer. Res. Appl. **8**, 130–141 (2009)
31. Lee, J., Kwon, K.N.: Consumers' use of credit cards: store credit card usage as an alternative payment and financing medium. J. Consum. Aff. **36**, 239–262 (2002)
32. Lydia, G.L., Maysami, R.C., Koh, H.C.: Singapore credit cardholders: ownership, usage patterns, and perceptions. J. Serv. Mark. **22**, 267–279 (2008)
33. Manzano, J., Navarre, C., Mafe, C., Blas, S.: Key drivers of internet banking services use. Online Inf. Rev. **33**, 672–695 (2009)
34. McDonald, R.P., Ho, M.-H.R.: Principles and practice in reporting structural equation analyses. Psychol. Methods **7**, 64–82 (2002)
35. Meier, A., Stormer, H.: eBusiness & eCommerce: Managing the Digital Value Chain. Springer, Berlin (2009)
36. Mitchell, V.W.: Consumer perceived risk: conceptualisations and models. Eur. J. Mark. **33**, 163–195 (1999)
37. Murray, K.B.: A test of service marketing theory: consumer information acquisition activities. J. Mark. **55**, 10–25 (1991)
38. Nunnally, J.C., Berstein, I.H.: Pschychometric Theory, 3rd edn. McGraw-Hill, New York (1994)
39. Ooi, K.B., Tan, G.W.: Mobile technology acceptance model: an investigation using mobile users to explorer smartphone credit card. Expert. Syst. Appl. **59**, 33–46 (2016)
40. Ozturk, A.B.: Customer acceptance of cashless payment systems in the hospitality industry. Int. J. Contemp. Hosp. Manag. **28**, 801–817 (2015)
41. Payam, H., Hamid, R.K.: The mediating role of the dimensions of the perceived risk in the effect of customers' awareness on the adoption of internet banking in Iran. Electron Commer. Res. **12**, 151–175 (2012)
42. Polasik, M., Wisniewski, T.P.: Empirical analysis of internet banking adoption in Poland. Int. J. Bank Mark. **27**, 32–52 (2009)
43. Roselius, T.: Consumer rankings of risk reduction methods. J. Mark. **35**, 56–61 (1971)
44. Roy, S.K., Balaji, M.S., Ankit, K., Sekhon, H.: Predicting internet banking adoption in India: a perceived risk perspective. J. Strat. Mark. **25**, 418–438 (2016)
45. Schumacker, R.E., Lomax, E.G.: A Beginner's Guide to Structural Equation Modeling, 3rd edn. Taylor and Francis Group, New York (2010)
46. Shih, H.P.: An empirical study on predicting user acceptance of e-shopping on the web. Inf. Manag. **41**, 351–368 (2004)
47. Simon, J., Smith, K., West, T.: Price incentives and consumer payment behaviour. J. Bank. Financ. **34**, 1759–1772 (2010)
48. Stone, R.N., Gronhaug, K.: Perceived risk: further considerations for the marketing discipline. Eur. J. Mark. **27**, 39–50 (1993)
49. Tan, G.W., Ooi, K.B., Chong, S.C., Hew, T.S.: NFC mobile credit card - the next frontier of mobile payment. Telemat. Inform. **31**, 292–307 (2014)
50. Teoh, W.M., Chong, S.C., Yong, S.M.: Exploring the factors influencing credit card spending behavior among Malaysians. Int. J. Bank Mark. **31**, 481–500 (2013)
51. Tseng, S.Y.: Bringing enjoy shopping by using CC: the antecedents of internal beliefs. J. Econ. Econ. Educ. Res **17**, 16–28 (2016)
52. Tu, T.T., Chang, H.H., Chiu, Y.H.: Investigation of the factors influencing the acceptance of electronic cash stored-value cards. Afr. J. Bus. Manag. **5**, 108–120 (2011)

53. Varaprasad, G., Chandran, K.S., Sridharan, R., Unnithan, A.B.: An empirical investigation on credit card adoption in India. Int. J. Serv. Sci. Manag. Eng. Technol. **4**, 13–29 (2013)
54. Vuong, D.H.Q., Trinh, H.N.: Perceived risk and the intention to use credit cards. Int. Res. J. Financ. Econ. **159**, 76–89 (2017)
55. Vuong, D.H.Q., Trinh, H.N.: Vietnam credit card market: much potential and ineffective. Econ. Stud. **57**, 47–56 (2017)
56. Wang, Y.M.: Determinants affecting consumer adoption of contactless credit card: an empirical study. Cyber Psychol. Behav. **11**, 687–689 (2008)
57. Wang, L., Wei, L., Jiang, L.: The impact of attitude variables on the credit debt behavior. Nankai Bus. Rev. Int. **2**, 120–139 (2011)
58. Yang, Q., Pang, C., Liu, L., Yen, D.C., Tarn, J.M.: Exploring consumer perceived risk and trust for online payments: an empirical study in China's young generation. Comput. Hum. Behav. **50**, 9–24 (2015)
59. Zhao, A.L., Lloyd, S.H., Ward, P., Goode, M.M.: Perceived risk and Chinese consumers' internet banking services adoption. Int. J. Bank Mark. **26**, 505–525 (2008)

Public Services in Agricultural Sector in Hanoi in the Perspective of Local Authority

Doan Thi Ta[1(✉)], Thanh Vinh Nguyen[1], and Hai Huu Do[2]

[1] Academy of Politics Region I, Ho Chi Minh City, Vietnam
doanta07@gmail.com, vinhthanhhv1@gmail.com
[2] Ho Chi Minh City University of Food Industry, Ho Chi Minh City, Vietnam
haidh1975@gmail.com

Abstract. The term of "*public service*" is being harshly debated in the scientific forums with respect to its connotation and denotation to realize public services in social and economic development. The typical research of Hanoi's agricultural sector is extremely essential to take proper interventions and state management of agricultural public services. The research has demonstrated the direct relationship between the influential factors and the agricultural development in Hanoi as follows: (i) market access has direct and positive impacts on Agricultural development in Hanoi; (ii) agricultural production information supports and production supports have direct and negative impacts on Agricultural development in Hanoi.

Keywords: Public services · Public investment · Agriculture
Economic development · Information supports
Production supports and market access

1 Overview of Public Services in Agricultural Development

1.1 Domestic and International Researchers

Addressing the issue of public services supporting agricultural development, the relevant documents may include:

Rainey (1973), "*Public Services in Rural Areas*", analyzes the provisions of public services such as education, water, public safety, roads, telephones and entertainment in rural areas. The research shows a gap in the quality, access and satisfaction of public services between rural and urban areas in the way that rural areas use lower-quality public services. Therefore, measures are proposed to improve the efficiency of public services and public investment in rural areas of the United States.

Luo and Wang (2009), "*Problems in Rural Public Service and its Countermeasures: Investigation on Rural Areas of Ningxia Hui Autonomous Region*

V. Kreinovich et al. (Eds.): ECONVN 2019, SCI 809, pp. 621–635, 2019.
https://doi.org/10.1007/978-3-030-04200-4_44

China", Asian Agricultural Research Journal 9, analyzes and evaluates public services in rural areas (Ningxia Hui) on three aspects, of educational, medical, and cultural services. The result shows that public service provision in rural areas is severe shortage which hinders fulfilment with actual needs of rural people, rural development and agricultural production.

Mogues et al. (2009), "*Access to and Governance of Rural Services: Agricultural Extension and Drinking Water Supply in Ethiopia*" considers extending access to agricultural encouragement and water supply services in. The result demonstrates that access to public services in rural Ethiopia is extremely low and there are differences among groups and regions. Although the level of satisfaction with public services in rural zones is quite high, only 8% of farmers use these services. Therefore, the study shows that to effectively provide public services for agriculture and rural areas, it is important to take into account needs, perceptions, conditions of farmers, besides improving the capacity of public services.

Ming and Junmin (2010), "*Equalizing Essential Public Service and Poverty Reduction under the Background of Development Mode Transformation*", Research Institute of Fiscal Science, Ministry of Finance analyzes the need to ensure fair public services in poverty reduction target. According to the authors, equity in access to essential public services will improve the quality and development of communities in disadvantaged areas, thereby poverty reduction process to be implemented sustainably. Essential public services will provide the infrastructure and conditions for production development in heavily poverty impacted areas. On the other hand, public services ensure basic living standards and minimize the risk of re-engage in the poverty cycle.

Hu et al. (2010), "*Effects of Inclusive Public Agricultural Extension Service - Results from a Policy Reform Experiment in Western China*", IFPRI research paper profess that the commercialization of China's rural public services began in the 1990s, which led to poor farmers' serious lack of access to public services. In recent years, the Chinese government has taken basic measures to meet farmers' access to public services based on three main perspectives: (1) considering farmers are target-centered, (2) determining the level of effectiveness of farmers' need for service extension, and (3) building a better and accountable system for public services.

Tervo (2011), "*Accessibility Analysis Of Public Services In Rural Areas Under Restructuring*", University Oulu, used the GIS approach to examine access to public services in rural Finland, particularly by population groups: adolescents, mature workers, the elderly in rural areas.

World Bank (2011), "*Strengthening the Management of Agriculture Public Services*", considers activities that support the redevelopment and improvement of agricultural support services in Haiti after the devastating earthquake, from which measures are proposed to strengthen the management and delivery of public agricultural support services.

Anh et al. (2011), "*Equitability in Access to Rural Public Services in Vietnam: An Outlook from the Red River Delta*", in International Business and

Management Vol. 2 deliver an idea on the importance of ensuring social equity in accessing public services as the foundation for promoting development. In the transition to a socialist-oriented economy, rural market and public services in Vietnam are moving towards decentralization. Public service users have to pay for the service fee, and charge-free services are no longer provided.

1.2 Understanding of Public Services in the Agricultural Sector

1.2.1 Definition of Public Service

Public services are closely associated with the category of public goods. In the economic sense, public goods have some basic characteristics. *Firstly*, non-excludability means that the kind of goods does not prevent anyone from its use. *Secondly*, non-rivalry means that when a good is consumed, it does not reduce the amount available for others. *Thirdly*, non-removability means that when goods are not consumed, public goods still exist. Generally, goods with all three characteristics are called pure public goods and those with none of characteristics are called non-pure public goods.

From the perspective of state authorities, administrative researchers suppose that public services are activities of state agencies in the performance of state administrative functions and assurance of supplying public goods for the common and essential needs of the society. This understanding emphasizes the role and responsibility of the state for supplying public goods and claims that the prime feature of public services is to fulfill the essential needs of society and community, the implementation of that activity can be undertaken by the state or private.

In Vietnam, regardingless of power functions, such as legislative, executive, judiciary, diplomacy, etc., the States functions of service provision for the society highly emphasizes on the States role in providing services to the community. It is important to distinguish public service activities (*called public welfare activities*) from the power administrative activities according to government policies in order to eliminate the bureaucracy, subsidy mechanism, offload the state apparatus, exploit all potential resources and improve the quality of public services. The State does not have the monopoly of providing public services, but the State can completely socialize some services, thereby needing to give the non-state sector (or the private sector) a part of the supply of some services, such as health, education, water supply and drainage, etc.

It can be seen that the concept and scope of public services are approached at various perspectives despite general characteristics of serving social essential needs and interests. Even if a part of the public service distribution is shifted by from the State to the private sector, the State still has a regulatory role to ensure fairness in the distribution of these services and to overcome market unsoundness. From the above characteristics, "*public service*" is understood in the following broad sense and the narrow sense:

In the broad sense, public services are goods and services that the Government intervenes in providing to ensure effectiveness and equitability. Therefore, public services are all activities carrying out the inherent functions of the Government, including from the issuing of policies, laws, courts, etc., to public health, education and transport activities.

In the narrow sense, public services are goods and services that directly serve the needs of organizations and citizens and the Government intervenes in providing to ensure effectiveness and equitability.

From the above different perspectives, the authors state that: *Public services are services that directly serve the essential common interests and needs of organizations and citizens; the State undertakes or authorizes the non-state establishments to supply these services in order to ensure the social order, equality and development.*

1.2.2 Analysis Framework of Public Services

1.2.2.1 Research Model

Public services in meeting the needs of agricultural production from the stages of the agricultural production chain: *(1) Production information supports*, refers to the supply of information of programs and development projects in agricultural production; construction and expertise of programs and projects; legal assistance (legal documents, procedures, contracts, relevant certificates, etc.); training plans and programs for human resources in agricultural production; preservation and storage of gene sources, gene funds of plants/animals; quality test and control of plants and animals; supply of seeds, gene sources, transplantation and genetic conservation (Fig. 1).

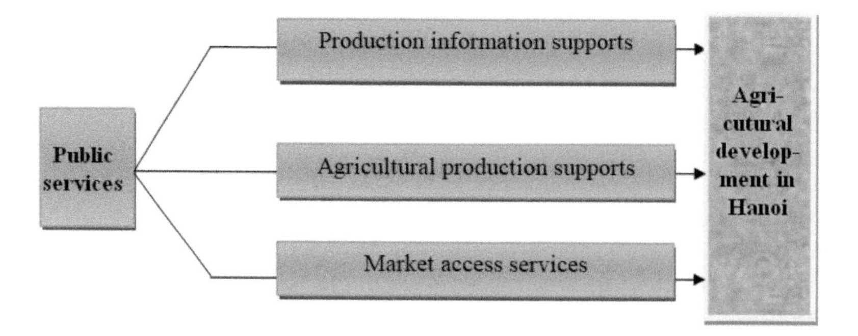

Fig. 1. Research model

(2) Agricultural production supports, refers to the application of scientific and technological achievements (production technology, biotechnology, information technology, etc.); agricultural promotion, financial and credit supports for agricultural production; supports for seeds and materials for agricultural production; rural environmental sanitation related to production and processing of agricultural products (garbage, emissions, waste water, etc.); supply of intra-field roads; dedication of staff and scientists in the agricultural and industrial

promotion units in agricultural production; disaster prevention, contract and legal right guarantees in agricultural production.

(3) Market access services, refers to the provision of market-oriented information of agricultural production; branding services, customer care in agriculture and trade promotion; the provision of information of quantity, quality and price through the trade exchange of agricultural products; disaster prevention, insurance and legal protection right for farmers.

1.2.2.2 Public Services in Agricultural Sector

Agriculture is a concept of industry or sector (the others: industry and service), including those that use soil, water and grassland as primary production materials and also a base of two other sectors. Based on the standard industrial classification of General Statistics Office of Vietnam, it can be seen that agriculture is one of the three sectors of the economy (agriculture, industry and service), comprising subsectors such as agriculture, forestry and fisheries (see Fig. 2 below). Agriculture produces basic materials of the society, using soil for cultivation and livestock and exploiting plants and animals as the main raw materials to produce food and a number of raw materials for industry. Agriculture is a large production industry covering many sub sectors: cultivation, livestock and primary processing of agricultural products, in a broader sense, it also includes forestry and fisheries. This industry is divided into two main types: (i) Subsistence farming features limited inputs and family self-sufficiency in outputs; (ii) Intensive farming is the field of agricultural production that is specialized in all stages of agricultural production, including the use of machinery in cultivation, livestock, or processing of agricultural products.

On the basis of a general approach, public services in agriculture promote the agriculture and rural development and relates to the agricultural production process from "*input*" to "*output*" (such as supplies of agricultural materials, seeds, agricultural promotion, technology transfer, plant protection, veterinary, quality control, irrigation, mechanization for agricultural production, agricultural credit, agricultural insurance, trade promotion and information, etc.) to serve agricultural production.

- *Production information supports*: The highest authority, a state administrative organ, is responsible for handling issues of organizations and citizens (such as providing administrative, judicial documents, certificates and licenses). Administrative and managerial services are carried out in accordance with the authority, order and procedures by law in content such as: (i) Expertise in programs, projects, schemes and plans; Preparation of guidelines, criteria, major standards, etc.; (ii) all kinds of licenses (such as import and export licenses of plants and animals, import licenses for plant protection and veterinary drugs, license to practice veterinary medicine and supply veterinary services, plant protection and animal feed, license of fishermen and fishing vessel, license of quarantine and field trials, etc.); (iii) relevant certificates (such as certificates of trading in animals, plants and fisheries, certificates of safe zones, certificate of practice of disinfection vaporization, certificates of origin of seeds, animal feed, certificate of product qualification, certificate of plant protection drug, etc.).

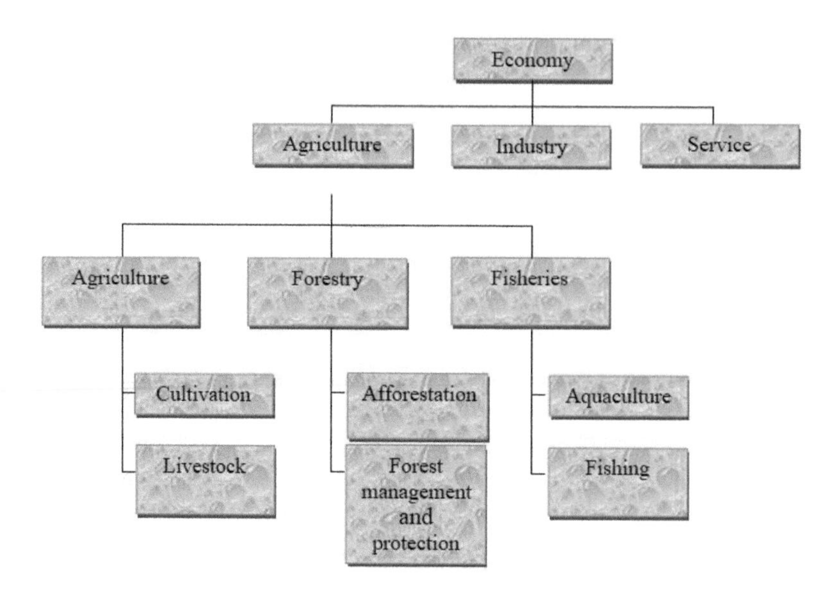

Fig. 2. Diagram of subsectors/sectors of the national economy. Source: Vietnam Statistical Yearbook 2017

- *Agricultural production supports*: With legal services associated to public administrative services, the Ministry of Agriculture and Rural Development is responsible for organizing and managing services in decentralization and authority of administrative units, public welfare units, specialized agencies and individuals to take charge of the functions and the provisions of law. Supports for agricultural production and business such as: (1) Supports for the construction of agricultural infrastructure (including electricity, roads, schools, stations, domestic water, planning and irrigation canals); (2) Technical supports (scientific research, technology transfer, agricultural production, agricultural promotion, irrigation and information services); (3) Testing services, agricultural promotion services, centers servicing agricultural production (from seeds, animals with original seeds, gene sources of plants and animals as well as gather of agricultural products and consumption of agricultural products); (4) Support and vocational training services in agriculture, rural areas and shift of employment structure; (5) financial services for agricultural production such as credit, business loans, savings and agricultural insurance; (6) forecasting services of agricultural calendar based on plants, seasons, weather and temperature, humidity and agro-forestry-fishery markets (price, quantity, etc.).

- *Market access services*: With the rapid development of the market-oriented economy, especially when Vietnam becomes the official member of World Trade Organization (WTO), information access is especially necessary not only for business enterprises, but also for the efficiency of agricultural productivity of famer households. Thank to useful information sources, households can decide what inputs, what price, and how many outputs, etc. In the context of international and domestic agro-market, there are more complex changes and changeable price, information updates become the necessary and vital condition to ensure that agricultural production sales with the best price. At the same time, trade promotion and brand of products through agricultural associations, trade fairs and sales activities are conducted; enhancement of promoting Vietnamese trademarks, access of good origin and geographical indications of industrial products to international and domestic consumers. In addition, it must be taken into account with prevention of risk factors and diseases in agricultural production; farmer support, production regulation and guidance of buying agricultural insurance, access to agricultural product markets, weather forecast information and prevention of natural calamities in agricultural production.

1.2.2.3 Assessment Criteria for Public Services in Agricultural Sector Development

Firstly, Assessment Criteria for Production Information Supports

- Approval of development strategy of agricultural industry/sector (q1.9a)
- Approval of programs and projects (q1.9b)
- Construction and appraisal services for programs, projects, plans (q1.9c)
- Legal assistance (legal documents, procedures, contracts, relevant certificates, etc.) (q1.9d)
- Training programs and plans of agricultural human resources (q1.9e)
- Preservation and storage of gene sources (q1.9f)
- Quality test and control of seeds of plants and animals (q1.9g)
- Research of transplantation of rootstock seeds, gene sources and conservation (q1.9h)

Secondly, Assessment Criteria for Agricultural Production Supports

- Application of scientific achievements (q1.10a)
- Agricultural promotion, financial and credit supports (q1.10b)
- National reservation for production and processing (q1.10c)
- Rural environmental sanitation services (waste, gas, waste water, etc.) (q1.10d)
- Supply of intra-field system (q1.10e)
- Serving quality of staffs and scientists (q1.10f)
- Disaster prevention (q1.10g)
- Insurance, contract and legal rights (q1.10h)

Finally, Assessment Criteria for Market Access Services

- Information of seeds, fertilizers and materials (q1.11a)
- Information supply, customer care in agriculture and trade promotion (q1.11b)
- Weather forecast (q1.11c)
- Supply of books, libraries and communication, etc. related to agricultural promotion (q1.11d)

1.2.2.4 Methods of Analysis

- Statistical and comparative, model and econometric methods: Database and forecasting the impact of factors, specifically:
  - Reliability analysis is used to measure the reliability and effectiveness of measures (criteria). Based on the Likert scale (from 1-Very bad to 5-Very good), the eigenvalue estimation method is applied and only those with eigenvalue >1 will be retained in the analytical model. The retention factor closely correlates with the remaining variables when the correlation coefficient in the Rotated Component Matrix is greater than 0.5.
  - Bi - variate correlation is used to testify relationships of control variable pairs and analyze multiple regression so that the relationship between dependent and independent variables can be determined. Based on the Likert scale (from 1-Very bad to 5-Very good), the reliability rating through the Cronbach Alpha coefficient is used to exclude unreliable variables of less than 0.6 and corrected item total correlation of less than 0.4.
  - Factor analysis is used to group variables through multiple regression analysis into groups of factors that need to be adjusted in the analytical model. The data analysis factor correlates closely with the remaining variables when the coefficient in the Rotated Component Matrix must be greater than 0.5, and based on Chi-square with P value less than 0.005, TLI and CFI must be greater than 0.8.
- *(Primary) Database*: The information was collected via delivering questionnaires for officials at local authorities. The research was conducted by a team of experts who analyzed and selected typical and reliable agricultural production in Hanoi such as Thach That District, Ba Vi District And Soc Son District. Besides, the discussion on agricultural production issues from beginning stage to the way of organizing production as well as finding markets for agricultural products also played a role in this research process. At the same time, we also pointed out the problems that have not been solved and what should be deployed in such a challenging agricultural production today. The 210 questionnaires were distributed in three local districts (*66 in Thach That District, 85 in Ba Vi District and 59 in Soc Son District*) within 5 months from March to August 2017 to collect information from agricultural activities.
- *Collection and processing process*: In the data processing method, encrypted data was input into SPSS software for descriptive statistics, plotting, regression analysis and kernel analysis. The study of factors based on correlated relationships in regression analysis and the statistical significance of variables in factor analysis aimed at finding solutions to adjust the variables of research that generate agricultural production improvement.

2 Survey Results on Public Services in the Field of Agriculture in Hanoi

Research results from 210 questionnaires in three local districts (*66 in Thach That District, 85 in Ba Vi District and 59 in Soc Son District*) were obtained through the quantitative model and the results of the study are presented as follows:

Firstly, we test the scale through the Cronbach's Alpha reliability coefficient to exclude unreliable variables in order that these variables cannot make up the dummy factor when we analyze the EFA-Exploratory Factor Analysis factor. Cronbach's alpha coefficient must be greater than 0.6 and the correlation coefficient of the total value of each scale must be greater than 0.3. The results of Cronbach Alpha analysis in Table 1 show that all scales are satisfactory and are used for factor analysis.

Table 1. Summary of reliability and relative minimum variables of scales- Source: SPSS extracted by the authors, 2017

No.	Scale	Cronbach's Alpha reliability coefficient	Number of observed variables
1	Information supports for production (q1.9a_q1.9h)	0,863	8
2	Agricultural production supports (q1.10a_q1.10h)	0,860	8
3	Market access (q1.11a_q1.11d)	0,828	4
4	Agricultural development in Hanoi (q1.9_q1.11)	0,931	20

After Cronbach's Alpha analysis, followed by EFA analysis with the Principal Component Analysis method and Varimax rotation to evaluate the uniqueness, convergence value and discriminant value of the scale. It is clear that KMO = 0.854 > 0.5, which is sufficient for factor analysis in the integration of variables, namely: With a sample size of 210, factor loading of observed variables must be greater than 0.5; the variables converge on the same factor and distinguish from the others. As a result of the analysis below, all factor loading of the observed variables are greater than 0.5; Bartlett (significance level = 0.000) has a coefficient of KMO = 0.912 with 16 variables (eliminated by variables of q1.11e to q1.11h; q1.10b, q1.10c, q1.10e, q1.10f, q1.10g; q1.9a, q1.9c, q1.9g, q1.9h;), after EFA analysis was extracted into 3 factors with wrong 67.82% (over 50%). In the end, we incorporate the data for the final result of the factor analysis in the model given in each rotation matrix correlation, as follows (Figs. 3 and 4):

Kaiser-Meyer-Olkin Measure of Sampling Adequacy.		.854
Bartlett's Test of Sphericity	Approx. Chi-Square	1064.723
	df	55
	Sig.	.000

Fig. 3. KMO and Bartlett's Test - Source: SPSS extracted by the authors, 2017

	Component		
	1	2	3
q1.11a	.889		
q1.11b	.767		
q1.11c	.724		
q1.9b	.575	.558	
q1.9e		.809	
q1.9f		.763	
q1.9d		.662	
q1.10d			.729
q1.11d			.713
q1.10h			.687
q1.10a			.639

Fig. 4. Rotated component matrix - Source: SPSS extracted by the authors, 2017 Extraction Method: Principal Component Analysis. Rotation Method: Varimax with Kaiser Normalization (Rotation converged in 7 iterations.)

Finally, the authors have applied CFA to test the suitability of the research model with the data collected after preliminary evaluation based on Cronbach's Alpha and EFA reliability coefficients in the previous section.

The results of the CFA analysis to test the suitability of the research model with the market data are as follows: Chi-square = 183,533; df = 71; P = 0.000 < 0.05; Chi-square/df = 2.585 < 5; GFI = 0.893, TLI = 0.907, CFI = 0.928 > 0.9; RMSEA = 0.08 = 0.08. Thus, the analytical indices satisfy the standard conditions and ensure that the research model is perfectly matched with to the market data (Fig. 5).

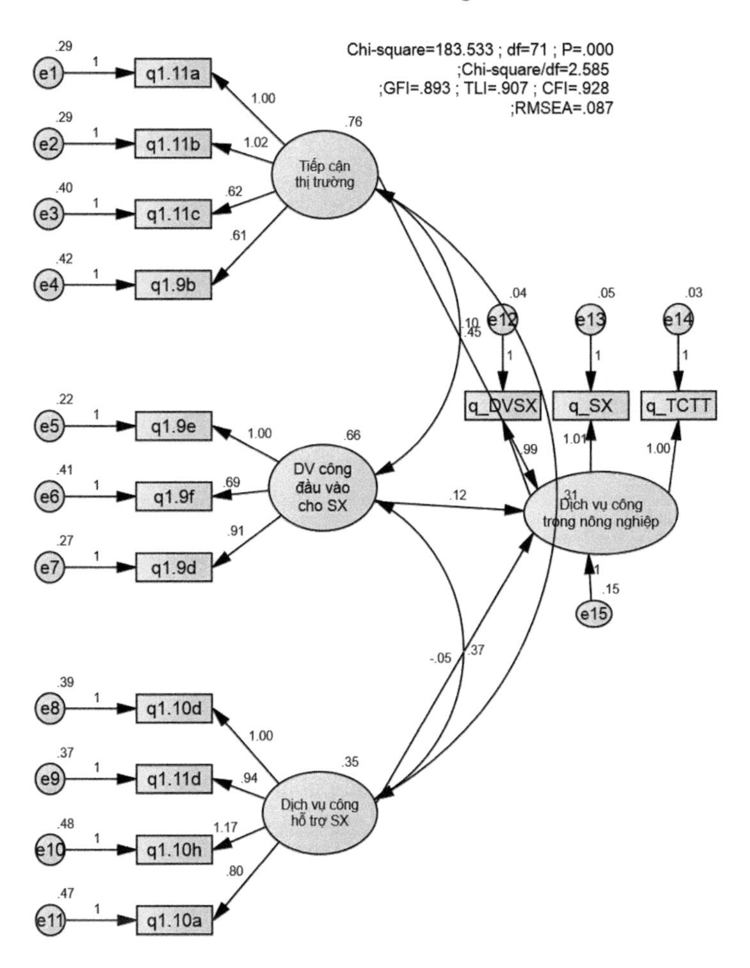

Fig. 5. Test research model - Source: By the authors, 2017

Structural Equation Modeling (SEM) Analysis shows testing results of the relation among three factors: Market access; production information supports; Public Services for production support and agricultural development in Hanoi (Fig. 6).

It can be seen from the observation of analysis results that, not all relationships in the theoretical model are significant at P-value < 0.05. Examining the direct relationship between the influencing factors (Market Access, Input Services for Production, Public Service for Production Support) and Hanoi Agricultural Development adds results: (i) Market access services, both directly and negatively impacting Hanoi's agricultural development (0,096); (ii) information for production (0. 117) and public support services (−0,047), both directly and positively impacting on Hanoi's agricultural development. From this, the study draws on the hypothetical test results in Table 2.

			Estimate	S.E.	C.R.	P	Label
Agricultural development in Hanoi	<---	Market access	.096	.051	1.897	.058	
Agricultural development in Hanoi	<---	Information supports	.117	.077	1.525	.127	
Agricultural development in Hanoi	<---	Production supports	-.047	.106	-.441	.659	

Fig. 6. Analysis results of Structural Equation Modeling (SEM) - Source: By the authors, 2017 *(Estimate: average estimated value; SE: standard error, CR: critical value, P: Probability level; ***: p< 0,001)*

Table 2. Hypothesis testing results - Source: By the authors, 2017

Hypothesis	Description	Result
H1	Market access has positive impacts on agricultural development in Hanoi City	Accepted
H2	Information supports have positive impacts on agricultural development in Hanoi City	Rejected
H3	Public services for production have negative impacts on agricultural development in Hanoi City	Rejected

1. *The hypothesis test results was accepted*, public services the most significantly impacted on agricultural development in Thach That, Ba Vi and Soc Son districts. Supply of Information, approval of programs and projects (q1.9b) associated with agricultural development; weather forecast information (q1.11c), information of seed, fertilizer and materials, information (q1.11a); customer information care in agriculture and trade promotion (q1.11b) ... will positively impact other groups of service factors.
2. *The hypothesis test results are rejected*, indicating that the public service in agriculture needs to be implemented in a uniform manner and confirming inverse correlation of the factors extracted from the model public services need to be invested in order to flourish in agricultural development such as: (i) Legal aid (legal documents, procedures, contracts, certification of documents ...) (q1.9d); Training programs and plans of agricultural human resource (q1.9e); Preservation and storage of gene sources original seed, gene fund of animals/plants (q1.9f); (ii) rural environmental sanitation (garbage, emission, waste water ...) (q1.10d); Application of scientific achievements (q1.10a); Insurance, contract and legal rights (q1.10h); Supply of book, library and communication, ... related to agriculture promotion work (q1.11d).

3 Recommendations on Public Services in Agricultural Development in Hanoi

Firstly, enhancement of implementation and development of 15 public services in the field of agriculture and rural development in a comprehensive manner, including: (1) Forest Protection and Development; (2) Conservation, rescue, restoration of forest ecology and resources; (3) Preservation of specimens in the forestry subsector; (4) Assay, test and quarantine livestock breeds, animal feeds and breeding environment; (5) Assessment and monitor the quality of breeding animals, animal feeds and biological products for environmental improvement in animal husbandry; (6) testing of pesticides; (7) survey for measures to prevent harmful organisms and diseases to protect production; (8) assessment of pests and diseases; (9) agricultural promotion; (10) sperm insemination, high-quality cows and high-yield pigs in the city; (11) assay of crop and forest plants, aquatic breeds; (12) preservation and storage of original seed, purebred seed; (13) assessment and certificate of agricultural products and materials in conformity with standards and technical regulations; (14) assessment and certificate of management process and system of production, preliminary processing and processing of agricultural products; (15) test and inspection of agricultural environment, agricultural materials.

Secondly, the entities involved in the coordination and implementation of public services have clear responsibilities and obligations: The Department of Agriculture and Rural Development actively advises, develops and submits to the People's Committee for approval of economic and technical norms and cost norms for public services using the state budget in the field of agriculture and rural development in the city. The Finance Department appraises to the City People's Committee price brackets or prices of public services in the field of agriculture and rural development of the city on the basis of the economic and technical norms and cost norms approved by the competent agencies. The right to promulgate and the roadmap shall be fully calculated according to regulations. Selection of public service units to offer public service in the field of agriculture and rural development in the form of mission assignment, order or bidding decentralized by the city and current regulations is conducted after consulting with the financial agency of the same level. Guidance of the public service units of the city which provide services in the field of agriculture and rural development is organized and implemented according to regulations.

Finally, enhancement of the capacity of state management at all levels in the agricultural sector should be done, including complete of the project aiming at strengthening the contingent of agricultural cadres at commune level; strengthening the management, supporting and improving quality and efficiency of business production of agricultural cooperatives; Strictly managing and inspecting the quality of breeds, agricultural materials and hygiene and food safety, specifically: (1) To strictly manage the production process according to regulations; To intensify the inspection and examination of the quality of plant seeds, livestock breeds, fertilizers, plant protection drugs, animal feeds - aquatic products, veterinary drugs and microorganisms in service of cultivation and husbandry, ... to well perform the epidemic prevention for plants, animals and aquatic products in the city; (2) To strengthen state management measures, analyze and certify quality; To coordinate with inter-branch inspection forces in inspecting hygiene and food safety and inspecting origins of agricultural products and food circulated and consumed in the capital; (3) To agricultural products of the provinces brought back to Hanoi, agriculture setting up a process, system of quality inspection, assessment and certificate for products brought to Hanoi for consumption; Inter-provincial links on administrative procedures related to slaughter management, food safety certificates, transport of agricultural products and foodstuffs to Hanoi and vice versa.

References

Anh, L.H., Giam, D.Q., Lam, B.T., Huyen, V.N., Cuong, T.H.: Equitability in access to rural public services in Vietnam: an outlook from the red river delta. J. Int. Bus. Manag. **2**, 209–218 (2011)

Decision No. 17/2012/QD-UBND, approving the master plan for agricultural development in Hanoi by 2020, vision to 2030, dated July 9, 2012 by the People's Council of Hanoi

Decision No. 3748/QD-BNN-KH, approving the development orientation of plant and animal varieties by 2020, vision to 2030, dated September 15, 2015 by the Ministry of Agricultural and Rural Development

Decision No. 27/2017/QD-UBND, guiding the implementation of Resolution No. 25/2013/NQ-HDND dated December 04, 2013 by the People's Council of Hanoi on the incentive policies for the development of specialized agricultural production areas in Hanoi for the period of 2014-2020, dated August 18, 2017 by the People's Committee of Hanoi

Decision No. 28/2017/QD-UBND, guiding the implementation of Resolution No. 03/2015/NQ-HDND dated July 08, 2015 by the People's Council of Hanoi on several policies for Hanoi Hi-tech Agricultural Development Program for the period of 2016–2020, dated August 7, 2017 by the People's Committee of Hanoi

General Statistics Office of Vietnam: Statistical Yearbook 2016, Vietnam (2016)

Hai, D.H.: Corporate Culture – An Intellectual Peak, Monograph, Transportation and Communication Publishing House, Hanoi (2016)

Hai, D.H.: Analysing the effects of the exporting on economic growth in Vietnam. Springer, Cham (2017)

Ha, D.T.H.: The State Management of Public Services. Scientific and Technical Publishing House (2007). Luo, Q., Wang, J.: Problems in rural public service and its countermeasures: investigation on rural areas of Ningxia Hui autonomous region China. J. Asian Agric. Res. (9) (2009)

Tervo, M.: Accessibility Analysis of Public Services in Rural Areas under Restructuring. University Oulu, The EU (2011)

National Assembly: Law on Government Organization, No. 76/2015/QH13 dated June 19, 2015 (2015)

Rainey, K.D.: Public Services in Rural Areas. Publication ERIC, The USA (1973)

Hu, R., Cai, Y., Chen, K.Z., Cui, Y., Huang, J.: Effects of Inclusive Public Agricultural Extension Service - Results from a Policy Reform Experiment in Western China, IFPRI discussion papers 1037, International Food Policy Research Institute (IFPRI) (2010)

Ming, S., Junmin, L.: Equalizing Essential Public Service and Poverty Reduction under the Background of Development Mode Transformation. Research Institute of Fiscal Science, Ministry of Finance (2010)
Thanh, C.V.: Public Service and Socialization of Public Services: Some Theoretical and Practical Issues. National Political Publishing House, Hanoi (2004)
Mogues, T., Cohen, M.J., Birnern, R.: Access to and Governance of Rural Services: Agricultural Extension and Drinking Water Supply in Ethiopia, ESSP II Discussion Paper 8, The Ethiopia (2009)
World Bank: Strengthening the Management of Agriculture Public Services. Publish World Bank, Washington, D.C. (2011)

Public Investment and Public Services in Agricultural Sector in Hanoi

Doan Thi Ta[1(✉)], Hai Huu Do[2], Ngoc Sy Ho[1], and Thanh Bao Truong[1]

[1] Academy of Politics Region I, Ho Chi Minh City, Vietnam
doanta07@gmail.com, ngocho.hvct@gmail.com, trbaothanh@yahoo.com
[2] Ho Chi Minh City University of Food Industry, Ho Chi Minh City, Vietnam
haidh1975@gmail.com

Abstract. The terms of "public investment" and "public services" are being harshly debated in scientific forums with respect to its connotation and denotation to implement the legal provisions of the Vietnam Law on Public Investment 2014 into social life and economic development. In order to take proper steps in policy intervention and state management of public investment in agricultural sector, a typical case study of Hanois agriculture is indeed needed. Research results show that: (i) market access directly and positively has impacts on Agricultural development in Hanoi City (0.096); (ii) research and development (0,117) and production and processing (−0.047) have direct and negative impacts on Agricultural development in Hanoi City.

Keywords: Public investment · Public services · Agriculture Economy · Economic development · Etc.

1 Introduction

1.1 Definition of Public Investment

Under the Vietnam Law on Public Investment 2014, from the perspective of the law: "*Public investment is the investment by the state in programs and projects to build socio-economic infrastructure as well as the investment in programs and projects serving socio-economic development.*", therefore, public investment can be understood as follows: Firstly, public investment involves all content related to the investment and use of state owned capital, including the investment or investment support of state owned capital in non-profit socio-economic development programs and projects; Secondly, it refers to the investment and business activities that use state owned capital, especially the management of investment activities of state-owned enterprises. For example, with regard to the resources of investment, the investments of any kind and for any purpose are all public investment if the capital is owned by the state, not by any individual or legal entity; however, in terms of investment purpose, public investment would only mean the investment in non-profit community service programs and projects. Hence,

V. Kreinovich et al. (Eds.): ECONVN 2019, SCI 809, pp. 636–659, 2019.
https://doi.org/10.1007/978-3-030-04200-4_45

public investment is the investment of state sector, including: investment from State budget (allocated to central ministries and localities); investment under national target program (normally non-profit); investment credit (at reasonable interest rates); and investment from SOEs (linked to profit targets). And, the objects of public investment are: (i) Investment in socio-economic infrastructure programs and projects; (ii) Investment in the operation of state agencies, non-business units, political organizations, and socio-political organizations; (iii) Investment and support of the supply of public utilities and services; (v) State investment in project implementation in the form of public-private partnerships.

1.2 Definition of Public Services

Public services are closely associated with the category of public goods. In the economic sense, public goods have some basic characteristics. Firstly, non-excludability means that the kind of goods does not prevent anyone from its use. Secondly, non-rivalry means that when a good is consumed, it does not reduce the amount available for others. Thirdly, non-removability means that when goods are not consumed, public goods still exist. Generally, goods with all three characteristics are called pure public goods and those with none of characteristics are called non-pure public goods.

In Vietnam, regardingless of power functions, such as legislative, executive, judiciary, diplomacy, etc., the States functions of service provision for the society highly emphasizes on the States role in providing services to the community. It is important to distinguish public service activities (called public welfare activities) from the power administrative activities according to government policies in order to eliminate the bureaucracy, subsidy mechanism, offload the state apparatus, exploit all potential resources and improve the quality of public services. The State does not have the monopoly of providing public services, but the State can completely socialize some services, thereby needing to give the non-state sector (or the private sector) a part of the supply of some services, such as health, education, water supply and drainage, etc.

1.3 Public Investment and Public Services in the Agricultural Sector

1.3.1 Agriculture and Agricultural Sector

Agriculture is a concept of industry or sector, including those that use soil, water and grassland as primary production materials (as foundation) of the second sector (industry), and the third sector (service). Based on the standard industrial classification of the Vietnam General Statistics Office in 2016, it can be seen that agriculture is one of the three sectors of the economy (agriculture, industry and service), comprising sub sectors such as agriculture, forestry and fisheries (see Fig. 1 below):

Agriculture produces basic materials of the society, using soil for cultivation and livestock and exploiting plants and animals as the main raw materials to

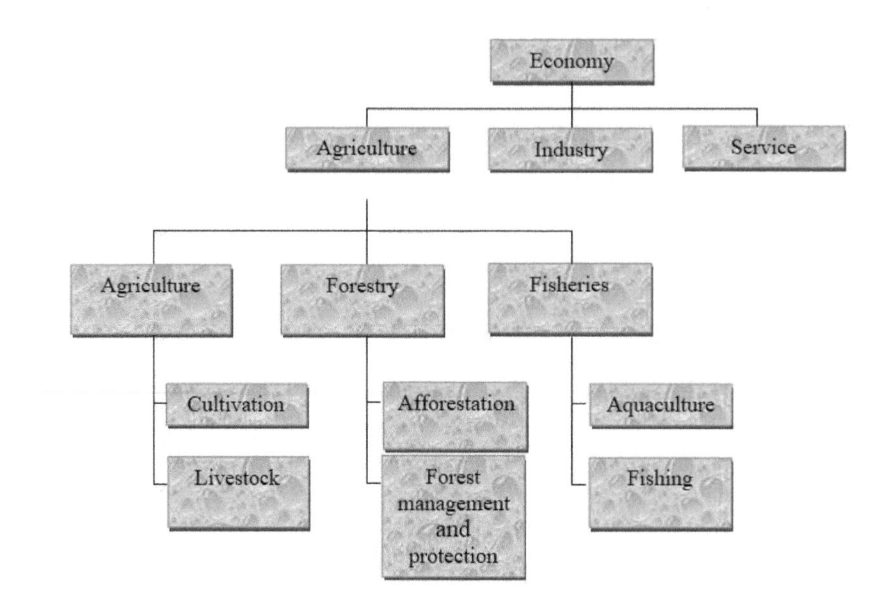

Fig. 1. Diagram of sub sectors/sectors of the national economy - Source: Vietnam Statistical Yearbook 2017

produce food and a number of raw materials for industry. Agriculture is a large production industry covering many sub sectors: cultivation, livestock and primary processing of agricultural products, in a broader sense, it also includes forestry and fisheries. This industry is divided into two main types: (i) Subsistence farming features limited inputs and family self-sufficiency in outputs; (ii) Intensive farming is the field of agricultural production that is specialized in all stages of agricultural production, including the use of machinery in cultivation, livestock, or processing of agricultural products.

1.3.2 Public Investment in Agricultural Sector

- *Investment in agricultural production planning*: Before implementing any project, one must base on the master plan and detailed plan to implement the designed items. Planning work in agriculture is understood as the process of research and design of plans associated with the specific nature of socio-economic conditions. Therefore, public investment in planning means the investment in overall and detailed planning to serve as a basis for the development of agriculture in the coming time. At the same time, planning work shall be carried out by practical programs and projects to facilitate the prediction of enterprises and farmers in a lasting and sustainable way.
- *Investment in agricultural infrastructure*: Development of agricultural infrastructure is to invest in the construction of systems of technical and material elements that are deployed on the agricultural production space, such as irrigation systems, infield canals, reservoir system and internal roads to serve the

production; building programs and projects on the construction of quarantine stations of animals, agricultural products and supplies, etc.; centers that provides information about legal issues, policies, prices and other information related to agricultural development; and even infrastructure system to attract domestic and foreign enterprises to invest in agricultural development, such as microbiology industry, processing industry, supply of seeds and breeds to farmers, etc.

- *Investment in research, transfer and application of science and technology in agricultural production*: The State will take the main functions and tasks, such as implementing the research, transfer and application of scientific and technological achievements at home and abroad to the fields of production and local life; Carrying out the contracts for technology transfer and scientific research; Organizing scientific research programs and projects, and transferring technology in the form of pilot models, and wide-scale replication in agricultural production; Selecting technical advances and new technological processes related to new plants and animals suitable with local conditions. Science and technology is an integral link in agricultural production to improve the quality of agricultural products. Science and technology changes farming practices, towards professionalism in agricultural production from seeds and breeds, production processes to supplying products to market.
- *Investment in trade promotion and advertising*: To well carry out the trade promotion and advertising, the State shall have to build programs and projects to promote Vietnams agricultural products to domestic and foreign customers. Public investment involves the trade promotion and advertising activities to expand the market, bring the brand to customers, provide geographical indication and origin, establish partnerships and organize exhibitions.

1.3.3 Public Services in Agricultural Sector

On the basis of a general approach, public services in agriculture promote the agriculture and rural development and relates to the agricultural production process from "input" to "output" (such as supplies of agricultural materials, seeds, agricultural promotion, technology transfer, plant protection, veterinary, quality control, irrigation, mechanization for agricultural production, agricultural credit, agricultural insurance, trade promotion and information, etc.) to serve agricultural production.

- *Supply of legal services related to agricultural production*: B The highest authority, a state administrative organ, is responsible for handling issues of organizations and citizens (such as providing administrative, judicial documents, certificates and licenses). Administrative services and administrative management are carried out in accordance with the authority, order and procedures by law in content such as: (i) Expertise in programs, projects, schemes and plans; Preparation of guidelines, guidelines, standards, industry standards, etc.; (ii) all kinds of licenses (such as import and export licenses of plant and animals, import licenses for plant protection and veterinary drugs, license to practice veterinary medicine and supply veterinary services, plant protection and animal feed, license of fishermen and fishing vessel, license of quarantine and field trials, etc.); (iii) relevant certificates (such as certificates of trading in animals, plants and fisheries, certificates of safe zones, certificate of practice of disinfection vaporization, certificates of origin of seeds, animal feed, certificate of product qualification, certificate of plant protection drug, etc.).

- *Regulation on business activities*: With legal services associated to public administrative services, the Ministry of Agriculture and Rural Development is responsible for organizing and managing services in decentralization and authority of administrative units, public welfare units, specialized agencies and individuals to take charge of the functions and the provisions of law. Support services for agricultural production and business such as: (1) support services for the construction of agricultural infrastructure (including electricity, roads, schools, stations, domestic water, planning and irrigation canals); (2) support services for technique (scientific research, technology transfer, agricultural production, agricultural promotion, irrigation and information services); (3) testing services, agricultural promotion services, centers servicing agricultural production (from seeds, animals with original seeds, gene sources of plants and animals as well as gather of agricultural products and consumption of agricultural products); (4) Support and vocational training services in agriculture, rural areas and shift of employment structure; (5) financial services for agricultural production such as credit, business loans, savings and agricultural insurance; (6) forecasting services of agricultural calendar based on plants, seasons, weather and temperature, humidity and agro-forestry-fishery markets (price, quantity, etc.).
- *Information support services, trade promotion (product consumption) and other prevention services*: With the rapid development trend of the market-oriented economy, especially when Vietnam becomes the official member of WTO (World Trade Organization), information access is especially necessary not only for business enterprises, but also for the efficiency of agricultural productivity of famer households. Thank to the source of useful information households can decide what inputs, what price, and how many outputs, etc. In the context of international and domestic agro-market, there are more complex changes and changeable price, information updates become the necessary and vital condition to ensure that agricultural production sales with the best price. At the same time, trade promotion and brand of products through agricultural associations, trade fairs and sales activities are conducted; enhancement of promoting Vietnamese trademarks, access of good origin and geographical indications of industrial products to international and domestic consumers. In addition, it must be taken into account with prevention of risk factors and diseases in agricultural production; farmer support, production regulation and guidance of buying agricultural insurance, access to agricultural product markets, weather forecast information and prevention of natural calamities in agricultural production.

2 Research Model of Public Investment in Agriculture

2.1 Research Model

Public investment in agricultural production are taken in the following steps (Fig. 2):

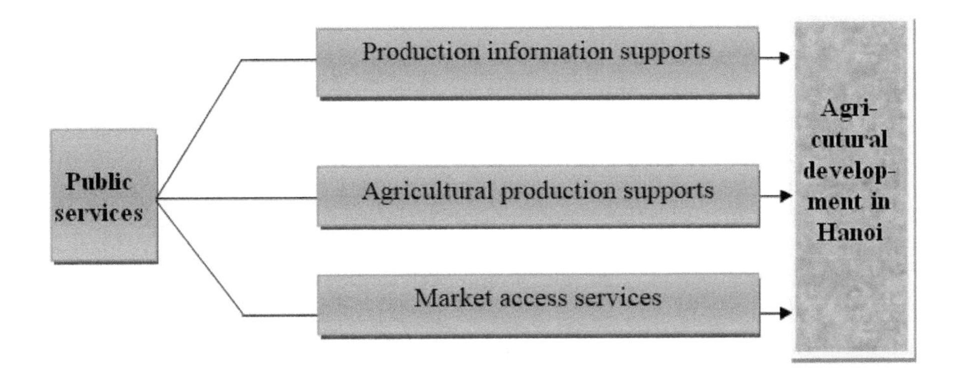

Fig. 2. Research model - Source: By the authors

(1) *Public investment in research and design* is research on seeds and issues of transplantation; agricultural development planning; policy development and instructional process associated with agricultural production; investment in the construction of laboratories and production experiments related to research, conservation and development of seed sources.
(2) *Public investment in production and processing*, is the investment and supporting process of agricultural production activities such as building, developing and protecting branch and trademark of agricultural products; building safe production processes in agriculture; ensuring conditions for the implementation of safe processes agricultural production; developing of agricultural production infrastructure (irrigation, interconnected transportation, storage and preservation); regionally linking in production; investing in the construction of infrastructure servicing agricultural production (inter-field connection and connected transport) as well as building a sample model in the direction of safety in agricultural production.
(3) *Public investment relating to the provision of information*, market and distribution is the process of investing and building input-market research centers (breed, process, technology, fertilizer, ...); output-market research centers (price information, market demand ...); trading platform of agricultural produce connecting information between buyers and sellers through contracts and legal information; the system of means connected to machines, equipment and electronic information ensuring the process of trading goods conveniently in the transaction of local agricultural products.

Public services in meeting the needs of agricultural production from the stages of the agricultural production chain:

(1) *Public services in research and design*: is the process of supply of information of programs and development projects in agricultural production; construction and expertise of programs and projects; legal assistance (legal documents, procedures, contracts, relevant certificates ...); training plan and programs for human resource in agricultural production; preservation and storage of gene sources, gene fund of plants/animals; quality test and control of plants and animals; supply of seeds, gene sources, transplantation and genetic conservation.

(2) *Supply services in production and processing of agricultural products* include the application of scientific and technological achievements (production technology, biotechnology, information technology); agricultural promotion, finance and credit for agricultural production; seed and materials for agricultural production; rural environmental sanitation related to production and processing of agricultural products (garbage, emissions, waste water and something like that); supply of intra-field roads; serving quality of staff and scientists in the agricultural and industrial promotion units in agricultural production; disaster prevention, contract and legal right guarantees in agricultural production.

(3) *Market and supply information services in agricultural production*: Market information services are information supply of plant/animal gene sources as well as preservation and storage of gene sources; information of seeds, fertilizers and materials for agricultural production; scientific and technological achievements and application consultancy in agricultural production; agricultural promotion, finance and banking; prices of inputs and raw materials for agricultural production; Market-oriented information for agricultural production; branding services, customer care in agriculture and trade promotion; disaster prevention, insurance and legal protection right for farmers.

2.2 Analysis and Research

In agricultural production, process of agricultural production are approached with product chains, under the influence of a variety of factors from economic and political institutions, legal environment, mechanisms, policies and technical factors (technology). Public investment and public services in the agricultural sector must also start from the agricultural production chain, as follows (Fig. 3):

2.3 Research Methodology and Data Processing Procedures

- Statistical and comparative methods, econometric models: Data system forecasting the impact of factors, specifically: + Preliminary assessment of reliability and value of the scale by Cronbach alpha reliability coefficient and

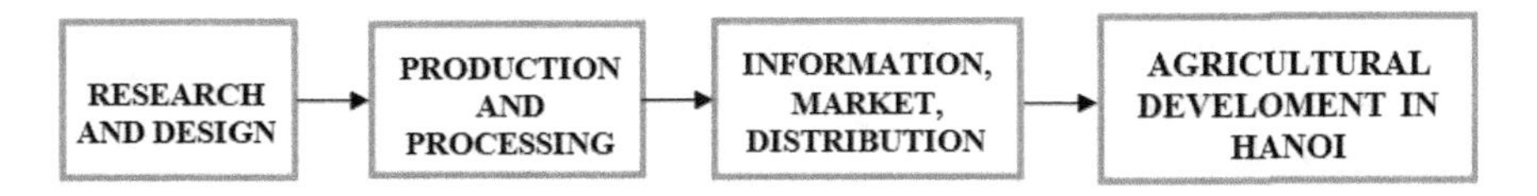

Fig. 3. Analysis and research

Exploratory Factor Analysis (EFA) through SPSS 23 is to assess the reliability of Scales, to eliminate observational variables not explaining research concepts (unsatisfactory) and to reconstruct the rest of the observed variables into appropriate elements (measurement components), as the basis for the modification of the research model, research hypotheses and further content analysis and tests. + Confirmatory Factor Analysis (CFA) is used to test suitability of scale models with market data; + Structural Equation Modeling (SEM) is used to test suitability of theoretical models and research hypotheses.

- Data sources (primary data): Information collected by interview questionnaire for opinions of officials in agricultural management departments and producers in agricultural enterprises and cooperatives. The research was conducted by a team of experts who analyzed, selected typically and ensured reliability in agricultural production in Hanoi such as Thach That district, Ba Vi district and Soc Son district. The seminar was aimed at exchanging agricultural production issues from input to way of production organization, as well as finding consumption markets for agricultural products from ministries and direct producers in agriculture sector. At the same time, outstanding problems and content need conducting with new opportunities and challenges in agricultural production today also were mentioned. The topic questionnaire was collected from 210 agricultural managers in three local districts (66 staffs from Thach That district, 85 staffs from Ba Vi district and 59 staffs from Soc Son) in five months from March to August 2017 to gather information from practitioners.
- Collection and processing: Processing method of data and research results was encoded data, the SPSS software was used for descriptive statistics, graphing, regression analysis and factor analysis. The study of factors was based on correlated relationships in regression analysis and the statistical significance of variables in factor analysis to find solutions of adjusting research variables and creating flourishing agricultural production in the direction of sustainable development. The impact assessment process was carried out in the following steps (Fig. 4):

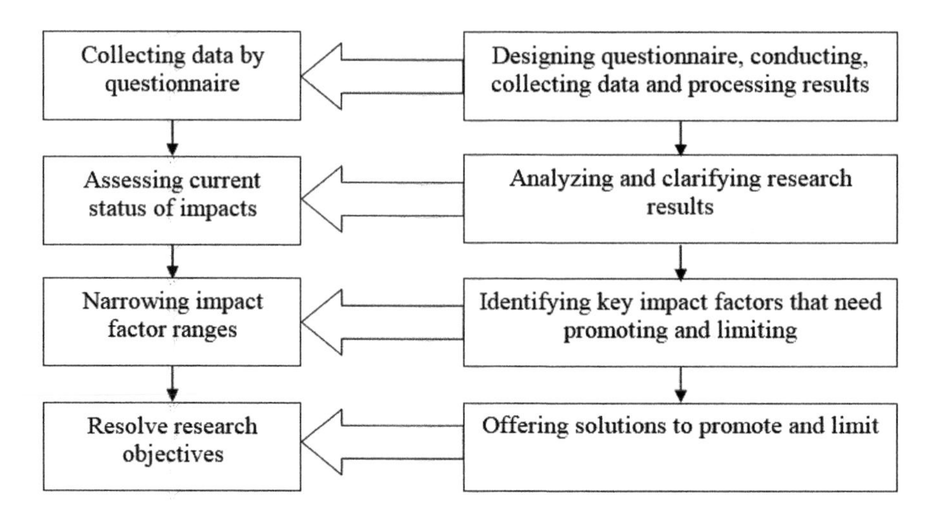

Fig. 4. The impact assessment process

3 Assessment Results of Public Investment in Agricultural Sector Development

3.1 Assessment Criteria of Public Investment

3.1.1 Assessment Criteria for Research and Design

- Assessment of investment in Hanoi's agriculture with the following contents:
  - Researching on seeds and issues of transplantation (q1.1a)
  - Developing and guiding the processes associated with agricultural production (q1.1b)
  - Planning agricultural development (q1.1c)
  - Developing Agricultural policies (q1.1d)
- Assessment of public investment in seeds conservation and issues of transplantation in agricultural production in Hanoi:
  - Conserving and transplanting original seeds and genetic resources of plants and animals (q1.2a)
  - Researching, conserving and developing seed and gene sources (q1.2b)
  - Building production laboratories and experiments (q1.2c)
  - Inspecting and controlling the process of transplantation original seed and gene sources (q1.2d)

3.1.2 Criteria for Production and Processing

- Public investment activities in the field of production planning and processing of agricultural products:
  - Building, developing, protecting of branches and trademarks of agricultural products (q1.3a)
  - Building up the safe process of agricultural production (q1.3b)

  - Ensuring conditions for the implementation of safe agricultural production process (q1.3c)
  - Developing of agricultural infrastructure (q1.4a)
  - Planning of agricultural production areas (q1.4b)
  - Regionally connecting in agricultural production (q1.4c)
- Public investment activities in supporting production and processing of agricultural products (agriculture, forestry and fisheries) in Hanoi:
  - Creating agricultural development environment (q1.5a)
  - Encouraging agricultural development (q1.5b)
  - Assisting agricultural development (q1.5c)
  - Public investment in model construction (q1.6a)
  - Building quality control system of agricultural product (q1.6b)
  - Building warning and forecast systems in agricultural production (q1.6c)
  - Investing infrastructure for agriculture (q1.6d)

3.1.3 Criteria for Assessing Market Access in Public Investment

- Establishment of input-market research center (q1.7a)
- Establishment of output-market research center (q1.7b)
- Establishment of trading platforms of agricultural product (q1.7c)
- Establishment of communications (equipment, information technology, etc.) (q1.7d)

3.2 Preliminary Assessment of the Scale

The research results obtained from 210 agricultural managers in three local districts (66 managers in Thach That District, 85 managers in Ba Vi District and 59 managers in Soc Son District) by the quantitative model are as follows:

The scale testing via the Cronbach's Alpha reliability coefficient is conducted in order to exclude garbage variables, avoid the event that these variables constitute the dummy factor when analyzing the EFA - Exploratory Factor Analysis factor. The testing standard is that Cronbach's Alpha coefficient must be greater than 0.6 and the correlation coefficient of total variable of each scale must be greater than 0.3. The analysis results of Cronbach Alpha in Table 1 show that all scales have satisfied the standards, reached the reliability and been used to analyze the next factors.

After Cronbach's Alpha analysis, EFA analysis using the Principal Component Analysis method and Varimax rotation is followed in order to assess the unidirectional nature, convergent validity and discriminant validity of the scale. With a sample size of 210, the factor loading coefficient of the observed variables must be greater than 0.5; the variables converge on the same factor and distinguish it from other factors. In addition, the KMO test coefficient must be in allowable range of 0.5 and 1.0. According to the following analysis results, all factor loading coefficients of the observed variables are greater than 0.5; Bartlett's test (Sig. = 0.000) indicates KMO = 0.798; all 25 variables after EFA analysis are extracted into 6 factors with Average Variance Extracted greater than 50% in Table 2.

Table 1. Reliability and correlation of minimum total variable of the scales - Source: By the authors, 2017

No.	Scale	Synthetic reliability coefficient Cronbach's Alpha	Number of observed variables
1	Assessment of investment in agriculture (q1.1a_q1.1d)	0,828	4
2	Public investment for seed conservation and acclimatization (q1.2a_q1.2d)	0,843	4
3	Planning on production and processing of agricultural products (q1.3a_q1.3c; q1.4a_q1.4c)	0,787	6
4	Support to production and processing of agricultural products (q1.5a_q1.5c; q1.6a_q1.6d)	0,860	7
5	Market approach (q1.7a_q1.7d)	0,871	4
6	Agricultural development in Hanoi City (q1.1_q1.7)	0,892	3

Table 2. Exploratory factor analysis results - Source: By the authors, 2017

No.	Name of factor group	Variable
1	Assessment of investment in agriculture (q1.1a_q1.1d)	From q1.1a to q1.1d
2	Public investment for seed conservation and acclimatization (q1.2a_q1.2d)	From q1.2a to q1.2d; From q1.3a to q1.3c
3	Planning on production and processing of agricultural products (q1.3a_q1.3c; q1.4a_q1.4c)	From q1.4a to q1.4c
4	Support to production and processing of agricultural products (q1.5a_q1.5c; q1.6a_q1.6d)	From q1.5a to q1.5c, q1.6a; q1.6d
5	Market approach (q1.7a_q1.7d)	From q1.7a to q1.7d
6	Agricultural development in Hanoi City (q1.1_q1.7)	q_NCTK; q_SX; q_TCTT

Kaiser-Meyer-Olkin Measure of Sampling Adequacy: 0,798
Bartlett's Test of Sphericity: Sig. = 0,000
Total Average Variance Extracted reached the value of: 68.55% > 50%

Therefore, after conducting Cronbach's Alpha and EFA, there are 25 observed variables to be extracted into 6 groups of factor in 3 groups of research element, namely in Table 3:

Table 3. Groups of factor following Cronbach's Alpha and EFA - Source: By the authors, 2017

Group of factor	Numerical order of the groups of factor according to EFA for the 1st time	The author renames the groups of factor
Research and design (q1.1a_q1.1d); (q1.2a_q1.2d)	Group 1: q1.1a to q1.1d	Assessment of public investment in agriculture (N1_NC)
	Group 2: q1.2a to q1.2d	Public investment for seed conservation and acclimatization (N2_NC)
Production and processing (q1.4a to q1.4c); (q1.5a_q1.5c; q1.6a_q1.6d)	Group 3: q1.4a to q1.4c	Production and processing (N3_SX)
	Group 4: q1.5a_q1.5c; q1.6a_q1.6d	Support to production and processing of agricultural products (N4_SX)
Market approach (q1.7a_q1.7d)	Group 5: q1.7a to q1.7d	Market approach to agricultural production (N5_TT)
Agricultural development in Hanoi City	Group 1: (q1.1a_q1.1d); (q1.2a_q1.2d)	Research and design (N1-N2)
	Group 2: (q1.4a to q1.4c);(q1.5a_q1.5c; q1.6a_q1.6d)	Production and processing (N3-N4)
	Group 3: (q1.7a_q1.7d)	Market approach (N5)

3.3 Model and Research Hypothesis Testing

The CFA method used in the Structural Equation Modeling (SEM) has more advantages than conventional methods. Thus, in this research, the group of authors applied CFA to test the suitability of the research model with the data obtained after a preliminary assessment using the Cronbach's Alpha and EFA reliability coefficients in the previous section.

The analysis results of CFA shall test the suitability of the research model with the market data as follows: Chi-square $= 617.71$; df $= 233$; P $= 0.000 < 0.05$; Chi-square/df $= 2.651 < 5$; GFI $= 0.806$, TLI $= 0.844$, CFI $= 0.868 > 0.8$; RMSEA $= 0.08 = 0.08$. Thus, the analytical indexes all satisfy the standard conditions and ensure that the research model is perfectly suited to the market data (Fig. 5).

Structural Equation Modeling (SEM) Analysis shows the testing results of the relationship between the five factors affecting: Market approach; Public investment; Assessment of investment level in agriculture; Planning on production and processing; Support to production and processing, and Agricultural development in Hanoi City in Fig. 6.

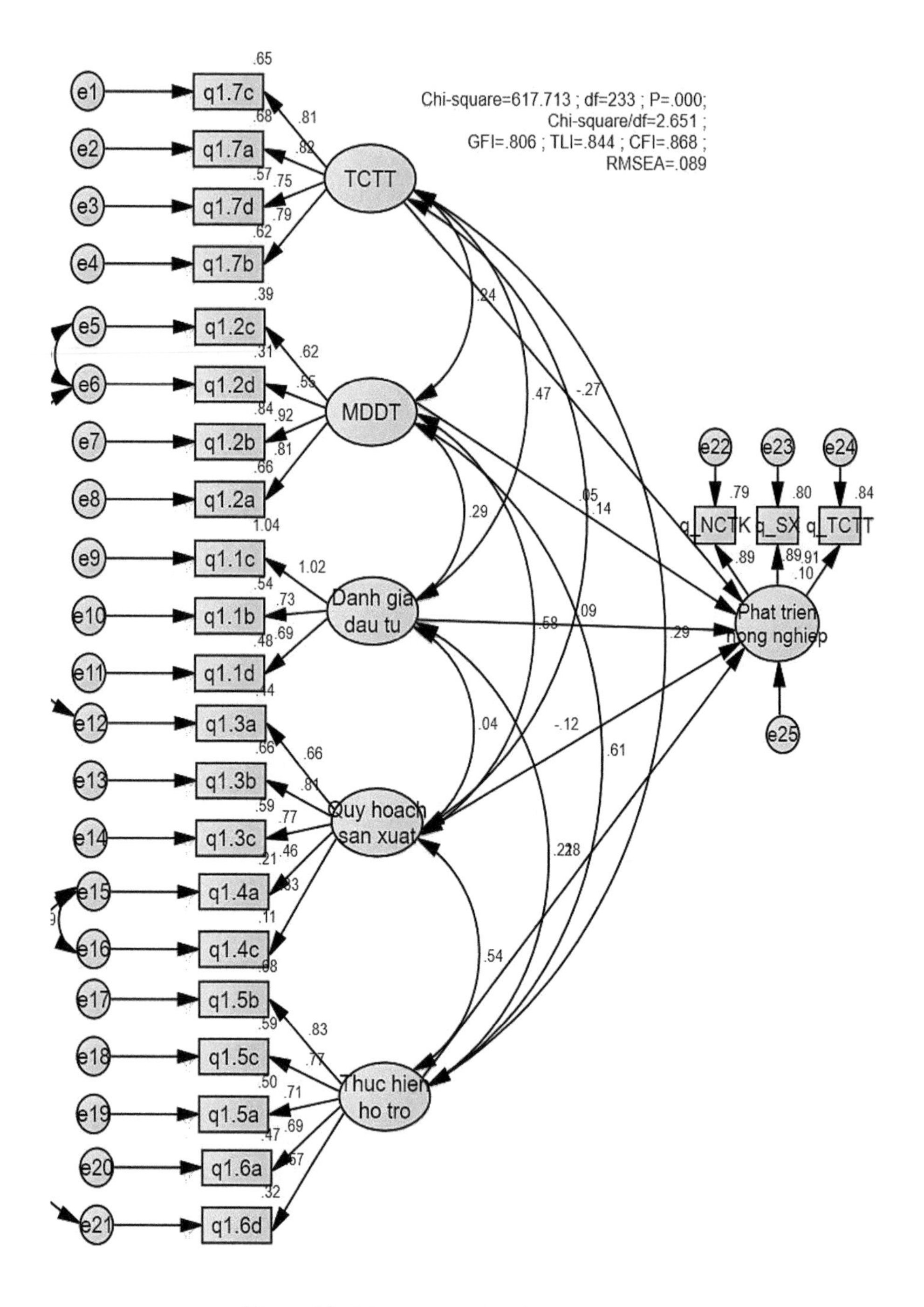

Fig. 5. Model testing - By the authors, 2017

			Estimate	S.E.	C.R.	P	Label
Agricultural development in Hanoi City	<---	Market approach	-.166	.055	-3.019	.003	
Agricultural development in Hanoi City	<---	Public investment	.042	.089	.470	.638	
Agricultural development in Hanoi City	<---	Assessment of investment in agriculture	.044	.039	1.121	.262	
Agricultural development in Hanoi City	<---	Planning on production and processing	-.086	.077	-1.110	.267	
Agricultural development in Hanoi City	<---	Support to production and processing	.163	.065	2.500	.012	

Fig. 6. Structural Equation Modeling (SEM) Analysis Results. Estimate: average estimated value; SE: standard error, CR: Critical Ratio, P: Probability, ***: $p < 0.001$

It can be seen from the observation of analysis results that, not all relationships in the theoretical model are significant at P-value <0.05. There is only a significant impact between Market approach, Support to production and processing and Agricultural development in Hanoi City. In which, Support to production and processing is the most influential factor in the Agricultural development in Hanoi City (0.163) and followed by the factor of Assessment of investment in agriculture (0.044). Meanwhile, the inspection of direct relationship between the influential factors (Market approach, Planning on production and processing, Public investment, Assessment of investment in agriculture) and Agricultural development in Hanoi City supplemented the following results: (i) Support to production and processing directly and positively has impacts on Agricultural development in Hanoi City (0.163); (ii) Market approach (−0,166) and Public investment (0.042), Assessment of investment in agriculture (0.044), Planning on production and processing (−0086) all have negative impacts on Agricultural development in Hanoi City. From there, the research draws on the hypothesis testing results in Table 4.

Table 4. Hypothesis testing results - Source: By the authors, 2017

Hypothesis	Description	Result
H1	Market approach has positive impacts on agricultural development in Hanoi City	Accepted
H2	Public investment has positive impacts on agricultural development in Hanoi City	Accepted
H3	Assessment of investment in agriculture has positive impacts on agricultural development in Hanoi City	Rejected
H4	Planning on production and processing has positive impacts on agricultural development in Hanoi City	Rejected
H5	Support to production and processing has positive impacts on agricultural development in Hanoi City	Rejected

4 Assessment Results of Public Services in Agricultural Sector Development

4.1 Assessment Criteria in Public Services

[1] **Assessment criteria in research and design** + Approval of development strategy of agricultural industry/sector (q1.9a) + Approval of programs and projects (q1.9b) + Construction and verification of programs and projects (q1.9c) + Legal aid (legal documents, procedures, contracts, certificates of documents, etc.) (q1.9d) + Training programs and plans of agricultural human resource (q1.9e) + Preservation and storage of gene sources (q1.9f) + Quality test and control of seeds and animals (q1.9g) + Research of transplantation of seed origin, gene sources and conservation (q1.9h) [2] **Criteria for production and processing** + Application of scientific achievement (q1.10a) + Agricultural promotion, financial and credit (q1.10b) + National reserves in service of production and processing (q1.10c) + Rural environmental sanitation (garbage, emissions, waste water ...) (q1.10d) + Supply of intra-field system (q1.10e) + Serving quality of staffs and scientists (q1.10f) + Disaster prevention (q1.10g) + Insurance, contract and legal rights (q1.10h) [3] **Criteria for market access** + + Information of seeds, fertilizer and supplies (q1.11a) + Supply of information, customer care in agriculture and trade promotion (q1.11b) + Weather forecast information (q1.11c) + Supply of book, library and communication ... related to agricultural promotion work (q1.11d).

4.2 Preliminary Assessment of the Scale

The research results obtained from 210 agricultural managers in three local districts (*66 managers in Thach That District, 85 managers in Ba Vi District and 59 managers in Soc Son District*) by the quantitative model are as follows:

The scale testing via the Cronbach's Alpha reliability coefficient is conducted in order to exclude garbage variables, avoid the event that these variables constitute the dummy factor when analyzing the EFA - Exploratory Factor Analysis factor. The testing standard is that Cronbach's Alpha coefficient must be greater than 0.6 and the correlation coefficient of total variable of each scale must be greater than 0.3. The analysis results of Cronbach Alpha in Table 5 show that all scales have satisfied the standards, reached the reliability and been used to analyze the next factors.

After Cronbach's Alpha analysis, EFA analysis using the Principal Component Analysis method and Varimax rotation is followed in order to assess the unidirectional nature, convergent validity and discriminant validity of the scale. With a sample size of 210, the factor loading coefficient of the observed variables must be greater than 0.5; the variables converge on the same factor and distinguish it from other factors. In addition, the KMO test coefficient must be in allowable range of 0.5 and 1.0. According to the following analysis results, all factor loading coefficients of the observed variables are greater than 0.5; Bartlett's test (Sig. = 0.000) indicates KMO = 0.854; all 20 variables after EFA analysis are extracted into 3 factors with Average Variance Extracted greater than 50% in Table 6.

Table 5. Reliability and correlation of minimum total variable of the scales- Source: By the authors, 2017

No.	Scale	Synthetic reliability coefficient Cronbach's Alpha	Number of observed variables
1	Research and design (q1.9a_q1.9h)	0,863	8
2	Production and processing (q1.10a_q1.10h)	0,860	8
3	Market access (q1.11a_q1.11d)	0,828	4
4	Agricultural development in Hanoi (q1.9_q1.11)	0,931	3

Table 6. Exploratory Factor Analysis Results - Source: By the authors, 2017

No.	Name of factor group	Variable
1	Research and design (q_TK)	q1.9d, q1.9e, q1.9f
2	Production and processing (q_SX)	q1.10a, q1.10d, q1.10h, q1.11d
3	Market access (q_TCTT)	q1.9b, q1.11a, q1.11b, q1.11c
4	Agriculture development in Hanoi (q1.9_q1.11)	q_TK; q_SX; q_TCTT

Kaiser-Meyer-Olkin Measure of Sampling Adequacy: 0.854
Bartlett's Test of Sphericity: Sig. = 0.000
Total Average Variance Extracted reached the value of: 67.82% > 50%

Therefore, after Cronbach Alpha and EFA has eliminated flowing variables: q1.9a, q1.9c, q1.9g, q1.9h; q1.10b, q1.10c, q1.10e, q1.10f, q1.10g; q1.11d, 11 observed variables are extracted into 3 groups in 3 groups of research element, namely in Table 7:

4.3 Model and Research Hypothesis Testing

The CFA method used in the Structural Equation Modeling (SEM) has more advantages than conventional methods. Thus, in this research, the group of authors applied CFA to test the suitability of the research model with the data obtained after a preliminary assessment using the Cronbach's Alpha and EFA reliability coefficients in the previous section.

The analysis results of CFA shall test the suitability of the research model with the market data as follows: Chi-square = 183.53; df = 71; P = 0.000 < 0.05; Chi-square/df = 2.585 < 5; GFI = 0.893, TLI = 0.907, CFI = 0.928 > 0.9; RMSEA = 0.08 = 0.08. Thus, the analytical indexes all satisfy the standard conditions and ensure that the research model is perfectly suited to the market data (Fig. 7).

Table 7. Groups of factor following Cronbach's Alpha and EFA - Source: By the authors, 2017

Group of factor	Numerical order of the groups of factor according to EFA for the 1st time	Renamed groups of factors
Research and design (q1.9a_q1.9h)	Group 1: q1.9d, q1.9e, q1.9f	Research and design (N1_NC)
Production and processing (q1.10a_q1.10h)	Group 2: q1.10a, q1.10d, q1.10h, q1.11d	Production and processing (N2_SX)
Market access (q1.11a_q1.11d)	Group 3: q1.9b, q1.11a, q1.11b, q1.11c	Market access for agricultural production (N3_TT)
Agricultural development in Hanoi City	Group 1: q1.9a_q1.9h	Research and design (N1-N2)
	Group 2: q1.10a_q1.10h	Production and processing (N3-N4)
	Group 3: (q1.11a_q1.11d)	Market access (N5)

Structural Equation Modeling (SEM) Analysis shows the testing the results of testing the relationship between three factors: market access; Research and development; Production and processing in Hanoi's Agricultural Development in Fig. 8.

It can be seen from the observation of analysis results that, not all relationships in the theoretical model are significant at P-value <0.05. There is only a significant impact between market access and agricultural development in Hanoi City. Meanwhile, the inspection of direct relationship between the influential factors (market approach, research and development, production and processing) and Agricultural development in Hanoi City supplemented the following results: (i) market access directly and positively has impacts on Agricultural development in Hanoi City (0.096); (ii) research and development (0,117) and production and processing (−0.047) have direct and negative impacts on Agricultural development in Hanoi City. From there, the research draws on the hypothesis testing results in Table 8.

1. *The hypothesis test results was accepted*, public services the most significantly impacted on agricultural development in Thach That, Ba Vi and Soc Son districts. Supply of Information, approval of programs and projects (q1.9b) associated with agricultural development; weather forecast information (q1.11c), information of seed, fertilizer and materials, information (q1.11a); customer information care in agriculture and trade promotion (q1.11b) ... will positively impact other groups of service factors.

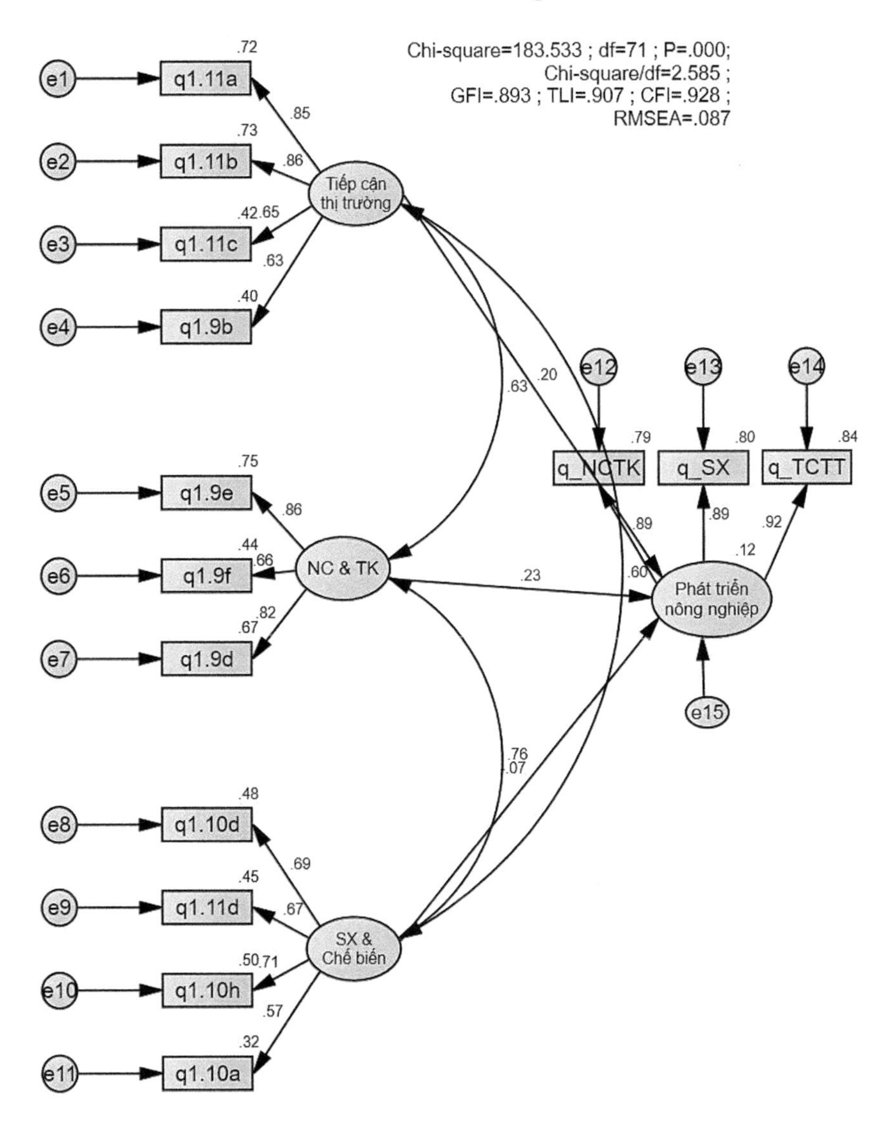

Fig. 7. Model testing

2. *The hypothesis test results are rejected*, indicating that the public service in agriculture needs to be implemented in a uniform manner and confirming inverse correlation of the factors extracted from the model public services need to be invested in order to flourish in agricultural development such as: (i) Legal aid (legal documents, procedures, contracts, certification of documents ...) (q1.9d); Training programs and plans of agricultural human resource (q1.9e); Preservation and storage of gene sources original seed, gene fund of animals/plants (q1.9f); (ii) rural environmental sanitation (garbage, emis-

		Estimate	S.E.	C.R.	P	Label
Agricultural development in Hanoi City <---	Market access	0.096	0.051	1.897	0.058	
Agricultural development in Hanoi City <---	Research and development	0.117	0.077	1.525	0.127	
Agricultural development in Hanoi City <---	Production and processing	-0.047	0.106	-0.441	0.659	

Fig. 8. Structural Equation Modeling (SEM) Analysis Results - Source: By the authors, 2017

Table 8. Hypothesis testing results - Source: By the authors, 2017

Hypothesis	Description	Result
H1	Market access has positive impacts on agricultural development in Hanoi City	Accepted
H2	Research and development has positive impacts on agricultural development in Hanoi City	Rejected
H3	Production and processing has positive impacts on agricultural development in Hanoi City	Rejected

sion, waste water ...) (q1.10d); Application of scientific achievements (q1.10a); Insurance, contract and legal rights (q1.10h); Supply of book, library and communication, ... related to agriculture promotion work (q1.11d).

5 Recommendations on Public Investment and Public Services in Agricultural Development in Hanoi

5.1 On Public Investment

Firstly, public investment in the master plan for agricultural production should be directed towards: (i) The master plan for agricultural development in Hanoi City in line with the master plan for agriculture development of the whole country and the Red River Delta; in line with the master plan for socio-economic development of Hanoi City until 2020, orientation to 2030; the general plan for the development of Hanoi City until 2030 with a vision to 2050 and other related plans; (ii) The investment in modern agricultural development on the basis of application of advanced scientific and technological advances and advanced production methods in order to raise productivity, quality, efficiency and competitiveness associated with processing and product consumption; The investment in agriculture in association with the labor restructuring in agriculture, the process of urbanization and development of newly civilized rural areas, the increase of land use efficiency and agricultural labor productivity, enhancing income and the life of farmers; (iii) The investment in agricultural development in the direction towards ecological urban agriculture, contributing to creating environmental landscapes, promoting the development of eco-tourism in harmonious and sustainable manner to the environment, step by step adapting to the global climate change; (iv) The investment in the planning for agriculture of Hanoi City in the socio-economic development plan of the Government, urban development planning and regional development planning; (v) Strengthening investment towards the formation of specialized commodity production areas, green belts, ecological and high-tech agricultural areas; (vi) The agricultural development in line with the market demand, contributing to ensuring national food security and advancing nutrition security; food safety and hygiene; (vii) The announcement of the plan after being approved; strengthening the management, inspection and supervision of the plan implementation in the districts; and the continuation of public investment to complete the detailed planning: breeding, fishery, forestry, irrigation, dykes; high-tech agricultural production zones, determination of planning sub-zones in order to form concentrated production areas.

Secondly, the public investment solution must ensure genetic resources and acclimatization to meet the needs of agricultural development in Hanoi City in the direction of sustainable development: (i) The development of plant seeds and animal breeds on the basis of maximum exploitation of genetic resources of domestic plants and animals; at the same time, expansion of the exchange of genetic resources with other countries in the world to select new varieties with genetic diversity, suitable for ecological regions; (ii) The development of plant and animal varieties on the basis of creating favorable conditions for all economic sectors to invest in agricultural development, especially those engaged in research, production and business of varieties; The priority of the State budget on research and development of varieties that other economic sectors have not, or less invested. Specifically: Firstly, invest to complete the system of agencies specializing in research, production and supply of plant seeds and animal breeds according to the plan already approved by competent authorities such as: the system of institutes and schools complying with the decisions of the Prime Minister and competent authorities, the State encourages enterprises to set up organizations for research and selection of varieties; complete the production and supply system of varieties under the plan approved by the Ministry of Agriculture and Rural Development in the Decision No. 1238/QD-BNN-TCLN dated 28/05/2012, Decision No. 1683/QD-BNN-CN dated 19/07/2012, Decision No. 1771/QD-BNN-TCTS dated 27/07/2012; Secondly, upgrade facilities, especially 52 Seed Research Centers, 42 Livestock breeding centers, which are business units managed by the Department of Agriculture and Rural Development of Hanoi City and coordinated in changing the mode of operation according to the enterprise model, then equitized in order to achieve autonomy in research and implementation.

Thirdly, Hanoi City continues promulgating the guidelines on agricultural production development in the period of 2015 – 2020: continue reviewing and revising the system of documents regulated by the central and city government regarding the legal mechanisms, policies and regulations related to agricultural production development; study and propose to amend, supplement and complete the inappropriate contents, mechanisms and policies; provide specific and preferential mechanisms and policies to attract and create conditions for all economic sectors to invest in agriculture and rural development; continue investing in the formulation and promulgation of mechanisms and policies to encourage enterprises, implement regrouping of lands and investment in agricultural production, focus on support for production of varieties and application of technological advances, training, infrastructure, development of agricultural commodity production... to improve the policies on investment in agriculture, specifically: Implement equitable investment programs such as program on encouraging the development of specialized agricultural production regions in Hanoi City for the period 2014-2020 (Decision No. 27/2017/QD-UBND of the People's Committee of Hanoi City); program on high-tech application in agricultural development in Hanoi City for the period of 2016-2020 (Decision No. 28/2017/QD-UBND of the People's Committee of Hanoi city); programs and projects adjusting and supplementing the list of works, the land recovery projects, the list of projects changing the purpose of planting rice and protective forests in the city of Hanoi in 2017 under the authority of the People's Council of Hanoi City (Resolution No. 03/NQ-HDND of the People's Council of Hanoi City); continue investing in a number of pilot policies to encourage agricultural development and rural infrastructure construction in Hanoi City (Decision No. 59/2016/QD-UBND of the People's Committee of Hanoi City).

Fourthly, the investment capital in agriculture in Hanoi City should be increased: Total investment capital for agricultural production development in Hanoi City is estimated at 60,168.9 billion VND by 2020 (Budget source: 22,941.4 billion VND, accounting for 38.1%), of which 30,184.9 billion VND for the period of 2011-2015 and 29,984 billion VND for the period of 2016-2020. Therefore, it is necessary to attract all investment capital sources for development of agriculture, forestry and fishery, including domestic investment capital (state budget, enterprises, credit institutions, people's capital,...), foreign investment by the policies and mechanisms, improvement of investment environment, priority for agricultural development, planning and focusing on agricultural production and promotion of chain linkage in development to ensure legality and traceability. Pay attention to and allocate funds to invest in the programs and schemes on agricultural development; especially high-tech application in agriculture; renovate the infrastructure system for agricultural production; infrastructure of specialized production areas; mechanize in agricultural production; pre-process, process and store products; take advantage of capital sources from the ministries and branches at the central level, domestic and foreign organizations (ODA, FDI, NGO, ...), bilateral cooperation programs between Hanoi City and other provinces and cities of our country in terms of agriculture and rural development.

Fifthly, it is required to invest and promote the production and restructuring of the agricultural sector in Hanoi City, specifically: (i) Develop and organize the effective implementation of key programs and schemes, focus on developing the plants and animals with high economic value such as the scheme on plant seeds and animal breeds, scheme on fruit trees, safe vegetables, ornamental plants, high-quality rice, development of tea areas and agricultural mechanization...; invest in the construction of infrastructure works for the production: irrigation, intra-field traffic, variety production establishments, pay attention to the infrastructure in the concentrated production areas; promote the application of technical and technological advances in cultivation, animal husbandry, and aquaculture in terms of harvesting, preserving and processing agricultural products. Apply new technologies such as biotechnology, new material technology, and information technology in the fields of production, processing and consuming products, at the same time apply technical advances in management and operation of commodity production, marketing techniques. (ii) Invest to restructure the agricultural sector of Hanoi City: Revise the structure of plants and animals; potential, advantages of land, labor force, technical infrastructure, market; select and put into production of plants and animals of competitive advantage, which can replace imported products and be capable of being exported; build and form specialized commodity production areas (vegetables, flowers, fruits, rice, pigs, dairy cows, beef, poultry, aquatic products, etc.); apply high-tech to increase productivity and quality of agricultural commodities, in which: Rice production focuses on high quality rice varieties and special rice varieties of Hanoi; produce vegetables and fruits on the basis of high-tech application to ensure food safety and hygiene, compete with imported goods in the domestic market and serve export such as cabbages, Kohlrabi, carrots, gourds, winter melons, cucumbers, tomatoes, peppers, onions, beans of all kinds, ... and spices of all kinds; produce the main categories of cut flowers with an application of high-tech (rose, daisy, gladiolus, lily, tuberose, lily, gerbera, dahlia, peony flower, peach blossom and orchids), valuable orchids and ornamental plants; produce specialty fruits of Hanoi (such as Dien grapefruit, Canh orange, late Ha Tay) and aromatic banana; produce clean tea, apply high technology in production and processing ...; Regarding breeding, artificial insemination and embryo transfer are used to crossbreed the local yellow cows with Zebu breeds (Sind, Brahman), Belgian cows; create the hybrid cows (HF1, HF2) by HF purebred bull semen of the Netherlands, ...; create lean pigs by the method of pig embryos imported from abroad with high-yield and lean (Yorkshire, Landrace, Duroc, Pietrain, etc.).

Finally, enhancing the market forecast capacity should be paid attention to: Promote trade promotion and market forecast. Build a network of wholesale markets and rural markets in order to consume products for farmers. Invest in building agricultural product processing establishments to create a stable output market. Develop the product brand from the application of standard production technology for high quality and accredited by the national and international accreditation agencies, to the promotion and supply of products to consumers. The Promotion Center for Investment, Trade and Tourism in the city shall preside and coordinate with the Departments of Industry and Trade, the Department of Agriculture and Rural Development, the Department of Science and Technology and concerned branches, agencies in boosting brand promotion activities, introducing products, promoting investment, exporting goods and agricultural products to develop production, improve investment efficiency and income of farmers.

5.2 On Public Services

Firstly, enhancement of implementation and development of 15 public services in the field of agriculture and rural development in a comprehensive manner, including: (1) Forest Protection and Development; (2) Conservation, rescue, restoration of forest ecology and resources; (3) Preservation of specimens in the forestry subsector; (4) Assay, test and quarantine livestock breeds, animal feeds and breeding environment; (5) Assessment and monitor the quality of breeding animals, animal feeds and biological products for environmental improvement in animal husbandry; (6) testing of pesticides; (7) survey for measures to prevent harmful organisms and diseases to protect production; (8) assessment of pests and diseases; (9) agricultural promotion; (10) sperm insemination, high-quality cows and high-yield pigs in the city; (11) assay of crop and forest plants, aquatic breeds; (12) preservation and storage of original seed, purebred seed; (13) assessment and certificate of agricultural products and materials in conformity with standards and technical regulations; (14) assessment and certificate of management process and system of production, preliminary processing and processing of agricultural products; (15) test and inspection of agricultural environment, agricultural materials.

Secondly, the entities involved in the coordination and implementation of public services have clear responsibilities and obligations: The Department of Agriculture and Rural Development actively advises, develops and submits to the People's Committee for approval of economic and technical norms and cost norms for public services using the state budget in the field of agriculture and rural development in the city. The Finance Department appraises to the City People's Committee price brackets or prices of public services in the field of agriculture and rural development of the city on the basis of the economic and technical norms and cost norms approved by the competent agencies. The right to promulgate and the roadmap shall be fully calculated according to regulations. Selection of public service units to offer public service in the field of agriculture and rural development in the form of mission assignment, order or bidding decentralized by the city and current regulations is conducted after consulting with the financial agency of the same level. Guidance of the public service units of the city which provide services in the field of agriculture and rural development is organized and implemented according to regulations.

Finally, enhancement of the capacity of state management at all levels in the agricultural sector should be done, including complete of the project aiming at strengthening the contingent of agricultural cadres at commune level; strengthening the management, supporting and improving quality and efficiency of business production of agricultural cooperatives; Strictly managing and inspecting the quality of breeds, agricultural materials and hygiene and food safety, specifically: (1) To strictly manage the production process according to regulations; To intensify the inspection and examination of the quality of plant seeds, livestock breeds, fertilizers, plant protection drugs, animal feeds - aquatic products, veterinary drugs and microorganisms in service of cultivation and husbandry, ... to well perform the epidemic prevention for plants, animals and aquatic products in the city; (2) To strengthen state management measures, analyze and certify quality; To coordinate with inter-branch inspection forces in inspecting hygiene and food safety and inspecting origins of agricultural products and food circulated and consumed in the capital; (3) To agricultural products of the provinces brought back to Hanoi, agriculture setting up a process, system of quality inspection, assessment and certificate for products brought to Hanoi for consumption; Inter-provincial links on administrative procedures related to slaughter management, food safety certificates, transport of agricultural products and foodstuffs to Hanoi and vice versa.

References

1. Decision No. 17/2012/QD-UBND, approving the master plan for agricultural development in Hanoi by 2020, vision to 2030, dated July 9, 2012 by the Peoples Council of Hanoi
2. Decision No. 27/2017/QD-UBND, guiding the implementation of Resolution No. 25/2013/NQ-HDND dated December 04, 2013 by the Peoples Council of Hanoi on the incentive policies for the development of specialized agricultural production areas in Hanoi for the period of 2014-2020, dated August 18, 2017 by the Peoples Committee of Hanoi
3. Decision No. 28/2017/QD-UBND, guiding the implementation of Resolution No. 03/2015/NQ-HDND dated July 08, 2015 by the Peoples Council of Hanoi on several policies for Hanoi Hi-tech Agricultural Development Program for the period of 2016-2020, dated August 7, 2017 by the Peoples Committee of Hanoi
4. Decision No. 3748/QD-BNN-KH, approving the development orientation of plant and animal varieties by 2020, vision to 2030, dated September 15, 2015 by the Ministry of Agricultural and Rural Development
5. General Statistics Office of Vietnam: Statistical Yearbook 2016, Vietnam (2016)
6. Behnassi, M., Shanhid, S.A.: Sustainable Agricultural Development: Recent Approaches in Resources Management and Environmentally-Balanced Production Enhancement. Spinger, Dordrecht (2012)
7. National Assembly: Law on Public Investment No. 49/2014/QH13, Vietnam (2014)
8. National Assembly: Law on Government Organization, No. 76/2015/QH13 dated June 19, 2015 (2015)
9. Resolution No. 03/NQ-HDND, on the amendment and supplement of the list of works and projects on land recovery; The list of projects changing the purpose of land for planting rice and protective forests in Hanoi, dated July 3, 2017 by the Peoples Council of Hanoi
10. Thanh, H.H.: Some key solutions to improve the environmental awareness of farmers to contribute to sustainable development of ecological agriculture in the Northern Delta. Academy-level scientific research project. Vietnam Academy of Social Sciences (2012)
11. The State Audit Office of Vietnam, Public investment: Actual state and solutions, Hanoi (2012). http://www.sav.gov.vn/1782-1-ndt/dau-tu-cong-%E2%80%93-thuc-trang-va-giai-phap.sav
12. The Vietnam Ministry of Planning and Investment: Strategic orientation towards sustainable development, Vietnam Agenda 21, Hanoi (2002)

Assessment of the Quality of Growth with Respect to the Efficient Utilization of Material Resources

Ngoc Sy Ho[1](✉), Hai Huu Do[2], Hai Ngoc Hoang[1], Huong Van Nguyen[3], Dung Tien Nguyen[4], and Tai Tu Pham[1]

[1] Academy of Politics Region I, Hanoi, Vietnam
ngocho.hvct@gmail.com, haihnhvkv1@gmail.com, ptt.kt14@gmail.com
[2] Ho Chi Minh City University of Food Industry, Ho Chi Minh City, Vietnam
haidh1975@gmail.com
[3] Hung Yen University of Technology and Education, Hanoi, Vietnam
vanhuong75hy@gmail.com
[4] National Economics University, Hanoi, Vietnam
dungnt@neu.edu.vn

Abstract. The Vietnam's economic growth has been high in the 30 years of Renovation and ranked the second of the world (after the China's one) for several years, which is highly appreciated by the international community. However, there are inadequacies and discrepancies between the rate and quality of the growth. The questions are whether such rate and quality of the growth are commensurate with the potential of Vietnam or not, what is the position of Vietnam in the region and what Vietnam should do to narrow the development gap in comparison with regional countries. This paper has given some answers to such questions.

Keywords: Economic growth · Quality of growth

1 The Quality of Economic Growth

1.1 Definition of the Quality of Economic Growth

According to R. Lucas and J. Stiglitz (who are American economists winning the Nobel Prize for Economics in 1995 and 2001), the quality of economic growth manifests itself in several following key criteria. (1) The economic growth is sustainable in the long term, capable of avoiding or standing against the external adverse changes; (2) The in-depth economic growth is reflected in the high and constantly increased Total Factor Productivity (TFP); (3) The growth must go ensure the economic efficiency and competitiveness; (4) The growth must reach the goal of improving welfare, reducing poverty and inequity; (5) The growth must go hand in hand with the sustainable environmental protection

V. Kreinovich et al. (Eds.): ECONVN 2019, SCI 809, pp. 660–677, 2019.
https://doi.org/10.1007/978-3-030-04200-4_46

and development; (6) The growth always supports and promotes the renewal of democracy and vice versa.

The study of “Quality of economic growth” by Vinod et al. (2000) has put forward arguments about the quantity and quality of economic growth. The authors did not directly study sustainable economic growth, but indirectly carried out from the quality of growth. From their point of view, keeping the growth in the long term and attaching great importance to the both quantity and the quality of growth are necessary to obtain the sustainable growth. The authors analyzed the factors influencing the quality of growth, and illustrated them by using analysis and data from many countries in the world. “*The quality of economic growth is the economic growth at a relatively high rate in a long time, and such growth must ensure the equity, social progress, and environmental improvements*”, said the authors. Based on research on development theory and practice, in 2000, Vinod et al. outlined three components of growth quality as follows:

(1) The high growth rate should be maintained in the long term; (2) The growth must contribute directly to the sustainable improvement of social welfare, equal distribution of achievements and poverty reduction; (3) The growth must assure not to degrade the environmental quality.

In the UNDP’s Human Development Report 1999 with combined theoretical and practical analysis, some concepts of growth were outlined as negative growth for the first time.

- *Rootless growth*: is the growth that causes people’s cultural identity to wither.
- *Ruthless growth*: is the growth in which its fruits mostly benefit the rich, while the poor gain little or even nothing, the poor increase, the gap between the rich and the poor becomes wider.
- *Jobless growth*: is the growth that does not expand the opportunities for employment, or refers to long hours and very low incomes in low productivity in traditional agriculture and informal sectors.
- *Futureless growth*: is the growth that undermines future generations by squandering resources, or destroying the environment.

Then, the researchers also outlined some of the negative growth concepts as:

- *Stagnant growth*: is the growth that is achieved in a short period of time, then decreases gradually, making the economy become stagnant. This growth is based mainly on traditional agriculture, handicraft, processing, assembly, and without in-depth investment. This model is common in early developing countries which are likely to get stuck in “middle income trap” when reaching this level.
- *Distorted growth*: is the growth that is largely based on over-exploitation of resources, overly-generous material subsides by various means such as tax exemption, tax debt, or state capital and credit incentives, while investment in human and basic and long-term engineering-technology are paid less attention. Compared to the above mentioned models, this model at first seems to be better for the poor and contribute to improving social welfare. The biggest

disadvantage of this model, however, is the socioeconomic burden on the Government, administerization of economic relations, leading to bias investment, over-emphasizing on physical capital investment by offering capital incentives, and increasing public investment that makes it unworkable and leads to the risk of widespread waste and corruption. Consequently, the growth can only be achieved as long as the Government is able to maintain the subsidies, otherwise the consequence is the public debt crisis and the disturbed economy.

The most common viewpoint of concepts is the close relation between the economic growth and its quality. The growth focusing on only high growth rate and ignoring the productivity, efficiency and competitiveness is considered as the unsustainable growth. The economic growth without improving living standards of the poor, promoting the democracy, or even degrading the social morality is defined as the rootless growth. The growth accompanied by the habitat destruction is the futureless growth. The growth must be associated with the quality and sustainable development and develop all economy, society and environment. Besides the quantity, it is necessary to pay more attention to the quality of the growth. However, there is no united definition of the growth quality and there are many approaches to this definition with both different strengths and shortcomings. Some most outstanding viewpoints are as follows:

- The economy with high growth quality must ensure that comparative advantages are promoted for rapid and efficient development and the growth must be attached to the scientific and technological advancement and environmental protection.
- The high growth quality is featured that the economic growth is accompanied with the forward shift of the economic structure; effective utilization of inputs, high total factor productivity, strong competitiveness of enterprises, poverty reduction and environmental protection.
- The growth quality should be demonstrated by the productivity and efficiency of investments, the competitiveness of the economy, including not only inputs such as management and allocation of resources in the reproduction process, but also outputs such as the improvement of living standards, equal distribution of outcomes, environmental protection and ecological stability.
- Other methods based on inputs and direct factors influencing on the growth to approach and evaluate the economic growth are the general production function, the relation between outputs and inputs.

The above-mentioned viewpoints more or less address the sustainable development. If broadly understood, the quality of growth can reach the conception of sustainable development. However, there is an outstanding difference between the growth quality and sustainable development. For the sustainable development, there are three must-be-considered aspects of economy, society and environment; but for the growth quality, it may be necessary to look into one of aspects. Nevertheless, it is also possible to see in the conception of quality of economic growth a fairly high consistency that is *quick, effective and long-term growth; economic growth must go hand in hand with improving quality of life and protecting environment*, and continued to be clarified by Vietnamese scholars. For example, according to Anh et al. (2005), the quality of economic growth is reflected in the sustainability of growth with the effective utilization of inputs in the production process, equal output distribution, life quality improvement, and ecological environment protection. According to Tran Tho Dat (2010), the quality of economic growth is quick, effective and sustainable development of economy and it is reflected in a number of indicators such as total factor productivity, improvement of people's living standards; shift of economic structure towards modernity; advancement, equality of society; and environmental protection.

Internationally, the economic growth quality has been mentioned and approached from economic growth for a long time. Most countries have concentrated all resources on economic growth. However, such countries could not achieve expected goals through the quick economic growth, and China is the most obvious example which achieved the short-term high economic growth, but could not maintain long-term sustainable growth. Specifically, the poverty is not reduced, and the environmental is degraded. It also does not ensure that poor countries can catch up with rich countries. Since the late 1990s, the sustainability and quality of the economic growth have begun to be addressed in studies and they all have focused on both rate and quality of the growth.

In Vietnam, researches in growth quality have just been taken into account. However, Vietnamese researches in growth quality often deeply understand theories of international scholars to analyze one or some conceptions of quality of economic growth such as: Aspects of economic restructuring, competitiveness of the economy, labor productivity, total factor productivity (TFP), incremental capital output ratio (ICOR), social or environmental aspects of quality of economic growth. The study of quality of economic growth has been going on for a long time. From the above analysis, the authors put forward a view on the economic growth quality as follows: ***Quality of economic growth is a long-term sustainable, effective and relatively high economic growth, and at the same time the growth must guarantee social equality and advancement, as well as environmental improvement.***

With the understanding of the quality of economic growth mentioned above, it is necessary to clarify some aspects as follows:

- Firstly, quick growth may not lead to high growth, but quick growth is the highest growth level as possible in terms of a particular economy.
- Secondly, the effectiveness of growth is reflected in many aspects such as the efficiency of inputs, effectiveness of investments, efficiency in foreign trade activities, etc.
- Thirdly, the advancement and social equality in the growth quality first and foremost is the distribution of achievements of the growth. In this aspect, quality growth must go hand in hand with poverty alleviation and life improvement of all population; the growth shall create more jobs, open up new opportunities for people to participate in the next growth process. In addition,

economic growth must contribute to improving social welfare through public goods and services, and better meet the needs of the people.
- Fourthly, in terms of environment, quality growth should be understood as the process of growth using economically and effectively natural resources, conserving biodiversity, and preventing environmental pollution. Environmental cost related ti the growth is required to be at minimum.

Therefore, while economic growth reflects the size of the economy, or changes in the size of growth, which aims at gradually fulfilling the human's basic physical needs such as food, water, shelter. Then, the quality of growth is step by step towards meeting the high level needs such as moral requirements like justice, peace in mind, joys, social status, etc. Moreover, the quality of economic growth reflects the nature and means of achieving economic growth. In case economic growth is achieved through capital, labor growth, and resources consumption, etc. with low effectiveness of capital and labor productivity utilization, the use of such resources is unreasonable, which will lead to slowing down economic growth, and with the result of social and environmental issues. Such economic growth is considered not to guarantee the quality and "deviate" from improving the quality of growth. If economic growth is mainly based on non-production factors such as science, management level, human resources, etc., such economic growth is in the direction of quality assurance. In brief, there are two following fundamental issues about quality of economic growth. *Firstly*, which is the economic growth based, on factors of capital, labor, resources, etc. or on ones of science, technology, management methods, quality of human resources, etc.? *Secondly*, how does economic growth affect human life on the social, economic, and environmental aspects?

1.2 Analytical Framework and Evaluation Indicators of Quality of Growth

Analytical Framework of Quality of Growth. There has not been a unified framework for quality of growth in the world so far. One of the most fundamental reasons may be the large gap in the development level among countries and the variation in the growth pattern that each country pursues. In the most general approach, the basis for analyzing and evaluating the quality of growth is often based on four contents complementing each other: (1) investment for the formation of capital assets involved in the value creation process; (2) growth pattern of a country; (3) distribution (both income and opportunity) during the growth process; and (4) effective management based on the quality of the Government's institution and policy.

The analysis of growth resources is the most commonly used method to evaluate a country's growth pattern. There are a lot of factors and agents involving in growth process, but only production factors of labor, physical capital, human capital, natural resources (and environment), and technological advances directly involves in growth process. Technological advances on the one hand affect the efficiency and productivity of the remaining factors, and on the other hand contribute to total factor productivity (TFP). Production factors contribute to the process of creating growth, forming a country's growth model, and thus contributing to welfare. To determine how input factors affect growth quality, the article uses the analytical framework as follows.

Factors affecting the speed and quality of growth include: capital, labor, technological advances, and resources, in which capital is an important factor, and the biggest contributor to Vietnam's growth over a long period of time. And at present, this factor is still playing a major role in the country's economic growth. However, the effectiveness of using investment capital and capital mobilization structure still has many inadequacies, and should have a specific assess in the period of 2006–2016, which is the milestone after Vietnam's WTO accession, to find out solutions for Vietnam's sustainable growth in the upcoming period (Fig. 1).

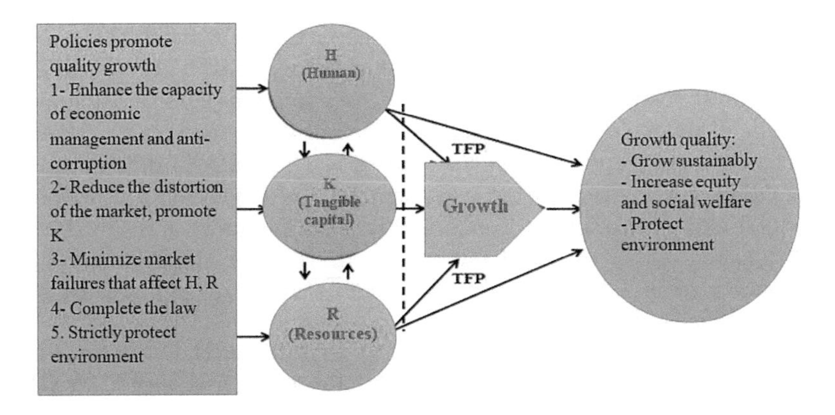

Fig. 1. Analytical framework of quality of growth - Source: Collected (2017)

Indicators Reflecting the Quality of Economic Growth

- *Labor productivity index*: Labor productivity is calculated by comparing 2 indicators of GDP and average labor in a year (or hours of labor) of the economy according to this formula:

$$\text{Labor productivity} = \frac{\text{Total GDP}}{\text{Total number of employees}}$$

In terms of principle, it is necessary to use GDP at a fixed price, at that time, the higher GDP per capita is, the higher labor productivity is. At the sector and enterprise level, GDP could be replaced by other indicators representing business results such as value added, revenue and profit. After calculating labor productivity in each sector and enterprise through each stage, it is possible to analyze and evaluate the development level of labor productivity and its development trend, thereby to assess the quality of growth under the angle of labor productivity growth.

- *The effectiveness of investment capital*: To evaluate the efficiency of investment capital, it is possible to use a variety of indicators. For example, the following indicators can be used to evaluate the utility of a production capital in general and a fixed capital in particular to produce Gross Domestic Product (GDP).

$$\text{The efficiency of capital} = \frac{\text{GDP per year of research}}{\text{Total amount of capital produced in the research year}}$$

However, it is difficult to define the "total production capital" at the macroeconomic level, especially when Vietnam has not yet recorded the "fixed assets" of the whole economy, sector or trading. Instead, the ICOR coefficient is used to assess the effectiveness of capital investment.

The ICOR coefficient is an economic indicator that how much the amount of investment capital has to increase to add up 1 dong of GDP. Investment made in the ICOR coefficient includes expenditures to increase fixed assets, mobile assets, and amounts of forming value added of economic sectors. The ICOR coefficient varies depending on the state of the economy in different periods, depending on the structure of investment and the efficiency of using physical products and services in the economy. With that in mind, the ICOR coefficient is considered one of the most important criteria for assessing the quality of economic growth. There are two commonly used methods of calculating the ICOR coefficient:

- The first method: ICOR = I1/(Y1 − Y0), where: I1 is the total investment of the research year, Y1 is GDP of research year and Y0 is GDP of the previous year. Indicators of investment capital and GPD for calculating ICOR by this method should be based on the same price (actual or comparative price).
- The second method: ICOR = (I/Y)/gy, where: I/Y is the ratio of investment capital comparing to GDP, gy is the growth rate of GDP. The ICOR coefficient in this method shows that: how much the capital rate comparing to GDP has to increase to add up 1% of GDP. The lower the ICOR coefficient is, the more effective the economic investment is. The lower ICOR coefficient means that capital investment rate needs to be lower than GDP to maintain the same economic growth rate. However, as the rule that if the rate of return decreases as the economy grows (GDP per capita increases), the ICOR coefficient will increase, it means that investment capital rate needs to be higher than GDP to maintain the same growth rate.

- *Total factor productivity (TFP)*: TFP reflects the contribution of "intangibles" such as: technical advances, technological innovation, management improvement, and enhancement of labor's skills and qualification. TFP is important in analyzing the quality of economic growth, mathematically TFP usually uses the growth function of Cobb-Douglas to calculate TFP:

$$Y = AxF(K^{\beta}xL^{\alpha}), \tag{1}$$

where: Y is output, K is capital, L is labor, A = TFP, β is the contribution factor of capital, ($\alpha = 1 - \beta$) is the contribution coefficient of labor. The way of calculating growth rate and contribution rate of TFP in increasing output is as follows:
+ TFP growth rate: The formula for calculating TFP is as follows: iTFP = iY − αiL − βiK, of which iY is growth rate of output (in term of amount or GDP); iK is the growth rate of fixed capital; iL is the growth rate of labor.

+ Contribution share of TFP increase to output: The formula of TFP share in GDP is as follows: % contribution of TFP = (iTFP/iY) × 100%, of which iTFP is the growth rate; iY is the growth rate of output (or GDP).
TFP reflects the efficiency of using inputs in the economy or determines the quality of economic growth. Thus, a quality growth economy is the economy that has a high contribution of TFP to growth. In contrast, a growth economy with low TFP contribution is the economy which does not guarantee the quality, or grows in width (in nature, growth in width is mainly based on the growth of capital and labor).

- *This indicator reflects the production capacity and technology of the economy.* This is an important indicator of production capacity, processing of the economy, thereby reflecting the production technology. This indicator is calculated by the ratio between GDP (Gross Domestic Product), which is defined as a monetary measure of the market value of all the final goods and services produced in a period (usually yearly) of time, and GO (Gross output), which is defined as a measure of total economic activity in the production of new goods and services in an accounting period (usually yearly).

$$\text{Production capacity} = \frac{\text{GDP}}{\text{GO}} \times 100\%$$

The higher the production capacity indicator is, the higher the value added ratio of the economy is; the lower the processing level is, the higher production capacity is. In contrast, when the production capacity index is low, the economy mainly depends on processing, production capacity, and low technology.

- *The index reflects the level of saving natural resources and environmental protection: Green GDP*
+ Green GDP is calculated as follows: Green GDP = Net GDP-Cost of natural resources and environment
+ Percentage of Green GDP comparing to net GDP = $\frac{\text{Green GDP}}{\text{GDP}} \times 100\%$
This indicator shows the cost of natural resources and environment destroyed in production. The higher the ratio of Green GDP comparing to GDP is, the lower natural resources is used and the lower the environment is affected in the economy.

2 Assessment of the Economic Growth Quality Based on Criteria

2.1 Labor Productivity in the Economic and Regional Scales

- *Outcomes*: Vietnam's labor productivity has been significantly improved over the years (see Table 1 and Fig. 2). The average productivity increased by 3.9% per year for the 2006–2015 period, of which the 2006–2010 period increased by 3.4% per year, and the 2011–2015 period increased by 4.2% per year. Compared to 2010, the labor productivity in 2015 increased by 23.6%, though it is still lower than the set target of 29%–32%, the labor productivity growth rate is higher than that of 2006–2010 period, contributing to narrowing its gap with other ASEAN countries.

Table 1. Labor productivity by industry - fixed price (million VND/year) - Source: Statistical Yearbook 2016

	Social labor Pro.GDP/Worker		Agriculture, forestry, fisheries		Industry, Capital construction		Services	
	General (mil./prs)	Rate (%)	Value (mil./prs)	Rate (%)	Value (mil./prs)	Rate (%)	Value (mil./prs)	Rate (%)
2011	45.53	3.49	17.41	2.70	82.05	4.08	64.73	0.15
2012	46.92	3.05	17.88	2.70	85.4	4.08	64.74	0.15
2013	48.72	3.84	18.33	2.52	88.72	3.89	66.77	3.14
2014	51.11	4.91	18.94	3.33	92.93	4.75	69.56	4.18
2015	54.38	6.40	19.72	4.12	81.32	−12.49	63.45	−8.79
2016	57.27	5.31	21.00	6.49	80.24	−1.33	66.19	4.32

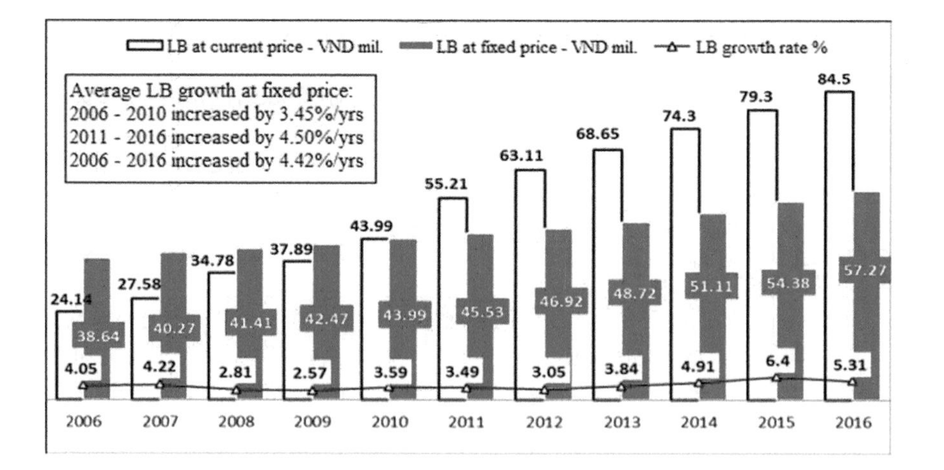

Fig. 2. Labor productivity (LB) in the 2006–2016 period - Source: Calculated from GSO data over the years

The contribution rate of labor force and labor productivity to the economic growth in the 2005–2007 period was 40% and 60%, respectively, then declined to the rate of 50% and 50%, respectively. From 2013 to 2016, the rate fluctuates at 30% and 70%, respectively. These are positive signs for the early improvement of growth quality, though it is not as robust and stable as those in other countries during the industrialization process of economy. According to General Statistics Office (GSO) report, in 2016, the labor productivity of the whole economy at current price reached VND 84.5 million per worker (equivalent to USD 3,853 per worker). At a constant price in 2010, the labor productivity of the whole economy in 2016 increased by 5.31% compared to 2015. By economic sector, the average labor productivity in the agriculture, forestry and fisheries sector was lowest with VND 32.9 million per worker; the industry and construction reached VND 112.0 million per worker; and the service sector reached VND 103.5 million per worker.

- *Weaknesses and shortcomings*: In general, Vietnam's labor productivity in the 2011–2015 period was still at a low level in comparison with the set target and regional countries, though it has improved significantly over the years. In the context of no comparison, it showed a "spectacular" progress, but the outlook is still at a low level compared to other regional countries. For example, in 2015, Vietnam's labor productivity at current price was USD 3,660, just equal to 4.4% of that of Singapore, 17.4% of Malaysia, 35.2% of Thailand, 48.5% of the Philippines, and 48.8% of Indonesia. In terms of the productivity of each individual worker, comparing to other countries, the labor productivity of Vietnamese workers is not lower (shown in various Asean skills competitions), but even higher in a number of fields. However, a poor overall labor productivity is caused by not only skills, workmanship, or qualifications of each worker, but also other factors. The social combination of labor, unsuitable labor restructuring, and a high rate of labor force have dragged Vietnam's labor productivity far behind many other countries in ASEAN region, even approximate to Laos, Cambodia, etc. (Fig. 3).

  It can be observed that the overall labor productivity has increased steadily with a relatively modest growth rate. For the 2006–2015 period, the average labor productivity increased by 3.9% per year, of which the 2006–2010 period increased by 3.4% per year, and the 2011–2015 period increased by 4.2% per year. Compared to 2010, the labor productivity in 2015 increased by 23.6%, though it is still lower than the set target of 29% to 32%, the labor productivity growth rate is higher than that of 2006–2010 period, contributing to narrowing its gap with other ASEAN countries. (As per Purchasing Power Parity (PPP) in 2005: Singapore's labor productivity in 1994 was 29.2 times higher than that of Vietnam, but in 2013, this gap was only 18 times; similarly, the labor productivity gap between Malaysia and Vietnam decreased from 10.6 times to 6.6 times; its gap with Thailand decreased from 4.6 times to 2.7 times; with the Philippines dropped from 3.1 to 1.8 times; and with Indonesia dropped from 2.9 to 1.8 times) (Fig. 4).

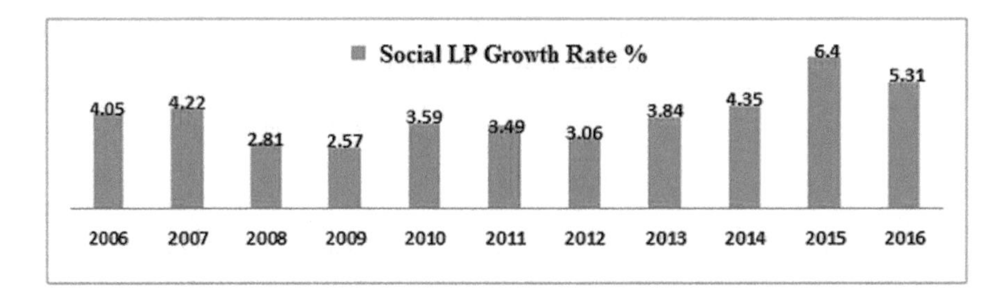

Fig. 3. Social labor productivity growth rate - Source: Statistical Yearbook over the years

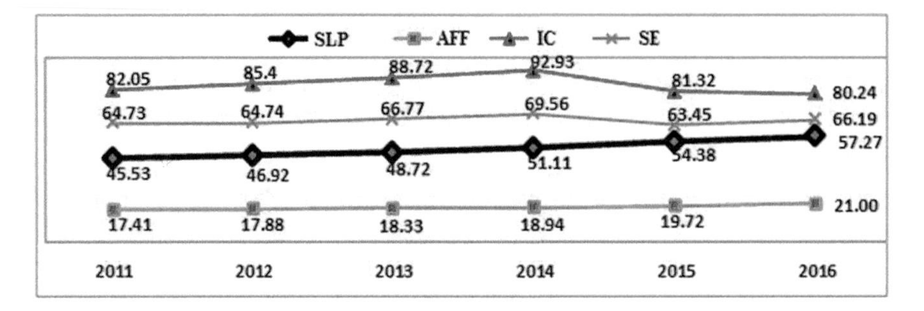

Fig. 4. Labor productivity by industry - fixed price in 2010 (VND million) - Source: Statistical Yearbook over the years

2.2 Capital Efficiency

- *Outcomes*: Capital is a key input factor for growth, especially in developing countries like Vietnam. Investment capital is a factor that has affected both the supply and demand, and also a main factor affecting Vietnam's economic growth in recent years. In the 2004–2010 period, the Rate of investment/GDP from a low level of under 30% of GDP increased to more than 30% and then over 40% of GDP; and since 2011 the Rate of investment/GDP has fallen gradually and fluctuated at over 30% of GDP (Table 2).
  The trend of reducing the proportion of social investment/GDP has lasted continuously since 2007 (of which 2009 is the year which has the highest rate of investment/GDP and also has the highest ICOR). Between 2012 and 2016, the Rate of Investment/GDP was over 30%, the lowest since 2000. Many economists say that it is suitable for the level of Vietnam's economy. If combined with technological innovation, and enhancement of micro and macro management level, it is for sure that the growth rate and growth quality will be improved in the coming time.
  Before 2000, the State capital was over 50% of total social investment, and in 2016 it reduced to 37.6%, which is a major effort in SEO reform and structural change in public expenditure. The Vietnam's public debt per capita has increased slightly, and reached USD 1,000 in 2015. Regarding public debt

Table 2. Total Social Investment (TSI)/GDP and ICOR - Source: Calculated from GSO data over the years

Year	GDP growth (at fixed price) (%)	GDP growth (at current price) (%)	TSI growth (at current price) (%)	Social In./GDP (at current price) (%)	ICOR (without latency) (times)
(1)	(2)	(3)	(4)	(5)	(6)
1991–1995	**Average five-year ICOR**			**3.50**	
1996–2000	**Average five-year ICOR**			**4.80**	
2000	6.80			34.20	5.03
2001	6.90			35.40	5.13
2002	7.10			37.40	5.21
2003	7.30			39.00	5.34
2004	7.80			40.70	5.21
2005	8.40			40.90	4.87
2001–2005	**Average five-year ICOR**			**5.15**	
2006	8.23			41.50	5.04
2007	8.46			46.50	5.50
2008	6.23			41.50	6.66
2009	5.32			42.70	8.03
2010	6.78	19.45	17.10	41.90	6.18
2006–2010	**Average five-year ICOR**			**6.28**	
2011	5.89	27.97	5.70	34.60	5.87
2012	5.03	16.37	7.00	32.10	6.38
2013	5.42	10.44	10.29	30.44	5.61
2014	5.98	9.86	11.30	31.00	5.18
2015	6.68	6.48	12.00	32.60	4.88
2011–2015	**Average five-year ICOR**			**5.58**	
2016	6.21	8.04	8.70	32.98	5.31

Note: Use the formula ICOR = (5)/(2) stated in part 1

per capita, Vietnam's public debt is at a relatively low rate compared to other Asian countries. Also in 2015, the country with the highest public debt per capita was Singapore with USD 56,000, followed by Malaysia with USD 7,696.9 USD, Thailand with USD 3,450.8. Vietnam, Indonesia, and the Philippines had the public debt per capita in 2015 of approximately USD 1,000 (Fig. 5).

– *Reviews*: The situation of unfocused investment has not been resolved: in 2010, the central and local ministries and agencies allocated the state budget to a total of 16,658 projects with an average capital allocated to each project of nearly VND 7 billion; in 2010, the average capital allocated to Project

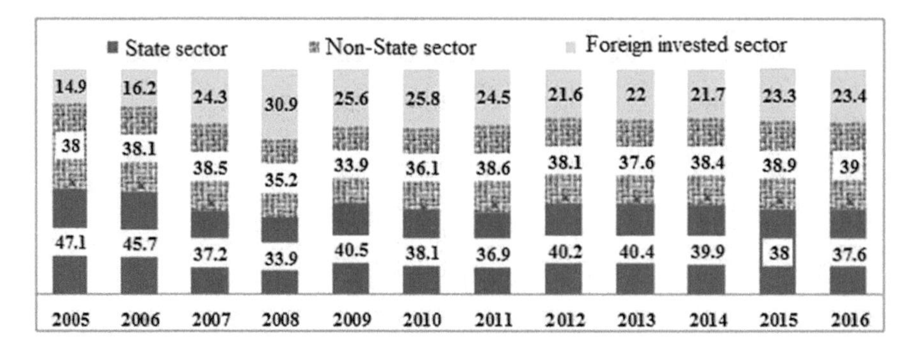

Fig. 5. Investment capital structure by economic sector over years (%) - Source: Calculated from GSO data over the years

Group A at the central level was approximately VND 115 trillion. In 2011, the average scale of an investment project was VND 11 million per project; in 2012, this amount increased to VND 17 million per project. Specifically, (1) Regarding the public debt, figures show that the relative percentage of public debt rose from 36.5% of GDP in 2001 to 62.2% of GDP in 2015, and 63.7% of GDP in 2016. In terms of scale, the figure in 2015 reached VND 2.68 quadrillion, 2.3 times more than that of 2010, and 14.8 times more than that of 2001. The public debt growth rate over the past 5 years was around 18.4%, three times higher the economic growth; (2) Vietnam's investment efficiency is low, reflected by the ICOR which is constantly increasing and at a high level. In the 1900s, the proportion of social investment/GDP was just 30%, and the ICOR was low (with an average value of 3.5 for the 1991–1995 period), which might give us a feeling that this period had a high investment efficiency. However, it was a fact that the economy was lack of capital and the agricultural sector (a sector with low ICOR value) was still an important driving force. As investment increased together with the effect of latency, the ICOR value rose to 4.8 for the 1996–2000 period, to 5.15 for the 2001–2005 period, and to a peak of 6.28 for the 2006–2010 period, which was the period in which most commentaries realized and sharply criticized the growth based on cheap labor and capital, and width development. The proportion of investment/GDP tended to decrease in the 2011–2015 period. And it has still been at a value of over 30% since 2011. Experts said that this is a reasonable rate in the context that Vietnam's economy has been still at a lower average level, so the ICOR value dropped to 5.58.

2.3 Quality of Labor Force

- *Abundant labor force*: According to GSO Socio-economic report, the country's labor force aged 15 and above in 2016 was 54.4 million persons, increasing by 455.6 persons compared to 2015, in which: Male workers were 28.1 million persons, accounting for 51.6%; female workers were 26.3 million persons, accounting for 48.4%. By area, the labor force aged 15 and above in urban area was 17.5 million persons, accounting for 32.1%, and those in rural area was 36.9 million persons, accounting for 67.9%. Labor force within the working age was estimated to be 47.7 million persons in 2016, increasing by 275.9 thousand persons in comparison with the previous year; in which, male workers were 25.8 million persons, accounting for 54.1%; female workers were 21.9 million persons, accounting for 45.9%. Labor force within the working age in urban area was 16.0 million persons, accounting for 33.4%; those in rural area was 31.8 million persons, accounting for 66.6%.

- *Low quality of human resources*: Human resources are the most decisive factor that determines the development of the country's economy and the existence of enterprises. However, in Vietnam, the quality of human resources has not been paid adequate attention and remained weak and poor. The high number of working-age people does not mean that the Vietnamese labor market could meet the labor demand of enterprises. The primary reason is that the number of skilled and qualified workers of Vietnam is still very limited. The percentage of trained workers is only 15.4%, while that of rural areas is only 10%. With an abundant labor force, Vietnam's unemployment rate has been constantly increased in recent years, while its enterprises have still complained of labor shortage. It is due to the fact that the Vietnamese workers meet only the demand for quantity, not the demand for quality. This not only makes it difficult for enterprises, but also causes the workers to lose their job opportunities. The unemployment rate within working-age people in 2016 was 2.30% (2.33% in 2015, and 2.10% in 2014), in which urban area was 3.18% (3.37% in 2015; 3.40% in 2014); rural area was 1.86% (1.82% in 2015; 1.49% in 2014). The unemployment rate among youth aged from 15 to 24 in 2016 was 7.34%, in which urban area was 11.30%; rural area was 5.74%. The quality of labor in Vietnam scored 3.9 points on the scale of 10. Among over 53.4 million employed workers aged 15 and above, there are only 49% of them being trained, of which those who attend three-month or over training courses only accounts for 19%. Vietnam's labor productivity has been rising in recent years, but it is still at a low level compared to other countries in the region and the world. Though it has improved significantly over the years, it is still lower than other countries in the region. In 2015, Vietnam's labor productivity at current price reached USD 3,660, just equal to 4.4% of that of Singapore, 17.4% of Malaysia, 35.2% of Thailand, 48.5% of the Philippines, and 48.8% of Indonesia.

2.4 Total Factor Productivity

- *Outcomes*: Total Factor Productivity (TFP) is a productivity measure of both "labor" and "capital" in a specific activity or for the whole economy. TFP reflects the advancement of science, engineering and technology, showing thereby that the increase in output depends not only on the increase in the amount of inputs (traditional method) but also on the quality of inputs, including labor and capital. At the economic level, TFP is evaluated based on two key indicators, the TFP growth rate and its contribution to the economic growth (GDP) (Table 3).

- *Reviews*: For the 2006–2010 period, Vietnam's TFP growth rate was −0.27% on average, it decreased in 2008 and 2009. The TFP has been increasing constantly and rapidly since 2010, and reached up to 3.18% in 2015. Data on the contribution of factors to GDP growth show a positive trend that the TFP' contribution to GDP is increasing gradually. In 2011, the TFP growth contributed 14.01%, and in 2015, it contributed up to 48.40% to the economic growth. According to calculations, the average TFP' contribution to Vietnam's economic growth for the 2011–2015 period was 30.33%. The contribution of TFP growth to Vietnam' economic growth has been increasing gradually. This shows that the inputs (capital and labor) are being used in an effective manner; the economy has a positive shift towards focusing on the growth quality, such as the application of technical and technological advances, and the concentration of resources on the fields with higher economic efficiency. Although there has been some criticism about the accuracy of qualifying the contribution of TFP and S&T to GDP, especially the difficulties in statistics and determination of the contribution of non-scientific and non-technological elements in TFP, it is generally agreed that the science and technology, and the coordinated management of factors play the most important role in TFP and GDP growth.

Table 3. TFP growth rate and its contribution to the economic growth for the 2006–2015 period

Year	GDP growth rate (%)	TFP growth rate (%)	Contribution of factors to the GDP growth (%)		
			Capital	Labor	TFP
2006	8.23	0.37	73.85	20.71	5.44
2007	8.46	0.36	69.79	24.96	5.25
2008	6.23	−1.03	90.54	28.26	−18.80
2009	5.32	−0.78	84.50	30.33	−14.82
2010	6.78	0.48	65.84	26.42	7.74
2011	5.89	0.85	60.61	25.37	14.01
2012	5.03	1.03	55.68	24.37	19.95
2013	5.42	1.73	50.46	17.04	32.50
2014	5.98	2.16	47.74	15.46	36.72
2015	6.68	3.18	49.90	1.70	48.40
Average 2006–2010	7.04	−0.27	78.60	27.13	−5.73

Source: Vietnam National Productivity Institute (VNPI) reports in 2014 and 2015

3 Some Recommendations Proposed to Improve the Quality of Vietnam's Economic Growth in the Near Future

Firstly, improve the quality of human resources: To promote the reforms of education and training towards improving the quality to meet the real demand of the economy; to force and/or encourage to develop the forms of associated training and cooperative training among universities, agencies, research organizations and enterprises to meet the requirements of human resources, and science and technology for production and business; to diversify the education and training resources, focusing the state budge on "elitist" education and training, combining with strengthening socialization, mobilizing resources for the development of education and training systems, vocational training, health care, promoting international cooperation; to encourage private-sector investment, foreign investment, and the social participation to effectively implement the national target programs on education, training, vocational training, and heath care. To ensure the equality and transparency among the forms of education and training.

Secondly, improve the efficiency of exploitation and use of resources: To minimize the negative impacts on the environment through appropriate mechanism, policies, and methods, especially tax policies, fees and charges, regime of inspection and supervision, and strict handling of violations. To ensure the equality and equity in the mechanism of exploitation and use of resources; For the purpose of improving the quality, efficiency and competitiveness of the economy in the context of deepening impact of climate change on the world economy, to carry out the Vietnamese Party's policy and State enforcement policy to transform to a green model with low resource consumption, "the growth model from mainly the width development to harmonious development between the width and depth" together with the effective exploitation and use of resources, requiring an instant change in the economy's structure, including the industry sector, territory sector, and the structure of economic sectors, towards giving priority to industries, regions, and economic sectors which show an efficient and economical use of resources.

Thirdly, the development of science and technology is a key driving force for the rapid and sustainable development: To shift the focus of scientific and technological activities to serve the industrialization and modernization, and depth development, thus contributing to the rapid growth of productivity, quality, efficiency, and enhancing the competitiveness of the economy. To perform the following tasks in a coordinated manner: raising the capacity and renewing the management mechanism, promoting the application of science and technology, enhancing international integration in relation to science and technology. To improve the capacity of science and technology in a focused and targeted manner. To develop synchronously and effectively use infrastructures and human resources. The State is expected to focus on investing in national key tasks, scientific and technological solutions for spearhead products, and at the same time enhancing socialization and mobilization of all social resources, especially from enterprises, for the development of science and technology. To attach the scientific and technological objectives and tasks to the socio-economic objectives and tasks at each level, sector, and locality.

References

Anh, N.T.T., Ba, L.X., et al.: The quality of economic growth: initial assessment of Vietnam. CIEM Statistics, Hanoi (2005)

Anh, V.T.: The Quality of Economic Growth of Hanoi. Publishing House of Vietnam National University, Hanoi (2015)

Chinh, N.T.: Enhancement of the growth quality, efficiency and competitiveness of the economy is associated with effective exploitation and use of natural resources. Institute of Strategy and Policy on Natural Resources and Environment, Hanoi (2013)

Perkins, D.H., et al.: Economics of Development, Sixth Edition, Copyrighted Material, Fulfilled by Amazon. Fung Kwan, Kengieng Mak (2008), Productivity Growth in the Greater Pearl Delta of China, Asia Business Council (2006)

General Statistics Office of Vietnam: Annual Statistics, Publishing House of Statistics, Hanoi (2010–2016)

General Statistics Office of Vietnam: Statistics of labour and employment over years, Information and Documents of GSO, Hanoi (2011–2016)

General Statistics Office of Vietnam: Summarized report "Status and solutions to improve labor productivity in Vietnam". Website of General Statistics Office of Vietnam published on 29 March 2016

National Institute of Finance: Macroeconomic Stability and Comprehensive Integration. Finance Publishing House, Hanoi (2015)

Lucas, R.E.: On the mechanics of economics of development. J. Monet. Econ. **22**(1), 3–42 (1988)

Solow, R.: A contribution on the theory of economics growth. Q. J. Econ. **70**, 65–94 (1965)

Saigon Website: Renovation of allocation of resources and distribution of benefits to motivate the development, Website of The Saigon Times, published on 21 October 2016. http://www.thesaigontimes.vn/Uploads/Articles03/114348/1d86e_Le-Dang-Doanh.doc

Thomas, et al.: Measuring Education Inequality. Working paper, World Bank Institute, Wasington D.C. (2000)

Transparency International: Corruption Perceptions Index 2013: Asia - Pacific, The annual report (2014)

Tuan, N.A.: Labor productivity in Vietnam 2015 - Noticeable figures. Vietnam. J. Sci. Technol., June 2016. Vietnam National Productivity Institute, Hanoi (2016)

UNFPA: Taking advantage of the population "gold" in Vietnam: opportunities, challenges and policy recommendations. Pulish by UNFPA, Hanoi (2010)

UNDP: Management initiatives on gender and economic policy in Asia and the Pacific. Centre for Asia - Pacific Publishing, Bangkok, Thailand (2012)

Vietnam National Productivity Institute: Vietnam National Productivity Report, National Centre for Socioeconomic Information and Forecast, Ministry of Planning and Investment, Hanoi (2014–2015)

Vinod, et al.: The Quanlity of Growth. Published for the World Bank, Oxford University Press (2000)

Yuan, Y.: Enabling Industry to Take-off and Up-grade in a Transforming Economy, the Case of Shenzhen SEZ Chine. Shenzhen Universsity, China, Center for Special Economic Zone Research (2011)

World Bank: Vietnam Development Report 2006–2016: Business, World Bank, Hanoi (2004–2016)

World Economic Forum: The Global Competitiveness Report, Oxford University Press New York, London (2007–2016)

Is Lending Standard Channel Effective in Transmission Mechanism of Macroprudential Policy? The Case of Vietnam

Pham Thi Hoang Anh(✉)

Banking Academy of Vietnam, 12 Chua Boc St., Dong Da, Ha Noi, Viet Nam
anhpth@hvnh.edu.vn

Abstract. This study aims to evaluate the effectiveness of lending standards in macroprudential policy (MaPP) transmission mechanism in Viet Nam during 2000–2016. By employing the regression model, the paper suggested that restriction on institutional entities that could borrow foreign currency denominated loans from banks was the most effective MaPP instrument in Viet Nam. However, the paper provides empirical evidence that setting higher risk weights on loans to securities and housing sectors than other types of loans has an unexpectedly reverse impact on credit growth. In spite of mix results among each MaPP measures, interestingly, the model suggested that overall macroprudential package expressed by the MaPP index was effective in Viet Nam.

Keywords: Macroprudential policy · Transmission mechanism
Lending standard channel · Viet Nam

1 Introduction

Macroprudential policy (MaPP) can be defined as application of a set of prudential instruments whose aim is to reduce systemic risks and the vulnerability of the financial system. Furthermore, macroprudential policy is expected to help predict and cope with sudden financial instability, and thereby minimize serious macroeconomic consequences. The study aims to analyze and evaluate the effectiveness of lending standards channel in macroprudential policy transmission mechanism in Viet Nam during 2000–2016.

The recent global financial crisis has shown financial regulation and supervision are based on not only the micro approach but also the macro approach. Although the term "macroprudential policy" has been used for a long time since the late 1970s (Clement 2010), but has become considerably more popular with the rapid increase of the policy research since the global financial crisis (Galati and Moessner 2011). However, there exist many points of view leading to a lack of consensus on issues related to macroprudential policy.

V. Kreinovich et al. (Eds.): ECONVN 2019, SCI 809, pp. 678–693, 2019.
https://doi.org/10.1007/978-3-030-04200-4_47

The general view of macroprudential policy is to limit systemic risks and costs of the system crises. According to the Bank of England (2009), the MaPP should aim at the more stable real economy in general, and in the more resilient financial system in particular. In addition, Borio and Drehmann (2009) argued that objectives of MaPP are to limit the risk of episodes of system-wide distress that have significant macroeconomic costs. In the time dimension, MaPP aims to limit the increase the systemic weaknesses over time by reducing the pro-cyclical reaction between asset prices and credit, and to curb an unsustainable increase in financial leverage and the funding instability. In terms of cross-sectional dimension, the MaPP aims to control increasing the vulnerability of the financial system caused by the interlinkages between the financial intermediaries and the essential role of financial institutions in key markets and can help individual institutions too important to fail. Three goals are due to the existence of three groups of systemic externalities generated by (i) trends amplify the negative shocks of the financial system, (ii) a feedback mechanism between financial system and macro-economy leading to an increase in risks of the financial systems from macroeconomic shocks, and (iii) the linkages in the financial system increases the vulnerability of financial systems from non-systemic and systemic shocks. IMF (2013) also noted that MaPP should aim to curb the systemic weaknesses rather than other goals of macroeconomic management as manage the level and composition of aggregate demand.

As regards instruments, IMF (2013) suggested the monetary authorities may impose such as preventive and counter-cyclical capital buffers, sectorial capital requirements, liquidity measures, Loan-to-Value (LTV) ratio, Debt-to-Income (DTI) ratio. However, MaPP still uses convetional tools of other policies such as reserve requirement (monetary policy), charge levied on wholesale loans (fiscal policy), etc. Before that, a research of the Bank of England (2011) had a different approach by dividing MaPP instruments into three groups including: (i) the instruments affecting the balance sheet of the financial institutions; (ii) the instruments influencing the conditions and terms of loans and other financial transactions, and (iii) the instruments influencing the market structure. Cerutti et al. (2015) relied on the impact objects to classify MaPP instruments into: foreign exchange related MaPP instruments, borrower targeted MaPP instruments, and financial institutions targeted MaPP instruments.

Other than the above approaches, the study by Zhang and Zoli (2016) classified MaPP instruments based on intermediate target of housing price. Accordingly, MaPP instruments including 2 groups: (i) housing related measures such as loan to value ratio (LTV), debt to income ratio (DTI), risk weights on mortgage, loan loss provisions on mortgage loans, allowance); and (ii) non-housing related measures, including instruments related to capital measures, credit limits, liquidity). There are many different approaches, but in general, MaPP instruments are quite varied and consensus with the IMF (2013). So far, most of instruments applied mainly to the banking system are micro-prudential instruments which are adjusted to suitable with macroeconomic objectives (Claessens 2014).

It seems that comprehensive studies of the MaPP transmission mechanism are still limited, most of which only refer to one or a few channels but unclear. Most of studies evaluate the effectiveness of macroprudential policy transmission mechanism to intermediate targets such as credit growth, and foreign exchange market stabilization which will minimize probability of systemic risks (Aiyar et al. 2014; Cerutti et al. 2015; Vandenbussche et al. 2015; Hartmann 2015; Tresel and Zhang 2016). The paper follows a new approach in classifying MaPP instruments into four groups including lending standards, foreign exchange, capital requirements and liquidity. Therefore, transmission mechanism of MaPP to intermediate and ultimate objectives may through different channels such as lending standard channel, foreign exchange related channel, capital requirement channel, and liquidity channel.

Tresel and Zhang (2016) have estimated the impact of the change in lending standards to credit growth in the region of European currency countries. Keys et al. (2010) showed that by imposing stricter laws on mortgages, the lending standards were less loosened by securitization in the United States (the US). Nadauld and Sherlund (2009) on the securitization business of subprime real estate mortgages in the US supposed that a stricter regulation on capital requirements could help to limit the growth of the bubble.

Caruana (2005) and Saurina (2009a) found empirical evidence that regulations on risk provision only has small impact on credit growth, while very useful in building counter-cyclical buffers, helps strengthen the solvency of the bank in Spain. Saurina (2009b) also found empirical evidence which shows that although the dynamic risk provisioning could not response to all of the credit losses in recession period, but was quite useful in increasing strengthen the resilience of individual banks and the banking system in Spain during the last financial crisis. Lim et al. (2011) pointed out that the various instruments of macroprudential policy reduced the cyclical nature of credit growth by reducing the correlation between credit growth and GDP growth. Especially, the minimum capital requirement and reserve requirement limits credit growth, but the effect is different in periods of boom and credit crunch. Aiyar et al. (2014) suggested that tightening minimum capital requirements will reduce the supply of credit, thereby reduce the risk of hot credit growth, and help the economy to achieve financial stability. The finding is also confirmed in studies of Labonne et al. (2016), Gambacorta and Shin (2016).

Main findings of this paper are as follows. First, in terms of lending standards, restriction on institutional entities that could borrow foreign currency (FC) denominated loans from banks was found to be the most effective MaPP instrument in Viet Nam during 2000–2016. In other words, empirical evidence suggested that regulations of only institutional entities having foreign currency revenue could borrow FC loans might lead to a significant decrease in credit growth rate in Viet Nam. Second, in contrast, the paper provides empirical evidence that activation of setting higher risk weights on loans to securities and housing sectors than other types of loan is found to have an unexpectedly reverse impact on credit growth. Third, in spite of mixed results among each

MaPP measures, interestingly, the paper suggested that overall macroprudential package expressed by the MaPP index was effective in reducing credit growth in Viet Nam.

The remainder of this paper is organized as follows. Section 2 reviews development of macroprudential policy, focusing on lending standard measures in Viet Nam during 2000–2016. In Sect. 3, the paper employs OLS model to assess the effectiveness of lending standards instruments in macroprudential policy transmission mechanism in Viet Nam. The final part makes some policy recommendations for the State Bank of Viet Nam to enhance the effectiveness of macroprudential policy.

2 Overview of Lending Standards in Macroprudential Policy in Viet Nam

Like other countries, macroprudential policy has been increasingly concerned by the State Bank of Viet Nam (SBV) since the global financial crisis. At that time, Viet Nam's banking system has experienced many serious difficulties including high non-performing loan ratio, cross ownership, shortage of liquidity, etc. However, all aspects of MaPP (including concept, legal framework, instruments as well as transmission mechanism) were dealt with at very simple level of knowledge. Interestingly, the SBV launched an actively and flexibly comprehensive package of measures including both economic and administrative ones to escape from the crisis in the financial market in general, and in banking system in particular. Most of them were conventional instruments of monetary policy, but some of them might be considered as MaPP instruments. In other words, these instruments were not activated on behalf of macroprudential department, but their ultimate objectives were to stabilize the financial market in general, and the banking system in particular. Under this view, the paper suggested that, MaPP instruments actually activated in cooperation with monetary policy instruments to obtain targets such as price stability, financial stability and economic growth in Viet Nam. During 2000–2016, the SBV actually activated a package of MaPP measures that could be classified into four types: lending standard, capital requirements, foreign exchange, and liquidity instruments. The paper, however, focused on reviewing development of lending standards instruments as a part of MaPP in Viet Nam that summarized in Table 1.

3 Evaluation of the Effectiveness of Transmission Mechanism of Macroprudential Policy Through Lending Standard Channel in Viet Nam

3.1 Model Specification

The paper applied the Ordinary Least Square (OLS) model suggested by Cerutti et al. (2015) to evaluate the effectiveness of transmission mechanism of macroprudential policy through lending standard channel. The evaluation framework is set

Table 1. Implementation of Lending Standards Instruments in Viet Nam during 2000-2016

	Instruments	Background	Time and degree of activation	Expected results
1	Ceiling credit growth rate for each or group of commercial banks	There were weaknesses/problems in Vietnamese banking system: (i) Non-performing loan rate was too high (around 17–18 %); (ii) Many banks faced with shortage of liquidity; (iii) Profit of Vietnamese banks decreased	- January, 2012: The SBV divided banks into four groups, which have maximum loan growth rates of 17 %, 15 %, 8 % and zero respectively in the year of 2012; - January 2013: The SBV divided banks into four groups, which have maximum loan growth rates of 12 %, 9 %, 5 % and 23% (e.g. Sacombank and Navibank) respectively in the year of 2013	- Maintain a reasonable credit growth rate in order to promote economic growth and ensure financial stability; - Restrict weak banks in over expanding lending activities
2	Restrictions on institutional entities that could borrow foreign currency denominated loans from banks (only for entities having revenues in foreign currencies)	- Foreign currency loan growth rate increased significantly at rate of 48.45% (in 2010) and 18.7% (year 2011) in compared with local current loan growth rate of 10.2% → high pressure on devaluation of local currency → unexpected fluctuations in foreign exchange market → financial instability	- Legal framework: + Circular 07, dated 24, March, 2011 (came into effect 9 May, 2011); + Circular 03/2012/TT-NHNN (dated 08 March, 2012); + Circular 37/2012/TT-NHNN (dated 28 December, 2012) + Circular 29/2013/TT-NHNN (dated 06 December, 2013); + Circular 24/2015/TT-NHNN (dated 08 December, 2015) (Replaced for Circular 29) + Circular 31/2016/TT-NHNN (dated 15 November, 2016)	- Decrease foreign currency loan growth rate → decrease loan dollarization → lower pressure on devaluation of local currency; → stabilize in foreign exchange market → stabilize financial market

(*continued*)

Table 1. (*continued*)

	Instruments	Background	Time and degree of activation	Expected results
3	Limits on loans per total deposit ratio (LDR)	There were weaknesses in Vietnamese banking system: (i) Non-performing loan rate was too high (around 17–18%) (ii) Many banks faced with shortage of liquidity; (iii) Profit of Vietnamese banks decreased	- Circular 36/2014/TT-NHNN (came into effect 1st February 2015)	- Maintain a reasonable credit growth rate in order to promote economic growth and ensure financial stability
4	Applying higher risk weights on loans to securities and housing sectors	- Loans to securities and housing sectors increased significantly → bubbles in securities and housing markets → financial instability	- Decision 03/2007/QD-NHNN (dated 19 January, 2007) applied the risk weight of 150% to securities loans; - Circular 13/2010/TT-NHNN (dated 20 May, 2010) applied the risk weight of 250% to securities and housing loans; - Circular 36/2014/TT-NHNN: The State Bank of Vietnam reduced the risk weights for credit facilities extended to the real estate and securities sectors from 250% to 150%	- Give directions to loans into priority fields; - Decrease credit growth rate in risky sectors such as securities and real estate market → eliminate systemic risks → ensure financial stability

(*continued*)

Table 1. (*continued*)

	Instruments	Background	Time and degree of activation	Expected results
5	Loan to value ratio (LTV)		Circular 41 (dated 31st December, 2016): The SBV imposed different risk weights for different LTV ratio, as follows (It will come into effect in 2020): a. Collateral is real estate for non-business purposes + LTV < 40%: risk weight is 30% + 40% < LTV < 60%: risk weight is 40% + 60% < LTV < 80%: risk weight is 50% + 80% < LTV < 90%: risk weight is 70% + 90% < LTV < 100%: risk weight is 80% + LTV > 100%: risk weight is 100% b. Collateral is real estate for business purposes + LTV < 60%: risk weight is 75% + 60% < LTV < 75%: risk weight is 100% + LTV > 75%: risk weight is 120%	
6	Debt to Income ratio (DTI or) DSC)		Circular 41 (dated 31st December, 2016): SBV should not impose a maximum of DTI ratio. SBV will impose different risk weights for different combination of DTI and LTV ratio (It will come into effect in 2020)	

Source: Author's compilation from www.sbv.gov.vn

as follows: (i) Through intermediate objectives: Lending standards ↑ (Tighten lending standard) → credit supply ↓ → mortgage loans ↓ → domestic credit growth ↓ (ii) Through ultimate objectives: Lending standards ↑ (Tighten lending standard) → credit supply ↓ → mortgage loans ↓ → housing price ↓ → risk of housing bubble ↓ risks and instability of financial system ↓. However, due to unavailability of historical housing prices in Viet Nam, the paper will focus on only the first stage of evaluation framework. In other words, it will assess impacts of activation of lending related MaPP instruments on credit growth (an intermediate objective). Therefore, my model is set into three equations:

$$DC_t = \alpha + \beta_1 DC_{t-1} + \beta_2 CPI_t + \beta_3 GDP_t + \beta_4 IR_t + \beta_5 MaPP1_t + \beta_6 MaPP2_t + \beta_7 MaPP3_t + \beta_8 MaPP4a_t + u_t \quad (1)$$

$$DC_t = \alpha + \beta_1 DC_{t-1} + \beta_2 CPI_t + \beta_3 GDP_t + \beta_4 IR_t + \beta_5 MaPP1_t + \beta_6 MaPP2_t + \beta_7 MaPP3_t + \beta_8 MaPP4b_t + u_t \quad (2)$$

$$DC_t = \alpha + \beta_1 DC_{t-1} + \beta_2 CPI_t + \beta_3 GDP_t + \beta_4 IR_t + \beta_5 MaPPIndex_t + \beta_6 CrisisEco_t + u_t \quad (3)$$

In which

DC: Domestic credit growth (in percent)
IR: Lending interest rate (in percent)
CPI: Inflation rate (in percent)
GDP: Gross Domestic Product growth rate (in percent)

The first two variables are collected from the International Financial Statistics (IMF) and is calculated based on q-o-q basis. The last two are collected from the General Statistics Office of Viet Nam. All variables are on a quarterly basis for the period Q1/2000 - Q4/2016, giving a total of 84 observations.

$CrisisEco$: A dummy variable representing the global financial crisis, so it is coded as 1 during Q3/2008 - Q4/2014), otherwise 0.

$MaPP$: Vector of macroprudential policy instruments related to lending activities, consisting of the following dummy variables based on the legal documents of the State Bank of Viet Nam (Table 1):

(i) $MaPP1$: Existence of the ceiling credit growth rate for each or group of commercial banks. The State Bank of Vietnam classified commercial banks into four groups (from 1 to 4) based on their performance and soundness. The monetary authority assigned different celling credit growth rate for different groups on the sense that the better the bank performance is, the higher the credit growth rate is assigned. This MaPP instrument was activated for the years 2012 and 2013. Therefore, MaPP1 will be coded 1 for these years and 0 otherwise.

(ii) $MaPP2$: Restrictions on institutional entities that could borrow foreign currency (FC) denominated loans from banks (Circular 07/2011, dated 24 March and came into effect 9 May, 2011):

$$Q1/2000\text{-}Q1/2011 : MaPP2 = 0$$

$$Q2/2011\text{-}Q4/2016 : MaPP2 = 1$$

(iii) $MaPP3$: Limits on loans per total deposit ratio (LDR) (Circular 36/2014, dated 20 November 2014 and came into effect 1 February, 2015)

$$Q1/2000\text{-}Q1/2015 : MaPP3 = 0$$

$$Q2/2015\text{-}Q4/2016 : MaPP3 = 1$$

(iv) $MaPP4$: Application of higher risk weights on loans to securities and housing sectors (Decision 03/2007). This variable could be coded in two ways:

Option 1:

$$Q1/2000\text{-}Q4/2006 : MaPP4a = 0$$

$$Q1/2007\text{-}Q4/2016 : MaPP4a = 1$$

Option 2:

$$Q1/2000\text{-}Q4/2006 : MaPP4b = 1$$

$$Q1/2007 - Q2/2010 : MaPP4b = 1.5$$

$$Q3/2010 - Q4/2014 : MaPP4b = 2.5$$

$$Q1/2015 - Q4/2016 : MaPP4b = 1.5$$

(v) $MaPPIndex$: A macroprudential policy index, calculated by adding all variables (applied for dummy variables only).

3.2 Findings and Comments

All variables are stationary at level (domestic credit growth and inflation), or first difference level (GDP growth rate and interest rate). All diagnosis tests are checked to illustrate that the OLS model is suitable for evaluating the effectiveness of transmission mechanism of macroprudential policy through lending standard channel (See Appendix 1 for more details). In order to check robustness of results, the paper will assess three models with same fundamental variables but MaPP variables are coded in different way.

Based on the OLS regression results as described in Table 2, the following key findings have been clarified:

First, the model reveals an interesting finding that restrictions on institutional entities that could borrow FC denominated loans from banks (as one of MaPP instruments) resulted in a significantly negative impact on credit growth at 1% of significant level in Viet Nam for period 2000–2016 (for both models) (Table 2). In other words, this could be considered as an effective instrument in decreasing pressure on "hot" credit growth in Viet Nam. This finding is consistent with those of Ostry et al (2011), Zhang and Zoli (2016).

Viet Nam experienced higher inflation rate compared with Asian countries in history (Kubo 2017). In order to curb, the country has to increase interest rate for dong deposit and loans in the market that led to a larger interest rate differential (between foreign and domestic currency loans). For example, in 2010,

Table 2. Effectiveness of Lending Standards Channel on Credit Growthin Transmission Mechanism of Macroprudential Policy

Variables	Expected sign	Model 1	Model 2	Model 3
C		42.833***	37.678***	53.671***
Domestic Credit	+	0.705***	0.773***	0.848***
GDP Growth	+	0.644	0.177	−0.123
Inflation	-	−0.325***	−0.338***	−0.454***
Interest Rate	-	0.349	0.159	0.784*
MaPP1	-	0.414	−0.995	
MaPP2	-	−9.395***	−9.478***	
MaPP3	-	1.111	4.956	
MaPP4a	-	4.507***		
MaPP4b	-		4.589**	
MaPP_Index	-			−1.754*
Crisis_Eco				2.652
R-Squared		0.898	0.864	0.808

Note: ***,**,* indicate coefficientsignificant at the 1%, 5% and 10% level, respectively. (Source: Author's calculation)

while interest rate for dong denominated loans was high (around 14–18% per year), those of the US dollar denominated loans was relatively low (about 6–7.5% per year). It, therefore, led to a fact that institutions preferred to borrow in foreign currency (especially in the US dollar) (Pham 2011). At the end of the first quarter in 2010, domestic currency denominated loans grew at rate of 0.57%, while the US dollar denominated loans reached at the top of 14.07%. That development could result in negative impacts on financial market in general, and in foreign exchange market in particular. First, sharp increase in FC loans would create a "bubble" supply of foreign currency (usually the US dollar) at the time of loan disbursement. In this case, exceed supply of US dollar led to revaluation pressure on dong and devaluation of the US dollar that could be harmful to Viet Nam's competitiveness for export. However, we could observe a reverse performance in the FOREX at the maturity date of loans (normally happened at the end of the year). At that time, borrowers should buy the US dollar back leading to exceed demand and causing pressure on devaluation of dong and unexpected fluctuations on the FOREX. Therefore, by imposing restrictions on institutional entities that could borrow FC denominated loans from banks, such negative consequences could be eliminated in Viet Nam. In addition, high FC loans means high loan dollarization ratio that could be harmful to the effectiveness of monetary policy and distort the financial market (Hauskrecht and Nguyen 2004; Alvares-Plata and Garcia-Herrero 2008; Kubo 2017).

Second, an instrument of applying risk weights on loans to securities and housing sectors at a higher rate than other sectors is found to have significantly positive impact on credit growth at 1% level. The finding is inconsistent with a conventional expectation from the monetary authority in the sense that the higher the risk weights imposed, the lower the domestic credit growth rate for a bank. In practice, even though the SBV imposed a higher risk weights on securities and housing loans (from 100% to 150% in 2007, and 250% in 2010), domestic credit still increased significantly. This finding could be explained by speculating psychology (demand side) and the retail banking strategy (supply side) among small commercial banks in Viet Nam as follows: (i) On the demand side: During 2007–2011, there were bubbles in securities and housing markets, and speculating psychology among individuals and institutions spread nationwide. However, both individuals and institutions seemed not to care realized seriously potential consequences of that bubbles on the Viet Nam economy in general, and in financial market in particular. They still wanted to borrow a lot of money to put in securities and housing market. (ii) On the supply side: After accession to the WTO in 2007, number of commercial banks increased significantly in Viet Nam that led to a fierce competition in providing financial services (e.g. mobilizing funds, making loans, etc). Some of them were rural banks in transformation process to "urban" or "standard" ones. In this case, new small banks with low reputation might not care about higher risk weights and loosen their lending standards to boost credit growth.

Third, other two instruments (ceiling credit growth rate for each or group of commercial banks and limits on loans per total deposit ratio) were found to have insignificant impacts on credit growth in Viet Nam during 2000–2016. It means that these instruments were not effective in transmitting to intermediate and ultimate objectives of macroprudential policy. In practice, in 2012 and 2013, the SBV classified the banking system into four groups with different maximum level of credit growth. This measure aimed at maintaining a reasonable credit growth rate in order to promote economic growth and ensure financial stability. In addition, it could prevent weak banks in over expanding lending activities. However, in practice, it could cause unexpected issues: (i) some commercial banks want to expand loan at a lower rate than their room; (ii) while other banks could expand their lending activities in spite of low ceiling credit growth rate. In order to solve these problems, the SBV should loosen the targeted credit growth rate.

Fourth, in spite of mix results among each MaPP measures, interestingly, the OLS model suggested that overall macroprudential package expressed by the index was effective in reducing credit growth in Viet Nam at 10%. This finding is consistent with those of Zhang and Zoli (2016), Cerutti et al. (2015). It implies that combination of MaPP instruments could be effective in stabilizing the credit market in the country.

Fifth, the OLS model reveals an empirical finding that there was a negative association between inflation rate and lending activities in Viet Nam during 2000–2016. In details, if inflation increases by 1%, domestic credit growth rate might decrease by 0.32% to 0.45% at 1 percent level of significance for different models. In conventional theory, inflation is a key determinant in the commercial bank lending volumes. Huybens and Smith (1999) and Boyd and Smith (1998).

asserts that inflation has adverse impact on long term lending and the movements in open market interest rates are fully and quickly transmitted to commercial loan to customers, furthering suggests that the amount of bank lending declines with inflation. Other variables such as GDP growth and lending interest rate are found to be positive but insignificant, showing a positive relationship between these variables and domestic credit growth. The positive sign of beta coefficient shows that an increase in GDP growth rate and lending interest rates determines a rise in credit growth. This result, however, is not in line with other studies in this field, showing that lower interest rates should promote credit to the private sector, implying a negative sign for this variable. The interestingly unexpected finding in Vietnam can be accounted for by several reasons. During 2007–2011, as mentioned above, there were bubbles in securities and housing markets. It, therefore, led to a fact that individuals and institutions still want to borrowed money to speculate in that two markets in spite of higher interest rate. Moreover, together with negative impacts from the global financial crisis, the country fell into an economic recession since late 2011. The SBV lowered lending rates to overcome economic difficulties, but it failed because of both lower supply and demand of loans. On one hand, due to high non-performing loan ratio, banks were not willing to make loans by tightening their own lending requirements. On the other side, enterprises were not willing to invest or not looking for effective projects during the economic recession.

4 Concluding Remarks

This paper analyzes and evaluates the effectiveness of lending standards instruments in macroprudential policy transmission mechanism in Viet Nam during 2000–2016. By employing a simple OLS model for quarterly data, the paper reveals a very interesting empirical evidence that restrictions on institutional entities that could borrow FC denominated loans from banks (as one of MaPP instruments) resulted in a significantly negative impact on credit growth at 1% of significant level in Viet Nam for period 2000–2016. The finding implies that the State Bank of Viet Nam should activate this instrument when the country faces with "hot" foreign currency credit growth rate or in the case of high loan dollarization background. By doing that, Viet Nam's monetary authority could reduce negative impacts of credit bubbles, and then stabilize the financial market.

The paper also provide interesting empirical evidence that imposing higher risk weights on loans to securities and housing sectors than other sectors is found to have an unexpectedly reverse impact on credit growth in Viet Nam during 2000–2016. This finding could be explained by psychological behavior among investors as well as retail banking strategy among small commercial banks. It means that, if the monetary authority could set stricter regulations and supervisions, this instrument could be activated in order to direct lending activities toward safer economic sectors in Viet Nam.

The other two MaPP instruments (e.g. ceiling credit growth rate for each or group of commercial banks and limits on loans per total deposit ratio) were found to have insignificant impacts on credit growth in Viet Nam during 2000–2016. It means that these instruments were not effective in transmitting to intermediate and ultimate objectives of macroprudential policy. This finding implies that the State Bank of Viet Nam should not intervene in banks' business strategy, especially setting ceiling credit growth rate for each bank. The bank itself should identify and set its own target to maintain and ensure its soundness and strength.

Appendices

Appendix A

See Tables 3 and 4.

Table 3. Descriptive statistics of all variables for lending standard channel

	DC_G	GDP_G	CPI2	IR	CRISIS_ECO	MaPP1
Mean	28.18277	6.687391	106.6198	10.82001	0.397059	0.117647
Median	28.38100	6.763924	105.7700	10.30000	0.000000	0.000000
Maximum	63.26933	9.238767	127.9000	20.10000	1.000000	1.000000
Minimum	3.744024	3.123255	97.60000	6.960000	0.000000	0.000000
Std. Dev.	13.34764	1.267280	6.617869	2.791163	0.492926	0.324585
Skewness	0.254546	−0.118683	1.323971	1.147799	0.420779	2.373464
Kurtosis	2.862115	3.009480	4.663267	4.388990	1.177055	6.633333
Jarque-Bera	0.788196	0.159893	27.70448	20.39735	11.42215	101.2476
Probability	0.674288	0.923166	0.000001	0.000037	0.003309	0.000000
Sum	1916.428	454.7426	7250.144	735.7610	27.00000	8.000000
Sum Sq. Dev.	11936.69	107.6020	2934.345	521.9696	16.27941	7.058824
Observations	68	68	68	68	68	68
	MaPP2	MaPP3	MaPP4	MaPP4A	MaPP_Index	
Mean	0.338235	0.117647	0.588235	1.551471	1.161765	
Median	0.000000	0.000000	1.000000	1.500000	1.000000	
Maximum	1.000000	1.000000	1.000000	2.500000	3.000000	
Minimum	0.000000	0.000000	0.000000	1.000000	0.000000	
Std. Dev.	0.476627	0.324585	0.495812	0.611700	1.204597	
Skewness	0.683837	2.373464	-0.358569	0.689437	0.512111	
Kurtosis	1.467633	6.633333	1.128571	1.866666	1.706278	
Jarque-Bera	11.95293	101.2476	11.38017	9.026263	7.714455	
Probability	0.002538	0.000000	0.003379	0.010964	0.021126	
Sum	23.00000	8.000000	40.00000	105.5000	79.00000	
Sum Sq. Dev.	15.22059	7.058824	16.47059	25.06985	97.22059	
Observations	68	68	68	68	68	

Table 4. Unit root test

Variable	ADF Test	Variable	ADF Test
Domestic Credit	−3.299908**	Inflation	−3.618409***
GDP Growth	−1.420738	Interest Rate	−3.618409
D(GDP Growth)	−5.337912***	D(Interest rate)	−6.874825***

Note: ***,**,* indicate coefficientsignificant at the 1%, 5% and 10% level, respectively.

Appendix B: Diagnosis Tests for Models of Evaluating the Effectiveness of MaPP Transmission Mechanism of Lending Standard Channel

Model 1:

Breusch-Godfrey Serial Correlation LM Test

F-statistic	2.833757	Prob. F(2,56)	0.0672
Obs*R-squared	6.157592	Prob. Chi-Square(2)	0.0460

Heteroskedasticity Test: Breusch-Pagan-Godfrey

F-statistic	1.806817	Prob. F(8,58)	0.0942
Obs*R-squared	13.36636	Prob. Chi-Square(8)	0.0999
Scaled explained SS	6.726627	Prob. Chi-Square(8)	0.5664

Model 2:

Breusch-Godfrey Serial Correlation LM Test

F-statistic	1.021981	Prob. F(2,56)	0.3665
Obs*R-squared	2.359340	Prob. Chi-Square(2)	0.3074

Heteroskedasticity Test: Breusch-Pagan-Godfrey

F-statistic	0.876541	Prob. F(8,58)	0.5416
Obs*R-squared	7.226718	Prob. Chi-Square(8)	0.5124
Scaled explained SS	4.450382	Prob. Chi-Square(8)	0.8144

Model 3:

Breusch-Godfrey Serial Correlation LM Test:

F-statistic	1.364178	Prob. F(2,58)	0.2637
Obs*R-squared	3.010125	Prob. Chi-Square(2)	0.2220

Heteroskedasticity Test: Breusch-Pagan-Godfrey

F-statistic	0.785109	Prob. F(6,60)	0.5850
Obs*R-squared	4.877309	Prob. Chi-Square(6)	0.5596
Scaled explained SS	5.601895	Prob. Chi-Square(6)	0.4692

References

Aiyar, S., Calomiris, C., Wieladek, T.: How does credit supply respond to monetary policy and bank minimum capital requirements? BoE Working Papers 508. Bank of England, London (2014)

Alvares-Plata P., Garcia-Herrero, A.: To dollarize or de-dollarize: Consequences for monetary policy. Discussion paper 842, DIW Berlin, German Institute for Economic Research (2008)

Bank of England: The role of macroprudential policy, Bank of England Discussion Paper, November 2009

Bank of England: Instruments of Macroprudential Policy, Bank of England Discussion Paper, December 2011

Boyd, H.J., Smith, B.D.: The evolution of debt and equity markets in economic development. Econ. Theory **12**, 519–560 (1998)

Borio, C., Drehmann, M.: Towards an operational framework for financial stability: 'fuzzy' measurement and its consequences, BIS Working Papers, no 284, June 2009

Caruana, J.: Monetary policy, financial stability and asset prices, Occasional Papers, 0507, Bank of Spain (2005)

Cerutti, E., Classens, S., Laeven, L.: The Use and Effectiveness of Macroprudential Policies: New Evidence, IMF Working Paper, WP/15/61 (2015)

Claessens, S.: An Overview of Macroprudential Policy Tools, IMF Working Paper WP/14/214 (2014)

Clement, P.: The term "macroprudential": origins and evolution, BIS Quarterly Review, March 2010

Galati, G., Moessner, R.: Macroprudential Policy-A Literature Overview, BIS Working Papers No 337 (2011)

Gambacorta, L., Shin, H.S.: Why bank capital matters for monetary policy, BIS working paper No 558 (2016)

Hartmann, P.: Real estate markets and macroprudential policy in Europe, Working paper series, European central bank, No. 1796 (2015)

Hauskrecht, A., Nguyen, T.H.: Dollarization in Vietnam. In: Paper prepared for the 12th Annual Conference on Pacific Basin Finance, Economics, Accounting, and Business, Bangkok, 10–11 August (2004)

Huybens, E., Smith, B.: Inflation, financial markets and long-rum real activity. J. Monetary Econ. **43**(2), 283–315 (1999)
International Monetary Fund: Key aspects of Macroprudential policies, IMF Staff Papers, June 2013
Keys, B., Mukkerjee, T., Seru, A., Vig, V.: Did securitization lead to lax screening? Evidence from subprime loans. Q. J. Econ. **125**(1), 307–362 (2010)
Kubo, K.: Dollarization and De-dollarization in Transitional Economies of Southeast Asia. IDE-JETRO Series, Palgrave MacMilan (2017)
Lim, C., Columba, F., Costa, A., Kongsamut, P., Otani, A., Saiyid, M., Wezel, T., Wu, X.: Macroprudential Policy: What Instruments and How to Use Them? Lessons from Country Experiences, IMF Working Paper No. 11/238 (Washington: IMF) (2011)
Nadauld, T., Sherlund, S.: The role of the securitization process in the expansion of subprime credit, Finance and Economics Discussion Series 2009-28. Board of Governors of the Federal Reserve System, Washington, April 2009
Ostry, J.D., Ghosh, A.R., Habermeier, K., Laeven, L., Chamon, M., Qureshi, M.S., Kokenyne, A.: Managing capital inflows: What tools to use? IMF staff discussion note, SDN/11/06, 5 April (2011)
Pham, T.H.A.: Assessing Vietnam's exchange rate policy in 2010, Banking Science and Training Review (2011)
Saurina, J.: Loan loss provisions in Spain. A working macroprudential tool, Bank of Spain Financial Stability Review, No. 17, pp. 11–26 (2009a)
Saurina, J.: Dynamic Provisioning. The experience of Spain, Crisis Response, Public Policy for the Private Sector. Note Number 7, July. The World Bank (2009b)
Tresel, T., Zhang, Y.S.: Effectiveness and Channels of Macroprudential Instruments: Lesson from the Euro Area, IMF Working Paper, WP/16/4 (2016)
Vandenbussche, J., Vogel, U., Detragiache, E.: Macro-prudential policies and housing prices: a new database and empirical evidence for Central, Eastern, and Southeastern Europe. J. Money Credit Banking **47**(S1), 343–377 (2015)
Zhang, L., Zoli, E.: Leaning against the wind: macroprudential policy in Asia. J. Asian Econ. **42**, 33–52 (2016)

Impact of the World Oil Price on the Inflation on Vietnam – A Structural Vector Autoregression Approach

Nguyen Ngoc Thach(✉)

Institute of Banking Research and Technology, Banking University of Ho Chi Minh City, 36 Ton that Dam Street, District 1, Ho Chi Minh City, Vietnam
thachnn@buh.edu.vn

Abstract. This paper aims to analyze the impact of the world crude oil price (hereinafter referred to as "the world oil price") on the inflation of Vietnam from the first quarter of 2000 to the fourth quarter of 2015 by using Structural Vector Autoregression (SVAR) method, Impulse Response Functions (IRFs) and Forecast Error Variance Decomposition (FEVD). The results show that the world oil price has positive effects on the inflation (measured by CPI). When the world oil price increases by one standard deviation, the inflation rises by 2.3416% in the first quarter and this uptrend continues to the fourth quarter. Concomitantly, the strongest impact of the world oil price on the inflation is observed in the fifth quarter, albeit diminishing after that. The results also indicate that in general, the world oil price has negative impact on Vietnam's real GDP growth. This paper provides some implications of domestic petroleum price regulation to improve the efficiency of the monetary policy.

Keywords: Monetary policy · World oil price · Inflation · Price regulation SVAR

1 Introduction

In most countries, petroleum and oil are strategic energy commodities, which have critical roles in the economy through their significant impacts on various industries and civilians' life. Petroleum is the input factor for manufacturing, daily activities, and national security; representing an important factor in boosting economic growth of a country. Therefore, petroleum, directly and indirectly, takes up a high proportion of the consumer goods basket. Petroleum is also among the non-monetary commodity group of inflation, which cannot be controlled by central banks. Therefore, if this inflation component accounts for a dominant proportion in the consumer commodity basket, the fluctuations of the world and domestic oil prices weaken the efficiency of monetary policy. This means that the main target of this policy is to control inflation. During the period of petroleum price fluctuations, price stabilization becomes a central issue of inflation control.

Vietnam imports a large proportion of finished petroleum products to serve domestic demands. During 2005-2014 period an average growth rate of petroleum

V. Kreinovich et al. (Eds.): ECONVN 2019, SCI 809, pp. 694–708, 2019.
https://doi.org/10.1007/978-3-030-04200-4_48

usage reached 8%. Currently, this demand is approximately 16.7 to 17.2 million tons of petroleum per year, among which transportation takes up 53% while agricultural and civil activities comprise around 8% [6]. Petroleum products, such as gasoline, fuel oil, heating oil, are derived from crude oil. Hence, the changes in crude oil price have direct impact on the cost of those products. In practice, the demand for petroleum in Vietnam is projected to continue the upward trend in conjunction with the rate of economic growth. Vietnam's Government has made significant efforts to renovate the mechanism of stabilizing petroleum prices. However, petroleum prices still fluctuate constantly, which causes difficulties for businesses and civilians, reduces the efficiency of monetary policy being implemented by the State Bank of Vietnam. Therefore, this paper aims to estimate the effects of the world oil price on the inflation of Vietnam and provide some implications of price regulation to achieve the monetary policy's high efficiency.

2 Theoretical Background

2.1 The World Oil Price

Oil is considered as "black gold" as it is a significant input of most economic activities. It is one of the materials for power production and transportation means. In petrochemistry, the material is used to produce plastics and many other products. Changes in the oil price can have substantial impacts on the economy.

Oil price can be understood as the price of a standard oil barrel, readily-delivered, typically Brent or WTI. The price of standard oil fluctuates wildly under the impacts of global political-economic events. In fact, oil industry categorizes crude oil under the area where it is originated like "West Texas Intermediate" (WTI) or "Brent"), generally based on its proportion, and relative density ("light", "medium", or "heavy"). Crude oil is also categorized as "sweet" or "sour" depending on the level of sulfur, in which sweet oil contains less than 0.5% of sulfur and oil contains about 1% of sulfur and requires more effort to produce it in accordance with current indicators. Heavier oil contains higher sulfur levels. Redundant sulfur is extracted from crude oil during the purification process because sulfur dioxide exhausting to the air during the burning process is a heavy pollutant.

A standard oil barrel is a commercial measurement unit of the quantity of crude oil. A barrel is 42 US gallon or 158.9873 litter; seven oil barrels is approximately 1.113 tons while one US gallon is around 3.785 cm^3 or 3.785 L.

The formula of calculating crude oil price is determined by the price of different standard oil types, which include:

West Texas Intermediate (WTI) oil: is a very high-quality type of oil, sweet and light. It is usually piped to Cushing oil center, Oklahoma (North America) before being processed and this is where the daily price of WTI crude oil is determined. WTI is the standard oil for pricing other types of crude oil around the world.

Brent oil: consists of 15 types of oil from Brent and Ninian fields which are mixed into Brent at Sullom Voe storage, Sheland island. This is also a high-quality type of oil, sweet, light in the North Sea which is located between the United Kingdom and North

Europe countries such as Sweden, Norway, Denmark…). Other types of crude oil around the world, including oil produced in Europe and oils imported to Europe from Africa and Middle East, are based on the price of Brent oil to form the price formula.

Dubai-Oman oil: is based sour oil. This type of oil is used as the standard oil forming the price of other types of crude oil in the Asia-Pacific of Near East and Middle East oil.

Tapis (Malaysia): is based on light sweet oil. This type of oil are used as the standard oil for calculating the price of different types of light oil in Asia-Pacific (Far East).

Minas oil (Indonesia): is used as a reference for heavy oil in Far East.

OPEC oil basket: includes a mixture of heavy and light crude oil which is extracted by OPEC countries and this type of oil is heavier than Brent and WTI oil. The word OPEC, in fact, stands for Organization of the Petroleum Exporting countries who are formed together to control the world oil prices. OPEC can adjust the oil production quota of its members. The energy prime minister summit of OPEC is held twice a year to evaluate the oil market and propose appropriate solutions to assure the oil supply. OPEC has tried to control the oil price between upper and lower levels by increasing or decreasing the supply. This is very important for market analysis.

2.2 Inflation and Inflation Measurement

Marx indicates that inflation is the situation in which bank notes are so abundant in distribution channels, exceeding the actual demand of the economy, devaluing the currency and redistributing the national income [17]. According to his perspective, inflation only appears when the supply of money surpasses the demand for money of goods distribution channels.

Keynesian theory claims that increasing money supply quickly will continually increase the goods' price at high percentage, thus causing inflation [16]. According to this viewpoint, only the aggregate demand which includes the money supply can cause high inflation. In other words, any events from the supply side are not the reason of high inflation.

Monetarists, among them Friedman [3] is the leader, conclude that inflation, in anywhere and at any time, is a monetary phenomenon. When money supply increases, inflation is inevitable. This theory is based on the neutrality of money in the long run, meaning that the increase in money supply does not affect the supply of goods and services as well as employment in the long term.

According to the supply-side theory, what causes inflation is the situation originated from the increase in production costs of companies. The increase is caused by budget deficit, making the government to raise tax as well as to levy more taxes on companies. In this case, cost-push inflation arises.

Generally, inflation is a phenomenon which occurs when the price level of the economy increases stably over a certain period of time. The price level is the average price of all goods and services in an economy which indicates volatility in the prices and purchasing power of the currency for other products. When the price level increases stably during a certain period, typically from a few months or more, it is considered as inflation.

Inflation is measured by the Consumer Price Index (CPI), Producer Price Index (PPI) and GDP deflator.

CPI is an indicator used to measure the price level of an economy, reflecting the trend of price volatility over time of goods in the consumer goods and service basket. The basket consists of representative goods and services, and it is updated regularly to maintain its suitability in each period. In the US, the basket comprises 265 primary products of 85 cities. In Vietnam, CPI consists of 572 primary products of 63 provinces and cities [4]. Vietnam's CPI calculated by Laspeyres formula is appropriate to the international rule:

$$I^{t\to 0} = \frac{\sum_{i=1}^{n} p_i^t q_i^o}{\sum_{i=1}^{n} p_i^o q_i^o} = \sum_{i=1}^{n} W_i^o * \left(\frac{p_i^t}{p_i^o}\right),$$

where $I^{t\to 0}$ is the CPI of the time t compared to time 0; p_i^t is the price of product i in the time t; p_i^o is the price of product i at time 0; W_i^o – fixed number in 2009.

Inflation is calculated by PPI in the same way with CPI approach. However, PPI is calculated on a larger number of products and by the trade price (the price in the first trade). In the US, PPI is measured on 3,400 products while the number in Vietnam is 1,800 products [5].

In addition, inflation is measured by GDP deflator (or implicit price deflator). This is the price level of all products and services in GDP used to determine inflation percentage. GDP deflator is determined by the percentage of the nominal GDP and the real GDP as in the following formula:

$$\text{GDP deflation in year t} = \frac{\textit{GDPt based on the current price}}{\text{GDPt based on the compared price}} \times 100 = \frac{\sum_i p_i^t q_i^t}{\sum_i p_i^0 q_i^t} \times 100,$$

where GDP_t is the gross domestic product in year t; p_i^0 is the base price of product i; q_i^0 is the quantity of product i in year 0 (base year).

2.3 Role of the Central Banks' Monetary Policy in Controlling Inflation

There are different approaches to controlling inflation. These approaches often fall into two categories: situational and strategic.

Situational approach requires temporary solutions used to quickly reduce inflation. When high inflation or hyperinflation arises, commonly-applied solutions include: fiscal restraint, monetary restraint, price control, income restriction.

Strategic approach aims to affect all aspects of the economy to maintain monetary stability. These solutions usually include social-economic development strategies, reforming public finance, improving market competition.

The mentioned inflation control strategies are implemented by not only central banks, but also other governmental authorities because central banks intervene only the core inflation, while other inflation components need other authorities' adjustment. For example, income restriction policy is implemented by Labour Ministry, fiscal restraint policy by Finance Ministry. Especially, price regulation policy also helps to prevent wild fluctuations in the price of such strategic goods as petroleum. To control inflation effectively, it is necessary for governmental authorities to collaborate in harmony. Central banks take a decisive role in solving this issue as they employ monetary policy tools to adjust CPI.

2.4 Empirical Evidence

Many empirical studies have been conducted to analyze the impacts of oil price on macroeconomic stability, especially the efficiency of monetary policy. These studies focused on investigating inflation in different countries.

For many European countries, Cunado and Garcia [2] pointed out that the increase of oil price could have long-term impact on inflation and GDP growth. For China, Qianqian [24] found that the increase of oil price reduced the output and increase CPI. In Japan, Rodriguez and Sanchez [25] claimed that the increase of oil price had negative impact on industrial production, but positive impact on inflation. Twenneboah and Adam [29] studied the impacts of shocks in the volatility of the world oil price on the monetary policy of Ghana in short and long term. By using Vector Error Correction Model (VECM), the study proved that, in the short-term, oil price shocks had positive effects on GDP, inflation (measured by CPI) and negative effects on interest rates. This result indicated that to minimize the effect of external shocks, which was caused by changes in the oil price, in the short-run, monetary policy should be adjusted toward cutting interest rates to prevent the price of goods from increasing. However, this policy could only be complemented in the short-term, but in the long-term, oil price shocks affected GDP negatively after four quarters, inflation and interest rates after two quarters. Meanwhile, oil price shocks always have positive effects on the exchange rate in short and long term. Kargi [10] investigated the impacts of oil price on inflation and economic growth in Turkey from the first quarter 1998 to the sixth quarter of 2013 by using the Granger causality test analysis. The result confirmed the existence of the Granger causality relationship between oil price, economic growth, inflation in the long-term. Specifically, oil price had positive effects on inflation, but negative effects on economic growth of Turkey. Yildirim, Ozcelebi and Ozkan [30] studied the impacts of the oil price's increase on the monetary policy of top oil import countries (The US, EU, China and Japan) during 2000–2013 period. The research used SVEC model to analyze the impacts of the oil price's increase on production, consumer prices and interest rates. The results showed that the oil price's increase leaded to the rise of consumer price, volatility in industrial production and interest rates, and after that affected monetary policy. The research concluded that oil price affected the prices of goods, inflation and interest rates through transmission channels. Therefore, these countries needed to construct an optimal monetary policy to eliminate the negative impacts of oil price's increase. Hesary and Yoshino [7] studied the impacts of oil price on GDP growth and inflation of China, Japan and the US from 1/2000 to 12/2013. To evaluate the impacts of oil price on macroeconomic variables (including GDP, CPI, monetary supply and exchange rates), the research used SVAR model. The results revealed that the impact of oil price on economic growth in China is by far more significant than in the US and in Japan. However, inflation in China was less affected by changes of oil price than the other two countries.

In Vietnam, there have not been many studies on the impacts of oil price on inflation. For example, Trung and Thong [23] examined macroeconomic factors affecting inflation in Vietnam during 1992–2012 period. VECM estimation on variables of CPI, GDP, monetary supply (M2), credit, interest rates, exchange rate, oil price and world rice prices showed that Vietnam's inflation was affected by expected inflation and exchange rate. In the short-term, monetary policy is not quick and effective responsive tools for controlling inflation in Vietnam. Anh, Lan, Ngoc, Phuong and Tung [22] investigated the volatility of the world oil price and its impacts on Vietnam's economy. Their result revealed that shocks of the aggregate demand and spare demand caused the volatility in the world oil price throughout 1975–2015 period. Oil supply shocks had minimal impact on the volatility of oil price, except 1976–1982 period, and this role diminished over time. The shock increasing in oil supply and total oil demand contributed to the economic growth of Vietnam while the shock increasing in spare demand hampered the economic growth. The shock increasing in oil supply reduced inflation while the shock increasing in total demand and spare demand increased Vietnam's inflation.

3 Methodology and Data

3.1 Methodology and Model

To estimate the impact of the world oil price on Vietnam's inflation, this paper employs SVAR model. SVAR model has been used in many studies on the impacts of oil prices on macroeconomic variables, which include inflation and money supply [7, 11–13, 20]. Also, SVAR model can be used to decompose oil price shocks into supply shocks and demand shocks [10, 20].

Based on [9], Hesary et al. [7] conduct studies in developing countries in recent years. Due to data limitation in Vietnam, we focus our research on the impact of the world oil price on inflation (measured by CPI) and GDP to assure that the estimation model is appropriate to the research purpose. The research model is as follow:

$$y_t = [\text{Lnoil}_t, \text{CPI}_t, GDP_t],$$

where Y_t are time series, Lnoil is the world oil price, CPI is inflation rate and GDP is real GDP growth rate.

Constraints remain the same as in the model of [7]. Constraint matrices A and B for a developing economy – Vietnam in SVAR equation is structured as follow:

$$\begin{bmatrix} 1 & 0 & 0 \\ a_{21} & 1 & 0 \\ a_{31} & a_{32} & 1 \end{bmatrix} \text{x} \begin{bmatrix} u_t^{Lnoil} \\ u_t^{CPI} \\ u_t^{GDP} \end{bmatrix} = \begin{bmatrix} b_{11} & 0 & 0 \\ 0 & b_{22} & 0 \\ 0 & 0 & b_{33} \end{bmatrix} \text{x} \begin{bmatrix} {}_t^{Lnoil} \\ {}_t^{CPI} \\ {}_t^{GDP} \end{bmatrix},$$

In this model, Lnoil is an exogenous variable, CPI and GDP are endogenous, and therefore they do not have causality impacts on foreign variables, in line with [9, 10, 26]. Thus, Lnoil will not consist of endogenous variables or elements in ($a_{12}, a_{13} = 0$).

In addition, according to economic theory and empirical studies [14, 19, 27], there exists a one-way relationship from CPI to GDP, but GDP does not affect CPI. This is an advantage of SVAR model for small economies because it helps to reduce the numbers of estimation variables in the model [15] (Table 1).

Table 1. Variable and data source description

Variable	Measurement	Source
Lnoil	Ln(the average price of oil during the period t)	IFS (IMF) [8]
CPI	(CPI t – CPI t-1)/(CPI t-1)	General Statistical Office of Vietnam
GDP	(GDP t – GDP t-1)/(GDP t-1)	IFS (IMF) [8] and Trading Economics

Source: The author collected.

3.2 Data

The research data is collected from three main sources: IMF, Trading Economics and General Statistical Office of Vietnam, from the first quarter of 2000 to the fourth quarter of 2015. Inflation and GDP growth of Vietnam are presented in Fig. 1.

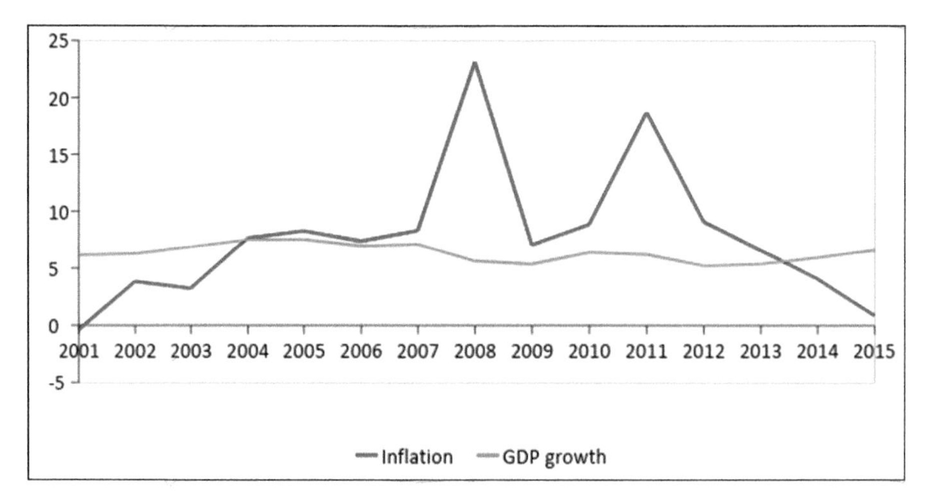

Fig. 1. Inflation and GDP growth during 2001–2015 period Source: [6, 28]

Vietnam's inflation of from 2001 to 2015 was complex. From 2001 to 2007, CPI was low and stable at below 10%. However, inflation rose significantly during 2007–2008 period and in 2008, it rocketed to 23.12%, the highest figure during the recent 10 years. After a downward trend of 2009, inflation increased significantly and surpassed 10% in 2011. From 2012 to 2014, inflation started decreasing and stabilizing.

In fact, wild fluctuations in oil price in the past led to high inflation and long depression in a number of countries. Oil price fluctuations with the peak increase emerged in the 1970s. After some adjustments in the following years, the strong increase continued from 4/1980 to 7/2008, following by significant decreases from 8/2008 to 02/2009. Oil price continued increasing and stabilizing from 3/2009 to 5/2014. However, it plummeted from 6/2014 to 4/2016.

Oil price volatility affected economic operations through different channels such as changes in the domestic oil price. Specifically, petroleum prices in Vietnam revealed a complex trend during 2001–2015 period, as shown in Fig. 2. In July 2008, the price of petroleum products rose, in which some products increased suddenly such as RON 92 (68%). In 2014, petroleum prices showed a complex trend again when it increased significantly before tumbling by 29.3% during the following six months.

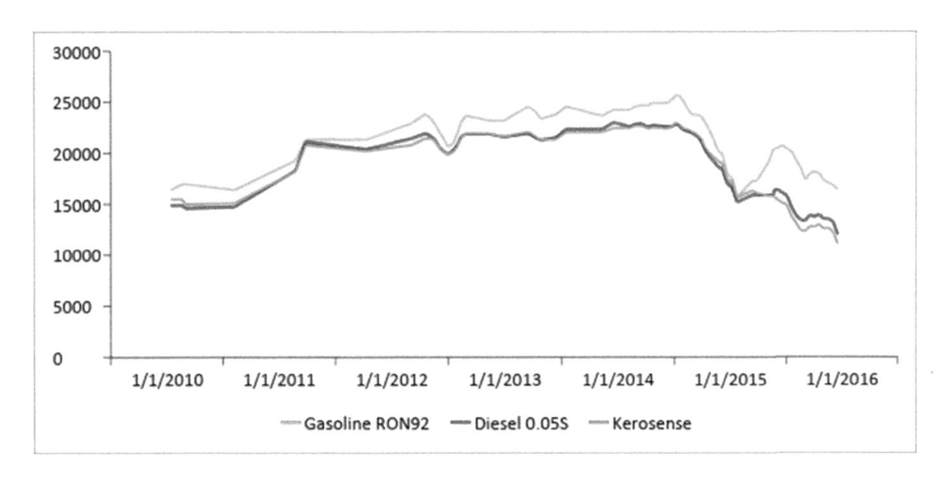

Fig. 2. Prices of RON 92, diezen 0.05S and kerosense during 2000–2015 period Note: Unit VND. Source: Petrolimex [21]

The world oil price's volatility was the reason of such volatility shown in Fig. 2. In 2008, the world oil price shown unusual fluctuations, as a result of the reduction in supply, which triggered the energy war. In 2014, the world oil price plummeted to below 60 USD/barrel because of the political conflict between Russia, the US, and Europe on Ukraine's issues. In addition, OPEC members with almost 40% of the world oil market did not reach an agreement on controlling the oil supply, leading to the price decrease. However, the main reason of the price fall was the shale oil revolution in the US.

Ineffective price regulation was the reason of unusual price volatility as well. In Vietnam, petroleum market was mostly monopolized by state-owned companies such as Petrolimex, PV Oil, as a result, petroleum prices were determined by these companies. This means the market prices of petroleum products did not reflect the supply and demand relationship.

4 Empirical Results

4.1 Test of Stationarity

SVAR estimation requires variables in the model to have the stationarity test. Augmented Dickey-Fuller (ADF) method is used to test the stationarity of data. The test results show that Lnoil and CPI variables are stationary when taking the first order difference, GDP variable is stationary at 0 lag with a significance level of 5%. Details of the variables' stationarity test are presented in Table 2.

Table 2. Stationarity test of variables

Variable	Lag's length	Statistics t	P-value
Lnoil	0	−1,521	0,5231
	1	−6,063	0,0000*
CPI	0	−1,836	0,3628
	1	−3,610	0,0056*
GDP	0	−2,996	0,0352*

Note: (*) significant at 1% level.
Source: The author's calculation.

4.2 Results of the Optimal Lag Selection

The purpose of this test is to select lag variables for the model and avoid missing important explanatory variables to obtain the optimal model. Criteria of the model's lag selection is determined by Log-Likelihood (LL), Likelihood Ratio (LR), in which the values are greater for good quality samples, and Final Prediction Error (FPE), Akaike Information Criteria (AIC), Hannan-Quinn Information Criteria (HQIC), Schwarz Bayesian Information Criteria (SBIC), in which the values are as small as possible. Based on Table 3, the selected optimal lag is 4.

Table 3. Lag selection criteria for the model

lag	LL	LR	df	P	FPE	AIC	HQIC	SBIC
0	−214.005				.314267	7.35611	7.39735	7.46175
1	−160.474	107.06	9	0.000	.0695	5.84658	6.01152	6.26913
2	−141.15	38.648	9	0.000	.049112	5.49661	5.78526*	6.23607*
3	−129.506	23.237	9	0.006	.045197	5.407	5.81936	6.46337
4	−120.137	18.739*	9	0.028	.045179*	5.39447*	5.93055	6.76776

Source: The author's calculation.

4.3 Estimation of Matrices A and B of the Structural Model

Estimation parameters of matrices A and B of each model are presented in Table 4. The results show that most parameters have statistical significance of 1%. LR test is statistically significant and therefore constraints of SVAR are appropriate.

Table 4. xxxx

Estimation results of matrix A			
$Lnoil_t$	$Lnoil_t$	CPI_t	GDP_t
$Lnoil_t$	1	0	0
CPI_t	−4.515*	1	0
GDP_t	−1.338**	0.071	1
Estimation results of matrix B			
$Lnoil_t$	0.131*	0	0
CPI_t	0	1.425*	0
GDP_t	0	0	0.619*

Note: (*) and (**) significant at 1% and 5% level, respectively.
Source: The author's calculation.

4.4 Model Stability Test

The test results in Fig. 3 show that the model is stable. The specific test results of the model indicate that all roots of the companion matrix stay within the unit boundary and SVAR model is stable.

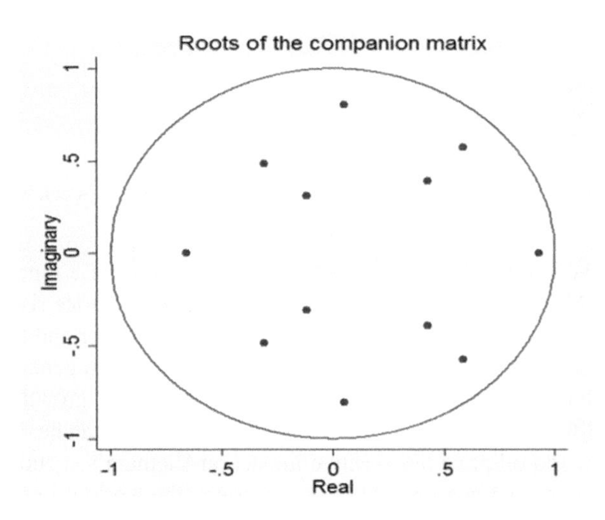

Fig. 3. Model stability test Source: The author's calculation

4.5 IRF Analysis and Variance Decomposition

4.5.1 IRF Analysis

The impulse response function is used to identify the impacts over time of the world oil price on CPI and real GDP growth and to analyze the volatility of these variables when shocks happen.

Impact of the World Oil Price on CPI. The push response function analysis of variables in SVAR estimation shows that when the world oil price increases to a standard deviation, CPI rises by 2.3416% in the first quarter and this increase in CPI lasts until the fourth quarter after the world oil price's increase. In addition, CPI is most affected by the world oil price in the fifth quarter and the effect diminishes after that. Based on these results, it can be concluded that the world oil price has positive impact on CPI of Vietnam and this is in line with [1, 29] (Fig. 4).

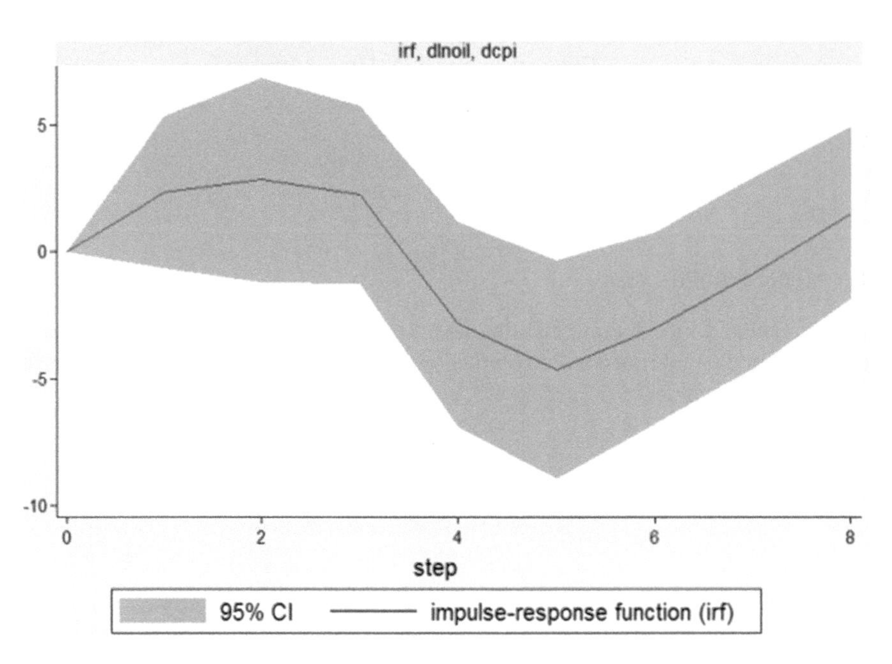

Fig. 4. Reactions of Dcpi to Dlnoil. Source: The author's calculation

Impact of the World Oil Price on GDP. The impulse response function analysis of variables in SVAR estimation shows that when the world oil price rises to a standard deviation, GDP growth increases by 1.7185% in the first quarter and this growth lasts only until the third quarter after the world oil price's increase. In general, the world oil price tends to have negative impact on Vietnam's real GDP growth. This result is statistically significant at two standard deviations. Therefore, it can be concluded that in short-term, the world oil price has positive impact on Vietnam's economic growth and the impact lasts over two quarters. After two quarters, the world oil price has negative impact on Vietnam's economic growth. This result is in line with [1, 18, 29] (Fig. 5).

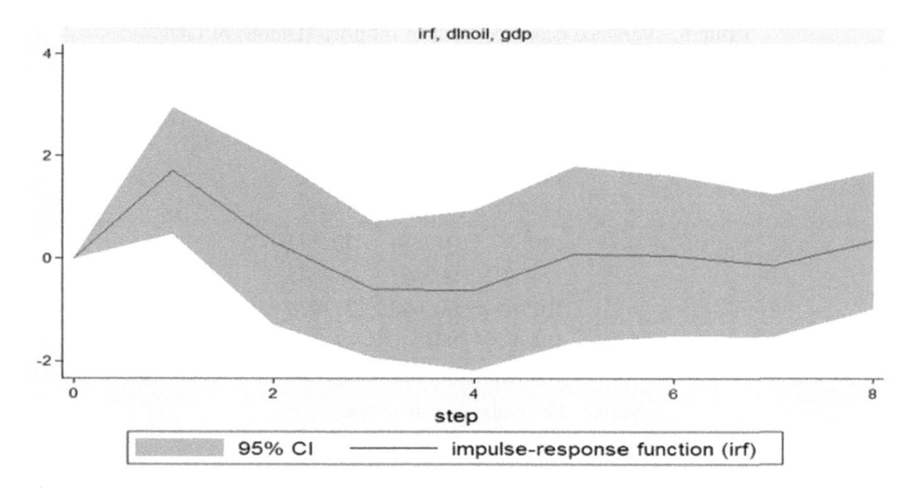

Fig. 5. Reactions of GDP to Dlnoil. Source: The author's calculation

To obtain detailed analysis on the impacts of the world oil price on inflation and real GDP growth of Vietnam, variance decomposition is conducted in the following sub-section.

4.5.2 Variance Decomposition

Estimating the reaction of variables resulted from structural shocks reveals directions and levels of the variables' reaction but not the role of the shocks for the variables' motions during the research period. Therefore, we conduct an analysis of variance decomposition of variables in SVAR model which analyzes the impact of the world oil price on Vietnam's inflation. The world oil price variable explains 14.79% of inflation rate and 4.36% of real GDP growth of Vietnam (Tables 5 and 6).

Table 5. Variance decomposition of oil price (Dlnoil) of CPI (DCPI)

Step	(1) fevd	(1) Lower	(1) Upper
0	0	0	0
1	.147926	−.019318	.315171
2	.217537	.004545	.430528
3	.258895	.018646	.499145
4	.243487	.017526	.469448
5	.248454	.032015	.464893
6	.299441	.044516	.554365
7	.320834	.048092	.593575
8	.321161	.0477	.594622

Table 6. Variance decomposition of oil price (Dlnoil) of GDP

Step	(1) fevd	(1) lower	(1) Upper
0	0	0	0
1	.043568	−.058312	.145449
2	.149193	−.030933	.32932
3	.139638	−.037635	.31691
4	.14612	−.010284	.302523
5	.124225	.003807	.244643
6	.107165	−.000648	.214978
7	.104592	−.004242	.213425
8	.104648	−.007041	.216336

Source: The author's calculation.

4.6 Empirical Result of the Impact of the World Oil Price on Vietnam's Inflation

This research has examined the impact of shocks in the world oil price on inflation (and real GDP growth) in Vietnam during 2000–2015 period. SVAR model, variables, and hypotheses in this research are selected to be appropriate with data conditions in Vietnam and are based on the research model of [7, 10]. The research results reveal that when the world oil price increases to a standard deviation, inflation rises by 2.3416% during the first quarter and inflation continues the upward trend throughout four quarters after the oil price's increase. In addition, inflation is most affected by the world oil price in the fifth quarter and after that the effect diminishes. Also, when the world price rises to a standard deviation, GDP growth increases by 1.7185% in the first quarter and the effect diminishes until the third quarter. Generally, the world oil price tends to have negative impact on the economic growth of Vietnam.

5 Conclusion and Policy Implications

Vietnam is a developing country with high energy intensity, especially petroleum products. The country's reliance on oil price is relatively high. Although exporting crude oil, this country is a big importer of petroleum products to serve its domestic manufacturing and consumer activities. Meanwhile, derivative markets, especially derivative oil products, have not grown much, hindering the ability of the economy to prevent risks from fluctuations in oil price. Therefore, researching into the impact of the world oil price on Vietnam's inflation (and real GDP growth) is very necessary.

The above analysis shows that Vietnam's inflation and real GDP react strongly to the world oil price shocks. To minimize the impacts of the world oil price on domestic prices, the solutions can be short-term and long-term. In this paper, the author focuses on the price regulation policy of Vietnam's Government. Strong fluctuations in the domestic petroleum prices over the recent years indicate that there are minuses in the

Government's policy of stabilizing petroleum prices. The ineffective regulation of petroleum prices can not handle to minimize their strong fluctuations. This requires to implement a more effective price regulation policy to ensure the efficiency of the monetary policy.

Therefore, petroleum price regulation policy needs to be directed toward creating a fair competitive environment for businesses. Companies will find it difficult to take advantage of the ownership of the infrastructure system and they will have to rent the systems as other companies. At the moment, as all of the ten leading importers are state-owned companies, the market competition is not strong enough. It is essential to shorten the gap in market share of companies to create a balance in the petroleum market. To intensify competition between companies, it's necessary to increase the number of private firms joining the market. Also, there is a need to raise businesses' awareness in preventing risks from oil price fluctuations by derivative tools. The Government needs to develop legal documents to specify derivative transactions because these transactions have not been well developed. Besides, companies need to acquire competent staffs who possess good understanding of preventing risk from oil price fluctuations because this practice is complex. Finally, companies need to be proactive in preventing risks, especially state-owned ones.

References

1. Cologni, A., Manera, M.: Oil prices, inflation and interest rates in a structural cointegrated VAR model for the G-7 countries. Energy Econ. **30**, 856–888 (2008)
2. Cunado, J., Gracia, P.D.F.: Do oil price shocks matter? Evidence for some European countries. Energy Econ. **25**(2), 137–154 (2003)
3. Friedman, M.: Inflation and unemployment. Economic Sciences, pp. 267–284 (1976)
4. General Statistical Office of Vietnam: Press release of updated calculation methods of the consumer price index during the 2009–2014 period (2009a). (in Vietnamese)
5. General Statistical Office of Vietnam: Vietnam's producer price index (PPI) (2009b). (in Vietnamese). https://www.gso.gov.vn/default.aspx?tabid=450&ItemID=12207. Accessed 02 Feb 2017
6. General Statistical Office of Vietnam: Social-economic updates (in Vietnamese). http://gso.gov.vn/default.aspx?tabid=621. Accessed 02 Feb 2017
7. Hesary, T.F., Yoshino, N.: Causes and remedies for Japan's long-lasting recession: lessons for the People's Republic of China, ADBI Working 554, Asian Development Bank Institute (2015)
8. IMF: International Financial Statistics (IFS). http://data.imf.org/?sk=5DABAFF2-C5AD-4D27-A175-1253419C02D1. Accessed 02 Feb 2017
9. Jayaraman, T.K., Lau, L.: Oil price and economic growth in small Pacific island countries. Modern Economy **2**(2), 153–162 (2011)
10. Kargi, B.: The effects of oil prices on inflation and growth: time series analysis in Turkish economy for 1988:01-2013: 04 period. Int. J. Econ. Res. **2**(5), 29–36 (2014)
11. Kilian, L.: Not all oil price shocks are alike: disentangling demand and supply shocks in the crude oil market. Am. Econ. Rev. **99**(3), 1053–1069 (2009)
12. Lee, K., Ni, S.: On the dynamic effects of oil price shocks. J. Monetary Econ. **49**, 823–852 (2002)

13. Lippi, F., Nobili, A.: Oil and the macroeconomy: a structural VAR analysis with sign restrictions. Center for Economic Policy Resarch, Working Paper 6830 (2008)
14. Mahmoud, L.O.M.: Consumer price index and economic growth: a case study of Mauritania 1990-2013. Asian Econ. Soc. Soc. **5**(2), 16–23 (2015)
15. Mala, R., Param, S.: Structural VAR approach to Malaysian monetary policy framework: evidence from the pre- and post Asian crisis periods. Department of Econometrics and Business Statistics Monash University, Australia (2007)
16. Mishkin, S.F.: The cause of inflation. National Bureau of Economic Research, pp. 2–16 (1984)
17. Moseley, F.: Marx's Theory of Money – Modern Appraisals, pp. 40–43. Mount Holyoke College, Palgrave Macmilan (2005)
18. Thanh, N.D., Trinh, B., Thang, D.N.: Impacts of the increase of oil prices: Initial quantitative analysis. Sci. J. Hanoi Natl. Univ. Econ. Bus. **25**, 25–38 (2008). (in Vietnamese)
19. Omoke, P.C.: Inflation and economic growth in Nigeria. J. Sustain. Dev. **2**(3), 159 (2010)
20. Peersman, G., Robays, I.V.: Oil and the Euro area economy. Econ. Policy **24**(60), 603–651 (2009)
21. Petrolimex: News letter (in Vietnamese). http://www.petrolimex.com.vn/nd/thong-cao-bao-chi.html. Accessed 02 Feb 2017
22. Anh, P.T.H., Lan, C.K., Ngoc, D.B., Phuong, N.M., Tung, T.H.: The volatility in the world oil price and its impacts on Vietnam's economy. Banking Academy (2015). (in Vietnamese)
23. Le Trung, Phan, Le Thong, Pham: Macroeconomic factors affect inflation in Vietnam. Bank. Technol. Rev. **102**, 17–29 (2014). (in Vietnamese)
24. Qianqian, Z.: The Impact of international oil price fluctuation on China's economy. Energy Procedia **5**, 1360–1364 (2011)
25. Rodriguez, R.J., Sanchez, M.: Oil-induced stagflation: a comparison across major G7 economies and shock episodes. Appl. Econ. Lett. **17**(15), 1537–1541 (2010)
26. Shaari, M.S., Hussain, N.E., Abdullah, H.: The effects of oil price shocks and exchange rate volatility on inflation: evidence from Malaysia. Int. Bus. Res. **5**, 9–16 (2012)
27. Shahzad, H.: Inflation and economic growth: evidence from Pakistan. Int. J. Econ. Finance **3**(5), 262–276 (2011)
28. Trading Economics. https://tradingeconomics.com/. Accessed 15 Feb 2017
29. Tweneboah, G., Adam, A.M.: Implications of oil price shocks for monetary policy in Ghana: A vector error correction model. University Library of Munich (2008)
30. Yildirim N., Ozcelebi, O., Ozkan, S.O.: Revisiting the impacts of oil price increases on monetary policy implementation in the largest oil importers. In: Zbornik Radova Ekonomskog Fakulteta U Rijeci-Proceedings of Rijeka Faculty of Economics, vol. 33, pp. 11–35 (2015)

The Level of Voluntary Information Disclosure in Vietnamese Commercial Banks

Tran Quoc Thinh[1(✉)], Ly Hoang Anh[2], and Pham Phu Quoc[3]

[1] Accounting and Auditing Department, Banking University of Ho Chi Minh City, Ho Chi Minh City, Vietnam
thinhtq@buh.edu.vn

[2] Banking University of Ho Chi Minh City, Ho Chi Minh City, Vietnam

[3] Research Institute of Ho Chi Minh City, Ho Chi Minh City, Vietnam

Abstract. The level of voluntary information for commercial banks (CBs) is important for users. There are a number of factors that affect to the level of voluntary information, in which corporate governance (CG) is great interest. In order to assess the impact of CG on the level of voluntary information of 30 Vietnamese CBs in the period of 2012–2016, the authors used quantitative method with panel data (OLS). The research results shown that there were two factors including the board size and the proportion of foreign shares that affected the same direction to level of voluntary information. So that, the authors suggested policies such as managers of CBs increase the number of board members, as well as considered to increase foreign ownership (within the limits set by the State Bank of Vietnam) to contribute the transparent information of Vietnamese stock exchange.

Keywords: Commercial banks · Voluntary information
Corporate governance

JEL Classification: G21

1 Background

In the process of regional and international integration, disclosure plays an important role for demand of investors. Information provided by commercial banks (CBs) should meet utility for users (Dhouibi and Mamoghli 2013). The level of voluntary information is influenced by many factors, including corporate governance (CG) and it is concerned in recent years (Mensah 2012; Hawashe 2015). CG will help the company to improve its ability to develop and mobilize capital from international markets as well as to build credibility and trust with stakeholders such as shareholders, investors (State Securities Commission and IFC 2010). As such, CG has helped to make the information of CBs more complete, reliable, public and transparent (Herwiyanti et al. 2015). However, the level of information disclosure of the CBs in Vietnam still has certain limitations. The disclosure of the information by the voluntary is still incomplete by Vietnam CBs (An 2016) and this will affect the decisions of investors.

V. Kreinovich et al. (Eds.): ECONVN 2019, SCI 809, pp. 709–718, 2019.
https://doi.org/10.1007/978-3-030-04200-4_49

Studies related to factors that affect to the level of voluntary information of CBs were used the panel data method (OLS). The authors found that it was the popular method for analyzing and evaluating the content of researchs over time. Therefore, the authors still use this method to carry out research. The previous researchs considered the factors affected the level of voluntary information of CBs and that was considered primarily as factors of the financial indicators (Mensah 2012; Ly 2015; An 2016), or the combination of financial indicators and CG factors (Hossain and Taylor 2007; Dhouibi and Mamoghli 2013; Hawashe 2015). There is currently no intensive research that focused on CG factors. Therefore, in this study, the authors focus on the impact of CG factors. The authors chose 30 Vietnam CBs and especially in the last five years because this period CBs had a lot of change in the acquisition, merger and restructuring.

The research structure consists of 5 parts. Section 2 presents theoretical structure. Part 3 will design the research. Section 4 is the result of research and discussion. Section 5 is a conclusion and suggest policies.

2 Theoretical Structure

2.1 The Concept

Corporate Governance (CG)

Charreaux (1997) considered CG as an organized control that governs the behavior of managers and determines their powers. CG was narrowly defined as tools to ensure maximum return on investment for its shareholders and creditors. The Organization for Economic Co-operation and Development (OECD) noted that the CG was internal measures to control the company, involve relationships between the board of directors and shareholders of a company (OECD, 2015).

Voluntary information

According to Merriam Webster Dictionary, voluntary is defined as acting of one's own free will, or another definition of English Oxford Dictionary that voluntary is proceeding from the will or from one's own choice or consent. Popa and Ion (2008) argued that disclosure was additional information that was intended to satisfy the needs of outside users such as financial analysts, consulting firms. Voluntary information was the choice of business, not mandatory. A voluntary disclosure was a company that may or may not need to disclose accounting information that was not required by law (Citro 2013). According to Hawashe (2015), voluntary disclosure was additional information that was intended to satisfy the needs of outside business users such as financial analysts, consulting firms, Investors.

2.2 Foundation Theories

This topic presents two theories as stakeholder theory and representation theory. Firstly, the stakeholders were those who could have a significant or significant impact on the success of the business. The stakeholder groups included customers, shareholders, suppliers, employees, the community. According to stakeholder theory, management decisions should be designed to please all stakeholders and acknowledge that negative actions could lead to adverse reactions from these stakeholders. Consequently, stakeholders had a significant impact on management decisions, which in turn affected the profitability of entity (Freeman and Mcvea 2001). Secondly, according to Jensen and Meckling (1976), there was a separation between ownership and control, conflicts arose between the owner and the operator. Both sides could want to maximize their benefits. Executives were expected to behave as desiration and bring the best benefit to owners (shareholders), and the executives pursued their own interests, because it did not always act in the best interests of shareholders. Theoretically, the main problem raised how the agent works in the best interest of the employer when they had an information advantage over the employers. These managers tended to make decisions that benefit themselves rather than the company's interests.

2.3 Overview of Previous Studies

There are studies to investigate the factors that affect the level of voluntary disclosure. Hossain and Taylor (2007) investigated the relationship between CG characteristics and voluntary levels in the annual report of 20 CBs in Bangladesh from 2000 to 2001. Characteristics to be tested is the size of the bank, audit firm and profitability. The results revealed that the size of the bank and the audit firm were the variables that influenced the level of voluntary information. Conversely, there was no relationship between voluntary information and profitable. Mensah (2012) conducted a empirical study about factors such as bank size, profitability, debt to equity ratio, liquidity and size of audit firm to the level of voluntary information in the 2009, with annual report of 21 CBs of Ghana. The author used the quantitative method. Research results indicated that profitability was positively correlated with the level of voluntary information, while debt to capital ratios, liquidity, bank size and audit firms did not have impact to the level of voluntary information. Dhouibi and Mamoghli (2013) examined the influencing of factors to the voluntary information of the 10 CBs in Tunisia for the period 2000–2011. The results shown that the size of the board, the degree of centralization, the ownership of the state reduces the level of voluntary information. However, the proportion of independent members in the board, concurrently board chairman and CEO, and reputation of the audit firm were not related to the level of voluntary information. Hawashe (2015) studied that bank characteristics influenced the level of voluntary disclosure in Libya CBs. By means of the OLS regression analysis, the study looked at seven factors and the results shown that the scale and status of listings were linked to the level of voluntary disclosure.

For Vietnam, there were some researches about the factors that influenced the level of voluntary information in CBs. Typically, Ly (2015) studied the influencing of factors to the level of voluntary information in the 25 annual reports of Vietnamese CBs in the period of 2012–2013. The study analyzed nine independent factors. The results shown that audit company, number of years of operation and listing status had a positive relationship to the level of voluntary information. Similarly, An (2016) conducted 30 CBs annual reports, from 2010 to 2015 and used quantitative methods to test the impact of factors on the level of voluntary information of CBs. The results shown that there were five factors including the size of the bank, the listing status and the auditing firm had a positive impact, while the financial leverage and return on assets had a negative impact on the level of voluntary information of the CBs.

It can be seen that the studies on the level of voluntary information in CBs have largely studied the financial characteristics interwoven with the CG, with some studies have considered that CG affected to level of voluntaty information at CBs. Especially in Vietnam, there is no specific research on the impact of CG factors to the level of voluntary information in CBs.

3 Research Design

3.1 Describe the Overall Pattern of the Study

At the time of the study, there were 30 CBs in Vietnam. Since then, the authors conducted the study of data in the annual report of 30 CBs, period of 2012–2016.

3.2 Research Models

Dependent variable

The Securities Disclosure of Interests (SDI) was first introduced by Alfaraih and Alanezi (2011) to measure the level of information disclosure. Based on that research, authors used the SDI to measure the level of voluntary information published by CBs. According to this measurement, the level of presentation would be calculated according to the following formula:

$$V_j = \frac{\sum_{i=1}^{n_j} d_{ij}}{n_j}$$

where:

V_j: voluntary information of CB j

n_j: number of items of voluntary information for CB j, nj $\leq$ 19 (The list of voluntary disclosure items of Vietnam CBs is set out in the appendix)

d_{ij}: 1 if item is disclosed, 0 if item is not disclosed for CB j

Independent variables

BSIZE: the number of board members.

INDEP: the percentage of independent members in the board on the total number of board members.

DIIFS the proportion of major shareholders (5% of the shares) of the total number of shareholders

CO.OWN: the proportion of organization shares in the total number of shares

ST.OWN: the proportion of state-owned shares in the total number of shares

FR.OWN: the proportion of foreign shares in the total number of shares

3.3 Research Models

Based on the model of Dhouibi and Mamoghli (2013), the authors have surveyed a number of experts in the banking sector to look at the factors the influence to Vietnam's condition. So the research model is made up of six variables such as board size, percentage of independent members, degree of ownership dispersion, the own of board, the own of state, the own of foreign. So, regression model:

$$\mathbf{VOLUNTARY} = \beta\mathbf{0} + \beta\mathbf{1} * \mathbf{BSIZE} + \beta\mathbf{2} * \mathbf{INDEP} + \beta\mathbf{3} * \mathbf{DIFFS} + \beta\mathbf{4} * \mathbf{CO.OWN} + \beta\mathbf{5} * \mathbf{ST.OWN} + \beta\mathbf{6} * \mathbf{FR.OWN} + \varepsilon$$

4 Results and Discussion

4.1 Descriptive Statistics Results

The data in Table 1 shown that the disclosure level was 13.03 points (minimum was 10 and maximum was 17). Thus, the voluntary level of publication was quite high. The average number of board members on the CBs was nearly eight. This shown that the number of board members were within the specified limits. The percentage of independent members in the board was 15.85%. Degree of ownership dispersion was about 31.34%. The proportion of organization ownership was 39.34%, followed by the state ownership of 9.63% and finally the foreign ownership of 6.51%. In general, the standard deviation of the variables was not significant.

Table 1. Descriptive Statistics

Descriptive statistics					
Variable	N	Minimum	Maximum	Mean	Std. Deviation
VOLUNTARY	150	10	17	13.0333	1.7587
BSIZE	150	5	11	7.1600	1.7839
INDEP	150	.00	1.00	.1585	.0949
DIIFS	150	.05	.96	.3134	.2710
CO.OWN	150	.00	.98	.3934	.3014
ST.OWN	150	.00	.96	.0963	.2480
FR.OWN	150	.00	.30	.0651	.1018
Valid N (listwise)	150				

Source: Analysis data from SPSS 22.0

4.2 Matrix Correlation Coefficient

Table 2 shown that the correlation between the level voluntary disclosure and independent variables, in which the correlation coefficient varied depending on the level of voluntary information. In addition, Table 2 also shown that the correlation coefficient between independent variables was less than 0.8. This partly demonstrated the non-existent multi-collinearity.

Table 2. Correlations

Variables		VOLUNTARY	BSIZE	INDEP	DIIFS	CO. OWN	ST. OWN	FR. OWN
Pearson Correlation	VOLUNTARY	1.000						
	BSIZE	.212	1.000					
	INDEP	.006	−.249	1.000				
	DIIFS	−.021	.208	−.276	1.000			
	CO.OWN	.014	.215	−.151	.718	1.000		
	ST.OWN	.099	.411	−.225	.772	.560	1.000	
	FR.OWN	.129	.331	−.084	.248	.340	.201	1.000
Sig.(1-tailed)	VOLUNTARY	.						
	BSIZE	.005	.					
	INDEP	.471	.001	.				
	DIIFS	.398	.005	.000	.			
	CO.OWN	.433	.004	.033	.000	.		
	ST.OWN	.114	.000	.003	.000	.000	.	
	FR.OWN	.008	.000	.153	.001	.000	.007	.

Source: Analysis data from SPSS 22.0

4.3 Conformity Assessment of the Model

Table 3 shown that the adjusted R^2 was 0.415, which meant that the independent variable accounted for 41.5% variation of the dependent variable.

Table 3. Model Summary

Model	R	R Square	Adjusted R Square	Std. Error of the Estimate	Change Statistics				
					R Square Change	F Change	df1	df2	Sig. F Change
1	.685	.581	.415	1.74524	.081	1.230	6	143	.000

Source: Analysis data from SPSS 22.0

4.4 Model Fit Testing

This test was to examine the linear relationship between the dependent variable and all independent variables.

Ho: βi = 0: The variables introduced into the model may not affect the level of voluntary information.

H1: βi # 0: Variables introduced into the model may affect the level of voluntary information.

The results from Table 4 shown that the sig value was .0000, so the hypothesis Ho is rejected. The linear regression model was consistent with the data set.

Table 4. ANOVA

Model		Sum of Squares	df	Mean Square	F	Sig.
1	Regression	37.459	6	3.746	1.230	.000
	Residual	423.374	143	3.046		
	Total	460.833	149			

Source: Analysis data from SPSS 22.0

4.5 Regression Results

The authors had performed regression based on selected variables, and specific results were the follows:

Based on the results from Table 5, the authors excluded p-value (Sig) of variables greater than 0.05. The regression model was defined as follows:

$$\mathbf{VOLUNTARY = 12.534 + 0.133 * BSIZE + 1.896 * FR.OWN}$$

Table 5. Coefficients

Model		Unstandardized Coefficients	Std. Error	Standardized Coefficients	t	Sig.	Coefficients[a] 95.0% Confidence Interval for B	
		B		Beta			Lower Bound	Upper Bound
1	(Constant)	12.534	1.117		11.217	.000	10.325	14.743
	BSIZE	.133	.099	.135	2.344	.011	−.063	.330
	INDEP	.872	1.624	.047	.537	.592	−2.339	4.083
	DIIFS	−1.276	1.061	−.197	−1.203	.231	−3.373	.821
	CO.OWN	−.046	.743	−.008	−.062	.950	−1.516	1.423
	ST.OWN	1.411	1.052	.199	1.341	.182	−.670	3.491
	FR.OWN	1.896	1.675	.110	2.132	.002	1.415	5.207

Source: Analysis data from SPSS 22.0

4.6 Discuss the Results

Research results shown that two factors including the board size (BSIZE) and the proportion of foreign shares (FR.OWN) and all affected the same direction to the level of voluntary information of Vietnam CBs. For BSIZE variable, the research results shown a positive relationship to level of voluntary information in CBs in Vietnam and the result was inconsistent with the results of Dhouibi and Mamoghli (2013). According to FR. OWN variable, the research results shown a positive relationship between the the proportion of foreign shares and the level of voluntary information in CBs in Vietnam and the result was consistent with the results of Dhouibi and Mamoghli (2013).

5 Conclusion and Suggest Policies

5.1 Conclusion

The disclosure is important to the users and this helps shareholders understand the financial situation, business of CBs. Specially, voluntary information is an opportunity for CBs to improve the quality of information, meeting the demand for information for users, especially investors. The research results shown that there were two factors and those influenced the level of voluntary information in CBs of Vietnam including board size and the proportion of foreign ownership. Since then, the authors suggest policies for managers of CBs in Vietnam such as increasing the number of board members to provide comments and discussions in order to have useful and complete information for users. At the same time, managers of banks need to consider increasing foreign ownership (within regulatory limits) to contribute to the health of the information environment. This contributes to transparent information of listing units, and gives more confidence to investors.

5.2 Suggest Policies

Board Size

The number of board members is important in contributing to the increased disclosure of CBs. There were still a few CBs with fairly modest members, according to the survey results of only 5 members. Therefore, managers of CBs need to increase the number of board, but also to the maximum under Law on Banking in 2010 and Law on Credit Institutions in 2010.

The Proportion of Foreign Ownership

Foreign ownership is important in raising the level of voluntary disclosure. This is appropriate in real terms as foreign investors require more stringent disclosure. In addition, foreign investors tend to require information transparency. Therefore, the CB has a high foreign ownership, the greater the level of voluntary information. Managers of CBs should consider to increase the foreign ownership (within the limits set by the State Bank of Vietnam, Law on Banking in 2010 and Law on Credit Institutions in 2010) to attract external capital as well as contribute to the health of the information environment. This contributes to transparent information of Vietnamese stock exchange.

Appendix

The list of voluntary disclosure items of Vietnamese commercial banks

No	The voluntary disclosure items
1	Income situation of employees
2	Status of fulfillment of obligations to the state budget
3	Assets, valuable papers for mortgage, pledge and discount, rediscount
4	Assets, valuable papers put to mortgage, pledge and discount, rediscount
5	Contingent liabilities and commitments
6	Commissioning activities
7	Trustee and agent activity
8	Other off-balance sheet activities that bear significant risks
9	Information on related parties
10	Events after the balance sheet date: The creditors explain the material events.
11	Geographic concentration of assets, liabilities and off-balance sheet items
12	Risk management policy related to financial instruments:
13	Credit risk
14	Interest rate risk
15	Currency risk
16	Payment risk
17	Other market price risks
18	Core segment report
19	Secondary segment report

References

Alfaraih, M.M., Alanezi, F.S.: Does voluntary disclosure level affect the value relevance of accounting information. Acc. Taxation **3**, 69–92 (2011)

Ly, B.N.: Factors influencing the level of voluntary disclosure in the annual report of the commercial banking system in Vietnam. Masters thesis, University of Economics (2015)

Charreaux, G.: Le gouvernementdesenterprises: Corporategovernance, théoriesetFaits. Economica, Paris (1997)

Citro, F.: Disclosure level evaluation and disclosure determinant analysis: a literature review. International Virtual Scientific Conference, University of Salerno Fisciano (2013)

Dhouibi, R., Mamoghli, C.: Determinants of voluntary disclosure in Tunisian bank's reports. Res. J. Finance Acc. **4**, 80–94 (2013)

Freeman, R.E., Mcvea, J.F.: A stakeholder approach to strategic management. SSRN **1**, 1–33 (2001)

Hawashe, A.A.: Commercial banks' attributes and annual voluntary disclosure: the case of Libya. Int. J. Acc. Financ. Reporting **5**, 208–233 (2015)

Herwiyanti, E., Ma, R.A.S.W., Rosada, A.A.: Analysis of factors influencing the islamic corporate governance disclosure index of Islamic Banks in Asia. Int. J. Humanit. Manag. Sci. **3**, 2320–4044 (2015)

Hossain, M., Taylor, P.J.: The empirical evidence of the voluntary information Disclosure in the annual reports of banking companies: the case of Bangladesh. Corporate Ownership Control **4**, 111–125 (2007)

Jensen, M.C., Meckling, W.H.: Theory of the firm: managerial behavior, agency costs and ownership structure. J. Financ. Econ. **3**, 305–360 (1976)

An, L.T.T.: Factors influencing the level of voluntary information of commercial banks in Vietnam. Masters thesis, University of Economics - Ho Chi Minh city (2016)

Mensah, A.B.K.: Association between firm-specific characteristics and levels of disclosure of financial information of rural banks in the Ashanti region of Ghan. J. Appl. Finance Banking **2**, 69–92 (2012)

OECD: G20/OECD Principles of Corporate Governance. OECD Report to G20 Finance Ministers and Central Bank Governors (2015)

Popa, A., Ion, P.: Aspects regarding corporate mandatory and voluntary disclosure. J. Fac. Econ. Econ. **3**, 1407–1411 (2008)

State Securities Commission and IFC: Corporate Governance Handbook. Agricultural Publishing House, Hanoi (2010)

Corporate Governance Factors Impact on the Earnings Management – Evidence on Listed Companies in Ho Chi Minh Stock Exchange

Tran Quoc Thinh[1(✉)] and Nguyen Ngoc Tan[2]

[1] Accounting and Auditing Department,
Banking University of Ho Chi Minh City, Ho Chi Minh City, Vietnam
thinhtq@buh.edu.un
[2] People's Committee of Ho Chi Minh City, Ho Chi Minh City, Vietnam

Abstract. In the trend of regional and international integration, many issues confront companies facing global challenges because of the competitive pressure on the international market. Many companies have had a lot of strategies to break in and take the lead for investors, but there are also some businesses in some ways for Earnings Management (EM). EM action has a great impact on investors and shareholder interests. Authors considered the quantitative model to examine the impact of corporate governance factors to EM behavior for 173 listed companies of Ho Chi Minh Stock Exchange (HOSE) in the period 2013–2017. The results shown that the professional experience had a positive effect, while the board size variable had the opposite effect to EM. Authors suggested some policies for listed companies that need to implement the monitoring mechanism, strengthen the internal control tools, internal audit to control the timely EM behavior.

Keywords: Earnings management · Listed company · Hose

JEL Classification: G21

1 Background

Economic integration will create the tremendous opportunities for the development of countries, one of the conditions for entry into the world economy. So every business has a target listing on the stock market. To be listed on the stock exchange, companies must be presented financial reporting in accordance with the standards and must provide useful information for decision-making investors. However, EM of companies has caused many unfavorable effects for investors in particular, users of accounting information in general Fuzi et al. (2015). For Vietnam, the situation of companies with profit-taking behavior will lead to bad effects, even bankruptcy of companies (Nhi and Trang 2013). Thus, the problem is that listed companies have EM actions to achieve the goals set earlier. EM action has a great impact on investors and shareholder interests. This has affected investors' confidence in attracting funds for the development of the business.

V. Kreinovich et al. (Eds.): ECONVN 2019, SCI 809, pp. 719–725, 2019.
https://doi.org/10.1007/978-3-030-04200-4_50

2 Theoretical Basis and Research Before

2.1 Related Concepts

Corporate Governance (CG)

Charreaux (1997) considered CG as an organized control that governs the behavior of managers and determines their powers. CG was narrowly defined as tools to ensure maximum return on investment for its shareholders and creditors. The Organization for Economic Co-operation and Development (OECD) noted that the CG was internal measures to control the company, involve relationships between the board of directors and shareholders of a company (OECD 2015).

Earnings Management (EM)

According to Investopedia, EM is the use of accounting techniques to produce financial reports that present an overly positive view of a company's business activities and financial position. Many accounting rules and principles require company management to make judgments following these principles. EM takes advantage of how accounting rules are applied and creates financial statements that inflate earnings, revenue, or total assets. Schipper (1989) proposed the definition of EM as a profit adjustment to achieve the manager's stated goal. It was a calculated intervention in the process of disclosing information on the financial statements to the outside with the aim of achieving some personal gain. Profit adjustment was a deliberate intervention in the process of providing financial information to achieve personal goals. Adjustment of profits reflected the actions of executives in choosing accounting methods to benefit them or increase the market value of the company (Scott 1997). Healy and Wahlen (1999) asserted that EM occured when managers used judgment in financial statements and in structured transactions to change financial statements or mislead certain stakeholders about financial statements. The underlying economic performance of the company or its impact on the outcome of the contract depended on the accounting data reported. EM was characterized as a negative mechanism and opportunity when managers deliberately relied on large gaps in their own discretionary financial reporting discrepancies.

2.2 Theoretical Basis

Freeman and Mcvea (2001) argued that stakeholders were shareholders, employees, creditors, suppliers, consumers, unions and regulators. Stakeholder theory raised a controversial issue among researchers as emphasizing the key role of managers to play the responsibility to stakeholders without mentioning how to keep their concerns. Therefore, managers were responsible for protecting the interests of the parties and maintaining a share of the respective interests of each holder.

Agency theory derived from economic theory, developed by Jensen and Mackling. The representative theory was considered the relationship between leaders such as shareholders and representatives such as corporate executives or corporate executives. In this theory, the shareholders were the owners or the head of the company, hiring others to do the work. The head of the company authorized the operation of the company for directors or managers, they were representatives for the shareholders. The representation problem arised from the separation of ownership and control in modern enterprises in which the shares were dispersed (Jensen and Meckling 1976).

2.3 Previous Studies

Studies about impact factors to the EM of companies have interested by many researchers in the world as well as in Vietnam. For typical foreign studies, Shah et al. (2009) studied 120 listed companies from a variety of sources, collecting data from the balance sheet and annual report of companies. The author concludes that the independent board had a positive effect on the EM. Ghazali et al. (2015) studied 389 listed companies in Malaysia from 2010 to 2012. The results shown that the company size had the opposite effect on RM but financial leverage and margins were in the same direction. Bassiouny et al. (2016) looked at 60 listed companies in Egypt from 2007 to 2011. Research results indicated that the company's financial leverage had a significant positive relationship with EM. Abbadi et al. (2016) tested 121 companies in Amman form 2009 to 2013. Research results shown that corporate governance had a significant negative impact on EM. Daghsni et al. (2016) carried out empirical including 70 companies. The study results shown that two variables had a positive effect on EM such as Board and CEO responsibilities, board activities. For Vietnamese studies, Van (2012) used the Modified Jones model to identify EM behavior of 60 listed companies on the Hanoi Stock Exchange. The results shown that companies had behaviors to adjust the profit. Lien (2014) reviewed the sample of 101 joint-stock companies listed on the HOSE for five years from 2009 to 2013. By quantitative method, the author pointed the separating the role of Board and CEOs, non-executive board members, independent board members, and board members had a positive effect on EM. Tu (2014) reviewed 100 listed companies of HOSE during the period 2009–2013. The results shown that the board of directors and company size had opposite effect to EM. Phuong (2014) reviewed 101 listed companies in Vietnamese stock market, from 2010 to 2013. The results shown that there were two variables including return on equity and company size affect EM.

Through previous research, the topic has examined the factors that influenced the EM of the companies. It can be seen that, the foreign studies have been currently no case research of the factors that affected the EM of Vietnam companies. In Vietnam, there has been no in-depth study of corporate governance to EM in recent times. Therefore, the authors's research has practical implications for the current situation in Vietnam.

3 Research Methodology

3.1 Describe the Research Sample

The authors collect 172 listed companies (excluding non-financial corporations) with sufficient information on corporate governance factors in the annual report of HOSE in the period of 2013–2017.

3.2 Research Models

The topic of choosing EM approach to the company's listing on the criteria focused on factors related to the corporate governance factors. Thus, the theme of Daghsni et al. (2016) is chosen for the study, incorporating unique elements of Vietnam, which has the same content of the EM metrics. Inheriting the above model, the topic considers only.

Based on the model of Daghsni et al. (2016), the authors have surveyed a number of experts to look at the factors the influence to Vietnam's condition. So the research model is made up of five variables such as Board size (BSIZE); The percentage of independent board members (INDEP); The Percentage of female members (FEMALE); Frequency of meetings (MEETS); Professional experience (EXPER). So, the regression model is expressed in terms of specific variables:

$$\mathrm{EM} = \beta 0 + \beta 1 * \mathrm{BSIZE} + \beta 2 * \mathrm{INDEP} + \beta 3 * \mathrm{FEMALE} + \beta 4 * \mathrm{MEETS} + \beta 5 * \mathrm{EXPER} + \varepsilon$$

Inside Earnings management (EM): is measured by DeAngelo (1986), which identifies the motivation for profit adjustment for each type of enterprise. There are studies using an improved model by Friedlan (1994) to facilitate the retrieval of metadata. The formula is specified:

Discretionary accrual can be adjusted yearly t = Discretionary accrual year t/Net revenue year t - Accrual accounting year t/Net revenue year t – 1.

In particular, Discretionary accrual year t = Profit after tax year t - Net cash flow from operating activities in year t.

BSIZE: the number of board members.

INDEP: the percentage of independent members in the board on the total number of board members.

FEMALE: the ratio of female members in the Board to the total number of board members.

MEETS: the number of board meetings in a year.

EXPER: the average number of years in the board's professional field of expertise.

4 Results

4.1 Evaluation of Model Fit

Table 1 shown that the adjusted R^2 was 0.416 which meant that the independent variable accounts for 41.6% variation of the EM dependent variable.

Table 1. Model summary

Model	R	R square	Adjusted R square	Std. error of the estimate
1	.589	.503	.416	.00102

Source: Analysis data from SPSS 22.0

4.2 Check the Suitability of the Model

This test examined the linear relationship between EM dependent variables and independent variables.

H0: βi = 0: Variables introduced into the model may not affect EM.
H1: βi # 0: Variables introduced into the model may affect EM.

The results from the above table shown that Sig value of EM was 0.000, so the null hypothesis was rejected. So the linear regression model matched the data set (Table 2).

Table 2. ANOVA

Model		Sum of squares	df	Mean square	F	Sig.
1	Regression	.195	5	.023	1.701	.000
	Residual	1.512	855	.043		
	Total	1.452	860			

Source: Analysis data from SPSS 22.0

4.3 Regression Results

Based on the results from Table 3, authors excluded the p-value value was greater than 0.05. So the regression model was defined as follows:

$$\mathbf{EM = -0.128 + 0.131 * EXPER - 0.102\,BSIZE}$$

The results of this study were similar to those of Daghsni et al. (2016). Research results shown that Professional experience variable (EXPER), the coefficient β was 0.131 greater than 0, indicated a positive relationship between the EM, while Board size variable (BSIZE), the coefficients was –0.102 is less than 0, inversely related to EM.

Table 3. Coefficients

Model		Unstandardized coefficients		Standardized coefficients	t	Sig.
		B	Std. error	Beta		
1	(Constant)	–.128	–.132		–2.738	.013
	BSIZE	–.102	–.212	–.163	–2.336	.006
	INDEP	–.019	–.029	–.081	–1.081	.128
	FEMALE	.102	.201	.042	.221	.826
	MEETS	–.019	–.029	–.081	–3.681	.238
	EXPER	.131	.121	.042	2.221	.002

Source: Analysis data from SPSS 22.0

5 Conclusions and Recommendations

5.1 Conclusion

EM is an important issue related to the quality of accounting information and this affects users in making business decisions. Control and management are one of the issues that need to be considered and sanctions to help information on the stock market be honest and reasonable. This also contributes to the health information to meet the trend of integration and development. Authors examined the sample survey of 173 listed companies by HOSE from 2013 to 2017, the result shown that Professional experience (EXPER) had a positive relationship to EM, while the Board size (BSIZE) had a negative relationship to EM. Since then, the authors suggested that the Board should implement the monitoring mechanism by the Supervisory Board, strengthen internal control tools, internal audit to check timely EM behaviors.

5.2 Policy Suggestions

Managers need to be aware of the long-term strategy of avoiding short-term goals by implementing EM to tailor information for personal gain. This leads to the prestige and position of the companies and the loss of investors and shareholders, who invest capital in production and business activities of enterprises. Managers should consider business ethics on the basis of providing transparent and honest information to the user. Besides, the Board should have strict sanctions when detecting instances involving EM by managers as well as Board should implement the monitoring mechanism by the Board of Supervisors, strengthen the internal control tools, internal audit to timely control EM behavior.

References

Abbadi, S.W., Hajazi, Q.F., Rahahled, A.S.: Corporate governance quality and earnings management: evidence from Jordan. Australas. Account. Bus. Financ. J. **10**, 55–75 (2016)

Bassiouny, S.W., Soliman, M.M., Ragab, A.: The impact of firm characteristics on earnings management: an empirical study on the listed firms in Egypt. Bus. Manag. Rev. **7**, 91–101 (2016)

Daghsni, O., Zouhayer, M., Mbarek, K.B.H.: Earnings management and board characteristics: evidence from French listed firms. Account Financ. Manag. J. **1**, 92–110 (2016)

DeAngelo, L.E.: Accounting numbers as market valuation substitutes: a study of management buyouts of public stockholders. Account. Rev. **61**, 400–420 (1986)

Freeman, R.E., Mcvea, J.F.: A stakeholder approach to strategic management. SSRN **1**, 1–33 (2001)

Friedlan, J.M.: Accounting choices of issuers of initial public offerings. Contemp. Account. Res. **11**, 1–31 (1994)

Fuzi, S.F.S., Halim, S.A.A., Julizaerma, M.K.: Board independence and firm performance. Procedia Econ. Financ. **37**, 460–465 (2015)

Charreaux, G.: Le gouvernement des enterprises: Corporate governance, théories et Faits. Economica, Paris (1997)

Ghazali, A.W., Shafieb, A.S., Sanusib, Z.M.: Earnings management: an analysis of opportunistic behaviour, monitoring mechanism and financial distress. Procedia Econ. Financ. **28**, 190–201 (2015)

Lien, G.: Study the relationship between corporate governance and profit-driven behavior of companies listed on the Ho Chi Minh City Stock Exchange, Master's thesis, University of Economics (2014)

Healy, P., Wahlen, J.M.: A review of the earnings management literature and its implications for standard setting. Account. Horiz. **13**, 365–383 (1999)

Jensen, M.C., Meckling, W.H.: Theory of the firm: managerial behavior. Agency Costs Ownersh. Struct. **3**, 305–360 (1976)

Phuong, N.T.: Test the relationship between the level of information disclosure on financial statements with EM of listed companies in Vietnam, Master's thesis, City University of Economics (2014)

OECD: G20/OECD Principles of Corporate Governance. OECD Report to G20 Finance Ministers and Central Bank Governors (2015)

Van, P.T.B.: Study on the model for identifying behavior of profit adjustment of listed companies on Hanoi Stock Exchange. Bank. Mag. **9**, 31–36 (2012)

Schipper, K.: Commentary on earnings management. Account. Horiz. **3**, 91–102 (1989)

Scott, W.: Financial Accounting Theory. Prentice-Hall, Upper Saddle River (1997)

Shah, S.Z.A., Zafar, N., Durrani, T.K.: Board composition and earnings management an empirical evidence form Pakistani listed companies. Middle East. Financ. Econ. **3**, 28–38 (2009)

Tu, T.T.M.: Analysis of the factors affecting AD behavior on the financial statements of joint stock companies listed on the Ho Chi Minh City Stock Exchange, Master's thesis, University of Economics (2014)

Nhi, V.V., Trang, H.C.: The behavior of profit adjustment and bankruptcy risk of companies listed on the Ho Chi Minh Stock Exchange. J. Econ. Dev. **276**, 23–29 (2013)

Empirical Study on Banking Service Behavior in Vietnam

Ngo Van Tuan[1(✉)] and Bui Huy Khoi[2]

[1] Banking University of Ho Chi Minh City, 36 Ton That Dam, Nguyen Thai Binh Ward, District 1, Ho Chi Minh City, Vietnam
tuannv@buh.edu.vn

[2] Industrial University of Ho Chi Minh City, 12 Nguyen Van Bao Street, Govap District, Ho Chi Minh City, Vietnam
buihuykhoi@iuh.edu.vn

Abstract. The aim of this research aimed to investigate the relationship emotional evaluation, rational evaluation and customer brand relationship in Vietnamese retail banking service. Survey data was collected from 450 customers some bank brands in HCM City. The research model was proposed from the study of emotional evaluation, rational evaluation and customer brand relationship of some authors in abroad. The reliability and validity of the scale were tested by Cronbach's Alpha, Average Variance Extracted (Pvc) and Composite Reliability (Pc). The analysis results of structural equation model (SEM) showed that the relationship emotional evaluation, rational evaluation and customer brand relationship had a relationship with each other.

Keywords: Smartpls 3.0 software · Emotional evaluation · Rational evaluation Customer brand relationship · Structural Equation Model · SEM Factors · Relationship

1 Introduction

Brand equity is one of the most important marketing concepts and has been an area of interest for marketing academics and practitioners as well. There are a numbers of models of brand equity in common marketing settings [1–3] or in financial service perspectives [4]. However, to my best knowledge, *there is no model of brand equity that particularly focuses on banking service.*

It might be worthwhile and necessary to build a brand equity model in banking service. Brand equity in banking service deserves elaboration in some regards. "First and foremost, unlike other financial firms, banks act as intermediaries between borrowers and lenders and, in so doing, they offer a unique form of asset transformation" [5]. Bank transactions usually involve a large sum of money and hence, trust and price (in terms of interest rates…) appear to be critical matters in the industry. Second, bank transactions, especially lending, are more complicated than transactions for other products and services. For example, before a loan is approved, it takes time and effort to get through an assessment process that is strictly regulated (by the State bank and/or by laws). Finally, most of the brand equity models are conceptualized by Western

V. Kreinovich et al. (Eds.): ECONVN 2019, SCI 809, pp. 726–741, 2019.
https://doi.org/10.1007/978-3-030-04200-4_51

authors and validated in developed countries. This poses the question of whether or not these models work well in a developing country like Vietnam. The aim of this research aimed to investigate the relationship emotional evaluation, rational evaluation and customer brand relationship in Vietnamese retail banking service.

2 Literature Review

Brand Equity

Building a strong brand involves creating brand equity. In a common sense, *brand equity is defined as the added value endowed by the brand to the product* [6]. In the last two decades, brand equity has become the most interesting research topic in marketing for both academics and practitioners. Despite the fact that brand equity is a potentially important marketing concept, it is not without controversy [4]. It is because brand equity is defined in different ways for different purposes [7]. However, in a general sense, the literature suggests that there have been two primary perspectives relating to studying brand equity [4, 8, 9]. The first approach is motivated by financial outcome for the firms. With this perspective, the brand is evaluated financially for accounting purpose and is usually manifested in the balance sheet. The second approach is based on the customer-brand relationship. This study adopts the later approach, customer-based brand equity (hereinafter referred to as CBBE).

There have been also debates on the importance of brand equity for products and services. Some researchers argue that branding (and thereby brand equity) is more important for services due to the intangible nature and the so-called 'credence' attributes of services, which makes it difficult for customers to examine the content and quality of a service before, during and even after the consumption of the service [10]. However, the findings of Krishnan and Hartline [10] do not support the contention that brand equity is more important in services than for products.

Aaker [1] defines brand equity as *"a set of assets and liabilities linked to a brand's name and symbol that adds to or subtracts from the value provided by a product or service to a firm and/or that firm's customers"*. Aaker conceptualizes a model of brand equity consisting of 4 main components: (1) brand loyalty, (2) brand awareness, (3) perceived quality, (4) brand associations (which are driven by brand identity: the brand as a product, the brand as an organization, the brand as a person and the band as a symbol). The fifth component is other proprietary brand assets such as patents, trademarks and channel relationships.

Keller [8] generalized the concept of brand equity by the CBBE model. He defines CBBE *"as the differential effect that brand knowledge has on consumer response to the marketing of that brand"*.

According to Keller [9], a brand is said to have positive CBBE if the consumer reacts more favorably to the marketing of the brand than they do to an unknown or fictitious version of the product or service in the same context. On the other side, a brand is said to have negative CBBE if the consumer reacts less favorably to the marketing of the brand under the same situation. This effect differs based on how favorable, strong and unique brand associations are evoked in the customer's mind.

Recently, Taylor et al. [4] proposed a model of brand equity (customer-based) for financial services. According to this model, brand equity is derived from the customer's perception of the quality and thereby the brand value. Other components of their brand equity construct are hedonic brand attitude, utilitarian brand attitude and brand uniqueness. According to the model, brand satisfaction and loyalty intention are the consequences, and positively relate to the brand equity.

However, the current study adopts the CBBE model developed by Martensen and Grønholdt [2]. This model captures aspects closely related to banking services. Martensen and Grønholdt [2] categorize brand associations into two types: (1) rational association and (2) rational and emotional association.

The rational associations are in connection with the customers' perceptions about the functional benefits, tangible aspects or the cost-value evaluation. These associations are very important in banking services. For example, price is a key factor that affects a customer's decision to stay with a bank [11, 12]. In other research, Gounaris et al. [13] suggest that, "with regard to financial services, consumers tend to become more involved, they develop the habit of 'shopping around' to find the best bargain".

The emotional associations related to either the intangible or tangible aspect. For example, a customer may feel confident or recognized (social approval) when she or he deals with a great bank's brand (emotional). This emotion, in turn, is the result of consuming excellent service offered by the bank (performance of the product and service). These associations will be discussed in details in the below.

Brand Associations

Rational Associations

Though the product quality is a component of the original model of CBBE; however, banking is a service-dominant industry and all banking products, as termed in the industry, are actually services or packages of service. Therefore, it is argued here that the *product quality* suggested by Martensen and Grønholdt [2] is not necessarily to be included in the research model which is only intended to apply in the banking service. Instead, this study focus on service quality as a component that speaks for the quality aspect of the model.

Service Quality

Service quality has become an increasingly important factor for success and survival in the banking sector [14]. It's a critical factor that affects an organization's competitiveness and an essential determinant that enable a company to differentiate itself from competitors [13].

Without doubt, service quality is a key driver of customer satisfaction and thereby loyalty. Olsen and Johnson [15] view service quality as "a key psychological reaction to the value that a service company provides". Same as with physical products, customers perceive service quality differently. This results from the difference between perceived quality and objective quality and can be expressed by an equation of performance and expectations, service quality = [performance – expectations] [16–18].

Martensen and Grønholdt [2] measure the service quality by three criteria: assurance, responsiveness and empathy. However, with a service-dominant industry like banking, it seems that service quality should be examined from a broader perspective. Thus, to have an insight into the consumer's perception about service quality in banking, the current study adapted the construct of Gounaris et al. [13] for measuring service quality in banking services.

Price

Price is one of the elements of the traditional marketing mix, and price is often stressed as a driver in customer satisfaction and loyalty models [2]. Keller [8] views price as a non-product-related attribute because it does not speak much for the product performance or service function. However, price is an important attribute association. In most cases, it is considered an important criterion for purchase.

In their model of CBBE, Netemeyer et al. [19] suggest that willingness to pay a price premium is a core/primary facet of CBBE. By testing and extending the Netemeyer et al. [19] CBBE model, Taylor et al. [4] confirm that willingness to pay a price premium is positively related to the brand value. They also argue that brand loyalty intention is positively related to the willingness to pay a price premium.

There is another approach to consider price premium. According to Aaker [1], price premium may be negative. Customers might expect a certain level of price advantage in a brand (for example 10% lower) compared to other higher-priced brands, and be willing to buy this brand if the advantage was greater 15% for instance. This negative price premium could reflect substantial brand equity for the lower-priced brand.

In banking service, price is indicated in terms of loan interest rate, credit interest rate and other charges and fees that customers pay to use the bank facilities. Price in banking service is a sensitive factor. Research into the small and medium sized businesses indicates that "pricing of a loan facility (e.g. an overdraft) has a strong impact on customer loyalty…" [11]. This is in line with Keaveney [20] that one of the three major factors for switching is a pricing problem, including non- competitiveness of the fee and interest rates, which capture 17% among other reasons. However, dissatisfied with this result, Colgate and Hedge suggest further research into the role of pricing [12]. Responding to this call, Bogomolova and Romaniuk carried out a study of the business banking industry and found that the top two reasons for switching to another bank are getting "better deal with the other bank" and the fees are too high [21].

Rational and Emotional Associations

Brand Promise

A brand is essentially a marketer's promise to deliver a predictable product or service performance [22]. Ambler [23] defines a brand as "the promise of the bundle of attributes that someone buys which provides satisfaction…. The attributes that make up a brand may be real or illusory, rational or emotional, tangible or invisible". This is in line with Kapferer (2008) who argues that "consumers don't just buy the brand name; they buy branded products that promise tangible and intangible benefits created by the efforts of the company" [9].

Why is brand promise important? It is widely agreed in the literature that one determinant of customer satisfaction is the gap between customers' experiences and their expectations [9, 17, 18, 24] and brand promise sets up this benchmark.

Brands thus become credible only through the persistence and repetition of their value proposition [9]. In other words, brand promise should be credible and deliverable. This is in line with Martensen and Grønholdt that "promise should be the hub of value creation for the customer [2]. The unique values should mirror meaningful promises to the consumer – promises that are credible and that the brand can fulfill".

Brand Trust and Credibility

Marketing literature has shown that "an essential and very important part of a brand is the trust consumers have in the brand living up to their expectations" [2]. There are different definitions about brand trust, for example, brand trust can be defined as "the confidence a consumer develops in the brand's reliability and integrity" [25, 26]. In this perspective, Delgado et al. believe that brand trust is uni-dimensional and driven by a consumer's overall satisfaction with the product and confident expectations of the brand's reliability and intentions in situations entailing risk to the consumer [27]. Trust is also viewed as a group of beliefs held by a person derived from his perceptions about certain attributes" [28]. In other words, trust implies that the customers believe that the brand can deliver both functional and emotional benefits.

Consumer trust is also important and sometimes is considered a prerequisite for the development of an attitude-based relation between the consumer and the company. From a consumer perspective, trust helps to reduce the perceived risk linked to the purchase or use of a company's products [29]. According to Martensen and Grønholdt [2], trust also provides assurance of quality, reliability, etc. and is thus a factor in providing the consumer with an experience of dealing with a credible and reliable company – a factor that is important in connection with the consumer's decision process. Hence the company should be aware of communicating values that they cannot deliver.

In the modern banking industry, internet banking is an indispensable and critically important part. Some studies have analyzed the importance of trust in internet relationships and suggested trust is habitually related to security and risk avoidance [28]. In internet banking, trust captures two different aspects: the customer's belief in banker goodwill and the reliability of the internet infrastructure.

Another dimension of this aspect is credibility. As mentioned previously, together with trust, credibility is especially important in the banking industry, as the bank brand is in fact the institution. Thus it is important for the bank to have high credibility. Empirical studies suggest that the consumers' perception of a company's credibility plays a central role in their perception of and attitude to the company, its products and communication, including [2].

In conclusion, an empirical study into the impact of trust on brand equity pointed out that "brand equity is best explained when brand trust is taken into account reinforces the idea that brand equity is a relational market-based asset" [27]. Martensen and Grønholdt argue "that being a credible company has a considerable influence on the consumer attitudes towards the brand and its ads, and eventually the consumers' intention to buy the company's products [2]. Therefore, the company should make a real effort to find out what they need to do to create high credibility among the consumers".

Brand Differentiation

The brand should differentiate itself from its competitors and offer the market something unique. "Uniqueness is defined as the degree to which customers feel the brand is different from competing brands—how distinct it is relative to competitors" [19]. However, the differences should be perceived as meaningful to the consumer [19]. Creating unique brand associations is in line with creating points of difference when positioning the brand. Besides addressing the distinctive benefits a brand will deliver to its consumers, target consumers must also find these benefits personally relevant and important [22].

Having a bank brand viewed as a corporate brand makes it possible for a bank to position itself in the minds of the consumers with a broader and more varied image than it does through a particular product or service. Keller [30] argues: "a corporate brand is distinct from a product brand in that it can encompass a much wider range of associations. A corporate brand thus is a powerful means for firms to express themselves in a way that is not tied into their specific products or services".

Brand Evaluations

Rational Evaluations

Brand Value

Brands should create value [2]. This value is perceived by comparing the benefit that the consumer expects to receive to their experience with a particular brand. This benefit is either functional or emotional [8]. If the benefit is less than expected, the consumer will be dissatisfied. Another words, the customer compares the quality that they perceived with their actual experience with the brand to evaluate the value they receive by consuming the brand.

Another way that value is created is based on the relationship between quality and perceived price [2]. In this regard, Zeitham [31] describes four consumers' perceptions about value: (1) value is low price, (2) value is the quality I get for the price, (3) value is what I get for what I give and (4) value is whatever I want in a product (this is in line with the previous perspective of value).

Regardless of what perspective is taken into account, value is a subjective term that totally depends on the perception of the customer. "It is the individual customer's preferences that determine whether the value is low or high" [2].

This evaluation is rational as the customer subjectively judges the value of a brand based on the benefit that they intentionally expected or the trade-off that they receive for what they give. According to Martensen and Grønholdt [2], there exists a strong relationship between perceived value and customer loyalty. They argue that, before buying a product or service, a customer usually seeks for possibilities and considers alternatives that live up to his/her requirements. The one with highest value will possibly be chosen.

Customer Satisfaction

Satisfaction does not always lead to loyalty; however, it is widely agreed in the literature that satisfaction is the key precursor to customer loyalty. According to Oliver (1999), "satisfaction is defined as pleasurable fulfillment. That is, the consumer senses that consumption fulfills some need, desire, goal, or so forth and that this fulfillment is pleasurable" [32].

The above-mentioned definition is in line with Parasuraman and Kotler [17, 18, 22] that satisfaction results from the difference between prior expectations and the actual performance of the product or service as perceived after the consumption.[1]

However, in banking services, with a variety of products and services, it is hard to evaluate the influence of satisfaction on the customer-brand relationship just through a single product or service. Thus, for satisfaction to have an affect on loyalty, individual satisfaction episodes should become aggregated or blended [32]. Therefore, satisfaction mentioned in this study is "overall satisfaction".

Emotional Evaluations

In most cases, customers buy a brand for not only functional benefits but also emotional and self-expressive benefit [1]. Martensen and Grønholdt [2] argue that "Brands should provoke excitement and evoke a higher experience than simply product-function. Brands should create positive feelings with us – we need to feel touched emotionally". According to these authors, a brand should also create intensive and fantastic experience to the customer. This feeling helps to consolidate the customer-brand relationship to "a point of connectedness that it is a rare experience for that customer to purchase anyplace else" (***feeling evaluation***).

In the CBBE model, Martensen and Grønholdt [2] also include "***self-expressive benefits and social approval***" as a sub component of brand evaluation. They argue that a brand can help a person to recognize himself or herself (or to be recognized) within a group that he or she thinks that he or she belongs to, and to show personal values and attitudes through the brands that that person buy and use.

However, unlike in physical products and other services, the similarity of products and services between banks may mean that self-expressive benefits are seen as less important than social approval. The argument is that, as mentioned previously, the customer may find it important to deal with a great bank brand in order to be recognized in a certain social status or to generate trust to their partners. With this perspective, the customer may wants to maintain their relation with the great bank as they can not find this kind of benefit with less well known brands.

Customer-Brand Relationship

Research on brand equity generally agrees that the final brand-building step is developing customer brand relationships or bonding, and that an important element in this connection is loyalty [2].

Aaker [1] views brand loyalty as a dimension of brand equity. In the CBBE pyramid developed by Keller [30], brand loyalty is at the top of the building blocks and is characterized in term of intensive relationship.

Despite its apparent benefits to any firm, loyalty is viewed quite differently from different perspectives. This might result from the variety of the customer's perceptions about the value that a brand delivers. Jacoby and Chestnut [33] define brand loyalty as a result of two components: "(1) A favorable attitude toward the brand, and (2) Repurchase of the brand over time."

One of the broadest definitions of loyalty is of Oliver [32] which describes loyalty as "a deeply help commitment to re-buy or re-patronize a preferred product/service consistently in the future, thereby causing repetitive same-brand or same brand set purchasing, *despite* situational influences and marketing efforts having the potential to cause switching behavior".

According to Oliver [32], customers become loyal in four phases. At the shallowest level, called *cognitive loyalty*, loyalty might result from the belief of the customer in the brand. The brand information is either retrieved from vicarious knowledge about the brand (from communication, word of mouth…) or current experience-based information. At this stage, if the satisfaction does not involve then the depth of loyalty is merely the brand performance.

Loyalty shifts to the next phase if satisfaction steps in. At this phase, attitudes toward the brand are formed basing on satisfaction, or pleasure accumulated through the consumption of the brand. Commitment at this episode is referred as *affective loyalty*. Though loyalty in this stage is at deeper level than cognitive loyalty is and not as easily dislodged, it's still venerable to switching.

It is desirable if loyalty moves to a deeper level, the *conative loyalty* (behavioral intention). The development of loyalty at this phase is based on repetitive positive experience with the brand. It reflects the customer favorable intention toward the brand such as deeply commitment to buy. However, Oliver argues that this desire is rather the repurchase intention and motivation, and may be "anticipated but unrealized".

The ultimate phase of loyalty proposed by Oliver [32] is *action loyalty* (other authors refer to as *behavioral loyalty* – Keller [30]). At this phase, not only the intention to re-buy is shifted to the action of re-buying (and "repeat purchases", Keller [30]) but also that desire engages in "overcoming obstacles"

Martensen and Grønholdt [2] adapt a more operational point of view: "Customer loyalty has two sides to it, which on the one hand results in an effective continuation and extension of the business partnership, and on the other hand in a recommendation of the supplier, the brand, the product or the services for other potential customers." According to them, customer loyalty takes place when the customer keeps on maintaining the relation with the company in terms of repurchases and purchase intention which can predict future behavior, and on the other hand, the loyalty will result in re-patronizing the company to purchase other products.

However, Martensen and Grønholdt [2] also agree that loyalty is also portrayed as certain attitudinal loyalty where the customer thinks that the company is distinctive and particularly attractive compared to its rivals. This is also in line with Oliver [34] that the customer's experiences with the company and its products are accumulated in a positive way as mentioned above (conative loyalty).

In the banking industry, research by Colgate [12] into the reasons that the customers switch or stay with their bank after a service failure shows that a majority of customers "who felt a strong sense of loyalty to their bank" decide to stay. According Colgate [35], this loyalty might result from the customer's confidence with the relationship they have shaped with the service provider.

Finally, all hypotheses, factors and observations are modified as Fig. 1.

"Hypothesis 1 (H1). Rational evaluation is positively related to price competitiveness

"Hypothesis 2 (H2). H2: Rational evaluation is positively related to brand promise

"Hypothesis 3 (H3). Rational evaluation is positively related to perceived service quality

"Hypothesis 4 (H4). H4: Rational evaluation is positively related to brand trust and credibility

"Hypothesis 5 (H5). Rational evaluation is positively related to brand differentiation

"Hypothesis 6 (H6). Emotional evaluation is positively related to price competitiveness

"Hypothesis 7 (H7). Emotional evaluation is positively related to brand promise

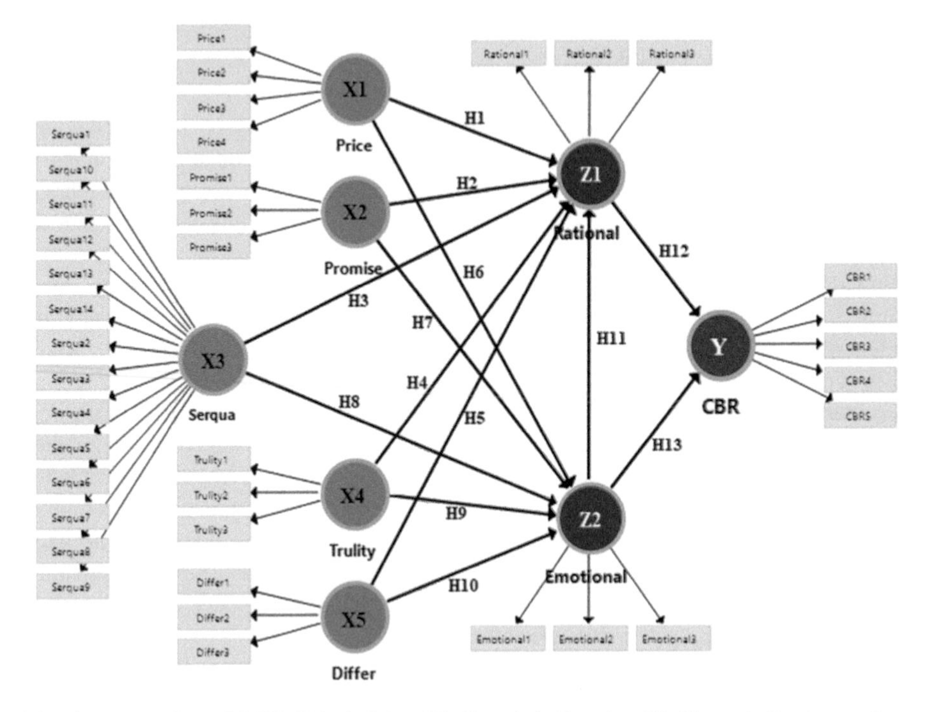

Fig. 1. Research model [X1 (Price): Price, X2 (Promise): Promise, X3 (Serqua): Service quality, X4 (Trulity): Trust and Credibility, Z1 (Rational): Rational Evaluation, Z2 (Emotional): Emotional Evaluation, Y (CBR): Customer-Brand Relationship, [Source: Designed by author.]

"Hypothesis 8 (H8). Emotional evaluation is positively related to perceived service quality

"Hypothesis 9 (H9). Emotional evaluation is positively related to brand trust and credibility

"Hypothesis 10 (H10). Emotional evaluation is positively related to brand differentiation

"Hypothesis 11 (H11). Rational evaluation is positively related to Emotional evaluation

"Hypothesis 12 (H12). Customer-brand relationship is positively related to rational evaluation

"Hypothesis 13 (H13). Customer-brand relationship is positively related to emotional evaluation

3 Research Method

This study was conducted in Ho Chi Minh city in Vietnam with two phases: a pilot test and the main study. The purpose of pilot test was to refine the questionnaire to help respondents to avoid problems in answering questions and to increase the quality of data recorded for the main survey. In the first phase, a qualitative approach was employed in order to explore whether the scale for measuring the constructs of brand equity were suitable in Vietnamese culture and the Vietnamese banking service. The first draft of questionnaire was developed in English. It was then translated into Vietnamese. Some amendments have been made where needed. This step was carried out by using group discussion techniques. Two mini group discussions were conducted. In the first discussion, four bank experts including two branch directors and two managers (all were male) from different banks were invited. The purpose of this step is to examine the clarity the instrument and to be sure that all survey questions were clear in meaning and sufficient to cover the research matter in reality, from the perspective of a banking professional. A quantitative approach was then used in the second phase. Data were collected by interviewing bank's customers. Respondents were selected by convenient methods with a sample size of 450 consumers bought retail banking service in Hochiminh City in Vietnam. There were 113 (25.1%) males and 334 (74.9%) females in this survey.

The questionnaire answered by respondents is the main tool to collect data. The questionnaire contained questions about Banking Service Behavior. A Likert-scale type questionnaire was used to detect Banking Service Behavior.

The survey was conducted on May 03, 2018. Data processing and statistical analysis software is used by Smartpls 3.0 developed by SmartPLS GmbH Company in Germany. The reliability and validity of the scale were tested by Cronbach's Alpha, Average Variance Extracted (Pvc) and Composite Reliability (Pc). Cronbach's alpha coefficient greater than 0.6 would ensure the scale reliability [36]. Composite Reliability (Pc) is better than 0.6 and Average Variance Extracted must be greater than 0.5 [37, 38]. Followed by a linear structural model SEM was used to test the research hypotheses [39].

Datasets

We validate our model on three standard datasets for Banking Service Behavior in Vietnam: Excel.csv and Smartpls.splsm. Dataset has eight variables: five independent variables, two intermediate variables and one variable. There are 450 observations and 38 factors in dataset. Excel.csv were used for descriptive statistics and Smartpls.splsm for advanced analysis.

4 Results and Findings

Structural Equation Modeling (SEM) is used on the theoretical framework. Partial Least Square method can handle many independent variables, even when multicollinearity exists. PLS can be implemented as a regression model, predicting one or more dependent variables from a set of one or more independent variables or it can be implemented as a path model. Partial Least Square (PLS) method can associate with the set of independent variables to multiple dependent variables [39].

4.1 Consistency and Reliability

In this reflective model convergent validity is tested through composite reliability or Cronbach's alpha. Composite reliability is the measure of reliability since Cronbach's alpha sometimes underestimates the scale reliability [39–41]. Table 1 shows that composite reliability varies from 0.807 to 0.918 which is above preferred value of 0.5. This proves that model is internally consistent. To check whether the indicators for variables display convergent validity, Cronbach's alpha is used (from 0.643 to 0.879). From Table 1, it can be observed that all the factors are reliable (Cronbach's alpha > 0.60) and Pvc > 0.5 (from 0.587 to 0.790). The **Serqua** has Pvc = 0.362 (<0.5) but Cronbach's alpha = 0.864 (<0.6) and Pc = 0.887 (>0.5) so it is supported.

Table 1. Cronbach's alpha, composite reliability (Pc) and AVE values (Pvc)

Factor	Cronbach's Alpha	Average Variance Extracted (Pvc)	Composite Reliability (Pc)	P	Findings
CBR	0.879	0.675	0.912	0.000	Supported
Differ	0.643	0.583	0.807	0.000	Supported
Emotional	0.866	0.790	0.918	0.000	Supported
Price	0.765	0.587	0.850	0.000	Supported
Promise	0.778	0.693	0.871	0.000	Supported
Rational	0.865	0.787	0.917	0.000	Supported
Serqua	0.864	0.362	0.887	0.000	Supported
Trulity	0.691	0.620	0.828	0.000	Supported

$$\alpha = \frac{k}{k-1}\left[1 - \frac{\sum \sigma^2(x_i)}{\sigma_x^2}\right] \quad \rho_C = \frac{\left(\sum_{i=1}^{p} \lambda_i\right)^2}{\left(\sum_{i=1}^{p} \lambda_i\right) + \sum_{i=1}^{p}\left(1-\lambda_i^2\right)} \quad \rho_{VC} = \frac{\sum_{i=1}^{p} \lambda_i^2}{\sum_{i=1}^{p} \lambda_i^2 + \sum_{i=1}^{p}\left(1-\lambda_i^2\right)}$$

k: factor, xi: observations, λ_i is a normalized weight of observation variable, σ^2: Square of Variance, i; $1 - \lambda i^2$ – the variance of the observed variable i. **Source:** *Calculated by Smartpls software 3.0*

Structural Equation Modeling (SEM)

SEM results showed that the model is compatible with data research: SRMR has P-value = 0.000 (<0.05) in Table 3 [41, 42].

In bootstrapping, resampling methods are used to compute the significance of PLS coefficients. Output of significance levels can be retrieved from bootstrapping option. Table 6 shows the results of hypotheses testing; all the t values above 1.96 are significant at the .05 level (Fig. 2).

Hypotheses H2, H3, H4, H6, H7, H8, H10, H11, H12 and H13 were supported in Table 2. The results indicated H1, H5 and H9 unsupported in Table 2 [39].

Table 2. Structural Equation Modeling (SEM)

Relation	Beta	SE	T value	P	Findings
Differ -> Emotional	0.123	0.050	2.453	0.015	Supported
Differ -> Rational (H5)	0.019	0.058	0.327	0.744	Unsupported
Emotional -> CBR	0.403	0.045	9.029	0.000	Supported
Emotional -> Rational	0.365	0.052	6.969	0.000	Supported
Price -> Emotional	0.124	0.046	2.709	0.007	Supported
Price -> Rational (H1)	0.064	0.049	1.309	0.191	Unsupported
Promise -> Emotional	0.203	0.045	4.513	0.000	Supported
Promise -> Rational	0.121	0.056	2.170	0.030	Supported
Rational -> CBR	0.399	0.045	8.934	0.000	Supported
Serqua -> Emotional	0.325	0.064	5.054	0.000	Supported
Serqua -> Rational	0.177	0.068	2.614	0.009	Supported
Trulity -> Emotional (H9)	0.108	0.055	1.950	0.052	Unsupported
Trulity -> Rational	0.123	0.059	2.091	0.037	Supported

Beta(r): SE = SQRT(1 – r2)/(n – 2);CR = (1 – r)/SE; P-value = TDIST (CR, n – 2, 2). **Source**: *Calculated by Smartpls software 3.0*

Table 3. Standard of model SEM

Standard	Beta	SE	T-value	P	Findings
SRMR	0.063	0.004	16.324	0.000	Supported

Source: *Calculated by Smartpls software 3.0*

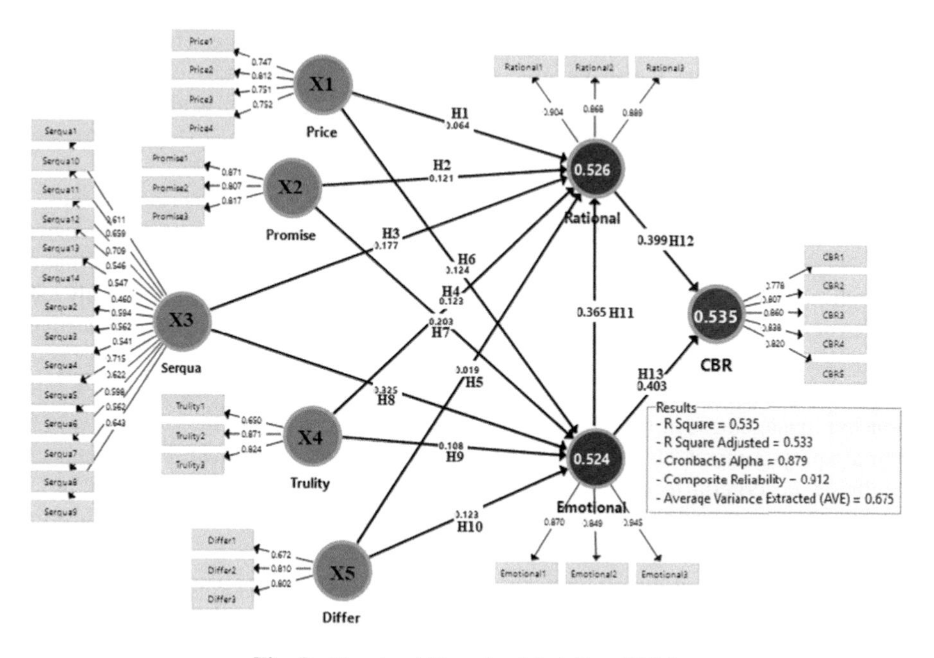

Fig. 2. Structural Equation Modeling (SEM)

4.2 Findings

SEM results have brought out some unexpected outcomes. "Brand differentiation" (H5) and "Price" (H1) were not supported in their relationship with "rational evaluation". Trust and credibility (H9) were not supported in their relationship with "Emotional evaluation". The contention might be that in the rational perspective, customers make their judgment based on what they see, feel or touch (service quality). It might be more important that a "real" rational evaluation, the cost/benefit evaluation (price) and/or what they expect to receive is satisfied by the bank (brand promise). In this perspective, emotion-based associations, like differentiation or trust and credibility, might not make sense. The results shown in Table 2 clearly support this argument.

Regarding the relationships between emotional evaluation and the brand associations, only nine hypotheses were confirmed Hypotheses H2, H3, H4, H6, H7, H8, H10, H11, H12 and H13.

5 Conclusion and Discussion

The aim of the current study is to test a general model of customer-based brand equity into the banking service perspective. The findings suggest that the theoretical model is not fully supported. However, the modified model can be used as a point of departure for those who intend to study the CBBE in the banking industry in Vietnam. As there is no specific model of CBBE for banking services in Vietnam so far, this model is the first that provides a clear image of the dimensions that contribute to the brand equity in banking service.

Secondly, this study contributes to the marketing literature a measurement scale as a useful instrument to measure the brand equity in banking service. The advantage of this instrument is that it can be used flexibly. For example, most of the observed variables presented in this study might also be useful for those who would follow Aaker's approach to measure brand equity of banking service in an emerging economy like Vietnam.

This study provides an insight into brand equity in banking industry. And thus, it can furnish bank managers a structured approach to formulate their branding strategies. The weight of the relationship between each of the brand equity dimensions or that of the sub components and the loyalty formation helps them to prioritize and allocate limited resources across brand equity dimensions/components to reach their objectives in a most efficient way.

The modified CBBE model can be also used as a guideline for customer relationship management in banking service. By better understanding the contributions of brand equity components to the customer attitude towards the brand, bank managers might set up criteria to classify customers into different groups, for instance "price sensitive" group or "rationally-based" group and "emotionally-based" group. Then the bank can have different policy for each group.

At the beginning of this study it was expected that the rational perspective (in terms of low price) would be most important for the banking industry. But interestingly the findings of this study do not support this view. Even though price competitiveness dominates the customer rational evaluation and is also involved in the emotional evaluation, the emotional evaluation finally contributes a larger proportion to the customer loyalty. On the one hand, it positively impacts on the rational perceptions, i.e. the more the customer's mentally prefer the brand, the higher they perceive the brand value and the greater their satisfaction with the brand. On the other hand, emotional perceptions play a larger role in forming customer loyalty to the brand.

Understanding the way that customer loyalty is formed, bank managers might need to inspire emotions in the customer's mind by offering superior service performance, differentiating the brand from competitors by providing customers with advantage that other banks would find hard to copy, generating trust from the target audience with consistent service quality and never communicating a value or service that the bank can not deliver.

Like any other research, this study has many limitations. The first one is in the sampling. Samples were selected by the convenience method. This is the least reliable form of non-probability sampling. Respondents were bank customers who are currently in transaction with the selected banks. Many of them are very familiar with the bank therefore they might over-rate the bank. In addition, the sample consists of individual customers only. This may not fully reflect all aspects of customer perception about bank operations, as others, for example business customers, may have very different views.

References

1. Aaker, D.A.: Measuring brand equity across products and markets. Calif. Manag. Rev. **38** (3), 102 (1996)
2. Martensen, A., Grønholdt, L.: Building brand equity: a customer-based modelling approach. J. Manag. Syst. **16**(3), 37–51 (2004)
3. Juga, J., Juntunen, J., Paananen, M.: Impact of value-adding services on quality, loyalty and brand equity in the brewing industry. Int. J. Qual. Serv. Sci. **10**(1), 61–71 (2018)
4. Taylor, S.A., Hunter, G.L., Lindberg, D.L.: Understanding (customer-based) brand equity in financial services. J. Serv. Mark. **21**(4), 241–252 (2007)
5. Havrilesky, T.M., Shelagh, H.: Operations Management, and Regulation Modern Banking. Wiley, Chichester (2005). England Standards and Pool's (in English)
6. Farquhar, P.H.: Managing brand equity. Mark. Res. **1**(3), 24–33 (1989)
7. Keller, K.L., Parameswaran, M., Jacob, I.: Strategic Brand Management: Building, Measuring, and Managing Brand Equity. Pearson Education India, Delhi (2011)
8. Keller, K.L.: Conceptualizing, measuring, and managing customer-based brand equity. J. Mark. 1–22 (1993)
9. Kapferer, J.-N.: The New Strategic Brand Management: Advanced Insights and Strategic Thinking. Kogan Page Publishers, London (2012)
10. Krishnan, B.C., Hartline, M.D.: Brand equity: is it more important in services? J. Serv. Mark. **15**(5), 328–342 (2001)
11. Burton, S., Lam, R., Lo, H.: Investigating the drivers of SMEs' banking loyalty in Hong Kong (2005)
12. Colgate, M., Norris, M.: Why customers leave or decide not to leave their bank. Univ. Auckl. Bus. Rev. **2**(2), 40–51 (2000)

13. Gounaris, S.P., Stathakopoulos, V., Athanassopoulos, A.D.: Antecedents to perceived service quality: an exploratory study in the banking industry. Int. J. Bank Mark. **21**(4), 168–190 (2003)
14. Chi Cui, C., Lewis, B.R., Park, W.: Service quality measurement in the banking sector in South Korea. Int. J. Bank Mark. **21**(4), 191–201 (2003)
15. Olsen, L.L., Johnson, M.D.: Service equity, satisfaction, and loyalty: from transaction-specific to cumulative evaluations. J. Serv. Res. **5**(3), 184–195 (2003)
16. Cronin Jr., J.J., Taylor, S.A.: Measuring service quality: a reexamination and extension. J. Mark. 55–68 (1992)
17. Parasuraman, A., Zeithaml, V.A., Berry, L.L.: SERVQUAL: a multiple-item scale for measuring consumer perceptions of service quality. J. Retail. **64**(1), 12–40 (1994)
18. Parasuraman, A., Zeithaml, V.A., Berry, L.L.: Reassessment of expectations as a comparison standard in measuring service quality: implications for further research. J. Mark. **58**(1), 111–124 (1988)
19. Netemeyer, R.G., Krishnan, B., Pullig, C., Wang, G., Yagci, M., Dean, D., Ricks, J., Wirth, F.: Developing and validating measures of facets of customer-based brand equity. J. Bus. Res. **57**(2), 209–224 (2004)
20. Keaveney, S.M.: Customer switching behavior in service industries: an exploratory study. J. Mark. 71–82 (1995)
21. Bogomolova, S., Romaniuk, J.T.: Why do they leave? An examination of the reasons for customer defection in the business banking industry. In: ANZMAC 2005 (2005)
22. Kotler, P.: Framework for Marketing Management. Pearson Education India, Delhi (2015)
23. Ambler, T.: Brand equity as a relational concept. J. Brand Manag. **2**(6), 386–397 (1995)
24. Oliver, R.L.: A cognitive model of the antecedents and consequences of satisfaction decisions. J. Mark. Res. 460–469 (1980)
25. Chatterjee, S.C., Chaudhuri, A.: Are trusted brands important. Mark. Manag. J. **15**(1), 1–16 (2005)
26. Filo, K., Funk, D.C., Alexandris, K.: Exploring the role of brand trust in the relationship between brand associations and brand loyalty in sport and fitness. Int. J. Sport. Manag. Mark. **3**(1–2), 39–57 (2008)
27. Delgado-Ballester, E., Munuera-Alemán, J.L.: Does brand trust matter to brand equity? J. Prod. Brand Manag. **14**(3), 187–196 (2005)
28. Cruz, P.P., Repeses, E.C., Laukkanen, T., Munoz, P.A.: Trust in E-bank: a cross-national study. In: ANZMAC Conference: Pricing and Financial Issues in Marketing (2005)
29. Feldwick, P.: Brand equity: do we really need it. How to use advertising to build strong brands, pp. 69–96 (1999)
30. Keller, K.L.: Building customer-based brand equity: a blueprint for creating strong brands (2001)
31. Zeithaml, V.A.: Consumer perceptions of price, quality, and value: a means-end model and synthesis of evidence. J. Mark. 2–22 (1988)
32. Oliver, R.L.: Whence consumer loyalty? J. Mark. 33–44 (1999)
33. Jacoby, J., Chestnut, R.W.: Brand Loyalty: Measurement and Management. Wiley (1978). Incorporated
34. Oliver, R.L.: Loyalty and profit: long-term effects of satisfaction (1997)
35. Colgate, M., Tong, V.T.-U., Lee, C.K.-C., Farley, J.U.: Back from the brink: why customers stay. J. Serv. Res. **9**(3), 211–228 (2007)
36. Nunnally, J.C., Bernstein, I.: The assessment of reliability. Psychom. Theor. **3**(1), 248–292 (1994)
37. Hair, J.F., Black, W.C., Babin, B.J., Anderson, R.E., Tatham, R.L.: Multivariate Data Analysis, vol. 6. Pearson Prentice Hall, Upper Saddle River (2006)

38. Hair Jr., J.F., Hult, G.T.M., Ringle, C., Sarstedt, M.: A Primer on Partial Least Squares Structural Equation Modeling (PLS-SEM). Sage Publications, Thousand Oaks (2016)
39. Khoi, B.H., Van Tuan, N.: Using SmartPLS 3.0 to analyse internet service quality in Vietnam. In: Anh, L.H., Dong, L.S., Kreinovich, V., Thach, N.N. (eds.) International Econometric Conference of Vietnam, pp. 430–439. Springer (2018)
40. Wong, K.K.-K.: Partial least squares structural equation modeling (PLS-SEM) techniques using SmartPLS. Mark. Bull. **24**(1), 1–32 (2013)
41. Latan, H., Noonan, R.: Partial least squares path modeling: basic concepts, methodological issues and applications. Springer (2017)
42. Henseler, J., Hubona, G., Ray, P.A.: Using PLS path modeling in new technology research: updated guidelines. Ind. Manag. Data Syst. **116**(1), 2–20 (2016)

Empirical Study of Worker's Behavior in Vietnam

Ngo Van Tuan[1(✉)] and Bui Huy Khoi[2]

[1] Banking University of Ho Chi Minh City, 36 Ton That Dam, Nguyen Thai Binh Ward, District 1, Ho Chi Minh City, Vietnam
tuannv@buh.edu.vn

[2] Industrial University of Ho Chi Minh City, 12 Nguyen Van Bao Street, Go Vap District, Ho Chi Minh City, Vietnam
buihuykhoi@iuh.edu.vn

Abstract. The aim of this research examines what factors motivate the worker involved in the construction industry firms in Ho Chi Minh City in Vietnam and their level of job satisfaction. Survey data was collected from 252 people in HCM City. The research model is proposed from the study of job satisfaction of some authors in abroad. The reliability and validity of the scale were tested by Cronbach's Alpha, Average Variance Extracted (Pvc) and Composite Reliability (Pc). The analysis results of structural equation model (SEM) showed that the job satisfaction and some factors have a relationship with each other. The finding of this study provides valuable insights for the management of construction industry firms understanding the factors effecting job satisfaction.

Keywords: Vietnam · Job satisfaction · Structural Equation Model SEM · Factors · Construction goods · Relationship · Smartpls 3.0 software

1 Introduction

The industry of construction and organizations of construction have had a long development progress quickly and steadily. The number of construction projects of high buildings, houses, sky-scrapers from foreign investors and domestic investors has been increased rapidly. Therefore, the relevant industries like bricks, cement, aluminium windows, glass etc. have been developed accordingly. Ultimately, the retention of competent employees is considered a key factor for success and sustainability growth, but it is a big concern to any business.

The industry of construction and organizations of construction have had a long development progress quickly and steadily. The number of construction projects of high buildings, houses, sky-scrapers from foreign investors and domestic investors has been increased rapidly. Therefore, the relevant industries like bricks, cement, aluminium windows, glass etc. have been developed accordingly. Ultimately, the retention of competent employees is considered a key factor for success and sustainability growth, but it is a big concern to any business.

The purpose of this study was to determine factors effected to job satisfaction amongst employees of construction industry firms in Ho Chi Minh City. A research

V. Kreinovich et al. (Eds.): ECONVN 2019, SCI 809, pp. 742–750, 2019.
https://doi.org/10.1007/978-3-030-04200-4_52

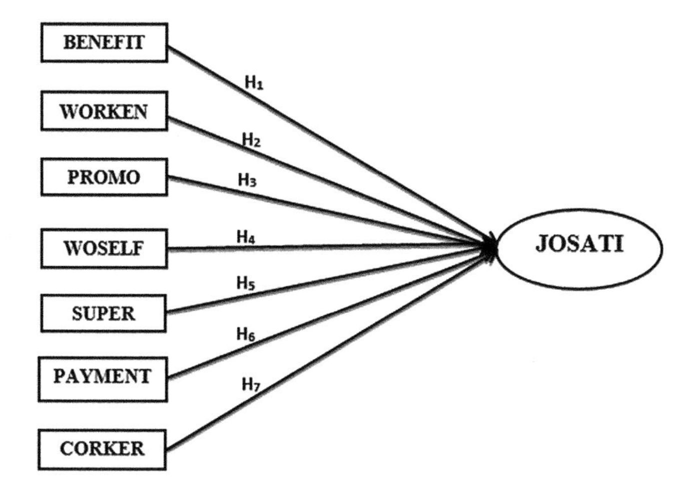

Fig. 1. Research model (**BENEFFIT**: *benefit*, **WORKEN**: *working environment*, **PROMO**: *promotion*, **WORKSELF**: *work itself*, **SUPER**: *supervisor*, **PAYMENT**: *payment*, **CORKER**: *co-workers*, **JOSATI**: *job satisfaction*)

3 Research Method

We followed the methods of Khoi and Van Tuan (2018). Research methodology is implemented through two steps: qualitative research and quantitative research. Qualitative research was conducted with a sample of 30 people. First periPod 1 is tested on a small sample to discover the flaws of the questionnaire. Second period of the official research was carried out as soon as the question was edited from the test results. Respondents were selected by convenient methods with a sample size of 252 people in 12 construction firms In Ho Chi Minh City in Vietnam. There were 151 (59.9%) males and 101 (40.1%) females in this survey. Their ages and qualification were in Table 1.

Table 1. Age groups and qualification

Age groups	Amount	Percent (%)	Qualification	Amount	Percent (%)
Under 25	39	15.5	High schools	4	1.6
From 25–34	196	77.8	Intermediate	21	8.3
From 35–44	13	5.2	College	59	23.4
From 45–54	4	1.6	University	166	65.9
			Master upwards	2	0.8
Total	252	100.0	Total	252	100.0

Their timeserving and working position were as Table 2:

The questionnaire answered by respondents is the main tool to collect data. The questionnaire contained questions about the position of the job satisfaction and factors

Table 2. Timeserving and working position

Timeserving	Amount	Percent (%)	Working position	Amount	Percent (%)
Employee	227	90.1	Upper 1 year	61	24.2
Manager	13	5.2	From 1–<3 years	114	45.2
Chief and Deputy of Bureau	12	4.8	From 3–<5 years	53	21.0
			From 5–<10 years	24	9.5
Total	252	100.0	Total	252	100.0

and their personal information. A Likert-scale type questionnaire was used to detect employment attitudes and job satisfaction.

The survey was conducted on July 03, 2017. Data processing and statistical analysis software is used by Smartpls 3.0. The reliability and validity of the scale were tested by Cronbach's Alpha, Average Variance Extracted (Pvc) and Composite Reliability (Pc). Followed by a linear structural model SEM was used to test the research hypotheses (Khoi and Van Tuan 2018).

4 Results

Structural Equation Modeling (SEM) is used on the theoretical framework. Partial Least Square method can handle many independent variables, even when multicollinearity exists. PLS can be implemented as a regression model, predicting one or more dependent variables from a set of one or more independent variables or it can be implemented as a path model. Partial Least Square (PLS) method can associate with the set of independent variables to multiple dependent variables (Khoi and Van Tuan 2018).

4.1 Consistency and Reliability

In this reflective model convergent validity is tested through composite reliability or Cronbach's alpha. Composite reliability is the measure of reliability since Cronbach's alpha sometimes underestimates the scale reliability (Wong 2013; Khoi and Van Tuan 2018). Table 3 shows that composite reliability varies from 0.824 to 0.918 which is above preferred value of 0.5. This proves that model is internally consistent. To check whether the indicators for variables display convergent validity, Cronbach's alpha is used. From Table 2, it can be observed that all the factors are reliable (>0.60) and Pvc > 0.5. The WORKEN has Pvc = 0.388 (<0.5) and Cronbach's alpha = 0.596 (<0.6) but Pc = 0.691 (>0.5) so it is supported.

Table 3. Cronbach's alpha, composite reliability (Pc) and AVE values (Pvc)

Factor	Cronbach's Alpha	Average Variance Extracted (Pvc)	Composite Reliability (Pc)	P	Findings
BENEFIT	0.717	0.540	0.824	0.000	Supported
CORKER	0.843	0.761	0.905	0.000	Supported
JOSATI	0.849	0.768	0.908	0.000	Supported
PAYMENT	0.880	0.737	0.918	0.000	Supported
PROMO	0.916	0.747	0.937	0.000	Supported
SUPER	0.861	0.591	0.896	0.000	Supported
WORKEN	0.596	0.388	0.691	0.000	Supported
WOSELF	0.801	0.546	0.856	0.000	Supported

$$\alpha = \frac{k}{k-1}\left[1 - \frac{\sum \sigma^2(x_i)}{\sigma_x^2}\right] \rho_c = \frac{\left(\sum_{i=1}^{p} \lambda_i\right)^2}{\left(\sum_{i=1}^{p} \lambda_i\right)^2 + \sum_{i=1}^{p} (1-\lambda_i)^2} \quad \rho_{vc} = \frac{\sum_{i=1}^{p} \lambda_i^2}{\sum_{i=1}^{p} \lambda_i^2 + \sum_{i=1}^{p} (1-\lambda_i^2)}$$

k: factor, xi: observations, λ_i is a normalized weight of observation variable, σ^2: Square of Variance, i; $1 - \lambda i^2$ – the variance of the observed variable i.

4.2 Structural Equation Modeling (SEM) in the First

SEM results in the first in the Fig. 2 showed that the model is compatible with data research. The job satisfaction is affected by WORKEN, PAYMENT, PROMO and WOSELF about 43.4. The BENEFIT, CORKER and SUPER are not relative with job satisfaction because their p-value is greater than 0.05 as Table 4.

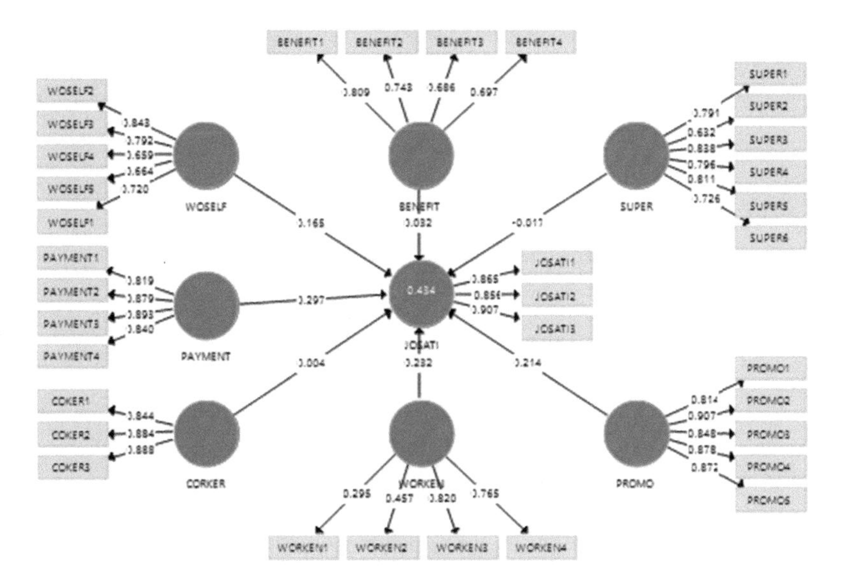

Fig. 2. Structural Equation Modeling (SEM) in the first

Table 4. Structural Equation Modeling (SEM)

Relation	Beta	SE	T-value	P	Findings
BENEFIT -> JOSATI	0.032	0.056	0.575	0.565	Unsupported
CORKER -> JOSATI	0.004	0.061	0.073	0.942	Unsupported
PAYMENT -> JOSATI	0.297	0.061	4.867	0.000	Supported
PROMO -> JOSATI	0.214	0.079	2.697	0.007	Supported
SUPER -> JOSATI	–0.017	0.074	0.237	0.813	Unsupported
WORKEN -> JOSATI	0.232	0.071	3.264	0.001	Supported
WOSELF -> JOSATI	0.165	0.056	2.918	0.004	Supported

Beta(r): SE = SQRT(1 – r2)/(n – 2); CR = (1 – r)/SE; P-value = TDIST (CR, n – 2, 2).

4.3 Structural Equation Modeling (SEM) in the Last

SEM results showed that the model is compatible with data research: SRMR, d_ULS and d_G has P-value = 0.000 (<0.05) (Henseler et al. 2016) (Fig. 3 and Table 5).

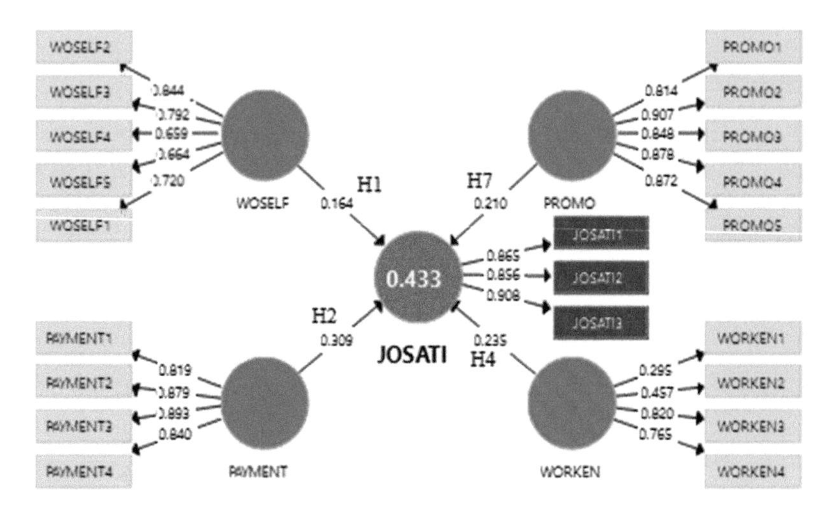

Fig. 3. Structural Equation Modeling (SEM) in the last

Table 5. Standard of model SEM

Standard	Beta	SE	T-value	P	Findings
SRMR	0.068	0.005	13.439	0.000	Supported
d_ULS	0.835	0.114	18.765	0.000	Supported
d_G	0.839	0.064	13.029	0.000	Supported

In bootstrapping, resampling methods are used to compute the significance of PLS coefficients. Output of significance levels can be retrieved from bootstrapping option. Table 6 shows the results of hypotheses testing; all the t values above 1.96 are significant at the .05 level. Hypotheses H1, H2, H4 and H7 were supported. The results indicated WORKEN, PAYMENT, PROMO and WOSELF affecting job satisfaction about 43.3%.

Table 6. Structural Equation Modeling (SEM)

Relation	Beta	SE	T-value	P	Findings
PAYMENT -> JOSATI	0.309	0.057	5.377	0.000	Supported
PROMO -> JOSATI	0.210	0.070	2.997	0.003	Supported
WORKEN -> JOSATI	0.235	0.062	3.812	0.000	Supported
WOSELF -> JOSATI	0.164	0.055	2.973	0.003	Supported

Beta(r): SE = SQRT(1 − r2)/(n − 2); CR = (1 − r)/SE; P-value = TDIST(CR, n − 2, 2).

Findings

The results of this study suggested that work itself, payment, promotion, and working environment related to Job Satisfaction. However, the research found insufficient evidence to suggest the relationship between benefit, promotion and co-worker with job satisfaction.

5 Conclusions

This study is aimed to quantitatively determine four factors impacting job satisfaction. However, some limitations of this study should be noted. Other variety, beyond the above-mentioned, have not been considered. The research has not been considered the impact of the workers' behavioral response, organizational culture, social factors, work-life balance, and economic crisis and market conditions influencing the impact of the predictors upon the outcome variety.

This study was undertaken because of the researchers' interest in determining the factors effecting employee's job satisfaction. It was also believed that managers need a more in-depth understanding of the relationships these independent variables have with a dependent. Although this study is limited in its generalization, it suggests that particular demographic characteristics can affect a person's level of satisfaction with a construction operation. Further, particularly dimensions of a furniture employee's job satisfaction can predict his or her commitment to that organization. It is suggested that more homogenous demographic traits in participants should be identified that moderate this relationship and that a larger sample of same industry operations be used.

References

Adams, R.: Work motivation amongst employees in a government department in the provincial government Western Cape. Doctoral dissertation, University of the Western Cape (2007)

Byars, L.L., Rue, L.W.: Human Resource Management, 8th edn. McGraw Hill, New York (2006)

Graham, G.H.: Understanding human relations: the individual, organization, and management. Science Research Associates (1982)

Henseler, J., Hubona, G., Ray, P.A.: Using PLS path modeling in new technology research: updated guidelines. Ind. Manag. Data Syst. **116**(1), 2–20 (2016)

Lee, X., Yang, B.: The influence factors of job satisfaction and its relationship with turnover intention: taking early-career employees as an example. Anales de Psicología **33**, 697–707 (2017)

Luddy, N.: Job satisfaction amongst employees at a public health institution in the Western Cape. Doctoral dissertation, University of the Western Cape (2005)

Pandey, P., Asthana, P.K.: An empirical study of factors influencing job satisfaction. Indian J. Commer. Manag. Stud. **8**, 96–105 (2017)

Schermerhorn, J.R.: Management for Productivity, 4th edn. Wiley, Canada (1993)

Schulz, S., Steyn, T.: Educators' motivation: differences related to gender, age and experience. Acta academica **35**(3), 138–160 (2003)

Smith, P.C.: The measurement of satisfaction in work and retirement: a strategy for the study of attitudes (1969)

Spector, P.E.: Job Satisfaction: Application, Assessment, Causes, and Consequences, vol. 3. Sage Publications, Thousand Oaks (1997)

Zheng, H.P., Wu, J.P., Wang, Y.M.A., Sun, H.P.: Empirical study on job satisfaction of clinical research associates in China. Ther. Innov. Regul. Sci. **51**, 314–321 (2017)

Wong, K.K.-K.: Partial least squares structural equation modeling (PLS-SEM) techniques using SmartPLS. Mark. Bull. **24**(1), 1–32 (2013)

Khoi, B.H., Van Tuan, N.: Using SmartPLS 3.0 to analyse internet service quality in Vietnam. In: Anh, L.H., Dong, L.S., Kreinovich, V., Thach, N.N. (eds.) Econometrics for Financial Applications, Studies in Computational Intelligence, vol. 760. Springer (2018)

Empirical Study of Purchasing Intention in Vietnam

Bui Huy Khoi[1(✉)] and Ngo Van Tuan[2]

[1] Industrial University of Ho Chi Minh City, 12 Nguyen Van Bao Street, Go Vap District, Ho Chi Minh City, Vietnam
buihuykhoi@iuh.edu.vn

[2] Banking University of Ho Chi Minh City, 36 Ton That Dam, Nguyen Thai Binh Ward, District 1, Ho Chi Minh City, Vietnam
tuannv@buh.edu.vn

Abstract. The aim of this research investigates if and how brand image, brand origin, country of origin, country of manufacture, brand awareness and corporate social responsibility has an impact on purchasing intention for imported goods in Ho Chi Minh City in Vietnam. Survey data is collected from 345 consumers in HCM City. The reliability and validity of the scale are tested by Cronbach's Alpha, Average Variance Extracted (Pvc) and Composite Reliability (Pc). The analysis results of structural equation model (SEM) shows that the purchase intention and some factors have a relationship with each other. The finding of this study provides valuable insights for the management of import goods firms understanding the factors effecting.

Keywords: Smartpls 3.0 software · Purchase intention
Structural equation model · SEM · Factors · Import goods · Relationship

1 Introduction

The majority of us have seen, when consumers decide to buy a product, the design, quality and features of the product are factors strongly influence their purchasing decisions. But deep down inside each customer always has one base, something called faith strong impact on the assessment of the quality and features of the product, which is brand origin. This concept is brand origin (BO), and can be defined as the place, region or country where the target customers table perceive the brand to belong to [1]. Indeed, BO images related positively to both dimensions of brand equity [2]—an important factor affecting purchase intention of consumer shopping behavior. However, BO has been little research on the impact of its main intentions of consumers shopping behavior, and even fewer and newer in the consumer market in Vietnam. Therefore, the effects of brand origin to the intention of the consumer shopping Vietnam is necessarily needed. Parallel with BO, another term is necessary to pay attention when we mentioned to brand origin, sometimes causing confusion or even differences when assessing the impact of a brand, especially in Vietnam, which is the COO. Country of origin (COO) can be defined as the country of manufacture or assembly. COO have been researched for a long time, even more than BO. On the other

V. Kreinovich et al. (Eds.): ECONVN 2019, SCI 809, pp. 751–764, 2019.
https://doi.org/10.1007/978-3-030-04200-4_53

hand, in recent time, Vietnamese is extremely sensitive in identifying a brand, accepting a brand through country of origin (COO), for example, they are firstly pay attention to the words "Made in …" when choosing a product, a brand. Therefore, COO will be an interesting factor, contributing a very important part in the process of researching the influence of the BO to the purchase intention of Vietnamese, thereby giving true conclusions about its effect to the purchase intention of Vietnamese.

In addition, the researchers strongly recommended to measure the impact of Brand Image to purchase intention beside studying the effects of BO [3, 4]. Because the assessment of a product (affecting to purchase intention) depends on the acceptance of the product brand (brand image). Moreover, according to the result of previous researches, brand image affects to customer purchase intention [5, 6] or the attitude of consumers to brand image have a positive impact on purchase intention [7]. Therefore, brand reputation (BREPI) is also an extremely important factor.

Brand awareness is significant factor impacts consumer decision—making. Brand awareness may be signal presence and substance because of high awareness for a long time, because the firm's products are widely distributed, and because the products associated with the brand are purchased by many other buyers.

Consumers' choices have implications for the whole society. Socially responsible corporations are more attractive to consumers. In addition, many academic researches confirm that CSR has a positive influence on consumer evaluations and purchase intensions of a company or product where consumer awareness is an independent variable, which is experimentally manipulated.

2 Literature Review

2.1 Purchase Intention

Understanding customer's need is an important issue to know their purchase intentions. So, enterprises need to research carefully the customer's inside problems. Consumers include many age, gender, personality, lifestyle, way of thinking, awareness … Each customer has its own feel and think in the process of purchasing and consumption. Prior to purchasing, consumers' collect product information based on personal experience and external environment. When the amount of information reaches a certain level, consumers start the evaluation process, and make a purchase decision after comparison and judgment. Therefore, purchase intention is often used to analyze consumer behavior in related studies [8]. The EKB model, developed by Engel et al. [9], describe the process used to evaluate consumers decision making. This model mentioned that consumer behavior is a continuing process, including awareness of a problem, information gathering, solution evaluation, and decision making. The process is also affected by internal and external factors such as information input, information process, general motives, environment … Among these factors, information gathering and environmental stimulation are two cardinal influential factors in the final decision making. According to Kotler, consumer behavior occurs when consumers are stimulated by external factors and come to a purchase decision based on their personal characteristics and decision making process [10]. These factors include choosing a product, brand, a retailer, timing, and quantity. This means consumers' purchase behavior is affected by their choice of product and brand.

There are many factors that affect to purchase intentions such as brand awareness, brand image, brand origin, country of origin, quality, prestige, ... In most cases, brand name is perceived as a key indicator of quality [11], and foreign brands generally help enhance a brand's perceived quality. Consumers reply on various quality to evaluate their perceptions of foreign brand quality. When consumers have experienced the product quality of a brand, they know how the product is and they will tend to consume more. Therefore, even when seeing a new product, even never used, as long as it is their favorite brand's product, they are still willing to buy.
So, we gave the proposed research hypotheses as following:

"**Hypothesis 2 (H2).** *There is an impact of brand origin on purchase intention.*"
"**Hypothesis 6 (H6).** *There is an impact of country of origin on purchase intention."*
"**Hypothesis 9 (H9).** *There is an impact of brand awareness on purchase intention."*
"**Hypothesis 12 (H12).** *There is an impact of brand reputation on purchase intention."*
"**Hypothesis 13 (H13).** *There is an impact of corporate social responsibility on purchase intention."*

2.2 Brand Origin and Country of Origin

Thakor and Kohli (1996) observed that literature have concentrated several aspects of brands that may affect purchase. One significant characteristic associated with many brands are the origin cues [1]. These cues have received little or no attention. He also defined brand origin (BO) as the place, region or country where the target customers perceive the brand to belong to. Now, a product is not produced, manufactured, designed, or assembled in its country of origin, it is itself country invented its product, may be the enterprises have realized the advantages from countries where they are going to choose countries of manufacture their products. However, it has been observed that many global firms position their brands with respect to their national origins [3, 12]. Brand origin has been found to affect consumers' quality perceptions, brand related attitudes, and purchase intentions, and it has resulted in brand origin stereotypes [13]

Following associative memory network theory [14], the strong brand association of BO is accessible to the consumer upon brand name activation. Many dimensions linked to BO are important for consumers' product. For example, BO images such as innovativeness, design, and prestige relate to both brand image and brand quality [15, 16].

The typicality of a brand as a representative of the BO, or the degree to which a brand represents a BO, may moderate the relationship between BO and brand equity. Research on memory association strength suggests that greater typicality enables consumers to categorize or recall brands faster after exposure to a brand or category cue [14]. By the way, it can effect to consumers' purchase intension from their easy identity about a brand.

So, we gave the proposed research hypotheses as following:

"**Hypothesis 1 (H1).** *There is an impact of brand origin on brand reputation.*"
"**Hypothesis 2 (H2).** *There is an impact of brand origin on purchase intention.*"
"**Hypothesis 3 (H3).** *There is an impact of brand origin on brand awareness.*"

If you choose a product or brand—that's mean, you make your purchase intention, what factor do you care? May be they are price, quality, and many others factor that influences them. One of these factors is the brand's country of origin (COO) [17].

Many customers often confuse between Brand origin (BO) and Country of origin (COO), they are different in many way as definition, meaning, image … but, they have relationship together and affect strongly to purchase intentions.

There are many definition of COO, on basic, country-of-origin as the country that conducts manufacturing or assembling [18], as IPhone is a Apple's mobile phone—a brand from USA, but they have IPhone "made in" China. Nike maybe another example, it is a brand of sport from USA, but Nike shoes are manufactured in Asia countries, such as Malaysia, Vietnam … Other researcher indicate that country-of-origin means the country that a manufacturer's product or brand is associated with; traditionally this country is called the home country [19]. For example, we are usually impressed by some well-knows brands in the world, the car of Germany or electronic equipment of Japan.

COO as an extrinsic information, a safe base which customer maybe firstly pay attention to evaluate a product before they make a decision to buy it, specially Vietnamese. Country image as an evaluation is important for consumer since consumer evaluation on product is not only based on value or quality of product but also based on what country that produced the product, how it produced and who made the product. COO as an item of evaluation is being a consumer consideration not only in developing countries but also in developed countries [20]. COO is consumers' perception toward country reputation that produced a product. A good country reputation such as a country that known has high technology capabilities is perceived that the country's product has a good products' quality.

In Vietnam, that maybe pay attention so much, because Vietnamese often concept that level of country's developing and national culture will strongly effect on production of the country. Therefore, it is proposed that:

"**Hypothesis 4 (H4).** *There is an impact of country of origin on brand awareness.*"
"**Hypothesis 5 (H5).** *There is an impact of country of origin on brand reputation.*"
"**Hypothesis 6 (H6).** *There is an impact of country of origin on purchase intention.*"
"**Hypothesis 7 (H7).** *There is an impact of country of origin on corporate social responsibility.*"

2.3 Brand Reputation

The purchase intention may indirectly depend on their perceived reputation [21]. Brand reputation would be a positive factor effects to corporate social responsibility, customers' purchase intention. So, we gave the proposed research hypothesis as following:

"**Hypothesis 12 (H12).** *There is an impact of brand reputation on purchase intention.*"

2.4 Brand Awareness

Keller refers to brand awareness as how easy it is for a consumer to remember a brand and defines brand awareness as "strong, favorable and unique brand associations in memory" [22]. Aaker defines it as "the ability of a potential buyer to recognize or recall that a brand is a member of a certain product category" [23].

Aaker also stated that brand awareness is measured according to different ways in which consumers remember a brand such as follows: brand recognition—when consumers have prior affirmation to a brand; brand recall—when consumers recall brands that meet a category need; top of mind—when consumers recall the first brand; dominant—when consumers recall the only brand [23].

Specifically, brand recognition is the lowest level of awareness and is related to the consumers' ability to confirm previous exposure to the brand when given the brand as a cue [22]. It is based upon an aided recall test, which finds the respondents' ability to identify brands in a certain product class when being provided with the names [23]. The second highest level of awareness is brand recall and is related to consumer's ability to retrieve the brand from memory when given a relevant cue [22]. In this level, they have not any unaided recall test, their respondents' ability to name brands in a certain product group without being provided with any names. According to Aaker, the brand recognition only deals with consumers' past exposure to a brand and not details about the place or the reason of the exposure and also emphasized that the only important issue in this respect is to remember the prior exposure. Brand recall was stated that a given brand plays the role of a stimulus and the need stands as the response [23].

Brand awareness is significant factor impacts consumer decision—making [24] occurs when consumers can recall or recognize a brand or simply when consumers know about a brand [22, 24]. Brand awareness may be signal presence and substance because of high awareness for a long time, because the firm's products are widely distributed, and because the products associated with the brand are purchased by many other buyers.

Thus, brand awareness effects to brand reputation, corporate social responsibility and consumers' purchase intension is a hypothesis should be supposed.

"**Hypothesis 8 (H8).** *There is an impact of brand awareness on brand reputation.*"
"**Hypothesis 9 (H9).** *There is an impact of brand awareness on purchase intention.*"
"**Hypothesis 10 (H10).** *There is an impact of brand awareness on corporate social responsibility.*"

2.5 Corporate Social Responsibility (CSR)

The concept of corporate social responsibility (CSR) calls for a lengthy discussion due to its varied history. There are many different definitions about CSR at the length of time. Carroll (1979) understands social responsibility of the business includes the expectations of society about economic, law, morality and charity for the organization at a given moment [25]. CSR can be defined as involves to all aspects of business behavior so that the impacts' activities are incorporated in every corporate agenda [26]. Researchers noted that CSR are the basic expectations of the company regarding initiatives that take the form of protection to public health, public safety, and the environment [27]. This can be inferred that traditional discussions on the issue of CSR have centered on economic, legal and ethical obligations and contemporary discussions on CSR focus on the use of CSR as a strategic tool [28]. Companies would advertise their CSR activities to communicate corporate image and build reputation, which benefits the company financially as an active to demonstrate for benefit not only the society but the business as well. Lantos also adds that a company's CSR activities designed to bring exposure for the company, improve the company's reputation and brand image which reflects positively on profits [29]. So, this is an indirect way which indicates involvement of consumers' purchase intension.

Consumers' choices have implications for the whole society. Socially responsible corporations are more attractive to consumers. In addition, many academic researches confirm that CSR has a positive influence on consumer evaluations and purchase intensions of a company or product where consumer awareness is an independent variable, which is experimentally manipulated. However, consumers usually have some knowledge about different firms' CSR, but quite limited. Consumers are not active information seekers of a firm's CSR. A main reason for consumer choice is whether they favor the product rather than the producer's CSR or price, value, brand image and trend are the most important factors that influence consumers' choice. However, consumers do state that a firm's CSR has an impact on their choices. Specially and environmentally responsible ways, event they claim that they are willing to pay a higher price for products of socially responsible firms.

So, by this way or by other way, CSR can be the potential factor has an influence on consumers' purchase intention.

"**Hypothesis 11 (H11).** *There is an impact of corporate social responsibility on brand reputation.*"
"**Hypothesis 13 (H10).** *There is an impact of corporate social responsibility on purchase intension.*"

Finally, all hypotheses, factors and observations are modified as Fig. 1.

3 Method

We followed the methods of Ly H. Anh, Le Si Dong, Vladik Kreinovich, and Nguyen Ngoc Thach [30]. Research methodology is implemented through two steps: qualitative research and quantitative research. Qualitative research was conducted with a sample of 30 people. Quantitative research is implemented two periods. First period 1 is tested on a small sample to discover the flaws of the questionnaire. Second period of the official research was carried out as soon as the question was edited from the test results with a sample of 345 people. Respondents were selected by convenient methods with a

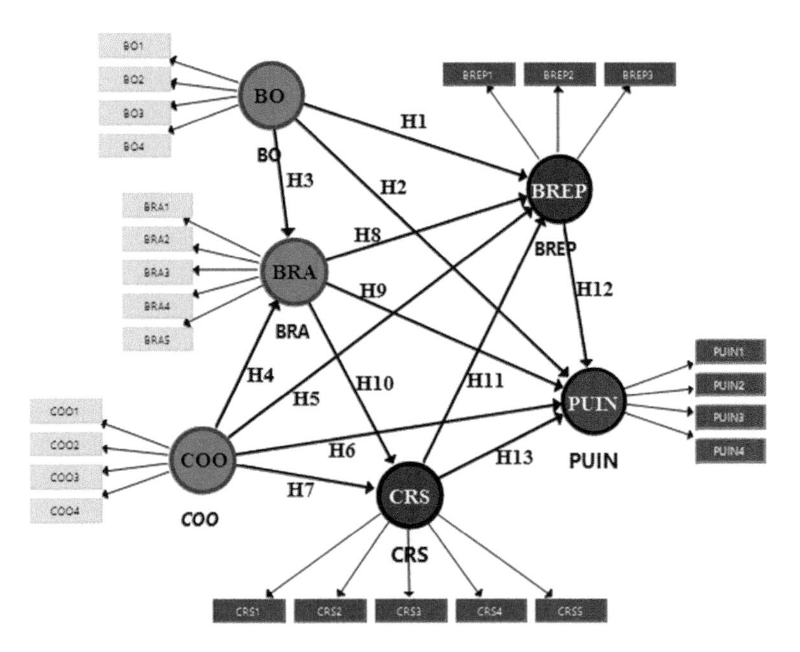

Fig. 1. Research model. **BO**: brand origin, **BRA**: brand awareness, **COO**: country of origin, **BREP**: brand reputation, **CRS**: corporate social responsibility, **PUIN**: purchase intension, **Source**: Designed by author.

sample size of 345 consumers bought import goods in Ho Chi Minh City in Vietnam from 10 companies in Table 3. There were 193 (55.9%) males and 152 (44.1%) females in this survey. Their ages and qualification were in Table 1. Table 2 was consumer's income in this survey in Vietnam. Table 3 was brand names from import goods that Vietnamese were favorable in this survey. There were 10 import goods such as: Nike, Adidas, Puma, Reebok, Iphone, Samsung, Honda, Philips, LG and Sony in this survey.

Table 1. Age groups and job

Age groups	Amount	Percent (%)	Job	Amount	Percent (%)
Under 18	19	5.5	Student	250	72.5
			Officer	47	13.6
From 18 to 30	284	82.3	Worker	19	5.5
From 30 to 40	25	7.2	Businessman	24	7.0
Upper 40	17	4.9	Freelance work	5	1.4
Total	**345**	**100.0**	**Total**	**345**	**100.0**

Source: Calculated by SPSS.sav and Excel.csv.

Table 2. Income

Income	Amount	Percent (%)
Under VND 5 million	244	70.7
From VND 5 to 10 million	60	17.4
From VND 10 to 20 millions	27	7.8
Under VND 20 millions	14	4.1
Total	**345**	**100.0**

Source: Calculated by SPSS.sav and Excel.csv.

Table 3. Brand name and country of origin

Brandname	Code	Amount	Percent (%)
Nike (American)	1	29	8.4
Adidas (Germany)	2	35	10.1
Puma (Germany)	3	22	6.4
Reebok (England)	4	44	12.8
Iphone (American)	5	26	7.5
Samsung (Korean)	6	63	18.3
Honda (Japan)	7	70	20.3
Philips (Netherland)	8	28	8.1
LG (Korean)	9	14	4.1
Sony (Japan)	10	14	4.1
Total	**10**	**345**	**100.0**

Source: Calculated by SPSS.sav and Excel.csv.

The questionnaire answered by respondents is the main tool to collect data. The questionnaire contained questions about the position of the purchase intension and factors and their personal information. A Likert-scale type questionnaire was used to detect purchase intension attitudes.

The survey was conducted in January 2018 in Ho Chi Minh City, Vietnam. Data processing and statistical analysis software is used by Smartpls 3.0 developed by SmartPLS GmbH Company in Germany. The reliability and validity of the scale were tested by Cronbach's Alpha, Average Variance Extracted (Pvc) and Composite Reliability (Pc). Cronbach's alpha coefficient greater than 0.6 would ensure the scale reliability [31]. Composite Reliability (Pc) is better than 0.6 and Average Variance Extracted must be greater than 0.5 [32, 33]. Followed by a linear structural model SEM was used to test the research hypotheses [30].

Datasets

We validate our model on three standard datasets for purchase intension in Vietnam: SPSS.sav, Excel.csv and Smartpls.splsm. Dataset has six variables: two independent variables, three intermediate variables and one variable. There are 345 observations and 30 factors in dataset. SPSS.sav and Excel.csv were used for descriptive statistics and

Smartpls.splsm for advanced analysis. We enclose a coded data in PDF file. It is exported from SPSS.sav (Excel.csv).

4 Measures

Structural Equation Modeling (SEM) is used on the theoretical framework. Partial Least Square method can handle many independent variables, even when multicollinearity exists. PLS can be implemented as a regression model, predicting one or more dependent variables from a set of one or more independent variables or it can be implemented as a path model. Partial Least Square (PLS) method can associate with the set of independent variables to multiple dependent variables [30].

4.1 Consistency and Reliability

In this reflective model convergent validity is tested through composite reliability or Cronbach's alpha. Composite reliability is the measure of reliability since Cronbach's alpha sometimes underestimates the scale reliability [30, 34, 35]. Table 3 shows that composite reliability varies from 0.793 to 0.887 which is above preferred value of 0.5. This proves that model is internally consistent. To check whether the indicators for variables display convergent validity.

Cronbach's alpha is used. From Table 4, it can be observed that all the factors are reliable (Cronbach's alpha > 0.60) and Pvc > 0.5. The COO has Pvc = 0.498 (<0.5) but Cronbach's alpha = 0.656 (<0.6) and Pc = 0.793 (>0.5) so it is supported.

Table 4. Cronbach's alpha, composite reliability (Pc) and AVE values (Pvc)

Factor	Cronbach's Alpha (α)	Average Variance Extracted (Pvc)	Composite Reliability (Pc)	*p* Value	Findings
BO	0.819	0.649	0.880	0.000	Supported
BRA	0.836	0.598	0.878	0.000	Supported
BREP	0.661	0.596	0.815	0.000	Supported
COO	0.656	0.498	0.793	0.000	Supported
CRS	0.841	0.612	0.887	0.000	Supported
PUIN	0.720	0.569	0.833	0.000	Supported

$$\alpha = \frac{k}{k-1}\left[1 - \frac{\sum \sigma^2(x_i)}{\sigma_x^2}\right] \rho_c = \frac{\left(\sum_{i=1}^{p}\lambda_i\right)^2}{\left(\sum_{i=1}^{p}\lambda_i\right)^2 + \sum_{i=1}^{p}(1-\lambda_i)^2} \quad \rho_{vc} = \frac{\sum_{i=1}^{p}\lambda_i^2}{\sum_{i=1}^{p}\lambda_i^2 + \sum_{i=1}^{p}\left(1-\lambda_i^2\right)}$$

k: factor, xi: observations, λ_i is a normalized weight of observation variable, σ^2: Square of Variance, i; $1 - \lambda i^2$ – the variance of the observed variable i. **Source:** Calculated by Smartpls software 3.0.

4.2 Structural Equation Modeling (SEM)

SEM results in the Fig. 2 showed that the model is compatible with data research. The purchase intension (**PUIN**) is affected by brand origin (**BO**), brand awareness (**BRA**), country of origin (**COO**), brand reputation (**BREP**) and corporate social responsibility (**CRS**) about 20.8%. The brand reputation (**BREP**) is affected by brand origin (**BO**), brand awareness (**BRA**), country of origin (**COO**) and corporate social responsibility (**CRS**) about 35.5%. The corporate social responsibility (**CRS**) is affected by brand awareness (**BRA**) and country of origin (**COO**) about 19.0%. The brand awareness (**BRA**) is affected by BRAND ORIGIN (**BO**) and country of origin (**COO**) about 2%.

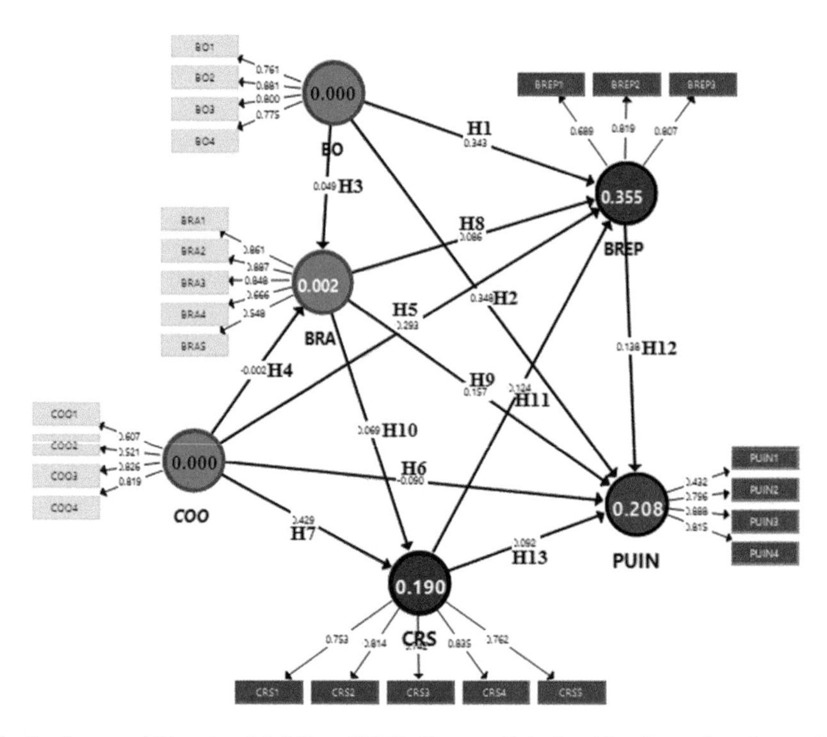

Fig. 2. Structural Equation Modeling (SEM). **Source**: Calculated by Smartpls software 3.0.

The BO and COO are not relative with BRA because their p-value is greater than 0.05 as Table 5. The other hypotheses are unsupported as: BRA → CRS, COO → PUIN, BRA → BREP and CRS → PUIN.

SEM results showed that the model is compatible with data research: SRMR has p-value ≤ 0.001 (<0.05) [35, 36] in Table 6.

Hypotheses H1, H2, H5, H7, H9, H11 and H12 were supported. The results indicated H3, H4, H6, H8, H10 and H13 unsupported.

Table 5. Structural Equation Modeling (SEM)

Relation	Beta	SE	T Value	*p* Value	Findings
BO → BRA	0.048	0.070	0.707	0.480	Unsupported
BO → BREP	0.343	0.058	5.945	0.000	Supported
BO → PUIN	0.355	0.086	4.059	0.000	Supported
BRA → BREP	0.088	0.052	1.644	0.101	Unsupported
BRA → CRS	0.071	0.052	1.320	0.187	Unsupported
BRA → PUIN	0.166	0.050	3.174	0.002	Supported
BREP → PUIN	0.136	0.068	2.024	0.043	Supported
COO → BRA	0.005	0.059	0.028	0.978	Unsupported
COO → BREP	0.292	0.066	4.426	0.000	Supported
COO → CRS	0.437	0.064	6.703	0.000	Supported
COO → PUIN	−0.092	0.068	1.325	0.186	Unsupported
CRS → BREP	0.122	0.056	2.217	0.027	Supported
CRS → PUIN	0.093	0.052	1.754	0.080	Unsupported

Beta(r): SE = SQRT(1 − r2)/(n − 2); CR = (1 − r)/SE; P-value = TDIST(CR, n − 2, 2).
Source: Calculated by Smartpls software 3.0.

In bootstrapping, resampling methods are used to compute the significance of PLS coefficients. Output of significance levels can be retrieved from bootstrapping option. Table 6 shows the results of hypotheses testing; all the t values above 1.96 are significant at the 0.05 level [30].

Table 6. Standard of model SEM

Standard	Beta	SE	*T*-Value	*p*	Findings
SRMR	0.067	0.005	12.674	0.000	Supported

Source: Calculated by Smartpls software 3.0.

5 Discussion

The results of this study suggested that **BO**, **COO** and **CRS** related to **BREP**; **COO** related to **CRS**; **BREP**, **BO** and **BRA** related to **PUIN**. However, the research found insufficient evidences to suggest the relationship between brand origin on brand awareness, country of origin on brand awareness, country of origin on purchase intension, brand awareness on brand reputation, brand awareness on corporate social responsibility and corporate social responsibility on purchase intension.

After exploring factor analysis test conducted SEM, the remaining hypotheses of the SEM model are hypotheses H1, H2, H5, H7, H9, H11 and H12. Let's go find out the H3, H4, H6, H8, H10 and H13 consider why these hypotheses did not affect relationships of consumers. Opinion was based on the ability to recognize and feel the personal perspective.

Next is the COO, as already mentioned in Literature review, there are a lot of factors that have been studied previously author (abroad) and most have concluded that the impact COO shopping intentions. However, In Vietnam, this factor is removed, so the possibility is not appropriate factors in the environment of Vietnam, or otherwise unrealized COO strongly influenced the purchase intentions of Vietnam's consumer.

And finally the elements of CSR. Can understand CSR-related business activities affect the social, environment, human … that carry humanitarian and high social responsibility. This variable also excluded from the study model, can explain as follows: first of all on the consumer side, they have no need or no need to learn about the social responsibility of import goods.

These are a few personal views about the independent variables with or without the intention to affect consumers shopping question.

5.1 Restrictions

The first drawback to the research question is that this study focuses only on one brand, in one country, and it cannot be generalized at the same time or to different brands in different markets at the same time. Moreover, as with time, the resources available for this research are limited, and it is difficult to perform a larger study, toward higher network generalization. We will try to improve this limitation in the future.

Some other factors may affect the generalizability of the study re the questions in the questionnaire, which were only related to how consumers feel in the Vietnam market, more specifically in the area only Ho Chi Minh city area, focusing largely on the ages between 18 and 30 should not be highly representative. Additionally, the study only collected data through the online survey form; thus, the accuracy of the information is not high, nor was the information gathered thoroughly, opinions of consumer product. The sampling method is convenient sampling, so not highly representative.

5.2 Future Orientation

From the results of the study, as well as realizing the above restrictions, we will attempt a more complete study at a later time.
Several groups proposed solutions to improve these limitations as follows:

- Expand the scope of research; not just focus on one city or one country that will try to expand the brand; add a few more to create a more complete study.
- Expanding the scope of the survey, the number of samples, and data collection methods. Can use additional methods of interviewing, direct surveys, etc., to improve the quality of data collected.

6 Conclusions

The research question was built from the legacy of questions from previous studies, and surveys in developing countries, more specific is Vietnam.

After studying the theoretical basis and the study of the variables in all, a questionnaire was sent to the respondents and focus mainly aged 18–30 in Ho Chi Minh City, Vietnam in the form of online surveys. The results from the data collected is then analyzed Smartpls 3.0.

Acknowledgments. This work was supported in part by Industrial University of Ho Chi Minh City, Vietnam.

Conflicts of Interest. The authors declare that there are no conflicts of interest regarding the publication of this paper.

References

1. Thakor, M.V., Kohli, C.S.: Brand origin: conceptualization and review. J. Consum. Mark. **13** (3), 27–42 (1996)
2. Hamzaoui-Essoussi, L., Merunka, D., Bartikowski, B.: Brand origin and country of manufacture influences on brand equity and the moderating role of brand typicality. J. Bus. Res. **64**(9), 973–978 (2011)
3. Balabanis, G., Diamantopoulos, A.: Brand origin identification by consumers: a classification perspective. J. Int. Mark. **16**(1), 39–71 (2008)
4. Batra, R.: Effects of brand local and nonlocal origin on consumer attitudes in developing countries. J. Consum. Psychol. **9**(2), 83–95 (2000)
5. Yu, C.-C., Lin, P.-J., Chen, C.-S.: How brand image, country of origin, and self-congruity influence internet users' purchase intention. Soc. Behav. Pers. Int. J. **41**(4), 599–611 (2013)
6. Tulipa, D., Muljani, N.: The country of origin and brand image effect on purchase intention of smartphone in Surabaya – Indonesia. Mediterr. J. Soc. Sci. **6**(5), 64–70 (2015)
7. Wu, C.-S.: A study on consumers' attitude towards brand image, athletes' endorsement, and purchase intention. Int. J. Organ. Innov. **8**(2), (2015)
8. Fishbein, M., Ajzen, I.: Belief, Attitude, Intention, and Behavior: An Introduction to Theory and Researched. Addison-Wesley, Reading (1975)
9. Engel, J., Blackwell, R., Kollat, D.: Consumer Behavior, 3rd edn. Holt, Rinehart and Winston, Inc., New York (1978)
10. Kotler, P., Keller, K.L.: Marketing Management. Pearson (2012)
11. Rao, A.R., Monroe, K.B.: The moderating effect of prior knowledge on cue utilization in product evaluations. J. Consum. Res. **15**(2), 253 (1988)
12. Oetzel, J., Doh, J.P.: MNEs and development: a review and reconceptualization. J. World Bus. **44**(2), 108–120 (2009)
13. Zhuang, G., Phau, I., Wang, X., Zhou, L., Zhou, N.: Asymmetric effects of brand origin confusion. Int. Mark. Rev. **25**(4), 441–457 (2008)
14. Collins, A.M., Loftus, E.F.: A spreading activation theory of semantic processing. Psychol. Rev. **82**, 407–428 (1975)
15. Chao, P.: Impact of country-of-origin dimensions on product quality and design quality perceptions. J. Bus. Res. **42**(1), 1–6 (1998)
16. Han, C.M., Terpstra, V.: Country of origin effects for uni national and bi national. J. Int. Bus. Stud. **19**(2), 235–255 (1988)
17. Mohd Yasin, N., Nasser Noor, M., Mohamad, O.: Does image of country-of-origin matter to brand equity? J. Prod. Brand Manag. **16**(1), 38–48 (2007)

18. Ahame, J.U., Johnson, J.P., Jang, X., Fatt, C.K.: Does country of origin matter for low-involvement products? Int. Mark. Rev. **21**(1), 102–120 (2004)
19. Samiee, S.: Customer evaluation of products in a global market. J. Int. Bus. Stud. **25**(3), 579–604 (1994)
20. Bilkey, W.J., Nes, E.: Country-of-origin effects on product evaluations. J. Int. Bus. Stud. **13** (1), 89–100 (1982)
21. Tangmanee, C., Rawsena, C.: Direct and indirect effects of perceived risk and website reputation on purchase intention: a mediating role of online trust. Int. J. Res. Bus. Soc. Sci. **5** (6), 1–11 (2016)
22. Keller, K.L.: Strategic Brand Management: Building, Measuring, and Managing Brand Equity. Pearson, Upper Saddle River (2013)
23. Aaker, D.A.: Measuring brand equity across products and markets. Calif. Manag. Rev. **38** (3), 102 (1996)
24. Huang, R., Sarigöllü, E.: How brand awareness relates to market outcome, brand equity, and the marketing mix. J. Bus. Res. **65**(1), 92–99 (2012)
25. Carroll, A.B.: A three-dimensional conceptual model of corporate performance. Acad. Manag. Rev. **4**(4), 497–505 (1979)
26. Orgrizek, M.: The effect of corporate social responsibility on the branding of financial services. J. Financ. Serv. Mark. **6**(3), 215–228 (2002)
27. Joyner, B.E., Payne, D., Raiborn, C.A.: Building values, business ethics and corporate social responsibility into the developing organization (2002)
28. Anim, P.A., Cudjoe, A.G.: The influence of CSR awareness on consumer purchase decision of a telecommunication network in ghana (a case of la nkwantanag madina municipality). Int. J. Sci. Technol. Res. **4**(2), 8–16 (2015)
29. Lantos, G.P.: The ethicality of altruistic corporate social responsibility (2001)
30. Khoi, B.H., Van Tuan, N.: Using SmartPLS 3.0 to analyse internet service quality in Vietnam. In: Anh, L.H., Dong, L.S., Kreinovich, V., Thach, N.N. (eds.) Econometrics for Financial Applications. Studies in Computational Intelligence, vol. 760, pp. 430–439. Springer (2018)
31. Nunnally, J.C., Bernstein, I.: The assessment of reliability. Psychom. Theor. **3**(1), 248–292 (1994)
32. Hair, J.F., Black, W.C., Babin, B.J., Anderson, R.E., Tatham, R.L.: Multivariate Data Analysis, vol. 6. Pearson Prentice Hall, Upper Saddle River (2006)
33. Hair Jr., J.F., Hult, G.T.M., Ringle, C., Sarstedt, M.: A Primer on Partial Least Squares Structural Equation Modeling (PLS-SEM). Sage Publications, Thousand Oaks (2016)
34. Wong, K.K.-K.: Partial least squares structural equation modeling (PLS-SEM) techniques using SmartPLS. Mark. Bull. **24**(1), 1–32 (2013)
35. Latan, H., Noonan, R.: Partial Least Squares Path Modeling: Basic Concepts, Methodological Issues and Applications. Springer (2017)
36. Henseler, J., Hubona, G., Ray, P.A.: Using PLS path modeling in new technology research: updated guidelines. Ind. Manag. Data Syst. **116**(1), 2–20 (2016)

The Impact of Foreign Reserves Accumulation on Inflation in Vietnam: An ARDL Bounds Testing Approach

T. K. Phung Nguyen(✉), V. Thuy Nguyen, and T. T. Hang Hoang

Banking University of HCM City, Ho Chi Minh City, Vietnam
{phungntk,thuynv,hanghtt}@buh.edu.vn

Abstract. This paper evaluates the short-term and long-term effects of foreign reserves accumulation on inflation in Vietnam between Q1/2004 and Q1/2017 in the context of a dollarized economy. The paper has been based on Irving Fisher's classic monetary theory and inherited Steiner's model (2009), by approaching the Autoregressive Distributed Lag Bounds Test (ARDL Bounds Test) and using the Error Correction Model (ECM). The results show that when the accumulation of foreign reserves and other macroeconomic variables change, the time for inflation to return to equilibrium in the long run is more than one year. The results of short-term and long-term analysis shows that accumulation of foreign reserves have an impact on inflation in Vietnam.

Keywords: Accumulation of foreign reserves · Inflation
Dollarization · ARDL bounds test

1 Introduction

The world economy has experienced very serious crises such as the financial crisis in East Asia in 1997 or the Global Financial crisis of 2008. The seriousness of crises and the dependence on external financial sector with its relevant conditions have led governments to increase insurance for their own country (Denbee et al. [6]). High foreign reserves can help reduce the effects of the crisis on growth in emerging markets (Moghadam et al. [21]). With that trend, foreign reserves continue to be highly valued in The Global Financial Safety Net (GFSN). According to IMF 2016 [10], the main objectives of the GFSN include: provide insurance for countries against a crisis; supply financing when crises hit and incentivize sound macroeconomic policies. In particular, foreign reserves are an important traditional component of the GFSN. This is the first tool against external liquidity shocks of each country. On the other hand, foreign reserves are a symbol of financial health which can help developing countries and emerging economies to penetrate the international market (Drummond et al. [7]; Nowak et al. [25]). In Vietnam, foreign exchange reserves have tended to increase in recent years (Fig. 1).

V. Kreinovich et al. (Eds.): ECONVN 2019, SCI 809, pp. 765–778, 2019.
https://doi.org/10.1007/978-3-030-04200-4_54

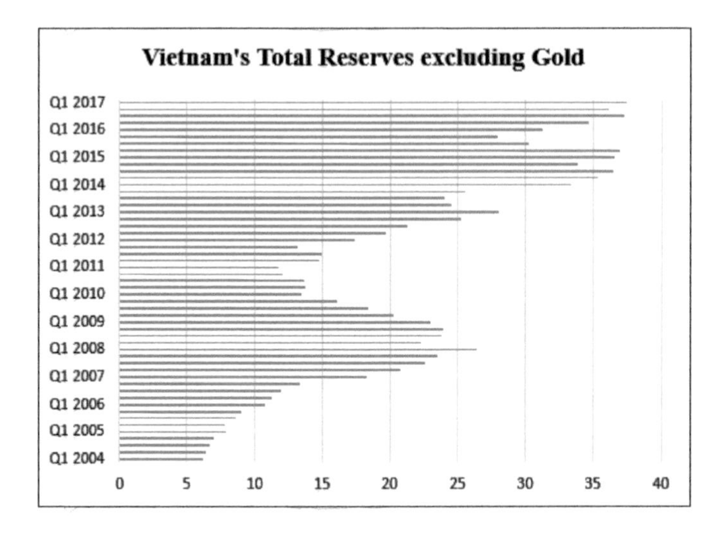

Fig. 1. Volatility of Vietnam's total foreign reserves from Q1/2004 to Q1/2017 Source: IFS (2018)

However, the accumulation of foreign reserves can increase the money base and the expansion of money supply if the sterilization of central bank is not sufficiently, lead to increased inflation of the economy (Heller [14]; Stenier [31]; etc.). And inflation is one of the macroeconomic indicators that need to be controlled and managed. So the central banks have to meet many difficulties and challenges when accumulating foreign reserves. They have to solve the problem of how to increase foreign reserves without increasing inflation. Thus, studying the impact of accumulating reserves on inflation is very necessary for Vietnam in the process of international economic integration.

There were many papers analyzing this relationship for different countries and group of countries, such as Heller [14], Khan [16], Lin and Wang [17], Elhiraika and Ndikumana [8], Steiner [30], Krušković and Maričić [3], Chaudhry et al. [4], Chen and Huang [5], Zhou et al. [33], Trinh [27], etc. However, no papers have investigated the characteristics of each country's economy, which characterizes the Vietnamese economy as a dollarized economy. Therefore, this study focuses on assessing the impact of accumulating reserves on inflation in a dollarized economy.

2 Theoretical Framework and Empirical Evidence

2.1 Theoretical Framework

The increase in foreign reserves leads to a change in the money supply of a country, and the increase in money supply affects the country's inflation. This mechanism is explained in two stages: The first one is the impact of accumulating foreign reserves on money supply. In order to investigate the impact of foreign reserves accumulation on money supply, we first examine the relationship between indicators on the central bank's balance sheet (Table 1).

Table 1. The central bank's balance sheet

Net Foreign Assets – NFA
Foreign Assets Foreign Liabilities
Net Domestic Assets – NDA
Net domestic credits + *Net Claims on Government* + *Claims on Other Depository Corporations* Net other Items
Monetary Bases
Currency in circulation Deposits of commercial banks

Source: Author's synthesis

In the central bank's balance sheet, the indicators are calculated as follows:

$$\text{Net Foreign Assets} = \text{Foreign Assets} - \text{Foreign Liabilities} \quad (1)$$

$$\text{Net Domestic Assets} = \text{Net domestic credits} + \text{Net other Items} \quad (2)$$

$$\text{Monetary Bases} = \text{Currency in circulation} + \text{Deposits of Commercial banks} \quad (3)$$

On the other hand, Monetary Bases can be expressed as:

$$\text{Monetary Bases (MB)} = \text{Net Foreign Assets (NFA)} + \text{Net Domestic Assets (NDA)} \quad (4)$$

From this formula, we can see that: When the central bank accumulates foreign reserves, Net Foreign Assets increases by an amount of ΔNFA, and so that, Monetary Base increases by an amount of ΔMB. On the other hand, money supply depends on two factors: the money multiplier (mm) and the Money Base (MB) (Mishkin) [20].

$$\text{Ms} = \text{mm.MB}. \quad (5)$$

where mm is the money multiplier which tells us what multiple of the monetary base is transformed into the money supply.

Thus, when the Monetary Base MB increases, the money supply Ms also increases. Confirming the positive relationship between foreign reserve accumulation and money supply is demonstrated in empirical studies by Heller [14], Khan [16], Steiner [30], Zhou et al. [33].

The second one is the impact of money supply on inflation. The impact of money supply on inflation is first explained by Irving Fisher's classic monetary theory. Fisher [9] examines the link between the money supply (Ms) and the total amount of spending on final goods and services produced in the economy (P*Y), where P is the price level and Y is aggregate output (income). This relationship is shown by the targets of velocity of money ($V = P * Y/M$). So, Eqs. (6) and (7) show the relationship between income and money amount and velocity of money.

$$\mathrm{Ms} * \mathrm{V} = \mathrm{P} * \mathrm{Y} \quad (6)$$

$$\text{It follows: } \mathrm{P} = \mathrm{Ms} * \mathrm{V}/\mathrm{Y} \quad (7)$$

In a short time, V is quite constant. In addition, wages and prices are fully flexible, so that Y is usually kept at the full employment level, so that Y can be considered not change in a short time. As a result, price movements are only the result of changes in the amount of money. When the Ms increases, P will rise and inflation will increase. In addition, there have been many empirical studies confirming the strong impact of money supply on inflation, such as McCandless and Weber [18], Nassar [22], Hossain [15], Nguyen [23] etc. In summary, the impact of foreign reserves accumulation on inflation can be summarized in the following diagram:

Foreign Reserves ↑ ⇨ Money Base ↑⇨ Money Supply ↑⇨ Inflation ↑

2.2 Empirical evidences

Empirical studies of the effect of foreign reserves accumulation on inflation are divided into three main approaches: worldwide, groups of nations or each nation approach.

With regard to the worldwide approach, the first paper exploring the relationship between foreign reserves accumulation and inflation is due to Heller [14]. The author uses data from 1951 to 1974 of the 126 member countries of the IMF and Switzerland. The result shows that the global reserves has a significant impact on the total money supply in the world. On the other hand, when directly estimating the global reserve to global consumer prices, the author has found a significant lagged link between these two important variables. Similarly, the accumulation of foreign reserves affects inflation with varying degrees, including the representative studies of Khan [16], Rabin and Pratt [28], Heller [14], Steiner [30] and Steiner [31].

Following the national groups approach, there are some typical studies such as Lin and Wang [17], Elhiraika and Ndikumana [8] and Krušković and Maričić [3]. Lin and Wang [17] examines the foreign exchange reserves and inflation in five East Asian countries. The results show that the relationship between exchange reserves and inflation is positive in Japan and negative in Korea and Taiwan. Next is the study by Elhiraika and Ndikumana [8]. The study used panel date form 21 countries in Africa (from 1979 to 2005) to explore the origins, dynamics and economic impact of foreign reserves accumulation. As a result, accumulation of foreign reserves has no significant impact on inflation, but leads to higher prices in the long run. Krušković and Maričić [3] analyzes the impact of foreign exchange reserves on economic growth in emerging economies including Brazil, China and Russia for the period 1993–2012. Using the ONK method and the Fixed Effect Model for quantitative analysis, the authors conclude that accumulation of foreign exchange reserves does not lead to inflation if the rate of accumulation of foreign reserves does not exceed the rate of economic growth.

As the paper concentrates on each nation approach, some relevant studies are reviewed here. Chaudhry et al. [4] analyze the relationship between foreign exchange reserves and inflation in Pakistan between 1960 and 2007. The results show that foreign exchange reserves are inversely related to inflation. Chen and Huang [5] uses a nonparametric model to analyze the transmission mechanism for accumulating foreign reserves against inflation in China. Analytical variables include reserves, money supply, consumer price index, nominal interest rates and GDP from January 1993 to March 2008. The authors concludes that the acceleration in foreign exchange reserves lead to an increase in the money supply, which in turn lead to an increase in inflation. Zhou et al. [33] use monthly data from January 2008 to December 2011 on foreign reserves and consumer price index to build a VAR model to explore the impact of foreign exchange reserves growth on the CPI in China. Granger causality tests show that foreign reserves are a cause for CPI increase. The foreign exchange reserves increases the CPI by 20% with a lag of one to eight months. In Vietnam, Trinh [27] uses the VAR model to measure the impact of accumulation of foreign reserves on inflation between Q1/2004 and Q1/2017. The results of Generalized Impulse Respond Function indicate that foreign reserves account for an increase inflation, start from the third quarter and reach a new equilibrium from the 7th quarter at 1.1% level.

In general, no matter what the underlying analysis, most studies show that accumulation of foreign reserves affects inflation.

3 Research Methodology

3.1 Conceptual Model

The model was originally derived from Steiner [30]. Steiner's model is based on Irving Fisher's classic monetary theory in conjunction with the theory of money supply and money multiplier. From Fisher's exchange equation:

$$\mathrm{Ms} * \mathrm{V} = \mathrm{P} * \mathrm{Y} \tag{8}$$

After taking natural logarithms and differentiating with respect to time this can be expressed in rates of change as:

$$\mathrm{d(lnMs) + d(lnV) = d(lnP) + d(lnY)}$$

It follows:

$$\frac{d(Ms)}{Ms} + \frac{d(V)}{V} = \frac{d(P)}{P} + \frac{d(Y)}{Y} \tag{9}$$

Where d is the differential operator.

In addition, we have:

$$Ms = mm.MB \text{ and } MB = NDA + NFA$$

It follows:

$$\frac{d(Ms)}{Ms} = \frac{d(mm)}{mm} + \frac{NFA}{MB}.\frac{d(NFA)}{FA} + \frac{NDA}{MB}.\frac{d(NDA)}{NDA} \tag{10}$$

Replace (10) into (9), we get:

$$\frac{d(P)}{P} = \frac{NFA}{MB}.\frac{d(NFA)}{NFA} + \frac{NDA}{MB}.\frac{d(NDA)}{NDA} + \frac{d(mm)}{mm} + \frac{d(V)}{V} - \frac{d(Y)}{Y} \tag{11}$$

Equation (11) shows the relationship between accumulation of foreign reserves and inflation. If the central bank accumulates foreign reserves and fully sterilizes the effects on the monetary base, the price level is unaffected. If, however, the central bank does not sterilize, the increase in foreign reserves causes the NFA to change $\frac{NFA}{MB}.\frac{d(NFA)}{NFA}$ and directly translates into an increase of the price level.

In addition, to consider the effect of accumulating foreign reserves on inflation in the context of a dollarized economy, the authors adds dollarization to the research model. Dollarization is the use of any other foreign currency in the domestic economy (Reinhart et al. [29]). Dollarization is still a common phenomenon in developing countries, including Vietnam (Hong [24]). The State Bank of Vietnam (SBV) SBV has taken many measures to prevent dollarization. This is a special feature of the structure of the Vietnamese economy, affecting the monetary management of the SBV in many aspects (Goujon) [12]. Considering in relation to inflation, studies on the impact of dollarization on inflation have contrasted findings. While Bahmani and Domac [1], Yeyati [32] show an increase in inflation, Gruben and Mcleod [13] and Berg et al. [2] find a decline in inflation as a result of dollarization. According to Mengesha and Holmes [19], this depends on the purchasing power. Normally, inflation weakens the purchasing power of the domestic currency. As a result, individuals tend to trade weaker currencies for a stronger currencies. If the exchange leads to fully dollarization, companies don't face currency mismatches. Because they can earn foreign currencies and pay off in foreign currencies. Moreover, the purchasing power of the new currency is strong. Therefore, inflation will fall in the economy. However, if currency swaps do not lead to fully dollarization, inflation tends to increase with the increase in dollarization. Thus, theoretical model is as follows:

$$P = f(NFA, NDA, mm, V, Y, DL) \tag{12}$$

Where DL is the rate of dollarization of the economy.

3.2 Description of Variables and Data Sources

We use data of Vietnam from the first quarter of 2004 to the first quarter of 2017. The data was collected in 2004 because this is the time before Vietnam's accession to the WTO. The economy has begun to change in order to prepare for the international economic integration trend, including the change of foreign reserves. Data sources are mainly from IFS 2018 [11] and Thomson Reuter Data Stream. The calculations of variables and sources of data are shown in Table 2.

Table 2. Variables and data sources

Variable	Symbol	Calculating method	Sources
Adjusted NFA (1)	NFA_t^*	$NFA_t^* = \frac{NFA_t - NFA_{t-1}\frac{e_t - e_{t-1}}{e_{t-1}}}{GDPn_t}$	IFS 2018
Adjusted NDA (2)	NDA_t^*	$NDA_t^* = (MB_t/GDPn_{t)} - NFA_t^*$	IFS 2018
Money multiplier	mm_t	M_{2t}/MB_t	IFS 2018
Inflation (3)	CPI_t	CPI is seasonally adjusted by Cenxus X12	IFS 2018
Output gap (4)	Y_t	$GDPr_t$ -$GDPp_t$	Thomson routers
Velocity of money	V_t	$GDPn_t/M_{2t}$	
Dollarization	DL_t	FD_t/M_{2t}	IFS 2018
		FD_t: Foreign Deposits	
		M_{2t}: Money Supply	

(1) We use the NFA adjusted for subtracting the increase in NFA due to exchange rate fluctuations. As the exchange rate change, the NFA value in VND as the foreign exchange difference is recorded at the end of the accounting period, this change increases the accumulation of foreign exchange reserves. However, it is just the value on the book rather than the actual increased value of foreign exchange reserves. Accordingly, NFA adjustment has the formula as above. Where: e_t and e_{t-1} are the VND/USD rates at the end of t and t-1 respectively; GDP is nominal GDP.
(2) Because the NDA is calculated in accordance with the NFA, the exchange rate difference is also adjusted as follows: $NDA_t^* = (MB_t/GDP_{nt}) - NFA_t^*$. Where: MB_t is Money Base
(3) The price leve P is representative by Inflation CPI.
(4) We use out put gap as the change of aggregate output of the economy. Output gap is calculated by the difference between the real output and the potential output. Where $GDPr_t$ is the real GDP; $GDPp_t$: Potential GDP is calculated using the Hodrick-Prescott filter with a smoothing factor of 1600 in the 9.0 Eviews software.

3.3 Estimation Method

The study uses the ARDL Bounds Test model developed by Pesaran et al. [26] to examine cointegration of variables. We then use the ECM error correction model to determine the rate of adjustment in the short term to return to the long-run equilibrium of inflation. Next, we assess the short-term impact and estimate the long-term effects of foreign reserves accumulation on inflation. We approach this model as it is an appropriate model for evaluating cointegration of variables in the small sample case. The ECM equation is as follows:

$$\Delta CPI_t = \alpha_0 + \lambda EC_{t-1} + \sum_{j=1}^{q0-1} \alpha_j \Delta CPI_{t-j} + \sum_{j=0}^{q1-1} \beta_1 \Delta NFA^*_{t-j} + \sum_{j=0}^{q2-1} \gamma_j \Delta NDA^*_{t-j} + \sum_{j=0}^{q3-1} \delta_j \Delta mm_{t-j} + \sum_{j=0}^{q4-1} \varphi_j \Delta Y_{t-j} + \sum_{j=0}^{q5-1} \mu_j \Delta V_{t-j} + \sum_{j=0}^{q6-1} \rho_j \Delta DL_{t-j} + \varepsilon_t \quad (13)$$

Where λ is the speed of adjustment (the short-term adjustment rate of the CPI to return to the long-run equilibrium when independent variables change). EC_{t-1} is the error in the CPI_{t-1} regression under independent variables. EC_{t-1} is defined as follows:

$$EC_{t-1} = CPI_{t-1} - \theta_0 - \theta_1 NFA^*_{t-1} - \theta_2 NDA^*_{t-1} - \theta_3 mm_{t-1} - \theta_4 Y_{t-1} - \theta_5 V_{t-1} - \theta_6 DL_{t-1} \quad (14)$$

Where: θ_0 is the intercept of the long –term equation

$\theta_1, \theta_2, \theta_3, \theta_4, \theta_5, \theta_6$ are the regression coefficients of the long-term equation.

Substituting Eq. (14) into Eq. (13), we get the ECM equation as follows:

$$\Delta CPI_t = \alpha_0 + \lambda CPI_{t-1} - \lambda\theta_0 - \theta_1 \lambda NFA^*_{t-1} - \theta_2 \lambda NDA^*_{t-1} - \theta_3 \lambda mm_{t-1} - \theta_4 \lambda \Delta Y_{t-j} - \theta_5 \lambda V_{t-1} - \theta_6 \lambda DL_{t-1} + \sum_{j=1}^{q0-1} \alpha_j \Delta CPI_{t-j} - \sum_{j=0}^{q1-1} \beta_1 \Delta NFA^*_{t-j} + \sum_{j=0}^{q2-1} \gamma_j \Delta NDA^*_{t-j} + \sum_{j=0}^{q3-1} \delta_j \Delta mm_{t-j} + \sum_{j=0}^{q4-1} \varphi_j \Delta Y_{t-j} + \sum_{j=0}^{q5-1} \mu_j \Delta V_{t-j} + \sum_{j=0}^{q6-1} \rho_j \Delta DL_{t-j} + \varepsilon_t \quad (15)$$

Equation (15) is the estimation equation for the cointegration test of variables and is the basis for determining the coefficients of the long-term equation. Long-term equation are defined as follows:

$$CPI_t = \theta_0 + \theta_1 NFA^*_t + \theta_2 NDA^*_t + \theta_3 mm_t + \theta_4 Y_t + \theta_5 V_t + \theta_6 DL_t + u_t \quad (16)$$

4 Estimation Results

4.1 Unit Roots Tests

According to Pesaran et al. [26], The ARDL Bounds test is based on the assumption that the variables are I(0) or I(1). So we first test the stationarity of all variables. Variables in the model were tested for stationarity with two methods: Augmented Dickey - Fuller (ADF) and Phillips - Perron (PP). The results are shown in Table 3.

The results of the stationarity test show that NFA*t, NDA*t are stationary at level but the remaining variables are stationary at first difference with a significance of 10%. Thus, the research data are eligible to use the ARDL Bound Test model.

Table 3. ADF and PP unit root tests of variables

Variable	ADF Test t-statistic	PP Test t-statistic	Variable	ADF Test t-statistic	PP Test t-statistic
CPIt	−0.57	−0.45	DL	−0.71	−0.28
ΔCPI_t	−4.02(***)	−3.04(***)	ΔDL	−6.66(***)	−13.95(***)
mm_t	−0.79	−0.97	V	−2.19	−6.85(***)
Δmm_t	−8.17(***)	−11.96(***)	ΔV	−2.88(*)	−31.76(***)
NDA^*_t	−4.03(***)	−4.21(***)	Yt	−3.24(**)	−2.56
NFA^*_t	−3.76(***)	−7.5(***)	ΔYt	−3.31(**)	−7.45(***)

Note: *Critical values of 1%, 5% and 10%, respectively, are −3.53; −2.90 and −2.59*
Δ is the first difference

4.2 ARDL Bounds Tests for Cointegration

Estimating the ARDL model with 9 Eviews software, the results show that the maximum lag order of variables in the Akaike information criteria (AIC) is ARDL (2, 1, 2, 3, 4, 1, 2). We conducted the Bound Test which results as follows (Table 4):

Table 4. Results from bound tests

Rank	F statistic	Critical value bounds							
K 6	9.42			2.5%		5%		10%	
		I(0)	I(1)	I(0)	I(1)	I(0)	I(1)	I(0)	I(1)
		2.88	3.99	2.55	3.61	2.27	3.28	1.99	2.94

Source: Authors' calculations

Thus, statistical value F is greater than the boundary value at significant levels from 1% to 10%. This proves that there is a long-term cointegration relationship of variables in the model.

To determine the reliability of the model, we continued the diagnostic tests for the ECM Eq. (15) including: Heteroskedasticity Test (Breusch-Pagan-Godfrey), Serial Correlation LM Test, Histogram Normality Test, CUSUM Test and CUSUM of Square Test. The results are shown in Table 5 and Fig. 2.

The results show that the residuals of the model are no autocorrelation, no heteroscedasticity, and normally distributed. In addition, the plot of the CUSUM and CUSUMSQ statistic fall inside the critical bands of the 5% confidence interval of parameter stability. This proves that the model is reliable and stable.

After testing diagnostic conditions, we continue to estimate the long run coefficients and the speed of adjustment coefficient by the difference Eq. (15). The results are show in Table 6:

The speed of adjustment coefficient $EC_{t-1} = -0.22$ shows that when inflation is out of equilibrium, the negative adjustment coefficient brings inflation to the

Table 5. Results of diagnostic tests

No.	Diagnostic tests	Probabilitiy
1	Heteroskedasticity Test: Breusch-Pagan-Godfrey	Prob(F21,27) = 0.87
2	Breusch-Godfrey Serial Correlation LM Test	Prob(F4.23) = 0,11
3	Histogram– Normality Test	Jarque Bera = 0.64
		Prob = 0.72

Source: Authors' calculations

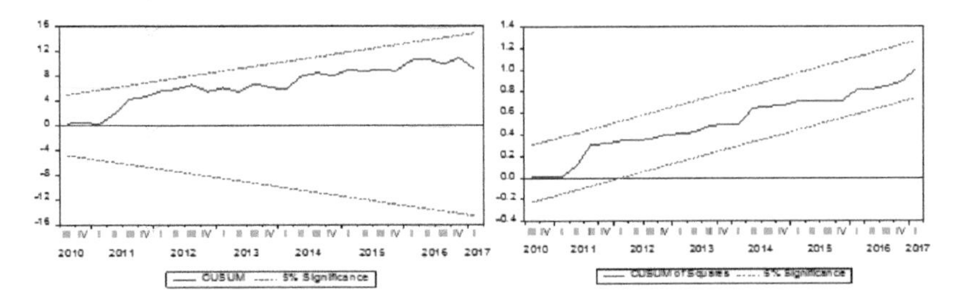

Fig. 2. Results of CUSUM test and CUSUM of squares test ***Source: Authors' calculations***

Table 6. Estimated long run coefficients and speed of adjustment coefficient

Variable	Coefficient	Standard error	t statistic
NFA^{*}_{t}	0.39(***)	0.125735	3.10
NDA^{*}_{t}	0.67(***)	0.178620	3.75
mm	0.18(***)	0.035651	5.09
GAP_{t}	0.34(*)	0.190366	1.78
V_{t}	0.39(***)	0.137086	2.86
DL_{t}	−0.42(***)	0.062725	−6.84
EC_{t-1}	**−0.22 (***)**	**0.023**	**−9.74**

Source: Authors' calculations

long-run equilibrium with a correction rate of 22% and takes time as $1/0.22 = 4.5$ times (over a year) to support equilibrium in the absence of other factors. The results also show that in the long term, NFA^{*}_{t}, NDA^{*}_{t}, mm_{t}, Y_{t} and V_{t} have a positive significant impact on CPI_{t} and DL_{t} is negatively signed.

In addition, to evaluate the short-term effects of accumulation of foreign reserves and dollarization to inflation, we continue to test the Wald Test of the coefficients of NFA^{*} and DL variables in the difference Eq. (15), the results are as follows (Table 7):

Thus, the results show that the null hypothesis H_0, the coefficients of the variables NFA^{*} and DL in the difference Eq. (15) are zero, are rejected. This shows that the coefficients are non-zero. Therefore, in the short run, both NFA^{*} and DL have an impact on CPI.

Table 7. Wald test results for ECM equations

Variable	Wald test		
	t- statistic	F-statistic	Prob
NFA_t^*	4.94	-	0.000
DL	-	25.42	0.000

Source: Authors' calculations

4.3 Robustness Test

To test the stability of the model and the estimated results, we re-estimate the model from the first quarter of 2005 to the first quarter of 2017. The results show that the variables in the model still have cointegrations after the Bounds Test. Estimated results for long-term coefficients shows that the NFA is positive for the CPI (coefficient of 0.39) and DL negative for the CPI (coefficient of −0.42). The EC_{t-1} adjustment coefficient is −0.225 and is statistically significant. We continue to make Wald test to determine the short-term effects of NFA and DL on CPI. The results reject the null hypothesis is that these coefficients are zero. Thus, like the original study, NFA and DL have short-term effects on CPI.

5 Discussion of Research Findings

The results of the study show that accumulation of foreign reserves has the same effect on inflation in the long term, meaning that the accumulation of foreign reserves will increase inflation. This result is similar to the conclusions of empirical studies in other countries and in the world, such as Heller [14], Steiner [30], Lin and Wang [17], Chen and Huang [5]. In Vietnam, from 2000 until now, money supply has always been one of the reasons mentioned to increase inflation. Particularly in 2007, when the SBV accumulates a large amount of foreign reserves but does not withdraw VND from circulation, the money supply in the economy increased, causing inflation in 2008 to 23%. In addition, the results are in contrast to Chaudhry et al. [4], with the conclusion that foreign exchange reserves has a negative impact on inflation in Pakistan. Because the case of Pakistan is different from developing countries, including Vietnam. According to Chaudhry and et al. [4], developing countries have higher incomes and more elastic imports. Pakistan's imports are based on food, crude oil, agricultural materials, machinery and medicines, etc and all imports are more or less based on foreign reserves. The decline in foreign exchange reserves in turn reduced the immediate importation of industrial and agricultural raw materials and created a shock that raises prices.

In terms of the dollarization and inflationary implications, the long-term coefficient estimates show that dollarization is counteracting inflation, or in other words, when dollarization is decline, the inflation increases. Dollarization has always been a concern of the SBV and the SBV has taken many measures to combat the dollarization of the economy. From the first quarter of 2004 to the first quarter of 2017, after 13 years, the rate of dollarization has fallen from 24% to 0.9%. However, the declining dollarization rate has fueled inflation (Fig. 3). Compared with the base year 2010, from 2004 up to now, the consumer price index has always tended to increase.

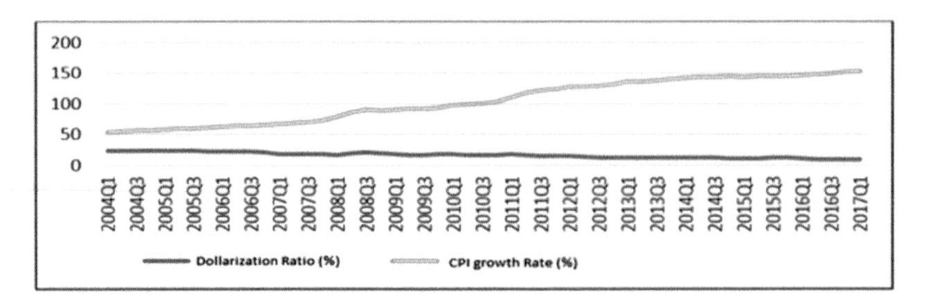

Fig. 3. Dollarization and consumer price index from Q1/2004 to Q1/2017 ***Source: IMF (2018) and Authors' calculations***

As the dollar decreases, the trend of people holding foreign currencies will decrease, so they sell foreign currencies to commercial banks, moving from the form of foreign currency reserves to VND (Viet Nam Dong). The commercial banks resell foreign currency to the SBV. Thus, the accumulation of foreign reserves by the SBV will increase. Foreign reserves accumulation increase causing inflation to increase due to the same effect of accumulation of foreign reserves on inflation.

6 Conclusions and policy implications

Studying the effect of accumulating foreign reserves on inflation in Vietnam in the context of a dollarized economy shows that in the long term, the accumulation of foreign reserves causes inflation and the dollarization have the opposite effect on inflation. The results have reflected the sterilization of the SBV is not effective when they intervene in the foreign exchange market. In the next time, along with the increase in foreign reserves and the implementation of anti-dollarization measures, the SBV should focus on sterilization policy. As a result, the SBV will limit the spillover effects of foreign reserves accumulation to make stabilization and development of economic.

References

1. Bahmani-Oskooee, M., Domac, I.: On the link between dollarisation and inflation: evidence from Turkey. Comp. Econ. Stud. **45**(3), 306–328 (2003)
2. Berg, A., Borensztein, E., Mauro, P.: Monetary regime options for Latin America. Financ. Develop. **40**(3), 24–27 (2003)

3. Krušković, B.D., Maričić, T.: Empirical Analysis of the impact of foreign exchange reserves to economic growth in emerging economics. Appl. Econ. Financ. **2**(1), 102–109 (2015)
4. Chaudhry, I.S., Akhtar, M.H., Mahmood, K., Faridi, M.Z.: Foreign exchange reserves and inflation in Pakistan: evidence from ARDL modelling approach. Int. J. Econ. Financ. **3**(1), 69 (2011)
5. Chen, L., Huang, S.: Transmission effects of foreign exchange reserves on price level: evidence from China. Econ. Lett. **117**(3), 870–873 (2012)
6. Denbee, E., Jung, C., Paternó, F.: Stitching together the global financial safety net. Bank of England Financial Stability Paper, (36) (2016)
7. Drummond, P., Mrema, A., Roudet, S., Saito, M.: Foreign Exchange Reserve Adequacy in East African Community Countries. International Monetary Fund (2009)
8. Elhiraika, A., Ndikumana, L.: Reserves accumulation in African countries: sources, motivations, and effects. Economics Department Working Paper Series, 24 (2007)
9. Fisher, I.: Purchasing Power of Money: Its Determination and Relation to Credit Interest and Crises. Rev AM Kelley, New York (1922)
10. IMF: Adequacy of the Global Financial Safety Net, March 2016
11. IFS: International Financial Statistic (2018)
12. Goujon, M.: Fighting inflation in a dollarized economy: the case of Vietnam. J. Comp. Econ. **34**(3), 564–581 (2006)
13. Gruben, W.C., McLeod, D.: Currency competition and inflation convergence. Center for Latin American Economics. Federal Reserve Bank of Dallas. Working Paper, 204 (2004)
14. Heller, H.R.: International reserves and world-wide inflation. IMF Staff Papers **23**(1), 61–87 (1979)
15. Hossain, A.A.: Monetary targeting for price stability in Bangladesh: how stable is its money demand function and the linkage between money supply growth and inflation? J. Asian Econ. **21**(6), 564–578 (2010)
16. Khan, M.S.: Inflation and international reserves: a time-series analysis. Staff Papers **26**(4), 699–724 (1979)
17. Lin, M.Y., Wang, J.S.: Foreign exchange reserves and inflation: an empirical study of five east Asian economies, pp. 1–18. Aletheia University, Taiwan, Chengchi University, Taiwan (2005)
18. McCandless Jr., G.T., Weber, W.E.: Some monetary facts. Federal Reserve Bank of Minneapolis. Q. Rev. Fed. Reserv. Bank Minneap. **19**(3), 2 (1995)
19. Mengesha, L.G., Holmes, M.J.: Does dollarization reduce or produce inflation? J. Econ. Stud. **42**(3), 358–376 (2015)
20. Mishkin, F.S.: The Economics of Money, Banking, and Financial Markets, vol. 422. Pearson education, London (2007)
21. Moghadam, R., Hagan, S., Tweedie, A., Viñals, J., Ostry, J.D.: The Fund's Mandate–Future Financing Role. IMF, 25 March 2010
22. Nassar, K.B.: Money demand and inflation in Madagascar (No. 5-236). International Monetary Fund (2005)
23. Nguyen, B.V.: Effects of fiscal deficit and money M2 supply on inflation: Evidence from selected economies of Asia. (2015). Browser Download This Paper
24. Hong, N.T.: Dollarization and monetary policy management in Vietnam. Retrieved from the State Bank of Vietnam database (2012)
25. Nowak, M.M., Hviding, M.K., Ricci, M.L.A.: Can higher reserves help reduce exchange rate volatility? (No. 4–189). International Monetary Fund (2004)
26. Pesaran, M.H., Shin, Y., Smith, R.J.: Bounds testing approaches to the analysis of level relationships. J. Appl. Econ. **16**(3), 289–326 (2001)

27. Trinh, P.T.T.: Impact of accumlation foreign exchange reserves on inflation : VAR approach. J. Econ. Develop. **26**(4), 46–68 (2015)
28. Rabin, A., Pratt, L.J.: A note on heller's use of regression analysis. Staff Papers (International Monetary Fund) **28**(1), 225–229 (1981)
29. Reinhart, C.M., Rogoff, K.S., Savastano, M.A.: Addicted to dollars (No. w10015). National bureau of economic research (2003)
30. Steiner, A.: Does the Accumulation of International Reserves Spur Inflation? A Panel Data Analysis. University of Osnabrueck, Osnabrueck (2009)
31. Steiner, A.: Does the accumulation of international reserves spur inflation? a reappraisal. North Am. J. Econ. Financ. **41**, 112–132 (2017)
32. Yeyati, E.L.: Financial dollarization: evaluating the consequences. Econ. Policy **21**(45), 62–118 (2006)
33. Zhou, L., Zhang, N., Chen, Q.Y.: Foreign exchange reserves, monetary policy and inflation: an empirical study from China. Adv. Inf. Sci. Serv. Sci. **5**(4), 920 (2013)

The Impact of Oil Shocks on Exchange Rates in Southeast Asian Countries - A Markov-Switching Approach

Oanh T. K. Tran[1], Minh T. H. Le[2], Anh T. P. Hoang[2(✉)], and Dan N. Tran[2]

[1] Banking University of Ho Chi Minh City,
36 Ton That Dam Street, District 1,
Ho Chi Minh City, Vietnam
kimoanhtdnh@gmail.com

[2] University of Economics Ho Chi Minh City,
59C Nguyen Dinh Chieu Street,
Ward 6, District 3, Ho Chi Minh City, Vietnam
{minhtcdn,anhtcdn,dantcdn}@ueh.edu.vn

Abstract. The study aims to investigate the impact of oil shocks on real exchange rates of ten Southeast Asian countries from 2005 to 2017. The results of the linear model show that only five out of ten countries the study has found the impact of the oil price shock on real exchange rates, but there is no evidence that supply shocks and aggregate demand shocks affect exchange rates in all these countries. However, the results of the Markov-switching model reveal that the supply shocks have same side impacts on some oil-exporting countries (Indonesia and Vietnam) and adverse impacts on oil-importing countries (Singapore, Thailand, Cambodia), while aggregate demand shocks only affect Indonesia, Brunei and Laos. It is also noted that oil price shocks have strong and persistent impacts on exchange rates in oil-exporting countries (except Vietnam). Additionally, the study results indicates evidences that support the presence of regime switching in oil-importing countries, namely Singapore and the Philippines.

Keywords: Oil shocks · Exchange rates · Markov-switching model

1 Introduction

Oil the most important source of energy - has always been regarded as an essenttial instrument for global economic development. With enormous marine potentials, of which reserves and crude oil production is regarded large worldwide, 10 developing economies in Southeast Asia (Timor Leste is not included because of lack of data) are step by step closely integrating into the world trade. However, it is unfortunate that these countries mostly export crude oil only and it is noticeable that they have to import most of the oil productsto serve their daily activities, productions and businesses. In the process, in order to boost exports

V. Kreinovich et al. (Eds.): ECONVN 2019, SCI 809, pp. 779–794, 2019.
https://doi.org/10.1007/978-3-030-04200-4_55

while improving trade balance, an important factor to consider is the exchange rates. Many studies on the relations between oil prices and exchange rates of oil-exporting and importing countries, show that flutuation in global oil prices could impact exchange rates through different transition channels; therefore, such fluctuations indirectly affect the economic development of these countries.

Most of these studies were conducted in major oil-exporting and importing countries such as the United States, Canada, and China, and only determined the impacts of oil price shocks on the exchange rates in a unique status. The studies use the linear regression model and only recognize the limited effects of oil price shocks on the exchange rates while the relations between these two variables is reciprocal and changeable at any stage. As a result, such linear models may not reflect all the goals that the authors need to study. Beckmann and Czudaj [7] have applied the Markov-switching model to study the relations between oil prices and exchange rates in different states. For that reason, they are not qualified enough to modelize the oil market as Kilian [25] nor able to isolate the origins of oil price shocks (supply shocks and demand shocks). For the above reasons, in this paper, we shall consider the impacts of oil price fluctuations on the economy by separating oil price fluctuations based on three major shocks including supply shocks, aggregate demand shocks and separate demand shocks. We will then assess the responses of the exchange rates of oil importers and exporters under the impact of supply shocks and demand shocks in the OLS linear regression model and Markov-switching model. The use of this Markov-switching model helps to identify non-linear impacts of the three oil shocks on the exchange rates that might be ignored if a linear regression model is applied only.

2 Literature Review

2.1 Transition Channels of Oil Price Shocks to Exchange Rates

From a theoretical point of view, an oil price shock can be transmitted to a country's exchange rates through two separate channels: trade conditions and asset effect channel. Many research have been conducted to study possible relations between exchange rates and oil prices through these two channels in which trading conditions affect both oil importers and exporters in different ways (Corden and Neary [15], Amano and Van Norden [2,3], Backus and Crucini [5], Chen and Rogoff [12], Cashin et al. [11]). It is understood that trade condition is the ratio between price index of export goods and that of import goods. It is also recognized that trade conditions relates to current accounts and balance of payments. When export commodity prices increase faster than the import's (in case both commodity prices increase), or perhaps both import and export commodity prices decrease, their value export prices decrease less than the import ones, national trade conditions are potentially beneficial for the increase of the sale from export activities. Following this, the nation's currency demand rises together with the local currency appreciation. If the export price increases by a smaller percentage than the import price, the currency value will decrease in relation to that country's business partners. To oil importers, rising oil prices often leads to a deterioration of the trade balance, followed by devaluation of the local currency (Fratzscher et al. [18]). Meanwhile, to oil exporters, a positive trade conditional shock could eventually lead to Dutch disease by pushing up the price of non-tradable commodity and appreciating the real exchange rates (Buetzer et al. [10]).

The role of oil shocks in changing trade conditions as well as their impacts on the exchange rates is of a great interest. Some studies assume the relation between oil prices and exchange rates in both oil importers and oil exporters. From a theoretical standpoint, Krugman [27] and Golub [19] have developed models in which the movement of oil prices generated asset transfer effects and portfolio redistributions, leading to some adjustments of exchange rates to balance the asset market. According to this viewpoint, the rise in oil prices is related to the transfer of assets from oil importers to oil exporters, leading to the underestimation (appreciation) of the exchange rates of the importing countries (exporting countries) through current account imbalances and portfolio redistributions (Rasmussen and Roitman [29], Buetzer et al. [10], Fratzscher et al. [18]). This means that as soon as oil prices rise, the exchange rates of oil exporters are appreciated and that of oil importers are underestimated. In case the importers take a small share of the oil market in comparison with a larger share of the oil exporters, assets transfer from oil importers to oil exporters improve trade balances of oil importers. In the long run, it would have a positive effect on exchange rates for oil importers (Corden [14], Grauwe [21]).

2.2 Previous Studies

Cheng [13] estimated the ECM model among commodity prices, the US dollars, world industrial outputs, federal funds rates, and commodity inventories and found that rising oil prices were associated with a devaluation of the US dollar and the strongest impact could last for several years.

Akram [1] estimated an SVAR model on the quarterly data of the OECD's industrial outputs, real short-term US interest rates, the exchange volumes of real dollar exchange rates and commodity prices (including oil prices) and found that the dollar devaluation was linked to a rise in commodity prices.

Coudert et al. [16] found a long-term volatility between commodity prices and exchange rates at 0.5 for commodity exporters and 0.3 for variation in oil exporters with uncertainty of the direction of causality between oil prices and asset ones (including exchange rates).

Fratzscher et al. [18] used the SVAR model and found causal relations between exchange rates and oil prices in both directions: oil prices increased by 10% leading to real exchange rates of the US dollar decrease 0.28% meanwhile, the US dollar dropped 1%, causing oil prices to rise 0.73%. Interestingly, their categorization of variance suggested that a negative correlation of exchange rates with oil prices was explained by risk shocks (the 2008–2009 global financial crisis and the financialization of the oil market). These results are consistent with the study carried out by Grisse [20], Beckmann and Czudaj [7], who all found that causal relations were in both directions.

In contrast to the above studies, Buetzer et al. [10] used data from 44 emerging countries, and assessed the impact of oil price shocks on exchange rates in two steps. First, they recorded various oil shocks identified by Kilian [25], then analyzed the effects on the nominal and real exchange rates as well as the return on equity. Contrary to their prediction, there was no link evidences between the currency of the oil exporters and the importers were found followed an oil price shock, particularly, an actual oil price increased. However, they noted that a surge in oil demand caused significant upward pressure on the currencies of oil exporters and they tended to counteract by accumulating foreign reserves. Basher et al. [6] expanded Kilian [25] on the crude oil market with other important macroeconomic variables and used less restrictive setup to analyze oil price shocks in the context of emerging markets. Recently, Atems et al. [4] applied the Kilian's method [25] to examine the impacts of oil shocks on exchange rates of the six developed countries. Their linear model showed that after an oil demand shock, the exchange rates fell and all recorded reactions were identical in oil-exporting and importing countries.

In addition, relatively few studies examine how different the effects of oil price shocks occur on oil exporters and oil importers. To oil exporters, Beckmann and Czudaj [7] identified a causal relation between exchange rates and oil prices for Brazil, Canada and Russia while to Mexico and Norway, rising oil prices has been correlated the decrease of their currencies compared with the US dollars. On the contrary, their results did not provide a clear pattern of causal relations for oil importers.

A study by Narayan [28] selected 14 Asian countries, including seven Southeast Asian countries - Vietnam, Cambodia, Indonesia, Myanmar, Malaysia, Singapore, the Philippines and Thailand. The author wondered whether the exchange rates fluctuations of these countries can be predicted by oil prices with a new GLS model based on regression model of time series prediction and these oil prices would serve as an important predictor of exchange rates for Bangladesh, Cambodia, Hong Kong and Vietnam. Major findings revealed that higher oil prices expectedly lead to higher real exchange rates or a depreciation of the Vietnam Dong in the future. The findings for other three countries (Bangladesh, Cambodia, Hong Kong) were totally opposite. Specifically, an increase of oil prices led to high appreciation of the currencies of Cambodia, Hong Kong and Bangladesh in the future. The findings also show that a prediction for the impacts of oil prices on exchange rates fluctuations in other sample countries such as Thailand and Malaysia was unclear and very limited. Huang and Feng [22] studied the effects of three oil shocks on China's economy by using the four-dimensional VAR model. First, empirical results indicated that with an increase of oil import demands, oil price shocks practically resulted in an undervalue of real exchange rates in the long term. It was owing toChina's less dependence on imported oil compared to its trading partners. Second, response models of real exchange rates with shocks structurally occur were in accordance with theoretical predictions. That was, the increase of supply shocks cause weakness while a demand shocks increase led to appreciation of the real exchange rates of China. Third, the analysis of variance revealed that real demand shocks were the most significant and strongest influencing factors on variations of the real exchange rates. Supply-side changes have arisen from stable productivity growth through technological advances and successive restructures of SOEs. Supply shocks were found to represent a greater impact in China than in the case of developing countries in other studies.

3 Research Methodology

3.1 Data Descriptions

To conduct the study, monthly data is gathered from ten Southeast Asian countries including: Vietnam, Indonesia, Laos, Cambodia, Thailand, Singapore, Philippines, Brunei, Malaysia and Myanmar for the period of 2005–2017 (Timor Leste is not included because of lack of data). Based on export turnover of crude oil, we classify the selected countries into two groups: oil-exporting countries consisting of Brunei, Malaysia, Vietnam and Indonesia and oil-exporting and importers of crude oil including Singapore, Laos, Cambodia, Philippines, Thailand and Myanmar to ensure a precise result and clear classification for each country's currency fluctuations.

The initial data to identify three oil shocks include oil supply, Brent oil prices and global real economic activities. Total oil supply (thousand barrels per day) and oil prices (USD per barrel) are available on the US Energy Information Administration[1]. Oil production figures are chosen from January 2005 to September 2017. Brent oil data are from January/2005 to October/2017. The Global Economic Performance Index variable representing the global aggregate demand is taken from Lutz Kilian's[2] from January 2005 to December 2017. Other items the nominal exchange rates variable (domestic currency per US dollar) and the CPI (for the calculation of the real exchange rates) of the ten Southeast Asian nations and the United States are taken from the IMF data[3].

3.2 Research Model

In this paper, we evaluate the response of exchange rates of importing and exporting countries under the impacts of supply shocks and demand shocks, through the linear regression model as follows:

$$\Delta fx_{i,t} = \beta_{0,i} + \beta_{1,i}\varepsilon^{s}_{i,t} + \beta_{2,i}\varepsilon^{d}_{i,t} + \beta_{3,i}\varepsilon^{p}_{i,t} + \beta_{4,i}\Delta fx_{i,t-1} + u_{i,t} \tag{1}$$

+ $\Delta fx_{i,t}$ is the first difference of the logarithm of the real exchange rates of nation i

[1] http://www.eia.gov/totalenergy/data/monthly/index.cfm.
[2] http://www-personal.umich.edu/~lkilian/paperlinks.html.
[3] http://data.imf.org.

+ ε^s supply shock
+ ε^d aggregate demand shock
+ ε^p demand shock

Apply the method of Kilian [25], the structural VAR model which is used to identify shocks is:

$$A_0 y_t = A(L)\, y_{t-1} + \varepsilon_t$$

where ε_t denotes the vector of serially and mutually uncorrelated structural innovations. We postulate that A_0^{-1} has a recursive structure such that the reduced form errors e_t can be decomposed according to $e_t = A_0^{-1}\varepsilon$:

$$e_t = \begin{pmatrix} e_{1t}^{\Delta prod} \\ e_{2t}^{rea} \\ e_{3i}^{rpo} \end{pmatrix} = \begin{bmatrix} a_{11} & 0 & 0 \\ a_{21} & a_{22} & 0 \\ a_{31} & a_{32} & a_{33} \end{bmatrix} \begin{pmatrix} e_{1t}^{\text{oil supply shock}} \\ e_{2t}^{\text{aggregate demand shock}} \\ e_{3t}^{\text{oil-specific demand shock}} \end{pmatrix}$$

The restrictions on A_0^{-1} may be motivated as follows: Crude oil supply shocks (referred to as *oil supply shocks* for short) are defined as unpredictable innovations to global oil production. Crude oil supply is assumed not to respond to innovations to the demand for oil within the same month. Innovations to global real economic activity that cannot be explained based on crude oil supply shocks will be referred to as shocks to the global demand for industrial commodities (or *aggregate demand shocks* for short). Finally, innovations to the real price of oil that cannot be explained based on oil supply shocks or aggregate demand shocks by construction will reflect changes in the demand for oil as opposed to changes in the demand for all industrial commodities (referred to as oil-specific demand shocks for short).

To explain the non-linear relations between real exchange rates and oil shocks, a Markov-switching model for the Eq. (1) is defined as follows:

$$\Delta fx_{i,t} = \beta_{0,i,s_t} + \beta_{1,i,s_t}\varepsilon_{i,t}^{s} + \beta_{2,i,s_t}\varepsilon_{i,t}^{d} + \beta_{3,i,s_t}\varepsilon_{i,t}^{p} + \beta_{4,i,s_t}\Delta fx_{i,t-1} + u_{i,t} \quad (2)$$

Hamilton's Markov switching model [23], or the state transition model, is one of the most popular non-linear time series models. This model includes many equations that can describe time series behaviors in different states. It also allows causal relation of time fluctuations across states rather than linear models with continuous and irreducible state variables. The Markov-switching model has an advantage of using estimated information on the probability of varying states in the same state, rather than a linear model which estimates information for a completely separate state.

4 Epirical Results

4.1 Stationary Test

Initially, we determine the stationarity for all oil variables as well as exchange rates fluctuations of selected countries with Unit root test by Dickey and Fuller extension method (ADF). As shown in Table 1, three oil variables, with p = 0.0286 < 0.05, Price stops at the base level. The other two variables, Supply and Demand do not stop at the root level but at the first level of difference.

By a similar method, stationary test's findings with exchange rates fluctuation in ten Southeast Asian countries show that all of variables stop at the base level.

Table 1. Stationary test of oil shocks

	Base string		Level 1 difference	
	p-value	t-statistic	p-value	t-statistic
Price	0.0286	−3.1050		
Demand	0.1884	−2.2547	0.0000	−8.5558
Supply	0.9079	−0.3808	0.0000	−12.1594

Source: calculated by authors

Table 2. Stationary test for exchange rates fluctuation of Southeast Asian countries

	Vietnam	Malaysia	Philippines	Singapore	
p-value	0.0000	0.0092	0.0000	0.0000	
t-statistic	−8.0303	−3.5112	−7.8512	−8.7652	
	Brunei	Laos	Myanmar	Thailand	Indonesia
p-value	0.0000	0.0000	0.0000	0.0000	0.0000
t-statistic	−8.6779	−6.9844	−10.9753	−7.5964	−6.8679

Source: calculated by authors

4.2 Optimal Lag Determination

To determine the optimal lag, we run Vector Autoregression Estimates with the temporary lag 2 and then VAR Lag Order Selection Criteria. As indicated in Table 3, lag 2 is the optimal lag because it is suitable with all criteria of LR, FPE, AIC, SC, HQ.

To check the accuracy and reliability of the optimal lag, we apply the unit circle test and find that lag2 is appropriate because all points are inside the circle.

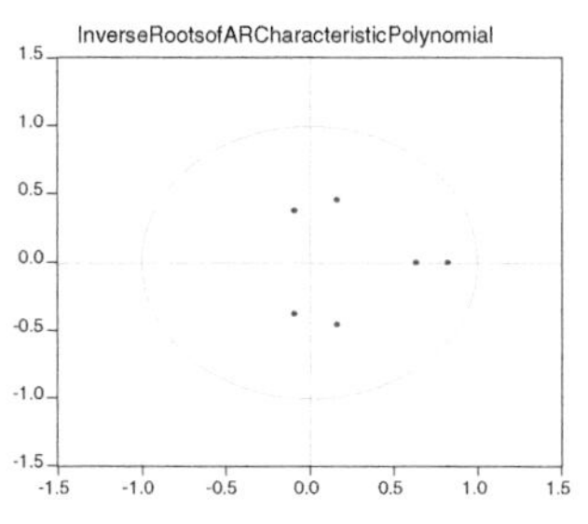

Table 3. Optimal lag determination

Lag	LogL	LR	FPE	AIC	SC	HQ
0	−49.15065	NA	0.000479	0.869177	0.938865	0.897478
1	114.0744	315.5684	3.66e−05	−1.701240	−1.422491	−1.588039
2	146.0219	60.16784*	2.50e−05*	−2.083699*	−1.595888*	−1.885596*
3	148.0124	3.649187	2.81e−05	−1.966873	−1.270000	−1.683870
4	155.0068	12.47337	2.91e−05	−1.933447	−1.027512	−1.565542
5	157.3510	4.063339	3.26e−05	−1.822517	−0.707521	−1.369712
6	165.7834	14.19439	3.30e−05	−1.813056	−0.488997	−1.275350
7	170.1102	7.067205	3.58e−05	−1.735170	−0.202050	−1.112563
8	176.8360	10.64910	3.75e−05	−1.697266	0.044916	−0.989758

Source: calculated by authors

4.3 Determine Three Types of Oil Shocks by SVAR Model

A, B matrix of SVAR is built as described:

A =				
1	0	0		
C(1)	1	0		
C(2)	C(3)	1		
B =				
C(4)	0	0		
0	C(5)	0		
0	0	C(6)		
	Coefficient	Std. Error	z-Statistic	Prob.
C(1)	204.0332	149.3965	1.365716	0.1720
C(2)	1.368145	1.043866	1.310652	0.1900
C(3)	-0.000678	0.000618	-1.097545	0.2724
C(4)	0.006212	0.000391	15.87451	0.0000
C(5)	10.41738	0.656233	15.87451	0.0000
C(6)	0.072256	0.004552	15.87451	0.0000
Log likelihood	139.6737			
Estimated A matrix:				
1.000000	0.000000	0.000000		
204.0332	1.000000	0.000000		
1.368145	-0.000678	1.000000		
Estimated B matrix:				
0.006212	0.000000	0.000000		
0.000000	10.41738	0.000000		
0.000000	0.000000	0.072256		

Following SVAR matrix, we determined three oil shocks by taking the residuals, (Supplyshock) defined by the Supply variable, Aggregate Demandshock, defined by Demand variable and Price shocks or spercific demand shocks, determined by Price variable followed by their stationarity for regression implementation checking. The findings indicate that all three oil shocks stop at the root level with p = 0.0000.

4.4 Linear Regression Results

We consider the impacts of oil price shocks on the exchange rates based on the linear regression model OLS estimating the variable of volatility of the exchange rates regressed on three oil shocks (Supplyshock, Demandshock and Priceshock). At the same time, a lag2 variable of the national exchange rates (LR) is also added as an explanatory variable. The results are presented in Table 4.

Table 4. Linear regression results

Nation	LR	Supplyshock	Priceshock	Demandshock
Vietnam	0.059916	−5.63547	0.686667	0.006997
	−0.5331	−0.7247	−0.6	−0.4451
Indonesia	−0.034045	31.84063	−19.75658***	0.03343
	−0.7018	−0.6468	−0.0011	−0.4214
Malaysia	0.78907***	−24.02548	−8.282366**	0.037988
	0	−0.5621	−0.02	−0.1223
Brunei	0.0056	−28.10039	−3.487744**	0.010808
	−0.9519	−0.16	−0.0457	−0.3687
Singapore	−0.025985	−30.0501	−2.53261	0.016764
	−0.777	−0.1993	−0.2016	−0.229
Thailand	0.105063	−27.23406	−5.44349***	0.003828
	−0.2458	−0.1912	−0.0025	−0.7552
The Philippines	−0.019165	1.621845	−1.883475	−0.009031
	−0.8375	−0.9441	−0.346	−0.5091
Cambodia	0.144788	−21.04878	−3.519725**	0.00732
	−0.1036	−0.2007	−0.0122	−0.4459
Laos	0.208811**	9.832188	−0.161223	0.005288
	−0.0273	−0.4603	−0.8923	−0.507
Myanmar	−0.007156	311.5407	−52.63875	0.139898
	−0.9389	−0.6586	−0.391	0.7408

*Source: calculated by authors; Symbol * * *, * *, * correspond with p-value 1 %, 5% and 10% .*

As indicated in Table 4, upon three oil shocks, all p-values are greater than 0.05 for supply shocks and aggregate demand shocks, which means that no effects of two types of shocks on the exchange rates are found in any of the listed countries. It is emphasized that oil price shocks (real demand shocks) is the only and significant factor that affects real exchange rates and has statistic meanings in Brunei, Malaysia, Indonesia, Cambodia and Thailand. However, one particular point is noted that the oil price shocks is negatively correlated with the real exchange rates in all of the above countries regardless of oil importers (Cambodia, Thailand) or oil exporters (Brunei, Malaysia, Indonesia). These results are in line with the estimations as Atems et al. [4] who found that the responses of real exchange rates to oil price shocks were identical for both oil-exporting and importing countries.

Thus, supply and demand shocks do not have any impacts on the exchange rates; oil price shocks have a negative relation with exchange rates in five countries regarding a linear model's results. Perhaps, this isalso true in practice owing to the non-linear relations between exchange rates and oil shocks, it is not detected by a linear regression frame. Potenitally, linear estimations miss the non-linear relations as well as not fully reflect the possible outcomes among the variables. Consequently, we will consider the results by using another non-linear model, the Markov-switching model.

4.5 Markov-Switching Model's Results

4.5.1 The Impacts of Supply Shocks

The Markov-switching model examines the impacts of oil price shocks on exchange rates in two Regime 1 and Regime 2 states. The results in Table 5 indicate that Vietnam and Indonesia are two oil exporters in which Supply shocks have the same effects on exchange rates. This result is consistent with the theory assuming that an increase of supply shocks would underestimate the national currency of oil exports, thereby increasing the exchange rates of the countries. This is due to the abundant supply following the cheaper price of oil. As a sequence, less income from the sale of oil results in a decrease of currency demands of the less oil-exporting countries. Eventually, this makes the national currency undervalue. On the other hands, for oil importers, a surge in supply shocks will cause the currency of oil-importing countries to appreciate and reduce their exchange rates. This implies a reverse relation between the supply shocks and the exchange rates of oil importers. Such experimental results are also true for oil importers of Laos, Thailand and Singapore.

As an oil exporter, Brunei, however, produces the opposite outcomes. It can be explained by other external factors affecting the exchange rates on which more explicit research should be done. In general, the results of the study provide a lot of evidences of the impacts of supply shocks on exchange rates in Southeast Asian countries consistent with previous theories and studies.

Table 5. Markov-switching model's results

Nation	Status	LR	Supplyshock	Priceshock	Demandshock
Vietnam	Regime 1	−0.062462	−16.5492	0.717856	0.003933
		−0.4858	−0.2475	−0.5382	−0.6103
	Regime 2	0.62029**	102.9491**	4.325632	0.006932
		−0.0179	−0.0368	−0.4463	−0.8733
Indonesia	Regime 1	−0.07832*	30.48757	−1.5639	−0.00412
		−0.0838	−0.3814	−0.6157	−0.8452
	Regime 2	0.000***	644.9465***	−130.3374***	−2.426928***
		0	−0.0006	0	0
Malaysia	Regime 1	0.04617	−26.47214	−5.315928*	0.028168
		−0.634	−0.4373	−0.0677	−0.1292
	Regime 2	0.280002**	26.34872	−14.39756**	−0.025896
		−0.0244	−0.6514	−0.0266	−0.6804
Brunei	Regime 1	−0.059577	2.971984	2.082535	0.034032**
		−0.5606	−0.8897	−0.2656	−0.0116
	Regime 2	−0.134476	−139.6775**	−17.29192***	−0.03456
		−0.6036	−0.0108	−0.0007	−0.2674
Singapore	Regime 1	−0.008917	−41.03659**	−6.979348**	0.021533
		−0.9147	−0.0462	−0.0109	−0.1054
	Regime 2	0.408265	420.5095	35.35748***	0.361051
		−0.6997	−0.2959	−0.0002	−0.1719
Thailand	Regime 1	−0.042998	−25.58755	-6.582604***	0.001912
		−0.6378	−0.1734	−0.0003	−0.8803
	Regime 2	−0.118557	−121.0294***	−5.887484*	0.010558
		−0.4106	−0.0036	−0.0685	−0.6013
The Philippines	Regime 1	−0.826979***	−11.21246	−13.73098*	−0.012845
		−0.009	−0.8304	−0.0606	−0.5752
	Regime 2	0.334538*	−3.293698	2.85528*	0.013053
		−0.071	−0.911	−0.0736	−0.4862
Cambodia	Regime 1	0.012774	−16.47598	−3.002343***	0.003843
		−0.8511	−0.2011	−0.008	−0.6088
	Regime 2	2.904167***	−42.82363	−7.355281*	0.033706
		0	−0.6316	−0.0825	−0.4225
Laos	Regime 1	0.427147**	−110.0857***	−4.203569	0.029433*
		−0.05	−0.0088	−0.2341	−0.0889
	Regime 2	0.094217	6.319454	1.371482	−0.00152
		−0.1953	−0.549	−0.1406	−0.8025
Myanmar	Regime 1	0.197403	−112.8218	−2987.596	−7.720435
		−0.9997	−0.9989	−0.9987	−0.9991
	Regime 2	0.005257	5.867496	−0.696552	0.009242
		−0.1908	−0.8468	0.7923	−0.6116

*Source: calculated by authors; Symbol * * * ,* * ,* correspond with p-value 1 % , 5% and 10% .*

4.5.2 Impact of Two Supply Oil Shocks

- To oil-exporting countries:

Except Vietnam, the other three oil exporters (Brunei, Indonesia, Malaysia) are all affected by oil price shocks. It is found that there is a negative correlation between the two variables in a state (Regime 1) in Brunei and Indonesia only. It is totally opposite to the case of Malaysia in which the stability and certainty of the relationship in both states (Regime 1 and Regime 2) is clearly determined. This is in line with studies by Bénassy-Quéré et al. [8], Kilian et al. [26], Bodenstein et al. [9].

The impacts of aggregate demand shocks on the exchange rates have the same side effects in Brunei. This is contrary to the theory that aggregate demand shocks affect currencies of oil exporters through a change of oil prices and of demands for other goods these countries export. Typically, this leads to an appreciation of the local currency, ie the exchange rates will decrease. However, depending on the export ratio of a country's total exports, the central bank has their incentives to actively cope with the upward pricing pressure by foreign exchange reserves (Buetzer et al. [10]). This effect may reduce the currency appreciation in oil exporters (as predicted by the theory) in response to an increase of an aggregate demand shock. That may be the reason why Brunei reveals opposite outcomes to the theory.

- To oil-importing countries:

There are a variety of evidences for the impacts of demand shocks on exchange rates in oi-importing countries. Not mention only one-dimensional effects, the outcomes clearly indicate a transition of states among the countries.

As shown in Table 5, the Philippines and Singapore are two countries that have the impacts of oil price shocks change in two states. Typically, Singapore and the Philippines are in the first state (Regime1) at which the adverse impacts of oil price shocks on the exchange rates are noticed while in state 2, it is shown that the oil price shocks and the exchange rates have same side relations. This implies that with an increase of oil price shocks, the exchange rates will change their state either appreciation or depreciation. Although it is quite difficult to pinpoint an exact reason for these results, changes can be derived from the economic foundation. Kaminsky [24] argues that if economic growth is related to exchange rates, the differences in business cycles among nations can lead to long-term change. In addition, Evans and Lewis [17] have shown that a fair trade forecast of future exchange rates may explain the change of exchange rates appreciation and depreciation.

Cambodia and Thailand are the two countries that produce similar results of impacts of oil price shocks which are opposite relations in both states. This outcome is consistent with Narayan [28].

Particularly, in the study results, Myanmar's variables do not show statistically significant results with a very high P-value in both linear and non-linear models. This means that oil prices do not seem to have any impacts on exchange rates in Myanmar, which is probably due to the intervention of their government on exchange rates policy. In details, 35 years prior to 2012, Myanmar government set up a fixed regime of exchange rates, anchoring the currency exchange rates into the International Monetary Fund's (IMF) and Special Drawing Rights (SDR), limiting initial capital inflows invested in Myanmar. However, April 2012 is a landmark period for Myanmar's exchange rates fluctuations at which their central bank, CBM, adopted a floating exchange rates regime, offering a daily reference rate, based on the supply and demand situations of the currency market. On the "black market" - unofficial market, an exchange of about 800 Kyat is for 1 US dollar and the US dollar is priced 120 times the official exchange rates. When the floating exchange rates policy is applied, the reference exchange rates that CBM offers will make the Kyat of Myanmar being officially quoted at a rate of 5–6 kyat for 1 USD growing up to the same levelof the black market's rate. For these reasons, from April 2012 onwards, the exchange rates of Myanmar has increased from over 800 Kyat in exchange for 1 USD. Research data chosen from 2005 to 2017 may not be sufficient enough to produce results as suggested by the theory.

All in all, whether under the impacts of demand shocks or supply shocks, the output results are sometimes incompatible with the theory as the exchange rates are clearly influenced by several factors which we shall consider and study more closely.

4.5.3 Estimated Time and Probability of Status Transition

The expected time of state 1 and state 2 is determined by the values Du1 and Du2 in Table 6. The Pij value is the probability converting from state i to state j.

Table 6. Estimated time and probability of status transition

	P11	P12	P21	P222	DU1	DU2
Vietnam	0.921929	0.078071	0.504363	0.495637	12.80886	1.9827
Indonesia	0.982681	0.017319	0.411697	0.588303	57.73961	2.428971
Malaysia	0.982025	0.017975	0.097295	0.902705	55.63395	10.278
Brunei	0.624697	0.375303	1.0000	0.0000	2.664516	1.0000
Singapore	0.929694	0.070306	1.000	0.000	14.22362	1.0000
Thailand	0.924338	0.075662	0.227037	0.772963	13.21672	4.404574
The Philippines	0.563956	0.436044	0.203147	0.796853	2.293348	4.922536
Cambodia	0.969211	0.030789	0.573206	0.426794	32.47936	1.744573
Laos	0.515241	0.484759	0.093	0.907	2.06288	10.75269
Myanmar	0.000204	0.999796	0.008334	0.991666	1.000204	119.9833

The results in Table 6 show that Singapore and the Philippines are the only two countries with status transitions. However, the expected time for two states of these countries is different. In Singapore, status 1 is expected for fourteen months and status 2 for one month while two months for status 1 and four months for status 2 in the Philipines. Furthermore, regarding time differences for two states, the Philippines is expected to be less than that of Singapore. For other countries, the expected time for Indonesia in status 1 is the highest with fifty seven months, the lowest is Myanmar and Laos from one to two months respectively. The results implies that the expected time varies across countries and regions, depending on other external factors.

The Pij state transition probability is considerably high in the majority of the selected countries, especially P11 and P22 values, which may be up to 99% (Myanmar) or 90% -98% in Vietnam, Singapore, Malaysia, etc. At the same time, the probability of transitions from one state to another (P12 and P21) is much smaller in the studied countries.

5 Conclusion

The objective of this paper is to examine the impacts of oil prices and oil price shocks on exchange rates. This is a critically important research topic because an oil shock can affect a country's trade conditions and possibly its competitiveness. The effects of oil shocks on exchange rates vary depending on whether or not a country is an oil net exporterfrom a net importer. By using data from ten Southeast Asian countries during the period of 2005–2017, the results indicate that:

Firstly, the linear regression model shows that only five out of ten countries give statistically significant results of oil price shocks on exchange rates and no evidences of supply shocks and demand shocks on exchange rates are found in all of chosen countries.

Second, the impacts of oil price shocks on the exchange rates is more statistically significant than the aggregate demand shocks. It is noticed that Myanmar is the only studied country that is not affected by oil price shocks. Moreover, aggregate shocks only affects Indonesia, Brunei and Laos; oil price shocks have a strong and persistent impact on exchange rates in oil-exporting countries (except Vietnam). A surge in oil prices has made the currencies of oil exporters highly appreciated, exchange rates fall in a state or repeated in two states under considerations. To oil importers, the Markov-switching model has found a reversal of states in Singapore and the Philippines. An increased oil price shock has reduced exchange rates of both countries in status 1 but increased in status 2. Explanations of it is challenging and one can be derived from objective factors that possibly influences exchange rates. In addition, the same opposite effects of oil price shocks on both Thailand and Cambodia are recorded in both states.

Third, the study finds evidences that supply shocks affect exchange rates of non-linear countries in a single state only. This has been recorded a same side effect in oil-exporting countries (Indonesia and Vietnam) and vice versa in oil-importing countries (Singapore, Thailand and Cambodia). Considered as an oil-exportinng country, however, Brunei's exchange rates fall when a surge in supply shocks occurs. Its cause may need to be explored in more comprehensive research, as exchange rates are in fact impacted by several other external factors.

References

1. Akram, Q.: Commodity prices, interest rates and the dollar. Energy Econ. **31**, 838–851 (2009)
2. Amano, R., Van Norden, S.: Oil prices and the rise and fall of the US real exchange rates. J. Int. Money Financ. **17**, 299–316 (1998)
3. Amano, R., Van Norden, S.: Exchange rates and oil prices. Rev. Int. Econ. **6**, 683–694 (1998)
4. Atems, B., Kapper, D., Lam, E.: Do exchange rates respond asymmetrically to shocks in the crude oil market? Energy Econ. **49**, 227–238 (2015)
5. Backus, D., Crucini, M.: Oil prices and the terms of trade. J. Int. Econ. **50**, 185–213 (2000)
6. Basher, S., Haug, A., Sadorsky, P.: Oil prices, exchange rates and emerging stock markets. Energy Econ. **34**, 227–240 (2012)
7. Beckmann, J., Czudaj, R.: Is there a homogenous causality pattern between oil prices and currencies of oil importers and exporters? Energy Econ. **40**, 665–678 (2013)
8. Bénassy-Quéré, A., Mignon, V., Penot, A.: China and the relationship between the oil price and the dollar. Energy Policy **35**, 5795–5805 (2007)
9. Bodenstein, M., Erceg, C., Guerrieri, L.: Oil shocks and external adjustment. J. Int. Econ. **83**, 168–184 (2011)
10. Buetzer, S., Habib, M., Stracca, L.: Global exchange rates configurations: Do oil shocks matter? European Central Bank, Working Paper Series No 1442 (2012). https://www.ecb.europa.eu/pub/pdf/scpwps/ecbwp1442.pdf. Accessed 01 May 2018
11. Cashin, P., Cespedes, L., Sahay, R.: Commodity currencies and the real exchange rates. J. Develop. Econ. **75**, 239–268 (2004)
12. Chen, Y.-C., Rogoff, K.: Commodity currencies. J. Int. Econ. **60**, 133–160 (2003)
13. Cheng, K.: Dollar depreciation and commodity prices. IMF World Economic Outlook, pp. 72-75 (2008)
14. Corden, W.M.: Booming sector and Dutch disease economics: Survey and consolidation. Oxford Econ. Pap. **35**, 359–380 (1984)
15. Corden, W., Neary, J.: Booming sector and de-industrialization in a small open economy. Econ. J. **92**, 825–848 (1982)
16. Coudert, V., Mignon, V., Penot, A.: Oil price and the dollar. Energy Stud. Rev. **15**(2) (2008). https://halshs.archives-ouvertes.fr/file/index/docid/353404/filename/papierPC7eng.pdf. Accessed 01 May 2018
17. De Evans, M., Lewis, K.: Do long-term swings in the dollar affect estimates of the risk premia? Rev. Financ. Stud. **8**, 709–742 (1995)
18. Fratzscher, M., Schneider, D., Van Robays, I.: Oil prices, exchange rates and asset prices. European Central Bank, Working Paper Series No 1689 (2014). https://www.ecb.europa.eu/pub/pdf/scpwps/ecbwp1689.pdf?7f2c60ad14bac661a7ce6465aa34f78b. Accessed 01 May 2018
19. Golub, S.: Oil Prices and Exchange rates. Econ. J. **93**(371), 576–593 (1983)
20. Grisse, C.: What drives the oil-dollar correlation? Unpublished manuscript (2010). https://www.aeaweb.org/conference/2011/retrieve.php?pdfid=650. Accessed 01 May 2018

21. Grauwe, D.P.: International Money: Post War - Trends and Theories. Oxford University Press Catalogue, Oxford (1989)
22. Huang, Y., Feng, G.U.O.: The role of oil price shocks on China's real exchange rates. China Econ. Rev. **18**(4), 403–416 (2007)
23. Hamilton, J.D.: A new approach to the economic analysis of nonstationary time seriesand the business cycle. Econometrica **57**, 357–384 (1989)
24. Kaminsky, G.: Is there a peso problem? evidence from the dollar/pound exchange rates, 1976–1987. Am. Econ. Rev. **83**, 450–472 (1993)
25. Kilian, L.: Not all oil price shocks are alike: disentangling demand and supply shocks in the crude oil market. Am. Econ. Rev. **99**, 1053–1069 (2009)
26. Kilian, L., Rebucci, A., Spatafora, N.: Oil shocks and external balances. J. Int. Econ. **77**, 181–194 (2009)
27. Krugman, P.: Oil shocks and exchange rates dynamics. In Exchange rates and international macroeconomics. University of Chicago Press (1983). http://www.nber.org/chapters/c11382.pdf. Accessed 01 May 2018
28. Narayan, S.: Foreign exchange markets and oil prices in Asia. J. Asian Econ. **28**, 41–50 (2013)
29. Rasmussen, T., Roitman, A.: Oil shocks in a global perspective: Are they really that bad?. IMF Working Paper WP/11/194 (2011)

Analysis of Herding Behavior Using Bayesian Quantile Regression

Rungrapee Phadkantha[1,2(✉)], Woraphon Yamaka[3], and Songsak Sriboonchitta[3]

[1] Center of Excellence in Econometrics, Chiang Mai University, Chiang Mai, Thailand
rungrapee.ph@gmail.com

[2] International College of Digital Innovation, Chiang Mai University, Chiang Mai, Thailand

[3] Center of Excellence in Econometrics, Faculty of Economics, Chiang Mai University, Chiang Mai, Thailand
woraphon.econ@gmail.com, Songsakecon@gmail.com

Abstract. The purpose of this paper is to analyse the herding behavior of eight stock sectors of the Stock Exchange of Thailand consisting of Banking (BANK), Commerce (COM), Communications (COMUN), Energy and Utilities (ENERG), Food and Beverage (FOOD), Personal Products and Pharmaceuticals (PERS), Petrochemicals and Chemicals (PETRO), and Property Development (PROP), for the period from January of 2014 to May of 2018. Conventionally, the model used for finding the herding behavior is linear regression, which is based on the mean of the distribution. However, the result obtained from the linear mean regression model may not be consistent with the heterogeneity of investor behavior. Thus, in this study, we employ Bayesian quantile regression to empirically estimate the daily stock returns. The innovation of this paper is to examine the data conditional on different quantiles and test the behavioral relation between stock returns and market movements with different quantile distributions. The results show that five out of the eight sectors have herding behavior at quantile 0.25 and the remaining three sectors have no herding behavior.

Keywords: Herding behavior · Quantile regression · Bayesian · SET

1 Introduction

As the stock market usually fluctuates due to the economic instability, uncertainty will be a matter of greater concern in the decision-making of investors. A substantial number of studies in behavioral economics have primarily focused on one key concept, namely, herding behaviour. Herding behaviour is defined as a situation whereby a group of stock market investors intentionally follow the actions of other investors by trading in the same direction over a period of time (e.g. [9,13,22]). This herding behaviour causes the trading price to deviate from

V. Kreinovich et al. (Eds.): ECONVN 2019, SCI 809, pp. 795–805, 2019.
https://doi.org/10.1007/978-3-030-04200-4_56

its fundamental value of the securities and thereby leading to an uncertainty and inefficiency in the market. Normally, individual investors are not sufficiently competent to make a rational investment decision based on information obtained as they, sometimes, need to follow a market leader, who may receive inside information about securities or get better information than other investors. This is to say, there exists an asymmetric information in the market. According to Christie and Huang [8], herding becomes more pronounced during extreme market period as investors follow the actions of the majority investors without relying on their private information during extreme market conditions. The work of Hwang and Salmon [14] mentioned that this herding behavior causes stock prices to move abnormally and contribute to a higher risk in the investment.

In the financial markets herding could mean that investors buy or sell securities regardless of their underlying fundamentals. It happens quite often in the financial markets, in particular stock markets of advanced and emerging markets. Many researches on herding behavior have been conducted for validating the existence or absence of it in specific stock markets. A summary of several results is given as follows: some studies have proven that the herding behavior is more likely to happen in an emerging market when compared to developed market ([1,2,4,6,11,17]). Furthermore, Lao and Singh [17] provided the evidence of the asymmetric effects in herding behavior patterns. They implied that investors tend to herd more intensively either an upward or a downward market. Chen, Demirer and Kutan [10] confirmed that the presence of herding behavior is most likely to occur during the extreme market movements, as investors are willing to follow the market consensus. This finding corresponded to the study of Chiang and Zheng [7] as they have suggested that there occurs an extreme movement persistently during the financial crisis. In addition, the market capitalization of firms also has the influence on herding. It is evident that herd behavior is more pronounced in small stocks than larger capital stocks. According to Shiller [21] herding behavior can push the stock prices away from their fundamental economic levels and possibly cause inefficiencies and speculative bubbles. Even though the motivations behind this herding phenomenon are explained from several perspectives, there is a general conformity concerning its effect in financial markets, referring to substantial movements in asset prices and increasing price volatility.

In herding behavior detection, it is generally done by linear regression which is based on the mean of the conditional distribution of stock return dispersions. However, this model specification is inconsistent with the behavior since the tail information has not been considered [14]. Thus, in this study, we consider using the quantile regression as an innovation tool for examining the data to detect the herding behavior through the relationship between stock return dispersions and aggregate market movements with different quantile distributions ([3,20]). Moreover, we use the Bayesian approach to estimate the parameter in quantile regression. This estimation has attracted increasing attention from many studies ([12,19]). Pastpipatkul, et al. [18] suggested that Bayesian estimation can escape possible misspecification errors and improve efficiency in estimation due to an identification of a stochastic or deterministic trend for the estimation of parameters. In a nutshell, this study introduces the quantile regression method for testing evidence of investors herd behavior in the eight sectors of the Stock Exchange of Thailand. To the best of our knowledge, the detection of herding behavior in the Stock Exchange of Thailand using quantile regression has not been explored in the literature yet and thus this is the first attempt in finding heterogeneity herding strategy in Thai stock market.

The other sections of the paper are as follows. Section 2 discusses the Bayesian quantile regression and strategy of herding behavior. Section 3 shows the data description. The empirical result and data analysis are shown in Sect. 4. Finally, Sect. 5 provides the conclusion.

2 Methodology

2.1 Herding Strategy: Cross Section Absolute Deviation

Chang et al. [6] said that the process of market investment decisions depends on overall market conditions. In normal market, the asset pricing model is reasonable. If the returns of the securities increase, individual investors will buy more securities. On the other hand, if the market is volatile, investors tend to ignore their own information and prefer herding investment. The return of securities usually depends on this behaviour. Chang et al. [6] suggested that the presence of herding behavior in the markets can be detected by cross-sectional absolute deviation (CSAD) which is measured by

$$CSAD_t = \frac{1}{n}\sum_{i=1}^{n} |R_{i,t} - R_{m,t}|, \tag{1}$$

where $R_{i,t}$ is the return of a security i and $R_{m,t}$ is the overall return calculated from the average return of n securities. Then test the herding behavior from the equation

$$CSAD_t = \gamma_0 + \gamma_1 |R_{m,t}| + \gamma_2 R_{m,t}^2 + \varepsilon_t \tag{2}$$

To determine whether there exists herding behavior, we can observe the coefficient γ_2. If γ_2 is negative significance, it indicates that the market has a herding behavior; otherwise herding behavior does not occur. As the market experiences a large price swing, market participants tend to suppress their private information and tend to herd using the information emerging from the consensus of all market constituents. The returns under these conditions tend to converge, causing the return dispersion to either decrease or increase at a decreasing rate.

2.2 Bayesian Quantile Regression

Quantile regression is a statistical method introduced by Koenker and Bassett [15]. Given a set of independent variables, the quantile regression models present the relationship between the independent variables at the specific quantiles (also called percentiles) of the dependent variable. The parameter estimates can change according to the specified quantile level. It is widely used in cases where a study seeks to estimate the different percentiles of a population of interest. In other words, it provides a detailed characterization of the data by estimating the differential effects of covariates on the entire conditional distribution of the response variable. Thus, it can be considered as a natural analogue in the regression analysis for providing a thorough and robust data analysis. For a comprehensive explanation of this methodology, see Buchinsky [5]. A simple approach of this model is suggested in Koenker and Bassett [15]. Consider the following quantile regression model:

$$y_t = \mu(x_t) + \varepsilon_t, \tag{3}$$

where $\mu(x_t)$ may be thought of as the conditional mean of y_t given the vector of regressors x_t and ε_t is the error term with mean zero and constant variance. It is not necessary to specify the distribution of the error term as it is allowed to take any form. Typically,

$$\mu(x_t) = x'_t\gamma \tag{4}$$

for a vector of coefficients γ. The τ-th $(0 < \tau < 1)$ quantile of ε_t is the value, q_τ for which $P(\varepsilon_t < q_\tau) = \tau$. The τ-th conditional quantile of y_t given x_t is then simply

$$q_\tau(y_t|x_t) = x'_t\gamma(\tau), \tag{5}$$

where $\gamma(\tau)$ is a vector of coefficients dependent on τ. In the conventional least squares method, the τ-th is defined as any solution for $\hat{\gamma}(\tau)$. Thus, the minimization problem is

$$\begin{aligned} \hat{\gamma}(\tau) &= \min_{\gamma(\tau)\in R^k} E\left[\rho_\tau(y_t - x'_t\gamma(\tau))\right] \\ &= \min_{\gamma(\tau)\in R^k} \sum\nolimits_{t=1}^{T} [\rho_\tau(y_t - x'_t\gamma(\tau))], \end{aligned} \tag{6}$$

where the loss function

$$\rho_\tau(u) = u(\tau - I(u < 0)). \tag{7}$$

2.3 Bayesian Estimation

The use of Bayesian inference in generalized linear and additive models is quite standard these days. Unlike the conventional methods, Bayesian inference provides one with the entire posterior distribution of the parameter of interest. In addition, it allows for parameter uncertainty to be taken into account when making predictions. However, in the area of quantile regression, there is very little in the literature along the Bayesian lines. Koop [16] said that the Bayesian method is based on conditional probability analysis which can be written as:

$$P(\gamma, \sigma^2|Y) = \frac{P(Y|\gamma, \sigma^2)P(\gamma, \sigma^2)}{P(Y)} \propto P(Y|\gamma, \sigma^2) \times P(\gamma, \sigma^2), \tag{8}$$

where $P(Y)$ is distribution of information, $P(\gamma,\sigma^2)$ is joint prior distribution of (γ,σ^2), $P(\gamma,\sigma^2|Y)$is joint posterior distribution of $(\gamma,\sigma^2|Y)$ and $P(Y|\gamma,\sigma^2)$ is likelihood function. In the Bayesian estimation, we need to derive the posterior distribution which consists of the likelihood and prior distribution. In this study, the likelihood can be written as

$$P(Y|\gamma,\sigma^2) = \begin{matrix} (2\pi)^{-NT/2}\left|\sigma^2\right|^{-T/2} \\ \exp\left\{ -\frac{1}{2}(Y_{i,t} - \sum_{p=1}^{P}\gamma_p Y_{i,t-p})\Sigma^{-1}(Y_{i,t} - \sum_{p=1}^{P}\gamma_p Y_{i,t-p})' \right\} \end{matrix}, \tag{9}$$

Then, multiplying with the prior we have the joint posterior as

$$P(\gamma,\sigma^2|Y) \propto P(Y|\gamma,\sigma^2) \times P(\gamma,\sigma^2), \tag{10}$$

where $P(\gamma,\sigma^2)$ is the prior distribution.

In this study, we adopt a Bayesian estimation for quantile regression which is quite different from what has been done previously in this context. Irrespective of the actual distribution of the data, Bayesian inference for quantile regression proceeds by forming the likelihood function based on the asymmetric Laplace distribution [15]. In general, one may choose any prior to produce the joint posterior, thus, in this study, we use normal prior for estimated coefficient γ and inverse gamma priors for σ^2 as we expect the behavior is normal and there may not exist an extremely swing.

3 Data Description

We study eight sectors of the Stock Exchange of Thailand consisting of Banking (BANK), Commerce (COM), Communication (COMUN), Energy and Utilities (ENERG), Food and Beverage (FOOD), Personal Products and Pharmaceuticals (PERS), Petrochemicals and Chemicals (PETRO) and Property Development (PROP). The daily data of these variables are collected from the period January, 2014 to May, 2018. All data are collected from Bloomberg. We then transform the data using the differences between its logarithmic closing index. The data description is provided in Table 1. All data series are stationary as shown by the Augmented Dickey-Fuller (ADF) test. Skewness is asymmetry in a statistical distribution, in which the curve appears distorted or skewed either to the left or to the right. If it is negative skew, the distribution is concentrated on the right. On the contrary, if it is positive skew, the distribution is concentrated on the left. We observe that the returns exhibit either negative or positive skewness, meaning that our returns may not have a normal distribution. Thus, we conduct the Jarque-Bera normality test and it is evident that stock returns reject the null hypothesis of normal distribution.

4 Empirical Result

Bayesian Quantile regression can be used to detect herding behavior of investors. By setting $\tau = 0.25$ we obtain quantile estimate for low $CSAD_t$ (small difference

Table 1. Data description

	BANK	COM	COMUN	ENERG	FOOD	PERS	PETRO	PROP	MARKET
Mean	0.0001	0.0005	−0.0001	0.0003	0.0001	−0.0003	0.0004	0.0003	0.0003
Maximum	0.075	0.038	0.089	0.049	0.042	0.111	0.062	0.057	0.045
Minimum	−0.044	−0.044	−0.165	−0.079	−0.041	−0.063	−0.066	−0.044	−0.048
Std. Dev.	0.01	0.008	0.013	0.011	0.008	0.012	0.014	0.009	0.007
Skewness	0.287	0.019	−1.77	−0.23	−0.274	0.358	0.032	−0.185	−0.263
Kurtosis	6.797	5.074	32.802	7.082	5.097	13.226	5.589	6.119	7.452
MBF Jarque-Bera	0.000	0.000	0.000	0.000	0.000	0.000	0.000	0.000	0.000
MBF Unit root	0.000	0.000	0.000	0.000	0.000	0.000	0.000	0.000	0.000

Table 2. Coefficients of the CSAD model of herding using Bayesian Quantile Regression

Coefficient	Quantile 0.25	Quantile 0.5	Quantile 0.75
BANK γ_0	−0.0011**	0.000002	0.0023**
γ_1	−1.2901**	−23.39**	−6.1179**
γ_2	−0.7704**	0.2838**	0.6277**
COM γ_0	−0.0033**	−0.000001**	0.0036**
γ_1	−8.3309**	−11.61**	−7.9999**
γ_2	−0.3353**	0.1205**	0.2951**
COMUN γ_0	−0.0022**	−0.00001**	0.0006**
γ_1	−3.3850**	−23.17**	−12.78**
γ_2	−0.7753**	0.3088**	1.32**
ENERG γ_0	−0.0043**	−0.000005**	0.0020**
γ_1	−22.4809**	−38.75**	−13.7952**
γ_2	−0.5448**	0.5111**	1.3994**
PERS γ_0	−0.0040**	0.0001**	0.0060**
γ_1	4.2192**	−1.49	−0.2487
γ_2	−0.3179**	0.0034	0.0412**
PETRO γ_0	−0.0042**	−0.0003**	0.0030**
γ_0	0.5284	−2.9843**	−11.5482**
γ_2	0.01183	0.14827**	0.5107**
PROP γ_0	−0.0049**	−0.00001	0.0053**
γ_1	0.0662	−0.2979	−4.0371
γ_2	−0.0094	0.0042**	0.1500**

Note: **95 % Bayesian credible interval

between sector return and stock return), $\tau = 0.5$ we obtain quantile estimate for median (intermediate difference between sector return and stock return) and $\tau = 0.75$, we obtain quantile estimate for high (large difference between sector return and stock return). The result of herding behavior by Bayesian Quantile regression is shown in Table 2. Remind that the herding behavior is determined by the coefficient (γ_2). If the coefficient (γ_2) is negative, it means that there exists the herding behavior in that sector of the Stock Exchange of Thailand. The results provide evidence of the herding behavior of investors for all sectors in the Thai stock market at quantile 0.25, except for PETRO and FOOD as indicated by the positive sign of γ_2. This result confirms that the herding behavior exists in the Thai stock market when the difference between o sector returns and market return are relatively small. However, the sign of the coefficients γ_2 becomes positive when $\tau \in (0.50, 0.75)$ for all sectors. This indicates that the rate of sector return dispersion increases nonlinearly in the upturn market. This implies that investors change their focus from the aggregate market consensus to stock fundamentals when the market is rising notably. To have a clear picture of the trading behavior, visual representation of the coefficients γ_2 ranging from quantile 0.1 to 0.9 for eight sectors are presented in Figs. 1, 2, 3, 4, 5, 6, 7 and 8. The horizon blue line indicates the zero value of γ_2 and if the coefficient γ_2 is lower than this line, herding behavior exists. The results confirm that the herding behavior takes place in the lower quantile, say less than quantile 0.45 in BANK, COM, PERS, COMUN, ENERG; less than quantile 0.3 in PROP; and less than 0.2 in PETRO and FOOD. Surprisingly, in the case of median quantile

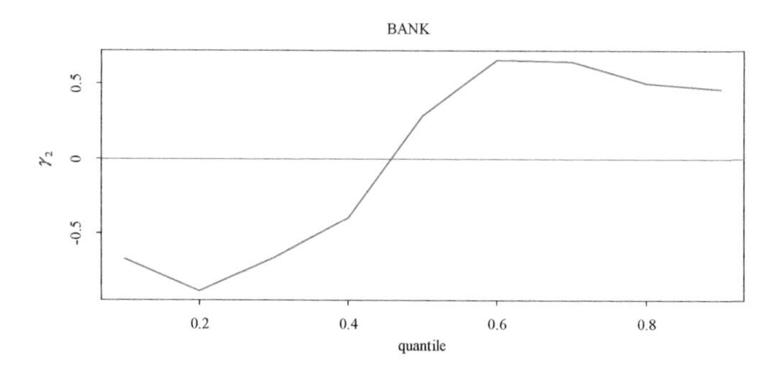

Fig. 1. Plot quantile in BANK

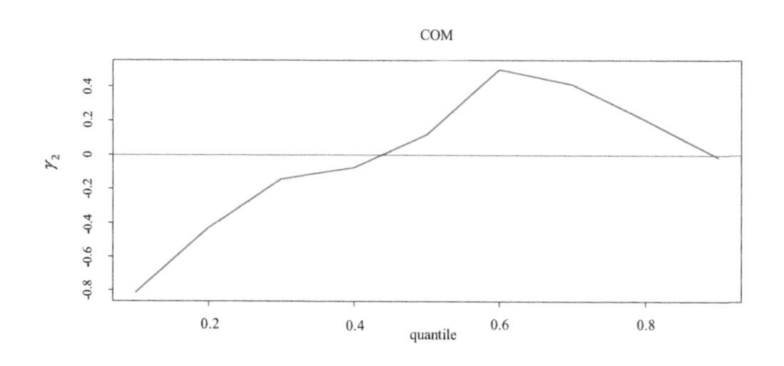

Fig. 2. Plot quantile in COM

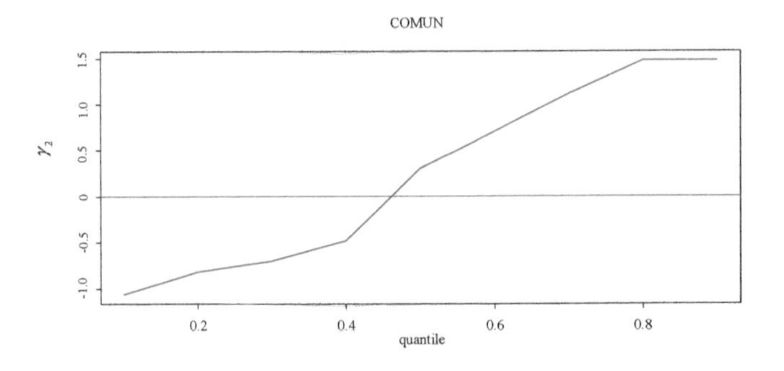

Fig. 3. Plot quantile in COMUN

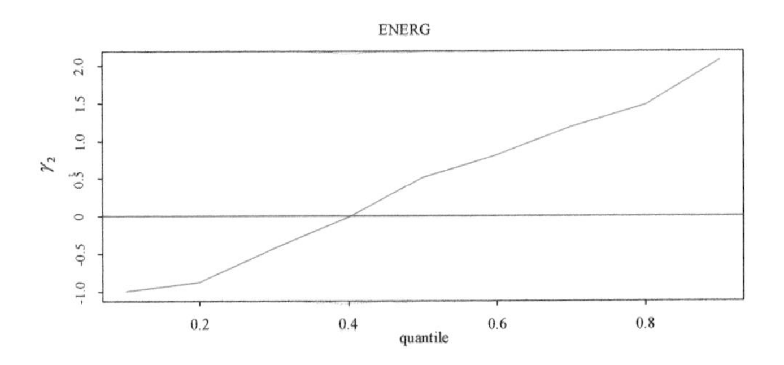

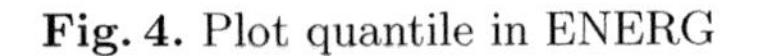
Fig. 4. Plot quantile in ENERG

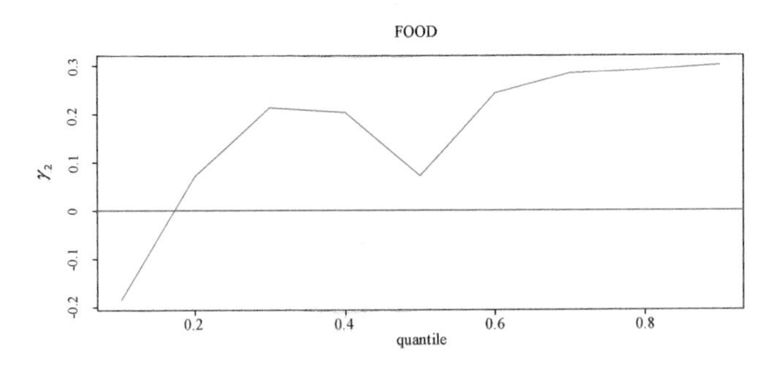

Fig. 5. Plot quantile in FOOD

up to the upper quantile, the results indicate no evidence of herding behavior of investors. The results can be supported by the existence of more informed investors in Thai stock market, particularly in the market upturn. The trading strategy of investors is based on public and private information that they hold during the market upturn.

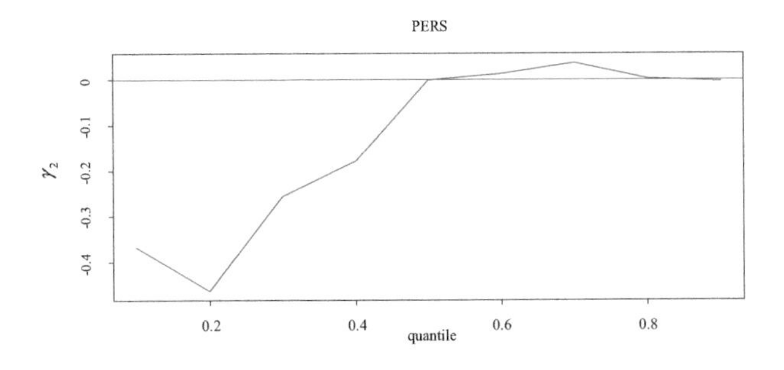

Fig. 6. Plot quantile in PERS

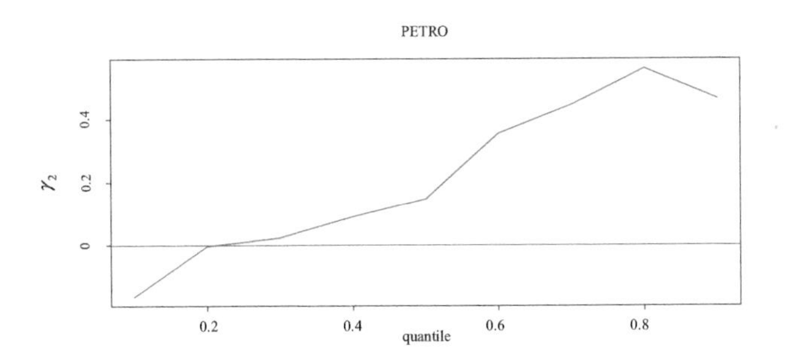

Fig. 7. Plot quantile in PETRO

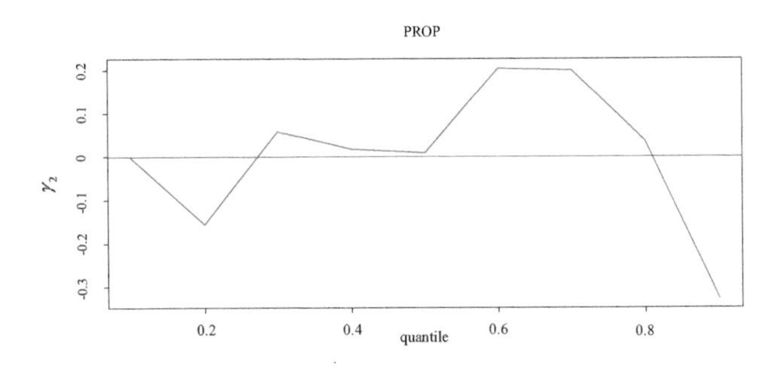

Fig. 8. Plot quantile in PROP

5 Conclusion

The main contribution of this paper is using Bayesian approach to Quantile regression to analyze herding behavior of investors in Thailands Stock Exchange market at the sector level, consisting of Banking, Commerce, Communications, Energy and Utilities, Food and Beverage, Personal Products and Pharmaceuticals, Petrochemicals and Chemicals, and Property Development. These eight sectors can be viewed as having high market capitalization in the Thai stock market. To detect the trading behavior, we use CSAD method of Chang et al.

[6] to detect herding behavior which is the process of market investment decisions depending on overall market conditions

We find that the coefficient γ_2 shows a negative sign in six out of eight sectors in the low quantile level indicating the presence of a herd behavior in the market downturn. However, the sign of this coefficient changes from negative to positive as the quantile level increases above 0.45. The possible explanation of this finding is that the investors might expect the potential loss in the market downturn when the prices of stocks tend to substantially decline and probably to the levels lower than their fundamental value. Thus, this uncertainty may spark the sell-offs in the market during this turmoil period. In brief, this might constitute the evidence of herding. However, the investors tend to focus and consider more on information of stocks rather than the aggregate market consensus when the stock market has a good condition and the risk that the market turning into downturn is low. For the other two sectors, Food and Beverage and Petrochemicals and Chemicals, the coefficient γ_2 presents the positive sign after quantile 0.25 to 0.90.

For future study, it should be possible to divide the range of each quantile into more frequencies. In addition if possible, study should be pursued on the behavior of the economic crisis and the forecast of economic crisis from the herding behavior to identify the potential trends in the future.

References

1. Ababio, K.A., Mwamba, J.M.: Test of herding behaviour in the Johannesburg stock exchange: application of quantile regression model. J. Econ. Financ. Sci. **10**(3), 457–474 (2017)
2. Al-Shboul, M.Q.: Asymmetric effects and the herd behavior in the Australian equity market. Int. J. Bus. Manag. **7**(7), 121 (2012)
3. Barnes, M.L., Hughes, A.T.W.: A quantile regression analysis of the cross section of stock market returns (2002)
4. Bikhchandani, S., Sharma, S.: Herd behavior in financial markets. IMF Staff Papers **47**(3), 279–310 (2000)
5. Buchinsky, M.: Recent advances in quantile regression models: a practical guideline for empirical research. J. Hum. Resour. **33**, 88–126 (1998)
6. Chang, E.C., Cheng, J.W., Khorana, A.: An examination of herd behavior in equity markets: an international perspective. J. Bank. Financ. **24**(10), 1651–1679 (2000)
7. Chiang, T.C., Zheng, D.: An empirical analysis of herd behavior in global stock markets. J. Bank. Financ. **34**(8), 1911–1921 (2010)
8. Christie, W.G., Huang, R.D.: Following the pied piper: do individual returns herd around the market? Financ. Anal. J. **51**(4), 31–37 (1995)
9. Demirer, R., Kutan, A.M.: Does herding behavior exist in Chinese stock markets? J. Int. Financ. Mark. Inst. Money **16**(2), 123–142 (2006)
10. Demirer, R., Kutan, A.M., Chen, C.D.: Do investors herd in emerging stock markets? Evidence from the Taiwanese market. J. Econ. Behav. Organ. **76**(2), 283–295 (2010)
11. Economou, F., Kostakis, A., Philippas, N.: Cross-country effects in herding behaviour: evidence from four south European markets. J. Int. Financ. Mark. Inst. Money **21**(3), 443–460 (2011)

12. Gerlach, R.H., Chen, C.W., Chan, N.Y.: Bayesian time-varying quantile forecasting for value-at-risk in financial markets. J. Bus. Econ. Stat. **29**(4), 481–492 (2011)
13. Gleason, K.C., Mathur, I., Peterson, M.A.: Analysis of intraday herding behavior among the sector ETFs. J. Empir. Financ. **11**(5), 681–694 (2004)
14. Hwang, S., Salmon, M.: Market stress and herding. J. Empir. Financ. **11**(4), 585–616 (2004)
15. Koenker, R.W., d'Orey, V.: Algorithm AS 229: computing regression quantiles. J. R. Stat. Soc. Ser. C (Appl. Stat.) **36**(3), 383–393 (1987)
16. Koop, G., Potter, S.M.: Bayesian analysis of endogenous delay threshold models. J. Bus. Econ. Stat. **21**(1), 93–103 (2003)
17. Lao, P., Singh, H.: Herding behaviour in the Chinese and Indian stock markets. J. Asian Econ. **22**(6), 495–506 (2011)
18. Pastpipatkul, P., Yamaka, W., Wiboonpongse, A., Sriboonchitta, S.: Spillovers of quantitative easing on financial markets of Thailand, Indonesia, and the Philippines. In: International Symposium on Integrated Uncertainty in Knowledge Modelling and Decision Making, pp. 374–388. Springer, Cham, October 2015
19. Reich, B.J., Fuentes, M., Dunson, D.B.: Bayesian spatial quantile regression. J. Am. Stat. Assoc. **106**(493), 6–20 (2011)
20. Saastamoinen, J.: Quantile regression analysis of dispersion of stock returns-evidence of herding. Keskustelualoitteita **57**, 1–19 (2008)
21. Shiller, R.J.: Behavioral economics and institutional innovation (2005)
22. Wermers, R.: Mutual fund herding and the impact on stock prices. J. Financ. **54**(2), 581–622 (1999)

Markov Switching Dynamic Multivariate GARCH Models for Hedging on Foreign Exchange Market

Pichayakone Rakpho[1,2](✉), Woraphon Yamaka[1], and Songsak Sriboonchitta[1]

[1] Center of Excellence in Econometrics, Faculty of Economics, Chiang Mai University, Chiang Mai, Thailand
pichayakone@gmail.com, woraphon.econ@gmail.com, Songsakecon@gmail.com
[2] International College of Digital Innovation, Chiang Mai University, Chiang Mai, Thailand

Abstract. Foreign exchange rates is a significant factor affecting foreign transactions such as trade and investment. Foreign exchange rates, especially EUR/USD and GBP/USD, have a high fluctuation in recent years and lead a severe risk to investors. In this study, we consider a hedging strategy as a tool for offsetting the potential losses of investors. We introduce two classes of Markov Switching correlation model, namely MS-CCC-GARCH and MS-DCC-GARCH to compute the optimal hedge ratios and portfolio weights in the foreign exchange rates (EUR/USD and GBP/USD) for the period of 2013–2018. We also compare the performance of these two models with CCC-GARCH, DCC-GARCH models. The results show that MS-DCC-GARCH perform better for EUR/USD and GBP/USD spot and futures pairs. We finally complement our analysis by computing the dynamic hedge ratio and optimal portfolio weight, the result shows that the hedge ratios for both currencies are mostly remaining closely to 1 over the sample periods. However, we notice that, in some periods, the hedge ratios are particularly low in the low volatility market regime.

Keywords: Markov switching model · CCC-GARCH model
DCC-GARCH model · Hedging model · Foreign exchange rates

1 Introduction

Foreign exchange rates is a significant factor in foreign transactions, trade and investment. Under the economic globalization regime, the exchange rates has changed over time and exhibited a large fluctuation during the recent decade. This uncertainty could bring a concern, regarding a substantial loss to exporters, importers and investors. To deal with this risk, the hedging strategy is introduced as a tool for offsetting potential losses. The aim of hedging is to use the futures contracts which is one of the most widely used techniques to manage the risk. This is to say that the investors or traders can reduce the foreign exchange

V. Kreinovich et al. (Eds.): ECONVN 2019, SCI 809, pp. 806–817, 2019.
https://doi.org/10.1007/978-3-030-04200-4_57

risk by buying the foreign currency futures. The problem is how many future contracts should be held for each unit of the underlying currency. Thus, in this study the hedge ratio is considered to find the optimal futures contracts for the investors and traders.

In the traditional computation, this hedge ratio can be calculated by the expected mean between the two returns using the regression technique. However, this method is inapplicable to estimate hedge ratios as it suffers from the problem of serial correlation and heteroscedasticity which often exists in the financial time series [5]. Therefore, the recent studies considered using the bivariate conditional volatility models to estimate time varying hedge ratio, e.g. constant conditional correlation (CCC)-GARCH introduced by Bollerslev in 1990 [4], dynamic conditional correlation (DCC)-GARCH of Engle and Sheppard [8], and copula-based GARCH model of Hsu et al. [11]. In CCC-GARCH model, it assumes that the correlation of CCC is constant over time, and only the conditional variance of GARCH process is time-varying. However, the assumption of constant correlation is not always reasonable and the hedge ratio can result in greater risk reduction than the static one [11]. Hence, in 2000, Engle and Sheppard [8] introduced the DCC-GARCH model which has a clear advantages in the correlation can be vary over time. Later, the copula-GARCH is proposed to relax the symmetric and linear correction in DCC and CCC-GARCH model. Although these models have been applied and perform well in measuring the hedge ratio [1], there are an evident that the ratio will become smaller when the market trend is stable, but it becomes larger when the market presents high volatility (see [6,13]). They suggested that the dependence or correlation of the financial time series often exhibit different structures depending on the regime of the market or economic conditions at that period. Therefore, quantifying the hedge ratio should be taken into account this structural change.

In this study, we aim to contribute to the literature by proposing the nonlinear model for measuring the optimal hedge ratio related to foreign exchange rates. To achieve our goal, a Markov switching approach is considered to extend into the DCC and CCC-GARCH models. Note that the copula-GARCH is not yet considered in this study and this issue will be left for the future study. We believe that these two models will become more realistic and flexible reflecting the variety of foreign exchange rates characteristics. In addition, this innovation provides an ideal alternative model for the construction of hedging portfolios.

The remainder of the paper is organized as follows. In Sect. 2 we briefly describe two regimes switching multivariate GARCH models, MS-CCC-GARCH and MS-DCC-GARCH, and the derivation of the OHR and hedging effective index. In Sect. 3 the data and the descriptive statistics are presented. Section 4 analyses the empirical estimates, and Sect. 5 presents the conclusions.

2 Methodology

2.1 MS-CCC-GARCH and MS-DCC-GARCH Models

The Markov switching GARCH model is a nonlinear specification model which reflects different states of the volatilities namely high and low volatilities. To have a better understanding, let we start with GARCH model. The GARCH model was put forward by Bollerslev [3] and it can be explained by the following equations

$$r_t = \mu_t + \varepsilon_t, \tag{1}$$

$$\varepsilon_t = \eta_t \sqrt{h_t}, \tag{2}$$

$$h_t = \omega + \sum_{i=1}^{q} \alpha_i \varepsilon_{t-i}^2 + \sum_{j=1}^{p} \beta_j h_{t-1}, \tag{3}$$

where r_t is log return of an asset at time t. μ_t is constant term at time t. ε_t is residual series at time t. h_t is the conditional variance at time t. η_t is independently and identically distributed (iid) standardized, random variables.

The Markov-switching (MS) model was introduced by Hamilton [9]. Changes in regimes can be estimated with the multiple regimes as (s_t) and are governed by a hidden Markov chain. Bauwens, Preminger, and Rombouts [2] explained that, the persistence in the estimated single regime of GARCH procedure could be considered as resulting from the misspecification and thus they recommend a way to control it using a MS-GARCH model. In the multivariate aspect, CCC and DCC approaches are used for joining the univariate MS-GARCH and obtaining the MS-CCC-GARCH model which can be written as

$$H_{t,s_t} = D_{t,s_t} R_{s_t} D_{t,s_t}, \tag{4}$$

$$h_{t,s_t} = \omega_{s_t} + \sum_{i=1}^{q} \alpha_{i,s_t} \varepsilon_{t-i,s_t}^2 + \sum_{j=1}^{p} \beta_{j,s_t} h_{t-j,s_t}, \tag{5}$$

where H_{t,s_t} is $n \times T$ matrix of conditional variances from univariate MS-GARCH model at regime s_t. $D_{t,s_t} = diag\,(h_{1t,s_t}, h_{2t,s_t}, ..., h_{nt,s_t})$, R_{s_t} is $n \times n$ regime dependent constant correlation parameters.

For the MS-DCC-GARCH model, it has the same structure as the MS-CCC-GARCH, but the conditional correlation is not restricted to be constant, therefore, the conditional correlation in Eq. (4) can be derived from

$$R_{s_t} = Q_{s_t}^{*-1} Q_{s_t} Q_{s_t}^{*-1} \tag{6}$$

$$Q_{s_t,t} = \ (1 - \theta_{1,s_t} - \theta_{2,s_t}) \bar{Q}_{s_t,t} + \theta_{2,s_t} Q_{s_t,t-1} + \theta_{2,s_t} \varepsilon_{s_t,t-1} \varepsilon'_{s_t,t-1}, \tag{7}$$

where $D_t = diag\,(h_{1t}, h_{2t}, ..., h_{nt})$, H_{t,s_t} is a $n \times n$ conditional covariance matrix at regime s_t, R_{t,s_t} is conditional correlation matrix at regime s_t, Q_{t,s_t} is conditional correlation matrix with time varying at regime s_t. This indicates that all parameters are governed by a state variable s_t which is assumed to evolve according to s_{t-1} with transition probability

$$Pr\left(s_t = j \left| s_{t-1} = i\right.\right) = p_{ij}, \sum_{j=1}^{j} p_{ij} = 1, \quad for \quad i = 1, ..., j, \tag{8}$$

where $\Pr\left(s_t = j \left| s_{t-1} = i\right.\right)$ presents probability in Markov switching process $(i = 1, 2)$.

2.2 Hedging Ratio Model

The optimal hedge ratio is defined as the holdings of futures which minimize the risk of the hedging portfolio and number of futures contracts (F) held to hedge against spot position (S). Johnson [12] determined the risk as the variance between the returns in the portfolio. If the joint distribution of the spot and futures returns remains the same over time, the conventional risk-minimizing hedge ratio δ^* will be

$$\delta^* = \frac{Cov\left(\Delta S, \Delta F\right)}{Var\left(\Delta F\right)} = \frac{Cov\left(r_S, r_F\right)}{Var\left(\Delta r_F\right)}. \tag{9}$$

Let ΔS and ΔF be the respective changes in the spot and futures prices, δ^*serves to estimate the number of futures contracts used to mitigate the risk of changes in the price of spot.

$$\delta^* = \frac{Cov\left(r_S, r_F\right)}{Var\left(r_F\right)} = \frac{\rho_{S,F} * h_S * h_F}{h_F^2} = \rho_{S,F} * \frac{h_S}{h_F} = \frac{\rho_{S,F}\sqrt{h_S}}{\sqrt{h_F}}. \tag{10}$$

The optimal portfolio weight of Foreign exchange rate spot/futures holding is given by:

$$w_{SF} = \frac{h_F - h_{SF}}{h_S - 2h_{SF} + h_F}, \tag{11}$$

where $w_{SF}^* \left(1 - w_{SF}\right)$ is the weight of the spot/futures of EUR/USD and GBP/USD spot/futures at time t. We can write the hedging ratio as h_{SF} is the conditional covariance between spot and futures returns divided by h_F is the futures returns' conditional variance.

2.3 Estimation

We can write the likelihood function of MS-CCC-GARCH (1,1) and MS-DCC-GARCH (1,1) with j regimes for two assets by

$$\mathbf{L}(\Theta_{s_t} \left| r_t^S, r_t^F\right.) = \sum_{j=1}^{J} \left(\prod_{t=1}^{T} f(\Theta_{s_t} \left| r_t^S, r_t^F\right.) \left(\Pr(s_t = j \left|\Theta_{S_t}\right.)\right) \right), \tag{12}$$

where $f(\Theta_{s_t} \left| r_t^S, r_t^F\right.)$ is the bivariate normal density distribution. Θ is all available information sets of the model, and $\Pr\left(s_t = j \left|\Theta_{s_t}\right.\right)$ is state's probabilities which are obtained from the Hamilton's filter algorithm Hamilton [9]. Assume that we consider two regimes model, thus we will obtain two regimes conditional correlation and variance. Thus, we measure the hedge ratio and also the optimal portfolio weight in terms of the expected value. Then, the expected hedge ratio and expected weight of foreign exchange rates spot/futures holding are as follows

$$E\delta^* = \sum_{s_t=1}^{2} \left[\frac{\rho_{S,F,(s_t)} \sqrt{h_{S,(s_t)}}}{\sqrt{h_{F,(s_t)}}} \Pr\left(s_t = j \left| \Theta_{s_t} \right.\right) \right], \tag{13}$$

$$Ew_{SF} = \sum_{s_t=1}^{2} \left[\frac{h_{F,(s_t)} - h_{SF,(s_t)}}{h_{S,(s_t)} - 2h_{SF,(s_t)} + h_{F,(s_t)}} \Pr\left(s_t = j \left| \Theta_{s_t} \right.\right) \right]. \tag{14}$$

3 Data

This paper uses daily closing prices data of spot and futures for foreign exchange rates series consisting of EUR/USD (Euro compared with United States Dollar) and GBP/USD (The British Pound compared with United States Dollar). The data are collected from September 2013 to May 2018, covering 1,196 daily observations. Both spot and futures daily settlement prices are obtained from Bloomberg. Each data series is transformed into log-difference ($\ln P_t - \ln P_{t-1}$), before it is conducted for analysis using the MS-CCC-GARCH and MS-DCC-GARCH models. Table 1 presents the descriptive statistics of the log-difference four data series, EUR/USD spot, EUR/USD futures, GBP/USD spot, GBP/USD futures. The four data series exhibit a negative average growth rate whereas EUR/USD-F performs the largest mean value. The skewness of these series shows different sign, the positive for the EUR/USD spot and EUR/USD futures returns and the negative for GBP/USD spot and GBP/USD futures returns. The kurtosis values are all higher than 3. This indicates that our data series have asymmetric distributions and heavy tails. The Jarque-Bera normality test is also provided to confirm this non-normality pattern. The result shows that MBF Jarque-Bera are equal to zero, indicating with decisive evidence of the normality hypothesis. In addition, the stationary test (Augmented Dickey Fuller), MBF unit root is stationary with decisive evidence result for all log-difference series. As we can see from Fig. 1, the daily data of spot and futures for foreign exchange rates series exhibit a stationary feature over the whole sample period. Note that the interpretation of MBF is referred to Vovk [14].

4 Empirical Results

4.1 Model Selection

In this study, we introduce two classes of regime switching correlation model thus it would be better to compare these models in order to find the best fit model. Note that the GARCH (1,1) specification is used in all models as it is

Table 1. Descriptive statistics

	EURUSD	EURUSD_F	GBPUSD	GBPUSD_F
Mean	−9.84E−05	−0.000129	−5.80E−05	−5.93E−05
Maximum	0.030158	0.029005	0.023404	0.022364
Minimum	−0.023821	−0.020671	−0.033753	−0.034308
Std. Dev.	0.005376	0.005017	0.005405	0.0052
Skewness	0.133174	0.130698	−0.091789	−0.188414
Kurtosis	5.363871	5.070712	5.305168	5.674128
Jarque-Bera	281.9983	217.0827	266.4838	363.4324
MBF Jarque-Bera	0.0000	0.0000	0.0000	0.0000
Unit root test	−35.22284	−35.04712	−36.01406	−35.71002
MBF Unit root test	0.0000	0.00000	0.0000	0.0000

Note: MBF is Minimum Bayes factor which computed by $eplogp$, where p is $p - value$ (see [10,14]).

Table 2. AIC and BIC for model selection

Model	EUR/USD		GBP/USD	
	AIC	BIC	AIC	BIC
CCC-GARCH (1,1)	20968.88	20999.4	20807.4	20837.92
DCC-GARCH (1,1)	18581.4	18622.09	18970.82	19011.51
MS-CCC-GARCH (1,1)	−18536.59	−18455.22	−18866.89	−18785.51
MS-DCC-GARCH (1,1)	**−19373.27**	**−19281.73**	**−19613.66**	**−19522.11**

simple and good enough to capture the volatility (see, [8]). In this section, we also compare the MS-CCC-GARCH and MS-DCC-GARCH with one regime models, namely CCC-GARCH, DCC-GARCH. The comparison criteria considered here is Akaike Information Criterion (AIC) and Bayesian Information Criterion (BIC) methods. According to the results in Table 2, it is found that MS-DCC-GARCH models are the best fit model for both EUR/USD and GBP/USD spots and futures as the lowest AIC and BIC are shown. Therefore, the further analysis of the hedge ratio is based on this model.

4.2 Estimation of MS-DCC-GARCH (1,1) Model

Tables 3 and 4 present the results of two regime MS-DCC-GARCH (1,1). The model provides regime dependent variance equation for two regimes. Let's consider these equations to interpret the meaning of each regime. Here, we consider the degree of volatility persistence in each regime which can be measured by the sum of ARCH and GARCH estimates, $(\alpha_{i,s_t} + \beta_{j,s_t})$, and the higher value of $(\alpha_{i,s_t} + \beta_{j,s_t})$ corresponds to the higher unconditional variance of the process. The result of volatility persistence of each hedge ratio is provided in

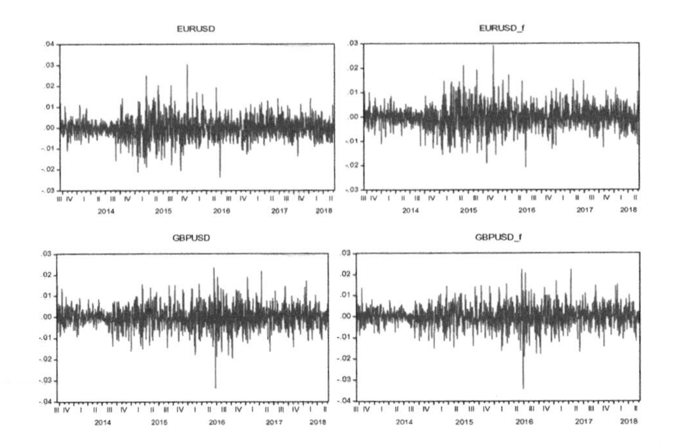

Fig. 1. Spot and futures daily returns of EUR/USD and GBP/USD during 2014–2018

Table 5. These findings lead us to interpret regime 1 as high volatility regime while regime 2 as low volatility regime. This indicates that regimes 1 and 2 tend to be associated with low and high volatility regime, respectively. It is interesting to compare the volatility of EUR/USD and GBP/USD. We find that hedged portfolio on EUR/USD of spot is more volatile than EUR/USD futures and GBP/USD spot is more volatile than GBP/USD futures. The estimates for $\theta_{1,(s_t=i)} + \theta_{2,(s_t=i)}$ across the regimes are different; the low and high volatility regimes are characterized by different dynamic correlation structures. We observe that the sum $\theta_{1,(s_t=i)} + \theta_{2,(s_t=i)}$ of EUR/USD spot and futures pair is 0.833(1) and GBP/USD for low and high volatility regime. While the sum $\theta_{1,(s_t=i)} + \theta_{2,(s_t=i)}$ are 0.823(0.963) for the low (high) volatility regime. This result corresponds to the study of Da Silva Filho [7] and Pathairatkul et al. [13] as they mentioned that the conditional correlation during high volatility market regime is less than that during low volatility market regime. In addition, Tables 3 and 4 also provide the result of the transition matrix. We can observe that the transition probability estimates P_{11} (P_{22}) are 0.95 and 0.8999 (0.95 and 0.9501) for EUR/USD and GBP/USD spot and futures pairs, respectively. Hence, the high volatility regime is more persistent for the EUR/USD market than for the GBP/USD market, while the low volatility regime is more persistent for GBP/USD market than EUR/USD market.

The estimated MS-DCC-GARCH (1,1) model also produces the probabilities of two regimes for the period from 2013 to 2018. In this section, we plot only the low volatility regime probabilities of EUR/USD and GBP/USD pairs in Fig. 2. Based on the results, it confirms that the low volatility regime probabilities regime is more persistent for the GBP/USD market than for the EUR/USD market as the GBP/USD spot and futures pair spends much of the time for the low volatility regime compared to the high volatility regime. We observe that the low probabilities mostly take place in all periods, except for the period from 2013 to 2014 which corresponds to the uncertainty of the US economy. For

Table 3. Estimation results of MS-DCC-GARCH for EUR/USD

Parameter	EUR/USD spot	EUR/USD futures
$\alpha_{0,(s_t=1)}$	0.0000(0.6065)	0.0000(0.6065)
$\alpha_{1,(s_t=1)}$	0.1101(0.7375)	0.0734(0.0709)
$\beta_{1,(s_t=1)}$	0.8155(0.0000)	0.5436(0.0000)
$\alpha_{0,(s_t=2)}$	0.0000(0.6065)	0.0000(0.6065)
$\alpha_{1,(s_t=2)}$	0.0131(0.9661)	0.0087(0.8433)
$\beta_{1,(s_t=2)}$	0.4977(0.0000)	0.3318(0.0035)
$\theta_{1,(s_t=1)}$	0.0047(0.2444)	
$\theta_{2,(s_t=1)}$	0.9952(0.000)	
$\theta_{1,(s_t=2)}$	0.0039(0.8096)	
$\theta_{2,(s_t=2)}$	0.8293(0.0000)	
P_{11}	0.95(0.0000)	
P_{22}	0.95(0.0000)	

Note: (), MBF is Minimum Bayes factor which computed by *eplogp*, where p is $p - value$ (see [10,14]).

Table 4. Estimation results of MS-DCC-GARCH for GBP/USD

Parameter	GBP/USD spot	GBP/USD futures
$\alpha_{0,(s_t=1)}$	0.0000(0.6065)	0.0000(0.6065)
$\alpha_{1,(s_t=1)}$	0.1491(0.000)	0.0893(0.8589)
$\beta_{1,(s_t=1)}$	0.8054(0.000)	0.5147(0.000)
$\alpha_{0,(s_t=2)}$	0.0000(0.6065)	0.0000(0.6065)
$\alpha_{1,(s_t=2)}$	0.1716(0.0001)	0.1144(0.8468)
$\beta_{1,(s_t=2)}$	0.4264(0.000)	0.2868(0.000)
$\theta_{1,(s_t=1)}$	0.1000(0.2239)	
$\theta_{2,(s_t=1)}$	0.5000(0.2197)	
$\theta_{1,(s_t=2)}$	0.0640(0.0022)	
$\theta_{2,(s_t=2)}$	0.7591(0.000)	
P_{11}	0.8999(0.000)	
P_{22}	0.9501(0.000)	

Note: (), MBF is Minimum Bayes factor which computed by *eplogp*, where p is $p - value$ (see [10,14]).

EUR/USD, the low volatility regime clearly exists during 2014–2016. After that, the probabilities are low along the period from 2016 to 2018. This indicates that there is high volatility in EUR/USD market after 2016. This period corresponds to Brexit, which is an abbreviation for "British exit," referring to the UK's decision in a June 23, 2016 referendum to leave the European Union (EU). The vote's result defied expectations global markets, causing the British pound to fall to its lowest level against the dollar in 30 years thereby pushing the high volatility in this regime.

Table 5. The volatility persistence measured by the sum $(\alpha_{(s_t=i)} + \beta_{(s_t=i)})$

Regime	EUR/USD spot	EUR/USD futures	GBP/USD spot	GBP/USD futures
1	**0.9256**	**0.6171**	**0.9545**	**0.604**
2	0.5108	0.3405	0.5981	0.4012

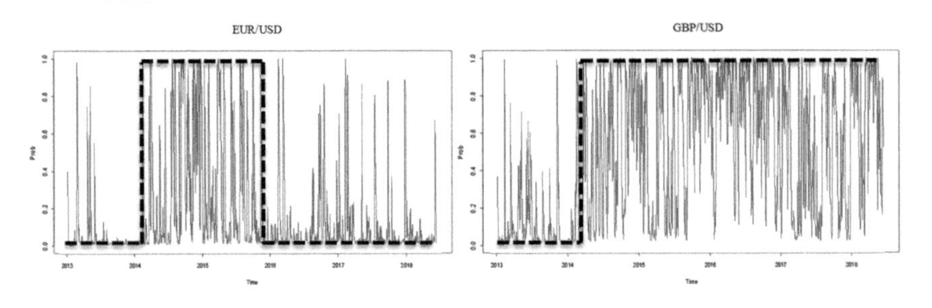

Fig. 2. Filtered probabilities of EUR/USD and GBP/USD

4.3 Volatility of Retunes

In this section, the volatilities of these returns are plotted in Fig. 3. As our model is two-regime MS-DCC-GARCH (1,1), two-regime conditional volatilities

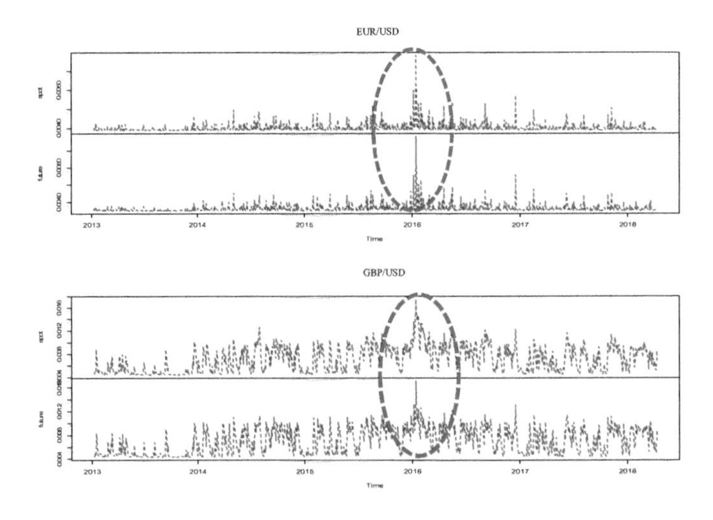

Fig. 3. Volatility of Retunes Series for EUR/USD and GBP/USD

are obtained from the GARCH process for each regime. We can observe that the expected conditional volatilities of EUR/USD, GBP/USD are relatively high during 2016, corresponding to the period of Brexit as we mentioned before.

4.4 Hedging Ratio and Optimal Portfolio Weight

Figures 4 and 5 show estimation of time varying hedging ratio and optimal portfolio weight for the foreign exchange rates series consist of (EUR/USD and GBP/USD) estimated using daily data by MS-DCC-GARCH (1,1) method. It can be seen that the time varying hedging ratios and optimal portfolios clearly change over the different phases of the economic situation. We observe that the highest average hedge ratio value is 0.98 for EUR/USD, meaning that if investors buy spot 100 contracts, they will long (buy) futures 98 contracts to prevent the risk while the average hedge ratio value of GBP/USD equal 0.96 meaning that, if investors buy spot 100 contracts, they will short (sell) futures 96 contracts to prevent the risk. Additionally, we notice that the hedge ratio of EUR/USD is low during 2014–2015 corresponding to the period of low volatility regime in the previous result. Thus, we can conclude that the hedge ratio will be low for the low volatility market regime. The investor should reduce their short (sell) position of EUR/USD futures contracts. Then, when we look at the optimal portfolio weight during the same period, the value of weight varies over the range 0.475–0.490. This suggests that the optimal holding of one EUR/USD spot/futures portfolio is 0.475–0.49 for spot and 0.525–0.51 for futures. In parts of GBP/USD spot/futures portfolios, the average hedge ratio remains at 0.96 throughout the entire period, except for the beginning of 2018. The portfolio holdings during this time are nearly 0.4, so the effect of an optimal GBP/USD hedge is negligible in this period.

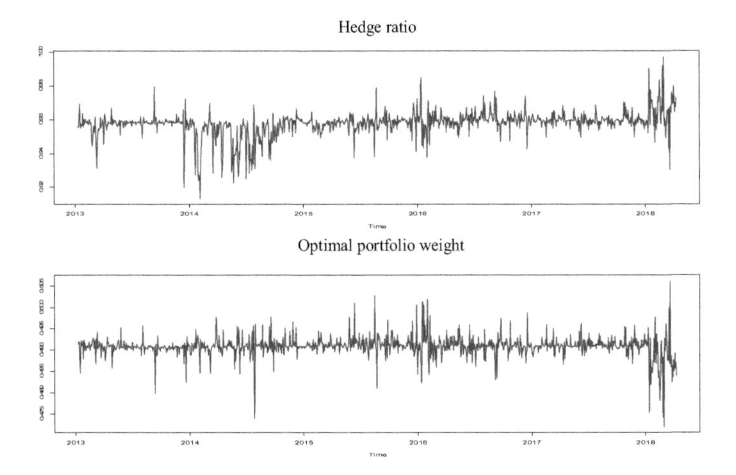

Fig. 4. Hedging ratio and optimal portfolio weight of EUR/USD

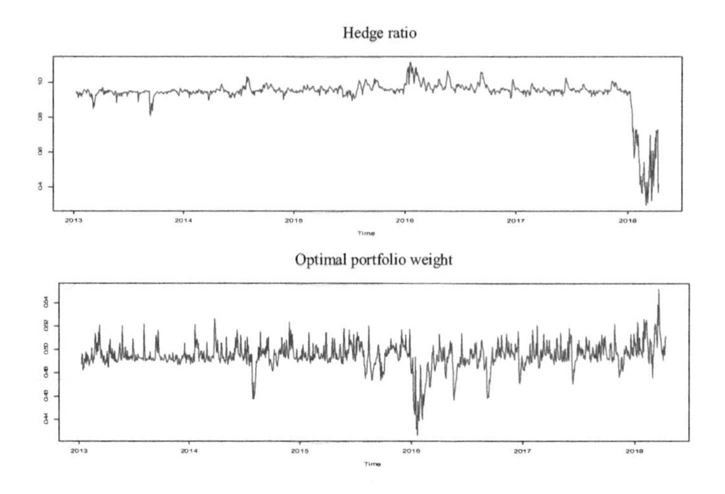

Fig. 5. Hedging ratio and optimal portfolio weight of GBP/USD

5 Conclusion

In this paper we employ nonlinear MS-DCC-GARCH (1,1) model of Billio and Caporin (2005) with two regimes for constructing the hedging strategy on foreign exchange rates series consisting of EUR/USD and GBP/USD. These two currencies are the world's main currency in terms of liquidity level as there are widely used in the international trade and market. The factor determining its popularity among traders is its high volatility and thereby providing an opportunity to make solid profit even on short-term transactions. In practice, the investor might exposure himself to the higher risk in the market, thus the hedging strategy is used to reduce the risk. In this study, we consider the spot and futures returns of EUR/USD and GBP/USD to calculate the optimal portfolios weight and optimal hedging ratios. The empirical results suggest that MS-DCC-GARCH is superior to one regime CCC-GARCH, DCC-GARCH and two regimes MS-CCC-GARCH model as the lower AIC and BIC are obtained. An interesting aspect of our regime switching model is that we can obtain a weak and strong persistence in the Markov chain, which produces both high and less volatility of dynamic correlations along the sample period. We further evaluate the optimal weights and dynamic hedging strategies for our foreign exchange rates. Our results mainly show significant volatility and time-varying hedge ratios of markets. These evidences are very important for investors because putting an investment in different period seems to face with a different market situation.

References

1. Ahmad, W., Sadorsky, P., Sharma, A.: Optimal hedge ratios for clean energy equities. Econ. Modell. **72**, 278–295 (2018)
2. Bauwens, L., Storti, G.: A component GARCH model with time varying weights. Stud. Nonlinear Dyn. Econ. **13**(2), 1–31 (2009)
3. Bollerslev, T.: Generalized autoregressive conditional heteroskedasticity. J. Econ. **31**(3), 307–327 (1986)
4. Bollerslev, T.: Modelling the coherence in short-run nominal exchange rates: a multivariate generalized ARCH model. Rev. Econ. Stat. **72**, 498–505 (1990)
5. Chang, C.L., González-Serrano, L., Jimenez-Martin, J.A.: Currency hedging strategies using dynamic multivariate GARCH. Math. Comput. Simul. **94**, 164–182 (2013)
6. Chodchuangnirun, B., Yamaka, W., Khiewngamdee, C.: A regime switching for dynamic conditional correlation and GARCH: application to agricultural commodity prices and market risks. In: International Symposium on Integrated Uncertainty in Knowledge Modelling and Decision Making, pp. 289–301. Springer, Cham, March 2018
7. Da Silva Filho, O.C., Ziegelmann, F.A., Dueker, M.J.: Modeling dependence dynamics through copulas with regime switching. Insur. Math. Econ. **50**(3), 346–356 (2012)
8. Engle, R.F., Sheppard, K.: Theoretical and empirical properties of dynamic conditional correlation multivariate GARCH (No. w8554). National Bureau of Economic Research (2001)
9. Hamilton, J.D.: A new approach to the economic analysis of nonstationary time series and the business cycle. Econometrica **57**, 357–384 (1989)
10. Held, L., Ott, M.: On p-values and Bayes factors. Annu. Rev. Stat. Appl. **5**, 393–419 (2018)
11. Hsu, C.C., Tseng, C.P., Wang, Y.H.: Dynamic hedging with futures: a copula-based GARCH model. J. Futur. Mark. **28**(11), 1095–1116 (2008)
12. Johnson, L.L.: The theory of hedging and speculation in commodity futures. Rev. Econ. Stud. **27**(3), 139–151 (1960)
13. Pastpipatkul, P., Yamaka, W., Sriboonchitta, S.: Analyzing financial risk and co-movement of gold market, and Indonesian, Philippine, and Thailand stock markets: dynamic copula with Markov-switching. In: Causal Inference in Econometrics, pp. 565–586. Springer (2016)
14. Vovk, V.G.: A logic of probability, with application to the foundations of statistics. J. R. Stat. Soc. Ser. B (Methodol.) **55**, 317–351 (1993)

Bayesian Approach for Mixture Copula Model

Sukrit Thongkairat[1,2(✉)], Woraphon Yamaka[1], and Songsak Sriboonchitta[1]

[1] Center of Excellence in Econometrics, Faculty of Economics, Chiang Mai University, Chiang Mai 50200, Thailand
sukrit415@gmail.com, woraphon.econ@gmail.com, Songsakecon@gmail.com
[2] International College of Digital Innovation, Chiang Mai University, Chiang Mai 50200, Thailand

Abstract. This paper aims to use the Bayesian estimation as an alternative method for formulating and estimating mixed copula models. This method has claimed to be more efficient than the conventional maximum likelihood estimator as it can deal with the high dimension copula and large parameter estimates under limited sample. In this study, we present various mixed copula functions constructed from both Elliptical and Archimedean copulas. We employ a simulation study to investigate the performance of this estimator for comparison with the maximum likelihood estimator. The results show that the Bayesian estimation is considerably more accurate than maximum likelihood estimator in various scenarios. Finally, we extend the Bayesian mixed copula to the real data and show that our approach perform well in this real data analysis.

Keywords: Mixed copula · Bayesian · Maximum likelihood
Financial markets

1 Introduction

The measurement of correlation is useful in explaining the relationship between the interested variables such as economic, environment, medicine, and social variables. However, recently, the presence of a correlation is quite difficult to detect properly as the nature of the data is more complex when compared to the behavior of the past data. In addition, the traditional correlation measures are insufficient to infer the presence of a causal relationship. Several conventional correlation measures are often expressed in the form of Pearson, Spearman, and/or Kendall tau, which are based on linear correlation and normal assumptions. However, these correlation measures might not be realistic in the real data applications, in particular financial study. It is found that the distribution of financial returns may not be normal (see, Jondeau and Rockinger [11]; Brooks et al. [3]). Also, these conventional measures might fail to verify the asymmetric correlation in the financial information, and thereby leading to an inaccurate

V. Kreinovich et al. (Eds.): ECONVN 2019, SCI 809, pp. 818–827, 2019.
https://doi.org/10.1007/978-3-030-04200-4_58

inference. Hence, these methods are not appropriate for measuring the correlation in the financial data series.

Recently, a powerful tool to describe the dependenciest between variables are the copula function of Sklar [17]. This function has been widely accepted and used to explain the relationship of two or more random variables. Thus, we might say that copula model is better than those traditional methods. Several classes of copulas have been proposed to join the random variables. These classes adequately describe some joint distributions, but not all of them can deal with complicated dependence structures. One way to cover more distributions is to consider convex combinations of distributions from different copula classes. Such combinations are known as mixed copulas. Nelson [14] introduced the convex combination approach to generalize the dependency structure of copula, allowing the dependence structure to have more than two copula functions.

In the estimation point of view, the development of estimation and statistical inferential methodology for copula models has been limited. Most of the researches employed a maximum likelihood estimator (MLE) to estimate the unknown dependence parameter in the copula model. The most recent literature using mixed copula with MLE can be found in (Hu [8]; Czado [5]; Nguyen [15]; Maneejuk [12]; Tansuchat and Maneejuk [19]). It is surprising that there has been only limited work on using Bayesian approach to estimate copula models. Bayesian is one of the measures in estimation theory and decision theory, moreover, it is also widely accepted as a good estimator. An alternative way to evaluate an estimator is the maximum posterior within Bayesian statistics. The previous Bayesian work on the copula model can be found in ([6,9,13,18]). These previous studies mentioned the advantage of Bayesian estimation over the MLE in two aspects. Firstly, Bayesian is the sample method, thus it can provide the reliable confidence interval for dependence parameters which are difficult to obtain by MLE [13]. Secondly, Bayesian can solve many difficult problems, especially when the number of parameters is large, which cannot be solved in ML framework. Smith, Gan, and Kohn [18] also suggested that MLE is also difficult when one or more of the margins is discrete, or if the one-step joint estimation of the copula and marginal parameters are used.

Although, the Bayesian estimation in copula model has been already proposed, none of them considered using the Bayesian estimation in mixed copula model. Therefore, this paper attempts to fill the gap by proposing the Bayesian estimation in various combinations of mixed copulas. We consider mixed copulas based on both Elliptical copulas and Archimedean copulas. Then, we apply our proposed method to measure the dependence among oil, gold and stock markets in the United States (US) and extend to do the portfolio optimization. The rest of this paper is organized into four parts as follows. Second section is about methodology, the basic concepts of mixed-copula model and Bayesian estimation. Third section describes the simulation study. Forth section presents our application study. Last section provides a conclusion.

2 Methodology

First, we briefly present the basic idea of the copula approach. Then, the Bayesian estimation procedures are explained.

2.1 Basic Concepts of Mixed Copula

The concept of copula is introduced in Sklar theorem [17], it is the function that is used to join the multivariate distribution functions. If $F_1(x_1), ..., F_n(x_n)$ are the marginal cumulative distribution function (cdf) of the random variable $x_1, ..., x_n$, the joint cdf can be written as

$$F(x_1, ..., x_n) = C(F_1(x_1), ..., F_n(x_n)) = C(u_1, ..., u_n), \tag{1}$$

where C is a copula function and u_i denotes the cumulative distribution value of x_i. Based on the viewpoint of density function, the above equation can be rewritten as:

$$h(u_1, ..., u_n \,|\phi,\, \theta) = \prod_{i=1}^{n} f(x_i \,|\phi_i) \cdot c(u_1, ..., u_n \,|\theta)\ , \tag{2}$$

where $f(x_i \,|\phi_i)$ is the marginal density of each variable x_i, $c(u_1, ..., u_n \,|\theta)$ is the copula density. The parameter ϕ_i and θ are the marginal and dependence parameters, respectively.

In this study, we consider two classes of copula, namely Elliptical and Archimedean copulas. The most used families of Elliptical copulas are the Gaussian (Normal) and Student-t copulas. This copula class has the advantage of being able to identify the correlation between the marginals. However, it is ineffective to identify the closed form expressions and is limited to a radial symmetry. Differently, the Archimedean class, the copula function allows us to capture the asymmetric nature of the data. The most used families of Archimedean copulas are Clayton, Gumbel, Frank, and Joe distributions. As we mentioned before, these copula functions still have a weakness in measuring dependence thus the mixed copula class is proposed by Nelsen [14] to make the copula function become more flexible to various dependence structure. Thus, we can consider various functions from Elliptical and Archimedean to further construct the mixed copula model. To construct the mixed copula function, Nelsen [14] proposed a convex combination approach and thus the mixed copula density can be written as

$$c_{mix}(u_1, ..., u_n \,|\theta) = wc_1(u_1, ..., u_n \left|\theta^1\right.) + (1-w)c_2(u_1, ..., u_n \left|\theta^2\right.)\ , \tag{3}$$

where w is the weight parameter of the interval data with values $[0, 1]$. θ^1 and θ^2 are the dependence parameters of first and second copula functions, respectively. The advantage of this method is a combination of a convex flexibility in determining the weight to calculate the appropriate copula function between these two functions. In the estimation aspect, we consider using two steps IFM (inference for margins) suggested by Patton [16] to obtain the Bayesian estimation. Thus, the marginal estimate parameters specified in the marginal density of each variable and dependence parameter in copula function are estimated separately.

2.2 Marginal Density Specification

Prior to measuring the dependence in copula function, the marginal distribution of each variable is computed using the univariate GARCH (I, J). the specific form of this model can be followed by

$$y_t = u_0 + \varepsilon_t, \tag{4}$$

$$\varepsilon_t = h\eta_t, \tag{5}$$

$$h_t^2 = \alpha_0 + \sum_{i=1}^{I} \alpha_i \alpha_{t-i}^2 + \sum_{j=1}^{J} \beta_j h_{t-j}^2. \tag{6}$$

Equations (4) and (6) are the conditional mean and variance equation, respectively. ε_t is the residual, consisting of standard variance, h_t, and standardized residual, η_t. The standardized residual will be given by the best-fit GARCH (I, J) specification, and then transformed into a uniform distribution $[0, 1]$.

2.3 Bayesian Estimation

Since we adopt a Bayesian approach to be an estimator of mixed copula approach, this statistical model must be completed by specifying the prior distributions for all model parameters. In this study, we specify a uniform prior for the dependence and weight parameters of mixed copula model since we have little prior information available. Note the two step IFM of Joe and Xu [10] is employed here, thus we estimate the marginal parameters ϕ in a first step by Bayesian GARCH model with student-t distribution of Ardia [1] and then estimate the association parameters Θ given $\widehat{\phi}$ IFM the following posterior.

$$f(\Theta, \widehat{\phi}\,|u)\alpha \left[c(u_1, ..., u_n\,|\theta^1, \theta^2, w) \cdot f(y_1, ..., y_n; \widehat{\phi}_1, ...\widehat{\phi}_n)\right] \cdot f(\Theta) \tag{7}$$

where Θ is all dependence and weight parameters. $c(\cdot)$ is the mixed copula density and $f(\Theta) = 1$ is the uniform prior density. Intuitively, our posterior distribution will match the likelihood function in MLE. Unlike the MLE, the estimation for Bayesian estimation is based on Markov chain Monte Carlo (MCMC) methods (see e.g. Chib and Jeliazkov [4]. for a comprehensive overview). In this estimation, we use a Metropolis-Hasting (MH) algorithm (see Hastings [7]) to approximate the posterior distribution of the joint parameter vector as the close from solution of our conditional posterior is not available here.

To sample these parameters, we run the Metropolis-hasting sampler for 10,000 iterations where the first 2,000 iterations serve as a burn-in period. For Metropolis-hasting algorithm, the estimation is based on the Monte Carlo iterates

$$\left\{\Theta^{(1)}, ..., \Theta^{(10000)}\right\}, \tag{8}$$

Note that each iteration is tested to obtain the updated parameter. In this test, we use the acceptance ratio which can be written as

$$r = \frac{f(\Theta^{(j)}, \widehat{\beta}\,|u)}{f(\Theta^{(j-1)}, \widehat{\beta}\,|u)}, \tag{9}$$

If $r <$ random $U(0,1) \rightarrow \Theta^{(j)} = \Theta^{*}$ otherwise $\Theta^{*} = \Theta^{(j-1)}$ The individual MH steps for each iteration $\Theta^{(j)}$ are performed using a symmetric truncated normal random walk proposal, corresponding to the lower and upper bound of copula and weighted parameter. In addition, the variances of the normal proposal parameters are tuned to achieve parameter acceptance rates between 20%–80% as suggested by Besag et al. [2].

3 Simulation Study

We assess the effectiveness of the Bayesian estimation by comparing Maximum likelihood estimator on the simulated data. To demonstrate and investigate the performance, we consider the four mixed copula functions, namely Normal-Student-t, Clayton-Gumbel, Joe-Clayton, Frank-Clayton copula functions. The true copula parameter of each function is provided in the second column of Table 1. In this simulation study, we generate 100 datasets with 100 and 500 observations. Then, these data are re-estimated and computed the absolute bias value of each parameter. In addition, there is another way to summarize the results over all parameters by computing the mean of the bias over the parameters in each scenario. Table 1 summarizes the estimated posterior parameters based on the mixed copula model. For comparison we also include the corresponding MLE. The result shows that our Bayesian estimation performs well in all cases and the bias value tends to be lower as the sample size increases, corresponding to the asymptotic properties estimator. By comparing each estimated parameter, we find that the Bayesian clearly outperforms the MLE only on the Normal-Student-t mixed copula with N = 500. For other cases, it is more accurate than MLE for most but not all parameters bias. So, we consider comparing each scenario by using the mean of the bias over the parameters; the results are provided in the Mean bias row. Based on this table, one can conclude that Bayesian is substantially more accurate than MLE in all cases. Overall, this estimator is more accurate than the MLE in more than half of the bias of parameters.

Table 1. Simulation results

Mix-copula	True value	N = 100		N = 500	
		Bayesian	MLE	Bayesian	MLE
Normal-Student-t	$\theta_N = 0.5$	0.0548	**0.0006**	**0.0077**	0.0885
	$\theta_T = 0.5$	**0.0081**	0.2128	**0.0025**	0.0361
	$df = 4$	**0.5406**	8.5209	**0.6110**	11.7142
	$\omega = 0.5$	**0.0978**	0.4000	**0.0171**	0.0425
	Mean bias	**0.1753**	2.2846	**0.1596**	2.9703
Clayton-Gumbel	$\theta_C = 2$	**0.2916**	0.4873	**0.0756**	0.1366
	$\theta_G = 2$	0.2608	**0.0726**	**0.1308**	0.1718
	$\omega = 0.5$	0.1037	**0.1031**	0.0099	**0.0079**
	Mean bias	**0.2187**	0.2210	**0.0718**	0.1054
Joe-Clayton	$\theta_J = 2$	0.0929	**0.0325**	**0.0592**	0.1989
	$\theta_C = 2$	**0.3855**	0.4421	0.1111	**0.0724**
	$\omega = 0.4$	**0.0013**	0.0046	0.0401	**0.0355**
	Mean bias	0.1599	0.1599	0.0701	**0.0426**
Frank-Clayton	$\theta_F = 2$	**0.1295**	0.5350	0.0869	**0.0384**
	$\theta_C = 2$	**0.0007**	0.7472	0.1386	**0.1239**
	$\omega = 0.4$	0.1645	**0.0595**	0.1204	0.1275
	Mean bias	**0.0982**	0.4472	0.1153	**0.0966**

Note: lower bias presents in bold number.

4 Application Study

This study uses daily time series (April 1, 1996 to March 31, 2016) of the stock index (S&P 500), gold prices (USD/Oz) and crude oil price (West Texas Intermediate). This data set is retrieved from DataStream, and then transformed into log return. Table 2 presents the descriptive statistics of the log return series. The sample mean of all the returns are very close to each other; however, the oil return has greater variation than the others. All data sets show non-normality with negative skewness and high kurtosis values (>3). In addition, the Jarque-Bera (J-B) test confirms that there is no normal distribution in all return series as the test is rejected for all returns. Also, the Augmented Dickey Fuller (ADF) test is conducted to show the stationarity of the data series, and the results are presented in Table 2. Table 3 presents the estimated coefficients and standard errors for marginal distributions of GARCH from Bayesian estimation. The coefficients of GARCH model are all statistically significant as the zero does not exist in the 95% Bayesian credible interval. We can observe that the ARCH parameters (α_1) exhibit significant, indicating that one-period lagged error affects the current conditional volatility. The volatility is quite persistent for all return series as shown by the significance GARCH components (β). The sum of (α_1) and (β) is close to one for all series, indicating a high level of persistence.

Table 2. Descriptive statistics

	Stock	Gold	Oil
Mean	0.0002	0.0002	0.0001
Median	0.0003	0.0000	0.0000
Maximum	0.1096	0.0889	0.1290
Minimum	−0.0947	−0.0982	−0.1444
Std. Dev.	0.0122	0.0110	0.0220
Skewness	−0.2296	−0.0958	−0.1213
Kurtosis	7.9901	6.8708	3.0689
Jarque-Bera	13898	10251	2055.3
Probability	0.0000	0.0000	0.0000
ADF-test	-17.304^a	-17.658^a	-15.195^a

Source: Calculation.
Note: "a" denotes a significance at 1% level.

Table 3. Marginal distributions for Bayesian Estimation on GARCH (1, 1) models

	Stock	Gold	Oil
α_0	0.00952 (0.0094, 0.0100)	0.00896 (0.0083,0.0100)	0.01007 (0.0096,0.0105)
α_1	0.783263 (0.1000, 0.9039)	0.76829 (0.1000,0.9690)	0.77238 (0.1000,0.9378)
β	0.044634 (0.0017,0.3433)	0.07743 (0.0002,0.4391)	0.04434 (0.0010,0.3110)
df	2.407798 (2.0063,2.0098)	2.44509 (2.0206,2.0439)	2.39013 (2.0054,2.0081)

We have created different combinations of the Normal, Student-t, Clayton, Gumbel, Frank and Joe copulas. As mentioned earlier, the mixed copulas allow us to capture different dependence structures among stock, oil and gold prices. The various combination mixed copula models are provided in Table 4. We also compare the results in Table 4. Different mixed copulas are estimated to find the best fit structure of our variables. We find that the best mixed copula for our returns is mixed Normal and Student-t copula, since it has the highest marginal log likelihood values. This indicates that stock, oil and gold markets exhibit the symmetric dependence. Therefore. We select the mixed Normal and Student-t copula as the inference model in our application study.

Tables 5 and 6 provide the result of parameter estimates from Bayesian and MLE, respectively. Apparently, similar results are obtained from these two estimations, but indicating the robustness of our Bayesian computation. Then, let us consider the result from our Bayesian estimation, Table 6 shows that the weight for Normal copula is 0.019 while the weight for Student-t copula is 0.981,

Table 4. Model selection

Copula	Marginal log likelihood
Gaussian - Student-t	35, 098.7457
Gaussian - Clayton	11, 025.2021
Gaussian - Frank	10, 107.6169
Gaussian - Gumbel	10, 392.7667
Gaussian - Joe	8, 021.0465
Student-t - Clayton	34, 398.8183
Student-t - Frank	3, 746.6765
Student-t - Gumbel	35, 006.9696
Student-t - Joe	35, 018.5256
Clayton - Frank	3, 662.3406
Clayton - Gumbel	9, 903.3234
Clayton - Joe	10, 636.6802
Frank - Gumbel	7, 829.1191
Frank - Joe	4, 806.8350
Gumbel - Joe	7, 491.0753

Source: Calculation.

Table 5. Parameter estimation under mixed Normal – Student-t copula from Bayesian

	coef	S.D.	2.5%	97.5%
θ_N	-0.03067	0.0003	-0.03141	-0.02996
θ_T	0.06417	0.0001	0.06400	0.06433
df	2.80553	0.0002	2.80480	2.80598
ω	0.01991	0.00003	0.01985	0.01996

Table 6. Parameter estimation under mixed Normal – Student-t copula from MLE

	coef	S.D.	t-stat	p-value
θ_N	-0.0312	0.0263	-1.1863	0.2355
θ_T	0.0643	0.0058	11.1020	0.0000
df	2.8049	0.0237	118.4786	0.0000
ω	0.0200	0.0027	7.4838	0.0000

Source: Calculation.

indicating that there is a high possibility of symmetric and heavy dependence among oil, gold, and stock volatilities in US market. In addition, we can observe that the copula dependence θ_T is 0. 06417, indicating that oil, gold, and stock volatilities are likely to move in the same direction, but the dependence is sparse.

5 Conclusion

This paper proposes to estimate the mixed copula model using the Bayesian approach. This method has been employed by only a few empirical analysts to date. However, they confirm the better computation performance than the maximum likelihood estimator in various contexts. Firstly, Bayesian can provide the reliable confidence interval for dependence parameters. Secondly, it can deal with many difficult problems, especially when the number of parameters is large, and the number of observations is small. To show the performance of this method in mixed copula model, we conduct a simulation study to examine its accuracy and also compare to the MLE on the simulated data. So, we consider using the mean of the bias over the parameters. The simulation results confirm the higher performance of the Bayesian estimation in various scenarios, especially when the simulated data is small.

Our method is extended to study in real data. In this application study, we apply the Bayesian Estimation on GARCH to model daily returns of oil, gold, and stock markets in the U.S., then the dependence structure between innovations of asset returns are captured by various classes of mixed copulas. We find that the mixed Normal and Student-t copula is the best fit model in capturing the dependence of our data. The result shows that Student-t copula has a greater weight than Normal copula, indicating that there is a high possibility of symmetric and tail dependence among oil, gold, and stock volatilities in US market. Finally, we can conclude that the volatility of oil and gold stocks dependence is sparse and tends to move in the same direction.

Acknowledgements. The authors are grateful to Puay Ungphakorn Centre of Excellence in Econometrics, Faculty of Economics, Chiang Mai University for the financial support.

References

1. Ardia, D.: bayesGARCH: Bayesian Estimation of the GARCH (1, 1) Model with Student-t Innovations in R (2007). http://CRAN.R-project.org/package=bayesGARCH
2. Besag, J., Green, P., Higdon, D., Mengersen, K.: Bayesian computation and stochastic systems. Stat. Sci. **10**, 3–41 (1995)
3. Brooks, C., Burke, P., Heravi, S., Persand, G.: Autoregressive conditional kurtosis. J. Financ. Econ. **3**(3), 399–421 (2005)
4. Chib, S., Jeliazkov, I.: Marginal likelihood from the Metropolis-Hastings output. J. Am. Stat. Assoc. **96**(453), 270–281 (2001)
5. Czado, C., Kastenmeier, R., Brechmann, E.C., Min, A.: A mixed copula model for insurance claims and claim sizes. Scand. Actuar. J. **2012**(4), 278–305 (2012)
6. Danaher, P.J., Smith, M.S.: Modeling multivariate distributions using copulas: applications in marketing. Mark. Sci. **30**(1), 4–21 (2011)
7. Hastings, W.K.: Monte Carlo sampling methods using Markov chains and their applications (1970)

8. Hu, L.: Dependence patterns across financial markets: a mixed copula approach. Appl. Financ. Econ. **16**(10), 717–729 (2006)
9. Huard, D., Évin, G., Favre, A.C.: Bayesian copula selection. Comput. Stat. Data Anal. **51**(2), 809–822 (2006)
10. Joe, H., Xu, J.J.: The estimation method of inference functions for margins for multivariate models (1996)
11. Jondeau, E., Rockinger, M.: Conditional volatility, skewness and kurtosis: existence, persistence, and comovements. J. Econ. Dyn. Control **27**(10), 1699–1737 (2003)
12. Maneejuk, P., Yamaka, W., Sriboonchitta, S.: Mixed-copulas approach in examining the relationship between oil prices and ASEAN's stock markets. In: International Econometric Conference of Vietnam, pp. 531–541. Springer, Cham, January 2018
13. Min, A., Czado, C.: Bayesian inference for multivariate copulas using pair-copula constructions. J. Financ. Econ. **8**(4), 511–546 (2010)
14. Nelsen, R.B.: An Introduction to Copulas (Springer Series in Statistics), p. 3. Springer New York Inc., Secaucus (2006)
15. Nguyen, C., Bhatti, M.I., Komorníková, M., Komorník, J.: Gold price and stock markets nexus under mixed-copulas. Econ. Model. **58**, 283–292 (2016)
16. Patton, A.J.: Modelling asymmetric exchange rate dependence. Int. Econ. Rev. **47**(2), 527–556 (2006)
17. Sklar, M.: Fonctions de repartition an dimensions et leurs marges. Publ. Inst. Statist. Univ. Paris **8**, 229–231 (1959)
18. Smith, M.S., Gan, Q., Kohn, R.J.: Modelling dependence using skew t copulas: Bayesian inference and applications. J. Appl. Econ. **27**(3), 500–522 (2012)
19. Tansuchat, R., Maneejuk, P.: Modeling dependence with copulas: are real estates and tourism associated? In: International Symposium on Integrated Uncertainty in Knowledge Modelling and Decision Making, pp. 302–311. Springer, Cham, March 2018

Modeling the Dependence Among Crude Oil, Stock and Exchange Rate: A Bayesian Smooth Transition Vector Autoregression

Payap Tarkhamtham[1,2(✉)], Woraphon Yamaka[2], and Songsak Sriboonchitta[2]

[1] International College of Digital Innovation, Chiang Mai University, Chiang Mai, Thailand

[2] Centre of Excellence in Econometrics, Faculty of Economics, Chiang Mai University, Chiang Mai, Thailand

payap.tar@gmail.com

Abstract. This study aims to investigate the relationship among West Texas Intermediate crude oil price which represents crude oil, Down Jones Industrial Index and exchange rate. We use smooth transition vector autoregression with Bayesian estimator to estimate the relationship among them due to the conventional VAR often faces with the over-parameterization problem and we can avoid using p-value in making the statistical inference. The results show that there is different relationship among variables in each regime. The impulse response indicates that most of variables will converge to their equilibrium within about 20 months. Finally, the spillover effect shows that three variables either influence or are influenced by other variables.

Keywords: Smooth transition VAR · Bayesian · Logistic function Spillover effects

1 Introduction

Oil is one of the most important commodities in the world. It is a vital input in production, transportation, and almost all human activities and considered as the lifeblood of the economy. According to International Energy Agency (2015), oil becomes the most widely consumed energy in the world. Over the past few decades, oil prices have turbulently bounced up and down over the time. For example, the oil price had sharply increased in 2007 and rapidly fell in 2008 (see Fig. 1). The fluctuation of oil price affects many sectors. For example, expensive fuel can cause higher transportation costs, as well as inflating prices of goods and services.

As suggested in the literature, oil price changes invariably have significant effects on the stock market and exchange rate. Nguyen and Bhatti [14] examined the dependence structure between oil price and the stock market in Vietnam

V. Kreinovich et al. (Eds.): ECONVN 2019, SCI 809, pp. 828–839, 2019.
https://doi.org/10.1007/978-3-030-04200-4_59

Fig. 1. Monthly crude oil spot prices

and China and the results indicate that if the oil prices decrease, Vietnam's stock market will also be down, but China's stock market seems independent from the crude oil price movement. Similarly, Ding et al. [6] and Zhu et al. [20] demonstrated that the impact of oil on stock markets is different across countries. Moreover, Diaz et al. [5] examined the relationship between oil price volatility and stock returns in the G7 economies and showed that the world oil price volatility is generally more significant for stock markets than the volatility of oil price given in local currency. Furthermore, many studies have been conducted on the spillover effect of the exchange rate and the stock market. For example, Naresh et al. [13] investigated the spillover effect of US dollar on the stock indices of BRICS and found that all the markets falling for a rise in the price of US Dollar and vice versa.

Theoretically, Golub [7] and Krugman [12] studied the role of oil prices in explaining exchange rate movements and revealed that oil exporting country may faced exchange rate appreciation when oil price rises and depreciation when oil price falls. In addition, Bloomberg and Harris [2] mentioned that the same relationship between these two variables can be explained by the law of one price. Since, particularly, oil is traded in US dollar and the fluctuation in the US dollar will change the purchasing power of the currencies of other countries. Thus, a depreciation of USD reduces the oil price and increases other countries purchasing power which in turn increases the oil demand that consequently pushes the oil price in USD, and this is the law of one price. Moreover, Aloui et al. [1] and Chang [3] examined the linkage between oil prices and exchange rates. The results revealed that this relationship is weak and almost zero in the pre-crisis, while during the crisis, the relationship became strongly positive. Turhan et al. [18] and Reboredo et al. [16] also investigated the linkage between oil prices and exchange rates and the results are like the work of Aloui et al. [1] and Chang [3], but the direction of correlation is negative during the crisis. According to these studies, the relationship between these variables can be different in each period.

In the literature, there are many methodologies used to investigate the relationship among variables. One of the most flexible and easy to use in multivariate analysis is Vector autoregressive (VAR) model. The VAR model was introduced by Sims [17] and has been widely used in literature. However, it still has some drawbacks as it cannot capture a nonlinear structure in the relationship between the data. Thus, Weise [19] proposed a logistic smooth transition vector autoregressive (LSTVAR) model, considering the structural change, the lag variable, and autoregressive slope parameter. Thus, in this study, we employ this model to investigate the relationship among crude oil price, the stock market, and exchange rate. This model allows us to characterize the relationship among these variables under different economic conditions, namely booms and busts. In the estimation of STVAR, our study considers the Bayesian estimation as it is an efficient tool for estimating the large parameters in the system equations. Pastpipatkul et al. [15] suggested that the Bayesian estimation can deal with the over-parameterization problem which often occurs in the VAR framework. Therefore, our study takes the advantage of this estimator in computing the parameters in STVAR.

The rest of paper is structured as follows. Section 2 describes the methodology; Smooth transition vector autoregression (STVAR) and Bayesian inference. Section 3 shows data description and the estimated results of STVAR, impulse response function and forecast error variance decomposition (spillover effect). Finally, Sect. 4 presents conclusion.

2 Methodology

2.1 Smooth Transition Vector Autoregression (STVAR)

The vector autoregressive (VAR) model was introduced by Sims [17] as a stochastic process model used to capture linear interdependence among multiple time series. Let $Y_t = (y_{1t}, y_{2t}, \ldots, y_{nt})$ as $(n \times 1)$ vector of variables at time t. The VAR model with lag-p is as follows:

$$Y_t = \sum_{i=1}^{p} \Pi_i Y_{t-i} + \varepsilon_t \tag{1}$$

where Π_i is a $n \times n$ autoregressive coefficient matrices of the lag terms Y_{t-i}, where $i = 1, .., p$ correspond to lags. The vector $\varepsilon_t = (\varepsilon_{1t}, ..., \varepsilon_{nt})$ is multivariate normally distributed error where $E(\varepsilon_t) = 0$ and $E(\varepsilon_{it}, \varepsilon_{jt}) = \Sigma$ for $i = j$ and 0 otherwise, hence $\varepsilon_t \sim NID\,(0, \Sigma)$, where Σ is regarded as non-negative variance-covariance matrix of the residuals. The STVAR model suggested by Weise [19] can be defined as follows:

$$y_t = \mu + \sum_{i=1}^{p} \Pi_i y_{t-i} + G\left(\gamma, c; z_t\right) \left[\mu + \sum_{i=1}^{p} \Pi_i y_{t-i}\right] + \varepsilon_t, \tag{2}$$

where y_t is $n \times 1$ vector of variables at time t, μ are $n \times 1$ intercept vector, Π_i is $n \times n$ coefficient matrix of the lag terms Y_{t-i}, where $i = 1, .., p$ correspond to lags. Then, we assume that the regime changes are captured by first order logistic smooth transition function, defined as

$$G(\gamma, c; z_t) = (1 + \exp\{-\gamma(z_t - c)\})^{-1}, \tag{3}$$

where $G(\gamma, c; s_t)$ is a logistic transition function. γ is non-negative speed of the smooth transition (smooth parameter). If $\gamma \to \infty$, it becomes a two-regime model. z_t is the transition variable which is one of the variables in y_t. c is the threshold parameter.

2.2 Posterior Estimation

In this study, we employ Gibbs Sampling and Metropolis Hastings to sample the parameters in the posterior distribution. This posterior consists of the estimation of likelihood function and the prior distribution whereas the likelihood can be written as

$$L(\theta|Y) = \prod_{t=1}^{T}\left(\frac{1}{\sqrt{|\Sigma|^T}} \times \exp\left[-\frac{1}{2}tr\varepsilon'_t\varepsilon_t\Sigma^{-1}\right]\right), \tag{4}$$

where ε_t is the M dimension errors derived from Eq. 2. θ is the all parameter estimates of STVAR equation. Then in setting the priors, we specify the prior density for all parameters in the model. We specify a standard normal prior for Π_i as $p(\Pi) \propto N(\tilde{\Pi}, a)$, where $\tilde{\Pi}$ and a are the mean and variance of Autoregressive parameters. The parameter Σ is specified to have an Invert Wishart prior with scale matrix $\varepsilon'_t\varepsilon_t$ and degree of freedom V, thus we have $p(\Pi) \propto IW(\varepsilon'_t\varepsilon_t, V)$. For the transition and smooth parameters, we use the uniform distribution and gamma distribution respectively. Note that the prior distribution for the threshold c is restricted as uniformly distributed between the upper and lower limits of the middle 80% of the observed transition variables to guarantee to have the threshold within the range of transition variable. Therefore, we can construct our posterior as follows

$$P(\theta) \propto L(\theta|Y) \cdot p(\Pi) \cdot p(\Sigma) \cdot p(c) \cdot p(\gamma). \tag{5}$$

Then, to estimate these all parameters, we employ Gibbs Sampling method to sample Π and Σ from the posterior distribution as we can find the close form solution of these parameters. The conditional posterior of Π and Σ can. be computed as in the following

Let $X = \{Y_{t-p}, G(\gamma, c; z_t) \cdot Y_{t-p}\}$, the conditional posterior distribution for Π is

$$\Pi^* = (X'X)^{-1}X'Y + \frac{(X'X)^{-1}}{a\Sigma^2}. \tag{6}$$

And the conditional posterior distribution for Σ is

$$\Sigma^* = IW(\varepsilon'_t\varepsilon_t/T, V). \tag{7}$$

Finally, for the parameters c and γ, there exists no close form solution to derive them directly in the Gibbs Sampling, thus the Metropolis Hastings within Gibbs is used to find these parameters. To do this, the acceptance ratio is conducted to evaluate the updated parameters in each draw [4]. Then, the Gibbs Sampling scheme for our posterior computation, therefore, takes the following form.

(1) Draw Π^i from $P(\Pi^{(i)} \left| \Pi^{(i-1)}, \Sigma^{(i)}, c^{(i)}, \gamma^{(i)}, Y_t \right.)$;
(2) Draw Σ^i from $P(\Sigma^{(i)} \left| \Pi^{(i)}, \Sigma^{(i-1)}, c^{(i)}, \gamma^{(i)}, Y_t \right.)$;
(3) Draw c^i from $P(\Sigma^{(i)} \left| \Pi^{(i)}, \Sigma^{(i-1)}, c^{(i)}, \gamma^{(i)}, Y_t \right.)$;
(4) Draw $[\gamma^i$ from $P(\gamma^{(i)} \left| \Pi^{(i)}, \Sigma^{(i)}, c^{(i)}, \gamma^{(i-1)}, Y_t \right.)$;
(5) Repeating step 1–4 for 10,000 iterations.

Finally, we discard the first 2,000 iterations as the burn-in and take the expectation of the parameter samplers to make the inference.

3 Data and Results

3.1 Data Description

In this study, we use the monthly data of West Texas Intermediate crude oil (OIL) spot price, Down Jones Industrial Average Index (DJI) and Trade Weight US Dollar Index (USD). The 360 observations are collected from Thompson Reuter and cover the periods from July 1988 to June 2018. Then, we transform the data into log-return $r_t = \ln(p_t/p_{t-1})$, where r_t is a log-return, p_t is the data at time t and p_{t-1} is the data at time t − 1. The log-return series are plotted in Fig. 2.

Table 1 reports the descriptive statistics of variables. The average returns for all variables are positive and close to zero. The skewness is negative for OIL and DJI, indicating that the data are more likely negative, and the skewness of USD is positive indicating that the data are more likely positive. The kurtosis of all variables is greater than 3, thus they are leptokurtic. Therefore, we use Jarque-Bera to test whether variables have a normal distribution and found that the variables depart from normal distribution. Finally, Augmented Dickey-Fuller test (ADF) was used to test that the variables are stationary. Finding that MBF shows decisive evidence against the null hypothesis with minimum Bayes factor (<1/300), indicating the variables are stationary.

3.2 Testing for a Threshold Effect

Using the non-linear type model needs the verification of the non-linear structure in the data. Our interest is on the existence of the threshold effect in the variables of interest. This concern is very important motivation for using STVAR models. If the real data generating process is nonlinear, the STVAR is acceptable. If, there is a linear relationship in the data, STVAR might fail to give the reasonable results and a linear VAR model may provide more robust estimation

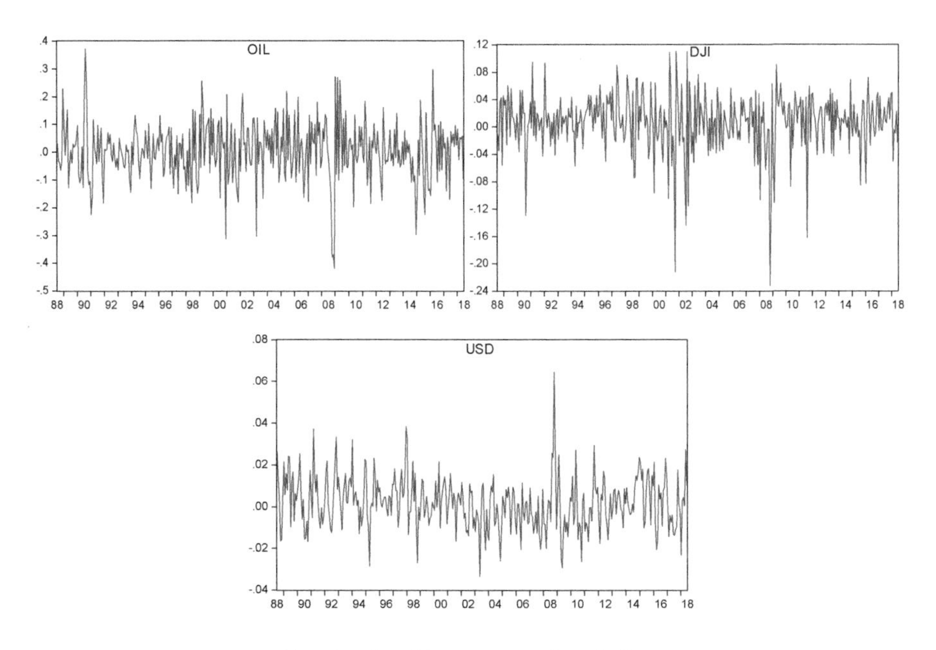

Fig. 2. Monthly returns of variables

Table 1. Descriptive statistics

Statistics	OIL	DJI	USD
Mean	0.004	0.0068	0.002
Median	0.0082	0.0109	0.0019
Maximum	0.373	0.1112	0.0647
Minimum	−0.4191	−0.2316	−0.0334
Std. Dev.	0.1068	0.0435	0.0125
Skewness	−0.27	−1.3196	0.4422
Kurtosis	4.4376	7.7103	4.45
Jarque-Bera	35.3767[0.0000]	437.2875[0.0000]	43.2664[0.0000]
ADF			
None	−17.9995[0.0000]	−19.4523[0.0000]	−12.2605[0.0000]
Intercept	−17.9967[0.0000]	−19.9291[0.0000]	−12.4351[0.0000]
Intercept and Trend	−17.9736[0.0000]	−19.9259[0.0000]	−12.5130[0.0000]

Note: [] shows the Minimum Bayes factor which can computed by *eplogp*, where p is p-value (see, [10]). The MBF can interpreted as follows: MBF between 1–1/3 is considered weak evidence for H_1, from 1/3–1/10 considered moderate evidence, 1/10–1/30 considered substanial evidence, 1/30–1/100 considered strong evidence, 1/100–1/300 considered very strong evidence, and <1/300 considered decisive evidence. (see [8,9])

if the threshold effect is trivial. To examine the threshold effect in the Bayesian framework, the Bayes factor is employed to select the best model for further analysis. Bayes factor is a useful tool for selecting a possible model which is presented in the ratio of the posterior under one model to another model. Here, the linear VAR model is specified to be a null hypothesis model, denoted by M_1 and the STVAR model to be an alternative model, denoted by M_2. The Bayes factor BF is computed by

$$BF = \frac{\int P(\theta_1 \,|M_1)}{\int P(\theta_2 \,|M_2)} \tag{8}$$

where $\int P(\theta_1 \,|M_1)$ and $\int P(\theta_2 \,|M_2)$ are the integral of posterior density of the null and alternative models, respectively. θ_1 and θ_2 are the vector of parameters of M_1 and M_2, respectively. For choosing the appropriate model, referred to the approach of Kass and Raftery [11].

In this test, we test the threshold effect in all variables, consisting of OIL, DJI, and USD. The result of this test is provided in Table 2. We can observe that there exists the threshold effect when DJI is determined to be transition variable as the $\log(BF)$ is less than 2, indicating the rejection of the linear VAR model. This evidence supports us to use LSTVAR model with transition variable DJI.

Table 2. Threshold effect testing

Transition variable	$\log(BF)$	Threshold effect
OIL	5.5118	No threshold effect
DJI	−0.1154	Threshold effect
USD	16.2188	No threshold effect

Note: $\log(BF) < 2$: supports M_2, $2 < \log(BF) < 6$: positive supports M_1, $6 < \log(BF) < 10$: strongly supports M_1, $\log(BF) > 10$: very strong supports M_1

3.3 Smooth Transition Vector Autoregression (STVAR)

To check whether the parameter estimates are significant or not, we consider the 95% Bayesian credible interval. If there exists zero in the interval 2.5% to 97.5%, we can confirm that this parameter is not significant. The result shown in Table 3 reveals that the autoregressive parameters are not significant as the zero values are exhibited in the 95% credible intervals, while threshold and smooth parameter provide a Bayesian significant. This result confirms the non-linear structure of our model, however the relationship among these three markets seem quite weak.

Table 4 provides the parameters estimated of the 2 regimes in the crude oil model. The first regime shows that most of the mean values in each equation is less than the second regime. Thus, this indicates that regime 1 is the low regime,

Table 3. Bayesian credible interval

Variables	Constant	y_{1-t}	y_{2-t}	y_{3-t}
Regime 1				
CRD	[−0.6484, 0.2047]	[−11.9600, 12.1100]	[−14.4400, 8.9660]	[−223.7000, 153.3000]
DJI	[−1.0480, −1.0310]	[−0.2260, 0.2446]	[−0.2147, 0.2420]	[−3.5620, 3.8320]
USD	[0.0428, 0.0888]	[−0.6020, 0.7053]	[−0.0278, 1.2430]	[−6.9150, 13.6100]
Regime 2				
CRD	[−0.3685, 1.2460]	[−23.0700, 22.7500]	[−16.8400, 27.3900]	[−295.5000, 426.4000]
DJI	[1.9920, 2.0250]	[−0.4653, 0.4292]	[−0.4564, 0.4052]	[−7.3430, 6.8170]
USD	[−0.1679, −0.0806]	[−1.3580, 1.1300]	[−2.3480, 0.0467]	[−25.3100, 13.9900]
Smooth	[4.9800, 5.0200]			
Threshold	[−0.0333, −0.0317]			

Note: [] is Bayesian credible interval 2.5% and 97.5%

Table 4. Estimation of STVAR

Variables	Constant	y_{1-t}	y_{2-t}	y_{3-t}
Regime 1				
CRD	−0.2220 (0.2183)	0.0249 (6.1240)	−2.7560 (5.9150)	−36.1100 (95.5300)
DJI	−1.0400 (0.0111)	0.0070 (0.1082	0.0130 (0.1047)	0.0542 (1.6920)
USD	0.0662 (0.0099)	0.0432 (0.2753)	0.6125 (0.2690)	3.2780 (4.3180)
Regime 2				
CRD	0.4398 (0.4136)	−0.0868 (11.6600)	5.2430 (11.1800)	66.8000 (183.0000)
DJI	2.0080 (0.0214)	−0.1036 (3.2420)]	−0.0252 (0.1979)	−0.0134 (0.2059)
USD	−0.1249 (0.0188)	−5.5290 (8.2730)	−1.1600 (0.5085)	−0.0984 (0.5240)
Smooth	5.0000 (0.0316)			
Threshold	−0.0319 (0.0004)			

while regime 2 is the high regime. The direction of all coefficients becomes inverse when they vary across the regimes.

According to this result, we can make the interpretation as follows: In regime 1, an increase in OIL return by 1% in this month can further increase its return by 0.0249% in next month. On the other hand, an increase in DJI and USD returns by 1% could decrease the oil return by 2.756% and 36.11%, respectively. Then, an increase in OIL, DJI, and USD returns by 1%, DJI will increase 0.007%, 0.013%, and 0.0542%, respectively. Besides, an increase in USD returns by 1% will appreciate an USD by 0.0432%, 0.6125%, and 3.278%, respectively.

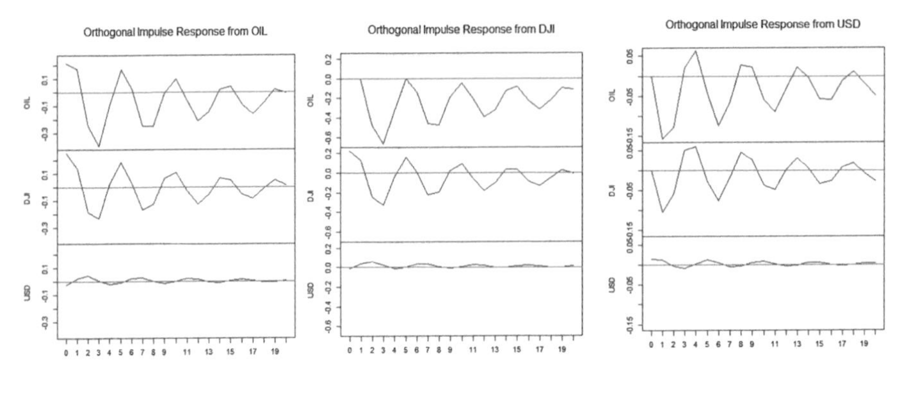

Fig. 3. Impulse response

In the second regime, we find that an increase in OIL return by 1% in this month can decrease the OIL return by 0.0868% in next month but contribute a positive effect to DJI and USD, by 5.243% and 66.8%, respectively. On the other hand, the homogenous effect of DJI return on the other variables is all negative. Likewise, the negative effect of USD return is also found in this regime for all variables.

3.4 Impulse Response Function Analysis

We then investigate the impulse response function of our model in order to see the reaction of each variable in response to some external shocks. Figure 3 illustrates the orthogonal impulse response of OIL, DJI, and USD returns. We can observe that these orthogonal impulse responses to shock have the same pattern bouncing up and down between positive and negative effect except the shock in USD as the response to this shock is quite low and stable. Let us consider the shock of oil returns, we can see that the shocks in OIL and DJI have a similar effect to the response in the other variables. The response of each variable to the shocks has both positive and negative direction and illustrates larger cyclical effects along the horizon line. However, the responses reach the equilibrium within 20 months. These results indicate that oil and gold shocks are likely to play an important role in the relationship among USD, oil and gold markets as it contribute a substantial effect to these markets including themselves.

3.5 Forecast Error Variance Decomposition: Spillover Effect

In this section, we present the spillover effects of our variables using forecast error variance decomposition from STVAR spillover framework. This spillover table therefore provides an approximate input-output decomposition of the total spillover of three variables. The result is reported in Table 5. The values in Panel A present the contribution of variance from each others. Panel B presents the overall contribution of variance for each variable. The sum of variances in a

Table 5. Forecasting results of error variance decomposition(Spillover effect)

	OIL	DJI	USD	SUM
Panel A				
OIL	76.178	0.012	23.81	100
DJI	51.387	41.783	6.83	100
USD	35.02	52.287	12.693	100
Panel B				
TO others	23.822	58.217	87.307	169.346
FROM others	86.407	52.299	30.64	169.346
NET spillovers	−62.585	5.918	56.667	

Note: The forecast variance decomposition for the three variables is based STVAR of order 1 (as determined by the BIC), identified

column, excluding the contribution to its own variance, is denoted as To others, indicating the spillover effect from one variable to the other variables. While the sum of variances in a row, excluding the contribution to its own variance, is denoted as From others, indicating the spillover effect from other variables to one variable. Finally, the NET spillovers presents the difference between To others and From others. If it is a positive value, it indicates that the variable is transmitter of spillover effects; whereas a negative value indicates the receiver spillover effects. According to this result, we can observe that there is quite a high level of spillover transmission among the OIL, DJI, and USD as the average total spillover is approximately 169.346%. Moreover, considering individual variable, the result shows both negative and positive sign of the net spillovers. OIL variable presents a negative net spillover, while there exhibits the positive sign for DJI and USD. This indicates that the net spillover of USD contributes the largest spillover to other variables while OIL represents the highest receiver. Thus, we can conclude that these three variables either influence or are influenced by other variables.

4 Conclusions

In this paper, we investigate the relationship between crude oil, stock market and exchange rate within the framework of STVAR model using Bayesian method as the estimator of the model. The results from STVAR model show that when the regime changes from regime 1 to regime 2 the direction of effect is inverse. However, the relationship among our variables are weak as the insignificant relationships are obtained. We further extend our analysis to study the spillover effect and impulse response function. The Impulse response reveals that if we shock each variable, the response of the markets is convergence to equilibrium within 20 months. We also find that the shock in USD would cause a small response in the other markets. However, the results of spillover effect among these three markets are quite different, we find that the spillover effect of USD is the largest. One possible reason is that the dynamics of the impulse responses is independent of the status of the USD when the shocks hit.

Acknowledgements. The authors are grateful to Puay Ungphakorn Centre of Excellence in Econometrics, Faculty of Economics, Chiang Mai University for the financial support.

References

1. Aloui, R., Assa, M.S.B., Hammoudeh, S., Nguyen, D.K.: Dependence and extreme dependence of crude oil and natural gas prices with applications to risk management. Energy Econ. **42**, 332–342 (2014)
2. Blomberg, S.B., Harris, E.S.: The commodity-consumer price connection: fact or fable? Econ. Policy Rev. **1**(3), 21–38 (1995)
3. Chang, K.L.: The symmetrical and positive relationship between crude oil and nominal exchange rate returns. N. Am. J. Econ. Financ. **29**, 266–284 (2014)
4. Chib, S., Greenberg, E.: Understanding the metropolis-hastings algorithm. Am. Stat. **49**(4), 327–335 (1995)
5. Diaz, E.M., Molero, J.C., de Gracia, F.P.: Oil price volatility and stock returns in the G7 economies. Energy Econ. **54**, 417–430 (2016)
6. Ding, H., Kim, H.G., Park, S.Y.: Crude oil and stock markets: causal relationships in tails? Energy Econ. **59**, 58–69 (2016)
7. Golub, S.S.: Oil prices and exchange rates. Econ. J. **93**(371), 576–593 (1983)
8. Goodman, S.N.: Toward evidence-based medical statistics. 1: The P value fallacy. Ann. Intern. Med. **130**(12), 995–1004 (1999)
9. Goodman, S.N.: Toward evidence-based medical statistics. 2: The Bayes factor. Ann. Intern. Med. **130**(12), 1005–1013 (1999)
10. Held, L., Ott, M.: On p-values and Bayes factors. Annu. Rev. Stat. Appl. **5**, 393–419 (2018)
11. Kass, R.E., Raftery, A.E.: Bayes factors. J. Am. Stat. Assoc. **90**(430), 773–795 (1995)
12. Krugman, P.: Oil shocks and exchange rate dynamics. In: Exchange Rates and International Macroeconomics, pp. 259–284. University of Chicago Press (1983)
13. Naresh, G., Vasudevan, G., Mahalakshmi, S., Thiyagarajan, S.: Spillover effect of US dollar on the stock indices of BRICS. Res. Int. Bus. Financ. **44**, 359–368 (2018)
14. Nguyen, C.C., Bhatti, M.I.: Copula model dependency between oil prices and stock markets: evidence from China and Vietnam. J. Int. Financ. Mark. Inst. Money **22**(4), 758–773 (2012)
15. Pastpipatkul, P., Yamaka, W., Wiboonpongse, A., Sriboonchitta, S.: Spillovers of quantitative easing on financial markets of Thailand, Indonesia, and the Philippines. In: International Symposium on Integrated Uncertainty in Knowledge Modelling and Decision Making, pp. 374–388. Springer, Cham (2015)
16. Reboredo, J.C., Rivera-Castro, M.A., Zebende, G.F.: Oil and US dollar exchange rate dependence: a detrended cross-correlation approach. Energy Econ. **42**, 132–139 (2014)
17. Sims, C.A.: Macroeconomics and reality. Econometrica J. Econ. Soc. **48**, 1–48 (1980)

18. Turhan, M.I., Sensoy, A., Hacihasanoglu, E.: A comparative analysis of the dynamic relationship between oil prices and exchange rates. J. Int. Financ. Mark. Inst. Money **32**, 397–414 (2014)
19. Weise, C.L.: The asymmetric effects of monetary policy: a nonlinear vector autoregression approach. J. Money Credit Bank. **31**, 85–108 (1999)
20. Zhu, H.M., Li, R., Li, S.: Modelling dynamic dependence between crude oil prices and Asia-Pacific stock market returns. Int. Rev. Econ. Financ. **29**, 208–223 (2014)

Effect of FDI on the Economy of Host Country: Case Study of ASEAN and Thailand

Nartrudee Sapsaad[1], Pathairat Pastpipatkul[2], Woraphon Yamaka[2(✉)], and Songsak Sriboonchitta[2]

[1] Faculty of Economics, Chiang Mai University, Chiang Mai, Thailand
nartrudee.supsaad@gmail.com
[2] Centre of Excellence in Econometrics, Faculty of Economics, Chiang Mai University, Chiang Mai, Thailand
ppthairat@hotmail.com, woraphon.econ@gmail.com

Abstract. The aim of this study is to investigate the effect of foreign direct investment (FDI) on the economy of host country by using the sample of ASEAN–FDI inflows to Thailand and Thailand's economic indicators during the year 2005–2017. In order to complete the better implication of FDI effect on the economy, this study incorporates other economic variables, namely stock market, exportation, GDP growth, consumer price index, importation and production of manufacturing index. Time series analysis and Bayesian method are employed in this study as they are useful estimators for making a prediction in various applied economic and econometric models. The basic background of Bayesian method refers to the Bayes theorem for computing the posterior distribution of interested parameters. Additionally, the posterior can be calculated by the prior distribution and the likelihood function through the joint distribution. This approach is based on multiple linear regression context which confirms the significant effect of FDI on the Thai economy. Nevertheless, there are FDI inflows from some countries in ASEAN which do not have any significant effect on Thai economy.

Keywords: Foreign direct investment · Thai economy
Bayesian multiple linear regression

1 Introduction

There have been many studies about the relationship between Foreign Direct Investment (FDI) and economic growth in developing countries. Some researchers fully realize that FDI plays role as a key driver to propel the economic growth because the effective FDI becomes the new challenge for the developing countries to accomplish sustainable development goals and it can enrich economy of the destination country through spillover effects and productivity advancement (see e.g., Berthélemy and Demurger [4], Borensztein et al. [5],

V. Kreinovich et al. (Eds.): ECONVN 2019, SCI 809, pp. 840–852, 2019.
https://doi.org/10.1007/978-3-030-04200-4_60

De Mello Jr. [10], Khalifah et al. [17]). Nevertheless, Bende-Nabende et al. [3] argued that FDI might not bring a host of benefits to every developing country but also cause many forms of costs to arise. Therefore, the effect between FDI and the host economy appeared to be mixed. FDI in theory and previous studies is believed to produce a higher productivity in the host country. In economics, Mankiw [19] defined the investment in such a way that "the investment means purchases of goods (such as capital equipment, structures and inventories) used to produce other goods; not financial investment, such as stock, bonds, and mutual funds". Therefore, the investment is one of the dynamic components of Gross Domestic Product (GDP). When the investment is increased, it gives a production boost, larger exportation, employment development; the latter is a rise in consumption and increased tax revenues by government spending. Finally, the GDP also rises; the host economy's opportunities have expanded. Unfortunately, in fact, FDI does not always boost the economy of every host country. FDI could be harmful to the host country as well since FDI might destroy the resource allocation and slow down the economic growth due to trade, price, financial and other factors [13] The study of Encinas-Ferrer et al. [12] also suggested that FDI could create the negative effect on the GDP growth in the host country due to a lower multiplier effect on the host country, when the proportion of FDI was smaller than the domestic investment.

Accordingly, the arguments are very interesting for this study to seek for the better implication of FDI effect on the economy of the host country. The study should also be concerned with various economic aspects, rather than GDP growth alone. The other studies have investigated FDI effect on the host economy in many ways. The details are as follows:

a. The effect of FDI inflow on the stock market in the host country The study of Claessens et al. [9] has interest in the effect of FDI on the 77 emerging stock markets which found that a positive effect in terms of financial market development in the host countries. The study of Sayek et al. [21] and Baker et al. [2] argued that FDI might have a negative effect to the stock market development since FDI could be the substitution of the short-term investment especially, the short-term equity and debt instruments; therefore they might be the speculator. Desai et al. [11] suggested that an increased FDI inflows with the form of intra-company loans or venture capitalist might cause poor stock market development in host country.

b. The effect of FDI inflow on the manufacturing production development in the host country Borensztein et al. [5] revealed that FDI played as a main channel of technology spillovers from the home country to the host country. Multinational enterprises (MNEs) provided the high level of technology with higher productivity for the host country.

c. The effect of FDI inflow on the inflation in the host country Trevino et al. [23] discovered the connection between FDI and inflation in the host country through Consumer Price Index (CPI). Their study found that increased FDI inflows rendered a decreasing in CPI rate, then a lower inflation rate. The results indicated a negative relationship between FDI and CPI and it also helped the host country to lessen inflation rate. The study of Reinhart and Rogoff [20] argued that the increasing FDI inflows might result in the price of goods and services and thereby inducing a higher inflation rate.

d. The effect of FDI inflow on the exportation and importation in the host country Zhang [25] researched the effect of FDI from the rest of the world in China's economy. His study indicated that FDI inflows had a positive relationship with China's export, that promoted high production in the local industry. The study of Chaisrisawatsuk, and Chaisrisawatsuk [7] on Thailand's economy found against that an increased FDI inflows could drop off the exportation in the host country to other trading partners because they exported the products from the host country to the home country and vice versa, that caused the crowding out export to the others. Furthermore, the host country had to import more raw materials and intermediate goods from the home country and/or the rest of the world because of the dependence on the production technology.

Regarding to the previous studies about the effect of FDI on the different countries. We found the important point that how about the effect of FDI inflows on Thailand economy with various economic aspects. In Southeast Asia, Chandprapalert [8] stated that Thailand economy was the second largest economy and its performance pretty good. Furthermore, Thailand was considered as one of important states on ASEAN. Keorite and Moubarak [16] also mentioned that Thailand was a popular destination for foreign investors and a newly industrialized country due to the abundance of resources and the policy with regarding foreign investors. Additionally, the magnitude impacts of FDI was remarkable for the intensive competition among ASEAN members. Most of the available investigations were interested in the effect between FDI inflows from ASEAN and Thailand economy. Nevertheless, they were not enough substance for considering FDI effect on the host economy because they have pointed out only the aspect of the GDP growth. Hence, the other economic aspects of the host country, for instance, the stock market, production improvement and inflation, should be addressed in the study. Moreover, only a few researchers have applied the Bayesian method in their studies, that is a straightforward idea to consider the applied economic problems. Simionescu [22] studied the relationship between FDI and EU economic growth during the economic crisis. Bayesian linear regression model was employed in his work in order to indicate the results regarding the relationship between FDI and EU. The study emphasised that FDI was the reciprocal effects on EU economic growth during economic crisis. Furthermore, Simionescu [22] recommended that Bayesian linear regression model was more appropriate on time series than a traditional linear regression because its coped with the incidence of the increased sample size or large number.

Ultimately, there is little study examining the relationship between FDI and the host economy through Bayesian method, especially FDI inflows from ASEAN to Thai economy. The aim of this study is to investigate the effect of FDI on the economy of host country by using the sample of ASEAN-FDI inflows to Thailand economy during the year 2005–2017. In order to complete the better implication the FDI effect on the host economy; we decide to study the effect of FDI inflows from ASEAN on Thai economy with various economic aspects - the stock market development, GDP, production improvement, importation, exportation and inflation. Time series analysis and Bayesian methods are employed to this study as they are useful estimators for making a prediction in the applied economic and econometric models. Thus, This study is designed to investigate "How can FDI inflows from ASEAN affect Thai economy?".

2 Methodology

2.1 Overview of Bayesian Regression Method

Canova [6] indicated that the Bayesian method was a useful estimator in many applied economic and econometric models. The basic background of this method was referred to the Bayes theorem which leads to computing the posterior of interested parameters. This posterior can be computed from prior and the likelihood. Let the vector of parameters of interest α lie in a set A and is the observed variable. Prior information is calculated by a density $g(\alpha)$. The sample of information is presented by a density $f(y\,|\alpha) \equiv L(\alpha\,|y)$, interpreted as the likelihood, thus we have

$$g(\alpha\,|y) = \frac{f(y\,|\alpha)p(\alpha)}{f(y)}, \tag{1}$$

$$g(\alpha\,|y) = L(\alpha\,|y)\;p(\alpha) \tag{2}$$

where $f(y) = \int f(y\,|\alpha)g(\alpha)$ is the marginal likelihood of y, $g(\alpha\,|y)$ is the posterior density of α and $p(\alpha)$ is the prior distribution of $alpha$. In this study, this Bayesian estimation is used to estimate the unknown parameters in the linear regression context.

$$y_t = \alpha x_t + \varepsilon_t \tag{3}$$

where $\varepsilon_t \sim i.i.d.N(0, \sigma^2)$ is the error term. y_t is the dependent variable at time t and x_t is the independent variable at time t. α is the parameter estimates. The model and method of this study are followed by the research of Kruschke et al. [18]. Thus Markov Chain Monte Carlo (MCMC) method with Gibbs sampling is used for estimating the simulation of posterior distribution in the context of regression model with Gaussian errors. The prior distribution incorporated in the posterior is specified as follows: a multivariate Gaussian prior on the vector α, and an inverse Gamma prior on the conditional error variance. Thus, $\alpha\tilde{}N(a_0, A_0^{-1})$, where a_0 is prior mean for α and A_0^{-1} is prior variance of α; and $\sigma^2 \sim \Gamma^{-1}(c/2, d/2)$ where c and d are shape and scale parameters for prior variance.

3 Data and Empirical Model

This study is designed to investigate "How can FDI inflows from ASEAN affect Thai economy?". From the evidence between ASEAN and Thailand, during the year 2005–2017. Time series analysis and Bayesian methods are employed in this study as they are useful estimators for making a prediction. This approach is based on multiple linear regression context which confirms the significant effect of FDI inflows from ASEAN to Thai economy.

The data sets are collected from the Bank of Thailand (BOT). We employ the time series data of quarterly data sets over the period from Q1,2005 to Q1,2017. The sample of FDI data are provided by ASEAN-FDI inflows to Thailand, which are classified by country, namely Indonesia (FDIID), Malaysia (FDIMY), the Philippines (FDIPH), Singapore (FDISG), Brunei Darussalam (FDIBN), Vietnam (FDIVN), Myanmar (FDIMM), Lao PDR (FDILA) and Cambodia (FDIKH). In order to complete the better implication for the effect of FDI on the economy in the host country; we decide to study the effect of FDI inflows from ASEAN to Thai economy with various economic aspects, namely stock market (SET_t), exportation ($THEX_t$), Gross Domestic Products growth ($THGDP_t$), inflation ($THINF_t$), production improvement (THP_t), and importation ($THIM_t$). Thus, we develop the empirical models as follows:

$$\begin{aligned} SET_t = &\alpha_{10} + \alpha_{11} FDIID_t + \alpha_{12} FDIMM_t + \alpha_{13} FDIPH_t + \alpha_{14} FDISG_t \\ &+ \alpha_{15} FDIBN_t + \alpha_{16} FDIVN_t + \alpha_{17} FDIMY_t + \alpha_{18} FDILA_t \\ &+ \alpha_{19} FDIKH_t + \varepsilon_t \end{aligned} \tag{4}$$

$$\begin{aligned} THEX_t = &\alpha_{20} + \alpha_{21} FDIID_t + \alpha_{22} FDIMM_t + \alpha_{23} FDIPH_t + \alpha_{24} FDISG_t \\ &+ \alpha_{25} FDIBN_t + \alpha_{26} FDIVN_t + \alpha_{27} FDIMY_t + \alpha_{28} FDILA_t \\ &+ \alpha_{29} FDIKH_t + \varepsilon_t \end{aligned} \tag{5}$$

$$\begin{aligned} THGDP_t = &\alpha_{30} + \alpha_{31} FDIID_t + \alpha_{32} FDIMM_t + \alpha_{33} FDIPH_t + \alpha_{34} FDISG_t \\ &+ \alpha_{35} FDIBN_t + \alpha_{36} FDIVN_t + \alpha_{37} FDIMY_t + \alpha_{38} FDILA_t \\ &+ \alpha_{39} FDIKH_t + \varepsilon_t \end{aligned} \tag{6}$$

$$\begin{aligned} THINF_t = &\alpha_{40} + \alpha_{41} FDIID_t + \alpha_{42} FDIMM_t + \alpha_{43} FDIPH_t + \alpha_{44} FDISG_t \\ &+ \alpha_{45} FDIBN_t + \alpha_{46} FDIVN_t + \alpha_{47} FDIMY_t + \alpha_{48} FDILA_t \\ &+ \alpha_{49} FDIKH_t + \varepsilon_t \end{aligned} \tag{7}$$

$$\begin{aligned} THP_t = &\alpha_{50} + \alpha_{51} FDIID_t + \alpha_{52} FDIMM_t + \alpha_{53} FDIPH_t + \alpha_{54} FDISG_t \\ &+ \alpha_{55} FDIBN_t + \alpha_{56} FDIVN_t + \alpha_{57} FDIMY_t + \alpha_{58} FDILA_t \\ &+ \alpha_{59} FDIKH_t + \varepsilon_t \end{aligned} \tag{8}$$

$$\begin{aligned} THIM_t = &\alpha_{60} + \alpha_{61} FDIID_t + \alpha_{62} FDIMM_t + \alpha_{63} FDIPH_t + \alpha_{64} FDISG_t \\ &+ \alpha_{65} FDIBN_t + \alpha_{66} FDIVN_t + \alpha_{67} FDIMY_t + \alpha_{68} FDILA_t \\ &+ \alpha_{69} FDIKH_t + \varepsilon_t \end{aligned} \tag{9}$$

4 Results and Discussion

4.1 Results

The empirical results based upon Bayesian regression context are presented in Tables 1, 2, 3, 4, 5 and 6. The results reveal that there are the mixed effects of ASEAN–FDI inflows on Thailand economy.

Table 1. The posterior mean coefficient of FDI inflows by ASEAN with 95% Highest Density Interval (HDI) effect SET index.

FDI inflow	FDI of ASEAN effect SET index		
	Posterior mean	Lower (2.5%)	Upper (97.5%)
Brunei	−9.25E−04	−3.32E−02	3.12E−02
Cambodia	8.21E−04	−1.66E−03	3.30E−03
Indonesia	2.11E−03	−3.29E−03	7.49E−03
Lao PDR	−1.72E−04	−3.41E−04	−2.44E−06
Malaysia	−1.24E−02	−6.10E−02	3.74E−02
Myanmar	−1.24E−03	−9.37E−03	6.89E−03
Philipines	−1.41E−03	−8.13E−03	5.36E−03
Sigapore	1.73E−03	−3.06E−03	6.49E−03
Viet Nam	−1.41E−03	−9.29E−03	6.59E−03

As shown in Table 1, the FDI inflow from Lao PDR seems to have a significant effect to SET index as the estimated coefficient of FDI from Lao PDR is −1.72E−04 with 95% HDI = [−3.441E−04, −2.41E−04], which indicates that the estimation of all 50,000 parameter posterior draws exclude zero from the 95% HDI and stay outside the region of practical equivalence (ROPE). Therefore, we decide to reject the null value. However, the FDI inflows from the other ASEAN neighbours do not have the significant effect to SET index as some estimated parameters in posterior draws include zero from the 95% HDI and stay inside the ROPE. Therefore, we decide to accept the null value.

As shown in Table 2, the FDI inflow from Lao PDR seems to have a significant effect to Thailand export sector as the estimated coefficient of FDI from Lao PDR is −1.78E−04 with 95% HDI = [−3.27E−04, −2.88E−05]. Moreover, the FDI inflow from Viet Nam might have a significant effect to Thailand export sector as the estimated coefficient of FDI from Viet Nam is −7.14E−03 with 95% HDI = [−1.41E−02, −9.87E−05]. As a result, both FDI inflows from Lao PDR and Viet Nam indicate that the estimation of all 50,000 parameter posterior draws exclude zero from the 95% HDI and stay outside the ROPE. Therefore, we decide to reject the null value. However, the FDI inflows from the other ASEAN neighbours do not have the significant effect to export as some estimated

Table 2. The posterior mean coefficient of FDI inflows by ASEAN with 95% Highest Density Interval (HDI) effect Thailand export.

FDI inflow	FDI of ASEAN effect Export		
	Posterior mean	Lower (2.5%)	Upper (97.5%)
Brunei	−1.38E−02	−4.22E−02	1.46E−02
Cambodia	9.07E−04	−1.28E−03	3.09E−03
Indonesia	3.63E−03	−1.12E−03	8.37E−03
Lao PDR	−1.78E−04	−3.27E−04	−2.88E−05
Malaysia	1.54E−02	−2.74E−02	5.93E−02
Myanmar	−5.40E−04	−7.70E−03	6.61E−03
Philipines	−1.44E−03	−7.35E−03	4.53E−03
Sigapore	1.76E−03	−2.46E−03	5.96E−03
Viet Nam	−7.14E−03	−1.41E−02	−9.87E−05

Table 3. The posterior mean coefficient of FDI inflows by ASEAN with 95% Highest Density Interval (HDI) effect Thailand GDP growth.

FDI inflow	FDI of ASEAN effect GDP growth		
	Posterior mean	Lower (2.5%)	Upper (97.5%)
Brunei	−2.86E−01	−6.89E−01	1.16E−01
Cambodia	1.40E−02	−1.70E−02	4.49E−02
Indonesia	6.24E−02	−5.09E−03	1.30E−01
Lao PDR	−1.97E−03	−4.09E−03	1.42E−04
Malaysia	−4.23E−01	−1.03E+00	1.98E−01
Myanmar	1.03E−01	1.75E−03	2.05E−01
Philippines	−5.51E−03	−8.95E−02	7.92E−02
Singapore	−1.96E−03	−6.19E−02	5.75E−02
Viet Nam	−1.73E−01	−2.71E−01	−7.30E−02

parameters in posterior draws include zero from the 95% HDI and stay inside the ROPE. Therefore, we decide to accept the null value.

As shown in Table 3, the FDI inflow from Myanmar seems to have a significant effect to Thailand GDP growth as the estimated coefficient of FDI from Myanmar is 1.03E−01 with 95% HDI interval = [1.75E−03, 2.05E−01], which indicates that the estimation of all 50,000 parameter in posterior draws are well above zero or the zero is excluded from the 95% HDI, and they stay outside the ROPE. Moreover, the FDI inflow from Viet Nam might have a significant effect to Thailand GDP growth as the estimated coefficient of FDI from Viet Nam is −1.73E−01 with 95% HDI interval = [−2.71E−01, −7.30E−02], which indicates that the estimation of all 50,000 parameter in posterior draws exclude zero and stay outside the ROPE. Therefore, we decide to reject the null value. However, the FDI inflows from the other ASEAN neighbours do not have the significant effect to GDP growth as some estimated parameters in posterior draws include zero from the 95% HDI and stay inside the ROPE. Therefore, we decide to accept the null value.

Table 4. The posterior mean coefficient of FDI inflows by ASEAN with 95% Highest Density Interval (HDI) effect Thailand CPI.

FDI inflow	FDI of ASEAN effect CPI		
	Posterior mean	Lower (2.5%)	Upper (97.5%)
Brunei	4.01E−04	−2.24E−02	2.31E−02
Cambodia	−1.62E−05	−1.77E−03	1.73E−03
Indonesia	−1.46E−05	−3.83E−03	3.78E−03
Lao PDR	−4.04E−06	−1.24E−04	1.15E−04
Malaysia	−5.45E−05	−3.44E−02	3.51E−02
Myanmar	1.85E−05	−5.72E−03	5.75E−03
Philippines	1.45E−04	−4.60E−03	4.93E−03
Singapore	4.45E−05	−3.34E−03	3.41E−03
Viet Nam	3.31E−05	−5.53E−03	5.68E−03

As shown in Table 4, all of FDI inflows from ASEAN neighbours do not have the significant effect to Thailand CPI as the estimation of all 50,000 parameter in posterior draws include zero from the 95% HDI and stay inside the ROPE. Therefore, we decide to accept the null value.

As shown in Table 5, all of FDI inflows from ASEAN neighbours do not have the significant effect to Thailand MPI as the estimation of all 50,000 parameters in posterior draws include zero from the 95% HDI and stay inside the ROPE. Therefore, we decide to accept the null value.

As shown in Table 6, the FDI inflow from Indonesia seems to have a significant effect to Thailand importation as the estimated coefficient of FDI from Indonesia is 5.07E−03 with 95% HDI interval = [2.97E−05, 1.01E−02], which indicates that the estimation of all 50,000 parameter in posterior draws are well above zero or the zero is excluded from the 95% HDI interval, and they stay outside the ROPE. Moreover, the FDI inflow from Lao PDR might have a significant effect to Thailand importation as the estimated coefficient of FDI from Lao PDR is −1.90E−04 with 95% HDI interval = [−3.48E−04, −3.19E−05], which indicates that the estimation of all 50,000 parameter in posterior draws exclude zero from the 95% HDI and stay outside the ROPE. Therefore, we decide to reject the null value. However, the FDI inflows from the other ASEAN neighbours do not have the significant effect to importation as some estimated parameters in posterior draws include zero from the 95% HDI and stay inside the ROPE. Therefore, we decide to accept the null value.

Table 5. The posterior mean coefficient of FDI inflows by ASEAN with 95% Highest Density Interval (HDI) effect Thailand MPI.

FDI inflow	FDI of ASEAN effect MPI		
	Posterior mean	Lower (2.5%)	Upper (97.5%)
Brunei	−3.69E−03	−2.80E−02	2.05E−02
Cambodia	5.33E−04	−1.33E−03	2.40E−03
Indonesia	1.00E−03	−3.06E−03	5.05E−03
Lao PDR	−1.17E−04	−2.45E−04	1.03E−05
Malaysia	−1.87E−03	−3.85E−02	3.55E−02
Myanmar	2.78E−04	−5.84E−03	6.39E−03
Philippines	6.17E−05	−4.99E−03	5.16E−03
Singapore	1.14E−03	−2.46E−03	4.72E−03
Viet Nam	−3.77E−03	−9.70E−03	2.24E−03

Table 6. The posterior mean coefficient of FDI inflows by ASEAN with 95% Highest Density Interval (HDI) effect Thailand import.

FDI inflow	FDI of ASEAN effect Import		
	Posterior mean	Lower (2.5%)	Upper (97.5%)
Brunei	−1.32E−02	−4.33E−02	1.68E−02
Cambodia	−4.25E−04	−2.74E−03	1.89E−03
Indonesia	5.07E−03	2.97E−05	1.01E−02
Lao PDR	−1.90E−04	−3.48E−04	−3.19E−05
Malaysia	−2.91E−02	−7.45E−02	1.73E−02
Myanmar	2.45E−03	−5.14E−03	1.00E−02
Philippines	7.32E−04	−5.54E−03	7.05E−03
Singapore	5.76E−04	−3.90E−03	5.02E−03
Viet Nam	−5.41E−03	−1.28E−02	2.05E−03

4.2 Discussion

The aim of this study is to investigate the effect of foreign direct investment (FDI) on the economy of host country by using the sample of ASEAN–FDI inflows to Thailand and Thailand economic indicators during the year 2005–2017. In order to complete the better implication of FDI on the economy, this study was incorporated by various economic indicators, namely stock market, exportation, GDP growth, consumer price index, importation and production of manufacturing index. Time series analysis and Bayesian method were employed to this study as they were useful estimators for making a prediction. Bayesian method referred to the Bayes theorem which leads to compute the posterior distribution of interested parameters. Additionally, the posterior can be calculated from the prior distribution and the likelihood function through the joint distribution. Moreover, in this study, we are interested in making a prognostication about Thai economy based on FDI inflows by ASEAN. Accordingly, this approach was based on the Bayesian multiple linear regression context which confirms the significant effect of FDI to Thai economy.

Regarding the results of this study, First, the FDI inflow from Lao PDR appears a significant negative effect on the stock market of Thailand, which implies that an increase of FDI inflow can inhibit the stock market development in the host country. According to the study of Sayek et al. [21] and Baker et al. [2] founded that FDI might have a negative effect to the stock market development since FDI could be the substitution of the short-term investment especially, the short-term of equity and debt instruments; therefore they might be the speculator. Desai et al. [11] suggested that an increase FDI inflows with the form of intra-company loans might affect poor stock market development in host country. In fact, the investment between Lao PDR and Thailand are rather the venture capitalist, which may lead to poor stock market. Our result indicates similar result. In contrast, the result of this study differed from Claessens et al. [9] where the effect was investigated in 77 emerging markets. They suggested that the effect of FDI on the host stock market were a positive effect. However, it does not appear on this study when investigated in Thailand stock market.

Second, the FDI inflows from Lao PDR and Viet Nam indicate the negative effects on the export. As the study of Chaisrisawatsuk and Chaisrisawatsuk [7], where was investigated in Thailand economy, found that an increase of FDI inflows could drop off the export on the host country to other trading partners because they exported the products from the host country to the home country and vice versa, whereas that caused the crowding out export to the others. According to BOT statistic, Thailand has been running a trade surplus and Baht has been appreciation. Therefore, these were the cause of reduction in export to the other countries. Regarding with the statistics, The top exports of Thailand to Lao PDR are oil, car, motor vehicles and parts, machinery and construction material, iron and steel, chemical products and so on. The top exports of Thailand to Viet Nam are vehicle, machinery, electronic equipment plastics, perfumes and cosmetics, other food preparations and so on. However, this study does not support Zhang [25] argument, where was researched in China. His study indicated that FDI inflows from the rest of the world were a positive relationship with China export.

Third, the FDI inflow from Myanmar shows a positive effect on the GDP growth. On the other hand, the FDI from Viet Nam shows a negative effect on the GDP growth. As the study of Simionescu [22] proved that the FDI inflows could generate the mixed effects on the economic growth. Adams [1] also found that FDI effects might be benefits and/or costs on the host country. As Mankiw [19] recommended that even if the host country productivity was produced by FDI with technology transfers and spillovers, FDI could be harmful to the host country as well since FDI might destroy the resource allocation and slow down in the economic growth due to trade, price, financial and other factors [13].

Encinas-Ferrer et al. [12] suggested that FDI could create the negative effect on the GDP growth in the host country due to a lower multiplier effect on the host country, when the proportion of FDI was smaller than the domestic investment. Regarding with the BOT statistic, FDI from Viet Nam was small proportion, it invested in the real estate.

Fourth, the FDI inflow from Indonesia shows a positive effect on Thailand importation. As Chaisrisawatsuk and Chaisrisawatsuk [7] perceptively stated that as more FDI inflows, the host country had to import more raw materials and intermediate goods from the home country and/or the rest of the world because of the dependence on the production technology. According to the statistic in 2017, Thailand's import from Indonesia amounted to 7.4 billion of US dollar, such as the mineral fuels including oil, vehicles, machinery, electronic equipment, copper and so on. Conversely, the FDI inflow from Lao PDR shows a negative effect on Thailand importation because Thailand was more abundant in resources and higher technology than Lao PDR. As a consequence, Thailand decided to import input and the production have done within the country, which caused a reduction in import sector. The explanation is supported by Gray [14], Hussin and Saidin [15] and Wadhwa [24].

Finally, the FDI inflows from ASEAN neighbours do not have any significant effects to Thailand CPI and MPI.

5 Conclusions

This study employed Bayesian multiple linear regression context to confirm the significant effect of FDI to Thai economy. Nevertheless, there were FDI from some countries in ASEAN did not have any significant effect on Thai economy, although the substantial inflow of FDI. By the theory, it believed that FDI generates the positive effect on the economic development of the host country. Moreover, other studies indicated a positive effect of FDI on economy. This study also confirmed the positive effect of FDI on the host economy. The more FDI inflow to the host country, which renders the stock market development, expansion of export, higher GDP growth and import reduction. On the other hand, this study founded the negative effect of FDI on the host country. The details are as follows: First, the poor stock market development. FDI could be the substitution of the short-term investment especially, the short-term of equity and debt instruments; therefore the foreign investment might be a speculation. In addition, FDI inflows with the form of intra-company loans or venture capitalist might affect poor stock market development in host country. Second, export reduction. An increase of FDI inflows could drop off the export on the host country to other trading partners because they exported the products from the host country to the home country and vice versa, whereas that caused the crowding out export to the others. Third, a reduction of GDP growth, even if the host country productivity was produced by FDI with technology transfers and spillovers, FDI could be harmful to the host country as well since FDI might destroy the resource allocation and slow down in the economic growth due to trade, price, financial, the insufficient FDI proportion and other factors. Fourth, a rise of import. The host country had to import more raw materials and intermediate goods from the home country and/or the rest of the world because of the dependence on the production technology.

This study is limited to the effect between FDI inflows by ASEAN and Thai economy and there are the weak effects on some countries due to lack of a strong data. Therefore, the FDI flows from other counties especially, the rest of the world, Japan, China, United States and other economics variables might be added because it is perhaps that some of them to have more significant effect and better inference than this study. Finally, The limitation of this study may be the benefit to develop the further study.

As a consequence, FDI does not always boost the economy of host country even though the country acquires more FDI. In other words, the contribution of FDI depends on both host country and foreign country - the policy, behaviour of government, the characteristic and proportion of foreign investment and trade policy. This study is important to imply that the host country government and policy maker should emphasize the policy or conditions that allow for maximizing benefits or for minimizing costs of FDI effect on the economy. They should exercise with the prudence for persuading more FDI inflows from ASEAN, that may be complementary or substitute to the Thailand production. For the stock market development, the government should monitor on the volume of short-term investment to prevent the stock market fluctuation by the speculation and set the conditions (including tax policy) for the intra-company loans and venture capitalist that are suitable. For the trade policy, the Thai Board of Investment (BOI) should promote the production aimed at export through the import tax exemption on raw materials, nevertheless they should beware of the exchange rate and inflation. Furthermore, they should negotiate with foreign investors for working conditions, technology transfer, training workers, resources access (i.e. financial resources, natural resources), then propose the policy and regulation are suitable.

Acknowledgements. The authors are grateful to Puay Ungphakorn Centre of Excellence in Econometrics, Faculty of Economics, Chiang Mai University for the financial support.

References

1. Adams, S.: Can foreign direct investment (FDI) help to promote growth in Africa? Afr. J. Bus. Manag. **3**(5), 178 (2009)
2. Baker, M., Foley, C.F., Wurgler, J.: Multinationals as arbitrageurs: the effect of stock market valuations on foreign direct investment. Rev. Financ. Stud. **22**(1), 337–369 (2008)
3. Bende-Nabende, A., Ford, J., Slater, J.: FDI, regional economic integration and endogenous growth: some evidence from Southeast Asia. Pac. Econ. Rev. **6**(3), 383–399 (2001)
4. Berthélemy, J.C., Demurger, S.: Foreign direct investment and economic growth: theory and application to China. Rev. Dev. Econ. **4**(2), 140–155 (2000)

5. Borensztein, E., De Gregorio, J., Lee, J.W.: How does foreign direct investment affect economic growth? J. Int. Econ. **45**(1), 115–135 (1998)
6. Canova, F.: Methods for Applied Macroeconomic Research, vol. 13. Princeton University Press, Princeton (2007)
7. Chaisrisawatsuk, S., Chaisrisawatsuk, W.: Imports, exports and foreign direct investment interactions and their effects. Asia-Pacific Research and Training Network on Trade Working Paper Series 45 (2007)
8. Chandprapalert, A.: The determinants of US direct investment in Thailand: a survey on managerial perspectives. Multinatl. Bus. Rev. **8**(2), 82 (2000)
9. Claessens, S., Klingebiel, D., Schmukler, S.: FDI and stock market development: complements or substitutes? World Bank Working Paper (2001)
10. De Mello Jr., L.R.: Foreign direct investment in developing countries and growth: a selective survey. J. Dev. Stud. **34**(1), 1–3 (1997)
11. Desai, M.A., Foley, C.F., Hines, J.R.: A multinational perspective on capital structure choice and internal capital markets. J. Financ. **59**(6), 2451–2487 (2004)
12. Encinas-Ferrer, C., Villegas-Zermeño, E.: Foreign direct investment and gross domestic product growth. Procedia Econ. Financ. **24**, 198–207 (2015)
13. Essays, UK.: Negative Effects of FDI in Host Countries Economics Essay. http://www.ukessays.com/essays/economics/negative-effects-of-fdi-in-host-countries-economics-essay.php?vref=1. Accessed Mar 2017
14. Gray, K.R.: Foreign direct investment and environmental impacts – Is the debate over? Rev. Eur. Comp. Int. Environ. Law **11**(3), 306–313 (2002)
15. Hussin, F., Saidin, N.: Economic growth in ASEAN-4 countries: a panel data analysis. Int. J. Econ. Financ. **4**(9), 119 (2012)
16. Keorite, M., Moubarak, M.: The impacts of China's FDI on employment in Thailand's industrial sector: a dynamic VAR (vector auto regression) approach. J. Chin. Econ. Foreign Trade Stud. **9**(1), 60–84 (2016)
17. Khalifah, N.A., Mohd Salleh, S., Adam, R.: FDI productivity spillovers and the technology gap in Malaysia's electrical and electronic industries. Asian Pac. Econ. Lit. **29**(1), 142–160 (2015)
18. Kruschke, J.K., Aguinis, H., Joo, H.: The time has come: Bayesian methods for data analysis in the organizational sciences. Organ. Res. Methods **15**(4), 722–752 (2012)
19. Mankiw, N.G.: Principles of macroeconomics, Cengage Learning, 2014. OECD, Foreign Direct Investment for Development Maximising Benefits, Minimising Costs. Organisation for Economic Co-operation and Development, France (2002)
20. Reinhart, M.C., Rogoff, M.K.: FDI to Africa: the role of price stability and currency instability (No. 3–10), International Monetary Fund (2003)
21. Sayek, S., Alfaro, L., Chanda, A., Kalemli-Ozcan, S.: FDI spillovers, financial markets and economic development, International Monetary Fund (2003)
22. Simionescu, M.: The relation between economic growth and foreign direct investment during the economic crisis in the European Union, Zbornik radova Ekonomskog fakulteta u Rijeci: časopis za ekonomsku teoriju i praksu **34**(1), 187–213 (2016)
23. Trevino, L.J., Daniels, J.D., Arbelaez, H.: Market reform and FDI in Latin America: an empirical investigation. Transnatl. Corp. **11**(1), 29–48 (2002)
24. Wadhwa, K.: Foreign direct investment into developing Asian countries: the role of market seeking, resource seeking and efficiency seeking factors. Int. J. Bus. Manag. **6**(11), 219 (2011)
25. Zhang, K.H.: How does FDI affect a host country's export performance? The case of China. In: International conference of WTO, China and the Asian Economies, pp. 25–26 (2005)

The Effect of Energy Consumption on Economic Growth in BRICS Countries: Evidence from Panel Quantile Bayesian Regression

Wilawan Srichaikul[1,2](✉), Woraphon Yamaka[1], and Songsak Sriboonchitta[1]

[1] Center of Excellence in Econometrics, Faculty of Economics, Chiang Mai University, Chiang Mai 50200, Thailand
srichaikul.w@gmail.com, woraphon.econ@gmail.com, Songsakecon@gmail.com

[2] International College of Digital Innovation, Chiang Mai University, Chiang Mai 50200, Thailand

Abstract. Energy consumption and economic growth relationship is an important global issue. Thus, in this study, we investigate the energy and economy nexus in the five major emerging national economies or BRICS countries, namely, Brazil, Russia, India, China and South Africa over the period 1990 to 2016 using Panel Quantile Bayesian regression approach. The results show that the effect of energy consumption on economic growth is positive and significant at the 95% Bayesian credible interval level for all quantiles. Meanwhile, other control variables are not statistically significant in affecting the energy consumption.

Keywords: Energy consumption · Economic growth
BRICS countries · Fixed effect panel quantile regression
Panel Quantile Bayesian estimation

1 Introduction

Energy is a prerequisite for economic development. No matter what the stage of economic development and the level of society's development. Energy is vital for present day's economic growth as it is a factor input in virtually all kinds of economic activities. The relationship between energy consumption and economic growth has been investigated in many countries ranging from developed economies to developing ones. For examples, Kraft and Kraft [11] studied this relation for both developing and developed countries, Akinlo [1] and Lee [13] focused only on developing countries while developed countries was investigated by Apergis and Payne [3]; Salim et al. [17]. However, studies on the relationship between energy consumption and economic growth are quite limited with respect to the BRICS countries (Brazil, Russia, India, China and South Africa). The review on the relationship between these two variables in BRICS countries can be found in Khobai, Abel and Le Roux [8].

V. Kreinovich et al. (Eds.): ECONVN 2019, SCI 809, pp. 853–862, 2019.
https://doi.org/10.1007/978-3-030-04200-4_61

In this study, our aim is to examine the relationship between energy consumption and economic growth in BRICS countries. These countries are recognized as the most developed among the emerging major economies. The economic growth of the BRICS countries is constantly increasing as they move to be the industrialized ones. As a result, the need for energy is also increasing in order to accommodate the demand of many sectors such as industry, transportation, economy, and others (Sasana and Ghozali [18]). The growth of BRICS countries is constantly enhancing the global energy demand (Khobai [7]). To achieve our goal, we adopt Panel Quantile regression (PQR) model to study the causal effect of energy consumption on economic growth as it allows us to explore a range of conditional quantiles, thereby exposing a variety of forms of conditional heterogeneity, and to control for unobserved individual effects. In addition, the Fixed Effect (FE) specification is used for controlling individual heterogeneity as it offers a more flexible approach to the analysis of panel data than that afforded by the random effects estimation (Kato et al. [6]). In the estimation point of view, we employ a Bayesian approach as the estimator for PQR model. The advantage of Bayesian estimator is that it can deal with the over-parameterization problem and also provide a convenient way of incorporating parameter uncertainty into predictive inferences. (Pastpipatkul et al. [16]).

Although, the rich literature on the relationship between energy consumption has already investigated in some papers, however, as we mentioned before, there is very little in the literature along the BRICS countries lines. Also, no work has been done under the FE-PQR with Bayesian framework. Therefore, our study is the first attempt in investigating the causal effect of energy consumption on economic growth using FE-PQR with Bayesian estimation. According to this model, we can easily analyze the impact factor of energy consumption for different growth levels.

The remainder of the paper is organized as follows. Section 2 briefly describes the methodology used in this study, Sect. 3 presents the data used in this study and model specification. In Sect. 4, we present the empirical results. Section 5 presents our conclusions.

2 Methodology

2.1 Fixed Effect Panel Quantile Regression

In this paper, we use a fixed effect panel quantile regression model to investigate the relationship between energy consumption and economic growth. The quantile regression technique was introduced in the seminal paper by Koenker and Bassett [10]. This method is a generalization of median regression analysis to other quantiles. The conditional quantile of y_{it} given x_{it} is as follows.

$$Q_{y_{it}}(\tau|x_{it}) = x'_{it}\beta(\tau). \tag{1}$$

Quantile regression is robust to outliers and heavy distributions. However, it does not take into account the unobserved heterogeneity of a country

Zhu et al. [19]. Therefore, some scholars Koenker [9] and Al-mulali et al. [2] studied the econometric theory of applying quantile regressions to panel data. In this paper, we employ a panel quantile method with fixed effects. Consider the following fixed effect panel quantile regression model:

$$Q_{y_{it}}(\tau|\alpha_i, x_{it}) = \alpha_i + x'_{it}\beta(\tau), \quad i = 1, ..., N; t = 1, ..., T \tag{2}$$

where α_i has a pure location shift effect on the conditional quantiles of the response. The effects of the covariates x_{it} are permitted to depend upon the quantile τ of interest. i is the index of individual and t is the index of time. N is the number of observations on the individual i. T is the number of observations on the time t.

Koenker [9] treated unobservable fixed effect as parameters to be jointly estimated with the covariate effects for different quantiles.

2.2 Panel Quantile Bayesian Estimation

In this study, the Bayesian estimation is used for finding the unknown parameters in the panel quantile model. According to the Bayes' theorem, the sample of the posterior distribution of this model can be formed as

$$Pr(\alpha(\tau), \beta(\tau)\,|y_t, x_t) \propto Pr(\alpha(\tau), \beta(\tau))L(\,y_t, x_t\,|\alpha(\tau), \beta(\tau)), \tag{3}$$

where $(y_t, x_t\,|\alpha, \beta)$ is the Asymmetric Laplace likelihood function of the model. The rest of the function is the prior distribution (α, β). First of all, according to Yu and Moyeed [20], the density of Asymmetric Laplace distribution is

$$L_A = \begin{cases} \dfrac{\alpha^T(1-\tau)^T}{\sigma}\exp\left[-\sum\limits_{t=1}^{T}\dfrac{(1-\tau)(y_{it} - x'_{it}\beta(\tau))}{\sigma}\right] & \text{if } y_{it} < \beta(\tau) \\ \\ \dfrac{\alpha^T(1-\tau)^T}{\sigma}\exp\left[-\sum\limits_{t=1}^{T}\dfrac{(-\tau)(y_{it} - x'_{it}\beta(\tau))}{\sigma}\right] & \text{if } y_{it} \geq \beta(\tau) \end{cases} \tag{4}$$

The posterior distributions depend on the interaction between likelihood and prior. With the influence of prior distribution on the posterior inference, it is necessary to conduct a prior sensitivity analysis, which becomes an important tool in Bayesian statistics. To perform sensitivity analysis on the priors, we estimate the posterior distribution with the normal prior $N(b, B)$ for parameter $\alpha(\tau)$ where b and B are respectively the prior mean and precision of $\beta(\tau)$. For the variance σ of the model, it is assumed to have inverse gamma prior, $\Gamma^{-1}(a, b)$, where and are shape and scale parameter, respectively, for σ. To sample the estimated parameter, we use Gibbs sampling method (Kozumi and Kobayashi [12]).

3 Data Analysis

3.1 Data Analysis

In this paper, we investigate the causal effect of energy consumption on economic growth by using data from BRICS countries (i.e Brazil, Russia, India, China and

South Africa) over the period 1990-2016 from the World Development Indicators. The energy consumption is expressed in terms of kilogram of oil used per capita and economic growth is expressed by GDP per capita current US$. Additionally, economic growth can be affected by others factors, we include some important variables in our model to avoid omitted variable bias problems. In the literature, there were various vital variables effecting the energy consumption, namely trade openness, population size and the industrial structure. In this study, we measure the trade openness as the share of trade openness in GDP. The population size is the total population of the country. The industrial structure is measured by the share of the industry value added to GDP. All variables are transformed into natural log. This study is conducted using an econometric framework as follows:

$$Q_{y_{it}}(\tau|\alpha_i, \xi_t, \chi_{it}) = \alpha_i + \xi_t + \beta_{1\tau}ENC_{it} + \beta_{2\tau}TRA_{it} + \beta_{3\tau}POP_{it} + \beta_{4\tau}IND_{it}, \tag{5}$$

where the countries are denoted by i and time by time t. y_{it} is the economic growth. The descriptions of variables are provided in Table 1.

Table 1. Variable definitions

Variable	Definition	Source
GDP	Economic growth (GDP per capita current US$)	World Development Indicators
ENC	Energy consumption (kg of oil equivalents per capita)	World Development Indicators
TRA	Trade openness (% of GDP)	World Development Indicators
POP	Total population	World Development Indicators
IND	The industrial structure (the share of the tertiary industry sector in GDP)	World Development Indicators

3.2 Descriptive Statistics

The summary of the descriptive statistics are illustrated in Table 2. We are clearly that the distributions of all of the variables are skewed, and high kurtosis, indicating that the five series distributions might not exhibit the normal distribution. Rather there exhibit the asymmetric and heavy tail distribution. To confirm this claim, The Jarque-Bera statistic also used to test the normality of the data. Here we apply the Minimum Bayes factor (MBF) approach to obtain the calibrated p-value of the test statistic. The MBF is interpreted following the Goodman [4] labelled intervals. MBF between 1 to 1/3 is considered weak evidence for H_1, form 1/3 to 1/10 considered moderate evidence for H_1, 1/10 to 1/30 is considered substantial evidence , form 1/30 to 1/100 Strong, form 1/100 to 1/300 Very strong, and <1/300 Decisive. According to the result, we find that all data series are decisive favor the alternative hypothesis H_1 or non-normal distribution.

Before estimating the Panel Quantile Bayesian regression models, we test whether the variables used are stationary. Therefore, in this study, we employ panel unit root tests according to Levin-Lin-Chu (LLC) (Levin, Lin and Chu [14])

Table 2. Summary statistics

Variable	GDP_{it}	ENC_{it}	TRA_{it}	POP_{it}	IND_{it}
Mean	6.5696	6.3223	7.559	19.4497	3.1235
Median	7.3608	7.0197	3.9334	19.0222	3.4813
Maximum	9.6808	8.6876	29.6783	21.0444	3.8785
Minimum	1.9732	2.115	2.752	17.4415	1.2436
Std. Dev.	2.4347	2.2523	7.8321	1.2705	0.9353
Skewness	−0.9426	−1.0535	1.7066	−0.003	−1.4071
Kurtosis	2.5182	2.6405	4.1979	1.5271	3.1370
Jarque-Bera	21.2960 (0.0000)	25.7012 (0.0000)	73.6021 (0.0000)	12.2026 (0.0000)	44.6518 (0.0000)

Note : ( ) MBF is Minimum Bayes factor which computed by $-e\,p\,log\,p$, where p is p-value (see, [5]).

Table 3. Panel unit root tests

Variable	Levin-Lin-Chu	
	Level	1st diff.
GDP_{it}	2.8243 (0.0185)	−3.9362 (0.0004)
ENC_{it}	−0.5160 (0.8753)	−4.9996 (0.0000)
TRA_{it}	1.4520 (0.3484)	−12.7190 (0.0000)
POP_{it}	−2.1519 (0.0987)	−2.0641 (0.1188)
IND_{it}	−2.0772 (0.1156)	−6.9630 (0.0000)

Note : ( ) MBF is Minimum Bayes factor which computed by $-e\,p\,log\,p$, where p is p-value (see, [5]).

According to Table 3, it presents the results of the panel unit root tests. The statistical test at level and 1st difference are considered here. The result show that the null hypothesis of the occurrence of a unit root could not be rejected for all of the variables at the level. However, the unit root null hypothesis for all of the variables at the 1st difference are decisive evidence for the stationary data. Therefore, we can conclude that our variable are stationary at the level and there are needed to transform in to the difference form.

4 Emprical Results

4.1 The Results of Panel Quantile Regression Estimation

In this section, we present the results of the Panel Quantile Bayesian regression estimation. The results are reported for the 5th, 10th, 20th, 30th, 40th, 50th, 60th, 70th, 80th, 90th and 95th quantiles of the conditional economic growth distribution. The estimation results are exhibited in Table 4 and Fig. 1 presents graphics of estimated coefficients. From the results, we can observe that the effect of energy consumption on the economic growth in BRICS countries has the same direction in all quantiles. The coefficient of ΔENC is highly significant and has a positive at lower, median, upper quantiles (5th, 10th, 20th, 30th, 40th, 50th, 60th, 70th, 80th, 90th and 95th quantile). At the 40th quantile, the coefficient of ΔENC is the highest among all quantiles. Moreover, our results show that the coefficient of ENC is positive and significant at the 95% confident interval level for all quantiles, implying that a higher energy consumption level is associated with the increase in economic growth in BRICS countries. The results imply that, as ΔENC increases by 1%, the level of economic growth increases by 2.6286%–3.1411% along the quantile levels, which is consistent with our

Table 4. Panel quantile Bayesian regression for BRICS countries

Variable	Quantile Bayesian				
	10th	20th	30th	40th	50th
ΔENC_{it}	2.6286*	2.6408*	2.5968*	2.5546*	2.5493*
	(0.8681,4.1171)	(1.4955, 3.7445)	(1.5879, 3.6378)	(1.6114, 3.6414)	(1.6795, 3.5252)
ΔTRA_{it}	−0.0079	−0.0104	−0.0127	−0.0135	−0.0145
	(−0.1740, 0.1524)	(−0.1310, 0.1053	(−0.0817, 0.07963)	(−0.0177, 0.0735)	(−0.1015, 0.0658)
ΔPOP_{it}	0.8952	0.8057	0.8397	0.5376	0.7836
	(−3.9401, 5.8783)	(−2.4240, 4.1337)	(−1.8522, 3.4776)	(−1.5853, 3.2974)	(−1.5859, 3.0687)
ΔIND_{it}	−2.3514	−2.3237	−2.2399	−2.1758	−2.1949
	(−7.1334, 1.3806)	(−5.7012, 0.8726)	(− 5.2632, 0.8362)	(−5.1132, 0.9134)	(−5.0326, 0.8337)
Variable	Quantile Bayesian				
	60th	70th	80th	90th	95th
ΔENC_{it}	2.5600*	2.6435*	2.7945*	2.9851*	3.1411*
	(1.5937, 3.7717)	(1.5844, 3.9903)	(1.4317, 4.3500)	(0.9386, 5.0566)	(0.2035, 6.2145)
ΔTRA_{it}	−0.0144	−0.0142	-0.0153	−0.0221	−0.0273
	(−0.0970, 0.0691)	(−0.0109, 0.0814)	(−0.1373, 0.1104)	(−0.1892, 0.1466)	(−0.2596, 0.2269)
ΔPOP_{it}	0.6096	0.324	−0.0149	−0.1057	−0.1142
	(−1.8583, 2.9971)	(−2.3961, 2.9471)	(−3.0663, 2.9478)	(−4.2407, 4.0802)	(−6.5024, 6.0445)
ΔIND_{it}	−2.4557	−2.9184	−3.4851	−3.7283	−3.4332
	(−5.3613, 0.7126)	(−5.6815, 0.2573)	(−6.2676, 0.3131)	(−7.3384, 0.3983)	(−8.8336, 2.9286)

Note: (1) Figures in parentheses are lower quantile and upper quantile respectively. (2) * is Bayesian significant

expectations because increases of ΔENC is expected to cause more economic growth. This is in consonance with the finding in a study by Khobai et al. [7]

Table 5. Bayes factor test for the equality of coefficient

ΔGDP_{it}	95% CI	Null Hypothesis
$\tau = 0.1$	(−0.032,0.031)	Accept
$\tau = 0.2$	(−0.038,0.015)	Accept
$\tau = 0.3$	(0.004,0.058)	Reject
$\tau = 0.4$	(0.047,0.101)	Reject
$\tau = 0.5$	(0.061,0.115)	Reject
$\tau = 0.6$	(0.041,0.095)	Reject
$\tau = 0.7$	(−0.043,0.012)	Accept
$\tau = 0.8$	(−0.196,−0.134)	Reject
$\tau = 0.9$	(−0.394,−0.318)	Reject
$\tau = 0.95$	(−0.559,0.462)	Reject

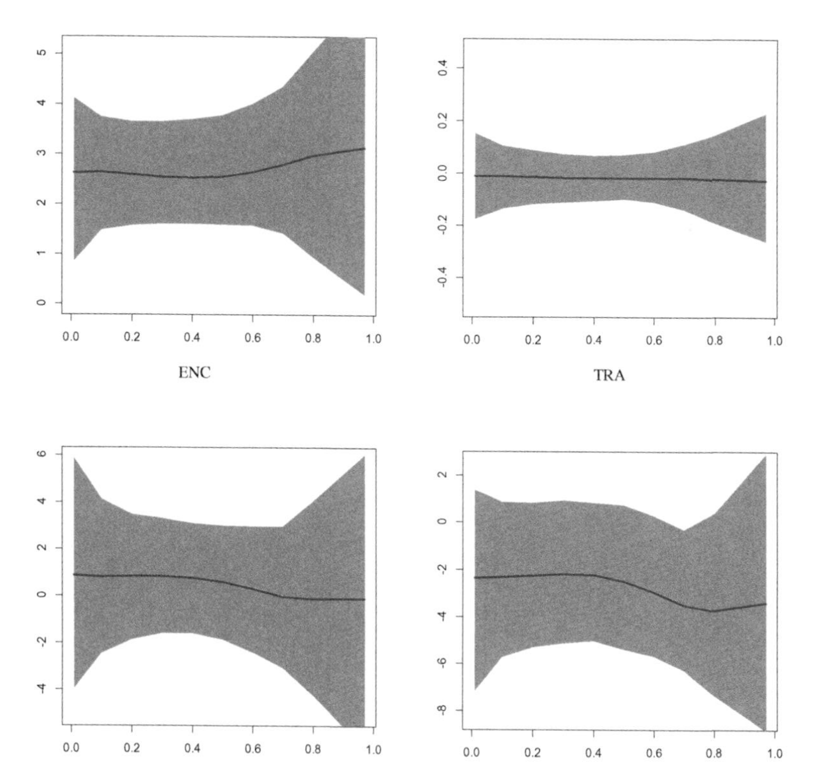

Fig. 1. Change in Panel Quantile Bayesian regression coefficients.

where they also suggested that there is a long run relationship between economic growth and energy consumption. However, the results of ΔTRA, ΔPOP, ΔIND are found not to have a significant effect on economic growth in BRICS countries. The corresponding panel quantile regression diagrams are provided in Fig. 1.

Finally, we can use inter-quantile tests to verify the heterogeneity of the obtained parameter at each quantile. According to the Morey and Rouder [15], Bayes Factor Approaches for testing interval null hypotheses is used to check the equality of estimated parameter across quantiles. The null hypothesis that the mean of the estimated parameter in quantile 1 is not different form the mean of the estimated parameter in other quantiles $\left[\bar{\beta}(\tau = 0.1) - \bar{\beta}(\tau = j) = 0\right]$, $j = 0.1, 0.2, ..., 0.95$ or not. In this test, we consider only the significant variable GDP and sample the difference between the $\beta^{ENC}(\tau = 0.1)$ and $\beta^{ENC}(\tau = j)$ from their corresponding posterior distributions. The result of 5,000 different mean samples are presented and illustrated in Table 5 and Fig. 2, respectively. In the interpretation, the null hypothesis of homogenous mean parameters is rejected if the 95% confident interval (95% *CI*) around the mean does not include zero. According the results, we find that the equality of the coefficients between

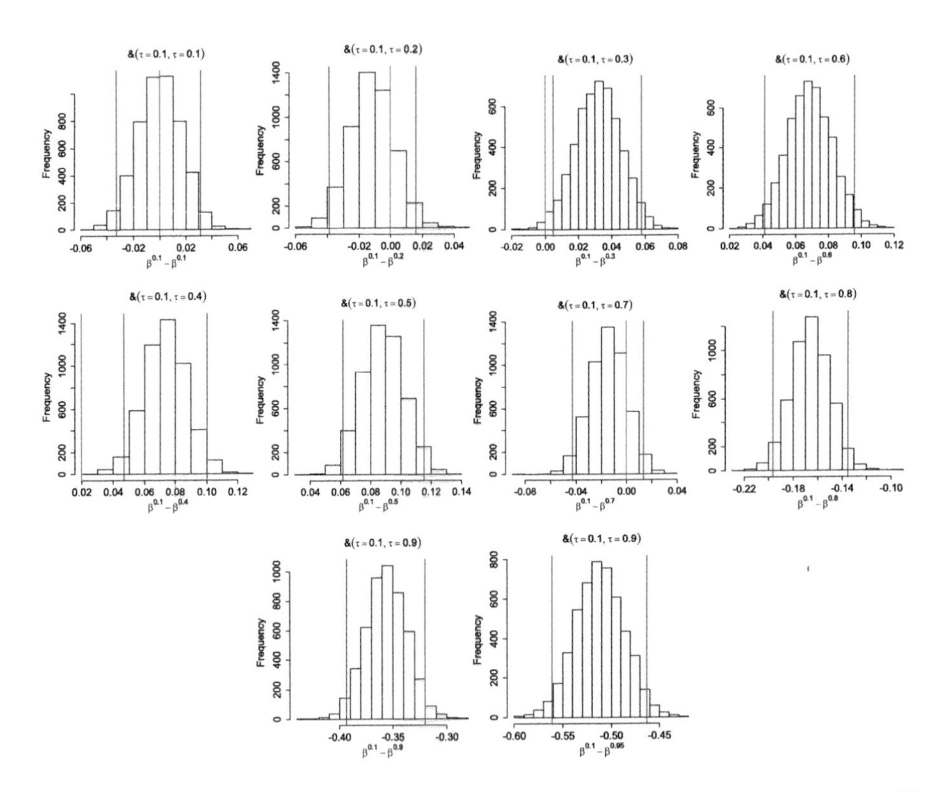

Fig. 2. Samples of the difference between mean parameter. The blue line presents 95% credible interval and red line presents zero difference between coefficient at quantile 0.1 and other quantiles.

($\tau = 0.1$) against ($\tau = j$) are rejected at all quantiles except ($\tau = 0.1, 0.2, 0.7$). Thus, this result confirms the used of quantile regression in investigating the effect of energy consumption on economic growth, indicating the heterogeneous effect of energy consumption on economic growth in BRICS.

5 Conclusion

In this paper, we employed Panel Quantile Bayesian regression approach to investigate the effect of energy consumption on economic growth in BRICS countries. In addition, to avoid an omitted-variable bias, certain related control variables are included in the model. We use a panel quantile regression model that takes into account an unobserved individual heterogeneity and distributional heterogeneity into consideration. Moreover, we employ a Bayesian approach as the estimator for panel quantile model. Our results suggest that the economic growth can exert significant effect at the 95% confident interval level for all quantiles on energy consumption in BRICS countries. By energy consumption increases by 1%, the level of economic growth increases by 2.6286%–3.1411% along the quantile levels. Meanwhile, other variables are not statistically significant to explain relationship on the energy consumption.

This study shows that a substantial portion of the economic expansion of the BRICS countries is energy consumption. However, the increase in energy consumption in light of economic growth may bring pollution problem from the elevating carbon emissions. Therefore, it is important that the government of BRICS countries should shift from fossil fuels, such as oil, to clean and renewable energy. Likewise, Government should invest more on technology for promoting development of new energy resources and renewable energy sources.

References

1. Akinlo, A.E.: Energy consumption and economic growth: evidence from 11 Sub-Sahara African countries. Energy Econ. **30**(5), 2391–2400 (2008)
2. Al-mulali, U., Fereidouni, H.G., Lee, J.Y., Mohammed, A.H.: Estimating the tourism-led growth hypothesis: a case study of the Middle East countries. Anatolia **25**(2), 290–298 (2014)
3. Apergis, N., Payne, J.E.: Energy consumption and growth in South America: evidence from a panel error correction model. Energy Econ. **32**(6), 1421–1426 (2010)
4. Goodman, S.N.: Toward evidence-based medical statistics. 1: the P value fallacy. Ann. Intern. Med. **130**(12), 995–1004 (1999)
5. Held, L., Ott, M.: On p-values and Bayes factors. Ann. Rev. Stat. Appl. **5**, 393–419 (2018)
6. Kato, K., Galvao Jr., A.F., Montes-Rojas, G.V.: Asymptotics for panel quantile regression models with individual effects. J. Econ. **170**(1), 76–91 (2012)
7. Khobai, H.: Electricity consumption and Economic growth: A panel data approach to Brics countries (2017)
8. Khobai, H., Abel, S., Le Roux, P.: A Review of the Nexus Between Energy consumption and Economic growth in the Brics countries (2017)

9. Koenker, R.: Quantile regression for longitudinal data. J. Multivar. Anal. **91**(1), 74–89 (2004)
10. Koenker, R., Bassett Jr., G.: Regression quantiles. Econ. J. Econ. Soc. 33–50 (1978)
11. Kraft, J., Kraft, A.: On the relationship between energy and GNP. J. Energy Dev. 401–403 (1978)
12. Kozumi, H., Kobayashi, G.: Gibbs sampling methods for Bayesian quantile regression. J. Stat. Comput. Simul. **81**(11), 1565–1578 (2011)
13. Lee, C.C.: Energy consumption and GDP in developing countries: a cointegrated panel analysis. Energy Econ. **27**(3), 415–427 (2005)
14. Levin, A., Lin, C.F., Chu, C.S.J.: Unit root tests in panel data: asymptotic and finite-sample properties. J. Econ. **108**(1), 1–24 (2002)
15. Morey, R.D., Rouder, J.N.: Bayes factor approaches for testing interval null hypotheses. Psychol. Methods **16**, 406–419 (2011)
16. Pastpipatkul, P., Yamaka, W., Wiboonpongse, A., Sriboonchitta, S.: Spillovers of quantitative easing on financial markets of Thailand, Indonesia, and The Philippines. In: International Symposium on Integrated Uncertainty in Knowledge Modelling and Decision Making, pp. 374–388. Springer, Cham, October 2015
17. Salim, R.A., Hassan, K., Shafiei, S.: Renewable and non-renewable energy consumption and economic activities: further evidence from OECD countries. Energy Econ. **44**, 350–360 (2014)
18. Sasana, H., Ghozali, I.: The impact of fossil and renewable energy consumption on the economic growth in Brazil, Russia, India, China and South Africa. Int. J. Energy Econ. Policy **7**(3), 194–200 (2017)
19. Zhu, H., Duan, L., Guo, Y., Yu, K.: The effects of FDI, economic growth and energy consumption on carbon emissions in ASEAN-5: evidence from panel quantile regression. Econ. Model. **58**, 237–248 (2016)
20. Yu, K., Moyeed, R.A.: Bayesian quantile regression. Stat. Probab. Lett. **54**(4), 437–447 (2001)

Analysis of the Global Economic Crisis Using the Cox Proportional Hazards Model

Wachirawit Puttachai[1(✉)], Woraphon Yamaka[2], Paravee Maneejuk[2], and Songsak Sriboonchitta[2]

[1] Faculty of Economics, Chiang Mai University, Chiang Mai, Thailand
wachirawitecon@gmail.com

[2] Centre of Excellence in Econometrics, Faculty of Economics, Chiang Mai University, Chiang Mai, Thailand
woraphon.econ@gmail.com

Abstract. This paper uses the Cox proportional hazards model to examine which of the structural characteristics could resist the US financial crisis survival countries. The dependent variable in this model is generated from GDP, and the Markov Switching Autoregressive (MS-AR) technique is used to detect the survival period as well as the crisis occurrence in each country. The survival of a country is found to be influenced by continents (Asia, Australia and Africa) and the higher development level. However, being the member of economic communities, APEC and WTO, increase the chance of the crisis occurrence.

Keywords: Cox proportional hazards model · Time to event data
US financial crisis · Markov Switching model

1 Introduction

The world economy has been significantly affected by the financial crisis since the Great Depression in 1930's. It was recorded that there are more than twenty crisis events occurring in the 21st century [20]. One of the most severe crises in this century is the financial crisis in 2008–2009, originating from the United States (US). This recent great recession did contribute to a large depression on the value of various assets in the market, leading to the collapse of financial sector in the world economy and deepening the contraction in economic activities [9], an event which took around 7 years for the world economy to recover [5].

Although the US financial crisis is viewed as one of the most severe for its effect has spread all over the world, it has not hit countries equally. Many studies of economic contagion and spillover effect found that the degree of financial and economic correlation among the countries are quite different and the effect seems to have different magnitude in various countries (see, Bekaert et al. [4], Kaminsky and Reinhart [14] and Rodriguez [17]). The particular propagation pattern of

V. Kreinovich et al. (Eds.): ECONVN 2019, SCI 809, pp. 863–872, 2019.
https://doi.org/10.1007/978-3-030-04200-4_62

the crisis clearly occurs to have allowed some of the countries with large asset prices and financial leverage to escape relatively lightly in terms of GDP loss. Conversely, countries with manufacture intensive economy appear to have been badly affected. With this finding, two questions arising from this observation are: What factors bring this inequality effect of crisis? and Which are the country's factors rendering its survival through a financial crisis?

To answer these questions, the Cox proportional hazards model (introduced in a seminal paper by Cox [6]) is considered in this study. Over many years ago this technique has been increasingly used in medical and social sciences, to predict and explain the characteristics causing something failure. Allison [1] showed that there are many reasons for this popularity and the foremost is probably the fact that the model itself does not need any information about some underlying distribution that we expect the survival time to follow. However, the application of this model in the economic study is limited as it is quite difficult to find or explore the dependent data of the model. In the conventional way working with survival time data or Time to Event data, the dependent variable is the time from when we start observing something until a specific event occurs which is quite difficult to quantify in economic study. In the recent literature, we found some applications of the model in banking and finance. Tveterås and Eide [21] employed a semi-proportional hazards model to analyze possible determinants of survival for ten plant cohorts during the period 1977 – 1992 in Norwegian Manufacturing. Shih and Giles [19] have studied the duration of bank rate spells under inflation targeting in Canada covering eight years. Danacica and Babucea [8] examined methodological and applicative problems in the factors influencing the duration of unemployment covering 2005 – 2006, examining three exogenous variables representing gender, age and educational level. Gunsel [11] adjusted the Cox proportional hazards model using a discrete-time logistic transformation to investigate the determinants of the timing of bank failure in North Cyprus over the period of 1984 – 2002. Giovannetti et al. [10] analyzed the likelihood of survival is invariant to firm size, international involvement and to technological intensity in Italian firms. Alves et al. [2], Pereira [16], Babajide et al. [3] and Cox et al. [7] conducted the Cox proportional hazards model to examine the operating and financial characteristics of bank failure. However, none of them use Cox model in the examination of the factors underlying the hits from economic crisis. Hence, this study attempts to fill the gap of applying Cox model in the economic crisis issue.

As we have mentioned that this technique differs from the other conventional regression as the dependent variables are the Time to Event data and the event occurrence of the data. To find the dependent crisis variable, we can consider the duration of failure event or crisis. And fortunately, we find that the Markov Switching model of Hamilton [12] seems to have an ability to measure the duration and detect the crisis in US financial crisis. The model has firstly been provided for measuring the duration of economic regimes, namely expansion and recession regime. Therefore, we apply this model to generate the dependent variable data in Cox model. In this study, we survey the influence of US financial crisis in 182 countries and investigate the crisis prospect in each country. This is to say, we employ the Markov Switching model to check whether there exists the US. financial crisis in each country as well as measure the duration of the crisis event.

The main contribution of this paper is applying the Cox proportional hazards model to analyze the structural characteristics rendering the global economic crisis during economic crisis 2008 – 2009. The results of this model will provide a great benefit for policymakers and government for proposing an appropriate economic development policy and the economic immunization policy against the crisis.

We organize the rest of the paper as follows. In Sect. 2, we briefly summarize the basic concepts and the baseline hazard function for the Cox model. Section 3 is on generating the dependent variables comprising the duration times and crisis status. In Sect. 4 we present our main results. Finally, the study ends with a conclusion in Sect. 5.

2 Methodology

In this section we provide short background of the methodology that will be used as a foundation in this study. We cover theory regarding survival analysis in general and the Cox proportional hazards model in particular.

2.1 Survival and Hazard Functions

Let the survival time T be a random variable with cumulative distribution function $P(t) = \Pr(T \leq t)$, and probability density function $p(t) = dP(t)/dt$. The survival function is the complement of the distribution function, $S(t) = \Pr(T > t) = 1 - P(t)$. And the relationship between $S(t)$ and $h(t)$, the hazard function, which estimates the sudden risk of failure at time t is expressed by:

$$h(t) = \lim_{\Delta t \to 0} \frac{Pr[(t \leq T < t + \Delta t)|T \geq t)}{\Delta t} = \frac{f(t)}{S(t)} \tag{1}$$

This function indicates the probability that an individual will experience an event (for example, death or bankruptcy) within a considered time period [22]. We note that the event in this study refers to the occurrence of US financial crisis in 2008 – 2009 thus the survival time is the period of time that individual country survive on this crisis event.

For example, Fig. 1 shows the features of variables for the survival data in this study. When the study ends, the country A still had not been affected by the event yet or it is the case of survival on crisis, while the others (B and C) experienced this crisis.

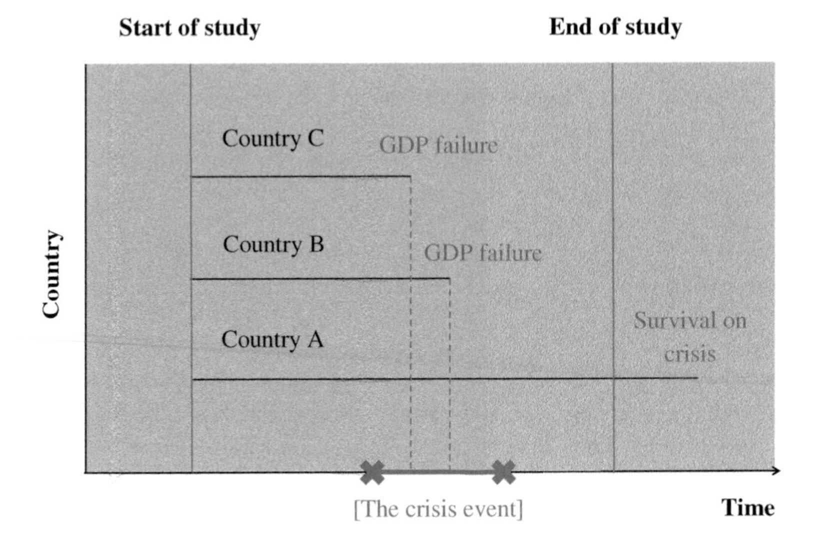

Fig. 1. The illustration of survival data

2.2 The Cox Proportional Hazards Model

Survival analysis typically examines the relationship of the survival distribution to covariates. The simplest way to examine this relationship is the linear regression specification.

$$\log h_i(t) \quad = \quad \alpha(t) + \beta_1 x_{i1} + \beta_2 x_{i2} + ... + \beta_k x_{ik}, \tag{2}$$

or, again equivalently,

$$h_i(t) \quad = \quad h_0(t) \exp(\beta_1 x_{i1} + \beta_2 x_{i2} + ... + \beta_k x_{ik}), \tag{3}$$

where $h_i(t)$ is the conditional hazard function of continuous random variable. Briefly, the hazard function can be interpreted as the crisis of country i and t is the crisis survivor time. If a country has a risk of crisis at 2007–2008, h_t is given as 1, otherwise $h_t = 0$. k is the number of independent variables, β_i are the partial regression coefficients, x_i are the values of the covariates for each individual observation, $h_0(t)$ is the hazard baseline. If a value of β_i greater than zero, or equivalently a hazard ratio greater than one, these indicates that the event hazard increases and the length of survival will decrease.

3 Data Generating Process

As already mentioned there were more than 20 crisis events in the 21st century but the largest one is the US financial crisis. Thus, this event and the Time to Event data are based on this crisis. To determine which structural characteristics respond to the crisis, the proxy of the crisis effect should be provided. In this study, we consider using the Gross Domestic Product (GDP) of the country as the representative of the economic condition. Therefore, the dependent variable in the Cox model is generated from GDP and the Markov Switching Autoregressive (MS-AR) technique is used to detect the survival period as well as the crisis occurrence in each country. Generally speaking, we define a certain GDP fall in the country (same for all countries) to be the event of interest (essential in survival analysis) in order to obtain each country's survival time.

The quarterly real GDP growth data between 1997–2016 of 182 counties are used in this study. The data is collected from International Financial Statistics (IFS). We estimate the dependent variable and the survival time of each country by using the MS-AR model which can be written as

$$Z_t = \begin{cases} \alpha_0 + \beta Z_{t-1} + e_t, & S_t = 0, \quad regime1, \\ \alpha_0 + \alpha_1 + \beta Z_{t-1} + e_t, & S_t = 1, \quad regime2, \end{cases} \tag{4}$$

where Z_t is the GDP growth at time t, $|\beta| < 1$ and ε_t are i.i.d. random variables with mean zero and variance are i.i.d. random variables with mean zero and variance σ_ε^2. With mean $\alpha_0/(1-\beta)$ this is a stationary AR(1) process when $S_t = 0$, and it switches to another with mean $(\alpha_0 + \alpha_1)/(1-\beta)$ when $S_t = 1$. Then provided that $\alpha_1 \neq 0$, this model accepts two dynamic structures at different levels, depending on the value of the state variable S_t [15]. In this case, z_t are dominated by two distributions with distinct means, and S_t determines the switching between these two regimes. In this study, we set the regime 1 as the crisis regime while regime 2 is given as normal regime. The obtained filtered probabilities from this model are further used to measure the survival time and detect the US financial crisis. If the probability of staying in the US financial crisis around 2008 – 2009 is greater than 0.5, the crisis is detected and the survival time is end and $h_t = 1$, otherwise $h_t = 0$. Note that the obtained dependent variables from 182 countries can be viewed as a cross-sectional data. After that, we assign the independent variables from all countries into 5 categories, as follows;

- Continents includes Asia and Australia, Africa, North America, South America and Europe
- Country classified by income level
- Country classified by economic development
- Country classified by human development index
- Economic community including OECD, EU, APEC, NAFTA, WTO and OPEC

These categories are sorted by World Bank, UN and OECD database in 2009. In this analysis, we set the dummy variable for all independent variables. The description of the data is provided in Table 1.

Table 1. Data description

Variable	Category	Total number of countries
Continents	Asia and Australia	52
	Africa	52
	North America	24
	South America	12
	Europe	42
Income	Low income countries	39
	Lower middle-income countries	50
	Upper middle-income countries	44
	High income countries	49
Economic development	Least developed countries	20
	Developing economies	122
	Developed economies	40
Human development	Low human development	20
	Medium human development	79
	High human development	43
	Very high human development	40
Economic community	OECD	30
	EU	27
	APEC	19
	NAFTA	3
	WTO	147
	OPEC	14
Total number of countries		182
Number of countries that responding to the crisis		114

4 Results

In this section, we present the main results obtained from the Cox regression model. Table 2 shows the partial regression coefficients of the structural characteristics or the independent variables for explaining the US. financial crisis.

According to Table 2, we observe the significant characteristics of the country for explaining the effect from US crisis to be Continents (Africa), Economic development, Human development, Economic community (APEC and OPEC), as there are found to have a moderate to strong evidence for $H_1 : \beta \neq 0$. We note that the Minimum Bayes factor(MBF) is used to make the statistic inference in this study, for more detail, see [13]. We find that the Africa continent presents a negative effect (p-value < 0.10 and MBF $= 0.453$) with hazard ratios

Table 2. Estimated coefficients

Independent variable		Beta	HR	$Pr(>\|Z\|)$
Continent	Asia and Australia	−0.95	0.92	0.059[0.453]
	Africa	−1.485	0.73	0.009 [0.115]
	North America	0.022	1.2	0.964 [1]
	South America	0.384	2.1	0.434 [1]
	Europe	−0.632	0.54	0.34 [0.997]
Income		−0.381	0.95	0.136 [0.737]
Economic development		−1.265	0.88	0.014 [0.162]
Human development		0.849	1	0.044 [0.373]
Economic community	OECD	−0.041	0.94	0.937 [1]
	EU	−0.479	0.66	0.513 [1]
	APEC	0.821	2.5	0.062 [0.468]
	NAFTA	0.236	2.6	0.808 [1]
	WTO	0.749	2	0.060 [0.458]
	OPEC	0.404	1.1	0.402 [1]

Note: [] MBF is Minimum Bayes factor which computed by $-e\,p\,log\,p$, where p is p-value (see, [13]).

0.73. From the estimation, being the country in African continent is associated with good economic performance because of the hazard ratio is 0.73, which means that the hazard of experiencing the crisis event reduce by 27%. This means that if the country is located in African continent, the risk of crisis is lower. Senbet and Otchere [18] suggested that although countries in the African continent have received investment from international investors, but with a very low level of such investment, they were far from suffering with crisis. Likewise, if the country is located in the Asia and Australia continent, the risk of crisis is lower as the negative beta coefficient is also obtained. Being the country in this continent could reduce the crisis risk by 5%. These two continents are viewed as the emerging economies which attract a substantial capital outflow from the US economy thus the chance of facing failure with the crisis is low when compared to the other continents.

For the factor of economic development, we also find a negative significant result indicating that the risk of crisis is low in the developed country. The reason why this structural characteristic produced the global GDP failure effect differently is that the financial markets of less developed countries (LDCs) are almost inefficient, so they fail to use the government policies to settle with economic crisis problem, then making them suffer more and also recover slowly. Meanwhile, the more developed countries (MDCs) have the strong economy and their financial markets are efficient. Thus, they can provide a potential policy to immunize their economy from the crisis. Conversely, the human development provides a positive significant effect and the estimated hazard ratio of 1 which

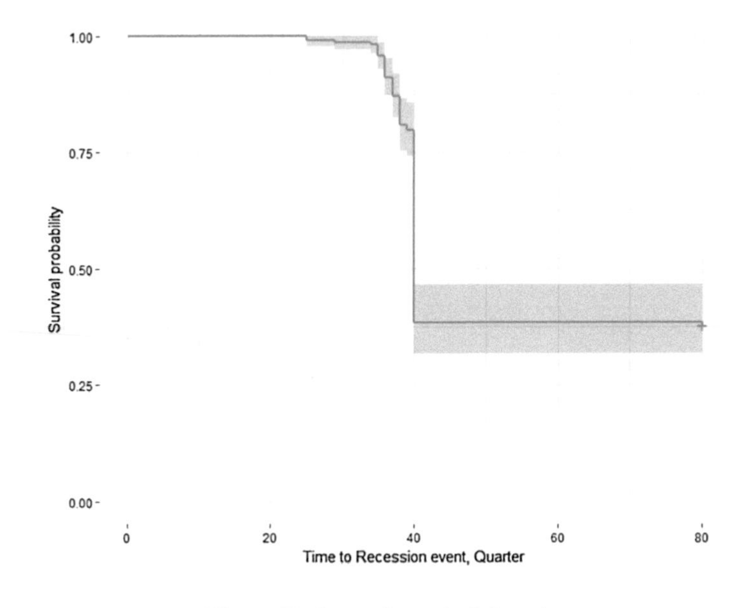

Fig. 2. Estimated survival function

means that a one-point increase or decrease has no impact on the survival in this model. Similar to the economic community variable, being the member in WTO and APEC cannot reduce the risk of US financial crisis.

In Fig. 2, the survival function is estimated from the Cox regression model. The time to crisis event data is discreet and the survival function has a step or stair shaped downward slope. We observe that the survival probability is quite high, almost 100% since the first quarter to the 25th quarter. This indicates that the risk of the crisis is low in those periods however the crisis events are mostly clumped together at the quarter 40, corresponding to the quarter 4 of 2007, which is around the US financial crisis. The survival probability in that period is dropped more than 60%, specifically around the 65%. Additionally, we find that the survival probability remains constant after quarter 40 as there are 68 countries that did not face the US financial crisis. Therefore, the survival function ends in a horizontal line in the last quarter.

5 Conclusion

Our study applies the Cox proportional hazards model to analyze which structural characteristics can be indicative of a country to be hit by the global GDP failure during the US financial crisis by using 182 countries. We generate the dependent variable as the Time to Event data by measuring the time from a starting point to the time at which a crisis event occurs. We find that among these countries, only 114 countries that were not immuned to the US financial crisis. The results from the Cox regression suggest that the continents

(Asia, Australia and Africa) and developed economies, the human development, and the economic community (APEC and WTO) factors provide a significant effect on the country's survival during 2008– 2009. In addition, the survival function is plotted to show survival probability from the crisis. We observe that the probability of survival from the crisis is gradually lower after 25th quarter and suddenly drops in the 40th quarter, which corresponds to the US financial crisis. We can say that there are unequal effects of the crisis to the global countries as the time to crisis event seems to be different.

Therefore, the financial system of each country should be improved for stabilizing the economy and protecting the effects of the economic crisis from outside country. And we should be mindful of the negative effects of economic community (APEC and WTO) as being the member in WTO and APEC cannot reduce the risk of US financial crisis. In other words, being the member in these two groups could have a higher chance to face with the crisis. We expect that there might be a contagion effect in these groups.

This study provides a new perspective in analyzing the global crisis. And the overall result clearly shows the characteristics that affect survival in crisis. This paper can be improved by considering other characteristics of interest in the next studies.

References

1. Allison, P.D.: Survival analysis. In: The Reviewer's Guide to Quantitative Methods in the Social Sciences, pp. 413–425 (2010)
2. Alves, K., Kalatzis, A., Matias, A.B.: Survival Analysis of Private Banks in Brazil (No. 21500002). EcoMod (2009)
3. Babajide, A.A., Olokoyo, F.O., Adegboye, F.B.: Predicting bank failure in Nigeria using survival analysis approach. J. South Afr. Bus. Res. (2015)
4. Bekaert, G., Ehrmann, M., Fratzscher, M., Mehl, A.: The global crisis and equity market contagion. J. Finan. **69**(6), 2597–2649 (2014)
5. Bordo, M.D., Haubrich, J.G.: Deep recessions, fast recoveries, and financial crises: evidence from the American record. Econ. Inq. **55**(1), 527–541 (2017)
6. Cox, D.R.: Regression models and life-tables. In: Breakthroughs in Statistics, pp. 527–541. Springer, New York (1992)
7. Cox, R.A., Kimmel, R.K., Wang, G.W.: Proportional hazards model of bank failure: evidence from USA. J. Econ. Finan. Stud. **5**(3), 35–45 (2017)
8. Danacica, D.E., Babucea, A.G.: Using survival analysis in economics. Survival **11**, 15 (2010)
9. Gilchrist, S., Schoenle, R., Sim, J., Zakrajšek, E.: Inflation dynamics during the financial crisis. Am. Econ. Rev. **107**(3), 785–823 (2017)
10. Giovannetti, G., Ricchiuti, G., Velucchi, M.: Size, innovation and internationalization: a survival analysis of Italian firms. Appl. Econ. **43**(12), 1511–1520 (2011)
11. Gunsel, N.: Determinants of the timing of bank failure in North Cyprus. J. Risk Finan. **11**(1), 89–106 (2010). https://doi.org/10.1108/15265941011012705
12. Hamilton, J.D.: A new approach to the economic analysis of nonstationary time series and the business cycle. Econ. J. Econ. Soc. **57**(2), 357–384 (1989)
13. Held, L., Ott, M.: How the maximal evidence of p-values against point null hypotheses depends on sample size. Am. Stat. **70**(4), 335–341 (2016)

14. Kaminsky, G.L., Reinhart, C.M.: On crises, contagion, and confusion. J. Int. Econ. **51**(1), 145–168 (2000)
15. Kuan, C.M.: Lecture on the Markov switching model, pp. 1–30. Institute of Economics Academia Sinica (2002)
16. Pereira, J.: Survival analysis employed in predicting corporate failure: a forecasting model proposal. Int. Bus. Res. **7**(5), 9 (2014)
17. Rodriguez, J.C.: Measuring financial contagion: a copula approach. J. Empirical Finan. **14**(3), 401–423 (2007)
18. Senbet, L.W., Otchere, I.: Financial sector reforms in Africa: perspectives on issues and policies. In: Annual World Bank Conference on Development Economics 2006: Growth and integration, pp. 81–120, January 2006
19. Shih, R., Giles, D.E.: Modelling the duration of interest rate spells under inflation targeting in Canada. Appl. Econ. **41**(10), 1229–1239 (2009)
20. Taylor, B.: Tiberius used quantitative easing to solve the financial crisis of 33 AD. Business Insider (2013)
21. Tveterås, R., Eide, G.E.: Survival of new plants in different industry environments in Norwegian manufacturing: as semi-proportional Cox model approach. Small Bus. Econ. **14**(1), 65–82 (2000)
22. Walters, S.J.: What is a Cox model? from University of Oxford Clinical School Information Management Services Unit (2008). www.jr2.ox.ac.uk/bandolier/painres/download/whatis/COX_MODEL.pdf

The Seasonal Affective Disorder Cycle on the Vietnam's Stock Market

Nguyen Ngoc Thach[1(✉)], Nguyen Van Le[2], and Nguyen Van Diep[3]

[1] Institute of Research and Technology, Banking University Ho Chi Minh City, Ho Chi Minh City, Vietnam
thachnn@buh.edu.vn
[2] SHB, Ha Noi, Vietnam
tuan.hq@shb.com.vn
[3] Faculty of Finance - Banking, Ho Chi Minh City Open University, Ho Chi Minh City, Vietnam
vandiep1302@gmail.com

Abstract. In this study, the authors used the TGARCH(1,1) model according to three different distribution patterns: normal distribution (Gaussian distribution), Student-t distribution, and generalized error distribution (GED) to analyze the effect of Seasonal Affective Disorder (SAD) on the Vietnam's stock market. With the data being the daily closing price of VN-Index collected from the Ho Chi Minh City Stock Exchange (HOSE) during the period from February 2002 to December 2017, the study results show that the SAD effect is confirmed to exist. Specifically, the SAD effect has had an asymmetric effect on stock returns, and the SAD effect also influenced the volatility of return. In particular, the SAD effect is more obvious in Ho Chi Minh City, whose latitude is lower than Hanoi's.

Keywords: Affect infusion model · Monday effect
Seasonal affective disorder · Stock returns · Volatility

1 Introduction

The Efficient Market Hypothesis (EMH) proposed by Fama [6] has been considered a central financial model for more than 40 years. According to Fama [6], an efficient market is the market in which stock prices reflect fully, immediately all available information on the market. EMH states that "it should be impossible to outperform the overall market", where all relevant information is captured by investors and reflected in market prices. At any time, the market also opportunely handle the latest information available.

Numerous studies have been conducted to check eligibility of the EMH and the results suggest that abnormalities have occurred in the stock market. Examples are the January effect (Rozeff and Kinney [21]), the Monday effect (Raj and Kumari [19]), and the Halloween effect (Bouman and Jacobsen [2]). In addition, a kind of mood disorder is determined by elements of weather (Floros [7]), motion of celestial bodies (Thach and Diep [22], Dichev and Janes [4], Krivelyova and Robotti [16]), and SAD (Murgea

V. Kreinovich et al. (Eds.): ECONVN 2019, SCI 809, pp. 873–885, 2019.
https://doi.org/10.1007/978-3-030-04200-4_63

[18], Kliger and Kudryavtsev [15], Kliger et al. [14], Dowling and Lucey [5], Garrett et al. [10], Kamstra et al. [12]) were also found on the stock market. These abnormalities have shown that capital markets are ineffective.

Many theories have been developed to find answers to the abnormalities in financial markets. In recent years, psychologists have confirmed that humans do not always behave appropriately, which can lead to inadequate decisions in financial market transactions. Behavioral finance enriches economic understanding by combining cognitive psychology into financial models and provides potential explanations for market abnormalities. Many studies in the field of behavioral finance have documented mood misattribution that affects investors' risk perception, which in turn influences their decision (i.e., investor behavior depends on the mood of the investor). For example, Akerlof and Shiller [1] argue that emotions and moods often represent crucial determinants and so prices are primarily determined by mood and emotions of investors.

Behavioral finance researchers have documented the influence of mood-proxy variables on global financial markets. Among the variables that represent mood, some researchers have argued that biorhythm disorders (such as SAD) have a strong influence on the mood of people. SAD is a term referring to the state of depression that tends to occur in the Fall and Winter due to lack of daylight (Melrose [17]). The term first appeared in the newspaper in 1985 and there are also other names such as Winter Depression, Winter Blues or the hibernation reaction. In addition, SAD is also believed to be the most important mood variable affecting the financial market (Murgea [18], Kliger and Kudryavtsev [15], Kliger et al. [14], Dowling and Lucey [5], Garrett et al. [10], Kamstra et al. [12]).

Studies on mood abnormalities have been performed in Vietnam stock market, such as the effects of super-moon (Thach and Diep [22]). However, research on the impact of SAD on the Vietnam's stock market has not been done. Therefore, this article focuses on a well-known psychological phenomenon, called SAD. Specifically, the article analyzes the impact of SAD on stock returns and their volatility to supplement the potential explanations for seasonal anomalies in the Vietnam stock market.

2 Literature Review

2.1 SAD, Mood and Human Behavior

In psychology, mood is defined as an emotional state. Mood is classified into two categories: good mood (positive mood) and bad mood (negative mood) (Frijda [9]). Human mood is influenced by many factors and it plays an important role in human decisions and behaviors, sometimes it leads to deviations from reasonable behavior (Kliger and Kudryavtsev [15]). Recently, the effects of psychological factors have attracted the growing concern of researchers in the field of economics and finance.

Many psychological and clinical studies documented the existence of SAD (Young [23], Melrose [17], Keller et al. [13], Dam et al. [3], Hawkins [11]). As a finding of these studies, 4–6% of the population has this disease and about 10–20% of the population has SAD but at a lesser level. SAD occurs primarily in the millennial and middle-aged (15–55 years) and occurs in both men and women, especially those who live in areas with poor lighting or a sudden change in light levels between seasons (Young [23]). Common symptoms of SAD include sleep problems, lethargy, depression, social problems, anxiety, feelings of despair, fatigue, difficulty concentrating, discomfort and sadness, and mood changes (Melrose [17]).

There are three causes of SAD. Firstly, according to Melrose [17], it is due to seasonal changes that can disrupt the balance of melatonin hormone. People with SAD have difficulty in producing excessive melatonin. Melatonin is a hormone produced by the pineal gland that responds to darkness by causing drowsiness. As Winter days get darker, melatonin production will increase and people with SAD will feel sleepy and lethargic. Secondly, people with SAD have difficulty regulating the neurotransmitter serotonin, a neurotransmitter that regulates mood. A lack of sunlight can cause a decrease in serotonin and can lead to depression (Melrose [17]). Thirdly, the combination of serotonin decrease and melatonin increase causes the changes in human circadian rhythms. Specifically, in the Winter, a longer night-time phenomenon occurs and, also, the Sun's shining time decreases, making the human body more difficult to adjust (leading to circadian disorder), leading to emotional disorder (Melrose [17]).

Thus, SAD has been shown to be one of the main factors affecting human circadian rhythms. The direct effect of SAD is to change the human biological balance, the adaptation mechanism is impaired and various mood disorders occur, especially depression and anxiety, thus has influenced the human decision mechanism.

2.2 Previous Empirical Studies

Psychology researchers have noted a strong association between depression and risk-taking. In particular, based on the influence of SAD on investor psychology, SAD changed investors' perceptions of risk and ultimately influenced the estimation of market risk. The findings of these studies have shown that SAD has a significant effect on mood, and mood has been shown to influence individual decision-making. Thus, SAD can be used as a mood-proxy variable to analyze mood effects to stock returns.

A pioneering study on the effect of SAD on the stock market was conducted by Kamstra et al. [12]. In particular, the authors studied the effects of SAD on stock markets in nine countries (the US, Sweden, Britain, Germany, Canada, New Zealand, Japan, Australia, and South Africa). Single regressions are made for each country to explain the different SAD effects in different countries. The findings show the impact of SAD on the US stock index (S&P 500, NYSE, NASDAQ, and AMEX), Sweden (Veckans Affärer), Britain (FTSE 100), Germany (DAX 30), New Zealand (Capital 40), Japan (NIKKEI 225), and South Africa (Datastream Global Index) in both cases: both before and after controlling well-known seasonal factors and other environmental factors. In particular, this study also shows that the SAD effect is stronger in countries located at higher latitudes than countries near the equator.

Garrett et al. [10] studied the SAD effect in the context of the equilibrium asset pricing model to determine whether seasonal factors can be explained using CAPM. The authors use daily and monthly data in six countries (the US, Sweden, New Zealand, the UK, Japan and Australia) to examine the relationship between SAD and stock returns). The results indicate that conditional CAPM allows the price of risk to change with respect to the seasonal variation in the length of day. This suggests that the

SAD effect is associated with increased risk aversion and seasonal depression, as reflected by a change in risk premium.

Dowling and Lucey [5] analyzed the effects of mood variables (weather, geomagnetic storms, SAD, daylight savings time) and lunar phases on stock prices in 37 countries. By using the GARCH (Generalized Autoregressive Conditional Heteroskedasticity) model, the authors have shown the strong influence of the SAD effect on stock returns and their volatility. In particular, the SAD effect is significant to small capitalization indices. Dowling and Lucey [5] argue that individual investors are more likely to be affected by the SAD in valuing stocks with small capitalization. Among all the variables that represent mood, SAD is the most important variable and the impact of SAD on the stock market is increasing with latitude.

The study of Kliger et al. [14] also shows the effect of SAD on efficiency of initial public offering (IPO). The authors examine the IPO first trading days and point out that in the short term, the stock return is lower when the day is shorter than the night. The authors suggest that the effect of SAD on investor mood has reduced demand for IPO. All the evidence found was consistent with the effect of SAD on the mood of investors and thus affected the rate of return on the stock market.

A number of later studies (Murgea [18], Kliger and Kudryavtsev [15]) also supported the SAD effect on the financial market.

3 Methodology and Data

3.1 Measurement of the SAD Effect

Based on the study of Garrett et al. [10], Kamstra et al. [12], the measure of the SAD effect is calculated in the following order:

First, the Sun's declination angle at time t (λ_t) is defined by the following formula:

$$\lambda_t = 0.4102 \sin[(\frac{2\pi}{365})(julian_t - 80.25)] \tag{1}$$

where "$julian_t$" is a variable running from 1 to 365 (366 in a leap year), representing the number of days in the year. $Julian_t$ gets a value of 1 on January 1st, 2 on January 2nd and so on.

Next, the length of the night (the hours of night) at t (H_t) is the time from sunrise to sunset at the latitude δ calculated by the following formula:

$$H_t = \begin{cases} 24 - 7,72 \text{arcos}[-\tan(2\pi\delta/360)\tan(\lambda_t)] \text{ in the Northern Hemisphere} \\ 7,72 \text{arcos}[-\tan(2\pi\delta/360)\tan(\lambda_t)] \text{ in the Southern Hemisphere} \end{cases} \tag{2}$$

with "arcos" as arc cosine.

Finally, the measure of the SAD effect at t (SAD_t) is calculated as follows:

$$SAD_t = \begin{cases} H_t - 12 \text{ (For trading days in the Fall and Winter)} \\ 0 \end{cases} \tag{3}$$

As SAD_t only changes through Fall and Winter, they are the seasons where medical evidence shows that SAD affects humans. By subtracting 12 (the average annual hours of nights at any given location), SAD_t reflects the length of the night over 12 h in the all and Winter. Fall begins at Fall equinox and ends at Winter Solstice. Meanwhile, the Winter begins with the Winter Solstice (the time when the Sun descends to the lowest point in the sky for later to return to the north, and the daytime become the shortest). Then, the days become longer and nights become shorter and this ends at Vernal equinox, this is the equinoctial point where day and night are equally long.

Vietnam is located in the Northern Hemisphere, so Fall is from 21 September to 20 December and Winter is from December 21st to March 20th (Kamstra et al. [12]). Hanoi and Ho Chi Minh City (HCMC) are the two cities used to calculate the length of the night and the SAD effect as these are the two largest economic and social centers in Vietnam.

The length of night in Hanoi and HCMC during the period from February 2002 to December 2017 is shown in Fig. 1. In particular, the length of the night begins to increase in the Fall period from September 21st and peak at the Winter Solstice (December 21st), then the length of the night begins to decrease and reach the bottom of the Summer Solstice (June 21st).

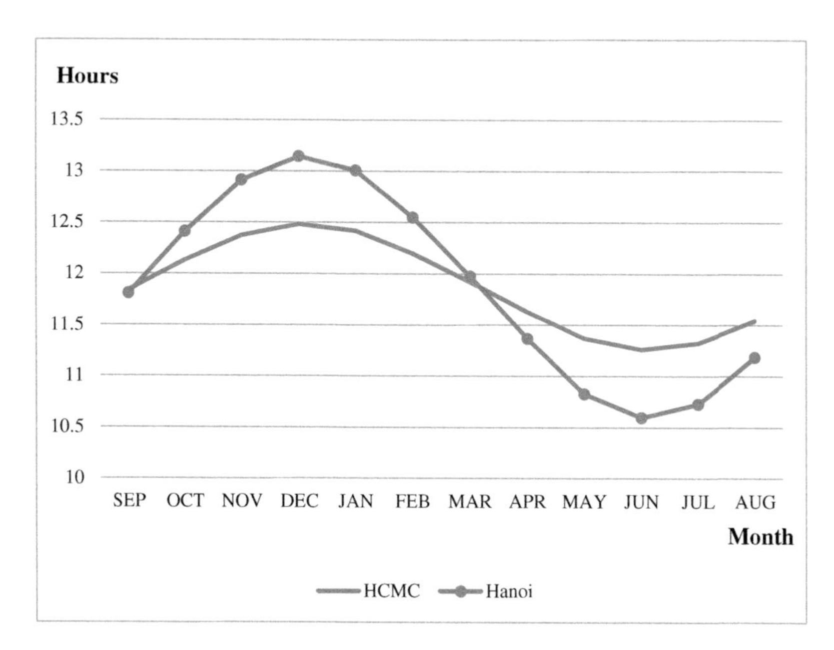

Fig. 1. Average length of night in Hanoi and HCMC

Table 1. Descriptive statistics for the length of the night and SAD effect

City	The length of night (H_t)		SAD effect (SAD_t)	
	Hanoi	HCMC	Hanoi	HCMC
Mean	11.85105	11.86272	0.68020	0.26012
Maximum	13.16958	12.49391	1.16958	0.49391
Minimum	10.57738	11.25302	−0.12131	−0.12403
Standard deviation	0.90422	0.43325	0.40562	0.19410
Observations	3,952	3,952	1,922	1,922

Notes: The latitude of Hanoi is 21°02′N and HCMC is 10°46′N.

For more details, Table 1 shows the length of the night (hours of the night) and the SAD effect (the length of the night over 12 h in the Fall and Winter) in Hanoi and HCMC. Observations of the entire sample were 3,952 and the number of observations in the Fall and Winter in the sample was 1,922. Specifically, the average daily length of the night in Hanoi is 11.85105 h, the standard deviation is 0.90422 h, and the large difference between the minimum value (10.57738 h) and the maximum value (13.16958 h) shows great volatility in the length of the night in Hanoi. In HCMC, the level of volatility in length of night is lower. Specifically, the average daily length of the night was 11.86272 h, the standard deviation of 0.90422 h, and the small difference between the minimum value (11.25302 h) and the maximum value (12.49391 h). The length of the night over 12 h in the Fall and Winter average of 0.68020 h (Hanoi), 0.26012 h (HCMC) and the standard deviation of 0.40562 h (Hanoi), 0.19410 h (HCMC) showed that in the Fall and Winter, the night was longer than the day. In particular, the SAD effect is stronger and fluctuates more in Hanoi as it has higher latitude than HCMC.

3.2 Data

The article uses the closing price of VN-Index for the period from February 2002 to December 2017. VN Index data is collected from the website of HOSE. VN-Index is a trend indicator of price volatility of all listed and traded stocks on HOSE. The return rate of the VN Index is calculated by the following formula:

$$R_t = \frac{P_t - P_{t-1}}{P_{t-1}} x100 \tag{4}$$

where R_t is the daily rate of return of VN-Index; P_t is the closing price of VN-Index on t and P_{t-1} is the closing price of VN-Index on t−1.

Table 2 summarizes the VN-Index's daily stock return. Specifically, the daily rate of return during the entire study period was 3,952 observations and 1,922 in Fall and Winter. Average daily return is 0.05099% (the entire sample), 0.07428% (in the Fall and Winter seasons). The standard deviation was 1.40320% (whole sample), 1.42814% (Fall and Winter) and the large difference between the minimum and maximum values of the return rate showed great volatility of stock return in this period. The stock return

Table 2. Descriptive statistics for the daily return on VN-Index

Period	02/2002–12/2017	Fall and Winter
Mean	0.05099	0.07428
Maximum	8.04888	8.04888
Minimum	−5.87174	−4.68216
Standard deviation	1.40320	1.42814
Skewness	0.04336	0.06503
Kurtosis	5.07324	5.14539
Jarque-Bera	709.0291***	369.9538***
Observations	3,952	1,922

Notes: *** Significant at 1% level.

curve with a skewness coefficient greater than 0 (Skewness > 0) should show the shape of a positive Skewness distribution (this indicates that the distribution has asymmetric sides), and the Kurtosis is greater than 3 (Kurtosis > 3) (this implies that the fat tailed profit chain) should have a more leptokurtic distribution. Also, Jarque-Bera test found that the hypothesis of normality was rejected at a statistical significance at 1%. All this suggests that the VN-Index's return rate data series does not follow the normal distribution rule.

3.3 Model and Methodology

Previous studies on the SAD effect (e.g., Murgea [18], Kamstra et al. [12]) used the Ordinary Least Squares (OLS) method to examine the effect of SAD on stock returns. However, this approach has some disadvantages including the heteroscedasticity of financial data cannot be taken into account, i.e. the volatility of stock prices at some point is much higher than usual. To overcome this problem, the conditional variance model of the Autoregressive Conditional Heteroskedasticity (ARCH) model is often used to simulate financial time series changing over time and be divided (i.e. there are very high changes/periods, followed by quiet periods). However, the basic ARCH model has some limitations: Too many lags may affect the results of estimation (because it significantly reduces the number of degrees of freedom in the model). The more delayed data series, the more variables are lost. Therefore, the Generalized Autoregressive Conditional Heteroskedasticity (GARCH) model is used instead of the ARCH model because of the higher degree of generalization. However, Rastogi [20] argues that the financial time series has three characteristics that differ from the normal time series (i.e., the financial time series have leptokurtic distributions, volatility clustering, and leverage effects). However, the GARCH model captures only two characteristics (leptokurtic distribution and volatility clustering) and ignores the leverage effect (i.e., asymmetric behavior). To incorporate the asymmetry of financial time series, Zakoian [24] proposed the TGARCH (Threshold GARCH) model. The main purpose of this model is to look at the asymmetry between negative shocks (bad news) and positive shocks (good news) as Zakoian [24] argues that negative shocks in the market have a stronger, more persistent effect than positive shocks and that they make investors pessimistic, depressed and even passively waiting for the signs of the market.

To overcome the disadvantage of the GARCH model, the article uses the TGARCH (1,1) model with three different distribution assumptions of standardized residuals: normal distribution, Student's-t distribution, and GED to examine the effect of SAD on returns and volatility in the Vietnam's stock market. The most appropriate model will be based on Akaike Information Criteria (AIC), Schwarz Information Criterion (SIC), and Hannan-Quinn Information Standards Criterion (HQC).

Based on the study of Garrett et al. [10], the TGARCH(1,1) model for the VN-Index's return rate will be as follows:

$$R_t = \mu_0 + \mu_1 SAD_t + \mu_2 Fall_t + \mu_3 Mon_t + \mu_4 Tax_t + \mu_5 R_{t-1} + \varepsilon_t \tag{5}$$

$$\sigma_t^2 = \omega + \alpha\varepsilon_{t-1}^2 + \gamma\varepsilon_{t-1}^2 d_{t-1} + \beta\sigma_{t-1}^2 + \upsilon SAD_t \tag{6}$$

Equation (5) is called the mean equation to test the effect of the SAD on the return rate, where R_t is the daily rate of return of VN-Index; SAD_t is defined by Eqs. (1), (2) and (3) with the latitude of Hanoi is 21°02′N and the latitude of HCMC is 10°46′N; $Fall_t$ represents the Fall effect ($Fall_t$ get the value of the SAD if it falls in Fall period, i.e., September 21st to December 20th and vice versa; Mon_t is the dummy variable that represents the Monday effect (Mon_t gets the value of 1 if t falls on Monday and 0 otherwise); Tax_t is the dummy variable representing the tax (Tax_t gets a value of 1 if it falls on the last trading day or the first five trading days of the tax year and 0 otherwise); R_{t-1} is the latency of the dependent variable; and ε_t is the error term.

Previous studies have shown that seasonally depressed investors tend to sell risky assets (stocks) and switch to investment in safe assets when daylight hours start decreasing in Fall. As the day becomes longer after the Winter Solstice at the end of December, the investor will recover from the SAD and return to the stock market, which will increase the stock price. Therefore, in Eq. (5) we have added the control variable $Fall_t$, we expect the value of the coefficients of the SAD_t variable to be positive and the coefficient of $Fall_t$ to be negative. The positive stock returns are due to the recovery from the SAD and in combination with negative return rates predicted by the Fall, suggested seasonal asymmetric effects of SAD on the volatility of return from Fall to Winter.

Equation (6) is called the conditional variance equation, where σ_t^2 is the conditional variance (or the volatility s of the daily rate of return of the VN-Index); dt−1 is a dummy variable (getting the value of 1 if $\varepsilon_{t-1} < 0$ and d_{t-1} is 0 if $\varepsilon_{t-1} \geq 0$). In the TGARCH model, good news/good mood ($\varepsilon_t \geq 0$) and bad news/bad mood ($\varepsilon_t < 0$) have different effects on conditional variance. In particular, good news (good mood) has an impact of α, whereas bad news (bad mood) has the effect of $\alpha + \beta$. If positivity of γ is statistically significant, the leverage effect exists and bad news increases volatility. In addition, coefficients ω, α and β must satisfy the conditions of the GARCH(1,1) model. Coefficient υ to measure the effect of the SAD effect on the volatility of the rate of return of the VN Index (The larger the value of SAD_t, the more confused investor mood, the SAD is expected to have negative effects on stock returns).

4 Empirical Results

4.1 Test of Stationarity

To avoid spurious regression, the VN-Index's return rate was put to the test for stationarity before analysis. This article examines the stationarity of the VN-Index's return rate in two cases: (i) with an intercept and (ii) with an intercept and trend). Two tests are used to test the stationarity of the VN-Index's return rate: Augmented Dickey Fuller (ADF) test and Phillips-Perron (PP) test. The results of the stationarity test shown in Table 3 show that the VN-Index's return rate is stationary at level in both cases of with an intercept and trend.

Table 3. Tests of stationarity

ADF test	Intercept	−14.77182***
	Trend and Intercept	−14.76981***
PP test	Intercept	−50.56206***
	Trend and Intercept	−50.55635***

Notes: *** significant at 1% level.

4.2 Estimation Result of the TGARCH(1,1) Model

Table 4 shows the estimation results of the TGARCH(1,1) model with three different distribution assumptions, namely normal distribution, Student's-t distribution, and GED. In the mean equation, the estimation coefficient of the SAD variable with latitude in Hanoi and HCMC was positive but not statistically significant at all 3 different distribution assumptions. The coefficients of Fall variable are negative but are only statistically significant at 10% in Student's-t distribution for both Hanoi and HCMC. The regression results show that the coefficients of the Mon variable are negative and statistically significant at 5% in all three distribution assumptions. Finally, the coefficient of Tax variable is positive but not statistically significant in all 3 distribution assumptions.

In the Variance Equation, the estimates of the three distribution assumptions show that: (i) the ARCH coefficient (α) and GARCH (β) coefficient are positive and significant at 1% (this suggests that the volatility shocks are quite persistent; (ii) the GARCH (β) coefficient is larger than the ARCH coefficient (α), so the volatility of the rate of return is influenced by past volatility rather than current volatility; (iii) the asymmetric coefficient (γ) in the three cases is positive, but the coefficient γ is only statistically significant at 10% in the case of the Student's-t distribution assumption (hence the leverage effect is not obvious). The coefficient of SAD with latitude in Hanoi and HCMC is positive and is statistically significant at 10% (in the case of normal distribution) and at 5% (in the case of Student's-t distribution and GED).

The TGARCH(1,1) model with Student's-t distribution is most suitable as it has the smallest AIC, SIC, and HQC values. Therefore, subsequent analysis will be based on the TGARCH(1,1) model estimation with Student's-t distribution. Estimation results in the mean equation show that the coefficient of SAD variable with latitude in Hanoi

Table 4. Estimation results of TGARCH(1,1) model

City	Hanoi			HCMC		
Distributions	Gaussian	Student's-t	GED	Gaussian	Student's-t	GED
Mean Equation						
Constant (μ_0)	0.015698	0.020577	0.017992	0.017676	0.021554	0.018887
SAD (μ_1)	0.060737	0.075996	0.064799	0.138508	0.175514	0.149729
Fall (μ_2)	−0.070986	−0.111145*	−0.097478	−0.178629	−0.267218*	−0.23684
Mon (μ_3)	−0.067629**	−0.067239**	−0.069787**	−0.067528**	−0.067054**	−0.069576**
Tax (μ_4)	0.052836	0.033189	0.014974	0.053571	0.034298	0.016076
R_{t-1} (μ_5)	0.200917***	0.201764***	0.199129***	0.201016***	0.201879***	0.199252***
Variance Equation						
Constant (ω)	0.027787***	0.015977***	0.021396***	0.028346***	0.016861***	0.02214***
ARCH effect (α)	0.209299***	0.219424***	0.215488***	0.209415***	0.219717***	0.215719***
Leverage effect (γ)	0.029163	0.050909*	0.04197	0.02941	0.050882*	0.042167
GARCH effect (β)	0.778614***	0.77624***	0.77619***	0.778211***	0.775778***	0.775721***
SAD (υ)	0.009258*	0.014378**	0.011881**	0.021421*	0.032111**	0.026691*
Information Criteria						
AIC	3.014995	2.989245	2.995027	3.015007	2.989364	2.995084
SIC	3.032484	3.008324	3.014106	3.032496	3.008443	3.014162
HQC	3.021198	2.996012	3.001794	3.021210	2.996131	3.001851

Notes: *, **, *** Significant at 10%, 5% and 1% levels, respectively.

and HCMC is positive but not statistically significant. In addition, the coefficient of Fall variable is negative and is statistically significant at 10% (Fall, as suggested by Garrett et al. [10], in the study model, is the SAD variable in Fall). It can be concluded that the SAD effect exists on the Vietnam stock market. This result supports the Affect Infusion Model (AIM) proposed by Forgas [8]. AIM proposes that individuals with negative mood avoid taking risks. On the other hand, individuals with positive moods tend to take more risks.

Specifically, when the Fall in Vietnam begins (i.e., the length of the night will increase), the number of people with SAD increased will lead to disappointment (negative mood), causing risk aversion, who will tend to stay away from risky assets and invest in less risky assets. This tendency increases as Fall begins to become more pronounced. SAD represents the sphere of influence and severity of seasonal affective disorder so that stock returns will decline during the Fall. Winter Solstice, the time of the shortest day of the year (the day beginning in the Winter), has led to a change in the psychology of people with seasonal affective disorder. As the length of the night begins to decrease, the frustration caused by the seasonal disorder begins to decrease (the mood of the investor becomes more positive) and the investor begins to return to investing in the risky assets on the stock market. As a result, the rate of return will increase in the Winter. As a result, the risk aversion caused by the SAD is asymmetric from Winter Solstice (bad effect in Fall and good effect in the Winter). The findings of the study show evidence of the effect of SAD on Vietnam stock market during the Fall. However, the findings do not provide clear evidence of the impact of the SAD on the Vietnam stock market in the Winter. The findings also show that Monday effect was present in the Vietnam stock market (the coefficient of Mon is negative and statistically significant at 5%), and the current rate of return depends on the rate of return of the previous day (coefficient of R_{t-1} is statistically significant at 1%). In addition, the effect of tax-loss selling has not been found on the stock market in Vietnam.

The variance of the TGARCH(1,1) model with Student's-t distribution shows that the conditions of the TGARCH(1,1) model are satisfied, where the coefficient γ is positive and is statistically significant at 10%, so the leverage effect is confirmed to exist (the effect of good news on the rate of return is about 0.22%, but the effect of bad news on the rate of return is about 0.27%). The coefficient of SAD is positive and statistically significant at 5%, indicating a decrease in the number of sunshine hours in the Fall that causes bad moods and an increase in the number of sunshine hours in the Winter to create a positive mood for the investors. This caused a volatility in the mood of investors and the consequent increase in the rate of return of the VN Index.

Comparing the effect of the SAD (with the latitude of Hanoi) and the effect of the SAD (with the latitude of HCMC) on the Vietnam's stock market, the results show that the impact of SAD on HCMC's latitude is more powerful (this means that the SAD effect is stronger in places with lower latitudes). This finding is not consistent with Dowling and Lucey [5], Kamstra et al. [12]. This may be because HOSE is located in HCMC, this is the largest economic center in Vietnam with the largest population, and therefore, there are many investors involved in trading on HOSE.

5 Conclusion

This study analyzes impact of the SAD effect (with latitude in Hanoi and HCMC) on the stock market in Vietnam. Sample data used are daily return rate of the VN-Index in the period from February 2002 to December 2017. In particular, the article uses the TGARCH(1,1) model with three different distribution assumptions (normal distribution, Student's-t distribution and GED) to examine impacts of the SAD effect on rate of return and its volatility. Based on AIC, SIC, and HQC, the TGARCH(1,1) model assuming Student's-t distribution is the most suitable and the results showed that there existed SAD effect on the Vietnam's stock market: (i) the SAD effect is seen as a factor behind the change in the rate of return on stock during the Fall; (Ii) the SAD effect is asymmetric between the Fall and Winter but the findings have not provided clear evidence of the effect of SAD on return rate in the Winter; (iii) the SAD effect has an impact on the return rate volatility of VN-Index; and (iv) the SAD effect is stronger at lower latitudes. In addition, the study results showed the presence of the Monday effect in Vietnam stock market. In general, the findings in this study strongly suggested that the SAD effect was caused by seasonal affective disorder. This led to a lower return rate where the length of the night becomes longer in the Fall (which implies that individual investors sometimes make investment decisions depending on their mood).

References

1. Akerlof, G.A., Shiller, R.J.: Animal Spirits: How Human Psychology Drives the Economy, and Why It Matters for Global Capitalism. Princeton University Press, Princeton (2009)
2. Bouman, S., Jacobsen, B.: The halloween indicator, 'sell in may and go away': another puzzle. Am. Econ. Rev. **92**(5), 1618–1635 (2002)
3. Dam, H., Jakobsen, K., Mellerup, E.: Prevalence of winter depression in Denmark. Acta Psychiatr. Scand. **97**(1), 1–4 (1998)
4. Dichev, I.D., Janes, T.D.: Lunar cycle effects in stock returns. J. Private Equity **6**(4), 8–29 (2003)
5. Dowling, M., Lucey, B.M.: Robust global mood influences in equity pricing. J. Multinational Financ. Manage. **18**(2), 145–164 (2008)
6. Fama, E.F.: Efficient capital markets: a review of theory and empirical work. J. Finance **25** (2), 383–417 (1970)
7. Floros, C.: On the relationship between weather and stock market returns. Stud. Econ. Finance **28**(1), 5–13 (2011)
8. Forgas, J.P.: Mood and judgment: the effect infusion model (AIM). Psychol. Bull. **117**, 39–66 (1995)
9. Frijda, N.H.: Emotion experience and its varieties. Emot. Rev. **1**(3), 264–271 (2009)
10. Garrett, I., Kamstra, M.J., Kramer, L.A.: Winter blues and time variation in the price of risk. J. Empir. Finance **12**(2), 291–316 (2005)
11. Hawkins, L.: Seasonal affective disorders: the effects of light on human behaviour. Endeavour **16**(3), 122–127 (1992)
12. Kamstra, M.J., Kramer, L.A., Levi, M.D.: Winter blues: a SAD stock market cycle. Am. Econ. Rev. **93**, 324–343 (2003)

13. Keller, C.M., Fredrickson, B.L., Ybarra, O., Cote, S., Johnson, K., Mikels, J., Conway, A., Wager, T.: A warm heart and a clear head - the contingent effects of weather on mood and cognition. Psychol. Sci. **16**(9), 724–731 (2005)
14. Kliger, D., Gurevich, G., Haim, A.: When chronobiology met economics: Seasonal affective disorder and the demand for initial public offerings. J. Neurosci. Psychol. Econ. **5**(3), 131–151 (2012)
15. Kliger, D., Kudryavtsev, A.: Out of the blue: mood maintenance hypothesis and seasonal effects on investors' reaction to news. Quant. Financ. **14**(4), 629–640 (2014)
16. Krivelyova, A., Robotti, C.: Playing the field: geomagnetic storms and international stock markets. Working Paper No. 5b, Federal Reserve Bank of Atlanta, Atlanta (2003)
17. Melrose, S.: Seasonal affective disorder: an overview of assessment and treatment approaches. Depression Res. Treat. **2015**, 1–6 (2015)
18. Murgea, A.: Seasonal affective disorder and the Romanian stock market. Economic Research-Ekonomska Istraživanja **29**(1), 177–192 (2016)
19. Raj, M., Kumari, D.: Day-of-the-week and other market anomalies in the indian stock market. Int. J. Emerg. Markets **1**(3), 235–246 (2006)
20. Rastogi, S.: The financial crisis of 2008 and stock market volatility - analysis and impact on emerging economies pre and post crisis. Afro-Asian J. Financ. Account. **4**(4), 443–459 (2014)
21. Rozeff, M.S., Kinney, W.R.: Capital market seasonality: the case of stock returns. J. Financ. Econ. **3**(4), 379–402 (1976)
22. Thach, N.N., Diep, V.N.: The impact of supermoon on stock market returns in Vietnam. In: Anh, L., Dong, L., Kreinovich, V., Thach, N. (eds.) Econometrics for Financial Applications, Studies in Computational Intelligence, vol. 760, pp. 611–623. Springer, Cham (2018)
23. Young, M.A.: Does seasonal affective disorder exist? A commentary on traffanstedt, mehta, and lobello (2016). Clin. Psychol. Sci. **5**(4), 750–754 (2017)
24. Zakoian, J.M.: Threshold heteroskedastic models. J. Econ. Dyn. Control **18**(5), 931–955 (1994)

Consumers' Purchase Intention of Pork Traceability: The Moderator Role of Trust

Nguyen Thi Hang Nga[1(✉)] and Tran Anh Tuan[2(✉)]

[1] Banking University of Ho Chi Minh City, 36, Ton That Dam Street, District 1, Ho Chi Minh City, Vietnam
hangnga.buh@gmail.com

[2] Ho Chi Minh City Institute of Development Studies, 28, Le Quy Don Street, District 3, Ho Chi Minh City, Vietnam
at7tran@gmail.com

Abstract. This study examines the effects of trust on consumers' intention to consume pork traceability based on Theory of Planed Behavior (TPB). Data are from a survey of 219 respondents from Ho Chi Minh city. Cluster analysis, regression, anova was used to analyze to the data. Results indicate that attitudes, social norms, perceived behavioral control have a positive effect on intention and these influences depend on the trust of consumer. When consumers have higher trust, they have more incentives to purchase and at the same time, the influence of attitudes towards purchasing intention is higher. Subjective norms only have a positive effect on intention to purchase when the trust of the consumer is low. In addition, consumer perceptions is quite positive for traceable pork. This results suggest that the managers and food management agency need to focus on consumers' trust in order to have effective communication strategies.

Keywords: Food · Intention · TPB · Traceability · Trust

1 Introduction

Pork is considered to be a popular dish in daily meals of Vietnamese families. Food consumption in general and pork in particular in the food market of Vietnam are facing great challenges related to the food safety issue. Recently, the food safety issue has received growing concern and attention from the public. Food safety incidents and scandals have occurred continuously in the animal husbandry and pork processing industry such as using tranquillizer before slaughtering, the use of beta-adrenergic agonists, or antibiotic residues are in excess of permitted level. Therefore, consumers are confused about whether they should buy pork for their daily meals, since they cannot determine which meat is of good quality and safe.

Traceability of meat products is one of food safety management methods, and it is also a communication channel which helps consumers purchase meat with clear origin. Food choice behaviour has been examined by many researchers in developed countries, they also mentioned the consumer trust in contexts associated with food safety risk in their studies, such as Lobb et al. (2007); Stefani et al. (2008). In general, previous studies suggest that consumer trust plays a vital role in explaining the intention of

V. Kreinovich et al. (Eds.): ECONVN 2019, SCI 809, pp. 886–897, 2019.
https://doi.org/10.1007/978-3-030-04200-4_64

consuming food. However, food consumption behaviour may vary with contexts from country to country due to cultural differences. In Vietnam, according to our knowledge, there hasn't been any research on the relationship between consumer trust and the behavior of choosing pork in general, and in particular, pork which can be traceability in the situation that involves food safety risks. Moreover, although traceability has been conducting in Vietnam, obstacles and limitations still exist.

To fill these gaps, firstly, this study aims to assess consumer sentiment about pork which can be traceable. The second objective is to segment customers on the level of their trust in pork with traceable in order to evaluate the difference in consumption intention between segments. The third one is to investigate the effect of trust on intention of buying pork with traceable and examine the regulatory role that consumer trust takes in the theory of planned behavior model. Based on the analysis of clustering techniques, regression analysis and ANOVA analysis, the results of this study would provide the basis for managers and state management agencies to establish appropriate policies.

2 Conceptual Framework

2.1 Theory of Planned Behavior

This study is based on the Theory of Planned Behavior (TPB) by Ajzen (1991). The theory suggests that behavior is influenced directly by behavioral intentions. Behavioral intentions are affected by attitudes, subjective norms and perceived behavioral control. A person's behavior is a combination of behavioral beliefs, normative beliefs, and control beliefs. Behavioral belief is the belief in the outcome of an action, which can lead to positive or negative behavior. Normative beliefs relate to social pressure to perform behavior. Control beliefs show the level of control by each individual to perform their behavior. The TPB theory is used as a general theoretical framework for predicting behavioral intentions in many types of areas, including the food consumption. Although using TPB in understanding human health behaviors is widely accepted, recent calls have been made to extend the TPB to include additional factors (Armitage and Conner 2001; Conner and Armitage 1998). In this study, we built a research model based on the TPB model and it includes trust influencing intention to consume meat in VietNam.

2.2 Attitude

Studies using the TPB theory have shown that attitude was one of the key factors explaining behavioral intention. Attitude is often defined as a psychological tendency that is expressed by evaluating a particular entity (food) with some degree of favor–disfavor, like–dislike, satisfaction–dissatisfaction or good–bad polarity (Eagly and Chaiken 1993). Attitudes indicate an individual's assessment of the degree of whether he or she likes or dislikes, satisfies or dissatisfied, good or bad when they perform action. If a person is aware of the consequence of a behavior is positive, they will have positive attitudes to do it and vice versa. When a person has positive attitudes, they are more likely to carry out action. Attitude is a positive factor affecting the intention in the food sector (Lobb et al. 2007; McCarthy et al. 2003; McCarthy et al. 2004; Tuu 2015). Based on the above discussions, the following hypothesis is proposed.

H1: Attitude has a positive effect on intention.

2.3 Subjective Norms

Subjective norms are normally supposed to capture the individual's perception being important to others in his or her social environment or expect him or her to behave in a certain way (Ajzen 1991). Subjective norms show the social pressure to do or not to do something (Thong and Olsen 2012). This pressure often comes from important people such as family, friends and colleagues. If the other involved people find that the behavior is positive (or negative) and the individual is motivated to meet the expectations of those related people, then there will be an affirmative subjective norm (or negative). In this study, subjective norms are defined as the approval of others' expectations, such as family norms (Olsen 2001). The results of previous studies suggest that subjective norms influence the intention positively (Thong and Olsen 2012). Based on the above discussions, the following hypothesis is proposed.

H2: Subjective norms has a positive effect on intention.

2.4 Perceived Behavioral Control

Perceived behavioral control refers to the perceiving of whether it is easy or difficult to perform action (Thong and Olsen 2012). As an individual has plenty of resources or opportunities they tend to think that there is less obstacles to doing things. Ajzen (1991) focused on perceived behavioral control as the person's beliefs as to how easy or difficult performance of the behavior is likely to be. He also suggested that control factors can be either internal to the person (e.g. skills, abilities, power of will, and compulsion) or external to the person (e.g. time, opportunity, and dependence on others). PBC is defined in this study as an integrated measure of internal and external resources that make it easy to act upon the motivation to consume (Tuu 2015). The results of previous studies also support the positive relationship between perceived behavioral control and purchase intention in the food sector (Lobb et al. 2007; Verbeke and Vackier 2005). The next hypothesis is thus proposed:

H3: Perceived behavioral control has a positive effect on intention.

2.5 Trust

Morrow et al. (2004) suggests that general trust is the extent to which one believes that others will not act to exploit one's vulnerabilities. Meanwhile, specific trust refer to beliefs about a particular object. According to Böcker and Hanf (2000) trust is recognized as a necessary way to reduce uncertainty about acceptable levels and to simplify decisions. The result research of Lobb et al. (2007) proposed that trust in source of information influences the intention to buy chicken in the UK. A recent study by Muringai and Goddard (2017) in Canada, the United States and Japan also indicates that trust affects the consumption of beef and pork. In the research on the intention to buy meat with traceability in Thailand. Buaprommee and Polyorat (2016) show that trust has a positive influence on buying decision. Vermeir and Verbeke (2007) also reported that the influence of attitudes, subjective norms, and perceived behavioral control on consumer intention is based on trust. The above discussions accordingly enable the following hypothesis to be suggested:

H4: Trust has a positive effect on intention.
H5: Trust has a positively moderates the (a) attitude, (b) subjective norms, (c) perceived behavioral control –intention relationship.

Based on the proposed hypotheses, the theoretical model is given in Fig. 1.

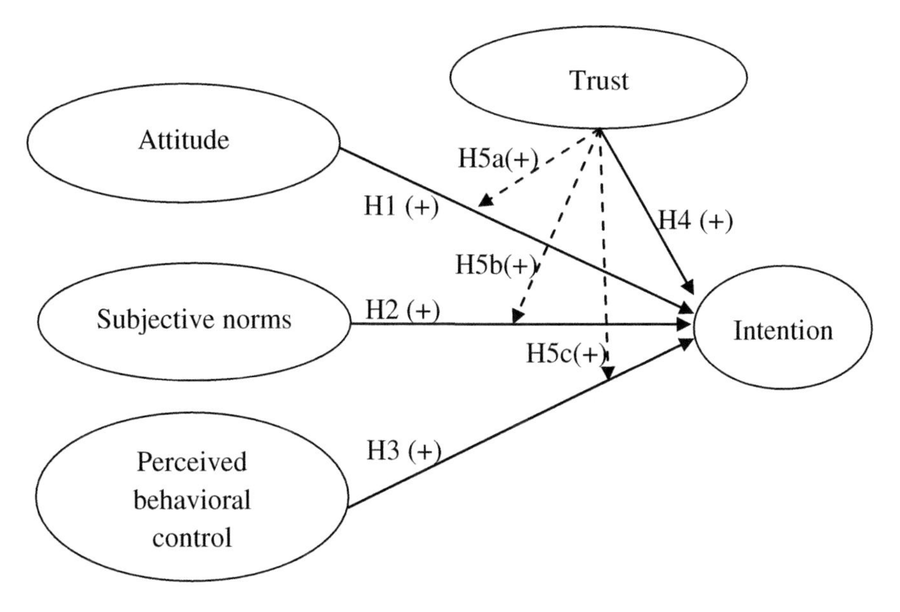

Fig. 1. Theoretical model

3 Research Methodology

The data in this study is collected by directly surveying 219 consumers in Ho Chi Minh City based on a 5- point Likert scale, in which 1 is completely disagree and 5 is completely agree. After obtaining data, the study performs analytical techniques such as descriptive statistics, testing the reliability of the scale, exploratory factor analysis (EFA), regression analysis, cluster analysis and ANOVA analysis using SPSS 16.0 software. EFA analysis is assessed using Barlett's test with KMO coefficient must be higher than 0,7; total variance explained is greater than 50% and factor loading is greater than or equal to 0,5 with the chosen significance level is 5%.

The criterion for measuring the scale reliability is the Cronbach's alpha coefficient is greater than 0,6 and the corrected item-total correlation is greater than 0,3. Testing the reliability of the scale: A useful coefficient for assessing internal consistency is Cronbach's alpha (Cronbach 1951). The formula is:

$$\alpha = \left(\frac{k}{k-1}\right)\left(1 - \sum \frac{s_i^2}{s_t^2}\right)$$

where k is the number of items, s_i^2 is the variance of the ith item and s_T^2 is the variance of the total score formed by summing all the items.

Cluster analysis: K-means method uses K prototypes, the centroids of clusters, to characterize the data. They are determined by minimizing the sum of squared errors (Ding and He 2004):

$$J_K = \sum_{k=1}^{K} \sum_{i \in C_k} (x_i - m_k)^2$$

Where $(x_1, \cdots, x_n) = X$ is the data matrix and $m_k = \sum_{i \in Ck} x_i / n_k$ is the centroid of cluster C_k and n_k is the number of points in C_k.

To test the mixed moderator role, the following M1, M2, M3 models are estimated using the OLS (Ordinary Least Square) through the three - step hierarchical regression model (Chaplin 1991).

$$\mathrm{Y} = \alpha + \beta \mathrm{X} + \varepsilon \qquad (\mathrm{M}_1)$$
$$\mathrm{Y} = \alpha + \beta \mathrm{X} + \gamma \mathrm{Z} + \varepsilon \qquad (\mathrm{M}_2)$$
$$\mathrm{Y} = \alpha + \beta \mathrm{X} + \gamma \mathrm{Z} + \delta \mathrm{X} * \mathrm{Z} + \varepsilon \qquad (\mathrm{M}_3)$$

In which:

- Y: Dependent variable
- X: Independent variable
- Z: Moderator variable

Based on previous researches, this research recommends the following model:

$$\mathrm{I} = \alpha + \beta_1 \mathrm{A} + \beta_2 \mathrm{S} + \beta_3 \mathrm{P} + \varepsilon \qquad (1)$$

$$\mathrm{I} = \alpha + \beta_1 \mathrm{A} + \beta_2 \mathrm{S} + \beta_3 \mathrm{P} + \beta_4 \mathrm{T} + \varepsilon \qquad (2)$$

$$\mathrm{I} = \alpha + \beta_1 \mathrm{A} + \beta_2 \mathrm{S} + \beta_3 \mathrm{P} + \beta_4 \mathrm{T} + \beta_5 \mathrm{A} * \mathrm{T} + \beta_6 \mathrm{S} * \mathrm{T} + \beta_7 \mathrm{P} * \mathrm{T} + \varepsilon \qquad (3)$$

In which:

- I: Intention (Dependent variable).
- A: Attitude.
- S: Subjective norms.
- P: Perceived behavioral control.

- T: Trust.
- ε: Random error

The questions of the scales used in this study are inherited from previous studies related to consumer's food choice behavior and qualitative research is applied concurrently to refine the questionnaire appropriately to the context of Vietnam. To be more specific, the behavioral attitude scale (A) in this study including six items that depict consumer attitudes when using pork in a family's daily meal were adapted from the study of Menozzi et al. (2015). The subjective norms scale (S) is measured according to three observational variables related to the opinions of those in the family that have impact on them, this scale is inherited from Tuu (2015). The perceived behavioral control scale (P) consists of six observational variables were adapted from previous studies, such as Menozzi et al. (2015). The trust scale (T) includes four observational variables inherited from the study of Buaprommee and Polyorat (2016); Menozzi et al. (2015). The consumer intention scale (I) has four observation variables that indicate the intention to consume pork in the near future, which are also inherited from Buaprommee and Polyorat (2016); Menozzi et al. (2015). Table 1 illustrates the question items in the shortened form of the scales.

4 Result

4.1 Reliability and Validity of the Measures

The study surveys 219 consumers in HCMC with 26.5% male and 73.5% female. People aged 18–24 account for 21.5%; people in the 25–34 age group make up 64.4% of the surveyed people; the 35–44 age group constitutes 12.3% and the percentage of people who are over 45 is 1.8%. In terms of educational background, the university level constitutes the majority of the group, which is 54.3% and the intermediate level accounts for 34.7%. People who earn from 5–10 million VND per month is 61.6% and those who are paid from 11 to 15 million VND a month make up 12.5%. The respondents are mostly office workers and civil servants, they all occupy 80% of the surveyed consumers. Of the 219 respondents, only 11% answer that they never hear of pork with traceable origin, 74% have heard about it and 15% say they have been told a lot about pork with traceable origin.

Results of descriptive statistics show that consumers rate relatively high in variables such as traceable pork is better for health (the mean of this variable is 3.75), safer (with an average of 3.74), better quality (the mean is 3.66), easier to control (about 3.68 on average) and also more expensive (with an average of 3.68) in the 5- point Likert scale. In general, consumers underestimate the perceived behavioral control variables (on average this scale is 3.1) in the 5- point Likert scale. Consumer confidence is also assessed as a moderate level (only 3.39). Consumers will be more confident if the meat is certified for traceable origins. The Cronbach's alpha coefficient illustrates that the scales meet the required reliability. Specifically, the Cronbach's alpha coefficients are greater than 0.6 and range from 0.851 to 0.927. The item-total correlation of the scales are greater than 0.3. This result is given in Table 1.

Table 1. Descriptive statistics of indicators and Cronbach'alpha

		Constructs and indicators	Mean	Std. Error	Factor loadings
		Cronbach'alpha: 0.889	**3.392**	**0.919**	
Attitude (A)	A1	Tastier	3.392	0.919	0.540
	A2	Healthier	3.753	0.809	0.909
	A3	Safer	3.739	0.824	0.867
	A4	More satisfying quality	3.657	0.891	0.887
	A5	More expensive	3.675	0.952	0.674
	A6	Guaranteed for being controlled	3.684	0.843	0.630
		Cronbach'alpha: 0.927	**3.735**	**0.874**	
Subjective norms (S)	S1	My family want me…	3.735	0.874	0.797
	S2	My family encourage me…	3.639	0.899	0.802
	S3	My family think that I should…	3.648	0.898	0.943
		Cronbach'alpha: 0.910	**3.237**	**0.980**	
Perceived behavioral control (P)	P1	Easy to look for this information	3.237	0.980	0.728
	P2	Feel confident when I look for it	3.164	0.948	0.818
	P3	Look for it without help from others	3.054	0.965	0.874
	P4	Easy to understand information	3.059	1.005	0.817
	P5	Confident that I'll understand it	3.086	0.941	0.722
	P6	Understand it without help from others	3.022	0.969	0.677
		Cronbach'alpha: 0.851	**3.383**	**0.913**	
Trust (T)	T1	I believe this pork can traceability	3.383	0.913	0.738
	T2	I trust the information provided	3.214	0.905	0.885
	T3	I trust it to be genuine	3.383	0.877	0.524
	T4	I trust the certified provided	3.502	0.831	0.626
		Cronbach'alpha: 0.887	**3.584**	**0.926**	
Intention (I)	I1	I have intention to buy and eat pork	3.584	0.926	0.550
	I2	I want to buy and eat pork	3.584	0.896	0.896
	I3	I will search for this pork to buy	3.570	0.907	0.718
	I4	I am willing to buy and eat pork	3.442	0.893	0.923

Source: Investigated by the author

The study performs exploratory factor analysis with Principal Axis Factoring and Promax rotation. The results show that the KMO of the Barlett's test is 0.897, which is greater than 0.6 with the Sig. = 0.000; all of the factor loadings are greater than 0.5 and the difference between factor loadings is less than 0.3; the cumulative variance explained is 73.17% and the Eigenvalues of the fifth factor were 1.062, which resulted in the exploratory factor analysis meets the requirements. The factor loadings of the exploratory factor analysis are shown in Table 1.

4.2 Cluster Analysis

After the results of verifying the reliability of the scales and performing the exploratory factor analysis are satisfactory, the study continue to conduct cluster analysis according to the belief variable. The cluster analysis procedure selected is non-hierarchical clustering (K-Means) using the optimal partitioning method. The result of cluster analysis according to the trust variable reveals that the two clusters were selected with Sig. of the F test in the observational variables is smaller than the significance level (5%), so it can be concluded that there is a statistically significance between clusters which have differences. As for cluster 1, there are 111 observations and cluster 2 had 108 observations. The mean of cluster 1 is 2.83 and for cluster 2, it is 3.96. Thus, cluster 1 is named as low trust and cluster 2 is cluster with high trust.

4.3 Regression Analysis

This study conducts regression analysis using the ordinary least squares method (OLS) with four independent variables: Attitude, subjective norms, perceived behavioral control, trust and the dependent variable is consumers' purchase intention. Results of the regression analysis indicate that the regression model is statistically significant at the 5% significance level. With the TPB model, the fit of the model is 43.1% and all variables have a positive effect on intention with the significance level of 5%. When extending the model by adding confidence variable, the result shows that the fit of the model is considerably improved, rising from 43.1% to 48.4%. However, the perceived behavioral control variable is not statistically significant in this situation. In the TPB model, attitude remains the most influential factor on intention, followed by subjective norms and perceived behavioral control. With the extended model, trust is the most influential factor on intention and the next is attitude, the last one is subjective norms. The test of the assumptions when conducting the OLS regression assures that all requirements are met. This result is illustrated in Table 2.

To examine the regulatory role of trust, the regression model was performed with two clusters of trust. Results of the two clusters suggest that, for the cluster with high trust, attitude, subjective norms, and perceived behavioral control have positive impact on purchase intention at a 5% significance level. In particular, the influence of subjective norms is greatest, followed by attitude and perceived behavioral control. As regards the cluster with low trust, the result reports that only attitude and perceived behavioral control affect intention positively at a 5% significance level, but the effect of subjective norms is not significant. Comparing the regression coefficients between the two clusters reveals that for the cluster with high trust, attitude influenced the intention

Table 2. Results of the regression analysis

Independent	Beta	Beta standard	P-value	R^2
Model TPB				
A	0.373	0.334	0.000	43.9%
S	0.255	0.272	0.000	
P	0.211	0.217	0.000	
Extended model				
A	0.276	0.247	0.000	49.3%
S	0.216	0.230	0.001	
P	0.077	0.079	0.191	
T	0.336	0.314	0.000	

more strongly (β = 0.496) than the cluster with low trust (β = 0.222). In contrast, perceived behavioral control has greater impact on intention in the low-trust cluster (β = 1.86) than in the high-trust cluster (β = 1.74). This result is presented in Table 3.

Table 3. Results of the regression analysis with two clusters of trust

Independent	Low trust			High trust		
	Beta	Beta standard	P-value	Beta	Beta standard	P-value
A	0.222	0.211	0.031	0.496	0.419	0.000
S	0.332	0.390	0.000	0.073	0.067	0.496
P	0.181	0.168	0.037	0.174	0.202	0.024

Therefore, the hypotheses H1; H2; H3; H4; H5a,b,c are all accepted. However, depending on the chosen model, the results will be different. In particular, the results of this study are consistent with the results of the previous studies when using the TPB model with the interpretation level of 43.9%. Meanwhile, attitude is considered as the main explanatory factor to purchase intention. But when expanding the model by adding a trust variable, trust plays the most important role in explaining purchase intention. When there is the presence of trust in the model, perceived behavioral control does not have the role of explaining intention to buy. This finding is similar with the results of several previous studies when addressing the role of perceived behavioral control. Perceived behavioral control is considered to be the least satisfying explanation for intention in the TPB model. If the attitude or norms are strong, the predictions of perceived behavioral control for intention would be low (Ajzen 1991). The findings of Verbeke and Vacackier (2005) also indicate a modest influence of perceived behavioral control when other factors were present.

In addition, our findings suggest that subjective norms influence purchase intentions depending on trust. With respect to those who have low trust, subjective norms play a vital role in explaining the intention to buy. Perhaps when consumers have low trust, the pressure of the involved people becomes important to them when performing action. However, when they have high trust, pressure is no longer important, thus consumers may not need opinions of the involved people anymore. This result will contribute to the explanation of why the outcome of the influence of subjective norms on intention is inconsistent among studies in the food sector. When reviewing previous studies, Ajzen (1991) also revealed failure of subjective norms when predicting intention in some studies. Further research has also focused on the role of subjective norms in explaining the intention of food consumption, and some authors even propose to remove subjective norms from the model, such as Yadav and Pathak (2016); Shin et al. (2016).

4.4 ANOVA Analysis

In order to test whether there is a difference in purchase intention between two clusters with high trust and low trust, the study conducted an ANOVA analysis. Before analyzing ANOVA, it is necessary to test whether the variances between the two groups are the same. The result presents that the variances between the high-trust cluster and the low-trust cluster are consistent at the significance level of 5% (Sig. = 0.175). Hence, the results of ANOVA analysis can be applied well. The result of ANOVA analysis shows that Sig. = 0.000, indicating a statistically significant difference in purchasing intention between the high-trust cluster and the low-trust cluster with a 5% significance level. The mean of purchase intention of the low-trust cluster is 3.2 and the corresponding figure for the high-trust cluster is 3.9. Therefore, the purchase intention of the high-trust cluster is firmer than the purchase intention of the cluster with low trust.

5 Conclusion, Implications, and Limitations

In this study, we examine the role of trust in explaining the intention to purchase pork with traceable origins of consumers in HCM City based on the theory of planned behavior. Trust plays a role as an independent variable and as a regulatory variable in the research model. The results of cluster analysis, regression analysis and ANOVA analysis show some primary results as follows:

First, variables in the TPB model such as attitude, subjective norms and perceived behavioral control explain 43.9% the change in purchase intention. In regard to the extended model, attitude, subjective norms, perceived behavioral control and trust explain 49.3% the change in purchase intention.

Second, when there is the presence of trust in the model, trust becomes the most influential factor and thus the impact of perceived behavioral control is not statistically significant. This result contributes to the explanation of the inconsistency of the influence that perceived behavioral control has in the TPB model related to the food sector.

Third, the results of the regression analysis for the two clusters, which present high and low trust, suggest that the contribution of subjective norms to the model is only meaningful for the cluster with low trust. People who have low confidence often undergo the pressure of those involved when intend to carry out action, whereas those with high confidence are not affected by this pressure. This result also helps to explain the inconsistency of the role of subjective norms in analyzing purchase intention of previous studies. Some researchers even propose to remove subjective norms from the model. Furthermore, the higher the trust is, the more likely consumers are to purchase.

Fourth, overall, most consumers think they have heard of traceable pork and they all have good thoughts about it, such as it would be better for health, it will has better quality, it is safer and easier to control. However, the perception of consumers' ability to control their behavior is still relatively low and the consumers' trust only remains at a moderate level. Moreover, the ANOVA analysis reveals that for the high- trust cluster, the purchase intention is higher than the one with low trust.

This result implies that managers and regulators in the food sector need to focus on the mass media in order to increase the positive perception of consumers. More importantly, consumer trust becomes a crucial factor in explaining consumers' purchase intention. The factors influencing purchase intention will change and depend on the different segments according to the consumer trust.

This study also has some limitations that the further studies need to improve, such as convenient sampling method, the scope of the survey is only limited in HCM City, and only using OLS as the regression analysis method. In addition, the intention to consume food in general and to consume pork in particular, both are affected by many other factors that this study has not mentioned. This means that, further studies should overcome the above limitations to increase the reliability of the study.

References

Ajzen, I.: The theory of planned behavior. Organ. Behav. Hum. Decis. Process. **50**, 179–211 (1991)

Armitage, C., Conner, M.: Meta–analysis of the theory of planned behaviour. Br. J. Soc. Psychol. **40**, 471–499 (2001)

Böcker, A., Hanf, C.H.: Confidence lost and—partially—regained: consumer response to food scares. J. Econ. Behav. Organ. **43**(4), 471–485 (2000)

Buaprommee, N., Polyorat, K.: The antecedents of purchase intention of meat with traceability in Thai consumers. Asia Pacific Manage. Rev. **21**(3), 161–169 (2016)

Conner, M., Armitage, C.J.: Extending the theory of planned behavior: a review and avenues for further research. J. Appl. Soc. Psychol. **28**, 1429–1464 (1998)

Cronbach, L.J.: Coefficient alpha and the internal structure of tests. Psychometrika **16**(3), 297–334 (1951)

Chaplin, W.F.: The next generation of moderator research in personality psychology. J. Pers. **59** (2), 143–178 (1991)

Ding, C., He, X.: K-means clustering via principal component analysis. In: Proceedings of the Twenty-First International Conference on Machine Learning, p. 29. ACM, July 2004

Eagly, A.H., Chaiken, S.: The psychology of attitudes. Harcourt Brace Jovanovich, Fort Worth (1993)

Lobb, A.E., Mazzocchi, M., Traill, W.B.: Modelling risk perception and trust in food safety information within the theory of planned behaviour. Food Qual. Prefer. **18**(2), 384–395 (2007)

McCarthy, M., de Boer, M., O'Reilly, S., Cotter, L.: Factors influencing intention to purchase beef in the Irish market. Meat Sci. **65**(3), 1071–1083 (2003)

McCarthy, M., O'Reilly, S., Cotter, L., de Boer, M.: Factors influencing consumption of pork and poultry in the Irish market. Appetite **43**(1), 19–28 (2004)

Menozzi, D., Halawany-Darson, R., Mora, C., Giraud, G.: Motives towards traceable food choice: a comparison between French and Italian consumers. Food Control **49**, 40–48 (2015)

Morrow, J.L., Hansen, M.H., Person, A.W.: The cognitive and affective antecedents of general trust within cooperative organisations. J. Manag. Issues **16**(1), 48–64 (2004)

Muringai, V., Goddard, E.: Trust and consumer risk perceptions regarding BSE and chronic wasting disease. Agribusiness, 1–27 (2017)

Olsen, S.O.: Consumer involvement in fish as family meals in Norway: an application of the expectance–value approach. Appetite **36**, 173–186 (2001)

Shin, Y.H., Hancer, M., Song, J.H.: Self-congruity and the theory of planned behavior in the prediction of local food purchase. J. Int. Food Agribusiness Mark. **28**(4), 330–345 (2016)

Stefani, G., Cavicchi, A., Romano, D., Lobb, A.E.: Determinants of intention to purchase chicken in Italy: the role of consumer risk perception and trust in different information sources. Agribusiness **24**(4), 523–537 (2008)

Thong, N.T., Olsen, S.O.: Attitude toward and consumption of fish in Vietnam. J. Food Prod. Mark. **18**(2), 79–95 (2012)

Tuu, H.H.: Attitude, social norms, perceived behavioral control, past behavior, and habit in explaining intention to consume fish in Vietnam. J. Econ. Dev. **22**(3), 102–122 (2015)

Verbeke, W., Vackier, I.: Individual determinants of fish consumption: application of the theory of planned behavior. Appetite **44**, 67–82 (2005)

Vermeir, I., Verbeke, W.: Sustainable food consumption among young adults in Belgium: theory of 8 planned behaviour and the role of confidence and values. Ecol. Econ. **64**(3), 542–553 (2007)

Yadav, R., Pathak, G.S.: Intention to purchase organic food among young consumers: Evidences from a developing nation. Appetite **96**, 122–128 (2016)

Income Risk Across Industries in Thailand: A Pseudo-Panel Analysis

Natthaphat Kingnetr[1(✉)], Supanika Leurcharusmee[1], Jirakom Sirisrisakulchai[1], and Songsak Sriboonchitta[1,2]

[1] Faculty of Economics, Chiang Mai University, Chiang Mai, Thailand
natthaphat.kingnetr@outlook.com, supanika.econ.cmu@gmail.com, sirisrisakulchai@hotmail.com, songsakecon@gmail.com

[2] Puey Ungphakorn Center of Excellence in Econometrics, Chiang Mai University, Chiang Mai, Thailand

Abstract. In this study, we investigate the labour income risk across industries in Thailand using the Labour Force Survey (LFS) data over 2008–2017 consisting of more than a million individuals. Two types of income risk are considered in this study: permanent and transitory. In order to estimate the risk, the LFS data is transformed into the pseudo-panel framework based on multiple labour characteristics. The results suggest that the transitory income risk is nearly twice as large as the permanent. In addition, we found that the top five industries facing strong income risks are transportation, agriculture, professional activities, manufacturing and financial and insurance activities.

Keywords: Income risk · Industry · Pseudo panel data · LFS Thailand

1 Introduction

There has been a growing literature investigating the way to estimate the income risk within labour market [9–11,13]. Krebs and Yao pointed out in [9] that the existence of income risk could potentially affect both a labour consumption pattern and welfare. The influence of income risk can be seen from [1] where it is shown that access to alternative income sources are required in order to smooth consumption whenever one is facing an unexpected income change. It was found in [6] that savings are the main shock buffer and it grows as the uncertainty in income increases. With that, such shock can mitigate an ability for workers to accumulate their physical and human capital, mitigating the opportunity to improve their incomes.

To the best of our knowledge, there is no existing work trying to estimate the income risk in the context of Thailand. This paper is the first attempt to investigate the dynamics of income risk inequality across different industries in Thailand. Data from the labour force survey (LFS) conducted by Thailand's National Statistics Office (NSO), covering the 2008 to 2017 period, are employed.

V. Kreinovich et al. (Eds.): ECONVN 2019, SCI 809, pp. 898–909, 2019.
https://doi.org/10.1007/978-3-030-04200-4_65

For the analysis, we follow the approaches suggested by [11]. The approaches allow one to take the shock of labour income into account and be able to decompose the income shock into two types of risk: permanent and transitory. By estimating both risks over time, one can see that the movement of such risk can provide a clue for the behaviour of labour regarding their saving and consumption in the future [8].

However, the LFS is not designed to be panel data which is crucial for the current approach of income risk estimation. This study, thus, attempts to create the pseudo-panel data based on several criteria that will be discussed later in a following session.

Our empirical results can be summarised as follows. First, we found that transitory income risk is larger than the persistent income risk for most of the industrial sectors considered in the study. Second, we found that overall the both types of risk are exhibiting a downward trend with the permanent risk being more stable overtime compared to the transitory. Third, the top five industry facing high income risk are transportation, agriculture, professional activities, manufacturing, and financial and insurance activities. We believe that this study would stimulate the concern about the income risk across industrial sectors as an alternative way to see the labour welfare. It also fills the gap in the literature which has been lacking a case study of Thailand.

The rest of this paper is organised as follows. Sections 2 and 3 provides discussion on the data and methodology of this study, respectively. Section 4 provides the empirical results. Finally, the conclusions are drawn in Sect. 5.

2 Data

In this study, the data regarding workers from Thailand are obtained from the Labour Force Survey (LFS) conducted by the National Statistics Office (NSO) over the period of 2008 to 2017. The data consisted of more than 800,000 individuals per year. Each individual is randomly interviewed based on NSO methodology across different regions in Thailand. In LFS, a questionnaire comprehensively covers important labour characteristics, such as hourly wage, salary, industrial classifications, age, sex, and education level. Due to how the survey is conducted, there are some observations containing missing data as well as those containing unrealistic values in certain fields. For instance, one may find individuals that receive zero salary, but he/she is employed and vice versa. Therefore, a data selection has to be made for this study.

We adapt the approach from [9,11] to prepare our data. First, individuals whose ages are between 25 and 70 are selected. This setting allows us to reduce the chance of having individuals who may be studying or out of labour force due to retirement. However, unlike other papers in literature which prefer the 25–60 setting, the inclusion of 61–70 people is to accommodate the fact that Thailand is entering the ageing society. It has become rather common to see these elderly people working in several occupations for the last five years [7]. Based on LFS, Fig. 1 shows that the share of elderly labour has been gradually increasing, while the opposite can be seen in the case of young employees.

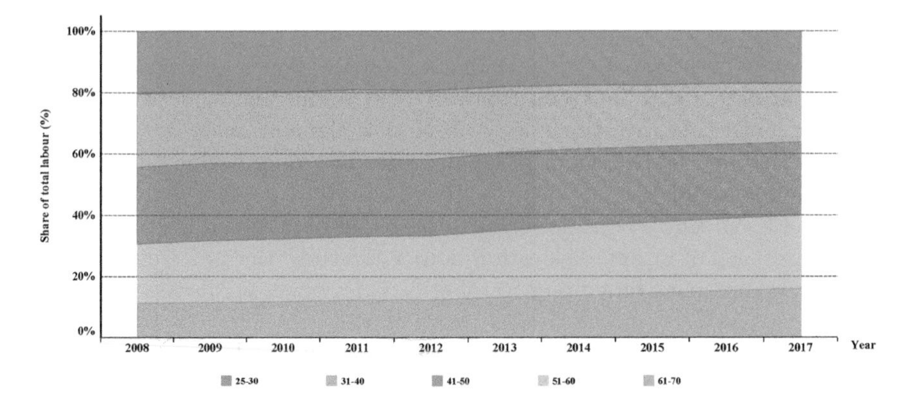

Fig. 1. The share of total labour by age groups from 2008 to 2017. **Source:** LFS 2008–2017, author's calculation.

Second, those whose monthly earnings are less than 2,250 baht are excluded. This allow us to get rid of individuals reporting unreliable figures. The criteria is based on the 2013 nationwide minimum daily wage law of 300 baht. Although such law exists, the actual earnings vary depending on type of job and bargaining power between employer and employee. Therefore, we decided to set the earning threshold to be a quarter of the minimum wage law.

Finally, we only consider individuals who are not self-employed in this study as their earnings are not reported in LFS. In addition, as pointed out by Kreb and Yao in [9], these types of people experience a different income process from employed people. They argued that the volatility of earnings in the case of the self-employed involve both risks from the labour market and the asset market.

Due to the these reasons, our data is trimmed down to approximately 180,000 individuals per year. Now, we are going to discuss about an overview and descriptive summary of the data after being trimmed down. Starting with average monthly income, Fig. 2 shows that the average monthly income has been increasing for the last decade. Interestingly, the female monthly income has taken over that of the male by 300 baht approximately in 2017 despite receiving 600 baht lower than males prior to 2015. In addition, there is a sharp increase in earnings in 2013 as a result of the nationwide minimum wage law [4].

Regarding the level of education attainment, three different levels are considered in this study: (1) the lower than high school including those who do not receive education, (2) the high school, and (3) the higher than high school. According to the share of educational attainment as shown in Fig. 3, it can be seen that females tend to have a higher education level than that of males and the percentage has been increasing since 2008. In 2017, Approximately 45% of females completed higher than the high school level of education, while the males remains less than 30%.

Figure 4 shows the share of total employment across 21 industrial classifications. It can be seen that manufacturing constitutes the most employment with

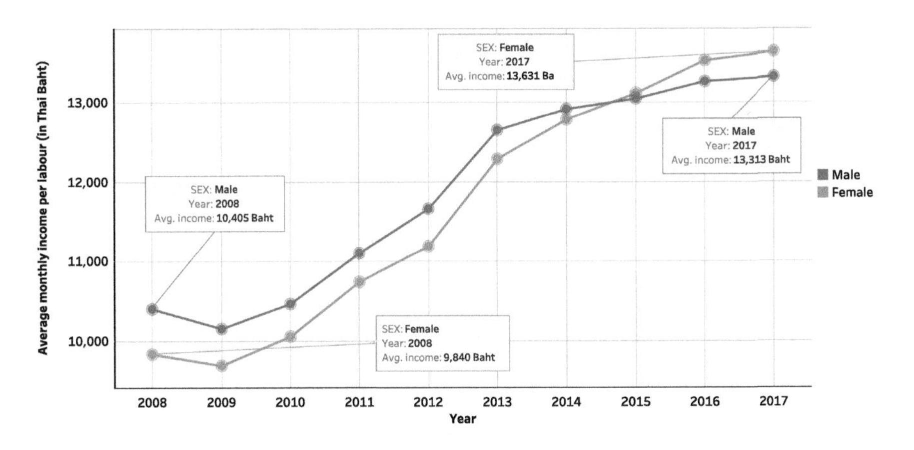

Fig. 2. The average monthly income from 2008 to 2017. **Source:** LFS 2008–2017, author's calculation.

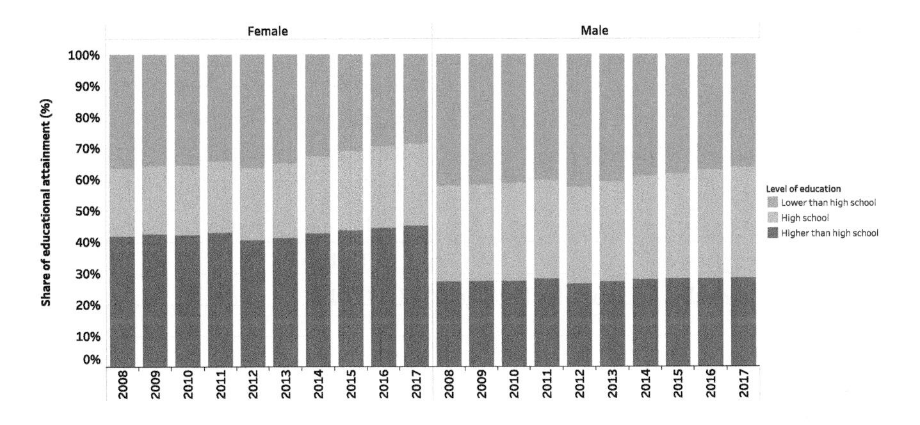

Fig. 3. The share of educational attainment by sex from 2008 to 2017. **Source:** LFS 2008–2017, author's calculation.

nearly 45% of the total. Next is the public administration and defence taking slightly more than a quarter of all employment. Wholesale and education show similar shares of employment of about 20% for each sector. Among these top 4 industries that cover nearly 90% of the total employment, the males takes more than a half except for the education sector. Most of interviewed workers are from the central region. The average age of workers is around 40 years old with average salary of 11,906 baht per month.

Finally, the further details regarding the composition and descriptive statistics of the data can be seen from Table 1. The majority of labour is male, taking over a half of the employment. In terms of education, 60% of Thai labours are well-educated with education equal to or greater than high school level.

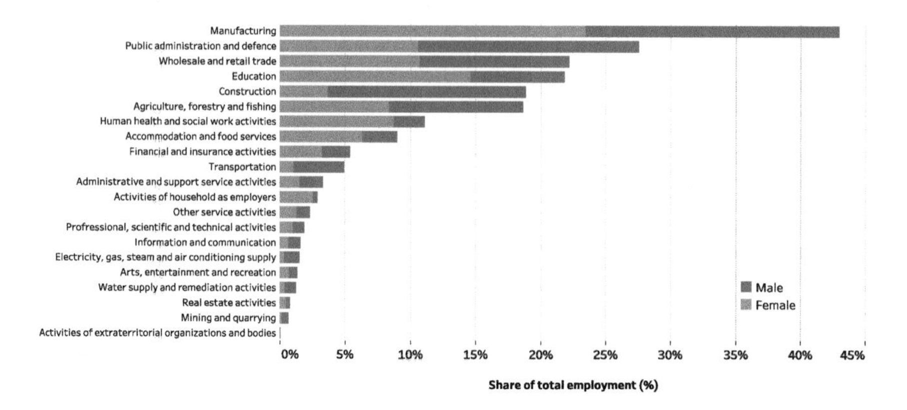

Fig. 4. The share of total employment by industrial classifications. **Source:** LFS 2008–2017, author's calculation.

Table 1. Composition of LFS data from 2008 to 2017

Number of observation	1.63 Million		
Male	54.61%	Bangkok	18.83%
Female	45.39%	Central	35.14%
Lower than high school	36.82%	North	13.82%
High school	30.84%	Northeast	19.79%
Higher than high school	32.34%	South	12.41%
Average age (year)	39.40		
Average monthly income (in Thai baht)	11906.45		

3 Methodology

In this section, we first describe briefly about how the income risk is estimated, followed by the criterion set for generating pseudo panel-data.

3.1 Estimating Income Risks

We define an individual income risk as 'unpredictable change of the individual income stream from its expected future income path'. The income risk measure used in this paper is the variance of changes of individual income. According to this measurement, the estimation involves two steps. We apply the same approach for estimating income risks as in the empirical works of [5,10]. The first step is to estimate the individual expected income from his/her characteristics such as age, gender, education, etc. From the first step, we get a stochastic change of individual income from its expected income. This change is not due to the changes in individual observable characteristics and can be used to measure the extent of income risk. The stochastic changes will be used in the second step to estimate the income risks for each industry.

Let y_{ijt} be the log of income for individual $i, i = 1, 2, ..., N$ from industrial sector $j, j = 1, 2, ..., J$ in time $t, t = 1, 2, ..., T$. The earning equation can be specified as

$$y_{ijt} = \alpha_{jt} + \beta_t \cdot x_{ijt} + u_{ijt}, \tag{1}$$

where α_{jt} and β_t denote time-varying parameters, x_{ijt} indicates a vector of individual observable characteristics, and u_{ijt} is the stochastic term. As discussed above, the changes in the stochastic terms u_{ijt} over time represent the unpredictable part of the individual income change.

Suppose that the stochastic term u_{ijt} can be separated into two unobserved components as

$$u_{ijt} = \omega_{ijt} + \eta_{ijt}, \tag{2}$$

where ω_{ijt} represents a permanent shock to income and η_{ijt} represents a transitory shock to income. The permanent component is persistent in the sense that it follows a random walk process:

$$\omega_{ijt+1} = \omega_{ijt} + \epsilon_{ijt+1}, \tag{3}$$

where $\{\epsilon_{ijt}\}$ is assumed to be independently distributed over t and identically and independently distributed across individuals i as $\epsilon_{ijt} \sim N(0, \sigma^2_{\epsilon jt})$. In the above specification, the transitory component has no persistence. We assume that $\{\eta_{ijt+1}\}$ is independently distributed over t and identically distributed across i, $\eta_{ijt} \sim N(0, \sigma^2_{\eta jt})$. The estimates of $\sigma^2_{\epsilon jt}$ and $\sigma^2_{\eta jt}$ give us the magnitudes of permanent and transitory income risks for each industrial sector j over time t.

To estimate $\sigma^2_{\epsilon jt}$ and $\sigma^2_{\eta jt}$, we consider the change in u_{ij} between period t and $t + n$:

$$\Delta_n u_{ijt} = u_{ijt+n} - u_{ijt} = \epsilon_{ijt+1} + \cdots + \epsilon_{ijt+n} + \eta_{ijt+n} - \eta_{ijt}. \tag{4}$$

The variance of these changes is given as

$$var(\Delta_n u_{ijt}) = \sigma^2_{\epsilon jt+1} + \cdots + \sigma^2_{\epsilon jt+n} + \sigma^2_{\eta jt} + \sigma^2_{\eta jt+n}. \tag{5}$$

The parameter $\sigma^2_{\epsilon jt}$ and $\sigma^2_{\eta jt}$ can be estimated using generalized method of moment (GMM) with the moment conditions in (5). The GMM estimator is specifically obtained by minimizing

$$\sum_{t,n} \{var(\Delta_n u_{ijt}) - \left(\sigma^2_{\epsilon jt+1} + \cdots + \sigma^2_{\epsilon jt+n} + \sigma^2_{\eta jt} + \sigma^2_{\eta jt+n}\right)\}^2. \tag{6}$$

Notice that we have a short time period (T = 10), hence we have a small sample for the estimation of $\sigma^2_{\epsilon jt}$ and $\sigma^2_{\eta jt}$. We then use equally weighted minimum distance (EWMD) estimation for small sample as suggested by [2,13]. [2] showed that using EWMD is superior to the two-stage GMM with optimal weighting matrix when taken into account the small sample bias.

3.2 Pseudo-Panel Data Preparation

Since the LFS is not designed to be a panel type, transforming data to a panel structure is required in order to apply the risk estimation technique discussed previously. We attempt to overcome this by creating a representative of a group of individuals exhibiting the same features under the following assumptions: firstly, individual's sex, educational attainment, type of employment, industrial sub-classification in which individual is employed as well as region must remain the same for all periods. Secondly, the individual's age must be available for all ten years since 2008 and increase by 1 for each consecutive year. The average value of individual residuals obtained from the first-step regression is then selected as the representative for a group of individuals with similar characteristics.

4 Empirical Findings

In this section, we start with the results from first-stage regression, followed by the estimates of overall income risk for each sub-sample considered in this study.

4.1 First-Stage Regression Results

Since this study seeks to estimate the income risk, the first step is to estimate relevant parameters for the earning equation. Even though we only employ the regression for the purpose of estimating income risk, we would still like to briefly discuss its results. According to Table 2, it can be seen that many coefficients exhibit expected signs. The coefficients of Age and Age^2 confirm the common finding in the literature of human capital theory [3,12]. In the case of other worker characteristics, those who are heads of households seem to generate higher earnings. The female workers experience a lower wage on average despite having a higher wage in 2015 as we showed in the previous section. Education still remains an important factor in driving earnings. It can be seen that the return on education increases as the level of education attainment increases. In term of marital status, in comparison to single individual, the married earn higher incomes on average while the rest face slightly lower incomes. Interestingly, those who work under public enterprise earn more income than the government employee while the private employee receives much less. Lastly, jobs in Bangkok provide higher incomes than the rest of the country, whereas the northern region faces the lowest monthly income on average.

4.2 Income Risk Estimates

After we obtain u_{ijt} from the first step, we will now move to the second step of the estimation of income risk using the approach discussed in the preceding section. Figure 5 shows the estimates of income risk using the whole sample. In can be seen that the value of transitory income risk is approximately twice as

Table 2. Regression results

Variable	Coefficient	Standard error
Constant	7.800	0.006
Age	0.052	0.000
Age^2	−0.042	0.000
Dummy variable		
Head of household	0.060	0.001
Female	−0.107	0.001
High school	0.388	0.001
Higher than high school	1.033	0.011
Married	0.040	0.001
Widowed	−0.111	0.002
Divorced	−0.025	0.002
Separated	−0.046	0.002
Public enterprise employee	0.229	0.002
Private employee	−0.199	0.001
Central	−0.269	0.001
North	−0.462	0.002
Northeast	−0.435	0.002
South	−0.373	0.002
N = 1,648,577	$R^2 = 0.546$	

Note: all coefficients are statistically significant at 5% level. Age^2 are divided by 100 prior to the estimation for readability purposes which does not affect our analysis. Although the similar results are also found in each sub-sample, they are omitted to conserve space.

large as the permanent income risk. In addition, there is a noticeable drop in the transitory income risk in 2013 which may be the result of the nationwide minimum wage law. On the other hand, permanent risk remains stably low overtime. Nevertheless, it is expected that employees working in each industrial sector would experience income risk differently.

The average industrial income risks are shown on Fig. 6. The findings are consistent with [10, 13] among others in which the transitory income risk tend to be higher than the permanent income risk. The top five industrial sectors facing high income risks are transportation, agriculture, professional activities, manufacturing, and financial activities. Whereas the bottom five are accommodation and food, construction, education, administrative activities, and real estate. Table 3 provides all industrial estimates of income risk. According to the Sect. 2, the

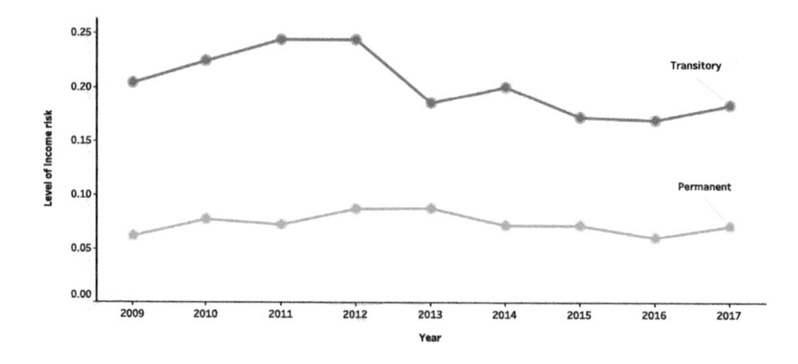

Fig. 5. Overall income risk from 2009 to 2017

agriculture, manufacturing constitute more than a half of the employees in this study. Therefore, we are going to further investigate and discuss their income risks overtime.

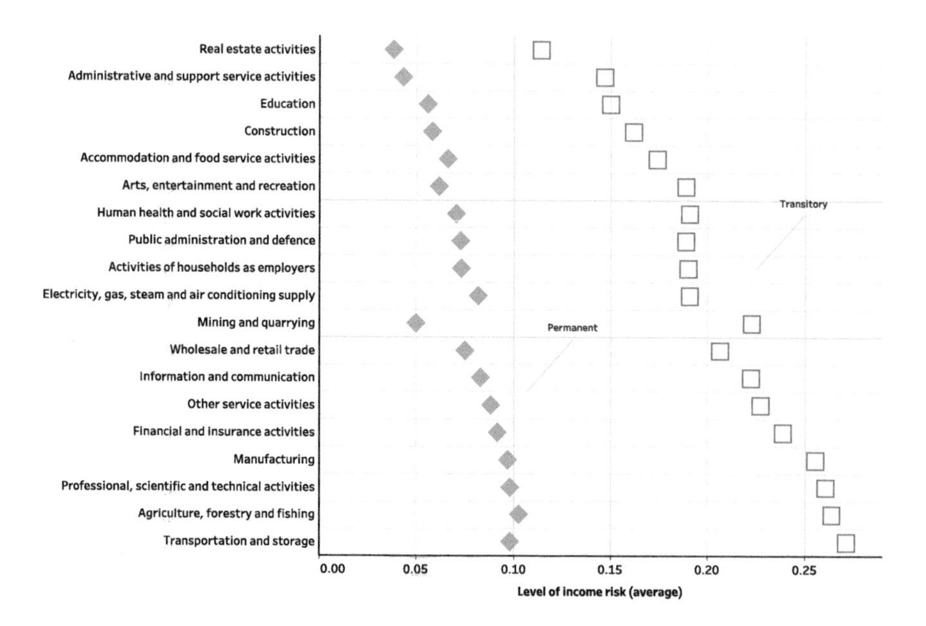

Fig. 6. Income risk by industrial classification (average over 2009 to 2017)

Figure 7 shows the income risks over time for both agriculture and manufacturing. Starting with the agriculture, we can see that both permanent and transitory income risks tend to decrease over time. On the contrary, the income risk both permanent and transitory seems to be going up in the case of manufacturing, especially in 2017.

Table 3. Estimates of industrial income risks from 2009 to 2017

Industrial classification	Permanent									Transitory								
	2009	2010	2011	2012	2013	2014	2015	2016	2017	2009	2010	2011	2012	2013	2014	2015	2016	2017
Real estate activities	0.016	−0.021	−0.127	0.246	−0.042	−0.026	0.279	−0.043	0.067	−0.013	0.053	0.400	0.027	0.004	0.127	0.160	0.093	0.174
Administrative and support service activities	−0.032	0.052	0.110	0.030	0.093	0.041	0.034	0.032	0.031	0.171	0.198	0.104	0.351	0.115	0.102	0.103	0.091	0.086
Education	0.024	0.061	0.048	0.065	0.058	0.067	0.045	0.036	0.097	0.148	0.150	0.150	0.176	0.135	0.163	0.145	0.112	0.171
Construction	0.038	0.091	0.059	0.074	0.073	0.044	0.044	0.053	0.048	0.212	0.169	0.186	0.203	0.171	0.136	0.123	0.128	0.129
Accommodation and food service activities	0.113	0.028	0.097	0.067	0.077	0.053	0.066	0.054	0.042	0.163	0.216	0.190	0.196	0.179	0.145	0.182	0.153	0.142
Arts, entertainment and recreation	0.078	0.013	0.049	0.163	0.074	0.062	0.041	0.030	0.045	0.114	0.251	0.495	0.193	0.145	0.154	0.145	0.094	0.110
Human health and social work activities	0.053	0.082	0.061	0.071	0.097	0.057	0.051	0.091	0.070	0.186	0.176	0.164	0.242	0.261	0.160	0.150	0.189	0.189
Public administration and defence	0.056	0.092	0.079	0.060	0.068	0.091	0.058	0.056	0.095	0.213	0.188	0.158	0.194	0.199	0.210	0.170	0.165	0.201
Activities of households as employers	0.065	0.049	0.108	0.075	0.039	0.076	0.096	0.025	0.128	0.194	0.239	0.192	0.169	0.162	0.155	0.234	0.148	0.218
Electricity, gas, steam and air conditioning supply	0.105	0.059	0.092	0.083	0.064	0.107	0.016	0.031	0.181	0.237	0.206	0.243	0.118	0.148	0.216	0.178	0.115	0.258
Mining and quarrying	−0.120	0.213	0.103	−0.095	0.195	0.099	0.090	0.199	−0.234	0.255	0.335	0.032	0.374	0.141	0.434	0.087	0.357	−0.011
Wholesale and retail trade	0.053	0.088	0.099	0.069	0.086	0.086	0.043	0.086	0.064	0.227	0.229	0.214	0.239	0.206	0.213	0.162	0.186	0.182
Information and communication	0.126	0.111	0.025	0.037	0.171	0.094	0.061	0.009	0.113	0.198	0.151	0.235	0.512	0.275	0.191	0.142	0.109	0.187
Other service activities	0.130	0.103	0.054	0.131	0.123	0.067	0.032	0.058	0.096	0.264	0.241	0.336	0.320	0.224	0.179	0.128	0.159	0.193
Financial and insurance activities	0.087	0.090	0.060	0.116	0.113	0.073	0.089	0.064	0.133	0.232	0.204	0.302	0.273	0.242	0.222	0.213	0.198	0.262
Manufacturing	0.047	0.131	0.103	0.101	0.106	0.095	0.070	0.069	0.153	0.299	0.258	0.248	0.311	0.238	0.248	0.216	0.204	0.279
Professional, scientific and technical activities	0.107	0.087	0.115	0.142	0.081	0.071	0.101	0.073	0.106	0.275	0.355	0.392	0.211	0.153	0.232	0.301	0.190	0.241
Agriculture, forestry and fishing	0.114	0.150	0.074	0.102	0.107	0.092	0.094	0.105	0.087	0.308	0.234	0.294	0.272	0.261	0.257	0.233	0.263	0.254
Transportation and storage	0.126	0.001	0.178	0.129	0.092	0.124	0.059	0.129	0.046	0.202	0.421	0.318	0.268	0.273	0.265	0.202	0.269	0.224

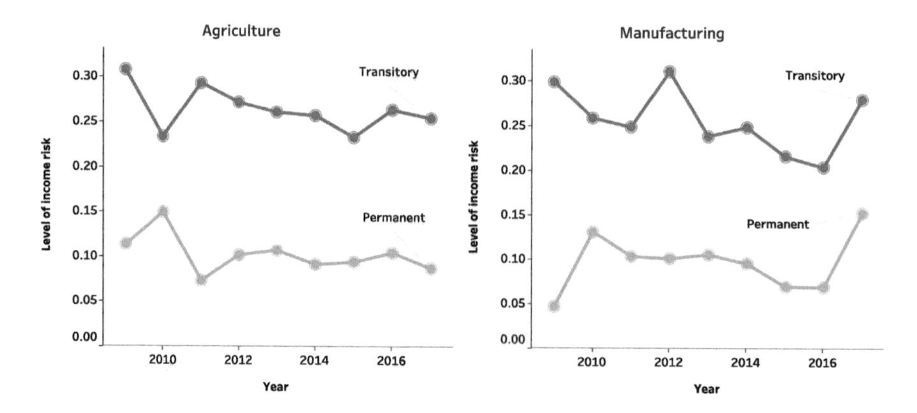

Fig. 7. Income risks for agriculture and manufacturing from 2009 to 2017

5 Conclusion

In this study, we investigated the income risk across 19 industrial sectors in Thailand using the LFS data from 2008 to 2017 consisting over a million individuals. To be able to employ the approach that could separate the income risk into permanent and transitory, we attempted to create the pseudo panel dataset out of the LFS based on multiple labour characteristics.

The results show that, overall, the transitory income risk is higher than the permanent income risk. In addition, findings from industrial sub-samples show that on average the transport sector experiences the income risk the most, followed by agriculture, professional activities and manufacturing. However, when taking the number of employees working in the sectors into account, we found that agriculture and manufacturing industry should receive greater attention in dealing with the income risk, especially the manufacturing industry since there is an increasing trend in risks in the case of both permanent and transitory.

To the best of our knowledge, this study was the first attempt to estimate an income risk across industries in Thailand over time using pseudo panel data. Therefore, there is room for future exploration such as an investigation on a channel through which the income risk is affected. Another recommendation is that one may consider other model specifications that relax the assumption of random walk process of the income risk.

References

1. Aiyagari, S.R.: Uninsured idiosyncratic risk and aggregate saving. Q. J. Econ. **109**(3), 659–684 (1994)
2. Altonji, J.G., Segal, L.M.: Small-sample bias in GMM estimation of covariance structures. J. Bus. Econ. Stat. **14**(3), 353–366 (1996)
3. Becker, G.: Human Capital: A Theoretical and Empirical Analysis, with Special Reference to Education, 2nd edn. National Bureau of Economic Research, Inc. (1975)

4. Bhaopichitr, K., Mala, A., Triratanasirikul, N.: Thailand economic monitor: December 2012 (English) (2012)
5. Carroll, C.D., Samwick, A.A.: The nature of precautionary wealth. J. Monetary Econ. **40**(1), 41–71 (1997)
6. Carroll, C.D., Samwick, A.A.: How important is precautionary saving? Rev. Econ. Stat. **80**(3), 410–419 (1998)
7. Fernquest, J.: Older people work longer in rapidly ageing Thailand. Bangkok Post (2016)
8. Hogrefe, J., Yao, Y.: Offshoring and labor income risk. ZEW Discussion Papers 12-025, ZEW - Zentrum für Europäische Wirtschaftsforschung/Center for European Economic Research (2012)
9. Kreb, T., Yao, Y.: Labor market risk in Germany. IZA Discussion Paper No. 9869 (2016)
10. Krebs, T., Krishna, P., Maloney, W.: Trade policy, income risk, and welfare. Rev. Econ. Stat. **92**(3), 467–481 (2010)
11. Krishna, P., Senses, M.Z.: International trade and labour income risk in the U.S. Rev. Econ. Stud. **81**(1), 186–218 (2014)
12. Mincer, J.: Schooling, Experience, and Earnings. National Bureau of Economic Research, Inc. (1974)
13. Palangkaraya, A.: Globalisation and the Income Risk of Australian Workers, chap. 7, pp. 165–196. No. 4 in Impact of Globalization on Labor Market. ERIA (2013)

Evaluating the Impact of Official Development Assistance (ODA) on Economic Growth in Developing Countries

Dang Van Dan[1(✉)] and Vu Duc Binh[2]

[1] Finance Faculty, Banking University HCMC, Ho Chi Minh City, Vietnam
dandv1978@yahoo.com

[2] Accounting - Financial Banking Faculty, Binh Duong Economic and Technology University, Binh Duong, Vietnam

Abstract. This paper is an empirical research study examining the impact of official development assistance (ODA) on economic growth in 60 developing countries covering Asia, Africa, Latin America and the Caribbean. Panel data analysis will be conducted for the period of 1996 to 2016, in order to examine the impact of ODA on economic growth. In addition to the static relationship framework, the Arellano-Bond Generalized Method of Moments econometric method will be applied to examine the dynamic framework between variables. The main findings of this paper suggest that ODA has positive and significant impacts on economic growth. In conclusion, the official development assistance (ODA) is a significant contributor to economic growth in developing countries.

Keywords: Official Development Assistance (ODA)
Economic growth · Developing countries

1 Introduction

Official Development Assistance (ODA) is defined by the Development Assistance Committee (DAC) of the Organization for Economic Cooperation and Development (OECD) as government aid that promotes and specifically targets the economic development and welfare of developing countries. ODA may be provided bilaterally, from donor to recipient, or channelled through a multilateral development agency such as the United Nations or the World Bank.

The official development assistance has a welfare enhancing effect, particularly when supporting consumption, capital investment, education and human capital development, entrepreneurship and poverty reduction efforts. As Gomannee, Girma and Morrissey [11] has stated "poor countries lack sufficient domestic resources to finance investment and the foreign exchange to import capital goods and technology, aid to finance investment can directly fill the savings-investment gap and, as it is in the form of hard currency, aid can

V. Kreinovich et al. (Eds.): ECONVN 2019, SCI 809, pp. 910–918, 2019.
https://doi.org/10.1007/978-3-030-04200-4_66

indirectly fill the foreign exchange gap". Papanek [16], Dowling and Hiemenz [9], Hansen and Tarp [8], Clemens et al. [7], Karras [14] and Asteriou [3] found evidence for positive impact of ODA on growth. On the other hand, some scholars have counter-argued that ODA can be harmful or ineffective when donors give complete control to the recipient country, which gives way to corruption. Instead of that, donors should direct the use of ODA to implement their own projects and programs. Griffen and Enos [12], Brautigam and Knack [17] and Ekanayake and Chatrna [10] found evidence for negative impact of ODA on growth. Meanwhile, others argued for the conditional effectiveness of aid, for instance, Burnside and Dollar [6] found that foreign aid is effective only in the presence of good macroeconomic policy environment; otherwise aid is ineffective. Mosley [15], Boone [5], Jensen and Paldam [13] found evidence to suggest that ODA has no impact on growth. The role of official development assistance (ODA) in the growth process of developing countries has been a controversial topic. There is still heated debate on whether or not ODA is effective in promoting economic growth in aid-recipient countries.

The contribution of ODA to economic growth of developing countries may be positive, negative or even non-existent, in statistical terms. The explanation for the inconclusive results remains unclear, many authors have suggested methodological and econometric causes. In the aid literature, various theoretical and empirical studies have been conducted in developing countries to determine the actual effects of ODA on economic growth. I have recorded a number of methodological and econometric weaknesses that may explain the inconsistent results of regression studies. Therefore, this paper investigates both the static and dynamic impact of the official development assistance (ODA) on economic growth in 60 developing countries, and proposes improvements to the methodological and econometric procedures found in studies of the relation between ODA and economic growth. Growth regressions, based on a large sample of developing countries covering a 21-year period, are estimated by using the generalized method of moments (GMM) suggested by Arellano and Bond [2].

2 Data Description and Sources

In order to test the implication of these models, a panel of aggregate data on ODA on a wide range of developing countries was collected. The study sample covers 60 developing countries from all regions classified by the World Bank from 1996 to 2016. The sample of countries consists of 15 low-income countries, 25 lower middle income countries, 20 upper middle income countries. The list of countries used in the empirical analysis is given in Table 1. The core data used in this study is taken from the World Bank's World Development Indicators (WDI) [18]. The variables of the study are the economic growth (GDP), the official development assistance (ODA), the gross capital formation as a proxy for domestic investment (INVEST), the labor growth (LABOR), the inflation rate (INF), OPEN as trade openness, and finally the INFRAST as the infrastructure index. Studied variables have been summarized in Table 2. Prior to empirical analyses it would be good step to present descriptive statistics of all series under consideration as can be found in Table 3. The correlation matrix has been shown in Table 4, it is depicted that official development assistance (ODA), investment, labor growth and trade openness is positively correlated with the economic growth (GDP). This indicates the fact that whenever there is an increase in these variables it will enhance the economic growth (GDP). It becomes more evident from Table 4 that ODA and GDP are positive, it looks like a contribution to the economy.

Table 1. List of developing countries included in the study

Upper middle income (3956 US$ $<$ GNI per capita in US$ $<$ 12235 US$)	Lower middle income (1006 US$ $<$ GNI per capita in US$ $<$ 3955 US$)	Low income (GNI per capita in US$ $\leq$ 1005 US$)
Albania	Angola	Burkina Faso
Algeria	Armenia	Benin
Argentina	Bangladesh	Haiti
Belize	Bolivia	Madagascar
Brazil	Cameroon	Malawi
Colombia	Congo, Rep.	Mali
Costa Rica	Côte d'Ivoire	Mozambique
Ecuador	Egypt, Arab Rep	Nepal
Fiji	El Salvador	Niger
Iran, Islamic Rep	Georgia	Rwanda
Kazakhstan	Ghana	Senegal
Macedonia, FYR	Guatemala	Tanzania
Malaysia	Honduras	Togo
Mexico	India	Uganda
Panama	Indonesia	Zimbabwe
Paraguay	Kenya	
Peru	Kyrgyz Republic	
South Africa	Moldova	
Thailand	Morocco	
Turkey	Nigeria	
	Pakistan	
	Philippines	
	Tunisia	
	Vietnam	
	Yemen, Rep	

Source: World Bank, 2016 [18]

Table 2. Variables of the study

Variable	Description of variables
GDP	Economic growth, GDP growth (% annual)
ODA	Net Official Development Assistance received (% of GNI)
INVEST	Investment, Gross capital formation as a percentage of GDP
LABOR	Labor growth, total labor force / total population
INF	Inflation, consumer prices (annual %)
OPEN	Trade openness, sum of Imports and Exports as a ratio of GDP
INFRAST	Infrastructure, fixed telephone subscriptions (per 100 people)

Table 3. Descriptive statistics

Variable	Mean	Std.Dev.	Min	Max
GDP	4.334874	3.920744	−28.09683	33.73578
ODA	4.722421	5.868301	−0.675395	50.07259
INVEST	22.93713	7.725353	1.523837	55.36268
LABOR	0.4170954	0.0696716	0.2312541	0.6033389
INF	33.109	701.2678	−7.113768	24411.03
OPEN	0.7287406	0.3466234	0.1563556	2.204074
INFRAST	8.09381	8.294625	0.0529789	38.33395

Source: Computed by the Researcher, 2018

Table 4. Correlation matrix

	GDP	ODA	INVEST	LABOR	INF	OPEN	INFRAST
GDP	1						
ODA	0.1494	1					
INVEST	0.1885	−0.0933	1				
LABOR	0.0489	0.0594	0.0951	1			
INF	−0.0618	0.0275	−0.0596	0.0312	1		
OPEN	0.0291	−0.0977	0.1965	0.1394	0.0228	1	
INFRAST	−0.0882	−0.4429	0.2047	0.1286	−0.0251	0.1535	1

3 Empirical Methodology

On the basis of considerations discussed above, in order to investigate the impact of official development assistance (ODA) on economic growth, the following static relation established using panel data:

$$GDP_{it} = \beta_0 + \beta_1 ODA_{it} + \beta_2 INVEST_{it} + \beta_3 LABOR_{it} + \beta_4 INF_{it} + \beta_5 OPEN_{it} + \beta_6 INFRAST_{it} + \alpha_i + \tau_t + u_{it}$$

Where GDP_{it} is economic growth of country i, in year t; ODA_{it} is official development assistance of country i, in year; other explanatory variable that are investment, labor force, inflation, trade openness, infrastructure. Country fixed-effects are represented by α_i; τ_t represents time period effects and finally u_{it} represents the error term.

In order to address any endogeneity issues of regression, and to capture persistence and potential mean-reverting dynamics in the economic growth, the estimations of dynamic panel data using Arellano-Bond's (1991) Generalized Method of Moments estimator where one period's lagged values of regressors are used as instruments. In this case the following equation is estimated:

$$GDP_{it} = \beta_0 + \beta_1 GDP_{it-1} + \beta_2 ODA_{it} + \beta_3 INVEST_{it} + \beta_4 LABOR_{it} + \beta_5 INF_{it} + \beta_6 OPEN_{it} + \beta_7 INFRAST_{it} + \alpha_i + \tau_t + u_{it}$$

The system GMM (SGMM) approach of Arellano and Bover (1995) and Blundell and Bond (1998) is used to control the endogeneity bias, unobserved country fixed effects, and other potentially omitted variables. Since the number of moment conditions increases with T which is a special feature of SGMM estimation in dynamic panel data, a Sargan test has to be performed to test the over-identification restrictions. There are many moment conditions cause bias while increasing efficiency. Therefore, a subset of these moment conditions could be used to take advantage of the trade-off between the reduction in bias and the loss in efficiency (see Baltagi) [4]. Sargan's J test for over- identification restrictions in a statistical model and AR(2) test for autocorrelation will be provided to support for the exogeneity of the instrument and the absence of autocorrelation, respectively.

4 Results

As suggested by Antonie et al. [1], in order to validate the Pooled OLS estimation, the poolability test detects with null hypothesis that all α_i are zero. The results suggest to reject the null hypothesis so that the Pooled OLS estimation is biased and not consistent. The presence of individual-specific effect is accepted.

Secondly, in order to decide between random-effects regression and OLS regression the Breusch-Pagan Lagrangian multiplier test rejects the null hypothesis, that means variances across entities are zero then the random-effects estimation will be appropriate.

Thirdly, the Hausman test employed with the null hypothesis of the preferred model is random-effects in comparison to the alternative hypothesis of the preferred model is fixed-effects estimation. The Hausman test checks whether the unique errors are correlated with the regressors, while null hypothesis are the unique errors which are not correlated. Since the Chi-Square probability of the test statistics accepts the null hypothesis, so the random-effects estimation must be preferred in order to analyze the functional relationship of the model.

Cross-sectional dependency, heteroskedasticity and serial correlation test will be carried out to gain a better understanding of the nature of the dataset.

In order to test cross-sectional dependency, as in contemporaneous correlation, a Pesaran CD test will be employed. The Pesaran CD test has rejected the null hypothesis of residuals which is not correlated with a probability that is less than 0.01. There is evidence of cross-sectional dependency as expected because of the cross-country observations which are influenced by common considerations such as similar political or economic issues.

In order to detect heteroskedasticity, a Breusch-Pagan Lagrangian test for groupwise heteroskedasticity in random-effects model has been implemented in Stata. It uses the null hypothesis as homoscedastic and in other words it is constant variance. The probability of the test statistics rejects the null hypothesis with a probability that is less than 0.01. This proves the presence of heteroskedasticity.

Finally, the Lagrange-Multiplier test for serial correlation will be carried out with the null hypothesis without serial correlation. Serial correlation causes the standard errors of the coefficients which are smaller than they actually are and it also causes a higher R-square. The Lagrange-Multiplier test for serial correlation will be carried out and has rejected the null hypothesis with a probability that is less than 0.01. There is evidence of first order auto-correlation. All diagnostic test are presented in Table 6.

In conclusion, there is heteroskedasticity, cross-sectional dependency and serial correlation problems in the model. Ignoring any of the cross-sectional dependency or the serial correlation or the presence of heteroskedasticityin the estimation of the panel models may induce a biased statistical result. In order to control heteroskedasticity and serial correlation, the model will be estimated with FGLS regression.

As it stated previously, a dynamic framework needed to be used to examine the relation between official development assistance (ODA) and economic growth due to the relation between the variables which occurs over time. Therefore, dynamic panel data estimation using the system GMM (SGMM) of Arellano and Bover (1995) and Blundell and Bond (1998) was carried out.

In Table 5, five regressions are reported. In the first four regressions, the static panel data estimations are carried out that are Pooled OLS, fixed-effects regression, random-effects regression and FGLS regression which controls heteroskedasticity and serial correlation. Finally, the fifth regressions includes dynamic panel data estimation using the System-GMM estimator. The statistically significant and positive relation of ODA found in the pooled-OLS, FEM, REM, FGLS and the dynamic panel data estimation indicates the significant positive impact of ODA on economic growth, which means the raise of the official development assistance will contribute to economic growth in developing countries. This is consistent with previous studies by Hansen and Tarp [8] and Clemens et al. [7], Karras [14] and Asteriou [3]. Another important point in the findings is the evidence of positive correlation between the domestic investment and the economic growth. The estimated coefficient for domestic investment is positive, suggesting that domestic investment is good for growth. The regression results give evidence of the positive impacts of ODA on economic growth,

Table 5. Estimation results

	(1) Pooled OLS	(2) FEM	(3) REM	(4) FGLS	(5) System-GMM
Dep.var.GDP					
ODA	0.080*** (0.020)	0.109*** (0.036)	0.082*** (0.025)	0.080*** (0.019)	0.042** (0.017)
INVEST	0.099*** (0.014)	0.143*** (0.022)	0.116*** (0.017)	0.099*** (0.014)	0.113*** (0.009)
LABOR	1.968 (1.521)	−1.879 (5.190)	1.580 (2.297)	1.968 (1.504)	1.763 (1.168)
INF	−0.001*** (0.000)	−0.001* (0.000)	−0.001** (0.000)	−0.001*** (0.000)	−0.001*** (0.000)
OPEN	−0.035 (0.305)	0.607 (0.891)	−0.143 (0.443)	−0.035 (0.302)	−0.290 (0.205)
INFRAST	−0.039*** (0.014)	−0.027 (0.036)	−0.040** (0.020)	−0.039*** (0.014)	−0.052*** (0.006)
GDP(-1)					0.126*** (0.018)
R square	0.158	0.174	0.1574		
Observations	1188	1188	1188	1188	1129
Wald test (p-value)	-	-	(0.000)	(0.000)	(0.000)
F test (p-level)	(0.000)	(0.000)	-	-	-
Hansen test (p-level)	-	-	-	-	0.553
AR(1) test (p-level)	-	-	-	-	0.000
AR (2) test (p-level)	-	-	-	-	0.845

Source: Computed by the Researcher, 2018

Note: (i) GDP denotes economic growth; ODA denotes official development assistance; INVEST denotes gross capital formation; LABOR denotes labor force; INF denotes inflation; OPEN denotes trade openness; INFRAST denotes infrastructure. (ii) *** and ** and * indicate rejection of null hypothesis at 1% and 5% and 10% significance level. (iii) AR-test is Arellano-Bond test.

Table 6. Hausman test and Diagnostics of the model

Other test/ Diagnostics	Test statistics	Probabilities
Hausman Test	3.14	(1.000)
Breusch-Pagan Lagrangian	129.43	(0.000)
Pesaran CD Test	19.888	(0.000)
Heteroskedasticity Test	169.45	(0.000)
Lagrange-Multiplier Test for Serial Correlation	71.54	(0.000)

Source: Computed by the Researcher, 2018

not only by direct channel, but also by indirect channel through improving the domestic investment. The main role of ODA in the promotion of economic growth has been considered impacting domestic source of finance such as savings, improving infrastructure and increasing domestic investment. In addition, inflation rate variable has the negative sign and it is statistically significant at the 1% of significance level. These findings are also consistent with the findings of previous studies.

5 Conclusion

This paper has sought to evaluate the impact of official development assistance on the economic growth of developing countries. One of the contributions of this paper is to input to the existing empirical literature on the impact of official development assistance (ODA) on economic growth of developing countries through its thorough analysis covering a large number of developing countries as well as in a longer time period. The study focuses on the period of 1996–2016 and 60 aid-receiving developing countries. The main results of this study suggest that ODA has a positive and statistically significant impacts on economic growth, that means ODA is a significant contributor to economic growth in developing countries. Indeed, a fair conclusion from this empirical evidence on ODA and economic growth shows that official development assistance appears to promote economic growth, ODA plays a role as a growth-enhancing factor. There are a number of mechanisms through which ODA can contribute to economic growth, including ODA for supplement domestic sources of finance, ODA for increasing investment, infrastructure; ODA for increasing the capacity to import capital goods and technology and; ODA for helping to stabilize the macroeconomics. The results of this paper indicate that the authorities in developing countries should pay more attention and they need to do this while attracting, managing ODA in order to maximize economic growth. There is need to implement appropriate policy, in order to achieve the positive impact of ODA on economic growth through increasing domestic investment, lower inflation rate.

References

1. Antonie, M.D., Cristescu, A., Cataniciu, N.: A panel data analysis of the connection between employee remuneration, productivity and minimum wage in Romania. In: Proceedings of the 11th WSEAS International Conference MCBE 2010, pp. 134–139 (2010)
2. Arellano, M., Bond, S.: Some tests of specification for panel data: Monte Carlo evidence and an application to employment equations. Rev. Econ. Stud. **58**(2), 277–297 (1991)
3. Asteriou, D.: Foreign aid and economic growth: new evidence from a panel data approach for five South Asian countries. J. Policy Model. **31**(1), 155–161 (2009)
4. Baltagi, B.H.: Econometric Analysis of Panel Data. Wiley, Chichester (2005)
5. Boone, P.: Politics and the effectiveness of foreign aid. Eur. Econ. Rev. **40**(2), 289–328 (1996)

6. Burnside, C., Dollar, D.: Aid, Policies, and Growth. World Bank Policy Research Working Paper No. 1777 (1997)
7. Clemens, M.A., Radelet, S., Bhavnani, R.R.: Counting chickens when they hatch: the short-term effect of aid on growth, Working Paper No. 6 44, Center for Global Development (2004)
8. Dalgaard, C.J., Hansen, H., Tarp, F.: On the Empirics of Foreign Aid and Growth. Centre for Research in Economic Development and International Trade Research Paper No. 02/08. **114**, 191–216 (2002)
9. Dowling, J.M., Hiemenz, U.: Aid, Savings and Growth in the Asian Region. Asian Development Bank Economic Office Report Series No. 3 (1982)
10. Ekanayake, E.M., Chatrna, D.: The effect of foreign aid on economic growth in developing countries. J. Int. Bus. Cult. Stud. **3**, 1–13 (2010)
11. Gomanee, K., Girma, S., Morrisey, O.: Aid and growth in Sub-Saharan Africa: accounting for transmission mechanisms. J. Int. Develop. **17**, 1055–1075 (2005)
12. Griffin, K., Enos, J.: Foreign assistance: objectives and consequences. Econ. Develop. Cult. Change **18**, 313–327 (1970)
13. Jensen, P., Paldam, M.: Can the two new aid-growth models be replicated? Inst. Econ. **127**(1), 147–175 (2003)
14. Georgios, K.: Foreign aid and long-run economic growth: empirical evidence for a panel of developing countries. J. Int. Develop. **18**, 15–28 (2006)
15. Mosley, P., Hudson, J.: Aid Effectiveness: A Study of the Effectiveness of Overseas Aid in the Main Countries Receiving ODA Assistance. University of Reading and University of Bath (1995)
16. Papanek, G.F: Aid, foreign private investment, savings, and growth in less developed countries. J. Polit. Econ. **81**(1), 120–130 (1973)
17. Knack, S., Brautigam, D.A.: Foreign aid institutions, and governance in Sub-Saharan Africa. Econ. Develop. Cult. Change **52**(2), 255–285 (2004)
18. World Data Bank (2016). https://data.worldbank.org/

The Effect of Macroeconomic Variables on Economic Growth: A Cross-Country Study

Dang Van Dan[1(✉)] and Vu Duc Binh[2]

[1] Finance Faculty, Banking University HCMC, Ho Chi Minh City, Vietnam
dandv1978@yahoo.com

[2] Accounting - Financial Banking Faculty, Binh Duong Economic and Technology University, Binh Duong, Vietnam

Abstract. This study examines the effect exacted by macroeconomic variables on the economic growth for selected 68 developing countries in a panel framework. Panel data analysis was conducted for the period 1996 to 2016, in order to examine the effect of macroeconomic variables on economic growth. The effect of macroeconomic variables was evaluated in a dynamic framework using system GMM (System - Generalized Method of Moments). The main findings of this paper indicated that high level domestic investment, labour and trade openess have positive and significant effect on economic growth. In contrast, inflation, money supply and interest rate have negative effect on growth in developing countries.

Keywords: Economic growth · Developing countries
Macroeconomic variables

1 Introduction

Macroeconomic management is one of the important concerns that held a highly prominent place in the literature. Monetary policy is a key factor of macroeconomic management in opened economy to stimulate economic stability and to promote economic development through its impact on economic variables. It is generally believe that monetary policy influences macroeconomic variables which include gross domestic product growth, inflation rate, money supply and interest rate in developing countrie (Anowor and Okorie [2]; Precious [16]). The accurate information on the effectiveness of the policy on the macro economy is the main issue of the policy maker to successfully implementation of any economic policy in general to achieve the sustainable economic growth, the authority and policy maker always targets on the intermediate variables include money supply and interest rate, which is considered as the most powerful instrument of monetary policy (Fasanya, Onakoya and Agboluaje) [10].

The relationship between macroeconomic variables and economic growth has been getting increasing attention in recent times for the important role it plays in

V. Kreinovich et al. (Eds.): ECONVN 2019, SCI 809, pp. 919–927, 2019.
https://doi.org/10.1007/978-3-030-04200-4_67

economic growth in developing countries. The effects of macroeconomic variables on economic growth have also been discussed by many scholars. For example, Fischer [11], in an empirical study of 73 countries for the period 1970–1985, found that high inflation negatively affect the growth rate of per capita income, and concludes that macro policies indeed matter for growth. Babatunde and Shuaibu [6] examined money supply, inflation and economic growth in Nigeria, the finding showed negative relationship between inflation and economic growth. Gul, Mughal and Rahim [12] pointed out that interest rate has negative impact on the output and they also found that money supply has strongly positive impact on the output. Ayub and Maqbool [5] stated that GDP is greatly affected by money supply, interest rate and inflation rate in Pakistan. Mensah and Okyere (2015) examined the impact of interest rate and inflation on real economic growth rate in Ghana concluded that interest rate has a negative influence on real growth rate. Alavinasab [1] examined the impact of monetary policy on economics growth in Iran by using time series data which appropriate with error correction model (ECM), the finding of regression showed that money supply and inflation had a long run significantly relationship on economic growth. Bukhari et al. [8] explored the public investment and economic growth in East Asian countries, they found that both public investment and private investment had a long run dynamic impact on economic progress in Asian countries. Rahman [17] explored the relationship of investment and economic progression in Bangladesh, they investigated the investment can impact on the Bangladesh economy growth, the empirical result explained that investment had a positive and significantly effect on gross domestic product. Dash [9] examined the public and private investment on economic growth in India, the result shows that public investment an optimistic and significance effect on gross domestic product.

On the other hand, some researchers found there is no relationship between monetary policies on economic growth. For example, Khabo and Harmse [14] studied the impact of monetary policy on the economic growth of a small and open economy of South Africa, the finding show that money supply and inflation are not significantly related to the change of economic growth. Babatunde and Shuaibu [6] also confirmed there is no relationship between money supply and economic growth.

In the growth literature, various theoretical and empirical studies have been conducted on developing countries to determine the actual effects of macroeconomic variables on economic growth. We have recorded a number of methodological and econometric weaknesses that may explain the inconsistent results of regression studies. Therefore, this paper investigates the dynamic effect of macroeconomic variables on economic growth in 68 developing countries, and proposes improvements to the methodological and econometric procedures found in studies of the effect of macroeconomic variables on the economic growth. Growth regressions, based on a large sample of developing countries covering a 21-year period, are estimated using the generalized method of moments (GMM) suggested by Arellano and Bond [4].

2 Data and Variables Description

The study sample covers 68 developing countries from all regions classified by the World Bank from year 1996 to 2016. The sample of countries consists of 15 low-income countries, 27 lower middle income countries, 26 upper middle income countries. The list of countries used in the empirical analysis is given in Table 1.

Table 1. List of developing countries included in the study

Upper middle income (3956 US$ $<$ GNI per capita in US$ $<$ 12235 US$)	Lower middle income (1006 US$ $<$ GNI per capita in US$ $<$ 3955 US$)	Low income (GNI per capita in US$ $\leq$ 1005 US$)
Albania	Angola	Burkina Faso
Algeria	Armenia	Benin
Argentina	Bangladesh	Haiti
Azerbaijan	Bhutan	Madagascar
Belize	Bolivia	Malawi
Brazil	Cameroon	Mali
Chile	Congo, Rep.	Mozambique
Colombia	Cote d'Ivoire	Nepal
Costa Rica	Egypt, Arab Rep	Niger
Ecuador	El Salvador	Rwanda
Fiji	Georgia	Senegal
Gabon	Ghana	Tanzania
Iran, Islamic Rep	Guatemala	Togo
Jamaica	Honduras	Uganda
Kazakhstan	India	Zimbabwe
Macedonia, FYR	Indonesia	
Malaysia	Kenya	
Mexico	Kyrgyz Republic	
Panama	Moldova	
Paraguay	Mongolia	
Peru	Morocco	
South Africa	Nigeria	
Thailand	Pakistan	
Tonga	Philippines	
Turkey	Tunisia	
Uruguay	Vietnam	
	Yemen, Rep	

Source: World Bank, 2016 [18]

The core data used in this study is taken from the World Banks World Development Indicators (WDI). The variables of the study are the economic growth (GROWTH), the inflation rate (INF), the money supply (MS), the interest rate (IR), the gross capital formation as a proxy for domestic investment (INV), the labor growth (LPR) and finally TRADE as trade openness. Variables of the study have been summarized in Table 2.

Table 2. Variables of the study

Variable	Description of variables
GROWTH	Economic growth, GDP growth (% annual)
INF	Inflation, consumer prices (annual %)
MS	Money supply, Broad money (% of GDP)
IR	Interest rate, Lending interest rate (%)
INV	Domestic investment, Gross capital formation as a percentage of GDP
LPR	Labor force participation rate, Total labor force/total population
TRADE	Trade openness, sum of Imports and Exports as a ratio of GDP

Prior to empirical analyses it would be good step to present descriptive statistics of all series under consideration as can be found in Table 3. The correlation matrix has been shown in Table 4, it shows that domestic investment, labor force and trade openness are positively correlated with the economic growth (GROWTH). This indicates that whenever there is an increase in these variables it will enhance the economic growth. In contrast, inflation, money supply and interest rate are negatively correlated with the economic growth (GROWTH).

Table 3. Descriptive statistics

Variable	Mean	Std.Dev.	Min	Max
GROWTH	4.330434	4.104396	−28.09683	34.5
INF	7.381157	11.31274	−18.10863	302.117
MS	44.06466	25.66518	6.546494	151.5489
IR	17.69932	14.3106	3.4225	217.875
INV	23.64789	8.545834	1.523837	67.9105
LPR	67.06528	10.79079	38.102	90.34
TRADE	0.7386433	0.3349047	0.1563556	2.204074

Source: Computed by the Researcher, 2018

Table 4. Correlation matrix

	GROWTH	INF	MS	IR	INV	LPR	TRADE
GROWTH	1						
INF	−0.0461	1					
MS	−0.1122	−0.1733	1				
IR	−0.1108	0.4317	−0.3110	1			
INV	0.1797	−0.0692	0.1987	−0.1998	1		
LPR	0.1393	−0.0085	−0.1761	0.1516	−0.0858	1	
TRADE	0.0224	−0.0643	0.3832	−0.1340	0.2217	−0.0083	1

Source: Computed by the Researcher, 2018

3 Model Specification

On the basis of considerations discussed above, in order to investigate the effect of macroeconomic variables on economic growth, the following static relationship was established using panel data:

$$GROWTH_{it} = \beta_0 + \beta_1 INF_{it} + \beta_2 MS_{it} + \beta_3 IR_{it} + \beta_4 INV_{it} + \beta_5 LPR_{it} + \beta_6 TRADE_{it} + \alpha_i + \tau_t + u_{it}$$

Here $GROWTH_{it}$ is economic growth of country i, in year t; other explanatory variable that are inflation, money supply, interest rate, domestic investment, labor force, trade openness. Country fixed-effects are represented by α_i; τ_t represents time period effects and finally u_{it} represents the error term. In order to address any endogeneity issues of regression, and to capture persistence and potentially mean-reverting dynamics in the economic growth, dynamic panel data estimations using Arellano-Bond's [4] Generalized Method of Moments estimator where one period's lagged values of regressors are used as instruments. In this case the following equation is estimated:

$$GROWTH_{it} = \beta_0 + \beta_1 GROWTH_{it-1} + \beta_2 INF_{it} + \beta_3 MS_{it} + \beta_4 IR_{it} + \beta_5 INV_{it} + \beta_6 LPR_{it} + \beta_7 TRADE_{it} + \alpha_i + \tau_t + u_{it}$$

The system GMM (SGMM) approach of Arellano and Bover (1995) and Blundell and Bond (1998) is used to control for the endogeneity bias, unobserved country fixed effects, and other potentially omitted variables. Since the number of moment conditions increases with T which is a special feature of dynamic panel data GMM estimation, a Sargan test has to be performed to test the over-identification restrictions. Too many moment conditions cause bias while increasing efficiency. Therefore, a subset of these moment conditions could be used to take advantage of the trade-off between the reduction in bias and the loss in efficiency (see Baltagi [7]). Sargan's J test for over-identification restrictions in a statistical model and AR(2) test for autocorrelation will be provided to support for the exogeneity of the instrument and the absence of autocorrelation, respectively.

4 Estimation and Empirical Results

As suggested by Antonie et al. [3], in order to validate the Pooled OLS estimation, the poolability test detects with null hypothesis that all α_i are zero. The results suggest rejecting the null hypothesis so that the Pooled OLS estimation is biased and not consistent. The presence of individual-specific effect is accepted. Second, in order to decide between random-effects regression and OLS regression the Breusch-Pagan Lagrangian multiplier test rejects the null hypothesis that variances across entities are zero therefore the random-effects estimation will be appropriate. Third, the Hausman test employed with the null hypothesis of the preferred model is random-effects versus the alternative hypothesis of the preferred model is fixed-effects estimation. The Hausman test checks whether the unique errors are correlated with the regressors, while null hypothesis are the unique errors which are not correlated. Since the probability of the Chi-Square of the test statistics rejects the null hypothesis, so the fixed-effects estimation must be preferred in order to analyze the functional relationship of the model. Heteroskedasticity and serial correlation test will be carried out to gain a better understanding of the nature of the dataset. In order to detect heteroskedasticity, a modified Wald test for groupwise heteroskedasticity in fixed-effects models that has been implemented in Stata. It uses the null hypothesis that is homoscedastic and in other words constant variance. The probability of the test statistics rejects the null hypothesis with a probability that is less than 0.01. This proves the presence of heteroskedasticity. Finally, the Lagrange-Multiplier test for serial correlation will be carried out with the null hypothesis that no serial correlation. Serial correlation causes the standard errors of the coefficients to be smaller than they actually are and it also causes a higher R-square. The Lagrange-Multiplier test for serial correlation will be carried out and has rejected the null hypothesis with a probability that is less than 0.01. There is evidence of first order autocorrelation. All diagnostic test are presented in Table 6. In conclusion, there is heteroskedasticity and serial correlation problems in the model. Ignoring any of the serial correlation or the presence of heteroskedasticity in the estimation of the panel models may induce a biased statistical result. In order to control the heteroskedasticity the model will be estimated with a robust fixed-effects estimation. Also, the model will be estimated by using a fixed-effects estimation with Driscoll and Kraay standard errors to control heteroskedasticity and serial correlation as suggested by Hoechle [13]. Fixed-effects within regression with Driscoll and Kraay standard errors assume the error structure to be heteroskedastic, serially correlated, and up to some lag possibly correlated between groups. As it stated previously, a dynamic framework needed to be used to examine the relationship between macroeconomic variables and economic growth due to the relationship between the variables which occurs over time. Therefore, dynamic panel data estimation using the system GMM (SGMM) of Arellano and Bover (1995) and Blundell and Bond (1998) was carried out.

In Table 5, six regressions are reported. In the first five regressions, the static panel data estimations are carried out that are pooled OLS, fixed-effects regression, random-effects regression and robust fixed-effects estimation which control

Table 5. Estimation results

	(1)	(2)	(3)	(4)	(5)	(6)
	Pooled OLS	FEM	REM	FE robust	FE dris~kraay	System GMM
Dep.var.GROWTH						
INF	0.002 (0.021)	−0.015 (0.024)	−0.003 (0.022)	−0.015 (0.046)	−0.015 (0.032)	−0.016** (0.006)
MS	−0.027*** (0.005)	−0.070*** (0.011)	−0.043*** (0.008)	−0.070*** (0.020)	−0.070*** (0.016)	−0.023*** (0.002)
IR	−0.049*** (0.012)	−0.082*** (0.017)	−0.063*** (0.015)	−0.082***(0.022)	−0.082*** (0.023)	−0.041*** (0.002)
INV	0.085*** (0.014)	0.112*** (0.022)	0.101*** (0.018)	0.112*** (0.034)	0.112*** (0.026)	0.073*** (0.002)
LPR	0.057*** (0.012)	−0.015(0.054)	0.046** (0.021)	−0.015 (0.069)	−0.015 (0.046)	0.032*** (0.003)
TRADE	0.340 (0.359)	3.382*** (0.958)	1.322** (0.593)	3.382* (1.990)	3.382* (2.077)	0.328*** (0.099)
GROWTH(-1)	-	-	-	-	-	0.287*** (0.009)
R square	0.0924	0.0903	0.1055	0.0903	-	-
Observations	1020	1020	1020	1020	1020	981
Wald test (p-value)	-	-	(0.000)	-	-	(0.000)
F test (p-level)	(0.000)	(0.000)	-	(0.000)	(0.000)	-
Hansen test (p-level)	-	-	-	-	-	0.599
AR(1) test (p-level)	-	-	-	-	-	0.000
AR (2) test (p-level)	-	-	-	-	-	0.461

Source: Computed by the Researcher, 2018

Note: (i) GROWTH denotes economic growth; INF denotes inflation; MS denotes money supply; IR denotes interest rate; INV denotes domestic investment; LPR denotes labor force participation rate; TRADE denotes trade openness. (ii) *** and ** and * indicate rejection of null hypothesis at 1% and 5% and 10% significance level. (iii) AR-test is Arellano-Bond test.

Table 6. Hausman test and Diagnostics of the model

Other test/ Diagnostics	Test statistics	Probabilities
Hausman Test	22.47	(0.001)
Breusch-Pagan Lagrangian	182.74	(0.000)
Heteroskedasticity Test	672.62	(0.000)
Lagrange-Multiplier Test for Serial Correlation	9.658	(0.002)

Source: Computed by the Researcher, 2018

heteroskedasticity, and fixed-effects estimation with Driscoll and Kraay standard errors to control heteroskedasticity and serial correlation, respectively. Finally, sixth regressions includes dynamic panel data estimation using the System-GMM estimator. The statistically significant and positive relationship of domestic investment found in the pooled-OLS, FEM, REM, fixed-effects estimation with Driscoll and Kraay and the dynamic panel data estimation indicates the significant positive effect of domestic investment on economic growth, which means that the raise in the domestic investment will contribute to economic growth in developing countries. This is consistent with previous studies by Bukhari et al. [8], Rahman [17] and Dash [9]. On the other hand, inflation is statistically significant only in System-GMM estimations and reduces the rates of economic growth according to expectations. In all the cases, one of the main findings surprisingly suggests the significant negative impact of money supply on economic growth, which means that developing countries pursue macroeconomic policies that result in high levels of money supply suffer low rates of economic growth. Another important point in the findings is the evidence of negative correlation between the interest rate and the rates of economic growth, parallel to the findings of Mughal and Rahim [12]; Ayub and Maqbool [5]; Mensah and Okyere [15] and among many others.

5 Concluding Remarks

This paper analyzes the effects of macroeconomic variables on the economic growth of developing countries. These effects are analyzed using panel data series for macroeconomic variables, while accounting for differences income levels: low income, lower middle income, upper middle income. The study focuses on the time period 1996–2016 and 68 developing countries. The major point emerging from this study is that domestic investment has a positive and statistically significant effects on economic growth, that is domestic investment is a significant contributor to economic growth in developing countries. Indeed, a fair conclusion from this empirical evidence on domestic investment and economic growth is that domestic investment appears to promote economic growth, domestic investment as a growth-enhancing factor. The results of this paper indicate that the authorities in developing countries should pay more attention and they need to do this while attracting, managing domestic investment in order to maximize economic growth. We also suggest that the government should keep inflation under control, the government can use the interest rate as a tool to promote domestic investment and economic growth by lowering the rate. The government and policy makers can use this information to guide monetary policy in developing countries to achieve macroeconomic goals, take steps to strengthen economic growth, solve problems on inflation.

References

1. Alavinasab, S.M.: Monetary policy and economic growth: a case study of Iran. Int. J. Econ. Commer. Manage. **4**(3), 234–244 (2016)
2. Anowor, O.F., Okorie, G.C.: A reassessment of the impact of monetary policy on economic growth: study of Nigeria. Int. J. Develop. Emerg. Econ. **4**(1), 82–90 (2016)
3. Antonie, M.D., Cristescu, A., Cataniciu, N.: A panel data analysis of the connection between employee remuneration, productivity and minimum wage in Romania. In: Proceedings of the 11th WSEAS International Conference on Mathematics & Computers in Business & Economics (MCBE) 2010, pp. 134–139 (2010)
4. Arellano, M., Bond, S.: Some tests of specification for panel data: Monte Carlo evidence and an application to employment equations. Rev. Econ. Stud. **58**, 277–297 (1991)
5. Ayub, S., Maqbool, S.F.: Impact of monetary policy on gross domestic product (GDP). Asian J. Bus. Manage. **3**(6), 470–478 (2015)
6. Babatunde, M.A., Shuaibu, M.I.: Money supply, inflation and economic growth in Nigeria. Asian-Afr. J. Econ. Econometrics **11**(1), 147–163 (2011)
7. Baltagi, B.H.: Econometric Analysis of Panel Data. Wiley, Chichester (2005)
8. Bukhari, I.A., Saddaqat, M.: The public investment and economic growth in East Asian countries. Int. J. Bus. Inf. **2**(1), 57–79 (2007)
9. Dash, P.: The impact of public and private investment on economic growth in India. J. Decis. Makers Indian Inst. Manage. **41**(4), 288–307 (2016)
10. Fasanya, I.O., Onakoya, A.B.O., Agboluaje, M.A.: Does monetary policy influence economic growth in Nigeria? African Econ. Bus. Rev. **12**(1), 635–646 (2013)
11. Fischer, S.: Growth, macroeconomics and development. In: NBER Macroeconomics Annual, pp. 329–364. The MIT Press, Cambridge (1991)
12. Gul, H., Mughal, K., Rahim, S.: Linkage between monetary instruments and economic growth. Univ. J. Manage. Soc. Sci. **2**(5), 69–76 (2012)
13. Hoechle, D.: Robust standard errors for panel regressions with cross-sectional dependence. Stata J. **7**(3), 281–312 (2007)
14. Khabo, V., Harmse, C.: The impact of monetary policy on the economic growth of a small and open economy: the case of South Africa. S. Afr. J. Econ. Manage. Sci. **8**(3), 348–362 (2005)
15. Mensah, A.C., Ebenezer, O.: Real economic growth rate in Ghana: the impact of interest rate, inflation and GDP. Glob. J. Res. Bus. Manage. **4**(1), 206–212 (2015)
16. Precious, C.: Impact of monetary policy on economic growth: a case study of South Africa. Mediterr. J. Soc. Sci. **5**(15), 76–84 (2014)
17. Rahman, A.: The impact of foreign direct investment on economic growth in Bangladesh. Int. J. Econ. Finance **7**(2), 178–185 (2015)
18. World Data Bank, World Development Indicators (2016). https://data.worldbank.org/

The Effects of Loan Portfolio Diversification on Vietnamese Banks' Return

Van Dan Dang[1(✉)] and Japan Huynh[2]

[1] Banking University of Ho Chi Minh City, Ho Chi Minh City, Vietnam
dandvl978@yahoo.com
[2] Vietnam Joint Stock Commercial Bank for Industry and Trade, Ho Chi Minh City, Vietnam
japanhuynh@gmail.com

Abstract. In this paper, the authors estimate the impact of loan portfolio diversification on bank return by using annual data from 25 commercial banks in Vietnam in the period of 2008–2017. In order to achieve the study objective, the author chooses the HHI measure to evaluate the loan portfolio diversification which is classified by economic sectors. The data used is unbalanced panel data while Pooled OLS, FEM and REM analysis methods are used for regression. The FEM regression is the most appropriate model to show that the diversification of the loan portfolio has the negative effect on bank return. Thus, in the banking market context in Vietnam, specialized banks have a slightly higher return than diversified banks during the research period.

Keywords: Bank return · Loan porfolio · Diversification · Commercial bank

1 Introduction

In the trend of banking modernization, the operations of the banking system are gradually moving to non-lending activities, diversifying their lucrative activities and reducing the risk from the traditional lending sector. However, with the function of capital rotation – both mobilizing and lending, it can be seen that with most commercial banks in Vietnam nowadays, lending is always having the most important role.

Recently, there has been a lot of difficulties in the operation of Vietnam's banking sector. The main reason is that the loans are given to the industries and firms which are run ineffectively. In other words, the loan portfolio is not profitable for the bank at the moment. More broadly with the global financial market, the question of whether banks should diversify or specialize their lending activities is very important to consider in the context of the consequences of the global crisis in 2008. This financial crisis has proven to be an excessive exposure for banks in the real estate market in the United States, which has since turned into a global financial crisis.

In Vietnam, commercial banks provide loans to many economic sectors. However, in the whole list of loans, only some basic industries make up a large proportion. This leads to the fact that the lending activities of Vietnam's commercial banks are quite risky. The risks of concentration can affect the business performance. It is obvious that

V. Kreinovich et al. (Eds.): ECONVN 2019, SCI 809, pp. 928–939, 2019.
https://doi.org/10.1007/978-3-030-04200-4_68

building an effective loan portfolio for banks is greatly concerned. In this paper, the author tries to answer one aspect of this problem by estimating the effect of the loan portfolio diversification on the bank return.

2 Theoretical and Empirical Literature

That theoretical frameworks which have been developed to argue whether the diversification or concentration of the loan portfolio will yield greater returns still have no consensus among scholars and professionals so far. On one hand, traditional banking theory advocated by Diamond (1984) and Marinč (2009) states that banks should pursue a diversification strategy and invest in various economic sectors to reduce the risks of concentration. Therefore, the banks can stay away from danger of financial shock. On the other hand, corporate finance theory with representative Mishkin (2013) shows that corporations should adopt a business strategy that focuses on the activities they know well and has a deep professional background.

The concept of efficient frontier was developed by Markowitz in 1952 and refers to the portfolio with the best expected return that can be obtained at a given level of risk. Investors can then, depending on their level of accepting risk, choose to move along the efficient frontier in an upward-sloping line, which shows a positive relation between risk and return. This implies that when all banks operate at their efficient frontiers, changes in portfolio diversification will not affect the performance of banks. However, according to Markowitz's study, it is easy to see that the risk index is measured by the deviation of future stock price against the expected price. Meanwhile, in the context of banking operations, the risk is determined by the likelihood of a loan loss, often measured by indicators such as bad debt ratios or the rate of risk provision. From these points, it can be difficult to apply this theory to banks which have very specific business backgrounds, plus the fact that in the banking market of other countries and even of Vietnam, we have no reason to believe that all the banks in the research sample operate at the efficient frontier.

Many empirical studies have been made to find out the diversification of the loan portfolio classified by industries and its impact on the bank's profitability. Typical studies can be referred to: Acharya et al. (2006) with 105 Italian banks during the period 1993–1998; a study by Deutsche Bundesbank, Hayden et al. (2006) with data from 983 German banks between 1996 and 2002; a study by Tabak, Lazio and Cajueiro (2010) with data of 96 commercial banks in Brazil between 2003 and 2009; most recently, Aarflot and Arnegård (2017) validated the efficiency of diversifying the loan portfolio to the performance of 112 banks in Norway for the period 2004–2013.

The results from these empirical studies seems to coincide with the view point of corporate finance theory of diversification, stating that corporations should concentrate on operations in which they possess expertize. Among the studies just mentioned, only research by Sigve Aarflot and Lars Arnegård at Norwegian banks has found a strong positive relationship between diversification and bank return. There is a clear trend that banks specializing in operations are more profitable than banks with more diversified loan portfolio. Moreover, the effectiveness of diversification strategies differs in some respects depending on the level of basic risk.

3 Methodology of Research

In this paper, in line with other comparable studies, the authors examines the effectiveness of diversifying the loan portfolio to bank return by estimating the general linear model as follows:

$$\mathbf{ROA_{bt} = \beta_0 + \beta_1 HHI_{bt} + \beta_2 Size_{bt} + \beta_3 Eq_{bt} + \beta_4 Per_{bt} + \varepsilon_{bt}} \quad (1)$$

(In which: ROA: return on total assets of bank, HHI: loan concentration index of the loan portfolio; Size: bank size; Per: personnel cost ratio; EQ: equity ratio; ε: model error term).

The model uses the HHI variable (measure of loan portfolio concentration) as the main explanatory variable. At the same time, the model also includes the control variables which are Size (bank size), Eq (the ratio of equity) and Per (personnel cost ratio) to have a more appropriate explanation that accurately reflects the true value and avoids the factors caused by the omitted variables. The return of Vietnam's commercial banks measured by the ROA (return on total assets) is the variable explained in the model. The data collected is included in all kinds of each bank's financial statements. Because there are periods when banks do not publish financial statement or financial statements do not have enough information, especially notes to financial statement; the data is unbalanced panel data. The dataset includes 230 observations.

4 Construction of Variables

4.1 Concentration Variables

- ***Hirschman-Herfindahl Index (HHI)***

The HHI is used to assess the concentration of the loan portfolio into certain industries, which can accordingly assess the diversification of the loan portfolio.

$$HHI_{bt} = \sum_{i=1}^{n} r_{bti}^2$$

The relative exposure of the bank b at time t to each economic sector i is defined as:

$$r_{bti} = \frac{\text{Nominal Exposure}_{bti}}{\text{Total Exposure}_{bt}}$$

The HHI is generally defined as the sum of the squares of each company's market share in an industry, which means that HHI equal to 1 represents a monopoly situation when one company dominates the entire industry. For the purpose of this study, using HHI as a measure of loan portfolio diversification, the HHI is calculated as the sum of the squares of the bank's relative exposures to the industries, while nominal exposure is easily considered as total amount of debt for each industry and total exposure stands for the whole size of loan portfolio. That the HHI equals to 1 represents a specialized bank where all loans are allocated to one sector, while the HHI equals to 1/n represents a complete diversification bank, where loans are equally distributed across n sectors.

According to Markowitz's portfolio theory, if all banks operate at the efficient frontier, adjusting the loan portfolio diversification will not affect the bank performance due to the existence of the positive relationship between risk and return. However, it is difficult to apply this theory to banks with very typical business backgrounds Moreover, in the banking market in other countries and even in Vietnam there is no reason to believe that all banks in the research sample operate at the efficient frontier as mentioned above and hence we expects β_1 to be non-zero and statistically significant. As presented, the theoretical framework on the correlation between the diversification of loan portfolio and bank return is not uniform, while based on the results of impirical studies which found the evidence of supporting specialization as well as the view point of supporting corporate finance theory, that a bank should concentrate its loan portfolio to optimize return. Therefore, the estimated coefficient for the HHI variable will be $\beta_1 > 0$.

Hypothesis 1: The diversification of loan portfolio has an negative impact on bank return, or in other words the concentration of loan portfolio has an positive impact on bank return ($\beta_1 > 0$).

4.2 Control Variables

- ***Bank size (Size)***

$$\text{Size} = \text{Ln}(\text{Total Asset})$$

The bank size can be accompanied by the economies of scale in the market. The bank size variable controls the potential effects of scale return. Larger banks are able to expand their operations in terms of both number and network of clients, utilizing cheap and copious capital sources. A larger customer base, a larger network of investment and growth potential help increase profitability. However, economic rule also indicates that larger organizations may also be affected by diseconomies of scale, when reaching a certain scale. The scale is no longer advantageous to the bank but will reduce return by exceeding the bank's control.

These assumptions and the current situation of Vietnamese banking, where groups of large banks still have many advantages of mobilization costs, operating history, networks and, possibly, the ownership structure over groups of smaller banks and the state dominates the stake of banks, have resulted in a number of other big advantages. Since, in Vietnam, the effect of bank size and return is expected to be positive.

Hypothesis 2: The bank size has an positive impact on bank return ($\beta_2 > 0$).

- ***Equity ratio (Eq)***

$$\text{Eq} = \frac{\text{Equity}}{\text{Total Asset}}$$

This variable measures and represents equity ratio of the bank, which reflects the capital structure of each bank. Bank equity is considered as a buffer to protect the bank against the risk of financial exhaustion. Therefore, the high value of equity will help banks and their managers to feel more secured about the operational risk for the bank, and this is also a firm basis for banks to expand their business, bring about higher return. Thus expectation in the model will result in $\beta_3 > 0$.

Hypothesis 3: The equity ratio has an positive impact on bank return ($\beta_3 > 0$).

- ***Personnel cost ratio (Per)***

$$\text{Per} = \frac{\text{Personnel Cost}}{\text{Total Asset}}$$

The study adds an additional control variable which is the ratio of personnel costs to total assets as a proxy for bank efficiency. In traditional operating costs of banks, personnel costs occupy a very important position. Recent developments in technology and particularly in digitization have contributed to the reduction of personnel-costs ratio in banks, in line with the term "digital bank". It can be seen from the banking point of view that banks with higher rate of costs for human resource or for a unit of assets means the banks have to spend more on personnel cost then they are running relatively inefficient. Thus expectation in the model will result in $\beta_4 < 0$.

Hypothesis 4: The personnel cost ratio has an negative impact on bank return ($\beta_4 < 0$).

5 Choice of Estimation Method

5.1 Pooled Ordinary Least Squares (Pooles OLS)

In this estimation, intercept and slope coefficients are constant across time and banks. The regression model is shown as follows:

$$\mathbf{ROA_{it} = \beta_0 + \beta_1 HHI_{it} + \beta_2 Size_{it} + \beta_3 Eq_{it} + \beta_4 Per_{it}} + \varepsilon_{it} \tag{2}$$

The orientation of the Pooled OLS model is to: (i) Pool cross sections to get bigger sample sizes; (ii) To investigate the effect of time; (iii) To determine whether relationships have changed over time.

However, with its simplicity, this model has a big disadvantage. It is likely that the model will be explained wrongly due to the merger of different individuals at different times, the model ignores heterogeneity that may exist among the studied banks. It is possible that the characteristic of each individual is included in ε_{it}.

5.2 Fixed Effects Model (FEM)

The fixed effects model considers the individual characteristics of each bank in the sample. Thus, the intercept will vary in each individual, but the slope coefficients are assumed to be constant for all individuals. The coefficient β_{0i} in the following formula (with "i" assigned to each bank) represents individual differences, such as size or culture of work:

$$\mathbf{ROA_{it} = \beta_{0i} + \beta_1 HHI_{it} + \beta_2 Size_{it} + \beta_3 Eq_{it} + \beta_4 Per_{it} + \varepsilon_{it}} \quad (3)$$

Equation (3) is known as the fixed effects regression model. The term "fixed effects" is because each intercept (of each bank) is different from those of other banks though, it does not change over time and is fixed through time. If we write the intercept as β_{0it} then that means the intercept of each bank will change over time. But note that in Eq. (3), we assume that the intercepts are constant over time.

We can perform a test to determine whether FEM is superior to Pooled OLS model. Because the Pooled OLS model tends to ignore heterogeneous factors that are included in the calculation of FEM, so Pooled OLS model is a restricted version of FEM (if there is no distinct identity of individuals, FEM is similar to Pooled OLS). Thus, we can use the restricted F test to test whether there is no difference between among individuals.

5.3 Random Effects Model (REM)

In the fixed effects model, we assume that the intercept β_{0i} is constant for each studied bank and it is constant over time. In the random effects model, we assume that β_{0i} is a random variable with mean β_0 (there is no "i" here) and the intercept of any individual is shown as: $\beta_{0i} = \beta_0 + \varepsilon_i$.

The differences of each intercept (of each individual) are reflected in ε_i. Thus with REM, we can write as follows:

$$\mathbf{ROA_{it} = \beta_0 + \beta_1 HHI_{it} + \beta_2 Size_{it} + \beta_3 Eq_{it} + \beta_4 Per_{it} + w_{it}} \quad (4)$$

In which $w_{it} = \varepsilon_i + u_{it}$, there are two components of the w_{it}: ε_i is the noise component of each bank and u_{it} is the mixed noise component of the cross subjects and the time series.

Since ε_i is a part of w_{it}, it is possible that w_{it} has correlation with one or more explanatory variables. If this occurs, the REM will lead to an inconsistent estimation of the regression coefficients. Hausmans test will show whether or not w_{it} correlates with the explanatory variables – that means REM is a more suitable model than FEM.

6 Descriptive Statistics and Correlation Analysis

The annual average ROA of banks shows that performance tended to go down in the period between 2009 and 2015, when it drops from 1.8843% to 0.4843% respectively. The scale of assets has increased steadily over the years, but the return has decreased. This is also understandable that bad debts in this stage have increased sharply. However, it can be seen that since 2016, the banks' profitability has tendeds to be improved when the average ROA starts to go up again. In the period of 2016–2017, the banking system has showed positive signs in the handling of bad debts (Fig. 1).

Table 1. Summary statistics for relevant variables in our regression analysis

	Min	Max	Mean	St.dev	Observations
ROA	0.0986%	6.0875%	1.0531%	0.8374%	230
HHI	0.1220	0.5098	0.2475	0.0714	230
Eq	1.3290%	46.2446%	10.2943%	5.8998%	230
Size	14.6987	20.9074	18.1781	1.2391	230
Per	0.3377%	1.6479%	0.8373%	0.2537%	230

According to Moody's, ROA $\geq$ 1% is satisfactory. It is the fact that the average ROA of banks in Vietnam was lower than 1% from 2012 to 2017. Based on this assessment criteria, Vietnam's banking system is in a state of using assets not effectively.

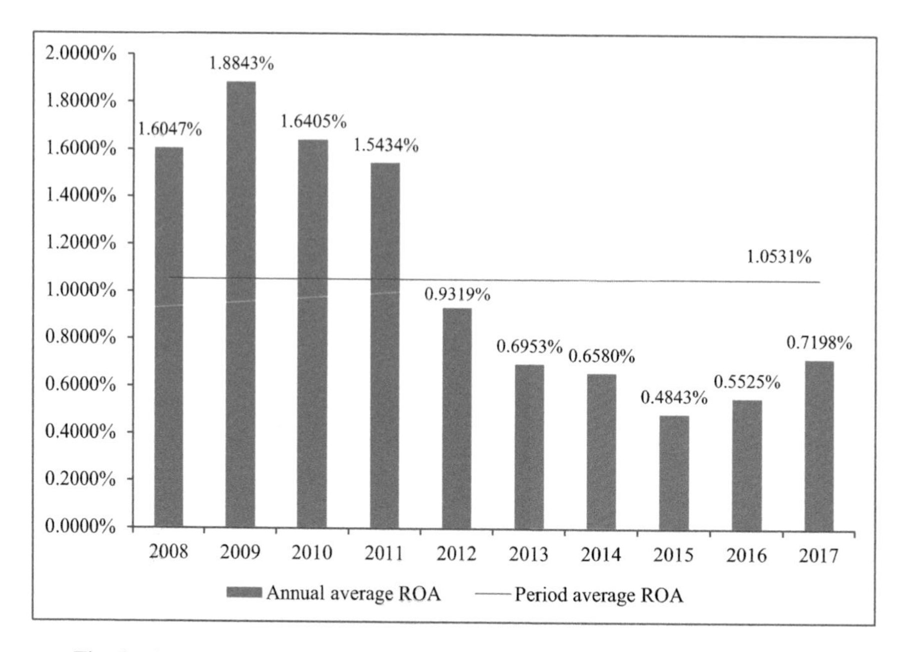

Fig. 1. Average ROA of Vietnam's banking system in the period of 2008–2017

Meanwhile, with the level of loan portfolio diversification of the banks in Vietnam during the study period, we can figure out the average HHI of the period is 0.2457. Comparing the level of diversification of the banking sector in Vietnam to other countries' according to previous empirical studies, it can be seen that the diversification of the loan portfolio in Vietnam is slightly higher than those of some banking systems, such as Brazil with average HHI of 0.3160 (Tabak et al. 2010), Norway with average HHI of 0.2892 (Aarflot and Arnegård 2017) or Germany with average HHI of 0.2910 (Hayden et al. 2006). On the opposite, if compared with the commercial banking system in Italy, the loan portfolio of banks here is a bit more diversified than in Vietnam, as the period average HHI is 0.2370 (Achayra et al. 2006).

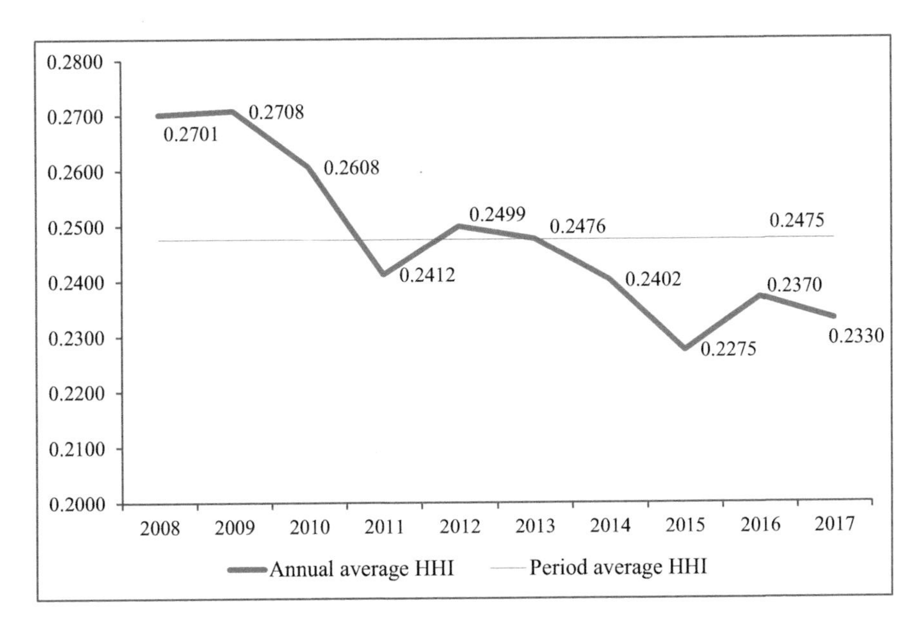

Fig. 2. Average HHI of Vietnam's banking system in the period of 2008–2017

As shown in Fig. 2, we can clearly see the volatility of HHI over time in Vietnam. Overall, the HHI has fallen from 2008 to 2017. The tougher competition has forced banks to expand lending to a wide range of industries, instead of just focusing on some specialized sectors as before. In addition, the move also noted that Vietnamese banks have been very cautious with the risk of loan-portfolio concentration and have restructured their loan portfolio in a more diversified way.

Table 2 shows the correlation matrix between the independent variables included in the model. It is clear that most correlations between independent variables are low, but the correlation coefficient between the bank size (Size) and the equity ratio (Eq) is relatively large at –0.6970.

Table 2. Correlation matrix

	HHI	Eq	Size	Per
HHI	1.0000	0.0226	–0.0807	–0.1878
Eq	0.0226	1.0000	–0.6970	0.1799
Size	–0.0807	–0.6970	1.0000	–0.0858
Per	–0.1878	0.1799	–0.0858	1.0000

This reveals that there might be the existence of multicollinearity in the regression model. Carry out the VIF test and the result shows that all VIF are less than 2. From these analysis results, the study showed no signs of multicollinearity between variables in the model.

7 Regression Results

After performing descriptive statistics and analyzing correlations among variables in the model, the paper continues to implement regression with the following models: Pooles OLS, FEM and REM (Table 3).

Table 3. Regression results from Pooled OLS, FEM and REM

		Pooled OLS	FEM	REM
HHI	Coefficient	0.008	0.032	0.023
	Standard error	0.007	0.011	0.009
	P-value	0.280	0.005***	0.011**
Eq	Coefficient	0.067	0.036	0.052
	Standard error	0.012	0.012	0.012
	P-value	0.000***	0.003***	0.000***
Size	Coefficient	0.001	–0.003	–0.001
	Standard error	0.001	0.001	0.001
	P-value	0.167	0.001***	0.212
Per	Coefficient	0.164	–0.012	0.122
	Standard error	0.207	0.252	0.232
	P-value	0.430	0.963	0.599

(***) Statistical significance at 1%; (**) Statistical significance at 5%; (*) Statistical significance at 1%; Variable explained is ROA.

Conduct F-tests resulting in p-value = 0.000 < 0.01 and Hausman's test for p-value = 0.000 < 0.01, the FEM was chosen to perform the final estimation and interpretation of the research results. Continue Modified Wald and Wooldridge tests in turn to identify heteroskedasticity and autocorrelation respectively in the FEM. The results show that there exists both heteroskedasticity and autocorrelation in the estimation model. To handle these, the paper runs the regression with cluster-robust standard errors (Hoechle 2007). Econometric software now integrates these special functions and particularly with Stata, the problem can be solved with the xtscc command.

As shown in Table 4, we have: (i) For HHI variable, the p-value = 0.008 should imply that statistical significance is at 1% and a regression coefficient of +0.032 with a positive sign indicates the positive effect on ROA variable; (ii) For Eq variable, the p-value = 0.015 should imply that statistical significance is at 5% and a regression coefficient of +0.036 with positive sign indicates the positive effect on ROA variable; (iii) For Size variable, the p-value = 0.033 should imply that statistical significance is at 5% and a regression coefficient of –0.003 bearing the negative sign indicates the negative effect on ROA variable; (iv) For Per variable, p-value = 0.957, this coefficient is not statistically significant.

Table 4. Regression results from modified FEM

	Coefficient	Standard error	P-value
HHI	0.032	0.009	0.008***
Eq	0.036	0.012	0.015**
Size	–0.003	0.001	0.033**
Per	–0.012	0.209	0.957
Obs	230		
Prob > F	0.0000		

(***) Statistical significance at 1%; (**) Statistical significance at 5%; (*) Statistical significance at 1%; Variable explained is ROA.

8 Result Discussions

Regression results show that the effect of HHI variable on ROA variable is positive. Based on the theory and practice of the Vietnam's banking system in the research period, this can be explained as follows:

- By focusing on certain sectors, lending banks may have better expertise in these sectors, as confirmed by Winton (1999) or Mishkin (2013). A bank with a more concentrated loan portfolio could benefit from monitoring and supervising more effectively due to better knowledge on industries and lower operating costs. On the other hand, diversification can reduce the efficiency of the bank, as it is more difficult for banks to keep track of their borrowers and they may face to adverse alternatives, which are derived from the edge of competition with other banks.
- Obviously, the trend of Vietnamese banks in the recent years shows that most banks tend to focus on lending in some high profitable sectors, typically in non-manufacturing sectors such as real estate, stock trading. In the context that these industries are still growing steadily and returns stay attractive, banks continuously grow credit in these industries. Additionally, there are some banks that have strengths in certain sectors such as import-export loans, industrial loans, construction loans, etc. They still prefer to concentrate their loans on these industries to maximize the competitiveness and bring about positive performances.
- The return of Vietnamese banks was strongly affected by bad debts during the study period. Bad debts of Vietnamese banks do not exist recently, but actually have accumulated many years ago. Especially in the period from 2009 to 2015, bad-debts boom strongly affected the bank's profitability. Dealing with bad debt has been actively implemented, in which the provisioning is considered as the leading tool.

As a result, banks have significantly decreased their return during this period (Fig. 1). There are many causes of bad debt, including the concentration of loans on high-risk industries or loans intended for large corporations on a large scale. It is clear that banks have diversified their loan portfolio as a result to attempt to reduce the level of concentration as before (Fig. 2). Based on these factors, it is possible to see, in the study period, there are two tendencies going together: the bank return has reduced and the loan portfolio has been diversified. These factors may add a basis for explaining why the diversification of the loan portfolio had the negative impact on bank return.

For bank size, the study found the negative effect on bank return. It can be explained for the Vietnam's banking market that large banks often lend large items, finance large projects of enterprises with poor marginality compared to smaller banks with smaller loans. Moreover, the risk of loan loss when financing large projects is be really heavy for big banks. Then we can see the positive effect of the equity ratio on bank return. This result is not surprising, because equity is a valuable buffer which help banks meet regulations and give more credibility to depositors, thereby reduce costs and successfully expand operation. Meanwhile, personnel cost ratio does not show any significant impact on bank return. This point can be explained that the salary culture of each bank is different in each period and the business direction can focus on different areas such as expanding the network, investing in modern technology or concentrating on human resources, attracting talents.

9 Conclusion and Future Work

The paper has found out that the diversification of the loan portfolio completely affects the return of Vietnamese commercial banks and this effect is negative. In other words, loan portfolio concentration seems to improve the performance of the bank return in Vietnam. By focusing only on a linear relationship between diversification and return, one could underestimate the importance of risk in the strategic decisions of banks. It is easy to see that the paper has simplified analysis and this should be extended to a more comprehensive assessment in the future.

References

Acharya, V.V., Hasan, I., Saunders, A.: Should banks be diversified? evidence from individual bank loan portfolios. J. Bus. **79**(3), 1355–1412 (2006)

Tabak, B.M., Fazio, D.M., Cajueiro, D.O.: The effects of loan portfolio concentration on brazilian banks' return and risk. J. Bank. Finance **35**(11), 3065–3076 (2010)

Diamond, D.W.: Financial intermediation and delegated monitoring. Rev. Econ. Stud. Ltd **51**(3), 393–414 (1984)

Hayden, E., Porath, D., von Westernhagen, N.: Does diversification Improve the Performance of German Banks? Evidence from Individual Bank Loan Portfolios, Deutsche Bundesbank, Discussion Paper Series 2: Banking and Financial Studies (5, 2006) (2006)

Hoechle, D.: Robust standard errors for panel regressions with cross-sectional dependence. Stata J. **7**(3), 281–312 (2007)

Marinč, M.: Bank monitoring and role of diversification. Trans. Stud. Rev. **16**(1), 77–91 (2009)

Markowitz, H.: Portfolio selection. J. Finance **7**, 77–91 (1952)

Mishkin, F., Matthews, K., Giuliodori, M.: The Economics of Money, Banking and Financial Markets (European edition). Pearson Education LTD, Harlow (2013)

Aarflot, S., Arnegård, L.A.: The effect of industrial diversification on banks' performance: A case study of the Norwegian banking market, SNF, Discussion Paper (9, 2017) (2017)

Winton, A.: Don't Put All Your Eggs in One Basket? Diversification and Specialization in Lending, Center for Financial Institutions Working Papers 00-16, Wharton School Center for Financial Institutions, University of Pennsylvania (1999)

An Investigation into the Impacts of FDI, Domestic Investment Capital, Human Resources, and Trained Workers on Economic Growth in Vietnam

Huong Thi Thanh Tran(✉) and Huyen Thanh Hoang

Statistics Division, Faculty of Accounting and Auditing,
Banking Academy, Ha Noi, Vietnam
{huongttt76,huyenht}@hvnh.edu.vn

Abstract. It is a general consensus that foreign direct investment (FDI) is a substantial source of capital, contributing to the total investment capital, as well as promoting economic growth of each country, especially when it comes to developing country as Vietnam. Vietnam offers attractive investment opportunities for foreign companies and has adopted a number of policies to attract foreign direct investment into the country. This paper examines the impact of FDI and other factors, including domestic investment capital (DIC), human resources (LB), and rate of trained workers (RTW) on the economic development of Vietnam, particularly with the view to considering the different among provinces. Data panel regression analysis were utilized to measure the relationship between independent (FDI) and dependent variables (GDP), with the data array obtained from 47 provinces and cities under central authority over the time period 2012 to 2015. The estimated result indicates that FDI, DIC and LB have a positive effect on the level of gross domestic product, while RTW has not affected the economic growth of Vietnam during the time period.

Keywords: Foreign direct investment · Domestic investment capital
Economic growth · Panel data regression model

1 Introduction

Over a 30-year period recently, the role of FDI in the Vietnamese economy has been increasingly important. Obviously, FDI is one of the essential sources for the domestic economic growth. FDI not only increases the supply of investment capital, but also encourages technology transfer, human capital accumulation for the economies with the view to openness and integration, especially in such developing countries as Vietnam, which is one of the countries with high economic growth rate in Asia. As a result, in recent years, Vietnam has attracted a considerable amount of FDI.

V. Kreinovich et al. (Eds.): ECONVN 2019, SCI 809, pp. 940–951, 2019.
https://doi.org/10.1007/978-3-030-04200-4_69

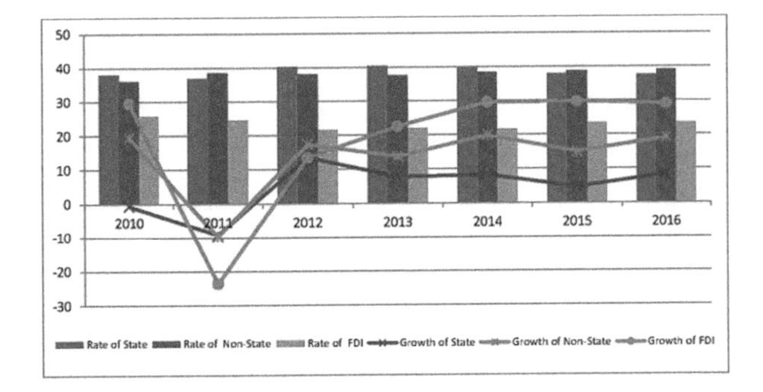

Fig. 1. Proportion and growth rate of investment capital in the economic sector in Vietnam, 2010–2016 Source: GSO of Vietnam

Figure 1 shows that over the last few years, the proportion of FDI is lower than that of both state and non-state sectors. However, it still has a higher growth rate. The growth rate of investment in all three regions declined sharply in 2011, of which FDI had the lowest growth rate. The remarkable decline of FDI in Vietnam in 20122 was influenced by such many factors as: (1) The weakness of the supporting industry. This is considered as a big "blockage" to prevent FDI inflows into Vietnam; (2) High inflation: in 2011, Vietnam's inflation was by far the highest level in Asia (23%), the price competitiveness of Vietnam is lost due to demand for staff increases, material costs, high interest rates, banks race to squeeze capital. As a result, foreign investors seem to behave differently to the growing domestic market of Vietnam and seek to move gradually to another countries. Although inflation has cooled down significantly, with only 6.81% in 2012, the resounding and results of high inflation in 2011 still seriously affected on activities of investors, who are concerned that inflation possibly recur at any time in Vietnam. Consequently, this would make investors worried about the attractiveness of the investment environment in Vietnam in the future.

Therefore, in 2012, despite a significant increase compared to the figure in 2011 in the FDI inflows registered of Vietnam, this increase is not sustainable with unexpected fluctuations. Over the period from 2012 to 2015, the number of FDI projects and the total registered capital witnessed a considerable increase. In 2016, with a series of Free Trade Agreement (FTA) in effect, there have been an ever-increasing trend in the FDI inflows. In general, the total registered capital of new projects, supplementary capital and investment in the form of capital contribution and share purchase in 2016 reached over 24.3 billion USD, an increase of 7.1% compared to the level in 2015. It is evident that the realized FDI in 2016 is estimated at 15.8 billion USD, reaching the peak level of FDI disbursement. (Source: Ministry of Planning and Investment).

In the period 2010–2016, the growth rate of FDI and GDP witnessed the same trend (shown in Fig. 2), which shows that FDI had a directly positive impact on the economic development of Vietnam.

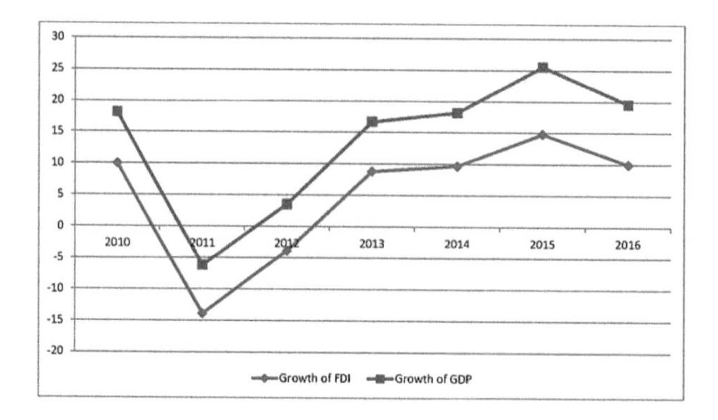

Fig. 2. Growth rate of FDI and GDP in Vietnam, 2010–2016 Source: GSO of Vietnam

2 Literature Review

There have been several studies concerning FDI and its influence on the output and growth of the economy. In previous studies, many economic researchers have concluded that there are positive effects of foreign direct investment on the economic development. Gudaro and Sheikh (2010), Rahman (2014) assess the impact of FDI, inflation rate, CPI on GDP growth in Pakistan during the period 1981–2010. They use multivariate regression and state that FDI had a positive effect, while inflation rate and CPI influenced negatively on the GDP of Paskistan between 1981 and 2010. Ali and Hussain (2017) apply a multivariate regression, using data array from 1991 to 2015 in order to investigate the effect of FDI on economic growth. In the model, GDP is the dependent variable, FDI, inflation rate, exchange rate and interest rate are the independent variables. The results showed that FDI, inflation rates, and exchange rates effect positively on Pakistan's GDP growth, while interest rates have a negative impact on this in the period 1991–2015. Agrawal and Khan (2011), by using a multiple regression, has investigated the relationship between FDI and the economic growth in India and China. In this model, GDP is the dependent variable, while FDI, investment capital, human resources are the explanatory variables. The time series data are derived from secondary sources such as: World Bank, the United Nations Conference on Trade and Development. They conclude that the variables of FDI, investment capital, labor force, human resources impact possitively on the GDP of China and India. The comparison between these two countries indicates that the impact of FDI on China's GDP growth is stronger than that in India in the period 1993–2009. According to the study done by Bhavan, Xu and Zhong (2011) on economic impact of FDI in South Asian countries, by taking time-series, cross-section analysis during the time period 1995–2008. Initially, they present a model to investigate the impact of factors on the potential of foreign investment. Subsequently, they use the growth model equation to assess the impact of FDI on economic growth, with the panel data regression model and the Arelano-Bond dynamic panel model. The results of this study suggest that: (1) sufficiency, promotion and cyclical factors are by far the most critical determinants of FDI in South Asian countries; (2) FDI has a positive impact on economic growth in South Asian. Aga (2014) highlights the effects of FDI on Turkey's economic growth throughout a time series approach during the period 1980–2012, with OLS and VAR estimation methods. In this model, GDP is the dependent variable, FDI, domestic investment (DIN) and trade liberalization (TL) are explanatory variables. They indicate that FDI and DIN have a positive impact on GDP growth, while TL affects negatively on GDP growth of Turkey over the time period.

As for Vietnam, the research into FDI and its impact on economic growth in Vietnam have been increasingly popular. Ha et al. (2017) use the time series method to analyze the effects of FDI on Vietnam's economic growth in the period 1990–2015. In the model, GDP is the dependent variable, FDI, total fixed capital, real exchange rate, real interest rate, inflation rate are the explanatory variables. They conclude that FDI, fixed capital, real exchange rates, real interest rates have positive effects, while inflation rate has a negative impact on GDP growth. Pham Anh and Ha (2012) assess the impact of FDI on economic growth in Vietnam by using the VAR model, with GDP being as the dependent variable, while disbursed FDI, social investment capital, total export, the number of labor force aged 15 and above, the number of students, and the Internet being as the explanatory variables. The result of the research show that FDI has stimulated exports and improved the quality of human resources, which is the premise for Vietnam's economic growth. Ly and Anh (2017) use panel data with a set data of 13 Asian countries from 1997 to 2006 to evaluate the effect of FDI on economic growth. In the research model, GDP growth is the dependent variable, while the lag of per capita GDP growth, domestic investment capital, labor force, FDI, exports, economic instability (rate inflation in countries) are the explanatory variables. From the estimation, they show that GDP per capita, FDI and domestic investment capital have had a positive impact on GDP growth. Son (2016) uses Cobb- Douglas expanded model with panel data regression to analyze the impact of FDI on the economic development in the Middle East of Vietnam. In this study, GDP growth is the dependent variable, while FDI, domestic investment, labor and human capital (the rate of increase of trained workers) are the explanatory variables. From the results, they highlight that FDI has a positive impact on the GDP growth of this region.

Accordingly, there have been few studies conducted on FDI and its impact on economic growth in Vietnam. Most of the studies show the positive impact of FDI on growth of the economy. However, in previous studies, there have been mainly use of time series and spatial data analysis such as OLS, VAR, etc. Therefore, they have not indicated the factor of inconsistency among provinces and cities in Vietnam and others, such as infrastructures, socio-economic policies, geographic location, natural resources, and so on. When researching on the whole economy of Vietnam, the authors find that in order to assess the impact of FDI and other factors on Vietnam's economic growth, particularly solving the different among provinces and cities in Vietnam, the panel data regression method has been suitable. This paper derived the panel data regression method with a data set of 47 provinces and cities under central authority in Vietnam during the time period 2012–2015.

3 Data Source and Methodology

In order to analyze the impact of FDI on economic growth, previous studies often used spatial or temporal cross-sectional analysis. Cross-sectional regression analysis are often unpredictable, and time series regression analysis are often less meaningful. With panel data regression analysis, the study has overcome the limitations of these two types. Therefore, in order to assess the impact of FDI and factors on Vietnam's economic growth, the authors select panel data regression method. This method would increase the sample size because of the calculated number of observations in both time and space dimensions. In addition, this method also allows us to study the dynamics of cross-over time units, as well as the inconsistency among provinces.

In the context of the study and actual data condition, in order to investigate the impact of FDI and factors on Vietnam's economic growth, the data regression model is as follows:

$$GDP_{ij} = \beta_0 + \beta_1\, LB_{ij} + \beta_2\, FDI_{ij} + \beta_3\, DIC_{ij} + \beta_4\, RTW_{ij} + c_i + u_{ij}$$

In which:

i: indicates a province
j: year
GDP: Gross domestic product (billion VND)
LB: Number of employees aged 15 and above working in economy (person)
FDI: Foreign direct investment (billion VND)
DIC: Domestic investment capital (billion VND)
c_i: characteristic of space
u_{ij}: random errors

Data sources for calculating these indicators are taken from the Statistical Yearbook of General Statistics Office of Vietnam and 47 provinces and cities under central authority in Vietnam.

There are three main methods commonly used for panel data regression, including: the POLS (Pooled OLS) model, the random effect model (REM), and the fixed effect model (FEM). As a result of the geographical inequality, the socio-economic factor between localities such as: infrastructures, socio-economic policies, geographic location, natural resources, and so on. Many factors are not observed or have no compatibility data. In such conditions, the application of the Regression Model for Panel Data Analysisis is the most appropriate choice to solve the geographic difference among provinces.

According to POLS model, all the coefficients, as well as intercept do not change over the time-series and space. This means that there will be no spatial or temporal characteristics, or in a way that c_i does not change over time, we can simply combine the crossover data and the time series in order to use OLS regression model.

As for REM, it is used in the case of heterogeneous spatial differences, or the difference in spatial variability, is not correlated with the independent variables in the model, which means that there is no correlation between c_i and the independent variables, and c_i is now considered as part of the random error, while u_ij is the spatial error and the time series combined. Assume that the spatial error components are not interrelated and do not self-correlate spatially and chronologically. When these assumptions are made, the obtained estimates will not converge to the overall parameter value, and then the FEM model then will be selected.

Regarding FEM, it is used when heterogeneous factors or particular characteristics of c_i space are correlated with independent variables. If cumulative and u_ij together into a composite random error then the estimation result would not be significant. At this point, we must approach the fixed-effects analysis method to control and isolate the effect of these particular characteristics from the explanatory variables in order to estimate the real effects of the explanatory variables.

The selection of the most suitable model is made through the following tests:

- Breusch-Pagan accreditation to choose POLS or FEM and REM models.

The hypothesis test is:

H_0: **No existence of random effect** $(\delta_u^2 = 0)$
H_1: **The existence of random effect**

If P value of the test $\chi^2 < 0,0000$ rejects the hypothesis H_0, it means that the model exists random effect, and then the model in the form of combined OLS should not be used and the model in the form of random effect shall be used.

- Hausman test is for selection of FEM or REM

The differences among provinces and cities is showed by c_i. If c_i is not correlated with independent variables in the model, then $v_ij = c_i + u_ij$ can be considered as a synthetic random error of the model (REM). On the contrary, if c_i correlates with independent variables, it cannot combine this element with random error factor, then the model is called a fixed effect model (FEM). In general, if the panel number is taken out or almost complete whole, the FEM is more suitable. When the panels are selected from the large whole, the REM may be appropriate.

The hypothesis test is:

H_0: **There is no correlation between** c_i **and independent variables.**
H_1: **There is correlation between** c_i **and independent variables.**

This test is guided as follows: If there is no significant difference between the estimated values from the two models of FEM and REM, it is a sign that c_i does not correlate with explanatory variables, then the REM is the appropriate choice. On the contrary, if there are significant differences between the estimated values from the two models of FEM and REM, it is a sign that there is a correlation between c_i and explanatory variables, then the FEM is the appropriate choice.

4 Estimated Results

Due to the actual data condition, on the purpose of investigation into the impact of FDI and other factors on the economic growth of Vietnam, the authors use a data set of 47 provinces and cities under the central government during the time period 2012–2015.

Step 1: Estimating the random effects model (REM)
In this model, dependent variable is gross domestic product (GDP), the explanatory variables are the number of laborers aged 15 and over working in the economy (LB), direct investment Foreign direct investment (FDI), domestic investment (DIC), and trained laborers (RTW). The result is shown in Table 1.

Table 1. REM Estimated results

```
Random-effects GLS regression                   Number of obs      =       186
Group variable: id                              Number of groups   =        47

R-sq:                                           Obs per group:
     within  = 0.1347                                          min =         2
     between = 0.2696                                          avg =       4.0
     overall = 0.2556                                          max =         4

                                                Wald chi2(4)       =     33.99
corr(u_i, X)   = 0 (assumed)                    Prob > chi2        =    0.0000
```

GDP	Coef.	Std. Err.	z	P>|z|	[95% Conf.	Interval]
LB	.0135619	.0059119	2.29	0.022	.0019748	.025149
FDI	.4255474	.2121401	2.01	0.045	.0097604	.8413344
DIC	.7423089	.3832835	1.94	0.053	-.008913	1.493531
RTW	759.8948	455.4093	1.67	0.095	-132.691	1652.481
_cons	8401.81	8900.672	0.94	0.345	-9043.187	25846.81
sigma_u	36657.88					
sigma_e	10016.613					
rho	.93052402	(fraction of variance due to u_i)				

Estimated results in Table 1 show that all four variables in the model are statistically significant. In particularly, LB and FDI have a positive impact on GDP at 5%, while DIC and RTW have a positive impact on GDP at 10%.

For the purpose of testing whether c_i being existing or not, as well as considering POLS model, we use the Breusch-Pagan test. If there is c_i existence, it is not suitable to choose the POLS model, and then the REM or FEM model will be selected (Table 2).

Table 2. Breusch-pagan test results

Breusch and Pagan Lagrangian multiplier test for random effects

GDP[id,t] = Xb + u[id] + e[id,t]

Estimated results:

	Var	sd = sqrt(Var)
GDP	1.97e+09	44377.5
e	1.00e+08	10016.61
u	1.34e+09	36657.88

Test: Var(u) = 0

chibar2(01) = 222.69
Prob > chibar2 = 0.0000

The result of Breusch-Pagan test indicates that the p-value of test was less than 5%. (P = 0.0000), so that we reject H_0. It means that the POLS model is inappropriate. Thus, this has the existence of c_i or there is difference of the geographical inequality c_i between provinces/cities in the whole economy.

Step 2: The selection of FEM or REM

In order to select FEM and REM, the authors use Hausman test (Table 3).

Table 3. Hausman test results

	—— Coefficients ——			
	(b) mohinhfe	(B) mohinhre	(b-B) Difference	sqrt(diag(V_b-V_B)) S.E.
LB	.0140245	.0135619	.0004626	.0015609
FDI	.3098243	.4255474	-.1157231	.0473348
DIC	.5994996	.7423089	-.1428092	.1616337
RTW	647.6882	759.8948	-112.2066	204.2874

b = consistent under Ho and Ha; obtained from xtreg
B = inconsistent under Ha, efficient under Ho; obtained from xtreg

Test: Ho: difference in coefficients not systematic

chi2(3) = (b-B)'[(V_b-V_B)^(-1)](b-B)
= 10.62
Prob>chi2 = 0.0140

The Hausman test results indicate that:

P = 0.01408, less than 5%, so rejecting H_0. It means that there is correlation between c_i and independent variables. Therefore, it is recommended to use FEM model.

Having determined the suitability of FEM, we then apply statistical hypothesis testing to see whether FEM have violated hypotheses or not through Wald and Wooldridge tests.

The results of these tests are represented in Tables 4 and 5.

Table 4. Wald test results

```
Modified Wald test for groupwise heteroskedasticity
in fixed effect regression model

H0: sigma(i)^2 = sigma^2 for all i

chi2 (47)  =    2.9e+33
Prob>chi2 =      0.0000
```

The Wald test result shows the existence of changed error variance of the model, with P_value being quite small (P = 0.0000 < 0.05). We reject H_0, which means the pattern of changed error variance exists.

In order to test the existing of the model, the Wooldridge test is used. The result of this test is shown in Table 5.

Table 5. Wooldridge test results

```
Wooldridge test for autocorrelation in panel data
H0: no first-order autocorrelation
    F( 1,      45) =    20.444
           Prob > F =     0.0000
```

The Wooldridge test results give the evidence for the existing of autocorrelation, with the p-value being less than 5%, so H_0 is rejected. Thus, this records the self-correlation between the random errors in REM model.

To overcome the above defects of the model, the authors select a panel data regression model with Robust option. The results of the estimation are represented in Table 6 below.

Table 6. Estimated results with robust option

```
Fixed-effects (within) regression               Number of obs      =       186
Group variable: id                              Number of groups   =        47

R-sq:                                           Obs per group:
     within  = 0.1361                                         min =          2
     between = 0.2545                                         avg =        4.0
     overall = 0.2417                                         max =          4

                                                F(4,46)           =       7.02
corr(u_i, Xb)  = 0.2544                         Prob > F          =     0.0002

                                      (Std. Err. adjusted for 47 clusters in id)
```

GDP	Coef.	Robust Std. Err.	t	P>|t|	[95% Conf. Interval]	
LB	.0140245	.0080694	1.74	0.089	-.0022183	.0302673
FDI	.3098243	.1211713	2.56	0.014	.0659192	.5537294
DIC	.5994996	.2412753	2.48	0.017	.1138376	1.085162
RTW	647.6882	407.6727	1.59	0.119	-172.9143	1468.291
_cons	12313.36	7705.57	1.60	0.117	-3197.147	27823.87
sigma_u	39100.342					
sigma_e	10016.613					

—more—

From the estimated results, we give some conclusions:

- From the estimated study, it is clear that all four explanatory variables have p-value less than 5%, then these factors are statistically equally significant. In particularly, LB, FDI have possitive impact on GDP at 5%.
- Breusch-Pagan test results indicate the existence of the inconsistency among provinces and cities in the whole economy. Thus, the model in the form of POLS should not be used, and the model in the form of random effect shall be used. In addition, there is correlation between ci and independent variables in the model, so that the selecting FEM is suitable.
- From Wald Test and Wooldridge Test results, it is obvious that there is changed error variances, as well as self-correlation between random errors in the model.
- Regression model with Robust option is selected in order to overcome the drawback of the model. From the estimated result, we conclude that: FDI, DIC affect positively on Vietnam's GDP growth in the period 2012–2015, of which DIC has a stronger impact. While LB also has a positive impact on GDP at 10% statistical significance. In addition, RTW has not really affected on the GDP of Vietnam at this stage.

5 Recommendations

From the empirical study, it is obvious that the increase of FDI is a substantial factor for Vietnam's economic growth. The result also highlights that its impact is nowhere near as strong as that of DIC. Therefore, in the coming time, in order to boost growth of economy, Vietnam needs to:

(1) Implement policies to attract FDI to create new impetus for economic growth. In order to attract FDI in Vietnam, it is necessary to continue on improving its investment environment, as well as creating favorable conditions for enterprises, especially FDI enterprises; Creating a truly competitive environment that values all forms of economy; Focusing on technology transfer and corporate governance; Continously encouraging education and training, with clearance of the spearheading knowledge sources in the hi-tech, financial and banking sectors as well, in order to attract FDI into Vietnam and to improve the quality of human resources;
(2) Continously keep the role of domestic investment capital as an important factor for GDP in economic growth, applying policies to promote internal accumulation of the economy;
(3) It is evident that qualifications of workers play an important role in economic growth, but this factor has not really had a positive impact on Vietnam's economic growth. Therefore, Vietnam should take measures to improve the quality of labor forces, through promoting training and re-training for laborers in order to meet the needs of society and international integration, improving the quality of training in educational institutions, including universities and colleges, in order to provide high quality labor forces to the economy.

References

Peneder, M.: Industrial structure and aggregate growth. WIFO, Structural Change anh Economic Dynamics **14**, 427–448 (2003)

Gudaro, A.M., Chahapra, I.U., Sheikh, S.A.: Impact of foreign direct of investment on economic growth: a case study of Pakistan. J. Manag. Soc. Sci. **6**(2), 84–92

Ur Rahman, Z.: Impact of foreign direct of investment on economic growth in Pakistan. J. Econ. Sustain. Dev. **5**(27), 251–255. ISSN 2222-1700 (paper), ISSN 2222-2855 (online)

Ali, N., Hussain, H.: Impact of foreign direc of investment on economic growth in Pakistan. Am. J. Econ. 163–170. p-ISSN: 2166-4951, e-ISSN: 2166-496X

Agrawal, G., Khan, M.A.: Impact of FDI on GDP: a comparative study of china and India. Int. J. Bus. Manag. **6**(10), 71–79

Bhavan, T., Xu, C., Zhong, C.: Determinants and growth effect of FDI in South Asian economies: evidence from a panel data analysis. Int. Bus. Res. **4**(1), 43–50

Aga, A.A.I.K.: The impact of foreign direct investment on economic growth: a case study of Turkey 1980–2012. Int. J. Economic Finance **6**(7), 71–84 (2014)

Ha, C.T., Wang, Y., Hu, X., Than, S.T.: The impact of foreign direct invesment on economic growth: a case in Viet Nam 1990 2015. Industrial Engineering Letters **7**(4) (2017). ISSN 2224-6096 (paper), ISSN 2225-0581 (online)

Anh, P.T.H., Thu, L.H.: An evaluation of relationship between foreign direct investment and economic growth in Vietnam. JED (220), 79–96, April 2014

Ly, T.D., Anh, L.H.: Impact of foreign direct investment on economic growth: the case of ASEAN +3. The result of Science and Technology Application, No. 10, September 2017. http://www.tapchicongthuong.vn/anh-huong-cua-dau-tu-truc-tiep-nuoc-ngoai-den-tang-truong-kinh-te-nghien-cuu-tai-cac-quoc-gia-asean-3-20171127021956748p0c488.htm

Son, T.: Evaluating the influence of foreign direct investment on economic growth: an empirical study in Central Region. In: National Conference of Statistics and Applied Informatics, pp. 253–260. Da Nang Press (2016)

The Impact of External Debt to Economic Growth in Viet Nam: Linear and Nonlinear Approaches

Lê Phan Thị Diệu Thảo(✉) and Nguyễn Xuân Trường

Faculty of Finance and Faculty of International Economics, Banking University, Ho Chi Minh City, Vietnam
{thaolptd, truongnx}@buh.edu.vn

Abstract. Analyzing the experimental of the impact of external debt on Vietnam's economic growth using the VECM model from a linear and non-linear perspective in the period from 2000 to 2013. Linear study results showed that external debt has a positive impact on economic growth in the long term. About external debt variable, a 1% increase in external debt would increase GDP by 1.29%. At the same time, the openness of the economy also positively impacts on economic growth at a rate of 1% increase in openness, increasing GDP by 0.5%. The study measured the debt threshold of 21.5% of GDP per quarter in the nonlinear model. Besides that external debt also has positive impact on economic growth in the long term. The results is the basis for giving some policy suggestions to the Government in planning the strategy of using external debt in short and long term for Vietnam in the future.

Keywords: External debt · Threshold debt · Economic growth Vietnam

1 Introduction

The external debt and growth economic's relationship is a topic which mentioned relatively in economic research literature. Theories focus on explaining this relationship based on the dynamic economic models of opened economies, with one side borrowing external debt for economic development, thereby using external saving to invest in the economy. This hypothesis becomes increasingly true for developing countries, with the use of abundant external resources modern technology shorten time use for developing in the hope of escape poverty, catching up with developed countries, increasing residents incomes. However, the other problem is when countries borrow heavily from abroad will lead to an accumulation of rising interest payments, leading to reduced investment and reduced social welfare. The question is whether increasing external debt will increase economic growth or vice versa as debt obligations increase. This reflects the existence of an optimal threshold for external debt, if they break through this level of debt, the countries will face increasing debt but negatively impact the growth of the economy. Thus, the borrowing countries need to pay attention to the debt threshold for optimal use of foreign resources in economic development.

V. Kreinovich et al. (Eds.): ECONVN 2019, SCI 809, pp. 952–967, 2019.
https://doi.org/10.1007/978-3-030-04200-4_70

Vietnam is a developing country and needs huge capital to build infrastructure and invest in development. Same as the other developing countries, Viet Nam has a high budget deficit, low foreign exchange revenues, leading to inadequate resources for investment therefore, abroad borrowing is one of the important resources to offset the development of the country, contributing to catch up with other countries in the region and the world. But how much external debt has a positive impact on the economic growth? Is there a nonlinear relationship between external debt and Vietnam's economic growth? These are the questions in this article aimed at finding answers to Vietnam, thus providing policy and institutional suggestions to motivate future economic growth.

2 Literature Review

Fosu (1999) studied the impact of factors such as labor growth, domestic investment, exports and external debt on economic growth in 35 African countries during the period from 1975 to 1994 by the OLS method. The results showed that external debt has a negative impact on economic growth and having the existence of the debt Laffer curve. Were (2001) also studied the effects of external debt on Keyna's economic growth during the period from 1970 to 1995 by the VECM method. The study indicated the same thing as in Fosu (1999).

Clements et al. (2003) studied relationship between external debt and economic growth in the 55 countries having low income in the period between 1970 and 1999 by using FEM and GMM models. He said the reduction of external debt in these countries would increase the growth rate of per capita income. In addition, Clements also determines the level of debt for low-income countries, about 30–37% of GDP. If this threshold is exceeded, the increase in external debt will reduce the per capita income.

Mohamed (2005) studied the impact of external factors such as external debt, exports, inflation on economic growth of Sudan from 1978 to 2001 using the OLS model. The results showed that external debt and inflation have a negative impact on economic growth.

The study was conducted by Frimpong and Oteng-Abayie (2006) to examine the impact of external debt on Ghana's economic growth over the period from 1970 to 1999. The results showed that external debt has a positive impact on economic growth in the long term.

Adegbite et al. (2008) investigating external debt in Nigeria for the period from 1975 to 2005 showed that external debt has a negative impact on national income. Sulaiman and Azeez (2012) also studied this problem in Nigeria for the period between 1970 and 2010.

Sha and Pervin (2012) studied the independence of the Bangladeshi economy against external debt of the government and guaranteed by the government in both short and long term. The authors used the time series data from 1974 to 2010. The authors conclude that the obligations of external debt in the public sector have a negative impact on economic growth in the short time, but the effect of total external debt in the public sector is unclear. However, increasing the external debt burden in the public sector will indirectly affect economic growth as it increases the debt obligations to the economy.

Dauda et al. (2013) examined the impact of external debt on economic growth in the period from 1991 to 2009. Analyzing quarterly time series data, the study had shown that increasing external debt contributes to Malaysia's growth in the long term.

Mohamed (2013) studied the effects of external debt in the short and long term on the economic growth of Tunisia from 1970 to 2010 using the VECM model. The results showed that debt has a negative impact on economic growth during the researched period, with 1% debt increasing bringing about reducing in economic growth by 0.15–0.17%. In the long term, every 1% increase in debt led to a 0.27% decrease in economic growth. In addition, the study also indicated the existence of a Laffer curve with a debt ceiling of 30% of GDP.

Korkmaz (2015) examined the relationship between external debt and economic growth in Turkey, according to quarterly data from Q1/2003 to Q3/2014. The results showed that external debt has a positive impact on Turkey's economic growth during the observed period.

Osinubi et al. (2010) studied external debt and the budget deficit affecting the economic growth of Nigeria in the period from 1970 to 2003. The research had shown the impact of budget deficits on the stability of external debt, a combination of this factor affecting Nigeria's economic growth in the short and long term. A stable external debt ratio is not a good policy. Good policy is to set up an external debt threshold for the country, contributing to the growth of economy following the debt Laffer curve and maintaining this threshold so that external debt does not exceed the threshold which has negatively impacts on the economy. The highlight of this study is finding the level of debt for Nigeria, approximately 60% of GDP, and it is affected by fiscal policy as well as the gap between real borrowing rates and economic growth rates.

Cechetti et al. (2011) examined the relationship between external debt and growth of economy in the 18 OECD countries 1980–2010. The authors had found that the rate of debt of government and household are around 85% of GDP and that of corporate is about 90% of GDP.

Mohd Daud (2016) studied the impact of government debt on Malaysian economic growth during the period form Q1/1996 to Q4/2011 by using the ARDL model. The results showed that there is a long-term nonlinear relationship between government debt and economic growth. In addition, the study also indicated the level of debt having positive impact on Malaysia's economic growth.

3 Methodology

This paper analyzes the impact of external debt on Vietnam's economic growth based on the VECM model. VECM model which developed by VAR model. The VECM model is considered to be an OLS regression equation between the present value of this variable (t) and its past value (t-1) and other variables in the model associated with the variable correct the error obtained from the cointegration. The VECM model is described as follows:

$$\Delta Y_t = \tau_1 \Delta Y_{t-1} + \tau_2 \Delta Y_{t-2} + \ldots + \tau_k \Delta Y_{t-k} + \Pi Y_{t-1} + \rho T + u_t$$

Where k is the latency of the model, $\tau_1, \tau_2, \ldots, \tau_k$ are square matrices level m with $\tau_i = \left(\sum_{j=1}^{k} \beta_j\right) - I_g$ and Π (g * g) is the square matrix that represents the long-term relationship between variables at equilibrium. Matrix Π is the product of the two matrices α (g * r) and β '(r * g) where r is the number of cointegration vectors that is also the order of the matrix Π. The matrix β 'is the cointegration vectors matrix that reflects the long-run relationship between the variables, α is the coefficient of the copper vector associated in the VECM model, T is the time trend.

The advantage of the VECM model is that it allows the measurement of the co-existence of multiple variables in the research model and allows for the measurement of the level of adjustment from the imbalance of the previous period. The data processing steps under the VECM model are as follows:

3.1 Unit Root Test Using Augmented Dickey Fuller (ADF)

First, test the cessation of the research data through Unit Root Test to determine the degree of integration of the time series. It then defines the appropriate latency structure for the VECM model based on the information standards such as LR (Likelihood Ratio), AIC (Akaike infor criterion), SC (Schawarz criterion), HQ (Hannan-Quinn criterion) and FPE (Final Prediction Error).

3.2 Structural Analysis – Granger Causality

Second, Granger Causality approach is employed to determine whether a specific variable or group of variables play any roles in the determination of other variables in the vector error correction (VEC) process (Johansen 1991, 1995). It tests whether an endogenous variable can be treated as exogenous and was done by examining the statistical significance of the lagged error-correction terms by applying separate t-tests on the adjustment coefficients. Thus, the variance decomposition provides information about the relative importance of each random innovation in the impact variables in the VEC.

4 Empirical Results

4.1 Analyze the Impact External Debt on Economic Growth by Linear Model

Research model based on the research conducted by Pattillo et al. (2002). However, there are some additional variables into the model in this study. The regression line is used to examine the impact of external debt on Vietnam's economic growth is:

$$\ln GDP = f \ln(EXD, OPE, M2, REER, DUM) \quad (1)$$

More specifically for the model:

$$\ln GDP = \alpha_0 + \alpha_1 \ln EXD + \alpha_2 DUM + \alpha_3 \ln OPE + \alpha_4 \ln M2 + \alpha_4 \ln REER + \alpha_5 T + u_t \quad (2)$$

Where:

Ln GDP is a dependent variable, taking the natural logarithm of GDP. This variable was used in the studies Clements (2005), Adegbite et al. (2008).

The independent variable ln EXD is the natural logarithm of the debt-to-GDP ratio in % GDP accordingly quarterly data. This variable is commonly used in foreign studies to assess the debt situation as well as the repayment capacity of countries. Studies conducted by Fosu (1996), Were (2001), Pattillo (2002), Clements (2005), Adegbite et al. (2008), Ayadi (2008), Tokunbo et al. (2010) and Korkmaz 2015) used this variable to assess the impact on economic growth. The results showed that external debt has the same or opposite direction to economic growth.

The independent variable ln OPE is the natural logarithm of the openness of the economy, calculated by taking the import-export value relative to GDP in % GDP accordingly quarterly data. This indicator was used in studies done by Clements (2005), Tokunbo et al. (2010), Daud et al. (2003).

Variable ln M2 is a logarithm of money supply in the economy, one of the macro variables affecting economic growth, representing the level of financial development of the economy. This indicator was used in Mohamed (2013) studies.

The ln REER variable is the logarithm of the real exchange rate in the economy. Exchange rates affect borrowing and debt repayments of borrowers as well as all activities in the economy. This indicator was used in the studies of Were (2001), Sulaiman and Azeez (2012).

DUM is a dummy variable, showing the impact of WTO integration on the openness of the economy, $\sigma = 0$ when Vietnam has not joined the WTO (from Q4/2006 and before) and $\sigma = 1$ when Vietnam became Official WTO membership (from Q1/2007).

T is the time trend and u_t is the residual of the model.

Research data is periodical data collected from various sources over the period from 2000 to 2013. The first data source is the Asian Development Bank (ADB). In addition, the GSO and World Bank (WB) data also are used. Data for subsequent years are not updated quarterly by these organizations.

In order to implement the VECM model, unit testing should be done by means of the ADF (Augmented Dickey-Fuller) method on the cropped data. The results show that the data variables are non-stop and the first-order difference equals 1% (Table 1). This is the basis and conducts the Johansen-Jesulius cointegration test.

Next, the study finds the optimal lag for the VECM model based on information standards. For different lag, the optimal lag according to the AIC standard varies, but the optimal lag according to the SBIC and HQIC standards is always 1. As a result, the optimal lag for the model is 1. Verify the cointegration relationship to demonstrate the

Table 1. Unit root test of stationary for variables

Variable	t-Statistic	1%	5%	10%	Conclutions
lnGDP	−2.603229	−3.555023	−2.915522*	−2.595565	Nonstationary
lnEXD	−1.509865	−3.557472	−2.916566	−2.596116	Nonstationary
lnOPE	−0.071092	−3.555023	−2.915522	−2.595565	Nonstationary
lnM2	−0.811474	−3.560019	−2.917650	−2.596689	Nonstationary
lnREER	−0.731572	−3.555023	−2.915522	−2.595565	Nonstationary
D(lnGDP)	−9.624394	−3.557472	−2.916566	−2.596116	Stationary
D(lnEXD)	−10.75118	−3.557472	−2.916566	−2.596116	Stationary
D(lnOPE)	−6.569152	−3.557472	−2.916566	−2.596116	Stationary
D(lnM2)	−5.194974	−3.560019	−2.917650	−2.596689	Stationary
D(lnREER)	−9.728703	−3.557472	−2.916566	−2.596116	Stationary

* *show significance at 5% respectively.*

Table 2. Results of cointegration tests

		Trace Test		Max-Eigen Test	
Hypothesized No. of CE(s)	Eigen value	Trace Statistic	Critical Values @ 5%	Max-Eigen Statistic	Critical Values @ 5%
None*	0.605655	123.7336*	88.80380*	50.24852*	38.33101*
At most 1*	0.475910	73.48503*	63.87610*	34.88895	32.11832
At most 2	0.359573	38.59608	42.91525	24.06351	25.82321
At most 3	0.215744	14.53257	25.87211	1.409524	19.38704

* *show significance at 5% respectively.*

long-term relationship between the variables in the model through the Trace and Max-Eigen tests with lag correspondingly to select the appropriate VECM model. The results show that Eq. 3 (with constant and without trend in the co-integration equation) and Eq. 4 (with constant and tended in the co-integration equation) (Table 2).

With Eqs. 3 and 4, the lag is 1, the study continues to select the number of co-integration for the model by checking Trace and Max-Eigen for the selected co-integration equations (Eqs. 3 and 4). The results show that there are two co-integration vectors under the Trace test and one co-integration under the Max-Eigen test between the variables in the model at the 5% significance level, reflecting the long-term correlation relationship of the VECM model and the selected equation is Eq. 4.

Conduct a regression estimation with VECM to determine the impact and correlation between the variables in the model corresponding to the two cointegration found in step 3. The results of the VECM model estimation show that the variables have effects same the expected impacts, but the lnM2(-1) variable is insignificant in the cointegration 1. The long equilibrium model shows in the Eqs. (5) and (6). The regression coefficients of external debt and the openness of the economy meet

Table 3. Estimated long-run model

Variable	Equation 1	t Statistic	Equation 2	t Statistic
LNGDP(−1)	1.0000		0.0000	
LNREER (−1)	0.0000		1.0000	
LNEXD(−1)	−1.164271	−1.28584	−0.414744	−3.98053
LNOPE(−1)	−6.310437	−6.43007	−0.604929	−5.35661
LNM2(−1)	0.398333	0.40107	0.367347	3.21425
T	0.203022	2.90389	0.004596	0.57132
C	27.41966		−4.824362	

Equivalent to the following cointegration equation:

Table 4. Estimated short-run model

Dependent Variable	D(LNGDP)		D(LNREER)	
Independent Variable	Coefficient	t Statistic	Coefficient	t Statistic
D(LNGDP(−1))	−0.45589	−3.36889*	−0.08782	−3.33489*
D(LNREER(−1))	−0.05043	−0.75606	−0.07086	−0.54582
D(LNEXD(−1))	0.04076	0.179052	−0.07368	−1.66734***
D(LNOPE(−1))	0.82896	2.27709**	0.28191	3.97912*
D(LNM2(−1))	0.25464	0.30152	0.37899	2.30589**
C	−0.11256	−1.95338***	−0.0254	−2.26511**
DUM	0.000188	1.92139***	−3.53E-0.5	−1.85171***
R^2	0.37926		0.5404	
Dependent Variable	D(LNEXD)		D(LNOPE)	
Independent Variable	Coefficient	t Statistic	Coefficient	t Statistic
D(LNGDP(-1))	0.04384	0.55699	0.07825	0.14838
D(LNREER(-1))	−0.4826	−1.24369	−0.66383	−2.5535**
D(LNEXD(-1))	−0.3720	−2.81626*	0.134707	1.52224
D(LNOPE(-1))	0.4062	1.91844***	0.01508	0.10633
D(LNM2(-1))	−0.4363	0.88808	−0.5668	−1.7223***
C	−0.0166	−0.5956	0.00649	0.28908
DUM	0.00013	2.29149**	0.000185	4.86141*
R^2	0.426990		0.44026	
Dependent Variable	D(LNM2)			
Independent Variable	Coefficient		t Statistic	
D(LNGDP(-1))	−0.0127		−0.58720	
D(LNREER(-1))	0.1037		0.97260	
D(LNEXD(-1))	−0.06907		−1.90249***	
D(LNOPE(-1))	0.02737		0.47041	
D(LNM2(-1))	0.66945		4.95821*	
C	0.035054		3.80501*	
DUM	−4.94E-0.5		−3.16094*	
R^2	0.444328			

*, **, *** *show significance at 1%, 5% & 10% respectively*

expectations and have a positive impact on economic growth but M2 money supply has the opposite effect (Table 3).

$$1 * \ln GDP_{(-1)} - 1.164271 * \ln EXD_{(-1)} - 6.310437 * \ln OPE_{(-1)} + 0.398333 * \ln M2_{(-1)} + 0.203022 * T + 27.41966 = 0 \quad (3)$$

$$\text{And}: \quad 1 * \ln REER_{(-1)} - 0.041474 * \ln EXD_{(-1)} - 5.35661 * \ln OPE_{(-1)} + 3.21425 * \ln M2_{(-1)} + 0.004569 * T - 4.824362 = 0 \quad (4)$$

Equations (1) and (2) are equivalent to the following:

$$\ln GDP_{(-1)} = 1.164271 * \ln EXD_{(-1)} + 6.10437 * \ln OPE_{(-1)} - 0.398333 * \ln M2_{(-1)} - 0.203022 * T - 27.41966 \quad (5)$$

$$\ln REER_{(-1)} = 0.041474 * \ln EXD_{(-1)} + 5.35661 * \ln OPE_{(-1)} - 3.21425 * \ln M2_{(-1)} - 0.004569 * T + 4.824362 \quad (6)$$

The short-term VECM estimate indicates that D(LNGDP) (which is dependent variable) have significant relationship with D(LNGDP(-1)), D(LNOPE(-1)) and dummy variable. Similarly, D(LNREER), dependent variable, is significant relationship with D (LNGDP (-1)), D (LNEXD (-1)), D (LNOPE (-1)), D(LNM2 (-1)) and DUM dummy variable. The equation with dependent variable is D(LNEXD (-1)) indicate that D(LNEXD (-1)), D (LNOPE (-1)) and DUM dummy variables have significant relationship with dependent variable. The equation with dependent variable is D(LNOPE) showing a significant relationship with D (LNREER (-1)), D (LNM2 (-1)) and dummy variables. Finally, as D(LNM2) is a dependent variable, independent variables LN (EXD (-1)), D (LNM2 (-1)) and DUM dummy variables are significant relationship with dependent variable.

In the next step, the Granger causality test is conducted to clarify the relationship between the variables in the VECM model. The results summarized qualitatively in Table 5 show that the null hypothesis where p is less than 10% is rejected which means that the alternative hypothesis is accepted.

As far as D(ln GDP) is concerned as a dependent variable, D(ln REER), D (ln EXD), and D (lnM2) do not cause D(ln GDP) at 5% levels. However, the combined test where both D(ln REER), D(ln EXD), D(lnM2) and D(ln OPE) are not the causal relations with D (GDP_SA) is rejected at 10% significance levels. This result shows that D (ln REER), D (ln EXD), D (ln M2) and D (ln OPE) cause D (ln GDP) which means that D(ln GDP) is the dependent variable.

When D(ln REER) is concerned as a dependent variable, either D(ln EXD) or D(ln M2) do not cause D(ln REER). However, the combined test where both

Table 5. Granger-causality test results.

Null Hypothesis	Chi-sq	P values
REER does not Granger cause GDP	0.571629	0.4496
EXD does not Granger cause GDP	0.032227	0.8575
OPE does not Granger cause GDP	5.185124	0.0228**
M2 does not Granger cause GDP	0.090912	0.7630
All variable does not Granger cause GDP	8.023949	0.0907***
GDP does not Granger cause REER	11.12151	0.0009*
EXD does not Granger cause REER	2.780032	0.0954***
OPE does not Granger cause REER	15.83343	0.0001*
M2 does not Granger cause REER	5.317143	0.0211**
All variable does not Granger cause REER	24.37559	0.0001*
GDP does not Granger cause EXD	0.310238	0.5775
REER does not Granger cause EXD	1.546765	0.2136
OPE does not Granger cause EXD	3.6804	0.0551***
M2 does not Granger cause EXD	0.788686	0.3745
All variable does not Granger cause EXD	9.076554	0.0592***
GDP does not Granger cause OPE	0.022018	0.8820
REER does not Granger cause OPE	6.520533	0.0107**
EXD does not Granger cause OPE	2.317211	0.1279
M2 does not Granger cause OPE	2.966491	0.0850***
All variable does not Granger cause OPE	9.238442	0.0554**
GDP does not Granger cause M2	0.344800	0.5571
REER does not Granger cause M2	0.945953	0.3308
EXD does not Granger cause M2	3.619470	0.0571***
OPE does not Granger cause M2	0.221281	0.6381
All variable does not Granger cause M2	4.460204	0.3473

*, **, *** *show significance at 1%, 5% & 10% respectively*

D (ln GDP), D(ln EXD), D(lnM2) and D(ln OPE) do not cause D(ln REER) is rejected at 1% significance levels. This result shows that D (ln GDP), D (ln EXD), D (lnM2) and D (ln OPE) are the causes of D (ln REER) which means that D(ln REER) is the dependent variable.

As far as D(ln EXD) is concerned as a dependent variable, D(ln GDP), D(ln REER) and D(ln M2) do not cause D(ln REER). However, the combined test where both D(ln GDP), D(ln REER), D(lnM2) and D(ln OPE) do not cause D(ln REER) is rejected at 10% significance levels. This result shows that D(ln GDP), D(ln REER), D(lnM2) và D (ln OPE)are the causes of D(ln EXD) which means that D(ln EXD) is the dependent variable.

When D(ln OPE) is concerned as a dependent variable, D(ln GDP) and D(ln EXD) do not cause D(ln REER). However, the combined test where both D(ln GDP), D(ln REER), D(ln M2) and D(ln OPE) do not cause D(ln OPE) is rejected at 10%

significance levels. This result shows that D(ln GDP), D(ln REER), D(lnM2) and D(ln OPE) are the causes of D(ln OPE) which means that D(ln OPE) is the dependent variable.

Finally, as far as D(lnM2) is concerned as a dependent variable, D(ln GDP), D(ln OPE) and D(ln REER) do not cause D(ln REER). However, the combined test where both D(ln GDP), D(ln REER), D(ln EXD) and D(ln OPE) do not cause D(lnM2) is rejected at 10% significance levels. This result shows that D(ln GDP), D(ln REER), D (ln EXD) and D(ln OPE) are the causes of D(ln M2) which means that D(ln M2) is the dependent variable.

Research conducted to measure the short-term impact of variables through impulse response Function and variance analysis. The results of the impulse analysis show that in the short run the variables respond to their own shocks primarily.

It can clearly be seen that the real exchange rate, external debt, openness of the economy and money supply are not affected by economic shocks. However, economic growth affects itself over two quarter after the shock. A real exchange rate shock affects economic growth in the first quarter and affect itself within the two-quarter period after the shock while external debt shocks affecting economic growth during three quarter and the real exchange rate within four quarter. In addition, external debt also reacts to itself in two quarter after the shock. Openness of economy shocks affects economic growth over the period of four quarter and external debt over the period of two quarter. Furthermore, Openness of economy also affects itself in within four quarter. Finally, money supply impacts on economic growth during three quarter and the exchange rate over two quarter.

To recognise the role of variables in shocks, the study performed the variance decomposition of variables in the model. D(GDP), D(lnREER), D(lnOPE) is mainly explained by itself in the short run. However, external debt has the greatest impact of openness and real exchange rates in the short term. Similarly, the M2 money supply is heavily influenced by external debt.

At this stage, some test will be run to fulfill the requirements of the VECM model. First of all, the inverse roots of AR test is conducted to ensure the stability of the model. The test results show that all inverse solutions are within unit circle. This shows that the VECM model ensures stability and sustainability.

Continuing research on the following assessments aims to ensure VECM requirements for residuals. Carrying out the Portmanteau test and the Lagrange factor (LM) to examine the sequence correlation of the residue. The results show that P values are greater than 10% so there is no basis to conclude that the residuals have autocorrelation. The standard normal distribution of the remainder indicates acceptance of the null hypothesis (there is no statistical basis for rejecting the residual hypothesis of the normal distribution model) with a significance level of 5%.

Empirical analysis shows that external debt has a positive impact on economic growth in the long run. For external debt, 1% increase in external debt make GDP increase by 1.16%. At the same time, the openness of the economy also positively impacts on economic growth with the coefficient 6.3 which means that when the openness of economy goes up by 1%, the growth of economy increases by 6.3% However, empirical analysis shows that the increase in M2 money supply has negative impact on economic growth. In particular, 1% increase in money supply reduce GDP by 0.39%. The findings of the empirical study respond to the hypothesis of the impact of external debt on Vietnam's economic growth during the researched period. In the short term, economic growth is mainly driven by exchange rate and economic openness, but the impact is not high. The empirical study of external debt in Vietnam during the period from 2000 to 2013 showed that positive impacts on the economy over the long run are consistent with previous studies such as Frimpong and Oteng-Abayi (2006), Dauda et al. (2013), Korkmaz (2015). In both short term and long term, external debt and economic openness have a positive impact on Vietnam's economic growth. Based on this result, the study will provide policy suggestions for Vietnam in the process of integration into the world economy in general and the financial market in particular.

4.2 Analyze the Impact External Debt on Economic Growth by Nonlinear Model

This part base on the study by Osinubi et al. (2010) to examine the nonlinear relationship between external debt and Vietnam's economic growth. The econometric study model illustrating the above relationship is of the form:

$$GDP = \alpha_0 + \alpha_1 EXD + \alpha_2 (EXD\text{-}EXD^*)*\sigma + \alpha_3 OPE + \alpha_4 T\ u_t$$

Where, GDP is dependent variable representing the growth rate of gross domestic product of Vietnam in %/quarter. This variable is used in the studies Clements (2005), Adegbite et al. (2008). …

EXD is the ratio of external debt to GDP in % GDP accordingly quarterly data. This variable is commonly used in foreign studies to assess the debt situation as well as the repayment capacity of countries. Studies conducted by Fosu (1996), Were (2001), Pattillo (2002), Clements (2005), Adegbite et al. (2008), Ayadi (2008), Tokunbo et al. (2010) and Korkmaz 2015) used this variable to assess the impact on economic growth. The results showed that external debt has the same or opposite direction to economic growth. EXD * is the optimal level of debt that the economy is aiming for so that external debt has a positive impact on economic growth. σ is dummy variable, equals 1 if EXD > EXD* and equals 0 if EXD < EXD*. OPE is the openness of the economy, calculated by taking the import-export value against GDP, in unit of GDP per quarter. This indicator is used in the studies of Clements (2005), Osinubi et al. (2010), Daud et al. (2003) … T is the time trend and u_t is the remainder of the model.

To assess the impact of external debt and external debt threshold on economic growth according to the research model, the first problem is to determine the level of external debt. The study uses the nonlinear curves of the quadratic equation to simulate external debt thresholds with economic growth through SPSS 20.0 software. This value is estimated based on the distribution of GDP growth with external debt variable in the quadratic curve and the maximum point. Simulating the debt Laffer curve is described in Fig. 1. The peak of this curve is considered to be the optimal level of debt for economic development, showing that Vietnam's external debt threshold is 21.5% per quarter. Use this result to evaluate the research model.

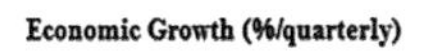

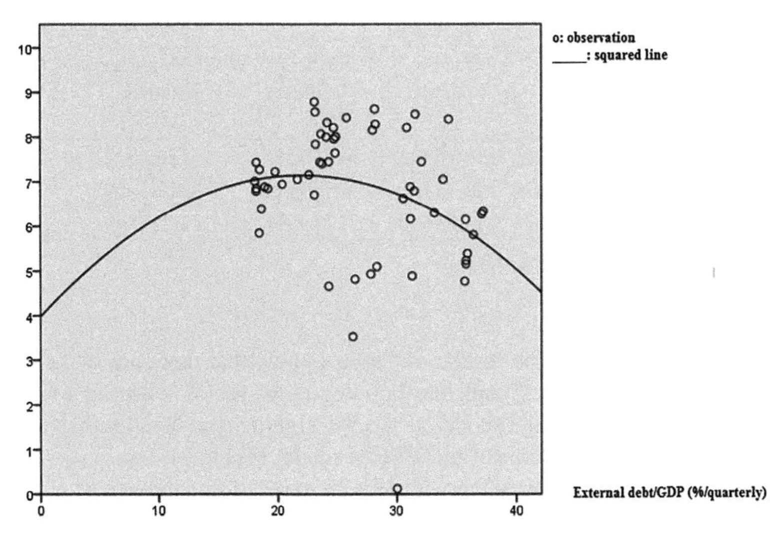

Fig. 1. Laffer curve of Việt Nam *Source: Author's calculations from SPSS 20 software*

Table 6. Unit root test of stationary for variables

Variable	t-Statistic	p-value	Conclutions
GDP_SA	−2.660527	0.0875*	Nonstationary
EXD_SA	−1.462006	0.545	Nonstationary
OPE_SA	1.161548	0.9975	Nonstationary
D(GDP_SA)	−10.28411	0.0000	Stationary
D(EXD_SA)	−10.52410	0.0000	Stationary
D(OPE_SA)	−4.857552	0.0002	Stationary

To implement the VECM model, unit testing should be done by means of the ADF (Augmented Dickey-Fuller) method on the cropped data. The results show that the data variables are non-stop and the level difference is 1%, (Table 6). This is the basis for the Johansen-Jesulius co-integration test.

With the optimal debt threshold found in Fig. 1, the study finds the optimal lag for the VECM model based on information standards using the Eviews 9.0 software. The optimal lag for the VECM model is 3.

Verification of the co-integration relationship was established to demonstrate the long-term relationship between the variables in the model through the Trace and Max-Eigen tests with the corresponding lag level found above. From there select the best VECM model. The results show that Eq. 4 (which is constant and tends to be in the co-ordinate equation). With Eq. 4, the delay is 3, the study continues to select the number of co-integration for the model by checking Trace and Max-Eigen for the selected

Table 7. Results of cointegration tests

		Trace Test		Max-Eigen Test	
Hypothesized No. of CE(s)	Eigen value	Trace Statistic	Critical Values @ 5%	Max-Eigen Statistic	Critical Values @ 5%
None*	0.912797	150.5823*	42.91525	126.8547	25.82321
At most 1*	0.350487	23.72750	25.87211	22.43966*	19.38704
At most 2	0.024462	1.287840	12.51798	1.287840	12.51798

** show significance at 5% respectively.*

co-integration equation. The results of Table 4 show that there are two co-integration vector under the Trace test and one co-integration vector with the Max-Eigen test between the variables in the model at the 5% significance level, reflecting the correlation relationship Long term of the VECM model (Table 7).

Conduct regression estimation with VECM to determine the impact and correlation between variables in the model corresponding to 2 cointegrations. The results of the VECM model estimation show that the variables have the same sign as expected. The regression coefficients of external debt are as expected and have a positive impact on economic growth (Appendix 2). The inclusion of dummy variables in the research model is statistically significant. The long-run equilibrium model is as follows:

$$\mathrm{GDP\_SA}_{(-1)} = 1.292369 * \mathrm{EXD\_SA}_{(-1)} - 0.068312 * \mathrm{T} - 26.17890$$

The study examined the impact of external debt on economic growth through the debt threshold. Sustainable fiscal policies require sound fiscal policy. The sustainability of fiscal policy is defined as maintaining a constant debt/GDP ratio. However, this is not the best policy. The best policy is to maintain the optimal debt ratio for the economy, boosting economic growth, corresponding to the peak of the debt Laffer curve. At the optimal level of debt, increasing external debt contributes to economic growth. Increasing debt is not a burden on the economy, contributing to fiscal sustainability. But if the economy has reached optimal levels of debt, debt stability becomes important.

Unstable debtors at optimal levels will face future debt burdens. Empirical research in Vietnam for the period 2000–2013 has shown a nonlinear relationship between external debt and economic growth. In addition, also confirmed the existence of debt Laffer curve. The results show that the long-run equilibrium relationship between GDP growth and the independent variables as well as the optimal debt range for Vietnam is 21.5%.

Compared to previous studies on Vietnam's debt threshold, this study also shows that the existence of nonlinear relationships of external debt to economic growth and debt range is 21.5% per annum in the period 2000–2013. The difference is in the new debt range compared to previous studies. In addition, it also shows the short and long term relationships of research variables.

5 Conclusion and Further Study

Study the impact of external debt on economic growth through linear and nonlinear models. The results of the study responded to the premise of the study that external debt has a positive impact on economic growth in the long run, as well as between external debt and economic openness. have a causal relationship. The empirical study of external debt and economic growth in Vietnam in the period 2000–2013 shows that there exists a debt laffer curve as well as a nonlinear relationship of external debt to economic growth. Calculating the debt threshold of 21.5% of GDP per quarter shows that if the debt ratio falls below the debt level, it will have a positive impact on economic growth and vice versa. However, the problem of capital use efficiency is still low and the management of external debt still has many issues that need to be changed in order to improve the efficiency of using external debt in Vietnam in the future.

One of the limitations of the study is the up-dating of quarterly external debt data. Currently, the World Bank and ADB do not update this data quarterly since 2014. In addition, much of Vietnam's macro data is updated yearly, with no quarterly analysis of data as direct investment. foreign currencies, foreign exchange reserves, foreign debts … Therefore, these variables have not been considered in the model impact on economic growth as well as the threshold of foreign debt of Vietnam. The sample size is small (53 observations) and not long enough for long-term analysis. In addition, the study does not mention the adverse effect of economic growth on the level of external debt.

The next step is to include more independent variables in the study in relation to external debt that affects Vietnam's economic growth. In addition, other models such as the MIDAS model can be considered to study this issue on the basis of mixed data.

References

ADB (2017). Vietnam: Macroeconomic and debt sustainability assessment 2017, truy cập tại <https://www.adb.org/sites/default/files/linked-documents/47293-001-sd-04.pdf>. truy cập ngày 10/10/2017

Adegbite, E.O., Ayadi, F.S., Felix Ayadi, O.: The impact of Nigeria's external debt on economic development. Int. J. Emerg. Markets **3**(3), 285–301 (2008)

Agénor, P.R., Montiel, P.J.: Development Macroeconomics. Princeton University Press (2015)

Ahokpossi, C., Allain, L., Bua, G.: A Constrained Choice? Impact of Concessionality Requirements on Borrowing Behavior (2014)

Chowdhury, A.: External debt and growth in developing countries: a sensitivity and causal analysis. WIDER-Discussion Papers (2001)

Clements, B., Bhattacharya, R., Nguyen, T.Q.: External Debt, Public Investment, and Growth in Low-Income Countries. IMF Working Paper No. 03/249 (2003)

Cohen, D.: Growth and external debt (No. 778). CEPR Discussion Papers (1993)

De Pinies, J.: Debt sustainability and overadjustment. World Dev. **17**(1), 29–43 (1989)

Deshpande, A.: The debt overhang and the disincentive to invest. J. Dev. Econ. **52**(1), 169–187 (1997)

Diallo, M.B.: Fiscal policy, external debt sustainability, and economic growth: Theory and empirical evidence for selected sub-Saharan African countries. Doctoral dissertation, New School University (2010)

Elbadawi, I., Ndulu, B., Ndungu, N.: Debt Overhang and Economic Growth in Africa. Iqbal and Kanbur, ed (1997)

Fosu, A.K.: The external debt burden and economic growth in the 1980s: evidence from sub-Saharan Africa. Canadian J. Dev. Stud./Revue canadienne d'études du développement **20**(2), 307–318 (1999)

Frimpong, J.M., Oteng-Abayie, E.F.: The impact of external debt on economic growth in Ghana: a cointegration analysis. J. Sci. Technol. (Ghana) **26**(3), 122–131 (2006)

Greene, J., Villanueva, D.: Private investment in developing countries: an empirical analysis. Staff Papers **38**(1), 33–58 (1991)

Hjertholm, P.: Debt relief and the rule of thumb: Analytical history of HIPC debt sustainability targets (No. 2001/68). WIDER Discussion Papers//World Institute for Development Economics (UNU-WIDER) (2001)

IMF: Theo External Debt Statistics: Guide for Compilers and Users 2003 (2003). <https://www.imf.org/external/pubs/ft/eds/Eng/Guide/index.htm>. Accessed 10 Oct 2017

Jeffries, I.: Vietnam: A Guide to Economic and Political Developments. Routledge (2007)

Kamin, S.B., Kahn, R.B., Levine, R.: External debt and developing country growth (No. 352) (1989)

Kaufmann, D., Kraay, A., Mastruzzi, M.: The worldwide governance indicators: methodology and analytical issues. Hague J. Rule Law **3**(2), 220–246 (2011)

Korkmaz, S.: The relationship between external debt and economic growth in Turkey. In: Proceedings of the Second European Academic Research Conference on Global Business, Economics, Finance and Banking (EAR15Swiss Conference) ISBN (pp. 978-1) (2015)

Krugman, P.: Financing vs. forgiving a debt overhang. J. Dev. Econ. **29**(3), 253–268 (1988)

Mensah, D., Aboagye, A.Q., Abor, J., Kyereboah-Coleman, A.: External debt among HIPCs in Africa: accounting and panel var analysis of some determinants. J. Econ. Stud. (2017). (just-accepted)

Mohamed, M.A.A.: The impact of external debts on economic growth: an empirical assessment of the Sudan: 1978–2001. Eastern Africa Soc. Sci. Res. Rev. **21**(2), 53–66 (2005)

Mohd Dauda, S.N., Ahmad, A.H., Azman-Saini, W.N.W.: Does external debt contribute to malaysia economic growth? Econ. Res.-Ekonomska Istraživanja **26**(2), 51–68 (2013)

Nakatani, P., Herrera, R.: The South has already repaid its external debt to the North: but the North denies its debt to the South. Monthly Rev. **59**(2), 31 (2007)

Nguyen, T.Q., Clements, M.B.J., Bhattacharya, M.R.: External debt, public investment, and growth in low-income countries (No. 3-249). International Monetary Fund (2003)

Pattillo, C.A., Poirson, H., Ricci, L.A.: What are the channels through which external debt affects growth? (2004)

Sachs, J.D.: Conditionality, debt relief, and the developing country debt crisis. In: Developing Country Debt and Economic Performance, The International Financial System, vol. 1, pp. 255–296. University of Chicago Press (1989)

Schclarek, A., Ramon-Ballester, F.: External Debt and Economic Growth in Latin America (2005). Paper not yet published. http://www.cbaeconomia.com/Debt-latin.pdf

Shah, M.H., Pervin, S.: External public debt and economic growth: empirical evidence from Bangladesh, 1974 to 2010 (2012)

Soludo, C.C.: Debt poverty and inequality in Okonjo Iweala, Soludo, and Muntar (eds.), The debt trap in Nigeria (2003)

Sulaiman, L.A., Azeez, B.A.: Effect of external debt on economic growth of Nigeria. J. Econ. Sustain. Dev. **3**(8), 71–79 (2012)

Swedish National Debt Office (2017). Organization, <https://www.riksgalden.se/en/aboutsndo/About-the-Debt-Office/Organisation/>. Accessed 9 Oct 2017

TFFS: Theo External Debt Statistics: Guide for Compilers and Users 2014 (2014). <http://www.tffs.org/pdf/edsg/ft2014.pdf>. Accessed 10 Oct 2017

Transparency International: Corruption Perceptions Index 2016 (2016). <https://www.transparency.org/news/feature/corruption_perceptions_index_2016?gclid=CjwKCAiA1O3RBRBHEiwAq5fD_CYFvCbqPjBVddwhwCeq6vAlmEL5dPGV24r3u1eJauOXCb2rX2uB7RoCnV0QAvD_BwE>. truy cập ngày 10/9/2017

Were, M.: The impact of external debt on economic growth in Kenya: An empirical assessment (No. 2001/116). WIDER Discussion Papers//World Institute for Development Economics (UNU-WIDER) (2001)

WB: CPIA debt policy rating (2013). <http://data.worldbank.org/indicator/IQ.CPA.DEBT.XQ?end=2015&locations=VN&start=2013>. Accessed 9 Oct 2016

WB: International Debt Statistics 2013 (2013). <https://openknowledge.worldbank.org/handle/10986/12226>. Accessed 9 Oct 2016

WB (2014). <http://siteresources.worldbank.org/INTDEBTDEPT/DataAndStatistics/20263218/HIPC-Fact-Sheet-web.pdf>. Accessed 10 Oct 2017

WB: International Debt Statistics 2014 (2014). Accessed <https://openknowledge.worldbank.org/bitstream/handle/10986/17048/9781464800511.pdf>. Accessed 9 Oct 2016

WB: International Debt Statistics 2017 (2017). <https://openknowledge.worldbank.org/bitstream/handle/10986/25697/9781464809941.pdf>. Accessed 9 Oct 2017

WB: Gross fixed capital formation (% GDP) (2017). <https://data.worldbank.org/indicator/NE.GDI.FTOT.ZS>. Accessed 9 Oct 2017

WB: International Debt Statistics 2018 (2018). <http://databank.worldbank.org/data/download/site-content/IDS-2018.pdf>. Accessed 22 Dec 2017

The Effects of Macroeconomic Policies on Equity Market Liquidity: Empirical Evidence in Vietnam

Dang Thi Quynh Anh[1,2(✉)] and Le Van Hai[1,2]

[1] Faculty of Finance, Banking University of Ho Chi Minh City, Ho Chi Minh City, Vietnam
{anhdtq,hailv}@buh.edu.vn
[2] Office of Educational Testing and Quality Assurance, Banking University of Ho Chi Minh City, Ho Chi Minh City, Vietnam

Abstract. This research assesses the impact of macroeconomic policies on the equity market liquidity in Vietnam, an attractive market in Southeast Asia. By using four different characteristics to measure equity market liquidity, this study employs a vector autoregressive (VAR) model to evaluate the influences of fiscal and monetary policies on the Vietnamese equity market in the period from January 2002 to December 2016. The findings show that both fiscal and monetary policies have relationships with the equity market liquidity. Based on the results, we recommend that investors and policymakers should make every effort to understand the simultaneous effects of both fiscal and monetary policies on the equity market liquidity rather than considering the effects of those policies separately.

Keywords: Monetary policy · Fiscal policy · Equity market liquidity Vietnam

1 Introduction

Liquidity is said to be the lifeblood of equity markets. It has prominent implications for traders, regulators, equity exchanges and the listed firms (Kumar and Misra) [20]. Liquidity also affects important decisions in corporate governance such as dividend policy, stock split, capital structure, company valuation... In addition, liquidity is used to assess the effectiveness and dynamics of the equity market.

In this study, our objective is to investigate the influence of monetary and fiscal policies on the liquidity of Vietnamese equity market. We examine whether the monetary policies of the central bank and the fiscal policies of the government are common determinants of equity liquidity. For example, when the central bank pursues an expansionary monetary policy, an increase in money supply could cause an increase in cash inflows to the equity market (Choi and Cook; Chordia et al.) [5,7]. Moreover, due to any systematic risk or information shock (e.g.

V. Kreinovich et al. (Eds.): ECONVN 2019, SCI 809, pp. 968–981, 2019.
https://doi.org/10.1007/978-3-030-04200-4_71

macroeconomic policy uncertainty) investors might change their asset holdings between equities and other financial securities.

Therefore, we observe the impact of standard monetary and fiscal policies on aggregate equity market liquidity in Vietnam. And then, we study the dynamic relationship between monetary and fiscal policies with equity market liquidity. As Choi and Cook [5] assert, the unpredictability of market liquidity is an important source of risk for investors, but it is clear that the risk is higher for emerging markets where investors generally have less opportunity to diversify their portfolios and face greater asymmetry of information. Consequently, identifying the macroeconomic determinants of the liquidity of these markets will help both local and international investors.

2 Literature Review

2.1 Monetary Policy and Equity Market Liquidity

There are many critical channels of monetary policy transmission. One of the main channels through which monetary policies affect the economy is the interest rate channel. This channel suggests that a change in interest rates will have an impact on the corporate cost of capital, which will eventually influence the present value of firms' future net cash flows. Consequently, higher interest rates lead to lower present values of future net cash flows, which, in turn, lead to lower stock prices (Mishkin) [25].

According to Fleming and Remolona [16] the expansion of monetary policy affects the liquidity of the stock market by reducing transaction costs and capital costs. For example, through open market operations, the purchase of securities by the central bank will increase the reserves of commercial banks and increase the money supply to the economy. Commercial banks can extend financing for margin trading on the equity market. Therefore, expansion of monetary policy will have a positive effect on the equity market liquidity.

Brunnermeier and Pedersen [4] have developed a model to study the relationship between market liquidity and funding liquidity. The central banks can help mitigate market liquidity problems by controlling funding liquidity. Central banks can also improve market liquidity by boosting speculator funding conditions during a liquidity crisis, or by simply stating the intention to provide extra funding during times of crisis, which would loosen margin requirements as financiers' worst-case scenarios improve.

2.2 Fiscal Policy and Equity Market Liquidity

The theories that address the link between the equity market with the fiscal policy can be subdivided into two opposite perspectives: The Keynesian positive effect hypothesis and the classical crowding out effect hypothesis. The Keynesian viewpoint centers on the use of automatic stabilizer and discretionary measures by the fiscal authority in the ways that support aggregate demand, boost the economy up and of course increase stock prices. The hypothesis believes that the effect of fiscal policy instrument on equity market is positive as fiscal policymakers can use budget deficit, tax and other discretionary measures to alter the interest rate thereby improving equity market performance. The classical crowding-out effect hypothesis: This Hypothesis centers on the negative impact of fiscal policy instruments on their sector and by extension the equity market. It explains that fiscal instruments have the potential to crowd out the loanable fund in the market and deter private sector activity, thereby having a negative impact on equity market liquidity.

2.3 Liquidity and Illiquidity Measures

Liquidity is an elusive concept and is not observed directly. In addition, equity market liquidity has multidimensional aspects that cannot be captured in a single measure (Amihud) [1]. Following the procedures suggested by Baker [2], Amihud [1], Goyenko and Ukhov [18] this study uses four different measures to capture the aspects of trading activity and price impact. The first proxy of liquidity for an asset that we use in this study is the turnover rate (TR), as suggested by Datar et al. [10]. The turnover rate of a stock is the number of traded shares divided by the number of shares outstanding in the stock. This is an intuitive metric of the liquidity of the stock.

$$\mathrm{TR_{iym}} = \frac{\sum_{\mathrm{d-1}}^{\mathrm{D_{iym}}} \mathrm{VO_{iymd}}}{\mathrm{NSO_{iym}}} \tag{1}$$

Where $\mathrm{TR_{iym}}$ is the turnover rate of stock i in month m of year y; $\sum_{\mathrm{d-1}}^{\mathrm{D_{iym}}} \mathrm{VO_{iymd}}$ is the monthly sum of the daily number of shares outstanding.

The second variable we use as a proxy for liquidity is traded volume (TV). Brennan et al. (1998) documents that a higher trade volume implies an increase in liquidity. The traded volume is calculated by using formula (2) below.

$$\mathrm{TV_{iym}} = \ln\left[\sum_{\mathrm{d-1}}^{\mathrm{D_{iym}}} \mathrm{VO_{iymd}}.\mathrm{P_{iymd}}\right] \tag{2}$$

In formula (2), $\mathrm{TV_{iym}}$ is the traded volume of stock i in month m of year y; $\mathrm{VO_{iymd}}$ is the number of daily traded shares and $\mathrm{P_{iymd}}$ is the daily price of each share. Therefore, the traded volume is calculated by taking the natural logarithm of the monthly sum of the daily product of the number of shares traded and their respective market price. Both TR and TV are based on trading activity and we can interpret them as liquidity proxies, as higher values are associated with more liquid assets.

Our third measure is Hui-Huebel Liquidity ratio by Lybek and Sarr [24], which relates the volumes of trades to their impact on prices and also to resiliency.

$$\mathrm{LR_t} = \frac{\sum_{\mathrm{d=1}}^{\mathrm{T}} \mathrm{P_{id}V_{id}}}{\sum_{\mathrm{d=1}}^{\mathrm{T}} |\mathrm{PC_{id}}|} \tag{3}$$

Where V_{id} is the traded volume of stock i in day d, P_{id} is the daily price of each share, PC_{id} is the difference between the price of stock i in day d and day d−1.

The fourth measure is of illiquidity by Amihud [1], which quantifies the response of returns to one dollar of trading volume. This illiquidity measure is very well established, particularly since studies such as Hasbrouck and Schwartz (1988), Goyenko and Ukhov [18], Lu-Andrews and Glascock [23].

$$ILLIQ_{iyd} = \frac{|R_{iyd}|}{TV_{iyd}} \quad (4)$$

where $ILLIQ_{iyd}$ is the illiquidity ratio of security i on day d of year y; R_{iyd} is the return on stock i on day d of year y and TV_{iyd} is the respective daily volume.

2.4 Macroeconomic Variables

The prime objective of this study is to investigate the effect of monetary policy (MP) and fiscal policy (FP) on the liquidity of Vietnamese equity market. To achieve this objective, we select several monetary and fiscal policy variables in line with previous studies, e.g. Chordia et al. [7], Goyenko and Ukhov [18].

We employ the aggregate money supply (M2) and the Inter-bank rate to be proxies for the monetary policy. For the aggregate money supply, we use the broad money supply (M2). Inter-bank rate (IBR) is relatively straightforward and is the rate at which the central bank offers credit to other financial institutions and thus acted as a control mechanism for market money supply.

In regard to fiscal policies, we consider two variables to capture the government's intervention in equity market liquidity. These variables are government borrowing from commercial banks (GB) and Treasury bill interest rate. Fisher [15] states that borrowing from commercial banks by the government can create a 'crowding out' effect and thus create competition for private savings where business firms may suffer from lack of credit opportunities.

The interrelationship between various macroeconomic variables and equity market liquidity is theoretically developed in Eisfeldt [13] and also empirically studied and documented in Söderberg [27], Næs et al. [26] and Fernández-Amadoret al. [14]. Based on these studies, we include the monthly growth rate of industrial production (IP) and monthly inflation rate (CPI) to capture inflation development.

3 Empirical Model

3.1 Vector Auto-Regression Analysis

To understand the relationship between equity market liquidity and macroeconomic variables, we utilize the vector auto-regression procedure employed in Chordia et al. [7] and Goyenko and Ukhov [18]. The vector auto-regression analysis can be mathematically expressed as follows:

$$X_t = c + \sum_{j=1}^{k} B_{ij} X_{tj} + u_t \quad (5)$$

where X_t is a vector that represents endogenous variables - liquidity, returns, industrial production, inflation, monetary and fiscal policy instruments; c is the vector of intercept, B is a 6 × 6 coefficients matrix ($i = 1$ to 6 monetary and fiscal policy variables, each with lag j), and u_t labels the vector of residuals. The lag order is estimated based on the Akaike information criterion and the Schwarz information criterion. If there is a difference in the lag orders of these two criteria, we use the shorter one for our model (see Chordia et al. [7]). The Augmented Dickey and Fuller test is used to check the non-stationarity of the variables.

3.2 Describe the Variables and Data Sources

We consider the following set of variables

$$X_t = [LIQ_t, GB_t, GIR_t, IP_t, INF_t, M2_t, IR_t, SR_t],$$

where LIQ_t is the dependent variable and represents the four (il)liquidity ratios (include TR, TV, LR and AILLIQ); GB_t the government bond, GIR_t the Treasury bill interest rate; IP_t the industrial production growth rate; INF_t the inflation rate; M2 the broad money growth rate, IR the inter-bank interest rate, SR monthly stock return.

The calculating method and data sources are summarized as follows (Table 1):

Table 1. Variables and calculating method

Variables	Symbol	Calculating method	Sources
Stock market liquidity (illiquidity)	TR	Formula (1)	HOSE, StoxPlus, Cafef
	TV	Formula (2)	
	LR	Formula (3)	
	Ailliq	Formula (4)	
Stock returns	SR	Monthly stock returns	
Inter - Bank Interest Rate	IR	Average of daily Inter - Bank Interest Rate in month t	SBV
Broad money	M2	M2 growth rate (monthly)	IFS
Industrial production	IP	Industrial production growth rate (monthly)	IFS
Inflation rate	INF	Consumer prices index (monthly)	IFS
Government interest rate	GIR	Treasury bill interest rate	IFS
Government Bond	GB	The growth of Government Bonds Volume	IFS

Table 2. The summarization of statistic variables

Variables	Mean	Median	Maximum	Minimum	Std. Dev.	Skewness	Observations
GB	3.425	2.9217	41.017	-13.104	5.815	1.857	180
GIR	10.927	10.335	20.250	6.960	3.000	1.061	180
M2	25.116	23.248	50.501	10.393	8.368	0.752	180
IR	6.729	6.511	18.651	0.541	3.512	1.080	180
INF	8.016	6.876	28.320	0.002	6.204	1.480	180
IP	12.021	10.104	27.718	(10.140)	7.716	1.035	180
SR	1.038	(0.108)	41.549	(26.105)	9.737	0.850	180
TR	3.209	2.736	13.366	0.106	2.030	1.738	180
TV	18,958	14,784	82,797	22	18,921	0.848	180
LR	196.382	128.554	908.946	1.063	223.057	1.137	180
AILLIQ	2.418	0.984	16.395	0.114	3.350	2.185	180

Our sample consists of the financial and macroeconomic data of the Vietnamese economy from January 2002 and ending in December 2016, which is 180 months. To compute the returns, liquidity and illiquidity measures stocks are included or excluded based on the criteria stated in (Chordia et al.) [7] and (Fernández-Amador et al.) [14].

The descriptive statistics of the data used in this study are presented in Table 2 above, the results from the Jarque-Bera, Skewness, and Kurtosis test show the normality of the data analyzed. The results from standard deviation with lower results also indicate that the data series is consistency over time.

4 Empirical Results

4.1 Unit Root Test

Theoretically, in order to avoid spurious regression, it is expected that time series data should be stationary for results validity to hold. In conducting a unit root test, traditional unit root tests techniques like Augmented Dickey-Fuller test (ADF) and the PP have been extensively used. If the time series data are not stationary, these will get the first difference to alter stationary.

The Table 3 shows that most time series have stopped at the level, except for money supply, interbank rates, government interest rate and industrial production. Thus, to ensure that the data series have a unit root, the first difference for these sequences is performed. The results show that these have a unit root at a first difference level with significance 1%.

4.2 The Impact of Monetary Policy Shocks

We estimate a total of 4 different VAR models for each of the four (il)liquidity measures and two monetary policy variables considered in our analysis. To the

Table 3. Unit root test of time series data

Variable	Level		1st difference		Result
	ADF	PP	ADF	PP	
LR	−14,0837*	−23,9088*			I(0)
TV	−10,2733*	−11,7968*			I(0)
TR	−6,1684*	−4,9772*			I(0)
Ailliq	−13,7046*	−14,0208*			I(0)
SR	−9,8265*	−9,8256*			I(0)
IR	−2,0751	−2,5923*	−11,7516*	−11,9321*	I(1)
M2	−1,2655	−2,6524***	−7,6133*	−10,7506*	I(1)
GB	−9,9223*	−9.7714*			I(0)
GIR	−2,8482***	−2,1067	−7,0323*	−8,9808*	I(1)
IP	−1,4527	−9,7858*	−5,0405*	−59,4559*	I(1)
INF	−2,6178***	−2,6421***			I(0)

Note: *,**,*** represent 1%, 5%; 10% respectively

understanding of the relation between (il)liquidity and monetary policy within the VAR system, we report the impulse response functions (IRFs) and variance decomposition as suggested in earlier studies, e.g. Chordia et al. [7], Goyenko and Ukhov [18], and Gagnon and Gimet [17].

The IRF traces the impact of a one-time, unit standard deviation, the positive shock to one variable on the current and future values of the endogenous variables. Results from the IRFs and variance decompositions are generally sensitive to the specific ordering of the endogenous variables. Therefore, in choosing an ordering, we rely on the prior evidence of Chordia et al. [7], Goyenko and Ukhov [18] and Fernádez-Amador et al. [14]. The order of our variables as follows: IP, INF, M2, IR and (il)liquidity. The liquidity and illiquidity at the end of the VAR ordering in our estimates to gain stronger statistical power (Goyenko and Ukhov) [18].

The accumulated responses of market (il)liquidity to one standard deviation of monetary policy shocks are shown in Fig. 1, traced forward over a period of 12 months. Here, responses are measured using the standard Cholesky decomposition of the VAR residuals. Most of the signs are in line with our hypothesis and significant. Vietnamese equity market liquidity increases (decreases) with easing (tightening) broad money supply.

Different from the money supply growth rate, the impulse response signs for the inter-bank rate are found not as expected. Higher (lower) inter-bank interest rate represents conservative (expansionary) monetary policy, thus generating no influences on liquidity market. The result indicates that market liquidity is more sensitive to the broad money supply growth than the inter-bank interest rate. Therefore, based on the overall impulse responses, we can conclude that equity market liquidity (illiquidity) tends to rise (decline) as the broad money

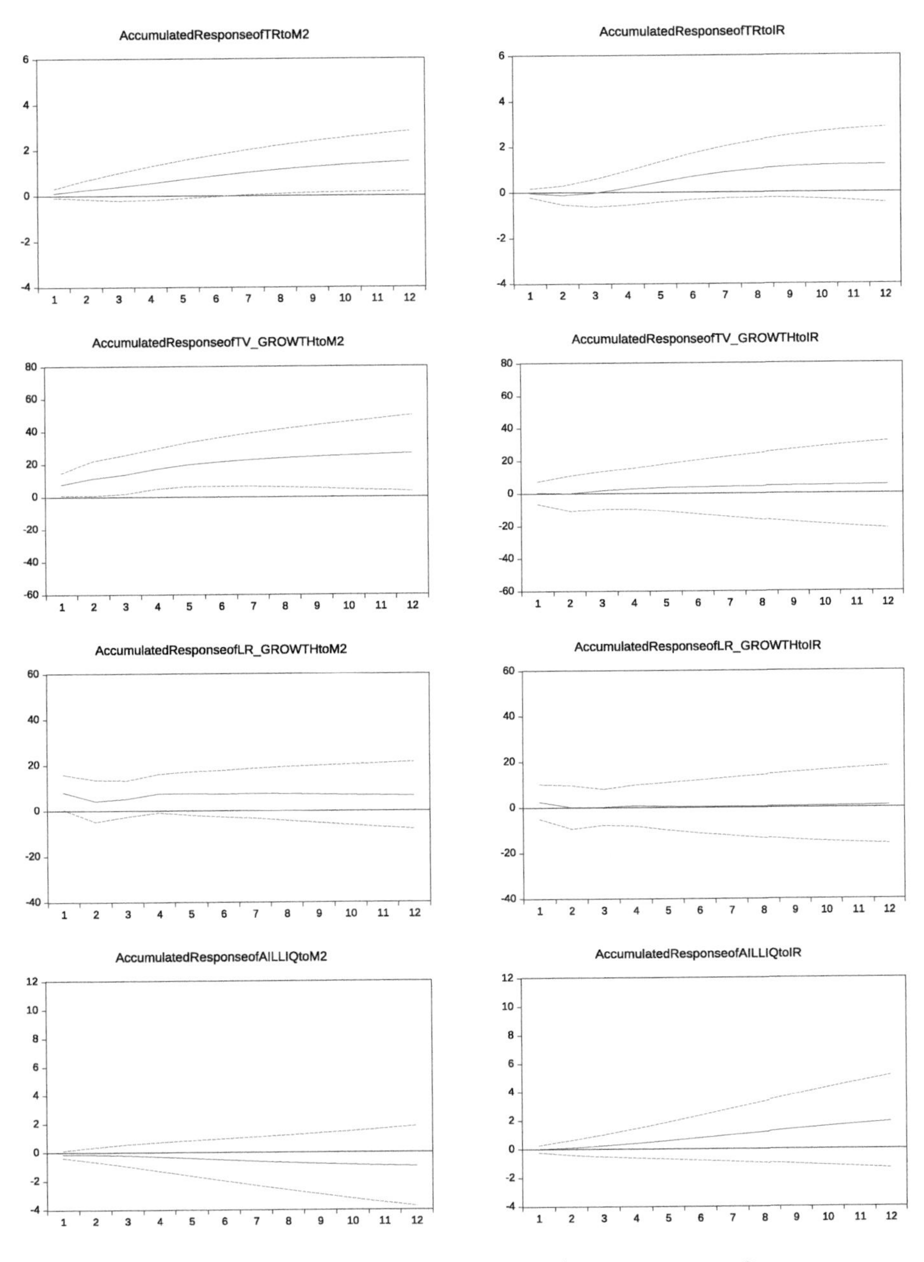

Fig. 1. Impulse response of Il(liquidity) to monetary policy

growth increases. That means expansionary (contractionary) monetary policy of central bank increases (decreases) the liquidity (illiquidity) in Vietnamese equity markets. The results are consistent with previous studies, such as Chordia et al. [7], Goyenko and Ukhov [18] and Fernández-Amador et al. [14].

Table 4. Variance Decomposition of Il(liquidity) market variables due to monetary policy shock

Variance decomposition of TV						
Period	IP	CPI	M2	IR	SR	TV
1	0.638	0.079	0.754	0.016	9.655	88.857
3	1.778	1.037	1.250	0.481	22.393	73.060
6	4.311	1.835	2.940	4.853	21.778	64.282
12	5.276	3.196	7.003	4.852	20.159	59.514
Variance decomposition of TR						
1	1.710	0.432	2.012	0.019	30.959	64.869
3	2.180	5.054	4.625	0.117	46.138	41.887
6	2.075	11.195	8.542	0.869	38.380	38.938
12	2.335	15.264	12.209	0.493	34.295	35.404
Variance decomposition of LR						
1	1.720	0.010	3.332	0.002	8.501	86.435
3	3.565	2.327	4.589	2.563	17.121	69.835
6	4.914	4.368	6.557	2.514	12.727	68.920
12	6.260	5.376	7.586	3.068	11.734	65.975
Variance decomposition of AILLIQ						
1	0.830	0.018	0.141	0.406	6.453	92.151
3	0.815	1.655	0.202	0.484	22.229	74.616
6	6.918	1.730	0.498	0.636	26.939	63.279
12	10.417	2.691	2.622	0.522	29.900	53.848

The variance decomposition of the liquidity measures to disentangle the information contributed by the monetary policy measures. The results (Table 4) indicate that the broad money growth and inter-bank interest rate respectively can explain more than a 12% variance of trading volume. Their effects also stabilize after 3 and 6 months respectively.

The inter-bank interest rate has a weak power to explain the variance of any of the four (il)liquidity measures. It can explain about 4.8% of the variance of TV and about 3.0% of the variance of LR. For TR and AIILIQ, IR cannot explain the volatility of these two variables, only under 0.5%. The Table 4 shows that the money supply shock is more likely to affect liquidity variables rather than inter-bank interest rate shock.

4.3 The Impact of Fiscal Policy Shocks

To evaluate the influence of fiscal policy variables on equity market (il)liquidity, we estimate a total of 4 different VAR models for each of the four (il)liquidity measures and two fiscal policy variables. Follow the recommendation of Gagnon and Gimet [17], we report the impulse response functions and variance decomposition to better understand the dynamics of fiscal policy within the VAR system. We order our variables as follows: macroeconomic variables - IP, INF and GB, GIR first, followed by SR and (il)liquidity ratios.

The signs of accumulated responses of market liquidity to a unit standard deviation innovation of fiscal policy shocks are presented in Fig. 2. In each group, the four variables represent the (il)liquidity responses to fiscal policy variables. The responses are estimated using a standard Cholesky decomposition of the VAR residuals and use the bootstrap 95% confidence bands to gauge the statistical significance of the responses.

The accumulated responses (Fig. 2) show that market liquidity increases following government borrowing shocks. The turnover ratio and the liquidity ratio give a positive response to government borrowing from the first month and reaches a new balance increase after 3 months. The response of Amihud [1] illiquidity is positive from the fourth month and reaches a new balance increase after 9 months.

Moreover, most of our liquidity variables are increased with the government interest rate shocks. In particular, trading volume and turnover rate increase over the 6-month period due to any changes in government interest rate. The effect of government interest rate on liquidity is weaker than that of government borrowing.

In addition, we estimate the variance decomposition of liquidity measures associated with fiscal policy variables. We can make several conclusions from these results. First, government borrowing and government interest rate explain up to 12% of the variation in liquidity.

The Table 5 shows that government borrowing and government interest rate can explain up to 6% of the variance in turnover rate and up to 10% in trading volume. But government borrowing and government interest rate only contribute around 3% of the information to Amihud [1] illiquidity ratio. However, equity market return explains most of the liquidity variances. SR contributes up to 40% of the volatility of TV and around 25% of TR. This is suitable for the equity market in Vietnam. As the stock return increases, investors will be encouraged to add more money to trading on the market to get more profit, thereby increasing the liquidity of the entire securities market.

The results of impulse responses and variance decomposition show that the liquidity measures of Vietnamese equity market react positively to a fiscal policy shock. This reaction of the liquidity to fiscal policy is not consistent with the crowding out hypothesis. The expansionary fiscal policy can increase firms and investors' access to credit and thus enhance market liquidity. Spilimbergo et al. [28], Blanchard et al. [3], Eggertsson and Krugman [11], Gagnon and Gimet [17] support this idea.

Moreover, other macroeconomic variables are also found to have statistically significant effects on equity market liquidity. The inflation shock has a negative effect on illiquidity market. That means when the consumer price index rise,

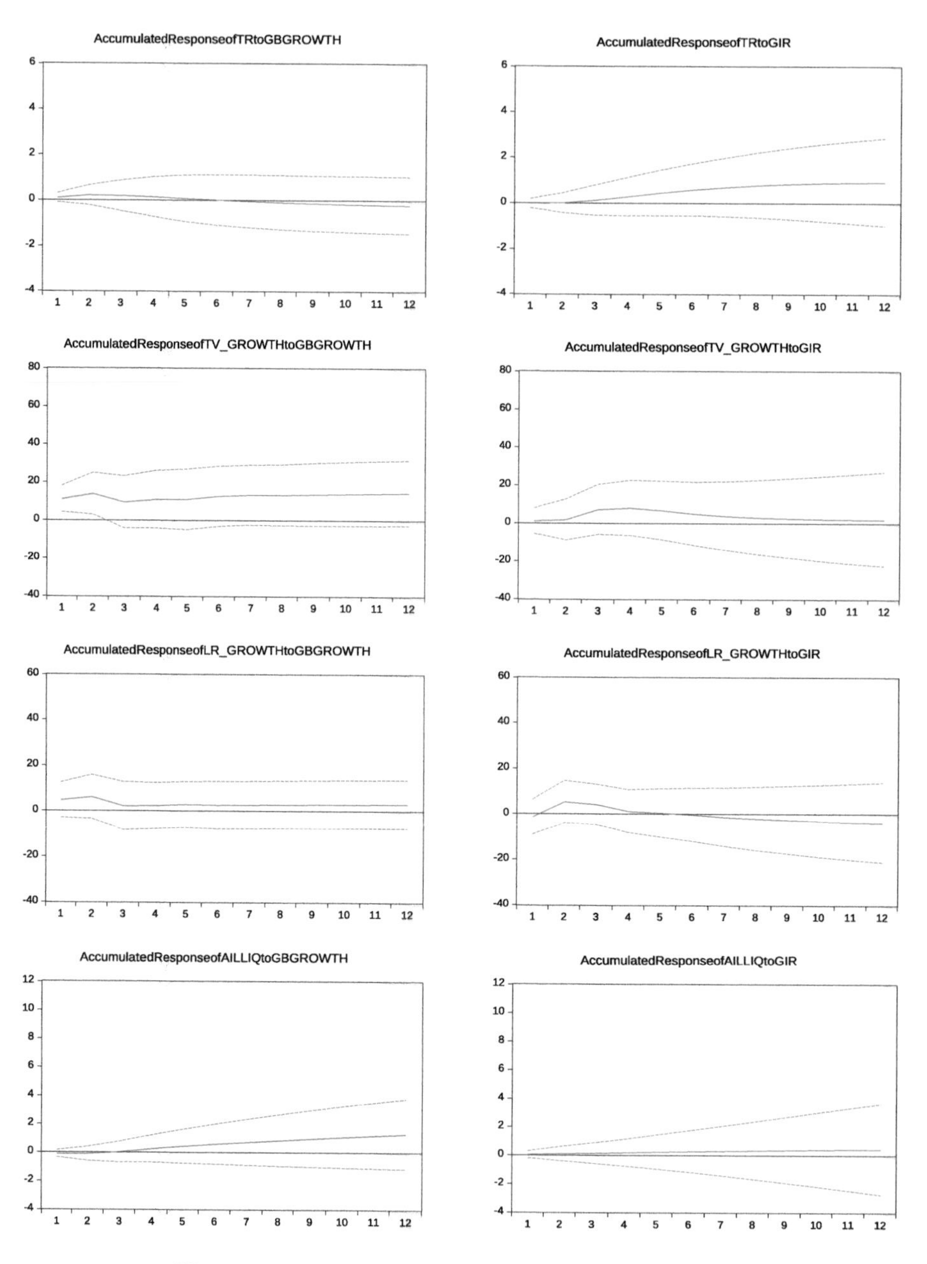

Fig. 2. Impulse response of Il(liquidity) to fiscal policy

the investor will shift to holding real assets, resulting in lower stock prices and transaction value.

The Table 6 shows that industrial production has significant positive impacts on the liquidity variables (especially TV and TR). We can explain that the stable macroeconomic environment creates opportunities for enterprises in production

Table 5. Variance Decomposition of Il(liquidity) market variables due to fiscal policy shock

Variance decomposition of TV						
Period	IP	INF	GB	GIR	SR	TV
1	2.214	0.001	2.413	0.192	18.680	76.500
3	5.806	6.515	1.916	0.670	41.842	43.250
6	8.401	6.283	2.163	0.918	40.527	41.707
12	8.779	6.337	2.193	1.017	40.229	41.446
Variance decomposition of TR						
1	0.535	0.000	0.312	0.029	0.028	99.095
3	1.906	0.333	0.502	0.536	20.722	76.001
6	2.592	0.310	1.109	4.882	24.285	66.821
12	2.972	0.415	1.155	5.901	24.128	65.429
Variance decomposition of LR						
1	0.299	0.001	0.079	0.393	0.005	99.223
3	0.766	1.690	0.492	0.859	15.114	81.079
6	1.064	2.767	1.329	0.939	15.001	78.900
12	1.076	2.803	1.377	1.304	15.196	78.243
Variance decomposition of AILLIQ						
1	0.056	1.722	0.017	0.154	0.242	97.810
3	0.034	2.366	1.802	0.169	15.959	79.671
6	5.577	1.916	2.624	0.282	23.070	66.530
12	8.836	1.505	2.116	0.307	27.669	59.567

Table 6. Summary of impulse response function signs to Industrial production and inflation shocks

	TV	TR	LR	AILLIQ
IP	+	+	ns	ns
INF	–	–	–	–

Note: + and – are positive and negative responses of four (il)liquidity measures to a unit standard deviation innovation in the monetary policy variables. 'ns' indicates no significant positive or negative response.

and business activities, raising revenue and accumulating profits. Since then, businesses have been actively reinvesting through the mobilization of capital from the society. This, in turn, makes the capital flow of social savings move constantly and create surplus value.

5 Conclusion

In this study, we use VAR models to investigate the effects of monetary and fiscal policy shocks on equity market liquidity in Vietnam. Using monthly data of the period from January 2002 to December 2016, we find evidence suggesting that both fiscal and monetary policies affect equity market liquidity.

Our major findings are as follows: first money supply growth, inter-bank interest rate. Government borrowing and government interest rate are receiving bidirectional causality from (il)liquidity measures. Second, the signs of impulse response functions are well in line with our hypothesis: expansionary monetary and fiscal policy increase overall market liquidity. Money supply can explain a large fraction of the error variance of the liquidity market. Third, different from the investigation of Chowdhury et al. [8] in Asia equity markets, our results of variance decomposition show no trail of the 'crowding out' effects. The government interest rate does not impact much on market liquidity. The reason is that in Vietnam the connection between government bonds market and the equity market is weak. So the increase in government interest rates does not affect corporate borrowing costs, as well as the cost of capital to investors.

Our findings are important for risk management officers and regulators. Regulators may use this study as evidence that (il)liquidity spirals are driven by monetary and fiscal policy variables. The impacts are not homogeneous and largely depend on the instruments used by the regulator. Therefore, they should be careful about applying their policy and consider the possible effects on the equity market while formulating those policies. This is because the ultimate impact of any policy changes on liquidity depends on the relative attractiveness of other asset markets: gold market, real estate market, forex market.

References

1. Amihud, Y.: Illiquidity and equity returns: cross-section and time-series effects. J. Finan. Markets **5**(1), 31–56 (2002)
2. Baker, H.K.: Trading location and liquidity: an analysis of US dealer and agency markets for common equitys (1996)
3. Blanchard, O., Giovanni, D.A., Mauro, P.: Rethinking macroeconomic policy. J. Money Credit Bank. **42**(9), 199–215 (2010)
4. Brunnermeier, M.K., Pedersen, L.H.: Market liquidity and funding liquidity. Rev. Financ. Stud. **22**(6), 2201–2238 (2009)
5. Choi, W.G., Cook, D.: Stock market liquidity and the macroeconomy: evidence from Japan. Natl. Bur. Econ. Res. **15**, 309–340 (2006)
6. Chordia, T., Roll, R., Subrahmanyam, A.: Market liquidity and trading activity. J. Financ. **56**(2), 501–530 (2001)
7. Chordia, T., Sarkar, A., Subrahmanyam, A.: An empirical analysis of stock and bond market liquidity. Rev. Financ. Stud. **18**(1), 85–129 (2005)
8. Chowdhury, A., Uddin, M., Anderson, K.: Liquidity and macroeconomic management in emerging markets. Emerging Markets Review (2017)
9. Darrat, A.F.: On fiscal policy and the stock market. J. Money Credit Bank. **20**(3), 353–363 (1988)

10. Datar, V.T., Naik, N.Y., Radcliffe, R.: Liquidity and stock returns: an alternative test. J. Financ. Mark. **1**, 2003–2219 (1998)
11. Eggertsson, G.B., Krugman, P.: Debt, deleveraging, and the liquidity trap: Fisher-Minsky-Koo approach. Q. J. Econ. **127**(3), 1469–1513 (2012)
12. Ehrmann, M., Fratzscher, M.: Taking stock: monetary policy transmission to equity markets. J. Money Credit Bank. **36**, 719–738 (2004)
13. Eisfeldt, A.L.: Endogenous liquidity in asset markets. J. Financ. **59**(1), 1–30 (2004)
14. Fernández-Amador, O., Gachter, M., Larch, M., Peter, G.: Does monetary policy determine stock market liquidity? New evidence from the euro zone. J. Empir. Financ. **21**, 54–68 (2013)
15. Fisher, D.: Monetary and Fiscal Policy, 1st edn. MaCmillan Press (1988)
16. Fleming, M.J., Remolona, E.M.: The term structure of announcement effects (2001)
17. Gagnon, M.-H., Gimet, C.: The impact of standard monetary and budgetary policies on liquidity and financial markets: international evidence from the credit freeze crisis. J. Bank. Financ. **37**(11), 4599–4614 (2013)
18. Goyenko, R.Y., Ukhov, A.D.: Stock and bond market liquidity: a long-run empirical analysis. J. Financ. Quant. Anal. **44**(1), 189–212 (2009)
19. Goyenko, R.Y., Holden, C.W., Trzcinka, C.A.: Do liquidity measures measure liquidity? J. Financ. Econ. 92(2), 153–181 (2009). Daily Data. J. Financ. 64 (3), 1445–1477
20. Kumar, G., Misra, A.K.: Closer view at the stock market liquidity: a literature review. Asian J. Financ. Account. **7**(2), 35–57 (2015)
21. Lavine, R., Zervos, S.: Stock markets, banks, and economic growth. Am. Econ. Rev. **88**, 537–558 (1998)
22. Lesmond, D.A.: Liquidity of emerging markets. J. Financ. Econ. **77**(2), 411–452 (2005)
23. Lu-andrews, R., Glascock, J.L.: Macroeconomic effects on stock liquidity (2010). SSRN 1662751
24. Lybek, M.T., Sarr, M.A.: Measuring liquidity in financial markets. International Monetary Fund (2002)
25. Mishkin, F.S.: The transmission mechanism and the role of asset prices in monetary policy. National Bureau of Economic Research (2001)
26. Næs, R., Skjeltorp, J.A., Ødegaard, B.A.: Stock market liquidity and the business cycle. J. Financ. **66**(1), 139–176 (2011)
27. Söderberg, J.: Do macroeconomic variables forecast changes in liquidity? An out-of-sample study on the order-driven stock market in Scandinavia. Working Paper. Vaexjoe University (2008)
28. Spilimbergo, A., Symanky, S., Blanchard, O., Cotelli, C.: Fiscal policy for the crisis. IMF Staff Position Note No. 2008/01 (2009)

Factors Affecting to Brand Equity: An Empirical Study in Vietnam Banking Sector

Van Thuy Nguyen(✉), Thi Xuan Binh Ngo, and Thi Kim Phung Nguyen

Banking University of HCM City, Ho Chi Minh City, Vietnam
{thuynv,binhntx,phungntk}@buh.edu.vn

Abstract. A strong brand is one of the most important assets of any company that want to be in a sustainable growing position within the context of international integration and tough competition today. Competitors of financial institutions in Vietnam are not only domestic but also foreign financial institutions. Unlike the manufacturing sector, the banking business is based on the trust of customers. A good bank brand is a prestigious brand, highly trusted by customers and influential decision to choose the use and maintenance of customers. This paper aims to provide empirical evidence of factors that impacted on bank brand equity in a specific context on Vietnam Banking Sector. The findings indicate strong support for bank brand equity coming form two factors among others: Brand Loyalty and Brand Association. The study is limited to 17 over 35 brands in Vietnam Banking Sector, and the survey was conducted with 378 banking customers in Ho Chi Minh Market only. However, the results provide deeply understanding of factors that impact on the brand equity, these may support banking managers to prepare properly strategy on brand equity investment for gaining customers' trust, generating purchase intentions, sales and financial values for the bank brand.

Keywords: Brand equity · Banking brand · Vietnam · Brand loyalty
Brand association · Brand satisfaction

1 Introduction

Traditionally, brand assessment was the field of marketing department. It is focus on the researching perception and behaviour of consumers toward to the brand. On other hand, the second approach of brand assessment is financial-based brand that rooted in traditional corporate finance theory and value a brand according to the common method as value a business and other commercial tangible assets.

Brand equity researchs are also conducted in two perspectives, consumer-based brand equity (CBBE) and financial based brand equity (FBBE). Simon and Sullivan [43] argues that the FBBE approach to value brand equity is based on the cash flow of branded products sales over those of non-branded products,

V. Kreinovich et al. (Eds.): ECONVN 2019, SCI 809, pp. 982–998, 2019.
https://doi.org/10.1007/978-3-030-04200-4_72

so the brand brings tangible financial value to the business. As the CBBE perspective, brand equity reflects the consumer's understandings, attitudes, emotions, and behavior toward a brand through their own experience to compare with those of competitors' brands (Aaker 1991 [1], 1996 [2]; Shocker et al. [42]; Mahajan et al. [26]; Ravald and Gronros [38]; Kotler [24]; Davis 2002 [13], Davis 2003 [14]; Keller 1992 [18], 1993 [19], 1998 [20], 2003 [21], 2008 [22]). Many empirical studies in the field of banking brand equity have been carried out to verify the CBBE approach, such as Pinar et al. [36], Severi and Ling [41], Rambocas and Kirpalani [37], Subramaniam et al. [46]. Keller 2008 [22] also insisted that an effective CBBE occurs "when consumers have a significant level of awareness and understanding about the brand. They also hold strong, salient and unique brand image in their mind".

This study attempts to verify the determinants that effect to the brand equity in financial services based on consumers' perception, specifically in Vietnam banking sector. We also analysis and evaluate the level impact of every determinants on the overall bank brand equity, the research results indicate several management implications that bank manager could consider to apply in their bank strategy to build brand equity. The remainder of this paper is decided into four sections. First, we will summarise the theoretical concepts underpinning our study, the research hypotheses are also identified in this section. Second, research methods will be discussed including the description of sample selection as well as statistical tools we use to validate our research model. Third, we will present the findings of this study. Finally, we will propose the managerial implications and discuss further research aims on bank brand equity.

2 Theoretical background

Branding is more than selling a product or service, such as a current account, a bank credit card or a life insurance policy. It is about finding out a target market and then designing a product and a brand personality that satisfied those needs of certain target markets. Branding is about discovering clear function needs among scattered and identifiable segments of customers, understanding the motivations and psychological needs of that segment and then integrating the these needs into a unique integrated selling offering. This is a challenge practice among financial services brands, where it is difficult to differentiate products practically. There are not to many ways financial services suppliers can provide savings, access to money, loans or insurance. Unlike other pure commodities, financial products can be easily duplicated by competitors, such as increasing the interest rate or adding same value added services to current account. They become like a commodity because of the very short period of time which our rivals can copy a successful business idea.

Theoretically, the approach of branding goods and services is similar. The ultimate goal is focused on building and leveraging the brand equity for a strong relationship between the brand and its customers. However branding service industry is different from branding in consumer manufacturing industry because it required high level of interaction between customers and staff or self-service technologies (Bitner et al. [8]). Through these touch points with services suppliers, customers earn experience and attitude against the service brand.

Brand Equity (BE)
As strong brands may reinforce market share, customer loyalty, generate sales and increase business profitability, they are valuable assets to a firm and therefore it is important factor in any business decisions. Aaker [1] defines brand equity as a set of assets and liabilities linked to a brand name and symbol that adds to or subtracts from the value provided by a product or service to a firm and that firm's customers. These assets can be grouped into five dimensions: brand awareness, brand associations, perceived quality, brand loyalty and other proprietary assets. Keller [21] also defines brand equity as differences in customer response to marketing activity. His customers brand equity model identifies 6 components including brand salience, brand performances, brand imagery, brand feelings, brand judgments and brand relationships. The concept of brand equity (Aaker [1]; Keller 2003 [21]) has both a financial and a marketing perspectives. From a financial point of view, brand equity must have a tangible financial value which is needed for a firm in case of merger, acquisition or investment purposes. Simom and Sullivan [43] as well as Biel [6] describe brand equity in terms of cash flow differences between where the brand name is added to a firm product and in other side where the same product does not have brand name. Estimating a financial value for the brand is necessary but it doesn't provide marketers tools to understand the process of building brand equity.

According to Yoo and Donthu [50], consumer-based brand equity approach is focusing on the steps of processing information and building confidence in the purchase decision by consumers. It also enhances efficiency and effectiveness of marketing mix decisions such as price, profits and brand extensions. Their study results indicated that the new brand equity scale is applicable, reliable, and relevant in different product categories in different cultures. These authors also pointed out that the four-dimensions brand equity model that comprise of brand awareness, brand association, perceived quality and brand loyalty are valid to identify brand equity.

Brand Awareness (BAW)
According to Netemeyer et al. [28], the extent to which consumers think about a brand when a product under that brand name mentioned is the ability to identify or recognize the brand by the consumer (Rossiter and Percy [39]). Aaker [1] defines brand awareness as the strength of a brand's presence in the consumer's mind. It refers to the ability that consumers recognize and recall a brand when thinking of a particular product. Keller [20] argues that brand awareness is developed from the familiar, frequency appearance of the brand in consumer mind through meeting their relevant needs and previous buying experience. He also suggests that brand awareness can influence customer buying decisions through strong brand associations.

Keller [22] affirms that brand equity increases when consumers have awareness and familiar with a particular brand, they have to hold a strong and positive association of that brand in mind. Aaker [2] points to the important role of brand awareness in brand equity valuation, as it measures the market share of customer minds or brand top-of-mind perceptions. He indicated that if the level of brand awareness is low then strong brand equity is insufficient built. This conclusion consistent with the view point of Vrontis and Papasolomou [48] that brand's strength is derived from high level of consumer awareness. Thus, brand awareness not only influences the consumer buying behavior but also affects the value of brand equity (Bird and Ehrenberg [7]). Consequently, brand awareness is a component of brand equity (Aaker 1991 [1], 1996 [2]; Keller [19]; Yoo et al. [51]; Yoo and Donthu [50]).

Based on the above discussion, our first research question is the level of brand awareness by customers of commercial banks affect the value of bank brand equity? Therefor, hypothesis H1 is proposed as follows:

Hypothesis H1: The level of brand awareness by customers of commercial banks has a positive impact on the brand equity bank brand equity.

Brand Association (BAS)

Aaker 1996 [2] emphasized that brand equity is supported in great part by the associations that consumers have with the brand, so brand association is any link in the consumer's mind of that brand. Brand association is considered as the consumer's perception of all forms, product attributes or only the particular product characteristics itself; is a picture of the brand that consumers have after recognizing that brand (Chen [11], Ramos and Franco [37]).

Keller 1998 [20] pointed out that brand association can be created through a combination of customer attitudes, product attributes and relevant benefits. In short, brand association is anything that consumers hold in their minds about a brand, these associations may relate to the results in terms of functionality, quality or symbolic meaning (O'Loughlin and Szmigin [32]). Brand association supports consumers in the process of collecting information, which is the basis for consumer choice because of having positive emotions and attitudes towards the brand (Aaker 1996 [2]). Lassar et al. [25] suggest that the brand association represents the relative brand's strength toward positive consumer perceptions of the brand, which enhance the brand equity.

From above mentioned points, brand association and brand equity are closely related. Our second research question is, in the context of banking industry in Vietnam, how is this relationship expressed? Hypothesis H2 is proposed:

Hypothesis H2: Brand association has a positive impact on the bank brand equity

Perceived Quality (PQ)

Perceived quality is subjective perception or evaluation by consumers about the overall quality and superiority of branded product when comparing with those of a competitor (Aaker 1991 [1], Zeithaml [52]).

Zeithaml [52], also indicated that perceived quality is a subjective assessment of product quality, it may or may not be the same as the actual quality of the product. Thus, the perceived quality is relatively, which may be different from one to another, given when consumers gaining consumption experience for products or services. This author asserts that perceived quality is a major factor influencing consumers' buying decision, When consumers have positive attitudes toward a brand, they tend to choose the brand due to the belief of the difference and outperform of that brand in comparing with competitor brands. In other words, perceived quality is the significant criteria for consumers to compare and choose between brands (Nguyen Van Thuy and Dang Ngoc Dai [29]). It can be accepted that perceived quality has positively correlation with brand equity (Motameni and Shahrokhi [27]; Yoo et al. [51]), which is an integral part of assessment the brand equity (Aaker 1991 [1]). Perceived quality is a determinant of brand equity which drive consumers choice among competing brands (Yoo et al. [51]). It is a resource for firms gaining competitive advantage.

Our third research question is, the consumer's perceived quality about a bank brand has any effect to the bank brand equity? Hypothesis H3 is proposed:

Hypothesis H3: The perceived quality has a positive impact on the bank brand equity.

Brand Satisfaction (SAT)

Customer satisfaction is a state happen when consumers satisfy their needs and desires (Oliver [33]; Olsen [34]), it is the result of perceived quality that consumers actual experience with a brand (Cronin et al. [12]). Therefore, the brand satisfaction reflects the overall consumer evaluation of a product after using it or from reference opinions before using it (Oliver [33]). Brand satisfaction represents consumer attitude and preferences toward a brand (Yasin et al. [49]).

Hong-Youl Ha [17] argues that brand satisfaction is a driven of brand loyalty, but it can also be considered as a important dimension to build brand equity. According to Bitner [8] and Oliver [33], customer satisfaction contribute to the economic efficiency, long-term financial performance and shareholder value (Hogan et al. [16]) increase market share and return on investment. Customer satisfaction is also associated with non-economic efficiency (Bloemer and Kasper [9]), enhance the brand loyalty via the relationship between brand with its consumer, in both terms behavior and attitude. Brand satisfaction influences current and future consumer behaviors, which may lead to customer buying intentions and repeated purchase behavior (Chang and Tu [10]). Beside that, customer satisfaction reinforces the firm's negotiation power with its stakeholders and facilitates stability growth in demand, increase brand investment and reduced overall costs. In short, it exist a positive relationship between customer satisfaction with the brand satisfaction and the brand equity (Pappu and Quester [35]). The fourth research question is how does consumer satisfaction toward a bank brand relate to the overall bank brand equity? Hypothesis H4 is stated as follows:

Hypothesis H4: Consumer satisfaction has a positive impact on the overall bank brand equity.

Brand Loyalty (BL)

Aaker1991 [1] defines brand loyalty as the customer's attachment to a brand; It is the customer's determination to use a brand when there is a need for a product or service (Sriram et al. [44]). According to Assael [4], there are two approaches to assess brand loyalty: (1) consumer behavior based approach (Oliver [33]) and (2) consumer attitudes approach (Oliver [33]; Yoo and Donthu [50]). Loyalty in terms of attitude expresses the consumer's emotion to the brand and the consumption trend, brand use intention permanently. Brand loyalty is a critical factor used to evaluate the firm's brand equity. Brand loyalty affects other aspects of consumer behavior. When consumers loyal to a brand, they often do not tend to compare their brands choice, they view their choice of a certain brand as no other substitution is better, they are not attracted by competing brands (Tong and Hawley [47]). This brings long-term profitability, building barriers to competition and creating firm competitive advantage .

According to Nguyen Van Thuy and Dang Ngoc Dai [29], the brand that build the higher customer loyalty the higher the profit and the higher the value of brand equity the firm earns. So, brand loyalty is the core value of brand equity (Aaker 1991 [1]) which is an important determinant that yields high brand equity value (Atilgan et al. [5]; Tong and Hawley [47]). Hypothesis H5 is proposed:

Hypothesis H5: Brand loyalty has a positive impact on the overall bank brand equity.

The linear regression equation and conceptual model are below:

$$\mathrm{BE} = \beta_0 + \beta_1{}^*\mathrm{BAW} + \beta_2{}^*\mathrm{BAS} + \beta_3{}^*\mathrm{PQ} + \beta_4{}^*\mathrm{SAT} + \beta_5{}^*\mathrm{BL} + \varepsilon \qquad (1)$$

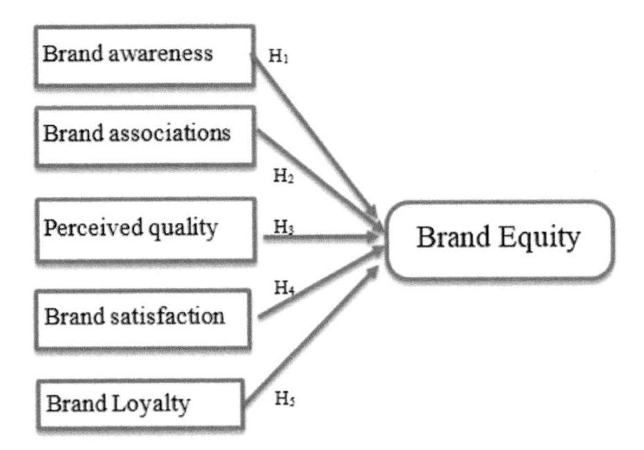

Fig. 1. Proposed theoretical research model

The research model illustraded in Fig. 1 and Eq. 1 summarizes the potential relationship between the proposal determinants of BE adoption.

3 Research Methodology

Scale Development

The study examined the relative effect of five variables (brand awareness, percieved quality, brand associations, brand satisfaction, brand loyalty) on banks brand equity. The proposed research model in Fig. 1 is specified as a structural equation model with 6 latent variables. Each of the latent variables in the model is operationalized by a set of indicators (measurement variables) which are adapted by the original scale of the previous authors then we adjusted to suit the context of the research in Vietnam banking sector. Table 1 shows the origin of the research scale measures.

Table 1. The origin of the research scale measures.

Variables/constructs	No. of items	Adapted form authors
Brand Awareness	5	Yoo et al. (2000); Hong-Youl Ha (2011)
Brand Associations	5	Aaker (1996), Buil et al. (2008), Keller (1993, 2008), Pappu et al. (2005), Hong-Youl Ha (2011)
Perceived quality	5	Buil et al., 2008, Pappu et al. (2006); Yoo et al., (2000); Tong and Hawley (2009) Kim and Kim (2005)
Brand Loyalty	5	Pinar, Girard and Eser, 2012; Yoo et al., 2000; Tong and Hawley, 2009; Kim and Kim 2005
Brand Satisfaction	5	Taylor et al. (2000); Rambocas, Kirpalani and Simms (2014)
Brand Equity	5	Lassar et al. (1995); Aaker (1996); Davis (2003)

Source: summarise for this study

Sample and Data Collection

This study was based on the development of a survey questionnaire that enabled the assessment of banks' consumer awareness, perceptions, attitudes and behaviour with respect to all aspects of bank brand equity which discussed in the literature review. The development of the survey followed a sequential three-stages process. Initially, having conducted a thorough literature review, secondly an exploratory research phase (which involved conducting ten in-depth interviews with banks' customers) was undertaken to solidify the conceptual framework, finally the survey conducted by quantitative method through direct interview with banks' customers. Data collected with banks' customers in Ho chi minh city from August to October 2016, in total 378 valid questionaire were obtained.

The sample statistics description show that the total banks in this study comprise 17 commercial banks such as Vietcombank, Vietinbank, BIDV, Argribank, Sacombank, Techcombank, MB, ACB, Eximbank, ... These commercial banks are considered having big business scale and long-term established brands. The details of sample charateristics is shown in Table 2.

Table 2. Sample characteristics

Indicators		Age					Sex		Sum	
		18–25	26–35	36–45	45–60	>60	Male	Female	N	%
Education	HighSchool	20	5	1	3	13	19	23	42	11.1
	University/College	99	76	40	56	11	120	162	282	74.6
	PostGraduate	15	14	13	9	3	26	28	54	14.3
Income/moth(VND)	<10 mil.	53	10	15	15	0	33	60	93	24.6
	10–20 mil.	68	42	27	39	19	77	118	195	51.6
	20–30 mil.	4	30	9	10	4	35	22	57	15.1
	>30 mil.	9	13	3	4	4	20	13	33	8.7
Sum	N	134	95	54	68	27	165	213	378	100.0
	%	35.4	25.1	14.3	18.0	7.1	43.7	56.3	100.0	

(Source: result of data analysis)

Data Analysis

Accoding to Hair et al. [15], to assess the initial reliability of the measures, Cronbach's alpha and the item-total correlation for all scales were applied. Cronbach's alpha for all the constructs was above 0.70 (Nunnally [31]). Furthermore, the item-to-total correlations were all above the threshold of 0.30 (Norusis [30]). Exploratory Factor Analysis is applicable to test to convergence validity of the scales, then Confirmatory Factor Analysis (CFA) is the next step used to test whether a prior theoretical model is the basis for a data set. CFA accepts the hypotheses of researchers, which are based on the relationship between each variable and one or more factors

Hypotheses were tested using structural equation modelling (SEM). The data was analysed using AMOS 20.0 software. Firstly, the psychometric properties of the scales were examined. Then, the structural model was evaluated by testing the research hypotheses that proposed above.

4 The Result and Findings

Reliability Confirmation

The Cronbach's alpha coefficient analysis for all the constructs was significant (from 0.806 to 0.908) which shown in Table 3.

We then performed exploratory factor analyses (EFA) using principal components analysis with Promax rotation. Results suggested that all factor loadings were above 0.5 and statistically significant which suggested the convergent validity of the scales (Steenkamp and Van Trijp [45]). The total variance explained exceeded 64.709% (>50%) of the variance of research sample with Eigenvalues

Table 3. Preliminary analysis - Cronbach's Alpha results

No.	Variables		Cronbach's coefficient alpha
1	Brand Awareness	BAW	0.868
2	Brand Associations	BAS	0.784
3	Perceived quality	PQ	0.876
4	Brand Loyalty	BL	0.864
5	Brand Satisfaction	SAT	0.813
6	Brand Equity	BE	0.821

greater than 1 (1.22). Barlett's test has significance level of Sig = 0. 000 and KMO = 0.884 (>0. 5).

In addition, the factor loading weights of all factors were satisfactory (> 0.5) and the convergence was highly validity. The results of the EFA analysis of brand equity show that the scale has a high convergence value. The loading factor is greater than 0.70 (from 0.759 to 0.853), Average Variance Extracted is 65.186% (>50%) and Eigenvalues is 2.607 (>1.00). Details results are presented in Table 4.

Table 4. Explorary factor analysis

	Independant variables					Dependant variables
	BAW	PQ	BL	BAS	SAT	BE
Factor loading weights	.842	.892	.870	.826	.826	.853
	.804	.863	.860	.747	.815	.831
	.800	.824	.812	.735	.798	.783
	.793	.736	.793	.701	.762	.759
	.768	.719	-	.527	-	-
Cronbach's Alpha	0.868	0.876	0.864	0.784	0.813	0.821

(Source: result of data analysis)

Confirmatory Factor Analysis

CFA used in the critical model to assess the distinction between research concepts cin the proposed model. The results indicate that the critical model consists of 305 degree freedom (df), χ^2=837.129 (p=0.000) and χ^2/df = 2.745 <3; The model fit indices were as follows: CFI = 0.913, TLI =0.900, IFI =0.914 (which are all greater than 0.90) and RMSEA = 0.068 < 0.07. Therefore, the research model is fit to this set of sample data. The values of λ_i is greater than 0.5 and statistically significant (all p values are equal to 0.000), so we can conclude that the observable variables used to measure the research concepts are in good convergent validity and in single dimensionality. CFA results for critical model is presented in Fig. 2.

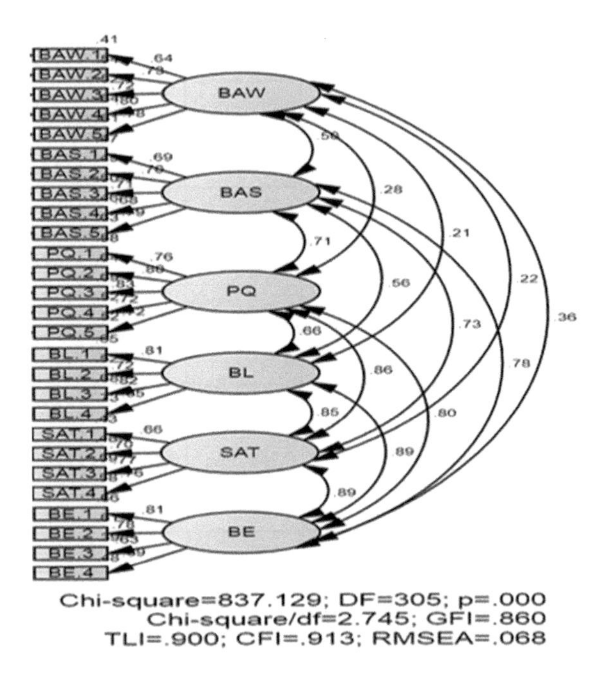

Fig. 2. The result of CFA (standardized Estimate) Saturated model

The results for validating discriminant between research concepts in the critical model are shown in Table 5. All correlation coefficients with the standard error between the factors are less than 1 and p value is 0.0000, so the research concepts are in good discriminant validity.

The model reliability test results in Table 6 provides a reliable coefficient of Cronbach's Alpha and overall reliability of all factors than greater 0.8 (from 0.812 to 0.879). As far as the average variance extracted of the concepts are all over 0.5. Therefore, the measure scales for all research concepts are qualified and reliability.

The CFA results assert the measurement model achieves a goodness-of-fit the data set, reliability, convergence validity, discriminant validity.

Results of the Model Estimation. A structural equations methodology using AMOS 20.0 was used to test the model (Arbuckle [3], Sanchez et al. [40]). The SEM results show that the model has df = 305 degrees of freedom, Chi-square = 837.129 (p = 0.000), $CMIN/df = 2.745 < 3$. The model fit ind-ices were as follows: GFI = 0.860, CFI = 0.913, TLI = 0.900 are both greater than 0.90 and $RMSEA = 0.068 < 0.07$, indicating that the conceptual model fits the set of sample data. Estimates show that all paths have positive relation and direct impact on the bank brand equity in Ho Chi Minh City. The results are shown in Fig. 3.

Table 5. The results of discriminant validity in the critical model

Correlation			Estimate	SE	CR	P-value
BAS	<->	PQ	0.710	0.022	13.344	0.000
BAW	<->	PQ	0.281	0.030	24.276	0.000
BAW	<->	BL	0.215	0.030	26.046	0.000
BAS	<->	SAT	0.734	0.021	12.691	0.000
PQ	<->	BE	0.795	0.019	10.951	0.000
BAW	<->	SAT	0.219	0.030	25.937	0.000
BAS	<->	BE	0.783	0.019	11.304	0.000
BAW	<->	BE	0.356	0.029	22.331	0.000
BL	<->	BE	0.885	0.014	8.004	0.000
BL	<->	SAT	0.854	0.016	9.093	0.000
SAT	<->	BE	0.895	0.014	7.628	0.000
PQ	<->	SAT	0.861	0.016	8.856	0.000
PQ	<->	BL	0.658	0.023	14.717	0.000
BAS	<->	BL	0.562	0.026	17.159	0.000
BAW	<->	BAS	0.504	0.027	18.609	0.000

(Source: Result of data analysis)

Table 7 shows that all hypotheses in the model are proven by SEM model testing. The estimation weights (unstandardized model) are positive (+) and statistically significant (p of SAT at 90% significance level) which demonstrate the concepts in the research models (brand awareness, brand association, perceived value, brand loyalty, brand satisfaction) have the positive impact on the bank brand equity. The conclude that the research concepts in the proposed model are supported theoretically.

In the regression analysis of the relationship between research concepts in the conceptual model, the greater value of the regression coefficients, the higher impact of the independent variable on the dependent variable. The SEM testing results also indicate that Heywood phenomenon does not occur during the estimation of the research model and the standard errors are smaller.

The standardized estimation results in Fig. 3 and Table 8 show that brand loyalty (BL) is the strongest contributor to the brand equity (standardized weighting is 0.567), followed by brand association (BAS) with weight = 0.309, the third is the perceived quality (PQ) with the weight = 0.165. Brand awareness (BAW) and brand satisfaction (SAT) have the same and least impact on the bank brand

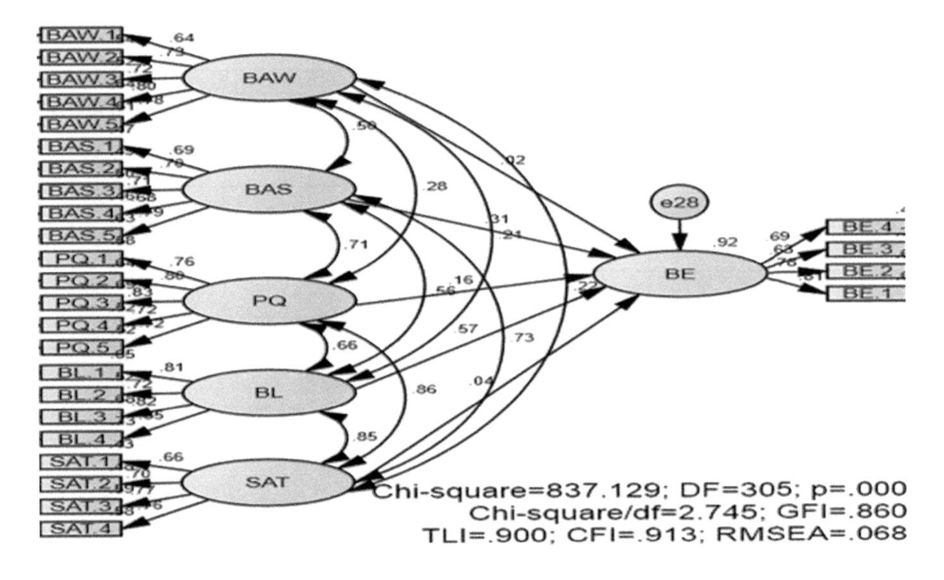

Fig. 3. Structural Equation Model

Table 6. Reliability of measure scales for all research concepts in the critical model

Constructs/Variables	Number of items	α	ρc	ρvc	Average λ
BAW	5	0.868	0.855	0.543	0.735
BAS	5	0.838	0.838	0.509	0.712
PQ	5	0.876	0.878	0.530	0.767
BL	4	0.864	0.879	0.646	0.802
SAT	4	0.813	0.812	0.520	0.720
BE	4	0.821	0.819	0.534	0.727

(Source: Result of data analysis)

equity. However, these impact level is not as strong as other factors. Therefore all the hypotheses H1, H2, H3, H4, H5 are accepted.

The linear regression of critical model:

$$BE = 0.025^{*}BAW + 0.309^{*}BAS + 0.165^{*}PQ + 0.037^{*}SAT + 0.567^{*}BL \quad (2)$$

Based on the results of the SEM analysis, the independent factors have a linear positive correlation with the bank brand equity. Here after are the hypotheses testing which is given in the research model.

- Hypothesis H1: The level of brand awareness has a positive impact on the value of commercial banks brand equity. The SEM results show that the standardised regression coefficient is positive ($\beta = 0.025$) and Sig. <0.005, indicating that the brand awareness is positively correlated with the brand equity value with a 95% confidence level.

Table 7. Unstandardized regression weights

Correlation			Estimate	S.E.	C.R.	P
BE	<-	BAW	0.020	0.038	0.517	0.005
BE	<-	BAS	0.258	0.067	3.844	***
BE	<-	PQ	0.140	0.087	1.606	0.008
BE	<-	BL	0.372	0.074	5.039	***
BE	<-	SAT	0.033	0.177	0.185	0.053

(Source: Result of data analysis)

Table 8. Results of the model estimation (Standardized)

Hypotheses	Correlation			Estimate	S.E.	C.R.	P
H1	BE	<-	BAW	0.025	0.038	0.517	0.005
H2	BE	<-	BAS	0.309	0.067	3.844	***
H3	BE	<-	PQ	0.165	0.087	1.606	0.008
H4	BE	<-	SAT	0.037	0.177	0.185	0.053
H5	BE	<-	BL	0.567	0.074	5.039	***

(Source: Study data analysis results)

- Hypothesis H2: Brand association has a positive impact on the of commercial banks brand equity. The SEM results indicate the positive relationship of brand association with the brand equity of 99% confidence level ($\beta = 0.309 > 0$ and Sig. <0.000).
- Hypothesis H3: Perceived quality has a positive impact on the commercial banksbrand equity. Based on the SEM analysis, the perceived quality factor had a standardized regression coefficient ($\beta = 0.165$) and statistically significant of 95% confidence (Sig. <0.008), indicating the positive impact on the perceived quality factor on the bank brand equity.
- Hypothesis H4: Brand satisfaction has a positive impact on the commercial banks brand equity. Results of the SEM analysis showed that the brand satisfaction factor had standardized regression coefficients $\beta = 0.037$ and Sig. <0.053, indicating a positive relationship between brand satisfaction and brand equity with 90% confidence.
- Hypothesis H5: Brand loyalty has a positive impact on the commercial banks brand equity . The SEM analysis showed that the brand loyalty factor had a standardized regression coefficient $\beta = 0.567$ and Sig. <0.000, indicating a positive relationship between brand loyalty and brand equity with a 99% confidence level.

5 Dicussion and Implications

The study on the bank brand equity conducted among customers of commercial banks in Ho chi minh city by quantitative method. The findings show that factors impact on the commercial banks brand equity including brand awareness, brand association, perceived quality, brand loyalty. These five factors can explains 91.6% of the change on bank brand equity with a 90% confidence level.

The impact extent of each factor on the bank brand equity is vary. Specifically, brand loyalty has the strongest impact on brand equity ($\beta = 0.567$), the second strongest impact on the brand equity is the brand associations ($\beta = 0.309$) followed by perceived quality ($\beta = 0.165$), and brand awareness ($\beta = 0.025$). In particular, the new finding of this study is about the brand satisfaction factor which is validity at 90% confidence even though the impact of this factor is not high ($\beta = 0.037$). These results lead banker to special attention to the brand loyalty factors, brand associations and perceived quality of banking service experience. As a result, banks need to invest in building strong brand equity in which the above critical factors need to be addressed:

- Enhancing customer loyalty to commercial banks. The increasingly competitive in financial market, maintaining customers loyalty will bring monetary value to banks business. The higher the brand loyalty by customer, the better profitability and the higher value adding to the bank brand equity. Therefore, solutions to increase customer loyalty should be seen as significant brand strategy of commercial banks. To enhance customer loyalty to the bank's brand, commercial bankers should focus on implementing several solutions such as: 1/ developing specialized products and services to create better competitive advantage; increasing level of interaction with bank's customers; 2/ building customer relationship management (CRM) system for the entire banking system, maintaining customer data and having appropriate policies for implementation of data marketing to meet customer personalised needs; 3/ The commercial bankers should try to increase the customers satisfaction through offering positive service experience to create values for their customers; 4/ Hiring and retaining high quality human resources are one of solutions that bankers can enhancing customer perceive quality toward every brand touch points; 5/ Together with the smart human resources policies, reinforcing the internal bank management system would help bankers improving consumer perceived quality.
- Amplifying the extent of consumer brand awareness. Many banks invest mainly on the level brand recognition which is externally perspective only. Bank need to re-think of the strategy to build positive brand associations as well to build brand internally through the corporate brand identity strategy. Diversified brand communication tools should be implemented strategically and consistency to both external and internal. Bankers can look into the brand extension strategies by either going to global market or joining international partnerships in local market. Setup subsidiary companies (real estate, securities trading, insurance or financial leasing....) allow bankers can package

their products and services into a whole that meet consumer needs efficacy. When a bank perform the brand extension strategy, it does not only mean to expand its operational market but also to prove the bank financial capability, the bank image and trust can be positively perceived by its customers.

- Building the positive brand associations toward the bank brand. Commercial banks now offer a wide range of financial products and services. However, these products are similarity in format or function benefits, that why consumers are unable to compare between competitive offers from different banks. This truth indicates that bank's marketing communication plan should be seriously considered to be more creative and effective for sending right messages in right places to right audiences to create strong brand associations and recalled communication effects through a variety means. Banks are in need to build a corporate culture within the bank environment, which emphasizes on transparency bank practices and customer-oriented attitudes. Bank's brand reputation is one of the first association that customers think of when referring to a bank. To enhance the prestige of the bank brand, bank should reassure that its brand promises and commitment to its customers is implemented daily by every single bank leaders and staff./.

References

1. Aaker, D.A.: Managing Brand Equity. The Free Press, New York (1991)
2. Aaker, D.A.: Building Strong Brands. The Free Press, New York (1996)
3. Arbuckle, J.L.: Amos^{TM} 17.0 User's Guide (2008)
4. Assael, H.: Consumer Behavior and Marketing Action. PWS-Kent Publishing, Boston (1992)
5. Atilgan, E., Aksoy, S., Akinci, S.: Determinants of the brand equity: a verification approach in the beverage industry in Turkey. Mark. Intell. Plan. **23**(3), 237–248 (2005)
6. Biel, A.: Discovering brand magic: the hardness of the softer side of branding. Int. J. Advert. **16**, 199–210 (1997)
7. Bird, M., Ehrenberg, A.S.C.: Non-awareness and non-usage. J. Advert. Res. **6**(4), 4–9 (1966)
8. Bitner, M.J.: Evaluating service encounters: the effects of physical surroundings and employee responses. J. Mark. **54**(2), 69–82 (1990)
9. Bloemer, J.M., Kasper, H.D.P.: The complex relationship between consumer satisfaction and brand loyalty. J. Econ. Psychol. **16**(2), 311–329 (1995)
10. Chang, C.H., Tu, C.Y.: Exploring store image, customer satisfaction and customer loyalty relationship: evidence from taiwanese hypermarket industry. J. Am. Acad. Bus. **7**, 197–202 (2005)
11. Chen, A.C.: Using free association to examine the relationship between the characteristics of brand associations and brand equity. J. Prod. Brand. Manag. **10**(6/7), 439–49 (2001)
12. Cronin, J.J., Brady, M.K., Hult, G.T.M.: Assessing the effects of quality, value, and customer satisfaction on consumer behavioral intentions in service environment. J. Retail. **76**(2), 193–218 (2000)
13. Davis, S.M.: Brand Asset Management: Driving Profitable Growth through Young Brands. Josey Bass, San Francisco (2002)

14. Davis, D.F.: The effect of brand equity in supply chain relationships, University of Tennessee dissertation., Knoxville, TN (2003)
15. Hair, J.J.F., Anderson, R.E., Tatham, R.L., Black, W.C.: Multivariate Data Analysis, 7th edn. Pearson Prentice Hall, Upper Saddle River (2010)
16. Hogan, J.E., Lehmann, D.R., Merino, M., Srivastava, R.K., Thomas, J.S., Verhoaf, P.C.: Linking customer assets to financial performance. J. Serv. Res. **5**(1), 26–38 (2002)
17. Ha, H.-Y., John, J., Janda, S., Muthaly, S.: The effects of advertising spending on brand loyalty in services. Eur. J. Mark. **45**(4), 673–691 (2011)
18. Keller, K.L.: Memory retrieval factors and advertising effectiveness. In: Mitchell, A.A. (ed.) Advertising Exposure, Memory and Choice, Hillsdale, NJ. Lawrence Erlbaum Associates (1992). in press
19. Keller, K.L.: Conceptualizing, measuring, and managing customer-based brand equity. J. Mark. **57**(1), 1–22 (1993)
20. Keller, K.L.: Strategic Brand Management: Building, Measuring and Managing Brand Equity. Prentice Hall, Upper Saddle River (1998)
21. Keller, K.L.: Strategic Brand Management, 2nd edn. Prentice Hall, Upper Saddle River (2003)
22. Keller, K.L.: Strategic Brand Management: Building, Measuring, and Managing Brand Equity, 3rd edn. Pearson Prentice Hall, Upper Saddle River (2008)
23. Kim, H., Kim, W.: The relationship between brand equity and firms performances in luxury hotels and chain restaurants. Tour. Manag. **26**, 549–560 (2005)
24. Kotler, P.: Marketing Management: Analysis, Planning, Implementation and Control, 10th edn. Prentice Hall, Upper Saddle River, Englewood Cliffs (2000)
25. Lassar, W., Mittal, B., Sharma, A.: Measuring consumer based brand equity. J. Consum. Mark. **12**(4), 4–11 (1995)
26. Mahajan, V., Rao, V.R., Srivastava, R.K.: An approach to assess the importance of brand equity in acquisition decisions. J. Prod. Innov. Manag. **11**, 221–235 (1994)
27. Motameni, R., Shahrokhi, M.: Brand equity valuation: a global perspective. J. Prod. Brand. Manag. **7**(4), 275–290 (1998)
28. Netemeyer, R.G., Balaji, K., Chris, P., Guangping, W., Yagci, M., Dean, D., Ricks, J., Wirth, F.: Developing and validating measures of facets of customer-based brand equity. J. Bus. Res. **57**(2), 209–224 (2004)
29. Thuy, N.V., Dai, D.N.: Fators impacting of service quality on customers satisfaction and loyalty: a case study of commercial EBanks in Ho Chi Minh City. J. Econ. Dev. Special, pp. 61 - 71 (2012)
30. Norusis, M.J.: SPSS for Windows: Advanced Statistics, Release 6.0, pp. 2–30. SPSS Incorporated, Chicago (1993)
31. Nunnally, J.C.: Psychometric Theory, 2nd edn. McGraw-Hill, New York (1978)
32. O'Loughlin, D., Szmigin, I.: Customer perspectives on the role and importance of branding in Irish retail financial services. Int. J. Bank Mark. **23**(1), 8–27 (2004)
33. Oliver, R.L.: A Behavioral Perspective on the Customer. McGraw-Hill, New York (1997)
34. Olsen, S.O.: Comparative evaluation and the relationship between quality, satisfaction and repurchase loyalty. J. Acad. Mark. Sci. **30**(3), 240–249 (2002)
35. Pappu, R., Quester, P.G.Q., Cooksey, R.W.: Consumer based brand equity: improving the measurement - empirical evidence. J. Prod. Brand. Manag. **14**(3), 143–154 (2005)
36. Pinar, M., Girard, T., Eser, Z.: Consumer-based brand equity in banking industry: a comparison of local and global banks in Turkey. Int. J. Bank Mark. **30**(5), 359–375 (2012)

37. Rambocas, M., Kirpalani, V.M.: Building brand equity in retail banks: the case of Trinidad and Tobago. Int. J. Bank Mark. **32**(4), 300–320 (2014)
38. Ravald, A., Grönroos, C.: The value concept and relationship marketing. Eur. J. Mark. **30**(2), 19–30 (1996)
39. Rossiter, J.R., Percy, L.: Advertising and Promotion Management. McGraw-Hill, New York (1987)
40. Sanchez, B.N., Budtz-Jorgensen, E., Ryan, L.M., Hu, H.: Structural equation models: a review with applications to environmental epidemiology. J. Am. Stat. Assoc. **100**(472), 1443–1455 (2005)
41. Severi, E., Ling, K.C.: The mediating effects of brand association, brand loyalty, brand image and perceived quality on brand equity. Asian Soc. Sci. **9**(3) (2013)
42. Shocker, A.D., Srivastava, R.K., Rueckert, R.W.: Challenges and opportunities facing brand management: an introduction to a special issue. J. Mark. Res. **31**, 149–158 (1994)
43. Simon, C.J., Sullivan, M.W.: The measurement and determinants of brand equity: a financial approach. Mark. Sci. **12**(1), 28–52 (1993)
44. Sriram, S., Balachander, S., Kalwani, M.U.: Monitoring the dynamics of brand equity using store-level data. J. Mark. **71**, 61–78 (2007)
45. Steenkamp, J.E.M., Van Trijp, H.C.M.: The use of LISREL in validating marketing constructs. Int. J. Res. Maket. **8**, 283–299 (1991)
46. Subramaniam, A., Mamun, A.A., Permarupan, P.Y., Zaino, N.R.B.: Effects of brand loyalty, image and quality on brand equity: a study among bank islam consumers in Kelantan, Malaysia. Asian Soc. Sci. **10**(14), 67 (2014)
47. Tong, X., Hawley, J.M.: Measuring customer-based brand equity: empirical evidence from the sportswear market in China. J. Prod. Brand. Manag. **18**(4), 262–71 (2009)
48. Vrontis, D., Papasolomou, D.: Brand and product building: the case of the Cyprus wine industry. J. Prod. Brand. Manag. **16**(3), 159–167 (2007)
49. Yasin, N.M., Noor, M.N., Mohamad, O.: Does image of country-of-origin matter to brand equity? J. Prod. Brand. Manag. **16**(1), 38–48 (2007)
50. Yoo, B., Donthu, N.: Developing and validating a multidimensional consumer-based brand equity scale. J. Bussiness Res. **52**, 1–14 (2001)
51. Yoo, B., Donthu, N., Lee, S.: An examination of selected marketing mix elements and brand equity. J. Acad. Mark. Sci. **28**(2), 195–211 (2000)
52. Zeilthaml, V.A.: Consumer perception of price, quality and value: a means-end model and synthesis of evidence. J. Mark. **52**(3), 2–22 (1988)

Factors Influencing to Accounting Information Quality: A Study of Affecting Level and Difference Between in Perception of Importance and Actual Performance Level in Small Medium Enterprises in Ho Chi Minh City

Nguyen Thi Tuong Tam(✉), Nguyen Thi Tuong Vy, and Ho Hanh My

Banking University of HCM City, Ho Chi Minh City, Vietnam
{tamnt,vyntt,myhh}@buh.edu.vn

Abstract. Ensuring reliability, consistency, honesty and understandability of accounting information is one of the most important responsibility of an enterprise. For the purpose of providing the bases for small and medium sized enterprises (SMEs) to enhance accounting information quality (AIQ), the research aims to discover factors affecting to AIQ in SMEs in Ho Chi Minh City (HCM City). In order to have deep explanation, the study goes on finding out whether there is difference in perception of importance of those factors to AIQ among job function. In spite of considering the importance of impact factors to AIQ, not all enterprises have actual performance to improve those factor in order to enhance AIQ. Hence, the research continues investigating whether there is a difference in perception of importance and actual performance level of above influencing factors to AIQ in SMEs in HCM City. Doing test basing on collected data from 138 surveys of SMEs in HCM City, the study's result indicates 18 factors influencing to AIQ. Among those 18 factors, there is a statistical significant difference in perception of importance of two factors "Accounting information system intuitive and easy to use" and "Physical environment" within various job function. The study also shows initial evidence about difference between perception of importance and actual performance level of influencing factors to AIQ.

Keywords: Accounting information quality
Accounting information · Influence factors · SMEs

1 Introduction

Enterprises need to use accounting information when making decisions about annual plans and business strategy (Susant [20]). Accounting information provides figures about enterprise's financial situation and performance's result as

V. Kreinovich et al. (Eds.): ECONVN 2019, SCI 809, pp. 999–1015, 2019.
https://doi.org/10.1007/978-3-030-04200-4_73

well as changes in management decision (Xu [25]). High accounting information quality is often considered as one of the main significant factors contributing to a company's success (Wang [24]). So, setting up a high quality AIQ plays an important part in an enterprise's operation. However, there is still a distance from perception of importance to actual performance level of impact factors to AIQ in practice (Xu [25]; Nguyen [16]). Although, being aware of importance of those factors' influences to AIQ, not many enterprises have enough activities in real performance in order to enhance them; and this leads to the following consequences:

Firstly, accounting information frauds in disclosing financial statements cause serious damages to users such as: diminishing trust in accounting information; decreasing the quality, and integrity of accounting information; causing enterprise's bankruptcy or insubstantial performance (Vlad et al. [22]).

Secondly, unreliable and inconsistent accounting information leads to inaccuracy effects of decisions (Salehi and Abdipour [18])

The result of Economic Survey in Ho Chi Minh City (2017) showed that SMEs accounted for 97.8% of total enterprises. However, only 37.81% enterprises have profit, 56.49% enterprises suffered losses, and the rest are at break-even point. The first reason is that most enterprises only focus on earning returns, and not pay enough attention to quality of accounting information displayed in financial statements. The second reason is that enterprises are not familiar with ensuring AIQ during their operation. There is a significant relationship between AIQ quality and operating's result (Susanto [21]). Hence, it is necessary to help SMEs in Ho Chi Minh City approach the concepts of AIQ, and set up AIQ system in order to enhance accounting information quality. For the purpose of guiding SMEs to approach AIQ, this study aims to detect factors that affecting accounting information quality in SMEs in HCM City, and find out the relationship between perception of importance and actual performance levels of those factors. The three specific objectives include:

+ Determining which factors influencing AIQ.
+ Testing whether there is difference in perception of importance of those factors to AIQ among job function.
+ Testing whether there is difference between perception of importance and actual performance level of those factors to AIQ.

This paper includes five sections: Sect. 2 provides literature review and research hypotheses; Sect. 3 presents research methodology and data collection; Sect. 4 shows research results and relevant discussions, and the last section is the conclusion.

2 Literature Review and Research Hypotheses

2.1 Quality of Accounting Information

Set of standards can be applied to measure quality of information as well as accounting information, because accounting information is a part of enterprise's information system. According to Wang and Strong [23], Delon and Mc Lean [5], and Ballou et al. [2], quality of information must be achieved following characteristics: accuracy, reliability, relevance, timeless, completeness, consistency, and understandability. Wang and Strong [23] also use dimension of consistent representation and accessibility, easy approach's operation to measure information quality. While Delon and McLean [5] indicated that information quality should be useful, sufficient, readable, clear and unique in their research.

However, setting up general standards to measure accounting information quality is really difficult (Klein and Kim [11]). Eppler and Wittig [7] displayed four attributes that used to measure the accounting information quality, including: (1) Providing a set of standards of accounting information quality that can be measured; (2) Providing a full standards' relationship description that can help scrutinize accounting information quality issue; (3) Providing a basis for accounting information quality's measurement and management; (4) These applied measurable standards are accepted widely.

In order to reach the convenience and consistency in assessing accounting information quality for enterprises from different countries, a set of international standards are set up. Nowadays, there are two sets of standards used to measure AIQ: Financial Accounting Standards Board (FASB) and The International Accounting Standards Board (IASB). FASB's measurement is basing on these following standards: matching, relevance, reliability, consistency, comparability, materiality, prudence (FASB, 1999). IASB's measurements are understandability, matching, reliability, comparability, timeliness (IASB, 2008). FASB and IASB have their own standards in measuring accounting information quality, so, global convergence of accounting standards is an inevitable tendency in the modern world.

Due to Carmona and Trombetta [3], the accounting information measurement satisfying standards in "convergence of IASB and FASB" is considered as "high-quality accounting". Hence, the research choses the convergence measures between IASB and FASB for the quality of accounting information displayed in Norwalk Agreement signed in 2004, which includes: matching, honesty, comparability, understandability, timeliness, materiality and verifiability.

2.2 Influence Factors to Accounting Information Quality

When mention about AIQ, enterprises tend to concentrate on achieving accuracy of accounting information, however, the quality of accounting information is not only measured by standard of accuracy, but also by other measures. Saraph, Benson, and Schroeder [19] figured out eight factors that influence data quality on view of Total Quality Management (TQM), which include: (1) Function of top management and quality policy; (2) Function of the management quality, (3) Training, (4) The design of product or service, (5) Management of supplier's quality, (6) The management and operation process, (7) Quality data and reporting, and (8) The relationship between employees. Motwani [14] offered several factors that affect data quality basing on TQM's view such as: Management of human resources and client, Strategic quality management, Group working for improvement, Planning of operation quality, Improvement of quality measurement systems, and Quality culture in corporation.

Xu [25], on his study about factors influencing to AIQ, suggested the main factors affect accounting information quality such as: (1) Top Management, (2) Training and Education, (3) Risk management, (4) Evaluate cost/benefit tradeoffs, (5) Middle management's commitment to DQ, (6) Clear AIQ vision for entire organization, (7) Nature of the Accounting Information Systems, (8) AIQ policies and standards, (9) Organisational structure, (10) AIQ controls, (11) Personnel competency, (12) Employee relations, (13) Management of changes, (14) Teamwork, (15) Understanding of the systems and AIQ, (16) Internal controls, (17) Audit and reviews,(18) Customer focus, (19) Performance evaluation & rewards.

Basing on the analysis of different AIQ standards of FASB, IASB, and researches of Xu [25], Motwani [14], Saraph, Benson, and Schroeder [19], the study sets up and considers a number of repeated influence factors in many researches when examining accounting information quality, including:

+ (B1) Management commitment to Accounting information quality (AIQ): recognising the role of AIQ and supporting AIQ activities
+ (B2) Training: suitable and update training programs are provided to staff so as to enhance the knowledge and skills in using AIS and gaining quality information. This factor is divided into two sub-factors depends on training purposes:
+ (B2.1) Primary training: supply basic training for new employees, new or upgrade systems
+ (B2.2) Advancing training: supply upgraded training for experienced employees and managers
+ (B3) Obvious AIQ vision for the organisation: Enterprise has suitable allocation of resources (capital, labor, technology) to ensure AIQ
+ (B4) Organisational structure: A streamlining, non-overlapping structure to increase AIQ.
+ (B5) AIQ policies and standards: standards, policies of AIQ are simple, matching, consistent, and measurable
+ (B6) Organisation have focus culture on AIQ: showing the deep concern about the role of high AIQ in all activities in enterprises
+ (B7) AIQ controls: setting up suitable policies and standards to assess and control the AIQ
+ (B8) Input controls: ensuring the accuracy of input information to prevent errors in operation system
+ (B9) Characteristics of the Accounting Information Systems(AIS): AIS is suitable, and measured by 3 sub-factors:
+ (B9.1) AIS-intuitive and easy to use
+ (B9.2) AIS-easy to upgrade
+ (B9.3) AIS-up-to-date
+ (B10) Employee relations: employees get on well with their colleagues, have opportunity in promotion and satisfy with enterprise's welfare

+ (B11) Management of changes: Ability of the managers in adapting any change in both internal and external environment
+ (B12) Going improvement: Ability of enterprise in catching up with the development of the operation system.
+ (B13) Teamwork: effective cooperation within and between departments in the enterprise
+ (B14) AIS and AIQ understanding of employee: people who work in AIS have comprehensive understanding about the system's operation and the AIQ's role. This factor is measured by two following sub factors:
+ (B14.1) Understanding the system's operation
+ (B14.2) Understanding the AIQ's importance, relations between AIQ and business objectives
+ (B15) Human resources capability: all levels, from top, middle management to employees, are well-trained, experienced, qualified, high skilled and knowledgeable in both technical and business area
+ (B16) Physical environment: Own capability of each employee from staff to managers (skills, knowledge, experience,...)
+ (B17) Internal controls: appropriate internal system and process controls on assessment management, security and human resource
+ (B18) Auditing and reviewing: Appropriate control in the system and the AIQ need to be ensured by conducting independent internal audit. AIQ should be reviewed regularly.

A accounting information may be useful in some cases while it is not in the other cases, depending on each difference type of users because information quality is a fitness for use. Hence, each survey participant may have difference views of the factors that affect accounting information quality. A next issue needed to solve is that there is any difference about the impact level of these factors among job function. This leads to the proposal of hypothesis H1:

H1: There is a significant difference in influence level of factors affect accounting information quality among interviewers' job function in SMEs.

However, there may have a gap between perception of importance and actual performance of factors influence to accounting information quality during organizations' operation. Thus, it is necessary to test whether that gap may exist or not through hypothesis H2:

H2: There is a significant difference between perception of importance and actual performance level of factors influence to accounting information quality in SMEs.

3 Research Methodology and Data Collection

3.1 Research Methodology and Questionnaire Structure

This study applies survey method with a set of questionnaire. The questionnaire is designed basing on the questionnaire in Hongjiang's study (Xu [25]). Statistical measurement is applied to process data collected from the questionnaire. There are three parts in this questionnaire:

Part A includes three questions of career, position and working experience.

Part B includes 18 factors (from B1 to B18) which verified into 22 detail factors. In part B questionnaire, there are three columns, including Column 1 stood for influence factors, which helps to find out factors affect to accounting information quality, Column 2 was for measuring the factors' importance, and Column 3 illustrated levels of performance. Likert scale score the responses for each factor (Table 1).

Table 1. Scale for influence, importance and performance

	Scale for influence		Scale for importance		Scale for performance
1	No influence	1	No important	1	No applicable
2	Little influence	2	Little important	2	Poor applicable
3	Average influence	3	Average important	3	Average applicable
4	Strong influence	4	Very important	4	Good applicable
5	Very strong influence	5	Very strong important	5	Very good applicable

According to Narli [15], the interval width of 5 Likert scale should be computed in order to set up the group boundary value for result discussions.

Interval Width = (Upper value - Lower value)/n = (5−1)/5 = 0.8.

Group boundary values are built that help to discuss research results basing on the above interval width, which are pointed in Table 2.

Table 2. Group boundary values of 5 Likert scale

Judgement scale for perception of influence		Judgement scale for perception of importance		Judgement scale for actual performance	
1.00−1.80	No influence	1.00−1.80	No important	1.00−1.80	No applicable
1.81−2.60	Little influence	1.81−2.60	Little important	1.81−2.60	Poor
2.61−3.40	Average influence	2.61−3.40	Average	2.61−3.40	Average
3.41−4.20	Strong influence	3.41−4.20	Very important	3.41−4.20	Good
4.21−5.00	Very strong influence	4.21−5.00	Very strong important	4.21−5.00	Very good Applicable

Part C was designed for the respondents to choose the most important and least important factors, which already displayed in part B. It helps to test again which are the critical factors that influence to accounting information quality.

The next applied method is hypothesis test. Several hypotheses are formed and tested in order to answer the research questions. Sample Paired T- test, ANOVA test, and Independent Sample T-Test are used to test the hypotheses of variation with a significance level of 5% and reliability of 95%.

Zikmund et al. [26] figured out the data need to be edited and coded to become secondary information. SPSS (Statistical Package for the Social Sciences) are applied to code and enter the completed questionnaire.

3.2 Data Collection

Through the survey questionnaire instrument, this study collects data of factors impacted on accounting information quality and interviewers' consideration about the importance of those factors and particular action level of those factors in practice. Basing on those data, the study can test the variations in perception of importance among survey interviewers as well as the variation between perception of importance and actual performance.

190 survey questionnaires were distributed to interviewers. There were 144 questionnaires returned; however, six questionnaires were eliminated due to not fulfilled answers, thus, the remaining 138 eligible questionnaires (rated 72.6% of the questionnaire is issued) were used for analysis.

Survey interviewers: According to Wang [24], in order to achieve high quality information, the research should collect data from people that have deep and actual knowledge of research problem. In order to find out the affect factors to accounting information quality in SMEs, the research selected target group of interviewers includes:

Accountants: are employees who join directly in the process and produce accounting information so they have deep knowledge in accounting system of the enterprise.

Internal Auditors: are independent people who inspect and evaluate the quality of accounting information, as well as honesty and reliability of financial statements.

Information staffs: are employees who have certain grasp about accounting software and accounting information systems.

Managers: are employers who make policies and regulations in order to support the activities related to accounting information quality in enterprises.

Sample size: In recent years, many researchers determined the sample size by the empirical formula. Generally, there are two directions proposed in the sample size selection:

Option 1: Take the absolute value for the sample set (N). This guide recommends that sample set size should be at least 100 samples (Comrey [4]; Gorsuch [9]).

Option 2: Take the ratio N/p, in which N is the number of population units and p is the number of observed units. This ratio should be at least 5 (Everitt [8]).

The research chose the approach of the sample size with the minimum number of 100 and the ratio of 5:1 in order to match two above directions of collecting sample size. Hence, with 22 factors, at least 110 interviews are selected. The actual sample size of 138 interviewers has met both conditions above.

4 Research Results and Discussions

4.1 Statistical Tests for Reliability of Scale and Normal Distribution

Reliability of Scale of influence factors to accounting information quality

It is not always easy to receive an honest answer from survey respondent. Thus, Cronbach's alpha is used to fix this problem and increase validity and reliability of data. A popular cited threshold for the Cronbach's Alpha is from 0.6 to 0.95 (Hoang and Chu [10]); factors which have Corrected Item-Total Correlation less than 0.3 will be deleted; in contrast, the factors are reliable if Corrected Item-Total Correlation is greater than 0.3.

It can be seen from Table 3 that, Cronbach's Alpha equals to 0.831 and all factors have Corrected Item-Total Correlation greater than 0.3, except factors "Management commitment to AIQ", "Employee relations" and "Internal controls". These three factors should be deleted out of the influence factors. After deleting these factors, Cronbach's Alpha increases to 0.834 and 19 remain factors all have Corrected Item-Total Correlation greater than 0.3 (Appendix A). This still satisfied the condition of reliability of survey scale, so that, these factors are kept to be analyzed the impact on accounting information quality.

Tests of Normality

Many of statistical analysis procedures, such as T-tests, analysis of variance, applying parametric tests, are based on the assumption that the data satisfies a normal distribution. However, the assumption of normality of the variables is not always supported. Elliott et al. [6] considered that parametric procedures could not be applied with un-normal distributed data. By using the Shapiro Wilk Test, the study examines whether data has a normal distribution.

The results from Table 4 show that data are normally distributed with the significant level at 5%. Hence, parametric statistics should be used for analyzing in order to test all the hypotheses which are proposed in the study.

4.2 Influence Factors to Accounting Information Quality in SMEs

The difference in influence level of factors affect accounting information quality among interviewers' job function (accountants, auditors, information staffs and managers) in SMEs

ANOVA test and ANOVA-Post hoc test are used to test the hypothesis H1 in order to find out the difference in influence level of accounting information quality among accountants, auditors, information staffs and managers in SMEs.

The results from Table 5 indicated that two factors "Physical environment" and "AIS intuitive and easy to use" have p-value < 0.05. This mean there is significant difference in these factors' influence levels to accounting information quality among accountants, auditors, information staffs and managers. For the remain factors, the lack of significance means that there is no difference in the perception of important of these factors influence to accounting information quality among various job function.

Table 3. Reliability of Scale

Position	Factors	Scale mean if item deleted	Scale variance if item deleted	Corrected item-total correlation	Cronbach's alpha if item deleted
1	**(B1) Management commitment to AIQ**	**78.30**	**69.539**	**0.142**	**0.833**
2	**(B10) Employee relations**	**78.43**	**66.656**	**0.292**	**0.829**
3	**(B17) Internal controls**	**78 .92**	**68.351**	**0.179**	**0.834**
	(B2) Training				
4	(B2.1) Primary training	78.29	66.499	0.368	0.825
5	(B2.2) Advancing training	78.36	64.525	0.505	0.819
6	(B3) Obvious AIQ vision for the organization	78.26	66.384	0.393	0.824
7	(B4) Organisational structure	78.43	65.736	0.395	0.824
8	(B5) AIQ policies and standards	78.57	65.546	0.417	0.823
9	(B6) Organisation have focus culture on AIQ	78.59	63.923	0.429	0.822
10	(B7) AIQ controls	78.54	66.046	0.371	0.825
11	(B8) Input controls	78.41	64.230	0.458	0.821
	(B9) Characteristics of the Accounting Information Systems				
12	(B9.1) AIS-intuitive and easy to use	78.34	66.839	0.342	0.826
13	(B9.2) AIS-easy to upgrade	78.52	65.784	0.421	0.823
14	(B9.3) AIS-up-to-date	78.36	64.158	0.504	0.819
15	(B11) Management of changes	78.69	66.304	0.375	0.825
16	(B12) Going improvement	78.59	66.111	0.372	0.825
17	(B13) Teamwork	78.30	65.758	0.445	0.822
	(B14) AIS and AIQ understanding of employee				
18	(B14.1) Understanding the system's operation	78.32	65.898	0.420	0.823
19	(B14.2) Understanding the AIQ's importance, and relations between AIQ and business objectives	78.40	66.811	0.371	0.825
20	(B15) Human resources capability	78.24	65.585	0.442	0.822
21	(B16) Physical environment	78.54	63.418	0.536	0.817
22	(B18) Auditing and reviewing	78.40	66.460	0.370	0.825

Table 4. Normality test of influence factors' distribution

Shapiro Wilk					
Position	Factors	Perception of Importance		Actual Performance	
		Statistic	Sig	Statistic	Sig
1	(B2.1) Primary training	0.82	0.00	0.89	0.00
2	(B2.2) Advancing training	0.86	0.00	0.88	0.00
3	(B3) Obvious AIQ vision for the organization	0.83	0.00	0.90	0.00
4	(B4) Organisational structure	0.86	0.00	0.89	0.00
5	(B5) AIQ policies and standards	0.86	0.00	0.90	0.00
6	(B6) Organisation have focus culture on AIQ	0.90	0.00	0.91	0.00
7	(B7) AIQ controls	0.86	0.00	0.88	0.00
8	(B8) Input controls	0.87	0.00	0.90	0.00
9	(B9.1) AIS-intuitive and easy to use	0.85	0.00	0.85	0.00
10	(B9.2) AIS-easy to upgrade	0.86	0.00	0.90	0.00
11	(B9.3) AIS-up-to-date	0.86	0.00	0.90	0.00
12	(B11) Management of changes	0.86	0.00	0.90	0.00
14	(B13) Teamwork	0.83	0.00	0.89	0.00
15	(B14.1) Understanding the system's operation	0.84	0.00	0.88	0.00
16	(B14.2) Understanding the AIQ's importance, and relations between AIQ and business objectives	0.88	0.00	0.85	0.00
17	(B15) Human resources capability	0.85	0.00	0.88	0.00
18	(B16) Physical environment	0.87	0.00	0.91	0.00
19	(B18) Auditing and reviewing	0.85	0.00	0.00	0.00

The results from Table 6 indicated the deep interpretation for each factor:

For "AIS-intuitive and easy to use", it is understandable that staffs need time to adapt with the change in enterprise's system. Hence, the system should be simple and easy to access. However, when there is any improvement of AIS, accountants, internal auditors will find it more difficult to update than information staffs because they are the one who directly produce accounting information in that system, while information staffs only control information, so they can not realize which difficulties that accountants and auditors may get during their activities.

For "Physical environment", according to Wang [24], information quality and work environment have a positive relation. A physical environment with modern

Table 5. ANOVA Test between group follows job function

Position	Factors	Job function				Asymp. Sig.
		Accountants	Auditors	Information staffs	Managers	
1	(B2.1) Primary training	3.88	3.92	4.00	3.89	0.845
2	(B2.2) Advancing training	3.84	3.58	3.95	3.82	0.351
3	(B3) Obvious AIQ vision for the organization	3.86	3.77	4.15	3.92	0.219
4	(B4) Organisational structure	3.68	3.81	3.95	3.75	0.604
5	(B5) AIQ policies and standards	3.73	3.31	3.65	3.61	0.142
6	(B6) Organisation have focus culture on AIQ	3.58	3.81	3.40	3.59	0.582
7	(B7) AIQ controls	3.59	3.38	3.85	3.64	0.970
8	(B8) Input controls	3.86	3.54	3.8	3.77	0.442
9	**(B9.1) AIS-intuitive and easy to use**	**3.8**	**3.58**	**4.30**	**3.84**	**0.015**
10	(B9.2) AIS-easy to upgrade	3.68	3.54	3.85	3.66	0.551
11	(B9.3) AIS-up-to-date	3.8	3.77	3.95	3.83	0.872
12	(B11) Management of changes	3.46	3.35	3.8	3.49	0.262
13	(B12) Going improvement	3.55	3.58	3.55	3.59	0.776
14	(B13) Teamwork	3.95	3.62	3.90	3.88	0.240
15	(B14.1) Understanding the system's operation	3.93	3.58	4.00	3.86	0.189
16	(B14.2) Understanding the AIQ's importance, and relations between AIQ and business objectives	3.92	3.73	3.60	3.78	0.084
17	(B15) Human resources capability	3.97	3.69	4.20	3.94	0.169
18	**(B16) Physical environment**	**3.73**	**3.58**	**3.85**	**3.64**	**0.037**
19	(B18) Auditing and reviewing	3.73	3.88	3.80	3.78	0.836

equipment has been found to be a factor that might influence accounting quality performance. The demand for enhancing computer systems of information staffs is higher than managers such as: accountants, information staffs because computing and the software is their main working facility. While managers do not pay attention to upgrade working infrastructure for staffs, because they consider that is not really a necessary problem. This makes the company cannot equip and update those facilities, and information staffs and accountants are likely to get more troubles during their work.

Table 6. ANOVA Post-hoc Test within group follows job function

Turkey HSD					
Dependent variable	Job function (I)	Job function (J)	Mean difference (I-J)	Std. error	Sig.
(B9.1) AIS Intuitive and easy to use	Accountants	Auditors	0.22	.172	.577
		Information Staff	**−0.50**	**.190**	**.045**
		Managers	−0.04	.198	.967
	Auditors	Accountants	−0.02	.172	.577
		Information Staff	**−0.72**	**.224**	**.009**
		Managers	−0.26	.231	.534
	Information staff	**Accountants**	**0.50**	**.190**	**.045**
		Auditors	**0.72**	**.224**	**.009**
		Managers	0.46	.245	.340
	Managers	Accountants	0.04	.198	.967
		Auditors	0.26	.231	.534
		Information Staff	−0.46	.245	.340
(B16) Physical environment	Accountants	Auditors	0.15	.198	.867
		Information Staff	**−0.12**	.219	.947
		Managers	0.09	**.228**	**.038**
	Auditors	Accountants	−0.15	.198	.867
		Information Staff	−0.27	.259	.717
		Managers	**−0.06**	.267	.303
	Information staff	Accountants	0.12	.219	.947
		Auditors	0.27	.259	.717
		Managers	**0.31**	**.282**	**.048**
	Managers	**Accountants**	**0.09**	**.228**	**.038**
		Auditors	0.06	.267	.303
		Information Staff	**−0.21**	**.282**	**.048**

4.3 The Difference Between Perception of Importance and Actual Performance Levels of Influence Factors to Accounting Information Quality in SMEs

The sample paired T-test is used to compared the paired sample's mean of the importance's perception and actual performance. The test discovered if there is the significant difference in the mean of the perception of importance and actual performance for each factor.

It can be seen from the Appendix B that there is difference between perception of importance and actual performance level of factors influence to accounting

Table 7. Mean importance and performance with ranking

Factors	Perception of importance		Actual performance level		Difference of mean
	Mean	Rank	Mean	Rank	
	(1)	(2)	(3)	(4)	(5) = (1)–(3)
(B9.3) AIS-up-to-date	**3.83**	7	**3.09**	12	**0.74**
(B3) Obvious AIQ vision for the organization	**3.92**	2	**3.19**	6	**0.73**
(B13) Teamwork	**3.88**	4	**3.17**	8	**0.71**
(B8) Input controls	**3.77**	11	**3.07**	13	**0.70**
(B2.1) Primary training	**3.89**	3	**3.20**	3	**0.69**
(B14.1) Understanding the system's operation	**3.86**	5	**3.20**	5	**0.66**
(B15) Human resources capability	**3.94**	1	**3.29**	1	**0.65**
(B7) AIQ controls	**3.64**	14	**2.99**	16	**0.65**
(B6) Organisation have focus culture on AIQ	**3.59**	17	**2.94**	18	**0.65**
(B2.2) Advancing training	**3.82**	8	**3.18**	7	**0.64**
(B14.2) Understanding the AIQ's importance, and relations between AIQ and business objectives	**3.78**	9	**3.14**	9	**0.64**
(B18) Auditing and reviewing	**3.78**	10	**3.14**	10	**0.64**
(B12) Going improvement	**3.59**	18	**2.95**	17	**0.64**
(B9.1) AIS-intuitive and easy to use	**3.84**	6	**3.21**	2	**0.63**
(B9.2) AIS-easy to upgrade	**3.66**	13	**3.06**	15	**0.60**
(B11) Management of changes	**3.49**	19	**2.91**	19	**0.58**
(B16) Physical environment	**3.64**	15	**3.07**	14	**0.57**
(B4) Organisational structure	**3.75**	12	**3.20**	4	**0.55**
(B5) AIQ policies and standards	**3.61**	16	**3.12**	11	**0.49**

information quality as all factors have p-value < 0.05. Thus, there is a significant difference between perception of importance and actual performance.

However, despite highly important consideration for all factors above with means running from 3.4 to 4.2, actual performance level of them is just considered as medium level. This means that there is a remarkable gap between the importance and performance of these factors. It can be seen from Table 7 that all these factors have the difference mean between the importance and performance of these factors nearly 0.5 and above. In more details, the 4-factors-on-top include "AIS-up-to-date", "Teamwork", "Obvious AIQ vision for the organization", and "Input controls" have a very highest difference of over 0.7. The next 11 factors have the difference of over 0.6. In practice, this big gap reflects the real situation of the relationship between perception of importance and actual performance of those factors in SMEs in Ho Chi Minh City. Although administrators have high perception of importance of the factors, they do not really have the appropriate activities to promote their implementation. For example, administrators recognize that the interaction between departments is both difficult and pressure. The lack of interaction and support between departments may cause delay, wrong and waste of time during working. This means that teamwork is very importance, however, in practice, not many enterprises have already established a clear and fair process in evaluating supporting level during working activities among departments and among employees. This may lead to benefit confliction among departments and among employee. If the employees are not satisfied and do not focus on working, they may make more mistakes during activities and this may cause bad information. Hence, it is necessary to get the gap between perception of importance and actual performance closer.

5 Conclusion

The study identifies a range of factors that affect the accounting information quality and indicates that there has a significant difference between perception of importance and actual performance of factors affect to accounting information quality. Most of the interviewers have remarkable perception about the importance of accounting information quality; yet, the actual performance is not as high as expectation. This difference is the main cause of the bad quality accounting information in SMEs in Ho Chi Minh City. The 4-factors-on-top include "AIS up-to-date", "Teamwork", "Obvious AIQ vision for the organization", and "Input controls". Administrators in SMEs should have actual actions in order to get the gap between perception and actual performance closer, such as: set up a fair reward regulations for teamwork, administrational actions follow company's objective, and willing update AIS.

This study also examines further about this difference in perception of importance in aspect of job function. In more details, it is shown that there is the difference in two influence factors ("AIS-intuitive and easy to use" and "Physical environment") to accounting information quality among job function: accountants, auditors, information staffs and managers.

Appendix A

Number of factors after deleting insignificant factors

Cronbach's Alpha = 0.834

Number of factors = 19

Position	Factors	Scale Mean if Item Deleted	Scale Variance if Item Deleted	Corrected Item-Total Correlation	Cronbach's Alpha if Item Deleted
1	B2.1a	67.40	55.818	.369	.829
2	B2.2a	67.47	54.032	.506	.822
3	B3a	67.37	55.855	.382	.828
4	B4a	67.54	55.039	.403	.827
5	B5a	67.68	54.890	.423	.826
6	B6a	67.70	53.571	.420	.827
7	B7a	67.65	55.645	.352	.830
8	B8a	67.52	53.770	.457	.825
9	B9.1a	67.45	55.899	.365	.829
10	B9.2a	67.63	55.096	.429	.826
11	B9.4a	67.46	53.842	.492	.823
12	B11a	67.80	55.491	.389	.828
13	B12a	67.70	55.802	.344	.830
14	B13a	67.41	54.856	.474	.824
15	B14.1a	67.43	55.137	.434	.826
16	B14.2a	67.51	56.193	.365	.829
17	B15a	67.35	55.002	.442	.825
18	B16a	67.65	53.119	.528	.821
19	B18a	67.51	56.135	.341	.830

Appendix B

Paired Sampler T-Test for Mean of importance and performance

		Paired Differences					t	df	Sig. (2-tailed)
		Mean	Std. Dev.	Std. Error Mean	95% Confidence Interval of the Difference				
					Lower	Upper			
Pair 1	B2.1b - B2.1c	0.696	0.825	0.070	0.557	0.835	9.907	137	0.00
Pair 2	B2.2b - B2.2c	0.638	0.887	0.076	0.488	0.787	8.442	137	0.00
Pair 3	B3b - B3c	0.732	0.956	0.081	0.571	0.893	8.998	137	0.00
Pair 4	B4b - B4c	0.551	1.011	0.086	0.380	0.721	6.397	137	0.00
Pair 5	B5b - B5c	0.493	0.898	0.076	0.342	0.644	6.446	137	0.00
Pair 6	B6b - B6c	0.652	0.877	0.075	0.505	0.800	8.739	137	0.00
Pair 7	B7b - B7c	0.652	0.851	0.072	0.509	0.795	8.999	137	0.00
Pair 8	B8b - B8c	0.703	0.947	0.081	0.544	0.862	8.721	137	0.00
Pair 9	B9.1b - B9.1c	0.630	0.846	0.072	0.488	0.773	8.750	137	0.00
Pair 10	B9.2b - B9.2c	0.601	0.940	0.080	0.443	0.760	7.516	137	0.00
Pair 11	B9.3b - B9.3c	0.732	0.956	0.081	0.571	0.893	8.998	137	0.00
Pair 12	B11b - B11c	0.587	0.808	0.069	0.451	0.723	8.533	137	0.00
Pair 13	B12b - B12c	0.638	0.896	0.076	0.487	0.788	8.364	137	0.00
Pair 14	B13b - B13c	0.710	0.906	0.077	0.558	0.863	9.210	137	0.00
Pair 15	B14.1b -	0.667	0.907	0.077	0.514	0.819	8.636	137	0.00
Pair 16	B14.2b -	0.638	0.943	0.080	0.479	0.796	7.942	137	0.00
Pair 17	B15b - B15c	0.652	0.834	0.071	0.512	0.793	9.186	137	0.00
Pair 18	B16b - B16c	0.565	0.879	0.075	0.417	0.713	7.552	137	0.00
Pair 19	B18b - B18c	0.645	0.809	0.069	0.509	0.781	9.368	137	0.00

References

1. Al-Hiyari, A., AL-Mashre, M.H.H., Mat, N.K.N.: Factors that affect accounting information system implementation and accounting information quality: a survey in University Utara Malaysia. Am. J. Econ. **3**(1), 27–31 (2013)
2. Ballou, D.P., Wang, R.Y., Pazer, H., Tayi, K.G.: Modeling data manufacturing systems to determine data product quality. In: Total Data Quality Management (TDQM) Research Program (1993)
3. Carmona, S., Trombetta, M.: The IASB and FASB convergence process and the need for 'concept-based' accounting teaching. Adv. Account. **26**(1), 1–5 (2010)
4. Comrey, A.L., Lee, H.B.: A First Course in Factor Analysis. Erlbaum, Hillsdale (1992)
5. Delone, W.H., McLean, E.R.: The DeLone and McLean model of information systems success: a ten-year update. J. Manag. Inf. Syst. **19**(4), 9–30 (2003)
6. Elliot, A.J., Maier, M.A., Moller, A.C., Friedman, R., Meinhardt, J.: Color and psychological functioning: the effect of red on performance attainment. J. Exp. Psychol. Gen. **136**(1), 154 (2007)
7. Eppler, M.J., Wittig, D.: Conceptualizing information quality: a review of information quality frameworks from the last ten years. In: Proceedings of the 2000 Conference on Information Quality (2000)
8. Everitt, B.S.: Multivariate analysis: the need for data, and other problems. Br. J. Psychiatry **12**(6), 237–240 (1975)
9. Gorsuch, R.L.: Factor Analysis, 2nd edn. Erlaum, Hillsdale (1983)

10. Hoang, T., Chu, N.M.N.: Phân tích dữ liệu nghiên cứu với SPSS [Analyzing data using SPSS]. Statistical Publishing House (2008)
11. Klein, H.J., Kim, J.S.: A field study of the influence of situational constraints, leader-member exchange, and goal commitment on performance. Acad. Manag. J. **4**(1), 88–95 (1998)
12. Kline, P.: Psychometrics and Psychology. Acaderric Press, London (1979)
13. Mintzberg, H.: The structuring of organizations: a synthesis of the research. University of Illinois at Urbana-Champaign's Academy for Entrepreneurial Leadership Historical Research Reference in Entrepreneurship (1979)
14. Motwani, J.: Measuring critical factors of TQM. Meas. Bus. Excel. **5**(2), 27–30 (2001)
15. Narli, S.: An alternative evaluation method for Likert type attitude scales: rough set data analysis. Sci. Res. Essays **5**(6), 519–528 (2010)
16. Nguyen, B.L.: Identify and control the factors that affect the quality of accounting information in ERF in Vietnam. Unpublished doctoral dissertation. University of Economics Ho Chi Minh City (2012)
17. Rapina: Factors influencing the quality of accounting information system and its implications on the quality of accounting information. Res. J. Financ. Account. **5**(2), 148–154 (2014)
18. Salehi, M., Abdipour, A.: A study of the barriers of implementation of accounting information system: case of listed companies in Tehran Stock Exchange. J. Econ. Behav. Stud. **2**(2), 76–85 (2011)
19. Saraph, J.V., Benson, P.G., Schroeder, R.G.: An instrument for measuring the critical factors of quality management. Decis. Sci. **20**(4), 457–478 (1989)
20. Susanto, A.: Accounting Information Systems: Concepts and Development of Computer-Based, Premium edition, First Printing, Linga Jakarta, Bandung (2008)
21. Susanto, A.: What factors influence the quality of accounting information. Int. J. Appl. Bus. Econ. Res. **13**(6), 3995–4014 (2015)
22. Vlad, M., Tulvinschi, M., Chiriţă, I.: The consequences of fraudulent financial reporting. USV Ann. Econ. Public Adm. **11**(1), 264–268 (2011)
23. Wang, R.Y., Strong, D.M.: Beyond accuracy: what data quality means to data consumers. J. Manag. Inf. Syst. **12**(4), 5–33 (1996)
24. Wang, R.Y.: A product perspective on total data quality management. Commun. ACM **41**(2), 58–65 (1998)
25. Xu, H.: Critical success factors for accounting information systems data quality. Unpublished Doctoral dissertation. University of Southern Queensland (2003)
26. Zikmund, W.G., Babin, B.J., Carr, J.C., Griffin, M.: Sampling Designs and Sampling Procedures, p. 411. Exploring Marketing Research, Thomson South-Western (1997)

Export Price and Local Price Relation in Longan of Thailand: The Bivariate Threshold VECM Model

Nachatchapong Kaewsompong[1](✉), Woraphon Yamaka[2], and Paravee Maneejuk[2]

[1] Faculty of Economics, Chiang Mai University, Chiang Mai 50200, Thailand
Nachat_flysky@hotmail.com
[2] Center of Excellence in Econometrics, Faculty of Economics, Chiang Mai University, Chiang Mai 50200, Thailand
woraphon.econ@gmail.com, mparavee@gmail.com

Abstract. There is an evidence of non-linear long-run equilibrium relationship between local longan price and all export longan price variables in Thailand. This paper examines the short run and long run relationship between the Thai local longan price and various export longan prices using a Threshold VECM Model. We used the monthly data for the period of 2005M1 to 2013M3. The threshold co-integration test confirms the non-linear long-run equilibrium relationship between local longan price and all export longan price variables. Also, we found the existence of long run equilibrium in local price relation with other export prices when the value of the error-correction is greater than the threshold level. Our results confirm that the longan market is cointegrated and it tends to be efficient and competitive when the longan market price is high.

Keywords: Longan · Price relation · Threshold VECM

1 Introduction

The total value of Thailand's exports has been more than 6 billion baht in recent years. Most exports are agricultural products which have continuously increased for the past decade. One of the most important agriculture product in term of volume and value of export is longan. It is consumed and marketed in the forms of fresh fruits, as well as dried, canned, and frozen products. With its enormous medicinal properties and values, it has recently been manufactured into different kinds of products. Among longan producing countries e.g. Thailand, China and Vietnam, Thailand is a densely populated agriculture country where longan is one of the major crops of the country. The value of longan export in Thailand is more than 20,000 million Baht in 2016, which is equivalent to Vietnam's value. However, this amount is still behind China, which is a leading producer.

Although Thailand has exported longan products substantially to the world market, especially to Hong Kong, China, Indonesia, Europe, and North America,

V. Kreinovich et al. (Eds.): ECONVN 2019, SCI 809, pp. 1016–1027, 2019.
https://doi.org/10.1007/978-3-030-04200-4_74

it has no control over the price as it still depends on the foreign markets. These situations have brought about a severe impact to longan producers in Thailand. The farmers cannot set the price higher to cover their costs. Although there are some innovation and technology introduced for developing their products, the farmers still face to lost revenue because they have been forced to set the price down by the foreign merchant and middleman. Moreover, they have been faced with some high costs for cultivating and harvesting. They are also faced with high competition in the world market. Thus, the price of longan has fluctuated and these fluctuations have occurred for quite some time in the long-term growth trend with regards to export. The fluctuation in longan price could influence the farmers' income. Thus, in this study, we will examine the co-movement of longan exported product prices and domestic price.

To this extent, proper monitoring and understanding of the cointegration in the longan prices in the Thai markets is critical to the government for the stability and management of longan price for farmers. In addition, As the market is cointegrated, it tends to be efficient, and competitive [4]. In the literature, there are many studies investigating the relationship and cointegration between the agriculture product prices using Vector Autoregressive model (VECM). Yu et al. [17] examined the cointegration and causality of crude oil prices and the vegetable oils prices. Saghaian [14] examined the link between corn prices, soybean prices, wheat prices, crude oil prices and ethanol prices. Mishra and Kumar [11] analysed the spatial integration of vegetable markets in Nepal market using weekly wholesale price and retail price data.

However, there has been less attention in the non-linear structure in the agricultural price which had been found in some studies ([12,13,15]). Rapsomanikis and Hallam [13] suggested that threshold behavior and discrete adjustment are exhibited in various asset prices including agricultural prices. Therefore, working with VECM regardless of the non-linear structure may provide biased inferences and hence misleading results. To accomplish our concept, we employ a nonlinear co-integration model or threshold Vector Autoregressive Error Correction model (TVECM) of Balke and Forby [2] to investigate the bivariate relationship between various exported product prices and domestic longan price of Thailand.

The remainder of the paper is organized as follows: Sect. 2 provides the model used in this study. In Sect. 3, we conduct a nonlinear and cointegration tests to examine the presence of a nonlinear co-integration. The data used in this study is presented in Sect. 4. Section 5 then provides the estimation result from the TVECM. Finally, the conclusion is provided in Sect. 6.

2 Methodology

The threshold vector error correction model was introduced by Balke and Fomby [2] to combine non-linearity and co-integration together. Balke and Fomby [2] pointed out that the long-run equilibrium relationship should be analysed by a co-integration test while assuming asymmetric adjustment. As we know, error-correction model (ECM) describes variables which respond to deviations from the equilibrium relationship when variables are co-integrated. This traditional approach can be focused on moving towards the long-run equilibrium for every time period. TVECM has been further extended by Hansen and Seo [7]. In this study, the bivariate two-regime TVECM is conducted which is written as:

$$\Delta Y_t = \begin{cases} \nu_1 + A_{11}\Delta Y_{t-1} + ... + A_{p1}\Delta Y_{t-p} + \beta'_1 w_{t-1} + u_{1t} & if \quad w_{t-1} \leq \gamma \\ \nu_2 + A_{22}\Delta Y_{t-1} + ... + A_{p2}\Delta Y_{t-p} + \beta'_2 w_{t-1} + u_{2t} & if \quad w_{t-1} > \gamma \end{cases} \tag{1}$$

where $\Delta Y_t = \{\Delta y_{2t}\, \Delta y_{1t}\}$ denotes an integrated order one, $I(1)$, vector of n-dimensional time series data, ν is a state dependent intercept term. A is $n \times n \times p$ regime dependent autoregressive parameter matrices with lag p. $\beta' w_{t-1}$ denotes $I(0)$ error correction term with a cointegrating vector. Note that w_{t-1} is the computed by the error term of the VAR model with integrated order zero, $I(0)$, dataset. As we consider a bivariate model, we need to impose some normalization on β to achieve identification. ε_t is error term with the assumed $N(0, \Sigma(S_t))$. In this model, the parameter estimates are allowed to switch between two regimes, according to the threshold value γ.

3 Testing for Threshold Co-integration with Asymmetric Adjustment

Balke and fomby [2] applied many univariate tests (e.g. [6,16]) to examine the co-integrating residual (i.e., the error-correction term) for testing a threshold effect in the model. Forbes et al. [5] developed a Bayesian estimation for financial arbitrage. Lo and Zivot [10] tested a threshold effect in the model by using a multivariate threshold co-integration model with a known co-integration vector. Hansen and Seo [7] examined an unknown co-integration vector. In particular, these authors developed a Lagrange multiplier (LM) test for the presence of a threshold effect based on the error-correction term.

For testing the threshold co-integration, Hansen and Seo [7] suggested the two heteroskedastic-consistent LM test statistics under linear co-integration in null against the alternative threshold co-integration. It means that no threshold co-integration under the null. So, the model (1) becomes the linear VECM. In case there is known priori of true co-integrating vector, the testing statistic is denoted as:

$$Sup\ LM^0 = \underset{\gamma_L \leq \gamma \leq \gamma_U}{Sup} \quad LM\,(\beta, \gamma) \tag{2}$$

where the known value β_0 is fixed at unity. And the other case that the test statistic of is unknown, the true co-integrating vector is denoted as:

$$Sup\ LM = \underset{\gamma_L \leq \gamma \leq \gamma_U}{Sup} \quad LM\left(\tilde{\beta}, \gamma\right),. \tag{3}$$

where $\tilde{\beta}$ is calculated from the null estimate of β In both tests, let $[\gamma_L, \gamma_U]$ be the search region, γ_L is the percentile at π_0 of $\tilde{w}_{t-1}$, and $\gamma_U$is the percentile at $(1 - \pi_0)$. Andrews [1] suggested that the good setting π_0 should belong to the interval 0.05 and 0.15. Finally, the 5000 simulation replications of the bootstrap methods procedure is applied to calculate the asymptotic critical values and p-values. We note that the Minimum Bayes factor(MBF) is used to make the statistic inference in this study. MBF is computed by $-eplogp$, where p is p-value, for more detail and interpretation, see [8].

4 Data

All the data in this study are from the monthly frequencies and cover the period from 2005M1 to 2013M3. To find the long run relation of Longan price variable, we use a monthly Longan price of Thailand data set that includes endogenous variables that are total price (TT), canned price (CAN), fresh price (Fr), dried price (Dry), and local price (Local). All of the variables are collected from the Office of Agricultural Economics. Before providing an analysis on the time series, we should determine an integrated order for each variable. We use univariate methods to test for the unit roots and identify the properties of stationarity in each time series variable. We investigate the order of integration of all variables by using the Augmented Dickey and Fuller [3] and KPSS [9] tests for examining the existence of unit roots. Note that in the ADF test, testing the hypothesis of unit root is carried out hard to ensure that the hypothesis is rejected. Then, we also used KPSS tests, which can distinguish variables that appear to be stationary and be integrated.

Table 1. Test for unit root

Variable	ADF		KPSS	
	Level	First difference	Level	First difference
TT	1.002528	−11.422***	0.189***	0.308
CAN	0.339005	−10.542**	0.228**	0.053
Fr	−0.479604	−11.771***	0.136***	0.106
Dry	2.618902	−3.696***	0.513***	0.285
Local	−0.477985	−10.596***	0.383***	0.5

Note: "***", "**" denote, respectively, very strong and decisive evidence support H_1: No unit root.

Table 1 shows the results of the stationary tests in the level and in first difference for all variables. We include a constant and a trend term in these tests. The optimal lag length of each case for the ADF tests is chosen by the Akaike information criterion (AIC) after testing for first and higher order serial correlation residuals. From Table 1, the results of testing for unit roots in the level variables with the ADF test were against the alternative with the KPSS test. For the ADF tests, all of them have indicated that accepting the null hypothesis contains a unit root in level, but it appears rejected when the null hypothesis or variables become stationary after taking the first differences. Moreover, the results of the KPSS, null hypothesis is stationary: tests show that the null hypothesis in the level reject all of the variables, which is a unit root, while the stationary under the null hypothesis is accepted after having the differencing once. Consequently, we suggest that all of the variables are integrated by order I(1). Then, we checked the order of lag of variables by LR test statistic (each test at 5% level) and Hannan-Quinn information criterion. The results are shown in Table 2.

Table 2. Hansen–Seo test of threshold co–integration

Model	lag	sup LM0	sup LM
TT vs Local	1		
Test satistic value		21.60099	66.07652
Bootstrap p-values		0.004***	0.06*
Threshold value		37.71496	4.11
Cointegrating vector estimate		1	0.722
CAN vs Local	3		
Test satistic value		42.40207	72.86081
Bootstrap p-values		0.05**	0.09*
Threshold value		-318.276	-3.687
Cointegrating vector estimate		1	1.364
Fr vs Local	1		
Test satistic value		32.43801	38.12623
Bootstrap p-values		0.07*	0.00***
Threshold value		28.30927	4.12
Cointegrating vector estimate		1	0.599
Dry vs Local	1		
Test satistic value		37.64151	75.74651
Bootstrap p-values		0.00***	0.07*
Threshold value		45.2627	-7.7
Cointegrating vector estimate		1	0.997

Note: Note: “***”, “**”, “*”, denote strong, very strong and decisive evidence support H_1: Exist threshold effect.

5 TVECM Estimation Results

5.1 Results of Asymmetric Threshold Co-integration Tests

Remind that the bivariate relationship between various exported product prices against domestic longan price of Thailand is investigated in this study. To examine whether there exists a non-linear cointegrated in each pair. We use $Sup\ LM^0$ and $Sup\ LM$ that was proposed by Hansen and Seo (2002) for the testing of threshold co-integration. We use 5000 simulation replications of bootstrap method to calculate p-values in both of the two tests. We check the lag length of the VECM model by the Akaike and Bayesian information criteria, where the results are lag 1 for the case of co-integration in total price (TT) vs local price (Local), fresh price (Fr) vs local price and dried price (Dry) vs local price. For cointegration in canned price (CAN) vs local price, lag 3 performs the best specification. The results are reported in Table 2. We find that results show that the null hypothesis of linear co-integration is rejected in all models. Consequently, we can further estimate TVECM for all pairs.

5.2 Results of the Two Regime Error Correction Models

From the results of the co-integration tests, we have to estimate two-regime VECM in order to investigate the asymmetric dynamic movement between local longan price and all export longan price variables. The results are given below. In each TVECM equation, ObsR1 and ObsR2 represent the percentages of sub-sample on total sample size when the error-correction term is below and above the certain threshold value, respectively. The p-value is reported in parentheses. Moreover, SSR is sum of square residual of the model. When we have no formal distribution theory for the parameter estimates and standard errors, these should be interpreted somewhat cautiously.

5.2.1 Model of Total Price vs. Local Price

$$\Delta TT_t = \begin{cases} \underset{(0.288)}{-2.807} + \underset{(0.019)}{0.104}\, trend + \underset{(0.06)}{0.518}\, w_{t-1} + \underset{(0.751)}{0.0466}\, \Delta TT_{t-1} + \underset{(0.22)}{0.255}\, \Delta Local_{t-1} + u_{1t}, & w_{t-1} \leq 4.11 \\ \underset{(0.03)}{7.383} \underset{(0.39)}{-0.053}\, trend \underset{(0.00)}{-0.846}\, w_{t-1} \underset{(0.7)}{-0.054}\, \Delta TT_{t-1} \underset{(0.653)}{-0.125}\, \Delta Local_{t-1} + u_{1t}, & w_{t-1} > 4.11 \end{cases} \tag{4}$$

$$\Delta Local_t = \begin{cases} \underset{(0.072)}{-2.912} + \underset{(0.13)}{0.041}\, trend + \underset{(0.09)}{0.281}\, w_{t-1} + \underset{(0.536)}{0.056}\, \Delta TT_{t-1} + \underset{(0.161)}{0.179}\, \Delta Local_{t-1} + u_{2t}, & w_{t-1} \leq 4.11 \\ \underset{(0.00)}{7.046} \underset{(0.44)}{-0.0295}\, trend \underset{(0.5)}{-0.082}\, w_{t-1} \underset{(0.5)}{-0.06}\, \Delta TT_{t-1} + \underset{(0.07)}{0.315}\, \Delta Local_{t-1} + u_{2t}, & w_{t-1} > 4.11 \end{cases} \tag{5}$$

ObsR1 = 73.2%; ObsR2 = 26.8%; SSR = 10293.22

The estimated VECM results between Total price and Local price are presented above. Let $w_{t-1} = \Delta^2 Y_{t-1}$ be the vector of the error-correction term. From the results, the error-correction effects or speed of adjustment, β, are both positive in the first regime. However, it appears a negative sign in the second regime. This indicates that there is no cointegration relationship between Total price and Local price in regime 1 or low longan price regime while Total price and Local price are found to have cointegrated relationship in the regime 2 or high longan price regime. Figure 1 plots the error-correction effect, the estimated regression response of ΔTT_t and $\Delta Local_t$ as a function of w_{t-1} in the previous period when holding the other variables constant. In Fig. 1, when the error-correction term is below the threshold value $w_{t-1} \leq 4.41$, we can see that there is a sharply error correction effect on the left side of the threshold in both total price and local price equations. The speed of adjustment in the error-correction to the response of both total price and local price would lead to a stronger positive response, indicating that there is no long run equilibrium in the low longan price regime. Nevertheless, on the right side of the threshold, the responses of both total price and local price will decrease when the error-correction exceeds over the threshold value, thus confirming a long run equilibrium.

These findings show that there exists a long-run relationship between total price and local price when the error-correction term exceeds a threshold, thus making efficiency with both prices. However, these prices seem to diverge from their equilibrium in the low price longan regime(error-correction term belows a threshold). Hence, the government should be careful and concentrate on the difference between these two prices and the appropriate policy should be used to manage the equilibrium price, when the longan market price turns to downturn.

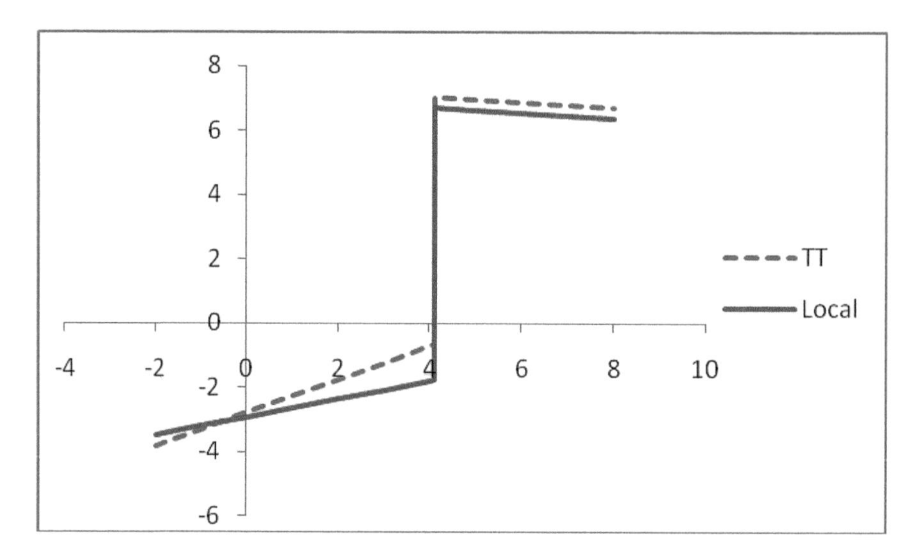

Fig. 1. Response of total price and local price to error correction (January 2005 to March 2013)

5.2.2 Model of Canned Price vs. Local Price

$$\Delta CAN_t = \begin{cases} \underset{(0.303)}{-0.062} trend \underset{(0.047)}{-0.2849} w_{t-1} \underset{(0.002)}{-0.5191} \Delta CAN_{t-1} \\ \underset{(0.007)}{-0.432} \Delta Local_{t-1} \underset{(0.189)}{-0.2555} \Delta CAN_{t-2} \underset{(0.052)}{-0.2590} \Delta Local_{t-2} \\ \underset{(0.617)}{+0.0742} \Delta CAN_{t-3} \underset{(0.077)}{-0.287} \Delta Local_{t-3} + u_{1t}, \quad w_{t-1} \leq -3.687 \\ \underset{(0.774)}{0.005} Trend \underset{(0.6931)}{-0.0349} w_{t-1} \underset{(0.490)}{-0.109} \Delta CAN_{t-1} \\ \underset{(0.104)}{+0.215} \Delta Local_{t-1} \underset{(0.021)}{-0.189} \Delta CAN_{t-2} \underset{(0.617)}{-0.058} \Delta Local_{t-2} \\ \underset{(0.355)}{+0.161} \Delta CAN_{t-3} \underset{(0.754)}{-0.035} \Delta Local_{t-3} + u_{1t}, \quad w_{t-1} \geq -3.687 \end{cases} \tag{6}$$

$$\Delta Local_t = \begin{cases} \underset{(0.609)}{0.0339trend} + \underset{(0.2458)}{0.1825w_{t-1}} - \underset{(0.011)}{0.4561\Delta CAN_{t-1}} \\ \underset{(0.8463)}{-0.0334\Delta Local_{t-1}} \underset{(0.001)}{-0.7474\,\Delta CAN_{t-2}} \underset{(0.073)}{-0.2629\,\Delta Local_{t-2}} \\ \underset{(0.057)}{-0.3159\Delta CAN_{t-3}} \underset{(0.2608)}{-0.2006\,\Delta Local_{t-3}} + u_{2t}, \quad w_{t-1} \leq -3.687 \\ \underset{(0.024)}{-0.449Trend} + \underset{(0.0039)}{0.2892w_{t-1}} + \underset{(0.811)}{0.0417\Delta CAN_{t-1}} \\ \underset{(0.228)}{+0.175\,\Delta Local_{t-1}} + \underset{(0.655)}{0.074\,\Delta CAN_{t-2}} \underset{(0.057)}{-0.247\,\Delta Local_{t-2}} \\ \underset{(0.605)}{-0.099\Delta CAN_{t-3}} + \underset{(0.895)}{0.016\,\Delta Local_{t-3}} + u_{2t}, \quad w_{t-1} \geq -3.687 \end{cases} \tag{7}$$

ObsR1 = 29.5%; ObsR2 = 70.5%; SSR = 4037.641

The estimated TVECM result between Can and Local price are presented above. From the results, there exists very strong evidence in the first regime of canned equation and the second regime of Local equation for error-correction effects. On the contrary, it does not appear at the evidence support of error-correction effect in first regime of local price and also in the second regime in the Can equation.

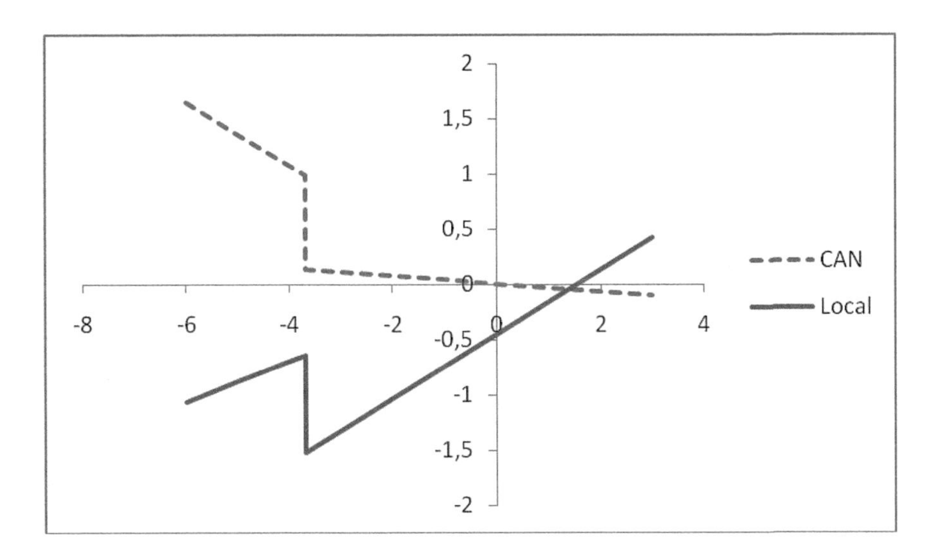

Fig. 2. Response of canned price and local price to error correction (January 2005 to March 2013)

Figure 2 plots the error-correction effect, the estimated regression response of ΔCAN_t and $\Delta Local_t$ as a function of w_{t-1} in the previous period when holding the other variables constant. We can see that the error correction effect β on the canned price equation is negative in both regimes, meaning that canned price has long run equilibrium in both regimes. Nevertheless, the local price equation diverges from the equilibrium as the positive error correction term is obtained in both regimes. These findings show that the error-correction terms respond more to canned price than local price and thereby making the canned price having more efficiency.

5.2.3 Model of Fresh Price vs. Local Price

$$\Delta Fr_t = \begin{cases} -\underset{(0.04)}{0.18}\, w_{t-1} + \underset{(0.88)}{0.0213}\, \Delta Fr_{t-1} - \underset{(0.8)}{0.0125}\, \Delta Local_{t-1} + u_{1t}, & w_{t-1} \leq 4.12 \\ \underset{(0.72)}{0.028}\, w_{t-1} - \underset{(0.001)}{0.624}\, \Delta Fr_{t-1} + \underset{(0.001)}{0.212}\, \Delta Local_{t-1} + u_{1t}, & w_{t-1} > 4.12 \end{cases} \tag{8}$$

$$\Delta Local_t = \begin{cases} \underset{(0.04)}{0.488}\, w_{t-1} + \underset{(0.21)}{0.463}\, \Delta Fr_{t-1} - \underset{(0.75)}{0.04}\, \Delta Local_{t-1} + u_{2t}, & w_{t-1} \leq 4.12 \\ \underset{(0.001)}{0.787}\, w_{t-1} - \underset{(0.00)}{1.22}\, \Delta Fr_{t-1} + \underset{(0.00)}{0.8}\, \Delta Local_{t-1} + u_{2t}, & w_{t-1} > 4.12 \end{cases} \tag{9}$$

The result of Fresh price and Local price are presented above. From the results, in both regimes of local price equation, error-correction effects are exist in all equations, except for error correction term of Fresh price in the second regime. Figure 3 plots the error-correction effect. In this figure, when the error-correction term is below the threshold value $w_{t-1} \leq 4.12$, we can see that there is a negative sign of error correction effect only on Fresh price equation in the first regime while, the others present all positive sign of error correction term. This indicates that the speed of adjustment in the error-correction to the response of Fresh price would lead to a price equilibrium in the low longan price regime.

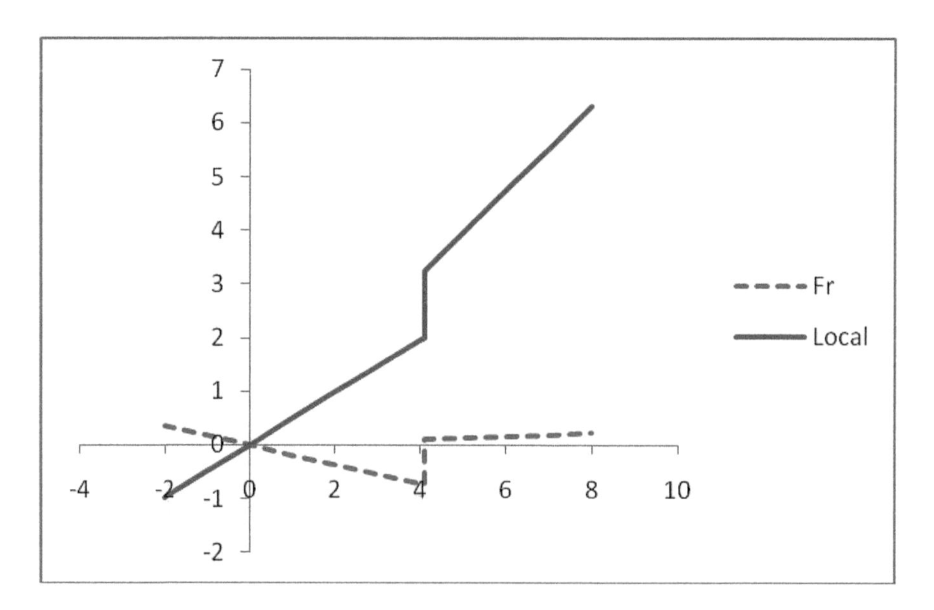

Fig. 3. Response of fresh price and local price to error correction (January 2005 to March 2013)

5.2.4 Model of Dried Price vs. Local Price

$$\Delta Dry_t = \begin{cases} \underset{(0.001)}{1.177} + \underset{(0.001)}{3.53}\, w_{t-1} - \underset{(0.001)}{4.18}\, \Delta Dry_{t-1} + \underset{(0.67)}{0.1}\, \Delta Local_{t-1} + u_{1t}, & w_{t-1} \leq -7.7 \\ \underset{(0.59)}{0.051} - \underset{(0.00)}{0.895}\, w_{t-1} - \underset{(0.81)}{0.022}\, \Delta Dry_{t-1} - \underset{(0.90)}{0.09}\, \Delta Local_{t-1} + u_{1t}, & w_{t-1} > -7.7 \end{cases} \tag{10}$$

$$\Delta Local_t = \begin{cases} \underset{(0.28)}{0.045} + \underset{(0.05)}{0.33}\, w_{t-1} + \underset{(0.67)}{0.1}\, \Delta Dry_{t-1} + \underset{(0.37)}{0.178}\, \Delta Local_{t-1} + u_{2t}, & w_{t-1} \leq -7.7 \\ \underset{(0.33)}{0.014} - \underset{(0.08)}{0.058}\, w_{t-1} - \underset{(0.4)}{0.013}\, \Delta Dry_{t-1} + \underset{(0.5)}{0.823}\, \Delta Local_{t-1} + u_{2t}, & w_{t-1} > -7.7 \end{cases} \tag{11}$$

ObsR1 = 33%; ObsR2 = 67%; SSR = 123990.5

Finally, the estimated results between Dried price and Local price are presented above. From the results, there exists a decisive evidence in error-correction effects and the negative effects are obtained for both Dried price and Local price equations in the second regime, indicating a cointegrated relationship between Dried price and Local price when the longan market price is high. Figure 4 plots the error-correction effects of these two variables. In this figure, when the error-correction term is below the threshold value $w_{t-1} \leq -7.7$, we can observe that there is a positive error correction effect both in Dried price and Local price equations. These show that there is no long run equilibrium for the first regime in both equations. After the threshold value $w_{t-1} > -7.7$, there exists a large response in the speed of adjustment of the error-correction β in Dried price and Local price equations. Moreover, it changes from the positive direction to negative direction, indicating a presence of the long run equilibrium in high longan price regime. This findings indicate that Dried price and Local price have more efficiency in high longan price regime.

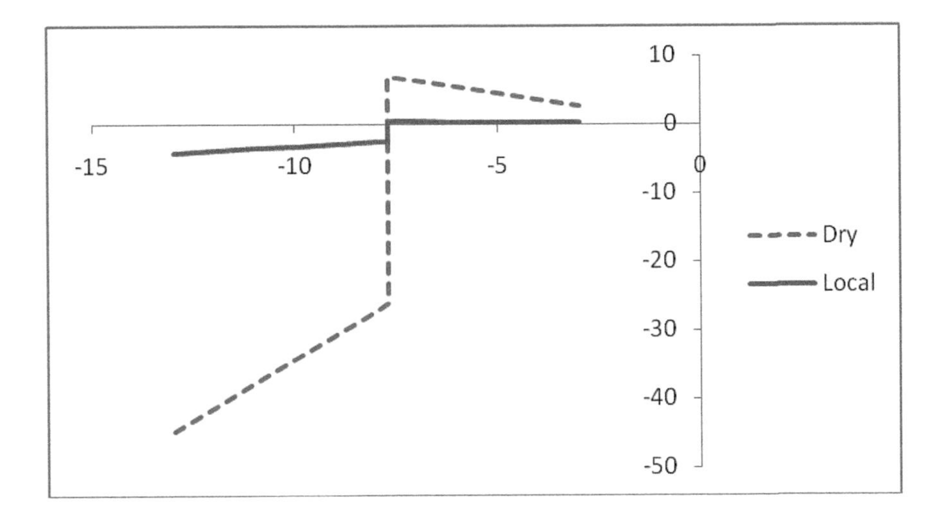

Fig. 4. Response of dried price and local price to error correction (January 2005 to March 2013)

6 Conclusion

This study examined the non-linear long-run equilibrium relationship between local longan price and all export longan price variables in Thailand, in order to exhibit unfavourable evidence of linear co-integration in the literature. Updated monthly data for the period of 1998:1 to 2013:3 were analysed. The threshold co-integration test was developed by the contribution of Hansen and Seo (2002), who concentrated on the possibility of an asymmetric adjusting process among time series variables.

The results from the test rejected the null hypothesis of linear co-integration between local longan price and all export longan price variables. The threshold co-integration model confirms that there is non-linear long-run equilibrium relationship between local longan price and all export longan price variables in Thailand. There exists an asymmetric dynamic adjusting processes between local longan price and all export longan price variables in Thailand. We can see that there are no long run equilibrium in local price, fresh price and total price when the error correction term is less than the threshold level. Moreover, it also has a non-equilibrium in fresh price in the second regime. Therefore, policy-makers should create an effective policy system to improve efficiency in this disequilibrium regime. The Thai economy is facing a challenge of making a sustainable management policy to become an equilibrium of these variables. Fiscal and monetary policy management is a possible tool to intervene and control this uncertain market.

To promote efficiency of adjusting with speed, it is necessary for the government to effectively and efficiently intervene in the configuration of the price of longan product. It should make directly at local price, fresh price and total price which have the size of the price gap less than threshold level and fresh price in the second regime for equilibrium in the long run. Especially in the local price, the government should emphasize this market because the local price in each regime has no equilibrium. On the contrary, there are dried, canned and total longan prices in the second regime that has a long run equilibrium. It had been less affected by the policy shock. So, the government should not emphasize in these variables but they should apply caution in the policy or monitor on these variables. Not only on the fiscal policy but also on creating the attitude towards working behavior for improving productivity because it is impossible for the government to intervene the longan price forever. The government has been exposed to expenses ranging within millions of baht just by intervening in the longan market each year. Therefore, investments in education, research and development in central longan market are the keys for increasing productivity of Thai farmers. Further study should be added on with the policy variables such as fiscal policy, monetary policy and trade policy for analysis of these shock effects to longan price.

Acknowledgements. We are grateful to Office of the Higher Education Commission for providing the scholarship and Faculty of Economics Chiang Mai University for providing economic tool for analysis of data. Moreover, we are grateful to Dr. Ravee Phoewhawm for providing the technical support to this paper.

References

1. Andrews, D.W.: Tests for parameter instability and structural change with unknown change point. Econometrica J. Econometric Soc., 821–856 (1993)
2. Balke, N.S., Fomby, T.B.: Threshold cointegration. Int. Econ. Rev., 627–645 (1997)
3. Dickey, D.A., Fuller, W.A.: Distribution of the estimators for autoregressive time series with a unit root. J. Am. Stat. Assoc. **74**(366a), 427–431 (1979)
4. Dwyer Jr., G.P., Wallace, M.S.: Cointegration and market efficiency. J. Int. Money Finance **11**(4), 318–327 (1992)
5. Forbes, C.S., Kalb, G.R., Kofhian, P.: Bayesian arbitrage threshold analysis. J. Bus. Econ. Stat. **17**(3), 364–372 (1999)
6. Hansen, B.E.: Inference when a nuisance parameter is not identified under the null hypothesis. Econometrica J. Econometric Soc., 413–430 (1996)
7. Hansen, B.E., Seo, B.: Testing for two-regime threshold cointegration in vector error-correction models. J. Econ. **110**(2), 293–318 (2002)
8. Held, L., Ott, M.: How the maximal evidence of p-values against point null hypotheses depends on sample size. Am. Stat. **70**(4), 335–341 (2016)
9. Kwiatkowski, D., Phillips, P.C., Schmidt, P., Shin, Y.: Testing the null hypothesis of stationarity against the alternative of a unit root: How sure are we that economic time series have a unit root? J. Econ. **54**(1–3), 159–178 (1992)
10. Lo, M.C., Zivot, E.: Threshold cointegration and nonlinear adjustment to the law of one price. Macroecon. Dyn. **5**(4), 533–576 (2001)
11. Mishra, R., Kumar, A.: The spatial integration of vegetable markets in Nepal. Asian J. Agric. Dev. **8**(1), 101 (2013)
12. Nazlioglu, S.: World oil and agricultural commodity prices: evidence from nonlinear causality. Energy Policy **39**(5), 2935–2943 (2011)
13. Rapsomanikis, G., Hallam, D.: Threshold cointegration in the sugar-ethanol-oil price system in Brazil: evidence from nonlinear vector error correction models. In: FAO Commodity and Trade Policy Research Working Paper, 22 (2006)
14. Saghaian, S.H.: The impact of the oil sector on commodity prices: correlation or causation? J. Agric. Appl. Econ. **42**(3), 477–485 (2010)
15. Tansuchat, R., Maneejuk, P., Wiboonpongse, A., Sriboonchitta, S.: Price transmission mechanism in the Thai rice market. In: Causal Inference in Econometrics, pp. 451–461. Springer, Cham (2016)
16. Tsay, R.S.: Testing and modeling multivariate threshold models. J. Am. Stat. Assoc. **93**(443), 1188–1202 (1998)
17. Yu, T.H., Bessler, D.A., Fuller, S.: Cointegration and causality analysis of world vegetable oil and crude oil prices. In: The American Agricultural Economics Association Annual Meeting, Long Beach, California, pp. 23–26, July 2006

Impact of the Transmission Channel of the Monetary Policies on the Stock Market

Tran Huy Hoang(✉)

University of Finance – Marketing, Ho Chi Minh City, Vietnam
th.hoang@ufm.edu.vn

Abstract. Developing since early 2000s only to date, the stock market of Vietnam has gradually become the most significant capital attraction channel of the enterprises, it is also the potential method for investment by the local and foreign investors accordingly. With the incessant innovation and development market, the Government has still applied the traditional monetary policies and the similar means. There aren't many reports on the scientific research which concentrate on the extensive research on these policies. The research topic of impact of the transmission channels of the monetary policies on the stock market includes the data collected from 2010 to 2017 for determining the appropriate policies for the current time of economy. The result is shown that the point of view is the same with the previously scientific researches.

1 Introduction

Mostly, the formerly experimental researches have shown the general results of the relationship between the monetary policies and the stock price, however, there aren't many presses which show in detail the correlation between the monetary policies and the stock price. With the different monetary regimes, the two above-mentioned factors contain the relationship with many various diversified degrees (Laopodis 2013), it is the same viewpoint as Fausch and Sigonius (2018) that the considerable and important responses of the stock market including the changes in the monetary policies can be observed when the net interest is negative. The research result shows that both the monetary policies and the stock market affect mutually, however, this relation isn't large enough for creating the remarkable changes (Laopodis 2013; Haitsma et al. 2016).

When implementing the establishment of one monetary policy, the authorities always consider the feedback of the stock market for adjustment. The large countries in the world admit that the stock market is one important factor for stabilizing the market economy and completing their new policies. Otherwise, the stock market is always the subject which is suffered from many impacts from many various channels of the monetary policies, specially, the interest and the interest ratio for compulsory reserves (Chatziantoniou et al. 2013). Therefore, the relationship between the monetary policies and the stock market, or the stock

V. Kreinovich et al. (Eds.): ECONVN 2019, SCI 809, pp. 1028–1051, 2019.
https://doi.org/10.1007/978-3-030-04200-4_75

price is the two-way relationship, both of sides must consider and respond with the movement of other party. Impact of changes for the monetary policies on the stock profit is inconsiderable in 1990s and such impact is improved, there are the significant statistical impacts since 2000a to date (Jansen and Zervou 2017), the research result supports the inconsiderable and poor important influences on the stock price during the period of economic bubble and these impacts are requested for maintaining when the bubble period is broken.

Actually, the policy planner has two options when making the decision to propose the new monetary policies: The expansionary monetary policy or the tight monetary policy, in spite of using through any channel for the market control, the purpose of the monetary policy is the market stability when the liquidity changes significantly (change of money supply in the stock market is the stock price) and interest. When the economy operates normally, the interest target is selected. It is supposed that the policy planner combines the expectations of the politicians as the standard for making the decision of policy, out of 70% countries, implementing the policy of interest use makes the stock price returns the long-term stock price (Hung and Ma 2017). In addition, the monetary policy has the differently impact for the bull market and the bear market or the various bull markets (Laopodis 2013).

The optimal expansionary monetary policy is made when having the price increase in the stock price, therefore, the transaction participants collect much profit when buying stock, or the more expensive assets in the present, it is divided equally the increase cost of the market participants in the economy (Zervou 2013). The research result of Fernández-Amador et al. (2013) shows that the expansionary monetary policy of ECB leads increase of liquidity in stock. The expansionary monetary policy leads increase in the stock profit (Haitsma et al. 2016). Zervou (2013) shows that the expansionary monetary policy makes increase of the (stochastic discount factor) because of increasing the output profit of transaction participants. Loss in the goods consumption for buying stock is less seriously, stock demand increases, stock price decreases. The restrictive expansionary monetary policy also creases or decreases the liquidity of individual shares, or, the more Company is small-scale, the more response for the liquidity impact of the monetary policy is significant. The foregoing findings are the general findings which are found that the developed regions, countries plan the clear, evident and periodic policy planning.

The tight monetary policy isn't notified in advance, it creates the negative information for the stock but the cash flow of owners (dividend) is put the price higher than the expected discount rate. This matter indicates that the unexpectation in the expected monetary policy shall make decrease for the profit, accordingly, it makes increase the liquidity for the stock market (Wu 2001). The tight monetary policy shall immediately make the bonds more attractive when comparing with the shares, accordingly, one part of impact for the monetary policy on the liquidity of the stock market shall be through the bond market (quality flight or liquidity flight) shown in the research of Haitsma et al. (2016). The tight policies which aren't notified in advance can increase the liquidity through impacts on the stock transactions. With new information in the market, the investors can make balance of their investment portfolio more positive between the shares and bonds, the cause is to promote the growth in the transaction quantity, Gospodinov and Jamali (2015).

Research of Tang et al. (2013) means the research on impact of the monetary policy on the monetary market and the stock market of China - the market has so many similar points with the market of Vietnam, the result shows that with the changes of monetary policy, it may considerably affect on the stock market. The stock profit shall increase after around 03 days since the date of happening the changes of the narrowed monetary policy in Shanghai stock market. Change of the stock market is determined by the shock of supply - demand and change on the monetary policy accordingly. Actually, in Vietnam, the monetary policy planners haven't properly approached on the response of the stock market. Although the stock expectations can create the liquidity for the stock market, the planners said that the liquidity of the stock market has the limit impact on the economy and it fails to make increase of effectiveness of the monetary policy. In addition, they don't think that determining the liquidity of the stock market belongs to their responsibility (Hung and Ma 2017). With the current situation, it is necessary for building the economic research project of the stock market.

2 The Research Method

Corresponding to many targets, the contents which are required for pay attention to, the used topic is diversified in the handling method, data analysis. Including the following analytical methods:

2.1 Method of Descriptive Statistics

Providing the mean statistics, maximum, minimum, and standard deviation. These statistics present the used data characteristics for research. This method of descriptive statistics is used for determining the nature of research subject, combines with the comparative method, we have the result required for evaluation.

2.2 The Quantitative Method Through Self-recurrent Model, Vector-VAR

This research topic uses the self-recurrent model, vector (VAR) for evaluating the trend and the relation degree between the time series. VAR model means VEC model, the recurrent variables, as the general kind of the single self-recurrent model when forecasting one set of variables. In the opinion of Sims (1980), VAR is considered as the valuable tool for surveying the dynamic effect of one shock for this variable on another variable. This is also the integrated method for checking the process with many time series when VAR provides the various criteria for proposing the optimal length for the variables. Furthermore, it includes the equation system that permits that the variables have the mutual relationship, and it is one system, accordingly, in which, including all variables which are considered as the endogenous variable. It is to say that this is the most model in the quantitative research for the monetary policies. Because the relationship between the economic variables doesn't purely impact one way only, in some cases, the independent variable (explanation variable) affects on the dependent variable and vice versa. Therefore, we should consider the mutual impact between these variables at the same time. In order to help to solve this matter, the optimal selection is to use VAR model, the model is rather flexible and easy for use in analysis with the multivariable time series. Analysis of VAR model includes three steps:

- ***The*** 1st ***step***: analyzing the forecast of the macro-variables, including the use of the vector-autoregression model (VAR model). This is the relative simple model using the data of time series, accordingly, the previous observation values are used for reaching the possible exact forecast. The difference between the forecast and result (the forecast mistake) for one specific variable is considered as one "shock", however, Sims finds that such forecasted mistakes have no the obvious meaning.
  For example, the interest is unexpectedly changed, it may the response of another shock, for example, unemployment or inflation, it also occurs "independently". Such independent change is called as "the basic shock".
- ***The*** 2nd ***step***: separates "the basic shock". This is prerequisite for the impact research of the "independent" interest change. Actually, one of the significant contributions of Sims is the demonstration for the comprehensive knowledge of operation method of the economy for reaching the recognition of "the basic shocks". Sims and his next researchers have developed the various methods for recognizing "the basic shock" in VAR model.
- ***The*** 3rd ***step***: Analyzing impulse – response [Temporary translation: Analyzing impulse and response]. This analysis illustrates the impact of basic shock for the macro-variables through the time.

Presently, VAR model is the integral tool of the Central Bank and the Ministry of Finance for analyzing the influence of many various shocks for the economy and influences of various policies for coping with the above-mentioned shocks.

VAR model in this research is detailed as follows:

$$Y_t = A_{0,1} + A_{1,t}Y_{t-1} + A_{p,t}Y_{t-p} + \varepsilon_t \quad (1)$$

In which:

- Y_t: is the vector, including: capitalization value of the stock market, VN-index, M2, exchange rate, the refinancing interest, credit limit and compulsory ratio.
- $A_{0,t}$: constant vector (intercept coefficient)
- $A_{0,t}$: as the coefficient matrix of other time (i = 1, ..., p)
- ε_t: as unit vector which satisfies some fixed conditions
- p: as the late value

Firstly, VAR model basically includes the endogenous variable: The capitalization value of stock market, VN-index, M2; and the variables representing the transmission mechanism of the monetary policy on the stock market. The sequence of variables is based on the assumption that change in the monetary policy shall be transmitted to the capitalization value of the stock market and VN-index. VAR model shall be expanded gradually through the various channels of the transmission mechanism of the monetary policy: Exchange rate, refinancing interest, credit limit and the compulsory reserves ratio in the model after terminating the inspection of basic model.

Data is handled by software "STATA 14"

3 Data Description and Variables

This research uses the data from the general department of statistics, from the annual reports, Decisions, Directives, Official letters of the State Bank, the database system, the financial criteria of International Monetary Fund (IFS-IMF); stage 2010–2017.

Model of Experimental Research: The topic aims at researching the influence of the monetary policy on the stock market of Vietnam through analyzing the transmission mechanism of the monetary policy, through the search efforts and knowledge of the author, many local and foreign research project on this matter have been done, the local research has been done actively for the last decade. From the pronounced result of research project, the author summarized some concepts which have discovered previously and tested separately by the researchers, including: The capitalization value of the stock market, stock index, M2, exchange rate, refinancing interest, credit limit. In which, opening the economy of Vietnam is considered, used the exchange rate VND/USD as the fixed exchange rate in the monetary policy of Vietnam. Besides, the direct tool which has been used by the State Bank of Vietnam in the monetary policy during the past is the compulsory reserves ratio, capital has been evaluated, considered in analyzing the impact of the macro-variable, the specific measurement of the impact degree and its transmission capacity in the operation of the monetary policy in Vietnam hasn't been researched in detail by any project.

4 Research Result

The research group uses the Augmented Dickey-Fuller test to check in all data of variables which exists the unit root. The testing result shows that the original data series have no stop (or including the unit root). After continuing testing with difference of Level 1 and Level 2 for all variables, with the meaning of 5%, the data series of the variables stop at the difference of Level 1. Therefore, the result for the integrated level of all variables is 1 or I (1). In order determine the effectiveness of the impact and evaluation of the monetary policy of the State Bank of Vietnam on the stock market, the research group divides the

Variables in model	Abbreviated	Time (month)	Source
Local area:			
Capitalization value of stock market	CV	2016T7–2017T11	UBCKNN
Stock index on HOSE	VN-index	2016T7–2017T11	UBCKNN
Money supply M2	M2	2010Q1–2017Q1	IFS-IMF
Transmission channels of the monetary policy			
Exchange rate	REER	2010Q1–2017Q2	IFS-IMF
Refinancing interest	INTEREST	2010Q1–2016Q4	IFS-IMF
Credit limit	TTTD	2010Q1–2017Q3	
Compulsory reserves ratio	DTBB	2016T1–2017T11	NHNN

variables which are considered into the group of 03 variables, then, carrying out the estimation of VAR model of each specific group. VAR model of the groups includes 03 variables, in consideration of the variable, in which, the representative variable for the transmission channels of the Central Bank is exogenous, the representative variables for the stock market is endogenous. It is supposed that change of monetary policy of the Central bank shall transmit and impact on the stock market. The testing result of VAR model is shown as follows:

Group 1: The bank interest and total money of M2 are the exogenous variable, VN-index is the endogenous variable.

Testing the unit root: Testing the unit root gives the result that all roots of the model is located within the unit circle, thus, VAR is stable. Stability of VAR model is the necessary condition for estimating VAR model (Fig. 1 and Table 1).

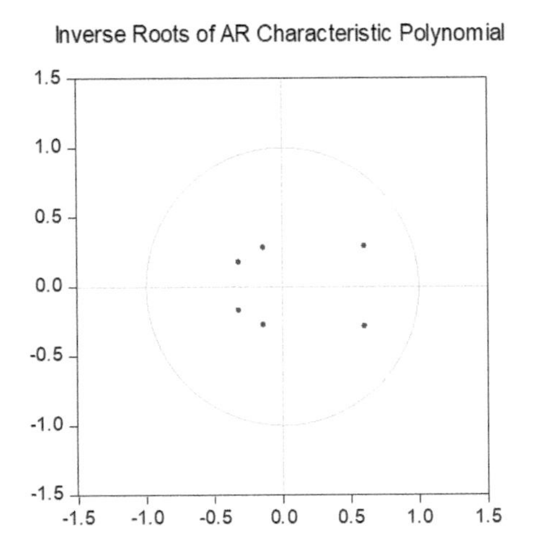

Fig. 1. Stability of model with the lag is 1

Table 1. The appropriate lag of model

Lag	LogL	LR	FPE	AIC	SC	HQ
0	−650.817	NA	1.04E+19	52.30533	52.4516	52.3459
1	−635.764	25.28824*	6.48e + 18*	51.82113*	52.40619*	51.98340*
2	−630.142	8.095769	8.82E+18	52.09137	53.11522	52.37534

Selecting the appropriate lag for VAR model

LR: Likelihood Ratio Test
PDE: Final Prediction Error
AIC: Akaike Information Criterion
SC: Schwarz Information Criterion
HQ: Hannan-Quinn Information Criterion

Basing on the criteria LR, PDE, AIC, SC and HQ of VAR model, the research group selects the appropriate lad as 1 (Table).

In order implement the research, the research group uses the lag in VAR for testing the existence of co-integration between the Interest variable and the two remaining variables. The achieved result of the values, Trace statistic and Max Eigenvalue show that there is at least 02 co-integrations between the variables in the model and there is existence for the long-term relationship of the considered variables (Table 2).

Table 2. Co-integration test

Hypothesized		Trace	0.05		Max-Eigen	0.05	
No. of CE(s)	Eigenvalue	Statistic	Critical value	*Prob.***	Statistic	Critical value	*Prob.***
None	0.439176	28.75258	29.79707	0.0656	13.88035	21.13162	0.3749
At most 1	0.34359	14.87222	15.49471	0.0619	10.10328	14.2646	0.2052
At most 2*	0.180209	4.768941	3.841466	0.029	4.768941	3.841466	0.029

Response of VN-index before the interest changes of the Commercial banks and total money

With the accumulative response graph of VN-index when the two bank interest variables and total money M2 change, the result is shown that when the interest of total money M2 change, as a result, there is the considerable change on the stock market. When the Central Bank implements the tight monetary policy, it causes the pressure of increase interest of the banks, it creates the impact immediately on VN-index. This can explain that when increasing the interest, the borrower trends on the loan reduction, the enterprises are more prudent in participating in the stock market and issuing the shares for capital attraction as well, in addition, the high interest also attracts the deposit need in the banks, it affects considerably on the stock market. Changing of increase total

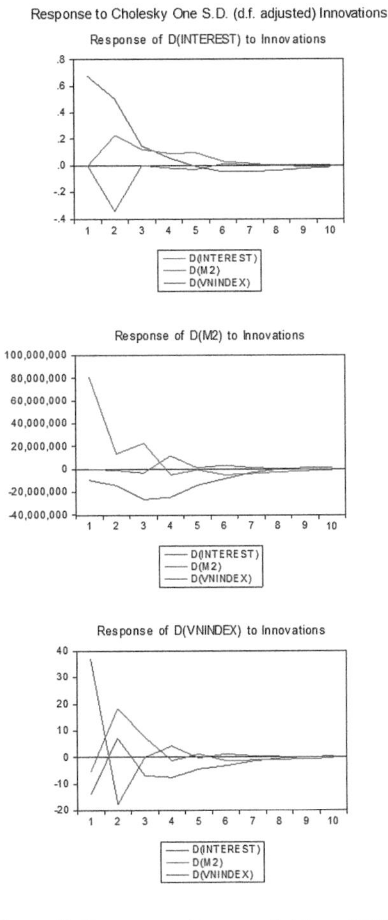

Fig. 2. Response function of the respective variables, INTEREST, M2, VN-index before changes of the two remaining variables.

money M2 and interest leads the significant decrease in VN-index, especially in the stage from Quarter 1 to half early Quarter 2. Since half Quarter 2 to half Quarter 4, when the Central Bank loosens the monetary policy, it impacts on the considerable increase of VN-index till half Quarter 3, after that, the growth is slow don and stops short. On the graph, the response functions of VN-index with INTEREST and M2 (Fig. 3), the breaking lines of responde line is put on or below the horizontal axis, therefore, it has the statistical meaning. Generally, the response line of VN-index for the bank interest trends the gradual increase toward the horizontal axis, the response function with M2 and INTEREST are advance toward 0 (Fig. 2).

Similar for VAR model, the group of the three remaining variables, the research group achieves the following result:

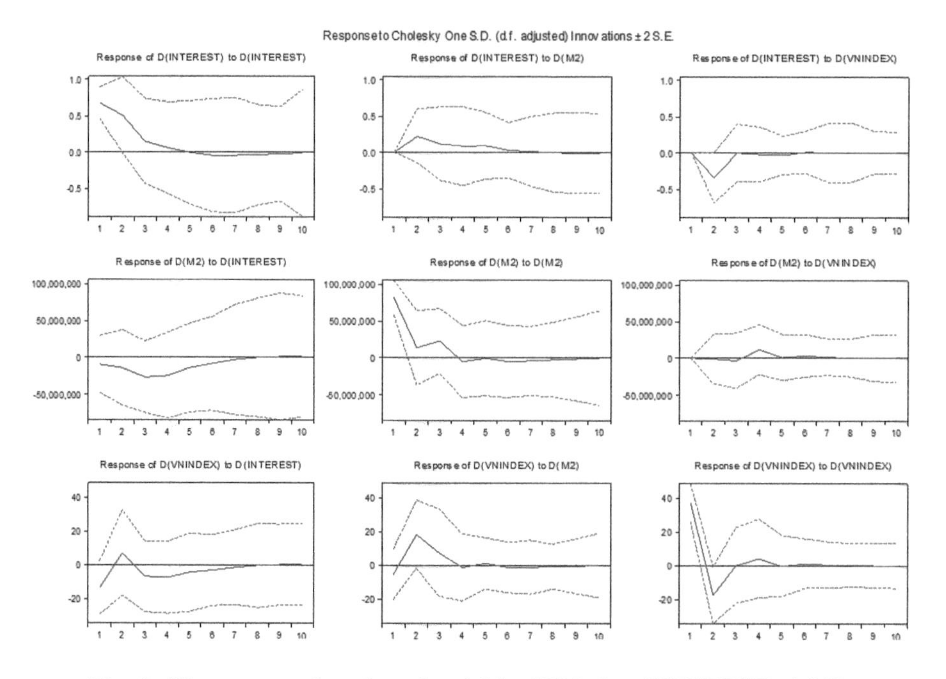

Fig. 3. The response function of variables VN-index, INTEREST và M2.

Group 2: Impact of the credit growth and total money M2 on VN-index.

Testing of the unit root shows the result that all roots of the model are located within the unit circle, therefore, VAT model is stable (Fig. 4). In order to find the optimal lag, the model is based on the criteria LR, PDE, AIC, SC and HQ of VAR model, the research group selects the appropriate lag 0. The result is shown in the following Table 3:

Table 3. Optimal lag

Lag	LogL	LR	FPE	AIC	SC	HQ
0	−684.673	NA^*	$1.57e+20^*$	55.01386*	55.16013*	55.05443*
1	−678.393	10.55167	1.96E+20	55.2314	55.81646	55.39368
2	−674.25	5.965989	3.01E+20	55.61996	56.64382	55.90393

The research group continues running the Granger Causality test for three variables, the achieved result is all p-value > α, therefore, VN-index is the dependent variable, the credit growth and total money M2 can impact VN-index (Table 4).

The result of co-integration test including the values of Trace statistic and Max Eigenvalue affirms that there is existence at least 1 co-integration vector

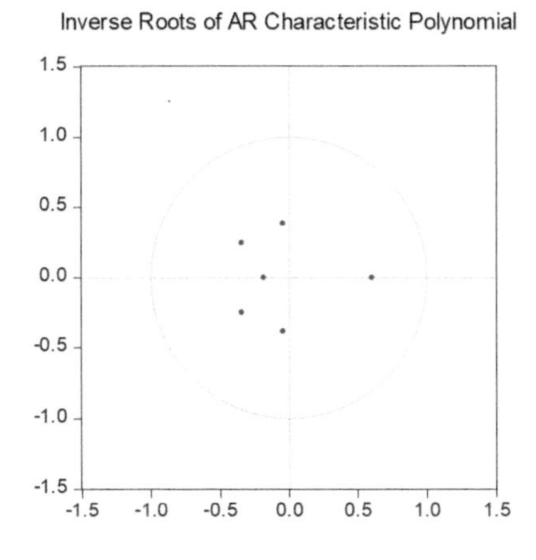

Fig. 4. Stability of model with lag 0.

Table 4. Optimal lag

Dependent variable:			D(VN-INDEX)
Excluded	Chi-sq	df	Prob.
D(TTTD)	0.114004	2	0.9446
D(M2)	5.966476	2	0.0506
All	7.878686	4	0.0961

with meaning of 5%. This shows that there is the considerable long-term relationship between the research variables (Table 5).

Observation of response function of VN-index with change of credit growth and total money M2 can come to the conclusion that the credit growth fails to impact obviously on VN-index, it only exists in the stage from Quarter 1 to half Quarter 2, there is considerable reduction of VN-index when the credit growth trends reduction. Immediately after, the response function of VN-index increases suddenly in Quarter 3, the fluctuation in Quarter 4, 5 and it is stable gradually then. Increase of money supply M2 leads the reduction in the response function of VN-index, the considerable amount of money is pumped into the share market and creates the bubble in the stock market. Actually, in the stage

Table 5. The co-integration test of the credit market, M2 and VN-index

Hypothesized		Trace	0.05		Max-Eigen	0.05	
No. of CE(s)	Eigenvalue	Statistic	Critical value	Prob.**	Statistic	Critical value	Prob.**
None*	0.588551	37.23993	29.79707	0.0058	21.31366	21.13162	0.0472
At most 1*	0.416641	15.92627	15.49471	0.043	12.93485	14.2646	0.0802
At most 2	0.117188	2.991423	3.841466	0.0837	2.991423	3.841466	0.0837

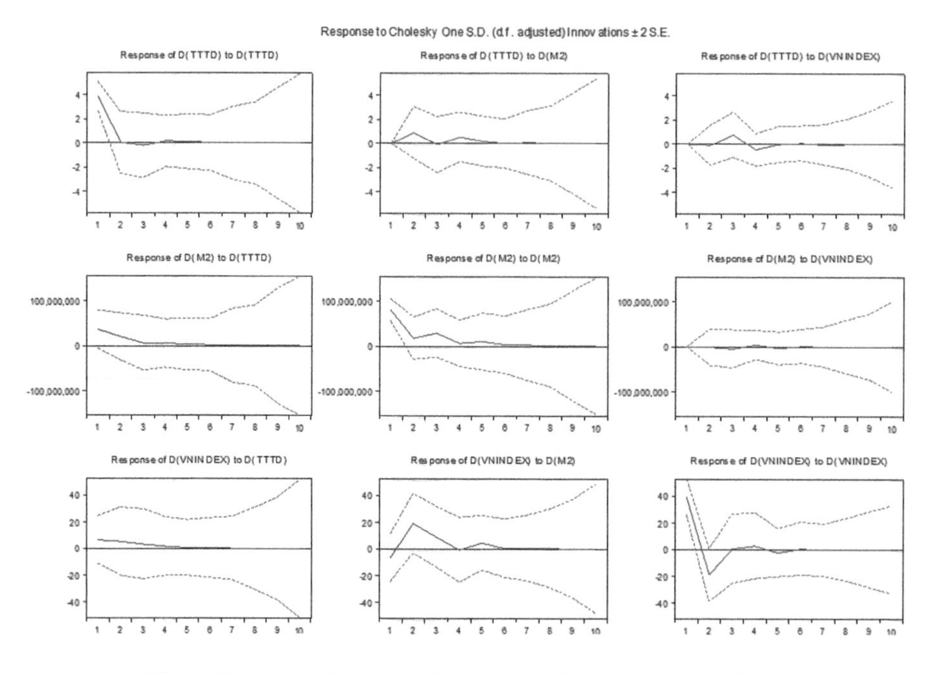

Fig. 5. Response function of variables VN-index, M2, TTTD.

of 2002–2007, at that time, VN-index increased more than fivefold till 1.000 points, immediately after, in the stage of 2008, because of the growth speed of money supply decreases considerably, it caused the race out of the share market since the scores of VN-index reduces quickly.

Group 3: VN-index is the dependent variable, total money M2 and the net exchange rate are the independent variable.

The result of lag is appropriate for VAR model with the standards LR, FPE, AIC, SC, HQ of the variables REER, M2 and VN-index is value 0 (Table 6). And the result of Granger Causality test shows that REER variable and total money M2 can impact on VN-index, with the meaning of 5% (Table 7).

LR: Likelihood Ratio Test
PDE: Final Prediction Error
AIC: Akaike Information Criterion
SC: Schwarz Information Criterion
HQ: Hannan-Quinn Information Criterion

The testing of unit root shows the result that all roots of model is located within the unit circle, VAR model is stable. The stable VAR model is the necessary condition so that VAR model can be estimated (Fig. 6).

In order to test the co-integration existence between the interest variable and two remaining variables, the research uses the values of Trace statistic and Max Eigenvalue. The result states that there is at least 02 co-integrations between

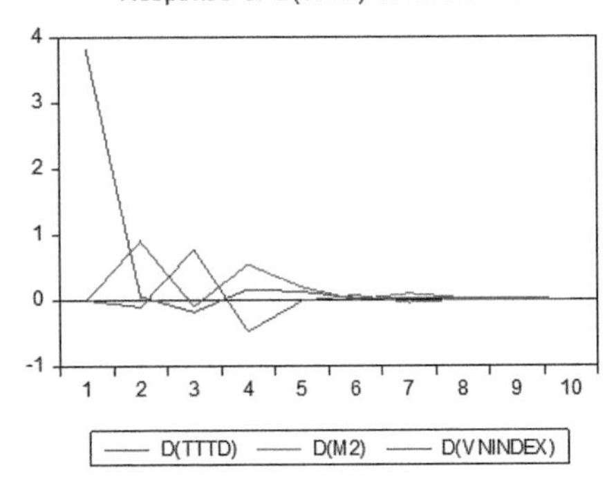

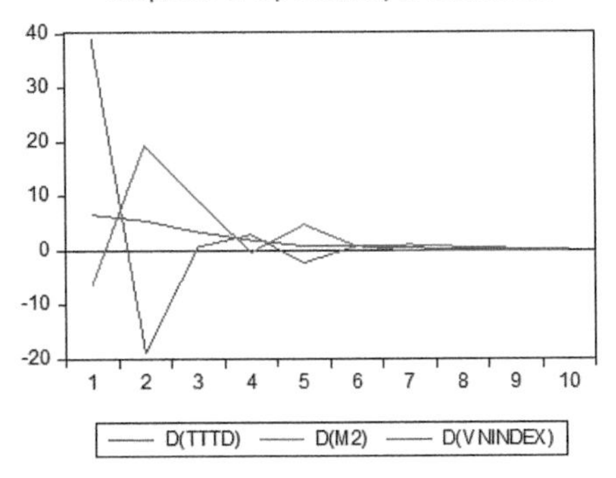

Fig. 6. The response function of respective variables of the credit market, M2, VN-index before changes of the two remaining variables.

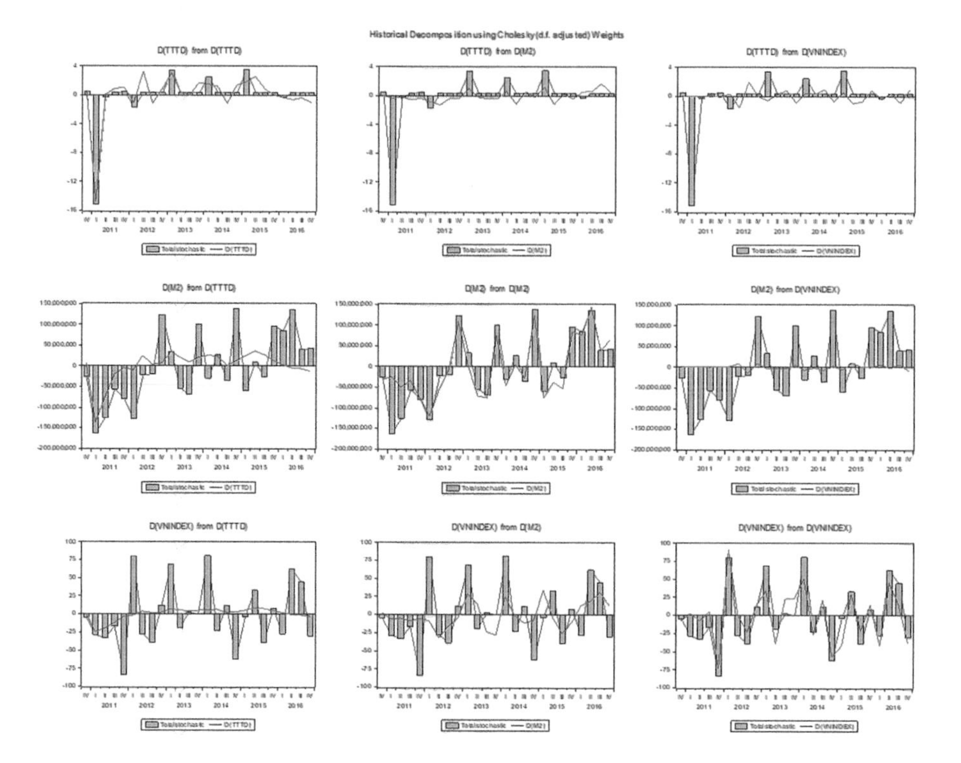

Fig. 7. The response function of respective variables of the credit market, M2, VN-index before changes of the two remaining variables.

Table 6. The appropriate lag of VAR model

Lag	LogL	LR	FPE	AIC	SC	HQ
0	−786.039	NA^*	$5.21e+23^*$	63.12314*	63.26940*	63.16371*
1	−778.996	11.83186	6.14E+23	63.27972	63.86478	63.44199
2	−769.494	13.68431	6.13E+23	63.23948	64.26333	63.52345

the variables in the model and there is existence of the long-term relationship of the considered variables (Figs. 7, 8 and Table 8).

The response of VN-index before changes of the net exchange rate and total M2.

Basing on the result graph, when the net exchange rate increases, there is decrease of VN-index from Quarter 1 to half Quarter 2. Since half Quarter 2 to half Quarter 3, the opposite trend of VN-index and REER still continue happening when REER reduces, accordingly, VN-index increases. When the exchange rate is considerably increased, value of local currency is reduced or the interest rate for the foreign currency is higher than the local currency, everyone trends to reserve the foreign currency or deposits at the bank. A large amount of people

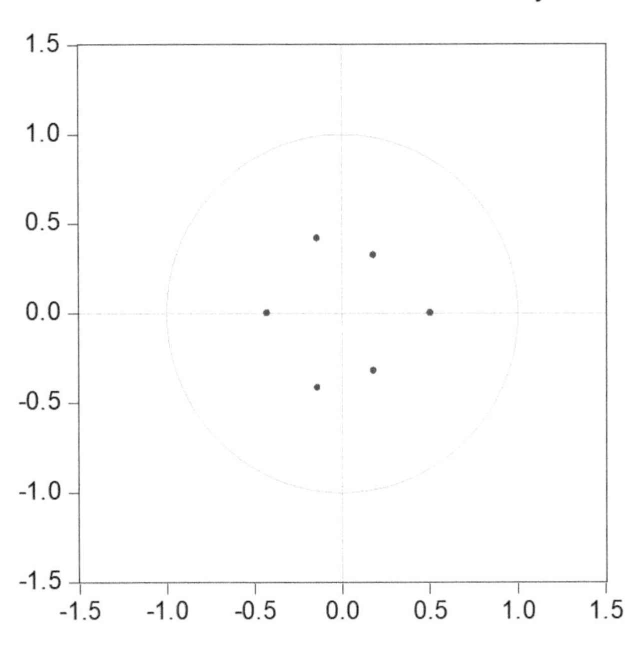

Fig. 8. Stability of model for the optimal lag is 0.

Table 7. Granger Causality test

Dependent variable:			D(VN-INDEX)
Excluded	Chi-sq	df	Prob.
D(TTTD)	2.412268	2	0.2994
D(M2)	3.947173	2	0.139
All	11.16212	4	0.0248

who buy shares leaves the stock market, when the State Bank increases the bank interest for the local currency for making balance of supply and demand for the foreign exchange. Since half Quarter 3 to half Quarter 4, REER and VN-index grows the same direction, after that, the stability is advanced forward to 0 (Figs. 9, 10, 11 and 12).

Table 8. Co-integration test

Hypothesized		Trace	0.05		Max-Eigen	0.05	
No. of CE(s)	Eigenvalue	Statistic	Critical value	Prob.**	Statistic	Critical value	Prob.**
None*	0.526611	31.51153	29.79707	0.0314	17.9481	21.13162	0.1318
At most 1	0.303292	13.56343	15.49471	0.0957	8.673319	14.2646	0.3144
At most 2*	0.184338	4.890113	3.841466	0.027	4.890113	3.841466	0.027

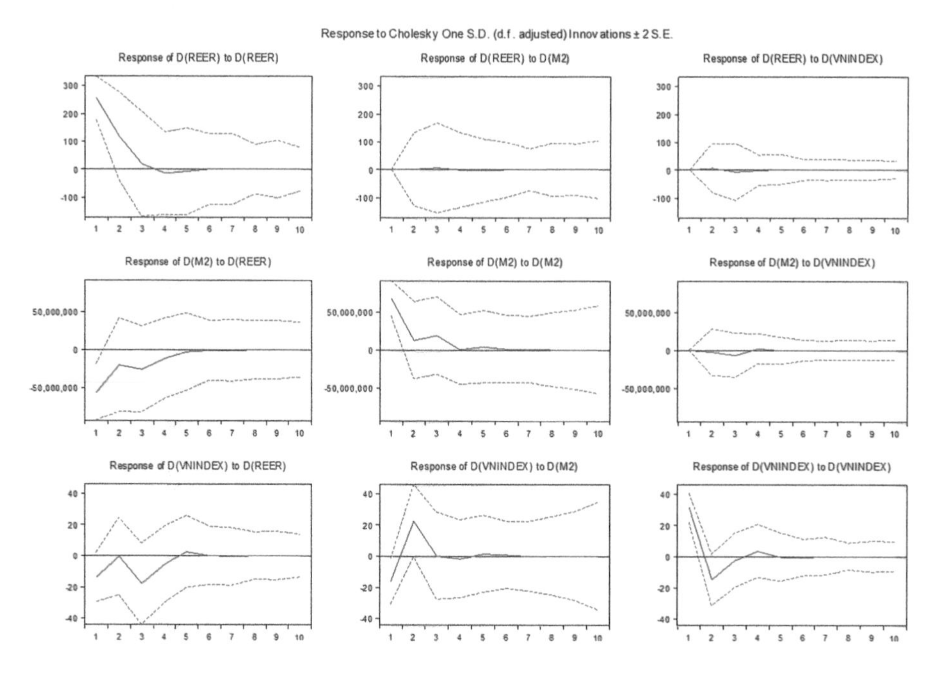

Fig. 9. Response of REER, M2 and VN-index.

Group 4: The capitalization value of the stock market (MARKETCAP) is the endogenous variable, the interest of Central Bank and total money M2 is the exogenous variable.

Testing of the unit root shows the result that all roots of model are located within the unit circle, thus, VAR model is stable. In order to find the optimal lag, the model is based on the criteria LR, PDE, AIC and HQ of VAR model, the research group selects the appropriate lag as 1. The result is shown in the following Table 9.

Table 9. The optimal lag

Lag	LogL	LR	FPE	AIC	SC	HQ
0	−837.173	NA	3.11E+25	67.21387	67.36013*	67.25444
1	−823.653	22.71383*	$2.19e + 25^*$	66.85226*	67.43732	67.01453*
2	−818.029	8.09915	2.97E+25	67.1223	68.14616	67.40628

LR: Likelihood Ratio Test
PDE: Final Prediction Error
AIC: Akaike Information Criterion
SC: Schwarz Information Criterion
HQ: Hannan-Quinn Information Criterion

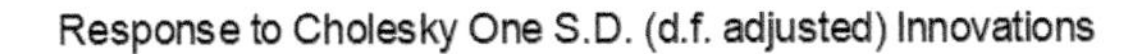

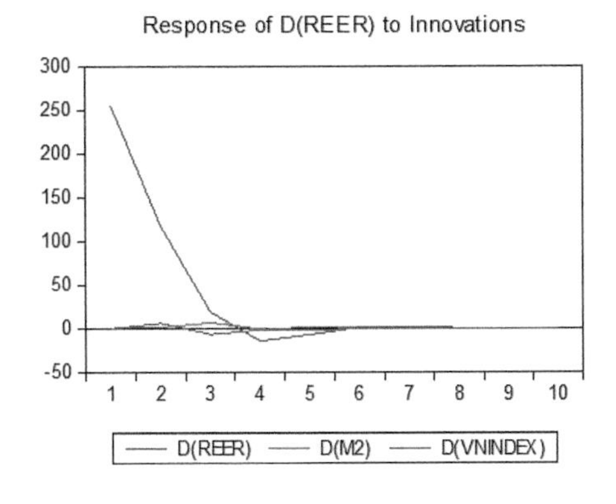

Response of D(M2) to Innovations

80,000,000
40,000,000
0
-40,000,000
-80,000,000
1 2 3 4 5 6 7 8 9 10

D(REER) D(M2) D(VNINDEX)

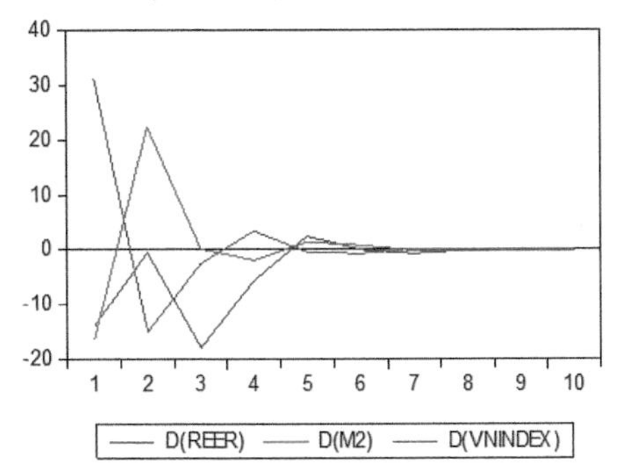

Fig. 10. Response function of respectively variables, REER, M2, VN-index before changes of 02 remaining variables.

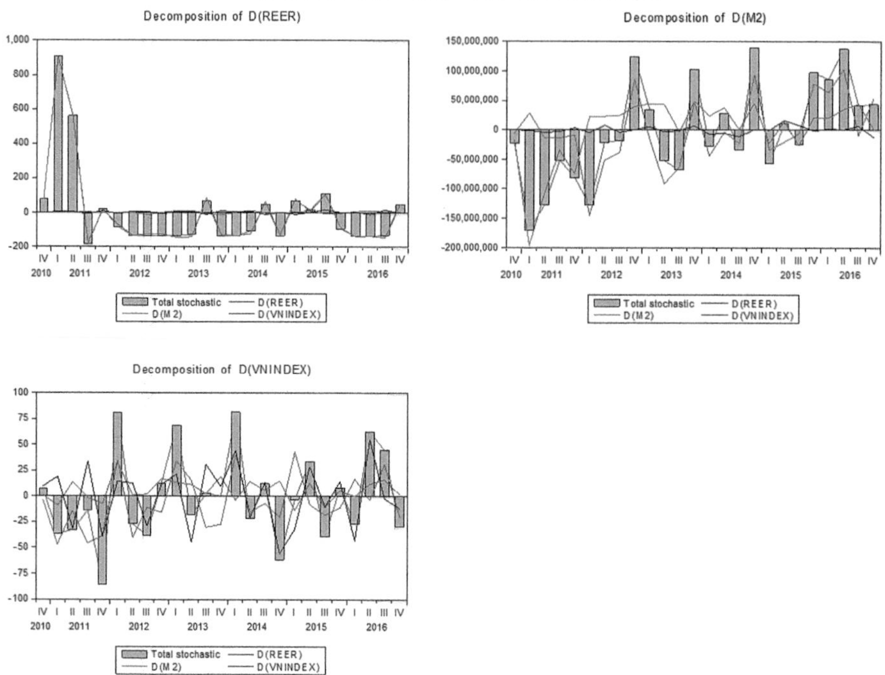

Fig. 11. .

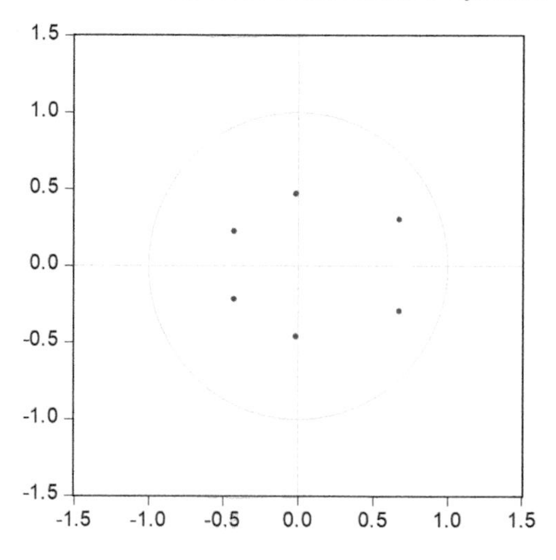

Fig. 12. Stability of model with the lag is 1.

The research group continues running the Granger Causality Test for three variables, the achieved result is p-value > α, therefore, the capitalization value of the stock market is the dependent variable on the interest of Central Bank and total money M2 which may impact on the capitalization value of the stock market (Table 10).

Table 10. Granger Causality test

Dependent variable:			D(MARKETCAP)
Excluded	Chi-sq	df	Prob.
D(INTEREST)	0.071369	2	0.9649
D(M2)	5.316642	2	0.0701
All		4	—

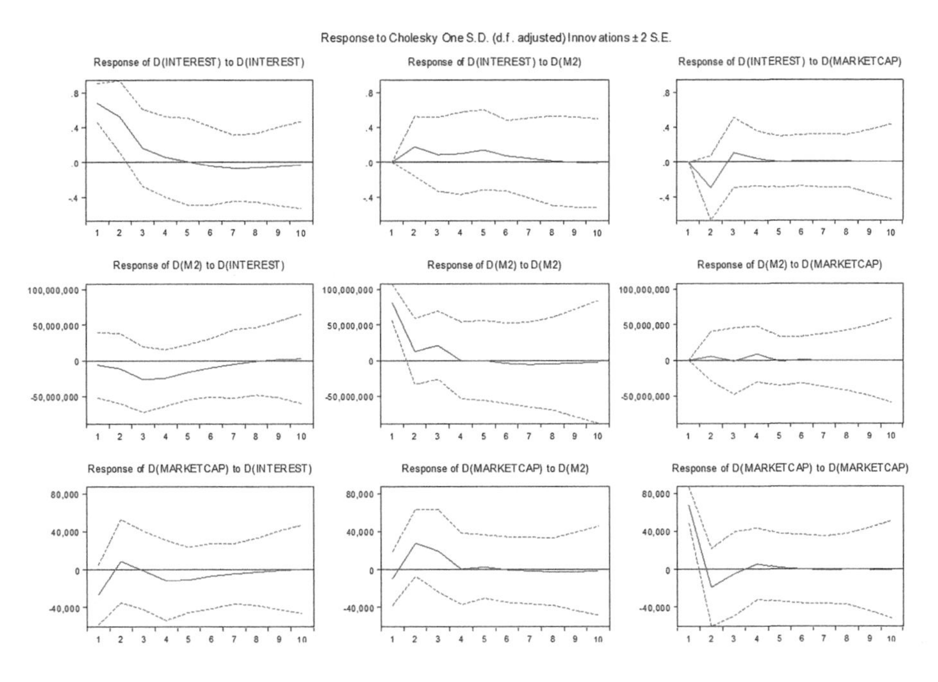

Fig. 13. The response function of values MARKETCAP, M2 and INTEREST.

The response function of the capitalization value of the stock market changes when the bank interest changes. In Quarter 1, the bank interest increases, as a result, the capitalization value of the stock market increases, in Quarter 2 to Quarter 4, when the interest decreases, there is also decrease of the capitalization value accordingly. Since Quarter 5 onwards, both bank interest and capitalization value trends increase and toward the horizontal axis, it shows the more exact impact. With the method of stock investment, the risk may be higher than other investment methods. When the Central Bank increases the interest, other stocks, such as, Government notes or bonds shall have the more attraction, normally, the earnings return of the stock market shall change the same direction with the market interest. However, when the interest inconsiderably increases and not enough competition with the earnings return of the stock market, the stock market still grows when the interest increases, in addition, the expectation is different and the accepted risk by the participants in the stock market is different and the potentiality of the present investment Company also contributes to create this phenomenon. When the non-risk interest increases, the share returns shall increase accordingly. Therefore, when the risk compensation is low, but total share returns doesn't increase or decreases, the investor shall think that investment to shares shall be more risk and they shall invest money to another channel (Table 11).

Group 5: The exchange rate and total money M2 is the exogenous variable, the capitalization value of the stock market is the endogenous variable.

The same as the previous VAR model, this model of Group 5 is stable (Fig. 13), the optimal lag is 0 (Table 12). Both variables, the net exchange rate and total money M2 can impact on the capitalization value of the stock market. The values of Trace statistic and Max Eigenvalue state that there are at least 02 strong co-integrations between the variables in the model and there is existence of the long-term relationship of the considered variables (Table 13).

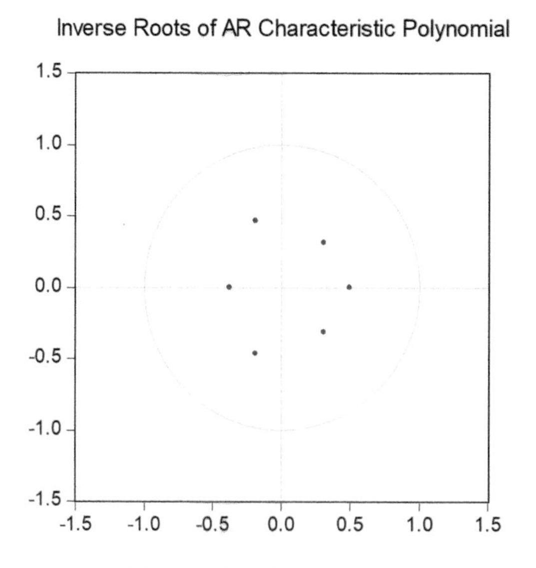

Fig. 14. Stability of model.

Granger Causality Test

LR: Likelihood Ratio Test
PDE: Final Prediction Error
AIC: Akaike Information Criterion
SC: Schwarz Information Criterion
HQ: Hannan-Quinn Information Criterion

Table 11. Granger Causality test

Dependent variable:			D(MARKETCAP)
Excluded	Chi-sq	df	Prob.
D(REER)	2.339443	2	0.3105
D(M2)	3.038919	2	0.2188
All	9.296025	4	0.0541

Table 12. Optimal lag

Lag	LogL	LR	FPE	AIC	SC	HQ
0	−972.864	NA^*	$1.61e+30^*$	78.06911*	78.21538*	78.10968*
1	−967.187	9.537637	2.12E+30	78.33494	78.92	78.49721
2	−960.516	9.606275	2.65E+30	78.52126	79.54511	78.80523

Table 13. Co-integration test

Hypothesized		Trace	0.05		Max-Eigen	0.05	
No. of CE(s)	Eigenvalue	Statistic	Critical value	Prob.**	Statistic	Critical value	Prob.**
None*	0.600188	35.65533	29.79707	0.0094	22.00224	21.13162	0.0377
At most 1	0.299908	13.65309	15.49471	0.093	8.557059	14.2646	0.3249
At most 2*	0.191306	5.096029	3.841466	0.024	5.096029	3.841466	0.024

In addition to response of the capitalization value of the stock market as explained above, the research group finds that there is the increasing variation of the capitalization value of the stock market with total money M2. The detail is from Quarter 1 to half Quarter 2, there is the concurrent increase of two variables, since Quarter 3 onwards, there is decrease and toward the horizontal axis. The expansionary monetary policy leads the expenditure increase in many fields, in which, the financial, stock investment is the method of most attention. Money supply increases, liquidity increases, it makes decrease of the market interest, the discount interest of the stock, as a result, the expectation price

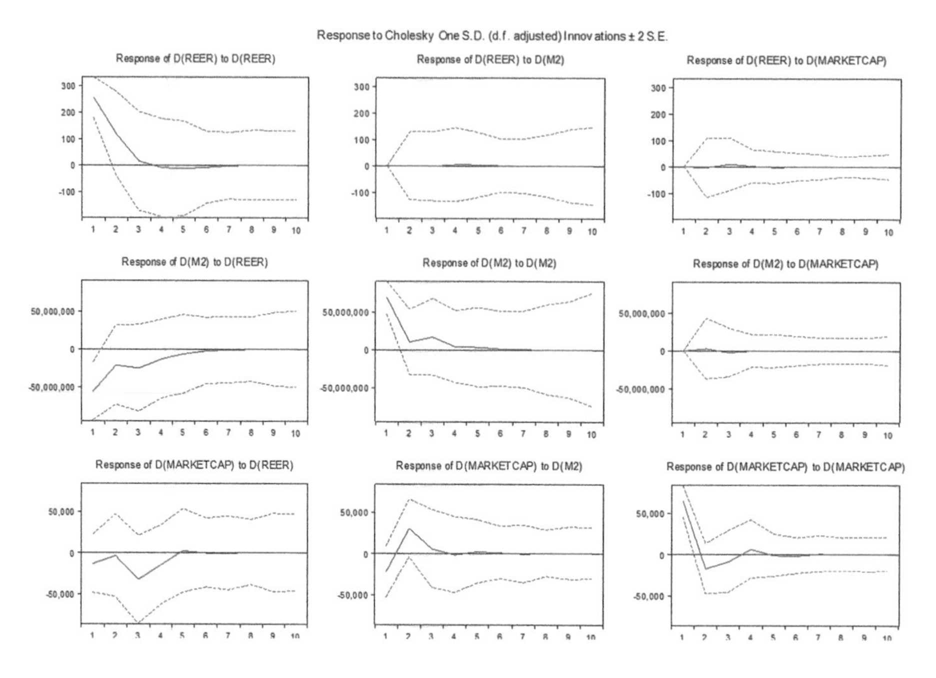

Fig. 15. Response function of variables MARKETCAP, M2 and REER.

increases and income increases. When implementing the tight monetary policy, it makes decrease of the capitalization value of the stock market, because: making decrease because of the high discount interest in the appraisal model; making decrease of the borrowing trend for investment to the stock; making increase of the enterprise's expenses, it affects considerably the company profit. And the stocks including "non-risk interest", such as, the Government paper credits, bonds are more attractive. Generally, increase of money supply shall make increase of the liquidity and credit for the investors of stocks, it makes increase of the stock price, the capitalization value of the stock market is higher, it leads the stable growth for the stock market (Figs. 14, 15 and Table 15).

Group 6: The capitalization value is the endogenous variable, the credit growth and total money M2 is the exogenous variable.

Table 14. Finding the optimal lag

Lag	LogL	LR	FPE	AIC	SC	HQ
0	−871.023	NA^*	$4.67e+26^*$	69.92182*	70.06809*	69.96239*
1	−865.9	8.606512	6.42E+26	70.23199	70.81705	70.39426
2	−862.959	4.234559	1.08E+27	70.71673	71.74059	71.00071

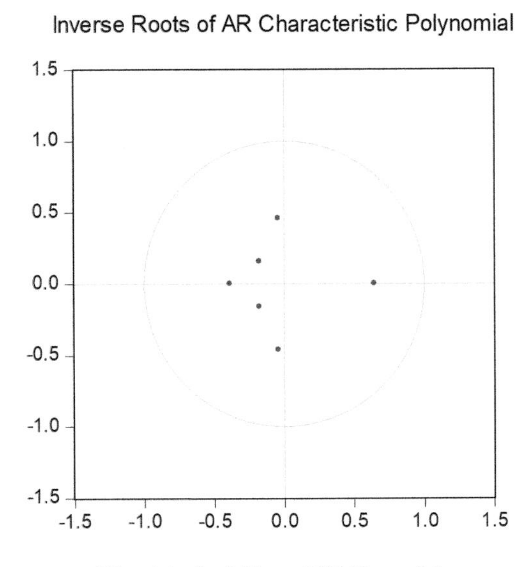

Fig. 16. Stability of VAR model.

As VAR model mentioned above VAR model of Group 6 is stable (Fig. 5), the optimal lag is 0 (Table 14). Both variables, the net exchange rate and total money M2 can impact the capitalization value of the stock market. Values of Trace statistic and Max Eigenvalue states that there is at least 01 strong co-integration between the variables in the model and there is existence of the long-term relationship of the considered variables (Table 16).

Table 15. Granger Causality test

Dependent variable:			D(MARKETCAP)
Excluded	Chi-sq	df	Prob.
D(TTTD)	0.291968	2	0.8642
D(M2)	5.433974	2	0.0661
All	6.548264	4	0.1618

Table 16. Co-integration test

Hypothesized		Trace	0.05		Max-Eigen	0.05	
No. of CE(s)	Eigenvalue	Statistic	Critical value	Prob.**	Statistic	Critical value	Prob.**
*None**	0.633754	39.97261	29.79707	0.0024	24.1068	21.13162	0.0185
At most 1*	0.427071	15.8658	15.49471	0.044	13.36784	14.2646	0.0689
At most 2*	0.098848	2.497961	3.841466	0.114	2.497961	3.841466	0.114

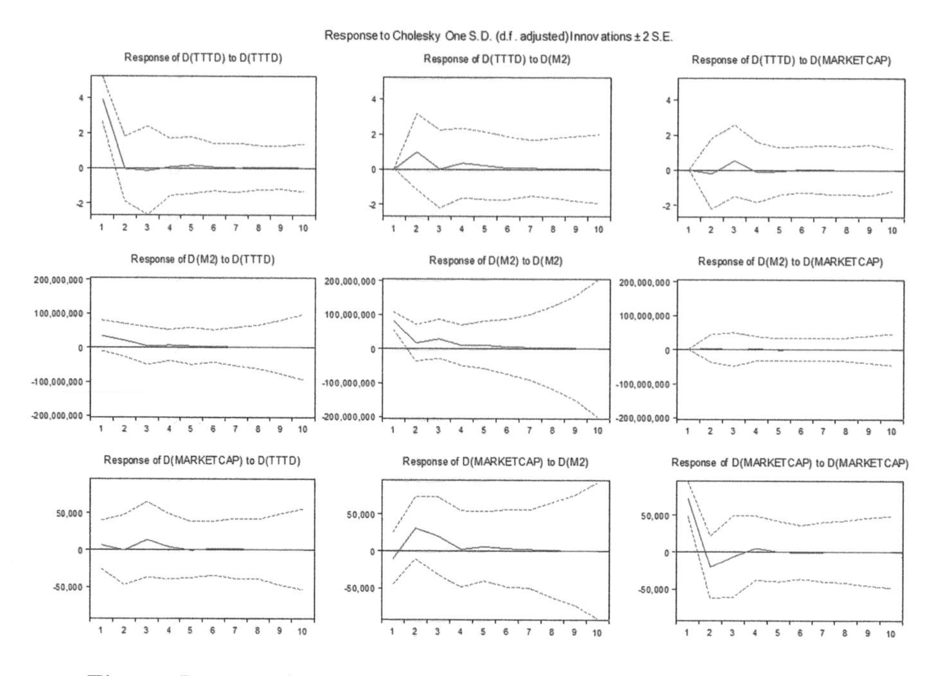

Fig. 17. Response function of the credit market, M2 and MARKETCAP

From Fig. 16, the research group comes to the conclusion that the capitalization value of the stock market has the same direction response with the credit growth, the most obvious is shown in the stage from Quarter 1 to Quarter 4 (Fig. 17).

5 Conclusion

Basing on the analytic result, it is found that the transmission channels of the monetary policy have the long-term relationship with the considered variables of stock market. The research group finds that although there is existence, this relationship doesn't truly have the considerable impact on the subject and it isn't really good for the stock market. Stimulating the stock market by using total money M2, the credit limit doesn't bring many significant results, because they are one of the main tools of the central bank for managing the market. This is to say that the monetary policy in Vietnam hasn't truly been effective in the long-term in the stock market, it mainly has impact on the inflation adjustment.

With the short-term relationship, the analytic result of Granfer test shows the impact of the bank interest, total money M2 and the net exchange rate on VN-index, impact only occurs in the average degree from Quarter 1 to Quarter 4, and it is stable then. In addition, the credit limit doesn't have the obvious impact on VN-index. The same as the capitalization value of the stock market, the bank interest, total money M2, and the net exchange rate is the transmission channels which impact on the capitalization value of the stock market. However, it lasts from Quarter 1 to Quarter 4, and the credit limit doesn't mostly cause the considerable impact on the capitalization value of the stock market. The result is also indicated through the response functions of the model.

With the above-mentioned result, it is found that the monetary policy of Vietnam only impacts in the first quarters, it doesn't last or maintain in the long-term. The monetary policies have the short-term impact only in the stock market and research group states that using the above-mentioned transmission channels doesn't really bring the effectiveness. This is rather correct with the foundation theory of economics in the monetary policy, basing on the finding of Mulldel- Fleming, one small, open economy and there is the exchange rate policy is less flexible the same as Vietnam, the monetary policy is very easy for counteracted and less uphold the effect, but on contrary, the fiscal policy is more appropriate with the conditions of above-mentioned economy.

The important point of the research is that research finds that the main transmission channels only impact temporarily on the stock market, thus, for adjusting the stock market, it is required for research through the new and more effective transmission channels, build the satisfactory policies of investment encouragements. In addition, it is required to pay attention to the research on building the foundation database for developing the ideology.

References

Laopodis, N.T.: Monetary policy and stock market dynamics across monetary regimes. J. Int. Money Financ. **33**, 381–406 (2013)

Fausch, J., Sigonius, M.: The impact of ECB monetary policy surprises on the German stock market. J. Macroecon. **55**, 46–63 (2018)

Haitsma, R., Unalmis, D., de Haan, J.: The impact of the ECB's conventional and unconventional monetary policies on stock markets. J. Macroecon. **48**, 101–116 (2016)

Chatziantoniou, I., Duffy, D., Filis, G.: Stock market response to monetary and fiscal policy shocks: multi-country evidence. Econ. Model. **30**, 754–769 (2013)

Jansen, D.W., Zervou, A.: The time varying effect of monetary policy on stock returns. Econ. Lett. **160**, 54–58 (2017)

Zervou, A.S.: Financial market segmentation, stock market volatility and the role of monetary policy. Eur. Econ. Rev. **63**, 256–272 (2013)

Fernández-Amador, O., Gächter, M., Larch, M., Peter, G.: Does monetary policy determine stock market liquidity? New evidence from the euro zone. J. Empir. Financ. **21**, 54–68 (2013)

Hung, K.C., Ma, T.: The effects of expectations-based monetary policy on international stock markets: an application of heterogeneous agent model. Int. Rev. Econ. Financ. **47**, 70–87 (2017)

Gospodinov, N., Jamali, I.: The response of stock market volatility to futures-based measures of monetary policy shocks. Int. Rev. Econ. Financ. **37**, 42–54 (2015)

Tang, Y., Luo, Y., Xiong, J., Zhao, F., Zhang, Y.C.: Impact of monetary policy changes on the Chinese monetary and stock markets. Phys. A Stat. Mech. Appl. **392**(19), 4435–4449 (2013)

Can Vietnam Move to Inflation Targeting?

Nguyen Thi My Hanh(✉)

Hanoi, Vietnam
mhanh105@gmail.com

Abstract. This study attempts to investigate the adoption of inflation targeting framework in Vietnam. This work is done by examining the satisfaction of one crucial prerequisite of inflation targeting: There is the existence of predictable and stable linkages between monetary policy instruments and inflation outcomes. The Johansen multivariate cointegration procedure and Vector Error Correction Model (VECM) approach are used to check the relationship between monetary policy instruments and inflation, and the findings point out that there exists the existence of stable and predictable linkage between monetary policy instruments and inflation in Viet Nam. However, the relationship is too weak. As a result, Viet Nam is yet a candidate for adopting inflation targeting framework.

Keywords: Monetary policy · Inflation targeting framework · Johansen test

1 Introduction

The State Bank of Vietnam focuses on many targets of monetary policy while other central banks in the world only focus on price stability. From 2012 until now, inflation of Vietnam is always controlled at one single digit. However, there is no pledge of The State Bank of Vietnam to stabilize prices in the long term, and as a result, inflation expectations which are important for price stability could not be anchored. It means that inflation might occur any time, which threaten the economy very badly like what happened in the period 2020–2025.

At the moment, Vietnam is choosing to manage exchange rate, or in other words, exchange rate is seen as a nominal anchor for Vietnam's monetary policy. According to the impossible trinity theory, it is impossible to have all three of the following at the same time: a fixed foreign exchange rate, free capital movement and an independent monetary policy. This theory has explain why many countries which chose exchange rate as a nominal anchor for monetary policy got into difficulties in conducting monetary policy such as Thailand, South Korea, Indonesia in 1997 Asian financial crisis. After that, these countries have moved to inflation targeting policy.

The World Bank's research group has suggested that Vietnam should choose only one nominal anchor for monetary policy: exchange rate or inflation because it seems impossible to achieve two targets at the same time. In order to modernize monetary policy framework, The State Bank of Vietnam is aiming to inflation targeting monetary policy in the period of 2011–2020. That's why the author will examine whether Vietnam could adoption inflation targeting policy in this paper.

V. Kreinovich et al. (Eds.): ECONVN 2019, SCI 809, pp. 1052–1061, 2019.
https://doi.org/10.1007/978-3-030-04200-4_76

2 Inflation Targeting

Inflation targeting was firstly adopted in New Zealand since 1990. Then it was soon followed by Canada, Australia, the United Kingdom, and Sweden. Recently, several emerging economies have also moved in this direction, including Chile, Mexico, and Brazil. Also, many other central banks are now considering the applicability of this regime to their countries because this scheme seems to lack some of the disadvantages of alternative monetary policy regimes. For example, monetary targeting policy has a number of benefits, but it is only applied if there is a correlation between the chosen aggregate and nominal income. However, in reality, this relationship does not exist. The adoption of monetary aggregates as the target variable for monetary policy is therefore incapable, and many countries decided to search for an alternative nominal anchor.

Exchange rate is considering as the other popular nominal anchor, which is often chosen by small, open economies. Such an anchor may take the form of a fixed exchange rate regime or a crawling peg with well-defined rules for the crawl. However, the recent financial crises in emerging economies have shown fixed exchange rate regime can suffer break since this regime can result in systemic banking and financial crises and broadly affect the level of economic activity. Finally, the failure of choosing monetary aggregate or exchange rate as a nominal anchor for monetary policy leads many countries consider adopting inflation targeting.

Nowadays inflation targeting framework has become popular in the world, however, it can be difficult to define what inflation targeting is (Rogoff 1985; Loayza and Soto 2001; Angeriz and Arestis 2008; Freedman and Laxton 2009a; Ncube and Ndou 2011). This study will use the definition of Mishkin (2007). He defined that inflation targeting monetary policy is a monetary policy strategies, including 5 elements: (1) the public announcement of medium-term numerical targets for inflation; (2) an institutional commitment to price stability as the primary, long-run goal of monetary policy and a commitment to achieve the inflation goal; (3) an information-inclusive approach in which many variables (not just monetary aggregates) are used in making decisions about monetary policy; (4) increased transparency of the monetary policy strategy through communication with the public and the markets about the plans and objectives of monetary policymakers; (5) increased accountability of the central bank for attaining its inflation objectives.

In practice, inflation targeting framework requires a quantitative statement as to what the inflation rate is consistent with the pursuit of price stability in the long run. Price stability is the primary goal of inflation targeting monetary policy. In addition, the target series needs to be defined and measured accurate, timely and easily understandable by the public. The central bank makes clear about the main tasks and criteria of monetary policy so that it makes monitoring inflation targeting simpler for the public. The more transparent the monetary policy become, the more credibility the central bank have. As a result, the central can easily pursue the inflation target.

Generally, one may distinguish between three types of inflation targeting: full-fledged inflation targeting (FFIT), inflation targeting lite (ITL), and electic inflation targeting (EIT) (Truman 2003). FFIT countries such as Sweden, the UK, Norway,

Czech Republic, Australia, New Zealand, and Canada concentrate on financial stability, the development of financial markets along with the level of credibility (medium to a high degree). These countries cannot attain and maintain low inflation rate without a clear commitment to inflation targeting so that they are forced to sacrifice output stabilization to the various degree. Meanwhile, EIT countries such as European Central bank, Denmark, Switzerland, Japan and the US can pursue the output stabilization objective together with price stability. However, ITL is at work in emerging economies that have a low credibility. It is regimes where central banks announce a broad inflation objective, but owing to their relative low credibility, they are not able maintain inflation as the foremost policy objective (Stone 2003).

3 Prerequisites for Inflation Targeting

The experience of successful inflation targeters points to several prerequisites for adoption of an inflation targeting framework. Firstly, instrument independence must be granted to central banks in order to fulfill their mandate of ensuring price stability. In other words, central banks must have the freedom to adjust its instruments of monetary policy, which helps achieve the objective of low inflation. Also, instrument independence means that the central bank is not constrained by the need to finance the government budget.

Secondly, the central bank must have an effective monetary policy instrument which has a relatively stable relationship with inflation. From the experience of countries which have success in adopting inflation targeting framework, they choose indirect instruments, such as short-term interest, rather than direct instruments, such as credit controls.

Thirdly, accountability and communication with the public are essential requirements for adopting inflation targeting framework. In particular, central banks publish periodically an inflation report, which updates its views of future prospects and explains its policy actions. That the public understand about how the central bank operates would help anchor inflation expectations. In case targets are breached, the central banks need to issue an open letter which explains causes of the breach and how and when inflation returns to tolerated levels. Moreover, to improve central banks' transparence, meeting minutes of monetary policy committees are published so that all decisions of central banks become completely clear to the public.

Fourthly, countries wishing to move to inflation targeting need to focus on several technical issues. For example, central banks need to make decisions to choose a suitable price index (underlying or core measure of inflation) and to specify the inflation target as a point, as a band, or as a medium-term average as well as the time horizon for meeting the target. In addition, forecast technique is a key element of a successful inflation targeting policy.

4 Is Vietnam Fulfilled the Prerequisites for Inflation Targeting?

4.1 Brief Theoretical Framework

Countries adopting inflation targeting framework normally use interest rates as the nominal anchor to control inflationary pressures. If inflationary exceeds the target rate, central banks would raise the interest rate to slow down the economy, and subsequently lower the inflation to get it closer to the target, vice versa. Meanwhile, real exchange rate incorporates external sectors in the domestic inflation process. It directly affects the cost of imported intermediate inputs to the production process on the inflation measure. Moreover, higher interest rate generally leads to a stronger currency that reduces international competitiveness, impacts export performance, and leads to relatively greater import penetrations. In other words, higher interest rate pushes up the value of currency and restrains economic activity, hence influences the aggregate demand and consumer price. Additionally, many studies have shown relationship between money supply and inflation (Lucas 1996). However, monetary targeting framework is not a good choice for several central banks such as the United States, Canada and the United Kingdom (Mishkin 2000). Then, these countries have switched to inflation targeting framework.

4.2 Model Specification and Data

Blejer and Leone (1999) have stated that one crucial prerequisite condition for inflation targeting is that there is a relationship between inflation and monetary policy instrument. As a result, this paper only focuses on the second prerequisite condition, or in other words, the question whether The State bank of Vietnam has an effective instrument policy will be answered.

To analyze the relationship between inflation and monetary policy instrument, Consumer Price Index (CPI) is chosen as a variable representing inflation and three variables such as Money supply (M2), Exchange rate (ER) and Interest rate (R) are chosen as variables representing monetary policy instruments. This data set is collected from International Financial Statistic. The paper employs quarterly data during 2000: Q1 to 2016:Q4. The Johansen cointegration test and The Vector Error Correction Model approach are applied on macroeconomic data. The model can be simplified as follows:

f(CPI) = f(M2, ER, R)

where CPI = Consumer Price Index (as proxy to measure inflation rate)

M2 = Money supply
ER = Exchange rate
R = Interest rate

4.3 Empirical Results

+ Unit root test:

Unit root test is performed to determine whether the individual series are stationary or not since the usage of non-stationary data can generate spurious regressions. In this case, Dickey-Fuller test is used. The null hypothesis is that each variable has a unit root, and the alternative hypothesis is that the series are stationary (or each variable has root outside unit circle).

In order to check for stationary condition of the above variables, this paper uses t-statistics critical values. The results suggest that all variables are stationary at level. Since these variables are integrated of order one, the cointegration test can be applied (Table 1).

Table 1. Dickey-Fuller test for unit root

Variables	Test-statistic	5% critical value	p-values
CPI	−3.863	−2.917	0.0023
R	−5.511	−2.917	0.0000
M2	−8.452	−2.917	0.0000
ER	−8.883	−2.917	0.0000

Source: Author's computation using Stata 11

+ Optimal lag selection:

Before running cointegration test, the optimal lag length should be decided because the weakness of Johansen test is sensitive to the lag length. The result suggests that the optimal lag length is 3 (Table 2).

+ Johansen tests:

Both Johansen's trace statistic test and Johansen's maximum eigenvalue statistic are applied. In particular, trace statistic test tests the hypotheses:

Table 2. Optimal lag selection

Selection-order criteria						Number of obs = 64		
Sample: 5–68								
Lag	LL	LR	df	p	FPE	AIC	HQIC	SBIC
0	−1266.21				2.0e+12	39.6941	39.7473	39.8291
1	−935.024	662.38	16	0.000	1.1e+08	29.8845	30.1103	30.5191
2	−904.879	60.29	16	0.000	7.0e+07	29.4025	29.8809	30.6168
3	−888.156	33.445	16	0.006	6.9e+07*	29.3799*	30.0709	31.134
4	−877.802	20.707	16	0.190	8.5e+07	29.5563	30.46	31.8501
Endogenous: CPI M2 ER R								
Exogenous: -cons								

Source: Author's computation using Stata 11

$$H_0(r) : r_0 \leq r \text{ and the alternative } H_1(r_0) : r_0 > r$$

Where r is the number of cointegrating vectors under the null hypothesis. Johansen's trace statistic is a join test where the null hypothesis H_0 examines that the number of cointegrating vectors is less than or equal to r against the alternative hypothesis H_1 which assumes that there are more than r.

Johansen also derives the Eigenvalue statistic for the hypotheses:

$$H0(r) : r_0 = r \text{ and the alternative } H_1 : r_0 =: r + 1$$

Where r is the number of cointegrating vectors under the null hypothesis H_0. The Johansen's maximum eigenvalue statistic examine tests its null hypothesis that the number of cointegrating vectors is r against an unspecified or general alternative which states that the number of cointegrating vectors is (r + 1) (Table 3).

Both test support for cointegration at a 5% significance level. Also, the presence of two cointegrations among the variables in Vietnam recommends a long-term relationship among them. The result of the below table will show the relationship (Table 4).

Table 3. Johasen tests for cointegration

Trend: constant				Number of obs = 65	
Maximum rank	parms	LL	Eigenvalue	Trace statistic	5% critical value
0	36	−932.56032		64.2383	47.21
1	43	−916.13772	0.39669	31.3927	29.68
2	48	−907.00589	0.24496	13.1290	15.41
3	51	−901.44838	0.15718	2.0140	3.76
4	52	−900.44138	0.03051		
Sample: 4–68				Lags = 3	

Source: Author's computation using Stata 11

The relationship is also presented by equations as follows:

$$e_{1t} = CPI + 0.028ER + 184R - 1244 \text{ or } CPI = 1244 - 0.028ER - 184R + e_{1t}$$

$$e_{2t} = M2 + 0.0029ER + 8.09R - 113 \text{ or } M2 = 113 - 0.0029ER - 8.09R + e_{2t}$$

Next, the study performs the impulse response function as an additional check. The impulse response test is to assess the responsiveness of the dependent variables in the VECM to shocks to each of the variables (Brooks 2008). There is a surprising result that CPI totally does not respond to shocks from M2, R and ER. It has found that though there exists a relationship between CPI and other variables, this relationship is too weak. This finding is also supported by variance decomposition analysis (Table 5). The changes of CPI are not explained by the changes of M2, ER and R (Fig. 1).

Table 4. Cointegrating equations

Cointegration equations

Equation	Parms	chi2	P > chi2
-ce1	2	25.75406	0.0000
-ce2	2	41.14545	0.0000

Identification: beta is exactly identified

Johasen normalization restrictions imposed

beta	Coef	Std. err	z	p>|z|	[95% confidence Interval]	
_ce1						
CPI	1	.	.	.	.	.
M2	(omitted)					
ER	.0283317	.016715	1.69	0.090	-.0044291	.0610925
R	184.5243	38.32858	4.81	0.000	109.4017	259.647
_cons	-1244.297	.	.	.	.	.
_ce2						
CPI	2.17e-19	.	.	.	.	.
M2	1	.	.	.	.	.
ER	.002972	.0007128	4.17	0.000	.001575	.004369
R	8.092902	.1.634428	4.95	0.000	4.889482	11.29632
_cons	-113.0033	.	.	.	.	.

Source: Author's computation using Stata 11

Table 5. Variance Decomposition analysis

Step	(1) fevd	(2) fevd	(3) fevd	(4) fevd	(5) fevd	(6) fevd
0	0	0	0	0	0	0
1	1	.153557	.261927	8.9e-06	0	.846443
2	.898911	.133765	.442961	.000375	.10082	.855074
3	.845566	.132758	.495815	.004324	.150102	.845631
4	.819083	.13479	.501762	.018166	.167543	.834235
5	.802622	.134363	.497942	.041073	.175314	.827964
6	.79072	.131925	.493172	.066939	.179956	.826216
7	.781533	.128396	.489283	.088708	.183	.827032
8	.774245	.124348	.486286	.1033984	.185136	.829023
9	.768269	.12017	.483808	.113854	.186775	.831367
10	.763242	.116903	.481597	.120136	.188109	.833663

(continued)

Table 5. (*continued*)

Step	(7) fevd	(8) fevd	(9) fevd	(10) fevd	(11) fevd	(12) fevd
Step	(7) fevd	(8) fevd	(9) fevd	(10) fevd	(11) fevd	(12) fevd
0	0	0	0	0	0	0
1	.019887	.002496	0	0	.718187	.018525
2	.080958	.062323	.000027	.01103	.473383	.0361
3	.157479	.091819	.000794	.019826	.345085	.037717
4	.202243	.105104	.001854	.024556	.291473	.035614
5	.22391	.110366	.002728	.027109	.267469	0.36367
6	.23511	.111405	.003635	.02823	.255861	.037169
7	.241925	.111739	.004648	.02846	.250045	.036964
8	.246781	.113347	.005706	.028273	.247017	.036139
9	.250673	.11689	.006735	.027937	.245316	.035132
10	.253998	.122242	.027565	.027565	.244244	.034149

Step	(13) fevd	(14) fevd	(15) fevd	(16) fevd
0	0	0	0	0
1	0	0	0	.978969
2	.000243	.000131	.002698	.901202
3	.003539	.001785	.001621	.86614
4	.011519	.00642	.004521	.841116
5	.019296	.010564	.010679	.812195
6	.02569	.013628	.015857	.784487
7	.030819	.016112	.018747	.762589
8	.034913	.018357	.019917	.74653
9	.03822	.020525	.020203	.734124
10	.040956	.02268	.020161	.723474

(1) Irfname = irf, impulse = CPI, and response = CPI
(2) Irfname = irf, impulse = CPI, and response = ER
(3) Irfname = irf, impulse = CPI, and response = R
(4) Irfname = irf, impulse = CPI, and response = M2
(5) Irfname = irf, impulse = ER, and response = CPI
(6) Irfname = irf, impulse = ER, and response = ER
(7) Irfname = irf, impulse = ER, and response = R
(8) Irfname = irf, impulse = ER, and response =M2
(9) Irfname = irf, impulse = R, and response = CPI
(10) Irfname = irf, impulse = R, and response = ER
(11) Irfname = irf, impulse = R, and response = R
(12) Irfname = irf, impulse = R, and response = M2
(13) Irfname = irf, impulse = M2, and response = CPI
(14) Irfname = irf, impulse = M2, and response = ER
(15) Irfname = irf, impulse = M2, and response = R
(16) Irfname = irf, impulse = M2, and response = M2

Source: Author's computation using Stata 11

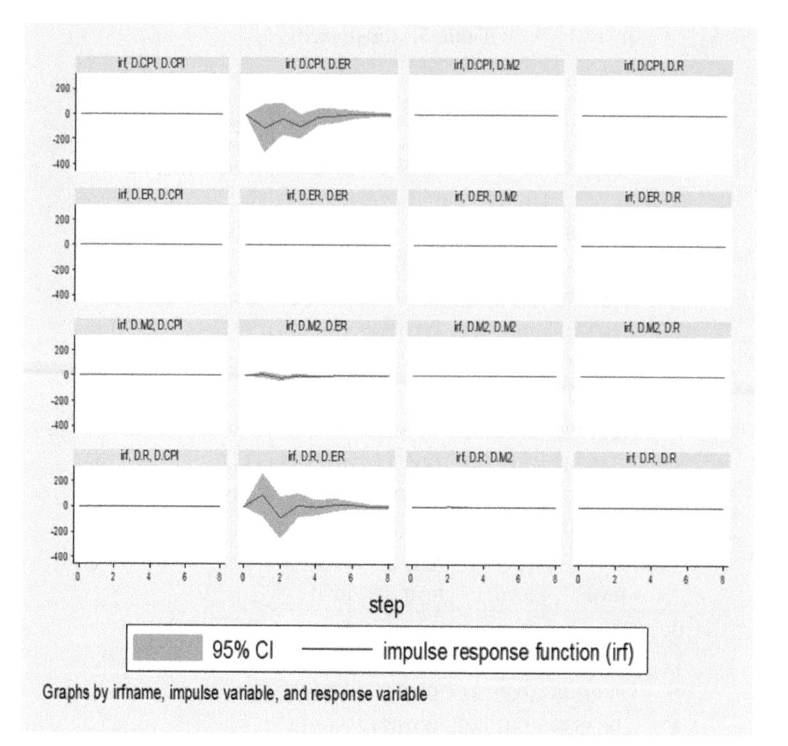

Fig. 1. Impulse response function (Source: Author's computation using Stata 11)

5 Conclusion

Although the above results have pointed that there exists a relationship between inflation and monetary policy instruments in Vietnam, the relationship is very weak. Moreover, from impulse response function analysis, it seems that monetary policy instruments have no influence on CPI. In other words, Vietnam has not yet satisfied the crucial condition for adopting inflation targeting framework.

According to IMF, Vietnam is pursuing the exchange rate targeting policy. This choice is suitable for many developing countries when their economies depend on export sector. These countries focus on keeping their currency stable so as to boost exports. However, this practice makes controlling inflation more complicatedly if the country cannot control foreign inflows. This situation happened in Vietnam during 2007–2011. At that time, Vietnam received a large amount of foreign capital after becoming a member of World Trade Organization. As a consequence, inflation rate hit double digits, which led to a difficult period for policy makers. Additionally, the impossible trinity theory indicated that one country cannot meet the three targets at the same time: a fixed exchange rate, a free capital movement and an independent monetary policy. Therefore, it is advisable that Vietnam should consider inflation targeting policy as a solution.

When moving to inflation targeting framework, there are many countries which have not satisfied all necessary conditions. Therefore, if Vietnam highly appreciates the advantages of inflation targeting framework, the country is still able to move to a new monetary policy framework in the future.

References

Angeriz, A., Arestis, P.: Assessing inflation targeting through intervention analysis. Oxf. Econ. Pap. **60**(2), 293–317 (2008)

Blejer, M.I., Leone, A.M.: Introduction and overview. In: Blejer, M.E., Ize, A., Leone, A.M., Werlang, S. (eds.) Inflation Targeting in Practice; Strategic and Operational Issues and Applications to Emerging Markets Economies. IMF (1999)

Brooks, C.: Introductory Econometrics for Finance. Cambridge University Press, Cambridge (2008)

Freedman, C., Laxton, D.: Why Inflation Targeting? IMF Working Papers, pp. 1–25 (2009a)

Johansen, S., Juselius, K.: Maximum likelihood estimation and inference on cointegration-with application to the demand for money. Oxford Bull. Econ. Stat. **52**, 169–210 (1990)

Loayza, N., Soto, R.: Ten years of inflation targeting: design, performance, challenges. Banco Central de Chile (2001)

Lucas, R.E.: Nobel lecture: monetary neutrality. J. Polit. Econ. **104**, 661–682 (1996)

Mishkin, F.S.: From monetary targeting to inflation targeting: lessons from the industrial countries. National Bureau of Economic Research (2000)

Mishkin, F.S.: Monetary Policy Strategy. MIT Press, Cambridge (2007)

Ncube, M., Ndou, E.: Working Paper 134-Inflation Targeting, Exchange Rate Shocks and Output: Evidence from South Africa (2011)

Rogoff, K.: The optimal degree of commitment to an intermediate monetary target. Q. J. Econ. **100**(4), 1169–1189 (1985)

Stone, M.R.: Inflation targeting lite. International Monetary Fund (2003)

Truman, E.M.: Inflation Targeting in the World Economy. Peterson Institute (2003)

Impacts of the Sectoral Transformation on the Economic Growth in Vietnam

Nguyen Minh Hai(✉)

Department of Economic Mathematics, Banking University of HCM City, Ho Chi Minh City, Vietnam
minhhai.nguyen77@gmail.com

Abstract. The objective of this paper is to research and analyze impacts of the sectoral transformation on the recent economic growth in Vietnam. At the same time, promoting economic growth towards sectoral restructuring is considered whether it is a critically viable direction for the Vietnam economy. The paper shows that the sectoral transformation during the period of 2000–2016 had significant impact to sustain economic growth. This has supported the direction of economic growth, moving towards a shift from a low-productivity sector to a high-productivity sector in the past is a right direction. Based on the analysis, the paper recommends a number of policies on promoting positive impacts to boost rapid and sustainable economic development.

Keywords: Sectoral transformation · Economic growth · Vietnam

1 Introduction

Sectoral transformation (ST) in Vietnam after the Doi Moi (Renovation) period is always one of the most popular topics, attracting much attention of researchers. Although there are a number of published studies but due to different perspective in different times. There is no comprehensive assessment of the ST in Vietnam economy after the renovation period. Recognizing Vietnam previous growth model, mainly focus on investments, cheap labor, raw materials extraction, export handicrafts, etc. is no longer suitable with the context of world economic integration. Moreover, experience from successful countries in the economic restructuring strategy by ST such as Korea, Indonesia, Singapore and Thailand shows that this strategy has contributed to bringing them from undeveloped economies to developed countries in the world, after decades of high growth and they have become Asia's brightest economies. Besides successful countries the economic restructuring, there are many unsuccessful countries, such as the sub-Saharan African countries, Latin America and the former Soviet Union, etc., which raises questions about the direction of ST impact on economic growth is really a right option for Vietnam. The Socio-Economic Development Strategy until 2020 has set the target: *"Economic restructuring towards industrialization and modernization is an indispensable way for Vietnam to quickly get out of backwardness, slow development and become a civilized and modern country"*. The Government has taken initiative in

V. Kreinovich et al. (Eds.): ECONVN 2019, SCI 809, pp. 1062–1072, 2019.
https://doi.org/10.1007/978-3-030-04200-4_77

restructure the economy, focusing on ST so that it is modern, effective and sustainable linked with the sectors following the nission: industry, agriculture, service plays the key role. Therefore, the review and re-assessment of the effectiveness of economic ST activities, especially whether ST has positive impacts on economic growth, whether it is a right decision for Vietnam at this period or not is very crucial. For all these reasons, the study will focus on analyzing impacts of ST on economic growth in Viet Nam over time based on data collected. Hopefully, the research results will be a scientific basis to guide the policy of sustainable growth for the following years.

2 Overview and Analytical Framework

2.1 Overview of Studies and Researches

The theory of economic transformation (ET) is one of the theories that has been studied from the early time. Smith (1776), Ricardo (1817) argued that the structural characteristics of economic components are closely related to level of development of a country and emphasize that ET is one of the prerequisites promote economic growth. Schumpeter's viewpoint (1939): ET is just a minor problem and the effectiveness of market allocation make considering ET as a natural outcome of the market development, not a crucial condition for economic growth. However, studies conducted by Chenery and Syrquin (1986), Kuznets (1966), Hoffman (1958) and recently by Rodrik (2008) expressed the view that support for classical economists, arguing that economic growth in semi-industrialized countries could not be explained by the neoclassical model and could only be improved by adding structural elements. The fact that all countries have prospered growth is a "*path-dependent*" process of sectoral transformation from the low-productivity sector into high-productivity ones. Emphasizing the importance of the ET to economic growth, classic theories such as "*take-off theory*" by Rostow (1960), "*dualism theory*" by Lewis (1954), "*balanced growth*" by Nurkse (1953) all assert that the transfer of resources from low productivity - to highproductivity areas is characteristic of economic development. At the same time, it is emphasized that the form of ET in developing countries is diverse and difficult to find a common for all countries, depending on the specific characteristics of each country.

Together with theoretical research, experiments on ST have also been conducted. One of significant research conducted by Chenery (1980), using an *input-output* model, applies to a number of semi-industrial countries to implement industrialization-led policy from export orientation to import substitution during the study period. The results showed that sustainable economic growth [1]requires a shift in the composition of production and the compatibility of both domestic demandand international trade opportunities. According to this method, Akita and Hermawan (2000), Hayashi (2005), show that Indonesia succeeded in implementing the ET during 1971–2000. In particular, the manufacturing industry (MI) expanded its production capacity, strong export

[1] Second sector: export expansion, manufacturing industry (MI).

potential and reduced dependence on imports. To measure the degree of interdependence among economic regions and Vietnam's economic growth with Indonesia, Malaysia in the 1996–2000 period, Akita and Hau (2006) affirmed that, similar to Malaysia, the source of Vietnam's economic growth was mainly based on the second sector[1]; different from Indonesia- the source of Indonesia's economic growth is the third sector[2]. However, the above results do not represent the characteristics of many developing countries, because the default of each country's technology level in the model is the same, there is no difference in each economy structure.

In addition, studies with array data have more realistic results. The studies conducted by Chenery (2000), Dani Rodrik (2012), Van Ark and Timmer (2003) in 39 economies are divided into three groups: developed, developing and planed economy in the regions: Africa, Latin America, Asia. The findings confirm that ST from low-productivity to high- productivity sector is a significant source of growth for developing countries rather than for developed countries. This occurs most strongly in the East and South Asia countries. The shift in the opposite direction occurs in sub-Saharan Africa, Latin American and former Soviet Union countries.

In summary, the differences in the results of ST research on economic growth are explained in different ways depending on the specific characteristics of the countries, in each stage of development and the degree of data accuracy. Therefore, in order to measure the impact of ST on economic growth, it is necessary to research thoroughly the microstructure of each country. This requires us to have a clear separation of the effects of the composition of the contribution component to economic growth. This study will focus on the impact of ST on Vietnam's economic growth in the period 2000–2016.

3 Research Methodology

3.1 Research Model

The author's research model is inherited from the basic model proposed by Sundrum (1990) and adjusted by Cornwall (1990) to explain the growth rate of labor productivity (LP) of the economy through both supply and demand approach. According to Cornwall, the economy is divided into four sectors: agriculture, industry, service and non-agriculture. Thus, the regression model assessing impacts of ST on economic growth is decomposed into equations based on the components of economic growth that affect economic growth. Specifically:

Equation 1: Impacts of sectoral input transformation on the economic growth:

$$Ln(GDP)_{it} = \alpha_1 LnK_{it} + \alpha_2 LnL_{it} + \alpha_3 K\_I_{it} + \alpha_4 K\_S_{it} + \alpha_5 L\_I_{it} + \alpha_6 K\_S_{it} + C_i + u_i \quad (3.1)$$

[2] Third sector: goods and service production.

Equation 2: Impacts of sectoral input transformation on the economic growth of the agricultural sector:

$$LnGDP\_A_{it} = \alpha_1 Ln(K\_A)_{it} + \alpha_2 Ln(L\_A)_{it} + \alpha_3 K\_I_{it} + \alpha_4 K\_S_{it} + \alpha_5 L\_I_{it} + \alpha_6 K\_S_{it} + C_i + u_i \quad (3.2)$$

Equation 3: Impacts of sectoral input transformation on the economic growth of the industrial sector:

$$Ln(GDP\_I)_{it} = \alpha_1 Ln(K\_I)_{it} + \alpha_2 Ln(L\_I)_{it} + \alpha_3 K\_A_{it} + \alpha_4 K\_S_{it} + \alpha_5 L\_A_{it} + \alpha_6 L\_S_{it} + C_i + u_i \quad (3.3)$$

Equation 4: Impacts of sectoral input transformation on the economic growth of the service sector:

$$Ln(GDP\_S)_{it} = \alpha_1 Ln(K\_S)_{it} + \alpha_2 Ln(L\_S)_{it} + \alpha_3 K\_A_{it} + \alpha_4 K\_I_{it} + \alpha_5 L\_A_{it} + \alpha_6 L\_I_{it} + C_i + u_{it} \quad (3.4)$$

Equation 5: Impacts of sectoral input transformation on the economic growth of non-agricultural sectors:

$$Ln(GDP\_NA)_{it} = \alpha_1 Ln(K\_NA)_{it} + \alpha_2 Ln(L\_NA)_{it} + \alpha_3 K\_I_{it} + \alpha_4 L\_I_{it} + C_i + u_{it} \quad (3.5)$$

When considering the economic growth of localities, it is important to notice the differences among the provinces in terms of geo-economics, resources, infrastructure, development level, socio-economic policies, etc. Therefore, in order to solve these heterogeneous problems, the Panel data models are most appropriate, as the difference between localities will be distinguished by the number C_i in the model.

3.2 Estimation Methodology

This research applies three common methods in data regression tables: (1) Pooled OLS model; (2) fixed effects model (FEM); random effects model (REM). After that, F tests, LM tests and Hausman tests will be used to select the appropriate regression model. In addition, in order to test the variance of the error, the self-correlation, the cross correlation among the panels, the Wald test, the Wooldridge test, the Pesaran test will be used.

3.3 Research Data

The experimental data used in this paper include: GDP data, capital (K) and labor (L) data of three sectors includes: agriculture, industry and service of 60 provinces across the country in the 2000–2015 period, which were standardized at constant prices in 2000 from the General Statistics Office (GSO).

4 Results and Discussion

4.1 Research Results

This section mainly presents regression results of Pooled OLS, FEM and REM for the entire research sample of each of the equations for the impacts of sectoral transformation (ST) on the economic growth. Estimates results of the model from Stata software are as follows:

Equation 1: Estimations of the model of impacts of sectoral input transformation on the economic growth:

Table 2 shows that the FEM model was selected, the obtained estimates were consistent estimates. Therefore the FEM model is applied for discussion. The estimated results from model (1) show that the variables K_I, L_I, L_S are statistically significant. The estimated coefficients of these variables are positive, which proves that each increase in the proportion of capital in industries and services has a stronger impact on economic growth than the increase proportion of capital in agriculture. This implies that the transformation of labor and other resources to high-productivity sectors has contributed to improving Vietnam's economic growth. The results also show that the attraction of labor and other resources in industry and services increases the proportion of labor and other resources in sectors, which play an important role in economic growth. Moreover, with the obtained results, the proportion of labor in the industry has the strongest impact (with 1.58%), followed by the proportion of industrial capital (with 0.65%) and finally the impact of increasing proportion of capital and service labor (with 0.49%). Thisalso asserts the argument that the ST is closely related to economic growth in Vietnam during the research period. In particular, the transformation of inputs from low-productivity to high-productivity sectors has a positive impact on growth.

Equation 2: Estimation results of the impacts of sectoral input transformation model on economic growth of the agricultural sector:

The estimation results of model (2) from Table 3 show that the FEM model was chosen. All structural variables in the model are statistically significant. Estimation coefficient of variables: K_I, K_S, L_I, and L_S have positive effects on GDP_A growth. This demonstrates that the proportion of capital and labor in industries and services have the promoting impact on agricultural growth. In addition, the comparative tests of estimation coefficient of the structural variables show that the coefficient of influence on GDP_A of variable K_I is stronger than the coefficient of K_S. This suggests that the increase in the proportion of capital in the service sector will have less impact on agricultural growth than that in the industry.

Equation 3: Estimation results of the model of impacts of sectoral input transformation on the economic growth of the industry:

The estimation results of model (3) from Table 4 show that the FEM model is chosen for the variables: K_S has positive effects on GDP_A growth, L_A has negatively impacts GDP_A; other variables: K_A, L_S are not statistically significant. This explains quite well the situation in Vietnam during the research period. The increase in the capital proportion of service sector (K_S) has a stronger impact on industrial growth than industrial sector in the capital structure, increasing the proportion of industrial labor has a stronger impact on industrial growth than that of agricultural labor in the labor structure. In addition, each of similar increase in the proportion of industrial and agricultural capital, or industrial and service labor, has the same effect on industrial growth.

Equation 4: Estimation results of the model of impacts of sectoral input transformation on economic growth of the service sector.

Results from Table 5 shows that the K_S, L_I variables and the structural variables K_A, K_I, L_I have positive effects on GDP_S growth. In which, the estimation

Table 1. Description of variables in the estimation model

Variables	Description of variables in the model
LnGDP	Logarithm GDP
LnGDP_A	Logarithm GDP in the agricultural sector
LnGDP_I	Logarithm GDP in the industrial sector
LnGDP_S	Logarithm GDP in the service sector
LnGDP_NA	Logarithm GDP in non-agricultural sector
LnK	Logarithm Capital
LnK_A	Logarithm capital in the agricultural sector
LnK_I	Logarithm capital in the industrial sector
LnK_S	Logarithm capital in the service sector
LnK_NA	Logarithm capital in the non-agricultural sector
LnL	Logarithm labor
LnL_A	Logarithm labor in the agricultural sector
LnL_I	Logarithm labor in the industrial sector
LnL_S	Logarithm labor in the service sector
LnL_NA	Logarithm labor in the non-agricultural sector
K_A	The proportion of agricultural capital in total capital
K_I	The proportion of industrial capital in total capital
K_S	The proportion of service capital in total capital
K_NA	The proportion of non-agricultural capital in total capital
L_A	The proportion of agricultural labor in total labor
L_I	The proportion of industrial labor in total labor
L_S	The proportion of service labor in total labor
L_NA	The proportion of non-agricultural labor in total labor

coefficients of K_A, K_I and L_I variables show that increase in the capital proportion of agricultural and industrial sectors will all have impact on the growth of the service sector and each increase in the proportion of service labor has a less impact on the growth of the service sector than the effect of a corresponding increase in the proportion of industrial labor in the labor structure. The results also show that the L_A variable has no effect on GDP_S growth in the entire sample. This implies that the same increase in the proportion of agricultural and service workers have the same impact on the growth of the service sector. This result has proved that sectoral transformation towards increasing the proportion of labor and other resources that are essential to boosting the growth of the service sector.

Equation 5: Estimation results of the model of impacts of sectoral input transformation on economic growth of non-agricultural production (Table 6).

Estimation results affirm the importance of industry in promoting the overall growth of non-agricultural sectors. This implies that each increase in the proportion of industrial labor has stronger effects on the growth of non-agricultural sectors than the corresponding increase in the proportion of industrial capital.

In conclusion, the results from Tables 1, 2, 3, 4 and 5 show that ST in Vietnam is closely related to economic growth during the research period. The results of the paper is the evidence affirm that ST is extremely crucial. This transformation is like a push to

Table 2. Regression results with the entire sample scale for the equation 1

Explanatory variables	Model (1)		
	Pooled OLS	FEM	REM
LnK	0.5383 (0.147)	0.3883 (0.237)	0.3883 (0.237)
LnL	0.3549 (0.463)	0.3349 (0.123)	0.3749 (0.123)
K_I	0.3563 (0.4517)	0.6533 (0.7817)	0.6533 (0.4187)
K_S	0.6219 (0.3678)	0.4959 (0.548)	0.6959 (0.678)
L_I	0.5523 (1.254)	1.5823 (0.254)	1.2223 (0.3554)
L_S	0.4605 (0.4093)	0.6905 (0.3093)	0.4605 (0.5093)
Independent coefficient	1.764 (2.575)	7.1664 (1.575)	2.3664 (1.575)
Number of observed variables	960	960	960
Number of provinces	60	60	60
Hausman (χ^2)		29.86	
Wald (χ^2)		1907.5	
Wooldridge		80.517	
Peasaran		26.381	

Table 3. Regression results with the entire sample scale for the equation 2

Explanatory variables	Model (2)		
	Pooled OLS	FEM	REM
LnK_A	0.8383 (0.447)	0.15555 (0.0237)	0.3883 (0.237)
LnL_A	0.5419 (0.523)	0.20596 (0.9523)	0.7749 (0.323)
K_I	0.5563 (0.1187)	1.2277 (0.1587)	0.6533 (0.4187)
K_S	0.7312 (0.6128)	1.0876 (0.1379)	0.8959 (0.3478)
L_I	0.8523 (1.354)	0.98733 (0.1309)	1.7223 (0.5154)
L_S	0.6005 (0.383)	0.8267 (0.4973)	0.5605 (0.4513)
Independent coefficient	1.3164 (2.375)	9.4623 (1.0775)	2.7664 (1.555)
Number of observations	960	960	960
Number of provinces	60	60	60
Hausman (χ^2)		26.89	
Wald (χ^2)		4895	
Wooldridge		28.83	
Peasaran		29.973	

boost and speed up the growth of sectors and the economy. In addition, the results support the important role of industry in attracting capital and labor to conduct ST and contribute to GDP growth.

5 Conclusions and Policy Recommendations

Based on the results of this analysis as well as current Vietnam economic contextVietnam, this research recommends some views and solutions to promote positively in recognizing characteristics of ST and economic growth in Vietnam as well as some recommendations for reasonable ST and promote the economy to grow rapidly in the coming years:

Firstly, the transformation of production factors from agriculture to industry and services has a positive impact, promoting the growth of sectors and the economy. Therefore, persisting in the orientation of transformation of labor structure from low-productivity to high-productivity sector should be thoroughly understood in Vietnam's strategic planning and development policy in the new phase. In the current context of Vietnam, especially in rural society, which is most clearly expressed, should accelerate structure transformation of rural households towards industrial, commercial and service households.

Table 4. Regression results with the entire sample scale for the equation 3

Explanatory variable	Model (3)		
	Pooled OLS	FEM	REM
LnK_I	0.3383 (0.147)	0.5475 (0.0237)	0.8113 (0.437)
LnL_I	0.4549 (0.463)	0.2515 (0.0624)	0.3419 (0.423)
K_A	0.6563 (0.4517)	0.22517 (0.2068)	0.7533 (0.1187)
K_S	0.5219 (0.4678)	1.2062 (0.1645)	0.5169 (0.777)
L_A	0.6123 (1.254)	- 0.9034 (0.3982)	1.2453 (0.3145)
L_S	0.6305 (0.3113)	- 0.1048 (0.6684)	0.6105 (0.4193)
Independent coefficient	1.264 (1.315)	7.7897 (0.8761)	2.7664 (1.9175)
Number of observed variables	960	960	960
Number of provinces	60	60	60
Hausman (χ^2)		16.29	
Wald (χ^2)		3333.5	
Wooldridge		48.827	
Peasaran		16.753	

Secondly, in order to promote economic growth and ST when allocating capital to industries in the economy, industrial capital needs to expand in both scale and density. Therefore, the government must create an enabling environment to attract investment in sectors that are encouraged to be competitive and provide good supporting services: infrastructure, capital, technology, and high quality human resources. In addition, it is important to increase the proportion of industrial labor to production growth of the non-agricultural sector in the economy.

Thirdly, policies must facilitate the flexible movement of resources into higher-productivity economic activities. Therefore, Vietnam should build a long-term roadmap, suitable with actual conditions and capabilities of the economy. The transformation of labor from agriculture to industry and services has recently played a very important role in Vietnam's economic growth. At present, agricultural labor still occupies a large proportion in the labor structure, so the transformation of labor to industry and services will continue to play an active role in Vietnam's economic growth in the future.

Table 5. Regression results with the entire sample scale for the equation 4

Explanatory variable	Model (4)		
	Pooled OLS	FEM	REM
LnK_S	0.5551 (0.247)	0.4527 (0.037)	0.3343 (0.337)
LnL_S	0.4549 (0.463)	0.2958 (0.1218)	0.4749 (0.123)
K_A	0.2336 (0.4517)	0.4159 (0.1091)	0.6533 (0.1177)
K_I	0.3217 (0.4178)	1.1532 (0.1415)	0.6959 (0.3178)
L_A	0.3523 (1.254)	0.63964 (0.4862)	1.3433 (0.4514)
L_I	1.605 (0.4093)	1.9664 (0.4619)	0.6605 (0.7518)
Independent coefficient	3.764 (2.575)	6.5969 (1.9615)	1.5674 (1.0375)
Number of observed variables	960	960	960
Number of provinces	60	60	60
Hausman (χ^2)		155.24	
Wald (χ^2)		3397.17	
Wooldridge		28.827	
Peasaran		39.397	

Table 6. Regression results with the entire sample scale for the equation 5

Explanatory variables	Model (5)		
	Pooled OLS	FEM	REM
LnK_NA	1.003 (0.2225)	0.5127 (0.0357)	1.3554 (0.237)
LnL_NA	0.2133 (0.3353)	0.2928 (0.1052)	0.3749 (0.2423)
K_I	0.5023 (0.2217)	0.2477 (0.091)	0.5526 (0.5187)
L_I	0.5119 (0.2071)	1.0512 (0.4165)	0.6959 (0.678)
Independent coefficient	0.7154 (1.154)	7.3402 (1.1326)	1.6444 (0.5550)
Number of observed variables	960	960	960
Number of provinces	60	60	60
Hausman (χ^2)		26.894	
Wald (χ^2)		4895.17	
Wooldridge		67.57	
Peasaran		28.459	

5.1 Limitations and Further Researchs

Limitations of the study were two important factors: latency and spatial error. Therefore, this research will be a prerequisite for subsequent studies using spatial economics models to research ST and economic growth in Vietnam.

References

Smith, A.: The Wealth of Nations. University of Chicago Press, Chicago (1776)
Akita, T., Hermawan, A.: The Sources of Industrial Growth in Indonesia, 1985–1995: An Input-Output Analysis, Working Paper, No. 4 (2000)
Akita, T., Hau, C.T.T.: Inter-sectoral Interdependence and Growth in Viet Nam: A Comparative Analysis with Indonesia and Malaysia. GSIR. Working Paper Series, EDP06-1 (2006)
van Ark, B., Timmer, M.: Asia's Productivity Performance and Potential: The Contribution of Sectors and Structural Change (2003). databases/10_sector/2007/papers/asia_paper4.pdf
Chenery, H., Syrquin, M.: Typical patterns of transformation. In: Chenery, H., Robinson, S., Syrquin, M. (eds.) Industrialization and Growth, A World Bank Research Publication. Oxford University Press, New York (1986)
Cornwall, J., Cornwall, W.: Growth theory and economic structure. Economica New Ser. **61** (242), 237–251 (1994)
Kuznets, S.: Modern Economic Growth: Rate, Structure and Spread. Vakils, Feffer and Sumons Private Limited, Bombay (1966)
Kuznets, S.: Two centuries of economic growth: reflections on US experience. Am. Econ. Rev. **67**, 1–14 (1977)
Hayashi, M.: Structural changes in Indonesian industry and trade: an input - output analysis. Dev. Econ. **43**(1), 39–71 (2005)
Hoffman, W.: The Growth of Industrial Economies. Oxford University Press, Manchester (1958)
Lewis, W.A.: Economic development with unlimited supplies of labor. Manchester Sch. Econ. Soc. Stud. **22**, 139–191 (1954)
Nurkse, R.: Problems of Capital Formation in Underdeveloped Countries. Oxford University Press, New York (1953)
Rodrik, D.: The real exchange rate and economic growth. Brook. Pap. Econ. Act. **2** (2008)
McMillan, M., Rodrik, D.: Globalization, Structural Change, and Productivity Growth, IFPRI Discussion Paper 01160 (2011)
Ricardo, D.: Principles of Political Economy and Taxation. Dent, London (1817)
Rostow, W.W.: The Stages of Growth: A Non-Communist Manifesto. Cambridge University Press, Cambridge (1960)
Schumpeter, J.A.: Business Cycles: Theoretical, Historical and Statistical Analysis of the Capitalist Process. McGraw Hill, New York and London (1939)

Bayesian Analysis of the Logistic Kink Regression Model Using Metropolis-Hastings Sampling

Paravee Maneejuk[1(✉)], Woraphon Yamaka[1], and Duentemduang Nachaingmai[2]

[1] Center of Excellence in Econometrics, Faculty of Economics, Chiang Mai University, Chiang Mai 50200, Thailand
mparavee@gmail.com, woraphon.econ@gmail.com

[2] Faculty of Economics, Chiang Mai University, Chiang Mai 50200, Thailand
dnachien@hotmail.com

Abstract. Threshold effect manifests itself in many situations where the relationship between independent variables and dependent variable changes abruptly signifying the shift into another state or regime. In this paper, we propose a nonlinear logistic kink regression model to deal with this complicated and nonlinear effect of input factors on binary choice dependent variable. The Bayesian approach is suggested for estimating the unknown parameters in the models. The simulation study is conducted to demonstrate the performance and accuracy of our estimation in the proposed model. Also, we compare the performance of Bayesian and the Maximum Likelihood estimators. This simulation study demonstrates that the Bayesian method works viably better when sample size is less than 500. The application of our methods with a birthweight data and risk factors associated with low infant birth weight reveals interesting insights

Keywords: Logistic kink regression model · Bayesian estimation
Maximum likelihood · Bayes factors

1 Introduction

In the discrete choice analysis, it is important to identify factor effecting of is a binary latent variable [8]. However, the traditional linear regression cannot be used to estimate this type of models as the left-hand side of equations takes values between 0 and 1, while the right-hand side may take any values of both discrete and continuous variables. Thus, the discrete models namely logistic regression have been proposed to handle this problem on the presupposition that individuals with a probability of 0.5 of choosing either of two alternatives are most sensitive to changes in independent variables. This assumption is imposed by the estimation technique because the logistic density function is symmetric about zero [16].

V. Kreinovich et al. (Eds.): ECONVN 2019, SCI 809, pp. 1073–1083, 2019.
https://doi.org/10.1007/978-3-030-04200-4_78

Recently, some empirical works related to the binary choice model have suggested the existence of the threshold effect in the model structure. Jiang et al. [11] analyzed Schizophrenia data using a nonlinear threshold logistic model and suggested there exhibits a complex effect of multiple genes in explaining the differences between low and high-risk groups. Fong, Di and Permar [5] studied the response of immune response against the infections like in the case of HIV disease and they found that the threshold effects are often plausible in a complex biological system. This is to say that the relationship between an outcome variable and independent variable changes as the independent value crosses a certain threshold or change point. Thus, in this study, we consider the nonlinear effect in the binary model as a novel tool for studies in the field of economic behavior. Rather than letting the logistic regression model be split into two (or more) groups based on a threshold value, we propose an alternative nonlinear model for the logistic regression. The model developed here is logistic kink regression model which is shown to be flexible where the kink effect or nonlinear effect can occur in some independent variables and the model structure does not change as there is only one covariate. Thus, the regression function is continuous, but the slope has a discontinuity at a threshold point or kink point [7]. Intuitively, the proposed model can investigate the slope change of the relationship between the outcome of interest and each independent variable at each side of the kink or threshold point. The kink approach has been proven to be successful in many recent studies [7,13–15].

The purpose of this study is to propose a modern nonlinear model for cross-section data analysis. In the estimation point of view, the classical methods, for example Maximum Likelihood estimator, are almost always used and perhaps the most commonly and widely accepted in parameter estimation in nonlinear model. However, these methods often face with the poor estimation from bias and inconsistency of the results when the sample size of the data is limited. Many researchers have concern about the use of limited data that can bring about an underdetermined, or ill-posed problem for the observed data, and the conventional estimators cannot easily serve to reach the global optimization. As it is widely understood, the larger sample size of data can bring the higher probability of finding a significant result [12]. Button et al. [3] suggested that when the samples are limited, it is often hard to get meaningful results.

To deal with this estimation problem, our study considers using a Bayesian estimation as the innovation tool for estimating the unknown parameters in logistic kink regression model. This estimation provides a flexible way of combining observed data with prior information from our knowledge thus we can construct the complete posterior to find the optimal parameter. In addition, the Metropolis-Hastings sampling is used as the tool for sampling the unknown parameters to avoid the difficulty reaching the global optimization. By using this sampler, we do not need to derive the closed form solution of the parameter estimates.

In this paper, our main objective is to propose a new class of nonlinear logistic models, to characterize the complex relationship between the independent and dependent variables in the economic context. Also, we introduced a Metropolis-Hastings based Bayesian approach to estimate the parameter of the model. To the best of our knowledge, this model has not been explored in the literature yet. Hence, this fact becomes one of our motivations to work on this paper.

The remaining of this paper is organized as follows: In Sect. 2, we will introduce the proposed logit kink regression model. The Bayesian estimation to estimate the unknown parameters in the model will be elaborated in Sect. 3. In Sect. 4, we conduct a simulation study to assess the accuracy and performance of the estimation and the model. Section 5 concludes this study.

2 Methodology

2.1 Review of Logit Regression Model

Prior to discussing about the nonlinear logistic regression model, let us briefly explain the conventional linear logistic regression model. The model can be in such form as

$$y*_i = X'\beta + u_i, \quad i = 1, ..., n \tag{1}$$

where y_i^* is a response variable with value 1 if $y_i^* > 0$, otherwise $y_i^* = 0$, X is an $n \times (k+1)$ data matrix with k independent variables and n observations. β is a $(k+1) \times 1$ vector of coefficients and u_i is an error term which is assumed to have logistic distribution. It is obvious that y_i^* is a binary choice and unobservable, thus according to the theory of probability model (e.g. [1]). Thus, the probability of $y_i^* = 1$ and that of $y_i^* = 0$ are, respectively,

$$\begin{aligned} prob(y*_i = 1) &= F(X'\beta), \\ prob(y*_i = 0) &= 1 - F(X'\beta), \end{aligned} \tag{2}$$

where $F(\cdot)$ is the logistic with distribution function

$$F(X'\beta) = \frac{exp(X'\beta)}{1 + exp(X'\beta)}. \tag{3}$$

2.2 Logistic Kink Regression Model

In this study, the kink regression model of Hansen [7] is extended to the logistic regression, thus we can modify the Eq. (1) as

$$\begin{aligned} y_i^* = \quad & \beta_0 + {\beta_1}^-(x'_{1,i} - \gamma_1)_- + {\beta_1}^+(x'_{1,i} - \gamma_1)_+ +, ..., \\ & + {\beta_k}^-(x'_{k,i} - \gamma_k)_- + {\beta_k}^+(x'_{k,i} - \gamma_k)_+ + \alpha Z_i + \varepsilon_i, \end{aligned} \tag{4}$$

where x_i is the regime-dependent independent variables of individual i, Z_i is the regime independent exogenous variables of individual i. β_0 is intercept term. $({\beta_1}^+, ..., {\beta_k}^+)$ and $({\beta_1}^-, ..., {\beta_k}^-)$ are the coefficients with respect to variable x_i for value of $x_{k,i} > \gamma_k$ and with respect to variable x_i for value of $x_{k,i} \leq \gamma_k$, respectively. α is the regime independent coefficient of Z_i. We use

$(x'_{k,i} - \gamma_k)_- = \min[x'_{k,i} - \gamma_k, 0]$ and $(x'_{k,i} - \gamma_k)_+ = \max[x'_{k,i} - \gamma_k, 0]$ as the indicator function for separating x_i into two regimes. Thus, the relationship between y_i^* and x_i is non-linear while there is a linear relationship between y_i^* and Z_i. In addition, the relationship of x_i with y_i^* changes at the unknown location threshold or kink point γ_k.

3 Bayesian Estimation and Bayes Factor for a Kink Effect

3.1 Bayesian Parameter Estimation

The Bayesian approach, it is characterized by the explicit use of probability distributions to draw inferences. It is different form the standard frequentist approach as it adds a prior probability distribution describing uncertainty about the underlying parameters of the sampling or data distribution. Therefore, this estimator consists of the likelihood of the data and the prior distribution of the parameter estimates. By using Bayes theorem, the combination of the likelihood of frequentist approach and prior distribution is possible, and this combination can give the posterior distribution of the parameter conditional on the data and prior. Let $\theta = \{\beta^-, \beta^+, \alpha, \gamma\}$, the standard form of the posterior distribution is

$$p(\theta\,|y, x, Z) \propto p(y, x, Z\,|\theta)p(\theta), \tag{5}$$

where $p\,(y, X, Z|\theta)$ is the likelihood probability function which includes the information of the data and $p(\theta)$ is the prior distribution. In this study, we consider the logistic distribution function as the likelihood function, thus the likelihood here can be written as

$$L(y^*\,|x, Z, \beta^-,\ \beta^+, \alpha, \gamma) = \prod_{i=1}^{n} F_i^{y_i^*}(1 - F_i)^{1-y_i^*}, \tag{6}$$

Let $\Psi{=}\beta_0 + {\beta_1}^-(x'_{1,i} \leq \gamma_1)_- + {\beta_1}^+(x'_{1,i} - \gamma_1)_+ +, ..., +{\beta_k}^-(x'_{k,i} - \gamma_k)_-$ $+{\beta_k}^+(x'_{k,i} - \gamma_k)_+ + \alpha Z_i$ and F_i is specified to be given by the logistic probability distribution evaluated at Ψ. Hence, the likelihood function of our model is

$$L(y^*\,|x, Z, \beta^-,\ \beta^+, \alpha, \gamma) = \prod_{i=1}^{n}\left[F_i(\Psi)^{y_i^*}(1 - F_i(\Psi))^{1-y_i^*}\right], \tag{7}$$

Suppose that our parameters follow an improper prior $p(\theta) = 1$, the posterior distribution become the likelihood function. Therefore, the posterior distribution of our model can be expressed as

$$p(\theta\,|y^*, x, Z) = \frac{exp\left[\sum_{i=1}^{n}(\Psi_i)y_i\right]}{\prod_{i=1}^{n}\left[1 + exp(\Psi_i)\right]}. \tag{8}$$

To sample all these parameters based on conditional posterior distribution, the study employs the Markov chain Monte Carlo, Metropolis-Hastings algorithm

for obtaining the sequence of parameter samples form from fully conditional distributions. Under the improper prior, there has a condition to use this algorithm as the posterior is not necessarily integrable. To deal with this problem the following is introduced **Hypothesis [H]**.
Given a sample with $n \geq 4$, we suppose there exists positive x_i and negative x_i associated with both of y_i^*=1 and 0.

Lemma 1. The posterior distribution of the Logit model $p(\theta|D)$ where D represents the observed data; is a true under [H]; i.e.

$$\int p(\theta\,|D)d\theta < +\infty \tag{9}$$

The proof of this Lemma 1 is referred to Altaleb and Chauveau [2]. In this study, we run the Metropolis-Hastings sampler for 20,000 iterations where the first 5,000 iterations serve as a burn-in period. For Metropolis-Hastings algorithm, the study applies it to find all unknown parameter including a value θ where the acceptance ratio is

$$r = \frac{p(\theta^*\,|y^*, x, Z)}{p(\theta_{b-1}\,|y^*, x, Z)}. \tag{10}$$

Then, the study sets

$$\theta_b = \begin{cases} \theta_{b-1} & U(0,1) < r \\ \theta_b^* & U(0,1) > r \end{cases}, \tag{11}$$

where θ_{b-1} is the estimated vector of parameter at $(b-1)^{th}$ draw and θ_b^* is proposal vector of parameters which randomly obtained from $N(\theta_{b-1}, 0.01)$. U is uniform (0,1). This mean that if the proposal θ_b^* looks good, keep it; otherwise, keep the current value θ_{b-1}. By using a Metropolis-Hastings algorithm, The study estimates the parameters using the average of the Markov chain on θ as an estimate of θ, when the posterior density is likely to be close to normal and the trace of θ_b looks stationary.

3.2 Bayes Factor for a Kink Effect

Since the nonlinear structure model has been proposed in this study, the Bayes factor test is used to examine the non-linear effect in the model. The purpose of this Bayes factor is to check whether a kink parameter is significantly exists or not. It can be used to assess the models of interest namely linear and kink logistic regression, so that the best fit model will be identified given a data set and a possible model set. In the other word, Bayes factor is a tool for model selection. Bayes factors is defined by the ratio of the posterior under one model to another model. In this study, the linear model is specified as a null model denoted by M_1 and the kink logistic model is specified as alternative model denoted by M_2. More specifically, Bayes factor BF is given by

$$BF = \frac{p(M_1, \theta_1\,|y^*, x, Z)}{p(M_2, \theta_2\,|y^*, x, Z)}, \tag{12}$$

where $p(M_1, \theta_1|y^*, x, Z)$ and $p(M_2, \theta_2|y^*, x, Z)$ are the posterior density of the null model and alternative model, respectively. θ_1 and θ_2 are the vector of parameters of M_1 and M_2, respectively. For choosing the appropriate model, we interpret Bayes factors following the Jeffreys' [10] labelled intervals. As a result, a between 1 and 3 was considered anecdotal evidence for H1 to H0, from 3 to 10 was considered substantial, from 10 to 30 strong, from 30 to 100 very strong and higher than 100 decisive.

4 Simulation and Application Studies

4.1 Simulation Results

In this section, we carry out the Monte Carlo simulation study to examine the performance of the Bayesian method in terms of coverage probability and estimation efficiency. The study focuses on the asymptotic properties of the posterior distribution in Eq. 8, to examine whether the likelihood is valid for posterior inference. Moreover, we compare the performance of Bayesian estimation with the conventional Maximum Likelihood (ML) estimation. The test based on bias and Mean Squared Error (MSE) is conducted for evaluating the performance of the final estimate. In the simulation, the following equation is used to generate the dataset y_i^*

$$y_i^* = \beta_0 + {\beta_1}^-({x'}_{1,i} - \gamma_1)_- + {\beta_1}^+({x'}_{1,i} - \gamma_1)_+ + \varepsilon_i, \tag{13}$$

where the true value for parameters α, β_1^-, and β_1^+ are $\alpha = 1$, ${\beta_1}^- = -2$, and ${\beta_1}^+ = 0.5$, respectively. The threshold value is $\gamma_1 = 3$. The covariate ${x'}_{1,t}$ is independently generated from the standard normal distribution $N(\gamma_1, 1)$ to guarantee that γ_1 is located in ${x'}_{1,t}$. For the binary y_i^* , we simulate from the binomial distribution conditional on the $F(\beta_0 + {\beta_1}^-({x'}_{1,i} - \gamma_1)_- + {\beta_1}^+({x'}_{1,i} - \gamma_1)_+)$. In this Monte simulation study, we simulate sample sizes $n = 200$, $n = 500$ and $n = 1,000$; and $R = 500$ data sets are simulated per sample size. Then, the performance of these two estimators are evaluated through the bias and Mean Squared Error (MSE) which are given as

$$Bias = \left| R^{-1} \sum_{r=1}^{R} (\tilde{\theta}_r - \theta_r) \right|, MSE = R^{-1} \sum_{r=1}^{R} (\tilde{\theta}_r - \theta_r)^2, \tag{14}$$

where $\tilde{\theta}_r$ and θ_r are, respectively, the estimated parameter and their true parameter values.

Table 1 reports the results of the sampling experiments for logistic kink regression model. We find that Bayesian does not completely outperform the MLE in all cases, when $n = 1000$ the value of Bias and MSE of parameter estimates obtained from Bayesian are higher than MLE. However, it still provides a good estimation and can be viewed as an alternative estimation for logistic kink regression model, especially when the sample size is small ($n = 200, 500$). We

Table 1. Simulation results

$n = 200$	MLE		Bayesian	
	Bias	MSE	Bias	MSE
β_0	0.0170	0.0770	**0.0095**	**0.0724**
β_1^-	0.0501	0.0844	**0.0042**	**0.0791**
β_1^+	0.1245	0.1604	**0.0495**	**0.1307**
γ_1	0.0200	0.0162	**0.0034**	**0.0153**
$n = 500$	MLE		Bayesian	
	Bias	MSE	Bias	MSE
β_0	0.0089	0.0289	**0.0083**	**0.0284**
β_1^-	0.0049	0.0359	**0.0039**	**0.0332**
β_1^+	0.0731	0.0693	**0.0295**	**0.0623**
γ_1	0.0003	0.0117	**0.0001**	**0.0109**
$n = 1000$	MLE		Bayesian	
	Bias	MSE	Bias	MSE
β_0	**0.0021**	**0.0178**	0.0028	0.0181
β_1^-	**0.0016**	**0.0203**	0.0026	0.0214
β_1^+	**0.0583**	**0.0282**	0.0205	0.0259
γ_1	**0.0003**	**0.0113**	0.0001	0.0114

Note: Bold numbers denote the lower Bias and MSE.

also examined the effect of sample size on estimator performance and found that the performance of Bayesian estimator is different when $n = 200$, $n = 500$ and $n = 1,000$. We can observe that the Bias and MSE tend to be close to zero as the sample size increase. This indicates that the performance is better when the sample size is increased conforming the asymptotic properties of this estimator.

4.2 Application Example

We empirically demonstrate the application of the proposed nonlinear model using real data based on a dataset from [9]. The birthweight data frame has 189 rows and 10 columns with information collected at Baystate Medical Center, Springfield, Mass during 1986. The data is used to investigate the risk factors associated with low infant birth weight. Thus, we consider the following equation:

$$low_i = f(age_i, lwt_i), \tag{15}$$

where low_i is the indicator of birth weight less than 2.5 kg. age_i is mother's age in years and lwt_i mother's weight in pounds at last menstrual period. Prior to estimating the logistic kink regression model, the nonlinear effect should be tested. Thus, Bayes factor test is used to determine whether our independent variable appears to have a nonlinear behavior. Using the Bayes factor formula, we test each covariate against low_i and the result is shown in Table 2. It provides the values of Bayes factor of these linear and nonlinear logistic regression models. We can observe that the value of BF of age_i is equal to 2.9028 while lwt_i is 4.5567. This indicates that Bayes factor tends to substantial favor the kink effect only in lwt_i while Bayes factor provides the anecdotal evidence for age_i. Thus, the relationship between low_i and lwt_i is non-linear while there is a linear relationship between low_i and age_i. As a result, the empirical equation in this example becomes

$$low_i = \beta_0 + \beta_1{}^{-}(lwt_i - \gamma_1)_{-} + \beta_1{}^{+}(lwt_i - \gamma_1)_{+} + \beta_2 age_i + \varepsilon_i, \quad (16)$$

Table 2. Bayes factor kink test

low_i	age_i	lwt_i
BF	2.9025	4.5567
Interpretation	Anecdotal	Substantial

To confirm the result of this Bayes factor, we are going to plot the Receiver Operating Characteristic Curve (ROC) curve and calculate the AUC (area under the curve) which are typical performance measurements for a binary classifier. The ROC is a curve generated by plotting the true positive rate (TPR) against the false positive rate (FPR) at various threshold settings while the AUC is the area under the ROC curve. The model with good predictive ability should have an AUC closer to 1 [6]. According to the result, the AUC of logistic kink regression is 0.6441 while the AUC of logistic regression is 0.6263. This result indicates that our proposed model is statistically superior to the conventional linear logistic regression in terms of prediction power. The plot of ROC is illustrated in Fig. 1.

Table 3. Simulation results

	Bayesian		MLE	
Coefficients:	Estimate	CI 95%	Estimate	p-value
β_0	0.0187	[0.0163, 0.0209]	0.0014	0.9985
β_1^-	−0.0216	[−0.0457, −0.0041]	−0.0461	0.1559
$\beta_1{}^+$	−0.0087	[−0.0191, 0.0012]	−0.0053	0.4878
β_2	−0.0883	[−0.1171, −0.0499]	−0.0394	0.2674
γ_1	124.0351	[115.512, 133.326]	114.871	0.0000

Note: CI is credible interval

Table 3 provides the result of parameter estimates from Bayesian and MLE. Apparently, similar results are obtained from these two estimations, but indicating the robustness of our Bayesian computation. In addition, convergence

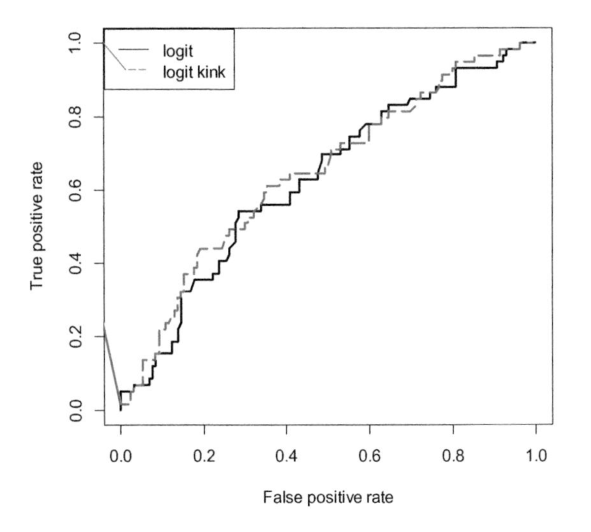

Fig. 1. ROC curves

diagnosis is considered to confirm the reliability of the MCMC. In the theory, it does not give a clear indication of whether it has converged. Thus, we depict the density based on the Metropolis-Hastings sampler draws, which gives some basic insight into the geometry of the posteriors obtained in this application analysis. The results show a fair good convergence behavior and it seems to converge to the normal distribution; thus, we can get an acceptable posterior inference for parameters that appear to have acceptable mixing (Fig. 2).

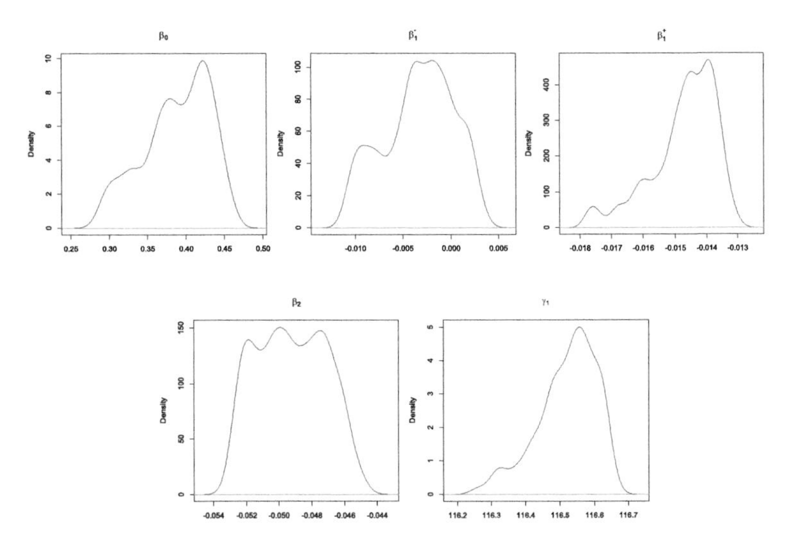

Fig. 2. Density of the logistic kink regression parameters coming from the M-H algorithm. The illustrations are for the 1000 last iterations of the chains.

5 Conclusion

In this study, we proposed a nonlinear logistic kink regression model in which a covariate has a different size effect on the binary choice variable. Intuitively, this type of nonlinear model can handle a nonlinearity in the relationship between two variables. To estimate this model, we employ a Bayesian estimation and use a Metropolis-Hastings algorithm to sample the parameters in the posterior. Moreover, the study also proposes to use a Bayes factor test to check whether logistic kink regression model is preferred relative to the linear logistic regression model.

We then conduct a simulation study to show the performance and accuracy of Bayesian estimation for our proposed model. The simulation results confirm that Bayesian estimator can give an accurate result for all unknown parameters. It performs well over a wide range of sample sizes and obtains higher accuracy as the sample size increases. When we compare the performance of Bayesian and MLE, we find that Bayesian outperforms the MLE when the observations are less than 500. Finally, our study gives an empirical example to show the performance of our model in real data. The result shows that our model is more accurate than the linear model in terms of AUC value.

References

1. Albert, J.H., Chib, S.: Bayesian analysis of binary and polychotomous response data. J. Am. Stat. Assoc. **88**(422), 669–679 (1993)
2. Altaleb, A., Chauveau, D.: Bayesian analysis of the Logit model and comparison of two Metropolis-Hastings strategies. Comput. Stat. Data Anal. **39**(2), 137–152 (2002)
3. Button, K.S., Ioannidis, J.P., Mokrysz, C., Nosek, B.A., Flint, J., Robinson, E.S., Munafó, M.R.: Power failure: why small sample size undermines the reliability of neuroscience. Nat. Rev. Neurosci. **14**(5), 365 (2013)
4. Chib, S., Albert, J.H.: Bayesian analysis of binary and polychotomous response data. J. Am. Stat. Assoc. **88**, 669–679 (1993)
5. Fong, Y., Di, C., Permar, S.: Change point testing in logistic regression models with interaction term. Stat. Med. **34**(9), 1483–1494 (2015)
6. Hanley, J.A., McNeil, B.J.: The meaning and use of the area under a receiver operating characteristic (ROC) curve. Radiology **143**(1), 29–36 (1982)
7. Hansen, B.E.: Regression kink with an unknown threshold. J. Bus. Econ. Stat. **35**(2), 228–240 (2017)
8. Hausman, J.A., and Wise, D.A.: A conditional probit model for qualitative choice: discrete decisions recognizing interdependence and heterogeneous preferences. Econometrica J. Econ. Soc. **46**, 403–426 (1978)
9. Hosmer, D.W., Lemeshow, S.: Applied Logistic Regression. Wiley, New York (1989)
10. Jeffreys, H.: Theory of Probability, 2nd edn. Oxford University Press, Oxford (1948)
11. Jiang, Z., Du, C., Jablensky, A., Liang, H., Lu, Z., Ma, Y., Teo, K.L.: Analysis of schizophrenia data using a nonlinear threshold index logistic model. PloS One **9**(10), e109454 (2014)

12. Peto, R., Pike, M., Armitage, P., Breslow, N.E., Cox, D.R., Howard, S.V., Mantel, N., McPherson, K., Peto, J., Smith, P.G.: Design and analysis of randomized clinical trials requiring prolonged observation of each patient. I. Introduction and design. British J. Cancer **34**(6), 585 (1976)
13. Pipitpojanakarn, V., Maneejuk, P., Yamaka, W., Sriboonchitta, S.: Expectile and quantile kink regressions with unknown threshold. Adv. Sci. Lett. **23**(11), 10743–10747 (2017)
14. Sriboochitta, S., Yamaka, W., Maneejuk, P., Pastpipatkul, P.: A generalized information theoretical approach to non-linear time series model. In: Robustness in Econometrics, pp. 333–348. Springer, Cham (2017)
15. Maneejuk, P., Yamaka, W., Sriboonchitta, S.: Analysis of global competitiveness using copula-based stochastic frontier kink model. In: Robustness in Econometrics, pp. 543–559. Springer, Cham (2017)
16. Walker, S.H., Duncan, D.B.: Estimation of the probability of an event as a function of several independent variables. Biometrika **54**(1–2), 167–179 (1967)

Analyzing Factors Affecting Risk Management of Commercial Banks in Ho Chi Minh City – Vietnam

Vo Van Ban(✉), Vo Đuc Tam, Nguyen Van Thich, and Tran Duc Thuc

Faculty of Foreign Language, Banking University of HCMC, Ho Chi Minh City, Vietnam
Vovanbanl@gmail.com

Abstract. Credit activities are considered as the most professional competences of commercial banks in Vietnam, which bring nearly 80–90% of profits to banks. However, credit risk at high level will affect directly to commercial banks' activities. Facing to challenges of international integration and improving competitive competence of commercial banks, risk management becomes very vital toward administration board of commercial banks in Ho Chi Minh City (HCMC), Vietnam. In this study, the author used quantitative method to investigate factors affecting to risk management in four commercial banks HCMC from evaluation of 120 managers and vice managers of branches in these banks. The findings of the survey presented that Clients' ability and Qualification of employees affecting most to risk management of Commercial Banks. Therefore, State Bank and governmental organizations have to have strict supervision and control toward commercial banks to limit risks. On the other hands, commercial bank itself has to improve risk management procedure, expertise of staff, limiting financial risk at low level as well as reduce risks for clients and the bank itself in financial service market in Vietnam.

Keywords: Ability · Environment · Liquidity · Qualification
Risk management · Vietnam

1 Introduction

Risk management of banking is considered as a serious matter discussed and studied by many researchers and financial organizations. According to these studies and analysis, risk culture was claimed for the recent global financial crisis in which many banking firms failed to understand the risk they were taking. The Financial Stability Board (2013) indicates the need of re-assessment of the three contributors to strengthen the financial organization's risk culture including the risk governance, the risk appetite and the compensation. Risk culture impacts deeply on the risk management framework toward the top of administrative board.

Vietnamese commercial banks nowadays are considered as crediting organizations, operating by company model such as Vietcombank, Vietinbank, BIDV, ACB, Tech-comBank, VPBank, Agribank, MB, Marintimebank, SHB, Eximbank, Navibank,

V. Kreinovich et al. (Eds.): ECONVN 2019, SCI 809, pp. 1084–1091, 2019.
https://doi.org/10.1007/978-3-030-04200-4_79

Sacombank, DongABank, Oceanbank, Kien Long Bank, Nam A Bank, HD Bank, MDB, Vietcapital Bank, SCB, TPBank, Lienviet Bank. In addition, there are foreign banks or branches of foreign banks in Vietnam as well as other financial organizations. The arisen risks to Vietnamese commercial banks are main subjects of risk management. The risks related to Vietnamese commercial banks are usually bad debt risks, risks of liquidity, and performance risks (Lang 2017).

As a result, risk management in the financial market becomes vital to Vietnam because the banking system has to shoulder the bad debt with high level as well as some weak commercial banks need to be processed.

Therefore, the author conducted this study to find out suitable solutions to improve the risk management for commercial banks in Ho Chi Minh City (HCMC) in Vietnam basing on analysis of risk management of commercial banks, clients' ability and outside environments in four selected commercial banks in HCMC such as ACB, BIDV, Sacombank and Techcombank.

2 Review of Literature

This section presents a brief review of recently published studies relating directly to risk management issues.

Many academics, practitioners and regulators accepted and regarded effective risk management as a major cornerstone of bank management. To deal with bank risk management, bankers and financial organizations have to acknowledge this reality and the need for a comprehensive approach, adopting the Basel I Accords, following by the Basel II Accords and recently by the Basel III. In addition, Sensarma and Jayadev (2009) found risk management as one of the determinants of returns of banks' stocks.

In the USA, Fan (2004) studied the sensitivity to risk of large domestic and recognized that profit efficiency was considered so sensitive to credit risk but not to risk of insolvency risk or to the mix of loan products. According to Halm (2004), in order to ensure successful financial liberalization, it is necessary to improve banking supervision and banks' risk management. Based on a study of interest rate and exchange rate exposure of Korean banks before the 1997 when the Asia Pacific economic crisis occurred, it was said that the performance of commercial banks was slightly related to their pre-crisis risk exposure.

Moreover, Fatemi and Fooladi (2006) investigated the practices of credit risk management in the largest US-based financial institutions and found that the single most important purpose served by the credit risk models utilized when identifying counterparty default risk. In the United Arab of Emirates (UAE), a comparative study of banks' risk management in locally incorporated banks and foreign banks was provided by Al-Tamimi and Al-Mazrooei (2007), identifying the three most important types of risks facing UAE commercial banks were foreign exchange risk, credit risk, and operating risk. In contrast, Al-Tamimi (2002) regarded the main risk facing UAE commercial banks as credit risk. For risk identification, they used financial statement as the main method when investigating branch managers while Al-Tamimi and Al-Mazrooei (2007) used the financial statement analysis and risk survey as the main methods for inspecting the bank risk manager, audits or physical inspections. The investigated banks were identified to become more sophisticated in managing their risk.

It is concluded that the locally incorporated banks become more efficient in managing risk when the variables such as risk identification, assessment and analysis were considered to have more influence in the risk management process. It is said that there was a significant difference between the UAE national banks and foreign ones in understanding risk and risk management, practicing risk assessment and analysis, as well as risk monitoring and controlling, but not in risk identification, credit risk analysis or risk management practices (Hameeda and Al-Aimi 2012).

Al-Tamimi (2008) found out the readiness to implement the Basel II Accord as well as the resources which are needed to accomplish in UAE commercial banks. It is said that commercial banks perceived the benefits, impact and challenges when associating with the implementation of the Basel II Accord. However, there were no positive tendency to implement Basel II of the UAE commercial banks and their impact of that implementation. There was also no difference between the UAE national and foreign banks that was found in the level of preparation for the Basel II Accord. Al-Tamimi (2008) concluded that employees' educational level was a significant difference in the level of the UAE banks when applying Basel II.

Hassan (2009) stated that Islamic banks were considered to identify a variety of risks when they offered the unique range of products. In addition, the staff working in the Islamic banks of Brunei Darussalam could understand remarkable types of risk and risk management as well as proving their ability to manage risk successfully. These banks had to face major risks such as foreign exchange risk, credit risk and operating risk. Therefore, to make their risk management practices more effective, the Islamic banks in Brunei had to pay attention to those variables and applied the Basel II Accord properly to improve the efficiency of risk management systems.

Greuning and Iqbal (2008) and Mirakhor (2011) suggested a comprehensive framework of risk management which could be applied to a conventional or Islamic bank. The findings of Hassan (2009) also supported to this tendency. Khan and Bhatti (2008) found that in order to improve risk management strategies and corporate governance of Islamic banks, they have to face another crucial challenge because of their adherence to Islamic law. The impact on the risk management of Islamic banks had certain applications, emphasis and inclusion or exclusion.

Pan (2016) stated that there were a close link between corporate governance of commercial banks and risk management. He also considered corporate governance as the basis of risk management, including improvement of ownership structure, a clear division of responsibilities, reasonable incentive mechanism, effective internal control system which present all aspects of corporate governance. They are also regarded as the fundamental guarantee for helping risk management techniques exerting its actual results.

Hameeda and Al-Aimi (2012) stated that Banks in Bahrain had a clear understanding of risk and risk management with efficient risk identification, risk assessment analysis, risk monitoring, credit risk analysis and risk management practices while credit, liquidity and operational risk relating to risk management were considered as the most important risks facing both conventional and Islamic banks. In

Vietnam, there were many studies toward risk management. Luc (2016) regarded risk management as an internal challenge in Vietnamese commercial bank system. Moreover, Mui (2014) suggested developing the Vietnamese banking system sustainably as a conception to cover risk management.

Generally speaking, the situation of risk in Vietnamese commercial banks is attached closely to bad debt, black credit, appropriated capitals, loss, fluctuation of monetary market (IDG, 2013) (Fig. 1).

Research framework

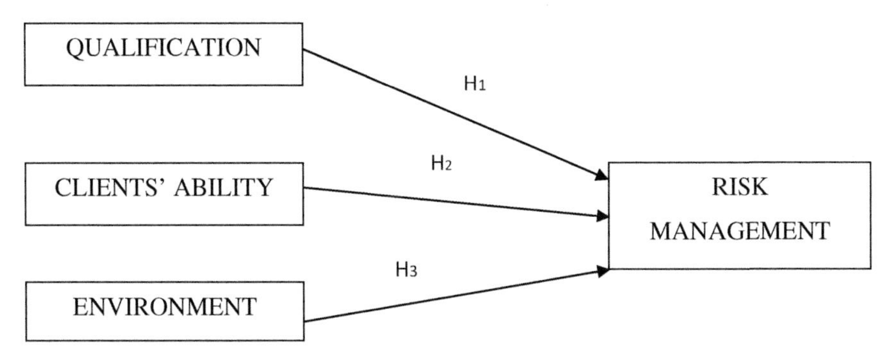

Fig. 1. The proposed theoretical framework of the study.

Research Questions and Hypothesis

Following the review of literature and objectives of this study, the author aimed to answer the following research question:

RQ1: What is the current status of risk management of commercial banks in HCMC?
RQ2: How do managers and vice managers of commercial banks evaluate risk management in term of Qualification of employees, Client's ability, and Reason from environment?
RQ3: What is the implication of this study toward business administration?

Hypothesis:

$\mathbf{H_1}$: There is a positive relationship between Qualification of employees and Risk management.
$\mathbf{H_2}$: There is a positive relationship between Clients' ability and Risk management.
$\mathbf{H_3}$: There is a positive relationship between Reason from environment and Risk management.

3 Research Methodology

In this study, the author used the quantitative approach and the technique of exploratory factor analysis to determine the validity and reliability of variables to find out the relationship between factors influencing risk management of commercial banks located in HCMC, Vietnam.

Basing on the rule of Bollen (1992), it requires at least 5 samples for an estimated parameter. Therefore, based on the number of parameters to be estimated (24), a sample size was 120 (24*5 = 120). The respondents are 120 managers and vice managers of 4 selected commercial banks located in HCMC such as ACB, BIDV, Sacombank, and Techcombank.

After collecting the data, the author used software SPSS 25.0 to analyze data, checking reliability of Cronbach's Alpha, Exploratory Factor Analysis (EFA) to explore factors affecting to risk management of four selected commercial banks in HCMC.

4 Findings

4.1 Evaluating the Suitability of the Multiple Regression Model of the Study

According to Trong (2008), the identified coefficient R^2 was proved to be the increased function by the independent factors input into the model. It is said that the more independent factors are put into the model, the more increasing level R^2 is. In this case, the adjusted R^2 coefficient is used to reflect the suitability of the multiple regression model. The summary of the model in this study is presented as following:

Table 1. Evaluating the suitability of the multiple regression model

Model summary[b]					
Model	R	R Square	Adjusted R Square	Std. Error of the Estimate	Durbin-Watson
1	,965[a]	,931	,930	,11365	1,021

[a]Predictors: (Constant), REASON FROM ENVIRONMENTS, CLIENTS' ABILITY, QUALIFICATION OF EMPLOYEES
[b] Dependent Variable: I am quite satisfied with my employee's expertise in risk management.

As shown in Table 1 above, the value of R was 0.931 > 0.5. Therefore, this model is suitable to evaluate the relationship between independent factors and dependent one. In addition, the value of R^2 was 0.930 and the multiple regression model were built to be suitable with 93%00 of data. In other word, 93% satisfaction of managers and vice managers of 4 commercial banks in HCMC were changed because of changing of independent factors of Qualification of employees (QOE), Clients' ability (CLA), and Reason from environment (REN). The 7% of remainder was due to other factors.

4.2 Testing Hypothesis About Meaning of Regression Coefficients...

The hypotheses of model were presented in Sect. 2.3 of this paper. In Table 2 below, when verifying t_{stat} and $t_{\alpha/2}$ of factors to measure the reliability, all independent factors (QOE, CLA, and REN) were passed because of $t_{stat} > t_{\alpha/2(0,552)} = 0.276$ (the smallest was 25.901) and value of Sig. was smaller than 0.05, presenting the high level of reliability.

Table 2. The result of regression weights.

Coefficients[a]

Model		Unstandardized Coefficients		Standardized Coefficients	t	Sig.	Collinearity Statistics	
		B	Std. Error	Beta			Tolerance	VIF
1	(Constant)	,044	,080		,552	,582		
	QUALIFICATION OF EMPLOYEES	,329	,012	,749	27,291	,000	,784	1,276
	CLIENTS' ABILITY	,323	,011	,755	28,866	,000	,863	1,159
	REASON FROM ENVIRONMENTS	,331	,013	,663	25,901	,000	,901	1,110

[a]Dependent Variable: I am quite satisfied with my employee's expertise in risk management.

4.3 Evaluating the Important Level of Factors Affecting to Clients' Satisfaction Towards the Service Quality of Commercial Banks in HCMC - Vietnam

Based on the Table 2 above, the multi-collinearity regression model of factors affecting to managers' and vice managers' satisfaction towards risk management of employees working in branches of four commercial banks located in HCMC, there was the standardized coefficients as followings:

SATISFACTION TOWARD RISK MANAGEMENT = 0.044 + **0.749** * QOE + **0.755** * CLA + **0.663** * REN

Therefore, all three factors Qualification of employees, Clients' ability, and Reason from environment had direct ration and plus effect to satisfaction of managers and vice managers of commercial banks toward risk management of four selected commercial banks in HCMC. When the banking employees perform and manage these variables well, the satisfaction of managers and vice managers become higher. In these 3 factors, the most affecting factor to managers' and supervisors' satisfaction towards expertise of employees in banking ranches of four commercial banks in HCMC was Clients' ability ($\beta = 0.755$), next was the Qualification of employees ($\beta = 0.749$), and final one was Reason from Environment ($\beta = 0.663$). In generally speaking, H_1, H_2, and H_3 for the theoretical research model were accepted. In conclusion, there were two thirds of managers and vice managers unsatisfying with their employees' performances in risk management of commercial banks in HCMC.

Therefore, managers and vice managers of commercial banks in HCMC should revise their clients' financial reports and investment projects carefully as well as re-train staff to apply the Basel II Accords and III to process risk management issues to prevent loss for the bank and clients. In addition, commercial banks in HCMC should connect and cooperate together to have a good connected e-banking system to solve risk management effectively.

5 Conclusions and Implication

All risks related to commercial banks in Vietnam are usually risks of bad debt, risks of liquidity, and risks of demonstrator. Each risk has an effective administration method. Therefore, managers of commercial banks in HCMC should follow these methods strictly to limit the risk in their own banks as following:

(1) Managing bad debt: Commercial banks have to decrease the percentage of bad debt under the safe range by restructuring weak banks, encouraging the merger of banks, re-buying weak commercial banks as well as the bad debts of commercial banks should be re-bought by Vietnam Asset Management Company to limit risks caused by bad debts.
(2) Managing liquidity: Risk of liquidity of commercial banks is not managed sustainably because of unbalance period. Vietnamese State Bank continues to decrease the top interest rates and encourages big commercial banks to support smaller ones in order to limit risk of liquidity of commercial banks. The signs of improvement toward liquidity risk were related to the increased interest rate for overnight from linked banks in short time, decreasing the overnight transactions, not having runs for interest rates as well as not having signs of decreasing deposits even the restructured banks. However, the risk of liquidity of HCMC commercial banks has been at high level as well as supervision toward liquidity risk of State Bank was not as good as expected.
(3) Operational risk: According to Basel II, risk of operation in banking system is damage of procedures, human resources, and the unfulfilled internal system, or outside factors. This risk might be also caused by informatics technology, internal fraud, operating structure, regulations, solving task process. These risks occur regularly in commercial banks due to human resources, the operational mechanism was not suitable with vision and mission, the improper business policy, risk of informatics system such as ATM out of order or links, or ethics of bank staff to have their own benefits. The main cause of operational risk is related to morals of banking staff and info-technology infrastructure, especially risks attached with new banking products based on the digital technology base.

Implication

Through this paper, the administration board of commercial banks in HCMC may have a better understanding of factors affecting most to risk management and pay attention to study clients' financial reports as well as examine clients' business project thoroughly before deciding to give loans. In addition, managers and credit managers of branches of commercial banks have to revise their human resources as well as training program to help their employees have good knowledge in risk management based on international standards to prevent loss for their own banks and improve the bank's reputation in the financial market in HCMC as well as in Vietnam.

References

Al-Tamimi, H.A.: Risk management practices: an empirical analysis of the UAE commercial banks. J. Financ. India **6**(3), 1045–1057 (2002)

Al-Tamimi, H.A., Al Mazooei, F.M.: Banks' risk management: a comparison study of UAE national and foreign banks. J. Risk Financ. **8**(4), 394–409 (2007)

Al-Tamimi, H.A.: Implementing Basel II: an investigation of the UAE banks' Basel II preparations. J. Financ. Regul. Compliance **16**(2), 137–187 (2008)

Bollen, A.K.: Tests for Structural of Equation Models. SAGE, Newbury Park (1992)

Fatemi, A., Fooladi, I.: Credit management: a survey of practices. Manag. Financ. **32**(3), 227–233 (2006)

FSB – Financial Stability Board. Risk management lessons from the Global Banking Crisis of 2008. Report in the Bank for International Settlements, Basel (2013)

Greuning, H.V., Iqbal, Z.: Banking and Risk Environment Islamic Finance: The Regulatory Challenge, pp. 11–39. Willey (2008)

Halm, J.H.: Interest rate and exchange rate exposures of banking institutions in pre-crisis Korea. Appl. Econ. **36**(13), 1409–1419 (2004)

Hasan, A.: Risk management practices of Islamic banks of Brunei Darrussalam. J. Risk Financ. **10**(1), 23–37 (2009)

Hameeda, A.H., Al-Aimi, J.: Risk management practices of conventional and Islamic banks in Bahrain. J. Risk Financ. London **13**(3), 215–239 (2012)

Lang, N.T.: Risk management in Vietnamese commercial banks and the current problems. J. Financ. Vietnam **2**(1), 1–5 (2017)

Luc, C.V.: Opportunity and Challenge for Vietnamese banks in the period 2016–2020. Scientific Conference of Banking Vietnam. Publisher National Economics University (2016)

Mirakhor, A.: Lesson from the recent crisis for Islamic finance. J. Econ. Manag. **16**(2), 132–138 (2011)

Pan, Z.: An empirical analysis of the Impact of Commercial Banks' Corporate Governance to Risk Control. Manag. Eng.: Bright. East **2**(2), 72–79 (2016)

Rosman, R., Abdul, R.: The practice of IFSB guiding principles of risk management by Islamic banks: International evidence. J. Islam. Account. Bus. Res. **6**(2), 150–172 (2015)

Sensarma, R., Jayadev, M.: Are bank stocks sensitive to Risk Management? J. Risk Financ. **10**(1), 7–22 (2009)

The Role of Market Competition in Moderating the Debt-Performance Nexus Under Overinvestment: Evidence in Vietnam

Chau Van Thuong[1], Nguyen Cong Thanh[1], and Tran Le Khang[2](✉)

[1] School of Accounting – Banking – Finance, Ho Chi Minh City University of Technology, Ho Chi Minh City, Vietnam
{cv.thuong,nc.thanh93}@hutech.edu.vn
[2] School of Economics, Erasmus University Rotterdam, The Hague, The Netherlands
khang.tl@vnp.edu.vn

Abstract. Previous studies suggested that the characteristics of an industry may play a significant role in the relationship between financial decisions and firm performance through the degree of concentration or competition. Therefore, this research aims to evaluate such a role in order to clarify the effect of industry competition on the relationship between debt and performance. Moreover, overinvestment is recently considered to be one of the causes leading to bad performance because it tends to worsen agency problems in enterprises. As a consequence, the paper is the first one to examine the different impacts of industry competition on the debt-performance relationship in companies with and without overinvestment. Collected from the financial statements of all listed firms on Vietnam's stock exchange, the dataset covers a wide range of 21 various industries over a seven-year period. The research methodology goes through two steps. Firstly, it calculates two alternative variables as the representatives of competition and overinvestment through different subequations. Secondly, it adds them to the main regression to estimate the results with the help of System-GMM estimator together with two instrumental variables, tangibility and non-debt tax shield, to deal with the endogenous problem. The findings show that debt ratio is positively related to firm performance and that the relationship might become stronger at the high level of industry competition. Nevertheless, the research indicates that the positive interaction between debt and competition gets weaker under overinvestment.

Keywords: Financial leverage · Industry competition · Over-investment Firm performance · Vietnam

JEL Classification: G32 · L11 · L25

1 Introduction

The debt-performance relationship attracts much attention and raises many debates in the global science community. Modigliani and Miller (1958) suggested that capital structure does not determine firm performance under some assumptions of a perfect

V. Kreinovich et al. (Eds.): ECONVN 2019, SCI 809, pp. 1092–1108, 2019.
https://doi.org/10.1007/978-3-030-04200-4_80

capital market. However, an enormous number of empirical studies have subsequently been conducted to confirm their relationship in reality. Surprisingly, almost all of their findings came to the consensus that capital structure is relevant to firm performance through the mechanism of the trade-off effect, limited liability effect, and discipling effect (Brander and Lewis 1986; Grossman and Hart 1983; Jensen 1986; Jensen and Meckling 1976; Khan 2012; Margaritis and Psillaki 2010; San and Heng 2011). Moreover, the sign of debt remains debatable among empirical studies. The trade-off between the costs and benefits of debt and equity (Jensen and Meckling 1976), the limited liability (Brander and Lewis 1986), and the discipling effect (Grossman and Hart 1983; Jensen 1986) support the positive sign. On the other hand, underinvestment (Myers 1977) and stakeholders' reactions the (Maksimovic and Titman 1991; Titman 1984) explain for the negative sign. Besides, the predation theory suggests that when a market is highly concentrated, it is easier for a company in the market to be swallowed by others in case it uses much debt in its capital structure (Bolton and Scharfstein 1990; Chavalier and Scharfstein Chavalier and Scharfstein 1996; Dasgupta and Titman 1998).

The predation theory emphasizes on the role of market product competition in regulating the real impact of debt on firm performance. In practice, some studies gave empirical evidence on such a role in the US and some emerging markets (Campello 2003, 2006; Chevalier 1995a, 1995b; Kovenock and Phillips 1997; Opler and Titman 1994). However, market product competition has not yet been researched in Vietnam since it was added to the list of emerging countries with its high economic growth, trade openness, as well as investment inflow within the two last decades. Moreover, because capital market in the country has not fully developed, its banking system has remained the major financing source for investment to Vietnamese enterprises. Thus, debt ratio in essence is a decisive factor of performance in the case of Vietnam (Fu-Min et al. 2014; Gueorguiev and Malesky 2012; Tran et al. 2015). Recently, Vietnam's global integration accompanied by its openness in various aspects of the economy and its implementation of privatization process among state-owned enterprises (SOEs) leads to the increased level of competition among companies in the market (Quy et al. 2014; Tran et al. 2015). In short, such a circumstance facilitates the moderation role of industry competition in capital structure of Vietnamese companies.

According to Agency Theory, the conflicts of interests between managers and shareholders force a firm to incur huge costs to settle them down (Gaver and Gaver, 1993; Jensen 1986; Jensen and Meckling 1976). Exposed to the free cash flow, managers enlarge sources under their control through carrying out a lot of investment projects, even those with negative net present value (NPV) in order to satisfy their personal gains. Hence, overinvestment signals a serious agency problem and makes firm performance worse. Generally, with high debt in the capital structure, enterprises operating in a concentrated market and experiencing overinvestment tend to perform inefficiently. In short, debt ratio, industry competition, and overinvestment should be collectively studied in the research.

The study aims at analyzing the role of industry competition on the debt-performance relationship under overinvestment to answer two questions: (1) Does the impact of debt on performance become better in a competitive market? and (2) Does overinvestment make the relationship worse? The answers to these questions partly contribute to the academic and practical world. For one thing, they provide empirical evidence on the critical role of competition and overinvestment in Vietnam, an emerging market, after the 2008 Global Financial Crisis. For another thing, they help investors set up a suitable investment portfolio and the government make appropriate policies in order to promote the freedom of the market as well as heighten the level of market competition in Vietnam.

The original data includes 699 companies listed on Vietnam's two stock market exchanges namely HOSE and HNX in the period of 2010–2016. However, after the data processing and missing removal, the final dataset covers 208 companies in a wide range of 21 industries from Thompson Reuters source. Overinvestment is measured by taking the estimated value of residual from the sub-equation model. Competition is calculated in two ways through the opposite HHI Index and the absolute value of coefficients of the sub-equation model (BI Index). Using System Generalized Method of Moments (SGMM) to handle the endogenous problem caused by the dynamic function, the study indicates the positive relationship between debt ratio and firm performance. Furthermore, in a competitive market, the use of debt is more likely to enhance its positive effect on performance. The result implies the existence of the predation theory in Vietnam's product market. Nevertheless, overinvestment among Vietnamese enterprises causes the interaction effect to be weaker due to high agency costs. The estimated results are robust with alternative proxies of both competition and performance.

The paper is divided into 5 sections. Section 2 presents the empirical reviews together with hypothesis development. Research methodology and estimation are explained in Sect. 4 and 5, respectively. The study ends with the conclusion in Sect. 6.

2 Literature Review and Hypothesis Development

2.1 Debt Ratio and Firm Performance

The relationship between debt and performance has raised much debates among various studies in corporate finance. Based on some assumptions of a perfect capital market without taxes, transaction costs, and asymmetric information, Modigliani and Miller (1958) propose Capital Structure Irrelevance Theory, supporting the idea that capital structure is irrelevant to operation effectiveness. In reality, the capital market has some imperfections. When relaxing the assumption of taxes, Modigliani and Miller (1963) demonstrate the beneficial role of tax shields in helping companies obtain the benefits of tax deduction. When relaxing the assumption of asymmetric information, Jensen and Meckling (1976) indicate the role of debt in solving the conflicts of interests based on Agency Theory. Asymmetric information makes managers more informed of business activities than shareholders. More seriously, with the free cash flow, managers attempt to establish a more extensive control over their companies through the pursuit of higher perquisites. It requires enterprises a certain amount of costs to align the interests between two sides. In this situation, an increase in debt ratio implies the reduction in the free cash flow and the participation of different partners in the capital market in the monitoring tasks with various disciplines as well as covenants (Grossman and Hart 1983; Harris and Raviv 1990; Jensen 1986). As a result, the use of debt is supposed to heighten firm performance by decreasing agency problems in an enterprise.

Hypothesis 1: financial leverage is positively related to firm performance.

2.2 Financial Leverage, Industry Competition, and Overinvestment

The association of debt with industry competition seems to be complicated. The limited liability effect suggests that companies with high debt ratio can aggressively compete with others in the market (Brander and Lewis 1986). Their aggressive behavior will lessen agency problems. However, the impact of such a behavior depends the level of competition as well as the characteristics of products in an industry (Wanzenried 2003). Thus, a company with high debt cannot increase their profits due to the limited liability effect. Particularly, when a market is completely competitive, this effect is more obvious because the limited liability causes production to become more aggressive and market prices to decrease. According to Cournot competition, the more substitutable market products are the lower profitability is.

The predation theory suggests that firms with high debt are more likely to be disadvantageous in terms of competitiveness compared to those with low debt in their capital structure. Fudenberg and Tirole (1986) show that the theory is even more obvious in markets with a high level of concentration. Preexisting companies are inclined to predate newcomers. The predation process will lower the profitability of entrant companies and put them into a gloomy prospect. Financially constrained enterprises are more vulnerable to the predation from other rivals in the market. Supporting this idea, Scharfstein (1990) explains the rivalry predation through debt covenants which are employed to mitigate agency problems. Once firms are exposed to restrictions from debt covenants, they are on the verge of liquidation and are forced to leave the market in case of failing to satisfy their obligations. Nevertheless, in a perfectly competitive market where every company only contributes a small part to the whole market's production, the rivalry predation tends to be negligible. The use of debt will lead to an increase in other competitors' market value; therefore, highly leveraged incumbent companies facilitate entrant ones to enter the market or expand their business activities (Chevalier, 1995a). Consequently, debt makes the product market more competitive. In his extensive analysis, Chevalier (1995b) suggests that in a concentrated market, companies with high debt are forced to charge high prices than those with low debt, which makes them making sensitive to their rivals' predation.

Chavalier and Scharfstein (1996) explain the debt-competition in another respect using the switching cost model. They argue that during a certain recession when the market is less competitive, highly leveraged enterprises are much inferior to their counterparts with low debt because they have to charge high prices in their products. Thus, business disadvantage is associated with product market competition. A economic recession will result in market concentration, making it possible for companies with low debt to predate those with high debt (Opler and Titman 1994). Economic downturns are highly correlated with market concentration. Thus, during a recession, debt ratio negatively effects firm performance (Campello 2003, 2006). Market concentration raises agency problems in an enterprise, so the characteristics of a competitive market can strengthen the discipling effect of debt and help reduce agency problems costs (Aghion et al. 1997; Grossman and Hart 1983).

According to Agency Theory, the interests between managers and shareholders are diverse (Jensen and Meckling 1976). Due to asymmetric information between two sides, managers often take advantage of management to satisfy their personal gains. In doing so, they expand assets under their control to achieve higher perquisites, secure the managing position, and build up an empire of their own (Brealey et al. 2008; Hail et al. Hail et al. 2014; Myers 1984), all of which leads to overinvestment, that is, investment in unprofitable projects. Hence, overinvestment is expected to worsen agency problems (Fu 2010; Liu and Bredin 2010; Titman et al. 2004; Yang 2005). In summary, overinvestment will exacerbate the debt-competition interaction, making firm performance inefficient.

Hypothesis 2: the effect of debt on performance is less severe in a competitive market than a concentrated one.

Hypothesis 3: the moderation impact of debt over industry competition is weaker under overinvestment.

3 Data and Methodology

3.1 Research Methodology

To test these three hypotheses, the study applies the following empirical model

$$\begin{aligned} \mathit{FirmPerformance}_{it} &= \alpha + \beta_1 \mathit{FirmPerformance}_{it-1} + \beta_2 \mathrm{LEV}_{it} + \beta_3 \mathrm{COM}_{jt} \\ &+ \beta_4 \mathrm{LEV}_{it} \mathrm{x}\, \mathrm{COM}_{jt} + \beta_5 \mathrm{LEV}_{it} \mathrm{x}\, \mathrm{COM}_{jt}\ \mathrm{x}\ \mathrm{Overinvestment}_{it} + \psi' \mathrm{x}_{it} + \varepsilon_{it} \end{aligned} \quad (1)$$

where $\mathrm{FirmPerformance}_{it}$ and $\mathrm{FirmPerformance}_{it-1}$ are respectively firm performance and the one-period lag of firm performance measured by alternative proxies of firm i at time t and t − 1; α is the constant; LEV_{it} is the ratio of total debt over total assets of firm i at time t; COM_{jt} is the proxy for the level of competition in industry j at time t, namely Herfindahl–Hirschman Index (HHI) and Boone Index (BI), which are described in detail below; $\mathrm{Overinvestment}_{it}$ is estimated by the error terms extracted from Eq. (4), those with positive signs are considered as overinvestment; x_{it} is a set of control variables described in the variable definition; ε_{it} is the error term[1].

The representatives for firm performance are return on assets (ROA) and return on equity (ROE), which are the earning before interests and tax (EBIT), earning before tax (EBT), earning after tax (EAT) divided by total assets and total equity respectively. Although these variables are thought to be affected by different accounting standards because their calculations are based on accounting books, they are considered to be better than Tobin's Q to be the representatives in this research. Demsetz and Lehn (1985) suppose that ROA and ROE reflect the present situation, while Tobin's Q shows future development. Demsetz and Villalonga (2001) emphasize that Tobin's Q is often affected by tangible assets whose depreciation is different from the real economic depreciation. Furthermore, the use of ROA and ROE helps mitigate the differences in

[1] See Appendix for the description of all variables.

firm size among companies in various industries. Debt ratio is measured by the ratio of total debt over total assets.

The study adds some control variables as determinants of performance to the regression including sale growth, firm size, and average return on assets. Sale growth (SGRO), the representative for growth opportunities (King and Santor 2008; Maury 2006), is measured by the differences between sale of firm i at time t and its sale at time t − 1 divided by sale at time t − 1. Firm size (Size) is the logarithm of total assets. According to Ghosh (2008), average return on assets (MROA) are the moving average of ROA in two consecutive years. The instrumental variables used to handle the endogenous problem in the regression model are tangibility (TANG) and non-debt tax shield (NDTS). TANG is the ratio of tangible assets over total assets. This variable plays a decisive role in a firm's access to financing capital (Booth et al. 2001; Campello 2006), especially in developing countries where the regulations to protect lenders and carry out loan contracts are loosely controlled. NDTS is the sum of research and development funds (R&D) and depreciation divided by total assets.

To examine the role of industry competition in the relationship between financial leverage and firm performance, the research has to identify the proxies for industry competition. In fact, there are two ways to measure industry competition: structural and non-structural (Lawton 1999). Structural approach evaluates market concentration using Herfindahl–Hirschman Index (HHI) (Campello 2006) or the level of concentration within four or five largest companies in a certain industry (CR4 or CR5) (Campello 2003; Chevalier 1995a, 1995b; Kovenock and Phillips 1997; Opler and Titman 1994). The degree of concentration (high HHI, CR4, or CR5) often accompanies lower competition and vice versa. Meanwhile, non-structural approach measures the level of competition from the market's behaviors. This measurement is appreciated more highly than structural approach because a high level of concentration does not imply lower competition in the market (Guzmán et al. 2012). In fact, the hypothesis on the relationship between market structure and the effectiveness shows that high concentration is simply the results of the market's effectiveness (Demsetz 1973). Some companies that are operating effectively can quickly expand their market shares, while those which are ineffective are smaller and smaller in size (Boone et al. 2004). Moreover, high concentration sometimes comes from the fierce competition of various companies in the market, leading to the fact that effective companies force ineffective ones to exit the market (Boone Boone 2008a). Thus, the level of concentration cannot correctly predict the level of competition in the market. To deal with such problems emerged from structural approach, Boone (2000) uses a new index to measure market competition, Boone Index (BI). The index measures the sensitivity of firm profitability to the ineffectiveness of the market. Because in a competitive market companies often have to suffer a big loss when they perform ineffectively, firm profitability will increase with how effective a firm performs, and such an increase will be higher in a competitive market (Boone 2008b).

Hence, BI is the proxy that is preferred in studies on industry competition and firm performance (Boone et al. 2013). However, to raise the reliability, the research will in turn employ these two alternatives to find out their impact on the relationship between financial leverage and firm performance. According to Beiner et al. 2011 (2011), HHI is measured as the total market shares of each firm in a certain industry.

$$HHI_{jt} = \sum_{i=1}^{Nj} \left(\frac{Sales_{ijt}}{\sum_{i=1}^{Nj} Sales_{ijt}} \right)^2 \quad (2)$$

In the formula, HHI_{jt} is HHI of industry j at time t; $Sales_{ijt}$ indicates the sales of firm i in industry j at time t. The higher HHI is the higher market concentration becomes (lower market competition).

BI is considered to be the index that helps directly evaluate the level of competition in the market. The index is based on the hypothesis of competition and effectiveness with the assumption that in a competitive market, if a firm does not operate effectively, it will incur losses (Boone 2008b; Boone et al. 2005; Boone et al. 2007). Therefore, an industry with high competition is expected to have a sharp decrease in variable profits due to the increase in the marginal costs. Then, BI is estimated through the following regression model:

$$VROA_{it} = \alpha + \beta_t lnMC_{ij} + \mu_{i,t} \quad (3)$$

where $VROA_{it}$ is the variable profits calculated by subjecting costs of goods from sales of firm i in industry j divided by total assets; $lnMC_{ij}$ is the logarithm of marginal costs which is costs of goods over sales of firm in in industry j; β_t is the coefficient of the model that is changing overtime. Its absolute value measures the degree of competition. The coefficient sign is expected to be negative. The higher the absolute value is he higher market competition is. Therefore, BI is the absolute value of β_t.

As pointed out in the hypothesis development, market competition is an important factor in analyzing the effect of financial leverage on firm performance. In order to catch such an impact, the interaction between financial leverage and industry competition is added to the regression model. Besides, the research also takes into account the problem of endogeneity in the model which are originated from three major reasons: simultaneity, measurement errors, and omitted variables. To mitigate the simultaneous effect between LEV and ROA, the study uses the lag of LEV due to the fact that financial leverage in the past often affects profits at present but the reserve relationship is impossible. However, in addition to simultaneity, the estimated results are partly affected by omitted variables and measurement errors. Therefore, GMM two-stage least square is used to deal with such a problem. Having the doubt that LEV is endogenous, the research decides to take TANG and NDTS as its instrumental variables.

These two instrumental variables are basically considered to be suitable. First, TANG is what the institutions use to evaluate the possibility of their customers' paying loans back so that they can make right decisions on lending capital (Booth et al. 2001; Campello 2006). Thus, the effect of this variable on firm performance is mainly through the financing capital to companies, showing that TANG is an appropriate instrumental variable for LEV (Campello 2006). Second, firms with higher non-debt tax shield are expected to have higher financial leverage (DeAngelo and Masulis 1980), and non-debt tax shield is not supposed to have the direct impact on earnings before tax and depreciation. This fact suggests that NDTS is an effective instrumental variable for financial leverage. Actually, Fama and French (2002) support the empirical evidence for the reverse relationship between non-debt tax shield and financial leverage. In short, the study uses both factors as instrumental variables.

Finally, overinvestment is measured through Eq. (4) using the fixed-effect technique. The estimated equation is generalized based on the ideas from previous studies (Bokpin and Onumah 2009; Carpenter and Guariglia 2008; Connelly 2016; Li and Zhang 2010; Malm et al. 2016; Nair 2011; Richardson 2006; Ruiz-Porras and Lopez-Mateo 2011). The explicit form of Eq. (1) is as follows:

$$NewInvestment_{i,t} = \alpha_0 + \alpha_1 CashFlow_{i,t} + \alpha_2 TobinQ_{i,t} + \alpha_3 FixCapitalIntensity_{i,t} + \alpha_4 FirmSize_{i,t} + \alpha_5 RevenueGrowth_{i,t} + \alpha_6 BusinessRisk_{i,t} + \alpha_7 Leverage_{i,t} + \omega_{i,t} \quad (4)$$

In the equation, $NewInvestment_{i,t}$ represents for the investment decision; $CashFlow_{i,t}$ reflects the cash available in a company after subtracting capital expenditures; $TobinQ_{i,t}$ is the representative of growth opportunity and market performance; $FixCapitalIntensity_{i,t}$ evaluates the ability to generate fixed assets through sales; $RevenueGrowth_{i,t}$ demonstrates the growth of the firm;$FirmSize_{i,t}$ shows a company's financial constraints; $BusinessRisk_{i,t}$ indicates the volatility of firm profitability; $Leverage_{i,t}$ is the capital structure of the company. The estimated error-term $\hat{\omega}_{i,t}$ taken from the above model is considered as the abnormalities in the investment decision. If the error term's value is positive, or $\hat{\omega}_{i,t} > 0$, $\hat{\omega}_{i,t}$ of firm i^{th} and year t^{th} is denoted as $Overinvestment_{i,t}$. This method of calculating overinvestment has been recently adopted by He and Kyaw (2018).[2]

3.2 Research Data

The research data is collected from Vietnamese listed companies in HOSE and HNX from 2010 to 2016. Based on the classification standard on Vietnam's Stock Exchange, the sample is classified into 21 different industries including durable goods, consumer goods, real estates, printing (except the Internet), transportation support, mining, professional contractors, electricity, basic metals, textiles, plastics an rubber, beverages and tobacco, paper, chemicals, non-metal minerals, food, electronic equipment, cultivation, sea transportation, heavy industry and civil construction, houses and buildings. Our data and correlation coefficients are summarized as Tables 1 and 2 below:

4 Results and Discussion

Table 3 presents the estimation results of Eq. (1). The SGMM method is used with the ratio of tangible assets to total assets and non-debt tax shield to total assets as instruments for debt ratio. The first six columns use the HHI index measuring the level of market concentration. Meanwhile, the last six columns use the BI index measuring market competition. Specifically, the lower the HHI, the lower the competition, while

[2] See Appendix for the description of all variables.

Table 1. Descriptive statistics

Variable	Obs.	Mean	Std. Dev.	Min	Max
EAT/TA	1,384	0.060576	0.054632	−0.041441	0.239514
EBT/TA	1,384	0.074129	0.065519	−0.041777	0.291967
EBIT/TA	1,384	0.094697	0.060779	−0.016461	0.303093
EAT/Equity	1,384	0.123358	0.090448	−0.128181	0.375616
EBT/Equity	1,384	0.151656	0.107330	−0.125376	0.447676
EBIT/Equity	1,384	0.219343	0.122889	−0.041824	0.561071
MROA	921.0	0.315095	0.232828	0.045080	1.109090
Size	1,384	27.08540	1.293908	23.95720	30.18850
Growth	1,384	0.108733	0.265465	−0.492889	1.131610
Leverage	1,383	0.512923	0.205921	0.103515	0.849271
Competition 1	1,388	−0.259280	0.083718	−0.670789	−0.135970
Competition 2	1,394	0.475526	0.496635	0.171184	2.278290

Source: Author's calculation

Table 2. Matrix correlation

	MROA	Size	Growth	Leverage	Competition1	Competition2
MROA	**1.0000**	−0.0876 (0.0095)	−0.0261 (0.4396)	−0.1467 (0.0000)	0.1452 (0.0000)	−0.0595 (0.0773)
Size	−0.0876 (0.0095)	**1.0000**	0.0993 (0.0003)	0.2130 (0.0000)	−0.1827 (0.0000)	0.1329 (0.0000)
Growth	−0.0261 (0.4396)	0.0993 (0.0003)	**1.0000**	0.0649 (0.0184)	−0.0021 (0.9399)	0.029 (0.2918)
Leverage	−0.1467 (0.0000)	0.2130 (0.0000)	0.0649 (0.0184)	**1.0000**	−0.0475 (0.0850)	0.1292 (0.0000)
Competition 1	0.1452 (0.0000)	−0.1827 (0.0000)	−0.0021 (0.9399)	−0.0475 (0.0850)	**1.0000**	0.2479 (0.0000)
Competition 2	−0.0595 (0.0773)	0.1329 (0.0000)	0.029 (0.2918)	0.1292 (0.0000)	0.2479 (0.0000)	**1.0000**

P-Values are given in the parentheses Source: Author's calculation

BI is in the opposite direction. Therefore, two competition variables, Competition 1 = (−HHI) and Competition 2 = BI, are generated to interpret the impacts of these two indicators on performance in the same direction.

The estimated results from the regression indicate that debt is positively associated with performance. Whereas, two representative variables for industry competition are negatively. These findings are suitable with the disciplining effect and Agency Theory (Berger and Di Patti 2006; Grossman and Hart 1983; Jensen 1986; Jensen and Meckling 1976; Weill 2008). Additionally, the significant positive impact of the two-variable interaction term between debt and competition is clearly shown in the estimation. This result demonstrates that when the market is highly competitive, the positive impact of debt on performance is stronger when firms are less likely be driven

Table 3. Regression Estimation

SGMM Estimations	Competition 1 = (-HHI)						Competition 2 = BI					
	EAT/TA	*EBT/TA*	*EBIT/TA*	*EAT/Equity*	*EBT/Equity*	*EBIT/Equity*	*EAT/TA*	*EBT/TA*	*EBIT/TA*	*EAT/Equity*	*EBT/Equity*	*EBIT/Equity*
Lag _ Performance	0.565***	0.583***	0.680***	0.256**	0.439***	0.559***	0.601***	0.685***	0.787***	0.431***	0.532***	0.532***
	(0.0887)	(0.0982)	(0.0776)	(0.121)	(0.0721)	(0.0615)	(0.0701)	(0.0777)	(0.0832)	(0.0502)	(0.0611)	(0.0686)
MROA	0.0143	0.0147	0.0195	0.0762*	0.0822***	0.0608*	0.0538***	0.0651***	0.0506**	0.104***	0.128***	0.110**
	(0.0194)	(0.0239)	(0.0154)	(0.0448)	(0.0290)	(0.0330)	(0.0177)	(0.0210)	(0.0219)	(0.0307)	(0.0407)	(0.0513)
Size	-0.00339	-0.00482	-0.00340	-0.0133	-0.00436	-0.0127**	-0.00358*	-0.00473**	-0.00623***	-0.00611**	-0.0101**	-0.0130***
	(0.00272)	(0.00341)	(0.00220)	(0.00939)	(0.00506)	(0.00598)	(0.00186)	(0.00218)	(0.00222)	(0.00281)	(0.00432)	(0.00500)
Growth	0.00518	0.00526	0.00841	0.00144	0.0112	0.0237*	0.00777**	0.00837**	0.00955**	0.0182**	0.0215***	0.0130***
	(0.00508)	(0.00588)	(0.00555)	(0.0139)	(0.0117)	(0.0136)	(0.00316)	(0.00347)	(0.00378)	(0.00714)	(0.00786)	(0.00419)
Leverage (Lev)	**0.243***	**0.329***	**0.248****	**0.897***	**0.419***	**0.851*****	**0.184****	**0.235****	**0.314*****	**0.377*****	**0.582*****	**0.738*****
	(0.142)	**(0.175)**	**(0.112)**	**(0.462)**	**(0.250)**	**(0.298)**	**(0.0884)**	**(0.103)**	**(0.103)**	**(0.134)**	**(0.204)**	**(0.226)**
Competition1 (Com1)	**-0.645****	**-0.819*****	**-0.483****	**-1.439***	**-0.649**	**-1.066***						
	(0.264)	**(0.310)**	**(0.228)**	**(0.837)**	**(0.473)**	**(0.593)**						
*Lev*Com1*	**1.480*****	**1.848*****	**1.136****	**3.512****	**1.842****	**2.316****						
	(0.538)	**(0.621)**	**(0.449)**	**(1.557)**	**(0.886)**	**(1.125)**						
*Lev*Com1*Overinvestment*	**-0.300***	**-0.339***	**-0.209**	**-1.073***	**-0.589***	**-0.208**						
	(0.172)	**(0.203)**	**(0.145)**	**(0.615)**	**(0.322)**	**(0.421)**						

(continued)

Table 3. *(continued)*

	(1)	(2)	(3)	(4)	(5)	(6)	(7)	(8)	(9)	(10)	(11)	(12)
Competition2 (Com2)							**-0.173****	**-0.192****	**-0.217****	**-0.308****	**-0.441****	**-0.540****
							(0.0871)	**(0.0952)**	**(0.0996)**	**(0.138)**	**(0.196)**	**(0.240)**
Lev*Com2							**0.392****	**0.435****	**0.470*****	**0.730*****	**1.029*****	**1.096*****
							(0.153)	**(0.167)**	**(0.177)**	**(0.250)**	**(0.356)**	**(0.420)**
Lev*Com2*Overinvestment							**-0.182****	**-0.194****	**-0.162***	**-0.281****	**-0.346***	**-0.319****
							(0.0740)	**(0.0882)**	**(0.0854)**	**(0.136)**	**(0.183)**	**(0.158)**
Observations	761	761	761	517	645	645	761	761	761	761	761	517
Number of instruments	22	22	30	23	32	30	26	26	26	32	26	24
Number of groups	164	164	164	160	163	163	164	164	164	164	164	160
F-Statistics	51.24	54.92	219.55	16.32	70.76	151.9	38.85	45.44	70.38	66.82	55.71	60.31
Prob.	0.000	0.000	0.000	0.000	0.000	0.000	0.000	0.000	0.000	0.000	0.000	0.000
Arellano-Bond test for AR(1)	-1.510	-1.600	-1.760	-2.590	-1.400	-1.520	-1.630	-1.690	-1.820	-1.530	-1.620	-1.780
Prob.	0.130	0.110	0.078	0.010	0.161	0.129	0.102	0.091	0.069	0.126	0.106	0.076
Arellano-Bond test for AR(2)	1.140	0.920	-0.080	0.950	1.080	1.320	0.350	-1.010	-0.770	1.090	1.120	0.480
Prob.	0.256	0.356	0.939	0.340	0.280	0.188	0.729	0.312	0.441	0.275	0.262	0.632
Hansen test of over-identification	14.3	13.3	19.65	13.94	29.02	25.9	15.62	16.22	14.12	31.81	25.98	15.77
Prob.	0.428	0.503	0.605	0.53	0.219	0.256	0.619	0.577	0.721	0.132	0.1	0.469

Standard errors in parentheses

*** $p<0.01$, ** $p<0.05$, * $p<0.1$

Source: Author's calculation

out of the market. The bad effect of debt in a concentrated market becomes less severe when the market becomes more competitive. The estimation is consistent with findings of a recent empirical study in Vietnam (Van Thuong et al. 2017). Interestingly, the influence of the debt-competition interaction becomes weaker overinvestment because overinvestment increases agency problems and weakens the good impact of debt. This evidence supports the hypothesis that the debt-performance nexus is conditional on market competition and overinvestment.

The estimation robustness is tested using alternative representatives for industry competition and firm performance. Industry competition is alternatively measured by the residual estimated from the subequation (BI Index) and Herfindahl-Hirschman Index (HHI Index). Furthermore, earnings before interests and taxes (EBIT), earnings before taxes (EBT), and earnings after taxes (EAT) over total assets and total equity are respectively employed to represent for performance. Consequently, the estimated coefficients of all different proxies reach the consistency in terms of signs and significance levels. All the relevant tests of System Generalized Method of Moments (SGMM) estimator appear to be comfortable in every single regression model in the research.

5 Conclusions

The debt-performance relationship is supposed to be moderated by industry competition and overinvestment because both factors are associated with agency problems within enterprises. Vietnam's current situation together with previous empirical studies have indicated the interdependence of debt, competition, and overinvestment and their combined impact on firm performance. With the use of System Generalized Method of Moments (SGMM), the research aims at identifying the role of industry competition in the debt-performance relationship under overinvestment. The paper clarifies that performance in Vietnamese listed companies are positively affected by debt ratio in the capital structure. Furthermore, the positive influence of debt gets stronger in an industry with high levels of competition, meaning that the bad effect of debt in this situation becomes less severe when entrants are less likely to be kicked out of the market by their competitors. However, the positive sign of the two-variable interaction becomes weaker when overinvestment is taken into consideration because overinvestment increases agency problems in companies.

Based on the estimated results, some recommendations are given to both the government and corporate companies. The government should heighten the level of competition through higher economic growth, better market regulations, and more transparent legal practices. Companies should limit the problem of overinvestment or mitigate agency problems by compensating managers with more benefits to increase their commitments toward acting in favor of shareholders' interests.

Appendixes

Table A1. Variable measurement for main econometric model

$$\begin{aligned} FirmPerformance_{i,t} = {} & \beta_0 + \beta_1 FirmPerformance_{i,t-1} + \beta_2 LEV_{i,t} + \beta_3 COM_{i,t} + \beta_4 LEV_{i,t} \times COM_{i,t} \\ & + \beta_5 LEV_{i,t} \times Overinvestment_{i,t} + \beta_6 COM_{i,t} \times Overinvestment_{i,t} \\ & + \beta_7 LEV_{i,t} \times COM_{i,t} \times Overinvestment_{i,t} + \psi' X_{i,t} + \varepsilon_{i,t} \end{aligned}$$

Dependent Variables

Variables	Denote	Definition
Firm Performance	EBIT/TA	EBIT (Earnings Before Interest & Tax) divided by Total Asset
	EBT/TA	EBT (Earnings Before Tax) divided by Total Asset
	EAT/TA	EAT (Earnings After Tax) divided by Total Asset
	EBIT/TE	EBIT (Earnings Before Interest & Tax) divided by Total Equity
	EBT/TE	EBT (Earnings Before Tax) divided by Total Equity
	EAT/TE	EAT (Earnings After Tax) divided by Total Equity

Explanatory Variables

Variables	Denote	Definition
Moving average of ROA	$MROA_{i,t}$	Moving average of EBITDA (Earnings Before Interest, Taxes, Depreciation and Amortization) over Total Asset ratios in three consecutive years
Company Growth	$Growth_{i,t}$	Growth rate of total sale
Company Size	$Size_{i,t}$	Natural logarithm of total asset
Leverage	$LEV_{i,t}$	Total liabilities over total asset
Competition	$COM^1_{i,t}$	HHI index
	$COM^2_{i,t}$	BI index

Table A2. Variable Measurement for subequation

$$NewInvestment_{i,t} = \alpha_0 + \alpha_1 CashFlow_{i,t} + \alpha_2 TobinQ_{i,t} + \alpha_3 FixCapitalIntensity_{i,t} + \alpha_4 FirmSize_{i,t} + \alpha_5 RevenueGrowth_{i,t} + \alpha_6 BusinessRisk_{i,t} + \alpha_7 Leverage_{i,t} + \omega_{i,t}$$

Dependent Variables		
Variables	**Denote**	**Definition**
New Investment	$NewInvestment_{i,t}$	Total investment including long-term and short-term investment divided by total asset
Explanatory Variables		
Variables	**Denote**	**Definition**
Cash Flow	$CashFlow_{i,t}$	The cash available in a company after subtracting capital expenditures
Market Performance	$TobinQ_{i,t}$	The Q ratio is calculated as the market value of a company divided by the firm's assets
Fix Capital Intensity	$FixCapitalIntensity_{i,t}$	The ability to generate fixed assets through sales measured by total fix asset divided by total sales
Firm Growth	$RevenueGrowth_{i,t}$	The growth rate of firm's revenue over year
Firm Size	$FirmSize_{i,t}$	Natural logarithm of total asset
Business Risk	$BusinessRisk_{i,t}$	Standard deviation of EBITDA (Earnings Before Interest, Taxes, Depreciation and Amortization) over Total Asset ratio in three consecutive years
Leverage	$Leverage_{i,t}$	Total liabilities over total asset

All the above variables are calculated by using financial data from ***Thomson Reuters Eikon Financial Analysis***

Table A3. List of 21 industries

Durable goods	Plastics and rubber
Consumer goods	Beverages and tobacco
Real estates	Paper
Printing (except the Internet)	Chemicals
Transportation support	Non-metal minerals
Mining	Food
Professional contractors	Electronic equipment
Electricity	Cultivation
Basic metals	Sea transportation
textiles	Heavy industry and civil construction
	Houses and buildings

References

Aghion, P., Dewatripont, M., Rey, P.: Corporate governance, competition policy and industrial policy. Eur. Econ. Rev. **41**(3–5), 797–805 (1997)

Beiner, S., Schmid, M.M., Wanzenried, G.: Product market competition, managerial incentives and firm valuation. Eur. Financ. Manag. **17**(2), 331–366 (2011)

Berger, A.N., Di Patti, E.B.: Capital structure and firm performance: a new approach to testing agency theory and an application to the banking industry. J. Bank. Financ. **30**(4), 1065–1102 (2006)

Bokpin, G.A., Onumah, J.M.: An empirical analysis of the determinants of corporate investment decisions: evidence from emerging market firms. Int. Res. J. Financ. Econ. **33**, 134–141 (2009)

Bolton, P., Scharfstein, D.S.: A theory of predation based on agency problems in financial contracting. Am. Econ. Rev. **80**, 93–106 (1990)

Boone, J.: Measuring product market competition. CEPR Discussion Paper (2636) (2000)

Boone, J.: Competition: theoretical parameterizations and empirical measures. J. Inst.Al Theor. Econ. JITE **164**(4), 587–611 (2008a)

Boone, J.: A new way to measure competition. Econ. J. **118**(531), 1245–1261 (2008b)

Boone, J., Griffith, R., Harrison, R.: Measuring competition. Paper presented at the Encore Meeting (2004)

Boone, J., Griffith, R., Harrison, R.: Measuring competition (Research Paper No. 022). Advanced Institute of Management (2005)

Boone, J., van Ours, J.C., van der Wiel, H.: When is the price cost margin a safe way to measure changes in competition? De Economist **161**, 1–23 (2013)

Boone, J., Van Ours, J.C., van der Wiel, H.: How (not) to measure competition (2007)

Booth, L., Aivazian, V., Demirguc-Kunt, A., Maksimovic, V.: Capital structures in developing countries. J. Financ. **56**(1), 87–130 (2001)

Brander, J.A., Lewis, T.R.: Oligopoly and financial structure: the limited liability effect. Am. Econ. Rev. **76**, 956–970 (1986)

Brealey, R.A., Myers, S.C., Allen, F.: Brealey, Myers, and Allen on valuation, capital structure, and agency issues. J. Appl. Corp. Financ. **20**(4), 49–57 (2008)

Campello, M.: Capital structure and product markets interactions: evidence from business cycles. J. Financ. Econ. **68**(3), 353–378 (2003)

Campello, M.: Debt financing: does it boost or hurt firm performance in product markets? J. Financ. Econ. **82**(1), 135–172 (2006)

Carpenter, R.E., Guariglia, A.: Cash flow, investment, and investment opportunities: new tests using UK panel data. J. Bank. Financ. **32**(9), 1894–1906 (2008)

Van Thuong, Chau, Le Khang, Tran, Thanh, Nguyen Cong: Captial structure and firm performance: the role of industry competition. J. Econ. Dev. **28–10**, 56–78 (2017)

Chavalier, J., Scharfstein, D.: Capital market imperfections and countercyclical markups. Amer. Econ. Rev. **86**, 703–725 (1996)

Chevalier, J.A.: Capital structure and product-market competition: Empirical evidence from the supermarket industry. Am. Econ. Rev. **85**, 415–435 (1995a)

Chevalier, J.A.: Do LBO supermarkets charge more? An empirical analysis of the effects of LBOs on supermarket pricing. J. Financ. **50**(4), 1095–1112 (1995)

Connelly, J.T.: Investment policy at family firms: evidence from Thailand. J. Econ. Bus. **83**, 91–122 (2016)

Dasgupta, S., Titman, S.: Pricing strategy and financial policy. Rev. Financ. Stud. **11**(4), 705–737 (1998)

DeAngelo, H., Masulis, R.W.: Optimal capital structure under corporate and personal taxation. J. Financ. Econ. **8**(1), 3–29 (1980)
Demsetz, H.: Industry structure, market rivalry, and public policy. J. Law Econ. **16**(1), 1–9 (1973)
Demsetz, H., Lehn, K.: The structure of corporate ownership: causes and consequences. J. Polit. Econ. **93**(6), 1155–1177 (1985)
Demsetz, H., Villalonga, B.: Ownership structure and corporate performance. J. Corp. Financ. **7** (3), 209–233 (2001)
Fama, E.F., French, K.R.: Testing trade-off and pecking order predictions about dividends and debt. Rev. Financ. Stud. **15**(1), 1–33 (2002)
Fu-Min, C., Wang, Y., Lee, N.R., La, D.T.: Capital structure decisions and firm performance of Vietnamese soes. Asian Econ. Financ. Rev. **4**(11), 1545 (2014)
Fu, F.: Overinvestment and the operating performance of SEO firms. Financ. Manage. **39**(1), 249–272 (2010)
Fudenberg, D., Tirole, J.: A "signal-jamming" theory of predation. RAND J. Econ. **17**, 366–376 (1986)
Gaver, J.J., Gaver, K.M.: Additional evidence on the association between the investment opportunity set and corporate financing, dividend, and compensation policies. J. Account. Econ. **16**(1–3), 125–160 (1993)
Ghosh, S.: Leverage, foreign borrowing and corporate performance: firm-level evidence for India. Appl. Econ. Lett. **15**(8), 607–616 (2008)
Grossman, S.J., Hart, O.D.: An analysis of the principal-agent problem. Econ. J. Econ. Soc. **51**, 7–45 (1983)
Gueorguiev, D., Malesky, E.: Foreign investment and bribery: a firm-level analysis of corruption in Vietnam. J. Asian Econ. **23**(2), 111–129 (2012)
Guzmán, G.M., Gutiérrez, J.S., Cortes, J.G., Ramírez, R.G.: Measuring the competitiveness level in furniture SMEs of Spain. Int. J. Econ. Manag. Sci. **1**(11), 09–19 (2012)
Hail, L., Tahoun, A., Wang, C.: Dividend payouts and information shocks. J. Account. Res. **52** (2), 403–456 (2014)
Harris, M., Raviv, A.: Capital structure and the informational role of debt. J. Financ. **45**(2), 321–349 (1990)
He, W., Kyaw, N.A.: Ownership structure and investment decisions of Chinese SOEs. Res. Int. Bus. Financ. **43**, 48–57 (2018)
Jensen, M.C.: Agency costs of free cash flow, corporate finance, and takeovers. Am. Econ. Rev. **76**(2), 323–329 (1986)
Jensen, M.C., Meckling, W.H.: Theory of the firm: managerial behavior, agency costs and ownership structure. J. Financ. Econ. **3**(4), 305–360 (1976)
Khan, A.G.: The relationship of capital structure decisions with firm performance: a study of the engineering sector of Pakistan. Int. J. Account. Financ. Report. **2**(1), 245 (2012)
King, M.R., Santor, E.: Family values: Ownership structure, performance and capital structure of Canadian firms. J. Bank. Financ. **32**(11), 2423–2432 (2008)
Kovenock, D., Phillips, G.M.: Capital structure and product market behavior: an examination of plant exit and investment decisions. Rev. Financ. Stud. **10**(3), 767–803 (1997)
Li, D., Zhang, L.: Does q-theory with investment frictions explain anomalies in the cross section of returns? J. Financ. Econ. **98**(2), 297–314 (2010)
Liu, N., Bredin, D.: Institutional Investors, Over-investment and Corporate Performance. University College Dublin, Dublin (2010)
Maksimovic, V., Titman, S.: Financial policy and reputation for product quality. Rev. Financ. Stud. **4**(1), 175–200 (1991)

Malm, J., Adhikari, H.P., Krolikowski, M., Sah, N.: Litigation risk and investment policy. J. Econ. Financ. **41**, 1–12 (2016)
Margaritis, D., Psillaki, M.: Capital structure, equity ownership and firm performance. J. Bank. Financ. **34**(3), 621–632 (2010)
Maury, B.: Family ownership and firm performance: Empirical evidence from Western European corporations. J. Corp. Financ. **12**(2), 321–341 (2006)
Modigliani, F., Miller, M.H.: The cost of capital, corporation finance and the theory of investment. Am. Econ. Rev. **48**(3), 261–297 (1958)
Modigliani, F., Miller, M.H.: Corporate income taxes and the cost of capital: a correction. Am. Econ. Rev. **53**(3), 433–443 (1963)
Myers, S.C.: Determinants of corporate borrowing. J. Financ. Econ. **5**(2), 147–175 (1977)
Myers, S.C.: The capital structure puzzle. J. Financ. **39**(3), 574–592 (1984)
Nair, P.: Financial liberalization and determinants of investment: a study of indian manufacturing firms. Int. J. Manag. Int. Bus. Econ. Syst. **5**(1), 121–133 (2011)
Opler, T.C., Titman, S.: Financial distress and corporate performance. J. Financ. **49**(3), 1015–1040 (1994)
Quy, V.T., Khuong, N.D., Swierczek, F.W.: Corporate performance of privatized firms in Vietnam (2014)
Richardson, S.: Over-investment of free cash flow. Rev. Acc. Stud. **11**(2–3), 159–189 (2006)
Ruiz-Porras, A., Lopez-Mateo, C.: Corporate governance, market competition and investment decisions in Mexican manufacturing firms (2011)
San, O.T., Heng, T.B.: Capital structure and corporate performance of Malaysian construction sector. Int. J. Humanit. Soc. Sci. **1**(2), 28–36 (2011)
Scharfstein, D.O.: Analytical performance measures for the miniload automated storage/retrieval system. Georgia Institute of Technology (1990)
Titman, S.: The effect of capital structure on a firm's liquidation decision. J. Financ. Econ. **13**(1), 137–151 (1984)
Titman, S., Wei, K.J., Xie, F.: Capital investments and stock returns. J. Financ. Quant. Anal. **39**(4), 677–700 (2004)
Tran, N.M., Nonneman, W., Jorissen, A.: Privatization of Vietnamese firms and its effects on firm performance. Asian Econ. Financ. Rev. **5**(2), 202 (2015)
Wanzenried, G.: Capital structure decisions and output market competition under demand uncertainty. Int. J. Ind. Organ. **21**(2), 171–200 (2003)
Weill, L.: Leverage and corporate performance: does institutional environment matter? Small Bus. Econ. **30**(3), 251–265 (2008)
Yang, Y.M.: Corporate governance, agency conflicts, and equity returns along business cycles (2005)

The Moderation Effect of Debt and Dividend on the Overinvestment-Performance Relationship

Nguyen Trong Nghia[1], Tran Le Khang[2], and Nguyen Cong Thanh[3](✉)

[1] Ho Chi Minh City University of Economics – Law, Ho Chi Minh, Vietnam
nguyentrongnghia@hcmup.edu.vn
[2] School of Economics, Erasmus University Rotterdam, The Hague, The Netherlands
khang.tl@vnp.edu.vn
[3] School of Accounting – Banking – Finance, Ho Chi Minh City University of Technology, Ho Chi Minh City, Vietnam
nc.thanh93@hutech.edu.vn

Abstract. Taking into consideration all Vietnam's non-financial companies listed on HOSE and HNX from 2006 to 2016, the research aims at clarifying the bad effect of overinvestment on firm performance as well as the moderation role of debt and dividend in mitigating agency costs caused by overinvestment. Utilizing two specific measurements of overinvestment via HP Filter and the positive error terms obtained from the subequation of Overinvestment Estimation, the study indicates the negative impact of overinvestment on profitability in Vietnamese enterprises. The harmful effect of overinvestment can be eased by the use of debt or the payouts of dividend. However, when they are combined, their separate influence of the two-variable interaction tends to be attenuated.

Keywords: Overinvestment · Debt · Dividend

JEL Classification: G31 · G35

1 Introduction

The most important triad of financial policies are regarded to be debt policy, dividend policy, and investment policy (Alli et al. 1993; Baker and Powell 2000). To clarify how much profit a company can earn, it is significant to figure out how much debt a firm should lever, how much dividend should be paid, and how much investment it should make. Within the scientific community, there have been many debates on the relationship among these three policies. Modigliani and Miller (1958) demonstrate that investment is independent of debt and dividend in the context of a flawless capital marketplace with the absence of taxes, transaction costs, liquidation costs, and asymmetric information. Nevertheless, these assumptions are no longer appropriate in an imperfect capital market. As a result, in this case, debt and dividend can affect investment policy in an enterprise.

V. Kreinovich et al. (Eds.): ECONVN 2019, SCI 809, pp. 1109–1120, 2019.
https://doi.org/10.1007/978-3-030-04200-4_81

A firm's future profitability partly depends on its own investment strategies within a constantly resilient setting overwhelmed with uncertainties (Kannadhasan and Aramvalarthan 2011). In the process of running a business, a manager is tasked with allocating capital resources so as to reach an optimal level of investment where there is an equality between the marginal benefit and marginal cost of capital investment. Once firm investment exceeds the optimal level, investment in projects would be unprofitable. Therefore, overinvestment causes inefficient firm operations. The interests between the principles and representatives, however, are not aligned. In order to harness as many private benefits as possible, managers are inclined to broaden financial possessions under their management whereas shareholders aim at maximizing profits. As a consequence, managers may make overinvestment in projects with negative net present value (NPV). Moreover, a firm has to pay an expensive cost to bring the interests of managers and shareholders to an agreement in order to solve such an urgent problem. Eventually, through raising agency problems and reducing firm profitability, it is likely that firm performance will be deteriorated by overinvestment (Grazzi et al. 2016; Gu 2013; Jensen 1986; Shima 2010).

Being conscious of the fact that a costly expenditure has to be incurred to handle the agency problem, a company often looks for another way to efficiently monitor managers' behaviour. This leads to the resolution of using debt in the capital structure and the pay-outs of dividends to shareholders. According to Agency Theory, both financial leverage and dividend payment can turn into beneficial devices in diminishing the free cash flow under the manager's control as managers are obliged to attain more profits to fulfil their commitments towards debt holders and shareholders. Through these two policies, additionally, stakeholders can share the heavy burden of monitoring responsibility in the capital market (Easterbrook 1984). In the course of time, it is assumed that the harmful effect of overinvestment on firm profitability is to be lightened, as the result of using debt along with dividend policy (DeAngelo et al. 2006).

This research aims at identifying the impact of overinvestment on firm profitability and enlightening the impact of debt and dividend on the overinvestment-performance relationship. Consequently, two main questions emerge: (1) Is overinvestment negatively related to firm profitability? (2) Is such a negative impact can be attenuated with the use of debt and dividend?

Due to the following reasons, Vietnam's listed companies are chosen in the study. The first reason is that with the presence of weak legal regulations and a high level of asymmetric information, Vietnam's financial market remains underdeveloped and financial resources are primarily taken from commercial banks. The second reason is the virtual neglect of overinvestment in this country where the interests between shareholders and managers are in serious conflicts. From the financial statements of all Vietnam's listed companies, the data is collected from 2012 to 2016. In addition, by using HP Filter and taking the positive residuals from the sub-equation including possible determinants of investment policy, two ways of measuring overinvestment existing in each enterprise are suggested. It is expected that these two blooming measures are better proxies for the problem of overinvestment compared with the old method of relying on Tobin's Q.

A tendency of overinvestment to worsen firm profitability can be seen from the estimated statistic from the model. Nevertheless, with the use of either debt or dividend policy, it is possible to minimize its negative relation with firm performance. More surprisingly, at the same time, the combination of these two policies can mitigate the constraining effects of the two-variable interactions. A substitution between financial leverage and dividend payment is implied. The negative sign of a single debt and dividend policy, as well as the positive coefficient of their interaction, also proves the preceding situation. Moreover, the consistence in signs and significant level across two alternative measures of overinvestment and six proxies of firm profitability shows that the results support the solidity of the regression model.

Ultimately, the authors anticipated that the findings of this study greatly contribute to the existing literature review in two aspects. Firstly, having taken into consideration the three-variable interaction of debt, dividend, and investment policy, it is likely to be the first paper to do so. Secondly, once again, it acknowledges the fundamental idea of the interdependence among debt, dividends, and overinvestment, which was introduced by Agency Theory. As a surplus, some recommendations for shareholders in dealing with the agency problem inside their businesses are also suggested.

Section two mentions the overall review of theories and empirical studies to develop research hypotheses. Data collection and description, measurement of overinvestment, and model specification are included in section three. Section four gives a clearer view of the estimated results. In Sect. 5, the research is summarized and some recommendations for corporate managers and shareholders are proposed.

2 Literature Review and Hypothesis Development

Based on some presumptions (Miller and Modigliani 1961; Modigliani and Miller 1958), it is believed that capital structure, dividend policy, and investment decisions are independent of one another. First, the absence of taxes, transaction costs, and bankruptcy costs all contribute to the perfection of the market. Second, information can be equally accessed by shareholders and managers, which means a market with two-way symmetric information. Third, the costs of debt are a burden both shareholders and debt holders have to bear. The relaxation of any assumption makes way for the imperfections of the capital market. Trade-off Theory, which expresses the benefits and costs of using debt in the capital structure, is formed by the existence of various kinds of taxes (Modigliani and Miller 1963), and the same thing is applicable to Tax Theory, which illustrates the reduction in dividend pay-outs (Litzenberger and Ramaswamy 1979). Pecking-order Theory signifies the hierarchy of financing sources, (Myers and Majluf 1984), Bird-in-hand Theory supports dividend payments to avoid future uncertainties (Gordon 1959, 1963), and Agency Theory expresses the interest conflicts between managers and shareholders (Jensen and Meckling 1976).

What makes their interests diverge is the separation between ownership rights of the principles and management rights of the agents (Jensen and Meckling 1976). Using their ability to access to internal information, managers often attempt to benefit themselves through getting higher salaries, securer jobs, and bigger properties under their control. These motivations are the reasons behind investment in unprofitable projects, which cause the problem of overinvestment. When a firm has a hard time attaining financing sources from the capital market to invest in projects with positive net present value, it implies that asymmetric information causes not only underinvestment (Brealey et al. 2008; Myers and Majluf 1984) but also overinvestment when shareholders find it hard to monitor business activities, which allows managers to possess more freedom to build up their own fortune (Hail et al. 2014). As a consequence, both the destruction of firm value and inefficiency in firm performance are the results of underinvestment and overinvestment (Fu 2010; Liu and Bredin 2010; Titman et al. 2004; Yang 2005).

Furthermore, through a wide range of empirical studies, the negative relationship between overinvestment and profitability is obviously illustrated. Shima (2010) emphasizes the negative effect of overinvestment on profitability. For Singapore's 360 listed firms from 2005 to 2011, Farooq et al. (2014) suggest three levels of investment which comprise of just-investment, overinvestment, and underinvestment. The research clarifies that only just-investment is effective for a firm, the others considerably reduce firm efficiency. Having analysed all Chinese listed companies in the period of 1998–2014, Guariglia and Yang (2016) find that it is a rare possibility that investment reaches the optimal level, as a result of limited financing resources and agency problems. As they claim it to be, agency problems are the main reason behind an enormous amount of investment, harmfully contribute to firm performance. Sharing the same similarity, Liu and Bredin (2010); Titman et al. (2004); Yang (2005) conclude that overinvestment bears a negative influence on firm performance. Ultimately, the research eventually develops the first hypothesis to shed light on the overinvestment-performance relation.

***Hypothesis 1*: Overinvestment negatively influences firm profitability**

Upon reaching its optimal level, the excess of the free cash flow creates an opportunity for managers to benefit themselves. By taking advantage of such situation, they can use such funds to broaden financial resources under their management or strengthen their position with the expansion of the business, all of which make up the reality of overinvestment. Thus, to prevent managers from expropriating compensations and making personal gains, deducting the free cash flow can be the solution to the problem (Jensen 1986); (Dyck and Zingales 2004; Nenova 2003); (Hope and Thomas 2008). Therefore, not only do the use of debt and the payment of dividends restraint the excessive free cash flow but they also pass the monitoring tasks from inside to outside partners (Alli et al. 1993; Biddle et al. 2009; Easterbrook 1984; Jensen 1986; Rozeff 1982). In addition, according to Richardson (2006) such an action is capable of lowering the free cash flow administered by managers. Lang and Litzenberger (1989) share the same idea that dividend and investment policy hold a mutual connection. A decrease in overinvestment, which subsequently improves a firm's market value, implies an increase in the distribution of dividend. Grossman and Hart (1982) emphasize that, by utilizing debt, firms can be experience the pressure of financial distress or worse bankruptcy. Moreover, through strict debt covenants, certain constraints are also established on managers' decisions by debt creditors. As a consequence, by continuing to overinvest in bad projects, managers will leave themselves at risk of losing perquisites or their own position in the company. Finally, for the moderate impacts of debt and dividend policy, the second hypothesis is proposed.

***Hypothesis 2*: debt and dividend tends to attenuate the negative influence of overinvestment on profitability.**

3 Data Methodology

3.1 Data Collection

From Thomson Reuters, all the financial statements of Vietnam's listed firms on HOSE and HNX from 2012 to 2016 are gathered for research data. Due to the enormous differences in the characteristics of products and their services, only non-financial companies are included in the sample data. After the processing operation, there remain 669 Vietnamese listed companies in the final dataset.

3.2 Model Specification

Following Chen et al. (2017); Altaf and Shah (2017), the dynamic model with the one-period lag of the dependent variable is taken into account in order to clarify the impact of overinvestment on firm performance together with the moderation effect of debt and dividend on the overinvestment-performance relationship.

$$\begin{aligned} Performance_{i,t} = {} & \lambda_0 + \lambda_1 Performance_{i,t-1} + \lambda_2 Size_{i,t} + \lambda_3 Growth_{i,t} + \lambda_4 Risk_{i,t} + \lambda_5 Liquidity_{i,t} + \lambda_6 Tangibility_{i,t} \\ & + \lambda_7 Dividend_{i,t} + \lambda_8 Debt_{i,t} + \lambda_9 Overinvestment_{i,t} \\ & + \lambda_{10} Dividend_{i,t} \times Debt_{i,t} + \lambda_{11} Debt_{i,t} \times Overinvestment_{i,t} + \lambda_{12} Dividend_{i,t} \times Overinvestment_{i,t} \\ & + \lambda_{13} Dividend_{i,t} \times Debt_{i,t} \times Overinvestment_{i,t} + \mu_{i,t} \end{aligned} \tag{1}$$

where $Performance_{i,t}$ and $Performance_{i,t-1}$ are respectively firm performance and one-period lag performance alternatively measured by earnings before interests and taxes (EBIT), earnings before taxes (EBT), and earnings after taxes (EAT) over total assets; $Dividend_{i,t}$ is cash dividend payments; $Debt_{i,t}$ is the ratio of total liabilities to total assets; $Overinvestment_{i,t}$ is the positive residual taken from Overinvestment Estimation in Appendix 2 or taken from HP Filter; $Size_{i,t}$ is firm size, the natural logarithm of total assets; $Growth_{i,t}$ is firm sales growth; $Risk_{i,t}$ is the variation in firm profitability; $Liquidity_{i,t}$ is measured by quick ratio; $Tangibility_{i,t}$ is the ratio of tangible fixed assets to total assets.[1]

Employing HP Filter and the subequation, overinvestment is measured by two different ways. Firstly, by subtracting the real to the fitted value of required investment to get the residual, over-investment can be calculated. This is believed to be unexpected investment. Amidst these residuals, the problem of overinvestment is implied (He and Kyaw 2018; Richardson 2006). Secondly, overinvestment is also measured using HP Filter technique (Hodrick and Prescott 1997). It is proposed to be the points above the trend line of the investment rate. Moreover, to avoid the possibility of lacking some important explanatory variables, some control variables are taken into account, which are firm size, growth, liquidity and tangibility (Altaf and Shah 2017; Chen et al. 2017; Fosu 2013).

Compared to those calculated on total assets, according to Table 1, it is observed that profitability measured on equity seems to have a stronger variation. The minimum values of profitability vary between –5.6% to –3.0%, and the maximum is from 25.7%

[1] See Appendix for the description of all variables in the main model and Overinvestment Estimation.

to 31.4%. In addition, from the sample measured by HP Filter and the sub-equation, overinvestment exists in roughly 35% and 42%.

Table 1. Summary statistics of all research variables

Variable	Obs.	Mean	Std. Dev.	Min	Max
EBIT/Total Asset	5,852	0.06303	0.05617	–0.05633	0.25661
EBT/Total Asset	5,852	0.07616	0.06644	–0.05505	0.30079
EAT/Total Asset	5,852	0.09405	0.06422	–0.02993	0.31353
Company Size	5,852	26.6709	1.28151	23.8265	29.8310
Risk	5,853	0.07834	0.06975	0.00407	0.37840
Liquidity	5,852	1.67759	1.17572	0.25585	6.92789
Tangibility	5,816	0.25386	0.19531	0.00478	0.79431
Dividend	3,996	0.53488	0.44375	–0.05574	3.22060
Debt	6,099	0.50751	0.22459	0.01251	0.94379
$Overinvestment^{REG}$	4,366	0.35044	0.47716	0.00000	1.00000
$Overinvestment^{HP}$	6,160	0.42305	0.49408	0.00000	1.00000

Sources: calculated by the author

4 Results and Discussions

The aftermath, as estimated in Table 2, reveals that both debt and dividend policy hold a negative influence on firm performance. Fascinatingly, compared with Pecking-order Theory and Tax Theory, it seems that these results share the same similarity (Litzenberger and Ramaswamy 1979; Myers and Majluf 1984). By displaying a positive sign, the interaction variable between financial leverage and dividend policy astonishingly signifies a substitution between these two policies; in other words, an increase in dividend payments will result in lessened financing resources of a firm, which obliges it to enter the capital market for funding new investments. This puts the company at the higher risk of getting more debt. To limit the negative effects of financial leverage, it is essential that the pay-outs of dividends are executed, as they allow firms to have more incentives to operate effectively in order not to be deep in debt. On the contrary, in the capital market toward business operations, both the establishment of debt covenants and the improvement of the monitoring partners can assist the use of debt in reducing the harmful impacts of dividend policy.

The study discovers that, without a doubt, overinvestment and firm performance are in a negative correlation. Being in harmony with Agency Theory and Free Cash Flow Hypothesis, the findings indicate that investing in projects with negative net present value, i.e. overinvestment, is presumed to bring about the deduction of firm profitability. Reliable proofs are also found, confirming that the harmful effects of overinvestment on firm performance can be moderated by the use of debt and the payment of dividends. Thus, this is true to the suggestion of cutting down on the excessive free cash flow by using financial leverage and dividends as the necessary devices. Nevertheless, it is possible that the constraining impacts that each single policy can have on

Table 2. Firm performance regression results

Firm Performance (FP)	Over-Investment measured by Sub-Equation			Over-Investment measured by HP-filter		
	FP = EBIT/Total Asset	FP = EBT/Total Asset	FP = EAT/Total Asset	FP = EBIT/Total Asset	FP = EBT/Total Asset	FP = EAT/Total Asset
Lag _ FP	0.883***	0.766***	0.748***	0.926***	0.830***	0.727***
	(0.0876)	(0.0811)	(0.0637)	(0.0796)	(0.0764)	(0.0676)
Company Size	0.00148**	0.00235***	0.00199***	0.00102*	0.00183***	0.00210***
	(0.000723)	(0.000684)	(0.000445)	(0.000599)	(0.000595)	(0.000426)
Growth	0.00219*	0.00170*	0.00155*	0.00235*	0.00214*	0.00158*
	(0.00113)	(0.00101)	(0.000894)	(0.00120)	(0.00114)	(0.000952)
Risk	–0.0917**	–0.0636	–0.106***	–0.125***	–0.120***	–0.0965***
	(0.0427)	(0.0407)	(0.0257)	(0.0334)	(0.0302)	(0.0226)
Liquidity	0.000414	0.00229**	0.00102**	0.000658	0.00125**	0.00109**
	(0.000492)	(0.000980)	(0.000432)	(0.000417)	(0.000506)	(0.000454)
Tangibility	–0.00340	0.00254	–0.00139	0.00331	0.000844	0.000846
	(0.00697)	(0.00623)	(0.00539)	(0.00577)	(0.00588)	(0.00491)
Dividend Policy	**–0.000364*****	**–0.000323*****	**–0.000281*****	**–0.000319*****	**–0.000287*****	**–0.000274*****
	(6.14e-05)	**(5.69e-05)**	**(4.70e-05)**	**(5.70e-05)**	**(6.23e-05)**	**(5.11e-05)**
Debt Policy	**–0.0528****	**–0.0956*****	**–0.0699*****	**–0.0395****	**–0.0678*****	**–0.0737*****
	(0.0213)	**(0.0231)**	**(0.0156)**	**(0.0168)**	**(0.0193)**	**(0.0144)**
Dividend × Debt	**0.000762*****	**0.000579****	**0.000421****	**0.000554*****	**0.000501*****	**0.000436*****
	(0.000271)	**(0.000237)**	**(0.000195)**	**(0.000138)**	**(0.000142)**	**(0.000112)**
Overinvestment	**–0.0219***	**–0.0219***	**–0.0144***	**–0.0153***	**–0.0146***	**–0.0136***
	(0.0118)	**(0.0114)**	**(0.00842)**	**(0.00906)**	**(0.00877)**	**(0.00750)**
Overinvestment × Dividend	**0.00437*****	**0.00440*****	**0.00339*****	**0.00460**	**0.00473**	**0.00370**
	(0.00127)	**(0.00130)**	**(0.00119)**	**(0.00343)**	**(0.00327)**	**(0.00253)**
Overinvestment × Debt	**0.0358***	**0.0375****	**0.0254***	**0.0267***	**0.0228**	**0.0212***
	(0.0192)	**(0.0186)**	**(0.0142)**	**(0.0146)**	**(0.0141)**	**(0.0121)**
Overinvestment × Dividend × Debt	**–0.00679*****	**–0.00667*****	**–0.00510*****	**–0.00681**	**–0.00727**	**–0.00562**
	(0.00193)	**(0.00197)**	**(0.00178)**	**(0.00535)**	**(0.00512)**	**(0.00396)**
Observations	2,269	2,269	2,269	2,318	2,318	2,318
Number of instruments	19	19	22	22	22	22

(continued)

Table 2. *(continued)*

Firm Performance (FP)	Over-Investment measured by Sub-Equation			Over-Investment measured by HP-filter		
	FP = EBIT/Total Asset	FP = EBT/Total Asset	FP = EAT/Total Asset	FP = EBIT/Total Asset	FP = EBT/Total Asset	FP = EAT/Total Asset
Number of groups	597	597	597	599	599	599
F-Statistics	1100.8	587.20	440.00	1120.7	565.30	422.50
Prob.	0.0000	0.0000	0.0000	0.0000	0.0000	0.0000
Arellano-Bond test for AR(1)	–6.0090	–5.9780	–6.3050	–6.5690	–6.3590	–6.3140
Prob.	1.86E-09	2.27E-09	2.88E-10	5.06E-11	2.03E-10	2.71E-10
Arellano-Bond test for AR(2)	0.5780	0.3950	0.5630	0.6620	0.5080	0.7280
Prob.	0.5630	0.6930	0.5730	0.5080	0.6120	0.4660
Hansen test of over-id.	3.9670	4.3610	8.2890	5.5820	7.2240	8.5180
Prob.	0.1380	0.1130	0.1410	0.3490	0.2040	0.1300

Standard errors in parentheses

*** $p < 0.01$, ** $p < 0.05$, * $p < 0.1$

Sources: calculated by the author

the overinvestment-profitability relationship can be diminished by the existence of financial leverage and dividend policy in the three-variable interaction. Once more, the substitution relation between financial leverage and dividend policy is emphasized. Undergoing various models utilizing different proxies for overinvestment and firm profitability, not only do the estimated results finally achieve the consistency in signs, but they also acquire the significance level, which allows the regression model to be further bolstered.

5 Conclusion

Conclusively, it is clear that the three variables including financial leverage, dividend payments, and investment policy are not independent of one another; however, to clarify the efficiency of business operations, they are literally weaved and combined. Having Agency Theory and Free Cash Flow held their implications, the study creates a new way of analysing the moderate effects of debt and dividend policy on overinvestment, for which the conflict of interest between shareholders and managers within a company is responsible.

With the dataset of all non-financial companies listed in Vietnam's stock exchange market from 2012 to 2016, the conclusion that overinvestment wields a negative impact on firm profitability is finally drawn. In a remarkable way, by subtracting the excessive free cash flow, the isolated usage of either dividend policy or debt policy can attenuate the adverse effect of overinvestment. Conversely, by virtue of the substitution effect between financial leverage and dividend payments, when combined, these two policies degenerate the overinvestment-performance relationship. With two substitute measures of overinvestment established under HP Filter technique together with various representatives for firm profitability, consisting of the positive residual taken from the sub-equation and the points over the trend line of the investment rate, the analysis of robustness is administered. Regardless of the replacement in proxies for both independent and dependent variables, not only do all estimated coefficients remain consistent in expected signs, but they also are consistent in the significance level, further bringing about the firmness of the model.

Judged from the outcome, some recommendations are proposed. Firstly, to mitigate the negative effect of overinvestment on firm profitability, financial leverage and dividend payments should be exploited for firms to limit the excess of free cash flow. Secondly, to deduct the possibility of overinvestment, managers should also take the enhancement of their governance into consideration for the agency problem to be lessened.

Appendix

Appendix 1: Variable Measurements

$$\begin{aligned} Performance_{i,t} = {} & \lambda_0 + \lambda_1 Performance_{i,t-1} + \lambda_2 Size_{i,t} + \lambda_3 Growth_{i,t} + \lambda_4 Risk_{i,t} + \lambda_5 Liquidity_{i,t} + \lambda_6 Tangibility_{i,t} \\ & + \lambda_7 Dividend_{i,t} + \lambda_8 Debt_{i,t} + \lambda_9 Overinvestment_{i,t} \\ & + \lambda_{10} Dividend_{i,t} \times Debt_{i,t} + \lambda_{11} Debt_{i,t} \times Overinvestment_{i,t} + \lambda_{12} Dividend_{i,t} \times Overinvestment_{i,t} \\ & + \lambda_{13} Dividend_{i,t} \times Debt_{i,t} \times Overinvestment_{i,t} + \mu_{i,t} \end{aligned}$$

Variables	Denote	Definition
Dependent Variables		
Firm Performance	EBIT/TA	EBIT (Earnings Before Interest & Tax) divided by Total Assets
	EBT/TA	EBT (Earnings Before Tax) divided by Total Assets
	EAT/TA	EAT (Earnings After Tax) divided by Total Assets
Explanatory Variables		
Risk	$Risk_{i,t}$	Standard deviation of ROA
Company Growth	$Growth_{i,t}$	Growth rates of total sales
Dividend	$Dividend_{i,t}$	Cash dividend payouts over earnings after taxes
Debt	$Debt_{i,t}$	Total liabilities/total assets
Liquidity	$Liquidity_{i,t}$	Quick ratio (current assets – inventories)/current liabilities
Tangibility	$Tangibility_{i,t}$	Tangible fixed assets/total assets
Overinvestment	$Overinvestment_{i,t}$	Investment residual measured by Overinvestment Estimation with $v_{i,t} > 0$ and Hodrick–Prescott Filter
Firm Size	$Size_{i,t}$	Natural logarithm of total assets

Appendix 2: Overinvestment Estimation

$$\begin{aligned} Investment^{NEW}_{i,t} = {} & \gamma_1 DebtRatio_{i,t} + \gamma_2 Risk_{i,t} + \gamma_3 CompanySize_{i,t} + \gamma_4 SaleGrowth_{i,t} \\ & \gamma_5 AssetTurnover_{i,t} + \gamma_6 GrowthOption_{i,t} + \gamma_7 CashFlow_{i,t} + v_{i,t} \end{aligned}$$

Variables	Denote	Definition
Dependent Variables		
New Investment	$Investment^{NEW}_{i,t}$	Total investment including long-term and short-term investment divided by total asset
Explanatory Variables		
Variables	Denote	Definition
Cash Flow	$CashFlow_{i,t}$	The cash available in a company after subtracting capital expenditures
Market Performance	$GrowthOption_{i,t}$	Growth option is Tobin's Q ratio calculated as the market value of a company divided by the firm's assets

(*continued*)

(continued)

Variables	Denote	Definition
Asset Turnover	$AssetTurnover_{i,t}$	The ability to generate fixed assets through sales measured by total fixed assets divided by total sales
Firm Growth	$SaleGrowth_{i,t}$	The growth rate of firm sales over year
Firm Size	$CompanySize_{i,t}$	Natural logarithm of total assets
Business Risk	$Risk_{i,t}$	Standard deviation of EBITDA (Earnings Before Interest, Taxes, Depreciation and Amortization) over Total Asset ratio in three consecutive years
Leverage	$DebtRatio_{i,t}$	Total liabilities over total assets

All the above variables are calculated by using financial data from ***Thomson Reuters Eikon Financial Analysis***

References

Alli, K.L., Khan, A.Q., Ramirez, G.G.: Determinants of corporate dividend policy: a factorial analysis. Fin. Rev. **28**(4), 523–547 (1993)

Altaf, N., Shah, F.: Working capital management, firm performance and financial constraints: empirical evidence from India. Asia Pac. J. Bus. Admin. **9**(3), 206–219 (2017)

Baker, H.K., Powell, G.E.: Determinants of corporate dividend policy: a survey of NYSE firms. Finan. Pract. Educ. **10**, 29–40 (2000)

Biddle, G.C., Hilary, G., Verdi, R.S.: How does financial reporting quality relate to investment efficiency? J. Account. Econ. **48**(2–3), 112–131 (2009)

Brealey, R.A., Myers, S.C., Allen, F.: Brealey, Myers, and Allen on real options. J. Appl. Corp. Finan. **20**(4), 58–71 (2008)

Chen, Y.-C., Hung, M., Wang, Y.: The effect of mandatory CSR disclosure on firm profitability and social externalities: evidence from China. J. Account. Econ. (2017)

DeAngelo, H., DeAngelo, L., Stulz, R.M.: Dividend policy and the earned/contributed capital mix: a test of the life-cycle theory. J. Financ. Econ. **81**(2), 227–254 (2006)

Dyck, A., Zingales, L.: Private benefits of control: an international comparison. J. Financ. **59**(2), 537–600 (2004)

Easterbrook, F.H.: Two agency-cost explanations of dividends. Am. Econ. Rev. **74**(4), 650–659 (1984)

Farooq, S., Ahmed, S., Saleem, K.: Impact of Overinvestment & Underinvestment on Corporate Performance: Evidence from Singapore Stock Market (2014)

Fosu, S.: Capital structure, product market competition and firm performance: evidence from South Africa. Q. Rev. Econ. Finan. **53**(2), 140–151 (2013)

Fu, F.: Overinvestment and the operating performance of SEO firms. Financ. Manage. **39**(1), 249–272 (2010)

Gordon, M.J.: Dividends, earnings, and stock prices. Rev. Econ. Stat. **41**, 99–105 (1959)

Gordon, M.J.: Optimal investment and financing policy. J. Finan. **18**(2), 264–272 (1963)

Grazzi, M., Jacoby, N., Treibich, T.: Dynamics of investment and firm performance: comparative evidence from manufacturing industries. Empirical Economics **51**(1), 125–179 (2016)

Grossman, S.J., Hart, O.D.: Corporate Financial Structure and Managerial Incentives. The Economics of Information and Uncertainty, pp. 107–140. University of Chicago Press (1982)

Gu, L.: Three Essays on Financial Economics. University of Illinois, Urbana-Champaign (2013)

Guariglia, A., Yang, J.: A balancing act: managing financial constraints and agency costs to minimize investment inefficiency in the Chinese market. J. Corp. Finan. **36**, 111–130 (2016)
Hail, L., Tahoun, A., Wang, C.: Dividend payouts and information shocks. J. Account. Res. **52** (2), 403–456 (2014)
He, W., Kyaw, N.A.: Ownership structure and investment decisions of Chinese SOEs. Res. Int. Bus. Finan. **43**, 48–57 (2018)
Hodrick, R.J., Prescott, E.C.: Postwar US business cycles: an empirical investigation. J. Money Credit Bank. **29**, 1–16 (1997)
Hope, O.K., Thomas, W.B.: Managerial empire building and firm disclosure. J. Accounting Res. **46**(3), 591–626 (2008)
Jensen, M.C.: Agency costs of free cash flow, corporate finance, and takeovers. Am. Econ. Rev. **76**(2), 323–329 (1986)
Jensen, M.C., Meckling, W.H.: Theory of the firm: Managerial behavior, agency costs and ownership structure. J. Financ. Econ. **3**(4), 305–360 (1976)
Kannadhasan, M., Aramvalarthan, S.: Relationships among Business Strategy, Environmental Uncertainty and Performance of firms Operating in Transport Equipment Industry in India (2011)
Lang, L.H., Litzenberger, R.H.: Dividend announcements: cash flow signalling vs. free cash flow hypothesis? J. Financ. Econ. **24**(1), 181–191 (1989)
Litzenberger, R.H., Ramaswamy, K.: The effect of personal taxes and dividends on capital asset prices: theory and empirical evidence. J. Financ. Econ. **7**(2), 163–195 (1979)
Liu, N., Bredin, D.: Institutional Investors, Over-investment and Corporate Performance. University College Dublin, Dublin (2010)
Miller, M.H., Modigliani, F.: Dividend policy, growth, and the valuation of shares. J. Bus. **34**(4), 411–433 (1961)
Modigliani, F., Miller, M.H.: The cost of capital, corporation finance and the theory of investment. Am. Econ. Rev. **48**(3), 261–297 (1958)
Modigliani, F., Miller, M.H.: Corporate income taxes and the cost of capital: a correction. Am. Econ. Rev. **53**(3), 433–443 (1963)
Myers, S.C., Majluf, N.S.: Corporate financing and investment decisions when firms have information that investors do not have. J. Financ. Econ. **13**(2), 187–221 (1984)
Nenova, T.: The value of corporate voting rights and control: a cross-country analysis. J. Financ. Econ. **68**(3), 325–351 (2003)
Richardson, S.: Over-investment of free cash flow. Rev. Acc. Stud. **11**(2–3), 159–189 (2006)
Rozeff, M.S.: Growth, beta and agency costs as determinants of dividend payout ratios. J. Financ. Res. **5**(3), 249–259 (1982)
Shima, K.: Lumpy capital adjustment and technical efficiency. Econ. Bull. **30**(4), 2817–2824 (2010)
Titman, S., Wei, K.J., Xie, F.: Capital investments and stock returns. J. Finan. Quant. Anal. **39** (4), 677–700 (2004)
Yang, W.: Corporate Investment and Value Creation. Indiana University Bloomington, Bloomington (2005)

Time-Varying Spillover Effect Among Oil Price and Macroeconomic Variables

Worrawat Saijai[1(✉)], Woraphon Yamaka[2], Paravee Maneejuk[2], and Songsak Sriboonchitta[2]

[1] Faculty of Economics, Chiang Mai University, Chiang Mai, Thailand
worrawat13@hotmail.com

[2] Centre of Excellence in Econometrics, Faculty of Economics, Chiang Mai University, Chiang Mai, Thailand
woraphon.econ@gmail.com

Abstract. A purpose of this study is to examine a dynamic relationship between oil price and macroeconomic variables namely consumer price index, interest rate, effective exchange rate, and broad money (M3). The rising prices of oil could affect producer's cost and, in turn, lead to rising average prices of all goods by theory. This situation is called inflation which can be observed from an increase in some macroeconomic indicators such as consumer price index. However, as the oil price are changing over time due to political and economic situations, the relationship between macroeconomic indicators should have more dynamic property. Therefore, this study employed the time-varying VAR model to examine this non-constant relationship. The estimated results show that the effect of oil price on some variables are time-varying while the other variable is constantly affected by the oil price.

Keywords: TV-VAR · Oil price · Macroeconomic variables Inflation · Volatility

1 Introduction

The characteristics of the oil markets affecting the economy has been of interested to many studies. The complexity between the relationship among oil price and economic variables made the empirical results for the causality still in doubt. However, there are enough evidences for supporting the relationship between oil price and macroeconomic factors. In the literature, Hamilton [12,14] remarked that United State economic recessions in 1960–61 was affected by an increasing in crude oil price. On the other hand, Blanchard and Gali [8], Kilian [15], Blanchard and Riggi [9], and Allegret et al. [2] presented that structure of economy and the way of using policy might be changed by the oil price transmission. Segal [21] found some interesting clue that the rising in oil price in 2008 didn't lead to high inflation and GDP growth because of the different context that demand driven became more important than supply driven which was vice versa in the

V. Kreinovich et al. (Eds.): ECONVN 2019, SCI 809, pp. 1121–1131, 2019.
https://doi.org/10.1007/978-3-030-04200-4_82

time before. Razmi et al. [20] found that the effects of oil price and macroeconomic indicators of 4 countries in Southeast Asia are different between before and after the oil crisis time in 2007–2008. These studies give an idea that the relationship and the size of effect between oil price and macroeconomic variable may change over time. There were some evidences that illustrate the relationship between oil price, oil price volatility, and macroeconomic factors by using the Time-varying Vector Autoregressive (TV-VAR) model [2,5]. This model is appropriate for capturing the structural change and its result is interesting. Allegret et al. [2] applied TV-VAR to study the effect of oil price on real effective exchange rate (REER) and confirmed that oil supply shocks and oil demand shocks contribute different impacts to REER. Oil demand shocks seem to produce greater effect, in contrary to the supply shocks which contribute only small effect. However, this study assumed that the structure of relationship between the variables is dynamic. The main purpose of this study is to examine the impacts of oil price shock on macroeconomic variables including CPI, Broad Money, Effective Exchange Rate (EER), and interest rate in Thailand using TV-VAR. This model is employed following the results of literature review suggesting that the context of relationship and transmission mechanisms between oil price and macroeconomic variables may change over time and that the model is characteristically able to capture the dynamism of oil price and macroeconomic variables nexus. In addition, this study focuses more specifically on the volatility among oil price and macroeconomic variables to trace the volatility spillover of these variables along the path of their relationship with EGARCH model. Thus, this study can contribute to the literature in two ways. First is the investigation of volatility using EGARCH model which allows us to quantify the conditional volatility of each variable. Second, the time varying relationship between these volatilities are investigated using TV-VAR model.

2 Review of the Literature

In the variables selection, there are many academic studies that focus on the relationship between oil price and macroeconomic variables mostly with the use of monthly and quarterly data. Hamilton [13] was among the pioneer studies that concentrated on oil price channels that affect economic output, GDP. According to Kilian [15], this research began to propose demand shock and supply shock in this field of research and other studies adopted this idea by replacing ordinary oil price by these shocks to provide more empirical studies in another aspect. The variables that those studies used could be found from. Kilian [16], Tuan and Nagata [22], and Aziz and Dahalan [3] which used oil price, economic output (GDP, GNP, import, export, etc.), and CPI; Abdullah and Masih [1] that used CPI, middle rate of base lending rate (BLR), 3 months Treasury bill discount rate (T-bill) and money supply (M2); and Razmi et al. [20] that used oil price, US industrial production, industrial production, CPI, aggregate money, interest rate, effective exchange rate, domestic credit, and stock price as variables. Many macroeconomic theories have been employed to explain the relationship of macroeconomic variables with oil price and its volatility. The results of some studies showed that oil price and its volatility play a significant role on CPI [3,16,19,20] and the positive oil price and oil demand shock could transmit a positive value effect to CPI shock, heightening inflation and its volatility. Some studies showed effect of the oil price shock on other macroeconomic variables. For example, Balke et al. [4] proved that an increase in oil price could produce huge effect on GDP, however, reduction in oil price caused small effect on the production factors. Some studies in this field [3,7–9,19,20] constructed VAR to study the transmission between variables. Some studies used different ways to study, for example, Tuan and Nakata [22] used VAR with Block Exogeneity and Basnet and Upadhyaya [6] employed Structural VAR. VAR is the most usual way to study the transmission between oil price and macroeconomic variables. The studies of relationship between oil price and macroeconomic variables were mainly conducted by using data from ASEAN5 including, Indonesia, Malaysia, Philippines, Singapore, Thailand, and Vietnam. For example, Razmi et al. [20] studied on ASEAN4 (-Vietnam). Aziz and Dahalan [3], Basnet and Upadhyaya [6] studied on ASEAN5 (Indonesia, Malaysia, Philippines, Singapore, Thailand), while Tuan and Nakata [22] studied about the ASEAN6 (+Vietnam). Razmi et al. [20] found that the impact of oil price volatility is important to explain the movement of macroeconomic indicators from the empirical study of 4 countries in ASEAN. Moreover, some countries have individual characteristic in the relationship between variables. This study also found that Thailand is the only one country that its production before 2007–2009 crisis got lower effect than the post-crisis time. Aziz and Dahalan [3] used the Panel VAR model to study Indonesia, Malaysia, Philippines, Singapore and Thailand and the results from the Impulse Response Functions showed that positive oil price shock caused negative effect on GDP, CPI, export, and import in the first and second quarters. The VAR estimation confirmed that oil price increase caused negative effect on GDP; however, Singapore got smaller effect than the others because of its smaller economic size compared to Malaysia and Indonesia that earn more revenue from oil trade. In the case of Thailand, Rafig, Salim, and Bloch [18] studied the impact of oil price on macroeconomic variables. The result of Granger causality test found that oil price volatility makes a one-way leading relationship to many macroeconomic variables, such as unemployment, interest rate, and etc. Furthermore, the result from Bivariate variance decomposition showed that the oil price fluctuation can bring a spillover effect to Thailand's GDP and inflation volatilities in massive size, 66–96 and 48–53 percent. Razmi et al. [19] studied the effect of oil price and US economy that was transmitted to Thailand's economy. The oil price showed substantial effect on Thailand macroeconomic factors, especially, before the economic crisis in 2007–2009, Thailand's CPI and Manufacturing production index were affected by global oil price in the pre-crisis higher than the post-crisis. However, oil did not generate a significant effect on interest rate, broad money, nominal effective exchange rate, and share prices. Moreover, the result from Variance Decomposition of CPI showed that the oil price and US industrial production index volatilities contributed up to 40 and 24 percent in the pre-crisis but this effect became lower in the post-crisis, 24 and 6 precent, respectively, on CPI and Manufacturing production index of Thailand. However, there are a few academic studies that focus on the spillover effects between the volatility of oil price and macroeconomic variables under the time varying context. Therefore, we attempt to fill the gap of the literature by using TV-VAR to study the time varying volatility relationship between oil price and macroeconomic variables in Thailand.

3 Methodology

3.1 Exponential Generalized Autoregressive Conditional Heteroscedasticity (EGARCH)

To measure the volatility of each series, our study considers using EGARCH which is the extension of GARCH model. This model is constructed under the structure of ARCH (Autoregressive Conditionally Heteroscedastic) which was first introduced by Engle [11]. The GARCH process is also included in order to achieve the uniqueness and stationarity, mean zero, and heavy tails of the volatility variable. In addition, it can account for leverage effects in the series. We can show the EGARCH(1,1) by this following equation

$$y_t = u + \varphi_p \sum_{p=1}^{P} y_{t-p} + \sigma_t \varepsilon_t, \tag{1}$$

$$\ln \sigma_t^2 = \omega + \alpha_1 \frac{\varepsilon_{t-1}}{\sqrt{\sigma_{t-1}^2}} + \beta_j \ln \sigma_{t-j}^2 + \gamma \left[\frac{|\varepsilon_{t-1}|}{\sqrt{\sigma_{t-1}^2}} - \sqrt{\frac{2}{\pi}} \right], \tag{2}$$

where t is time, $\omega > 0, \alpha_i \geq 0$, and $\beta_j \geq 0$, the innovation sequence $\{\varepsilon_k\}_{k=-\infty}^{\infty}$ has independent property and their distribution are $E\{\varepsilon_0\} = 0$ and $E\{\varepsilon_0^1\} = 1$. Under EGARCH assumption, all shocks could make the same impact on volatility. The main point of EGARCH is that the conditional variance (σ_t^2) depends on its own in the previous time (σ_{t-j}^2), ARCH process (ε_{t-1}^2) and the asymmetry or the leverage effect in the last term of volatility Eq. 2. Here, the mean Eq. 1 is assumed to follow the AR(p) process.

3.2 Vector Autoregressive Model

VAR is an effective model used to study the causality between variables, as well as economic variables, and could be conducted to see the transmission between those factors by lag augmentation. The VAR(p) model could be shown like this.

$$y_t = a_0 + \sum_{j=1}^{p} A_j y_{t-j} + \varepsilon_t. \tag{3}$$

We suppose that y_t is a matrix $MT \times 1$ which contains T, time, observations which are related to each dependent variable. The original VAR modeling comes along with lag p and which means that dependent variable could be employed with p lags. A_j is a matrix of coefficients with $M \times M$ dimension, a_0 is the deterministic component's vector that contains a constant term or/and dummy variables, while ε_t is a zero-mean white noise with positively definite-contemporaneous covariance matrix Σ_ε and zero covariance matrices. We can conclude that, Y is a $T \times M$ matrix, T is a full observation of each variable. The error terms of y_t and Y could be shown in ε, where $\varepsilon \sim N\left(0, \Sigma \otimes I_T\right)$.

3.3 Time Varying Parameter-Vector Autoregression

In this study, we stimulate transmission between variables through the time varying model. The model allows the coefficient to change over time t, according to Kalman's filter. We prefer to employ this model to study the changing of the relationship that could be changed over time with flexible and robust manner (see Nakajima [17]; and Del Negro and Primiceri [10]), the TV-VAR model could be shown in the following equation.

$$Y_t = c_t + \sum_{i=1}^{P} \beta_{i,t} Y_{t-p} + A_t^{-1} \varepsilon_t \quad , \varepsilon_t \sim N(0, \Sigma_{t,1}), \tag{4}$$

Thus, Time Varying equation can be shown as

$$\beta_{i,t} = F\beta_{i,t-1} + u_t \quad , u_t \sim N(0, \Sigma_2) \tag{5}$$

Y_t is a endogenous variable's vector; c_t is constant term's vector, $\beta_{i,t}$ is a time varying parameter's matrix; A_t is a lower triangular matrix, $\Sigma_{1,t}$ is a vector of time varying standard deviation; ε_t is a vector of error term; F is vector of time varying's coefficients $\beta_{i,t-1}$; u_t is a vector of error term in time varying equation.

4 Data and Empirical Model

The data used in this study are Thailand macroeconomic variables including Consumer Price Index (CPI) provided by Division of Trade Information and Economic Indices; Broad Money (B), Effective Exchange Rate (EER), and interest rate (R) which are provided by International Financial Statistics; the data of Crude oil price (OP): US dollars per Barrel are provided by Index Mundi database, the data cover the range between 1988;M7; 2017:M3. This research uses TV-VAR to prove the existence of the dynamic relationship between the volatilities. Firstly, we conduct unit root test to see the stationary property of each variable that is very important before using those variables in the next step which does not require non-stationary property. Secondly, we employ the EGARCH(1,1) with normal innovation to quantify the volatility of each variable. Then we conduct VAR and TV-VAR models to study the spillover effect of variables in both static and dynamic perspectives. Note that TV-VAR provides information to help us understand the dynamic relationship over time. The performance of these two models is also compared in this study to confirm reliable result of our study. Moreover, this study constructs an Impulse Response function which is based on TV-VAR to see the time-adjustment after the shock events occurred and then we introduce a Variance Decomposition to forecast the impact of volatility of independent variables in the future. The economic model in this study can be shown as

$$(hCPI_t, hB_t, hEER_t, hR_t, hOP_t) = f(hCPI_{t-p}, hB_{t-p}, hEER_{t-p}, hR_{t-p}, hOP_{t-p})$$

$hCPI$ = Volatility of Core Consumer Price Index of Thailand
hB = Broad Money of Thailand
$hEER$ = Effective Exchange Rate of Thailand
hR = Policy interest rate of Thailand
hOP = Global Crude oil price
t = time
p = number of lag,

where lag in the above function is selected by AIC. $hCPI$, hB, $hEER$, hR, and hOP are obtained from EGARCH process.

5 Empirical Results

Firstly, unit root test is employed to check the stationary property of each variable, and the result shows that those variables have no unit root problem. So, the variables are granted for our empirical model. This study conducts EGARCH model to see volatility of each variable. Table 1, it provides the result of EGARCH model, the summation of alpha and beta coefficient can be considered as representative of the market situation. All parameters, ω, α, β, and γ, are shown the significant level at 0.05-0.10, except in the model of Broad Money. The coefficient that shows the sensitivity over period is α. If $\alpha > 1$, it indicates that the volatility is highly dependent on the market shock. The result shows that the volatilities of CPI and EER tend to depend on market shock, as α values are larger than 0.1, and very sensitive to bad news. However, EGARCH allows α value to be negative and we observe that the α values of Crude Oil price and Broad Money tend to be more sensitive to good news. The result from $\beta < 0$ indicates that GARCH models of each generated variable are stationary. The result can imply that B, EER, and R volatilities are persistent because the value of Beta (β) ranges between 0.745–0.966. To study the leverage effect, the result from γ shows that volatilities of Broad Money, Crude Oil price, and effective exchange rate have the highest leverage effect (0.923, 0.312, and 0.221) and can imply that negative shocks create less volatilities than positive shocks, in the same manner, the volatility is very sensitive to good news.

This study constructs VAR and TV-VAR models to estimate the volatility spillover effect of macroeconomic variables. The results of static parameter VAR

Table 1. AR(1)- EGARCH(1,1) result

Variable	CPI	B	EER	R	OP
Intercept	0.001051***	−0.024854***	0.001536	−0.002528	−0.000604
AR(-1)	0.374971***	−0.057685	0.145010***	0.182074***	0.282784***
ω	-0.424483***	−0.876097***	−1.921221***	0.000045	−0.852535**
α	0.124667***	−0.031294	0.221089**	0.000000***	−0.154468**
β	0.966054***	0.764706***	0.745973***	0.917937***	0.84011***
γ	0.229769***	0.923824***	0.221373**	0.158197**	0.312738***
Log Likelihood	1628.435	217.0315	846.8306	537.1036	376.4288

Notes: *** and ** represent significant level at 0.01 and 0.05, respectively.

and time varying VAR are also compared to confirm the performance of TV-VAR over VAR model. The result from VAR(1) shows that CPI(−1), B(−1) and OP(−1) have significant impact on CPI (see Table 2). It can be interpreted that the spillover effect from CPI(−1) causes a negative effect on CPI but different from B(−1) and OP(−1) that the positive shocks in the volatilities raise positive shock which can be supported by IS and LM theory. Next, the volatility of Policy Interest Rate plays a significant role to Effective Exchange Rate's volatility by providing positive spillover effect. It means that after the Bank of Thailand unexpectedly decreased policy interest rate, the Policy Interest Rate provides negative effect to Thailand's exchange rate's volatility. Furthermore, three variables have significant impact on Policy Interest Rate's volatility, including CPI(−1), EER(−1), and R(−1), the direction of spillover effect from these variables is positive. It means that when the shock events happened to the volatility of Consumer Price Index, Effective Exchange Rate, and Policy Interest Rate in the previous month, the Policy Interest Rate of Thailand tend to increase to make the stability for Thailand's economy. However, the magnitude effect of the spillover effect from CPI(−1) is very high, compared to other variables. The estimated coefficients obtained from VAR and TV-VAR are provided in Table 2. The result shows that there are not much difference between coefficients from these two models. The sign and magnitude of the coefficients are similar confirming the robust result from these two models. However, we can make the comparison between these two models using Bayesian Information criterion (BIC). We find that the BIC value of TV-VAR is less than VAR model, indicating the higher performance of TV-VAR model.

We then illustrate the time varying coefficients obtained from TV-VAR model in Fig. 1, and the results show that the spillover effect of oil price to the macroeconomic variables consisting of Consumer Price Index, Broad Money, and interest rate. In the case of EER, we find that the effect of volatility of oil to this variable is constant over time, so we do not plot this result in this study. According to Fig. 1, we can observe the effect of oil price to other variables a changed over time, especially the effect of oil price to CPI. The positive effect of oil to CPI has substantially decreased and reached the minimum around 1997, corresponding

Table 2. VAR and TV-VAR model estimation results on the volatility of CPI

Variable	CPI		B		EER		R		OP	
	VAR	TVP-VAR	VAR	TVP-VAR	VAR	TVP-VAR	VAR	TVP-VAR	VAR	TVP-VAR
CPI(−1)	−0.115***	−0.156	−0.085	−0.096	0.035	0.048	0.083*	0.044	0.022	0.022
B(−1)	0.104*	0.068	0.02	-0.056	0.007	0.01	0.019	0.037	0.089	0.07
EER(−1)	0.076	0.044	0.029	0.053	−0.023	−0.01	0.08*	0.102	0.067	0.075
R(−1)	0.083	0.069	−0.03	−0.025	0.140*	0.106	0.264***	0.028	−0.014	−0.012
OP(−1)	0.088*	0.089	−0.053	−0.059	−0.052	−0.057	0.032	0.077	−0.032	−0.017
Constant	0.174***	0.198	0.099*	0.048	−0.049	−0.027	0.076	−0.024	0.004	−0.005
BIC of VAR = −420.154										
BIC of TV-VAR = −444.255										

Notes: lag selection was applied by considering AIC and HQ criterions, and the results show that lag 1 is the most appropriate; *** and ** represent significant level at 0.01 and 0.05, respectively; Beta of TVP-VAR is represented by mean of Beta

Fig. 1. TV-VAR Coefficients

Table 3. Forecast error variance decomposition

CPI's variance					
T	CPI	B	EER	R	OP
2	0.9761	0.0008	0.0106	0.0011	0.01131
3	0.9757	0.0008	0.0108	0.0011	0.01141
4	0.9758	0.0008	0.0108	0.0012	0.01141
B's variance					
T	CPI	B	EER	R	OP
2	0.00526	0.9831	0.0046	0.0000	0.0069
3	0.00535	0.9829	0.0048	0.0000	0.007
4	0.00535	0.9829	0.0048	0.0000	0.007
EER's variance					
T	CPI	B	EER	R	OP
2	0.01673	0.0033	0.9767	0.003	0.0002
3	0.01675	0.0033	0.9766	0.003	0.0002
4	0.01676	0.0033	0.9766	0.003	0.0002
R's variance					
T	CPI	B	EER	R	OP
2	0.0016	0.0017	0.0074	0.9852	0.004
3	0.0019	0.0017	0.0074	0.9851	0.0041
4	0.0019	0.0017	0.0074	0.9851	0.0041
OP's variance					
T	CPI	B	EER	R	OP
2	0.0032	0.0073	0.0087	0.0034	0.9774
3	0.0032	0.0073	0.009	0.0034	0.9772
4	0.0032	0.0073	0.009	0.0034	0.9772

to the Asian financial crisis. However, the effect of oil price has increased after 2000 and reached the maximum at 2013, before dropping again in 2014–2017. The result of oil price on the other two variables seems to suggest a less fluctuation when compared to CPI. The effects of oil volatility to these two volatilities are totally different. The effect of oil to Broad Money is negative and the effect tends to increase along our sample period. In contrast, the effect of oil price to Interest rate tends to decrease over time. In addition, we provide the result of Forecasting Variance Decomposition for error constructed from TV-VAR model. The result is provided in Table 3. The result shows that the variance in the macroeconomic variables are significantly explained by their own variance. Then, we consider the share of oil shock to the macroeconomic variables and the results show a small share of oil price shock to macroeconomic variables, accounting approximately 0.02 to 1.131 percent. According to this result, we can conclude that there is a small effect of oil shock to the macroeconomic variables.

6 Conclusion

In this study, we investigate the dynamic volatility relationship between oil and macroeconomic variables using the TV-VAR model. Firstly, the volatility of the variable is quantified by the volatility EGACRH model. The result from EGARCH model shows that EER has the highest volatilities compared to other variables, indicating that EER is very sensitive to economic events. It is mainly controlled by foreign exchange markets and intervened by the central bank to keep the direction of the price that benefits the nation. Then all volatilities series are applied to both VAR and TV-VAR models to investigate the relationship. The result from VAR model shows that there exists a significant volatility relationship among macroeconomic variables and oil, indicating that those variables exhibit a long run relationship. The result from TV-VAR model also supports the result of VAR model as coefficients from the TV-VAR model are close to those estimated from the VAR model. We observe that there exists a time varying effect of oil volatility on consumer price index, broad money, and interest rate over time, while the impact of oil price on exchange rate is rather constant. Finally, the Forecast Error Variance Decomposition gives an idea that the error of Crude Oil Price will increase and almost reach the maximum effect around 4 months.

References

1. Abdullah, A.M., Masih, A.M.M.: The impact of crude oil price on macroeconomic variables: new evidence from Malaysia. In: INCEIF 16th Malaysian Finance Association Conference (paper ID: MFA-FM-118), 4–6 June 2014, Kuala Lumpur, Malaysia (2014)
2. Allegret, J.P., Couharde, C.C., Mignon, V., Razafindrabe, T., et al.: Oil currencies in the face of oil shocks: What can be learned from time-varying specifications? Technical report (2015)

3. Aziz, M.I., Dahalan, J.: Oil price shocks and macroeconomic activities in Asean-5 countries: a panel VAR approach. Eurasian J. Bus. Econ. **8**(16), 101–120 (2015)
4. Balke, N.S., Brown, S.P.A., Yucel, M.: Oil price shocks and the U.S. economy: Where does the asymmetry originate? Energy J. **23**(3), 27–52 (2002)
5. Bashar, O.H., Wadud, I.M., Ahmed, H.J.A.: Oil price uncertainty, monetary policy and the macroeconomy: the Canadian perspective. Econ. Model. **35**, 249–259 (2013)
6. Basnet, H.C., Upadhyaya, K.P.: Impact of oil price shocks on output, inflation and the real exchange rate: evidence from selected ASEAN countries. Appl. Econ. J. **47**(29), 3078–3091 (2015)
7. Bernanke, B., Gertler, M., Watson, M.: Systematic monetary policy and the effects of oil price shocks. Brookings Pap. Eco. Ac. **1**, 91–157 (1997)
8. Blanchard, O.J., Gal, J.: The macroeconomic effects of oil price shocks: Why are the 2000s so different from the 1970s? In: NBER International Dimensions of Monetary Policy, pp. 373–421 (2007)
9. Blanchard, O.J., Riggi, M.: Why are the 2000s so different from the 1970s? A structural interpretation of changes in the macroeconomic effects of oil price. J. Eur. Econ. Assoc. **11**(5), 1032–1052 (2013)
10. Del Negro, M., Primiceri, G.E.: Time-varying structural vector autoregressions and monetary policy: a corrigendum. FRB of New York Staff Report, 619 (2013)
11. Engle, R.F.: Autoregressive conditional heteroskedasticity with estimates of the variance of United Kingdom inflation. Econometrica **50**, 987–1007 (1982)
12. Hamilton, J.D.: What is an oil shock. J. Econometrics **113**, 363–98 (2003)
13. Hamilton, J.D.: Are the macroeconomic effects of oil-price changes symmetric? A comment. Carnegie-Rochester Conf. Ser. Public Policy **28**(1), 369–378 (1988)
14. Hamilton, J.D.: Oil and the macroeconomy since World War II. J. Polit. Econ. **91**, 228–48 (1983)
15. Kilian, L.: Not all oil price shocks are alike: disentangling demand and supply shocks in the crude oil market. Am. Econ. Rev. **99**(3), 1053–1069 (2009)
16. Kilian, L.: The economic effects of energy price shocks. J. Econ. Lit. **46**(4), 871–909 (2007)
17. Nakajima, J.: Time-Varying Parameter VAR Model with Stochastic Volatility: An Overview of Methodology and Empirical Applications. Institute for Monetary and Economic Studies (No. 11-E-09), Bank of Japan (2011)
18. Rafig, S., Salim, R., Bloch, H.: Impact of crude oil price volatility on economic activities: an empirical investigation in the Thai economy. Resour. Policy **34**, 121–132 (2009)
19. Razmi, F., Mohamed, A., Lee, C., Habibulah, M.S.: The effects of oil price and US economy on Thailand's macroeconomy: the role of monetary transmission mechanism. Int. J. Econ. Manage. **9**(S), 121–141 (2015A)
20. Razmi, F., Mohamed, A., Lee, C., Habibullah, M.S.: The role of monetary policy in macroeconomic volatility of association of Southeast Asian Nations-4 countries against oil price shock over time. Int. J. Energy Econ. Policy **5**(3), 731–737 (2015B)
21. Segal, P.: Oil price shocks and the macroeconomy. Oxf. Rev. Econ. Policy **27**(1), 169–185 (2011). https://doi.org/10.1093/oxrep/grr001
22. Tuan, K.V., Nakata, H.: The macroeconomic effects of oil price fluctuations in ASEAN countries: Analysis using a VAR with Block Exogeneity, Discussion Paper Series A No.619 (2014)

Exchange Rate Variability and Optimum Currency Areas: Evidence from ASEAN

Vinh Thi Hong Nguyen(✉)

Faculty of International Economics, Banking University, Hochiminh City, Vietnam
vinhnth@buh.edu.vn

Abstract. The aim of this study is to analyze determinants of exchange rate variability and calculate the Optimum Currency Area (OCA) indexes based on OCA theory for ASEAN region. Applying the exchange rate variability approach, we examine the possibility of currency integration by using cross-sectional data-set for ten Southeast Asian countries over the period 2005Q1 to 2016Q4. The results indicate that the monetary convergence is significantly impacted by output volatility, dissimilarity of export, trade linkages, and inflation difference between two countries, while the size of the economy and financial development do not contribute to OCA criteria for ASEAN regions. The results of OCA indexes suggest that the integration process should start with Singaporean Dollar and Malaysian Ringgit, followed by Philippines Peso, Indonesian Rupiah, Laotian Kip, Vietnam dong, and then Cambodian Riel, bath Thai, Brunei Dollar and Myanmar Kyat.

Keywords: ASEAN · Monetary integration · Optimum currency area
Exchange rate variability

1 Introduction

The establishment of the ASEAN Economic Community (AEC) in 2015 is a crucial milestone in the regional economic integration. The vision of AEC 2025 is to create a deeply integrated ASEAN economy with objectives to support members sustain high economic growth when they face of global shocks and volatilities. Although monetary integration is concerned in the political arena, there are few empirical studies addressed on forming monetary union for ASEAN. To fill this gap, the study analyzes the possibility of currency integration among ten ASEAN countries.

Table 1 shows in broad terms the structure of the ASEAN (Association of Southeast Asian Nations) economies. The structures of ten countries are diverse but similar with manufacturing contributing a large part of GDP and exports. This study analyses the linkage of these factors and their effect of Optimum Currency Area (OCA) index for ASEAN countries. This paper applies the core implications of the theory of optimum currency areas to cross-country data. We investigate determinants of exchange rate variability based on OCA theory to find empirical support. This will also help countries of ASEAN area to prepare for participating ASEAN's monetary union.

V. Kreinovich et al. (Eds.): ECONVN 2019, SCI 809, pp. 1132–1141, 2019.
https://doi.org/10.1007/978-3-030-04200-4_83

Table 1. Structure of the Economy for ASEAN countries, 2016

Share of GDP (%)	BRU	CAM	IND	LAO	MAL	MYR	PHIL	SING	THAI	VIET
Trade	87.2	127	37.4	75.1	128.6	39.1	64.9	310.3	121.7	184.7
Agriculture	1.2	24.7	13.5	17.2	8.7	25.5	9.7	0	8.5	16.3
Manufacturing	11.5	16	20.5	7.8	22.3	22.8	19.6	17.7	27.4	14.3
Trade in Services	43	30.4	5.8	9.2	25.5	10	18.2	103.4	27	14.6
Share of export (%)										
Food	0.2	4.6	22.5	30.6	11.6	36.9	8.5	3	13.9	12.9
Agricultural raw materials	0	2.1	4.8	3.2	1.9	2.5	0.8	0.6	3.9	1.4
Manufactures	11.4	93.2	47.7	27.3	68.5	28.7	85.3	79.1	78.2	82.8
Import (%)										
Food	18.6	7.3	11.7	13.5	8.8	18.8	11.6	4.3	7	8.7
Agricultural raw materials	0.2	2.1	3.1	0.3	1.7	0.4	0.7	0.5	1.7	2.9
Manufactures	70.9	80.2	67	69.8	73.8	68.6	75.9	73.5	74.2	80.1

Source: World Development Indicators (2018)

The rest of the paper is structured as follows. Section 2 looks at previous research on the OCA theory and determinants of exchange rate variability. Section 3 provides the method that is used in this research, and describes the data that are used. Empirical results are presented in Sect. 4. Finally, Sect. 5 contains concluding remarks.

2 Literature Review

In the literature, Horvath and Komarek (2002) mentions that there are two main streams of theory of optimum currency areas. The first stream focus on economic characteristics to determine where the borders of exchange rates should be drawn from 1960s to 1970s, and the second stream from 1970s till now focus on the costs and benefits of OCA for single country participating in a currency area. The theory of OCA has developed since crucial contributions of Mundell (1961); McKinnon (1963); Kenen (1969). Mundell (1961) defines optimum currency area as an area with internal factor mobility and external factor immobility. His OCA framework emphasizes the importance of following determinants: first, an OCA area requires high degree of internal factor mobility and a low degree of external factor mobility, second, it should be a stable wage and price, and third, it has a simple labor mobility with national limitations.

Previous studies dealing with testing OCA theory employ exchange rate variability approach and calculate OCA index. One of the pioneers research is Bayoumi and Eichengreen (1997). To analysis the possibility of currency integration for European countries, Bayoumi and Eichengreen (1997) use exchange rate variability approach and calculate OCA index for European area. They employ variables such as output volatility, trade intensity, size of economics, dissimilarity of export commodity structure. In this paper, they also use nominal year-end bilateral rate because the results for real exchange rates, constructed from nominal rates using GDP deflators, are very similar (Bayoumi and Eichengreen 1997). Horváth and Komarek (2002) calculate OCA-indexes for the Czech Republic, EU, Germany and Portugal and compare the structural similarity of the Czech Republic and Portugal to the German economy and find that the Czech economy is closer. The results are reversed when the EU economy is considered as a benchmark country. Horváth and Kuâerová (2005) examine the determinants of bilateral real exchange rate variability for 20 developed countries. The finding shows that OCA criteria such as trade linkages, openness and size of economy or financial development explain a substantial part of real exchange variability.

Alvarado (2014) analyzes the optimum currency area (OCA) for ASEAN and ASEAN + 3 for the period of 2003–2012. The paper finds that nearly half of the country members have moved symmetrically, the monetary convergence is significantly influenced by the output disturbances and the trade linkages in both regions; while the size of the economy only becomes significant in ASEAN + 3 and the synchronic advantage is not contributing and even insignificant for ASEAN + 3. Achsani and Partisiwi (2010) use the exchange rate variability based on OCA index and hierarchical clustering analysis to analyze the possibility of currency integration among ASEAN + 3. The result shows that Singapore Dollar was the most stable currency in the region. ASEAN + 3 should start with Malaysia and Singapore, followed then by Japan, Thailand, South Korea and China. Kawasaki (2012) applies generalized purchasing power parity (G-PPP) model into an up-to-date non-linear econometric model and considering the adoption of the Asian monetary unit (AMU) into East Asian countries—ASEAN5, China, Korea, and Japan. The paper provides positive empirical results which suggest for forming a common currency in this area. Ogawa and Kawasaki (2007) adopt a multi-step process toward forming a common currency in East Asian and imply that ASEAN + 3 should launch a policy dialogue related to exchange rate as well as adopt a managed floating exchange rate system. Similarity, Lee et al. (2003) investigate the prospects of a currency union in East Asia and explore that two of the most important determinants of business cycle synchronizations are intra-region trade share and trade structure similarity. Calderón et al. (2002) imply that if countries have closer international trade links and more symmetric business cycles, they should join a currency union. Chaudhury (2009) examines the possibility of monetary integration for the seven ASEAN members (Singapore, Brunei, Philippines, Malaysia, Indonesia, Thailand and Vietnam) by calculating OCA indexes and using yearly data for 1980–2007. The result indicates that these seven countries are most likely to form the OCA. It is, however, early to make a final conclusion about the benefits to be gained if the union is formed.

This study contributes to test OCA theory by linking OCA criteria with bilateral exchange rate variability using cross-country data. We examine determinants of exchange rate variability that folow OCA theory and find empirical support for the possibility of currency integration ASEAN region.

3 Methodology

Following the model proposed by Bayoumi and Eichengreen (1998) and research on more determinants of exchange rate variability, the OCA indexes approach is adopted in order to examine the monetary integration for ASEAN area. The relationships between determinants and exchange rate variability can be specified as follows:

$$Vol(e_{ij}) = \alpha + \beta_1 BCS_{ij} + \beta_2 DISSIM_{ij} + \beta_3 TRADE_{ij} + \beta_4 SIZE_{ij} + \beta_5 FIN_{ij} + \beta_6 INF_{ij} + \varepsilon$$

In this regression, observation *ij* corresponds to two economies pair *i* and *j*. Where $Vol(e_{ij})$ is the exchange volatility between country *i* and country *j*, i.e., the OCA index BCS_{ij} is the business cycles synchronization, *DISSIMij* is the dissimilarity of export commodity structure, *TRADEij* is the trade intensity, *SIZEij* is the economic size, *INFLij* is the inflation differential, *FINij* is financial growth rate, and ε is the estimation error.

The variables of the above model are calculated as follows. To measure the OCA index, we draw the deviation standard of the nominal exchange rate movements i.e.,

$$Vol(E_{ij}) = SD(\Delta\, log e_{ij})$$

The business cycles synchronization is calculated by the following formula

$$BCS_{ij} = SD(\Delta Y_i - \Delta Y_j)$$

where ΔY_i and ΔY_j are the growth of the real GDP of country i and j. The trade intensity is adopted the following approach:

$$TRADE_{ij} = \frac{1}{T}\sum_{t=1}^{T}\left(\frac{ex_{it}}{y_{it}} + \frac{ex_{jt}}{y_{jt}}\right)$$

Where ex_{it} and ex_{jt}. are the ratio of bateral exports from country i to country j. The dissimilarity of export commodity structure is exploited as the following formula

$$DISSIM_{ij} = \frac{1}{T}\sum\nolimits_{t=1}^{T} \left(|A_{it} - A_{jt}| + |B_{it} - B_{jt}| + |C_{it} - C_{jt}|\right)$$

where A_{it} and A_{jt} are the share of agricultural trading of country i and j, B_{it} and B_{jt} are the share of mining trading, and C_{it} and C_{jt} are the share of manufacture trading. The inflation differential is determined by using the flowing formula:

$$INF_{ij} = \frac{1}{T}\sum_{t=1}^{T}(\pi_{it} - \pi_{jt})$$

where π_{it} and π_{jt} are the customer price index (CPI) of country i and j. The variable of economic size is determined as follows:

$$SIZE_{ij} = \sum_{t=1}^{T} (log y_{it} - \log y_{jt})$$

The financial development trade is computed by the following formula:

$$FIN_{it} = \frac{1}{T} \sum_{t=1}^{T} \left(\frac{M2_{it}}{y_{it}} + \frac{M2_{jt}}{y_{jt}} \right)$$

where $M2_{it}$ and $M2_{jt}$ are the circulated money in country i and j and y_{it} and y_{jt} is the current price GDP in country i and j.

Table 2 reports the summary and descriptions of variables used in this research. The main approach to test the theory of OCA is to analyze the determinants of exchange rate variability. The above variables represent OCA criteria, and it is considered the lower the volatility of exchange rates is among countries, the more well prepared they participate the monetary union (Horváth and Kuâerová 2005). The expected sign for dissimilarity of the structures of the two economies is positive because economies with less similar structures are more likely to undergo asymmetrical shocks and thus mean volatility of exchange rate are expected to increase. The trade links between the two economies expected to have negative sign. This means economies which trade with each other more intensively are less likely to undergo asymmetric shocks and thus mean exchange volatility are, ceteris paribus, expected to be less intensive (Frankel and Rose 1998). Bayoumi and Eichengreen (1998) suggest that smaller economies the more benefit from the services provided by a stable exchange rate. Therefore, the SIZE variable is expected to have positive sign with exchange rate volatility. The research also includes correlation of economic cycles in the two economies on the grounds that economies with less correlated cycles are more likely to undergo asymmetrical shocks and thus the BCS has positive sign with exchange rate movement. Similarity, financial development (FIN_{ij}) and inflation difference (INF_{ij}) are expected to have positive effect on volatility exchange rates.

4 Descriptions of Variables and Data Sources

In the paper, we use the original data are of quarterly frequency, and take average or standard deviation of all the variables over. As a result, the final data matrix is cross-sectional data. Because the given data are bilateral, the combination of bilateral relationships among 10 ASEAN countries leads to 45 observations. This study analyzes a cross sectional countries comprising ten countries, i.e., Brunei, Cambodia, Indonesia, Laos, Malaysia, Myanmar, Singapore, Philippines, Thailand, Vietnam, which consist of quarterly data from 2005Q1 till 2016Q4. All the data compiled from the International Financial Statistics (IFS) and the CEIC Database.

Table 2. Summary of explanatory variables

Classification	Variable	Descriptions
Independent variable	$Vol(e_{ij})$	The standard deviation of the nominal exchange rate movements
	BCS_{ij}	The standard deviation of the difference in the logarithm of real output between i and j,
Dependent variables	$DISSIM_{ij}$	The sum of the absolute differences in the shares of agricultural, mineral, and manufacturing trade in total merchandize trade
	TRADEij	The mean of the ratio of bilateral exports to domestic GDP for the two countries
	SIZEij	The mean of the logarithm of the two GDPs measured in U.S. dollars
	INFLij	The mean of inflation difference is obtained by the customer price index (CPI) of country *i* and j
	FINij	The mean of ratio of M2 to current price GDP in countries i and j

Table 3 reports the summary of statistics for the maximum, minimum, average and standard deviation of the variables used to estimate determinants of exchange rate variability. The minimum of exchange rate volatility is zero because this data is the exchange rate volatility of Singapore and Brunei pair. The Brunei dollar is interchangeable with the Singapore dollar at par according to Currency Interchangeability Agreement in 1967. From these figures, it can be seen the difference in trade intensity, economic size, financial growth and inflation difference among ASEAN countries.

Table 3. Descriptive statistics of variables

Variable	Obs.	Mean	SD	Min	Max
$Vol(e_{ij})$	45	0.175078	0.280194	0.00000	0.735409
BCS_{ij}	45	0.034509	0.010521	0.014931	0.055265
$TRADE_{ij}$	45	0.024758	0.036442	.0000127	0.218523
$DISSIM_{ij}$	45	.9484365	.4925421	.1962194	1.85446
$SIZE_{ij}$	45	−0.46563	0.976175	−2.01527	1.895223
FIN_{ij}	45	−0.34074	1.879241	−3.74724	3.779043
$INFL_{ij}$	45	−0.8007	4.698001	−10.7798	9.050208

Source: IFS and CEIC, author's own estimations.

Table 4 shows the correlation coefficients between variables which are relatively low. Most of variables have positive sign with exchange rate volatility, except trade intensity. All the signs of correlation matrix of variables are in the line with expected signs in the research model.

Table 4. Correlation matrix of variables

	VOL(e)	BCS	TRADE	DISSIM	SIZE	FIN	INFL
VOL(e)	1.0000						
BCS	0.1260	1.0000					
TRADE	–0.2769	0.0511	1.0000				
DISSIM	0.0947	0.5588	–0.0182	1.0000			
SIZE	0.0437	–0.4194	0.0964	–0.1295	1.0000		
FIN	0.0729	0.0277	0.1293	–0.2072	0.1446	1.0000	
INFL	–0.0186	–0.2132	0.0754	0.1105	0.3145	–0.5221	1.0000

Source: IFS and CEIC, author's own estimations.

5 Empirical Results

The estimation results are presented in Tables 5. The table reports an estimation of determinants of exchange rate variability by the OLS method. The findings shows that all coefficients yield the expected signs and four variables are significant. The variability of business cycles synchronization has a statistically and positively effect on exchange rate variability. Thus, the cyclical fluctuations of output had an adverse impact on exchange rate movements in the studied period. This finding also suggests that economic cooperation between the countries is very important to precede large exchange rate fluctuations.

Table 5. OLS regression result

Variable	Coefficient	Standard error
BCS	5.21539*	5.609947
TRADE	–2.47337**	1.199602
DISSIM	0.00772*	0.110781
SIZE	0.03225	0.053933
INF	0.00495**	0.012176
FIN	0.02071	0.029602
_cons	0.075051	0.153773
R2	0.28147	
Obs.	45	

Source: IFS and CEIC, author's own estimations.
***, **, * * *and* ** *denote significance at the 10%, 5% and 1% levels, respectively 5% and 10%.*

In this model, trade linkages variable has a significant effect on exchange rate variability with 5% level. The negative relation is consistent with the finding in Bayoumi and Eichengreen (1997); Alvarado (2014). The results confirm that higher bilateral trade together with high similarity of exports led to lower exchange rate volatility in ASEAN countries. The dissimilarity of the structures of the two economies is positive and significant effect on exchange rate variability. This implies economies with less similar structures are more likely to undergo asymmetrical shocks and lead volatility of exchange rate increase. The inflation difference is have a significant impact on exchange rate variability. With increasing inflation difference between two economics the exchange rate variability increase.

Both the size and financial development variables are not significant in the case of ASEAN countries. In other words, the exchange rate volatility is not affected by the size of the members as well as financial development.

Based on the regression result, we calculate the OCA indexes. The result for OCA indexes is presented in Table 6 respectively from smallest to largest. According to the calculated OCA indexes, during this period, the Singaporean Dollar is the most stable currency with the lowest OCA index followed by Malaysian Ringgit, Philippines Peso, Indonesian Rupiah, Laotian Kip, Vietnam dong, and then Cambodian Riel, bath Thai, Brunei Dollar and Myanmar Kyat.

Table 6. OCA Indexes for ASEAN calculated for the period 2005–2016

Pairs	OCA index	Pairs	OCA index	Pairs	OCA index
MlSg	**−0.238740**	LaSg	0.147071	BrVn	0.220201
LaVn	0.000108	LaPh	0.147074	InMy	0.229053
LaTh	0.023209	SgVn	0.147287	CaMy	0.233093
PhSg	0.077992	PhTh	0.156452	InLa	0.235276
CaVn	0.082665	InMl	0.157210	MyVn	0.242440
MlTh	0.099483	CaSg	0.168075	BrSg	0.245859
InVn	0.102360	CaIn	0.175128	MySg	0.250745
BrIn	0.104032	CaPh	0.180445	BrTh	0.255420
LaMl	0.107324	InPh	0.194722	MyTh	0.271724
PhVn	0.113344	MlPh	0.194885	BrMy	0.276645
CaMl	0.125241	CaTh	0.201769	MlMy	0.278546
ThVn	0.125765	InTh	0.202398	BrPh	0.291882
InSg	0.132527	BrMl	0.204253	MyPh	0.295423
MlVn	0.140180	SgTh	0.208193	BrLa	0.337025
LaMy	0.143340	CaLa	0.215525	BrCa	0.375864

Source: IFS and CEIC, author's own estimations.
Br = Brunei, Ca = Cambodia, In = Indonesia, La = Laos, Ml = Malaysia, My = Myanmar, Ph = Philippines, Sg = Singapore, Th = Thailand, Vn = Vietnam

The research consider indices for bilateral rates against Singapre because that country is widely viewed as the core member of ASEAN. Besides, the Brunei dollar is interchangeable with the Singapore dollar at par. Using Singapore Dollar as a benchmark, Fig. 1 shows that the countries with high similarity are Singapre – Malaysia,

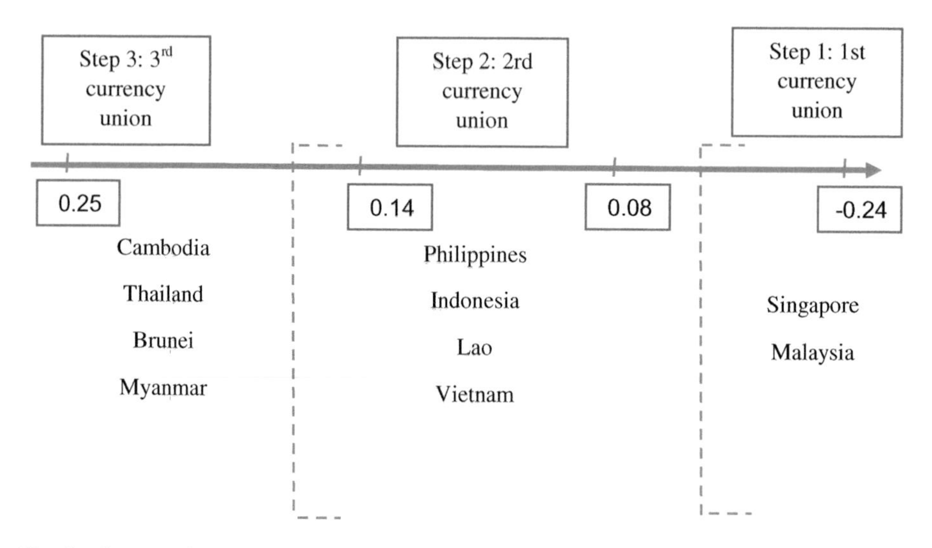

Fig. 1. Currency integration process among ASEAN countries. Source: IFS and CEIC, author's own estimations.

Philippines – Indonesia-Lao – Vietnam. Meanwhile, Cambodia – Thailand – Brunei – Myanmar have different characteristics. The findings also suggest that the integration process can be undertaken by respectively unifying Malaysian Ringgit and Singaporean Dollar (step 1), followed by Philippines Peso, Indonesian Rupiah, Laotian Kip, Vietnam dong (step 2), and then Cambodian Riel, bath Thai, Brunei Dollar and Myanmar Kyat (step 3).

6 Conclusions

This study estimate the impact of the determinants on exchange rate variability and OCA indexes based on sample of 45 cross-sectional data-set for ten Southeast Asian countries over the period from 2005Q1 to 2016Q4. This paper contributes to test OCA theory by linking OCA criteria with bilateral exchange rate variability. The empirical results provide some evidences to confirm that OCA criteria such as business cycles synchronization, trade linkages, dissimilarity of export and inflation difference explain a substantial part of exchange variability. We do not find clear evidence that variability of size and financial development have an effect on the nominal exchange variability in the analyzed countries. The results of OCA indexes suggested that the integration process should be started by unifying Singaporean Dollar and Malaysian Ringgit (step 1), followed by Philippines Peso, Indonesian Rupiah, Laotian Kip, Vietnam dong (step 2), and then Cambodian Riel, bath Thai, Brunei Dollar and Myanmar Kyat (step 3).

References

Achsani, A., Partisiwi, T.: Testing the feasibility of ASEAN + 3 single currency comparing optimum currency area and clustering approach. Int. Res. J. Financ. Econ. **37**, 79–84 (2010)

Alvarado, S.: Analysis of the optimum currency area for ASEAN and ASEAN + 3. J. US-China Public Admin. **11**(12), 995–1004 (2014)

Bayoumi, T., Eichengreen, B.: Ever closer to heaven? An optimum-currency-area index for European countries. Eur. Econ. Rev. **41**(3–5), 761–770 (1997)

Bayoumi, T., Eichengreen, B.: Exchange rate volatility and intervention: implications of the theory of optimum currency areas. J. Int. Econ. **45**(2), 191–209 (1998)

Calderón, C., Chong, A., Stein, E.: Trade intensity and business cycle synchronization: Are developing countries any different? Chilean Central Bank, Working Papers, 195. 12/2002 (2002). Accessed http://www.bcentral.cl/eng/studies/working-papers/195.htm

Chaudhury, M.R.: Feasibility and Implications of a Monetary Union in Southeast Asia, Middlebury College (2009)

Frankel, J., Rose, A.: Is EMU more justifiable ex post than ex ante? Eur. Econ. Rev. **41**(1998), 563–570 (1998)

Horváth, R., Komarek, L.: Optimum currency area theory: an approach for thinking about monetary integration. Warwick economic research papers No.647 (2002)

Horváth, R., Kuâerová, Z.: Real exchange rates and optimum currency areas: evidence from developed economies. Czech J. Econ. Finance, **55**(5–6), 253–266 (2005)

Kawasaki, K.: Are the ASEAN plus three countries coming closer to an OCA (2012). Accessed http://www.rieti.go.jp/en/publications/summary/12050008.html

Kenen, P.: The theory of optimum currency areas: An eclectic view. In: Mundell, R.A., Swoboda, A.K. (eds.) Monetary Problems of the International Economy, pp. 41–60. University of Chicago Press, Chicago, IL (1969)

Lee, J., Park, Y., Shin, K.: A currency union in East Asia. ISER Discussion Paper, 571 (2003). Accessed http://ssrn.com/abstract=396260

Mckinnon, R.: The theory of optimum currency area. Am. Econ. Rev. **53**(1963), 717–725 (1963)

Mundell, R.A.: A theory of optimum currency areas. Am. Econ. Rev. **51**(4), 657–665 (1961)

Ogawa, E., Kawasaki, K.: East Asian Currency Cooperation, p. 16 (2007). Japan. Accessed http://aric.adb.org/pdf/seminarseries/SS10paper_East_Asian_Currency.pdf

The Firm Performance – Overinvestment Relationship Under the Government's Regulation

Chau Van Thuong[1], Nguyen Cong Thanh[1], and Tran Le Khang[2](✉)

[1] School of Accounting – Banking – Finance, Ho Chi Minh City University of Technology, Ho Chi Minh City, Vietnam
{cv.thuong, nc.thanh93}@hutech.edu.vn

[2] School of Economics, Erasmus University Rotterdam, The Hague, The Netherlands
khang.tl@vnp.edu.vn

Abstract. With the purpose of identifying the negative association between overinvestment caused by agency problems and firm performance measured through profitability, the research employs the dataset of Thomson Reuters covering 669 Vietnamese non-financial listed companies in the period of 2012–2016. Applying the fixed effect technique to measure overinvestment and run the main econometric model, the paper finds that overinvestment is the cause leading to ineffective operations because it decreases firm profitability. The negative association between firm performance and overinvestment can be regulated with the intervention of the government through state ownership. The regulation impact of state ownership will be stronger in companies that has state ownership rate lower than 50%.

Keywords: State ownership · Overinvestment · Firm performance

JEL: G31 · G35

1 Introduction

Due to the harmful effects of state ownership on firm operations, a process of privatizing state-owned into private-owned enterprises, or so-called "privatization", is increasingly gaining its popularity in the modern world (Djankov and Murrell 2002; Peng, Buck and Filatotchev 2003; Rodríguez, Espejo and Cabrera 2007; Sheshinski and López-Calva 2003). Apparently, emerging countries where the government still holds much control of business activities through its large ownership rates, where market laws as well as regulations remains weak and loose, and where the financial market is only in its embryonic stage with so many legal cracks facilitate a breeding ground for state-owned enterprises. Therefore, the governments in these countries pay attention to the privatization process in order to foster economic growth and create a transparent business environment. In Vietnam, the privatization of government-owned companies has been conducted since 1986 when the Congress passed the economic reform program to drive this country towards a market-oriented economy. The adoption

V. Kreinovich et al. (Eds.): ECONVN 2019, SCI 809, pp. 1142–1153, 2019.
https://doi.org/10.1007/978-3-030-04200-4_84

of this practice has brought some noteworthily positive signals to Vietnam's economy through higher foreign direct investment, lower poverty rates, and better business environments. Additionally, privatization has resulted in the emergence of Vietnam's two market exchanges namely Ho Chi Minh Stock Exchange (HOSE) and Hanoi Stock Exchange (HNX). Although an absolute dependence of the government (100% state-owned) has been lifted in all companies listed on these two exchanges, a certain rate of state ownership still remains in quite a large number of firms, indicating the involvement of the government in business activities. State ownership is thought to exacerbate agency conflicts between managers and shareholders due to two major reasons. First, besides profit maximization, the government also have other non-profitable purposes to pursue. So, if a manager is often forced to make disadvantageous financial decisions, he, in some respect, will do harm to firm performance. Second, based on Agency Theory, managers often take advantage of the discretionary funds to benefit themselves but not shareholders, causing companies to incur agency costs to align shareholders and managers' interests (Jensen 1986). Also, managers in state-owned enterprises act in their personal gains. Whether state ownership will change agency problems in a positive or negative way is an unanswered question. With the dominance of state ownership, the government will pay a closer look at a firm's operation and management, reducing its agency costs and raising its profitability (Bos 1991). On the other hand, managers in state-owned enterprises are forced to hire more employees than necessary, to appoint politically connected but not well-qualified candidates to job positions, and to focus on social and political objectives (Boycko, Shleifer and Vishny, 1996; Dewenter and Malatesta 2001; Krueger 1988). All of these actions worsen existing agency problems. There is a belief that different rates of state ownership will exert discrepant impacts on how effectively a firm perform its operations. Thus, this research takes into account two levels of state ownership namely upper 50% and lower 50% with the purpose of identifying their differences. The combination of both Agency Theory and Free Cash Flow indicate that when facing the excess in the discretionary funds, managers try to do every way possible to enlarge financial sources under their control to gain more compensation and private benefits (Gaver and Gaver 1993). They even invest in projects with negative net present value (NPV) and make the problem of overinvestment worse. Hence, overinvestment is a problem rooted in the agency conflicts between managers and shareholders. In this case, whether state ownership will help decrease or increase the problem of overinvestment remains unanswered. Again, the study aims to identify whether state ownership can lessen the negative impact of overinvestment and whether its moderation effects vary between companies with SOE rate higher and lower than 50%.

The research data is derived from Thompson Reuters' financial statements for 669 non-financial companies listed on Vietnam's two biggest stock exchange markets for five consecutive years from 2012 to 2016. Two different proxies for overinvestment using the sub-equation (Richardson 2006; He and Kyaw 2018) and HP Filter (Hodrick and Prescott 1997) together with three various representatives of firm profitability are employed to test the robustness of the regression model. From the variable estimations, the study demonstrates the negative relationship between overinvestment and profitability. State ownership with its advantages in accessing financing resources and regulatory protections is supposed to reduce the harmful effect of overinvestment.

However, when SOE rate is classified into two types with state ownership over and under 50%, the estimated result is in support of a more beneficial impact of companies with SOE rate lower than 50% on the overinvestment-performance relationship. In short, this paper contributes a great part to empirical studies in the world, in general, and in Vietnam, in particular. For this first time, the problem of overinvestment is examined in the case of Vietnam with two alternative approaches. Besides, the study evaluates the moderation effect of state ownership on overinvestment. It, eventually, come with the suggestion that a certain percent of state-owned shares should be maintained in a company, and this percent should be lower than 50%.

The study is structured as followings. Section two reviews relevant theories and empirical studies to develop research hypotheses. Section three shows the method used to clarify these hypotheses. It includes data collection, variable description, models, and techniques. Section four displays the regression results, and section five concludes the study as well as gives some policy recommendations.

2 Literature Review and Hypothesis Development

2.1 Overinvestment and Firm Performance

If properly made following financial objectives, investment policy will make a great contribution to a firm's profit maximization. Grazzi, Jacoby and Treibich (2016) show that suitable investment policy will foster economic growth and widen the labour market. Licandro, Maroto and Puch (2004) and Nilsen, Raknerud, Rybalka and Skjerpen (2008) suggest that the increase in investment means the application of modern technologies, the replacement of old machines to new ones, the expansion of manufacturing lines that can help raise firm productivity. Moreover, the investment in research and development (R&D) is found to move in the same direction with firm productivity and profitability. In the decision-making process of investment strategies, managers are considered to be a key element. There will be no serious problem if the conflicts of interests between managers, the representatives, and shareholders, the company owners, are effectively negotiated. However, managers often act on their own interests with the aim of expanding their managing power over the company's property but not maximizing shareholders' benefits (Gaver and Gaver 1993; Jensen 1986; Jensen and Meckling 1976). If tightly controlled by shareholders and stakeholders, managers will hesitate to carry out investment projects or even forgo projects with positive net present value, causing the problem of underinvestment. If loosely managed, managers will easily take advantage of discretionary funds to broaden financial resources under their management as much as possible by investing in projects with negative net present value, leading to the problem of overinvestment (Brealey and Franks 2009). Guariglia and Yang (2016) give clearer evidence that underinvestment is due to financial constraints, overinvestment is because of the excessive free cash flow, both of them are originated from agency problems. Thus, except for the optimal investment which is effective for firm performance, overinvestment and underinvestment are harmful to business operations (Farooq, Ahmed and Saleem 2014; Jensen 1986; Liu and Bredin 2010; Titman, Wei, and Xie 2004; Yang 2005). Although the impact of overinvestment has been researched in other markets. It is not taken into consideration in the case of Vietnam. Thus, the study comes out with the first hypothesis.

Hypothesis 1: overinvestment has a negative effect on firm performance in Vietnamese enterprises.

2.2 State Ownership and the Overinvestment-Profitability Relationship

The effect of state ownership on firm performance has been arguably examined. The relationship with the government has both advantages and disadvantages in a company's operations (Fan, Wong and Zhang 2007; Hellman, Jones and Kaufmann 2000; La Porta, Lopez-de-Silanes, Shleifer and Vishny 2000a, b; Shleifer and Vishny 1994). La Porta, Lopez-de-Silanes, Shleifer and 2000a, b hold the view that the presence of state ownership accompanies the absence of the monitoring incentives toward managers' decisions, making agency problems more serious in state-owned enterprises. Boubakri, Cosset and Saffar (2008) stress that the pursuit of other goals including ensuring employment and wages of workers, promoting regional development, increasing national security, and producing affordable goods and services instead of profit maximization has led state-owned enterprises (SOEs) to perform ineffectively. It is this agency problem that distort investment efficiency and deteriorates firm performance.

Privatization, the reduction in the number of state-owned shares, plays a vital role in moderating the efficiency of investment (Megginson and Netter 2001). The transformation of SOEs to private investors are thought to cause serious agency problems (Boubakri et al. 2008; Denis and McConnell 2003; Guedhami, Pittman and Saffar 2009). Although privatization is associated with better governance and management, these improvements may disappear if the government continues to hold a major control over privatized companies (Boubakri and Cosset 1998; D'souza and Megginson 1999; Guedhami et al. 2009; Megginson and Netter 2001).

On the contrary, the benefits of having a close relation with the government cannot be ignored. Some previous studies prove the positive impact of closer connections with the government on firm performance (Boubakri et al. 2008; Boubakri and Cosset 1998; Faccio 2006). Such a good relationship can help companies easily obtain certain resources such as lands, capital, and licenses. Moreover, sometimes the management of the government will bring a close monitoring to the company's operations, reducing agency problems between managers and shareholders (Bos 1991). The government can also manipulate laws and regulations, taxes, and decisions to borrow loans. Therefore, its participation will bring political support to firms, enhancing firm performance (Lu 2000; Sun, Tong and Tong 2002). Finally, the research gives the second hypothesis.

Hypothesis 2: firm performance will be worse in companies with the high level of state control.

3 Data and Methodology

3.1 Data Collection

In this study, we examine the sample data of 669 non-financial firms listed on Vietnam's stock exchange market in the period of 2012–2016 from the data source of Thompson Reuters. Initially, overinvestment is measured through Eq. (1) using the fixed-effect technique. The estimated equation is generalized based on the ideas from previous studies (Bokpin and Onumah 2009; Carpenter and Guariglia 2008; Connelly 2016; Li and Zhang 2010; Malm, Adhikari, Krolikowski and Sah 2016; Nair 2011; Richardson 2006; Ruiz-Porras and Lopez-Mateo 2011). The explicit form of Eq. (1) is as follows:

$$\begin{aligned} NewInvestment_{i,t} = {} & \alpha_0 + \alpha_1 CashFlow_{i,t} + \alpha_2 TobinQ_{i,t} + \alpha_3 FixCapitalIntensity_{i,t} \\ & + \alpha_4 FirmSize_{i,t} + \alpha_5 RevenueGrowth_{i,t} + \alpha_6 BusinessRisk_{i,t} \\ & + \alpha_7 Leverage_{i,t} + \omega_{i,t} \end{aligned} \tag{1}$$

Where $NewInvestment_{i,t}$ represents for the investment decision; $CashFlow_{i,t}$ reflects the cash available in a company after subtracting capital expenditures; $TobinQ_{i,t}$ is the representative of growth opportunity and market performance; $FixCapitalIntensity_{i,t}$ evaluates the ability to generate fixed assets through sales; $RevenueGrowth_{i,t}$ demonstrates the growth of the firm; $FirmSize_{i,t}$ shows a company's financial constraints; $BusinessRisk_{i,t}$ indicates the volatility of firm profitability; $Leverage_{i,t}$ is the capital structure of the company. The estimated error-term $\hat{\omega}_{i,t}$ taken from the above model is considered as the abnormalities in the investment decision. If the error term's value is positive, or $\hat{\omega}_{i,t} > 0$, $\hat{\omega}_{i,t}$ of firm i^{th} and year t^{th} is denoted as $Overinvestment_{i,t}$. This method of calculating overinvestment has been recently adopted by He and Kyaw (2018). Besides, the research also employs Hodrick–Prescott (HP) Filter to be another alternative measure of overinvestment. This method helps to identify the fluctuation in firm investment, and the positive excess values compared to the fitted investment line are regarded as over-investment (Hodrick and Prescott 1997).[1]

3.2 Model Specification

In order to find evidence proving the aforementioned hypotheses, the paper carries out the regression model as Eq. (2) following several previous studies (Altaf and Shah 2017; Chen, Hung and Wang 2017; Fosu 2013). Besides major explanatory variables, this study adds all possible control variables to reduce the likelihood of endogeneity due to omitted regressors, which helps measure firm performance effectively (Wooldridge 2015). Thus, the main regression model is clearly shown as follows:

[1] See Appendix for the description of all variables.

$$\begin{aligned} Performance_{i,t} = & \beta_0 + \beta_1 Dividend_{i,t} + \beta_2 Leverage_{i,t} + \beta_3 AssetGrowth_{i,t} \\ & + \beta_4 ProfitVolatility_{i,t} + \beta_5 Liquidity_{i,t} + \beta_6 Tangibility_{i,t} \\ & + \beta_7 Overinvestment_{i,t} + \beta_8 SOErate_{i,t} + \beta_9 Overinvestment._{i,t} \times SOErate_{i,t} \\ & + \beta_{10} Overinvestment_{i,t} \times SOErate_{i,t} \times SOE_{i,t}^{<50\%} + \xi_{i,t} \end{aligned} \quad (2)$$

In Eq. (2), firm performance $Performance_{i,t}$ is measured by earnings before interests and taxes (EBIT), earnings before taxes (EBT), and earnings after taxes (EAT) over total assets as the dependent variables respectively. The primary explanatory variables include ownership rate $SOErate_{i,t}$ and overinvestment in the discrete form represented by the error-term of the subequation and in the continuous form calculated by HP Filter. The control variables are cash dividend payouts divided by the number of share outstanding $Dividend_{i,t}$, total liabilities over total assets $Leverage_{i,t}$, growth of total assets $AssetGrowth_{i,t}$, standard deviation of ROA $ProfitVolatility_{i,t}$, the company's quick ratio $Liquidity_{i,t}$, tangible fixed assets divided by total assets $Tangibility_{i,t}$. The three-variable interaction term is taken into consideration in order to test how the two-variable interaction terms react with the value of SOE rate, lower and higher than 50%. Therefore, $SOE_{i,t}^{<50\%}$ takes the value of 1 for the SOEs that have state ownership rates lower than 50%, otherwise. Tables 1 and 2 display the description and correlation of both dependent and independent variables in the research (see Footnote 1).

Table 1. Descriptive statistics

Variable	Observations	Mean	Variance	Min	Max
$EAT_{i,t}/Asset_{i,t}$	3,115	0.05426	0.00298	−0.05630	0.25565
$EBT_{i,t}/Asset_{i,t}$	3,109	0.06705	0.00424	−0.05500	0.29894
$EBIT_{i,t}/Asset_{i,t}$	3,118	0.08646	0.00394	−0.02990	0.31229
$Dividend_{i,t}$	2,073	1.30290	0.87132	0.00050	4.50000
$Leverage_{i,t}$	3,273	0.49956	0.04931	0.01311	0.94375
$AssetGrowth_{i,t}$	3,123	0.11273	0.05752	−0.23740	1.81471
$ProfitVolatility_{i,t}$	3,111	0.06833	0.00410	0.00407	0.37320
$Liquidity_{i,t}$	3,149	1.64385	1.29759	0.25585	6.86174
$Tangibility_{i,t}$	3,077	0.24305	0.03755	0.00488	0.79431
$SOErate_{i,t}$	1,558	41.0825	349.164	5.00000	96.7200
$Over-Investment_{i,t}^{REG.}$	2,980	0.32919	0.22090	0.00000	1.00000
$Over-Investment_{i,t}^{HP}$	3,345	0.01888	0.00299	0.00000	0.60553

Source: authors' estimation

Table 2. Matrix correlation of all explanatory variables

	$Dividend_{i,t}$	$Leverage_{i,t}$	$AssetGrowth_{i,t}$	$ProfitVolatility_{i,t}$	$Liquidity_{i,t}$	$Tangibility_{i,t}$	$SOErate_{i,t}$	$Over-Invest^{REG.}_{i,t}$
$Dividend_{i,t}$	**1.0000**							
$Leverage_{i,t}$	0.2386	**1.0000**						
$AssetGrowth_{i,t}$	−0.1340	0.1682	**1.0000**					
$ProfitVolatility_{i,t}$	−0.0383	0.2170	0.0285	**1.0000**				
$Liquidity_{i,t}$	0.1224	0.7330	0.1542	0.1806	**1.0000**			
$Tangibility_{i,t}$	0.0558	−0.0612	−0.0478	0.0933	−0.4100	**1.0000**		
$SOErate_{i,t}$	0.0251	0.0944	−0.0440	−0.0575	−0.0046	0.0719	**1.0000**	
$Over-Invest^{REG.}_{i,t}$	−0.0801	−0.0543	0.0227	−0.0505	−0.1411	−0.1133	0.0684	**1.0000**
$Over-Invest^{HP}_{i,t}$	−0.0415	−0.1572	0.0138	0.0367	−0.1236	−0.0980	−0.0086	0.4472

Source: authors' estimation

4 Results and Discussions

The regression results clarify that overinvestment does harm to firm performance. Such a finding implies the agency problem caused by the managers' investment in projects with negative net present value. Additionally, SOE rate is supposed to positively contribute to firm profitability. With the close relationship with the government, SOEs often receive

Table 3. Regression results using over-investment measured by sub-equation

$Performance_{i,t}$	$EBIT_{i,t}/Asset_{i,t}$		$EBT_{i,t}/Asset_{i,t}$		$EAT_{i,t}/Asset_{i,t}$	
Constant	0.0982***	0.0981***	0.109***	0.109***	0.0855***	0.0854***
	(0.00691)	(0.00689)	(0.00680)	(0.00678)	(0.00562)	(0.00562)
$Dividend_{i,t}$	−0.0401***	−0.0399***	−0.0405***	−0.0404***	−0.0333***	−0.0332***
	(0.00132)	(0.00132)	(0.00130)	(0.00130)	(0.00107)	(0.00107)
$Leverage_{i,t}$	−0.129***	−0.129***	−0.184***	−0.184***	−0.150***	−0.150***
	(0.00925)	(0.00924)	(0.00910)	(0.00909)	(0.00753)	(0.00753)
$AssetGrowth_{i,t}$	0.000676	0.000747	0.000842	0.000911	0.000666	0.000714
	(0.000572)	(0.000572)	(0.000563)	(0.000563)	(0.000466)	(0.000466)
$ProfitVolatility_{i,t}$	−0.00227	−0.00208	−0.0301	−0.0299	−0.0382*	−0.0381*
	(0.0261)	(0.0260)	(0.0257)	(0.0256)	(0.0212)	(0.0212)
$Liquidity_{i,t}$	−0.000358	−0.000308	0.000759	0.000808	0.000773	0.000807
	(0.000904)	(0.000903)	(0.000890)	(0.000888)	(0.000736)	(0.000736)
$Tangibility_{i,t}$	0.0273***	0.0285***	0.00915	0.0104	0.0123*	0.0132**
	(0.00781)	(0.00781)	(0.00768)	(0.00769)	(0.00636)	(0.00637)
$Over-Inv.^{REG}_{i,t}$	**−0.0216****	**−0.0354*****	**−0.0206****	**−0.0342*****	**−0.0176****	**−0.0270*****
	(0.00903)	**(0.0107)**	**(0.00889)**	**(0.0106)**	**(0.00736)**	**(0.00875)**
$SOErate_{i,t}$	**0.00027****	**0.000268****	**0.000278*****	**0.000276*****	**0.000231*****	**0.00023*****
	(0.000108)	**(0.000108)**	**(0.000106)**	**(0.000106)**	**(8.78e-05)**	**(8.77e-05)**
$Over-Inv.^{REG}_{i,t} \times SOErate_{i,t}$	**0.000261**	**0.000431****	**0.000371****	**0.000536*****	**0.000332****	**0.00045*****
	(0.000192)	**(0.000204)**	**(0.000189)**	**(0.000201)**	**(0.000156)**	**(0.000166)**
$Over-Inv.^{REG}_{i,t} \times SOErate_{i,t} \times SOE^{<50\%}_{i,t}$		**0.000478****		**0.000465****		**0.000323****
		(0.000201)		**(0.000198)**		**(0.000164)**
Observations	1,323	1,323	1,323	1,323	1,323	1,323
R-squared within	0.558	0.56	0.623	0.625	0.619	0.620
R-squared overall	0.554	0.556	0.621	0.623	0.617	0.618
R-squared between	0.026	0.031	0.338	0.349	0.258	0.268
F-Statistics	183.3	166.2	240.4	217.6	236.2	213.4
Prob.	0.000	0.000	0.000	0.000	0.000	0.000

*, **, *** corresponding to significance level of 10%, 5% and 1%

Source: authors' estimation

certain benefits in terms of political and financing support from the state, bringing more profitable opportunities to these companies (Adhikari, Derashid and Zhang 2006; Claessens, Feijen and Laeven 2008; Johnson and Mitton 2003; Lu 2000; Sun et al. 2002). Interestingly, based on the positive sign of the two-variable interaction between SOE rate and overinvestment, the study observes that state ownership tends to mitigate the negative impact of overinvestment on firm performance. The control of the government can help reduce agency costs within a business owing to the state's manipulation of policies, regulations, laws, and credit allocation decisions (Lu 2000). What's more, the beneficial moderation of SOE rate toward overinvestment are inclined to be stronger among companies with SOE rate under 50% and vice versa (Tables 3 and 4).

Table 4. Regression results using over-investment measured by HP-filter

$Performance_{i,t}$	$EBIT_{i,t}/Asset_{i,t}$		$EBT_{i,t}/Asset_{i,t}$		$EAT_{i,t}/Asset_{i,t}$	
Constant	0.0994***	0.0992***	0.110***	0.110***	0.0878***	0.0877***
	(0.00662)	(0.00656)	(0.00649)	(0.00643)	(0.00535)	(0.00531)
$Dividend_{i,t}$	−0.0396***	−0.0395***	−0.0402***	−0.0400***	−0.0330***	−0.0329***
	(0.00131)	(0.00130)	(0.00129)	(0.00128)	(0.00106)	(0.00105)
$Leverage_{i,t}$	−0.131***	−0.130***	−0.185***	−0.185***	−0.152***	−0.152***
	(0.00929)	(0.00921)	(0.00911)	(0.00903)	(0.00751)	(0.00745)
$AssetGrowth_{i,t}$	0.000461	0.000429	0.000640	0.000609	0.000489	0.000463
	(0.000568)	(0.000563)	(0.000557)	(0.000552)	(0.000459)	(0.000455)
$ProfitVolatility_{i,t}$	0.00309	0.00875	−0.0266	−0.0210	−0.0346	−0.0300
	(0.0260)	(0.0258)	(0.0255)	(0.0253)	(0.0211)	(0.0209)
$Liquidity_{i,t}$	−0.000407	−0.000414	0.000686	0.000679	0.000694	0.000689
	(0.000900)	(0.000892)	(0.000882)	(0.000875)	(0.000728)	(0.000722)
$Tangibility_{i,t}$	0.0303***	0.0312***	0.0110	0.0119	0.0134**	0.0141**
	(0.00774)	(0.00767)	(0.00759)	(0.00752)	(0.00626)	(0.00621)
$Over-Inv._{i,t}^{HP}$	**−0.488*** **	**−0.944*** **	**−0.505*** **	**−0.954*** **	**−0.488*** **	**−0.857*** **
	(0.111)	**(0.144)**	**(0.109)**	**(0.142)**	**(0.0896)**	**(0.117)**
$SOErate_{i,t}$	**0.000196** **	**0.000187** **	**0.000233** **	**0.000223** **	**0.000185** **	**0.000178** **
	(9.47e-05)	**(9.39e-05)**	**(9.29e-05)**	**(9.21e-05)**	**(7.66e-05)**	**(7.60e-05)**
$Over-Inv._{i,t}^{HP} \times SOErate_{i,t}$	**0.0109*** **	**0.0169*** **	**0.0121*** **	**0.0181*** **	**0.0115*** **	**0.0163*** **
	(0.00244)	**(0.00272)**	**(0.00239)**	**(0.00266)**	**(0.00197)**	**(0.00220)**
$Over-Inv._{i,t}^{HP} \times SOErate_{i,t} \times SOE_{i,t}^{<50\%}$		**0.0140*** **		**0.0138*** **		**0.0113*** **
		(0.00289)		**(0.00283)**		**(0.00234)**
Observations	1,323	1,323	1,323	1,323	1,323	1,323
R-squared within	0.561	0.569	0.629	0.635	0.627	0.633
R-squared overall	0.558	0.566	0.627	0.634	0.625	0.632
R-squared between	0.046	0.066	0.299	0.316	0.236	0.256
F-Statistics	186.1	172.7	246.3	227.9	244.4	226.1
Prob.	0.000	0.000	0.000	0.000	0.000	0.000

*, **, *** corresponding to significance level of 10%, 5% and 1%

Source: authors' estimation

5 Conclusion

The relationship between overinvestment and firm performance is regulated by the intervention of the government in a company's business activities. The research collects the dataset of financial statements of 669 Vietnamese non-financial listed companies in Ho Chi Minh and Hanoi Stock Exchange from 2012 to 2016, the study employs two different ways of measuring overinvestment with the use of the subequation and HP Filter. In the study, overinvestment is found to be negatively related to firm performance. Additionally, state ownership can regulate the negative effect of overinvestment on profitability. The regulating impact is supposed to become stronger when SOE rate is lower than 50%, meaning that in the process of privatization companies should reduce SOE rate to below 50% to capture the moderation effects of state ownership on the detrimental influence of overinvestment.

Appendixes

Table A1. Variable Measurement for main econometric model

Dependent Variables		
Variables	**Denote**	**Definition**
Firm Performance	EBIT/TA	EBIT (Earnings Before Interest & Tax) divided by Total Asset
	EBT/TA	EBT (Earnings Before Tax) divided by Total Asset
	EAT/TA	EAT (Earnings After Tax) divided by Total Asset
Explanatory Variables		
Variables	**Denote**	**Definition**
Profit volatility	$MROA_{i,t}$	Moving average of EBITDA (Earnings Before Interest, Taxes, Depreciation and Amortization) over Total Asset ratios in three consecutive years
Company Growth	$AssetGrowth_{i,t}$	Growth rates of total assets
Dividend	$Dividend_{i,t}$	Cash dividend payments
Leverage	$Leverage_{i,t}$	Total liabilities over total assets
Liquidity	$Liquidity_{i,t}$	The ratio of short-term assets to short-term debt
Tangibility	$Tangibility_{i,t}$	The sum of fixed assets, real estates, and inventory divided by total assets
Overinvestment	$Overinvestment_{i,t}$	The error terms of Equation (1) with $\hat{\omega}_{i,t} > 0$
State ownership rate	$SOErate_{i,t}$	State ownership rates
SOE dummy	$SOE_{i,t}^{<50\%}$	Dummy variable equal 1 if SOE rates < 50%

Table A2. Variable measurement for subequation

$$NewInvestment_{i,t} = \alpha_0 + \alpha_1 CashFlow_{i,t} + \alpha_2 TobinQ_{i,t} + \alpha_3 FixCapitalIntensity_{i,t} + \alpha_4 FirmSize_{i,t} + \alpha_5 RevenueGrowth_{i,t} + \alpha_6 BusinessRisk_{i,t} + \alpha_7 Leverage_{i,t} + \omega_{i,t}$$

Dependent Variables

Variables	Denote	Definition
New Investment	$NewInvestment_{i,t}$	Total investment including long-term and short-term investment divided by total asset

Explanatory Variables

Variables	Denote	Definition
Cash Flow	$CashFlow_{i,t}$	The cash available in a company after subtracting capital expenditures
Market Performance	$TobinQ_{i,t}$	The Q ratio is calculated as the market value of a company divided by the firm's assets
Fix Capital Intensity	$FixCapitalIntensity_{i,t}$	The ability to generate fixed assets through sales measured by total fix asset divided by total sales
Firm Growth	$RevenueGrowth_{i,t}$	The growth rate of firm's revenue over year
Firm Size	$FirmSize_{i,t}$	Natural logarithm of total asset
Business Risk	$BusinessRisk_{i,t}$	Standard deviation of EBITDA (Earnings Before Interest, Taxes, Depreciation and Amortization) over Total Asset ratio in three consecutive years
Leverage	$Leverage_{i,t}$	Total liabilities over total asset

All the above variables are calculated by using financial data from ***Thomson Reuters Eikon Financial Analysis***

References

Adhikari, A., Derashid, C., Zhang, H.: Public policy, political connections, and effective tax rates: Longitudinal evidence from Malaysia. J. Account. Public Policy **25**(5), 574–595 (2006)

Altaf, N., Shah, F.: Working capital management, firm performance and financial constraints: empirical evidence from India. Asia Pac. J. Bus. Adm. **9**(3), 206–219 (2017)

Bokpin, G.A., Onumah, J.M.: An empirical analysis of the determinants of corporate investment decisions: evidence from emerging market firms. Int. Res. J. Financ. Econ. **33**, 134–141 (2009)

Bos, D.: Privatization: a theoretical treatment. OUP Catalogue (1991)

Boubakri, N., Cosset, J.-C., Saffar, W.: Political connections of newly privatized firms. J. Corp. Financ. **14**(5), 654–673 (2008)

Boubakri, N., Cosset, J.C.: The financial and operating performance of newly privatized firms: evidence from developing countries. J. Financ. **53**(3), 1081–1110 (1998)
Boycko, M., Shleifer, A., Vishny, R.W.: A theory of privatisation. Econ. J., 309–319 (1996)
Brealey, R., Franks, J.: Indexation, investment, and utility prices. Oxf. Rev. Econ. Policy **25**(3), 435–450 (2009)
Carpenter, R.E., Guariglia, A.: Cash flow, investment, and investment opportunities: new tests using UK panel data. J. Bank. Financ. **32**(9), 1894–1906 (2008)
Chen, Y.-C., Hung, M., Wang, Y.: The effect of mandatory CSR disclosure on firm profitability and social externalities: evidence from China. J. Account. Econ. (2017)
Claessens, S., Feijen, E., Laeven, L.: Political connections and preferential access to finance: the role of campaign contributions. J. Financ. Econ. **88**(3), 554–580 (2008)
Connelly, J.T.: Investment policy at family firms: evidence from Thailand. J. Econ. Bus. **83**, 91–122 (2016)
D'souza, J., Megginson, W.L.: The financial and operating performance of privatized firms during the 1990s. J. Financ. **54**(4), 1397–1438 (1999)
Denis, D.K., McConnell, J.J.: International corporate governance. J. Financ. Quant. Anal. **38**(1), 1–36 (2003)
Dewenter, K.L., Malatesta, P.H.: State-owned and privately owned firms: an empirical analysis of profitability, leverage, and labor intensity. Am. Econ. Rev. **91**(1), 320–334 (2001)
Djankov, S., Murrell, P.: Enterprise restructuring in transition: a quantitative survey. J. Econ. Lit. **40**(3), 739–792 (2002)
Faccio, M.: Politically connected firms. Am. Econ. Rev. **96**(1), 369–386 (2006)
Fan, J.P., Wong, T.J., Zhang, T.: Politically connected CEOs, corporate governance, and Post-IPO performance of China's newly partially privatized firms. J. Financ. Econ. **84**(2), 330–357 (2007)
Farooq, S., Ahmed, S., Saleem, K.: Impact of Overinvestment & Underinvestment on Corporate Performance: Evidence from Singapore Stock Market (2014)
Fosu, S.: Capital structure, product market competition and firm performance: evidence from South Africa. Q. Rev. Econ. Financ. **53**(2), 140–151 (2013)
Gaver, J.J., Gaver, K.M.: Additional evidence on the association between the investment opportunity set and corporate financing, dividend, and compensation policies. J. Account. Econ. **16**(1–3), 125–160 (1993)
Grazzi, M., Jacoby, N., Treibich, T.: Dynamics of investment and firm performance: comparative evidence from manufacturing industries. Empir. Econ. **51**(1), 125–179 (2016)
Guariglia, A., Yang, J.: A balancing act: managing financial constraints and agency costs to minimize investment inefficiency in the Chinese market. J. Corp. Financ. **36**, 111–130 (2016)
Guedhami, O., Pittman, J.A., Saffar, W.: Auditor choice in privatized firms: empirical evidence on the role of state and foreign owners. J. Account. Econ. **48**(2–3), 151–171 (2009)
He, W., Kyaw, N.A.: Ownership structure and investment decisions of Chinese SOEs. Res. Int. Bus. Financ. **43**, 48–57 (2018)
Hellman, J.S., Jones, G., Kaufmann, D.: Seize the state, seize the day: State capture, corruption and influence in transition (2000)
Hodrick, R.J., Prescott, E.C.: Postwar US business cycles: an empirical investigation. J. Money Credit. Bank. 1–16 (1997)
Jensen, M.C.: Agency costs of free cash flow, corporate finance, and takeovers. Am. Econ. Rev. **76**(2), 323–329 (1986)
Jensen, M.C., Meckling, W.H.: Theory of the firm: managerial behavior, agency costs and ownership structure. J. Financ. Econ. **3**(4), 305–360 (1976)
Johnson, S., Mitton, T.: Cronyism and capital controls: evidence from Malaysia. J. Financ. Econ. **67**(2), 351–382 (2003)

Krueger, A.O.: The political economy of controls: American sugar: National Bureau of Economic Research Cambridge, Mass., USA (1988)
La Porta, R., Lopez-de-Silanes, F., Shleifer, A., Vishny, R.: Investor protection and corporate governance. J. Financ. Econ. **58**(1), 3–27 (2000a)
La Porta, R., Lopez-de-Silanes, F., Shleifer, A., Vishny, R.W.: Agency problems and dividend policies around the world. J. Financ. **55**(1), 1–33 (2000b)
Li, D., Zhang, L.: Does Q-theory with investment frictions explain anomalies in the cross section of returns? J. Financ. Econ. **98**(2), 297–314 (2010)
Licandro, O., Maroto, R., Puch, L.A.: Innovation, investment and productivity: evidence from Spanish firms (2004)
Liu, N., Bredin, D.: Institutional Investors, Over-investment and Corporate Performance. University College Dublin (2010)
Lu, X.: Booty socialism, bureau-preneurs, and the state in transition: Organizational corruption in China. Comp. Polit., 273–294 (2000)
Malm, J., Adhikari, H.P., Krolikowski, M., Sah, N.: Litigation risk and investment policy. J. Econ. Financ., 1–12 (2016)
Megginson, W.L., Netter, J.M.: From state to market: a survey of empirical studies on privatization. J. Econ. Lit. **39**(2), 321–389 (2001)
Nair, P.: Financial liberalization and determinants of investment: a study of indian manufacturing firms. Int. J. Manag. Int. Bus. Econ. Syst. **5**(1), 121–133 (2011)
Nilsen, Ø.A., Raknerud, A., Rybalka, M., Skjerpen, T.: Lumpy investments, factor adjustments, and labour productivity. Oxf. Econ. Pap. **61**(1), 104–127 (2008)
Peng, M.W., Buck, T., Filatotchev, I.: Do outside directors and new managers help improve firm performance? An exploratory study in Russian privatization. J. World Bus. **38**(4), 348–360 (2003)
Richardson, S.: Over-investment of free cash flow. Rev. Account. Stud. **11**(2–3), 159–189 (2006)
Rodríguez, G.C., Espejo, C.A.D., Cabrera, R.V.: Incentives management during privatization: an agency perspective. J. Manag. Stud. **44**(4), 536–560 (2007)
Ruiz-Porras, A., Lopez-Mateo, C.: Corporate governance, market competition and investment decisions in Mexican manufacturing firms (2011)
Sheshinski, E., López-Calva, L.F.: Privatization and its benefits: theory and evidence. CESifo Econ. Stud. **49**(3), 429–459 (2003)
Shleifer, A., Vishny, R.W.: Politicians and firms. Q. J. Econ. **109**(4), 995–1025 (1994)
Sun, Q., Tong, W.H., Tong, J.: How does government ownership affect firm performance? Evidence from China's privatization experience. J. Bus. Financ. Account. **29**(1–2), 1–27 (2002)
Titman, S., Wei, K.J., Xie, F.: Capital investments and stock returns. J. Financ. Quant. Anal. **39**(4), 677–700 (2004)
Yang, W.: Corporate investment and value creation. Indiana University Bloomington (2005)

Author Index

V. Kreinovich et al. (Eds.): ECONVN 2019, SCI 809, pp. 1155–1157, 2019.
https://doi.org/10.1007/978-3-030-04200-4

W

Y

Z